Composite Construction in Steel and Concrete IX

Composite Construction in Steel and Concrete IX

Proceedings of the Ninth International Conference on Composite Construction in Steel and Concrete 27–29th July 2021

Edited by
Markus Knobloch,
Ulrike Kuhlmann,
Wolfgang Kurz, and
Markus Schäfer

Ernst & Sohn
A Wiley Brand
WILEY

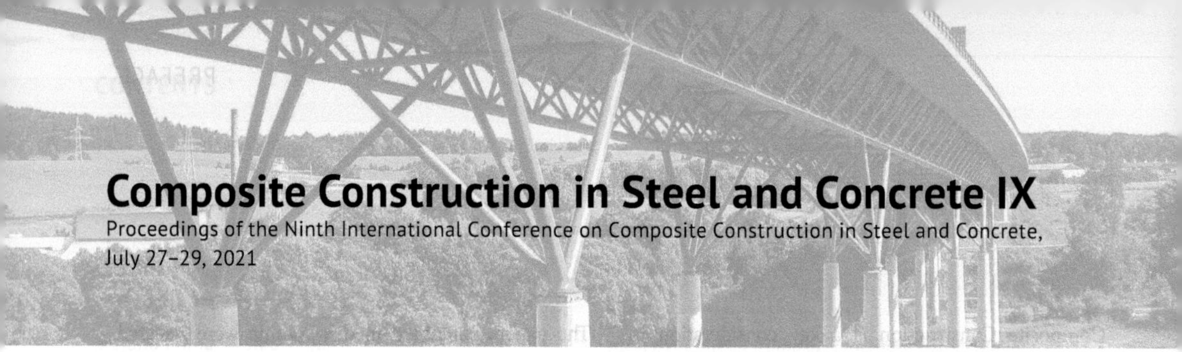

Composite Construction in Steel and Concrete IX

Proceedings of the Ninth International Conference on Composite Construction in Steel and Concrete, July 27–29, 2021

About this conference

The highly successful International Conference series on Composite Construction in Steel and Concrete is a major forum for researchers, practitioners, and engineers to share and discuss their research, practical experience and innovations related to composite constructions in steel and concrete with their peers of the global composite construction community. One of the principles of the conference series is that it should represent a forum where the latest research and case studies are presented. Papers are therefore submitted only a few months before the conference and may be adapted based on the outcome of the discussions during the conference, which ensures that only the most current work is published. The CCIX was hosted by the Ruhr-Universität Bochum RUB, University of Stuttgart, RPTU Kaiserslautern-Landau and University of Luxembourg in July 2021. As a result of the global COVID-19 coronavirus pandemic, it is the first Composite Construction Conference to be held completely online.

Review Process

All papers published in this issue were peer-reviewed by the Scientific Board of the conference series.

Committees:

Local Organizing Committee

Ulrike Kuhlmann	University of Stuttgart	Germany
Markus Knobloch	Ruhr-Universität Bochum RUB	Germany
Wolfgang Kurz	RPTU Kaiserslautern-Landau	Germany
Markus Schäfer	Universität Luxembourg	Luxembourg

Scientific Committee

Roland Bärtschi	Bärtschi Ingenieure	Switzerland
Alain Bureau	Centre Technique De La Construction Metallique (CTICM)	France
Adrian Ciutina	Polytechnic University of Timisoara	Romania
Graham Couchman	Steel Construction Institute (SCI)	UK
W. Samuel Easterling	Iowa State University	USA
Jerome Hajjar	Northeastern University	USA
Stephen Hicks	University of Warwick	UK
Markus Knobloch	Ruhr-Universität Bochum RUB	Germany
Venkatsh Kodur	Michigan State University	USA
Ulrike Kuhlmann	University of Stuttgart	Germany
Wolfgang Kurz	RPTU Kaiserslautern-Landau	Germany
Dennis Lam	University of Bradford	UK
Jean-Paul Lebet	École Polytechnique Fédérale de Lausanne	Switzerland
Roberto Leon	Virginia Polytechnic Institute and State University	USA
Matti V. Leskela	University of Oulu	Finland
Richard Liew Jat Yuen	National University of Singapore	Singapore
Renata Obiala	ArcelorMittal Global R&D	Luxembourg
José Oliveira Pedro	Instituto Superior Técnico of Lisbon	Portugal
Gian Andrea Rassati	University of Cincinnati	USA
Manuel L. Romero	Universitat Politècnica de València	Spain
Markus Schäfer	Universität Luxembourg	Luxembourg
Brian Uy	University of Sydney	Australia
Milan Veljkovic	Delft University of Technology	Netherlands
František Wald	Czech Technical University in Prague	Czech Republik
Rebekka Winkler	Ruhr-Universität Bochum RUB	Germany
Ben Young	The Hong Kong Polytechnic University	China

Organized by

Ruhr-Universität Bochum RUB

RPTU Kaiserslautern-Landau

Bauingenieurwesen
Prof. Dr.-Ing. Wolfgang Kurz
STAHLBAU

University of Stuttgart
Institute of Structural Design

Université du Luxembourg

Partners

Ernst & Sohn
A Wiley Brand

gruner

NPS® by Tecnostrutture®
your GREEN BUILDING SYSTEM

Sponsors

HRA

Ingenieurbüro Prof. Dr.-Ing Ulrike Kuhlmann

Leonhardt, Andrä und Partner

peikko®

EIFFAGE SEH

Support

Federal Ministry of Transport and Digital Infrastructure

CODIFICATION

9TH INTERNATIONAL CONFERENCE ON COMPOSITE CONSTRUCTION IN STEEL AND CONCRETE
M. Knobloch, U. Kuhlmann, W. Kurz, M. Schäfer. (Eds.)
Online, Germany, July 27-29, 2021

ENGINEERING MODEL FOR THE VERTICAL SHEAR CAPACITY OF COMPOSITE SLABS WITH ADDITIONAL REINFORCING STEEL

Nicole Schmeckebier*, Wolfgang Kurz**

* KREBS+KIEFER Ingenieure GmbH Köln
e-mail: schmeckebier.nicole@kuk.de

** Technische Universität Kaiserslautern
e-mail: wolfgang.kurz@bauing.uni-kl.de

Keywords: Composite Slabs, Shear Capacity, Engineering Model, Design, Standards, Experimental Investigations.

Abstract. *Up to now, no independent model for the shear design of composite slabs under the consideration of the two types of longitudinal reinforcement – metal sheet and reinforcing steel – exists. Therefore, extensive investigations were conducted at Technische Universität Kaiserslautern. The research project was completed recently. A new engineering model, which was calibrated on tests, was developed. Furthermore, the engineering model was transferred into a design model. In this paper the experimental investigations are presented as well as the engineering model.*

1 INTRODUCTION

Eurocode 4 [1] says that the shear capacity of composite slabs should be calculated according to Eurocode 2 [2] by using the empirical supported formula of concrete slabs without shear reinforcement. This formula refers on the one hand to the tooth model from Reineck [3] and on the other hand to the analysis of an extensive data base of shear tests on concrete specimen [4]. This model is based on some assumptions that are not valid for composite slabs. The metal sheet has its own shear capacity and the dowel effect of the slab is weaker than the dowel effect of the reinforcing steel. Furthermore, the bond stiffness of metal sheets is significantly lower than that of reinforcing steel. In addition, there is the fact that the empirical formula of Eurocode 2 [2] was determined by analysis of an extensive database of shear tests only on concrete specimen.

Furthermore, a lack of security for the calculation of the shear capacity of composite slabs made of lightweight concrete was noticed in [5]. In this paper tests on composite slabs are described, whereby the aim of this research was the investigation of the longitudinal bond action. The observed failure mode was a combination of longitudinal shear and transverse shear failure. Therefore, a new design model for the shear capacity of composite slabs without additional reinforcement [6] was developed at Technische Universität Kaiserslautern. This design model is valid for composite slabs made of lightweight concrete as well as normal concrete, but the main focus of this research project were the studies on specimens with lightweight concrete. The model contains three mechanisms that act additive. The shear capacity of the metal sheet and the bearing capacity of the uncracked compression zone together are described as the basic value because they act permanently. A third mechanism, the tensile bearing effects in the crack tip, can be added for re-entrant profiles. This shape guarantees an effective bond action so that critical crack width can be avoided.

In practice, composite slabs are mostly built with additionally reinforcement. But the significant influence of the reinforcing steel cannot be considered entirely in the model of [6]. Therefore, further research was necessary.

2 EXPERIMENTAL INVESTIGATIONS

2.1 General

In general, shear failure doesn't occur in composite slabs. So, the enforcement of shear failure in tests is very difficult. Therefore, the specimens were designed as small stripes of slabs and with uncommon dimensions. Two cross sections are shown as an example in figure 1. To avoid longitudinal shear failure an effective end anchorage was used. As reinforcing steel threaded anchor bars made of high strength steel [7] with a diameter of 15 mm were used to increase the bending capacity compared to the shear capacity disproportionately. In accordance with the rules, the reinforcing steel was placed as mesh reinforcement directly on the top of the metal sheet. In each rib one rebar was located and normal concrete C30/37 was used. The test setup was realised as a three-point-bending-test with an offset arrangement of the load. The distance between the support and the load introduction was set to three times of the effective height d_m. The width of the specimens corresponded to twice of the ribs of the metal sheet.

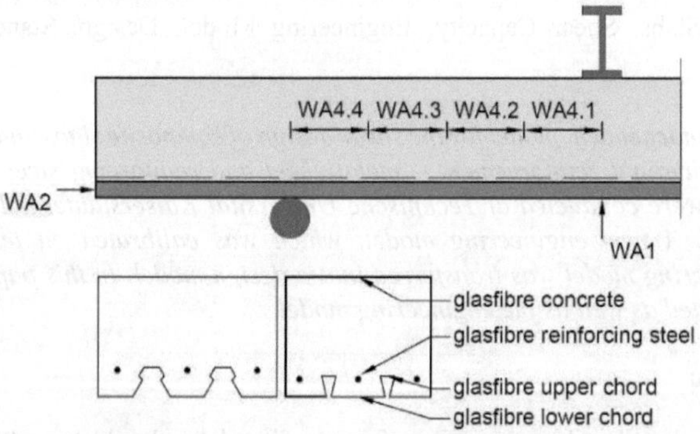

Figure 1. Schematic representation of cross sections of specimens and arrangement of measurement.

Table 1: Overview of combination of parameters of tests and failure loads of tests.

combination [-]	metal sheet [-]	thickness of metal sheet [mm]	height [mm]	width [mm]	f_{ctm} [N/mm²]	f_{cm} [N/mm²]	reinforcement [-]	V_{test} [kN]
1	SHR [8]	1.00	300	370	2.1	24.4	Ø15/15	168.0|137.2
2	C70 [9]	1.00	300	460	2.1	24.4	Ø15/18	172.2|185.6
3	Hody [10]	1.00	280	500	2.0	24.9	Ø15/20	178.9|164.0
4	ComFlor 80	1.00	280	690	2.0	24.9	Ø15/30	248.9|249.6
5	Hody [10]	0.75	280	500	2.4	27.8	Ø15/20	209.4|180.3
6	SHR [8]	0.75	300	370	2.4	27.8	Ø15/15	139.9|157.3
7	SHR [8]	1.25	300	370	2.4	27.8	Ø15/15	151.4|181.3
8	SHR [8]	1.00	300	370	2.2	25.0	Ø20/15	144.7| -
9	Hody [10]	1.00	280	500	2.2	25.0	Ø20/20	225.2|215.8
10	SHR [8]	1.00	300	370	3.3	56.1	Ø15/15	224.2|226.8
11	Hody [10]	1.00	280	500	3.3	56.1	Ø15/20	259.3|
12	C70 [9]	1.00	300	460	2.6	31.0	Ø15/18 u	- |190.7
13	ComFlor 80	1.00	280	690	2.6	31.0	Ø15/30 u	270.4|244.0
14	SHR [8]	0.75	250	370	2.5	30.4	Ø15/15	168.7|120.3
15	C70 [9]	1.00	250	460	2.5	30.4	Ø15/18	146.2|163.1

u ≙ location of reinforcing steel in through with c_{nom} = 20 mm, - ≙ outlier

Four different metal sheets were tested, whereby re-entrant profiles were included as well as profiles with an open shape. Furthermore, parameters were varied specifically to investigate their influence on the bearing behaviour. The parameters are the thickness of the metal sheet, the ratio of longitudinal reinforcement, the concrete strength, the location of the reinforcing steel and the height of the slab. Table 1 gives an overview over the combinations of the parameters. Two tests were performed with each combination. The specimens were named using the following principle: shape of metal sheet – thickness of metal sheet – height of specimen – concrete compressive strength – additional longitudinal reinforcement.

2.2 Test Results

The shear failure of composite slabs with additional reinforcing steel is very different to the bearing behaviour according to shear of concrete slabs. To explain that, the observations during the test procedure are described on the example of test C70-1,00-300-C30/37-Ø15/18h-1. Therefore, the force-deflection-diagram is shown in figure 2 and in figure 3 different crack patterns are presented. At a shear force of 30 kN the first bending cracks occurred, which results in the first change of the stiffness in the force-deflection-curve. With increasing load further formation of bending cracks could be observed. At a shear force of about 110 kN the diagram shows a significant load drop, which is followed by the second change of stiffness. The picture in the middle of figure 3 shows that a new crack appeared. This crack reaches the uncracked compression zone and runs inclined, so it could be identified as the shear crack. Because the shear crack wasn't critical, the cylinder load could be increased. During the load increase the crack propagation was still stable and a third area with a constant gradient could be seen in the force-deflection-diagram. The last picture in figure 3 shows that the further crack development was concentrated near to the support, because it was the horizontal part of the shear crack. At a shear force of about 160 kN the deflection of the specimen increased and the failure occurred a short time later. The failure was characterised by the opening of the shear crack. In summary the force-deflection-curve could be divided into four areas. The first one is the uncracked condition, followed by the area characterised by the formation of bending cracks. The third area describes the formation of the horizontal part of the shear crack and the last area is the failure of the specimen. Compared to concrete specimens which fail suddenly the behaviour at failure is different. After the shear crack reaches the concrete compression zone, the crack formation is still stable and the load could be increased significantly.

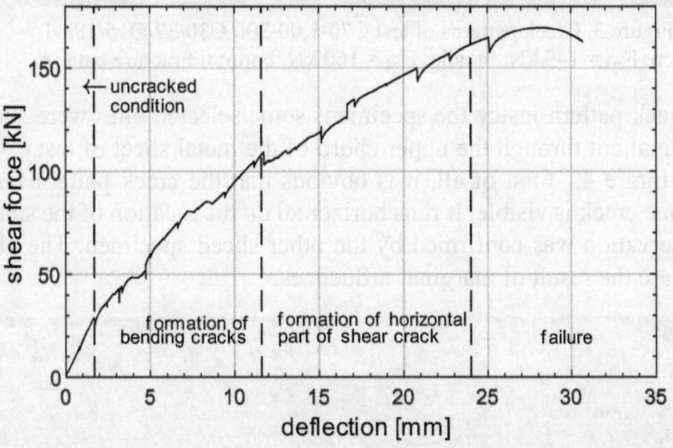

Figure 2. Force-deflection-diagram of test C70-1,00-300-C30/37-Ø15/18h-1.

Figure 3. Crack patterns of test C70-1,00-300-C30/37-Ø15/18h-1
top: F_{cyl} = 140 kN, middle: F_{cyl} = 160 kN, bottom: Fracture pattern.

To investigate the crack pattern inside the specimens some selected ones were sliced. As an example, a picture of the longitudinal cut through the upper chord of the metal sheet of test C70-1,00-300-C30/37-Ø15/18h-1 is shown in figure 4. First of all, it is obvious that the crack pattern inside and outside the specimen differs. Only one crack is visible. It runs horizontal on the location of the smallest concrete width to the support. This observation was confirmed by the other sliced specimen. The oblique cracks on the outside of the specimen are the result of marginal influences.

Figure 4. Longitudinal cut through the upper chord of specimen C70-1,00-300-C30/37-Ø15/18h-1.

The strains of the metal sheets were continuous measured with glass fibre cables on the upper chord as well as on the lower chord. Thus, it could be determined, that at the location of the shear crack the strains of the lower chord reached their maximum where the strains on the upper chord reached their minimum. Contrary to the expectation the maximum strains in the metal sheet are not at the location of the maximum bending moment of the specimen. Furthermore, the concrete compressive strains on the top of the specimen were measured. Following the Bernoulli Hypothesis, the concrete strains should increase constantly from the support to the load introduction. On the contrary a steady increase could only be observed at the location of the shear crack. Then, the concrete compressive strains increased significantly with their maximum at the load introduction. These observations lead to the conclusion that the Bernoulli Hypothesis is not valid for the shear force bearing behaviour of composite slabs. When the shear crack occurs, the bearing behaviour changes to a tied arch model with two compression struts. The concrete in the location of the shear crack supports itself on the metal sheet, which results in the high measured strains. The second compression strut runs directly to the support.

3 NEW ENGINEERING MODEL

3.1 Basic description of the design model

From the observations in the tests and the continuous strain measurement by glasfibres a simplified truss model was developed. It is presented in figure 5. After the shear crack occurs the concrete supports itself on the metal sheet which transfers the shear force through the crack. Then, the dowel effect of the longitudinal reinforcement induces the forces back into the concrete. This is characterised by the tension strut in the truss model that brings the forces of the direct compression strut into the support. The horizontal tie which is necessary for the equilibrium is the longitudinal reinforcement.

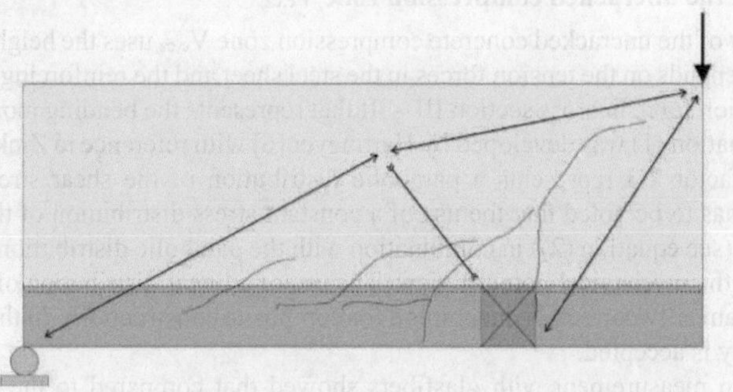

Figure 5. Truss model of shear transfer in composite slabs.

On the basis of the described truss model the engineering model for the design of the shear capacity of composite slabs with additional reinforcing steel was developed. The model is shown in figure 6 and contains four mechanisms that act additive. The shear transfer in the concrete compression zone $V_{c,cz}$, the shear transfer in the crack propagation zone $V_{c,ct}$ and the shear capacity of the tension strut that fails by kinking of the reinforcement or spalling of the concrete in the ribs $V_{c,ks}$ establish the equilibrium in the shear crack. In addition the vertical component of a strut that transfers the load to the support $V_{c,cs}$ is considered. The failure occurs when one of the mechanisms fails. The failure of the dowel effect of the reinforcement was identified as the kinematic condition for the overall failure of the specimen. Due to that the opening of the shear crack was possible which leads into the loss of the tensile bearing effects in the crack tip. The remaining two mechanisms cannot compensate these forces.

The strain measurement in the tension area showed that the metal sheet always reached the maximum tension bearing capacity in section II-II. Therefore, it is proposed to allocate the fully anchored tensile force to the metal sheet following the partial bond theory. The rest of the tension force, which is necessary for the equilibrium of moments, should be referred to the reinforcing steel. Furthermore, the high utilisation of the metal sheet is the reason that the own shear capacity of the metal sheet is not considered in the described model.

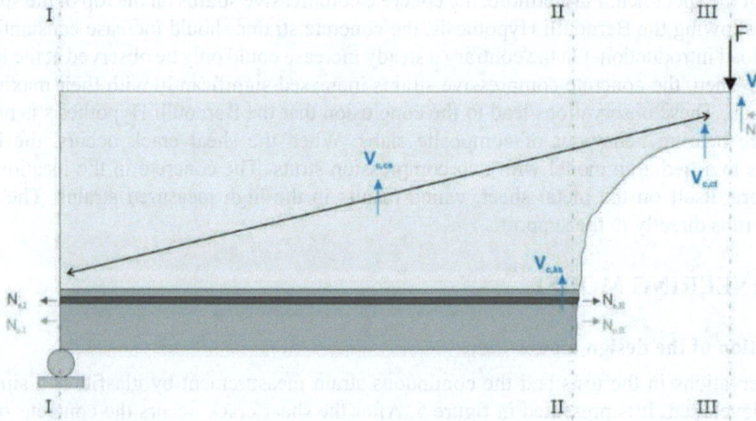

Figure 6. New shear design model for composite slabs with reinforcing steel.

3.2 Shear capacity of the uncracked compression zone $V_{c,cz}$

The shear capacity of the uncracked concrete compression zone $V_{c,cz}$ uses the height of the compression zone z_{pl}. This height depends on the tension forces in the steel sheet and the reinforcing steel in cross section II-II and the compression force in cross section III – III that represents the bending moment present in cross section III-III. The equation (1) was developed by Hartmeyer [6] with reference to Zink [11] and is assumed for this model. The factor 2/3 represents a parabolic distribution of the shear stresses in the concrete compression zone. It has to be noted that the use of a constant stress distribution of the normal stresses in the compression zone (see equation (2)) in combination with the parabolic distribution of the shear stresses is not consistent from the mechanical point of view. The use of a linear distribution of the normal stresses, which would be mechanically correct, is uncommon for composite constructions. In the interest of the Ease of Use this discrepancy is accepted.

Because the strain measurement with glasfibers showed that compared to the reinforcing steel the strains in the metal sheet were significantly higher, it is suggested to use the full anchored normal force of the metal sheet. The remaining forces necessary for the realisation of the bending moment should be assigned to the reinforcing steel.

$$V_{c,cz} = \frac{2}{3} \cdot z_{pl,III} \cdot b \cdot f_{ctm} \quad (1)$$

$$z_{pl,III} = \frac{N_{s,II}+N_{p,II}}{b \cdot f_{cm}} \quad (2)$$

3.3 Shear capacity of crack propagation zone $V_{c,ct}$

Hillerborg [12] introduced the critical length as a factor to describe the crack length that is able to transfer tensile stresses perpendicular to the crack. Hartmeyer [6] and Zink [11] used this approach to determine the shear capacity depending on the fracture energy G_f of the concrete. In this model the approach from Hartmeyer [6] was used and enhanced, s. equation (3).

$$V_{c,ct} = \alpha \cdot \beta \cdot l_{ch} \cdot b \cdot \frac{d_s}{d_{s,0}} = 0{,}12 \cdot \frac{G_f \cdot E_{cm}}{f_{ctm}} \cdot b \cdot \frac{d_s}{d_{s,0}} \tag{3}$$

The fracture energy is determined by the rules of Model Code 2010 [13]. The factors, which are necessary for the description of the characteristic length, were assumed from Remmel [14] and Grimm [15] ($\beta = 0{,}4$) and Reinhardt et al. [16] ($\alpha = 0{,}3$). The length of the shear crack with small crack width is not only dependent on the material behaviour but also on the total length of the crack. Good results were received by using the basic formula on very thick slabs. For the application in thinner slabs a linear dependence of the effective crack length on the effective height d was assumed. The basic value $d_{s,0}$ limits this effect to the maximum height d_s for which experimental data exists. The linear proportionality is a simplified assumption.

The investigation of the crack patterns in the test showed that the shear crack runs nearly horizontal when reaching the uncracked compression zone. Therefore, the consideration of the angle of the shear crack is not necessary.

3.4 Shear capacity inducing kinking and spalling $V_{c,ks}$

Two failure modes has to be considered, concrete failure $V_{c,ks,1}$ and steel failure $V_{c,ks,2}$, where the one with the lowest shear capacity should be used for the mechanism $V_{c,ks}$ according to equation (4).

$$V_{c,ks} = min \begin{cases} V_{c,ks,1} \\ V_{c,ks,2} \end{cases} \tag{4}$$

For the shear capacity inducing spalling the approach from Baumann and Rüsch [17] was enhanced according to equation (5). In figure 4 a horizontal crack can be found. As explained this crack occurs in all test specimen and it is located at the minimum width between shear connectors in the webs of the steel sheets. The width $b_{min,bar}$ has to be taken as the minimum width described above minus the diameter of the longitudinal reinforcement. Baumann and Rüsch [17] introduced a length d_{eq} in their formula which was adapted here to the height h_{pc}. For all types of steel sheets with shear connectors in the webs the height of the lowest connector above the bottom chord of the sheet is taken as h_{pc}. It is assumed that in other cases with re-entrant profiles h_{pc} is the middle between the upper and lower chord which both introduce shear by contact forces perpendicular to the span. The geometry values b_{min} and h_{pc} are explained in figure 7. In [17] the third square root of the concrete compressive strength was used to take the concrete tensile strength into account. Equation (5) shows that the concrete tensile strength is here considered directly with f_{ctm}.

$$V_{c,ks,1} = 1{,}64 \cdot b_{min,bar} \cdot h_{pc} \cdot f_{ctm} \tag{5}$$

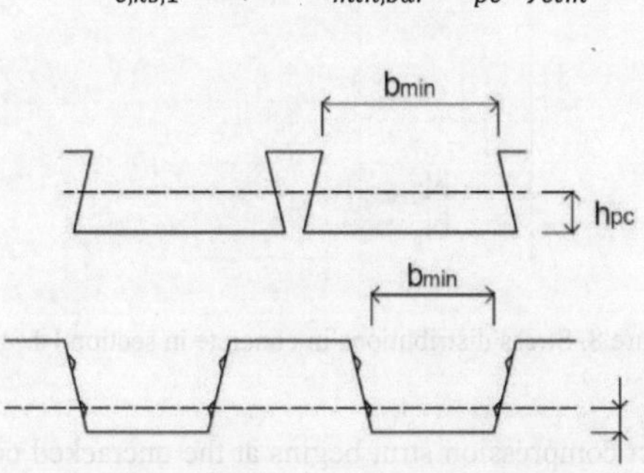

Figure 7. Geometry values b_{min} and h_{pc}.

The equations for the kinking resistance of the longitudinal reinforcement were taken from Model Code 2010 [13], see equation (6). Factor η_s is the utilization ratio of the reinforcing steel. As mentioned before, in all tests we observed that the utilization of the steel sheet was significantly higher than that of the reinforcement. Therefore, we propose that first the full utilization of the steel sheet in the decisive section is assumed and then the remaining tensile force is used to determine the utilization of the reinforcement. Furthermore, the approach in [13] contains a formulation dependent on the displacement of the crack origins. With a displacement of 0,2·d_s the maximum dowel effect can be activated. This formulation represents a criterion of serviceability. Because on the one hand the design of the shear force bearing behaviour belongs to ULS and on the other hand in the tests crack displacements in the range of 0,2·d_s could be observed the expression depending on displacement could be neglected.

$$V_{c,ks,2} = 1{,}6 \cdot A_s \cdot \sqrt{f_{sk}} \cdot \sqrt{f_{cm}} \cdot \sqrt{1-\eta_s^2} \qquad (6)$$

3.5 Shear capacity of the direct compression strut $V_{c,cs}$

The vertical component of the direct compression strut is another component of the shear capacity of composite slab in those cases where a strut can directly anchor on a direct support. The direct compression strut is at equilibrium with the anchored tension force in section I-I, so the vertical component has to be calculated according equation (7). In correspondence to concrete members the inclination angle of the compression strut is named θ.

$$V_{c,cs} = \left(N_{s,I} + N_{p,I}\right) \cdot \tan\theta \qquad (7)$$

In all test specimen the slabs were longer than the span so an overhanging length of about 500 mm was used to anchor the tension forces in the steel sheet and in the reinforcement. Due to the fact, that the bending moment at the support is about zero (just a small hogging moment out of dead weight of the slab) it is absolutely necessary that there has to be some bending moment in the concrete slab to reach this equilibrium. In some tests some minor cracks have been detected at to top side of the composite slab in the region of the support at ULS. Therefore, it is assumed as a simplification that the stresses in the concrete have a triangular shape with a concrete stress about zero at the top edge of the slab, see figure 8. This leads to the simplified assumption that the position of the direct compression strut is located at 1/3 of the height of the slab. For rectangular members this approach is correct. But in composite slabs the concrete in the area of the metal sheet is constricted, which has to be considered by determining z_{SP} as position of the centroid of compression stresses.

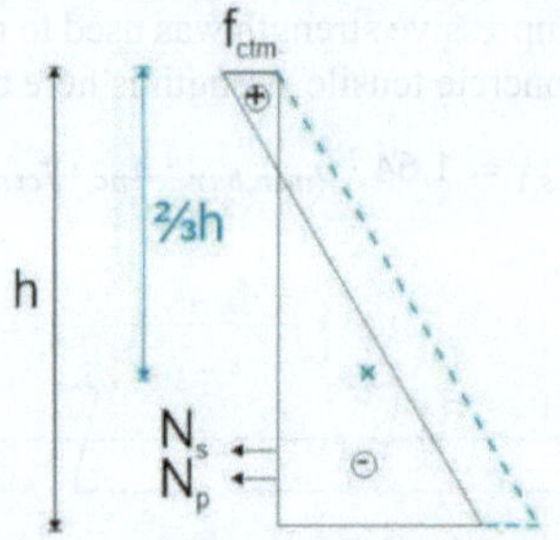

Figure 8. Stress distributions in concrete in section I-I.

At its opposite end the direct compression strut begins at the uncracked compression zone above the shear crack. In all test specimen this strut started at the load introduction into the specimen. So, the inclination angle could be determined according to equation (8), where z_{pl} has to be determined with the tension forces in section II – II of figure 6.

$$\tan\theta = \frac{z_{SP} - 0{,}5 \cdot z_{pl,III}}{3 \cdot d_m} \qquad (8)$$

Because of the dependency of the shape of the metal sheet the calculation of tan θ is time consuming. Therefore, a parameter study was done to provide a simplified assumption of tan θ. For the parameter study common cross sections of decks were used. The applied loads were chosen that high that the cross sections reached their bending capacity. Due to that the height of the uncracked compression zone increases and the inclination angle of the compression strut decreases. Figure 9 shows the results of the parameter study. For the simplified assumption of tan θ the value 0.21 was chosen as the minimum value. In general, for thinner decks in combination with higher concrete compressive strength greater values could be observed. Because higher concrete compressive strength are rarely used in composite slabs and the use of formula (8) is time-consuming, tan θ = 0.21 is a satisfied assumption.

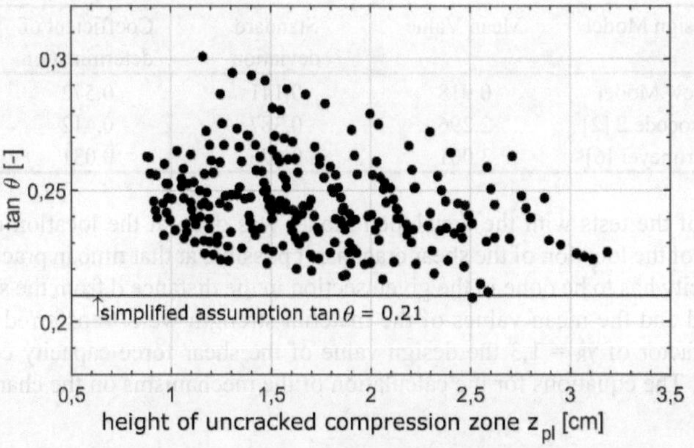

Figure 9. Results of parameter study to tan θ.

4 VERIFICATION AND STATISTICAL ANALYSIS OF THE DESIGN MODEL

With the developed model the shear force capacity could be calculated depending on the bending load. So, the model is very different to the current model [2] which describes the shear force capacity of a cross section of a bending member. Like described above three of the mechanisms are depending on the utilisation ratio of the longitudinal reinforcement. Thus, an independent consideration is not possible.

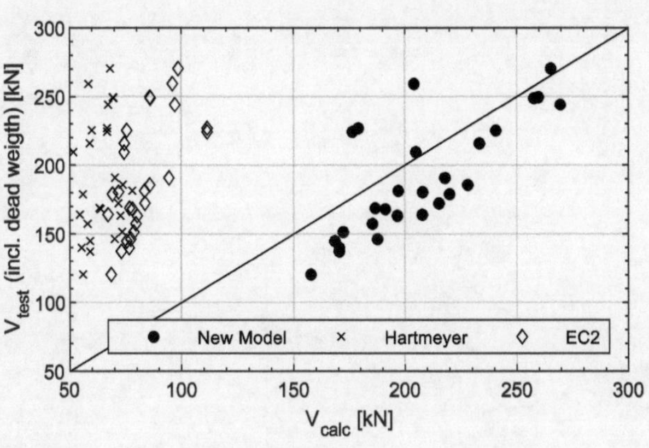

Figure 10. Comparative presentation of test results with calculated shear forces as mean values.

Figure 10 shows the plot of ultimate load in the tests V_{test} vs. the ultimate load according the developed design model V_{calc}. It shows a very good accordance between the tests and the model. Furthermore, the accordance is independent from the shape of the metal sheet. Also plotted in figure 10 are the comparisons between the test results and the shear forces calculated with the model of the current EN 1992 [2] and the model of Hartmeyer [6]. It becomes clear that with these models the shear force capacity of composite slabs cannot be described reliable. Even though the capacities are on the safe side, only with the new developed model the capacities can be calculated economical. Table 2 contains the values describing the mathematical reliability of the models. The new developed model is considered in the draft of Eurocode 4 [18].

Table 2: Comparison of test results with calculated shear forces as mean values.

Design Model	Mean Value	Standard deviation	Coefficient of determination
New Model	0,918	0,141	0,572
Eurocode 2 [2]	2,296	0,387	0,412
Hartmeyer [6]	3,001	0,676	0,031

The recalculation of the tests with the developed model was done at the location of the shear crack. Because the prediction of the location of the shear crack isn't possible at that time in practice the calculation of the shear force capacity has to be done in the given section in the distance d from the support. The model was statistical analysed and the mean values of the material strength were transferred into characteristic values. With a safety factor of $\gamma_R = 1{,}3$ the design value of the shear force capacity could be calculated according equation (9). The equations for the calculation of the mechanisms on the characteristic level are summarised in table 3.

$$V_{Rd} = \frac{0{,}694}{\gamma_R} \cdot \left[V_{c,cz} + V_{c,ct} + V_{c,ks} + V_{c,cs}\right] \quad (9)$$

Because with the developed model the calculation of the normal forces in different sections is necessary for the calculation of the shear force capacity the question arises whether the model could be simplified for the Ease of Use in practice. Therefore, a simplified design check is proposed. Only the tensile forces in the section of the support has to be calculated and used for the calculation of the different mechanisms. The tension force at support is at equilibrium with the compression force in the known distance d from the support. A verification of this approach was done with different statistical analysis. More information can be found in [19].

Table 3: Summary of equations for the calculation of the design shear force capacity.

Mechanism	Calculation
Compressive zone	$V_{c,cz} = \frac{2}{3} \cdot z_{pl} \cdot b \cdot f_{ctm}$ with: $z_{pl} = \frac{N_{s,II} + 1{,}15 \cdot \tau_{rk} \cdot b \cdot L_{II}}{1{,}33 \cdot b \cdot f_{ck}}$ z_{pl} = height of uncracked compression zone f_{ctm} = mean value of cylinder tensile strength of concrete $N_{s,II}$ = normal force in reinforcing steel in section II-II τ_{rk} = characteristic value of longitudinal shear stress of metal sheet L_{II} = horizontal distance between section I-I and II-II f_{ck} = characteristic cylinder compressive strength of concrete
Crack propagation zone	$V_{c,ct} = 0{,}12 \cdot \frac{G_f \cdot E_{cm}}{f_{ctm}} \cdot b \cdot \frac{d_s}{d_{s,0}}$ with: $G_f = 73 \cdot f_{cm}^{0{,}18}$ fcm [N/mm²] G_f = fracture energy of concrete [N/m] E_{cm} = mean value of Young`s modulus d_s = effective height of reinforcing steel $d_{s,0}$ = basic value of effective height of reinforcing steel = 270 mm
Dowel effect	$V_{c,ks} = \min\{V_{c,ks,1}; V_{c,ks,2}\}$ with: $V_{c,ks,1} = 1{,}64 \cdot b_{min} \cdot h_{pc} \cdot f_{ctm}$ b_{min} = minimum width of concrete inside of the metal sheet h_{pc} = vertical distance between bottom edge of metal sheet and the lowest shear connector $V_{c,ks,2} = 1{,}6 \cdot A_s \cdot \sqrt{f_{sk}} \cdot \sqrt{f_{ck}} \cdot \sqrt{1 - \eta_s^2}$ A_s = cross section of reinforcing steel f_{sk} = characteristic value of yield strength of reinforcing steel η_s = utilisation ratio of reinforcing steel
Direct compression strut	$V_{c,cs} = \tan\theta \cdot (N_{s,I} + N_{p,I})$ with: $\tan\theta = \frac{z_{SP} - 0{,}5 \cdot z_{pl}}{3 \cdot d_m}$ $N_{p,I} = 1{,}15 \cdot \tau_{rk} \cdot b \cdot L_I$ θ = inclination angle of compression strut $N_{s,I}$ = normal force of reinforcing steel in section I-I $N_{p,I}$ = normal force of metal sheet in section I-I z_{SP} = location of compression strut from the top edge of deck d_m = mean value of effective heitgh L_I = bond length of metal sheet behind the support

REFERENCES

[1] DIN EN 1994-1-1 2010, *Eurocode 4: Bemessung und Konstruktion von Verbundtragwerken aus Stahl und Beton – Teil 1-1: Allgemeine Bemessungsregeln und Anwendungsregeln für den Hochbau, Deutsche Fassung EN 1994-1-1:2004 + AC:2009*, Beuth, Berlin.

[2] DIN EN 1992-1-1 2010, *Eurocode 2: Bemessung und Konstruktion von Stahlbeton- und Spannbetontragwerken – Teil 1-1: Allgemeine Bemessungsregeln und Regeln für den Hochbau, Deutsche Fassung EN 1992-1-1:2004 + AC:2010*, Beuth, Berlin.

[3] Reineck, K.-H., *Ein mechanisches Modell für den Querkraftbereich von Stahlbetonbauteilen*, Dissertation, Universität Stuttgart, 1990.

[4] Reineck, K.-H., Kuchma, D.A. und Fitik, B., "Erweiterte Datenbanken zur Überprüfung der Querkraftbemessung für Konstruktionsbauteile mit und ohne Bügel", *DAfStb-Heft 597*, 2012.

[5] Brüdern, A.-E., Mechtcherine, V., Kurz, W. and Jurisch, F., "Self-Compacting Lightweigth Aggregate Concrete for Composite Slabs", *Proceedings of Advanced Concrete Materials International Conference*, Stellenbosch, South-Africa, 2009.

[6] Hartmeyer, S., *Modell zur Beschreibung des Querkrafttragverhaltens von Stahlverbunddecken aus Leicht- und Normalbeton*, Dissertation, Technische Universität Kaiserslautern, 2014.

[7] Z-12.5-96, *Allgemeine bauaufsichtliche Zulassung Z-12.5-96 – Ankerstabstahl St 900/1100 mit Gewinderippen AWM 1100 Nenndurchmesser: 15 und 20 mm, Stahlwerk Annahütte Max Aicher GmbH & Co. KG, 01.10.2015 – 01.10.2020,* Berlin, 2015.

[8] Z-26.1-45, *Allgemeine Bauartgenehmigung Z-26.1-45 – SUPER-HOLORIB SHR 51-Verbunddecke, Montana Bausysteme AG, 13.03.2018 – 13.08.2023*, Berlin, 2018.

[9] Z-26.1-22, *Allgemeine bauaufsichtliche Zulassung Z-26.1-22 – COFRASTRA Verbunddecken, ArcelorMittal Construction Deutschland GmbH, 28.07.2016 – 28.07.2021*, Berlin, 2016.

[10] Z-26.1-52, *Allgemeine Bauartgenehmigung Z-26.1-52 – Verbunddecke Hody SB 60, REPPEL b.v. Bouwspecialiteiten, 01.07.2019 – 01.07.2024*, Berlin, 2019.

[11] Zink, M., *Zum Biegeschubversagen schlanker Bauteile aus Hochleistungsbeton mit und ohne Vorspannung*, Dissertation, Universität Leipzig, 2000.

[12] Hillerborg, A. "Analysis of one Single Crack", *Fracture Mechanics of Concrete*, 1983.

[13] Model Code *Model Code 2010 – First complete draft, Volume 1*, Lausanne: International Federation for Structural Concrete (fib), 2010.

[14] Remmel, G., "Zum Zug- und Schubtragverhalten von Bauteilen aus hochfestem Beton", *DAfStb-Heft 444*, 1994.

[15] Grimm, R., "Einfluß bruchmechanischer Kenngrößen auf das Biege- und Schubtragverhalten hochfester Betone", *DAfStb-Heft 447*, 1997.

[16] Reinhardt, H.-W., Cornelissen, H., Hordijk, D.A., "Tensile Tests and failure Analysis of Concrete", *Journal of Structural Engineering*,112(11), 2462-2477, 1986.

[17] Baumann, T., Rüsch, H., "Versuche zum Studium der Verdübelungswirkung der Biegezugbewehrung eines Stahlbetonbalkens" *DAfStb-Heft 210*, 1970.

[18] prEN 1994-1-1, *Eurocode 4: Design of composite steel and concrete structures – Part 1-1: General rules and rules for buildings – SC4.T6: Draft prEN 1994-1-1:042020 + comments*. CEN European Committee for Standardization.

[19] Schmeckebier, N. *Ein Ingenieurmodell zur Berechnung der Querkrafttragfähigkeit von Stahlverbunddecken des Hochbaus mit praxisrelevanter Betonstahlbewehrung*, Dissertation, Technische Universität Kaiserslautern, 2021.

9TH INTERNATIONAL CONFERENCE ON COMPOSITE CONSTRUCTION IN STEEL AND CONCRETE
M. Knobloch, U. Kuhlmann, W. Kurz, M. Schäfer. (Eds.)
Online, Germany, July 27-29, 2021

EUROPEAN CODE DEVELOPMENTS

Stephen J. Hicks*, Markus Schäfer, Graham Couchman*****

* School of Engineering, University of Warwick, Coventry, UK (Vice-chair of CEN/TC250/SC4 and member of project team CEN/TC250/SC4.T3, SC4.T5 and SC4.T6)
e-mail: Stephen.J.Hicks@warwick.ac.uk

** University of Luxembourg, Department of Engineering, Structural Engineering & Composite Structures (Convenor of CEN/TC250/SC4.T6 and member of project team CEN/TC250/SC4.T1)
e-mail: markus.schaefer@uni.lu

*** SCI, Ascot, UK (Chair of CEN/TC250/SC4)
e-mail: G.Couchman@steel-sci.com

Keywords: Eurocode 4, Second generation of Eurocode 4, Codes and Standards, Composite.

Abstract *The second generation of Eurocode 4 has been developed through several project teams that report to CEN TC250 Subcommittee 4 (CEN/TC250/SC4) 'Design of composite steel and concrete structures', which is chaired by Dr Graham Couchman. Given that work on the revised version of Eurocode 4 is nearing completion, this paper presents a selection of the changes that will be included, together with some of the technical challenges that needed to be overcome. Finally, further enhancements that might be considered worthy for inclusion within future editions of this standard are presented.*

1 INTRODUCTION

Work on the Eurocodes commenced in 1975 when the European Commission decided on 'an action programme in the field of construction' based on article 95 of the Treaty of Rome, which was aimed at 'the elimination of technical obstacles to trade and harmonisation of technical specifications'. The Commission of European Communities (CEC) published eight European codes, or 'Eurocodes', for the design and execution of buildings and civil engineering structures. From these eight documents, the code for composite steel and concrete structures was published as Eurocode No. 4 in 1985 [1], which was based on the: ECCS Model Code [2]; international studies; together with Eurocode No. 1 (Common unified rules for different types of construction and material), No. 2 (Common unified rules for concrete structures), and No. 3 Common unified rules for steel structures). The European Commission transferred the preparation and the publication of the Eurocodes to the European Committee for Standardization (CEN) in 1989 through a series of mandates, in order to provide them with the status of a European Standard (EN). In the same year, the Construction Products Directive (CPD), was issued which introduced the concept of CE Marking for all construction products permanently incorporated into construction works [3].

Under the direction of Technical Committee CEN/TC250, the Eurocodes were published by CEN in 1992 as European pre-standards (ENV). Due to difficulties in harmonizing all aspects, the ENV Eurocodes included "boxed values" which permitted Member States to choose values for use within their territory through the publication of National Application Documents (NADs). Subcommittee 4 (CEN/TC250/SC4) was responsible for the ENV Eurocode 4, which was published in three Parts *viz*. ENV 1994-1-1 [4], ENV 1994-1-2 [5], and ENV 1994-2 [6]. To avoid repetition of information, and reduce potential ambiguity, values and properties are only given in one Eurocode. Because of this, ENV Eurocode 4 provided extensive cross-referencing to the ENV Eurocode 2 and Eurocode 3.

The EU mandate to CEN required that the content of the final ENs should be limited to the ENV versions modified in response to national comments; this became challenging in the development of the EN

Eurocode 4 due to changes in Eurocode 2 and Eurocode 3 that had been made to address these comments which, *inter alia*, included increasing the maximum yield strength of structural steel from 355 MPa to 460 MPa. Moreover, from national comments relating to the ease of use of ENV 1994-2 for bridge designers, the Eurocode 4 project team were given permission to repeat the 'general' Part 1-1 rules within Part 2 [7]. For the design of composite steel and concrete structures, the EN Eurocode 4 was published in the following three parts:

- EN 1994-1-1, Part 1-1: General rules and rules for buildings [8].
- EN 1994-1-2, Part 1-2: General rules – Structural fire design [9].
- EN 1994-2, Part 2: General rules and rules for bridges [10].

To enable the EN Eurocodes to be used within a particular territory, National Standards Bodies (NSBs) have published National Annexes (NAs) which contain: Nationally Determined Parameters (NDPs) (values of partial safety factors and classes applicable to that country, country specific data, and values where only a symbol is given in the EN); decisions on the status of informative annexes; and references to non-contradictory complementary information (NCCI). After a coexistence period, the EN Eurocodes replaced the former national standards in 2010 in countries that are members of CEN. The Construction Products Regulation (CPR) [11] replaced the CPD in 2011, which resulted in CE Marking becoming mandatory from 1st July 2013. The above provides a brief overview of the history of Eurocode 4 up to the EN version; a much more comprehensive review of the development from 1970 to 2010 is presented by Johnson [7]. More recently, the Eurocodes were adopted as national standards in Singapore [12], and it is anticipated that other countries may soon be implementing them, such as Hong Kong, Macau, Malaysia, Vietnam, Sri Lanka and Indonesia [13].

Following the publication of Mandate M/515 by the European Commission [14], work on the second generation of the Eurocodes commenced in 2015. Given that the work programme is nearing completion, the revised version of Eurocode 4 will soon become available. This paper presents a selection of the changes that will be included within the second generation of Eurocode 4, Part 1-1 (hereafter referred to as prEN 1994-1-1) [15], as well as highlighting future areas of improvement which may be considered worthy for future revisions.

2 SECOND GENERATION OF EUROCODE 4

A response to Mandate M/515 was prepared by CEN/TC250 [16], which set-out an ambitious and detailed work programme where discrete tasks are undertaken under the direction of one of TC250's existing subcommittees, working groups or horizontal groups. The mandate, *inter alia*, requires: extension of the Eurocodes in terms of new materials, products and construction methods; reduction in the number of NDPs (thereby leading to an alignment of safety levels); enhancing 'ease of use' for users; adoption of relevant ISO standards to supplement the Eurocodes (which implicitly recognizes the CEN-ISO Vienna agreement); and incorporation of recent results from scientific and technical associations, together with new research results. The revision can be broadly divided into two activities:

- General revisions and maintenance of the Eurocodes following the receipt of comments from the industry through a "systematic review" undertaken by NSBs.
- Technical enhancements of the Eurocodes within the scope of Mandate M/515.

For cases where there was insufficient agreement to develop a new EN, European Technical Specifications (CEN/TS) are also under development, which will complement and enlarge the suite of Eurocodes. A graphical representation of the structure for the second generation of the Eurocodes is presented in Figure 1.

The CEN/TC 250 work programme has been split into four overlapping phases, as follows:

- Phase 1: 25 Tasks (125 technical experts), 2015-2018
- Phase 2: 22 Tasks (88 technical experts), 2017-2020
- Phase 3 & 4: 26 Tasks (104 technical experts) 2018-2022

Each Task is the responsibility of a Project Team, which consists of a maximum of five or six members. The project team members were selected through a competitive tender and are contracted to the Royal

Netherlands Standardization Institute (NEN), which manages the work programme on behalf of CEN. For Eurocode 4, CEN/TC250/SC4 identified eight tasks within the work programme (see Table 1). At the time of writing, work is nearing completion with the last three project teams shown in Table 1 concluding in February 2022.

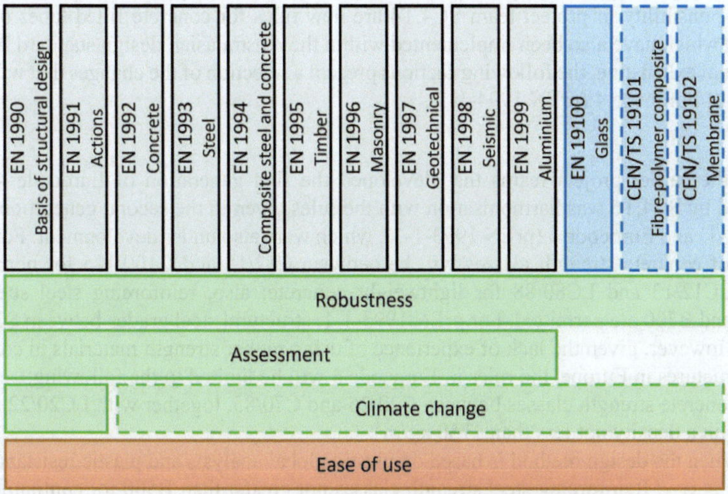

Figure 1. Structure of the second generation of the Eurocodes

Table 1: Tasks relating to Eurocode 4 within the CEN/TC250 work programme[16]

Project Team	Phase	Tasks
SC4.T1	1	Development of revisions to EN 1994-1-1, EN 1994-1-2 and EN 1994-2 in response to feedback from industry during the systematic review, including needs for harmonization with EN 1992 and EN 1993.
SC4.T2	1	Development of a new annex in EN 1994-1-1 for the design of composite beams with large web openings.
SC4.T3	1	Development of revised rules in EN 1994-1-1 for shear connection in the presence of modern forms of profiled sheeting.
SC4.T4	1	Development of new rules in EN 1994-1-2 for composite columns (concrete filled tubes) in fire.
SC4.T5	2	Development of new rules for EN 1994-1-1 and EN 1994-1-2 covering shallow floor construction, and other flooring types using precast concrete elements.
SC4.T6	3-4	Development of revised EN 1994-1-1
SC4.T7	3-4	Development of revised EN 1994-1-2
SC4.T8	3-4	Development of revised EN 1994-2

For the design of composite steel and concrete structures, it is anticipated that the second generation of Eurocode 4 will result in the following documents:

1. EN 1994-1-1, Part 1-1: General rules and rules for buildings.
2. EN 1994-1-2, Part 1-2: General – Structural fire design.
3. EN 1994-2, Part 2: Additional rules for bridges.
4. CEN/TS 1994-1-101 Design of composite steel and concrete structures — Design of double and single skin steel concrete composite (SC) structures.
5. CEN/TS 1994-1-102 Design rules for the use of Composite Dowels.
6. CEN/TS 1994-1-103 Design rules for composite columns comprising high performance materials.

As well as the new technical specifications, to reduce repetition and avoid any ambiguity, it has been decided to remove general rules from EN 1994-2 (as reflected in the change in title above) and provide cross-references to EN 1994-1-1; this change is consistent with Eurocode 3 and effectively results in EN 1994-2 becoming an "application part" [17]. Moreover, one of the important revisions to EN 1994-1-2 that was the responsibility of project team SC4.T4 are new rules for concrete filled tubes (CFTs) in fire conditions [18], which have also been implemented within the Australasian design standard AS/NZS 2327 [19]. Although not exhaustive, the following sections present a selection of the changes that will be included within Part 1-1 of Eurocode 4 (prEN 1994-1-1).

2.1 Materials

Similar to the earlier project teams that developed the first generation of Eurocode 4, one of the challenges faced by SC4.T6 was harmonisation with the rules given in the second generation of Eurocode 2 (prEN 1992-1-1) and Eurocode 3 (prEN 1993-1-1), which were also under development. For prEN 1992-1-1, the range of concrete strength classes will be between C12/15 and C100/115 for normal concrete, together with LC12/13 and LC80/88 for lightweight concrete; also, reinforcing steel strength classes between B400 and B700 are permitted. For prEN 1993-1-1, structural steel grades between S235 and S700 are supported. However, given the lack of experience of using higher strength materials in composite steel and concrete structures in Europe, the rules in Eurocode 4 will be limited to the following:

- Concrete strength classes between C20/25 and C70/85, together with LC20/22 and LC60/66 with a density not less than 1750 kg/m^3.
- When the design method is based on plastic global analysis and plastic resistance:
  - Reinforcing steel strength classes not greater than B500 for continuous composite beams, columns, and slabs.
  - Structural steel with a nominal yield strength not more than $f_y = 460$ MPa.
- When the design method is based on elastic or strain-limited design principles (see Section 2.6), the full range of reinforcing steel strength classes and structural steel grades permitted in prEN 1992-1-1 and prEN 1993-1-1, respectively may be applied.

Note that the concrete strength class of C70/85 has increased from C60/75 due to push test evidence suggesting that that is no adverse effects on the ductility of headed studs (see Section 2.3). Notwithstanding this, it is anticipated that a larger nominal yield strength for structural steel and higher concrete strength classes will be supported within the forthcoming CEN/TS on composite columns, which is consistent with other international standards and design guides where $f_y \leq 690$ MPa and $f_{ck} \leq 100$ MPa are permitted (see Section 2.10).

2.2 Composite beams with web-openings

Over the last 20 years, many long-span composite systems have been developed which permit mechanical services to be integrated within the structural depth of the floor. A common method of achieving this service integration is by means of openings cut within the webs of composite beams using I- or H-sections. The two main configuration of openings that are used are:

- isolated large openings at positions where the interaction between the openings is minimised; and
- regular openings, to form what are sometimes known as 'cellular beams'.

Whilst a draft amendment to the ENV version of Eurocode 3 was prepared in 1998 [20], which covered non-composite beams with web-openings, the committee draft was never published. Improvements to existing industry design guidance on composite beams with web-openings was developed through two major European research programmes [21],[22]. In the UK, this led to a design guide [23] that can be used alongside Eurocode 3 and 4.

From work undertaken by project team SC4.T2 (see Table 1), prEN 1994-1-1 will include a new Annex, which complements the new EN 1993-1-13 [17], and supports the design of composite beams with web-openings in sagging moment regions. For cases where the flexural stiffness of the concrete slab above the hole is significant, supplementary design rules [24] are given in a separate Annex which considers tension in the shear connectors and shear in the concrete.

2.3 Headed stud shear connectors embedded in solid concrete slabs and encasements

Internationally, higher strength concrete and a wider range of stud diameters are frequently being supported by composite design standards [19],[25]. Due to a greater body of push test data being available than when the existing Eurocode 4 was developed, several studies have investigated the design resistance of headed stud shear connectors to extend the scope of the current rules, as follows:

- From conducting an EN 1990 reliability analyses based on 101 push tests, Hanswille and Porsch [26] showed that some small modifications should be made the existing Eurocode 4 design model to justify the use of the target value of γ_V = 1.25 for the existing concrete strength classes of between C20/25 and C60/75, stud diameters of 16 mm $\leq d \leq$ 25 mm, and a specified ultimate tensile strength of the stud material of 460 $\leq f_u \leq$ 620 MPa.
- The database by Hanswille and Porsche was expanded to 242 tests by Hicks [27], who investigated what changes would need to be made to several design models to support the use of higher concrete strengths and larger stud diameters. From analyses according to EN 1990, it was found that by making small modifications to the Eurocode 4 design model, it could be safely extended to include concrete between C12/15 and C90/105, 12.7 mm $\leq d \leq$ 31.75 mm, and 400 $\leq f_u \leq$ 500 MPa (in addition, an enhancement to the Eurocode 4 model according to Döinghaus [28] for classes greater than C50/60 was also proposed). It was also recommended that the design model should be limited to a stud height to diameter ratio of $h_{sc}/d \geq 4$.
- Konrad *et al.* [29] developed a completely new design model. From reliability analyses considering 140 push tests, this design model was validated for concrete between C20/25 and C100/115, 16 mm $\leq d \leq$ 25 mm, and a characteristic tensile strength of the stud material $f_{uk} \leq$ 740 MPa. It was also recommended that $h_{sc}/d \geq 4$.

The design resistances predicted by Eurocode 4 compared with the above design models are presented graphically in Figure 2.

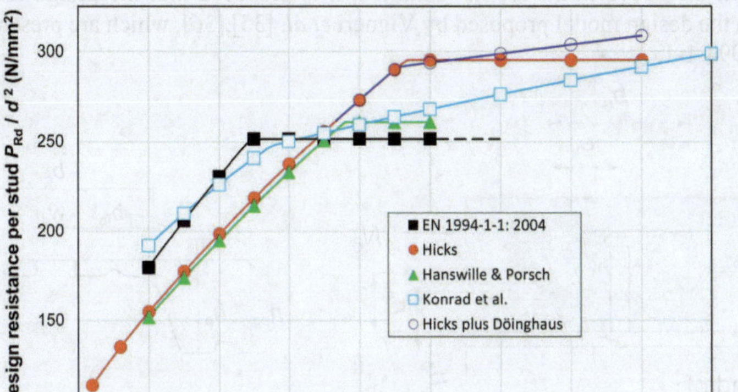

Figure 2. Design resistance of headed studs versus characteristic compressive cylinder strength of concrete with f_u = 450 MPa and $h_{sc}/d \geq 4$ for Eurocode 4 compared with other proposed design models (extension for prEN 1994-1-1 shown dotted).

Due to concerns that the ductility of headed studs may reduce in more modern concretes than those included within the push test databases discussed above [30], coupled with the fact that increasing the design resistance of headed studs might have adverse implications when fatigue is considered, the current rules will be maintained within prEN 1994-1-1. However, the concrete compressive strength will be increased to strength class C70/85 (shown by the dotted line in Figure 2), and the rules limited to $h_{sc}/d \geq 4$. The latter limit is consistent with the requirements in the AISC Specification [31].

2.4 Headed studs used with profiled steel sheeting in buildings

For studs welded within trapezoidal sheeting with the ribs transverse to the supporting beam the current Eurocode 4 provides equations to determine a reduction factor k_t, which is applied to the design resistance of a stud embedded in a solid concrete slab (see Section 2.3). The existing reduction factor equations are based on push test data from specimens that used sheeting types which were common in the 1980s. However, more recent research [32],[33] has shown that the reduction factor equations do not perform well for modern trapezoidal profiled steel sheeting. To remedy this situation, two new design models have been proposed for incorporation within Eurocode 4 *viz.*: new reduction factor equations by Konrad *et al.* [29], which need to be used with a new design model for headed studs embedded in solid concrete slabs (see Section 2.3); and a mechanical cantilever model that considers failure of a concrete failure cone in combination with plastic hinges developing within the shank of the stud, which is based on the work by Odenbreit and Nellinger [34] and extended by Vigneri *et al.* [35],[36].

From considering the advantages and disadvantages of the proposed models, it was felt that neither was able to completely replace the existing design rules in Eurocode 4 because they: reduced the current scope; required changes elsewhere (e.g. fatigue); or resulted in different design resistances for long established products that are known to perform well. On this basis, it was decided to retain the existing Eurocode 4 reduction factor equations, but reduce their scope for sheeting with transversely orientated ribs to:

- Trapezoidal sheets with an embedment depth $h_A \geq 2.7d$ and $e_k > 60$mm, where $h_A = (h_{sc}-h_p)$ (see Figure 3(a)).
- Re-entrant sheets with $h_A \geq 2d$.
- Reinforcement placed underneath the head of the stud.
- For trapezoidal sheets with a top re-entrant stiffener, $d_{ef} \leq 15$ mm and $b_{fp} \geq 25$ mm (see Figure 3(b)).

For cases when the above limits are not satisfied, it is proposed that the design resistances may be evaluated through the design model proposed by Vigneri *et al.* [35],[36], which are presented within a new Annex to prEN 1994-1-1.

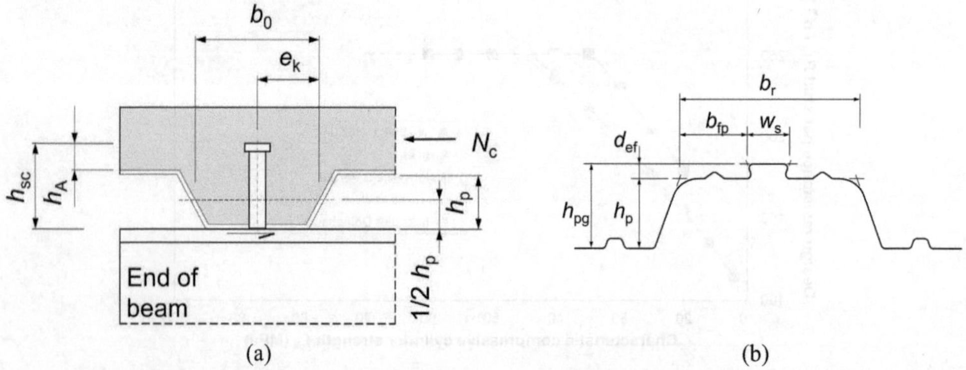

Figure 3. (a) Headed stud welded centrally within transversely orientated rib of trapezoidal profiled steel sheeting (b) cross-section of trapezoidal sheeting with top re-entrant stiffener [15].

2.5 Minimum degree of shear connection in beams for buildings

The existing Eurocode 4 rules for partial shear connection are based on two independent studies [37],[38]. These studies assumed that, in solid concrete slabs and composite slabs using profiled steel sheets prevalent in the 1980s, the characteristic slip capacity of 19 mm diameter studs was approximately δ_{uk} =

6 mm. The required slip was determined from numerical analyses of composite beams using various spans, cross-sections, and degrees of shear connection η. The rules for partial shear connection in Eurocode 4 were limited to situations where the required slip did not exceed 6 mm. Studs were deemed to be "ductile" in those situations.

From full-scale composite beam and companion push tests with trapezoidal sheeting and transversely orientated ribs, it has been found that a characteristic slip capacity of δ_{uk} = 10 mm can be achieved [32], [33]. Moreover, in practical design of composite beams in buildings, the utilization of the plastic bending resistance ρ_m (= M_{Ed} / [0.95$M_{Rd}(\eta)$]) can be low because the design is often limited by serviceability considerations. Finally, the earlier numerical analyses that formed the basis for the existing Eurocode 4 rules conservatively assumed propped construction, which is unusual in current practice. From a major European research programme [33], numerical analyses have been undertaken to investigate the sensitivity of these variables on the required slip. In the UK, this led to a design guide that can be used alongside Eurocode 4 [39]. From project team SC4.T3, the following changes are proposed within prEN 1994-1-1:

- Two ductility categories for ductile shear connectors are introduced: D2 for δ_{uk} = 6 mm; and D3 for δ_{uk} = 10 mm.
- Utilization of the plastic bending resistance of between $0.8 \le \rho_m \le 1.0$.
- Consideration of whether propped or unpropped construction is used which, for the latter, is related to the utilization of the bending resistance of the steel section from the self-weight loads before composite action $M_{a,Ed}$ cf. bending resistance of the composite beam with full shear connection ($\rho_{up} = M_{a,Ed} / M_{pl,Rd}$).

The proposed minimum degree of shear connection rules for composite beams in bending are presented graphically in Figure 4.

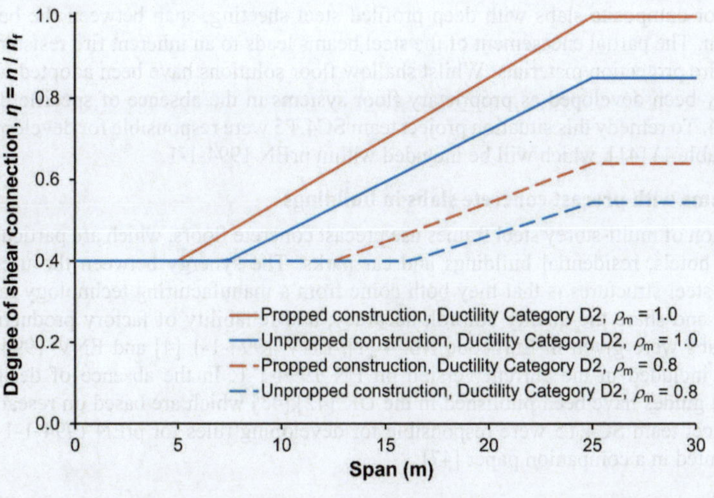

Figure 4. Minimum degree of shear connection versus span for S355 symmetric steel sections, Ductility Category D2 shear connectors and plastic bending resistance utilization of ρ_m = 1.0 or ρ_m = 1.0 (with ρ_{up} = 0.15 for unpropped construction).

2.6 Strain limited design of composite beams in bending

For composite beams where the plastic neutral axis is deep within the cross-section (such as may occur in slim floor beams, or for beams when steel grades S420 or S460 are used), the plastic bending resistance can overestimate the actual resistance. In these circumstances, the bending resistance should be evaluated from non-linear theory using the stress-strain relationships of the concrete, structural steel and reinforcement. Given that such a calculation is often impractical in normal design, Schäfer *et al.* [40]

developed a simplified method for evaluating the non-linear bending resistance which will be included in prEN 1994-1-1 with the following scope:

(i) concrete strength class between C20/25 and C50/60;
(ii) ratio of the overall depth of the concrete slab ($h_c + h_{pg}$) to the depth of the steel section h_a is between $0.15 \leq (h_c + h_{pg})/h_a \leq 0.7$, where h_c is the thickness of concrete flange above the ribs of the sheeting;
(iii) the depth of the profiled steel sheeting $h_{pg} \leq 135$ mm;
(iv) the effective width of the concrete flange $b_{eff} \geq 1000$ mm; and
(v) the cross-sectional area of the top flange A_{ft} in relation to the bottom flange area A_{fb} satisfies $A_{ft}/A_{fb} \leq 1.0$.

When conditions (i) to (v) are satisfied, the non-linear bending resistance may be taken to be:

$$M_{Rd} = \beta M_{pl,Rd} \tag{1}$$

where β is a reduction factor (which is related to the steel grade and the depth of the plastic neutral axis in relation to the overall depth of the beam, expressed by the ratio z_{pl}/h) and $M_{pl,Rd}$ is the design value of the plastic bending resistance of the composite cross-section with full shear connection.

2.7 Slim floor beams in buildings

Shallow floor, or slim-floor construction, has become popular throughout Europe as it provides a shallow structural zone, reduced number of beams, and flexibility in the layout of mechanical services. The key feature of slim-floor construction is the steel beams, which are integrated within the floor depth and possess a wide bottom flange to support the floor slabs. The floor slabs, consisting of either precast concrete hollow core slabs or composite slabs with deep profiled steel sheeting, span between the beams in the orthogonal direction. The partial encasement of the steel beams leads to an inherent fire resistance without the application of fire protection materials. Whilst shallow floor solutions have been adopted widely, they have predominantly been developed as proprietary floor systems in the absence of specific design rules given in Eurocode 4. To remedy this situation project team SC4.T5 were responsible for developing generic design rules (see Table 1) [41], which will be included within prEN 1994-1-1.

2.8 Composite beams with precast concrete slabs in buildings

A high proportion of multi-storey steel frames use precast concrete floors, which are particularly suited to sectors such as hotels, residential buildings and car parks. The synergy between the use of precast concrete slabs and steel structures is that they both come from a manufacturing technology rather than a site-based activity, and share the quality control, accuracy, and reliability of factory production. Whilst rules on precast slabs were given in Eurocode No. 4 [1], ENV 1994-1-1 [4] and ENV 1994-2 [6], this information is not included in the current version on EN 1994-1-1. In the absence of design rules in Eurocode 4, design guides have been published in the UK [42],[43] which are based on research by Lam [44],[45],[46]. Project team SC4.T5 were responsible for developing rules for prEN 1994-1-1 (see Table 1), which are presented in a companion paper [47].

2.9 Composite slabs

Whilst the *m-k* method has been widely used in the design of composite slabs for longitudinal shear, it has been known for some time that there are several shortcomings with the approach [48]; because of these shortcomings, the European industry has generally adopted the partial connection method, which has a much wider scope. To reflect current practice, it is proposed that prEN 1994-1-1 will no longer support the *m+k* method [49].

When the vertical shear resistance of composite slabs is considered in the current Eurocode 4, reference is made to the Eurocode 2 rules for slabs without shear reinforcement. From a comprehensive test programme on composite slabs with a wide variety of profiled steel sheeting cross-sections, it has been found that the current design approach is very conservative. To remedy this situation, a new design model for composite slabs has been developed [50], which is proposed to be included in prEN 1994-1-1.

2.10 Composite columns using high strength concrete and structural concrete

Internationally, higher strength concrete and structural steel is being introduced in design standards for concrete and steel structures (see Section 2.1). This is recognized in the Australasian composite steel and concrete design standards for buildings and bridges AS/NZS 2327 [19] and AS/NZS 5100.6 [25], where $f_y \leq 690$ MPa and $f_{ck} \leq 100$ MPa are permitted in the design of composite columns. Following the introduction of the Eurocodes in Singapore, a design guide was published [51] that extends the scope of the existing Eurocode 4 simplified design method to concrete strength class C90/105 and S550 steel, which is essentially considered as an NCCI. However, given that it was felt that a broader range of composite columns with high strength materials should be included, prEN 1994-1-1 supports their use through an extension to the general design method [49]. Nevertheless, it is anticipated that simplified design methods will be included in the companion CEN/TS (see Table 1).

More recently, from considering a database of more than 3200 tests [52], structural reliability analyses have been undertaken which investigated the performance of the current simplified design methods for concrete filled tubes (CFTs) given in a variety of international design standards when extended to $f_y = 850$ MPa and $f_{ck} = 185$ MPa. From this investigation [53], it was found that the Eurocode 4 simplified method for CFTs could be extended beyond its current limits, whilst still maintaining the existing reliability index β level (see Figure 5).

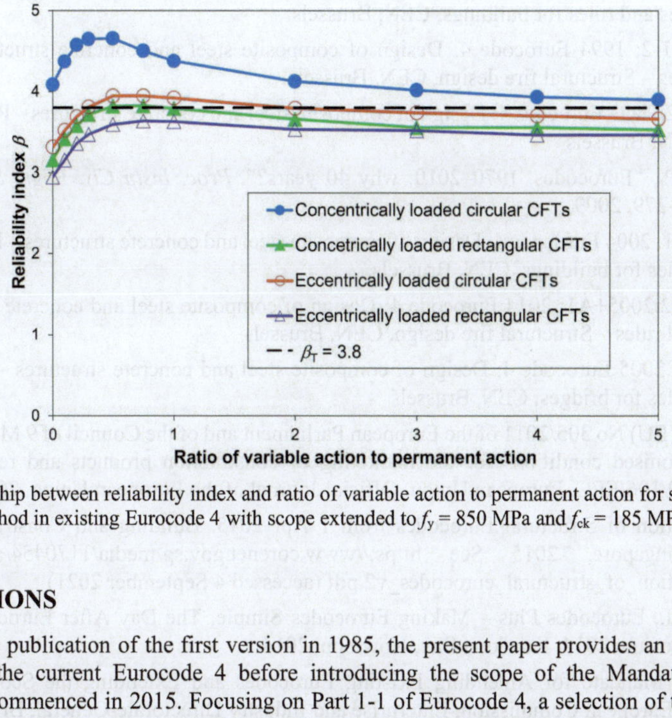

Figure 5. Relationship between reliability index and ratio of variable action to permanent action for simplified design method in existing Eurocode 4 with scope extended to $f_y = 850$ MPa and $f_{ck} = 185$ MPa

3 CONCLUSIONS

Following the publication of the first version in 1985, the present paper provides an overview of the development of the current Eurocode 4 before introducing the scope of the Mandate M/515 work programme that commenced in 2015. Focusing on Part 1-1 of Eurocode 4, a selection of the changes that will be included within the second generation of this Eurocode are presented, together with some of the technical challenges that needed to be overcome. Work undertaken within the Phase 4 project teams will be concluded in early 2022, before the CEN Enquiry draft for Eurocode 4 is completed. Due to the CEN process, together with the need to translate the drafts from English, it is anticipated that the complete second generation of the Eurocodes will be published in 2027.

ACKNOWLEDGEMENTS

The present authors have prepared this paper according to what they consider to be the significant changes to the existing EN 1994-1-1. However, the draft is still being prepared for CEN Enquiry and, because of this, may be subject to change. Any views expressed in this paper may not necessarily reflect those of the other members of CEN/TC250 Subcommittee 4. The first and second author acknowledge with thanks the financial support provided by the European Commission/European Free Trade Association under Grant Agreement SA/CEN/GROW/EFTA/515/2017-08 for Project Team SC4.T6 of the CEN/TC250 work programme.

REFERENCES

[1] Eurocode No. 4: Common unified rules for composite steel and concrete structures, CEC, Report EUR 9886 EN, 1985, https://op.europa.eu/s/oSuM

[2] Composite Structures. European Convention for Constructional Steelwork (ECCS) – Joint Committee for Composite Structures, Construction Press, London, 1981

[3] Directive 89/106/EEC Council Directive of 21 December 1988 on the approximation of laws, regulations and administrative provisions of the Member States relating to construction products. European Community, Official Journal of the European Union, 1989

[4] ENV 1994-1-1: 1992 Eurocode 4: Design of composite steel and concrete structures – Part 1-1: General rules and rules for buildings, CEN, Brussels.

[5] ENV 1994-1-2: 1994 Eurocode 4: Design of composite steel and concrete structures – Part 1-2: General rules – Structural fire design, CEN, Brussels.

[6] ENV 1994-2: 1997 Eurocode 4: Design of composite steel and concrete structures – Part 2: Composite bridges, CEN, Brussels

[7] Johnson R.P., "Eurocodes, 1970–2010: why 40 years?", *Proc. Instn Civ. Engrs Structs & Bldgs*, 162(6), 371-379, 2009

[8] EN 1994-1-1: 2004 Eurocode 4: Design of composite steel and concrete structures – Part 1-1: General rules and rules for buildings, CEN, Brussels.

[9] EN 1994-1-2:2005+A1: 2014 Eurocode 4: Design of composite steel and concrete structures – Part 1-2: General rules – Structural fire design, CEN, Brussels

[10] EN 1994-2: 2005 Eurocode 4: Design of composite steel and concrete structures – Part 2: General rules and rules for bridges, CEN, Brussels

[11] Regulation (EU) No 305/2011 of the European Parliament and of the Council of 9 March 2011 laying down harmonised conditions for the marketing of construction products and repealing Council Directive 89/106/EEC. European Union, Official Journal of the European Union, 2011.

[12] Implementation of Structural Eurocodes from 1 Apr 2015. Building and Construction Authority (BCA), Singapore, 2015 See https://www.corenet.gov.sg/media/1170454/appbca-2015-05-implementation_of_structural_eurocodes_v2.pdf (accessed 4 September 2021)

[13] Tomlinson J., Eurocodes Plus – Making Eurocodes Simple, The Day After Eurocodes, BSI Hong Kong Symposium 2014, Kowloon Bay, Kowloon, 2014

[14] M/515 EN. Mandate for Amending Existing Eurocodes and Extending the Scope of Structural Eurocodes. European Commission, Enterprise and Industry Directorate-General, Brussels, 2012

[15] prEN 1994-1-1: 2021 Eurocode 4: Design of composite steel and concrete structures – Part 1-1: General rules and rules for buildings, Draft March 2021, CEN/TC250/SC4 N2189.

[16] Response to Mandate M/515 EN 'Towards a second generation of EN Eurocodes', CEN/TC250 – N993, Brussels, 2013

[17] Kuhlmann, U., Schmidt-Rasche, C., Jörg, F., Pourostad, V., Spiegler, J. and Euler, M., "Update on the revision of Eurocode 3", *Steel Construction*, 14, 2-13, 2021.

[18] Albero, V., Espinos, A., Romero, M.L., Hospitaler, A., Bihina, G., Renaud, C. "Proposal of a new method in EN1994-1-2 for the fire design of concrete-filled steel tubular columns", *Engineering Structures*, 128, 237-255, 2016.

[19] AS/NZS 2327: 2017. Composite steel-concrete construction for buildings, Standards Australia/Standards New Zealand, Sydney/Wellington.

[20] ENV 1993-1-1: 1992/A2: 1998 Eurocode 3: Design of steel structures - Part 1-1: General - General rules and rules for buildings, Amendment A2, CEN, Brussels

[21] Large web openings for service integration in composite floors, Final Report for ECSC Research Contract 7210-PR-315, European Commission, 2006, https://op.europa.eu/s/o0Lh

[22] Large web openings for service integration in composites floors (LWO+), Final report for RFCS Project RFS-C2-05037, European Commission, 2009, https://op.europa.eu/s/o0Lo

[23] Lawson R.M., Hicks S.J. *Design of Composite Beams with Large Web Openings: In Accordance with Eurocodes and the UK National Annexes*, SCI P355, SCI, Ascot, 2011

[24] Classen, M., Kurz, W., Schäfer, M., Hegger, J. "A mechanical design model for steel and concrete composite members with web openings", *Engineering Structures*, 197, 2019

[25] AS/NZS 5100.6: 2017. Bridge Design, Part 6: Steel and composite construction, Standards Australia/Standards New Zealand, Sydney/Wellington.

[26] Hanswille, G., M. Porsch, M. "Zur Festlegung der Tragfähigkeit von Kopfbolzendübeln in Vollbetonplatten in DIN 18000-5 und EN 1994-1-1", Schriftenreihe des Instituts für konstruktiven Ingenieurbau, Ruhr-Universität Bochum, Festschrift Prof. Kindmann, Bochum, 2007.

[27] Hicks, SJ. "Design shear resistance of headed studs embedded in solid slabs and encasements", *Journal of Constructional Steel Research*, 139, 339-352, 2017.

[28] Döinghaus, P. Zum Zusammenwirken hochfester Baustoffe in Verbundträgern, Institut für Massivbau, RWTH Aachen, Dissertation, 2001.

[29] Konrad, M., Eggert, F., Kuhlmann, U., Schorr, J. "New approach for the design shear resistance of headed studs in profiled steel sheeting with ribs transverse to supporting beam", *Steel Construction*, 13, 252-263, 2020.

[30] Jähring, A. Zum Tragverhalten von Kopfbolzendübeln in hochfestem Beton, Lehrstuhl für Massivbau, Institut für Baustoffe und Konstruktion, Technische Universität München, Dissertation, 2008

[31] AISC 360-16. Specification for Structural Steel Buildings. American Institute of Steel Construction, Chicago, IL, 2016

[32] Hicks, S.J., Smith, A.L. "Stud Shear Connectors in Composite Beams that Support Slabs with Profiled Steel Sheeting". *Structural Engineering International*. 24(2), 246-253, 2014

[33] Development of Improved Shear Connection Rules in Composite Beams (DISCCo), Final report for RFCS Project RFSR-CT-2012-00030, European Commission, 2017, https://op.europa.eu/s/o2Ca

[34] Odenbreit, C., Nellinger, S. "Mechanical model to predict the resistance of the shear connection in composite beams with deep steel decking", Steel Construction, 10, 248-253, 2017

[35] Vigneri, V., Odenbreit, C., Schäfer, M., Hicks, S., Lam, D., Hanus, F. "Characterization of the load-slip behaviour of headed stud shear connections in narrow profiled sheeting". In M. Knobloch, U. Kuhlmann, W. Kurz, & M. Schäfer (eds.) Proceedings of the 9th international Conference on Composite Construction in Steel And Concrete, July 27 – July 29, 2021, Online, Germany

[36] Vigneri, V., Hicks, S.J., Taras, A, Odenbreit, C. "Design models for predicting the resistance of headed studs in profiled sheeting", Steel and Composite Structures, 42(5), 633-647, 2022.

[37] Johnson RP, Molenstra N. "Partial shear connection in composite beams for buildings", *Proc. Instn Civ. Engrs,* 91(2), 679–704, 1991

[38] Aribert J-M. "Dimensionnement de poutres mixtes en connection partielle" Mixed Structures Including New Materials. IABSE Symposium Brussels, Report 60, 215–220, 1990.

[39] Couchman, G.H. Minimum degree of shear connection rules for UK construction to Eurocode 4. SCI Publication 405, Steel Construction Institute, Ascot, 2015.

[40] Schäfer, M., Zhang, Q., Braun, M., Banfi, M. "Limitations of plastic bending resistance for composite beams deviated from strain-limitation", *Journal of Constructional Steel Research*, 180, 2021.

[41] Hicks, S.J, Braun, M., Markovic, Z., Way, J. "New Eurocode 4 design rules for shallow floor construction", *ce/papers*, 4(2-4), 1016-1020, 2021

[42] Hicks, S.J. and Lawson, R.M. Design of composite Beams using precast concrete slabs. SCI Publication 287, Steel Construction Institute, Ascot, 2003

[43] Couchman, G.H. Design of composite beams using precast concrete slabs in accordance with Eurocode 4. SCI Publication 401, Steel Construction Institute, Ascot, 2014

[44] Lam, D., Elliott, K.S., Nethercot, D.A. "Experiments on composite steel beams with precast concrete hollow core floor slabs", *Proc. Instn Civ. Engrs Structs & Bldgs*, 140, 127-138, 2000.

[45] Lam, D., Elliott, K.S., Nethercot, D.A. "Push-off tests on shear studs with hollow-cored floor slabs", *The Structural Engineer*, 76(9), 167-174, 1998.

[46] Lam, D. "Capacities of headed stud shear connectors in composite steel beams with precast hollowcore slabs", *Journal of Constructional Steel Research*, 63(9), 1160-1174, 2007.

[47] Hicks, S.J, Braun, M., Markovic, Z., Way, J. "New Eurocode 4 design rules for composite beams with precast concrete slabs". In M. Knobloch, U. Kuhlmann, W. Kurz, & M. Schäfer (eds.) Proceedings of the 9th international Conference on Composite Construction in Steel And Concrete, July 27 – July 29, 2021, Online, Germany

[48] Bode, H., Sauerborn, I. "Modern design concept for composite slabs with ductile behaviour", Composite Construction in Steel and Concrete II, Easterling, W.S. and Roddis, W.M.K. (eds), ASCE, NY, 1993, 125-141

[49] Schäfer, M. "European design code for composite structures in steel and concrete", *Steel Construction*, 12, 70-81, 2019

[50] Schmeckebier, N., Kurz, W. "Engineering model for the vertical shear capacity of composite slabs with additional reinforcing steel". In M. Knobloch, U. Kuhlmann, W. Kurz, & M. Schäfer (eds.) Proceedings of the 9th international Conference on Composite Construction in Steel And Concrete, July 27 – July 29, 2021, Online, Germany.

[51] BC4: 2021 Design guide for steel-concrete composite columns with high strength materials – An extension of Eurocode 4 method to C90/105 concrete and S550 steel. BCA, Singapore See https://www1.bca.gov.sg/docs/default-source/docs-corp-regulatory/building-control/bc4-2021.pdf (accessed 4 September 2021)

[52] Thai, H.-T., Thai, S., Ngo, T., Uy, B., Kang, W.-H., Hicks S.J. "Concrete-filled steel tubular (CFST) columns database with 3,208 tests", Mendeley Data, 2020 http://dx.doi.org/10.17632/j3f5cx9yjh.1

[53] Thai, H.-T., Thai, S., Ngo, T., Uy, B., Kang, W.-H., Hicks S.J. "Reliability considerations of modern design codes for CFST columns", *Journal of Constructional Steel Research*, 177, 2021

9TH INTERNATIONAL CONFERENCE ON COMPOSITE CONSTRUCTION IN STEEL AND CONCRETE
M. Knobloch, U. Kuhlmann, W. Kurz, M. Schäfer. (Eds.)
Online, Germany, July 27-29, 2021

EVOLUTION OF USA COMPOSITE CODES: CHANGES IN CHAPTER I (COMPOSITE CONSTRUCTION) OF THE AISC 360-22

Roberto T. Leon

* Virginia Tech, Blacksburg (VA), USA
e-mails: rleon@vt.edu

Keywords: codes, composite beams, composite columns, shear transfer, connectors.

Abstract. *This paper briefly reviews the evolution of composite codes in the USA in the context of important changes for the 2022 edition of the American steel design specification for buildings Major changes include the incorporation of composite walls as a new type of composite member, the addition of new provisions for shear capacity of filled members, and the introduction of a performance-based procedure for qualifying new types of anchors. An appendix also allows for the use of high strength materials for concrete filled rectangular members. The paper then describes the write's opinion on (1) important issues that need to be addressed on the direction of future codes, (2) areas where more research work is needed to provide more consistent and simpler design rules and (3) the desirability of including additional topics into the main specification.*

1 INTRODUCTION

The 9th International Conference on Composite Construction in Steel and Concrete 2021 provides an excellent opportunity to update colleagues around the world on changes to the upcoming 2022 version of AISC 360-22 [1], the American specification for steel building design. While the final, official version of the specification will not be ready for about another year, the version of Chapter I (Design of Composite Members) is almost complete at the time of this writing and no significant additional changes are expected.

The new version of Chapter I is about 32 pages long, about 5 more pages than the previous edition [2], and contains the following major changes, roughly in the order in which they appear:

- External to AISC, a major change has occurred in American specifications for composite construction, as the American Concrete Institute, in its 2019 version of ACI 318 [3], has eliminated all references to composite columns. Thus, the AISC 360 specification is now, de facto, the only comprehensive document addressing composite construction in the USA and some editorial and small technical changes have been made in the 2022 of AISC 360 to accommodate that transition.
- Wherever possible, all requirements related to detailing of longitudinal reinforcement, stirrups, ties, hoops and similar in AISC 360 are incorporated by reference from ACI 318. The intent was to promote the use of composite section by eliminating any confusion that might arise from perceived conflicts between ACI and AISC documents, particularly insofar as transverse reinforcement amounts and detailing are concerned. The only exception is for encased columns, where the existing AISC requirements have been kept because those were the ones used in the calibration studies. Materially, these differences are small for non-seismic applications. While no formal mechanism has been established to maintain this compatibility as ACI make changes to its future editions, it is hoped that such compatibility can be maintained.

- Major changes have occurred in the USA in the material specification for tubes. New versions of the ASTM A500 Grade C [4] and the newer A1085 [5] documents aim for a minimum yield strength for both circular and rectangular tubes of 345 MPa. This is a significant departure from the traditional values of 315MPa for rectangular and square tubes and 290 MPa for circular tubes. These values have an impact on design not only insofar as strength but also for local buckling, but local buckling provisions for those members have remained unchanged. Changes in these limits are tied to the opening of the largest and most modern tube rolling facility in the world, capable of producing square sections of up to 560 mm on a side, rectangular sections of up to 865 mm x 305 mm, and circular sections up to 710 mm, all with wall thicknesses up to 25 mm. Wide local availability of large tube sections is envisioned to increase significantly the use of these members.
- A new Appendix 2 that includes equations for determining the available strength of rectangular filled composite members with either (1) a specified minimum yield stress of structural steel exceeding 525 MPa but less than 690 MPa, and/or (2) a specified compressive strength, f'c, exceeding 69 MPa but less than 100 MPa, has been added. It is still possible that provisions for circular tubes may be added before the final version for 2022 is finished, but it is more likely that those will not be introduced until the 2028 edition. The decision to put the material in an Appendix rather than in the specification is due to the fact that it applies to a subset of the composite member types addressed in Chapter I and the need to avoid confusion as the equations in Appendix A do not match exactly those in Chapter I.
- The addition of composite walls as a separate type of structural member in Sections I1.4 though 6, including specific definitions for the effective area, moment of inertia and shear stiffness of those sections. These additions are described in detail in Section 3 of this paper.
- The addition of new equations for the determination of the shear strength of filled composite members. This is a significant improvement to Section I4 of the existing specification as current provisions are assumed to be extremely conservative. These changes are described in Section 4 of this paper.
- Last, but potentially a most important change, is the addition to Section I8 of a performance-based procedure to qualify any type of anchor (shear connector) for use in composite beams. The details of this addition are given in Section 5 of this paper.

2 CHAPTER I OF THE AISC 360-22 STEEL SPECIFICATION

The development of the provisions for composite construction for the 2022 edition of AISC 360 began even before the cycle formally started as AISC Technical Committee 5 (TC5 - Composite Construction) had a number of technical issues yet to address from the previous edition [2]. These included, among many others:

1. Coordinate with ACI to determine whether AISC or ACI should be the leader in developing and maintaining composite provisions on a topic by topic basis.
2. Clarification may be needed regarding application of direct bond provisions to tension hangers.
3. Improve provisions for composite column: extend slender column equations to compact, develop stability reduction factors, and revise resistance factors for members under combined axial and bending forces.
4. Allow higher strength material properties for encased and concrete-filled columns.
5. Develop provisions to address long term deflection and creep.

6. Coordinate on development of AISC 341 – Seismic Provisions [6] and ASCE 41 – Minimum Loads [7].
7. Develop guidelines for force transfer from a composite column to a base plate
8. Develop provisions for transverse reinforcement that address large cracks over girders.
9. Refine Chapter I Commentary exceptions for checking ductility/slip capacity of shear connectors.
10. Extend provisions to cover deeper deck profiles and larger diameter stud anchors along with developing a group reduction factor.

To one degree or another, items (1) through (6) in the above list were addressed in the present cycle. Items (7) through (10) are still under development and will not be addressed until the 2028 edition.

In addition, due to a change in how seismic provisions were to be addressed in the 2016-22 cycle by AISC, TC5 was also put in charge of developing and maintaining the seismic provisions for systems containing composite members (Chapters G and H of AISC 341 [8]). The idea behind this change was that both non-seismic and seismic design provisions need to be developed in a consistent fashion, and rather than having two committees (composite and seismic) that shared common members but separate development processes, it would be best to have all composite design provisions under the purview of one committee. In the case of the American steel and composite construction specifications, this is particularly important because the non-seismic provisions (AISC 360) are focused on member design, while the seismic ones (AISC 341) are focused on system performance. For the design of many structural member types, the existing AISC 341 defaults to the AISC 360 provisions; thus, to obtain the system performance desired, it is best if changes can be made in the AISC 360 document. The inclusion of composite walls as a separate member type in AISC 360-22 is a good example of this synergy as the it began as a seismic structural system in AISC 341. On the other hand, for encased members, the amounts and detailing of longitudinal and transverse bars differs substantially for the non-seismic and seismic provisions. As a result, some member detailing requirement need to be present in AISC 341.

Finally, and although external to TC5, AISC started a technical committee on evaluation and retrofit of existing structures (AISC TC 7 – Evaluation and Repair) in order to provide an update to the ones present in ASCE 41-16 [9]. Until recently, ASCE 41 was conceived as a stand-alone document for the evaluation and repair of all types of existing structures. As part of a wider trend in USA codes, the decision was made that ASCE should limit itself to the loading portion of the design process and that material organizations (AISC, ACI, etc.) should take over the resistance portion of the evaluation and repair processes. The work of TC7 did not directly impact that of TC5 in this cycle as the work of TC 7 was limited primarily to "translating" the existing ASCE 41 to AISC 360 language and creating a new, stand-alone set of provisions (AISC 342-22) for evaluation and repair [10]. However, it is expected that much more interaction will take place in the next code cycle as the composite provisions in ASCE 41 were outdated and there is a very large inventory of existing older building with composite members. In most cases those buildings were analysed using only the strength and stiffness of the steel members and connections; for proper assessment of the seismic behaviour of those structures, the influence of unintended and intended composite action needs to be considered. This will require an enormous effort for the 2022-28 cycle for TC7 and TC5 is expected to interact extensively in the process.

3 COMPOSITE WALLS

The new Chapter I adds composite shear walls, defined as members composed of of two steel plates connected by tie bars and infilled with concrete (Figure 1). These members have been used in high-rise construction [11], and their used is expanding to low and moderate height buildings.

Much of the original work on this type of members was done as part of developments in the nuclear industry (Varma, 2018) but are now being seen as a simple and economical substitute to traditional reinforced concrete walls, particular for areas where seismic design dominates and where drift and not strength are the governing parameter.

Chapter I provides some simple expressions (Equations 1-3) for the basic structural properties for these members for cases where the engineer chooses not to do a more detailed analysis:

$$(EI)_{eff} = E_s I_s + 0.35 E_c I_c \quad (1)$$

$$(EA)_{eff} = E_s A_s + 0.45 E_c A_c \quad (2)$$

$$(GA)_{eff} = G_s A_{sw} + G_c A_c \quad (3)$$

where,

A_c = area of concrete, mm^2

A_s = area of steel section, mm^2

A_{sw} = area of steel plates in the direction of in-plane shear, mm^2

E_c = modulus of elasticity of concrete = $0.043 w_c^{1.5} \sqrt{f_c'}$, MPa

G_s = shear modulus of steel = 76,880 MPa

G_c = shear modulus of concrete = 0.4 E_c

I_c = moment of inertia of concrete section about the elastic neutral axis of the composite section, mm^4

I_s = moment of inertia of steel shape about the elastic neutral axis of the composite section, mm^4

E_s = modulus of elasticity of steel = 200,000 MPa

w_c = weight of concrete per unit volume ($1500 \le w_c \le 2500$ kg/m^3)

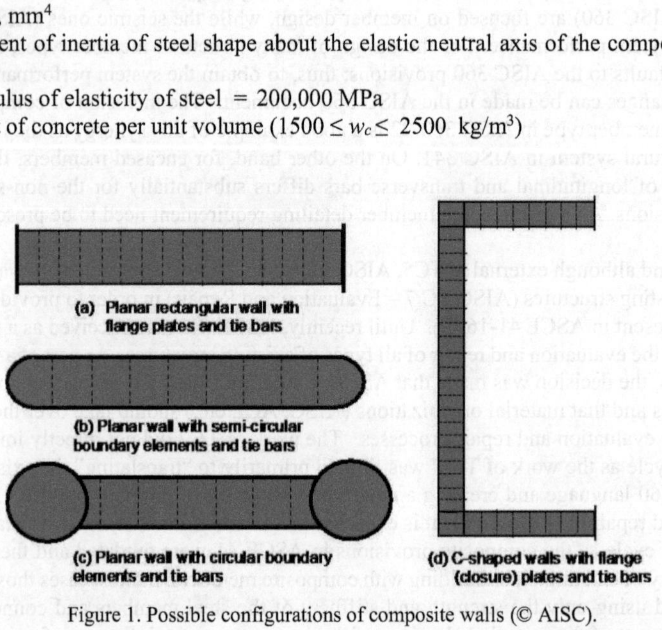

Figure 1. Possible configurations of composite walls (© AISC).

It is interesting to note that Equation (1) corresponds to the case of a regular filled composite rectangular column section where the area of steel sections and steel reinforcement is small compared to the overall area of the member. Thus, in a generic sense, walls and columns provide the equivalent flexural rigidity. Equation 2 is intended to be consistent with the amount of cracking assumed in the derivation of Equation (1). There are not corresponding expressions to Equations (2) and (3) for regular square and rectangular columns in the current version of AISC 360, although designers of those members typically will assume a value similar to that given by Equations (2) and (3) by default. The design requires that the wall aspect ratio be greater than 3, which results in shear stiffnesses close to the uncracked values, and Equation (3) reflects this assumption. Because most of these walls will be controlled by drift, and thus behave elastically under non-seismic design loads, the flexural stiffness to be used in the direct design method of analysis is not reduced ($\tau_b = 1.0$) as it is for regular columns ($\tau_b = 0.8$).

The minimum reinforcement ratio is set at 1% to be consistent with general reinforced concrete design where such limit is used to limit problems with creep and shrinkage under service loads; an upper limit of 10% is used as little research is available beyond this point and its is unlikely that the system would be economical beyond this limit.

A unique feature of the system is the use of tie bars to ensure that local buckling does not occur before the steel section reaches its yielding in compression. These tie bars, of course, also serve to stabilize the two thin walls during the fabrication, transportation and casting operations as the walls are shop fabricated. If additional shear transfer between the steel walls and the concrete is needed, shear studs or similar can be used.

The slenderness (b/t) of the plate is limited to:

$$\frac{b}{t} \leq 1.2\sqrt{\frac{E_s}{F_y}} \tag{4}$$

The spacing of tie bars must be less than the wall thickness, t_{sc} and the tie bar spacing to plate thickness ratio, s_t/t, must be less than:

$$\frac{s_t}{t} \leq 1.0\sqrt{\frac{E_s}{2\alpha+1}} \tag{5}$$

$$\alpha = 1.7\left[\frac{t_{sc}}{t} - 2\right]\left[\frac{t}{d_{tie}}\right]^4 \tag{6}$$

where ,

s_t = largest clear spacing of the ties, mm
t = plate thickness, in. mm
t_{sc} = thickness of composite plate shear wall, mm
d_{tie} = effective diameter of the tie bar, mm

In general, the tie bar requirements are governed by the buckling stress in the fabrication and transportation process, which is limited to 1000 psi in the equation:

$$\sigma_{cr} = \left(\frac{E}{\left(\frac{S}{t}\right)^2}\right)\frac{1}{(2\alpha+1)} \tag{7}$$

The provisions for composite walls in the non-seismic specification (AISC 360) were originally introduced primarily to support their use in the seismic provisions (AISC 341). It was deemed inconsistent to have a structural member in the seismic provisions that was not present in the conventional specification. As the cycle progressed, it became obvious that the members had potential for use in non-seismic areas and thus fairly comprehensive provisions were introduced into AISC 360. It is likely that the design provisions introduced in this cycle will be significantly extended and/or modified in the next one.

3 SHEAR CAPACITY OF CONCRETE FILLED TUBES

An area where the existing provisions were known to be very conservative and inconsistent is in the determination of the shear capacity of filled members. Currently, the choices are to (a) use the shear strength of the steel section alone given by the AISC specification, (b) use the shear strength of the concrete section provided by the ACI 318 rules, or (3) use the combined strength of the steel section and any transverse reinforcement. As (1) the shear design of composite walls was the subject of quite a bit of work as part of the changes in both the conventional and seismic provisions during this cycle, and, as noted in the previous section, (2) the design of composite walls seem to share many commonalities with that of conventional filled tubes, new provisions for the shear capacity of concrete-filled members have been introduced. As these members are very ductile, the resistance factors assumed are high ($\phi_v = 0.90$ (LRFD) and $\Omega_v = 1.67$ (ASD)).

For concrete-filled tubes, the member capacity is given by:

$$V_n = 0.6 A_v F_y + 0.06 K_c A_c \sqrt{f_c'} \tag{8}$$

where,

A_v = Shear area of the steel portion of a composite member. For circular members $A_v = 2A_s/\pi$; for rectangular members it is the sum of the area of webs in the direction of in-plane shear, mm²

A_c = Area of concrete in filled composite member, mm²

K_c = 1 for members with shear span-to-depth, $(M_u/V_u)/d$, greater than or equal to 0.7, where M_u and V_u are equal to the maximum required flexural and shear strengths, respectively, along the member length, and d is equal to the member depth in the direction of bending

K_c = 10 for members with rectangular compact composite cross sections and $(M_u/V_u)/d$ less than 0.5

K_c = 9 for members with round compact composite cross sections and $(M_u/V_u)/d$ less than 0.5

K_c = 1 for members having other than compact composite cross sections, irrespective of $(M_u/V_u)/d$

Linear interpolation between the above K_c values is used for members with compact composite cross sections and $(M_u/V_u)/d$ between 0.5 and 0.7. M_u, V_u and d are the moment, shear and effective section depth at the critical section, so $(M_u/V_u)/$ is the shear span ratio. It is expected that K_c =1.0 will be by far the most common cases as low shear span-to-depth ratios are not common except in connection areas.

For concrete-filled shear walls, the member shear capacity is given by:

$$V_n = \frac{K_s + K_{sc}}{\sqrt{3K_s^2 + K_{sc}^2}} A_{sw} F_y \tag{9}$$

where,

A_{sw} = area of steel plates in the direction of in-plane shear, in.² (mm²)

$K_s = G_s A_{sw}$

$$K_{sc} = \frac{0.7(E_c A_c)(E_s A_{sw})}{4E_s A_{sw} + E_c A_c} \tag{10}$$

4 PERFORMANCE-BASED DESIGN OF ANCHORS

Since the introduction of the Nelson shear stud into conventional USA construction in the 1950s, that type of anchor has dominated the American composite floor market. The determination of the strength of those anchors has traditionally been carried out through pushout tests that have never been standardized by ASTM, ANSI or similar American organizations. In addition, in the USA, the specifications for shear studs reside within the American Welding Institute (AWS) and thus outside the purview of AISC. Stud material is governed by the provisions of ASTM 108 for Grades G10110 through G10200 for either semi-killed or killed deoxidation material [11] and installation is governed by AWS D1.1 [12]. Traditionally the yield and tensile strength of the material used for shear studs has been taken as 345MPa and 415MPa, respectively. In the last decade the actual strengths measured in production have given values consistently above the nominal ones. As shear stud provisions in the American specification were considered to be somewhat generous, design values for studs by AISC have not been adjusted to reflect improved material properties.

As the design of anchors shifts from purely a strength consideration to one in which ductility and total deformation capacity play a role, TC5, along with similar committees of the American Institute of Iron and Steel (AISI) that work in the areas of composite decks, decided to develop a performance-based specification for such anchors in composite beams. The intent is to provide a quick and reliable qualification test for new types of anchors. Those provisions are now in Section I8.4 of the new specification, which require use of the AISI S923 [13] standard or an alternative test approved by the authority having jurisdiction. Note that S923 shares many similarities to Annex B of EN-1994-1-1 [14] and other international standards.

AISI 923 (Figure 2) is an equivalent pushout test now certified by the American National Standards Institute (ANSI) with the axial load applied in two steps:

1. Apply load, under load control, in increments of about 5% of the expected peak capacity of the specimen, but not more than 9kN until 80% of the expected capacity is reached.
2. After reaching 80% of the expected ultimate strength, apply loads in displacement control until slip increases by a fixed increment. Average slip increment should not exceed 1.25

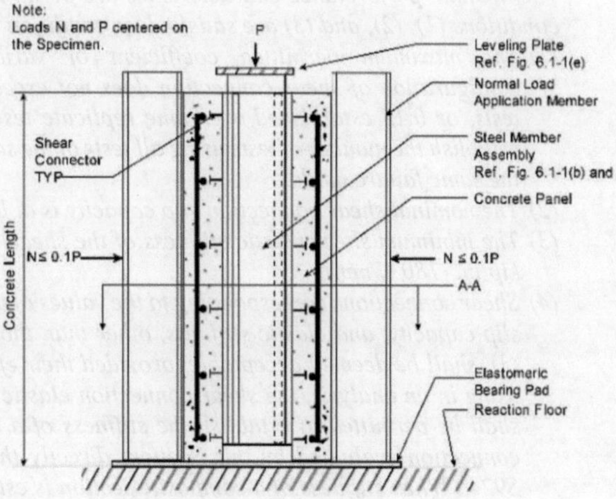

mm and the loading should continue until at least 30% of the ultimate strength has been lost.

(a) Side view of S923 specimen

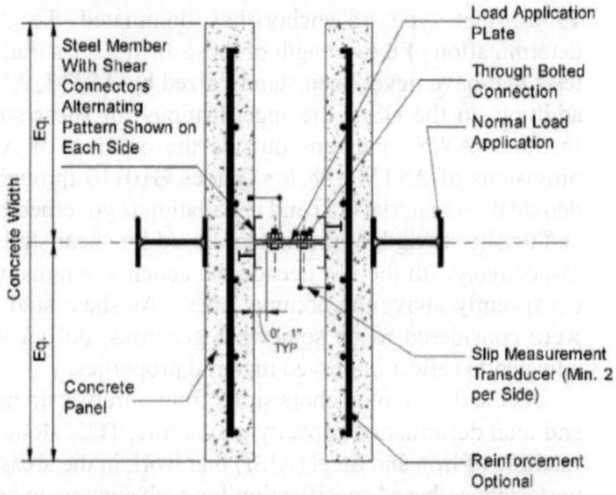

(b) Top view of S923 specimen

Figure 2 – Specimen for anchor qualification test AISI S923 (© AISI)

The requirements for the test setup, too long to describe here, include advanced test control, data acquisition, and document archival protocols. The ultimate slip capacity is taken at no less than 95% of the post-peak strength. The qualification limits and associated resistance factor, shall be determined as described in Section K2.1 of AISI S100 [15]. The procedure therein is a simple first moment approach.

AISC chose to include the following qualification limits, are taken verbatim from the proposed specification [1] language, as its qualification limits:

The nominal performance characteristics are permitted to be used in design provided either conditions (1), (2), and (3) are satisfied, or condition (4) is met:

(1) The maximum permitted coefficient of variation corresponding to each tested configuration of shear connection does not exceed 0.09 established over four replicate tests, or 0.15 established over nine replicate tests. It is permitted, for this purpose, to establish the number of tests using all tests of the same type of shear connection that exhibit the same failure mode.

(2) The nominal shear connection slip capacity is at least 0.25 in. (6 mm).

(3) The minimum shear elastic stiffness of the shear connection shall not be less than 2,000 kip/in. (180 N/mm).

(4) Shear connections corresponding to the values of coefficient of variation, shear connection slip capacity, and elastic stiffness, other than those stipulated in conditions (1), (2), and (3), shall be deemed acceptable, provided their effect is captured in the design. In lieu of using in an analysis, the shear connection elastic stiffness determined per this Section, it shall be permitted to establish the stiffness of a composite section, incorporating shear connection evaluated by this Section, directly through testing in accordance with AISI S924. When stiffness of a composite section is established in accordance with AISI S924, it shall be a mean tested value established based on at least three tests.

Within the context of AISI, the stiffness of shear connections is now proposed to be addressed through a separate set of tests that subjects the specific configuration (i.e., a composite beam with some particular steel deck and type of anchor) to the load history shown in Figure 3. This addresses the issue that the connection stiffness is influenced by many more parameters than those present in the pushout test. The test can be continued into a strength portion to determine the actual system strength (Figure 3(b)). Taken together, S923 and 924 provide a more rational and efficient path for the qualification of new anchor systems in composite beams.

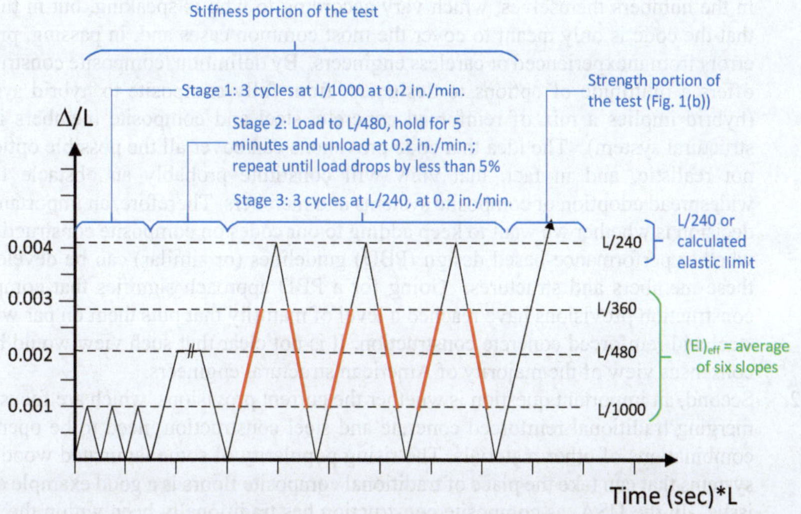

(a) Loading protocol for stiffness portion of S924 (© AISI)

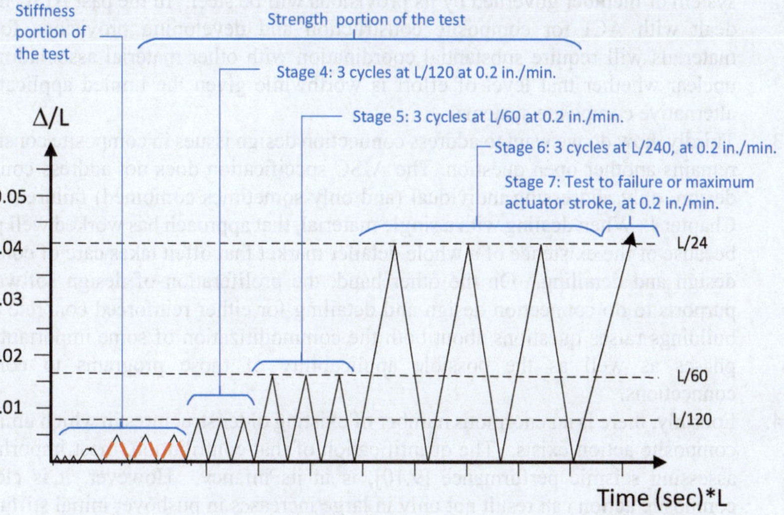

(a) Loading protocol for stiffness portion of S924 ((©) AISI)

Figure 3 – Load history used under AISI S923 (© AISI)

5 DECISIONS FOR FUTURE CODES

As the use of composite members becomes more widespread in the USA, and as this type of construction begins to aim for the low- to moderate rise market (5 to 20 stories), four significant decisions need to be made about the role of building codes in this process:

1. First, in writing codes in the USA, the assumption is often made that their provisions are aimed at about "70 of the designs that 90% of the engineers do." The point is not in the numbers themselves, which vary according to who is speaking, but in the fact that the code is only meant to cover the most common cases and, in passing, prevent errors from inexperienced or careless engineers. By definition, composite construction offers a multitude of options to designers, from fully composite to hybrid systems (hybrid implies a mix of reinforced concrete, steel and composite members in the structural system). The idea that code provisions can cover all the possible options is not realistic, and in fact, that view will constitute probably an obstacle to the widespread adoption of composite systems and members. Therefore, an important first decision is whether we want to keep adding to our codes on composite construction, or whether performance-based design (PBD) guidelines (or similar) can be develop for these members and structures. Going for a PBD approach signifies that composite construction provisions have reached a level of maturity that puts them on par with all steel and reinforced concrete construction. It is not clear that such view would be the consensus view of the majority of American structural engineers.
2. Second, an important question is whether the current provisions, which are focused on merging traditional reinforced concrete and steel construction, need to be opened to combinations of other materials. The rising popularity of some laminated wood floor systems that can take the place of traditional composite floors is a good example of this issue. In the USA, as composite construction has traditionally been within the AISC specification, so it is clear that the most important material involved in any composite system or member governed by its provisions will be steel. In the past AISC has only dealt with ACI for composite construction and developing provisions for other materials will require substantial coordination with other material associations. It is unclear whether that level of effort is worthwhile given the limited applications of alternative composite systems
3. Thirdly, how do we want to address connection design issues in composite construction remains another open question. The AISC specification does not address connection design, only addressing individual (and only sometimes combined) failure modes in Chapter J. When dealing with a single material, that approach has worked well perhaps because of the existence of a whole detailer market that often takes care of connection design and detailing. On the other hand, the proliferation of design software that purports to do connection design and detailing for either reinforced concrete or steel buildings raises questions about both the commoditization of some important design phases as well as the possible applicability of those programs to composite connections.
4. Fourthly, there is an enormous number of existing older structures in which unintended composite action exists. The quantification of that contribution, most importantly in assessing seismic performance [9,10], is in its infancy. However, it is clear that composite action can result not only in large increases in pushover initial stiffness and ultimate resistance but also in the misidentification of governing failure modes. The effort necessary to address this issue, both from the need for fundamental experimental data and from the need to develop new analytical tools, is large.

The decisions above are tied to both market forces and possible technical disruptors in the construction sector. In the writer's personal view, there is no consensus about whether any of the potential work outlined above is worth doing and where the necessary financing form this work will come from.

6 TECHNICAL NEEDS

While there are any number of important decisions that could be made to push composite codes in non-traditional directions, it is clear that American codes will need significant improvements in their future editions. A short listing of them, broken down by member types and limited to the non-seismic cases, follows.

6.1 Composite Beams

1. Distribution of shear studs – current provisions do not provide sufficient guidance for designers for cases where the design of the anchors is governed by high local shears. Traditionally, designers have calculated a total number of shear studs and distributed them uniformly along the span based on limit analysis theory. This approach is so common that many designers fail to identify situation were those assumptions do not work and take appropriate action. In addition, even for the cases of distributed loads, as span get longer as slip demands on the anchors increase, it may be a good idea to require that at least two studs be used in the last three flutes (or something similar) to address anchor ductility demand in composite beams with low degrees of shear connection.
2. Revaluation of current limits on exceptions on stud ductility check – the last version of the AISC specification introduced a requirement for anchor ductility but added three conditions in which those an explicit check was not needed. Those exceptions, and how they will be handled with deeper decks and larger diameter anchors, need clarification.
3. Deeper deck and longer studs – current AISC Specifications are keyed to experimental work done more than 50 years ago and is limited in most cases to decks with trapezoidal profiles, a maximum depth of 75 mm. and a unpropped construction process. In many practical situations today, use of different deck profiles, deeper decks and propped construction will probably provide a more economical and technically efficient solution. Extensive experimental research is needed to develop these provisions; reliable analytical tools exist but need calibration and verification data.
4. Use of high strength materials – the databases on which most current design provisions are based contain tests in which the materials used are well below the strength available in the market today. While use of such high strength materials seems obvious for the case of columns and elements carrying lateral forces, it is not clear if there is a practical need for developing design procedures for high strength materials for floor systems, where stiffness (i.e., vibrations and deflections) govern the design.
5. Improvement in serviceability – there are a number of practical situations where serviceability behaviour, and in particular floor cracking and deflections, can be improved by simple detailing provisions such as the addition of some minimum reinforcement over beam and girder lines. In the past, the specification and the commentary have stayed away from providing any requirements in this area, leaving it to the designer's discretion. In more complicate loading situations, such as occurs with the torsional forces in edge beams, some addition guidance is probably needed.
6. Coordination with other organizations – although AISC works with the Steel Deck Institute on developing consistent specifications, the same cannot be said for composite joists and trusses which fall within the purview of Steel Joist Institute. Additional

cooperation is addressing conflicting design requirement and improving guidance on serviceability for those members would be welcomed by many designers.

6.2 Composite Columns

1. Standardize column design provisions – because the column and beam-column design provisions for compact and non-compact composite columns were developed by different teams and at different times, and because the addition of Appendix 2 on high strength materials has added more equations, there is currently a lack of uniformity on the design equations for composite columns. It will be ideal ifs a "grand unified theory" (GUT, as in physics) could be developed and all design provisions be derivable from the same set of original equations and principles. This will also help engineers dealing with unsymmetrical sections and similar other configurations not explicitly covered by the specification today.
2. Definitions of stiffness for different analysis types – a major issue currently, particularly for seismic design, is that there is a not a clear definition of what effective flexural, shear and axial stiffnesses should be used for the different types of analysis (elastic, inelastic, pushover, time history, modal) permitted by the seismic design specification. Ideally, a set of principles could be derived to guide the engineer (i.e., when to use uncracked properties and when to use ultimate strength section properties).
3. Incorporation of new member types – there are number of types of members, such as double-skinned circular columns, concrete filled tubes with solid steel cores and encased composite columns with eccentrically placed steel sections, for which proposed design rules exist in the Japanese, Chinese, European and Australasian codes. Those provisions could be incorporated into the American codes easily if items (1) and (2) are completed.
4. Biaxial design – currently American design rules only address the strength of symmetrical columns along their principal axes and simplified interaction rules are adopted for biaxially loaded columns. Given that one of the most important advantages for composite columns over steel ones is the designer's ability to tailor their biaxial properties to match demand, provisions in the specification directly addressing the strength and stiffness values to be used in design will be extremely helpful.

6.3 Composite Walls

The introduction of composite walls into their new version of the specification signifies that AISC and TC5 will probably receive many questions and suggestions from designers over the next few years as they begin to utilize the provisions in both AISC 360 and 341. If experience is our guide, it is likely that there will be significant additions and modifications to the current provisions in the next cycle. Several talks at this conference address those issues directly, so they will not be elaborated upon herein.

6.4 Composite Connections

As noted earlier, composite connection design requires that both concrete and steel failure modes be considered. Thus, it is pertinent to ask if Chapter J addresses all the necessary failure criteria for composite connections, and more importantly, if there are interactions between failure modes that need to be considered.

7 CONCLUSIONS

This review article described the main technical changes that will appear in the upcoming edition of the American steel design specifications. The major changes include the addition of composite shear walls and provisions of shear capacity of filled sections, as well as the incorporation of a performance-based design approach to qualify new types of anchors for providing shear connection in composite beams. These changes are described in sections 3-5 of the paper. The rest of the paper presents the writer's personal view on some future issues, both from a procedural and technical standpoint, that will need to be addressed in the next editions of the specification.

REFERENCES

[1] AISC, *Specification for Structural Steel Buildings (ANSI/AISC 360-22)*, American Institute of Steel Construction, Chicago (IL), 2022 (unpublished).

[2] AISC, *Specification for Structural Steel Buildings (ANSI/AISC 360-16)*, American Institute of Steel Construction, Chicago (IL), 2016.

[3] ACI, *Building Code Requirements for Structural Concrete (ACI 318-19)*, American Concrete Institute, Farmington Hills (MI), 2019.

[4] ASTM, *A500/A500M-21: Standard Specification for Cold-Formed Welded and Seamless Carbon Steel Structural Tubing in Rounds and Shapes*, Philadelphia (PA), 2021.

[5] ASTM, *A1085/A1085M-15: Standard Specification for Cold-Formed Welded Carbon Steel Hollow Structural Sections (HSS)*, American Society for Testing and Materials, Philadelphia (PA), 2015.

[6] AISC, *Seismic Provisions for Structural Steel Buildings (ANSI/AISC 360-16)*, American Institute of Steel Construction, Chicago (IL), 2016.

[7] ASCE, *ASCE/SEI 7-16: Minimum Design Loads and Associated Criteria for Buildings and Other Structures*, American Society of Civil Engineers, Reston (VA), 2016.

[8] AISC, *Seismic Provisions for Structural Steel Buildings (ANSI/AISC 341-22)*, American Institute of Steel Construction, Chicago (IL), 2022 (unpublished).

[9] ASCE, *ASCE/SEI 41-17: Seismic Evaluation of Existing Buildings*, American Society of Civil Engineers, Reston (VA), 2017.

[10] AISC, *Seismic Provisions for Evaluation and Retrofit of Existing Structural Steel Buildings (ANSI/AISC 342-22)*, American Institute of Steel Construction, Chicago (IL), 2022 (unpublished).

[11] Morgen, B., Yang, H.J., Klemencic, R. and Hooper, J., *Rainer Square – Concrete Filled Composite Plate Shear Wall Tower Seattle*, Proceedings of the Eighth International Conference on Composite Construction in Steel and Concrete (Rassati et al., eds.), AISC/Virginia Tech, 2020.

[12] ASTM, *A108-18: Standard Specification for Cold-Formed Welded and Seamless Carbon Steel Structural Tubing in Rounds and Shapes*, Philadelphia (PA), 2018.

[13] AWS, D1.1/D1.1M:2015 Structural Welding Code—Steel, American Welding Society, Miami (FL), 2015.

[14] AISI, *AISI S923-20: Test Standard for Determining the Strength and Stiffness of Shear Connections of Composite Members*, American Iron and Steel Institute, Washington (DC), 2020.

[15] AISI, *AISI S924-20: Test Standard for Determining the Effective Flexural Stiffness of Composite Members*, American Iron and Steel Institute, Washington (DC), 2020.

[16] EN, *Eurocode 4: Design of composite steel and concrete structures - Part 1-1: General rules and rules for buildings*, European Committee for Standardization, Brussels, 2004.

[17] AISI, S100-16: North American Specification for the Design of Cold-Formed Steel Structural Members, American Iron and Steel Institute, Washington (DC), 2016.

9TH INTERNATIONAL CONFERENCE ON COMPOSITE CONSTRUCTION IN STEEL AND CONCRETE
M. Knobloch, U. Kuhlmann, W. Kurz, M. Schäfer. (Eds.)
Online, Germany, July 27-29, 2021

NEW EUROCODE 4 DESIGN RULES FOR COMPOSITE BEAMS WITH PRECAST CONCRETE SLABS

Stephen J. Hicks*, Matthias Braun, Zlatko Markovic*** and James Way******

* School of Engineering, University of Warwick, Coventry, UK
e-mails: Stephen.J.Hicks@warwick.ac.uk

** spannverbund Luxembourg S.A, Luxembourg
e-mail: matthias.braun@spannverbund.lu

*** University of Belgrade, Belgrade, Serbia
e-mail: zlatko@grf.bg.ac.rs

**** Kloeckner Metals UK, UK
e-mail: james.way@kloeckner.com

Keywords: Eurocode 4, Mandate M/515, codes and standards, composite beams, precast concrete.

Abstract. *As members of Project Team SC4.T5, the authors of this paper present the main outcomes from the work that is intended for implementation within the second generation of Eurocode 4. General design and application rules for composite design of precast concrete elements will be presented, with a particular focus on: the effective width of flanges using precast concrete elements for shear lag; design resistance to longitudinal shear; and the design resistance of headed stud connectors in the presence of precast hollow core slabs.*

1 INTRODUCTION

Steel construction has achieved a high market share in buildings and is often used in conjunction with various types of precast concrete floors. A high proportion of multi-storey steel frames use precast concrete floors, which are particularly suited to sectors such as hotels, residential buildings, and car parks. The synergy between the use of precast concrete slabs and steel structures is that they both come from a manufacturing technology rather than a site-based activity, and share the quality control, accuracy, and reliability of factory production.

Whilst industry guidance for designing composite beams with precast concrete slabs is available [1],[2], this form of construction is not currently covered within the scope of EN 1994-1-1 (Eurocode 4) [3]. As a consequence of this, there is a general need to develop harmonized design rules for Europe. This need was identified in the CEN/TC250 response to Mandate M/515 [4],[5] in the context of developing the second generation of the Eurocodes.

The CEN/TC250 work programme for developing the second generation of the Eurocodes commenced in 2015 and is split into four overlapping phases, which will conclude in 2022. For Eurocode 4 only a single task within the Phase 2 work was undertaken, entitled 'Development of rules covering shallow floor construction, and other flooring types using precast concrete elements'. Project Team SC4.T5 was made responsible for the task, which was broken down into the following work-packages:

- Development of design solutions for composite beams incorporating precast concrete units.
- Development of design solutions for shallow floor construction.
- Compilation of solutions into codified rules.
- Production of background documentation.

The main deliverable from the task was a new part for Eurocode 4, or proposed revisions to the existing EN 1994-1-1 and EN 1994-1-2 [6].

The membership of Project Team SC4.T5 consisted of the authors of the present paper (with the second author assigned the role of Project Team Leader). This paper presents an overview of the proposed design rules for composite beams incorporating precast concrete slabs within Eurocode 4. A companion paper presents the development of the rules for shallow floor construction [7].

2 COMPOSITE BEAMS WITH PRECAST CONCRETE SLABS

The rules are limited to use in buildings. The structural steel beam should be an open cross-section, which may be hot-rolled or welded. The precast concrete slabs may consist of solid composite slabs, together with hollow core slabs or composite hollow core slabs that are square or chamfered-ended (see Figure 1).

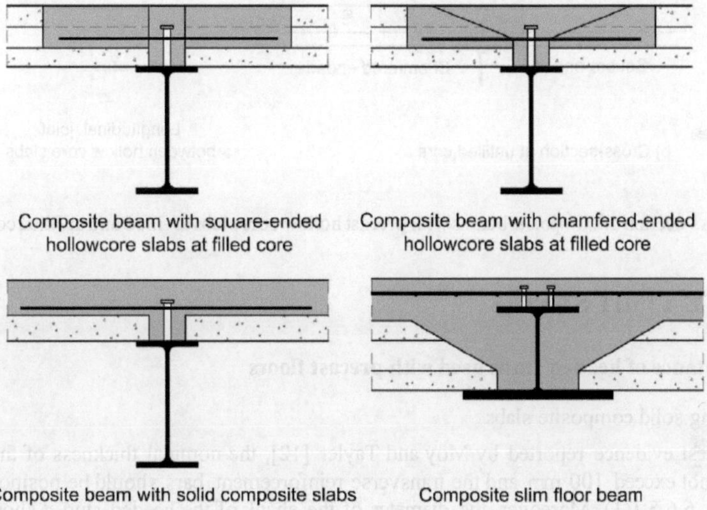

Figure 1. Typical cross-sections of composite beams with precast slabs

The steel beam and precast concrete slabs should be interconnected by mechanical shear connectors, to limit the longitudinal slip between concrete and steel and the separation of one component from the other. A particular design consideration is when precast hollow core slabs are used. In this special case the steel beams may provide 'flexible supports', which result in a reduction to the shear resistance of the slab from the development of transversal shear stresses [8],[9],[10]. For this special case, it is proposed that the designer makes reference to EN 1168 [11].

3 STRUCTURAL ANALYSIS

3.1 Effective width of flanges for shear lag

For floors with solid composite slabs, the effective width of the flanges in the composite beam may be determined using the existing rules given in EN 1994-1-1, 5.4.1.2. From full-scale composite beam tests [13] on floors with hollow core slabs or composite hollow core slabs, the effective width of the flanges may also be determined using EN 1994-1-1, but with the geometric width b_i taken as the smaller of the length of the infill within an open core ℓ_f (see Figure 2), or the distance from the outstand shear connector to a point mid-way between adjacent webs to the supporting steel beams, measured at mid-depth of the concrete flange [1],[2]. In addition, b_0 should be taken as the gap b_g between the ends of the hollow core slabs.

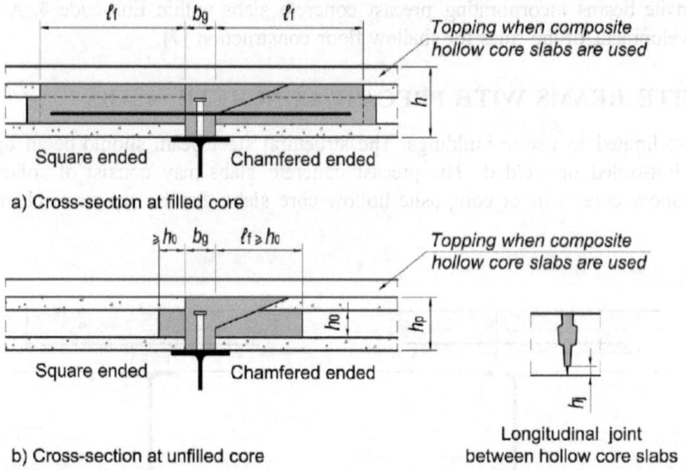

Figure 2. Cross-section of composite beams with precast hollow core slabs at filled and unfilled core positions

4 ULTIMATE LIMIT STATES

4.1 Design resistance of headed studs used with precast floors

4.1.1 Floors using solid composite slabs

From push test evidence reported by Moy and Tayler [12], the nominal thickness of the precast floor plate h_p should not exceed 100 mm and the transverse reinforcement bars should be positioned according to EN 1994-1-1, 6.6.5.1(1). Moreover, the diameter of the shank of the headed stud d should not be less than 19 mm. In these situations, the headed stud shear connector may be deemed to be ductile according to EN 1994-1-1, 6.6.1.1(5), and the design shear resistance should be taken as the resistance in a solid slab given by EN 1994-1-1, Equation (6.18) and (6.19).

For cases when the minimum dimensions given above are not satisfied, the design resistance of the shear connectors should be evaluated from specific tests according to EN 1994-1-1, Annex B.

4.1.2 Floors using precast hollow core slabs

From full-scale composite beam and companion push test evidence [13],[14], the nominal thickness of the hollow core slab excluding topping h_p should not exceed 265 mm and the transverse reinforcement bars should be positioned according to EN 1994-1-1, 6.6.5.1(1). The slabs may be square or chamfered-ended. For the latter case, the depth of the chamfer a_h should not exceed 85 mm, whilst its breadth a_b should not exceed 235 mm (see Figure 3). The diameter of the shank of the headed stud d should be within the range 19 mm $\leq d \leq$ 22 mm.

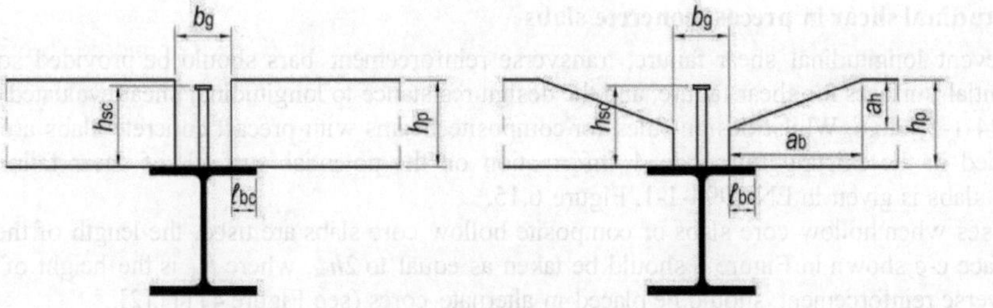

Figure 3. Detailing requirements for precast hollow core slabs.

In these situations, the design shear resistance of headed stud connectors should be taken as the resistance in a solid slab (calculated according to EN 1994-1-1, Equation (6.18) and (6.19)) and multiplied by the reduction factor k given by [15]:

$$k = \beta\varepsilon \leq 1.0 \tag{1}$$

with

$$\beta = \frac{b_g + 70}{140} \qquad \text{for } 70 \text{ mm } \geq b_g \geq 50 \text{ mm}$$

$$\varepsilon = \frac{\phi + 20}{40} \qquad \text{for } 20 \text{ mm } \geq \phi \geq 8 \text{ mm}$$

where b_g is the nominal distance between the ends of the hollow core slabs (mm) and ϕ is the diameter of the transverse reinforcing bars (mm).

Open cores should be provided to receive transverse reinforcement bars, which should be positioned below the heads of the stud shear connectors according to EN 1994-1-1, 6.6.5.1 (see Figure 4). Where the minimum dimensions above are not satisfied, the design shear resistance of the shear connectors should be evaluated from specific tests for precast hollow core slabs according to the proposed push test arrangement given in Annex B (see Section 7).

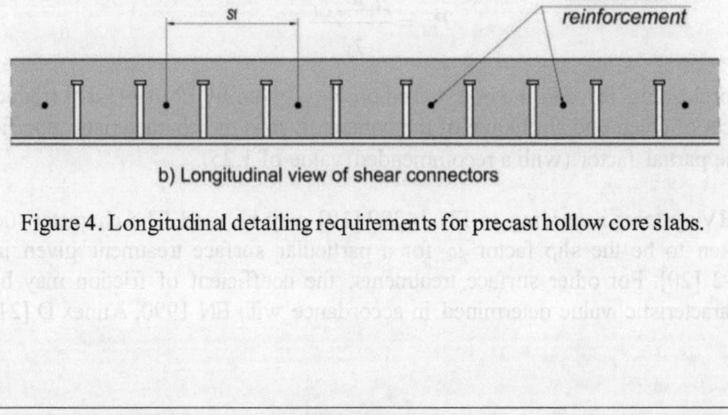

Figure 4. Longitudinal detailing requirements for precast hollow core slabs.

4.2 Longitudinal shear in precast concrete slabs

To prevent longitudinal shear failure, transverse reinforcement bars should be provided so that they cross potential surfaces for shear failure, and the design resistance to longitudinal shear evaluated according to EN 1994-1-1, 6.6.6. Whilst design rules for composite beams with precast concrete slabs are generally not provided in the existing Eurocode 4, information on the potential surfaces of shear failure in solid composite slabs is given in EN 1994-1-1, Figure 6.15.

For cases when hollow core slabs or composite hollow core slabs are used, the length of the potential shear surface c-c shown in Figure 5 should be taken as equal to $2h_{sc}$, where h_{sc} is the height of the studs. The transverse reinforcement should be placed in alternate cores (see Figure 4) [1],[2].

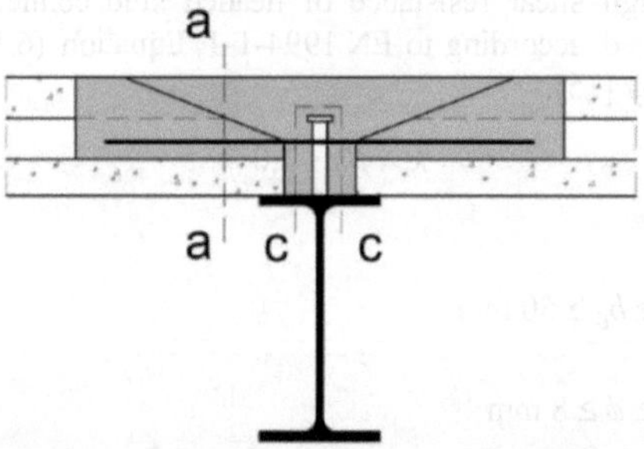

Figure 5. Typical potential surfaces of shear failure where precast hollow core slabs are used.

The transverse reinforcement should be provided in accordance with EN 1994-1-1, 6.6.6 and fully anchored within each open core, with the length of the concrete infill ℓ_f not being less than 500 mm (see Figure 2). As shown in Figure 2(b), cores without transverse reinforcement should be filled with concrete such that $\ell_f \geq h_0$.

5 BOLTED SHEAR CONNECTORS

In the interest of improving the sustainability of composite beams through encouraging re-use, design rules for bolted shear connectors are also proposed. Two types of bolted shear connector are considered *viz.* pre-stressed bolts and non-preloaded bolts.

5.1 Preloaded bolts

Given their use in Switzerland [16] and the UK [17], it is proposed that the earlier design rules previously given in ENV 1994-1-1 [18] for class 8.8 bolts are reintroduced. At the ultimate limit state, the design shear resistance per bolt should be taken as:

$$P_{Rd} = \frac{\mu_k F_{pr,Cd}}{\gamma_V} \tag{2}$$

where $F_{pr,Cd}$ is the preloading force in the bolt, based on $F_{p,Cd}$ given by EN 1993-1-8 (reduced to take account of the effects of creep and shrinkage of the concrete), μ_k is the characteristic coefficient of friction and γ_V is the partial factor (with a recommended value of 1.25).

Both HR and HV systems according to EN 14399 [19] may be used. The characteristic coefficient of friction may be taken to be the slip factor μ_k for a particular surface treatment given in the execution standard EN 1090-2 [20]. For other surface treatments, the coefficient of friction may be based on test results, with the characteristic value determined in accordance with EN 1990, Annex D [21].

The reduction in the preloading force in the bolt due to creep and shrinkage of the concrete should either be determined by long-term tests, or should be assumed to be not less than 40% of $F_{p,Cd}$.

5.2 Non-preloaded bolts

Non-preloaded bolts with embedded nuts may also be used as shear connectors. Preloaded bolts according to EN 14399 [19] may be used, provided that the preloading force is not higher than $F_{p,C}$ according to EN 1993-1-8 [22], and that the preloading is only applied within the region b between the embedded and outer nut (see Figure 6).

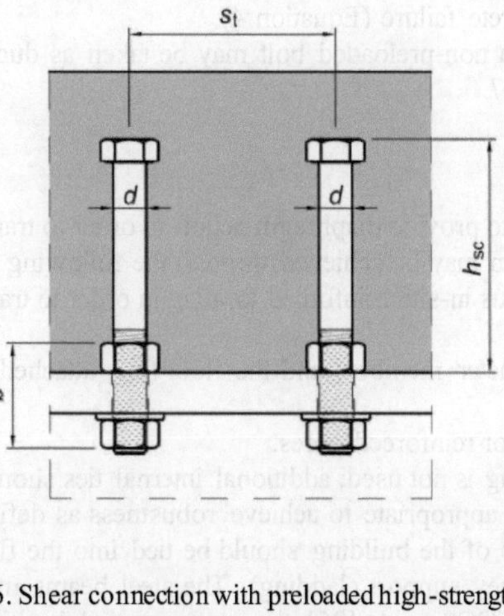

Figure 6. Shear connection with preloaded high-strength bolts

The design shear resistance of a non-preloaded bolt, with one or two embedded nuts, should be determined from [23]:

$$P_{b,Rd} = \frac{\alpha_b A_s f_{ub}}{\gamma_V} \quad (3)$$

or

$$P_{c,Rd} = \frac{55\alpha_c d^{1.9}\left(f_{ck}\dfrac{h_{sc}}{d}\right)^{0.4} + 22000}{\gamma_V} \quad (4)$$

whichever is smaller, with:

$$\alpha_b = 0.6\left(\frac{34}{d}\right)^{0.23} \quad (3)$$

$$\alpha_c = \frac{22.5}{d+3} \le 1.0 \quad (4)$$

where d is the bolt diameter in mm (12 mm $\le d \le$ 24 mm), A_s is the tensile stress area of the bolt in mm² ($A_s \approx 0.785\pi d^2/4$), h_{sc} is the overall nominal height of the non-preloaded bolted shear connector above the flange in mm, f_{ub} is the ultimate tensile strength in N/mm² for bolt classes 8.8 and 10.9, f_{ck} is the characteristic cylinder compressive strength of concrete at the age considered, of density not less than

1750 kg/m³ that is cast in-situ around the shear connector and γ_V is the partial factor (with a recommended value of 1.25).

5.2.1 Slip capacity

The characteristic slip capacity for non-preloaded bolts may be calculated to be:

$$\delta_{uk} = 0.56e^{2.45P_{b,Rd}/P_{c,Rd}} \tag{5}$$

where $P_{b,Rd}$ is the design value of shear resistance from bolt failure (Equation (3)), and $P_{c,Rd}$ is the design value of shear resistance to concrete failure (Equation 4).

According to Equation (7), a non-preloaded bolt may be taken as ductile according to EN 1994-1-1, 6.6.1.1(5), when $P_{b,Rd}/P_{c,Rd} > 0.97$.

6 TYING SYSTEMS

A floor is normally required to provide diaphragm action in order to transfer wind loads to braced walls or concrete core walls. This action may be achieved through the following measures:

- Provision of a continuous in-situ reinforced topping in order to transfer the in-plane forces in both orthogonal directions.
- Ties between the perimeter members and the floor (e.g. attached by headed stud connector and looped bars, etc.).
- Ties to the shear walls or reinforced cores.
- Where an in-situ topping is not used, additional internal ties should be provided.

The same measures are also appropriate to achieve robustness as defined by EN 1991-1-7 [24]. The steel beams around the perimeter of the building should be tied into the floor plate for diaphragm action, and for torsional resistance (if they support cladding). The steel beams may also be considered to act as peripheral ties. From the IPHA/ECCS guide [25], the recommended location of these ties for hollow core floors is shown in Figure 7.

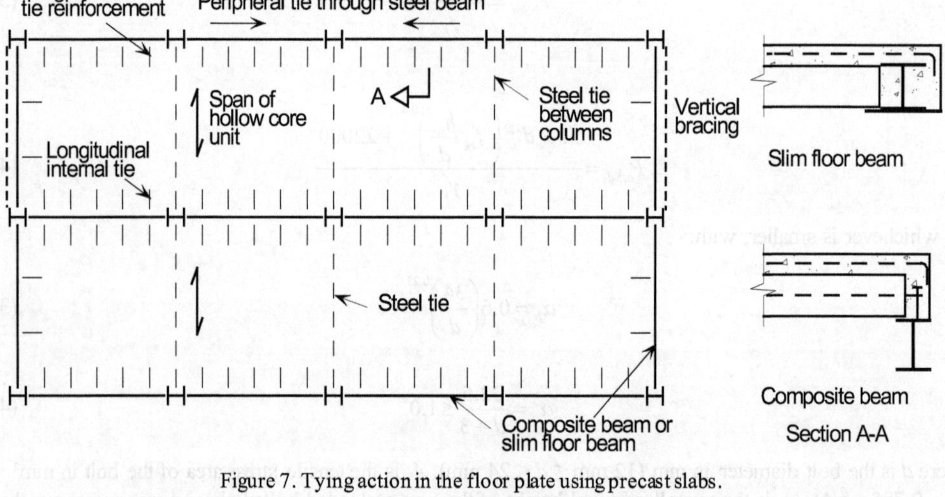

Figure 7. Tying action in the floor plate using precast slabs.

7 TESTS ON SHEAR CONNECTORS IN HOLLOW CORE SLABS

To ensure that the experimental resistance of shear connectors in hollow core slabs is determined consistently, a specific test is proposed for inclusion within EN 1994-1-1, Annex B (see Figure 8). As can be seen from Figure 8, due to the practicalities of testing shear connectors with hollow core slabs using the standard push specimen, a one-sided arrangement is proposed. This arrangement has been successfully used by Lam *et al.* [14], where the results from tests using this type of specimen formed the basis of Equation (1).

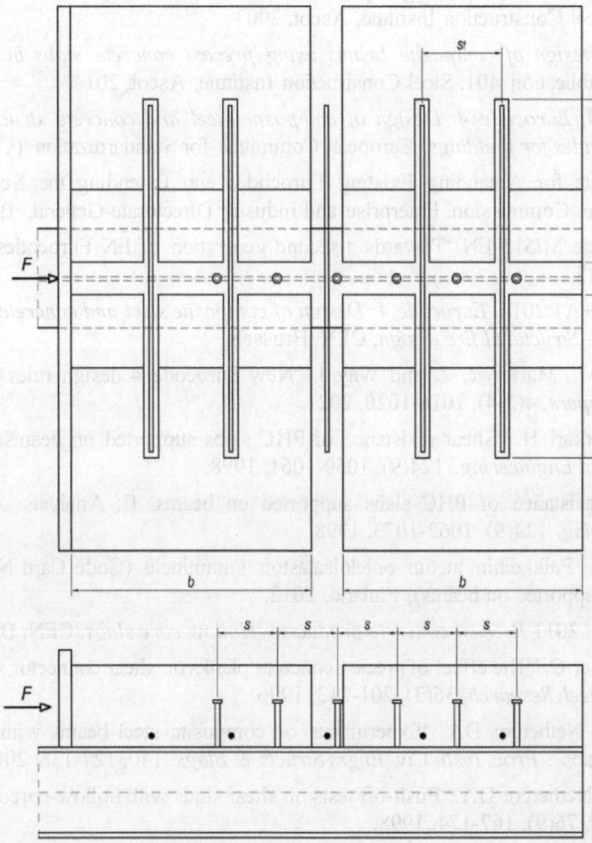

Figure 8. Test specimen for headed stud connectors within hollow core slabs or composite hollow core slabs.

8 CONCLUSIONS

This paper presents the proposed design rules for composite beams with precast concrete slabs, which were developed by Project Team SC4.T5 for inclusion within the second generation of Eurocode 4. The rules are intended to complement the existing industry guidance that has been used widely across Europe. The proposed design rules for composite beams using precast concrete slabs are currently being circulated for comment to the 34 National Standardization Bodies (NSBs) that make up the CEN membership and, as a consequence of this, may be subject to change. Any views expressed in this paper may not necessarily reflect those of the other members of CEN/TC250 Subcommittee 4 (CEN/TC250/SC4).

ACKNOWLEDGEMENTS

The authors acknowledge with thanks the funding received from the European Committee for Standardization (CEN) through the Grant Agreement SA/CEN/GROW/EFTA/515/2016-02 for Project Team SC4.T5 of CEN/TC250 work programme.

REFERENCES

[1] Hicks, S.J. and Lawson, R.M., *Design of composite Beams using precast concrete slabs*. SCI Publication 287, Steel Construction Institute, Ascot, 2003

[2] Couchman, G.H. *Design of composite beams using precast concrete slabs in accordance with Eurocode 4*. SCI Publication 401, Steel Construction Institute, Ascot, 2014

[3] EN 1994-1-1: 2004, *Eurocode 4: Design of composite steel and concrete structures - Part 1-1: General rules and rules for buildings*, European Committee for Standardization (CEN), Brussels.

[4] M/515 EN. Mandate for Amending Existing Eurocodes and Extending the Scope of Structural Eurocodes. European Commission, Enterprise and Industry Directorate-General, Brussels, 2012

[5] Response to Mandate M/515 EN 'Towards a second generation of EN Eurocodes', CEN/TC250 – N993, Brussels, 2013

[6] EN 1994-1-2: 2005+A1:2014. *Eurocode 4: Design of composite steel and concrete structures - Part 1-2: General rules – Structural fire design*, CEN, Brussels.

[7] Hicks S.J., Braun M., Markovic, Z. and Way J. "New Eurocode 4 design rules for shallow floor construction", *ce/papers*, 4(2-4), 1016-1020, 2021.

[8] Pajari M. and Koukkari H. "Shear resistance of PHC slabs supported on beams. I: Tests", *ASCE Journal of Structural Engineering*. 124(9), 1050-1061, 1998.

[9] Pajari, M. Shear resistance of PHC slabs supported on beams. II: Analysis. *ASCE Journal of Structural Engineering*, 124(9), 1062-1073, 1998.

[10] Normikortti N:o 18. Palkkeihin tuetun ontelolaataston suunnittelu (Code Card No. 18: Design of hollow core slabs supported on beams), Finland, 2012.

[11] EN 1168: 2005+A3: 2011 *Precast concrete products – Hollow core slabs*. CEN, Brussels.

[12] Moy S.S.J. and Tayler C. "The effect of precast concrete planks on shear connector strength", *Journal of Constructional Steel Research*, 36(3), 201-213, 1996.

[13] Lam D, Elliott KS, Nethercot DA. "Experiments on composite steel beams with precast concrete hollow core floor slabs", *Proc. Instn Civ. Engrs Structs & Bldgs*, 140, 127-138, 2000.

[14] Lam D, Elliott KS, Nethercot DA. "Push-off tests on shear studs with hollow-cored floor slabs", *The Structural Engineer*, 76(9), 167-174, 1998.

[15] Lam D. "Capacities of headed stud shear connectors in composite steel beams with precast hollowcore slabs", *Journal of Constructional Steel Research*, 63(9), 1160-1174, 2007.

[16] SIA 269/4: 2011 *Maintenance des structures porteuses - Structures mixtes acier-béton*. Swiss Society of Engineers and Architects (SIA), Zurich

[17] Design Manual for Roads and Bridges: Highway Structures & Bridges Inspection & Assessment: CS 457: *The assessment of composite highway bridges and structures (formerly BD 61/10)*, Highways England, 2019

[18] ENV 1994-1-1: 1992. *Eurocode 4: Design of composite steel and concrete structures Part 1-1: General rules and rules for buildings*. CEN, Brussels.

[19] EN 14399-1: 2005 *et seq. High-strength structural bolting assemblies for preloading*. CEN, Brussels

[20] EN 1090-2: 2008+A1: 2011 *Execution of steel structures and aluminium structures Part 2: Technical requirements for steel structures*. CEN, Brussels.

[21] EN 1990: 2002+A1: 2005 *Eurocode – Basis of structural design*. CEN, Brussels

[22] EN 1993-1-8: 2005. *Eurocode 3: Design of steel structures Part 1-8: Design of joints*. CEN, Brussels

[23] Pavlović M.S. *Resistance of bolted shear connectors in prefabricated steel-concrete composite decks*. PhD Thesis, University of Belgrade, Belgrade, 2013

[24] EN 1991-1-7: 2006+A1: 2014. *Eurocode 1: Actions on structures – Part 1-7: General actions – Accidental actions*, CEN, Brussels

[25] Guidelines for the application of prestressed hollow core slabs supported on built-in beams, ECCS No.103, European Convention for Constructional Steelwork/International Prestressed Hollowcore Association (ECCS/IPHA), Brussels/Tamworth, 1998

9TH INTERNATIONAL CONFERENCE ON COMPOSITE CONSTRUCTION IN STEEL AND CONCRETE
M. Knobloch, U. Kuhlmann, W. Kurz, M. Schäfer. (Eds.)
Online, Germany, July 27-29, 2021

POTENTIALLY UNSAFE STRUCTURAL CONSEQUENCES IN THE DESIGN OF COMPOSITE BEAMS SHEAR CONNECTORS

(A look at some of the Eurocode 4-1-1 design backgrounds)

Dr. Ioan Diacu
Omega Structure® - Paris – France

Keywords: Structural design, Structural safety, Composite steel-concrete beams, Shear connectors, Eurocode 4-1-1, Unsafe design resistance.

Abstract. *The paper investigates the design of composite - steel-concrete - beams, with emphasis on the shear connectors, as specified in the Eurocode 4-1. A structural security inaccuracy is identified and analyzed. Improvements for the future design code are proposed.*

1 SCOPE

The successful use of composite beams is fundamentally dependent on the design accuracy of the longitudinal shear connection between the concrete slab and the steel beam.

The ductile shear connection concept, together with the critical length idea, used in the current Eurocode 4-1-1[1], as well as in its second-generation code project, can produce some structural unsafety consequences that deserves a detailed investigation.

A principle elementary analysis is preliminarily discussed, in order to outline and define the problem, followed by a specific assessment and practical conclusions and proposals.

The paper points mainly on the § 6.2.1.2 and § 6.2.1.3 of [1] – positive bending, plastic analysis. Common notation used and explained in the Eurocodes are used whenever possible.

The paper is not primarily intended for direct engineering practice; but rather as a discussion un the future development of Eurocodes.

2 INTRODUCTION AND PRELIMINARY MATTERS

a) The equilibrium proof is a major test for the design analysis coherence.

In the old EN 1994-1-1 code [1] as well as in the new project, it seems that the equilibrium exigences are not minded enough comprehensively. An investigation is carried out in this paper, in several analysis steps.

b) According to the Structural Mechanics theories, in the case of a beam submitted to bending and shear (Fig.1), the longitudinal shear force ($\mathbf{F_{LS}}$) is proportional to the transversal shear force (**V**) – duality of shear stresses. The shear force is the gradient of the bending moment diagram. This means that, for a rigid connection hypothesis, only in the case of concentrated loads, the constant connector's distancing is fitted (Fig.2).

c) According to the classical (elastic) theory, the connections (welded, bolted or riveted) between the flanges and the web, for an **I** section steel beam, are sized with respect to the real shear forces (Fig. 2) and can be settled with a variable distancing on the beam length or can be conservatively sized at the maximum shear. The beam is often divided in some zones with uniform connectors spacings.

d) For simplicity, there is considered a **hypothetical case** of a **simply supported I section beam, uniformly loaded** (Fig. 4), and there is assumed that only the upper flange of this **I** section (Fig.3) is designed with a ductile connection to the lower **T** section. In this particular case, using an elastic analysis, the total longitudinal shear force ($\mathbf{A_{1f} \cdot f_y}$) is uniformly transferred over the critical length (**L/2**) (Fig.4).

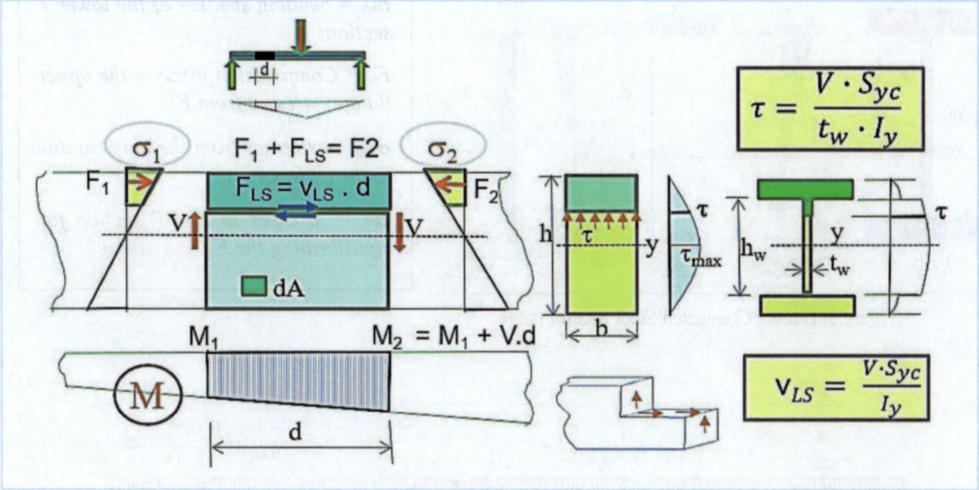

$$\Sigma F_{LS} = t_w \cdot \tau \cdot d = \int_{z_i}^{z_{max}} \frac{M_y + V \cdot d}{I_y} \cdot z \cdot dA - \int_{z_i}^{z_{max}} \frac{M_y}{I_y} \cdot z \cdot dA = \frac{V \cdot d}{I_y} \cdot \int_{z_i}^{z_{max}} z \cdot dA = \frac{V \cdot d}{I_y} \cdot S_{yc}$$

Figure 1. Elastic Shear Stresses in Beams

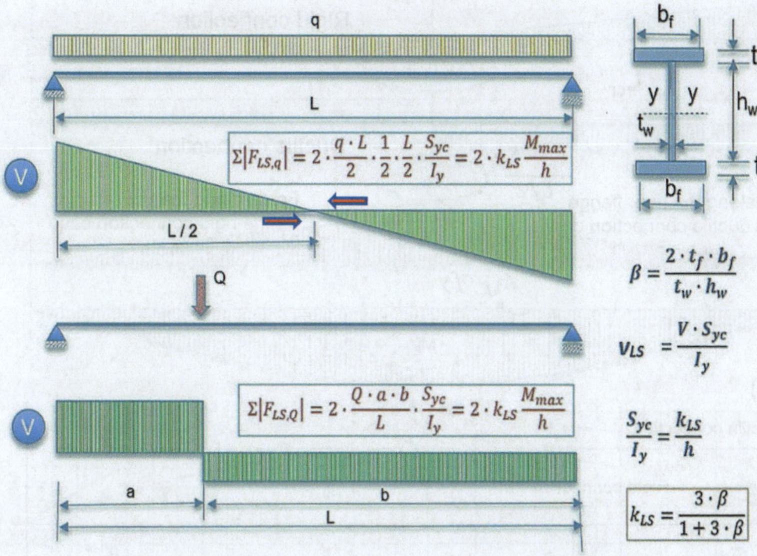

Fig. 2. Simply Supported Beam . Longitudinal Shear Forces.

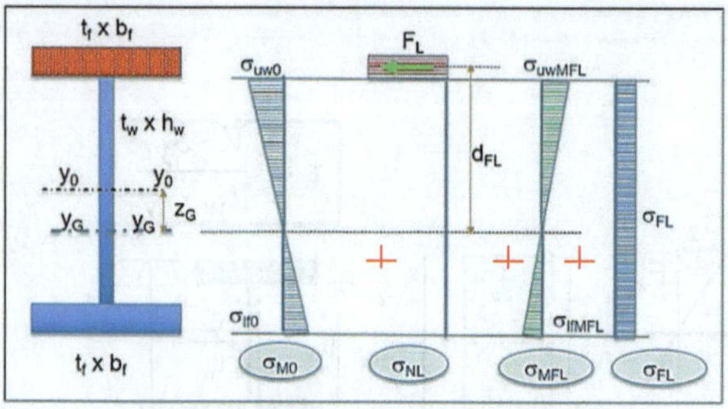

***σ_{M0}** = bending stresses of the lower T section;*

***F_L** = Compression force in the upper flange; stresses from F_L;*

***σ_{MFL}** = stresses from the moment due to the F_L force;*

***σ_{FL}** = stresses in the T section for equilibrating the F_L;*

Figure 3. Ductile Connected Steel Beam Flange

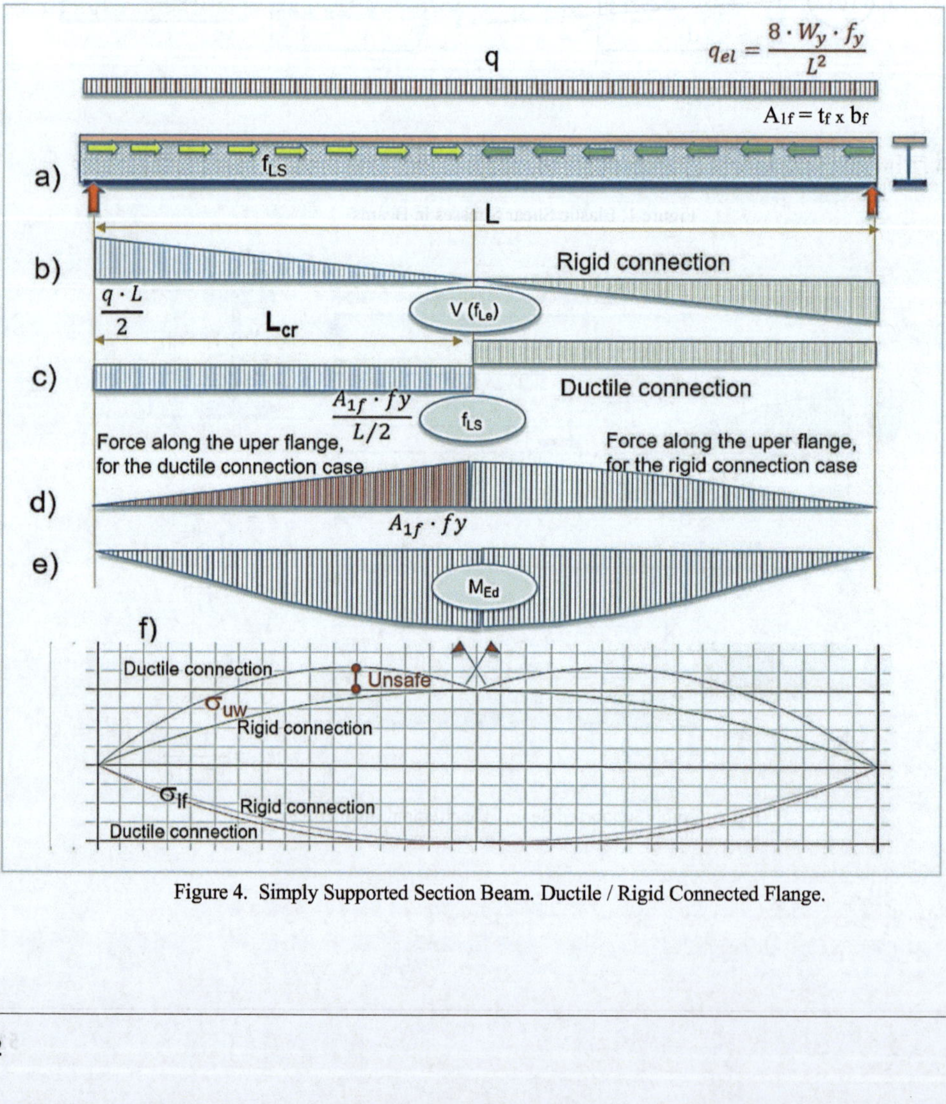

Figure 4. Simply Supported Section Beam. Ductile / Rigid Connected Flange.

e) In this context, the hypothesis that the whole section is adequately active all over the beam's length is no longer valid. Even if the beam has a constant section, because the longitudinal shear force - $\mathbf{f_{LS}}$ - is constant instead of being linearly variable, the maximum $\boldsymbol{\sigma}$ stresses are no longer in the same place and don't have the same values as in the case of a full rigid connection (Fig.4; Fig.5)

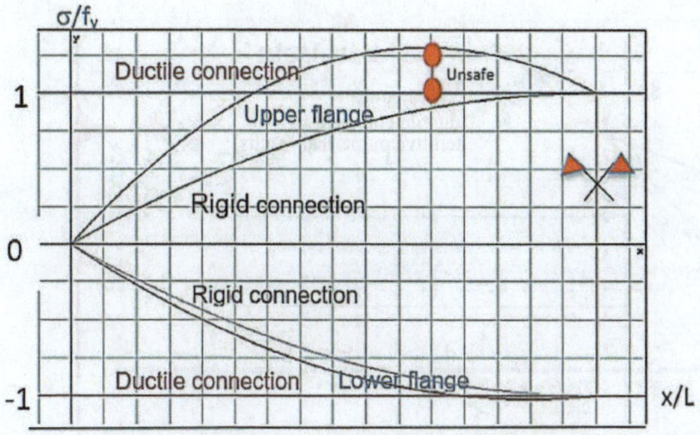

Fig. 5. Unsafe Effects of Ductile Connection

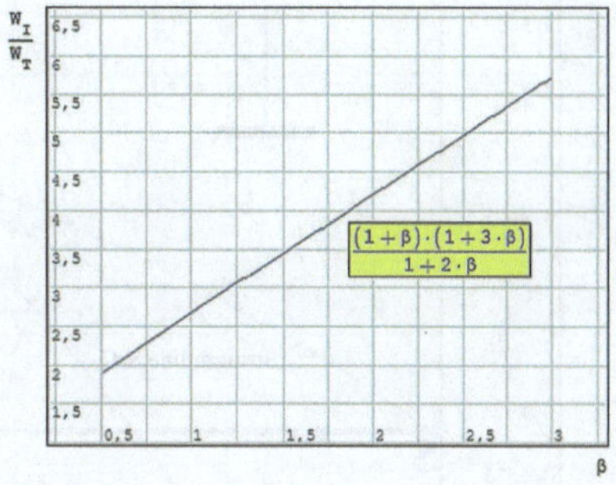

Figure 6. I and T Sections Resistance Modulus

f) In this theoretical case (Fig.4), using an elastic analysis (Fig. 3), the stress difference between the rigid and the ductile connection is high; the two limit cases are a **T** section versus an **I** section, with strength ratios varying between 2 and 6 (Fig. 6), whereas for the composite beams, the ratio is usually less than 4 (currently, less than 3); when the strength ratio is over 2,5 (in plastic) some additional (unknown) checks are required by the EN 1994-1-1 [1] .

3 THE CASE OF STEEL – CONCRETE COMPOSITE BEAMS.

a) In the case of a composite beam, the plastic (positive) bending resistance is heavily dependent on the connector's behavior. However, the connectors effect on the bending resistance is practically inside the range between the resistance of the steel section alone – **S** point on the Fig. 7 and the resistance of the fully active composite section– **C** point.

b) In the case of composite beams, aiming to place the connectors with a uniform spacing, over a critical length (between two successive critical cross-sections [1]), the introduction of the concept of ductile connectors leads to the same load on each connector - its full resistance. In order to establish the necessary number of connectors on a critical length, the total maximal longitudinal shear (considered as the minimum between the compression strength of the concrete slab and the traction strength of the steel beam) is divided by the minimum strength of a ductile connector. This strategy is widely found in the literature [2; 3; 4]. The objective is implicitly attainable for the case of beams with constant shear force along the critical length (Fig.2; concentrated force case), without the exigency of ductility

c) By adopting the ductile connector's strategy, in the case of a beam having a non-constant shear force along a critical length (e.g.: for the case of a simply supported beam, uniformly loaded), the available section bending resistance is diminished, following the proportionality (Fig. 2) between the transversal shear force - **V** - and the longitudinal shear force – $\mathbf{F_{LS}}$. In fact, in the frame of these assumptions, the hypothesis of plane sections conservation after the deformation is no more respected. In such case, the compressive forces on the concrete slab increase only linearly over the critical length, instead of parabolically, as for the $\mathbf{M_{Ed}}$ diagram.

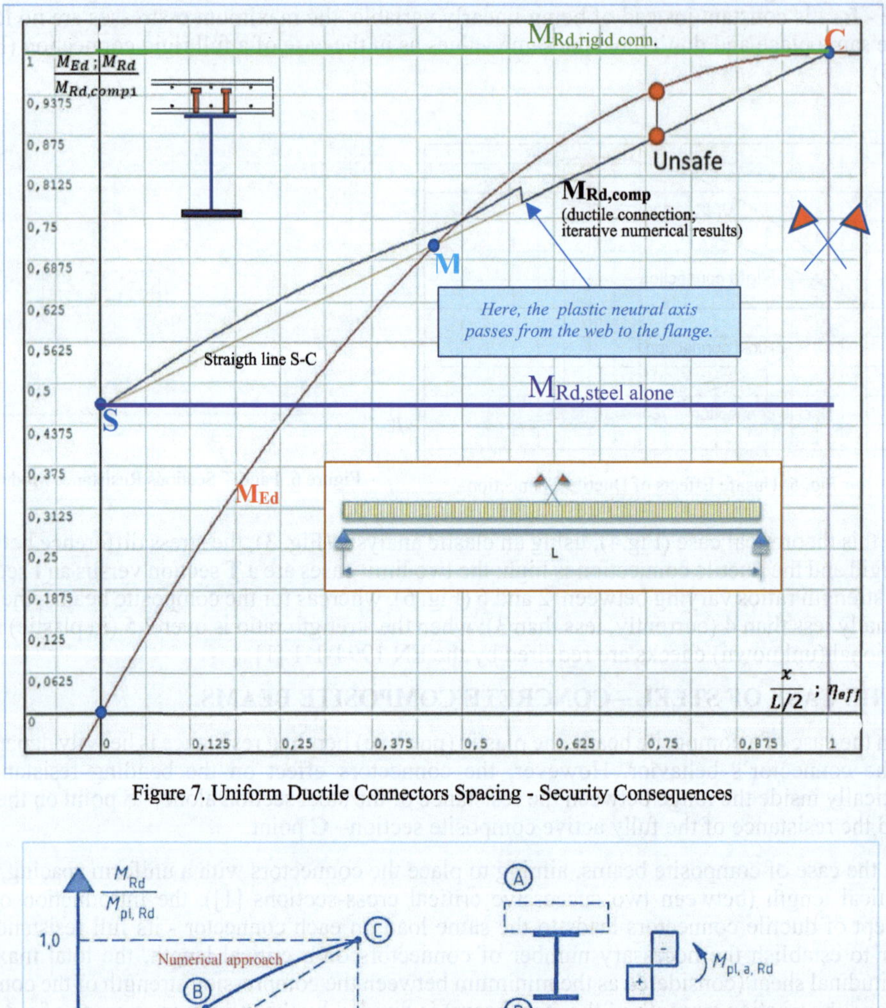

Figure 7. Uniform Ductile Connectors Spacing - Security Consequences

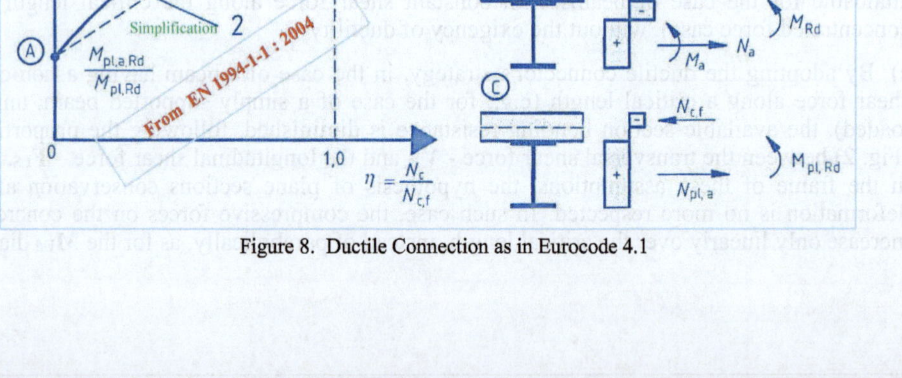

Figure 8. Ductile Connection as in Eurocode 4.1

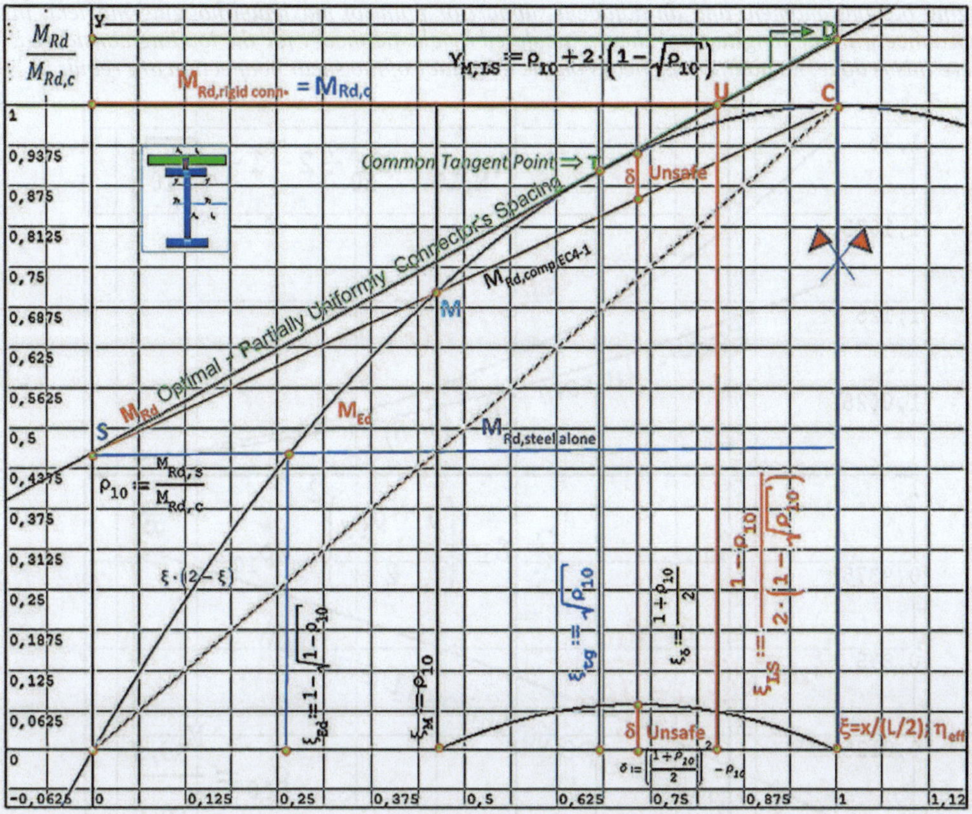

Figure 9. Simplified Figure 7 Information

d) The Eurocode assumption ("hope") (Fig. 8) is that all over the critical length, $\mathbf{M_{Rd}} \geq \mathbf{M_{Ed}}$, if the full $\boldsymbol{\eta}$ connection is achieved, point **C** in Fig. 7 and Fig. 9, where $\mathbf{M_{Rd,C}} = \mathbf{M_{Ed,C}}$. This assumption is demonstrated in Fig. 7 as being false. The quasi-straight line of $\mathbf{M_{Rd}}$ (deducted by iterative numerical procedures, built with [5] as [6]) and the parabola of the M_{Ed} are intersected, after the **S** point, at another two points: firstly, in the **M** point (Fig. 7) and secondly in the **C** point. Between the **M** and **C** points (on an enough long distance) (Fig. 9): $\mathbf{M_{Ed}} > \mathbf{M_{Rd}}$

e) In the context of a more accurate analysis, the consequent question is: "if there is a security discrepancy, is it a significant one and, if so - what has to be done in that case?".

f) Regarding the total number of connectors for a full connection, if EN 1994-1-1 is clear enough, concerning the connectors spacing, the specifications given at the paragraph 6.6.1.3 are ambiguous enough. The use of the "critical length" for connectors uniformly spacing is suggested:

"(3) Ductile connectors may be spaced uniformly over a length between adjacent critical cross sections..."

At 6.6.1.3 (5) in the EN 1994-1-1 it is mentioned this idea:

"*(5) The required number of shear connectors* ***may*** *be distributed between a point of maximum sagging bending moment and an adjacent support or point of maximum hogging moment, in accordance with the longitudinal shear calculated by elastic theory for the loading considered. Where this is done, no additional checks on the adequacy of the shear connection are required.*"

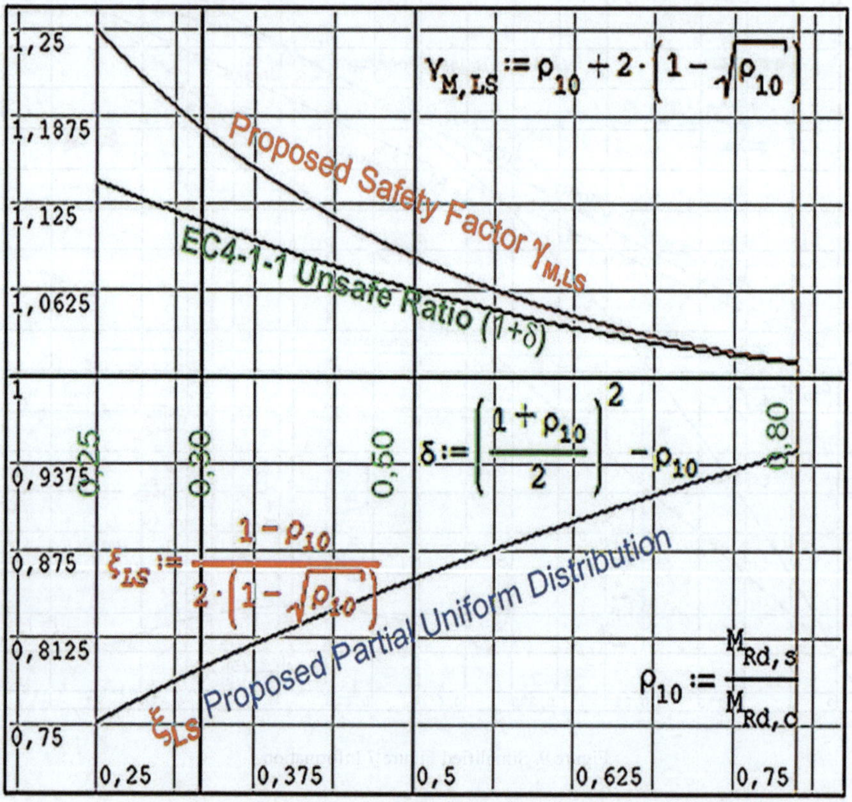

Figure 10. Maximum EC4-1 Unsafe Ratio Assessment. Proposed Safety Factor

g) It is obvious that the Eurocode 4 keeps a certain vagueness ("may") on the correct connectors' distribution. The "allusion way" for specifying generates uncertainties and hence an uncontrolled practice.

h) If the shear connection is uniform, the compression force in the slab is linearly augmented, as if the η coefficient varies linearly over the critical length (Fig. 7) and (Fig. 9). If the connection is rigid, the full resistance is available all over the beam's length; the force in the slab is augmented parabolically - with a linearly variable shear force as gradient. In the case of a ductile connection (Fig. 7), there is an unsafe zone on the critical length (between the **M** and **C** points), where the $\mathbf{M_{Ed}} > \mathbf{M_{Rd}}$, although at the beam's center $\mathbf{M_{Ed,C}} = \mathbf{M_{Rd,C}}$.

i) In the case of an **I** steel section alone (Fig.5 and 6), the ratio between the **I** section and the **T** steel section (resulted from an inactive flange) resistances, according to the elastic analysis, is a large value ($\mathbf{M_{Rd,I}/M_{Rd,T} = 2\text{-}6}$), the mixity steel concrete plays a lesser contribution compared to the simply steel beam alone ($\mathbf{M_{Rd,comp}/M_{Rd,steel} = 1{,}2\text{-}3}$). The safety consequences for the composite beam case are approximately evaluated in Fig. 10, based on the approximation from

Fig. 9; there are also proposed as well a general solution for the partial distribution of connectors as an equivalent additional security coefficient if the Eurocode's solution is kept.

j) For the case of combined loads (concentrated and distributed), (Fig. 11), the idea to size the longitudinal shear connection to a maximum value over each critical length is abusive. Fig. 11 shows that between $\mathbf{l_{cr,a}}$ and $\mathbf{l_{cr,b}}$ there is an important difference on the total longitudinal shear forces to be transferred. Probably, it could be a better idea to adapt the longitudinal shear connection dimensioned to the total real shear diagram proportions on each critical length, as mentioned in the Eurocode. Such a distribution could be as in Fig. 11 and 12.

The ratio between the contribution of the concentrated loads versus the distributed loads in the moments diagram values is important to be also taken into account.

A connector's distribution correction, as for the case of distributed forces, must also be considered.

k) Another way for designing the composite steel-concrete beams is by concentrating the entire forces transfer between the concrete slab and the steel beam (Fig. 13), near the ends supports. For information, the concrete slab longitudinal shortening is about 1…2 mm/m.

$$\Sigma F_{LS,Lcr,Ed,i} = \frac{V_{l,Ed,i} + V_{r,Ed,i}}{2} \cdot \frac{k_{LS}}{h} \cdot L_{cr,i}$$

$$n_{icr,conn} = \frac{(V_{l,Ed,i} + V_{r,Ed,i}) \cdot L_{cr,i}}{\sum_{k=1}^{n}((V_{l,Ed,k} + V_{r,Ed,k}) \cdot L_{cr,k})}$$

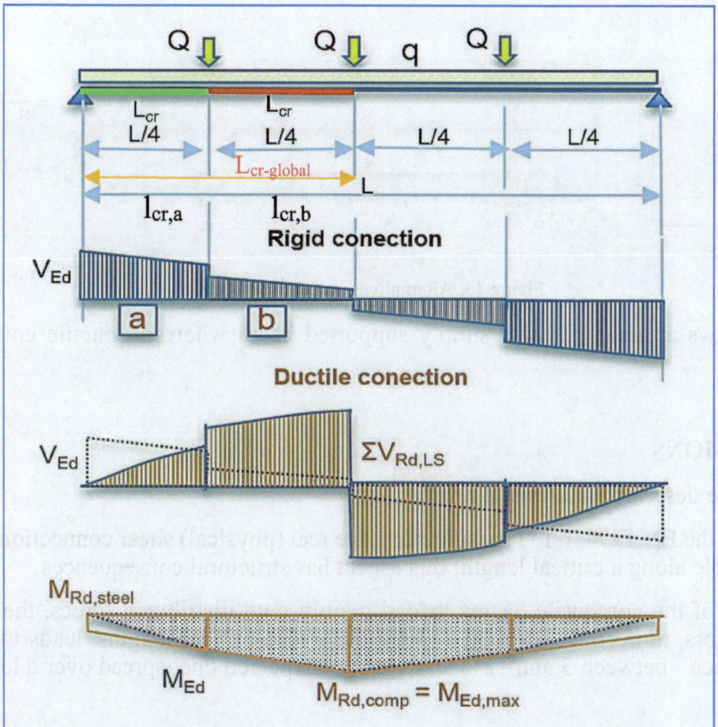

Fig. 11 Example of a Simply Supported Beam with Complex Loading

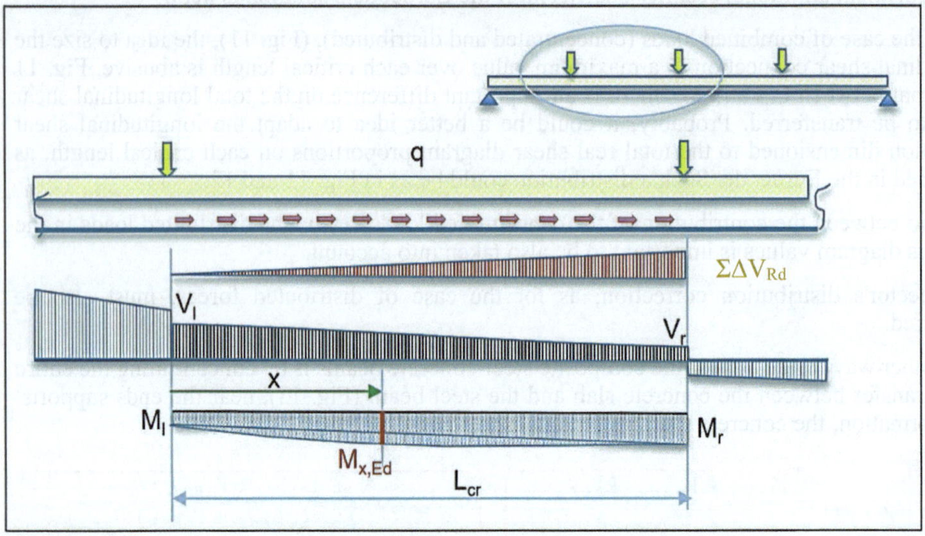

Figure 12. A Critical Length from Fig. 16

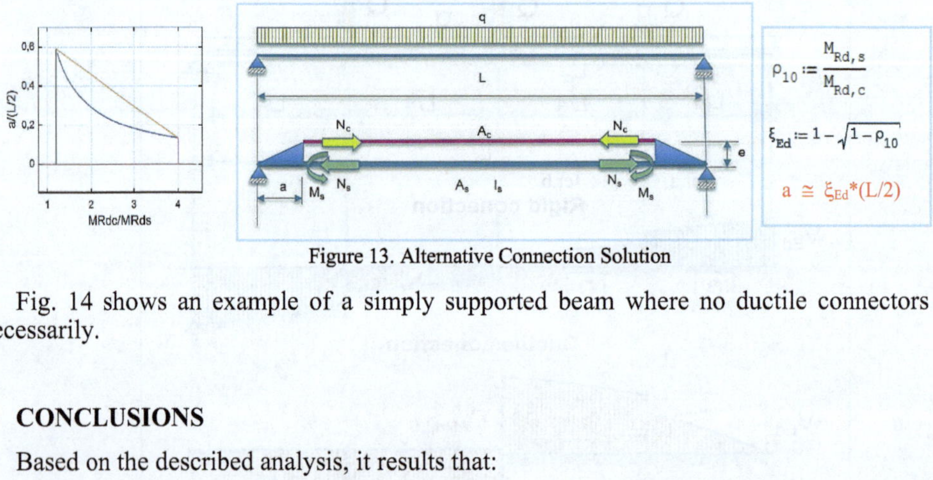

Figure 13. Alternative Connection Solution

l) Fig. 14 shows an example of a simply supported beam where no ductile connectors are necessarily.

4 CONCLUSIONS

Based on the described analysis, it results that:

Contrary to the EN 1994-1-1 [1] assumption, the real (physical) shear connection coefficient η is also variable along a critical length; this aspect has structural consequences.

In the case of the composite beams loaded mainly with distributed forces, the concept of ductile connectors, uniformly distributed along the whole critical lengths, leads to **an unsafe design resistance** - between 3 and 12% less than the expected one, spread over a long distance (Fig. 9 and 10).

The distribution of the entire necessary number of connectors, only on the first zone (75...80%) of each critical length, is able to compensate the aforementioned lack of security. Some detailed information is in the Fig.11.

Another solution could be the use of a uniform connectors distribution over the critical length, as in [1], accompanied by the use of a supplementary security factor - $\gamma_{M,SL}$ (Fig.11)

For the complex loadings, the uniform connector's distribution, between two successive points of null on the shear forces diagram, over the first 80% of each critical length, proportionally to the total longitudinal shear force on each critical length, obtained from an elastic analysis, is a reasonable secure strategy.

The aforementioned proposals can be easily implemented in the future second generation of the Eurocode 4-1.

ACKNOWLEDGMENTS

My full gratitude to Dr. Robert Statica, New Jersey Institute of Technology, Newark, NJ, USA, for the exigent and constructive lecture of the paper.

Thanks to the Late Prof. Nicolae Patraniche (UTCB) for his important support during the past hard times.

Thanks to Eng. Hema Aliche – Afnor Compétences, France - for the didactic opportunities.

My gratefulness to Maria - my beloved wife. My gratefulness to my Late Parents.

REFERENCES

[1] Eurocode 4-1 (EN 1994-1-1): Design of composite steel and concrete structures - Part 1-1: General rules and rules for buildings. CEN/TC250/SC4. 2005.

[2] R-P Johnson – Designers' Guide to Eurocode 4: Design of Composite Steel and Concrete Structures. Thomas Telford - ICE Publishing 2012.

[3] Dujmović / Androić / Lukačević - Composite Structures according to Eurocode 4. Worked Exemples. 2015 Wilhelm Ernst & Sohn.

[4] CSTB - France : Guides Eurocodes : Constructions mixtes acier-béton. Vérification des poutres de planchers mixtes acier-béton. Par : Jean-Marie Aribert ; Hoàng-Tùng Vu ; Mohammed Hjiaj ; Samy Guezouli . 2011.

[5] SMath Studio : https://en.smath.com/view/SMathStudio/summary

[6] Ioan Diacu; Ménad Chenaf - Le Compagnon de l'Eurocode 3 – CSTB Editions, France, 2015

[7] Diacu, I. (2021) Potential unsafe structural consequences in the design of composite beam shear connectors – A look at the background to some of the Eurocode 4-1-1 design rules Steel Construction. https://doi.org/10.1002/stco.202000057

COMPOSITE BEAMS

9TH INTERNATIONAL CONFERENCE ON COMPOSITE CONSTRUCTION IN STEEL AND CONCRETE
M. Knobloch, U. Kuhlmann, W. Kurz, M. Schäfer. (Eds.)
Online, Germany, July 27-29, 2021

DEFORMATION CAPACITY AND DUCTILITY OF SHEAR CONNECTORS

Roland Baertschi*, Samuel Garcia, Robert Kroyer** and Pascal Sutter****

* Baertschi Partner Bauingenieure AG
e-mail: roland.baertschi@baertschipartner.ch

** ETH Zurich, Chair of Steel and Composite Structures
e-mail: garcia@ibk.baug.ethz.ch, kroyer@ibk.baug.ethz.ch, pasutter@student.ethz.ch

Keywords: Composite Beams, Partial Shear Connection, End-Slip Method, Deformation Capacity.

Abstract. *Deformation capacity and slip range (eventually called "ductility") of shear connectors are key parameters for the design of shear connectors in composite beams, especially in partial shear connection. Existing design rules require a deformation capacity of at least 6 mm for the use of shear connectors with plastic design methods. This rule is combined with the assumption of rigid-ideal plastic behaviour of the shear connectors. Still, the real characteristics of most of today's usual shear connectors differ considerably from these assumptions, even more for shear connectors used in combination with metal decking. The paper describes the results of a literature research conducted in order to obtain reliable values for the deformation capacity and the slip range of headed stud shear connectors. Based on these results, advanced design methods can be used to realise composite beams with more realistic lower limits for the degree of partial shear connection and thus better economic performance.*

1 INTRODUCTION

Composite construction combines sustainable, quick lightweight steel construction with inexpensive concrete slabs. In buildings, partial shear connection is often used in order to accelerate the construction progress even further by casting the concrete slabs without propping the steel structure and thus reducing the forces to be transferred through the shear connectors. By consequence, a reduced number of shear connectors is sufficient to obtain the required load-bearing resistance for the beam.
In this state of partial shear connection, deformation capacity and slip range (often called "ductility") of the shear connectors define another criterion for the minimum degree of partial shear connection.

The regulations concerning the minimum degree of partial shear connection in today's codes are conservative but very crude and neglect most of the governing parameters, such as the construction process and the actual deformation capacity and slip range of the shear connectors used.

Shear connectors are considered "ductile" in today's Eurocode 4 [1] if they exhibit a characteristic deformation capacity of at least 6 mm. No explicit conditions are defined for the ascending branch of the shear connector's load-slip-curve, even though it is known from [2] that "soft" shear connectors reaching their characteristic load bearing capacity only at large slip (e.g. shear connectors in trapezoidal steel sheeting perpendicular to the beam) are strongly on the unsafe side if designed after Eurocode 4 [1].

Shear connectors exhibiting a characteristic deformation capacity of less than 6 mm must not be used for plastic design methods. On the other hand, shear connectors with much larger deformation capacities may be used, but without profiting from the favourable influence of their increased deformation capacity.

The "end-slip method" (also known as "theta-method") [2], [3] allows for much more accurate determination of the minimum degree of partial shear connection based on a consistent kinematic model using the actual deformation capacity and slip range of the shear connectors. Still, its application has been limited until today since the governing input parameters for most shear connectors were unknown.

Roland Baertschi*, Samuel Garcia**, Robert Kroyer** and Pascal Sutter**

2 THE MINIMUM DEGREE OF PARTIAL SHEAR CONNECTION

2.1 Overview

In partial shear connection the shear interface governs failure of the beam. In order to avoid premature failure of the composite beam, the shear connectors must exhibit both sufficient slip range and deformation capacity. The physical meaning of the initial slip δ_i, the slip range δ_s and the deformation capacity δ_u is explained in figure 1. The initial slip δ_i is the amount of slip at which a shear connector exhibits at least the characteristic shear resistance P_{Rk} for the first time. Deformation capacity δ_u corresponds to the slip value at which the shear force falls below the characteristic shear resistance P_{Rk} for the last time.

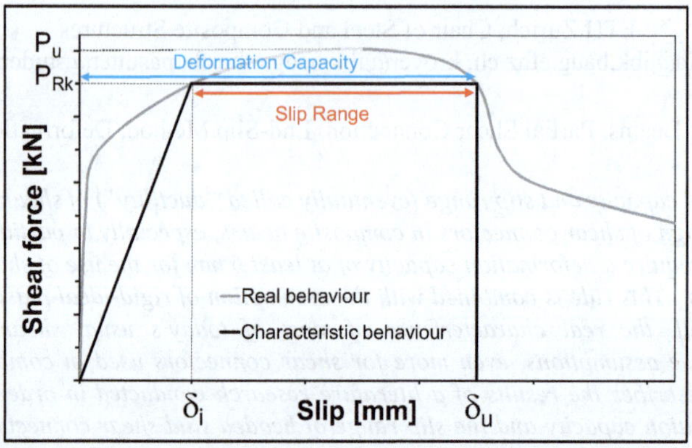

Figure 1. Real behaviour and characteristic load-slip behaviour of a shear connector in parameters P_{Rk} (shear resistance), δ_i (initial slip) and δ_u (deformation capacity). Slip range is $\delta_s = \delta_u - \delta_i$.

For headed studs and other connectors with similar load-slip behaviour, the minimum degree of partial shear connection can be determined using the formulas given in today's relevant codes.

For other connectors, the end-slip method [2] can be used for the design of the shear connection. The end-slip method allows the use of a wide variety of shear connectors for the plastic design of composite beams in partial shear connection. Shear connectors of limited deformation capacity can be used in composite beams as well with this method. For regular headed studs, the minimum degree of partial shear connection can be reduced considerably in many cases, even more with unpropped construction.

For plastic design of composite beams in partial shear connection the first design criterion is that the shear interface does not fail before the steel beam or the concrete slab are fully plastified (deformation capacity criterion). Second, all shear connectors must be at their full shear resistance at the same time (slip range criterion). Both criteria must always be respected.

Therefore, the maximum slip occurring in ultimate limit state (ULS) must not be greater than the deformation capacity of the connectors, and the minimum slip must not be smaller than the slip necessary for the connectors to reach their design shear resistance. These criteria are explained in figure 2. The red curve "*FE: δ_u = 6 mm, 2 concentrated loads@L/3*" shows the finite element simulation of a beam fulfilling both criteria. All connectors are at ULS at the same time and the steel beam is fully plastified at the same time. Thus, the beam reaches the bending resistance predicted using plastic design.

The blue curve "*FE: δ_u = 2 mm, 2 concentrated loads@L/3*" shows a beam with connectors of sufficient slip range (all connectors are at ULS at the same time), but insufficient deformation capacity (the connectors fail before the beam is fully plastified). Therefore, the beam does not reach the bending resistance predicted by plastic design. The orange curve "*FE: δ_u = 2 mm, uniform load*" shows a beam where both slip range and deformation capacity criterion are not fulfilled. In this case the connectors fail before the beam is fully plastified and before all connectors reach ultimate limit state. By consequence, the beam's bending resistance is lower than that obtained from plastic design as well.

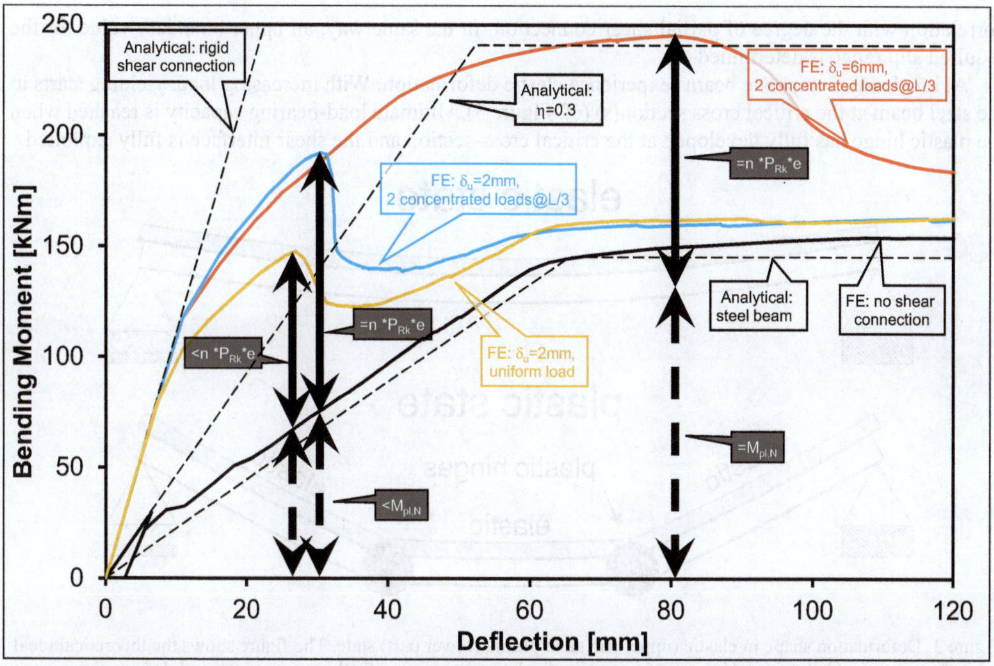

Figure 2. Load-deformation behaviour of a simply-supported composite beam with partial shear connection η.
η = 0.3, L = 7.2 m, steel section IPE 270, f_y = 300 N/mm².
Shear connectors with δ_u (0.9 * P_u) = 2 to 6 mm, evenly distributed over the whole length of the beam.

2.2 Existing design methods

The deformation capacity and slip range criteria are usually verified implicitly by respecting a minimum degree of partial shear connection. In Eurocode 4 [1] and other relevant codes, these criteria are met implicitly for connectors showing similar properties as headed studs with formulas as given in equation (1) or similar.

$$\eta = \frac{N}{N_f} \geq \frac{355N/mm^2}{f_y} \cdot (0.75 - 0.03 \cdot L_e) \geq 0.4 \quad (1)$$

With this strongly simplified approach simple and safe design is possible. Still, non-standard connectors cannot be used and many important parameters such as e.g. whether or not the beam is propped during concrete casting are not considered.

This method is allowed for use with "ductile" shear connectors only. Shear connectors are considered "ductile" if their characteristic deformation capacity amounts to at least δ_{uk} = 6 mm.

2.3 End-slip method

The end-slip method (theta-method) explicitly compares deformation capacity and slip range required in a composite beam at the desired degree of partial shear connection to that shown by the shear connectors used. The required deformation capacity and slip range (also called "ductility") are determined based on a mechanical model according to the rules given in [2]. The mechanical model calculates an upper boundary value for the slip occurring in a composite beam under assumption of an infinitely soft shear connection (no shear connection). Then it assumes that slip in a composite beam is reduced approximately in linear correlation with the degree of partial shear connection. In the same way, an upper boundary value for the required slip range is determined.

At low loads, composite beams experience elastic deformation. With increasing load, yielding starts in the steel beam at the critical cross section(s) (cf. Figure 3). Ultimate load-bearing capacity is reached when the plastic hinge has fully developed at the critical cross-section and the shear interface is fully exploited.

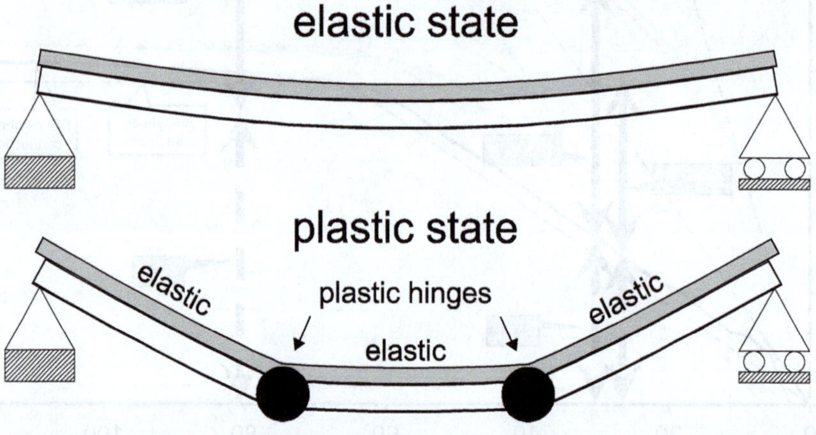

Figure 3. Deformation shape in elastic (upper part) and plastic (lower part) state. The figure shows the theoretical ideal case of symmetrical behaviour. However, in practice, the beam is never perfectly symmetrical, and deformation occurs in an asymmetrical way.

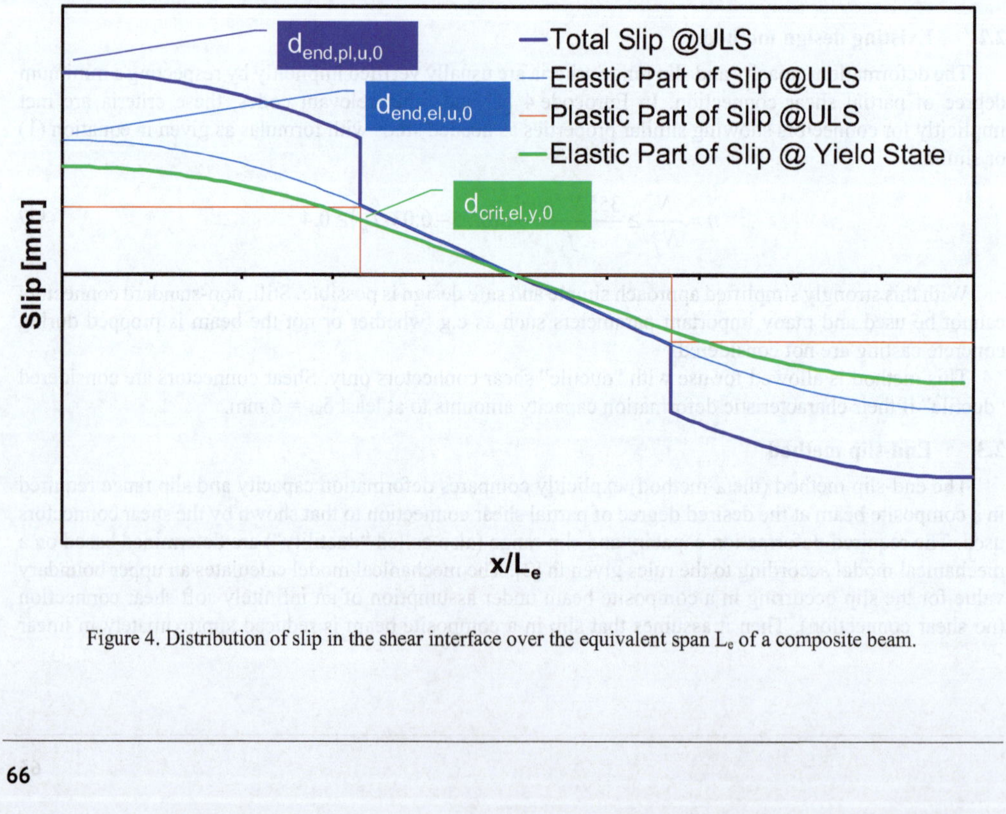

Figure 4. Distribution of slip in the shear interface over the equivalent span L_e of a composite beam.

Slip in the shear interface is shown in Figure 4 and consists of the elastic part well-described by the elastic rules (differential equation or approximation methods such as Möhler's γ-method [4]) and a plastic part which is constant between the critical cross-section and the support (or the location of zero bending moment in continuous beams).

The degree of partial shear connection shall be greater than the maximum of the values obtained from the two subsequent formulas:

$$\eta_{\min} \geq 1 - \frac{\delta_u}{d_{end,pl,u,0}} \quad \text{(deformation capacity criterion)} \qquad (2)$$

$$\eta_{\min} \geq 1 - \frac{\delta_u - \delta_i}{d_{end,el,u,0} - d_{crit,el,y,0}} \quad \text{(slip range criterion)} \qquad (3)$$

In [2] both criteria were combined in one formula:

$$\theta = \min\left(\frac{\delta_u}{d_{end,pl,u,0}}; \frac{\delta_u - \delta_i}{d_{end,el,u,0} - d_{crit,el,y,0}}\right) \text{ and thus } \eta_{\min} \geq 1 - \theta \qquad (4)$$

With:

δ_u Deformation capacity of the shear connector (6 mm for "regular headed studs")

δ_i Initial slip (slip at which the shear connector reaches its nominal shear resistance)

$d_{end,pl,u,0}$ Maximum slip at ultimate load without shear connection (cf. figure 4)

$d_{end,el,u,0}$ Maximum elastic slip at ultimate load without shear connection (cf. figure 4)

$d_{crit,el,y,0}$ Elastic slip in the critical cross-section at yield without shear connection (cf. figure 4)

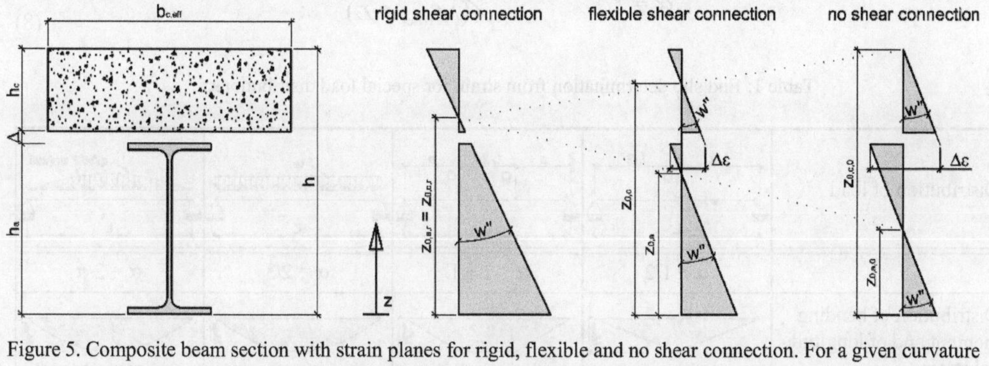

Figure 5. Composite beam section with strain planes for rigid, flexible and no shear connection. For a given curvature w''(x) the difference in slip at the shear interface Δε(x) depends on the distance z_{0c}-z_{ac}. Slip is the integral of Δε(x) over the beam length.

Slip in the shear interface under assumption of no shear connection is determined as follows:

$$d_{end,pl,u,0} = \alpha \cdot \frac{L_e}{2} \cdot \frac{1{,}5 \cdot f_{y,a} - 0{,}8 \cdot \sigma_{a,c}}{E_a} \cdot \left(1 + \frac{h_c + \Delta}{(h_a - z_{0,a,0})}\right) \quad \text{total end slip at ultimate limit state} \qquad (5)$$

$$d_{end,el,u,0} = \alpha \cdot \frac{L_e}{2} \cdot \frac{\frac{W_{pl}}{W_{el}} \cdot f_{y,a} - 0{,}8 \cdot \sigma_{a,c}}{E_a} \cdot \left(1 + \frac{h_c + \Delta}{(h_a - z_{0,a,0})}\right) \quad \text{elastic part of end slip at ultimate limit state} \qquad (6)$$

$$d_{crit,el,y,0} = \beta \cdot \frac{f_{y,a} - 0{,}8 \cdot \sigma_{a,c}}{E_a} \cdot \left(1 + \frac{h_c + \Delta}{(h_a - z_{0,a,0})}\right) \quad \text{elastic slip at yield at } x_{crit} \qquad (7)$$

Roland Baertschi*, Samuel Garcia**, Robert Kroyer** and Pascal Sutter**

With:

α	Integration factor after equation (8) and table 1.
L_e	Equivalent span (equals L: Span for simply supported beams).
$f_{y,a}$	Yield strength of steel beam.
$\sigma_{a,c}$	Stress on top of upper flange at construction stage (usually 0 for propped construction, between 0 and $f_{y,a}$ for unpropped construction, between 0 and $-f_{y,a}$ for precambered construction). Lower values are conservative. Therefore, for precambered construction a factor of 1.2 instead of 0.8 must be used.
E_a	Elastic modulus of steel beam.
h_c	Depth of concrete slab.
Δ	Height of an eventual gap between concrete slab and steel beam.
h_a	Depth of steel beam.
$z_{0,a,0}$	Neutral axis of the steel section without shear connection ($=h_a/2$ for symmetrical profiles).
β	Coefficient depending on the load arrangement. For 2 concentrated loads at positions x_{crit} and (L_e-x_{crit}) equation (9) applies. For one single load at mid-span, the same equation may be applied using $x_{crit} = L_e/2$. For uniform load equation (10) applies.
x_{crit}	Location of critical cross-section (distance from support for simply-supported beams, distance from point of zero bending moment for continuous beams). For mainly concentrated loads, the critical cross-section is located at first load from support, for mainly uniform loading, as a simplified approach: $x_{crit} = 0.4\ L_e$ may be used.

$$\alpha = \left(\int_0^L M(x)dx\right) \Big/ \left(M_{\max} * L\right) \tag{8}$$

Table 1: End slip determination from strain for special load distributions.

Distribution of load		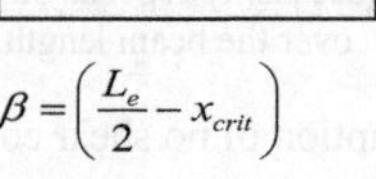		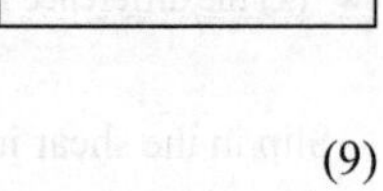
α	$\alpha = 1/2$	$\alpha = 1\text{-}a/L$	$\alpha = 2/3$	$\alpha = 2/\pi$
Distribution of bending moment and of longitudinal strain	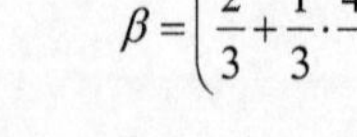	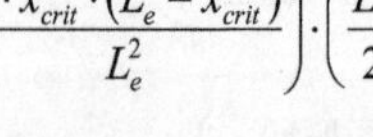	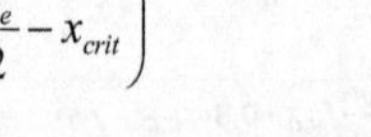	

$$\beta = \left(\frac{L_e}{2} - x_{crit}\right) \tag{9}$$

$$\beta = \left(\frac{2}{3} + \frac{1}{3} \cdot \frac{4 \cdot x_{crit} \cdot (L_e - x_{crit})}{L_e^2}\right) \cdot \left(\frac{L_e}{2} - x_{crit}\right) \tag{10}$$

Equations (5, 6, 7) show that the construction process (whether or not the beam is propped during concrete casting) has a great influence on the design of the shear interface. With propped construction $\sigma_{a,c}$ is about zero. If the beam is not propped, $\sigma_{a,c}$ can be up to almost as much as yield strength. In a third possible case, temporary props are used to pre-camber the steel beam. In that case $\sigma_{a,c}$ can be of opposite sign as the yield stress reached at ultimate limit state.

After unpropped construction the required deformation capacity in the shear interface is lower than after propped construction. If the beam was pre-cambered by means of temporary props, the design is likely to be critical. In that (rare) case, the existing formulation for the minimum degree of partial shear connection (cf. equation (1)) is on the unsafe side.

In composite beams that are not propped during concrete casting, shear connectors of reduced deformation capacity can rather be used than if the beam is propped. With shear connectors of sufficient deformation capacity, the degree of partial shear connection (and thus the number of shear connectors) can be reduced further for unpropped construction than if the beam is propped during concrete casting.

Slip at the critical cross-section is determined using equation (7). Shear connectors must not need more initial slip before they reach their characteristic shear resistance. Otherwise, the shear connectors are not able to bear all at the same time, and the shear resistance of all connectors in a shear length combined is too small.

2.4 Existing gaps in codes and knowledge

Application of the existing code formulas is very restricted, and that of the end-slip method requires information on initial slip and deformation capacity of the shear connectors used. The existing code formulas require a minimum characteristic deformation capacity of at least $\delta_{uk} = 6$ mm and mutually assume that initial slip is sufficiently small, which may be correct for many cases. Still, large initial slip is suspected in certain high-strength steel applications and even more with steel sheeting. Therefore, a data base was established from tests in literature and shall be extended in the future.

3 PROCEDURE

3.1 Collection of load-slip data

In the last decades, plenty of experimental tests on the structural behaviour of headed stud shear connectors have been conducted. Recently, a literature review was carried out at ETH Zurich in order to collect load-slip curves from standard pushout tests in a database [5]. So far, the literature considered covers mainly the testing campaigns on headed shear studs conducted in Central Europe from the last decades such as by Hanswille [6], but the database is supposed to be complemented by further test results from all over the world and more connector types in the near future.

3.2 Selection of load-slip data

While the vast majority of data collected concerns tests without steel sheeting, only limited data was found on tests with several types and configurations of steel sheeting. Therefore, only data from tests without steel sheeting was selected for further evaluation.

3.3 Determination of initial and deformation capacity values

With the aim of determining deformation capacity and slip range of the shear connectors, deformation capacity δ_u and initial slip δ_i as well as slip range δ_s were extracted from the selected data – both at the shear connector's characteristic shear resistance $P_{uk} = 0.9\ P_u$.

Initial slip δ_i was determined from the connector's load-slip curve as the slip, when a connector's shear resistance exceeds P_{uk} for the first time, while deformation capacity δ_u was determined as the slip at which a connector's shear resistance becomes permanently lower than P_{uk}. Slip range was then calculated as $\delta_s = \delta_u - \delta_i$ for each test.

A 95 % percentile value was determined as a characteristic value for initial slip, while a 5 % percentile value was used for deformation capacity, shear resistance and slip range.

The influence of selected parameters such as connector's geometry and steel or concrete strength to deformation capacity δ_u, initial slip δ_i and slip range δ_s was analysed systematically. The results are presented in the sequel.

4 RESULTS

4.1 General

Figure 6 shows the slip characteristics derived from selected push-out test results. In both plots, only data from tests where steel failure was governing is shown. Slip range is calculated as the difference of ultimate and initial slip $\delta_s = \delta_u - \delta_i$ for each data set. Both deformation parameters as well as their correlation is shown in figure 6 on the right. Initial slip values are all smaller than 5 mm and show a concentration at about 1 mm with a mean value of about $\delta_{i,m} = 1.8$ mm. Deformation capacity values start a little below 5 mm and range up to 35 mm with a mean value of about $\delta_{u,m} = 10$ mm, resulting in the slip range scatter illustrated on the left side in figure 6.

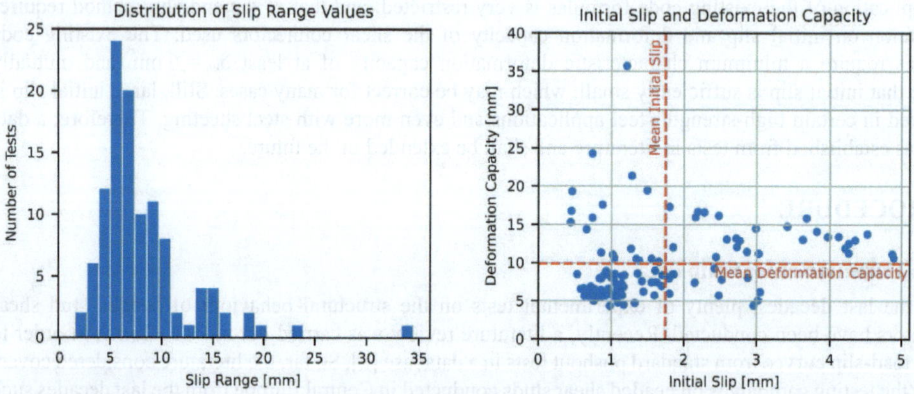

Figure 6. Distribution of slip range δ_s (left). Mean initial slip δ_i and deformation capacity δ_u (right).

Although the data shown in these two plots was already filtered by the expected failure mechanism, the characteristics of the corresponding specimens are not homogeneous in terms of stud diameter and height as well as in terms of steel and concrete strength. The data set was thus filtered again by these parameters.

4.2 Influence of stud geometry on the slip values

Although the data shown in figure 6 have already been filtered for the expected failure mechanism of the studs, the data still shows considerable scatter. For this reason, the data was further filtered by stud geometry, whereas stud diameters of 19, 22 and 25 mm were distinguished as well as stud heights of 75-80, 100, 120-125 and 200 mm.

Figure 7 shows the distribution of initial slip δ_i and deformation capacity δ_u among all values from the selected and filtered data set. The vertical axes describe the absolute number of test results in the given range, whereas the whole set of all test results is represented by the bars, and the markers represent the contribution of the subsets according to the filtering criterions. A general tendency towards large scatter is observed for the deformation capacity δ_u and even more for initial slip δ_i.

For the total of all selected data, the histogram for ultimate slip (deformation capacity) shows a shape similar to a Weibull- or lognormal distribution. For each single diameter value, the shape of the histogram looks much more irregular.

For initial slip, the situation is similar to that for ultimate slip (deformation capacity), whereas the distribution tends to be even more irregular for initial slip than for ultimate slip.

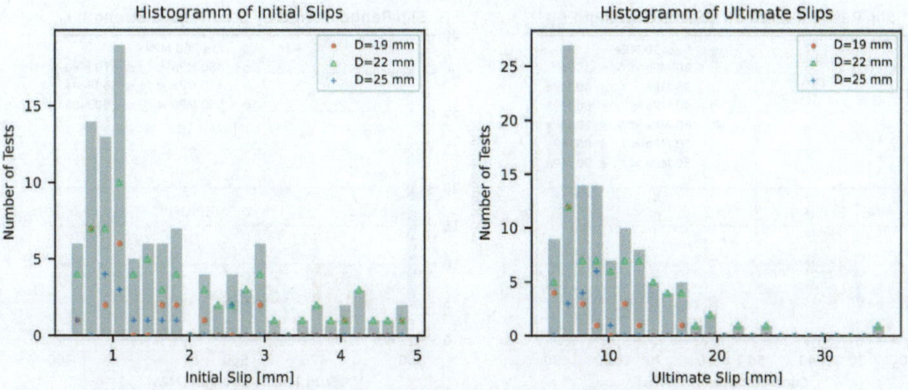

Figure 7. Initial slip δ_i (left) and deformation capacity δ_u (right) by stud diameter.

A deeper analysis of the stud geometry's influence on its load-slip-behaviour is shown in figure 8.

Figure 8 shows the slip range segmented by stud diameter (left) and by stud height (right). The scatter widths for different diameters and heights correlate to some extent with the number of data sets. The largest scatters are observed for 22 mm stud diameter and for 125 mm stud height.

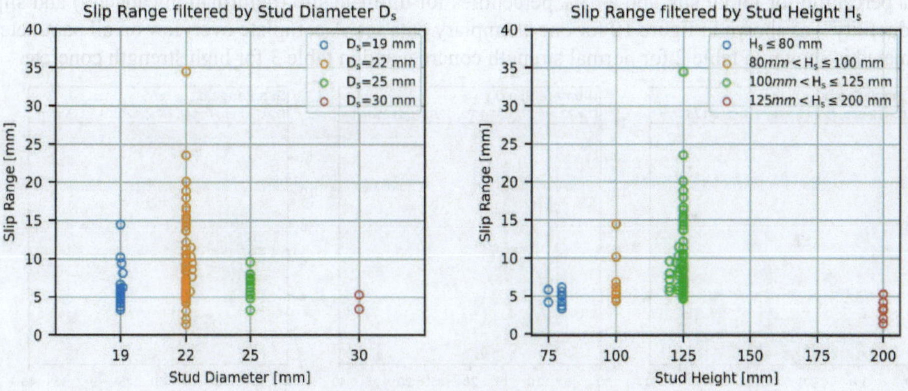

Figure 8. Slip range by stud diameter (left). Slip range by stud height. (right)

4.2 Influence of material properties on the slip values

Further analysis was conducted in order to obtain reliable information on the influence of material properties on the slip range. Figure 9 (left) shows that with high strength concrete (approximately $f_{ck} > 45$ MPa) slip range becomes much smaller than for normal strength concrete. In an analogous manner figure 9 (right) shows that for studs made of high-strength steel (approximately $f_u > 560$ MPa) slip range is much lower than for studs made of lower-strength steel, whereas the reliability of this information is limited by the low number of data sets with studs made from high-strength steel.

This confirms the application limits in today's codes such as [1] in terms of material properties. Further, this information is plausible from the mechanical model which states that a considerable amount of the slip range is established by concrete crushing in the lower region of the stud shaft. With high strength concrete, the concrete in the lower region of the stud shaft will not crush at all or only in combination with very small additional slip before failure, such that slip will merely occur from shear failure of the stud.

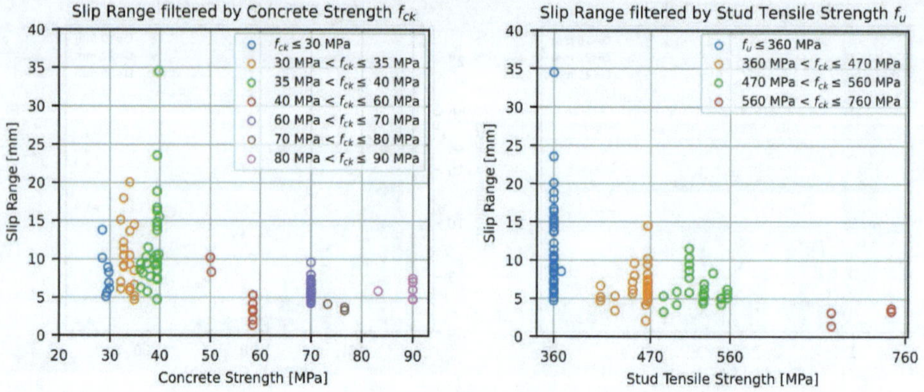

Figure 9. Slip range by concrete strength (left). Slip range by stud tensile strength (right)

4.3 Data fitting

Based on the previous observations, characteristic values for deformation capacity (ultimate slip) δ_{uk}, initial slip δ_{ik} and slip range $\delta_{sk} = (\delta_u - \delta_i)_k$ were determined for many cases using the parameters stud diameter, stud height, concrete strength and stud tensile strength. The characteristic values were determined as a 95% percentile for initial slip and as 5% percentiles for ultimate slip (deformation capacity) and slip range ("ductility") as shown in figure 10 for one exemplary data set. A complete overview on all parameter sets evaluated is shown in table 2 for normal strength concrete and in table 3 for high strength concrete.

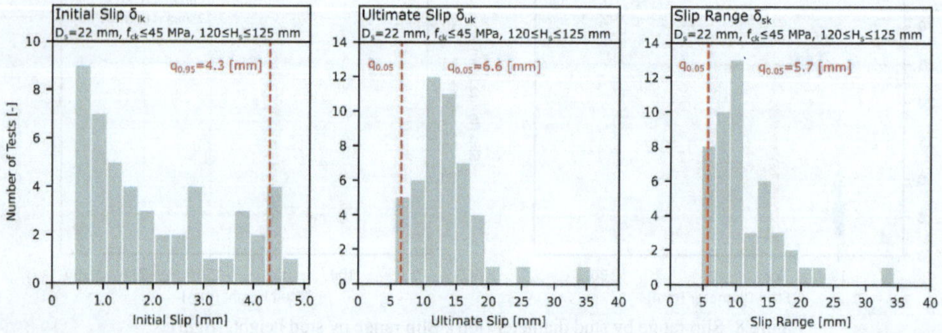

Figure 10. Distribution fit for initial slip δ_i (left), deformation capacity δ_u (middle) and slip range δ_s (right)

The evaluation shows that characteristic deformation capacity is greater than $\delta_{uk} = 6$ mm for all cases analysed with normal strength concrete, but smaller than $\delta_{uk} = 6$ mm for most cases with high strength concrete. Additionally, the data show a tendency towards larger characteristic deformation capacities for larger stud diameters.

Initial slip is much smaller for high strength concrete than for normal strength concrete with values of 4-5 mm for δ_{ik} with normal strength concrete and about half of that for high strength concrete.

Slip range values δ_{sk} between 4 and almost 6 mm result from the evaluation, whereas no obvious differences can be recognised between high strength concrete and normal strength concrete.

Table 2: Characteristic initial slip, deformation capacity and slip range for steel $f_u \leq 560$ N/mm² and concrete $f_{ck} \leq 45$ N/mm².

Stud Ø	Stud Height	# Tests	Initial Slip δ_{ik} (95%)	Deformation capacity δ_{uk} (5%)	Slip Range δ_{sk} (5%)
19 mm	75-80 mm	0			
	100 mm	1	-	-	-
	120-125 mm	6	4.6 mm	6.3 mm	4.8 mm
	200 mm	0	-	-	-
22 mm	75-80 mm	0	-	-	-
	100 mm	0	-	-	-
	120-125 mm	48	4.3 mm	6.6 mm	5.7 mm
	200mm	0			

Table 3: Characteristic initial slip, deformation capacity and slip range for steel $f_u \leq 560$ N/mm² and concrete $f_{ck} > 45$ N/mm².

Stud Ø	Stud Height	# Tests	Initial Slip δ_{ik} (95%)	Deformation capacity δ_{uk} (5%)	Slip Range δ_{sk} (5%)
19 mm	75-80 mm	8	1.6 mm	5.2 mm	4.2 mm
	100 mm	1			
	120-125 mm	6	1.1 mm	5.9 mm	4.8 mm
	200 mm	0	-	-	-
22 mm	75-80 mm	0			
	100 mm	6	2.4 mm	5.4 mm	4.5 mm
	120-125 mm	12	2.0 mm	5.8 mm	4.8 mm
	200 mm	2			
25 mm	75-80 mm	0	-	-	-
	100 mm	0			
	120-125 mm	12	1.7 mm	6.5 mm	5.3 mm
	200 mm	2			

4.4 Specimen with steel sheeting

Presence, orientation, type, geometry and thickness of an eventual steel sheeting have a significant influence on the load-slip behaviour of shear connectors. Current codes distinct between longitudinal and perpendicular orientation as well as between open-trough and dove-tail sheet types in order to predict shear capacity. The influence of trapezoidal steel sheeting on both deformation capacity and slip range of shear connections is not treated in today's codes. Also, no general rules concerning the load-slip behaviour of shear connectors in slabs with trapezoidal steel sheets could be established so far.

While a considerable amount of data was found on tests with steel sheeting spanning perpendicular to the steel beam, these data is from tests with many different sheeting geometries and connector arrangements, and only few data were found on tests with steel sheeting longitudinal to the beam. For each parameter set the number of tests is insufficient for reliable statistical evaluation. More data is needed for proper analysis.

5 CONCLUSIONS

The analyses performed confirm that usual headed stud shear connectors in combination with normal strength concrete fulfil the deformation capacity criterion $\delta_{uk} \geq 6$ mm required in today's codes for the plastic design of composite beams. Still, it was found that with high strength concrete ($f_{ck} \geq 45$ MPa) most shear connectors exhibit insufficient deformation capacities. Therefore, as a conclusion, composite beams with high strength concrete must not be designed with plastic methods, even more in partial shear connection.

In today's construction practice this limitation is of critical importance, even more as concrete suppliers must usually guarantee for certain minimum concrete strength values, but there is no limitation in terms of maximum strength. Even more, today's usual quality management routines do not comprise any consideration of excessive concrete strengths.

Today's codes contain only the deformation capacity criterion, assuming that initial slip is small enough that no unfavourable effect is expected from that. Still, especially with low-strength concrete and trapezoidal steel sheeting, the data analysed raises the suspicion that initial slip becomes of considerable importance for the design of composite beams both in full and partial shear connection, whereas partial shear connection in combination with mostly uniform loading is considered a worst case. Still, it is suspected that the considerable reduction of shear capacity for headed studs in combination with trapezoidal steel sheeting covers up for many of the deficiencies of today's codes in terms of initial slip. However, in order to prove this suspicion, more data is needed from tests with trapezoidal steel sheets in both directions.

Based on the mechanical model in [2] and [3] and on the data analysed, the introduction of an initial slip criterion in today's codes as a prerequisite condition for the application of shear connectors for both elastic and plastic design of composite beams is recommended.

Knowing both actual deformation capacity and slip range of a wide range of shear connectors in different configurations will allow safe and economic use of today's code and widespread application of the end-slip method. Building a reliable database is therefore of importance.

So far, only limited data from literature could be evaluated and transferred to the database. More data shall be obtained from literature in order to complement the number of configurations and to obtain even more reliable data.

REFERENCES

[1] CEN, *EN 1994-1-1:2010 Eurocode 4 - Design of composite steel and concrete structures Part 1-1: General design rules and rules for buildings*, Brussels, 2010.

[2] Bärtschi R., *Load-Bearing Behaviour of Composite Beams in Low Degrees of Partial Shear Connection*, PhD thesis (ETH Dissertation 15895), Institute of Structural Engineering, ETH Zurich, 2005, https://doi.org/10.3929/ethz-a-004906133.

[3] Bärtschi R., Fontana M., Taras A., Tobler R., *Zum Mindestverdübelungsgrad von Verbundträgern*, *Stahlbau*, 89(7), 2020.

[4] Möhler, K., *Über das Tragverhalten von Biegeträgern und Druckstäben mit zusammengesetztem Querschnitt und nachgiebigen Verbundmitteln*, Karlsruhe, 1956

[5] Sutter P., *Load-Deformation behaviour of shear connectors and its effect to the minimum degree of partial shear connection of composite beams*, Master's Thesis, Institute of Structural Engineering, ETH Zurich, 2021.

[6] Hanswille H.-G., *Versuchsbericht über die Durchführung von 77 Push-Out-Versuchen im Rahmen des Forschungsprojekts "Modellierung von Schädigungsmechanismen zur Beurteilung der Lebensdauer von Verbundkonstruktionen aus Stahl und Beton"*, Bergische Universität Wuppertal, 2007.

9TH INTERNATIONAL CONFERENCE ON COMPOSITE CONSTRUCTION IN STEEL AND CONCRETE
M. Knobloch, U. Kuhlmann, W. Kurz, M. Schäfer. (Eds.)
Online, Germany, July 27-29, 2021

HIGH-DEFINITION MODELING OF COMPOSITE BEAMS

Samiran Adhikari*, Gian A. Rassati*, James A. Swanson* and Rachel A. Chicchi*

* University of Cincinnati
e-mails: adhikasn@mail.uc.edu, gian.rassati@uc.edu, james.swanson@uc.edu, rachel.chicchi@uc.edu

Keywords: Composite Beams, Partial Composite Action, Nonlinear Analysis, Shear Connectors.

Abstract. *This paper presents a high-definition modelling approach for the simulation of the flexural behaviour of composite beams that accounts for the principal mechanisms associated with strength, stiffness, and ductility. The goal of the modelling approach presented is to enable parametric studies on the ultimate capacity of composite beams involving different amounts of shear transfer provided, various placement configurations of shear connectors, variable geometry of deck and shear connectors, and different materials. The study presented begins with the simulation of push-out tests that are validated against experimental results in literature. A recent experiment performed by NIST is used as the validation platform for the study. After validation, some parametric analyses are presented, and it is concluded that the modelling approach discussed herein is appropriate to predict the initial and post-peak flexural behaviour of composite beams. Recommendations are also made regarding the placement of shear connectors.*

1 INTRODUCTION

In 2019, the National Institute of Standards and Technology (NIST) performed a series of experimental tests on one-bay, one story composite frames focusing on the fire response of the composite beams [1]. Pre-testing analyses for the fire loads were presented at the 8th International Conference on Composite Construction in Steel and Concrete [2], however, the focus of this paper will be on the results of the baseline test that was conducted without fire loading, but just with applied vertical forces. The beam was designed to be 82% composite with the lightweight concrete slab, and the shear connectors were placed at uniform spacing along the length of the member, in the troughs of the ribbed metal deck, in a strong position. The test specimen, consisting of a beam with a composite slab simply connected to two columns, failed to develop the design moment capacity of the composite member, due to an unexpected failure of a shear connector at one of the beam's ends. This event created some concern due to the fact the specimen complied with the letter of the 2016 AISC Steel Design Specification [3]. This failure stirred some renewed interest in the evaluation of the behaviour of composite beams, especially for partially composite cases, and in the consideration of solutions outside of common practice, such as the concentration of shear connectors toward the ends of the beams, as well as the use of larger diameter, longer shear connectors, associated with the use of deeper metal decks, which would allow to span longer lengths economically. In the aftermath of the failure in the experiment, it was postulated that the uniform spacing of the shear connectors along the entire length of the beam, although meeting global equilibrium requirements, caused the outmost shear connectors to experience an excessive demand in terms of strength or ductility, resulting in a premature failure of the specimen. Questions have also been raised as to the quality of the weld for the shear connector that failed, which also pointed at the lack of a redundancy in this configuration, leading to a decreased reliability in case of understrength welds or connectors.

Considering the elevated cost associated with the fabrication and experimentation of such a specimen, the obvious recourse is to consider numerical simulations to perform parametric analyses investigating the numerous configurations that can be envisioned when considering the variation of parameters such as shear connector layout, material properties, shear connector diameter and length, depth of deck, impact of percentage of composite action, and influence of span length. Owing to the complex nature of strength, stiffness, and ductility development in steel-concrete composite sections, high-definition modelling was considered as the most promising approach for a comprehensive simulation of most, if not all, relevant mechanisms. Using the NIST experiment as the validation platform offers a unique opportunity for exercising the proposed modelling approach's prediction capabilities and to explore some configurations that could hardly be investigated experimentally.

This paper presents a brief literature review focusing on modelling issues for composite structures, followed by a brief summary of the main features of the NIST experiment. The paper then proceeds to present the modelling study that was performed, outlining the sensitivity studies and the results showing the influence of specific details on the overall structural response. The proposed modelling approach is validated on the NIST experimental results, and then some results of ongoing parametric studies are presented and discussed.

2 LITERATURE REVIEW

This brief review is organized by general topic, touching on a few of the most important details that affect the modelling choice for a steel-concrete composite beam.

2.1 Tests on composite beams

Considering the complexity of full-scale tests, experimental projects on composite beams are relatively few in literature. A few of the seminal works that are germane to this paper comprise the work of Grant et al. [4] where a reduction factor was defined for shear connectors in ribbed decks. This prompted the work of Easterling et al. [5] that, based on the results of four tests on composite beams, established the current approach recognizing that shear connectors can provide different capacities depending on their location within a rib of the metal decking and on the direction of the shear force they are resisting (i.e., the so-called "strong position" versus "weak position'). Leon and Ammermann [6] tested a two-bay composite frame focusing on the beam-column connection behaviour, demonstrating the benefits of semi-rigid connections. A bi-directional composite connection tests was reported by Green et al. in [7], leading to recommendation on the detailing of the slab in the vicinity of the column when cyclic reversal loads are expected. Nie and Cai [8] presented a set of experimental results on three composite beam configurations (one simple span, and two two-span continuous beams) with partial shear connection. Hicks and Smith [9] published results of testing of composite beams with ribbed metal decks, investigating the effects of different configurations of shear connectors and Ranzi et al. [10] reported the results of two full-scale tests, also focused on different configurations of shear connectors. To date, the attempts of characterizing the full range of response of an entire composite beam using high-definition finite element models are scant in literature. For example, Tahmasebinia et al. [11] performed a probabilistic study based on a Monte Carlo simulation using a set of 3D finite element simulations to reproduce the tests in [8], using contact surfaces, embedment constraints, and nonlinear materials. Katwal et al. [12] developed a model investigating the interaction between various components, the fracture of shear studs and of the profiled steel sheeting, as well as tensile and compression damage, also referring to the experimental tests in [8].

2.2 Push-out tests for shear connectors

As noted from the previous review of composite beam tests, headed shear stud connectors are the most prevalent means of transferring horizontal shear forces between steel and concrete components in composite beams. The traditional experimental testing approach is the so-called push-out test, in which the shear force versus shear deformation at the interface of a composite beam subassemblage is tested monotonically. Many such tests have been reported in literature [13, 14, 15, 16] and several researchers have focused on the numerical simulation of this type of tests. Lam and El-Lobody [17] published a set of experimental push-out tests paired to a finite element study using a high-definition modelling approach focused on reproducing the experimental results in a solid slab. The researchers showed that their high-definition model was capable of reproducing the different failure modes of the configurations that were tested. Qureshi et al. [18] studied the effect of shear connector spacing and layout on the shear capacity of a composite beam by means of another high-definition finite element model, which proved to be able to reproduce the associated experimental push-off tests accurately.

2.3 Modelling plasticity in concrete

One of the most complex hurdles in finite element modelling of composite beams is the definition of the material model for the reinforced concrete. The Concrete Damaged Plasticity (CDP) model is likely one of the best available choices [16, 18, 19] if a response to highly nonlinear conditions is necessary for the simulation. This model is based on a modified Drucker-Prager approach, as investigated in [20, 21]. A comprehensive study of the parameters affecting the CDP model is presented in [22], while [23] discusses how to establish CDP parameters based on laboratory tests on concrete.

3 THE NIST EXPERIMENT

A set of long-span composite floor beams subjected to combined gravity and fire loads were tested at the National Fire Research Laboratory to provide the necessary data to advance the performance-based design of structures under fire conditions [1, 2]. This experimental campaign focused on 12.8m (42 ft)-long composite beams connected through simple beam-column connections ("gravity connections") under fire conditions. The first of the five tests performed was considered as a baseline test, in which the structure was only subjected to vertical forces, but no fire conditions. This will be the focus of this paper. The composite slab had a width of 1.83m (6 ft), with a full thickness of 8.26cm (3.25 in.) over 7.62cm (3 in.)-deep ribs. The metal deck was installed with the ribs perpendicular to the longitudinal direction of the W460×52 (W18×35) ASTM A992 steel beam. The beam material average yielding stress was measured at 354MPa for the flange and 370MPa for the web. Nineteen mm (0.75 in.)-diameter, 121 mm (4.75 in.) tall headed shear studs were welded through the metal deck at a 30.5cm (1 ft) spacing in every 121-mm wide flute, providing a calculated 82% composite action. The shear connectors were placed in a strong position within the ribs of the metal deck. The beams were connected to W310×158 (W12×106) columns by means of two L127×127×9.5 (L5×5×3/8) angles, with one leg welded to a sacrificial plate to be connected to the column, in order to allow reuse of the columns for all tests, and the other bolted to the web of the beam with three 19mm (0.75 in.) ASTM F3125 grade A325 bolts. The resulting gap between the sacrificial plate and the steel beam was 12.7mm (0.5 in.). Welded wire mesh of 3.4mm diameter spaced at 152mm in both directions (6×6 W1.4×1.4) was provided in both longitudinal and transverse directions at the mid-depth of the topping slab. The concrete slab was poured in two casts, one with a concrete of mean compression strength 49MPa (7 ksi) and the other 54MPa (8 ksi). The columns were braced longitudinally, rendering them virtually rigid for the purposes of frame action, and the slab was not developed around the columns.

Forces were applied at six equally spaced locations along the length of each beam at a rate of 1.1 kN/min.(0.25 kips/min.). The edges of the concrete slab were braced laterally at the locations of loading to prevent twisting of the slab. During testing, when the applied loads reached 63kN (14.2kips), a failure of the shear connection at one end of the beam caused the test to be stopped, and the beam was unloaded to approximately 40kN (9 kips). After examination of the specimen, testing was resumed but the beam could never carry more than 61kN (13.7kips), at which point approximately one half of the beam had lost shear connectivity to the slab, with visible delamination of the steel deck units. At this point, the specimen was unloaded once again, and after investigation it was reloaded, reaching a peak load of 57kN (12.8 kips), at which point the weld of the angle connection on the same side of the beam that experienced the shear failure fractured at a measured specimen displacement of 360mm (14.2 in.). Figure 1 shows the curve of the average of the six applied loads versus the displacement at midspan, showing the two unloading events and the successive reloading. It should be noted that the six point loads were only approximately equal to each other during the testing, due to relative vertical displacements of their points of application as the beam deflected differentially along its length. Comparatively, the load that would have produced the design moment in the beam is 82.2 kN (18.5kips); thus, the beam in the experiment failed at approximately 77% of its design capacity. The failure was tied to the unzipping of the shear connection between the steel beam and the slab, due to either failure of the shear connector or to concrete failure, starting with the extreme shear connector and ultimately proceeding all the way to midspan.

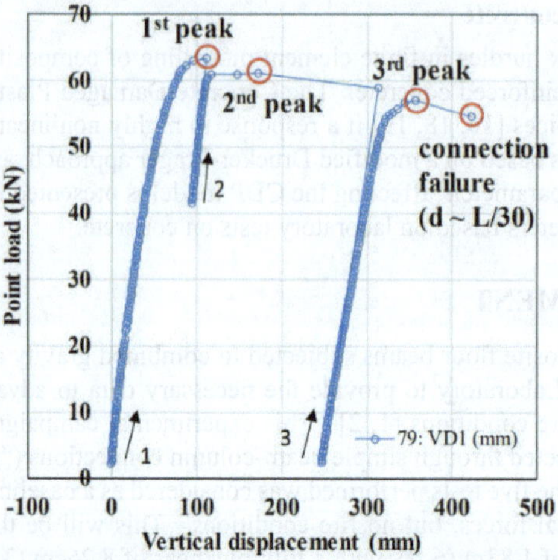

Figure 1. Average of point loads versus vertical displacement at midspan for the composite beam tested [2].

Figure 2. Detail of shear stud fracture surface (left) and view of the end length of the beam after concrete removal [2].

After most composite action was lost, one of the beam-column connections failed and the test was ended. After the test, it was postulated that the premature failure of the beam was due to the insufficient horizontal shear capacity at the ends of the beam, due to the uniform spacing of the shear connectors. In fact, if one were to calculate the shear demand over the length of the beam, one would find that for the first 107cm (42 in.) of the beam, there is a constant vertical shear equal to $3P$, where P is one of the six point loads applied to the specimen. At the load corresponding to the first failure, i.e., at P = 63kN (14.2kips), the corresponding horizontal shear demand would be approximately 339kN (76.2kips) over a 91.4cm (36 in.) length, where three headed studs where installed, providing a horizontal shear capacity of 283kN (63.6 kip), which is insufficient to resist that demand. This happens more often than one would think, and usually the ductility of the headed studs is relied upon to redistribute the excess horizontal shear demand to regions of the beam that have spare capacity to resist it. Consequently, it was further postulated that there might have been an issue with the installation of the last headed stud, resulting in a weld of reduced capacity, limiting the ability of that shear connector to deform sufficiently to redistribute demand further along the beam. Figure 2 shows a detail of the failure location of the last headed stud and the last 120cm of the beam after concrete removal. It can be seen that the last shear connector was the only one to have failed, whereas the others, while significantly deformed, were still positively connected to the top flange of the beam, indicating that concrete failure caused the complete progressive loss of shear connection, as opposed to progressive headed stud fracture.

4 MODELING STUDY

4.1 Component and material modelling

The greatest challenges related to the goal of simulating in great detail most or all of the mechanisms that influence the response of a steel-concrete composite beam are the post-elastic modelling of the materials, and in particular their damage initiation and evolution. It is relatively straightforward to successfully predict the elastic and post-elastic behaviour of steel, at least until fracture prediction is desired. Even then, criteria for damage initiation and damage evolution can be established [24] that allow for a reasonably accurate prediction of the initiation of fracture in steel. When considering the concrete material, however, the marked differences in tension and compression render these predictions more complex. For example, the finite element analysis software ABAQUS [25] accounts for detailed tension and compression behaviour of concrete. Figure 3 shows the uniaxial stress-strain response of concrete material, including softening damage to unloading branches, depending on the extent of tension damage of the material. Work by Hillerborg et al. [26] provides criteria for the establishment of fracture energy as a means of reducing mesh sensitivity when concrete cracking is simulated. The concrete material's softening behaviour can be simulated using a linear, a bilinear, or an exponential decay function, with the latter providing the best adherence to experimental results.

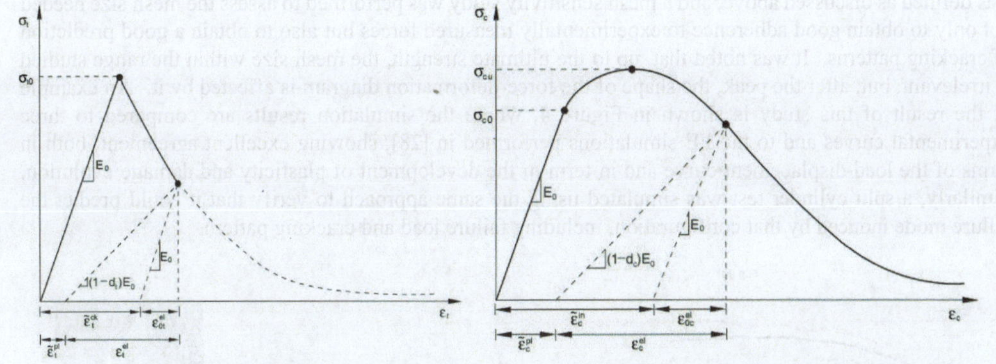

Figure 3. Stress-strain curves for concrete in tension (left) and in compression (right) [25].

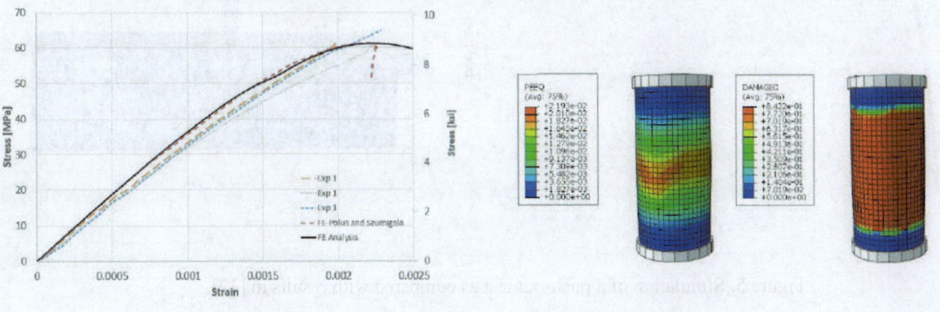

Figure 4. Verification of compressive response of concrete material compared to [28].

In compression, the concrete material is assumed to remain elastic until the stress reaches 40% of the ultimate strength; the plastic region is characterized by nonlinear hardening followed by strain-softening after the ultimate strength. Upon unloading, the linear response degrades. For this study, the stress-strain behavior reported in Majewski [27] was used. In particular, the model used the work in [27] to establish the strain corresponding to the peak capacity ε_{c1} and a strain at failure ε_{cu}, as a function of the average compressive strength of concrete f_{cm}, as shown in the equations (1) and (2):

$$\varepsilon_{c1} = 0.001\left[2 - \exp(-0.024 f_{cm}) - \exp(-0.140 f_{cm})\right] \tag{1}$$

$$\varepsilon_{cu} = 0.004 - 0.0011\left[1 - \exp(-0.0215 f_{cm})\right] \tag{2}$$

A linear response is assumed up to 40% of f_{cm}. Additionally, to define the post-peak compression damage, the plastic strain ε_c^{pl} is given from the relationship between the compressive inelastic strain ε_c^{in} and the damage parameter d_c, and is defined by equation (3), assuming an average Young's modulus for the material E_{cm}, as shown in Figure 3:

$$\varepsilon_c^{pl} = \varepsilon_c^{in} - \frac{d_c}{(1-d_c)}\frac{\sigma_{co}}{E_{cm}} \tag{3}$$

A series of component simulations were performed to evaluate the fitness of the material models chosen to simulate the expected experimental behaviours. Concrete cylinder testing reported in [28] were simulated in ABAQUS by modelling the concrete cylinder with solid 8-node reduced integration elements and the two steel loading plates with 4-node shell elements and hard contact surfaces. The concrete material was defined as discussed above, and a mesh sensitivity study was performed to assess the mesh size needed not only to obtain good adherence to experimentally measured forces but also to obtain a good prediction of cracking patterns. It was noted that, up to the ultimate strength, the mesh size within the range studied is irrelevant, but, after the peak, the shape of the force-deformation diagram is affected by it. An example of the result of this study is shown in Figure 4, where the simulation results are compared to three experimental curves and to the FE simulations performed in [28], showing excellent agreement, both in terms of the load-displacement curve and in term of the development of plasticity and damage evolution. Similarly, a split cylinder test was simulated using the same approach to verify that it could predict the failure mode induced by that configuration, including failure load and cracking pattern.

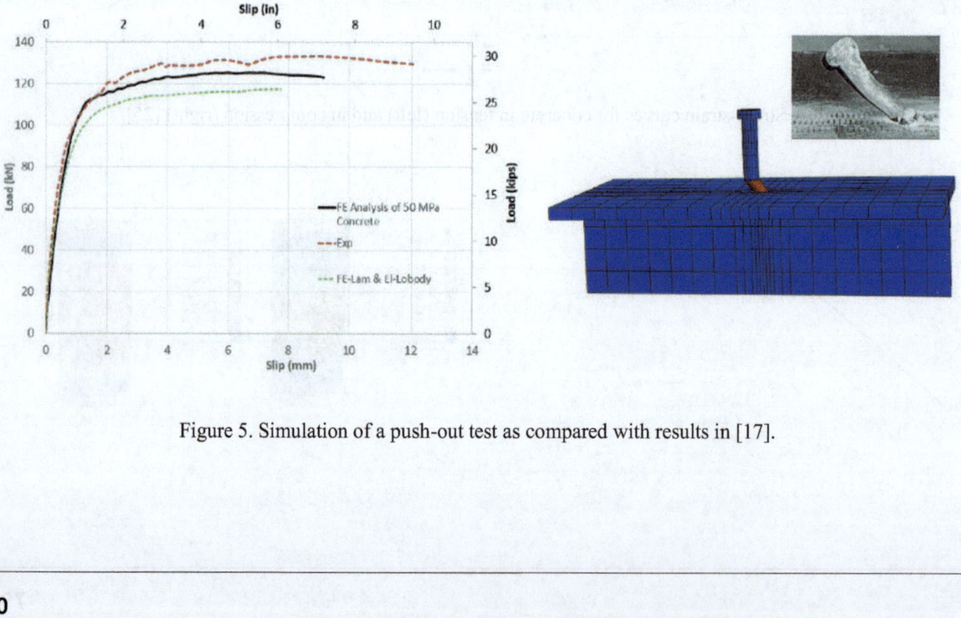

Figure 5. Simulation of a push-out test as compared with results in [17].

Furthermore, a push-out test from [17] was modelled and simulated to verify the prediction capability of the modelling approach considered. For such a model, an elastic-plastic with hardening model was used to model the shear stud material and for the reinforcing bars, while an elastic-perfectly plastic model was used for the steel beam. Once again, solid 8-node elements with reduced integration were employed for all components, with the exception of the reinforcing bars which were modelled as embedded 3-dimensional truss elements. Contact surfaces were modelled explicitly using hard contact in the normal direction and a penalty function formulation to account for tangential behaviour. A small friction coefficient of 0.01 was used. The connection of the studs to the beam flange was effected by means of multi-point constraints, to avoid the explicit meshing of the annular weld. The concrete mesh was refined in the vicinity of the shear stud, while the largest working mesh spacing of 25mm (1 in.) that was found in the previous sensitivity study was used for the rest of the concrete slab. A similar size mesh was used for the steel beam. Figure 5 shows the comparison of the simulated load versus slip behaviour with the experimental data and with the FE work by the original authors in [17], for a 50MPa (7.25 ksi) concrete material. The figure shows an excellent performance, resulting in an error on the peak force of 4.56% and on the slip at peak force under 10%. Similar results were obtained when simulating a push-out test on 35MPa (5 ksi) concrete, where the peak force was predicted within 5.88% of the experimental value, and the slip at peak force was within 9% of what was measured in the laboratory. Qualitatively, the predicted deformations match the experimental observations, even with the use of a relatively coarse mesh. In perspective, the good performance of a relatively coarse mesh is important, when considering the goal of modelling an entire composite beam to this level of detail.

4.2 Full-scale modelling

The NIST specimen was modelled in its entirety, but symmetry constraints were used so that only one half of the beam, cut along the vertical plane through its longitudinal axis, was considered. This allowed for a reasonably refined mesh around the studs and within the flutes of the metal deck, while maintaining an acceptable run time for the analyses. The explicit dynamic solver available in ABAQUS was used for the analysis, considering that the goal was to reproduce and ultimately predict the post-peak behaviour of the structure, including softening phenomena and damage initiation and evolution within the materials. Overall, the model has just over 1.2 million degrees of freedom. The first goal was to reproduce the NIST test, up to and past the measured peak strength, to validate the modelling approach and to gain confidence on its predictive power. Due to the configuration of the test specimen, it was decided to forego the modelling of the columns, considering that they were quite rigidly braced against in-plane bending, and to assume a fixed condition for the sacrificial plate to which the connection angles were welded. The weld was not modelled explicitly, but multi-point constraints tying the contact surfaces of the plate and of the leg of the angles were used in their stead. Pretensioned bolts through the outstanding leg angles, connecting them to the beam web, were modelled explicitly. The steel beam was modelled by means of solid 8-node reduced integration elements, while the steel deck was modelled explicitly using 4-node shell elements. Shear studs were modelled using 8-node solid elements following the same strategy as the push-out component test that was described in the previous section. The shear studs were tied to the deck through a multi-point constraint instead of explicitly modelling the annular weld. In turn, the metal deck is connected to the beam at the same location and for the same footprint of the shear stud through a surface-to-surface tie constraint. The concrete slab was also modelled using solid 8-node reduced integration elements, and the material was defined as described in the previous section. Slab reinforcement was modelled by means of 3-dimensional truss elements embedded in the slab material. Figure 6 shows an overall view of the model, and Figure 7 shows (a) detailed views of the assembly of beam, deck, and studs, (b) a specific detail showing the strong position for the installation of the shear studs, (c) a detail of the beam-column connection, showing the angles, bolted to the beam web and tied to the sacrificial end plate that is externally restrained, and (d) a detail of the steel reinforcement in the slab.

Figure 6 also shows the locations at which the loads are applied into the beam, modelled as rigid plates to simulate a point load as a distributed load over a small area avoiding the convergence difficulties that are often associated with a more direct approach. A parametric study was conducted to investigate the impact of externally fixing the connection plate at each end of the beam (i.e., the sacrificial plate that was connected to the columns in the experiment) as opposed to pinning it, or to assigning a rotational stiffness of the same order of magnitude as that of a pinned-pinned column loaded with a moment at midspan and braced at that location. Results showed that the semi-rigid and the fixed solution provided very similar results, which is supported by the observation that experimentally the columns were solidly braced against in-plane bending to the point of behaving almost rigidly. The loads were applied in a smooth step to prevent unwanted dynamic responses of the structure over an adequate time duration.

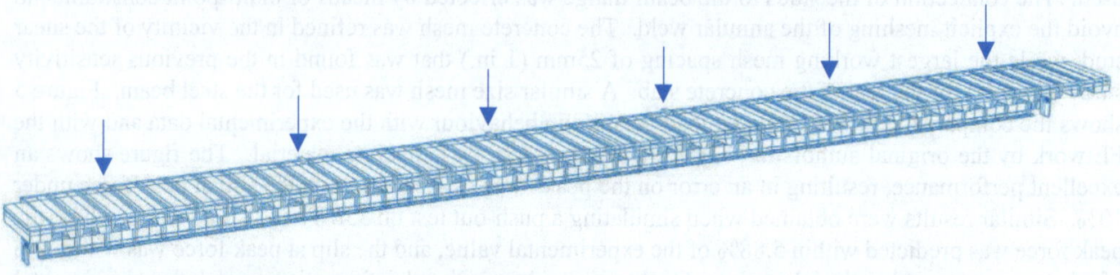

Figure 6. Overall view of the finite element model of the NIST beam showing the load locations.

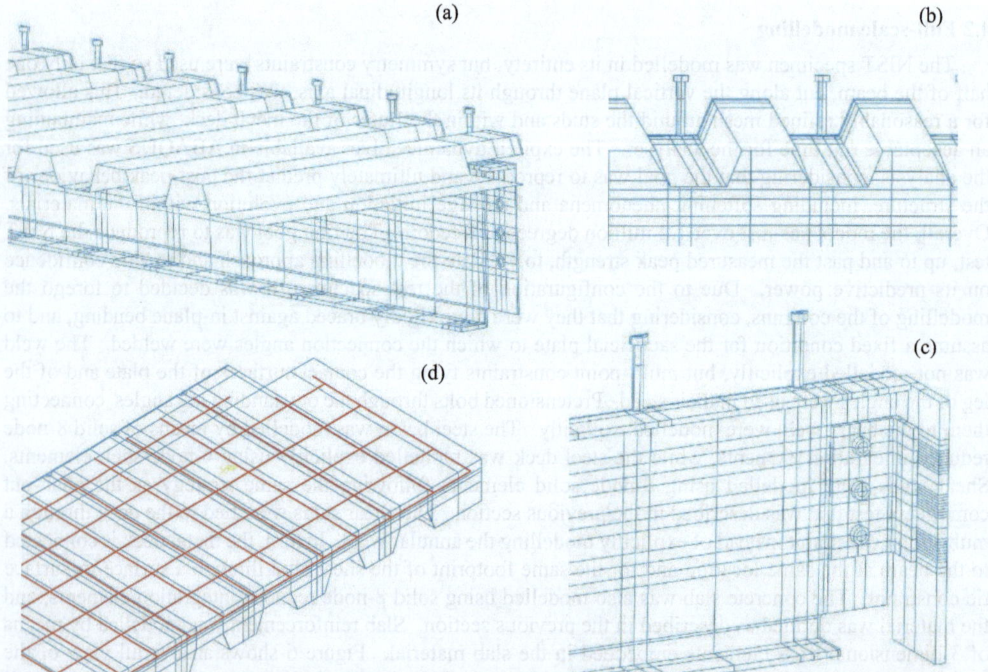

Figure 7. (a) Overall detail without slab and beam-column connection, (b) Detail view of stud position in the ribs of the metal deck, (c) Detail of beam-column connection, and (d) Detail of slab reinforcement.

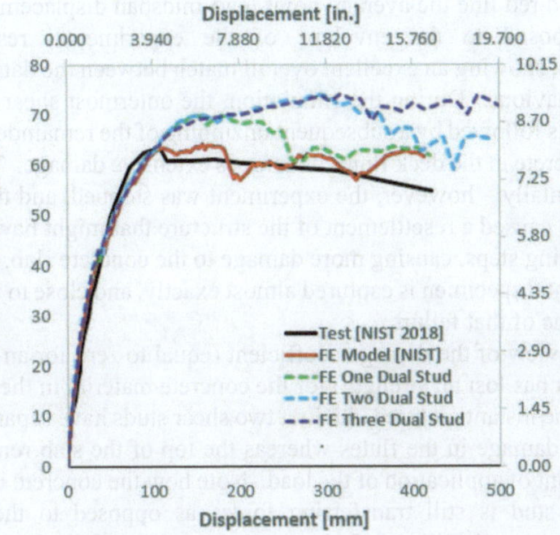

Figure 8. Load-displacement response comparison to the experimental results in [1].

Figure 9. Concrete damage coefficient at one end of the beam after fracture of the first two studs.

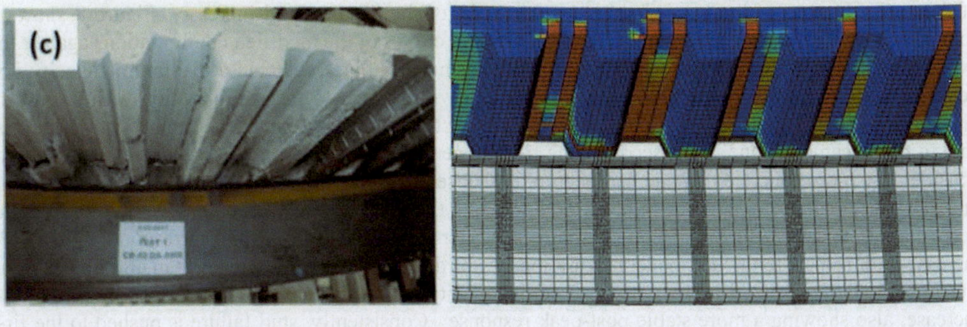

Figure 10. Comparison of damage pattern 1m from midspan at the end of the simulation with a picture from [1].

Figure 8 shows in a bold red line the average point load-midspan displacement chart of the simulated composite beam superimposed to the envelope of the experimental results (not including the unloading/reloading curves), showing an excellent overall match between the datasets, in terms of stiffness, strength, and post-peak behaviour. During the simulation, the outermost shear stud fails first, detaching from the deck. That failure is followed by a subsequent unzipping of the remainder shear connectors toward the mid-span, while the concrete in the deck flutes undergoes extensive damage. The unzipping of the studs was not observed experimentally – however, the experiment was stopped, and the structure unloaded and reloaded, which would have caused a resettlement of the structure that might have helped in preserving the studs in the subsequent loading steps, causing more damage to the concrete slab. It is encouraging that the failure load of the experimental specimen is captured almost exactly, and close to the actual beam deflection that was measured at the time of that failure.

Figure 9 shows a detail view of the damage coefficient (equal to zero for an undamaged material, and equal to 1 for a material that has lost all strength) for the concrete material in the slab near the shear studs. It is possible to see that, at the instant pictured, the first two shear studs have separated from the metal deck, as well as the compression damage in the flutes whereas the top of the slab remains substantially elastic with the exception of the point of application of the load. Note how the concrete in the flute is considerably more damaged where the stud is still transferring forces as opposed to the relatively low damage corresponding to the outermost stud. Figure 10 shows a comparison of the damage coefficient in the slab at 1m (3 ft) from the midspan, showing an excellent match with the experiment. Overall, based on these observations, it can be concluded that the proposed model can predict with encouraging accuracy the failure modes that were observed experimentally both quantitatively and qualitatively.

4.2 Parametric studies

Considering the excellent results in predicting the experimental failure strength, stiffness, and damage patterns, a series of parametric analyses were run focusing on the presumptive reasons for the premature failure of the experimental specimen at a load that did not reach the design level. The first of these analyses, on which this paper focuses, entails a configuration where the same number of shear connectors is used over the length of the beam. Instead of using a uniform distribution of studs, the trial configurations entail removing the stud nearest the midspan of the beam and placing it in the outermost flute of the deck, in a twin stud configuration. Further configurations were investigated repeating this operation for the second and the third flute of the deck, removing more studs from the midspan region. The rest of the shear connectors were still installed in each rib, along the centerline of the beam. This way, a 33%, 66%, and 100% increase in horizontal shear strength was provided at the ends of the beam (where the shear demand is greatest – and as discussed above where the shear demand was greater than the resistance provided by three headed studs), respectively. Simultaneously, no shear resistance was supplied where the theoretical shear demand is null or nearly zero. The rest of the model was maintained identical to the original, in order to provide good grounds for comparison. It is expected that by relying a little less on plastic redistribution of the shear demand of the headed studs, the capacity of the beam would increase. Figure 8 shows a comparison of the average point load-midspan deflection curves of the modified configurations against the experimental data. The simulation with twin studs in the first flute predicted a maximum capacity just below 70kN (15.7 kips), an 11% increase in capacity with respect to the configuration with a uniform placement of studs, which still was not sufficient to reach the design capacity of the beam. In fact, the first stud to fail became the one in the second flute from the end, and in general, by the end of the simulation, considerably more concrete damage had occurred. Consequently, another model was created eliminating one more shear connector near the midspan and doubling the headed studs within the second flute from the beam end. In this case, the predicted maximum capacity rose to 73kN (16.4 kips), corresponding to a 16% increase with respect to the original configuration. A third simulation, with twin studs in all three extreme flutes of the deck, resulted in a slight increase in capacity to 74kN (16.6 kips), corresponding to a 18% increase, also showing a more stable post-peak response. Consistently, stud failure is pushed to the first isolated stud along the length of the beam and increasing concrete damage can be seen as studs are doubled in the end flutes. Even in the configurations studied, the simulation cannot reach but 90% of the nominal moment capacity for the beam, which incidentally corresponds to its design capacity. Concrete in the flutes becomes considerably damaged toward the end of the simulation, corresponding to near loss of composite action. Simultaneously, as it took place experimentally, the outstanding legs of the angles in the beam-column connections undergo large plastic deformations, indicating impending failure. Overall, the change in configurations of the stud placement along the length of the beam produces the expected effect, although the full capacity of the beam is approached but not reached. Further studies are underway to better understand these phenomena.

5 CONCLUSIONS

This paper presents a high-definition finite element modelling approach that has proven capable of predicting with noteworthy accuracy the failure strength, stiffness, and damage distribution of a steel-concrete composite beam with partial shear interaction. This approach shows great promise for extensive parametric studies, as shown in one example in this paper. The reconfiguration of the headed shear connectors investigated shows a change in strength that is akin to what would be expected by simplified hand calculations. Further investigations along this line would enable researchers to propose changes to the current design specifications in order to improve the reliability of long span, partially composite beams. Additionally, the proposed modelling approach lends itself well for further investigations, involving multiple span beams with continuous slabs, and configurations with larger diameter studs and with longer studs in deeper decks, which might be more suitable for long spans.

REFERENCES

[1] NIST Technical Note 254, *Compartment Fire Experiments on Long-Span Composite-Beams with Simple Shear Connections. Part I: Experimental Design and Beam Behavior at Ambient Temperature,* https://doi.org/10.6028/NIST.TN.2054, NIST, Washington, DC, USA, 2019.

[2] Main J.A., Sadek F., Weigand J.M., Jiang J., Choe L., Selvarajah R., Hoehler M.S., Seif M., and Gross J. "Performance of Composite Beams in Fire: Pre-Test Analysis of Full-Scale Experiments", in Rassati G.A., Hajjar J.F., and Leon R.T., Eds. *Composite Construction in Steel and Concrete VIII*, July 29-August 2, 2017, Spring Creek Ranch, Jackson Hole, Wyoming, AISC, Chicago, IL 2020

[3] AISC 360-16, *Specification for Structural Steel Buildings*, American Institute of Steel Construction, Chicago, Illinois, 2016

[4] Grant J.A., Fisher J.W., and Slutter R.G., "Composite beams with formed steel deck", *Engineering Journal* 14(1):24-43, 1977

[5] Easterling W.S., Gibbings D.R., and Murray T.M., "Strength of shear studs in steel deck on composite beams and joists," *Engineering Journal*, 30(2):44–55, 1993.

[6] Leon R.T. and Ammerman, D. "Semi-Rigid Composite Connections for Gravity Loads", *Engineering Journal* 27(1):1-11, 1990

[7] Green T.P., Leon R.T., Rassati G.A. "Bidirectional tests on partially restrained, composite beam-to-column connections", *Journal of Structural Engineering* 130(2):320-327

[8] Nie J., Fan J., and Cai C., "Experimental study of partially shear-connected composite beams with profiled sheeting," *Engineering Structures*, 30(1):1–12, 2008.

[9] Hicks S.J. and Smith A.L., "Stud shear connectors in composite beams that support slabs with profiled steel sheeting", *Structural Engineering International* 24(2):246-253, 2014

[10] Ranzi G., Bradford M.A., Ansourian P., Filonov A., Rasmussen K.J.R., Hogan T.J., and Uy B. "Full-scale tests on composite steel-concrete beams with steel trapezoidal decking", *Journal of Constructional Steel Research*, 65(7):1490-1506, 2009

[11] Tahmasebinia F., Ranzi G., and Zona A., "Probabilistic three-dimensional finite element study on composite beams with steel trapezoidal decking," *Journal of Constructional Steel Research*, 80:394–411, 2013.

[12] Katwal U., Tao Z., and Hassan M.K., "Finite element modelling of steel-concrete composite beams with profiled steel sheeting," *Journal of Constructional Steel Research*, 146:1–15, 2018.

[13] Rambo-Roddenberry M.D. *Behavior and strength of welded stud shear connectors*, PhD Dissertation, Virginia Polytechnic Institute and State University, Roanoke, Virginia, 2002.

[14] Sublett C.N., Easterling W.S., and Murray T.M., *Strength of welded headed studs in ribbed metal deck on composite joists*, Report No. CE/VPI-ST 92/03, Virginia Polytechnic Institute and State University, Blacksburg, Virginia, 1992

[15] An L. and Cederwall K., "Push-out tests on studs in high strength and normal strength concrete", *Journal of Constructional Steel Research* 36(1):15-29, 1996

[16] Mirza O. and Uy B., "Behaviour of headed stud shear connectors for composite steel– concrete beams at elevated temperatures," *Journal of Constructional Steel Research*, 65(3):662–674, 2009.

[17] Lam D. and El-Lobody E., "Behavior of headed stud shear connectors in composite beam," *Journal of Structural Engineering*, 131(1):96–107, 2005.

[18] Qureshi J., Lam D., and Ye J., "Effect of shear connector spacing and layout on the shear connector capacity in composite beams," *Journal of Constructional Steel Research*, 67(4):706–719, 2011.

[19] Szumigala M., Chybinski M., and Polus L., "Stiffness of composite beams with full shear connection," in *IOP Conference Series: Materials Science and Engineering*, 471:52-83, IOP Publishing, Bristol, UK, 2019.

[20] Lubliner J., Oliver J., Oller S., and Oñate E., "A plastic-damage model for concrete," *International Journal of Solids and Structures*, 25(3):299–326, 1989.

[21] Lee J. and Fenves G.L., "Plastic-damage model for cyclic loading of concrete structures," *Journal of Engineering Mechanics*, 124(8):892–900, 1998.

[22] Kmiecik P and Kamiński M, "Modelling of reinforced concrete structures and composite structures with concrete strength degradation taken into consideration," *Archives of Civil and Mechanical Engineering*, 11(3):623–636, 2011.

[23] Jankowiak T. and Lodygowski T., "Identification of parameters of concrete damage plasticity constitutive model," *Foundations of Civil and Environmental Engineering*, 6(1):53–69, 2005.

[24] Wurzelbacher K.P., *A model for prediction of failure initiation in finite element analysis of bolted steel connection*, MS Thesis, University of Cincinnati, Cincinnati, Ohio, 2012

[25] Dassault Systémes, *ABAQUS Documentation User's Guide*, Simulia Corp., Providence, Rhode Island, 2019

[26] Hillerborg A., Modéer M., and Petersson P.-E., "Analysis of crack formation and crack growth in concrete by means of fracture mechanics and finite elements," *Cement and Concrete Research*, 6(6):773–781, 1976.

[27] Majewski S., *The mechanics of structural concrete in terms of elasto-plasticity*, Publishing House of Silesian University of Technology, Gliwice, Poland, 2003.

[28] Polus Ł. and Szumigała M., "Laboratory tests vs. FE analysis of concrete cylinders subjected to compression," in *AIP Conference Proceedings*, 2078:20-89, AIP Publishing LLC, Melville, New York, 2019

9TH INTERNATIONAL CONFERENCE ON COMPOSITE CONSTRUCTION IN STEEL AND CONCRETE
M. Knobloch, U. Kuhlmann, W. Kurz, M. Schäfer. (Eds.)
Online, Germany, July 27-29, 2021

NON-LINEAR ANALYSIS OF COMPOSITE BEAMS WITH MINIMAL MODELLING AND CALCULATION EFFORT FOR STRAIN-LIMITED DESIGN

Philipp Hauser*, Wolfgang Kurz*

* Technische Universität Kaiserslautern, Germany
e-mails: philipp.hauser@bauing.uni-kl.de, wolfgang.kurz@bauing.uni-kl.de

Keywords: Composite Beams, Strain-Limited Design, Design Model

Abstract. *In the following, an analysis method is presented that enables the modelling of composite beams in commercial programs with minimum effort, considering non-linear material behaviour and nonlinear flexibility of the composite joint. A particular focus is placed on the deformations, especially those of the composite joint in horizontal and vertical directions. A detailed insight into the necessary accuracy of the material modelling is presented. The model is subsequently verified by means of comparative analyses of real beam tests.*

1 INTRODUCTION

The construction industry has an important contribution to limiting global warming through efficient load-bearing structures and the associated savings in raw materials. However, the safety level of the structures must not suffer as a result. The effectiveness of composite beams made of concrete and steel is beside the performance of both composite materials mainly characterised by the load bearing and deformation behaviour of the shear connectors. In order to enable further developments with regard to safety and economic efficiency, calculation methods are required which, in addition to the load-bearing capacity, also consider the deformations of the beam and the deformability of the shear connectors.

In the current standard of EN 1994-1-1 [1], for the strain-limited design it is assumed that the total cross-section is kept plane, i.e. the flexibility of the composite joint is neglected. Since this flexibility of the composite joint, especially in beams with low shear connection ratio, plays a decisive role on the load-bearing and deformation behaviour of composite beams, an analysis model is presented which allows the realistic modelling of composite beams by using software that is frequently used in structural design offices. In the model the non-linear material behaviour of steel and concrete as well as a non-linear load-slip behaviour of the composite joint is considered.

In addition to the strain-limited design, for which detailed investigations regarding reliability are still required, the analysis model is already suitable for use in research and could be a very suitable complement to complex scientific programmes. For example, post-calculations of test results as well as investigations of the influence of individual parameters on composite beams are very easily possible and with high accuracy. By using commercial programs, adjustments in the model, for example a change of the positions of the shear connectors, as well as the direct output of the results, such as strains and internal forces, are much more easily possible. Moreover, the use of a high-performance computer is not necessary here.

In the following, the structure of the model is explained first. Subsequently, the material modelling and the influence on the results are described. To verify the models, comparative analyses of real beam tests and the results of numerical simulations are presented. Based on the analysis model, parameter studies on the influence of the load-deformation curve of the beam elements are presented.

2 GENERAL

In addition to the Gamma-Method, framework models have long been the standard method in timber construction for the calculation of composite timber cross-sections or timber-concrete composite structures. Due to the almost linear elastic material behaviour of timber, a linear elastic analysis in timber engineering normally is sufficient [2, 3, 4, 5]. In steel-concrete composite construction, due to the high ductility of the composite beams, consideration of the non-linear material behaviour is essential for modelling real experiments and for economic design. For this reason, framework models have only been used to a limited extent in composite constructions made out of steel and concrete.

However, due to increasingly powerful commercial programs for structural design and increasing computer performance, composite beams can now be analysed with framework models considering the non-linear material behaviour of the composite members as well as the non-linear load-slip relation of the composite joint. That means little modelling and computing effort compared to scientific numerical simulations. Moreover, the method can easily be adopted for practically relevant design problems due to the use of common engineering programmes.

As mentioned, the calculation model is a two-dimensional framework structure in which the upper chord corresponds to the concrete slab and the lower belt chord to the steel profile. The beams are assigned the respective cross-sectional properties of the real composite components. For the steel profiles, it is usually possible to use profile databases. Since the programme does not consider load propagation perpendicular to the model plane, it is essential to consider the effective widths of the concrete slab according to EN 1994-1-1 [1]. Since the calculation effort increases enormously due to the resulting trapezoidal concrete cross-sections, the effective width also can be modelled with areas of constant width. The reinforcement layer is modelled as one rectangular bar with the diameter of the reinforcement bar as height and a width resulting from the total area of the reinforcement laying in the effective width of the concrete. The reinforcement is placed on the axis of the concrete cross-section and the exact height is set by using a bar eccentricity. As usual in numerical simulations, a rigid bond between reinforcement and concrete is assumed. Based on the framework model given in [6] a horizontal and a vertical spring is arranged in the composite joint in place of each connector. These springs are connected to the steel beam and the concrete slab with vertical rigid bars. The beam elements lay on the centroid of the concrete slab and the steel profile. The model is shown in figure 1. Even though the present study only focuses on headed studs as shear connectors, if the load and deformation behaviour of other shear connectors are available, these can also be implemented in model.

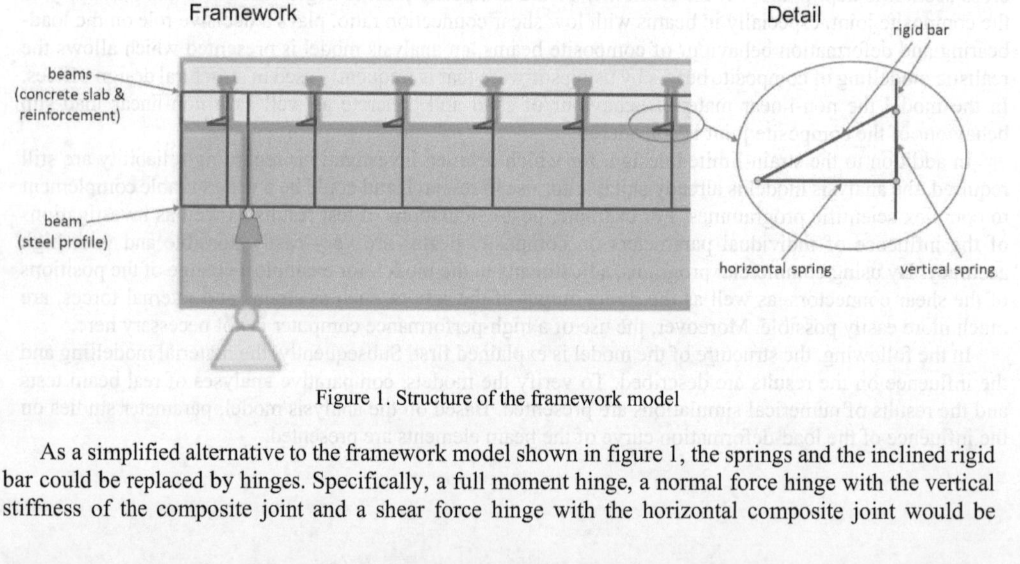

Figure 1. Structure of the framework model

As a simplified alternative to the framework model shown in figure 1, the springs and the inclined rigid bar could be replaced by hinges. Specifically, a full moment hinge, a normal force hinge with the vertical stiffness of the composite joint and a shear force hinge with the horizontal composite joint would be adopted. However, since in further researches the influence of the slip on the uplift is to be modelled via an inclined position of the "horizontal" spring, the set-up from figure 1 is preferred. The magnitude of the uplift can be specified via the angle of the spring, so research is currently still being carried out into a general parameter for the position.

In addition to the framework models, in which the concrete slab and steel profile are modelled as a beam section, shell models were also investigated. Here, the concrete slab is modelled as one shell and the steel profile as three shells, separated for the flanges and the web. Using shell elements, the strains and curvatures of concrete and steel could be modelled a more precise, especially locally. In contrast, the determination of the internal forces of the composite materials proved to be disadvantageous. While the internal forces of beam elements can be computed directly, the internal forces of shells must be determined either by manually setting sections in the programme or by integrating the stresses over the cross-section.

For the beams investigated here, steel profile with symmetrical cross-section and small ratio of plastic to elastic bending resistance, both methods achieve similar results. For general cross-sections, the more complex method using shell-elements seems to provide the more precise results. Since the modelling and calculation effort of the shell models is significantly higher, this paper will concentrate only on the framework models out of beam elements.

3 MATERIAL

If a powerful commercial software such as RFEM [7] from Dlubal is used to analyse the framework models, the material behaviour of the components is defined very detailed in tabular form so that plastic effects of the materials as well as cracking of concrete can be considered.

Based on results from push-out tests, simplified load-slip relationships of the shear connectors are determined. This simplified relationship is used in the model as stiffness of the horizontal springs. Figure 2 shows on the left the results of push-out tests of various headed studs with different concrete strengths from [8] and the simplified load-slip relationship of one headed stud on the right. To determine the horizontal spring stiffness for the model, numerous results of push-out tests of different types of shear connectors and concrete strengths can be integrated from various literature sources.

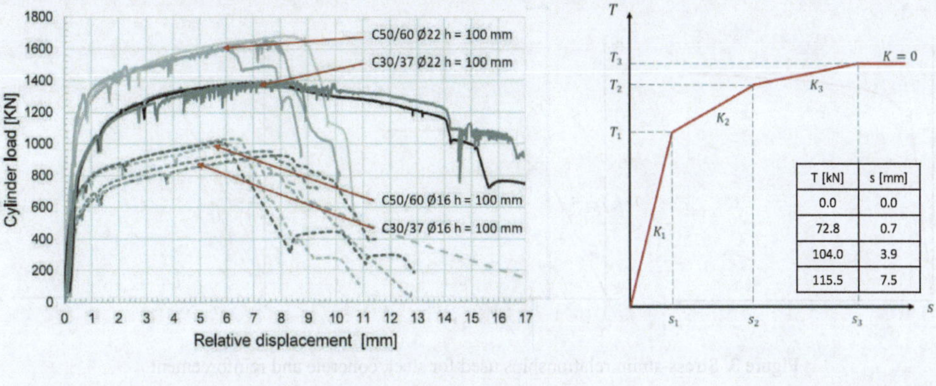

T [kN]	s [mm]
0.0	0.0
72.8	0.7
104.0	3.9
115.5	7.5

Figure 2. left: load-slip-relationship of headed stud from push-out tests from [8]; right: simplified load-slip relationship

Unlike the horizontal springs, the vertical springs in the model are defined separately for tensile and compressive forces. Since in the real structure the vertical compression forces are transmitted between concrete and steel via contact of both components, the vertical spring stiffness for compression forces is assumed to be almost unlimited rigid. In a model to determine the vertical stiffness of composite joints of headed studs proposed in [9] the stiffness of the concrete surrounded by the headed stud is applied while the headed stud is assumed rigid. First an equivalent circular concrete surface is determined. Due to the small number of available test results for headed studs under axial tensile forces specifying the connections stiffness, the equation for determining the equivalent concrete diameter is only calibrated on one test series. For the beams investigated below, consisting of a C30/37 concrete and 16 mm diameter headed stud, numerical simulations have shown that a major part of the uplift results from the elongation of the bolt shaft and that the concrete compression only has a minor contribution. For this reason, the tensile stiffness of the bolt shaft is modelled as vertical spring stiffness under tensile load. This assumption is also frequently made in fastening technology. Since the distribution of uplift varies depending on the stiffness of the connector and the concrete, further investigations are necessary to enable general statements. A serial arrangement of the bolt stiffness and the stiffness of the surrounding concrete is aimed. However, in order to be able to determine the stiffness of the concrete with confidence, further investigations are necessary to apply the model from [9] to general situations. Comparative calculations show that the vertical spring stiffness only influences the size of the uplift, but not the other results. If the uplift is not of interest, as it is usually the case, a rigid connection can be used in vertical direction. In the model, rigid bars with only vertical springs are arranged in the composite joint at the supports as well as under concentrated loads, comparable to load introduction stiffeners in real steel profiles. If no connectors are arranged at these points, the springs only transmit compressive forces.

Figure 3 shows the material models used for steel, concrete and the reinforcement. To consider the real load-bearing-behaviour of the composite beams the analysis is done with the average of the tested material properties. The stress-strain behaviour of the steel was modelled from data of tensile tests with consideration of the yield plateau and resolidification. The stress-strain behaviour of the concrete in compression was modelled parabolic according to EN 1992-1-1 [10] without modelling the descending range using the test results of the average compressive strength and the modulus of elasticity. Since it is not possible to consider the falling branch in the stress-strain-curve after reaching the compressive strength in the programme, the graph is constant up to a maximum strain $\varepsilon_{cu,1}$ of 3.5‰. This results in a slight overestimation of the concrete at higher compression strains. The concrete under tensile forces is modelled linearly elastic up to the tensile strength. In the programme, the function "crack" is activated in the material modelling, so that stress drops after once reaching the tensile strength.

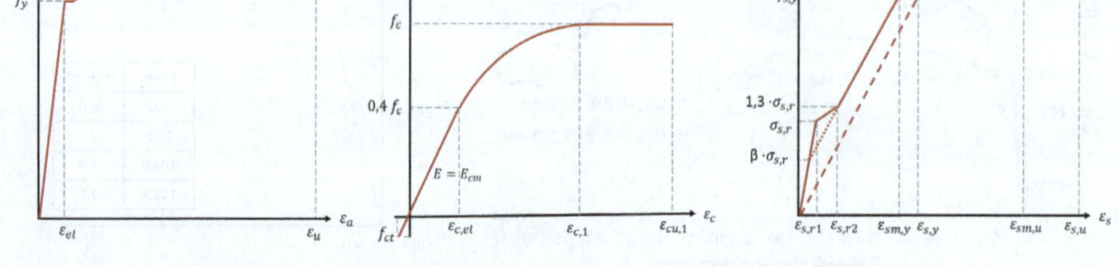

Figure 3. Stress-strain relationships used for steel, concrete and reinforcement

In reality, tensile stresses in reinforced concrete do not drop immediately after exceeding the tensile strength, but forces can still be transmitted across the cracks. To consider the effect of tension stiffening in nonlinear analysis, there are two different methods. In numerical simulations tension stiffening is usually considered via a stress-strain relationship or a stress-crack opening relationship with a falling range after exceeding the tensile strength. As an alternative, the tension stiffening may be considered via a modified stress-strain relationship of the reinforcement according to [11]. The stiffness of the reinforcement is increased depending on the cracking in the concrete in order to consider the concrete's effect. In figure 3 (right) the solid line corresponds to the course from [11]. As a further simplification and if repeated loading and unloading is to be expected, [12] suggests straightening of the peak in the range of $\sigma_{s,r}$. This approach differs only in the range from $\beta\ \sigma_{s,r}$ to $1.3\ \sigma_{s,r}$ and is shown in figure 3 with the dotted line. This simplification is used for the following analyses.

4 INFLUENCE OF MATERIAL MODELLING

Since material modelling can be complex and not all required parameters are accessible, a parameter study is carried out in addition to the "exact" material properties from figure 3 in order to investigate the influence of material modelling. Minimum requirements for the modelling are derived. For this purpose, the three single-span beams described in section 5 were analysed with different material modelling accuracies. The modelling of the stress-strain behaviour of steel and concrete, the yield strength of the steel, a neglect of tension stiffening and a neglect of the reinforcement were varied. One parameter was varied per calculation, all others remained constant according to figure 3. The most important results of the investigation are shown below in figure 4 using the example of the beam "SSB 70%" of section 5.

For steel, in addition to the material behaviour described above, a bilinear curve with experimentally determined tensile strength and associated strain as well as bilinear curves according to EN 1993-1-5 Annex C [13] were applied once with and once without steel hardening. The results of the "exact" material modelling and those with bilinear behaviour without considering the steel hardening are shown in the top left of figure 4. The influence of the yield strength is shown at the top right. Here, large deviations occur in the calculations. In addition to the material model from figure 3, the concrete was modelled with a trilinear material behaviour, in which the parabolic course according to EN 1992-1-1 [10] was approximated, as well as with a bilinear course, in which the slope up to the compressive strength constantly corresponds to the modulus of elasticity of the concrete.

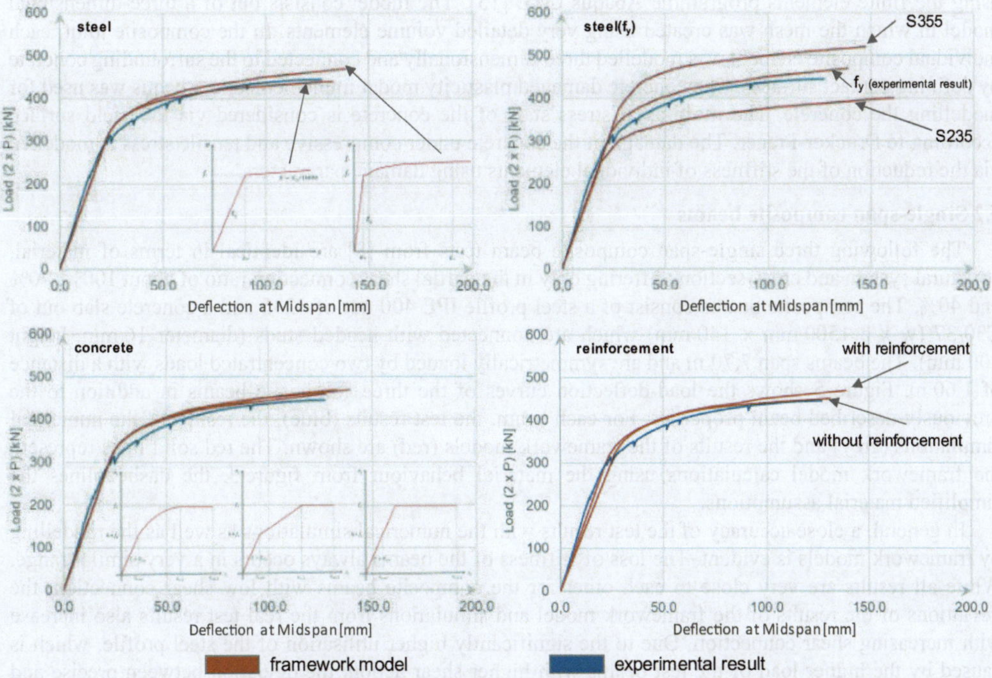

Figure 4. Influence of the accuracy of the materials modelling

Modelling of the concrete has no visible influence here. Since the concrete in the composite beams essentially only acts in compression, a model that neglects the reinforcement also achieves reliable results. Deviations only occur in the elastic range due to the neglect of tension stiffening.

The parameter study on the single-span beam with concrete in compression zone clearly shows that variations in the material behaviour of the steel have a much more significant influence on the results than the material behaviour of the concrete. This agrees with the results of safety-parameter investigations on single-span composite beams from [14].

It is shown that the application of the real yield strength of the steel has the greatest influence on the accuracy of the calculation. Due to overstrength of the steel, it is therefore essential to carry out tensile tests to know the precise start of plastification. If such results are not available, a rough estimate can only be made based on the strength classes. It is evident that the steel can also be approximated bilinearly according to EN 1993-1-5 [13]. In addition, it has been shown in figure 4 that the concrete can also be bilinearly approximated and the reinforcement can be neglected.

5 VERIFICATION OF THE MODEL

5.1 Introduction

To verify the analysis model described above, comparative calculations with existing tests on single-span composite beams were carried out. Furthermore, the application of the framework models to composite beams with negative moments, i.e., with concrete in the tension zone, is to be checked. For this purpose, comparative calculations are carried out on symmetrical double-span beams.

In addition, scientific numerical simulations were carried out for the investigated single-span beams using the finite elements programme Abaqus CAE [15]. The model consists out of a three-dimensional model in which the mesh was created using very detailed volume elements. In the composite joint, each individual composite element was modelled three-dimensionally and connected to the surrounding concrete by defining contact surfaces. The concrete damaged plasticity model implemented in Abaqus was used for modelling the concrete. The multi-axial stress state of the concrete is considered via the yield surface according to Drucker-Prager. The damage of the concrete under compressive and tensile stress is modelled via the reduction of the stiffness of individual elements using damage parameters.

5.2 Single-span composite beams

The following three single-span composite beam tests from [8] are identical in terms of material, structural system and cross-section, differing only in the partial shear connection ratio of about 100%, 70% and 40%. The composite beams consist of a steel profile IPE 400 out of S235 and a concrete slab out of C30/37 (w x h 1500 mm x 140 mm) which are connected with headed studs (diameter 16 mm, height 100 mm). The beams span 7.70 m and are symmetrically loaded by two concentrated loads with a distance of 1.60 m. Figure 5 shows the load-deflection curves of the three composite beams in addition to the previously described beam properties. For each beam, the test results (blue), the results of the numerical simulations (grey) and the results of the framework models (red) are shown. The red solid lines represent the framework model calculations using the material behaviour from figure 3, the dashed lines the simplified material assumptions.

In general, a close accuracy of the test results with the numerical simulations as well as the modelling by framework models is evident. The loss of stiffness of the beams always occurs in a very similar range. While all results are very close to each other for the composite beams with low shear connection, the deviations of the results of the framework model and simulations from the real test results also increase with increasing shear connection. Due to the significantly higher utilisation of the steel profile, which is caused by the higher load of the test beams with higher shear action, the deviation between precise and simplified material modelling also increases. It seems, that the flexibility of the composite joint is more accurately represented in the model than the material behaviour of the beam elements. To investigate the influence of the overestimation of the concrete at compression strength, the maximum concrete compression strains were determined from the framework models. Concrete compression strain above $\varepsilon_{c,1}$ does not occur with the beams "SSB 40%" and "SSB 70%", thus no overestimation of the concrete occurs here. Only with beam "SSB 100%" a compression strain of $\varepsilon_{c,1}$ exceeded from a load of 520 kN and thus the load-bearing capacity of the concrete is exceeded. This effect can be seen in figure 5 above this load level.

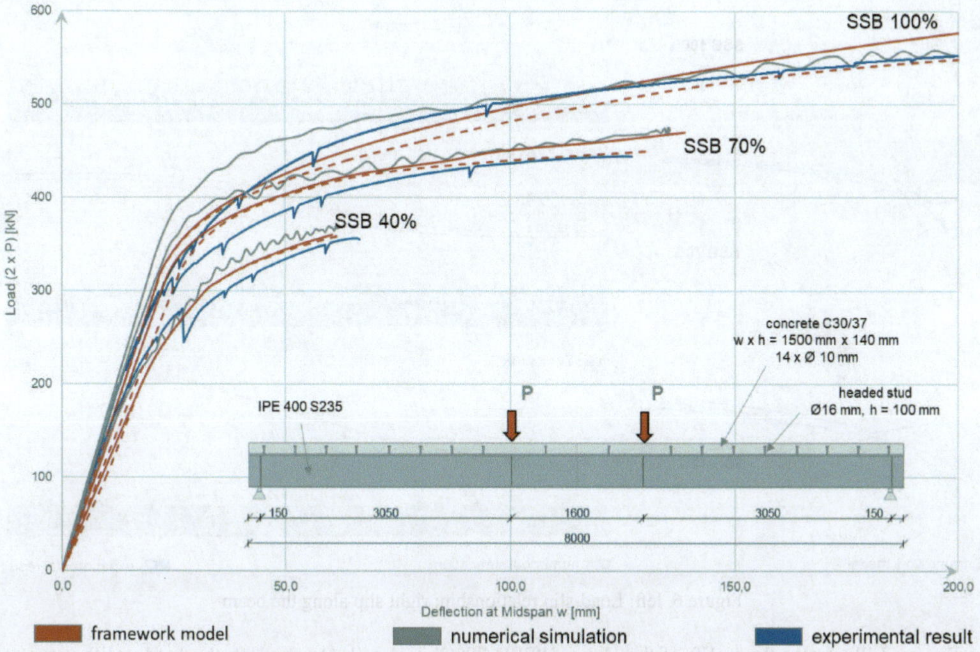

Figure 5. Load-deflection-curve of the three single-span beams

According to [8], the failure of the three test beams occurs after huge deformations due to a sudden fracture of the composite joint, in which the concrete slab separated over the entire length of one beam half. This failure is indicated in the load-end slip curves in figure 6 by a horizontal deformation curve at the end of the test. The sudden failure cannot be represented in the framework models. However, an almost horizontal course of the slip at beam end can also be seen in the results of the framework model, which is comparable to a failure of the bonded joint.

Figure 6 shows the load end slip curves of the three beams on the left. Since in the real tests, unlike in the model calculations, there was no absolute symmetry in the beams, the results from both beam-ends are shown. It can be seen that in the test, despite preloading cycles, it was not possible to eliminate the adhesive bond between the concrete and the steel. At the beginning of the tests, the load increases without slip occurring. Only at higher load levels the bond failed, and slip occurred.

While the load-slip curves of the framework models in the range of bearing capacity show very good accuracy with the real tests, clear differences occur at the lower load levels. This is due to the simplified trilinear load-slip characteristic of the framework models. Since the push-out tests succeeded better in loosening the adhesive bond through preloading cycles, this also results in a deviation between the dowel characteristic curve and the real component behaviour. In general, the load-slip behaviour can be represented well by means of a framework model; with more precise input values, the test results would be even more accurate. On the right in Figure 6 the slip curves of the three beams are shown over the beam length. It is shown that the curves of the framework models (red) correspond very precisely to the individual measuring points of the beam tests (blue dots).

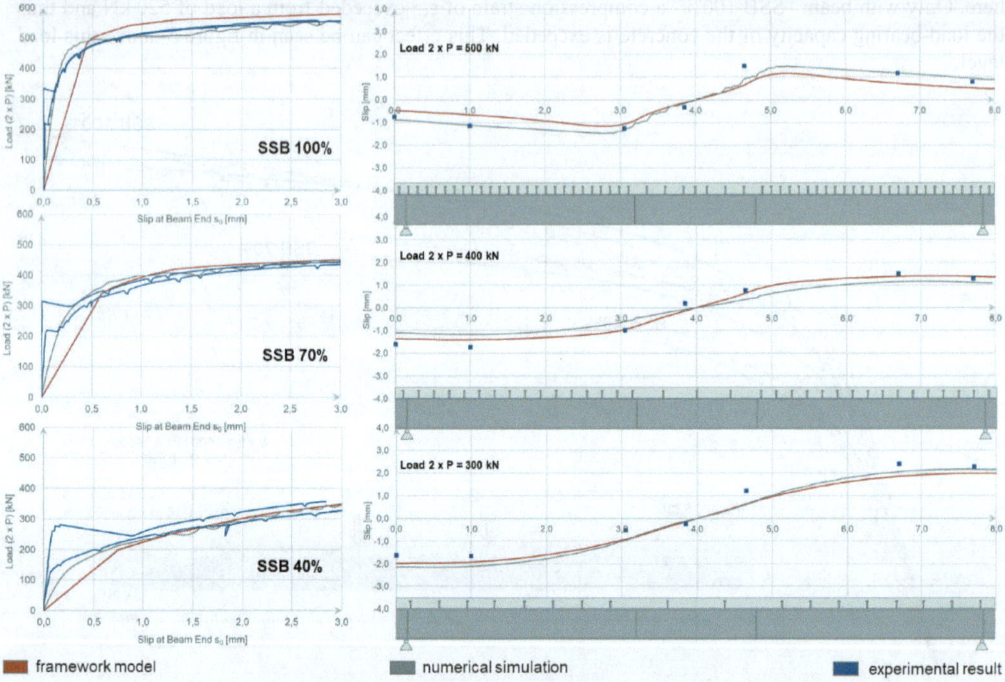

Figure 6. left: Load-slip relationship; right slip along the beam

Figure 7 illustrates the uplift of the slab of "SSB 70%" in detail. On the left, the load uplift curve at beam-end is shown. It is evident that there is good accuracy between the test and the simulations although there is a less stiff behaviour in the plastic range of the simulation. The simulations have shown a difference between local and global uplift effects. Whereas locally at the headed stud the uplift decreases to zero due to the compression, globally, i.e. between the edge of the steel profile and the concrete slab, uplift occurs almost over the entire length of the beam. Only the global uplift is presented in the following.

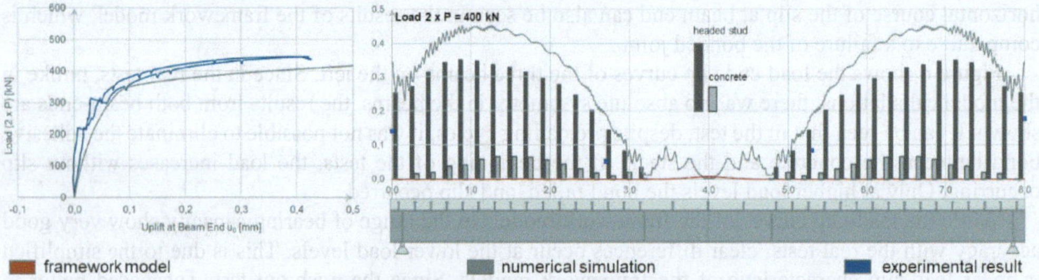

Figure 7. Uplift of the slab of "SSB 70%"

On the right, the grey curve shows the course of the uplift along the beam length at the load level of 400 kN as a result of the simulation. The four blue points represent measurements from the test. In red the course the uplift from the framework model with horizontal spring arrangement is shown. Even though slight uplift occurs, it is clear that it is of a much smaller order of magnitude than in the real tests. It is obvious that due to the modelling and the resulting deformation of the springs, a realistic modelling of the uplift effects was not possible. As mentioned at the beginning, an inclined arrangement of the "horizontal" spring could generate an uplift initiated by the slip as it occurs in the real construction, however here the uplift would be predetermined in the model.

Below the uplift curve, the magnitude of the longitudinal deformation of the headed stud as well as the magnitude of the concrete compression, which in total lead to the uplift, are shown by a bar chart. It is visible that at each point the uplift results mainly from the elongation of the headed stud. The assumption of the axial stiffness of the bolt shaft is therefore justified for the present composite joint. It is visible from the load-uplift curve that a non-linear behaviour of the bonded joint also occurs in vertical direction. The vertical stiffness would therefore have to be at least a multi-linear curve considering the plasticising of the headed stud.

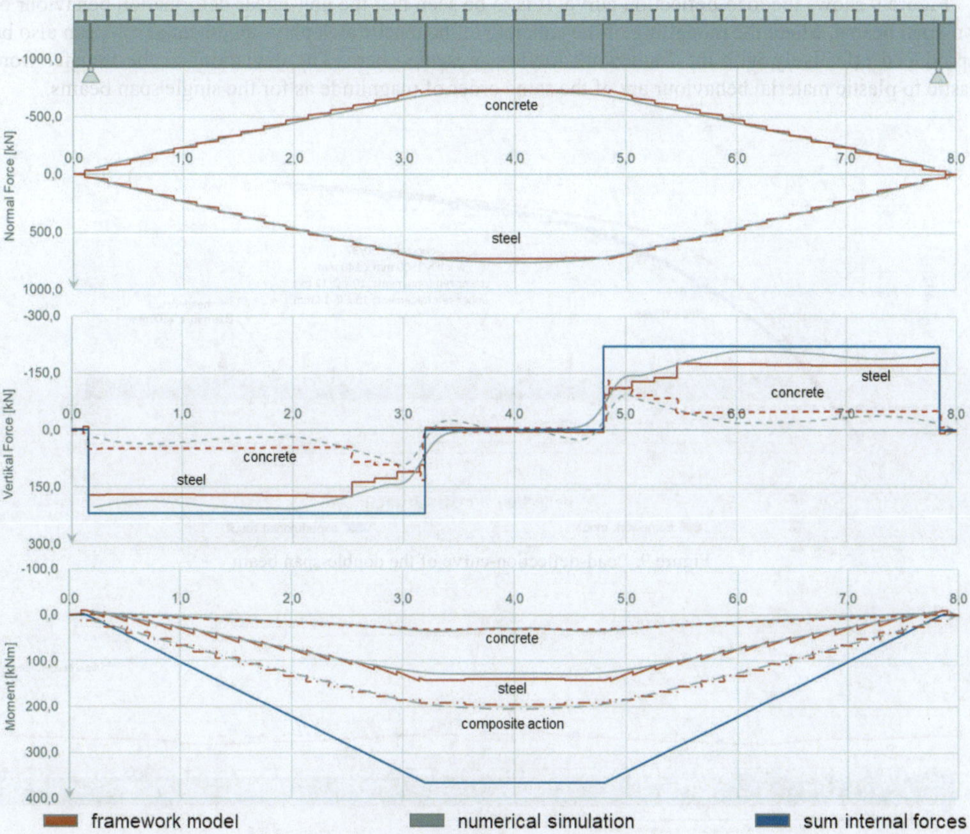

Figure 8. Internal forces of the beam "SSB 70%"

Figure 8 shows an example of the internal forces of the beam "SSB 70%" at the load level 2 x P = 440 kN, i.e., a load level far in the non-linear range of the load-deformation curve. The results of the partial internal forces of concrete and steel determined by means of a framework model and numerical simulations show very good consistency. Due to the framework model and the transmission of the forces at the defined nodes, the result curves show step changes at these nodes. While the curves of normal forces and moments show very good accuracy, where even the plastification of the steel is well represented, deviations occur in the distribution of shear forces. In general, it is visible that with the simplified modelling using framework models, a realistic distribution of the internal forces to the partial cross-sections is achieved and thus a realistic load-bearing behaviour of the model is assumed.

5.3 Double-span composite beams

In addition to the single-span composite beams, the framework models are also applied to multi-span composite beams or other beams with a concrete slab in tension. Here, comparative analyse is made for a 14 m long symmetrical composite beam with two spans with a partial shear connection ratio of about 40%. In the cross-sections, the beam has the similar properties as the previously investigated single span beams, except for different reinforcement. The exact properties are shown in figure 9. As the tests were just carried out, the results of the framework models are only compared with the experimental results of the beam, not with numerical simulations.

Figure 9 shows the load-deflection curve. It is to be seen that the non-linear deformation behaviour of two-span beams, where the modelling of the concrete in the tensile area plays an essential role, can also be reproduced effectively with the framework models presented here. The deviations in the transfer from elastic to plastic material behaviour are of the same order of magnitude as for the single span beams.

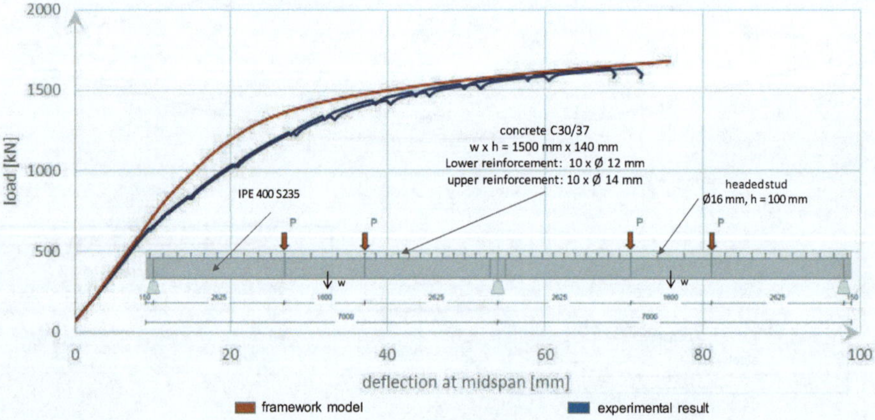

Figure 9. Load-deflection-curve of the double-span beam

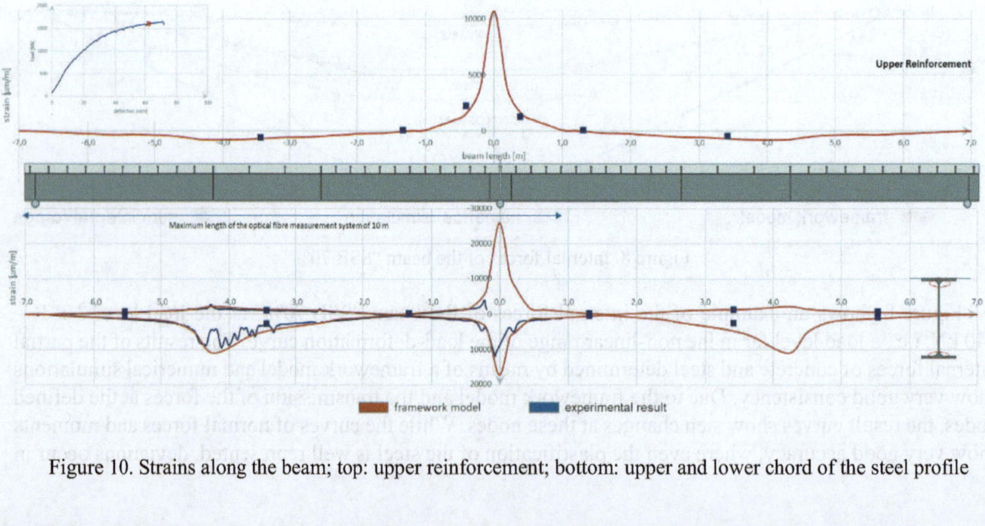

Figure 10. Strains along the beam; top: upper reinforcement; bottom: upper and lower chord of the steel profile

Figure 10 presents a comparison of the framework model's steel strains with the results of the experiment at a load of 1600 kN. The upper course shows the results of the upper reinforcement and the lower courses represents the strains at steel profile measured at the upper and lower chords. The blue squares represent the measured results of the strain gauges and the two blue curves represents the results of the fibre optical measurement system measured continously over the length at the steel profile. Due to the maximum length of this system, the strains could not be measured at the whole length. Here, aswell for the steel profile as for the reinforcement a high accuracy of the framework modell even for the non-linear material behaviour could be seen.

Figure 11 shows the load-slip-relations at the marked sections. Each red ellipsis represents the position of a displacement transducer for slip measurement in the beam test. Due to the use of symmetry in the diagram, at each measuring point two experimental results are shown. It is visible that the result of the framework model corresponds very accurately with the test results at every measurement point.

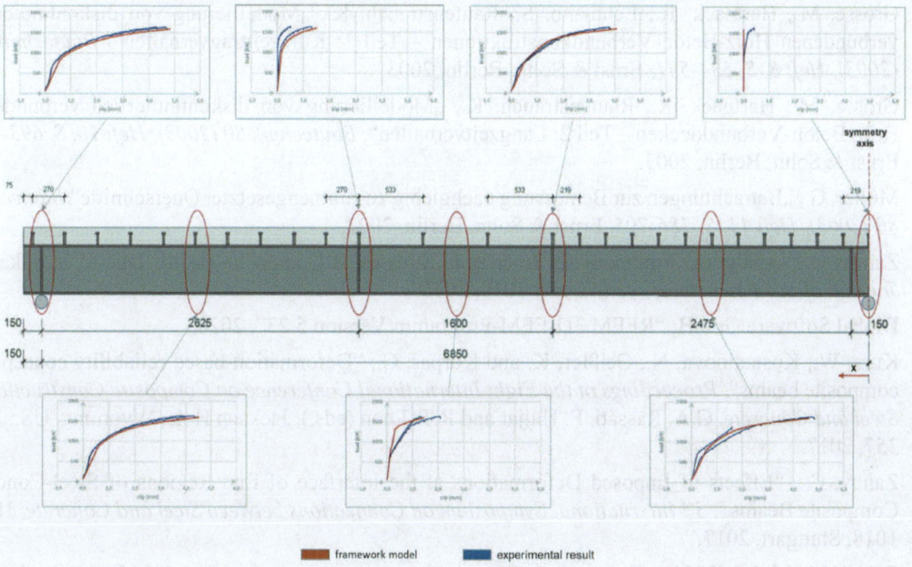

Figure 11. Load-slip relationship of the double-span beam at various sections

6 CONCLUSIONS

Based on the presented comparisons of the framework models with the test results of the beam tests of various static systems and partial shear connection ratios, it is demonstrated that the load bearing and deformation behaviour of composite beams can be modelled with good accuracy by framework models considering the non-linear material behaviour as well as the non-linear behaviour of the composite joint. While there are slight deviations in the load-deflection curves, the slip curves could also be reproduced very well, especially its non-linear behaviour. In addition, the good consistency of the curvatures and the internal force distribution shows clearly that a realistic load-bearing behaviour is modelled. Further investigations are necessary to determine the uplift behaviour, especially regarding the determination of the non-linear load-uplift curve as vertical spring stiffness as well as the arrangement of the spring in the model. Due to the high level of correlation between model calculations and real tests, it is possible to use the beam models in research as a complement to complex numerical simulations.

Since the real beam tests could be represented with high consistency, the framework model seems to be generally suitable for non-linear, strain-limited designs. While the non-linear material properties of steel, concrete and reinforcement could be derived from the respective standards, improved information would have to be developed for the representation of the composite joint depending on the type of shear connector as well as the concrete strength.

REFERENCES

[1] DIN EN 1994-1-1 (2010)., *Eurocode 4: Design of composite steel and concrete structures – Part 1-1: General rules and rules for buildings*, Beuth-Verlag, Berlin, Germany.

[2] Kneidl, R.; Hartmann, H., "Träger mit nachgiebigem Verbund. Eine Berechnung mit Stabwerksprogrammen", *Bauen mit Holz 97 (4),* 285-290, Bruderverlag, Köln, 1995.

[3] Grosse, M., Hartnack, R., Lehmann, S., Rautenstrauch, K., „Modellierung von diskontinuierlich verbundenen Holz-Beton-Verbundkonstruktionen – Teil 1: Kurzzeittragverhalten", *Bautechnik 80 (2003), Heft 8, S. 534-541,* Ernst & Sohn, Berlin, 2003.

[4] Grosse, M., Hartnack, R., Rautenstrauch, K., „Modellierung von diskontinuierlich verbundenen Holz-Beton-Verbunddecken – Teil 2: Langzeitverhalten", *Bautechnik 80 (2003), Heft 10, S. 693-701,* Ernst & Sohn, Berlin, 2003.

[5] Möller, G., „Betrachtungen zur Bemessung nachgiebig zusammengesetzter Querschnitte" *Bautechnik 80 (2003), Heft 11, S. 786-795,* Ernst & Sohn, Berlin, 2003.

[6] Zanuy, C., "Analytical Equations for Interfacial Stresses of Composite Beams Due to Shrinkage", *Journal of Steel Structures,* 15 (4), 999-1010, 2015.

[7] Dlubal Software GmbH; "RFEM 3D-FEM-Programm Version 5.23", 2020.

[8] Kurz, W., Kostadinova, N., Geißler, K. and Korpas, G., "Deformation-based reliability concept for composite beams", *Proceedings of the Eight International Conference on Composite Construction in Steel and Concrete*, G.A. Rassati, F. Hajjar and R.T. Leon (eds.), Jackson Hole, Wyoming, US., 346-357, 2017.

[9] Zanuy, C.; "Effects of Imposed Deformations at the Interface of End Regions of Steel-Concrete Composite Beams", *3rd International Symposium on Connections between Steel and Concrete,* 1007-1018, Stuttgart, 2017.

[10] DIN EN 1992-1-1 (2011)., *Eurocode 2: Design of concrete structures – Part 1-1: General rules and rules for buildings*, Beuth-Verlag, Berlin, Germany.

[11] CEB-FIP „Model Code 1990 – Design Code", London, 1993

[12] Deutscher Ausschuss für Stahlbeton "Heft 525 – Erläuterungen zu DIN 1045-1, 2. Überarbeitete Auflage", Berlin, 2010.

[13] DIN EN 1993-1-5 (2019)., *Eurocode 3: Design of steel structures – Part 1-5: Plated structures*, Beuth-Verlag, Berlin, Germany.

[14] Geißler, K.; Heyde, S., "Zum zuverlässigkeitstheoretischen Konzept der Verbundträgerbemessung", *Stahlbau 77 (2008), Heft 4,* 274-283, Ernst & Sohn, Berlin, 2008.

[15] Dessault Systems Simulia Corp. „Abaqus/CAE 2021" 2021.

9TH INTERNATIONAL CONFERENCE ON COMPOSITE CONSTRUCTION IN STEEL AND CONCRETE
M. Knobloch, U. Kuhlmann, W. Kurz, M. Schäfer. (Eds.)
Online, Germany, July 27-29, 2021

PROBABILISTIC DESIGN OF COMPOSITE GIRDERS CONSIDERING THE DEGRADATION OF THE SHEAR CONNECTION AND THE REDISTRIBUTION OF FORCES

Kevin Wolters*, Georgios Christou and Markus Feldmann***

* RWTH Aachen University, Institute of Steel Construction
e-mails: k.wolters@stb.rwth-aachen.de, feldmann@stb.rwth-aachen.de

** RWTH Aachen University, Institute of Structural Concrete
e-mail: gchristou@imb.rwth-aachen.de

Keywords: Composite bridges, fatigue, crack propagation, residual capacity, probabilistic design

Abstract. *Fatigue design of composite girders and in particular shear connectors according to Eurocode 4 is based on fatigue strength curves. Therefore, the lifetime of a girder can be limited by failure of a critical shear connector. The failure probability of a component is thus determined by the failure probability of one shear stud or composite dowel, although a composite girder is an internally highly statically indeterminate system with the possibility of redistributing forces in the shear connection. This paper outlines the positive effects on the lifetime of a girder by considering crack propagation and residual capacity of shear connectors with cracks as well as redistributing forces towards less stressed and damaged connectors. The failure probability of a girder is determined by a Monte Carlo simulation using a lamella model with incremental calculation of the degradation of the shear connection. The results show the economic potential of a future global safety concept for composite girders especially with composite dowels.*

1 INTRODUCTION

Due to their age and the increased transport weights in the last decades, many bridges in Germany and throughout Europe are currently in bad condition. Germany's bridge statistics show that almost 12 percent of all bridges on German highways will need to be extensively refurbished or replaced in the short to medium term [1]. The current and future demand for safe and economical bridge structures is expected to increase. In recent years, new bridges of medium span have preferably been built using composite construction.

The shear connection of composite girders and bridges is usually realized by shear studs according to Eurocode 4. Since the introduction of their German technical approval in 2013 [2] composite dowels are also used more and more often, mostly in bridge construction [3]. In addition to the static load-bearing capacity, the fatigue design of shear connectors is of particular importance. Here, fatigue resistance of shear studs and composite dowels is determined using fatigue strength curves, which indicate the 95% survival probability determined from load-controlled model tests (with constant shear forces) with only one or a few connectors. So, the fatigue design of the whole shear connection with dozens to hundreds of connectors is broken down to the design of one decisive stud or dowel. This completely disregards the fact that

- other shear connectors are still intact or have longer service lives.
- excessive crack growth in a connector leads to a decrease in stiffness and thus to a decrease in the transmitted shear force, provided that surrounding dowels show less damage (path-controlled behaviour). Excessively damaged dowels are thus unloaded and their service life is extended.

These positive effects apply in particular for composite dowels, which have a pronounced phase of crack initiation as well as crack propagation and a high residual load-bearing capacity even in case of long cracks. This is shown schematically in Figure 1a) for a single dowel. After crack initiation in the hot spot of the dowel - decisive for fatigue design so far - a distinct crack propagation phase follows with a sufficiently high residual capacity of the composite dowel. The acting shear force reaches the load capacity of the damaged composite dowel only in cases of larger crack lengths. Further to be considered and shown in the figure are increasing live loads over the service life of the structure and a higher scatter of the residual load-bearing capacity, because now also position, geometry and rate of crack propagation must be considered. Figure 1b) shows this development for shear studs. Due to the large notch effects of the weld collars of shear studs, recent research projects assume crack propagation almost from the first load cycle [4,5]. The lifetime prognosis shown is based on the consideration of the mean values and thus forms the 50% quantile. If the distributions of impact and resistance are known, forecasts for any other safety level can be derived accordingly.

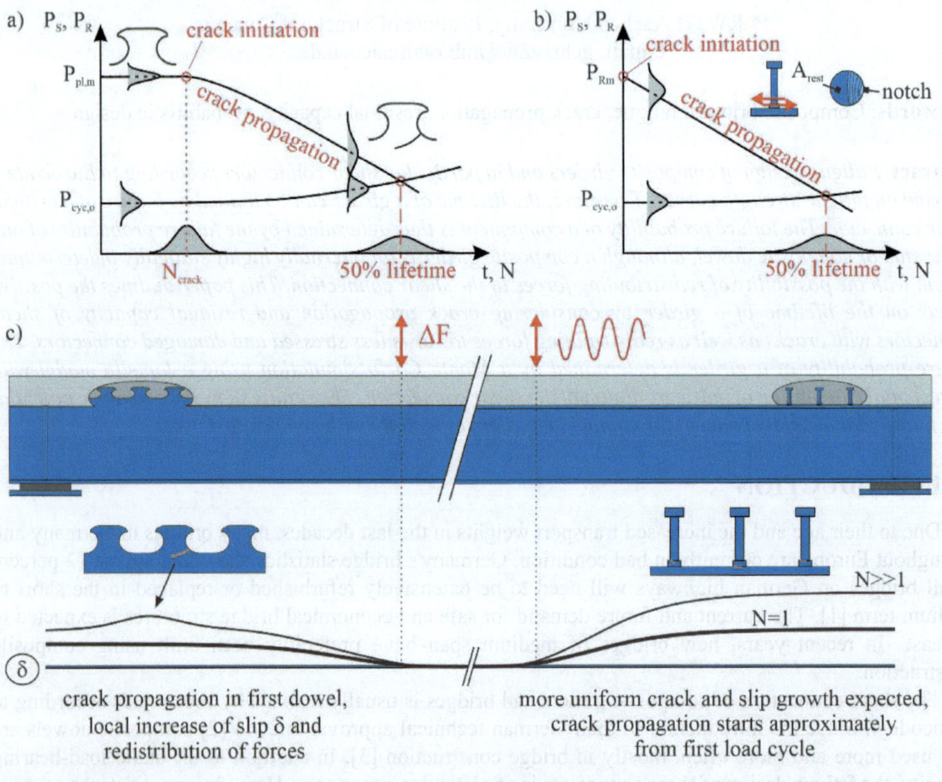

Figure 1. Exemplary lifetime prediction of a) one composite dowel and b) one shear stud with fatigue crack growth; c) slip δ in the shear connection and crack propagation with force redistribution especially using composite dowels.

Assuming a normal distributed point of crack initiation of composite dowels in a composite girder, crack propagation will start in a certain dowel while surrounding dowels are still uncracked. With progressing crack growth, the stiffness of the connector decreases and the slip of the shear connection increases locally, see Figure 1c). In contrast to Figure 1a), the shear force of the cracked dowel is redistributed to adjacent dowels and therefore decreases.

Consideration of a global fatigue behaviour of the whole shear connection could lead to a more favourable design level closer to the mean values of fatigue resistance in the future. This paper deals with the influencing parameters and methods of a probabilistic design of the shear connection using a lamella model.

2 STATE OF THE ART

2.1 Current fatigue design of shear studs and composite dowels

International fatigue studies on shear studs from the 1960s to the 1980s were used in [6] to derive a fatigue strength curve for Eurocode 4 and thus to apply the nominal stress concept from EN 1993-1-9 (detail category 90). The Wöhler curve for shear studs - determined by load cycle numbers, where connectors were cut off - has the advantage of a simple design, but it is extremely sensitive to the assumptions of the applied shear stress range due to Wöhler exponent m = 8. An error of 10% in the shear stresses assumed, for example, leads to an error of 114% in the theoretical service life [7]. The decreasing residual load-bearing capacity of increasingly damaged shear studs as shown in Figure 1b) is considered in Eurocode 4 by limiting the maximum cyclic load to 0.75 times, according to the national annex of Germany to 0.6 times the static load-bearing capacity. However, due to the decoupled consideration of the fatigue limit state and the ultimate limit state, no preliminary damage is considered when determining the maximum ultimate loads.

No new detail category for fatigue design was derived for composite dowels. Instead, the nominal stress range of a dowel is determined according to [2], see Figure 2, using stress concentration factors k_f, internal forces and cross-section values. For design of the steel dowels, the detail categories for machine flame cutting are used, category 125 without and 140 with post-treatment of the cutting line. Furthermore, the maximum cyclic load is limited to avoid cyclic concrete failure in the form of a steady increase of inelastic slip growth up to concrete pry-outs, see [2].

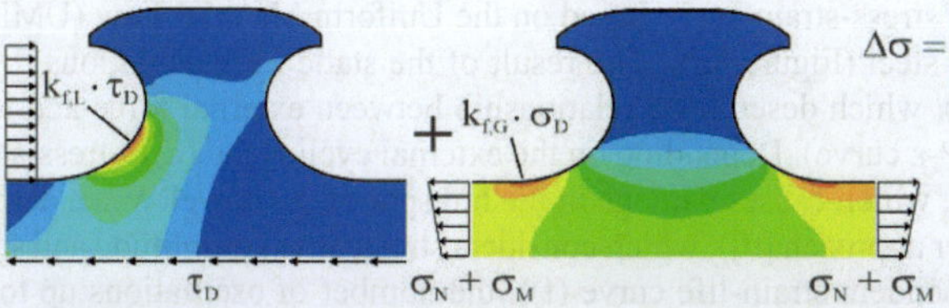

$$\Delta\sigma = \left| k_{f,L} \cdot \frac{\Delta V \cdot S_y}{I_y \cdot t_w} \right| + \left| k_{f,G} \cdot \left(\frac{\Delta N}{A} + \frac{\Delta M}{I_y} \cdot z_D \right) \right|$$

	$k_{f,L}$	$k_{f,G}$
clothoid shape	7,3	1,5
puzzle shape	8,6	1,9

Figure 2. Principal tensile stress distributions due to local shear and global forces for clothoid shaped dowels, stress concentration factors k_f for different dowels shapes and determination of the nominal stress range Δσ according to [2].

For both shear studs and composite dowels, the fatigue resistance was determined by push-out tests with only a few connectors and the fatigue check is performed for the decisive connector. The design level of the whole shear connection is therefore equal to the design level of a single connector.

2.2 Crack initiation, crack propagation and residual load-bearing capacity

2.2.1 Concepts for determination of crack initiation

While the lifetime of a shear stud is dominated by its phase of crack propagation due to the sharp notch of the weld collar as described earlier, composite dowels resist many load cycles until a crack is initiated. This depends among other things on material parameters, the quality of the cutting line and the stress amplitudes. The simplest way to assume the crack initiation point is using the fatigue resistance curve and the nominal stress concept, chapter 2.1. A more sophisticated method, supported by FE calculation, is the strain-life approach, using cyclic stress-strain (Ramberg-Osgood approach, [8], Eq. (1)) and strain-life curves (Manson, Coffin & Morrow, [9], Eq. (2)). Here, the parameters K', n', σ_f, ε_f', b and c can be determined in incremental step tests on the base material of the composite dowels [10].

$$\varepsilon_a = \varepsilon_{a,elastic} + \varepsilon_{a,plastic} = \frac{\sigma_a}{E} + \left(\frac{\sigma_a}{K'} \right)^{1/n'} \tag{1}$$

$$\varepsilon_a = \varepsilon_{a,elastic} + \varepsilon_{a,plastic} = \frac{\sigma_f}{E} \cdot (2N)^b + \varepsilon_f' \cdot (2N)^c \tag{2}$$

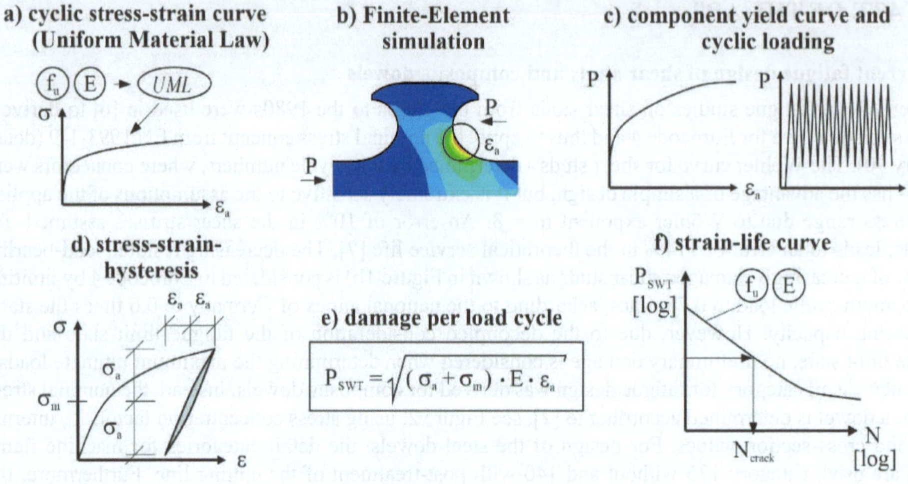

Figure 3. Summary of the strain-life approach.

Static FE simulations are first used to determine the relationship between external loads and local stresses or strains at the base. The cyclic stress-strain curve based on the Uniform Material Law (UML) is implemented as the material law for the steel (Figure 3a)). The result of the static FE calculations (b)) is the so-called component yield curve (c)), which describes a relationship between external force and local strain at the location of crack initiation (P-ε curve). Depending on the external cyclic load (c)), stress-strain hysteresis can be derived at the base (d)), which causes damage in the hot spot of the dowel. Via a damage parameter P_{SWT} (Smith, Watson & Topper approach [9]), which considers stress, stress amplitude and strain amplitude (e)), as well as a material dependent strain-life curve (f)), the number of oscillations up to the crack can be determined. The concept has been used for composite dowels in [11].

2.2.1 Crack propagation

Crack propagation in structural steel due to high cycle fatigue can usually be described via linear elastic fracture mechanics, neglecting small regions of plastic deformation at the crack tip. Therefore, stress intensity factors K are taken from literature for standard cases or are determined numerically. The stress intensity factor K describes the increase of elastic stresses near the crack tip, see Figure 4a). The loading of a crack tip can be divided in three components – tension, shear in plane and shear perpendicular to the crack plane. For composite dowels only the first two modes apply, Figure 4b). The crack propagation direction is dependant on the combination of both modes. In most cases stress intensity factors are calculated via analytical or empirical expressions of Eq. (3), where σ is the nominal stress, "a" is the crack length and Y and M_k are functions for different geometries of cracks and components.

$$K = \sigma \cdot \sqrt{\pi \cdot a} \cdot Y \cdot M_k \quad (3)$$

Using cyclic stresses, cyclic stress intensity factors can also be determined, Figure 4 c). Equations (4) and (5) were derived in [12,13] for composite dowels based on numerical parameter studies [14,15], where nominal stresses were replaced by shear stresses, considering the amplitude of dowel forces ΔP, the dowel spacing e_x, the web thickness t_w and one function Y both for crack and component geometry. Possible values for Y_m and Y_b are given in chapter 3.

$$\Delta K = \frac{\Delta P}{e_x \cdot t_w} \cdot \sqrt{\pi \cdot a} \cdot Y(a) \quad (4)$$

$$Y(a) = Y_m \cdot \ln(a) + Y_b \quad (5)$$

Finally, the crack propagation rate da/dN can be calculated via the material parameters C and m and the Paris Law [9] in Eq. (6), which is valid in a medium load range, Figure 4d).

$$da/dN = C \cdot \Delta K^m \tag{6}$$

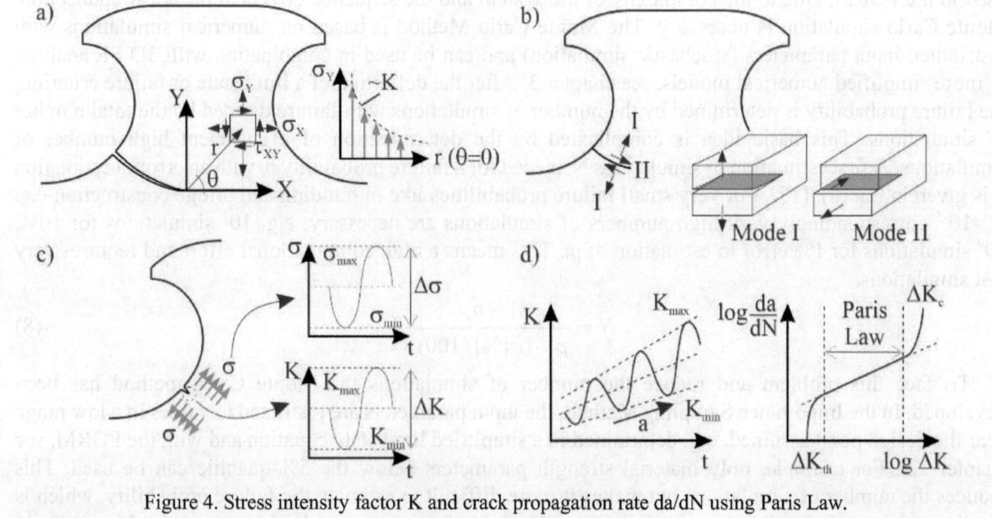

Figure 4. Stress intensity factor K and crack propagation rate da/dN using Paris Law.

2.3 Probabilistic design and First Order Reliability Method

Reliability theory is usually classified into five levels 1 to 4 depending on the degree of accuracy and complexity. The semi-probabilistic safety concept of the Eurocodes with the application of partial safety factors and combination factors corresponds to Level 1 of this classification. More complex and exact methods are Level 2 (probabilistic approximations), Level 3 (probabilistic "exact" solutions) and Level 4 methods (with additional optimization methods). For this purpose, the problem is usually transformed into a limit state equation Z, which contains any number of randomly distributed input variables, in the simplest case these are two normally distributed variables action (S) and resistance (R). Several methods have been developed for the evaluation of the limit state equation, whereby the First Order Reliability Method (FORM) with a limit state equation linearized at the design point is probably the most common method. Simple approaches exist for chain systems, where the failure probability of one link means failure of the chain. Like mentioned before, does this not apply to the behaviour of the shear connection. More complex approaches exist for brittle and ductile parallel systems. Often the estimation of failure probability is limited to determining upper and lower boundaries.

In literature, probabilistic approaches exist for the calculation of residual lifetime of steel bridges. An example is shown in Eq. (7), considering distributed values for loading (Weibull distribution with collective H and shape coefficient ν), fatigue resistance ($\Delta\sigma_D$, N_D, m) and damage parameter (Miner rule with failure at D_{limit}) [16].

$$Z = D_{\text{limit}} - \frac{\Gamma(m/\nu)}{\ln(H)^{m/\nu}} \cdot \frac{H}{N_D} \cdot \left(\frac{\Delta\sigma_{\max}}{\Delta\sigma_D}\right) \tag{7}$$

Failure probabilities p_f and safety levels β can be calculated with the FORM, provided there is a valid limit state equation. Eq. (7) can be adopted for one connector of the shear connection, but it is not possible to consider a redistribution of forces and sequence effects. Therefore, incremental calculations considering the whole shear connection are used instead in the following chapters.

2.4 Monte Carlo Method

For a composite girder with a high number of connectors with scattering crack initiation and propagation behaviour, redistributing forces and an ultimate limit state depending on the degradation of the shear connection, there is no closed solution for failure probability or a limit state equation Z, which can be used in the FORM. Due to the complexity of the system and the sequence effects in the shear connection, Monte Carlo simulation is necessary. The Monte Carlo Method is based on numerical simulations with distributed input parameters (stochastic simulation) and can be used in combination with 3D FE analysis or more simplified numerical models, see chapter 3. After the definition of a limit state or failure criterion, the failure probability is determined by the number of simulations with failure divided by the total number of simulations. This basic idea is complicated by the determination of a sufficient high number of simulations. A first estimation of simulations N needed for a failure probability p_f with an error in estimation α is given in Eq. (8), [17]. For very small failure probabilities like in building and bridge construction e.g. $p_f = 10^{-5}$ correspondingly very high numbers of simulations are necessary, e.g. 10^7 simulations for 10%, 10^9 simulations for 1% error in estimation of p_f. This means a high computational effort and requires very fast simulations.

$$N = \frac{1 - p_f}{p_f \cdot (\alpha[\%]/100)^2} \tag{8}$$

To face this problem and reduce the number of simulations the Monte Carlo method has been developed. In the Importance Sampling Method, the input parameters are restricted to values in a low range near the design point assumed, e.g. determined in a simplified limit state equation and with the FORM, see chapter 2.3. For example, only material strength parameters below the 5%-quantile can be used. This reduces the number of simulations but makes it more difficult to estimate the failure probability, which is not the quotient of simulations with failure to all simulations anymore. In this paper, the Monte Carlo Method is used exemplarily for an estimation of load cycles until a failure criterion is reached, see chapter 4. The Importance Sampling Method will be part of future studies.

3 LAMELLA MODEL AND INPUT PARAMETERS

For investigations on the complex interactions between steel girder, concrete slab and shear connectors in combination with the degradation of the shear connection due to fatigue loading, a framework or lamella model has been developed for MATLAB [21]. Steel girder and concrete slab are discretized as beam elements, both can be divided into an arbitrary number of lamellas, which has an impact on the computational time. The lamellas are connected over the length with stiff coupling elements, see Figure 5. Each composite dowel or row of shear studs can be modelled by a spring element, which has, amongst others, information about current crack length "a", stiffness, residual capacity and inelastic concrete deformation. Several input parameters are summarised in Figure 5.

The code translates the basic input about girder length, cross-section and positions of shear connectors into the beam-spring model shown. Therefore, nodes with two translational and a rotational degree of freedom are introduced and connected by 1D elements. For each beam, spring and coupling element an element stiffness matrix is created and built into a global stiffness matrix $\underline{K}$. The external loads are input as nodal forces into a load vector p. The deformation vector u is calculated considering the boundary conditions, Eq. (9).

$$\underline{K} \cdot u = p \Leftrightarrow u = \underline{K}^{-1} \cdot p \tag{9}$$

The deformation vector u contains all translational and rotational node deformations. Using element stiffness matrices, the internal forces in beam elements can be calculated. Furthermore, slip and shear force of each shear connector is calculated. More information about this method can be found in [18]. In contrast to FE simulations with volume elements e.g. using Abaqus, the calculation is done in less than a second, depending on the discretization. This allows for incremental calculations each ΔN, where crack propagation as well as a redistribution of forces can be considered.

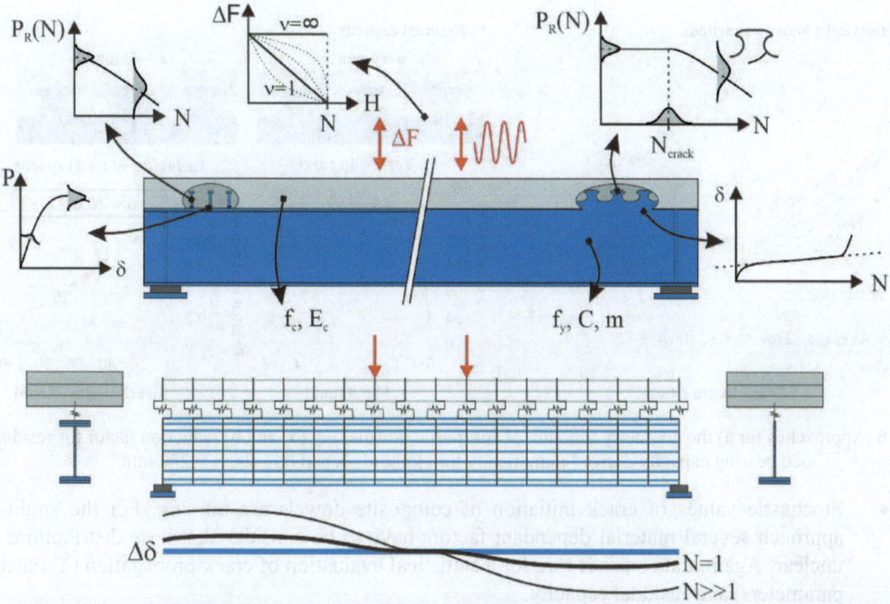

Figure 5. Input parameters to be considered in the beam-spring lamella model of the composite girder.

In each increment, the dowel forces lead to a certain damage or degradation due to ΔN load cycles. Important aspects that are or will be implemented in the model are shown in Figure 5. The external loading can be described by an arbitrary number of single or distributed line loads, with fixed values or as a collective. For composite dowels, the Palmgren-Miner rule is used until crack initiation. Here, the strain-life approach will be implemented in the future. After crack initiation, Eq. (4)-(6) are used to calculate the incremental crack length. Dependant on the current crack length of each dowel, stiffness and residual capacity are calculated. The calculation continues until an abort criterion is reached. This could be defined as a critical crack length, reaching the residual capacity in one dowel or even reaching the shear capacity of the whole shear connection or the moment capacity of the composite girder (partial shear connection).

The relations between forces, crack lengths, stiffness, residual capacity and inelastic slip were determined in further numerical studies with 3D solid element models in Abaqus or taken from literature. For composite dowels, Figure 6a) shows the geometry functions derived in [12]. The range of geometry functions is due to different crack initiation locations, web thicknesses, global loadings and influences of cracks in adjacent dowels. On the safe side, the upper limit formula can be used for Eq. (5). The residual capacity in Figure 6b) was determined using volume models with elastic concrete and ideal elastic-plastic steel behaviour [13]. Here, a more precise estimation will be done in the future. This information can be implemented in the lamella model. Approaches for shear studs can be found in [4,5].

For probabilistic calculations, any of the input parameters can be considered as a random variable with mean value, standard deviation and type of distribution, see Figure 5, either as a global random variable (e.g. yield strength of steel girder f_y) or as a local variable varying for each dowel (e.g. stiffness or load-bearing capacity). The biggest challenge is not the implementation of all these effects, the code can be developed quite easily. But further studies are needed to quantify realistic types of distributions, mean values and coefficients of variances:

- A database for the static load-bearing capacity of headed studs is good, but information about distributions of crack propagation rate, stiffness and residual capacity of cracked studs is rare.
- Only approaches with constant load amplitudes exist for inelastic slip behaviour using composite dowels [11,19]. Decreasing dowel forces can not be considered so far.

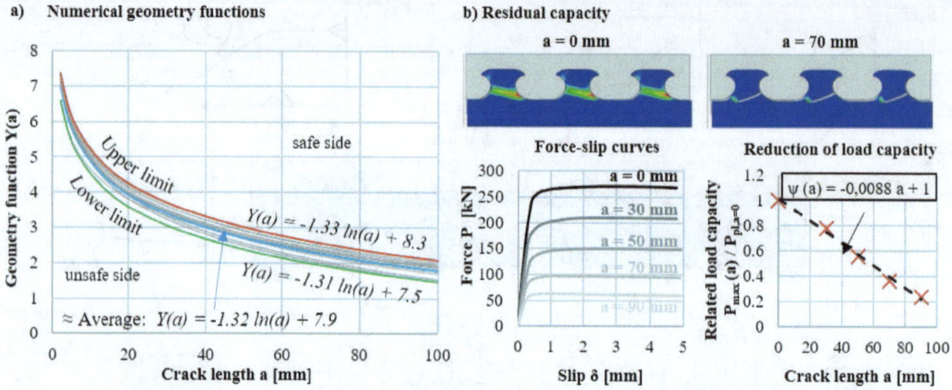

Figure 6. Approaches for a) the geometry function of composite dowels, Eq. (5) and b) reduction factor for residual load bearing capacity derived numerically for clothoid shaped dowels, $e_x = 200$ mm.

- Stochastic values of crack initiation of composite dowels are missing. For the strain-life approach several material dependant factors have to be considered, whose distributions are unclear. Again, data basis is rare for a statistical evaluation of crack propagation (Y function parameters) and residual capacity.
- For the material parameters C and m for the Paris Law, Eq. (6), several different assumptions can be found in literature, leading to a huge span of possible crack rates.
- Additionally, the correlation of different parameters, for example between stiffness and residual capacity depending on the crack length or between material parameters (C and m; parameters for the strain-life-approach, chapter 2.2.1) is unknown.

One the one hand there are still lots of uncertainties about input parameters which require further experimental and theoretical studies and therefore, results of the calculations are preliminary. But on the other hand, the lamella model is a powerful tool to check possible effects of the influence parameters on the behaviour of the girder. The method shown in chapter 4 will definitely help to gain more experience and better understanding of the fatigue behaviour of the shear connection.

4 PROBABILISTIC DESIGN WITH MONTE CARLO SIMULATIONS

As shown in the previous chapters, there are still lots of open questions regarding the distributions of input parameters within probabilistic calculations. So, the method used for simulations in this paper are based the Monte Carlo Method described in chapter 2.4, but not in its original sense to determine failure probability values p_f. Instead, a simplified model is used to gain experience about the influence of individual parameters on the fatigue behaviour of the shear connection and its interactions. The result of each simulation is the number of load cycles N_{ab} until an abort criterion is reached. Using randomly distributed input parameters, N_{ab} is also a random value. By thousands of simulations, the mean value, coefficient of variance and type of distribution of N_{ab} can be calculated. The influence of one individual parameter can thereby be estimated, setting all other parameters to fixed values. In conclusion, the most important parameters on the fatigue behaviour can be found this way. Future studies should focus on these parameters.

The following assumptions and boundary conditions apply to the various Monte Carlo simulations in this paper:

- According to experimental tests described in [12,13], a composite girder with composite dowels (dowel spacing $e_x = 200$ mm, equal to the distance between coupling elements and spring elements, respectively), a static length of 6.8 m, a halved HEM 500 and a concrete slab 100 x 18 cm is used for the parameter studies. The numerical model was validated using this beam tests.

- The external loading is fixed and not a random variable. According to Figure 5 a four-point bending load is applied, with a ΔF of 120 kN per load.
- Fatigue of the steel girder is neglected and is only considered in the shear connection. The focus is on fatigue of the steel teeth especially on crack propagation. Concrete failure is not part of the studies.
- In the experimental girder tests, three dowels were notched on each side to initiate crack propagation. The corresponding dowels are also "notched" in the model. So, crack propagation starts at the beginning. In the remaining dowels, the shear force is converted into a hot spot stress and a damage parameter (Miner rule) is introduced. If this parameter reaches 1.0 in a dowel, another crack is initiated.
- A crack length of 100 mm is used as an abort criterion. Maximum shear forces in the example calculations are about one third of the characteristic shear resistance per dowel. Since other dowels are still intact, there will be no complete failure of the shear connection and of the girder due to a low residual capacity. The focus is on the crack propagation in the steel teeth.
- One incremental calculation until N_{ab} is reached takes about 2-3 seconds. However, this means a high computational effort for Monte Carlo simulations with thousands of incremental calculations for each individual parameter. To save time, only one lamella is used for the steel girder and one for the concrete slab, leading to the time effort mentioned. $\Delta N = 10^5$ is used as incremental load cycle number.
- Input parameters concerning the geometry of the girder are fixed. Possible random parameters are
    - the material parameters C and m, Eq. (6), as random but global parameters for the whole girder
    - and the geometry function Y with values Y_m and Y_b, random for each dowel Eq. (4) and (5),
    - the hysteretic dowel stiffness C_{hyst} and the rate of degradation C_{deg} of the stiffness depending on the crack length "a" and
    - the resistance against crack initiation $\Delta\sigma_D$ (random fatigue strength curve).

As a reference, Figure 7 shows the exemplary girder model following the experimental tests, mean values of the input parameters and results for slip, dowel forces and crack development.

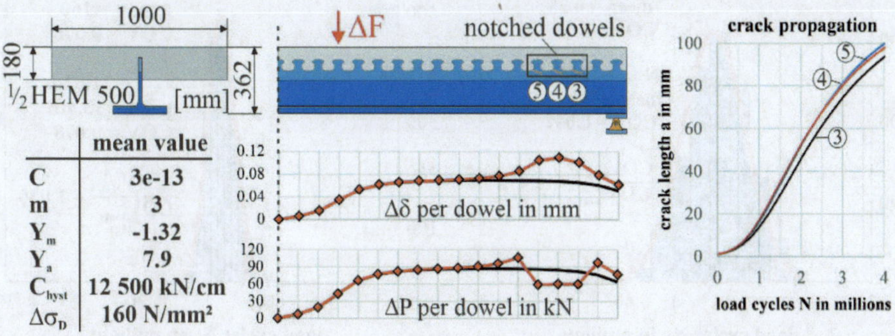

	mean value
C	3e-13
m	3
Y_m	-1.32
Y_a	7.9
C_{hyst}	12 500 kN/cm
$\Delta\sigma_D$	160 N/mm²

Figure 7. Girder with composite dowels for parameter studies of length 6.8m under four-point bending loading and with three notched dowels, mean values for input parameters leading to hysteretic slip $\Delta\delta$ and shear forces ΔP (black: first load cycle; red: last load cycle) and development of crack lengths for the three dowels.

Using the mean values as fixed parameters, the chosen abort criterion - a crack length of 100 mm - is reached in dowel 4 after four million load cycles. This is the reference value for N_{ab} in the following. The crack propagates similarly in dowels 4 and 5, a little bit slower in dowel 3. The slip increases locally around dowels 3 to 5, the shear forces in as well as the stiffness of cracked dowels decrease with the crack length. Shear forces are redistributed to adjacent dowels 1, 2, 6 and 7.

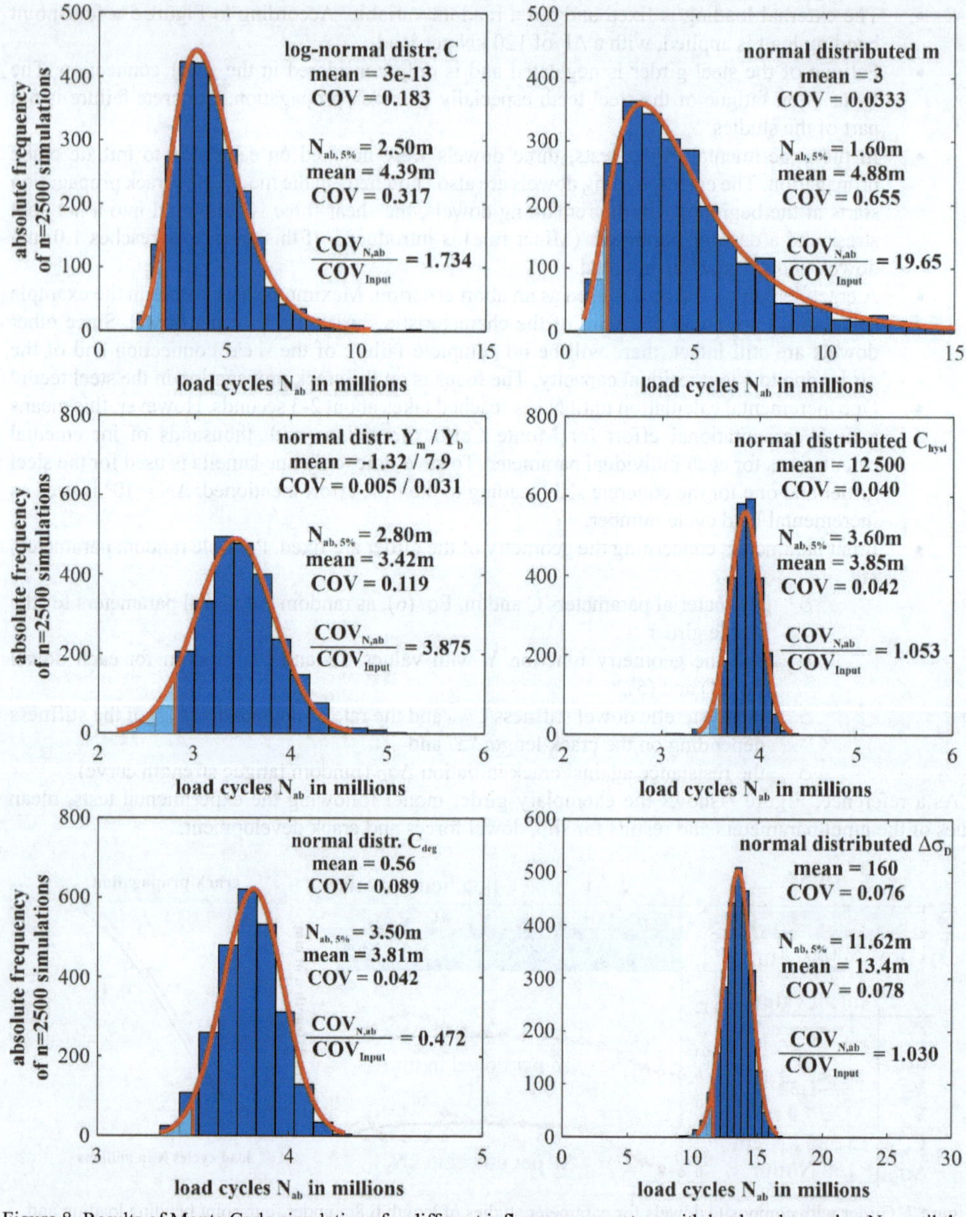

Figure 8. Results of Monte Carlo simulations for different influence parameters with load cycle number N_{ab} until the abort criterion "critical crack length 100mm" is reached, comparison of coefficients of variance (COV) of N_{ab} and input parameter to show effects of distributed input parameters.

Figure 8 summarizes the results of several Monte Carlo simulations. For each individual input parameter, 2500 incremental simulations with random distributed values were evaluated. It should be noted that the absolute frequency is dependant on the number of bars for the diagrams and is a qualitative value. Furthermore, different values for the coefficients of variance (COV) were used, a direct comparison of the influence of single parameters is difficult. Therefore, the COV's of input and result distributions are compared. A quotient of 1.0 shows that the distribution of the input parameter leads to a similar distribution of load cycle numbers until "failure". Quotients greater than 1.0 show an increase of effects of random input parameters. The results show a small effect of hysteretic dowel stiffness C_{hyst} on N_{ab} and even a decreasing effect of the degradation parameter C_{deg}, which is considered as the percentage of stiffness degradation at crack length 100mm compared to initial stiffness. For the Monte Carlo simulation "$\Delta\sigma_D$" no initial cracks were considered, leading to higher N_{ab}. Since results are dependant on a value for translation of shear force into hot spot stress, the increase (N_{ab} almost tripled) should not be taken quantitatively. A hand calculation according to [2] would lead to crack initiation after 2 million load cycles.

The random values Y_m and Y_b of the geometry function Y were considered as fully correlated. The upper and lower bonds of Figure 6a) were taken as 5% and 95%-quantile, respectively. The quotient of COV 3.88 is very high, so small changes in the input data have a stronger effect on the load cycle number. The material parameters of the Paris Law C and m, Eq. (6), were considered as individual and uncorrelated parameters. Here, the assumptions in literature differ. In the Probabilistic Model Code [20] m is chosen as a deterministic parameter and C as log-normal distributed. Other authors assume high correlations between C and m. In this paper, both parameters are checked individually. The simulations show a huge influence on the assumption of m. With a small COV a great span of possible N_{ab} is determined, due to the exponential influence of m in Eq. (6). The influence of C is smaller, but significant.

Summarized, the first Monte Carlo simulations show big influences of the material parameters C and m and the chosen geometry function Y, whereas other parameters might be neglectable. In future studies, realistic approaches of random parameters are to be developed as well as correlations between these parameters and interaction effects. Furthermore, the strain-life approach will be implemented for crack initiation and the abort criterion will be developed considering decreasing residual capacities of the dowels. Failure probabilities p_f will be determined by Monte Carlo simulations for different cross sections and static systems to develop a global concept of fatigue design of shear connections.

5 CONCLUSIONS & OUTLOOK

The paper shows first results of ongoing studies on the fatigue behaviour of the shear connection. The focus of these studies is on composite dowels as shear connectors due to their good fatigue behaviour. The method can be adapted for shear studs. The method of probabilistic Monte Carlo simulations using a lamella model for time efficiency was shown. Since realistic distributions of input parameters are not cleared yet, their qualitative effect on the distribution of load cycle numbers until "failure" was examined for each parameter. The results show big influences of the material parameters C and m assumed in calculations. Also, the geometry function for the crack propagation Y has a strong effect on the results. These are the most important parameters in future studies.

The lamella model will be extended by the strain-life approach to determine crack initiation points, inelastic behaviour of concrete slip, as well as the effects of cracked concrete, with combined spring-gap elements and abort criteria considering the residual load-bearing behaviour of cracked shear connectors. Therefore, more experimental, numerical (with 3D volume elements) and theoretical studies are necessary. The aim is to sophisticate the fatigue design, which is based on one decisive shear connector so far, to a global design of the whole shear connection. This will lead to a better understanding of the real behaviour in a composite girder and will exploit the economic potential of composite constructions. The lifetime of the exemplary shear connection, which was designed for 2 million load cycles according to [2], can be doubled to tripled if the crack propagation phase is considered.

REFERENCES

[1] Bundesanstalt für Straßenwesen, "Zustandsnoten der Brücken", (https://www.bast.de/BASt_2017/DE/Statistik/Bruecken/Zustandsnoten.pdf?__blob=publicationFile&v=15).

[2] Deutsches Institut für Bautechnik, *Stahlverbundträger mit Verbunddübelleisten in Klothoiden- und Puzzleform*, German technical approval Z-26.4-56, 2018.

[3] Kopp M., Wolters K., Feldmann M. et al, "Composite dowels as shear connectors for composite beams – Background to the design concept for static loading", *Journal of Constructional Steel Research*, 147, 488–503, 2018.

[4] Feldmann M., Gesella H., "Zur Vorhersage des Rißfortschritts bei einbetonierten Kopfbolzendübeln in Verbundträgern unter nicht ruhender Belastung", *Stahlbau*, 74, 767-779, 2005.

[5] Hanswille G., Porsch M., Üstündag C., "Studies of the lifetime of cyclically loaded steel-concrete composite bridges", *Steel construction*, 3, 140-147, 2010.

[6] Roik K., Hanswille G., *Background Report on Eurocode 4 - Limit State of Fatigue for headed Studs*, Federal Office for Building and Regional Planning, Bonn, 1989.

[7] Mensinger M., *Zum Ermüdungsverhalten von Kopfbolzendübeln im Verbundbau*, Dissertation, Kaiserslautern, 2001.

[8] Ramberg W., Osgood W. R., "Description of stress-strain curves by three parameters", Technical note / National Advisory Committee for Aeronautics, Washington D.C, 902, 1943.

[9] Radaj D., *Ermüdungsfestigkeit. Grundlagen für Leichtbau, Maschinen- und Stahlbau*, Springer Berlin, 2003.

[10] Bissing H., Knobloch M., "Data-based fatigue assessment – application of the Incremental Step Test and the strain-life approach", *Workshop SimFat - Large-Scale Testing & Advanced Numerical Simulation in Fatigue and Fracture*, Publication Series of the Institute for Structural Engineering, Ruhr-University Bochum, 20-25, 2021.

[11] Gallwoszus J.: *Zur Ermüdung von Verbundkonstruktionen mit Verbunddübelleisten*, Dissertation, Institute of Structural Concrete, RWTH Aachen University, 2015.

[12] Feldmann M., Wolters K., "Tragverhalten eines Verbundträgers unter Berücksichtigung des Risswachstums in Verbunddübelleisten", *Berichte aus dem Konstruktiven Ingenieurbau 03/2020 – Festschrift Prof. Mangerig*, 227-234, 2020.

[13] Wolters K., Christou G., Feldmann M.: "Crack propagation and residual load-bearing behavior of composite dowels – effects on the global behavior of composite girders under cyclic loading", *Proceedings Eurosteel Sheffield 2021* (paper accepted).

[14] Wolters K., Feldmann M.: "Fatigue and crack propagation investigations on composite dowels in an angled single push-out test", *Proceedings ASCCS Valencia*, 725-732, 2018.

[15] Wolters K., Feldmann M.: "Inkrementelle Rissfortschrittsberechnungen in Verbunddübelleisten", *Workshop SimFat - Large-Scale Testing & Advanced Numerical Simulation in Fatigue and Fracture*, Publication Series of the Institute for Structural Engineering, Ruhr-University Bochum, 62-67, 2021.

[16] Al Shamaa D., Geißler K.: "Grundlagen zur Festlegung der Sicherheitselemente des Betriebsfestigkeitsnachweises", *Stahlbau*, 86, 753–764, 2017.

[17] Klingmüller O., Bourgund U., *Sicherheit und Risiko im Konstruktiven Ingenieurbau*, Vieweg & Sohn Verlagsgesellschaft mbH, Braunschweig/Wiesbaden, 1992.

[18] Wolters K.; Feldmann M., "Ermüdungsuntersuchungen an Verbundträgern mit Verbunddübelleisten und nachgiebiger Verbundfuge mittels Stabwerkmodellen", *Bauingenieur*, 93, 11, 429-437, 2018.

[19] Christou G., Ungermann J., Wolters K. et al., "Ermüdung von Verbunddübelleisten – Analyse und Modellentwicklung", *Beton- und Stahlbetonbau*, 115, 5, 355-363, 2020.

[20] JCSS, "Probabilistic Model Code", (https://www.jcss-lc.org/publications/jcsspmc/JCSS-3-12-Fatigue.pdf).

[21] MATLAB and Statistics Toolbox Release 2020a, The MathWorks, Inc., Natick, Massachusetts, United States.

9TH INTERNATIONAL CONFERENCE ON COMPOSITE CONSTRUCTION IN STEEL AND CONCRETE
M. Knobloch, U. Kuhlmann, W. Kurz, M. Schäfer. (Eds.)
Online, Germany, July 27-29, 2021

PROGRESSIVE COLLAPSE MODELING OF A STEEL STRUCTURE WITH COMPOSITE SLAB USING COMMERCIAL FINITE ELEMENT SOFTWARE

Trent J. Phillips*, Gian A. Rassati*, James A. Swanson*, Nadia Baldassino and Riccardo Zandonini****

* University of Cincinnati
e-mails: phillth@mail.uc.edu, gian.rassati@uc.edu, james.swanson@uc.edu

** University of Trento
e-mails: nadia.baldassino@unitn.it, riccardo.zandonini@unitn.it

Keywords: Progressive Collapse, Composite Slab, SAP2000, Nonlinear Analysis.

Abstract. *This paper presents a study on the modelling approaches for column loss scenarios using software that is widely available to structural engineers worldwide, with the goal of providing practical guidance to designers faced with the need or the opportunity to perform progressive collapse analyses. The presented study leverages recent experimental results on column removal in a composite frame obtained at the University of Trento. Comparisons between experimental data, results of high-definition finite element analyses, and the simulation results obtained using commercially available software are performed and conclusions are drawn based on the modelling approach used. Furthermore, a parametric study on various column loss scenarios in a prototype building is conducted and the results are compared with high-definition model results in literature. It is concluded that the proposed modelling approach can be successfully used in practical cases, with results that are quite satisfactory both at the local and at the global level.*

1 INTRODUCTION

Ever since the 1968 partial collapse of the Ronan Point tower in the United Kingdom [1], the topic of progressive collapse of buildings has commanded a great deal of attention. The complexity of the issue, paired with the limited computational facilities available, has prevented much progress in the state-of-the-art until the 1990s and early 2000s, spurred by the catastrophic events of the Oklahoma City bombing in 1995 and the World Trade Center attacks in 2001. Since then, great strides have been made on the topic of progressive collapse, most notably establishing how the role of floor slabs is paramount to the robustness of building structures, since they help developing membrane action for floors upon large displacement, as well as contribute to redistribute loads among undamaged locations in a structure. The focus of this paper is on reinforced concrete slabs made composite with the steel flooring system, investigating how composite action can further improve the resiliency of a building.

This paper follows a series of experimental tests that were performed at the University of Trento as part of the RobustImpact project, funded by the Research Fund for Coal and Steel of the European Commission to investigate means of increasing resiliency in building structures. The results of these experimental tests are used in that project as a validation platform for a series of high-definition numerical models that are aimed at predicting the response of steel-concrete composite prototype buildings in column loss scenarios. The focus of this paper is to investigate practical modeling approaches for the simulation of progressive collapse of steel-concrete composite buildings using commercial finite element analysis software that might be readily available to an average design firm, as opposed to using advanced finite element analysis software that might be more geared toward research and academic endeavors but might not be readily available or even practically useable by practicing engineers.

For this project, based on a 2016 survey among structural engineers conducted by the Committee on Digital Design of the Structural Engineering Institute of the American Society of Civil Engineers jointly with the Council of American Structural Engineers [2], the structural analysis platform that was chosen as one of the most used in practice is SAP2000 [3].

This paper, following a brief review of the most relevant literature and a summary of the experimental and analytical work performed at the University of Trento, presents a series of modeling approaches and considerations to simulate the loss of a column in a steel-concrete composite building frame. The modeling approach that is developed as part of this project is then validated against the experimental results, and finally a prototype building is simulated with the goal of providing some general considerations.

2 LITERATURE REVIEW

Several publications in literature offer a summary of past seminal events that contributed to mold the current approach to progressive collapse design. Pearson and Delatte [1] discuss the effects of the Ronan Point collapse on design codes following the event, outlining how building regulations were affected by imposing requirements that a building structure over four stories in height be capable of resisting collapse under reduced safety factors in the event of a defined structural member or a portion of the structure being removed. Another publication [4] in the 25th anniversary of the collapse discusses the impact of the event on the contemporary building codes, outlining specifically the structural integrity clauses introduced in model building codes in the 1980s. Ellingwood et al. [5] provide a thorough analysis of past events and develop a set of guidance for design professionals to reduce the likelihood of progressive collapse in buildings. The observations and recommendations presented in [5] include the following: (i) slab on metal decks can provide full lateral support to beams; (ii) the use of shear connectors to effect composite action can greatly increase the redistribution capacity; (iii) reinforcing bars are recommended in lieu of welded wire fabric to ensure the ductility needed for redistribution; (iv) top- and bottom- continuous reinforcement in full-depth slabs is recommended for robustness; (v) if splicing is necessary, lap splices are preferred to mechanical devices; (vi) slabs should be designed to carry at least their own self weight in the event of a column loss; and (vii) although light weight concrete is often used to limit dead loads in buildings, the use of normal weight concrete can provide significant benefits in membrane behavior and catenary effects.

Numerous research projects in the last two decades focused on numerical advances that might allow a more in-depth analysis of progressive collapse, with the common goal of providing design and detailing recommendations for practical applications. For example, the research group led by El-Tawil at the University of Michigan [6, 7] focused their work on high-definition progressive collapse analyses, focusing on the effect of the concrete slab presence (and of the approaches for modeling it) on the progressive collapse prediction. Among the main results that are relevant to this work, those publications highlight the influence of the concrete slab on collapse-arresting mechanisms and generally on the system robustness, highlighting how an appropriate modeling of the floor slabs can prove to be one of the most difficult tasks, and how an adequate definition of its stiffness and strength may markedly affect the prediction results.

Other authors focused their attention on the influence of the structural configuration and of the selection of the point of collapse on the overall robustness of a building structure. Gerasimidis [8] studied the effect of building height and structural configuration on the likelihood of progressive collapse, concluding that in buildings of 15 stories or less only two failure mechanisms can be expected at the location of a column loss, namely the excessive yielding of a connecting beam, or buckling of an adjacent column. Also, the author explicitly considered the influence of a concrete slab, showing that beam capacities can be increased by the presence of a slab composite with them. Finally, the author showed that, as a building becomes taller, the influence of beam yielding becomes more and more negligible as the added weight will render the buckling of an adjacent column as the main failure mechanism. In a different paper, Gerasimidis et al. [9] focused more explicitly on the influence of the building height on global stability, showing the potential for the formation of hat-truss-like mechanisms in tall buildings that might function as hangers for the structure below. Kim and Kim [10], for steel moment frames, studied the effect of the column loss location on the robustness of the building structure, concluding that the removal of an edge or a corner column is more critical, due to the absence of stabilizing and redistributing mechanism on one or more sides of the column, whereas the removal of an interior column is less of a concern. The authors, in accordance with the results in [8] also showed that a taller building would be less prone to progressive collapse due to the better likelihood of the formation of a truss action at the upper, undamaged floors, providing hanging support to adequately connected structures.

Structural connectivity has always been recognized as a fundamental requirement for the development of a robust structure to resist progressive collapse. The availability of ductility and strength in connections to resist actions other than what would be expected of them during the normal operation of a building has always been seen as a necessity for resilient structures. Even in the 1980s it was suggested that buildings with lateral force resisting systems designed for earthquake resistance would prove to be much more resilient than buildings where seismic demand would not control the design, due in large part to the larger size of the members used and the greater attention provided to connection details, ensuring their strength and ductility. A considerable body of work has been developed since the 1990s on the modeling of beam-column connections, especially for what concerns the study of semi-rigid connections. Research performed on composite connections leading into the 21st century is summarized in [11], where a simplified and an extended component model is presented for composite connections. Component modeling has also been incorporated into the Eurocodes for steel and composite structures [12, 13], providing guidance on developing simplified curves to account for the actual stiffness and strength of beam-column connections.

Eurocode 1, part 1-7 [14] provides designers with tools to assess the probability of occurrence of accidental actions, the consequences of failure, and the level of acceptable risk. Eurocode 1 permits local failures to occur, as long as no key elements are involved, and allows the use of both horizontal and vertical ties to mitigate progressive collapse. The U.S. Department of Defense released in 2009 design document UFC 4-023-03, focused on the design of buildings to resist progressive collapse [15]. The approach of UFC 4-023-03 uses threat independent loading, focusing on the consequences of a major event rather than on the causes. Two design strategies are considered – an indirect method that addresses robustness by specifying minimum levels of strength, continuity, and ductility in a structure, and a direct method that is meant to achieve progressive collapse resistance through either the alternate path (AP) approach or the specific local resistance (SLR) approach. The AP method guides the designer to provide sufficient strength and stiffness in structural floors and framing to allow bridging of a failed column without undue consequences: bridging is attained by flexural resistance, catenary action, truss action, and any other mechanisms based on the structural configuration. The SLR method requires the building or parts thereof to be designed for a specific load or threat. In 2013, the U.S. General Services Administration published another document focusing on alternate load paths and progressive collapse resistance [16]. This document focuses on ground-level threats, considering number of stories above ground, type of occupancy, use of the facility, and strategic value of the building. This document is more explicitly focused on federal buildings and employs a risk-based approach to the design against progressive collapse.

3 SUMMARY OF WORK PERFORMED AT THE UNIVERSITY OF TRENTO

Researchers at the University of Trento, as part of a broader consortium of European researchers, within the RobustImpact project, funded by the Research Fund for Coal and Steel, designed, built, and tested experimentally a set of steel-concrete composite substructures with the goal of investigating the progressive collapse behavior of frames with composite slabs, focusing on alternate mechanisms of resistance in the slab and beam-column connections [17]. The full details of the experimentation and the ancillary numerical studies are reported by Roverso [18] and highlighted in [19]. The researchers designed two five-story prototype buildings, 18m tall, with six equal bays in the longitudinal direction and two bays in the transverse direction. The transverse direction bays were equal in one building (symmetric configuration) and of unequal lengths in the other (unsymmetric configuration). The buildings were otherwise identical. A 11.4m × 12.8m substructure was extracted from the ground floor of each of the buildings, encompassing a two-bay by two-bay portion of the whole structure, taken from one of the ends of the building. These two substructures were constructed in the laboratory for materials and structural testing at the University of Trento and tested to failure by simulating the removal of the central column of the subassemblies. A rendering of the experimental test is shown in Figure 1, while Figure 2 shows the naming scheme for beams and joints in both substructures.

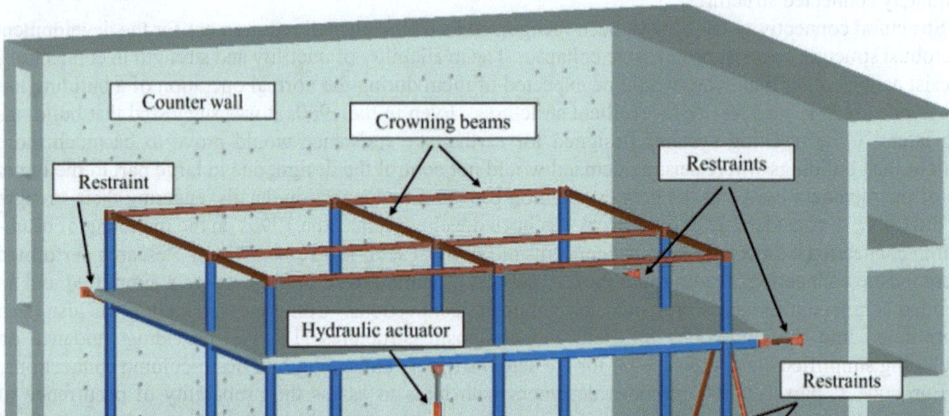

Figure 1. Rendering of experimental setup [18].

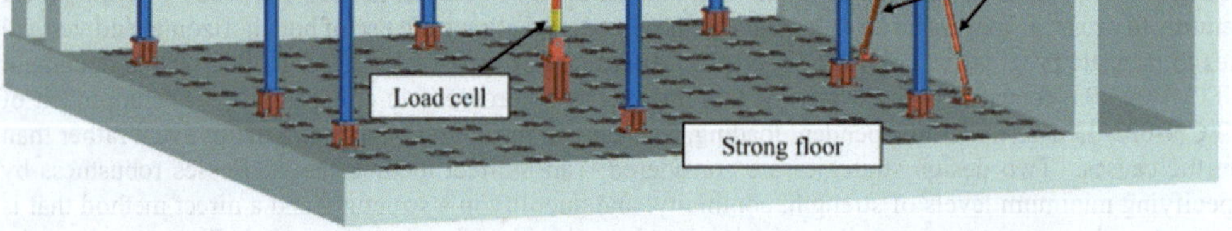

Figure 2. Nomenclature of beams and columns [18].

Column B2, which was the one to be removed to investigate the robustness of the frame, was replaced by a hydraulic actuator. During the initial loading phase, which accounted for the dead and live loads on the slab by means of sandbags placed uniformly across the structure, the actuator was set to hold position, in order to provide a rigid reaction force to the structure. The second phase of the test involved the slow depressing of the actuator to reach equilibrium, representing the actual flexibility of the column. The actuator was then further slowly depressed in order to simulate a quasi-static loss of the center column in the structure. Upon total loss of the column (represented by the actuator force reaching a zero value), the actuator was then put into tension, simulating the effect of the loads above the substructure after column loss. The load-deflection curves for both specimens are shown in Figure 3, where the response to the different load steps is shown explicitly. Failure of the specimen corresponded to a beam-column connection failure, which took place by fracture of two bolts in an end-plate connection. The instances where the first and the second bolt failed are also marked on those graphs. The two structures had a similar response to the loss of the central column, and the symmetric structure behaved somewhat better than the asymmetric structure, as it might be expected.

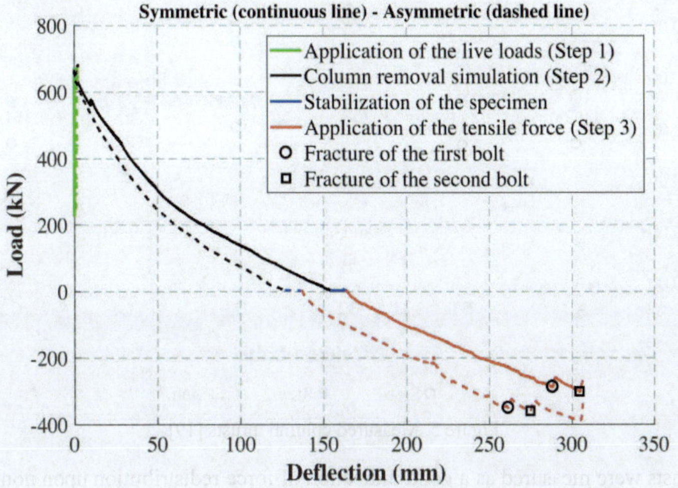

Figure 3. Load-deflection response of symmetric and asymmetric substructures [18].

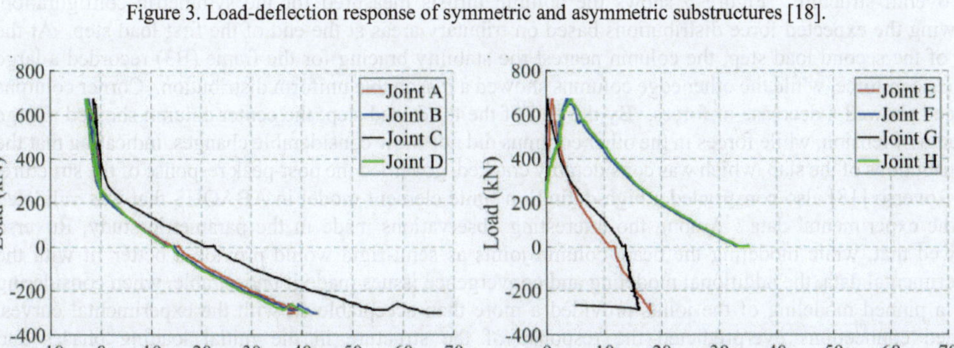

Figure 4. Load-rotation response of beam-column connections for interior (ABCD) and exterior joints (EFGH) [19].

Local behavior was also measured during the experiments, including load-rotation curves for the beam-column connections and thrusts at the column bases. Beam-column joints were identified by letters A through G, where the interior, cruciform joints, connecting to the middle column were marked A through D and the exterior joints were marked E through H. Figure 4 shows the connection responses. The difference in response between interior and exterior joints is evident: the interior joints exhibit a moment reversal in the first two stages of loading, while remaining substantially elastic up to the third loading step. Joint C, connecting to the weak axis of the column, exhibited a slightly less strong and more ductile behavior than the others. As for the exterior joints, the strong axis connections exhibited a considerably stiffer response than the weak axis joints. Additionally, joint G showed higher initial stiffness due to the proximity to one of the stability braces for the structure, which ultimately led the connected beam to undergo local buckling, limiting the overall rotation to 17 mrad.

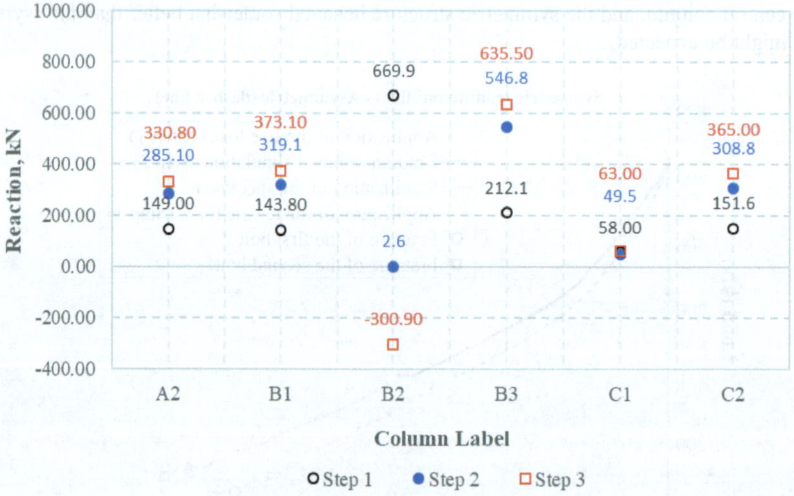

Figure 5. Measured column thrusts [19].

Column thrusts were measured as a good indication of force redistribution upon nonlinear response of the overall structure. Figure 5 shows the column thrusts measured for the symmetric configuration, showing the expected force distributions based on tributary areas at the end of the first load step. At the end of the second load step, the column nearest the stability bracing for the frame (B3) recorded a large increase in force, while the other edge columns showed a reasonable uniform distribution. Corner columns instead showed a decrease in forces. By the end of the third load step, the center column showed a large amount of tension, while forces in the other columns did not show considerable changes, indicating that the contributions of the slab, which was considerably cracked, governed the post-peak response of the structure.

Roverso [18] also constructed a high-definition finite element model in ABAQUS that was validated on the experimental data. Among the interesting observations made in the parametric study, Roverso showed that, while modeling the beam-column joints as semi-rigid would provide a better fit with the experimental data, the additional modeling and convergence issues made it undesirable, when considering that a pinned modeling of the joints provided a more than acceptable fit with the experimental curves. Pinned connections overpredicted the response of the structure in the initial loading phases and underpredicted the response after plasticity phenomena started to become important. However, the error in the prediction throughout was limited. The high-definition modeling also showed the necessity of careful modeling of slab reinforcement, explicitly modeling reinforcing bars where they were located experimentally, as opposed to considering a smeared reinforcement approach. Post peak response of the slab, in fact, is greatly affected by the placement and quantity of reinforcement. The numerical results on column thrusts showed a good match, once again showing that the columns directly adjacent to the location of the column loss were the most affected, whereas corner columns showed smaller changes. As for beam-column connections, the numerical predictions, especially using semi-rigid joints, matched the experimental data well. The researchers finally investigated the response of a full-scale building, modeled using the same approaches as the sub-assemblies, allowing for some general conclusions on the behavior of a full building. The conclusions support observations in literature that the presence of the rest of the building is actually beneficial to the robustness of the structure, showing better responses to a column loss than the isolated subassembly.

4 MODELING STRATEGIES IN SAP2000

SAP2000 v19 was chosen as one of the most pervasive structural analysis software among structural engineering firms in the United States. For the sake of simplicity, this study focused on the symmetrical configuration tested in [18], in order to limit the sources of possible variability from the experimental setup. The modeling approach presented herein is the result of many iterations that were used to explore and evaluate different solutions that could be employed within the element library of SAP2000.

4.1 Generalities and materials

In a first approximation, beams and columns of the appropriate sizes were modeled using standard beam elements in SAP2000, including bracing locations and lateral supports to reproduce the experimental setup. Beams were pinned to columns, and the columns were extended above the slab, similar to the experimental setup, with their free ends constrained as a rigid diaphragm, again to represent the constraining effect of the crowning beams that were installed in the specimen. Beams and columns were defined so that their nonlinear response would be captured by means of fiber hinges at discrete locations along the length of each element. Fiber hinges have the advantage of autonomously accounting for the interaction between axial and bending stresses, thus considerably simplifying the post-elastic modeling of the structural members. Materials were defined with the use of fiber hinges in mind for steel and concrete, based on the material testing data that was obtained experimentally. One specific modification to the standard definition of a nonlinear fiber was incorporated in the model for concrete fiber hinges. In fact, the default concrete stress-strain curve in SAP2000, is limited to the first and third quadrants, axes excluded. SAP2000 does not contain an explicit element deletion facility to account for damage initiation and evolution, and thus the software, once a point reaches the end of the defined stress-strain curve, continues to attempt to find a solution based on the previous slope of the curve. This can create considerable convergence issues – the remedy that is suggested based on the work herein is to continue the definition of both tensile and compressive branches of the material with a horizontal, zero-stress line both in tension and compression to facilitate convergence, virtually introducing element deletion into the analysis.

4.2 Slab modeling

Slab modeling, as indicated by multiple sources in literature, revealed itself to be the most challenging and simultaneously the most important aspect of the modeling for progressive collapse simulation. SAP2000 offers a family of nonlinear shell elements that would appear upon first examination to be ideal for the purposes of slab modeling. However, when subjected to large displacements, convergence proved to be quite challenging for these elements, thus limiting their usefulness for this particular application. Since the inelastic response, and in particular the post-peak behavior, of the slab is fundamental in progressive collapse simulations, a different modeling approach was envisioned for this project. A lattice structure made by beam elements, consisting of beams directed in both in-plane directions, was envisioned to reproduce the reinforced concrete slab. Mesh sensitivity analyses were performed, allowing to recommend a ten-by-ten pattern of subdivisions to reproduce the slab in each bay of the frame. Sixteen different beam sections were modeled to match the varying reinforcement layout in different regions of the slab, as was done in [18]. It was decided to take advantage of the Section Designer facility in SAP2000 to simplify the definition of each beam section – however, this limits the definition of reinforcement to the longitudinal direction of the beam element, which implies that different beam sections needed to be defined for X-direction and Y-direction reinforcement. Coincidentally, the ten-by-ten pattern corresponded to a distribution of 4 reinforcing bars at 150 mm spacing, which matched the experimental setup. Since beam sections are only defined longitudinally, there was by necessity an overlap of members, which required eliminating self-weight from one set of beams, in order to match the original loads. Additionally, the torsional constant for these beam members was set to zero to eliminate redundant stiffnesses – this resulted in the redistribution of torsional effects to the orthogonal connecting member in the form of strong axis bending.

However, using a lattice of crisscrossing beams for the slabs makes the application of distributed loads challenging. For this purpose, linear shell elements were employed to fill the gaps within the lattice of beams, with both flexural and membrane stiffnesses set to zero, in order to provide a simply defined surface for the application of distributed loads, without impacting the flexural and axial stiffnesses of the slab itself.

4.3 Modeling of the composite action

The first approach that was explored for the definition of composite action in the model was based on the experimental observation that, even at failure, the shear connectors had not yet shown any sign of yielding. For a fully composite structure, thus, it was considered that rigid links could be used to connect slab elements (represented as discussed above, by a lattice of beam elements), and steel beam elements at discrete locations along the length of beams and girders. This approach, however, resulted in large axial forces developing in concrete members between link locations, leading to a premature termination of the analysis due to fiber hinges reaching their capacity. This is not unexpected, although it may have been overlooked had fiber hinges not been used for the steel and concrete elements: rigidly linking two line elements in bending, acting at a finite distance from each other, naturally leads the model to develop a flexural resistance as the result of equal and opposite axial forces developing in each of the members, multiplied by their relative distance, rather than resisting flexure by activating curvatures of a compositely acting section. Due to this observation, this approach was quickly abandoned for what resulted in an overall simpler modeling technique, once again leaning on the Section Modeler facility in SAP2000. Considering the negligible influence of the deformation of the shear connectors, it was decided to approximate the composite slab behavior using a lattice of individual beam elements, this time incorporating both the steel section and the concrete section into a single composite section. The Section Designer would calculate all the appropriate mechanical quantities for the definition of the composite section, and fiber hinges would be defined for both steel and concrete across the entire composite section.

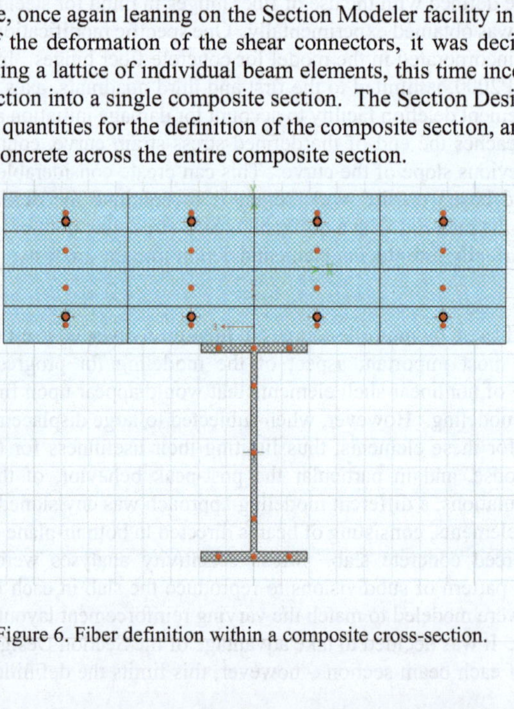

Figure 6. Fiber definition within a composite cross-section.

This resulted in a much simpler modeling approach that still incorporated all the desirable characteristics of the slab as described above. The only limitation of this approach is the discounting of any interface slip, which is justified in this case by dealing with a section that was fully composite.

4.4 Fiber hinge definitions

As a consequence of this approach, two sets of fiber hinges had to be defined, one for the composite elements, and the other for the concrete slab elements away from the steel beams. These hinges are P-M2-M3 hinges, i.e., they account for the interaction of axial stress and both strong- and weak-axis bending stresses. Fiber hinges also leverage the SAP2000 analytical engine, by accounting for the changing moment-rotation curve at that location at each increment of the analysis. The obvious drawback of using fiber hinges is the increased computational time, however analysis times experienced throughout this projects were tolerable. Fiber hinges are defined over a finite length of a member. A parametric analysis was conducted to establish an appropriate definition length, resulting in the observation that hinge length did not greatly affect the overall response of the structure. For this project, hinge lengths of 50mm were used for the composite members and of 65mm for the concrete slab members. The fiber layout that was used in these analyses was a relatively coarse layout, for the sake of computational efficiency, as shown in Figure 6. Hinges were located at the mid-length of each individual beam element used.

4.5 Connection modeling

Nonlinear links were selected to simulate the beam-column connection response at the ends of all beams. As compared to the initial trial runs with ideal pinned and fixed connections, the definition of a nonlinear moment-rotation response of the connection resulted in a much better prediction of the experimental behavior. Additionally, the local behavior prediction was also affected by the connection definition, showing much better adherence to the experimental values when the connections' nonlinear moment-rotation responses were incorporated. Multilinear Plastic links were selected, in order to be able to properly define the hysteretic behavior of the connections, considered that the experiment showed a full reversal of the moment demand on some of the connections. The links were defined in the M3 direction, accounting for the strong axis flexural effects, while all other degrees of freedom, including weak axis bending, torsion, shear in both directions, and axial forces, were modeled as rigid. Degradation was included in the moment-rotation definition of the links to more closely simulate the experimental behavior, and at the same time to avoid excessive hardening and late stage load concentration due to link stiffening. Degradation accounts for concrete crushing, rebar buckling, yielding of end-plates, and other phenomena that might take place. An alternate model was also create to explore the difference with the case with perfectly pinned beam-column connections. For this case, the links were made fully rigid, and the connecting composite beam element was released in the strong axis bending direction. The moment-rotation definition for the joints is based on the component method in [12,13], with the addition of an third linear branch to transition from the elastic region to the perfectly plastic region of the response, as was suggested in [11].

4.6 Loading

The Staged Construction option in SAP2000 was used to effect the removal of the center column in the simulated subassembly, after the system had reached equilibrium under the applied dead and live loads. Once the column removal was completed, an additional loading step was added where a force is applied at the location of the central column to continue pulling down on the structure, reproducing the last stage of the experimental test.

5 DISCUSSION OF RESULTS

Figure 7 shows a comparison of the load-deflection curves between the experimental specimen, the simulated substructure with semi-rigid connections and that with perfectly pinned beam-column connections. It can be readily noted how the simulation using semi-rigid connections, other than consistently slightly overpredicting the strength of the specimen, follows the post-peak slopes of the experiment very closely throughout the entire range of behavior. Conversely, the simulation with pinned connections matches the experiment remarkably well until the force in the column is zero. After the column starts picking up tension, the simulated data diverges, predicting a much smaller stiffness for the response.

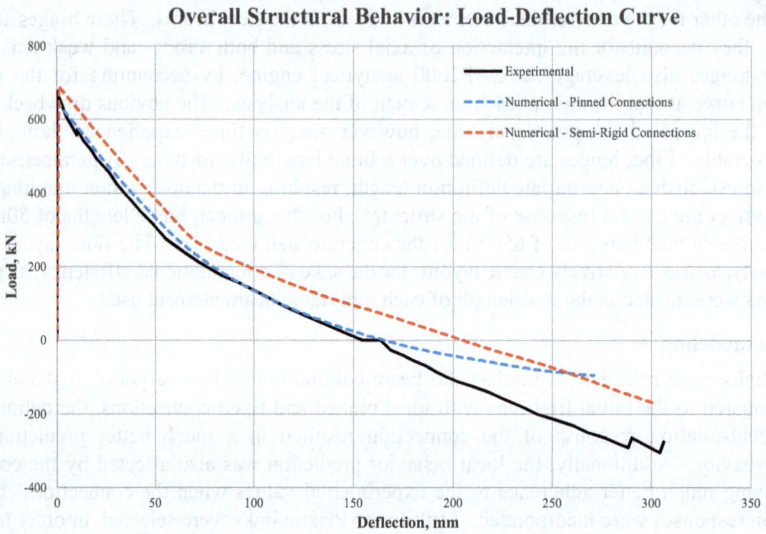

Figure 7. Comparison of load-deformation responses.

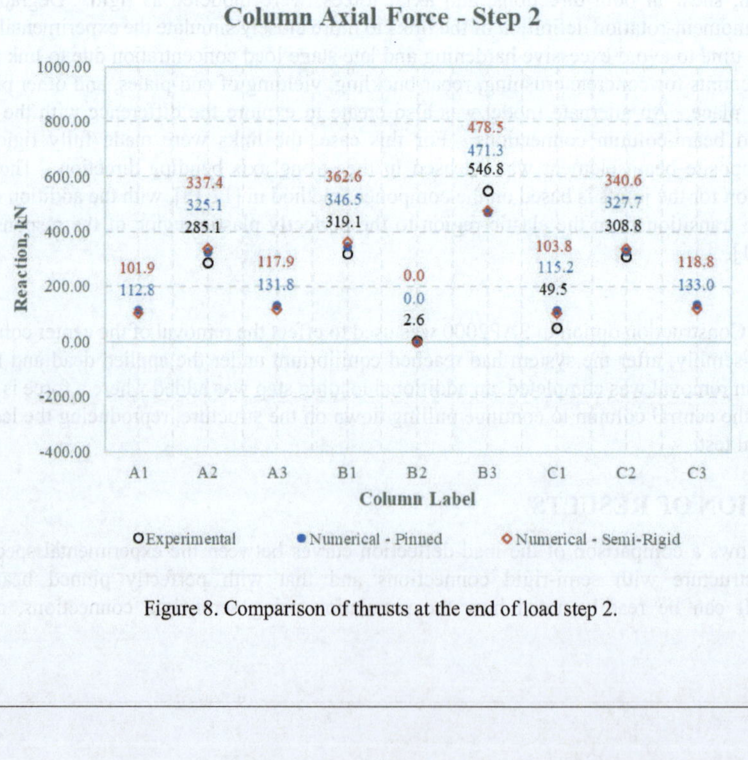

Figure 8. Comparison of thrusts at the end of load step 2.

It is clear how the influence of the joint response is felt particularly in this stage of the simulation, after the column has completely unloaded and large displacements are involved. Considering the level of simplification of the analytical model, when compared to a high-definition model, one could conclude that the results obtained are extremely promising, allowing to use a 'simpler' model to predict a very complex phenomenon quite closely, at least in terms of global response quantities.

Figure 8 shows a comparison of the column thrusts in the same three cases, at the end of the second load step, when the column is completely lost. The largest deviation from the experimental results is notable for column B3 – the reader will recall that this is the location in the experiment where the specimen was braced against the strong wall in the laboratory, thus causing a local stiffening effect. Trends seen in the experiment are replicated by the model, including the relatively small demand on corner columns after the middle column loss, and the redistribution of forces toward the edge columns. Similar trends can be seen for the other two load steps, not reported here for brevity.

Figure 9 shows some select comparisons between moment-rotation responses of some of the connections. In general, the simulated curves are in general agreement with the experimentally measured data. The semi-rigid model appears once again to provide a better prediction of the general response of the system throughout all the phases of loading.

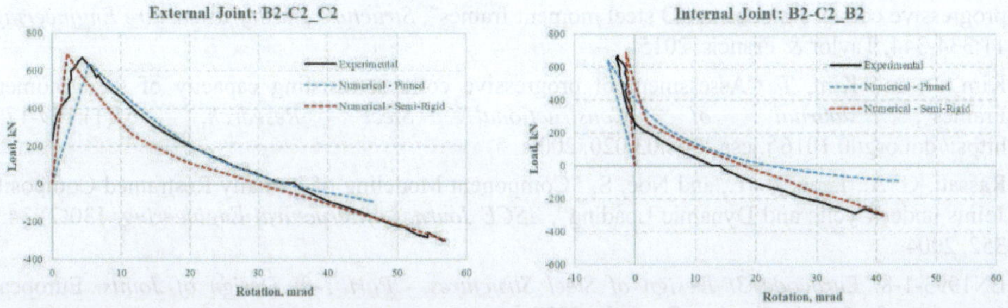

Figure 9. Comparison of load-rotation curves for an exterior and an interior joint.

6 CONCLUSIONS

In conclusion, the proposed modeling approach has proven capable of providing quite an accurate prediction of both global and local responses of a steel-concrete composite frame subjected to a column loss scenario. The use of SAP2000 allowed for the prediction of complex phenomena both at the global and at the local level, while limiting the model complexity. It is conceivable that SAP2000 models can be used on larger structural systems to investigate their robustness, without the need for extremely complex high-definition modeling approaches, simultaneously maintaining a reasonable demand for computational resources. The modeling approach proposed herein can be simplified by leveraging the APIs provided with SAP2000, automating to a large extent the model definition steps, freeing the analyst to focus on the definition of the connections and of the material properties. Additionally, while it was shown that explicitly modeling semi-rigid connections provides better overall results, the approach using pinned connections can also provide workable results without too much loss of accuracy in the prediction. Further comparisons are needed to pinpoint the limitations of the use of SAP2000 to simulate complex problems such as progressive collapse, but these initial results show great promise.

REFERENCES

[1] Pearson, C. and Delatte, N. "Ronan Point Apartment Tower Collapse and its Effect on Building Codes", *Journal of Performance of Constructed Facilities*, 19(2):172-177, 2005

[2] SEI-CASE Committee on Digital Design, *Survey on the use of analysis and design software,* 2016. www.seibim.org/resources/ Accessed on 3/23/2021.

[3] CSI America. *CSi Analysis Reference Manual*, Berkeley, California, 2017

[4] Longinow, A. and Ellingwood, B.R. "The impact of the Ronan Point collapse – 25 years after", *Structural Engineering World Wide 1998,* Paper OP312-2, Elsevier Science, New York, New York

[5] Ellingwood, B. R., Smilowitz, R., Dusenberry, D. O., Duthinh, D., Lew, H. S., and Carino, N. J., *Best practices for reducing the potential for progressive collapse in buildings,* NIST Interagency/Internal Report (NISTIR), National Institute of Standards and Technology, Gaithersburg, Maryland, 2007.

[6] El-Tawil, S., & Li, H. H. "Progressive Collapse Research: Current State and Future Needs.", *Advanced Materials Research,* 639-640(1), 3-12, 2013

[7] Li, H. and El-Tawil, S. "Three-dimensional effects and collapse resistance mechanisms in steel frame buildings", *ASCE Journal of Structural Engineering,* 140(8):1-11, 2014.

[8] Gerasimidis, S. "Analytical assessment of steel frames progressive collapse vulnerability to corner column loss", *Journal of Constructional Steel Research,* 95:1-9, 2014

[9] Gerasimidis, S., Deodatis, G., Kontoroupi, T., and Ettouney, M. (2015). "Loss-of-stability induced progressive collapse modes in 3D steel moment frames", *Structure and Infrastructure Engineering,* 11:334-344, Taylor & Francis, 2015

[10] Kim, J., and Kim, T. "Assessment of progressive collapse-resisting capacity of steel moment Frames", *Journal of Constructional Steel Research*, 65(1):169-179, https://doi.org/l0.1016/j.jcsr.2008.03.020, 2009

[11] Rassati, G. A., Leon, R. T., and Noé, S. "Component Modeling of Partially Restrained Composite Joints under Cyclic and Dynamic Loading", *ASCE Journal of Structural Engineering,* 130(2):343-352, 2004

[12] EN1993-1-8. *Eurocode 3: Design of Steel Structures - Part 1-8: Design of Joints.* European Committee for Standardisation, Brussels, 2007

[13] EN1994-1-1. *Eurocode 4: Design of Composite Steel and Concrete Structures -Part 1-1: General Rules and Rules for Buildings.* European Committee for Standardisation, Brussels, 2011

[14] EN1991-1-7. *Eurocode 1 -Actions on Structures - Part 1-7: General Actions -Accidental Actions.* European Committee for Standardisation, Brussels, 2011

[15] UFC 4-023-03. *Design of Buildings to Resist Progressive Collapse.* Department of Defense, Washington, District of Columbia, 2009.

[16] General Services Administration (GSA). *Alternate Path Analysis & Design Guidelines for Progressive Collapse Resistance.* General Services Administration, Washington, District of Columbia, 2013

[17] Zandonini, R., Baldassino, N., and Freddi, F. *Moment Resisting Steel-Concrete Composite Frames Under the Column Loss Scenario: Design of an Experimental Study.* Research Report, Department of Civil, Environmental and Mechanical Engineering, University of Trento, https://doi.org/10.15168/11572_219133, 2018

[18] Roverso, G. *Progressive Collapse Assessment of Steel and Concrete Composite Structures Subjected to Extreme Loading Conditions.* PhD Dissertation, Department of Civil, Environmental, and Mechanical Engineering, University of Trento, 2018

[19] Zandonini, R., Baldassino, N., Freddi, F., and Roverso, G. "Steel-concrete frames under the column loss scenario: An experimental study", *Journal of Constructional Steel Research*, 162:105527, https://doi.org/10.1016/j.jcsr.2019.02.036, 2019

COMPOSITE BRIDGES

9TH INTERNATIONAL CONFERENCE ON COMPOSITE CONSTRUCTION IN STEEL AND CONCRETE
M. Knobloch, U. Kuhlmann, W. Kurz, M. Schäfer. (Eds.)
Online, Germany, July 27-29, 2021

COMPOSITE DOWELS FOR BRIDGES: TRENDS AND CHALLENGES FOR NEW EUROPEAN DESIGN RULES

W. Lorenc*, G. Seidl, R. Kühne****

* Wrocław University of Science and Technology
e-mail: wojciech.lorenc@pwr.edu.pl

** Potsdam University of Applied Science
e-mail: guenter.seidl@fh-potsdam.de, robert.kuehne@fh-potsdam.de

Keywords: Composite dowels, Shear connection, Composite bridges

Abstract. *The advent of composite dowels (CDs) in Europe at the onset of the previous decade gave rise to the development of innovative composite bridge forms where a T-profile has replaced the conventional I-beam. Both the construction techniques and design approaches are unprecedented. Insights gained from the erection of new bridges and ensuing discussions have led to important findings that are imperative to incorporate into the forthcoming European approval (CEN-TS). The challenge concerning the design equations for concrete dowels is laid out.*

1 INTRODUCTION

The evolution of the composite dowel strip continues to be a work in progress. Reference [1] provides an overview of the current technological advancements. It underscores the transformation of the dowel strip, evolving from a shear connector to a novel hybrid cross-section. In the realm of bridge design, two subjects have emerged as highly significant:

1. The **shearing criterion** should include the confining effect in a new way.
2. The **pry-out cone criterion** for reinforced concrete is needed.

1.1 New definition of pry-out cone criterion for reinforced concrete

The pry-out cone failure mechanism of concrete part in shallow embedded CDs with the decisive parameter c_u (see Fig. 1a) in [2]:

$$P_{po,k} = \chi_x \cdot \chi_y \cdot 90 \cdot \frac{h_{po}{}^{1,5}}{e_x} \cdot \sqrt{f_{ck}} \cdot (1 + \rho_{Di})\ [\text{kN/m}] \qquad (1)$$

where:

$P_{po,k}$	shear capacity of composite dowel strip per length
χ_x, χ_y	reduction factors due to overlapping of the concrete cones
h_{po}	height of the pry-out cone
f_{ck}	compressive strength of the concrete (char.)
$\rho_{D,i}$	ratio between area of reinforcement and concrete in the dowel core weighted by the Young's modulus
e_x	distance between the steel dowel in longitudinal direction

During the design of a composite bridge next to city of Elbląg in Poland [3] it became evident, that the pry-out cone criterion according to Equation (1) doesn't appear in the case of large and reinforced concrete cover c_u if the guiding of the reinforcement is appropriate. The design rules in [2] are currently on the save side for the pry-out criterion. At this point, a novel approach would be proposed:

First and foremost, it's worth noting that the pry-out cone criterion doesn't seem to be applicable for many current bridges and should be replaced with a novel criterion designed for reinforced concrete in confined conditions.This proposed methodology is fitting for situations involving shallowly embedded composite dowels or non-confined concrete elements without stirrups. The state of confinement is actualized through the implementation of vertical stirrups (see Fig. 1). The interaction of overlapping pry-out cones, as delineated in [4], results in the underestimation of load-bearing capacity for scenarios with deeply embedded CDs exhibiting a high c_u value. In conclusion neither the pry-out cone for a single dowel without pry-out cone interaction nor the cone interaction itself is likely to occur in the case of deeply embedded CDs. Equation (1) doesn't take it into consideration. The direction for future research should aim to bridge the gap between the Equation (1) and a mechanical model for CDs under the specific conditions of reinforced concrete boundaries.

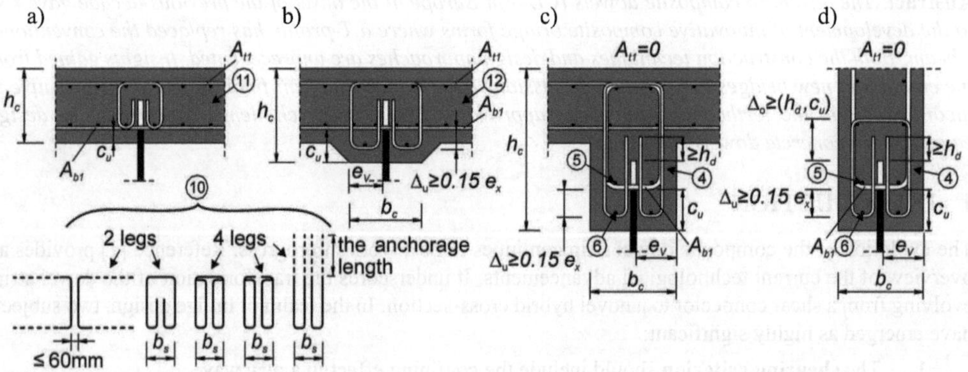

Fig. 1: Shallow (a) and deeply (b-d) embedded composite dowels (outline of [5])

1.2 The shearing criterion

The concept of the perfobond strip was initially proposed by Leonhardt and Andrä [6], aimed at introducing a continuous shear connector that utilizes the compressive strength of concrete in the design formula, as outlined in Equation (2). Subsequently, Oguejiofor and Hosain introduced formulas which is using the tensile strength [7] as outlined in Equation (3).

$$D_{\mathrm{du,b}} = 1.4 d^2 \beta_{\mathrm{WN}} \tag{2}$$

$$q_{\mathrm{u}} = \beta_1 h t f_{\mathrm{c}}' + (\beta_2 A_{\mathrm{cd}} + \beta_3 A_{\mathrm{cc}})\sqrt{f_{\mathrm{c}}'} + \beta_4 A_{\mathrm{tr}} A_{\mathrm{yr}} \tag{3}$$

where:

$D_{\mathrm{du,b}}$, q_{u}	shear capacity per concrete dowel
β_{WN}, f'_{c}	compressive strength of concrete
d	diameter of a hole
A_{cd}	shear area of concrete dowel
A_{cc}	shear area of concrete per connector
A_{tr}, f_{yr}	total area and yield strength of transvers reinforcement

β_1 to β_4	regression coefficients (determined by numerical data)

In Equations (1) and (2), β_{WN} and f_c' represent the compressive strength of concrete. In the context of the perfobond strip [6], a simple correlation was assumed between the shear strength of concrete in the hole and the compressive strength of the concrete. In reference [7], a combination of the action of concrete dowels and local compression was employed because the concept of a continuous shear connection was discarded for push-out tests and beam tests: the boundary conditions differ significantly compared to those in the perfobond strip tests [6][5] (see Fig. 5), where the dowel strip surface was screened. Utilizing contemporary notations, where f_{ck} denotes the concrete compressive strength and f_{ct} signifies the tensile strength, formula (2) can be simplified to (4) and formula (3) can be condensed to (5):

$$P_{\mathrm{Rk},1} = k_1 \cdot A_{\mathrm{D}} \cdot f_{\mathrm{ck}} \tag{4}$$

$$P_{\mathrm{Rk},2} \cong k_2 \cdot A_{\mathrm{D}} \cdot f_{\mathrm{ct}} \tag{5}$$

Here, A_D represents the area of the concrete dowel, while k_1 and k_2 are the respective coefficients. Proper calibration of these coefficients enables us to derive the same outcome from both formulas (4) and (5). It is noteworthy that in Japan (where PBL shear connectors are extensively utilized), the compressive strength of concrete is used in the design of concrete dowels [8], which contrasts with practices in Europe. In Europe, the tensile strength is employed [9] to characterize the shear strength of concrete dowels, and the approximation of the tensile strength is utilized as the square root of the compressive strength according to Equation (6):

$$P_{\mathrm{sh,k}} = \eta_{\mathrm{D}} \cdot e_{\mathrm{x}}{}^2 \cdot \sqrt{f_{\mathrm{ck}}} \cdot (1 + \rho_D) \tag{6}$$

where:

$P_{\mathrm{sh,k}}$	shear capacity of composite dowel strip per length
η_{D}	geometry-dependent reduction factor of steel dowel
e_{x}	distance between the steel dowel in longitudinal direction
f_{ck}	compressive strength of the concrete (char.)
ρ_{D}	ratio between area of reinforcement and concrete in the dowel core

2 DISCUSSIONS ON CONCRETE SHEARING

2.1 Historical background and design methods

A thesis is proposed to simplify further considerations: There is no fundamental difference in the functioning of concrete dowels as a joint between the steel flange and the concrete plate, as developed by Leonhardt (PBL) [5], and the composite dowels (CDs) currently in use. Since the inception of the perfobond strip, numerous researchers have endeavoured to construct a theoretical model describing PBLs and CDs. In Poland, a closed composite dowel joint (CCD) has recently been introduced into bridge engineering (see Fig. 2). This innovation prompted an analysis of the similarities and differences between PBL and CD, with concentrated research on this subject carried out in [10]. The study scrutinized design approaches employed in various countries for the design of PBL. The use of PBL shear connection is widespread in Japan (see Fig. 3).

Fig. 2: Composite dowels and closed composite dowels of the Elbląg bridge in Poland [14]

Fig. 3: CCD in Japanese bridges (2019), in cooperation with Osamu Ohyama and Nobuhito Okubo

The authors of [5] proposed a straightforward approach wherein the design formula (Equation (2)) was developed, depending on the concrete compressive strength and the coefficient derived from tests. A similar design strategy is currently employed in Japan [8].

$$V_{\mathrm{psud}} = 1.6 d^2 f'_{\mathrm{cd}} / \gamma_{\mathrm{b}} \tag{7}$$

where:

V_{psud}	design shear strength of a concrete dowel
f'_{cd}	design compressive strength of concrete
d	diameter of a hole
γ_{b}	member factor (usually 1.3)

In Canada experiments were conducted on beams equipped with welded sectional fasteners in the form of steel flat plates with perforations [11]. Based on numerical analyses, the formula (Equation (3)) was proposed [7]. This formula assumed that the load capacity of the connector is dependent on the square root of the concrete's compressive strength, which is linked to the concrete's tensile strength, and is adjusted with an empirical factor. Following this, many researchers sought to refine the aforementioned approach, and their efforts typically resulted in various magnitudes of the proposed coefficients. In [12], a model separating into zones A and B was proposed (see Fig. 4a). This approach was further developed in [13] (see Fig. 4b). In [4] and [14], the body of knowledge was expanded. In [4], while adopting the failure mechanism proposed in [13], a formula grounded in MC90 [15] was proposed. Although the core of this formula is relatively simple (Equation (8)), the final formula for the shear force transmitted by a composite dowel with reinforcement $T_{\mathrm{sh,re,k}}$ is rather complex (refer to [4], p. 109). It's noteworthy that this proposal accounted for compression perpendicular to the connection (refer to [4]: Equation 9-7), designated as σ_{s1} in Equation (8). This approach would allow for the inclusion of prestressing:

$$P_{\mathrm{sh,k}} = 0.40 \cdot f_{\mathrm{c}}^{2/3} \cdot (\sigma_{\mathrm{s1}} + \rho_{\mathrm{D}} \cdot f_{\mathrm{y}}) \tag{8}$$

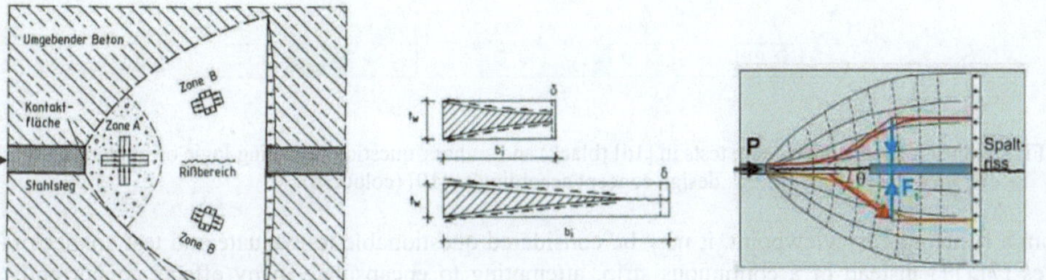

a) Model of Wurzer [20] b) Model of Zapfe [13] c) Model of Heinemeyer [14]

Fig. 4: Model of the bearing behaviour of concrete dowels

[14] presents analyses that incorporate all the previously mentioned works. The design concept eventually adopted in Germany takes on a slightly modified version of the concept proposed in [13], where the load capacity primarily hinges on the concrete tensile strength, the dowel area, and a coefficient. In conclusion, despite numerous efforts, it has not been possible to achieve the same consistency seen with the concrete shear failure model as with steel dowels, where the design formula, mechanical model, numerical model, and test results form a logical, consistent structure [16]. In the case of concrete, the test results, numerical model, mechanical model, and design formula do not form such a coherent whole - this is justified by the fact that the phenomenon is significantly more complicated than in the steel model scenario. While the research findings, the numerical model, and the design formula interact satisfactorily, the fact that the design formula does not mirror the results (experimental investigations and numerical analyses) is difficult to rationalize other than by the absence of a suitable mechanical model - despite many attempts to construct it. The justification for researchers (including the authors) is simply extremely complicated nature of the phenomenon.

However, it's worth noting that in Japan, a simple formula is employed for PBLs, essentially an extension of [6], and many bridges with PBL connections are designed (see Fig. 2). Therefore, it can be observed that the formula using concrete's compressive strength (in Japan) is seemingly as good as using concrete's tensile strength (in Europe), and vice versa. This isn't particularly surprising, as other formulas would also work, as would a purely empirical approach, such as proposing that for a given geometry (commonly used), the load capacity is, for instance, 500kN - or any other value derived from experimental investigations. A very similar concept was introduced in the National Annex to 1994-2 in the Czech Republic, where the load-bearing capacity of the PBL strip is a simple summation of three components.

Upon analyzing the method of shaping the fasteners according to [11] (see Fig. 5) and the structure of the formula in [7] (Equation 3), it remains unclear why the load capacity between the connectors (part of Equation 3 with β_1) should be determined by the compressive strength, and why the load capacity in concrete dowels (part of Equation 3 with β_2) should be determined by the tensile strength. This issue remains essentially the same when the fasteners are in close proximity (see Fig. 5, lower illustration).

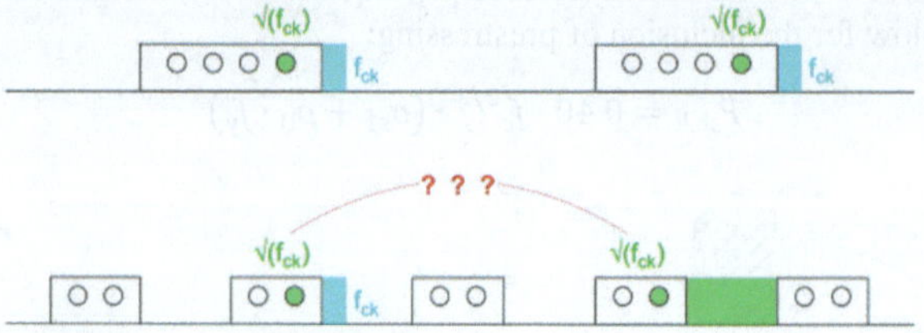

Fig. 5: Shear connection in beam tests in [16] (black) and authors' question regarding logic of implementation of design concept according to [10] (coloured)

From a contemporary viewpoint, it may be considered questionable to evaluate and test connectors with holes [7],[11] instead of a continuous strip, attempting to encapsulate many effects concurrently [7] - although it's important to emphasize that this critique is made from today's perspective: at the time, it was pioneering work. With our current knowledge, we understand that the solution presented in Fig. 5 is not optimal in terms of economic factors, production technology, and fatigue parameters. This raises the question of whether the absence of styrofoam at the front of the steel plate in the push-out tests in [7] (in contrast to [5]) was intentional or accidental. This query holds significance because it impacts all subsequent actions and the origins of the existing formulas for composite dowels.

2.2 Items to consider

2.2.1 Narrow concrete webs and confinement effect

The continual optimization of bridges has led to questions about the minimal viable thickness for a concrete web. A key aspect to understand is how the confined concrete effect in the direction perpendicular to the plane of CDs impacts the load-bearing capacity. It's crucial to underscore that this influence may be classified as either passive or active. The passive effect can be quantified in terms of the thickness of the concrete web where the CDs are embedded, the amount of reinforcement, or a combination of both. The active influence, on the other hand, can be expressed as an action of a force perpendicular to the plane of CDs.

Currently, the passive influence is somehow included in the formula ([9], Equation 1) but the mechanical model does not quantify it. The active influence has been examined exclusively for the tensile force [17], revealing that tension in the perpendicular direction doesn't substantially impact the ultimate load-bearing capacity - yet the mechanical model does not explain why this is so. The influence of compressive force, on the other hand, has not been analysed anywhere. Meanwhile, structures are being built in Poland where compression occurs perpendicular to the dowel, and it is evident that this influence must be significant and advantageous - currently, it is conservatively disregarded in design.

Furthermore, there has been no analysis of how thin the webs can be in bridges, a factor that is currently being optimized - specifically, the thickness of the concrete web. This brings us to the issue of passive influence. Therefore, for the new versions of Eurocodes (or CEN-TS) [5], the ad-hoc concept of e_v (Fig. 6) has been introduced, alongside a conservative notation.

"*The thickness* b_c *of the concrete web should be not less than* $b_{c,\min} = 250\text{mm}$. *The nominal concrete cover* e_V *from the side of the haunch or concrete web to the connector should be not less than* $(b_{c,\min} - t_w)/2$ *and* h_{po} *according to equation D.2(4) should be not greater than* $e_V + 0{,}13 \cdot e_x$."

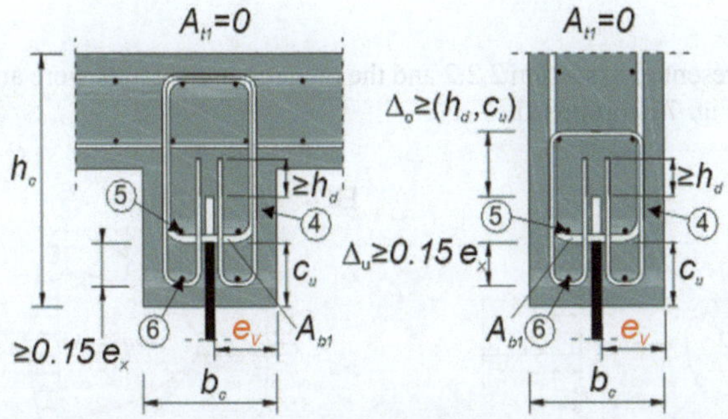

Fig. 6: Concept of e_V [5]

This temporary ad-hoc solution aims to prevent the design of excessively thin webs and should be substituted in the future with more refined guidelines. It is based on current experience from tested elements to (i) ensure the confinement effect and (ii) to prevent possible pry-out-cone failure at the vertical surface of the web, as was demonstrated in [18]. In fact, webs below 250mm are already under consideration, and their design is backed by research, although clear design rules are lacking. For now, to shorten the guidelines [5], the shear formula has been substituted with boundary conditions [1]. This approach also needs to be reviewed, especially considering that the guidelines are set to be updated as CEN-TS.

2.2.2 The shear failure mechanism

The failure mechanism suggested in [13] appears viable to describe. An attempt utilizing a simplified approach following the Model Code has been proposed in [4], and a suitable Finite Element Method (FEM) image confirming this mechanism is provided in [19]. It appears feasible to propose a mechanical model and derive the relevant formulas grounded in the Drucker-Prager failure criterion. This model would allow us to account for the influence of the axial force perpendicular to the plane, as well as the active and passive impacts of the concrete confinement.

In the context of the connector scale, the friction mechanism at elevated pressures should be the governing mechanism that describes the behaviour of the dowels [18], as depicted by the Drucker-Prager model (D-P), rather than the Strain-Tensor model (S-T) which is specifically designed for assessing cracked concrete on a distinct scale. The approach undertaken in [14], as seen in Fig. 4c, presents a valuable starting point for the analysis of this issue. With these considerations in mind, it is important to revisit and verify the model proposed in [12], taking into account the mechanism suggested in [13] and the current state of knowledge.

2.3 The model proposal

The previous proposals presented in section 2.2.2 and the own research results were analysed. On this basis, the concept presented in Fig. 7 is proposed:

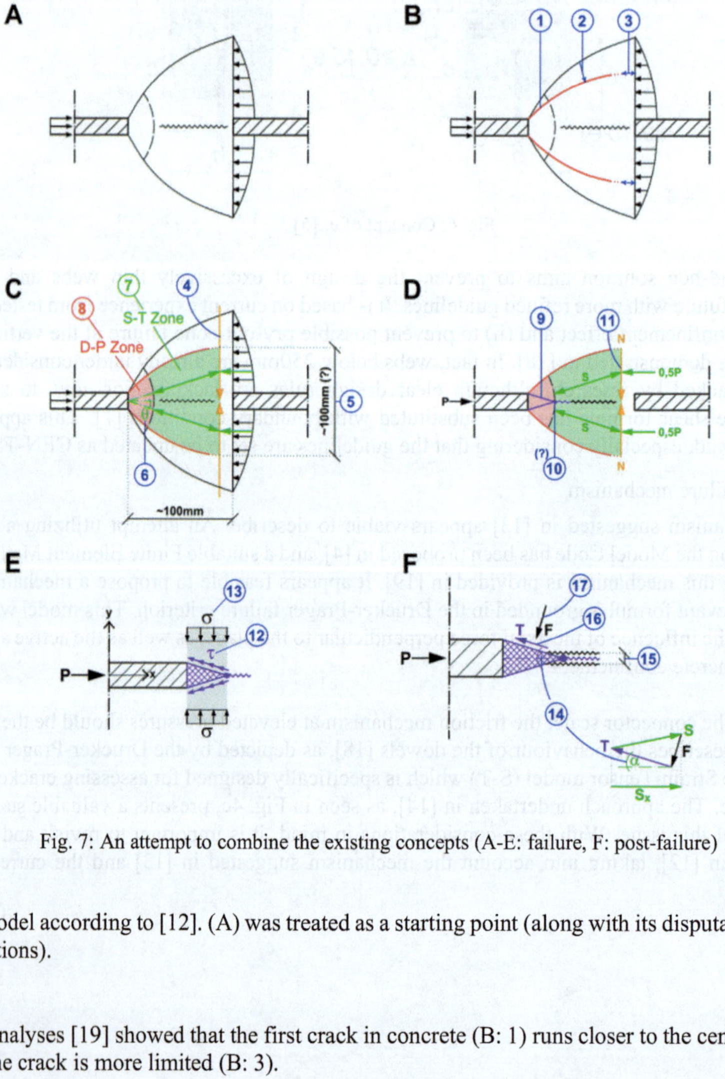

Fig. 7: An attempt to combine the existing concepts (A-E: failure, F: post-failure)

Step A:

The basic model according to [12]. (A) was treated as a starting point (along with its disputable –but needed – simplifications).

Step B:

Numerical analyses [19] showed that the first crack in concrete (B: 1) runs closer to the center of the dowel (B: 2) and the crack is more limited (B: 3).

Step C:

At this point, one can make a quantitative assessment (C: 5) and, based on the latest works from 2021 at Wrocław University of Technology (tests of beams with 20 cm concrete web, not published yet), state that with a standard dowel size, it is possible to close the force system in a 100mm space (C: 5). The first zone of the [19] model is attributed to the Drucker-Prager mechanism (C: 8) and the second to the S-T mechanism (C: 7). The geometry determines the angle Θ (C: 6) and Θ is small because the dimension (C: 5) is small. Of course, the boundary between the D-P and S-T regions is blurry and physically absent. The force in the reinforcing bar (C: 4) ensures that the system of forces is closed and from now on, only the D-P area can be considered because it determines the load capacity of the concrete dowel.

Step D:
At this point, reference is made to the shear plane proposed in [13] (D: 9), while it is noted that the position of the point where the wedge ends (D: 10) is at-hoc and requires discussion.

Step E:
The corresponding large σ (E: 13) is the confinement effect and it provides a correspondingly large τ (E: 12). One can notice that σ can come from the force in the rebar (D: 11), pre-stresses, etc. The integral of τ gives P. Thus, the limit load capacity of the dowel is conditioned and cohesion and σ play a key role here. After reaching the ultimate capacity, the structure degrades inside the D-P area and a wedge with a different geometry appears (F14).

Step F:
It appears mechanism based on friction and this is so called "post-failure" region (F: 14). A significant gap appears (F: 15) and geometry and kinematics allow to describe the phenomenon. The force N (D: 11) determines the force F (F: 17) and the friction angle determines the force T (F: 16). This creates the force S_x (F: 18). It is difficult to judge whether the angle Θ changes on the transition from E (failure) to F (post-failure) but it is worth noting that it does not have to, because the change goes inside the area D-P. In conclusion: Figures A and E (Fig. 7) are well known concepts, as is the idea of applying the S-T model. Combining them into a coherent whole is an original proposition. It is presented for the purpose of a broader discussion.

3 DISCUSSION ON PRY-OUT CONE

3.1 Historical background and design methods

In [20] it was generally stated that

"the confined reinforcement is important in order to avoid a spalling of the concrete coverage (brittle failure) for concrete-webs with external reinforcement. With the arrangement of $A_{s,conf}$ the maximum load bearing capacity can be risen significantly and ductile behaviour is provided."

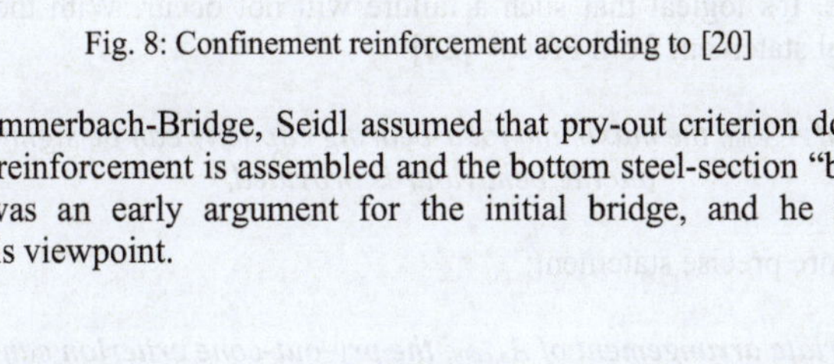

Fig. 8: Confinement reinforcement according to [20]

When designing the Simmerbach-Bridge, Seidl assumed that pry-out criterion does not apply because a sufficient confinement reinforcement is assembled and the bottom steel-section "blocks" the pry-out cone [20]. However, this was an early argument for the initial bridge, and he subsequently presented counterarguments to this viewpoint.

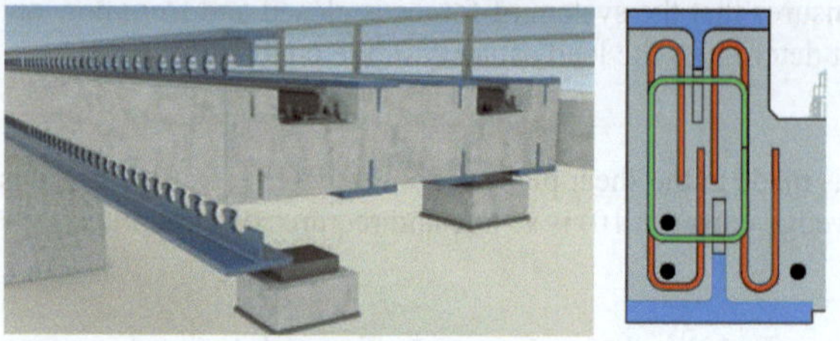

Fig. 9: Reinforcement according to [20] in the bridge concept [21]

This assumption is repeated in [9]:

"*The design against concrete pry-out can be neglected if the concrete is supported by steel flanges and at the same time confinement stirrups are installed (e.g. beam-type sections with external reinforcement).*"

3.2 Items to consider

Given our current understanding and the insights gained from bridge design experience [1], it seems that the early stages did not consider the simultaneous use of stirrups, which would have made the assumption of the potential for a pry-out-cone type mechanism no longer valid - even without any steel flanges. It logically follows that implementing a large value for Δ (Fig. 8) would prevent the occurrence of a pry-out-cone mechanism, thereby replacing the tensile strength of the concrete with the load-bearing capacity of the rebars. In other words, ensuring a spalling of concrete coverage criterion with a sufficiently large cover ensures that no pry-out-cone appears. Notably, this also calls the significance of the χ_x reduction factor in Equation 7 in [5] into question. By the same logic, the Ψ factor also appears questionable under such circumstances.

$$p_{\mathrm{Rd}} = 90 \cdot \frac{h_{\mathrm{po}}^{1,5}}{e_x} \cdot \sqrt{f_{\mathrm{ck}}} \cdot (1 + \rho_{\mathrm{Di}}) \cdot \chi \cdot \psi_{\mathrm{cr}} / \gamma_{\mathrm{V}} \ [\mathrm{kN/m}] \qquad (7)$$

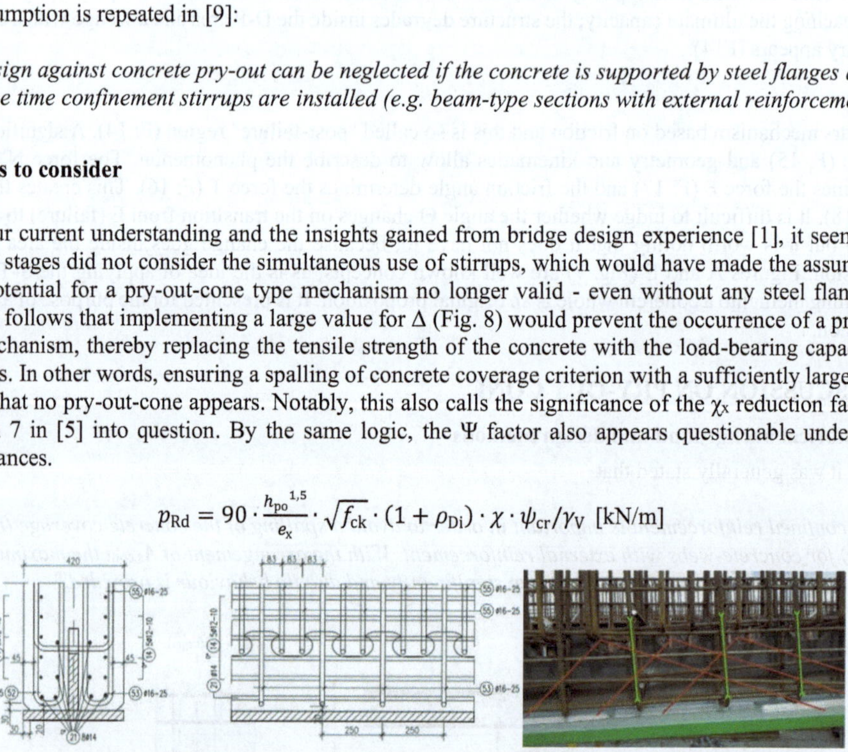

Fig. 10: Shear connection of Elbląg bridge [14] at the pillars (with virtual pry-out cones)

The issue is illustrated in Fig. 10. One might conceive of a pry-out-cone failure scenario assuming the absence of a steel flange. It's logical that such a failure will not occur. With the proper (confinement) reinforcement, the general statement from Preco+ [20]

"*With the arrangement of $A_{s,conf}$, the maximum load-bearing capacity can be significantly increased, and ductile behaviour is provided,*"

can be replaced with a more precise statement:

"*With the appropriate arrangement of $A_{s,conf}$, the pry-out-cone criterion can be eliminated.*"

No steel flange is required. This difference is of fundamental importance to bridge designers. At this point, to define what "appropriate" means, it is proposed (ad-hoc) that the confinement stirrups are provided densely, even more so than what is presented in Fig. 11, and that a conservative angle of 45 degrees is used. This would result in a factor of 0.5 instead of 0.3 in formula (7) of [9]. Of course, this must be confirmed through research and perhaps the current design rules using a 0.3 factor can be maintained. In summary, it is possible to eliminate the pry-out cone even for the ratio of small covers, as illustrated in Fig. 11.

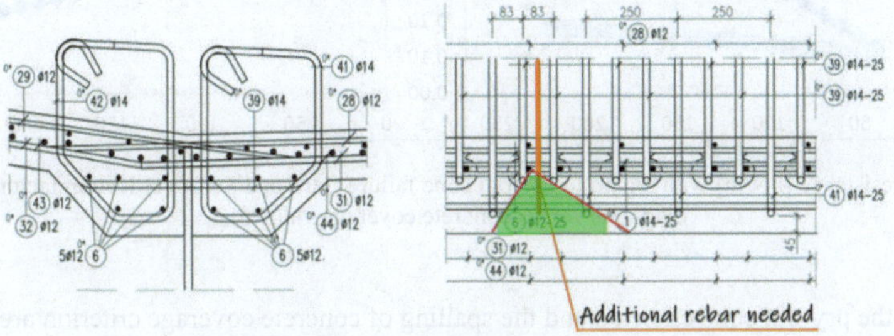

Fig. 11: Shear connection of Elbląg bridge at the field (with virtual pry-out cone)

Fig. 12 illustrates the parameters of the shear connection depicted in Fig. 9, considering a bottom coverage of 120mm. The corresponding design resistances are: $v_{L,Rd,pl} = 2760$ kN/m, $v_{L,Rd,sh} = 2512$ kN/m and $v_{L,Rd,po} =$ 1409 kN/m. Fig. 13 on the left shows how the pry-out cone resistance (vertical axis: $v_{L,Rd,po}$) varies with changes in coverage (horizontal axis: c_u), while Fig. 13 on the right shows how the reduction factor (vertical axis: $\chi = \chi_x$) changes with variable coverage (horizontal axis: c_u). As the reinforced concrete cover increases, the reduction in load capacity due to the χ factor also increases - this is established knowledge. However, it should be clarified that this applies to a low value of unreinforced cover in slabs, not the high value of reinforced cover in beams.

Steel: S 460		$E_a =$	210000	[N/mm²]			Reinforcement in dowels				
Concrete: C 40/50		$E_{cm} =$	37000	[N/mm²]	50/60					inside	outside
									d [mm]	12	12
									n	2	2
		Klothoide	1								
f_{yk}	f_{ck}	e_x	e_v	h_D	t_w	c_o	c_u	h_c	h_{po}	A_b	A_t
[N/mm²]	[N/mm²]	[mm]	[mm]	[mm]	[mm]	[mm]	[mm]	[mm]	[mm]	[mm²]	[mm²]
460	50	250	10000	100	30	1000	120	1220	152.5	226.08	226.08
A_{sf}	A_D	$A_{D,i}$	η_D	χ_x	χ_v	ρ_D	$\rho_{D,i}$				
[mm²]	[mm²]	[mm²]	[-]	[-]	[-]	[-]	[-]				
452.16	12500	305000	1.611	0.364	1.000	0.103	0.008				

Fig. 12: Parameters of typical (see Fig. 9) shear connection (c_u = 120mm).

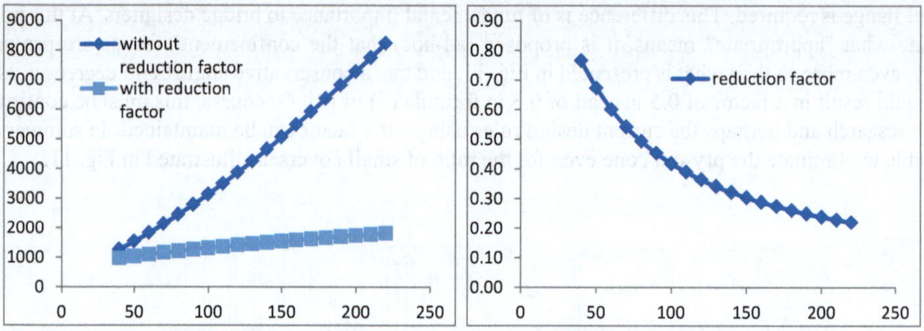

Fig. 13: Shear resistance [kN/m] with regard to pry-out cone failure (left) and Seidl's reduction factor (right): both diagrams versus concrete coverage value c_u

Fundamentally, the pry-out cone criterion and the spalling of concrete coverage criterion are reflections of the same phenomenon. In the context of bridge construction, it appears logical to evade circumstances where these criteria must be applied using concrete tensile strength, instead utilizing suitable reinforcement. Guidance on this is illustrated for beam-type sections in Fig. 10. Concurrently, it's beneficial to re-evaluate why the pry-out cone failure is considered a ductile failure mechanism, and to recognize that different types of pry-out cone failure exist, each with varying ductility and dependent on reinforcement systems. After the concrete cover has split and no reinforcement remains, it can no longer contribute to the load-bearing capacity, which is based on tensile strength (indicating brittle behaviour). There is a degree of ductility in the post-failure region of the dowel characteristic due to the main reinforcement in the dowel, but this should not be mixed with a formula based on the tensile strength of concrete.

3 SUMMARY

The present discussion addresses prevalent challenges associated with the design of composite dowels in bridge structures, with a particular focus on the concrete component of shear connections. The necessity for accurately considering the confinement effect has been underscored. The implementation of thin concrete webs and cross-prestressed dowels has drawn attention to the limitations of the current design formulas regarding shear. Concurrently, through thoughtful construction, it's not only feasible but also justified to eliminate the pry-out criterion. At present, it seems rational to design bridge structures such that the reinforcement prohibits pry-out cones, with 100mm appearing to be a reasonable coverage in bridges to ensure this.

The evolution and optimization of technology that occurred post-2017 in Poland, in conjunction with new insights, offers a fresh perspective on the necessities and solutions required for bridge construction, particularly in relation to thin concrete webs and pre-stressing. In this context, the stipulations of [5] are conservative (which is positive), but there is a need to augment them with provisions that reflect new knowledge and experiences.

Historically, the design has been consistently dictated by the potential for pry-out cone failure, which is why in the case of [5], the decision was made to replace the shear formula with boundary conditions. However, considering the arguments presented in this paper, it's proposed that the "shear design"-formula be reintroduced, albeit with the inclusion of appropriate boundary conditions (providing a confinement effect). This can be achieved by referencing existing research, without the need for substantial changes to the formula. Introduction of new formulas would require further research and some provisions (of new CEN-TS) should be left open.

REFERENCES

[1] Lorenc, W.: Composite dowels: the way to the new forms of steel-concrete composite structures. IABSE Symposium 20-22.05.2020, Wroclaw 2020.

[2] Allgemeine bauaufsichtliche Zulassung der Verbunddübelleiste, No. Z-26.4-56, Deutsches Institut für Bautechnik, Berlin, 2013. (and later version from 2018).

[3] Lorenc W., Kołakowski T., Hukowicz A., Seidl G. Verbundbrücke bei Elbląg - Weiterentwicklung der VFT-WIB-Bauweise. Stahlbau 86 (2017), Heft 2.

[4] Seidl, G.: Behaviour and load bearing capacity of composite dowels in steel-concrete composite girders, Dissertation, Raport serii PRE nr. 4/2009, Wrocław University of Technology, 2009.

[5] CEN/TC 250/SC 4 N 1817 Composite dowels: Annex D final draft (2017), not published.

[6] Leonhardt, F., Andrä, W., Andrä H-P., Harre. W.: Neues, vorteilhaftes Verbundmittel für Stahlverbund-Tragwerke mit hoher Dauerfestigkeit, Beton- und Stahlbetonbau, 12/1987.

[7] Oguejiofor, E.C., Hosaint U.M.: Numerical analysis of push-out with perfobond rib connectors. Computers and Structures Vol 62. No 4. pp 617-624., 1997.

[8] Basic Design and Theory of Hybrid Structures, Japan Society pof Civil Engineers (JSCE), 2019.

[9] Feldmann M., Kopp M., Pak D.: Composite dowels as shear connectors for composite beams – background to the German technical approval, Steel Construction 9 (2016), No. 2, pp. 80-88, Berlin 2016.

[10] Kozioł P.: Nośność łączników stalowych w strefie połączenia elementu zespolonego z elementem betonowym. Raport serii PRE nr 3/2018. Praca doktorska. Politechnika Wrocławska, 2018.

[11] Oguejiofor E. C., Hosai M. U.: Behaviour of perfobond rib shear connectors in composite beams: full-size tests. Can. I. Civ. Eng. 19, 224-235, 1992.

[12] Wurzer O.: Zur Tragfähigkeit von Betondübeln. Dissertation, Institut für Konstruktiven Ingenieurbau, Universität der Bundeswehr, München 1997.

[13] Zapfe, C.: Trag- und Verformungsverhalten von Verbundträgern mit Betondübeln zur Übertragung der Längsschubkräfte, Dissertation am Institut für Konstruktiven Ingenieurbau, Universität der Bundeswehr, München 2001.

[14] Heinemeyer, S.: Zum Trag- und Verformungsverhalten von Verbundträgern aus ultrahochfestem Beton mit Verbundleisten. Dissertation, Institut für Massivbau, RWTH Aachen, Aachen 2011.

[15] Comité Euro-International du Béton, Model Code 1990, Thomas Telford Services, Lausanne 1993

[16] Kożuch. M., Lorenc, W.: The behaviour of clothoid-shaped composite dowels, Experimental and numerical investigations, Journal of Constructional Steel Research 167 (2020)

[17] Kopp M., Pak D., Feldmann M.: Effects of transverse tension load on the static capacity and fatigue resistance of composite dowels, Steel Construction 9 (2016), No. 2, p. 115-121.

[18] Lorenc, W., Seidl G.: Anwendung von Verbunddübelleisten im Brückenbau. Münchener Stahlbautage 2017.

[19] Lorenc W.: Nośność ciągłych łączników otwartych w zespolonych konstrukcjach stalowo-betonowych (The design concept for the steel part of composite dowel shear connection in steel–concrete composite structures). Publishing Office of Wrocław University of Technology, 2011.

[20] Seidl G., Popa N., Zanon R., Lorenc W., Kożuch M., Rowiński S., Franssen J.-M., Fohn T., Hermosilla C., Farhang A., Nüsse G.: RFCS dissemination knowledge project PRECO+: Prefabricated Enduring Composite Beams based on Innovative Shear Transmission, RFS2-CT-2011-00026.

[21] Seidl G., Mensinger M., Koch E., Hugle F.: Eisenbahnüberführung Simmerbach – Pilotprojekt in VFT-Rail-Bauweise mit externer Bewehrung, Stahlbau 81 (2012), Heft 2, S. 100-107

9TH INTERNATIONAL CONFERENCE ON COMPOSITE CONSTRUCTION IN STEEL AND CONCRETE
M. Knobloch, U. Kuhlmann, W. Kurz, M. Schäfer. (Eds.)
Online, Germany, July 27-29, 2021

DEVELOPMENT OF AN ORTHOTROPIC COMPOSITE SLAB SYSTEM FOR ROAD BRIDGES

R. Stroetmann*, C. Karge* and T. Mansperger**

* TU Dresden, Institut für Stahl- und Holzbau, Dresden, Germany
e-mail: richard.stroetmann@tu-dresden.de, caecilia.karge1@tu-dresden.de
** Leonhardt, Andrä und Partner Beratende Ingenieure VBI AG, Dresden, Germany
e-mail: tobias.mansperger@lap-consult.com

Keywords: bridge deck system, orthotropic composite slabs, composite dowel strips, experimental tests.

Abstract. *Large increases in traffic loads led to extensive fatigue damage to existing orthotropic steel roadway slabs of bridges. Reinforced concrete bridges and conventional composite bridges are less susceptible to fatigue but considerably heavier. With the so-called ortho-composite-slab (OC-slab) a lightweight and robust construction is currently being developed as an alternative to existing construction methods. The basic ideas for the design are explained and an overview of the objectives of the AiF-FOSTA research project P1265 [1] is given. An overview of the experimental test program is given in section 3.3 It is used to determine the load-bearing and fatigue behavior of the OC-slab with composite dowel strips. Detailed information on tests on dowel strips in longitudinal and transverse direction are presented in [9]. Furthermore, application examples in practice are shown in section 4.*

1 INTRODUCTION

Orthotropic steel roadway slabs in bridge construction are characterized by high load-bearing capacity with low dead weight. Notch sharpness in details and strongly increased traffic loads led to extensive fatigue damage to existing orthotropic steel roadway slabs. With the so-called ortho-composite-slab (OC-slab), which consists of a longitudinally stiffened deck plate, a relatively thin layer of reinforced concrete and dowel strips as shear connectors, a lightweight and robust construction is being developed as an alternative. Similar types of roadway slabs were first used by Leonhardt [2] and Robinson [3] in the 1940s and 1950s. Due to the success of the orthotropic steel roadway slabs at that time, the lightweight composite construction became less and less important. The basic idea is taken up again in the OC-slab. However, the concrete is to be considered as statically contributing. In the earlier designs, it was mostly arranged structurally [4, 5].

2 SUPERSTRUCTURES IN LARGE BRIDGE CONSTRUCTION

Bridge superstructures are available in a wide variety of designs. There is the pure steel construction in form of the orthotropic steel roadway slab, the solid construction and composite constructions made of steel and concrete. The different superstructure types can be seen in Figure 1. The composite steel construction can be divided into two further variants. On the left, the construction with open steel box girder and in-situ concrete is shown, on the right the variant with closed steel box girder, semi-precast elements and in-situ concrete supplement.

Orthotropic steel roadway slabs are characterized by a high load-bearing capacity combined with a very low dead weight. However, strongly increasing traffic loads have resulted in extensive fatigue damage to existing orthotropic steel roadway slabs. Most affected is the connection of the trough stiffener to the deck plate. According to DIN EN 1993-2/NA [6], no computational fatigue check (F) is required for the deck plate and longitudinal stiffeners if they are designed in accordance with the structural implementation recommendation of EN 1993-2, Annex C [7]. The minimum deck plate thickness is specified as 14 mm. If the fatigue check were carried out for this minimum design, a calculated service life of only 30 – 50 years would result, taking the covering into account. Thicker deck plates are therefore used for newer bridge superstructures with orthotropic steel roadway slabs [5].

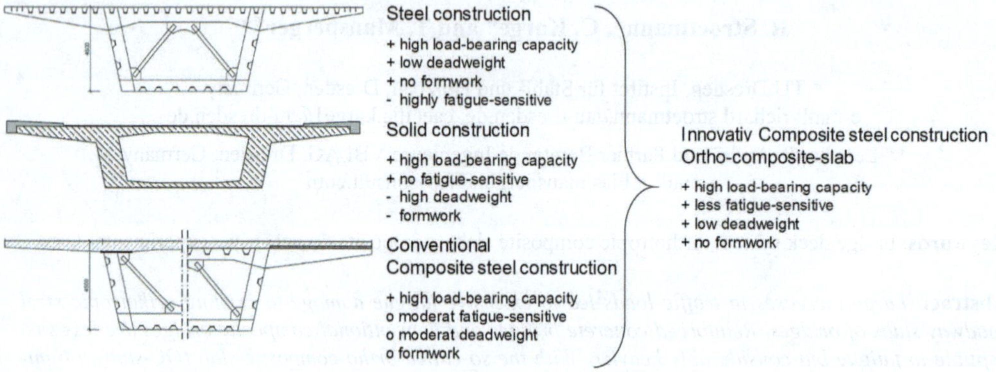

Figure 1. Overview of the main advantages and disadvantages of superstructures.

Solid constructions and conventional composite bridges are less susceptible to fatigue but considerably heavier. They also have a high load-bearing capacity, but are significantly more extensive in terms of fabrication. In solid constructions and composite constructions with open steel box girder, additional formwork work is required. In composite constructions with closed steel box girder, the formwork work has already been eliminated. However, semi-precast elements (SPE) have to be install.

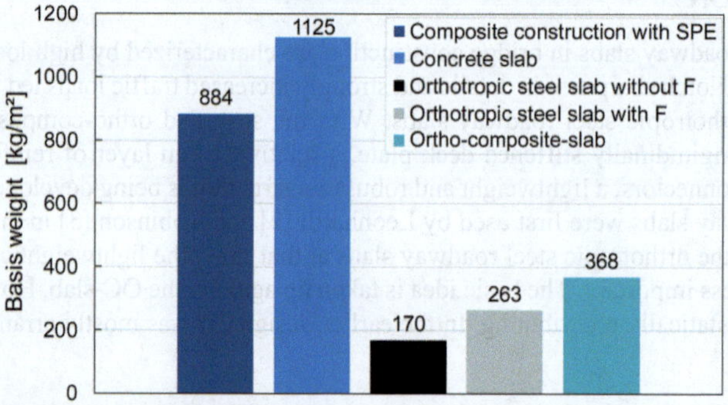

Figure 2. Basis weight of roadway slabs.

With the OC-slab a construction was developed that has a low dead weight like the orthotropic steel roadway slab, but at the same time is just as robust as the other constructions. In addition, due to a thin layer of concrete, the construction is significantly less susceptible to fatigue than the orthotropic steel roadway slab and, despite the concrete, there is no need for additional formwork work or the laying of semi-precast elements. Figure 2 illustrates the differences in the dead weight of the various roadway slabs. The conventional composite construction and the solid construction are the heaviest one with a weight per unit area of the roadway slab of 900 and 1100 kg/m² respectively. The lightest construction with 170 kg/m² is the orthotropic steel roadway slab, but it has already been mentioned, this is not designed for the current traffic loads. Orthotropic steel roadway slabs with thicker plates have a weight per unit area of about 260 kg/m². The OC-slab has a comparable low dead weight of 370 kg/m².

3 AIF-FOSTA RESEARCH PROJECT P 1265 – DEVELOPMENT OF OC-SLABS

3.1 Motivation and objective

The objective of the AiF-FOSTA research project P 1265 is to establish the fundamentals for the design and construction of OC-slabs with dowel strips for road traffic. This will create the essential requirements for the construction to establish itself as an economical and durable alternative to reinforced concrete slabs and orthotropic steel roadway slabs. The focus of the experimental, analytical and structural-mechanical investigations is to determine the load-bearing effects of the roadway slab under local wheel loads as well as the composite actions. The load-bearing capacity and fatigue strength of the dowel strips under loading in the transverse direction and under combined loading in the longitudinal and transverse directions are investigated.

3.2 Construction of the OC-slab

The construction of the OC slab consists of a modified orthotropically stiffened steel plate and a thin, single-layer reinforced cast-in-place concrete layer (Figure 3). The steel plate has a thickness of 10 mm and rests on trough stiffeners with a plate thickness of 6 mm. The stiffeners are spaced at 1050 mm. Composite dowel strips in clothoid form (CL) are provided as shear connectors in the new structure. They are used to transfer the shear forces between the steel plate and the concrete layer. The dowel strips are welded on edge to the steel plate as flat steel with recesses open at the top. In Germany, they are currently regulated by the general design approval (GDA) Z-26.4-56 [8]. Both the strips and the stiffeners are connected with fillet welds that are no thicker than 5 mm, so that single-pass welding is possible.

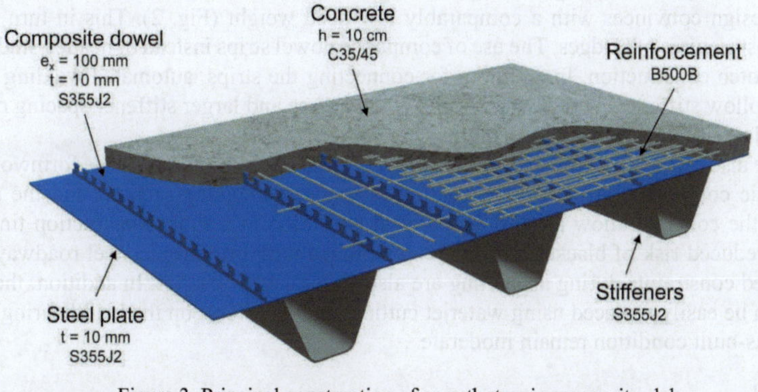

Figure 3. Principal construction of an orthotropic composite slab.

The composite dowel strips also serve as spacers for the reinforcement. Assembly bars are placed in every eighth or tenth recess. The longitudinal reinforcement and the transverse reinforcement of the concrete layer can be laid on top of them. The transverse reinforcement of the slab also serves as dowel reinforcement for the strips.

The thin top concrete layer can eliminate the weak points of the orthotropic steel roadway slab. Orthotropic steel roadway slabs showed many fatigue damages in the past, partly due to the design, but mainly due to the strong increase of traffic loads. Most of the damage occurred at the transition of the steel plate to the stiffener as a result of local wheel loads. This was caused by the insufficient stiffness of the steel plate. To illustrate the effect of the thin concrete layer, Figure 3 shows a comparative calculation between the two designs. The orthotropic steel roadway slab is shown on the left, the OC-slab on the right. Both were loaded with a single wheel load. The results show the transverse stresses in steel, first at the top and second at the bottom. Due to the thin concrete layer, the stresses could be reduced considerably. Accordingly, critical notch details, such as the connection of the stiffeners to the steel plate, are relieved. This leads to a drastic increase in fatigue strength.

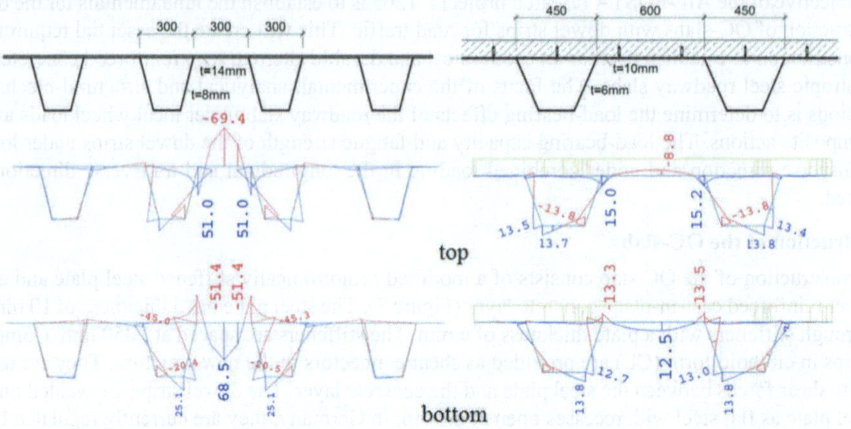

Figure 4. Comparison of stresses in steel between an orthotropic and ortho-composite slab.

In addition to the significantly higher fatigue strength compared to conventional steel roadway slabs, the new robust design convinces with a comparably low dead weight (Fig. 2). This in turn allows cost savings on the substructures of bridges. The use of composite dowel strips instead of headed stud connectors results in a low notch construction. In addition, for connecting the strips, automated welding is possible. The trapezoidal hollow stiffeners with the lower plate thicknesses and larger stiffener spacing compared to orthotropic steel decks reduce the welding volume.

The steel plate also serves as formwork for the concrete slab. The elimination of the formwork carriages required for classic composite bridges, the execution of longer concreting sections and the single-layer reinforcement of the concrete allow a more economical execution in a short construction time. Another advantage is the reduced risk of black ice formation compared with orthotropic steel roadway slabs. The temperature-related constraints during asphalting are also significantly smaller. In addition, the reinforced concrete layer can be easily replaced using waterjet cutting and the effects on the load-bearing capacity of the bridge in the as-built condition remain moderate.

3.3 Test program

Some design features of the OC-slab do not comply with current standards and guidelines. In contrast to conventional composite bridges, the composite dowel strip in OC-slabs must create a two-dimensional bond. The load-bearing behavior of the strips in transverse direction as well as biaxial has not yet been researched. The planned thin concrete layer of only 10 cm is also unusual. In addition, no shear reinforcement is provided in the thin slab. All this has to be investigated both statically and cyclically with the following test program.

Figure 5 shows the experimental matrix. The experimental test program includes static push-out tests on composite dowel strips in longitudinal and transverse direction, static and cyclic beam tests as well as a static and cyclic segment test.

The General design approval (GDA) Z-26.4-56 [8] includes the determination of the longitudinal shear capacity and the verification against fatigue for strips with a dowel spacing e_x of 150 to 500 mm. The dowel height depends on this distance. Due to the thin layer of concrete, it is necessary to arrange a strip with a low height and consequently a small dowel spacing. With a concrete layer of 10 cm, a dowel spacing e_x of

100 mm results for the clothoid geometry, deviating from the GDA, taking into account the required concrete cover. In order to extend the GDA by this dowel spacing, static push-out tests are carried out in longitudinal direction (Series 1 in Figure 5).

Series	Specimen	Loading	Objective of the study	Quantity	Material
Series 1		static	Shear and deformation behaviour in longitudinal direction to be extended by the geometrical limits of the current aBG Z-26.4-56	4	
Series 2		static	Shear and deformation behaviour in transverse direction to determine the stiffness and the design value of the shear load capacity	4	Steel S355J2 Concrete C35/45 B500B
Series 3.1	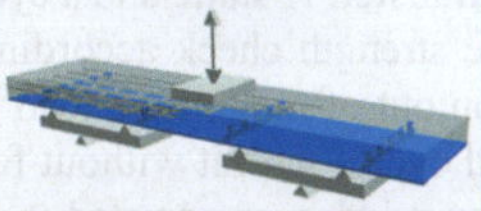	static	Load-bearing and deformation behaviour under load transfer in transverse direction and corresponding fatigue behaviour	1	
Series 3.2		cyclic		9	
Series 4.1	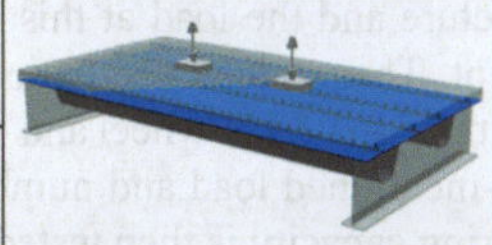	static	Load-bearing and deformation behaviour at local and global load transfer and corresponding fatigue behaviour	1	
Series 4.2		cyclic		1	

Figure 5. Test matrix.

Furthermore, unlike in conventional composite bridge construction, the composite dowel strips create a laminar bond. As a result, it is subjected not only to longitudinal but also to transverse shear loads. So far, neither the amount of load-bearing capacity nor the load-bearing mechanism in transverse direction has been investigated. The second push-out test is used to investigate the behavior of the dowel strips under transverse shear loading. Different from the tests before, the strip is arranged perpendicular to the direction of force. The design value of the transverse shear capacity is determined and, at the same time, the spring stiffness of the shear connector. The results of series 1 and 2 are explained in detail in [9].

While the push-out tests are used to test the composite dowel strip on an unyielding underground, the beam and segment tests take into account further influences from the loading and the load-bearing effect of the roadway slab. These include deflection from the plate and the bearing from the plate on the stiffeners. The transfer of local wheel loads is the main focus here.

The beam test is performed on a 0,6 m wide section of the roadway slab (Figure 5). The load-bearing and deformation behavior as well as the fatigue strength in transverse direction under a single wheel load are investigated. In the OC-slab, this beam rests on two trapezoidal hollow stiffeners. The load transfer is comparable to that of a beam on elastic support. In the test, this is simulated by two seesaw bearings. The support position of the seesaws based on numerical pre-calculations. The stiffeners webs close to the load take up significantly more load than the outer ones. The seesaw pivot point is therefore very close to the inner support points. Figure 6 shows the load distribution and the positioning of the seesaws. A static test is performed to estimate the linear elastic range for the following fatigue tests. The objective is to obtain a Wöhler-line for the connection of the composite dowel strips to the steel plate under load transfer in transverse direction. The aim is to find out whether the crack first develops at the seam transition to the strip or at the seam transition to the cover plate.

Both in the specimens of the beam tests and in the specimens of the segment tests, cracks with a width up to 0,3 mm are pre-stamped into the concrete before the tests are carried out. This creates a pre-damaging of the concrete.

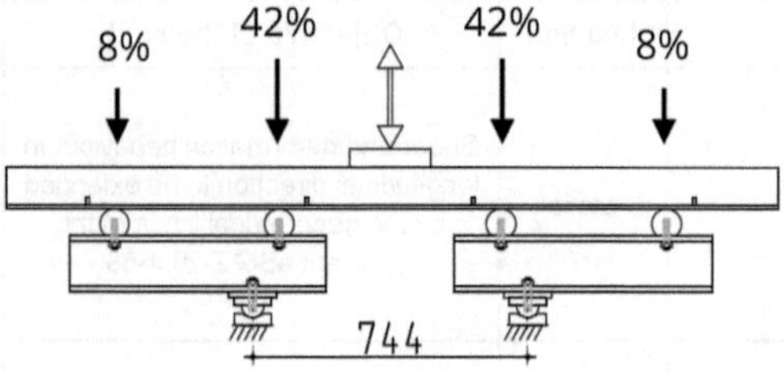

Figure 6. Seesaw bearing and load distribution of the beam test (series 3).

The structure of the segment test consists of two longitudinal stiffeners and the steel plate spanning between two cross girders. The entire design principle of the orthotropic composite slab is illustrated. The segments are 2,1 m wide and 4,2 m long. Loading is applied with two concentrated loads representing two wheel loads. The support on four outer points results in a biaxial load transfer. A static and a cyclic test are performed. For the cyclic test, the approach of the simplified fatigue strength check according to EC 3-2/NA NDP to 9.1.2 (1) was chosen. The largest wheel load oscillation out of the fatigue load models according to EC 1-2 is 75 kN. The structure must endure 5 million load cycles with it without fatiguing. In order to keep the test duration as short as possible, the load oscillation width was adjusted to 0,5 million cycles. There is a linear relationship between the stresses in the structure and the load at this load level. Therefore, the Wöhler-lines equation can be applied to this adjustment. Thus, a load oscillation width of 75 kN/wheel and 5 million load cycles, becomes a load oscillation width of 161,5 kN/wheel and 0,5 million load cycles. If the test specimen shows no signs of fatigue at the last-mentioned load and number of load cycles, fatigue strength has been demonstrated. The residual load-bearing capacity is then tested and compared with the load from the static test.

3.4 Critical notch details

The relevant notch details to be investigated with series 3 and 4 and accompanying numerical calculations are summarized in Figure 7. Possible fatigue cracking points are located at the connection of the stiffeners to the steel plate, the flame-cut edge of the composite dowel strip and its connection to the steel plate.

Under transverse shear loading, the dowel strip base is mainly subjected to bending. However, for the determination of the fatigue strength, the notch case catalogue of EN1993-1-9 do not list a fillet weld connection of a T-joint subjected to bending. Since one weld is in compression and the other in tension, the nominal stress notch cases 80 and 36* (numbers 2 and 3 in Figure 7), which apply to a T-joint constantly in tension or compression, are selected for this purpose. In notch case 80, the fatigue crack originates at the weld foot and continues through the strip. Notch case 36* describes a crack that originates in the weld root. Likewise, a crack could originate at the weld foot and continue through the steel plate. This notch detail is also categorized as notch case 80 (No. 4 in Figure 7). In test series 3, the fatigue behavior of the strip under transverse loading and consequently these three notch cases will be investigated. The aim is to be able to exclude notch case 36* and to determine a Wöhler-line for the connection of the strip under transverse shear loading.

Number 1 shows the notch case 125 or 140. This would cause a fatigue crack at the flame-cut edge of the composite dowel strip. This occurs mainly due to the cyclic longitudinal shear stresses of the dowels. At the connected webs of the stiffeners, a fatigue crack may occur at the transition of the fillet weld to the stiffener (notch case 50, No. 5 in Figure 7). These two notch details are additionally considered in the segment tests. It is necessary to find out which is the most stressed notch detail. Furthermore, it must be shown on the basis of the experimental tests, but above all also through numerical accompanying calculations, that the decisive notch detail nevertheless has sufficient fatigue strength.

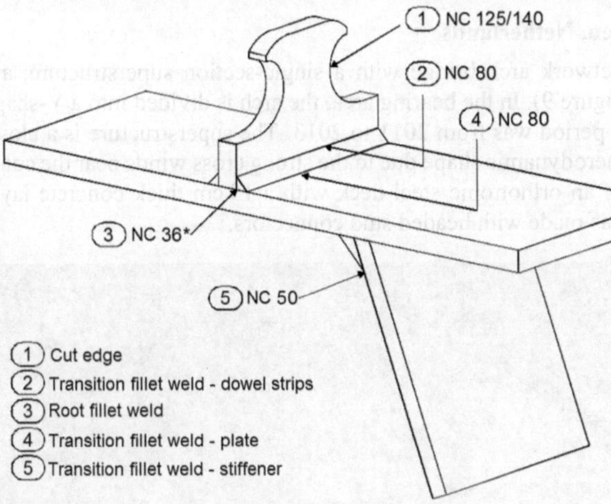

Figure 7. Relevant notch cases.

4 EXAMPLES OF APPLICATION

4.1 Elbebridge Wittenberge, part of BAB A14, Germany

Near Wittenberge in Germany, a new construction of a highway bridge over the Elbe River with a single cross-section for both directions of traffic is planned. The spans of the river bridge are 126, 160 and 126 m. The superstructure is a steel structure with a wave-like shaped central top-girder above the roadway slab. Due to the special cross-sectional shape, the roadway slab is mainly located in the area of the axis of gravity or in the compression area of the overall cross-section. Thus, no longitudinal tensile forces from the global load-bearing effect act on the roadway slab. The competition design originally proposed a conventional reinforced concrete slab. In a preliminary investigation, an OC-slab proved to be the most economical solution, partly because of its low weight and the high load-bearing capacity of the cross-section during construction (Figure 8). The roadway slab consists of a 10 mm thick steel plate supported by trapezoidal stiffeners with a spacing of 600 mm. A 15 cm thick, single-layer reinforced cast-in-place concrete slab is placed on top. Headed studs are acting as shear connectors.

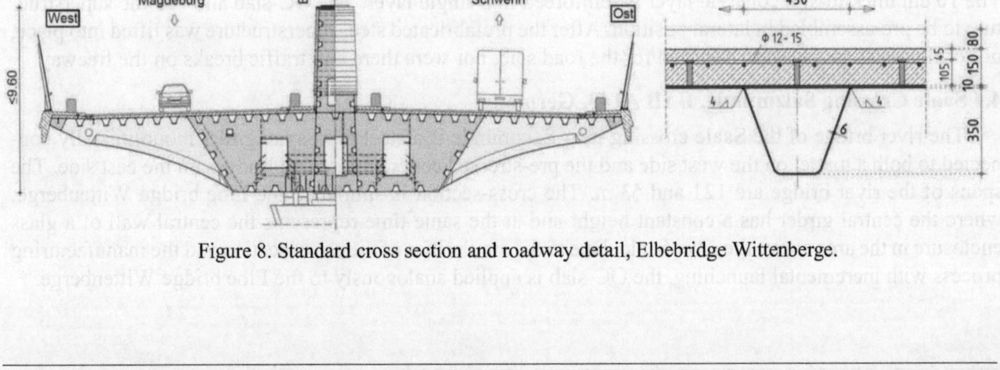

Figure 8. Standard cross section and roadway detail, Elbebridge Wittenberge.

In 2012, component tests were carried out in individual cases as part of the approval process. These tests already showed the characteristic punching shear failure at wheel loads in the range of 1 MN and thus a sufficient load-bearing capacity. The start of construction work is scheduled for spring 2022.

4.2 De Oversteeg, Nijmegen, Netherlands

"De Oversteeg" is a network arch bridge with a single-section superstructure, approx. 30 m wide, spanned by a mono arch (Figure 9). In the bearing area, the arch is divided into a Y-shape. The span length is 285 m. The construction period was from 2011 to 2013. The superstructure is a closed box girder with cross beams, which has an aerodynamic shape due to the strong cross winds near the coast. The bridge deck is an OC-slab consisting of an orthotropic steel deck with a 15 cm thick concrete layer and a two-layer reinforcement. The bond was made with headed stud connectors.

Figure 9. "De Oversteeg" in Nijmegen, Netherlands (Picture © LAP).

4.3 Cyclist bridge building 780b over A6 near Schwabbach, Germany

The BW 780b cycle path bridge in the course of the A6 freeway near Schwabbach in Germany crosses a six-lane freeway equipped with high gabion walls as noise barriers. The structure is an integral frame bridge with struts. The span is approx. 50 m. Before the highway was extended, there was a tunnel for the crossing cycle path. For the new bridge construction (Figure 10), long ramps had to be filled in and corresponding land acquisition had to be made. In order to keep this as low as possible, the gradient of the cycle path had to run as low as possible above the freeway. By using an OC-slab, the construction height was reduced by 20 cm. The OC-slab consists of a stiffened steel plate and puzzle-shaped composite dowel strips. The 10 cm thick in-situ concrete layer is reinforced in a single layer. The OC-slab allowed the superstructure to be pre-assembled in lateral position. After the prefabricated steel superstructure was lifted into place, no formwork elements were required for the road slab, nor were there any traffic breaks on the freeway.

4.4 Saale Crossing Salzmünde, BAB A143, Germany

The river bridge of the Saale crossing near Salzmünde is a steel twin-span girder monolithically connected to both a tunnel on the west side and the pre-stressed concrete foreland bridges on the east side. The spans of the river bridge are 121 and 53 m. The cross-section is similar to the Elbe bridge Wittenberge, where the central girder has a constant height and at the same time represents the central wall of a glass enclosure in the area above the river Saale. Due to the similarities of the cross-section and the manufacturing process with incremental launching, the OC-slab is applied analogously to the Elbe bridge Wittenberge.

Figure 10. Cyclist bridge building 780b over A6 near Schwabbach, Germany (Picture © LAP).

5 SUMMARY AND OUTLOOK

The OC-slab is a robust and lightweight alternative to existing steel and composite roadway slabs. Key design details, such as the thin concrete layer without shear force reinforcement or the composite dowel strips, which creates a two-dimensional bond between steel and concrete, are investigated using the test program presented here and supplemented by numerical calculations. The test program includes static push-out tests on composite dowel strips in their longitudinal and transverse directions, a static and several cyclic beam tests, and a static and cyclic segment test. While the OC-slab is mainly considered as a new construction option in the P 1265 project, there is also the possibility to use the basic principles of the OC-slab as a retrofit measure for existing orthotropic steel roadway slabs. The advantages presented speak for themselves. The large number of application examples also shows that there is great interest in implementing the new construction.

ACKNOWLEDGMENTS

The IGF project 20255 BR / FOSTA P1265 " Wirtschaftliche und dauerhafte Orthoverbundfahrbahnplatten mit Dübelleisten für Straßenbrücken" of FOSTA – Forschungsvereinigung Stahlanwendung e. V., Düsseldorf, was funded by the German Federal Ministry for Economic Affairs and Energy via the AiF as part of the program for the promotion of Collective industrial research (IGF) on the basis of a resolution of the German Bundestag. The project was carried out at the Institute of Steel and Timber Construction of the Technische Universität Dresden. The authors would like to thank the BMWi for financial support, FOSTA for taking over the sponsorship, organization and handling, as well as all members of the committee accompanying the project for their active participation and support in the form of provision of materials, production of test specimens and support constructions for carrying out the tests.

REFERENCES

[1] FOSTA-Forschungsprojekt P 1265, "*Wirtschaftliche und dauerhafte Orthoverbundfahrbahnplatten mit Dübelleisten für Straßenbrücken*", Research report (in progress), 2022.

[2] Leonhardt, F., "*Die neue Straßenbrücke über den Rhein von Köln nach Deutz*", Die Bautechnik, Jahrgang 26, Heft 7, 193-199, 1949.

[3] Fauchart, J.; Sfintesco, D., "*The development and use of the Robinsons composite deck in France*", IABSE Rep. Work. Commissions *2*, 155-164, 1968.

[4] Mansperger T., Jung R., "*Neue Wege im Verbundbrückenbau*", Stahlbau 87, Heft 8, 752–758, 2018.

[5] Jung, R.; Mansperger, T., "*Die Orthoverbund-Fahrbahnplatte*", Stahlbau 89, Heft 2, 129–137, 2020.

[6] DIN EN 1993-2/NA Nationaler Anhang - Eurocode 3, "*Bemessung und Konstruktion von Stahlbauten – Teil 2: Stahlbrücken*", 2014.

[7] EN 1993-2 Eurocode 3, "*Design of steel structures – Part 2: Steel Bridges*", 2010.

[8] Deutsches Institut für Bautechnik, „*Allgemeine Bauartgenehmigung für Stahlverbundträger mit Verbunddübelleisten in Klothoiden- und Puzzleform*", Zulassung Z 26.4-56, 2018.

[9] Karge C.; Stroetmann R., Wisnewski J., "*Experimental and numerical investigation of dowel strips for longitudinal and transverse loading*", 9th international conference on composite construction in steel and concrete, 2021.

9TH INTERNATIONAL CONFERENCE ON COMPOSITE CONSTRUCTION IN STEEL AND CONCRETE
M. Knobloch, U. Kuhlmann, W. Kurz, M. Schäfer. (Eds.)
Online, Germany, July 27-29, 2021

EFFECT OF CROSS-SECTION BRACING IN STEEL-CONCRETE COMPOSITE BRIDGE DECKS USING THE GENERALIZED BEAM THEORY

Luís Vieira*, José Oliveira Pedro*, Rodrigo Gonçalves and Dinar Camotim***

* *CERIS, DECivil, Instituto Superior Técnico, Universidade de Lisboa, Portugal*
e-mails: luisjfvieira@tecnico.ulisboa.pt, jose.oliveira.pedro@tecnico.ulisboa.pt, dcamotim@civil.ist.utl.pt

** *CERIS, DEC, Faculdade de Ciências e Tecnologia, Universidade Nova de Lisboa, Portugal*
e-mail: rodrigo.goncalves@fct.unl.pt

Keywords: Steel-concrete bridges, Flexible cross-section bracing, Distortion, Generalized Beam Theory (GBT), Box-girder, Twin-girder

Abstract. *This paper explores the structural analysis of steel-concrete bridge decks with explicit consideration of flexible cross-section bracing, using a novel procedure based on the Generalized Beam Theory (GBT). The effect of the bracing is included in a GBT-based bar finite element, which is used to analyze two numerical examples involving standard deck cross-sections and bracing systems. It is shown that the proposed finite element leads to extremely accurate solutions when compared with those obtained with full shell finite element models, with a huge economy in terms of DOFs. Moreover, it is shown that the modal features of GBT provide a very valuable tool for the structural interpretation of the results.*

1 INTRODUCTION

Steel-concrete composite bridge decks generally include intermediate cross-bracings to provide adequate stiffening. These bracing systems are important to restrain distortion effects in box-girder decks, improve the distribution of the live loads between multiple girders (if more than two), and transfer lateral loads from the girders to the deck slab [1].

Although the bridge structural behavior can be accurately assessed with shell finite element models, such models are often too demanding in terms of computational effort and require careful handling of the results. Simplified methods based on standard beam or grillage models [2,3] usually offer a simple approach, but they may lead to insufficiently accurate results especially if flexible intermediate bracings are present and their effect needs to be properly considered. Alternatively, Generalized Beam Theory (GBT) can be employed [4-10]. Originally proposed and developed by Schardt [11], GBT is currently well established as a powerful tool for the analysis of thin-walled prismatic members with deformable cross-sections, since it (i) retains (and improves on) the structural interpretation of a beam formulation and (ii) provides very accurate results with just a fraction of the computational effort required by full shell finite element models.

The objective of this paper is to present a new GBT approach that can explicitly simulate the effect of intermediate flexible cross-section bracing and predict, very accurately, torsion-distortion effects in steel-concrete composite bridges with such bracing. In order to demonstrate the capabilities and potential of the proposed approach, two commonly used twin-girder and box-girder decks with intermediate bracing are analyzed. The results obtained with the proposed GBT-based approach are compared with those provided by shell finite element models.

2 GENERALIZED BEAM THEORY WITH FLEXIBLE BRACINGS

2.1 Overview of the GBT formulation

According to fundamentals of Generalized Beam Theory (GBT), the beam kinematics are described through the superposition of pre-calculated cross-section deformation modes (e.g. major and minor axis bending, shear, torsion and higher-order – see Section 2.2) whose amplitudes along the beam longitudinal axis constitute the problem unknowns.

Consider the arbitrary thin-walled prismatic member depicted in Figure 1. Using Kirchhoff's assumption for thin plates, the displacement vector $\boldsymbol{U}$ can be expressed in terms of the wall mid-surface local axes as

$$\boldsymbol{U}(x,y,z)=\begin{bmatrix}U_x\\U_y\\U_z\end{bmatrix}=\begin{bmatrix}0 & \bar{\boldsymbol{u}}^t\\ \bar{\boldsymbol{v}}^t & 0\\ \bar{\boldsymbol{w}}^t & 0\end{bmatrix}\begin{bmatrix}\boldsymbol{\phi}\\ \boldsymbol{\phi}_{,x}\end{bmatrix}-z\begin{bmatrix}0 & \bar{\boldsymbol{w}}^t\\ \bar{\boldsymbol{w}}_{,y}{}^t & 0\\ 0 & 0\end{bmatrix}\begin{bmatrix}\boldsymbol{\phi}\\ \boldsymbol{\phi}_{,x}\end{bmatrix}=\boldsymbol{\Xi}_U\begin{bmatrix}\boldsymbol{\phi}\\ \boldsymbol{\phi}_{,x}\end{bmatrix}, \tag{1}$$

where the comma indicates a differentiation, $\boldsymbol{\phi}=\boldsymbol{\phi}(x)$ is a column vector containing the deformation mode amplitude functions, $\boldsymbol{\Xi}_U$ is an auxiliary modal matrix and $\bar{\boldsymbol{u}}$, $\bar{\boldsymbol{v}}$ and $\bar{\boldsymbol{w}}$ are column vectors containing the deformation mode wall mid-line displacement components along the x,y and z local axes, respectively.

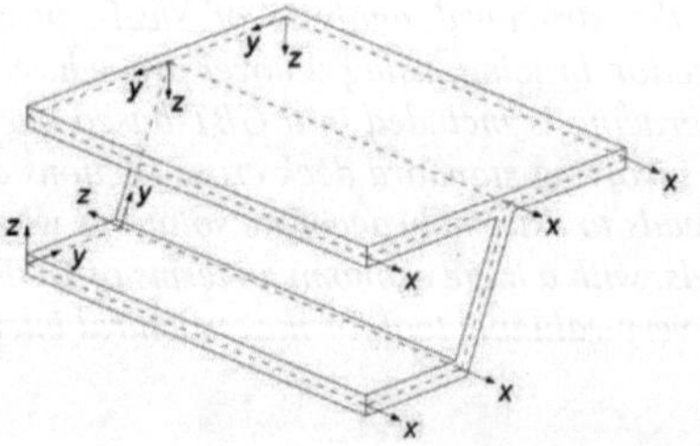

Figure 1. Arbitrary thin-walled member geometry and local coordinate systems.

The non-null components of the strain vector $\boldsymbol{\varepsilon}^t=[\,\varepsilon_{xx}\;\varepsilon_{yy}\;\gamma_{xy}\,]$, which can be separated into membrane (M) and bending (B) terms, are obtained from the displacements by means of

$$\boldsymbol{\varepsilon}=\boldsymbol{\varepsilon}^M+\boldsymbol{\varepsilon}^B=(\boldsymbol{\Xi}_\varepsilon^M(y)+z\boldsymbol{\Xi}_\varepsilon^B(y))\begin{bmatrix}\boldsymbol{\phi}\\ \boldsymbol{\phi}_{,x}\\ \boldsymbol{\phi}_{,xx}\end{bmatrix} \tag{2}$$

$$\boldsymbol{\Xi}_\varepsilon^M(y)=\begin{bmatrix}0 & 0 & \bar{\boldsymbol{u}}^t\\ \bar{\boldsymbol{v}}_{,y}{}^t & 0 & 0\\ 0 & (\bar{\boldsymbol{u}}_y+\boldsymbol{v})^t & 0\end{bmatrix},\;\boldsymbol{\Xi}_\varepsilon^B(y)=-\begin{bmatrix}0 & 0 & \bar{\boldsymbol{w}}^t\\ \bar{\boldsymbol{w}}_{,yy}{}^t & 0 & 0\\ 0 & 2\bar{\boldsymbol{w}}_{,y}{}^t & 0\end{bmatrix} \tag{3}$$

A plane stress state is assumed in each wall and the non-null stress components, assembled in Voigt-notation as $\boldsymbol{\sigma}^t=[\,\sigma_{xx}\;\sigma_{yy}\;\tau_{xy}\,]$, can be related to $\boldsymbol{\varepsilon}$, in the elastic case, through [4]

$$\boldsymbol{\sigma}=\boldsymbol{\sigma}^M+\boldsymbol{\sigma}^B=\begin{bmatrix}\sigma_{xx}^M\\ \sigma_{yy}^M\\ \tau_{xy}^M\end{bmatrix}+\begin{bmatrix}\sigma_{xx}^B\\ \sigma_{yy}^B\\ \tau_{xy}^B\end{bmatrix}=\begin{bmatrix}E & 0 & 0\\ 0 & 0 & 0\\ 0 & 0 & G\end{bmatrix}\boldsymbol{\varepsilon}^M+\begin{bmatrix}\frac{E}{1-\nu^2} & \frac{\nu E}{1-\nu^2} & 0\\ \frac{\nu E}{1-\nu^2} & \frac{E}{1-\nu^2} & 0\\ 0 & 0 & G\end{bmatrix}\boldsymbol{\varepsilon}^B \tag{4}$$

The GBT equilibrium equation system is obtained from a variational principle, such as the Principle of Virtual Work. Assuming that the loads are applied at the wall mid-planes, for simplicity, one has

$$\delta W_{int} = -\int_V \begin{bmatrix} \delta\boldsymbol{\phi} \\ \delta\boldsymbol{\phi}_{,x} \\ \delta\boldsymbol{\phi}_{,xx} \end{bmatrix}^T \left(\boldsymbol{\Xi}_\varepsilon^{M\,t} \begin{bmatrix} E & 0 & 0 \\ 0 & 0 & 0 \\ 0 & 0 & G \end{bmatrix} \boldsymbol{\Xi}_\varepsilon^{M} + \boldsymbol{\Xi}_\varepsilon^{B\,t} \begin{bmatrix} \frac{E}{1-\nu^2} & \frac{\nu E}{1-\nu^2} & 0 \\ \frac{\nu E}{1-\nu^2} & \frac{E}{1-\nu^2} & 0 \\ 0 & 0 & G \end{bmatrix} \boldsymbol{\Xi}_\varepsilon^{B} \right) \begin{bmatrix} \boldsymbol{\phi} \\ \boldsymbol{\phi}_{,x} \\ \boldsymbol{\phi}_{,xx} \end{bmatrix} dV \tag{5}$$

$$\delta W_{ext} = \int_\Omega \begin{bmatrix} \delta\boldsymbol{\phi} \\ \delta\boldsymbol{\phi}_{,x} \end{bmatrix}^t \boldsymbol{\Xi}_U{}^t\, \boldsymbol{q}\, d\Omega \tag{6}$$

where $\boldsymbol{q}^t = [\, q_x\, q_y\, q_z \,]$ is the applied load vector.

Performing a cross-section integration leads to

$$\delta W_{int} = -\int_L \begin{bmatrix} \delta\boldsymbol{\phi} \\ \delta\boldsymbol{\phi}_{,x} \\ \delta\boldsymbol{\phi}_{,xx} \end{bmatrix}^T \begin{bmatrix} \boldsymbol{B} & \boldsymbol{0} & \boldsymbol{D}_2 \\ \boldsymbol{0} & \boldsymbol{D}_1 & \boldsymbol{0} \\ \boldsymbol{D}_2{}^t & \boldsymbol{0} & \boldsymbol{C} \end{bmatrix} \begin{bmatrix} \boldsymbol{\phi} \\ \boldsymbol{\phi}_{,x} \\ \boldsymbol{\phi}_{,xx} \end{bmatrix} dx \tag{7}$$

where $\boldsymbol{B}$, $\boldsymbol{C}$, $\boldsymbol{D}_1$ and $\boldsymbol{D}_2$ are "GBT linear stiffness matrices", associated, respectively, with (i) transverse extension/flexure (ii) primary/secondary warping (iii) plate shear distortion/torsion and (iv) membrane/flexural Poisson effects [12]. These matrices are essential to the process of calculation of the deformation modes and are also quite useful in the identification and characterization of the physical phenomena underlying each deformation mode.

2.2 Cross-section deformation modes

The cross-section deformation modes are characterized through the mid-line displacement functions $\overline{\boldsymbol{u}}$, $\overline{\boldsymbol{v}}$ and $\overline{\boldsymbol{w}}$ (recall Eq.1) and are calculated through the so-called "GBT cross-section analysis", which consists mainly of two steps:

1) The cross-section is reduced to its wall mid-surfaces and discretized by means of cross-section nodes with the purpose of establishing an initial deformation mode basis related to the imposition of unit nodal in-plane and warping displacements (cross-section DOFs). Besides the "natural nodes", which are located at wall intersections or outstands, it is possible to consider "intermediate nodes" that add DOFs and thus improve the ability of the deformation modes to reproduce complex displacement fields (i.e., the accuracy of the formulation is enhanced).
2) The cross-section deformation modes are calculated from the initial basis through appropriate change of basis procedures involving the GBT linear stiffness matrices (recall Eq. 7), leading to a set of physically meaningful deformation modes.

For arbitrary flat-walled cross-sections, this can be done through the GBTUL software [13], freely available at http://www.civil.ist.utl.pt/gbt, which implements the theory laid out in Refs. [12,14,15]. However, GBTUL does not allow for the consideration of the rigid link elements used in this work to model the offset at the connection between the steel and concrete slab. For this reason, the GBTUL procedure was modified to account for this offset (the details are presented in a forthcoming journal paper). The proposed procedure splits the deformation modes into three main categories:

(i) *Vlasov modes*, which comply with Vlasov's null membrane shear strain hypothesis ($\gamma_{xy}{}^M = 0$)

(ii) *Shear Modes*, non-Vlasov modes with null membrane transverse extension ($\gamma_{xy}{}^M \neq 0,\ \varepsilon_{yy}{}^M = 0$)

(iii) *Transverse Extension Modes*, non-Vlasov modes with non-null transverse extension ($\varepsilon_{yy}{}^M \neq 0$)

For illustrative purposes, the deformation modes obtained with the proposed procedure for a composite I-girder are depicted in Figure 2.

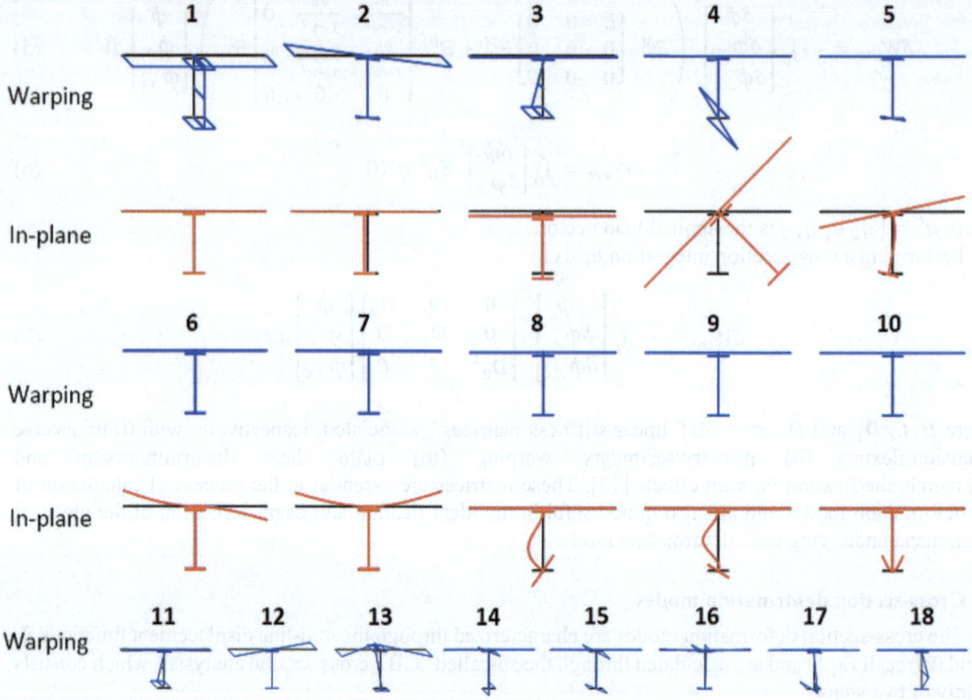

Figure 2. In-plane and warping configurations of the Vlasov (1-10) and shear (11-18) deformation modes of a composite I-girder cross-section with one intermediate node in the web.

As in conventional beam theories, axial extension and major/minor-axis bending are included in the Vlasov modes. For open cross-sections, the torsion mode is also a Vlasov mode, whereas for cross-sections with closed cells it lies in the shear mode space due to the presence of circulatory membrane shear flow. The Vlasov mode family also contains additional local-plate and distortional modes.

Membrane shear effects are captured by the shear modes which can be described, in open sections, using only warping functions. Membrane transverse extension is usually negligible and therefore transverse extension modes are discarded from the analysis, leading to important computational savings.

2.3 Finite element implementation

The finite element implementation is similar to that employed in Ref. [5], interpolating the deformation mode amplitude functions through $\boldsymbol{\Phi} = \boldsymbol{\Psi}\, \boldsymbol{d_e}$, where matrix $\boldsymbol{\Psi}$ contains the approximation functions and vector $\boldsymbol{d_e}$ contains the unknown nodal values, which constitute the problem unknowns. Hierarchic linear and quadratic Lagrange polynomials are employed for pure warping modes ($u_k \neq 0, v_k = w_k = 0$) and Hermite cubic polynomials for the remaining modes. This leads to a finite element with two end nodes having $2N - N_w$ DOFs each and one intermediate node with N_w DOFs, where N is the total number of modes and N_w the number of warping modes.

The GBT element stiffness matrix $\boldsymbol{K_e}$ and force vector $\boldsymbol{F_e}$ are determined by inserting the interpolation in Eqs. (7) and (6), respectively, leading to

$$\boldsymbol{K_e} = \int_{L_e} \begin{bmatrix} \boldsymbol{\Psi} \\ \boldsymbol{\Psi}_{,x} \\ \boldsymbol{\Psi}_{,xx} \end{bmatrix}^T \begin{bmatrix} \boldsymbol{B} & \boldsymbol{0} & \boldsymbol{D_2} \\ \boldsymbol{0} & \boldsymbol{D_1} & \boldsymbol{0} \\ \boldsymbol{D_2}^t & \boldsymbol{0} & \boldsymbol{C} \end{bmatrix} \begin{bmatrix} \boldsymbol{\Psi} \\ \boldsymbol{\Psi}_{,x} \\ \boldsymbol{\Psi}_{,xx} \end{bmatrix} dx \quad (9)$$

$$\boldsymbol{F_e} = \int_{\Omega_e} \begin{bmatrix} \boldsymbol{\Psi} \\ \boldsymbol{\Psi}_{,x} \end{bmatrix}^t \Xi_U{}^t\, \boldsymbol{q}\, d\Omega_e \quad (10)$$

The longitudinal discretization is defined so that the bracing locations coincide with element end nodes. By doing so, the bracing can be simulated as a set of nodal elastic constraints that directly relate to nodal DOFs.

In a shell finite element model, the kinematic boundary conditions are enforced at the nodes, whereas in a GBT model they are applied to the deformation modes. To ensure consistency, constraint equations are set up in the GBT model (following the Lagrange multiplier method [16]), to replicate exactly the boundary conditions of the shell model. The finite element procedure was implemented in Matlab [17].

2.4 Stiffness contribution of cross-section bracing

The stiffening effect of cross-section bracing is determined through a standard sub-structuring procedure, in which the cross-section bracing is analyzed separately by means of a local shell finite element model. The displacements of the beam-bracing interface associated with each DOF are imposed as settlement displacements in ABAQUS [18] using S4R and S3R elements (see Figure 3).

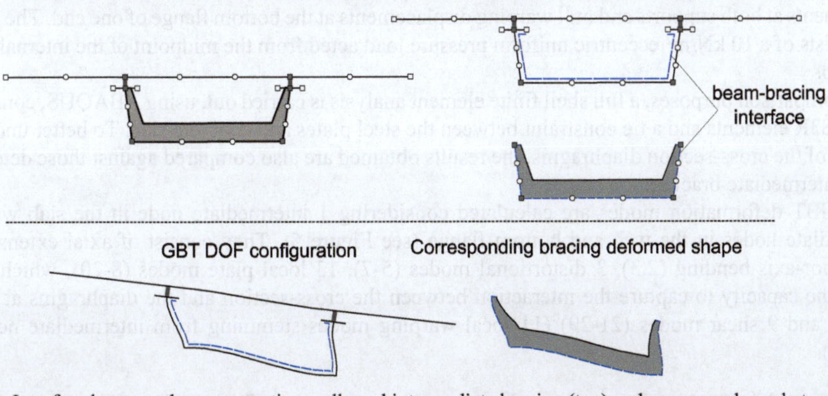

Figure 3. Interface between the cross-section walls and intermediate bracing (top) and correspondence between GBT DOF and bracing deformed configurations (bottom)

The contribution of cross-bracing to the element stiffness matrix is calculated by determining the work performed by the reaction forces F_{BS} on the beam-bracing interface displacements U_{BS},

$$\boldsymbol{K_{BS_{ij}}} = \int_{\Omega_{BS}} U_{BS_i} F_{BS_j}\, d\Omega_{BS} \quad (13)$$

where $i,j = 1, \dots, 2N - N_w$ refer to the nodal DOFs.

The same bracing system is usually repeated throughout the deck span and, therefore, its contribution to the element stiffness matrix is determined only once. Note also that this procedure is very versatile since it is not specific to any deck cross-section or cross-bracing configuration.

3 ILLUSTRATIVE EXAMPLES

3.1 Box-girder with flexible plate diaphragms

The first example deals with a slightly asymmetric box-girder with intermediate cut-out plate diaphragms – it is based on the replacement of two bridge decks in Luxembourg [19]. Additional stiffening is provided to the diaphragm through orthogonal "flange plates" along the internal edges, thus forming a T-section. The cross-section geometry, material properties and loading are displayed in Figure 4.

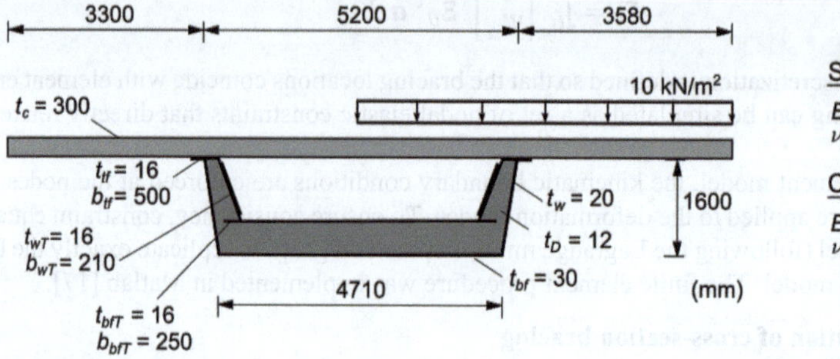

Figure 4. Deck cross-section and intermediate diaphragm geometry, loading and material properties.

A simply supported span of 35 meters is considered, having 5 intermediate diaphragms with a longitudinal spacing of about 5.83 meters. The beam support conditions consider null cross-section in-plane displacements at both supports and null warping displacements at the bottom flange of one end. The external load consists of a 10 kN/m² eccentric uniform pressure load acted from the midpoint of the internal portion of the slab.

For comparison purposes, a full shell finite element analysis is carried out, using ABAQUS, considering S4R and S3R elements and a tie constraint between the steel plates and concrete slab. To better understand the effect of the cross-section diaphragms, the results obtained are also compared against those determined without intermediate bracing.

The GBT deformation modes are calculated considering 1 intermediate node in the slab walls and 2 intermediate nodes in the web and bottom flange (see Figure 5). They consist of axial extension (1), major/minor-axis bending (2,3), 3 distortional modes (5-7), 13 local-plate modes (8-20), which largely improve the capacity to capture the interaction between the cross-section and the diaphragms at bracing locations, and 9 shear modes (21-29) (11 local warping modes stemming from intermediate nodes are discarded).

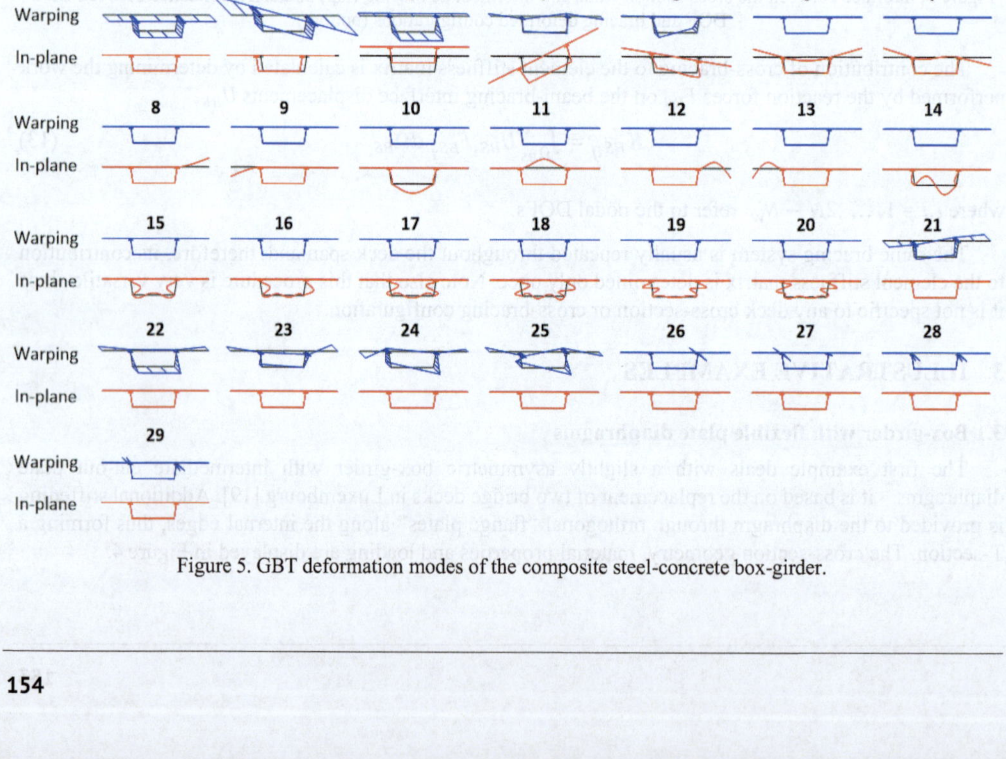

Figure 5. GBT deformation modes of the composite steel-concrete box-girder.

The GBT longitudinal discretization adopted consists of 24 equal-length finite elements, i.e., 4 elements between braced cross-sections, leading to a total of 1440 DOFs.

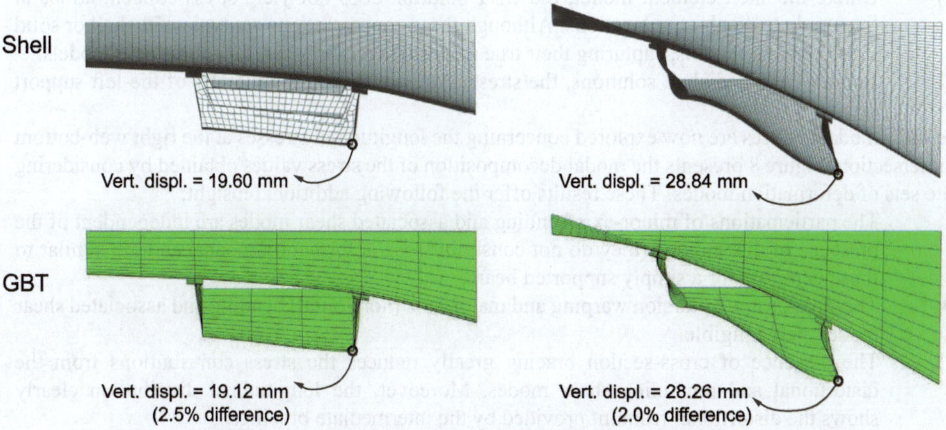

Figure 6. Deformed configurations (x100) of half of the girders with (left) and without (right).

The deformed configurations are displayed in Figure 6 and it is possible to observe an excellent agreement between the GBT and shell results. The significant reduction of cross-section distortion due to cross-section bracing is very clear. Moreover, GBT is able to capture the slight cross-section distortion that still occurs, even in braced cross-sections, which confirms the presence of relatively flexible diaphragms.

The distortional effects in box-girders produce longitudinal warping stresses which can be quite relevant when compared to the bending stresses [1]. Figure 7 presents the variation along the span of the longitudinal membrane stresses at the two web-bottom flange intersections, with and without intermediate diaphragms.

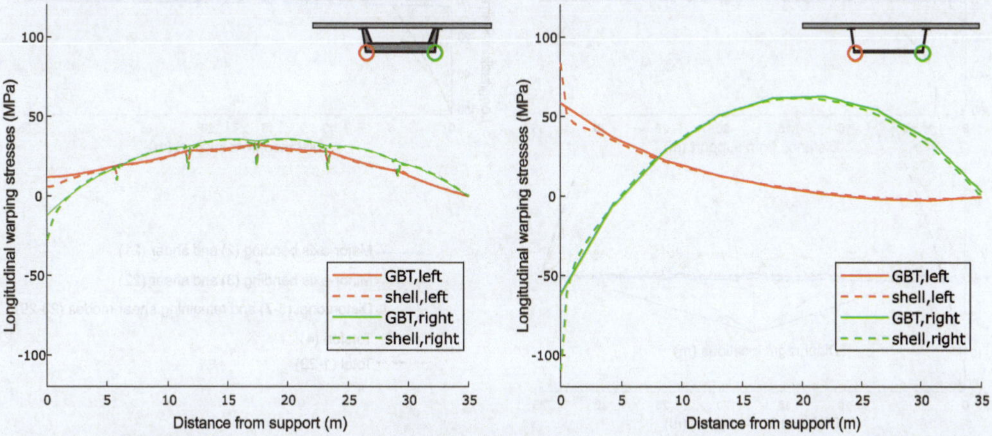

Figure 7. Longitudinal membrane warping stresses at web-bottom flange intersections with (left) and without (right) intermediate diaphragms.

The observation of these stress distributions prompts the following remarks:

(i) The inclusion of the cross-section diaphragms significantly reduces the longitudinal warping stresses (the maximum tensile stress reduces from 63 MPa to 36 MPa).

(ii) The difference between the stresses at the left and right bottom flange nodes is an indicator of the distortion warping stress magnitude. Naturally, this difference substantially decreases with the presence of intermediate bracing.

(iii) Unlike the shell element model, the GBT solution does not yield stress concentrations at intermediate diaphragm locations. Although these concentrations are typical in shell or solid finite element models, capturing their true magnitude requires careful and refined models.

(iv) In both GBT and shell solutions, the stresses are not null in the vicinity of the left support section.

The GBT modal features are now explored concerning the longitudinal stresses at the right web-bottom flange intersection. Figure 8 presents the modal decomposition of the stress values obtained by considering adequate sets of deformation modes. These results offer the following additional insight:

(i) The participations of minor-axis bending and associated shear modes are independent of the presence of diaphragms (they do not constrain these deformation modes) and are similar to those expected for a simply supported beam.

(ii) The stresses due to torsion warping and major-axis (horizontal) bending and associated shear modes are negligible.

(iii) The presence of cross-section bracing greatly reduces the stress contributions from the distortional and remaining shear modes. Moreover, the longitudinal distribution clearly shows the distortional restraint provided by the intermediate bracings.

(iv) The non-null support section stresses stem from a clamping effect provided to distortion. This is due to the warping restraint in the bottom flange of the left support section, which prevents the in-plane rotation of this plate and, therefore, acts as a fixed-end support for any deformation mode associated with bottom-flange lateral displacements (namely the distortional mode 5).

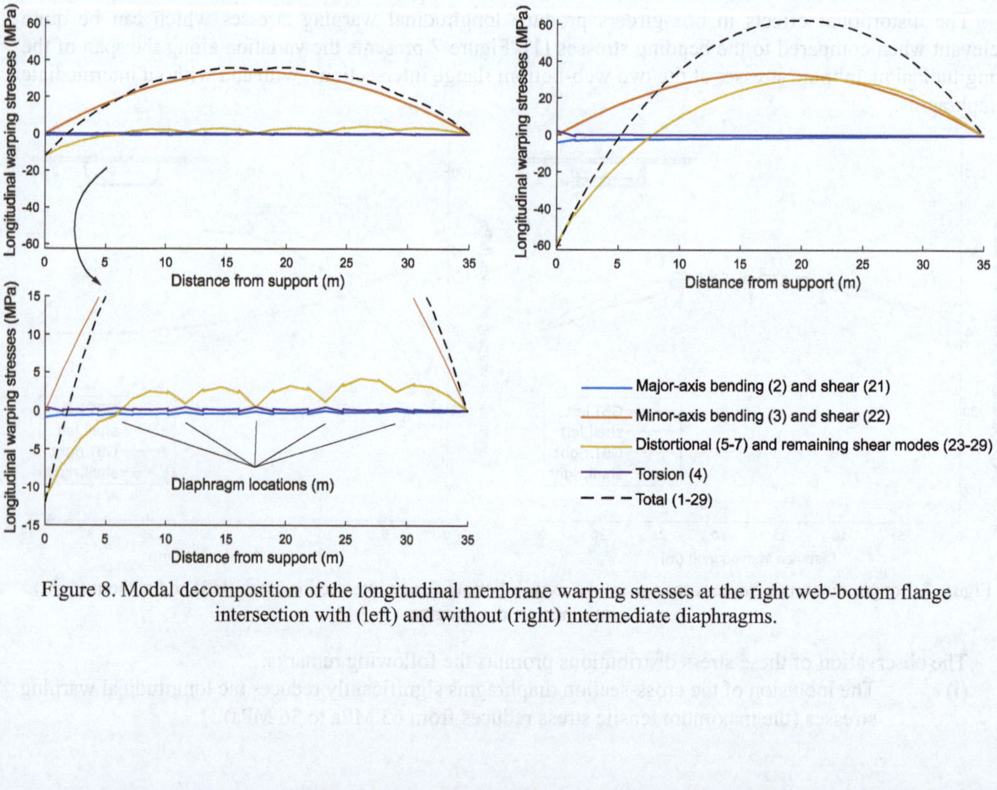

Figure 8. Modal decomposition of the longitudinal membrane warping stresses at the right web-bottom flange intersection with (left) and without (right) intermediate diaphragms.

3.2 Twin-girder with intermediate cross-beams

The second example concerns a twin-girder composite steel-concrete deck from the OPTIBRI research project [20]. The deck cross-section geometry, material properties and loading are displayed in Figure 9.

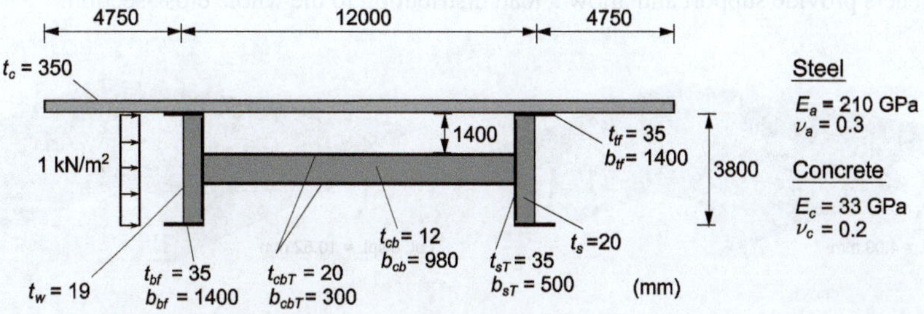

Figure 9. Deck cross-section and intermediate cross-beam geometry, loading and material properties

A simply supported span of 35 meters is again considered, with the girders connected through 5 equally spaced H cross-beams that are connected to internal vertical web-stiffeners with a T-section. The beam support conditions are similar, i.e., warping of the bottom steel flanges is restrained at one end and all in-plane displacements are prevented at both ends. The external load consists of a 1 kN/m^2 uniform pressure load acting on the web representing lateral wind action. The GBT deformation modes are obtained considering 2 intermediate nodes in the loaded web, leading to the Vlasov deformation modes displayed in Figure 10 (all shear modes are neglected in this example). The previous longitudinal discretization is kept, involving 24 finite elements and leading to a total of 799 DOFs.

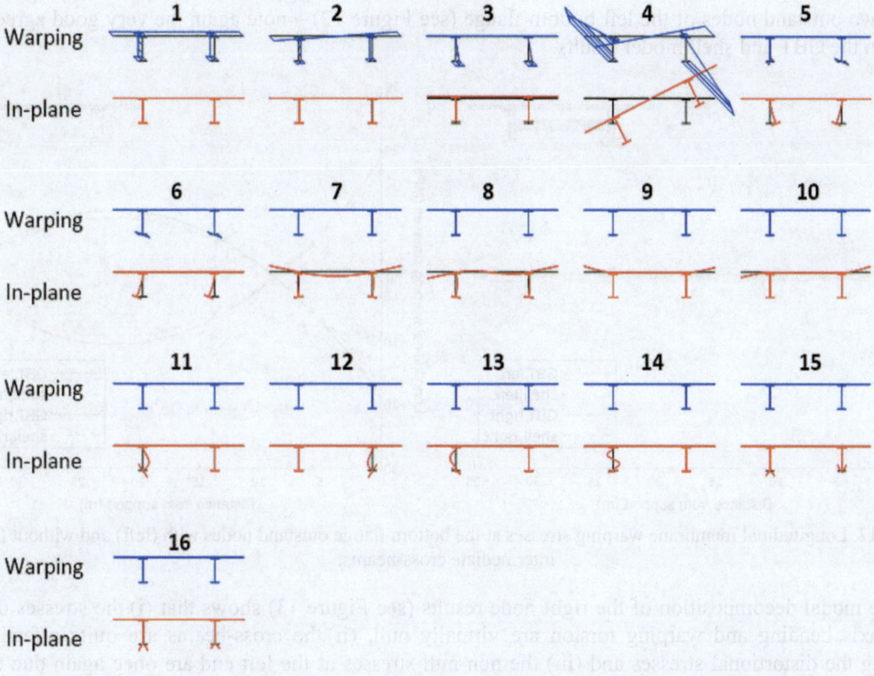

Figure 10. GBT deformation modes for the composite steel-concrete twin-girder.

Figure 11 presents the deformed configurations obtained with the shell and GBT finite element models, which are, once again, virtually coincident. If no bracing is present, the load causes lateral bending of the left web and bottom flange, whereas in the braced case it is clear that the cross-beams and connecting vertical stiffeners provide support and allow a load distribution to the whole cross-section.

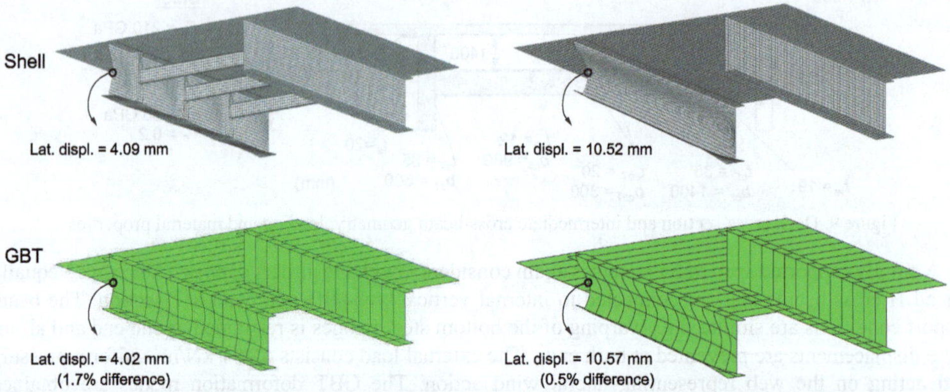

Figure 11. Deformed configurations (x100) of about half of the deck with (left) and without (right) intermediate cross-beams (cross-beams are not represented in the left GBT deformed configuration).

The stiffening action is also clear from the observation of the longitudinal membrane warping stresses at the two outstand nodes of the left bottom flange (see Figure 12) – note again the very good agreement between the GBT and shell model results.

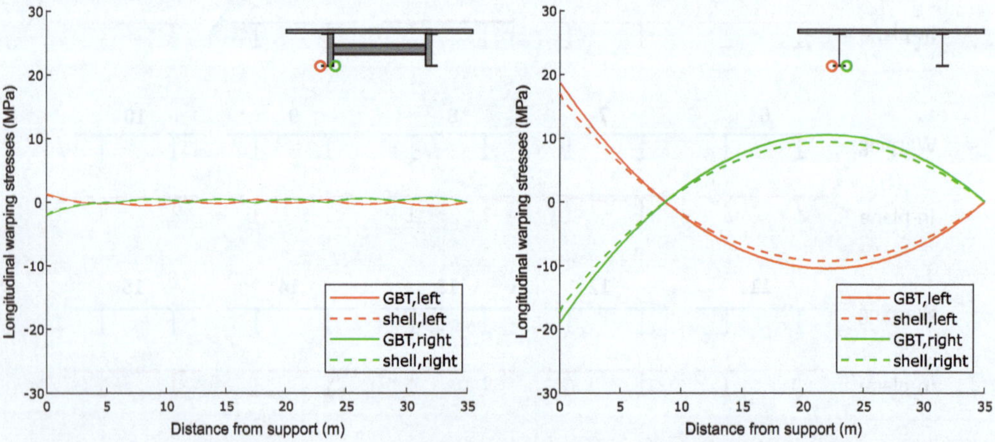

Figure 12. Longitudinal membrane warping stresses at the bottom flange outstand nodes with (left) and without (right) intermediate cross-beams.

The modal decomposition of the right node results (see Figure 13) shows that (i) the stresses due to major-axis bending and warping torsion are virtually null, (ii) the cross-beams are quite effective in reducing the distortional stresses and (iii) the non-null stresses at the left end are once again due to the warping displacement restraint provided at the bottom flanges that act as a fixed support for distortion.

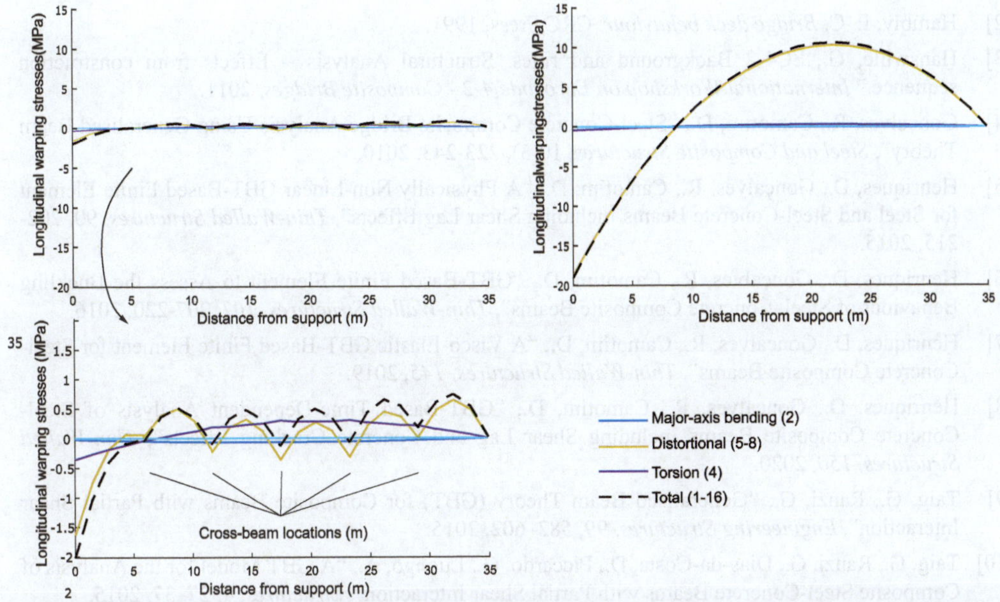

Figure 13. Modal decomposition of longitudinal membrane warping stresses at the bottom flange right outstand node with (left) and without (right) intermediate cross-beams.

4 CONCLUSIONS

This paper proposed a procedure that can explicitly consider the effect of flexible intermediate cross-bracing in the GBT structural analysis of steel-concrete composite bridge decks. Essentially, the method involves the use of a GBT-based bar finite element model in which the deck cross-section bracings are simulated through a set of elastic constraints. The respective stiffness values are determined by imposing the displacements corresponding to the various GBT deformation modes on the bracing substructure, which is modeled using shell finite elements.

The accuracy and capabilities of the proposed approach was demonstrated by means of two numerical examples involving a box-girder deck with intermediate plate diaphragms and a twin-girder deck with intermediate cross-beams – refined shell finite element models were used for comparison purposes. An excellent agreement was observed regarding both displacements and longitudinal warping stresses. Moreover, the modal features of GBT were used to provide structural insight regarding the influence of the bracing system on the bridge deck behavior. The results provided clear evidence concerning the merits and potential of the proposed procedure to perform serviceability limit state analyzes of steel-concrete composite bridge decks.

ACKNOWLEDGMENTS

The first author gratefully acknowledges the financial support of FCT (Fundação para a Ciência e Tecnologia, Portugal), through the doctoral scholarship SFRH/BD/136750/2018.

REFERENCES

[1] Kuhlmann, U., et al., *COMBRI Design Manual-Part I: Application of Eurocode rules-Part II: State-of-the-Art and Conceptual Design of Steel and Composite Bridges.* 2008.

[2] Hambly, E. C. *Bridge deck behaviour*. CRC Press, 1991.

[3] Hanswille, G.,"EC4-2 Background and rules. Structural Analysis – Effects from construction sequence." *International Workshop on Eurocode 4-2 - Composite Bridges,* 2011.

[4] Gonçalves, R., Camotim, D., "Steel-Concrete Composite Bridge Analysis Using Generalised Beam Theory", *Steel and Composite Structures*, 10(3), 223-243, 2010.

[5] Henriques, D., Gonçalves, R., Camotim, D., "A Physically Non-Linear GBT-Based Finite Element for Steel and Steel-Concrete Beams Including Shear Lag Effects", *Thin-Walled Structures,* 90, 202-215, 2015.

[6] Henriques, D., Gonçalves, R., Camotim, D., "GBT-Based Finite Element to Assess the Buckling Behaviour of Steel–Concrete Composite Beams", *Thin-Walled Structures, 107*, 207-220, 2016.

[7] Henriques, D., Gonçalves, R., Camotim, D., "A Visco-Elastic GBT-Based Finite Element for Steel-Concrete Composite Beams", *Thin-Walled Structures, 145*, 2019.

[8] Henriques, D., Gonçalves, R., Camotim, D., "GBT-Based Time-Dependent Analysis of Steel-Concrete Composite Beams Including Shear Lag and Concrete Cracking Effects", *Thin-Walled Structures, 150*, 2020.

[9] Taig, G., Ranzi, G., "Generalised Beam Theory (GBT) for Composite Beams with Partial Shear Interaction", *Engineering Structures,* 99, 582–602, 2015.

[10] Taig, G., Ranzi, G., Dias-da-Costa, D., Piccardo, G., Luongo, A., "A GBT Model for the Analysis of Composite Steel-Concrete Beams with Partial Shear Interaction", *Structures*, 4, 27–37, 2015.

[11] Schardt, R. *Verallgemeinerte Technische Biegetheorie*, Springer-Verlag, Berlin, 1989.

[12] Bebiano, R., Camotim, D., Gonçalves, R., "A Cross-Section Analysis Procedure to Rationalise and Automate the Performance of GBT-Based Structural Analyses", *Thin-Walled Structures,* 92, 29-47, 2015.

[13] Bebiano, R., Camotim, D., Gonçalves, R., "GBTUL 2.0 − A second-generation code for the GBT-based buckling and vibration analysis of thin-walled members", *Thin-Walled Structures,* 124, 235-257, 2018.

[14] Gonçalves, R., Ritto-Corrêa, M., Camotim, D., "A New Approach to the Calculation of Cross-Section Deformation Modes in the Framework of Generalized Beam Theory" *Computational Mechanics*, 46(5), 759-781, 2010

[15] Gonçalves, R., Bebiano, R., Camotim, D.,"On the Shear Deformation Modes in the Framework of Generalized Beam Theory", *Thin-Walled Structures*, 84, 325-334, 2014.

[16] Bathe K.J., *Finite Element Procedures*, Prentice-Hall Inc., New Jersey, USA, 1996.

[17] MATLAB, release R2018b, The MathWorks Inc., Massachusetts; 2018.

[18] ABAQUS, v2018, Dassault Systemes Simulia Corporation, Providence, 2018.

[19] Pedro, J.J.O. and Salmon, A., "New launching method of composite steel-concrete decks for replacing two bridges in Luxembourg", *IABSE Symposium, Guimaraes 2019: Towards a Resilient Built Environment Risk and Asset Management – Report*, 1116-1123, 2019.

[20] Habraken, A., et al, *Optimal use of high strength steel grades within bridge (OPTIBRI)*, EU publication, EUR 29546 En., 2019.

9TH INTERNATIONAL CONFERENCE ON COMPOSITE CONSTRUCTION IN STEEL AND CONCRETE
M. Knobloch, U. Kuhlmann, W. Kurz, M. Schäfer. (Eds.)
Online, Germany, July 27-29, 2021

INTEGRAL SHEET PILING ABUTMENTS OF MODULAR COMPOSITE BRIDGES FOR A TIME EFFICIENT CONSTRUCTION

Mike Tibolt*, Dennis Rademacher*, Oliver Hechler Kevin Wolters***, Nils Rittich*** and Stoyan Ivanov******

* ArcelorMittal Steligence®
e-mail: mike.tibolt@arcelormittal.com, dennis.rademacher@arcelormittal.com

**ArcelorMittal Sheet Piling – Luxembourg
e-mail: oliver.hechler@arcelormittal.com

*** RWTH Aachen
email: k.wolters@stb.rwth-aachen.de, n.rittich@stb.rwth-aachen.de

**** Schimetta Consult ZT GmbH
e-mail: Stoyan.Ivanov@Schimetta.at

Keywords: Composite bridges, integral abutments, modular construction, sheet pile abutments

Abstract. *Integral bridge structures with a high degree of prefabrication comply with all requirements of modern bridge structures. Composite bridges with sheet pile abutments using modular elements are such a solution although not yet a standardized solution with wide acceptance by bridge authorities and owners. In the present paper, two design concepts of a modular connection between superstructure in steel composite construction using hot-rolled profiles and integral steel sheet pile abutments are presented: A flexible concept with sheet piles as abutments for bridge spans 25-30m and a more rigid concept applying combined sheet pile walls for bridge spans up to 30-45m. The concepts are referencing the Knife Edge-Support (KES) as basis of design. The load paths are analyzed numerically, and the influencing parameters are identified. Finally, preliminary experimental results to proof the suitability of the connections will be confirmed presented.*

1 MODULAR COMPOSITE AND INTEGRAL BRIDGES: FUTURE SOLUTIONS

A large portion of the today's 50-60 years old bridges are aged and in need of refurbishment. Further, the traffic volume increased significantly, contributing to further degradation of the bridge infrastructure. As a result, many of those bridges have to be replaced by new structures before the end of the intended service life.

For a long time, solely the construction costs of the bridge were the governing criteria for the assessment of the economic efficiency and the selection of the bridge solution. Over the last years, bridge authorities and owners tend to a holistic approach to assess the economic efficiency of a bridge structure [1], [2], [3]. This approach accounts for not only the construction costs but also the construction time and the maintenance costs of the bridges as the main criteria for the decision for the solution finally projected. A limitation of the traffic disturbance during the construction of the bridge and of the maintenance activities during the in-service time reduces the related costs for the bridge owners and the society.

Integral composite single span bridges as shown in Figure 1 a) are typically used up to a span length of 50-55 m. But also, for smaller span ranges they are an efficient and economical alternative to concrete and prestressed concrete frame bridges. Generally, the abutments are either made of in-situ concrete on construction site or they are prefabricated in the workshop in parallel to the segmented superstructure and transported to the construction site where they are connected to the superstructure.

The use of steel sheet pile abutments offers extra benefits to composite and integral bridges. The sheet piles serve as load bearing elements as foundation, in addition to acting as a temporary construction pit retaining system and replacing expensive pile foundations (Figure 1 b)).

Figure 1. a) BAB A8 bridge structure 5 near Harlaching, Germany (© RWTH), b) Underpass at Swarzedz, Poland (©ArcelorMittal)

With sheet piles it is easily possible to construct a tight load bearing wall. Moreover, the superstructure and the abutment can be cast simultaneously, saving construction time. Sheet pile abutments thus comply with the principle of modular bridge construction. An independent study shows that the use of sheet pile abutments reduces the construction costs by 7% and the construction time by 10% compared to standard concrete solutions [2], [3]. Also, the environmental impact is reduced significantly using steel of latest production technologies using 100% scrap and renewable energy in production.

In Germany, first pilot projects for modular composite and integral bridges have already been realized, showing different concepts for the design of the superstructure. For a high degree of prefabrication and construction time reduction, the composite superstructure can be made of prefabricated concrete elements. These elements are positioned on the already installed steel beams and the composite action is activated by group of shear studs in pockets which are subsequently concreted (Figure 2 a)). If required, the prefabricated elements can be prestressed longitudinally and/or transversally to realize a force-locked or frictional connection between the slab elements. Other concepts dispense with a shear connection between the main girders and the slab to ensure the possibility of replacing precast elements, which are intended to be fit for traffic without an additional asphalt layer (Figure 2 b)).

Figure 2. a) Modular bridge Hammacher Strasse (Heitkamp), b) Segment bridge Bögl (© Max Bögl)

Alternatively, VFT-girders with additional in-situ concrete layer on top can be applied (Figure 3 a)). VFT-girders are composite elements formed out of a steel beam and a 10-12 cm prefabricated concrete slab. On the construction site the individual structural elements are joined together with the in-situ concrete supplement so that a monolithic load-bearing effect is achieved. The shear forces between the concrete and the steel beam are accomplished by headed shear studs using short studs for the prefabricated slab and longer studs for the in-situ concrete layer. VFT-girders are transported by heavy cargo trucks to the construction site and can be lifted by mobile cranes. The VFT technology allows spans up to 60m for integral bridges without middle supports.

Figure 3. a) VFT-girder with a thin precast concrete slab; Horlofftalbridge in Hungen, Germany (ArcelorMittal)

Most of the above-mentioned concepts are based on welded beams from standard steel plates in S355. For small and medium span bridges, the use of plates with higher steel strength is not common. On the other hand, the use of hot-rolled sections in S460M is well established in the European traffic network and advantageous for road bridges, since the higher steel strength can lead to weight and cost savings in the steel structure. Furthermore, the environmental impact can be reduced by using 100% recycled steel such as thermo-mechanically rolled high strength profiles. The corresponding EPD [4] shows significant lower CO2 emission, reduced by more than 50% during production stage, compared to the standard mix of structural steels based on World Steel Association database. The low carbon equivalent of such steel beams results in an improved weldability of the material without any preheating. Thereby valuable energy can be saved.

Modular composite and integral bridges with hot-rolled beams and sheet piles would combine the outstanding performance of composite construction with the advantages of integral bridges. The resulting benefits are manifold:

- Low construction and maintenance costs
- Driving quality is increased and noise level is reduced due to missing expansion joints
- No essential interference of the traffic under the bridge by elimination of the middle support
- Earthquake resistance - falling of the superstructure from the supports is eliminated, also damping for longitudinal modes is increased due to soil-abutment interaction.
- Added redundancy and capacity for all types of unforeseen events like impacts, explosions etc.
- The high degree of prefabrication significantly reduces the construction time
- Prefabrication in the workshop ensures a high quality of the modular elements
- The life-cycle-costs are low compared to standard bridge solutions [1]
- Very low environmental impact

Composite and integral bridges thus comply with the holistic approach and they are an attractive solution for the replacement of old bridge structures.

2 INTEGRAL SHEET PILE ABUTMENTS FOR COMPOSITE BRIDGES

For road bridges of an average span length and standard concrete abutments, 40% of the total erection costs of the bridge are dedicated to the realization of the abutments. This share increases up to 55% if additional temporary built-ins are required [5]. The advantage of sheet piles consists in their double function. In addition to the role as pit protection system, they also act as a permanent bearing system for the foundation. An additional concrete pile foundation is not required in this case. During the driving of the sheet piles into the soil, the piles are connected by interlocks to an integral wall. For a quick erection, the bridge superstructure and the sheet pile abutments can be cast at the same time. Compared to the standard concrete abutments, the cost for sheet pile abutments is reduced by 7% and the construction time by 10% using sheet piles [2], [3]. "EcoSheetPile™ Plus" steel sheet piles are made from 100 % recycled scrap steel and with 100 % renewable electricity reducing the environmental impact even further compared to standard sheet pile abutment already having 36% lower GWP emission compared to alternative concrete solutions over lifetime [2]. They are part of the "XCarb™ recycled and renewably produced" label which is audited and certified (incl. electricity origin) by an independent third-party. The result in [3] holistically verifies that, especially for bridges which need to have deep foundations (pile foundations), steel sheet pile abutments must be assessed to ensure the final design decision is the most cost-efficient. In addition, steel sheet piles have the advantage of being mostly underground where the corrosion rates are almost negligible, e.g. 2.4 mm over 100 years as specified in Eurocode 3–5 (Piling) [6]. If the structure is subject to more aggressive environments, several options are available: accounting for corrosion loss in design, organic coatings, hot dip galvanizing, or special "weathering" steel grades. By far the most used solution is to account for the forecasted steel loss over design life in the specification of the sections. In consequence, sheet pile abutments comply with modular construction.

There are two principal ways how sheet piles can be used in bridge abutments. Cantilever walls rely on their embedment depth to provide lateral support for the soil and to carry the weight of the bridge superstructure (Figure 4 a)). The retained height is generally restricted to about 4.5m to limit the lateral deformation in the in-service state and the wall section size. The sheet piles generally have a concrete capping beam to transfer the load from the bridge superstructure into the sheet pile foundation and to compensate tolerances from pile installation. Cantilevers allow for spans of 25-30 m. Anchored retaining walls are similar to cantilever walls with an additional support close to the top of the sheet piles (Figure 4 b)). The anchor limits the lateral displacement of the wall and allows for an increased retaining height (up to 15m). Spans in the range of 30-45 m can be achieved. Wing walls using sheet piles maybe projected to further provide rigidity respectively account for longitudinal load bearing capacity.

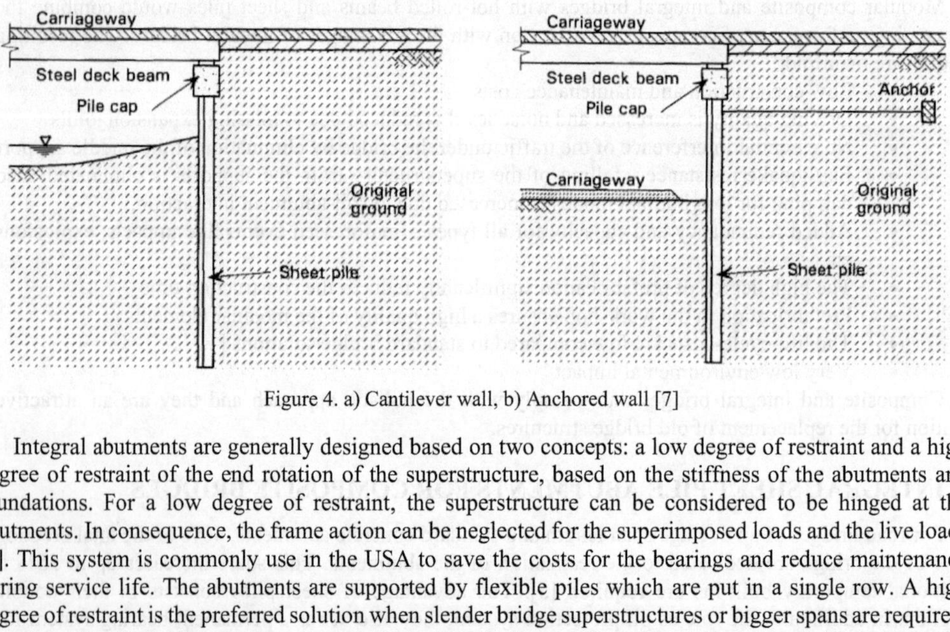

Figure 4. a) Cantilever wall, b) Anchored wall [7]

Integral abutments are generally designed based on two concepts: a low degree of restraint and a high degree of restraint of the end rotation of the superstructure, based on the stiffness of the abutments and foundations. For a low degree of restraint, the superstructure can be considered to be hinged at the abutments. In consequence, the frame action can be neglected for the superimposed loads and the live loads [7]. This system is commonly use in the USA to save the costs for the bearings and reduce maintenance during service life. The abutments are supported by flexible piles which are put in a single row. A high degree of restraint is the preferred solution when slender bridge superstructures or bigger spans are required. In order to increase the negative corner moment at the abutments and to shift up the field moment, the rotation of the horizontal member (the superstructure) is partly restraint by the stiff vertical members at the ends. This is for instance the classic solution for integral bridges in Germany. Both concepts can also be realized with sheet pile abutments. Low restraints (25-45%) can be achieved with AZ sheet piles (Figure 5 a)) and a high restraint (75%) can be realized with combined walls (Figure 5 b)).

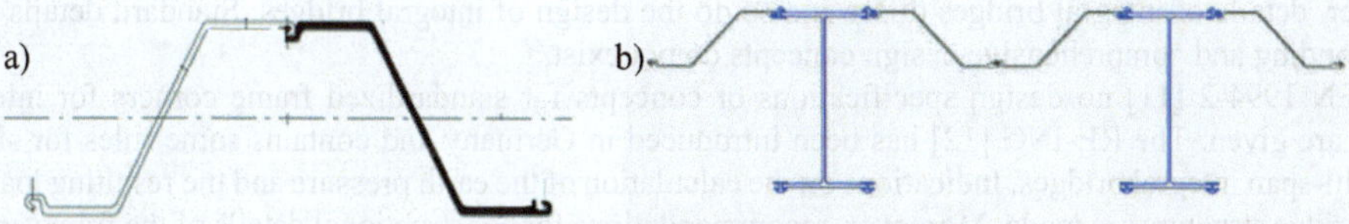

Figure 5. a) Sheet pile (AZ sections), b) combined walls (HZ and AZ sections)

In the European funded RFCS project INTAB [8], an innovative frame corner connection between the composite superstructure of the bridge made of hot-rolled profiles and the integral sheet pile abutments has been developed (Figure 6 a)). The composite girders (e.g. VFT-girders) are installed on IFB-beams which are welded on top of the sheet piles. The frame corner as well as the girder can be cast in parallel. The hogging moment is transferred via corresponding reinforcement bars. The degree of restraint achieved with the so called "INTAB"-connection was about 50%. It is important to note that the sheet piles beyond the outer edges of the abutment significantly contribute to the stiffness of the system. They need to be considered in the design accordingly.

The "INTAB"-system is based on the Knife Edge Support (KES) (Figure 6 b), which was was derived by a research programme of ArcelorMittal in collaboration with the University of Darmstadt as well as Hegger + Partner Ingenieure Aachen, Germany. The KES consists in an innovative form of a concrete capping beam installed on the sheet pile head [9]. The support can transfer static and cyclic vertical loads and static horizontal loads form the concrete cap into the sheet pile and subsequently into the soil. During installation on construction site, first the sheet piles are driven into the soil, secondly a platform is fixed to the sheet pile for the positioning of the prefabricated reinforcement cage and to serve as formwork and finally the cap is cast and the platform is removed. The KES is covered by a general construction technique permit (aBG) for Germany [10].

Capping beam designs based on KES method are able to transfer vertical loads up to 3834 kN/m [10] into a simple sheet pile wall (AZ 50), which is sufficient for the average range of small and medium bridge spans.

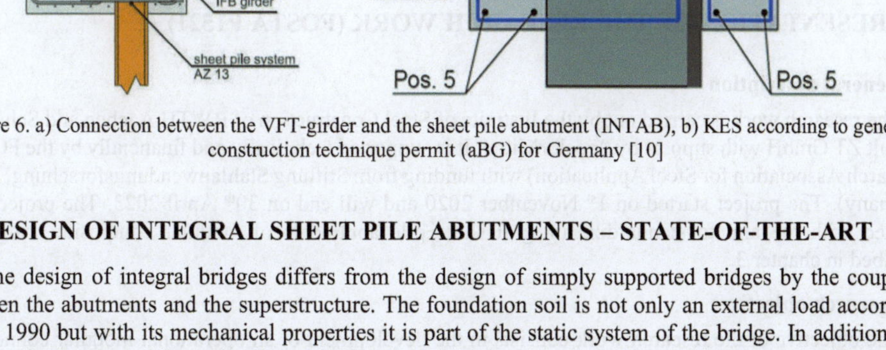

Figure 6. a) Connection between the VFT-girder and the sheet pile abutment (INTAB), b) KES according to general construction technique permit (aBG) for Germany [10]

3 DESIGN OF INTEGRAL SHEET PILE ABUTMENTS – STATE-OF-THE-ART

The design of integral bridges differs from the design of simply supported bridges by the coupling between the abutments and the superstructure. The foundation soil is not only an external load according to EN 1990 but with its mechanical properties it is part of the static system of the bridge. In addition, the degree of restraint and the acting forces in the connection between the deck and the abutments need to be quantified. Of course, the stiffness of the abutments as well as the forces depend on the connection design.

However, details of integral bridges differ and so do the design of integral bridges. Standard details with corresponding and comprehensive design concepts do not exist.

In EN 1994-2 [11] no design specifications or concepts for standardized frame corners for integral bridges are given. The RE-ING [12] has been introduced in Germany and contains some rules for single and multi-span integral bridges. Indications on the calculation of the earth pressure and the resulting loading on the bridge structure are made. Moreover, recommendations for constructional details of the frame corner can be found in [12]. The recommendations and rules are however very general and do not allow for a complete static design of integral bridges. Finally, sheet pile abutments are not tackled at all in the document.

In [8] an innovative connection between the composite superstructure and the integral sheet pile abutments has been developed (Figure 6 a)). However, a holistic design approach for the verification of the connection has not been proposed and the experimental test campaign did not include cyclic horizontal loading. Nevertheless, the feasibility of an enduring moment bearing connection under cyclic loading from traffic loads or temperature was shown. In addition, the horizontal displacements due to temperature and the interaction between the soil and the structure were preliminary investigated.

In [13] a new connection between the abutments and H-piles was proposed. The H-pile is loaded around its weak axis. The experimental test campaign was limited to static loading for the derivation of the stiffness and the resistance of the connection. The potential of sheet pile abutments for bridges with low truck circulation is highlighted in [14] but no consistent design rules are delivered. Design recommendation for integral bridges in the US are given in [15]. The recommendations are limited to H-pile foundations for the abutments which is the most common integral bridge system in the USA. Sheet pile abutments are not covered, and the engineer should rely on his experience from former projects for the design of integral bridges. In the UK, design guidance for sheet pile abutments are given in [7]. The abutment can be composed of single sheet piles (e.g. AZ sections) or combined walls (e.g. HZ-M walls) and the rules are valid up to spans of 35 m. However, the design guide is only valid for joint- or simply supported bridges with sheet pile abutments. Integral bridges are not covered.

From what precedes, it becomes clear that a significant lack of design rules and standardized solution for the connection between the bridge superstructure and the abutments exists. Additionally, a standard solution for the frame corner with a well-defined load path for the introduction of the negative moment from the superstructure into the sheet pile abutments is currently missing. In consequence, some key-points as for instance the load path in the corner frame, the horizontal cyclic loading from traffic loading in the sheet pile abutments and the compensation of tolerances need extensive investigations.

Integral sheet pile abutments for modular composite integral bridges have a huge potential, although the lack of design rules and missing standardized solutions for the modular connection currently prevents their widespread application.

4 PRESENTATION OF THE RESEARCH WORK (FOSTA P1521)

4.1 General description

The research work is carried out by the Institute of Steel Construction of RWTH Aachen and Schimetta Consult ZT GmbH with support by ArcelorMittal. It is supported technically and financially by the FOSTA (Research Association for Steel Application) with funding from Stiftung Stahlanwendungsforschung, Essen (Germany). The project started on 1st November 2020 and will end on 30th April 2023. The project was initiated with the aim to standardize integral sheet pile abutments of modular composite bridges as described in chapter 3.

4.2 Research objective

The objective of the research work consists in the development of an operational modular connection between the integral sheet pile abutments and the modular composite bridge deck. The focus is to put on single-span composite road bridges made of hot-rolled profiles up to steel grade S460M to allow for a time efficient and economic construction of the bridge. Two different concepts for the connection will be worked out: a stiff connection (high restraints) and a flexible connection (low restraint) to comply with the two philosophies for the design of integral abutments and to cover a large range of spans. The main requirements for the connection are a high modularity, a quick construction, a low degree of maintenance and simplicity. The solution should be a standard solution which small and medium size companies can easily reproduce and apply for their bridge projects. An important point is the acceptance of the solution by the authorities. Therefore, the development will be done in close collaboration with bridge authorities, bridge owners and contractors. Finally, a design concept for the connection will be developed. The concept will be based on preliminary scientific work and will be presented in a comprehensive way. It will give answers to the lack of knowledge as described in chapter 3. The design concept will be implemented in a design and practical guide and will be made available in the public domain.

4.3 Description of the scientific work

The research project has been divided into 4 work packages, which are summarised in the following chapters.

4.3.1 Requirements for the connection

In the first work package, the requirements to be met by the connection are identified. Valid design rules (e.g. EC, ZTV-ING, RE-ING) are investigated to consider their demands on the connection at an early stage of development. Special attention is given to the driving tolerances of the sheet piles since they have a significant impact on the design of the connection. The slenderness, the span and the concept of the bridge superstructure strongly influence the moment distribution, the forces and moments to transfer in the connection and finally the geometry of the connection. Therefore, lower- and upper limits for the slenderness and the span of the bridge superstructure are defined for which the developed connection can be applied. The limits cover the ranges of slenderness and spans of integral bridges in practice.

4.3.2 Conceptual design of the connection

In the second work package, two different concepts for the connection are developed:

- A connection with a low degree of restraint between 25% and 45%. This is most likely achieved with simple sheet piles of type AZ. The sheet piles transfer all the vertical and horizontal loads.
- A connection with a high degree of restraint of about 75%. The high restraints are realized with combined walls (HZ and AZ sections) as integral abutments. The intermediate piles transfer the horizontal loading from earth and water pressure while the bearing piles transfer the vertical and horizontal loads from the superstructure.

The focus of the development is put on a modular solution. Application of prefabricated concrete elements is investigated. They can be used as lost formworks to reduce the construction time. Methods to fix the prefabricated elements to the bridge superstructure, the sheet pile abutments and/or the soil are investigated. Crucial point is a feasible fixation on construction site. The transfer of the forces resulting from the hogging bending moments in the corners is a major point of investigation. Possible solutions are the load transfer by means of a corresponding reinforcement arrangement like in [8], using composite dowels or by clamping. Variants will be worked out and discussed with bridge authorities, design offices and contractors during the project meetings. The influence of the stiffness of the connections on the horizontal displacement and rotation of the sheet piles due to temperature loads and the resulting cracks in the concrete are analysed. Possible anchorage of the sheet pile is considered. Figure 7 shows different preliminary ideas for the conceptual design of the modular connection from the project proposal. The concepts are based on the principle of the Knife Edge-Support. The final determination of the connection lay-out is done together with bridge authorities, design offices and contractors. In this way, a comprehensive connection lay-out with close relation to practical use is developed. Results of the first project meeting are shown in chapter 5.

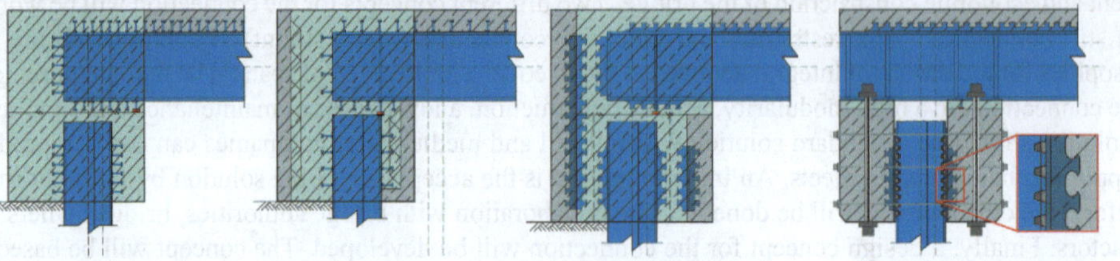

a) In-situ concrete b) Formwork (reduced moment capacity) c) Formwork (high moment capacity) d) maximum modularity

Figure 7. Ideas for conceptual design of sheet pile abutments from a) minimum to d) maximum modularity (from the project proposal)

4.3.3 Numerical and experimental investigations

In the third work package, the connection concepts developed in work package 2 are numerically and experimentally investigated to proof their feasibility. In a first step, the loading from self-weight, traffic and temperature on the abutments is determined in commercial frame programs with consideration of the construction phases. In a second step, numerical models of the connections are generated in a FE-program. The 3D-models account for geometrical and material non-linearities and all components of the connections are considered. Prior to calibration with test data, the models allow to consider the following points for an optimal planning of the experimental campaign:

- The suitability of the concepts is numerically clarified. Deficits are identified and allow for an adaption of the concept at an early stage. After this iteration step, the concepts for the connection are finally specified for more thorough numerical and experimental investigations.
- The force paths in the connection are identified. The position and the form of the reinforcement can be adopted to the force flow and the shear connectors can be further optimized (e.g. form, number, position…).
- The influencing parameters are identified. The load-deformation behaviour of the connection certainly depends on the material strength, the dimensions of the steel and concrete components, the type of shear connectors and the type of loading. The qualitative influence of those parameters is identified, and the most influencing parameters are considered in the tests.
- The measurement equipment for tests is determined. Failure modes are identified in the numerical simulations to adapt the measurement technique accordingly and to be able to measure important failure indicators during the tests.

The numerical models will be calibrated once the test campaign has been concluded and an extensive parameter study to quantify the influence of the identified parameters will be performed. These results are considered in the derivation of the design rules. The feasibility of the integral connection between the composite deck and the sheet pile abutments is finally confirmed in further experimental testing. The dimensions and assembly of the test specimen are defined by means of the preliminary numerical results. This allows to generate meaningful test results with only few large-scale tests. The final lay-out of the test specimens is only defined once the conceptual design of the connection has been fixed.

The load bearing resistance of the individual components of the connection is determined in static and cyclic small-scale tests. The load bearing resistance for vertical loads is regulated in [10] for standard sheet piles. Therefore, the focus is put on combined walls which are not covered in [10]. It is assumed that the vertical load bearing resistance will not be the governing failure mode of the connection. Since the knowledge for cyclic horizontal loading is limited, static as well as cyclic tests with horizontal loading are foreseen, Figure 8 a). In these tests, the shear resistance of the connection is determined.

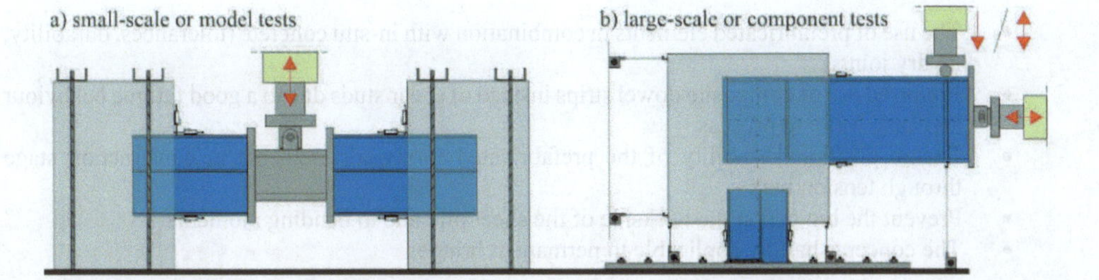

Figure 8. Preliminary test set-ups for static and cyclic experimental investigations using a) model or b) component tests

Finally, the load-deformation behaviour of the connection is tested in large-scale tests. Here, static tests with vertical loading and different moment-shear ratios, as well as cyclic tests under vertical, horizontal or combined loading are provided, Figure 8 b). With the cyclic tests, it is not intended to derive new fatigue detail categories. Therefore, no large number of tests is necessary. The forecast of the load bearing and fatigue behaviour of the modular connection will be based on the combination of existing approaches for fatigue strength curves. The tests will show that the connection offers the required load bearing resistance and that it can safely transmit the loads from the superstructure into the integral abutments. The test results are used to calibrate the numerical models to be used in a quantitative parameter study.

4.3.4 Design guideline

In the fourth and last work package, design and construction rules are developed based on the outcomes from work packages two and three. The rules allow a simple design of integral sheet piles and they will be discussed with design offices and bridge authorities to ensure their acceptance. It is important to note that the concept should meet current normative regulations and it shall not require additional approvals. A design example finally shows the application of the rules. The design rules will be prepared for a future implementation in the corresponding standards (i.e. EN 1994-2).

5 CONCEPTS FOR INTEGRAL SHEET PILE ABUTMENT – FIRST RESULTS

5.1 Discussion on the proposed preliminary concept ideas

The preliminary conceptual design for sheet pile abutments shown in Figure 7 were discussed with bridge authorities, design offices and contractors during the first coordination meeting of the project. Main bridge actors prefer the two concepts b) and c) with partial prefabricated elements acting as integrated formwork. The tolerances and durability can be guaranteed by the in-situ concrete in between. Solution d) with only prefabricated elements is a good concept for temporary bridges due to its high degree of modularity and possibility of disassembling the elements. The durability of solution d) could be an issue for permanent bridges. In the framework of the project, it was decided to develop the concepts b) and c).

Horizontal shear studs cannot be welded automatically and need to be welded by hand (e.g. fillet welds), which is – for instance – generally not allowed by the German road authorities. Alternatively, they could be welded on steel plates in the workshop and the steel plates are subsequently welded on the sheet piles on construction site. This however requires additional welding on construction site. The filigree prefabricated elements of solution b) could be an issue under fatigue loading. Concept c) is the preferred solution with regards to its fatigue behaviour. Composite dowels [16] transfer high forces with an acceptable fabrication effort of the prefabrication elements.

The spalling and cracking of the concrete at the sheet pile and a pull-out of the sheet pile from the concrete as observed in [8] must be avoided. The test specimen in [8] have no rebars going through holes in the sheet piles or any connectors applied on the sheet piles. This explains an observation in [8], where big gaps between sheet pile and concrete occurred at the rear. From what precedes, the demand from practice on the concept of the connection can be summarized as follows:

- The use of prefabricated elements in combination with in-situ concrete (tolerances, durability, no dry joints)
- Preferred use of composite dowel strips instead of shear studs due to a good fatigue behaviour and easy welding process
- Ensure positional stability of the prefabricated formwork elements in construction stage through tension bars
- Prevent the big gap at the backside of the sheet pile due to bending moments
- The concept shall by applicable to permanent bridges

5.2 Proposed solution

Two aspects are important in the design of the bridge abutment. First, the method of transferring forces and moments from the superstructure to the sheet pile foundation must be clarified. Secondly, the modular construction of the abutment as well as the construction process are to be designed. Figure 9 a) shows two different concepts for the transfer of the corner moment. In the Strut and Tie model the bending moment is transferred primarily through the horizontal forces, with tension in the top reinforcement and compression at the bottom flange of the steel girders. In this case, it is not necessary to include the steel beam far into the abutment. The equilibrium in the node is ensured by the inclined strut of concrete. In the shear model the bending moment is transferred primarily through vertical forces, compression at the front part of the node and tension (actually shear in the composite dowels) at the end of the steel girders. In that case, the steel girder is encased into the node. A single concrete strut is hard to develop due to the small angle and the equilibrium is ensured primarily by shear transfer between the composite dowels.

The shear concept is chosen for the modular construction shown in Figure 9 b), a sophisticated concept based on Figure 7 c) as a stiff solution with a combined wall. In the construction process, the combined wall is built first. The sheet piles are driven to a lower level than the HZ-profiles to avoid cutting holes for the tension bars. The backfill is placed and compacted, composite dowels are fixed (either by welding or with friction-grip connections) on the HZ-profiles and temporary columns are placed for the precast concrete elements in the front. The rear concrete elements encase steel plates with composite dowels welded on both sides for the transfer of shear forces. Front and rear elements are placed, aligned and connected through tension bars, which take the hydrostatic forces due to the in-situ concrete filled in between. In the next step, the main girders (here as VFT-girders with a precast slab up to the edge of the abutment) are lifted in the recesses provided for this purpose in the front elements and placed on the HZ-profiles. Additional reinforcement such as stirrups constricting the zone of composite dowels are installed. The upper reinforcement of the slab is connected via tension sleeves to the reinforcement in the rear precast elements. Finally, the abutment and the slab are concreted. Vertical forces and moments are thus transferred mostly to the HZ-profiles while the sheet piles take horizontal forces due to the backfill. Tolerances of sheet piles are compensated by the in-situ concrete.

The concept guarantees a fast construction time, a good handling of tolerances, a clear load transfer in the axes of the main girders with a high moment load-bearing capacity. The applicability and the fatigue behaviour will be checked in the experimental component tests. Since composite dowels have a favourable fatigue behaviour, a good fatigue behaviour is also expected for the frame corner.

A similar concept can be derived for the flexible solution with sheet piles only (AZ sections). Here, a transversal steel profile arranged on the top of the sheet piles is required to distribute the vertical loads of the main girders over a larger width. The sheet piles will take vertical and horizontal forces and a pull-out of the rear sheet pile due to moments must be avoided. Therefore, tension bars could be arranged through cutting holes or shear connectors are welded or clamped on the sheet pile. Anyway, the stiff solution with combined walls shown in Figure 9 b) seems to be the favourable solution for further studies.

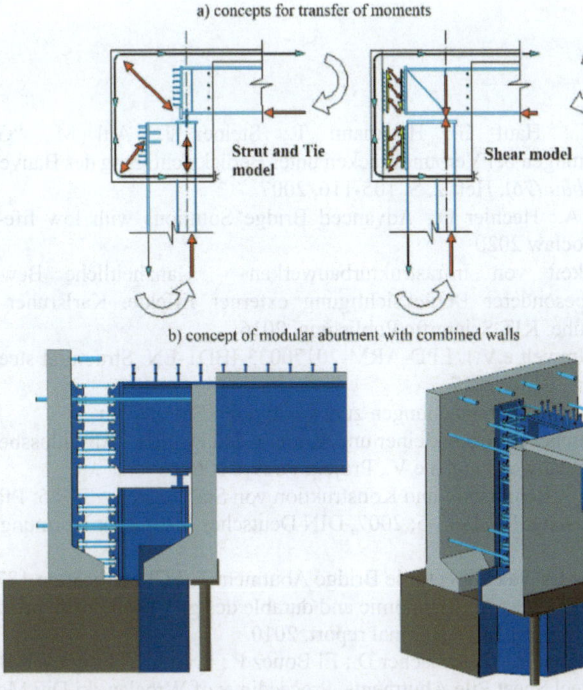

Figure 9. a) Concepts of load transfer in the abutment, b) concept for a stiff solution of sheet pile abutment based on the shear model using precast formwork elements, VFT-girder with shear studs

6 CONCLUSIONS

Thousands of highway bridges will have to be replaced in the medium term. Therefore, new concepts for a short erection time minimized traffic interference are to be developed. This paper shows the advantages of modular, composite and integral bridges, especially using sheet piles for the abutments. A new concept of modular abutments, connecting sheet piles with the superstructure, for a short erection time and with a favourable load bearing and fatigue behaviour will be developed in a German research project funded by the FOSTA. The objectives and the work packages of the project are summarised.

A sophisticated design concept of the modular connection was developed after a first project meeting, where preliminary concepts of the research proposal were discussed with companies, which are active in bridge construction, as well as bridge authorities. The concept shown in chapter 5 will be investigated further in numerical and experimental studies. Preliminary numerical results are presented in [17]. At the end of the project, a design guide for modular integral bridges with sheet pile abutments will be developed and published.

The concept of modular, composite bridges with steel sheet pile abutments is a promising alternative to reduce construction costs, traffic interference and overall life-cycle costs. Construction times and maintenance of the bridge structure are reduced to a minimum and construction quality is sustainably managed. With steels produced under the XCarb®, recycled and renewably produced label of ArcelorMittal, the environmental impact of the structure is further limited to a minimum catering responsible sourcing and promoting sustainable public procurement.

REFERENCES

[1] Kuhlmann U.; Pelke E.; Hauf G.; Herrmann T.; Steiner J.; Aul M., "Ganzheitliche Wirtschaftlichkeitsbetrachtungen bei Verbundbrücken unter Berücksichtigung des Bauverfahrens und der Nutzungsdauer", *Stahlbau (76)*, Heft 2, S. 105-116, 2007

[2] Rademacher D.; Barceló A.; Hechler O., Advanced Bridge Solutions with low life-cycle costs, Proceedings of IABSE Wroclaw 2020

[3] ZINKE, T.: Nachhaltigkeit von Infrastrukturbauwerken – Ganzheitliche Bewertung von Autobahnbrücken unter besonderer Berücksichtigung externer Effekte. Karlsruher Institut für Technologie (KIT). Karlsruhe: KIT Scientific Publishing, 2016

[4] IBU (Institut Bauen und Umwelt e.V.). EPD-ARM-20170033-IBD1-EN: Structural steel sections in HISTAR grades. Berlin: s.n., 21.02.2017

[5] Schmitt V.; Seidl G.; Vogel C., "Untersuchungen zum verstärkten Einsatz von Stahlverbundkonstruktionen bei Brücken kleiner und mittlerer Stützweiten" Abschlussbericht, Forschungsvereinigung Stahlanwendungen e.V., Projekt P629, 2006

[6] DIN EN 1993-5, Eurocode 3, Bemessung und Konstruktion von Stahlbauten – Teil 5: Pfähle und Spundwände, Deutsche Fassung: EN 1993-5: 2007, DIN Deutsches Institut für Normung e.V., Juli 2007 + AC: 2009.

[7] Yandzio, E., "Design Guide for Steel Sheet Pile Bridge Abutments", SCI Publication 187.

[8] Feldmann M.; Naumes J.; Pak D. et al, "Economic and durable design of composite bridges with integral abutments". RFCS Project INTAB, Final report, 2010.

[9] Barceló A.; Hechler O.; Zillgen H.; Rademacher D.; El Bouez P.; Prüm C., Design and Benefits of Composite Bridges with Steel Sheet Pile Abutments, Proceedings of Wroclawski Dni Mostowe, Wroclaw 2019

[10] Allgemeine Bauartgenehmigung, Stahlbetonholm mit Schneidenlagerung zur Einleitung von Vertikal- und Horizontalkräften in Stahlspundwandbohlen der Firma ArcelorMittal nach DIN EN 1992-1-1 mit DIN EN 1992-1-1/NA. Z-15.6-235, DIBt Deutsches Institut für Bautechnik, November 2017.

[11] DIN EN 1994-2, Eurocode 4, Bemessung und Konstruktion von Verbundtragwerken aus Stahl und Beton – Teil 2: Allgemeine Bemessungsregeln und Anwendungsregeln für Brücken, Deutsche Fassung: EN 1994-2: 2005, DIN Deutsches Institut für Normung e.V., Juli 2006.

[12] RE-ING, Richtlinien für den Entwurf, die konstruktive Ausbildung und Ausstattung von Ingeniuerbauten – Teil 2: Brücken – Abschnitt 5: Integrale Bauwerke, Bundesministerium für Verkehr und digitale Infrastruktur, Dezember 2017.

[13] Ahn J.-H.; Yoon J.-H.; Kim J.-H.; Kim S.-H., "Evaluation on the behavior of abutment-pile connection in integral abutment bridge", Journal of Constructional Steel Research, (67), p.1134-1148, 2011.

[14] White J. D.; Klaiber W., "Modified Sheet Pile Abutments for Low-Volume Road Bridges", Center for Earthworks Engineering Research, Iowa State University, 2012.

[15] Wasserman E.P; Walker J.H., "Integral Abutments for Steel Bridges" Vol. II, Chap. 5, Highway Structures Design Handbook, American Iron and Steel Institute, 1996

[16] Allgemeine Bauartgenehmigung, Stahlverbundträger mit Verbunddübelleisten in Klothoiden- und Puzzleform. Z-26.4-56, DIBt Deutsches Institut für Bautechnik, Mai 2018.

[17] Tibolt M.; Rittich N.; Wolters K.; Ivanoy S., "Modular composite bridges with integral sheet piling abutments for a time efficient construction", Proceedings of IABSE Symposium Prague 2022, Challenges for Existing and Oncoming Structures, Prague 2022

9TH INTERNATIONAL CONFERENCE ON COMPOSITE CONSTRUCTION IN STEEL AND CONCRETE
M. Knobloch, U. Kuhlmann, W. Kurz, M. Schäfer. (Eds.)
Online, Germany, July 27-29, 2021

NEW TYPE OF COMPOSITE ARCH ELEMENT USING COMPOSITE DOWELS: APPLICATION IN RAILWAY NETWORK ARCH BRIDGE

R.Sęk *, B.Pilujski **, D.Sobala **and W.Lorenc ***

* RS Projekt
e-mails: radoslawsek@rs-projekt.pl

** STRABAG Sp. z o.o..
e-mails: boguslaw.pilujski@strabag.com, dariusz.sobala@strabag.com

*** Wrocław University of Science and Technology
e-mails: wojciech.lorenc@pwr.edu.pl

Keywords: Composite dowels, Railway bridges, Network arch, Arch bridges.

Abstract: *This paper presents a new type of composite element using composite dowels that was developed for the purposes of the railway bridge over the Vistula River in Krakow, Poland. The structure is the world's first arch bridge using cold-bent HD rolled sections; hence the idea of a new type of composite connection between arch and tie is new as well, thus a new type of composite arch emerges. The concept, design and construction experience are presented herein.*

1 INTRODUCTION

The new network arch bridges along the E30 railway line crossing the Vistula River in the center of Krakow represent a significant step forward in bridge engineering [2]. The entire bridge crossing consists of three independent bridges (two single-track bridges and one double-track bridge), each consisting of three arch spans 49.5, 116 and 63.5m (Fig. 1). In 2016 PKP (Polish National Railway) issued a call for tenders in accordance with a formula that allows the contractor to modify the solutions detailed in the tender documents (i.e. design-build formula). The "concept of three arches" was proposed by CEO Bogusław Pilujski from Strabag Sp. z o.o. (the Contractor) within the framework of the design-build contract, to achieve superior aesthetical and economical effects [2]. According to this concept, the designer Radosław Sęk (RS Projekt) developed the final design. ArcelorMittal delivered the cold-bent steel elements (HD sections and T-sections) and Mostostal Krakow (sub-contractor of Strabag) fabricated steel structure.

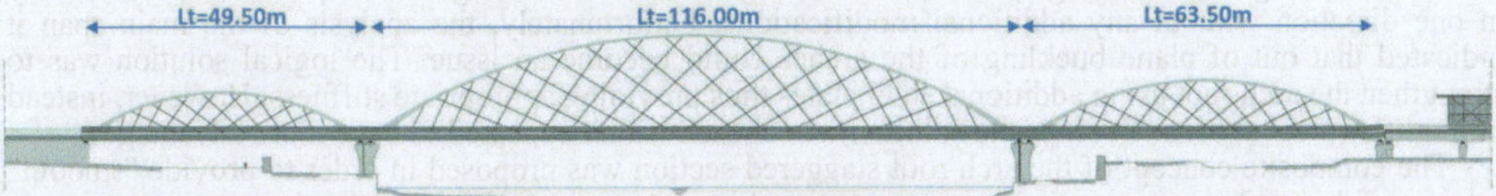

Figure. 1. Longitudinal section of a single-track bridge - the concept by Bogusław Pilujski.

The bridges are typical network arch constructions. Steel arches are made of S460 Histar I-rolled profiles (HL and HD section types), while the remaining elements are made of S355 and S460 steel. The hangers are S460 tension rods. Bridge deck and concrete parts of arch roots are of C50/60 concrete grade with B500 grade reinforcing bars.

In 2019-2020 the first of three bridges – the M2 – was constructed and opened to traffic. The M1 and M3 bridges are currently under construction.

Figure 2. Network arch bridge crossing the Vistula river in Krakow: internal span of M2 bridge during first-load tests (picture: http://krakow-rudzice.pl/).

The structure is in many respects first-in-the-world. The superstructure of each span is a new form of network arch, whose basic elements are cold-bent HD rolled sections [4] and new composite elements using T-sections connected to the concrete using composite dowels [1]. The solution obtained is characterized by a very slender arch and deck, and the innovativeness of the span construction manifests itself as follows:

(1). it is the first railway bridge using the cold-bent HD profiles (the idea was proposed by Tveit many years ago and up until now it has only been implemented for roadway bridges in Poland [4]),

(2). a completely new type of structural element has been used in the construction (scope of this paper), namely a composite element from two curved T-sections connected with concrete filling in the space between them using a composite dowel shear connection,

(3). for the first time a T-section (that was cut out from a rolled I-beam to obtain composite dowels) was cold-bent in the belt plane,

(4). the composite dowels that were utilized are the biggest and thickest yet.

For the first time composite dowels which are normally used in beams [3] have been implemented in a composite arch [1]. The application of a new type of composite element (Figs. 3-5) in the arch helped achieve the desired rigidity of the portal arch frame; the same refers to the smooth transition of the steel structure of the arch into a longitudinally prestressed concrete deck. The first bridge was built between 2019 and 2020 and it is a single-track structure. The next bridge which is a two-track structure is now under construction, where other cutting-edge solutions are being implemented (etc. anchoring hangers in a concrete deck using a new type of plates with composite dowels).

2 DESIGN IDEA

The early stages of the design process of the bridges assumed the usage of HD profiles cold-bent only in one direction without any additional modifications. Unfortunately, the analysis of the main span it indicated that out of plane buckling of the arches could become an issue. The logical solution was to strengthen the arch root using additional steel plates thus providing appropriate stiffness. However, instead of modified steel section, a composite solution was chosen which appeared to have more advantages.

The composite concept of the arch root staggered section was proposed in order to provide "smooth" transfer of the steel structure into the concrete one, taking into account the following issues:

1) ensuring adequate stiffness in terms of portal frame stability

2) ensuring the reliable and robust connection between steel arch and prestressed concrete deck

3) ensuring the reliability and resistance of the structure in terms of accidental derailment and rolling stock impact when entering the bridge

4) achieving high fatigue strength

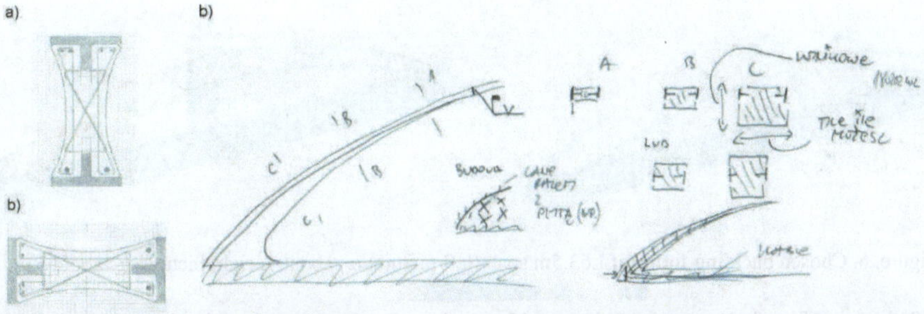

Figure 3. The idea of implementation of double composite section in arch bridges using cold-bent steel elements and composite dowels by Lorenc in 2018: a) idea for beam section proposed for beams by Seidl in 2007, b) 90 deg. rotation of section, c) implementation of transition zone between steel arch and prestressed concrete deck in 2018.

At the beginning of the design process of composite arch root section a number of challenges became apparent, the first difficulty being imperfections. Unfortunately, there was no clear design approach to this issue since Eurocode 4 does not provide any straightforward guidance on the design of composite elements with two separated steel parts. It was therefore decided to adopt a conservative approach and use an elastic design concept implementing buckling factor appropriate for steel I-sections (for steel checks). The usage of the "b" curve has been opted for according to EN 1993-1-1 for the buckling factor. For concrete parts, the usual concrete design procedure was applied. For the purpose of calculation, a relevant FE model was created using shell elements for concrete parts and beam elements for steel parts (Fig. 4). It is worth mentioning that even a plastic design approach would have been viable for out-of-plane bending (and limited stress design for certain) which is a decisive criterion for N+M design of this element, especially due to the fact that steel is outside of the concrete cross section.

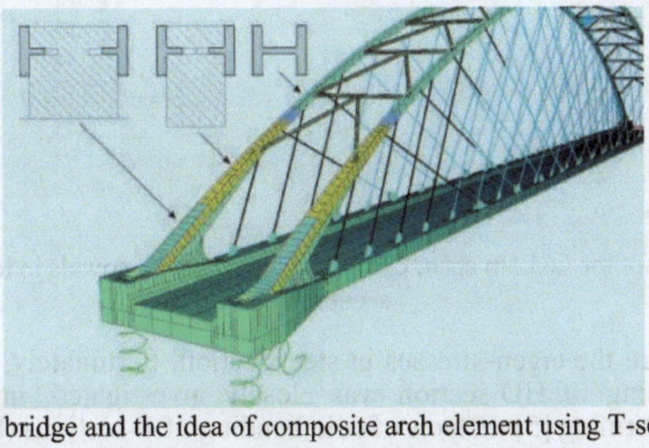

Figure 4. FE model of the M2 bridge and the idea of composite arch element using T-sections and composite dowels.

Hence, steel design concept governed dimensioning of steel and composite arch elements. The usage of composite arch root greatly reduced out-of-plane buckling and only typical buckling forms, characteristic for network arch constructions appeared in 116m span (α_{cr} results range close to 4). In smaller spans, the values were larger (and of course more diverged). Due to the length of the small spans and arch rise out-of-plane buckling also appeared. The buckling factors have been calculated using appropriate α_{cr} (Fig. 6 right) for composite arch elements.

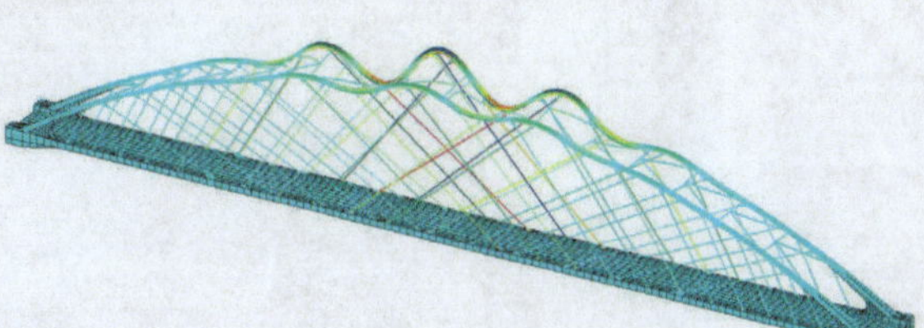

Figure. 5. In-plane buckling form of L116 span (α_{cr} about 4).

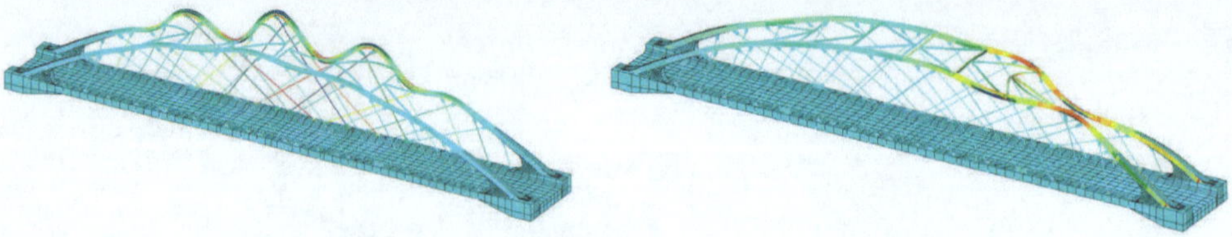

Figure. 6. Chosen buckling forms of L63.5m span (left picture: α_{cr} about 7, right picture: α_{cr} about 5).

Out-of-plane surface behavior was thoroughly analyzed, which resulted in the decision to provide transfer of forces between steel and concrete using composite dowels – a fairly straightforward solution requiring as little as creating a proper shape in the webs. Moreover, the composite dowels show bigger load bearing capacity and better fatigue capacity. Additional welded studs have been used only in the upper part of the composite element where there are relatively small loads and this is a kind of introducing part of the main element.

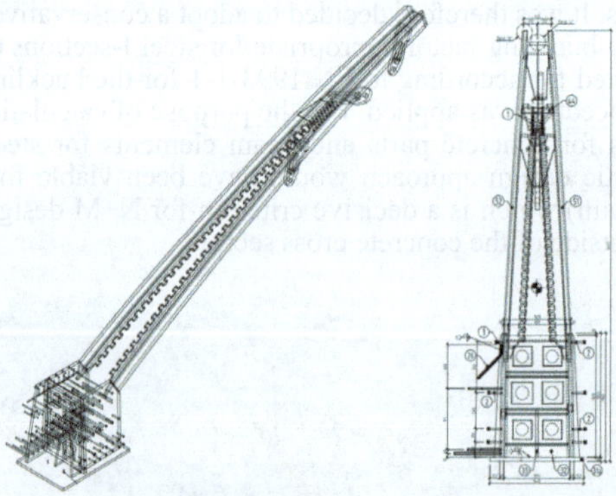

Figure 7. An example of arch root for L63.5m span. Clear visible composite dowels in lower part and studs in upper section.

Another complication were the eigen-stresses in steel section. Fortunately, the research was ongoing and the effect of cold forming of HD-section was closely investigated, including the redistribution of eigen-stresses due to cold forming [5]. Effects of cold bending of T-sections (both strong and weak axis) were also experimentally studied in [5]. The redistribution of eigen-stresses during cutting of I-section into T-section is still under discussion and R&D works are ongoing (collaborative works of ArcelorMittal and Wroclaw University of Technology). While it is a multi-level issue, the general conclusions are known. The exemplary cutting of HL1100 section out of Histar 460 steel is illustrated in fig. 8.

Figure 8. Separation of HL1100 profile (Histar 460 steel) during preparations tests for [5]; cutting effects on redistribution of eigen stresses

2 CONSTRUCTION & MANUFACTURING

2.1 Steel structure

Manufacturing of the T-sections elements was a well thought out process. Previous experience has showed possible problems in achieving proper geometry after bending (possible torsion) of T-sections, so the initial idea was to weld starting arch segments from straight elements. Unfortunately, it would entail using additional connection joints and produce inferior aesthetical appearance. Taking into account these disadvantages, it was decided to apply cold bending, similar to the main part of the arch made of HD profiles. The mutual cooperation of ArcelorMittal (AM) and Wroclaw University of Technology (PWr) led to the implementation of a new solution in the form of bent T-sections: PWr performed the first bending tests of temporarily connected T-sections (Fig. 9), and AM developed and implemented the target technology
(Fig. 10).

Figure 9. New technology tests by Wroclaw University of Technology in Poland [5].

Figure 10. Production of elements by ArcelorMittal in Luxembourg (pictures: ArcelorMittal).

The technology development works coincided with the research conducted by Marcinczak [5] in cooperation with AM, and thus it was possible to discuss exhaustively the many complex issues related to cold bending, such as ductility of steel after bending, changing the profile of residual stresses, changing the yield strength, etc. The T-profiles were cut from HL profiles made of Histar 460 steel. On account of the thick flanges and curves of the elements and the combination of the design processes (for a graphical representation of the elements taking into account the curves and preparation for welding, see fig. 11), the production process was complicated and proved to be an interesting challenge.

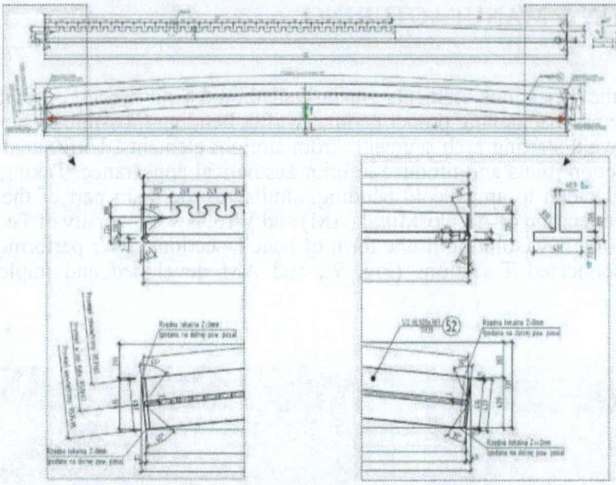

Figure 11. A fragment of the drawing showing the T-profile for the L63.5m span.

Cold-bent T-sections were delivered to Poland from Luxembourg, where steel elements of the bridge structure were connected (T-sections were welded together, the base was made, as well as the gusset plates of hangers and additional welded studs under the web: Figs. 12, 13).

Figure 12. Production of elements by Mostostal Krakow in Poland.

Figure 13. Steel element of composite arch (left) to be connected with HD profile (right).

The biggest and the thickest MCL composite dowels ever made (factor 1.28 for basic MCL100x250 shape and thickness 60.5mm) have been used in 116m span arch roots. Specification of composite dowels and arch root I-sections for 3 spans of M2 bridge is listed below (dimensions in mm):

- L49.5: 1/2HL920x725 (t_w/t_f=38.1/68.1) and f/L=314/10663, m=13.2 tons, CDs: MCL100x250
- L63.5: 1/2 HL920x787 (t_w/t_f=40.9/73.9) and f/L=294/11675, m=14.6 tons, CDs: MCL100x250
- L116: 1/2 HL920x1194 (t_w/t_f=60.5/109) and f/L=238/14197, m=22.6 tons, CDs: MCL128x320

After the fabrication of the arch root, it was transported to the construction site, where it was set in the right position with great precision. Subsequently, the formwork, reinforcement and ducts of the prestressing tendons were prepared. The detailed process of bridge construction is presented in [2].

Figure 14. Steel elements and reinforcing bars of concrete deck (left) and typical hanger anchorage with optical fiber attached for onsite researches (right)

2.2 Concrete part

After the steel elements were precisely positioned, the slab was concreted and tendons were tensioned (Fig. 14). The concreting of the area between the T-profiles above the slab was performed in the next step (Fig. 15). Due to the fact that arch roots are strongly reinforced with reinforcing bars and the limited access during concreting, it was necessary to use self-compacting concrete in this area.

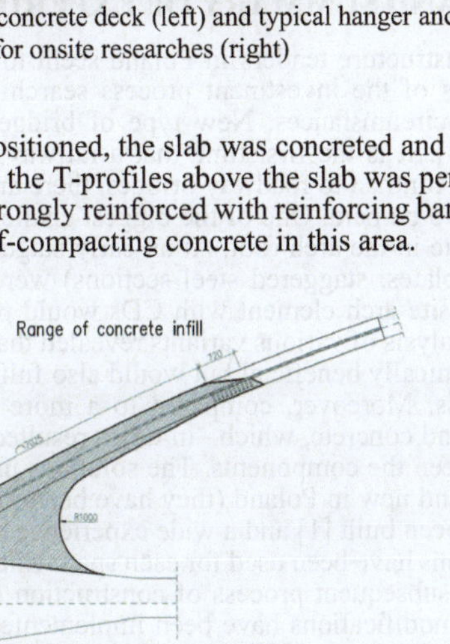

Figure 15. L63.5m span element: range of the concrete infill

In the area where the composite element is connected to the concrete deck, in addition to longitudinal prestressing, transversal prestressing also occurs (Fig 16).

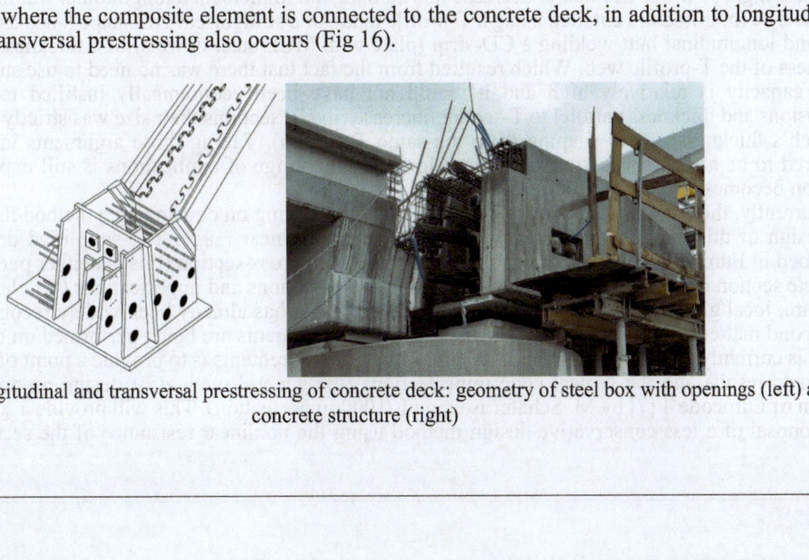

Figure 16. Longitudinal and transversal prestressing of concrete deck: geometry of steel box with openings (left) and concrete structure (right)

The composite elements of the arch above the deck were concreted in two stages (Fig. 17).

Figure 17. The arch root during construction, partially concreted (right)

2 CONCLUSIONS AND SUMMARY OF EXPERIENCES

The present-day infrastructure tenders in Poland seem to be dominated by a design-build approach, which makes participants of the investment process search for newer/more optimal solutions suitable depending on particular circumstances. New type of bridges using cold-bent HD sections have been developed in this vein [4]. It is the first time that a railway network arch bridge of this type has been constructed in Poland. In contrast to roadway bridges, there are definitely more requirements to fulfil and a greater attention needs to be paid. One of the biggest challenges was the design of a proper connection between steel and concrete in the arch root. At an early stage of the design, other conventional solutions (such as welding cover plates, staggered steel sections) were considered but due to their limitations it became clear that composite arch element with CDs would provide plenty of opportunity to create new forms [1]. An in-depth analysis of various variants revealed that implementing this state-of-the-art solution would not only be economically beneficial but would also fulfill the requirements of loadbearing capacity and of sufficient stiffness. Moreover, compared to a more traditional approach, it enabled a smooth transition between steel and concrete, which –in turn - resulted in a superior aesthetic effect: the structure shows the harmony between the components. The solutions using composite dowels (CDs) are perceived as modern rather than brand new in Poland (they have been in use for 10 years now). A number of girder bridges [3] have already been built [1] and a wide experience has been gained.

The composite solutions have been used for each span which amounts to twelve elements in total. After the design stage and the subsequent process of construction of the M2 bridge, valuable experience was gained and a number of modifications have been implemented for the next two bridges, as a result. The biggest change was the replacement of HL sections by HD sections in the arch root. This was due to the following reasons: first of all, it was observed that material waste after HL cutting (Fig. 10 left) was relatively high, so it was decided to decrease this amount and implement a longitudinal welding connection instead. The modification entailed using a T-profile made of a straight-cut HD shape (such as the adjoining one) and longitudinal butt welding a CD-strip (plate with MCL steel dowels) of a thickness less than the thickness of the T-profile web. Which resulted from the fact that there was no need to use such large CDs. Their capacity is relatively high and it would not have been economically justified to increase the dimensions and thickness parallel to T-section increase (in HL sections their size was strictly dependent on the web's thickness due to shaping from the same T-section). Taking those arguments into account, it appeared to be a very successful approach. Moreover, the range of applications is still expanding as the solution becomes more ubiquitous.

Currently, the Wroclaw University of Technology is working on developing a method that will enable the design of this new type of complex elements in a non-linear range. The simplified design method, described in Eurocode 4, is not applicable to the considered cross-section (this is neither partially encased concrete section nor prismatic). Residual stresses in the T-sections and imperfections (for elements and for structure, local/global) are of key importance. The first issue has already been widely recognized [5] and the second matter is now subject to analysis: precise measurements are being performed on the M1 bridge which is currently under construction. The aim of these measurements is to provide a point of reference for the onsite results and the typical code imperfections (the general method of design proposed in a new version of Eurocode 4 [7] by M. Schäfer is using L/1000 imperfection). This will provide a sound basis for the proposal of a less conservative design method using the nonlinear resistance of the section; a strain-limited design needs to be considered [8]. As experience from bending tests shows, the deformation of concrete under compression in the case of the considered cross-section is relatively large (owing to the application of steel flange). A similar outcome is expected for elements under compression and bending.

Figure 18. General view of new elements in M2 bridge (picture by P. Hamarnik)

REFERENCES

[1] Lorenc, W. Composite dowels: the way to the new forms of steel-concrete composite structures. *IABSE Symposium* 20-22.05.2020, Poland.

[2] Sęk, R., Szewczyk K., Pilujski, B., Sobala, D., Lorenc, W., Kożuch, M. Bridge over Vistula river in Cracow: the first railway network arch bridge using cold-bent HD sections and composite dowels. *IABSE Symposium* 20-22.05.2020, Poland.

[3] Lorenc, W., The model for a general composite section resulting from the introduction of composite dowels. *Steel Construction* 10 (2017):154-167.

[4] Kaczmarek, T.; Galewski, T.; Topolewicz, K. ; Sęk, R..; Radziecki, A. ; Ochojski, W. ; Kożuch, M.; Lorenc, W. Polish experience with network arch bridges using cold-bent HD sections. *Steel Construction* (2020). https://doi.org/10.1002/stco.202000034

[5] Marcinczak K., Zmiana właściwości wytrzymałościowych stali i układu naprężeń własnych po formowaniu na zimno dwuteowych kształtowników walcowanych ze stali S460M, Wydział Budownictwa Lądowego i Wodnego Politechniki Wrocławskiej, Raport serii PRE nr 1/2021, *rozprawa doktorska* (Change in quality and eigen-stress stress system after cold forming of S460M steel rolled I-sections, *PhD Dissertation*, Wrocław University of Technology, 2021)

[6] Zanon R., Assad M., Rademacher D., Lorenc W., Tied-arch bridges with jumbo shapes as arch member : state of the art and developments. *Costruzioni Metalliche.* 2020, nr 2, s. 66-82. (also: World Steel Bridge Symposium, April 21-24, 2020, virtual conference : conference proceedings, https://www.aisc.org/globalassets/nsba/conference-proceedings/2020/wsbs2020-rzanon-tied-arch-bridges-using-structural-shapes_20200123.pdf)

[7] Banfi M., The next generation of Eurocode 4. *EUROSTEEL 2017*, September 13–15, 2017, Copenhagen, Denmark.

[8] Schäfer M., Limits of plastic design for composite beams. Requirements for slim and compact composite sections. *EUROSTEEL 2017*, September 13–15, 2017, Copenhagen, Denmark.

9TH INTERNATIONAL CONFERENCE ON COMPOSITE CONSTRUCTION IN STEEL AND CONCRETE
M. Knobloch, U. Kuhlmann, W. Kurz, M. Schäfer. (Eds.)
Online, Germany, July 27-29, 2021

NEW TYPE OF TRANSITION ZONE FOR STEEL-CONCRETE HYBRID BEAMS IN BRIDGES

P. Kozioł*, W. Lorenc*, M. Kożuch*, W. Kosecki and A. Stempniewicz*****

* Wrocław University of Science and Technology
e-mails: piotr.koziol@pwr.edu.pl, wojciech.lorenc@pwr.edu.pl, maciej.kozuch@pwr.edu.pl

** Europrojekt Gdańsk S.A.
e-mail: w.kosecki@europrojekt.pl

*** FASYS MOSTY Sp. z o.o.
e-mail: stempniewicz-adam@tlen.pl

Keywords: hybrid beam, perfobond connector, destructive tests, composite structures.

Abstract. *A new kind of composite connection is implemented in bridge engineering in Poland, which leads to a new type of steel-concrete beams. First innovative bridges are constructed based on in-depth scientific works. Perfobond shear connector (PBL) concept has been combined with standard geometries of composite dowels and in this way, efficient shear connection has been created that is used in combination with composite dowels. It enables a new type of beams that have more general forms than standard composite beams – they are named hybrid beams. There are intensive R&D works going on towards optimization of the transition zone concept. The article widens the already executed experimental tests, summarises the development of the hybrid beams, and presents the application of the hybrid beams in bridges.*

1 INTRODUCTION

Over the last few years, Poland has reached significant development of its transport infrastructure. A well-developed road infrastructure enables the growth of industrial activity, improves the mobility of people, and is an important factor in the economic development of the country. Most of the investments are realized in the "Design and Build" delivery system, in which high quality and profitability requirements for the construction generate innovative design solutions. These factors stimulates designers to combine code check, FEA design and R&D works.

The search for new solutions based on the use of structural steel and concrete in the optimization of bridge spans has a great impact on the carried out research work. The basic knowledge, gained in many of research and scientific projects, supplemented, appropriately developed, and combined, is transformed into technical solutions with a wide range of application possibilities. As one of many examples, the solution of anchorage zone of a composite element into a concrete element can be used [1].

When typical PreCoBeam beams with MCL-type shear connectors are used in the span area, with high bending moments, it is possible to use its high bearing capacity. However, when using as a continuous beam or in the frame structure, prefabricated composite beams can be still optimized. In such static schemes, negative bending moments are created near the support parts, and the structural steel continued from the span to the support is in the compression zone. It is therefore advantageous to differentiate the type of beam cross-section along the beam, and use reinforced concrete near support parts as a more reasonable one. Beams with both concrete-type sections, and steel and concrete composite-type sections are named hybrid beams. The crucial part of the beams is a smooth transition zone with anchorage of the steel part in concrete.

The first viaduct, where the structural steel ends before the support zone, was built in 2009 in Poland. The WD4 viaduct carries S10 expressway between Stryszek and Białe Błota, and access road over two railway tracks linking Tczew and Chorzów Batory (Fig. 1). It is designed as three independent single-span

 Composite Construction in Steel and Concrete IX: Proceedings, Eds. M. Knobloch, U. Kuhlmann, W. Kurz, M. Schäfer.

frames. The load-bearing structure of each frame consists of four prefabricated VFT-WIB girders with steel T-sections embedded in concrete as external reinforcement. Composite precast elements cover only the span area over two railway tracks (Fig. 2). Near the support zone, the reinforced concrete is used. The bridge design and construction are described in detail in [7].

a)

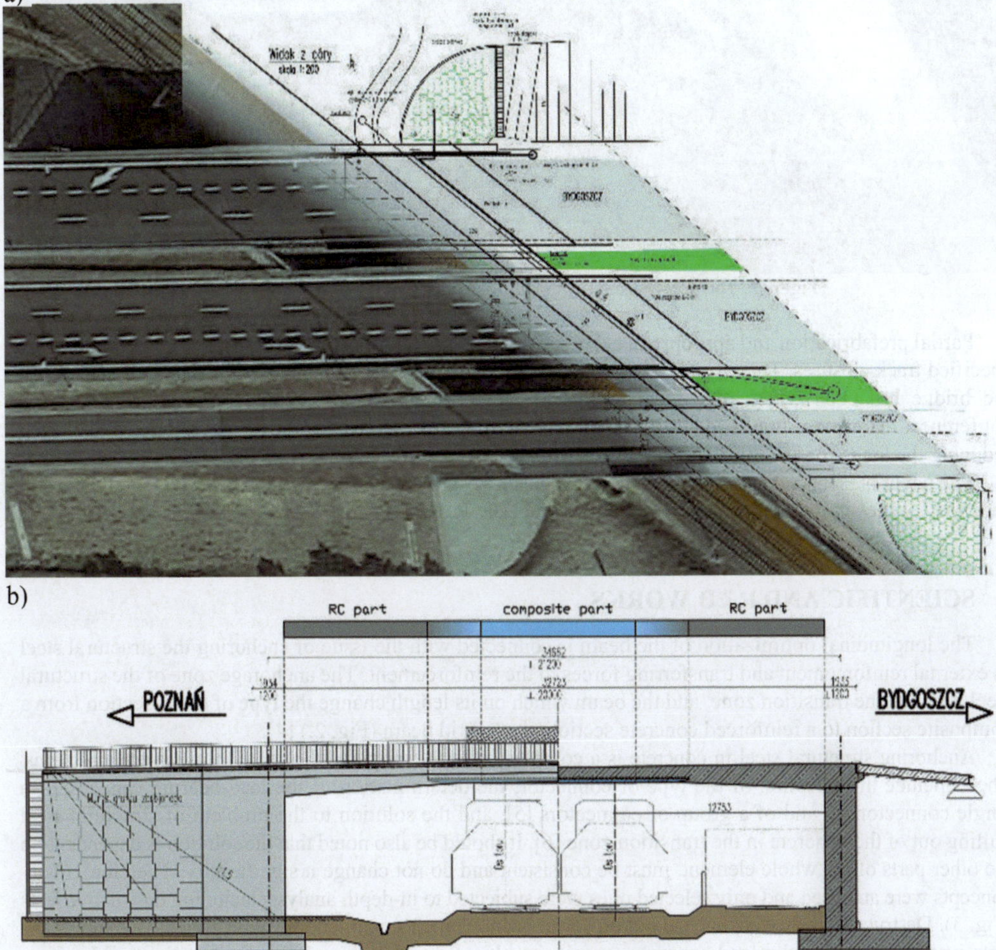

Figure 1. WD4 viaduct – the idea of the hybrid beams: a) plan and google map [9] top view on the bridge, b) cross-section of the one out of three spans.

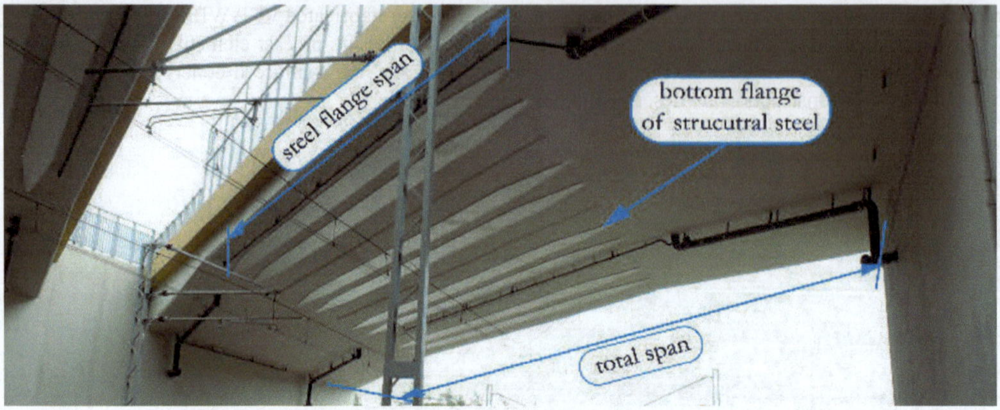

Figure 2. The WD4 viaduct with hybrid beams – view on superstructure.

Partial prefabrication and appropriate execution technology shortened the construction time to meet the specified track closures. The cross-section of the span is similar to the original design. The construction of the bridge has been widely discussed at different practitioners' and scientists' meetings as well as conferences. However, due to doubts and lack of tests' results, research aimed to provide the solution reducing the length of anchoring the structural steel has been conducted.

The paper presents different aspects of hybrid beam design, the research which has been conducted and summary of the application of hybrid beams in bridges in Poland.

2 SCIENTIFIC AND R&D WORKS

The longitudinal optimization of the beam is connected with the issue of anchoring the structural steel as external reinforcement and transferring forces to the reinforcement. The anchorage zone of the structural steel is called the transition zone, and the beam which on its length change the type of cross-section from a composite section to a reinforced concrete section, the hybrid beam (Fig. 2.) [4].

Anchoring structural steel in concrete is a complex issue that requires the solution of several problems. These include the selection of the type of connector, the determination of the load-bearing capacity of a single connector [8] and of a group of connectors [5], and the solution to the problem of structural steel pulling out of the concrete in the transition zone [6]. It should be also noted that the solution is dependent on the other parts of the whole element, must be consistent and do not change it significantly. Therefore, many concepts were analysed and only selected ones were subjected to in-depth analysis including destructive tests (Fig. 3). Destructive tests were carried out on seven beam elements, and ten transition zones. The aim of the tests was to determine the load-carrying capacity at ultimate and serviceability limit states, to identify a criterion for evaluating the behaviour of the connection in the transition zone, and to determine the failure mechanisms of the tested element.

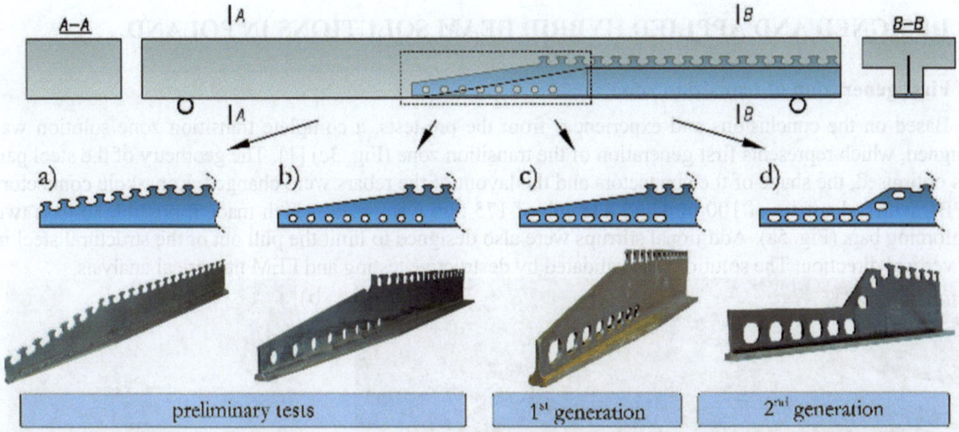

Figure 3. Evolution of the steel part in transition zone: a, b) pre-tests, c) first generation, d) second generation.

The first destructive tests of various types of transition zone solutions were carried out on beams on a semi-technical scale (Fig. 3a). The type of connectors, the distribution and shape of reinforcement in the transition zone were varied. As a result, satisfactory load capacity of the transition zone was proven, i.e. higher than the characteristic plastic load capacity of the composite and reinforced concrete section [3]. The results confirmed the validity of using closed connectors PBL (perfobond type) in this zone. The circular hole connector has similar load capacity, regardless of the direction of the resultant force, and ensures a durable and ductile connection. Tests set up with the failure of beam specimens are shown in Fig. 4.

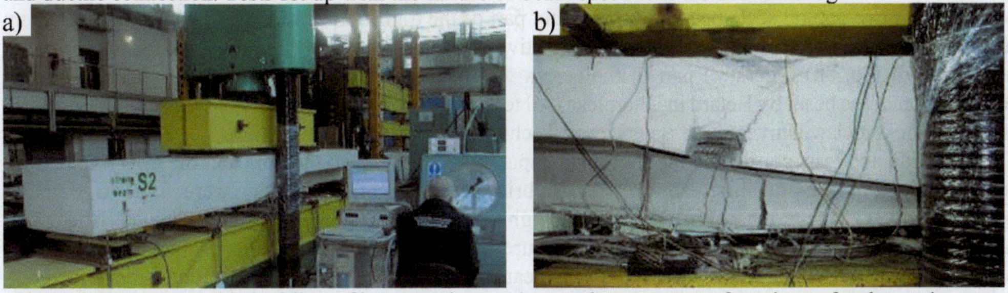

Figure 4. Tests set up with the failure of beam specimens: a) general arrangement of specimen after destructive tests, b) part of transition zone after experimental tests.

The moment resistance of the transition zone is determined, among other things, by the resistance of the connectors and the lever arm of internal forces on which the closed connectors operate. The dowels should therefore be placed as low as possible, i.e. close to the steel flange. However, placing the connectors as close to the edges of the concrete element reduces the load bearing capacity of the connectors by design, with the criterion of failure by pry-out cone.

In the final stage of conducted tests, plastic hinge was obtained resulting in exceeding the ultimate compressive strains in top concrete, what lead to its brittle cracking (Fig. 4b) [3]. At the same time, the bottom external reinforcement in transition zone was pulled out of the concrete in vertical direction. The pry-out failure criterion did not occur. The correct shape of the reinforcement ensured a high load-bearing capacity of the connectors, which was further confirmation of the validity of their use.

3 DESIGNED AND APPLIED HYBRID BEAM SOLUTIONS IN POLAND

3.1 First generation of transition zone

Based on the conclusions and experiences from the pre-tests, a complete transition zone solution was designed, which represents first generation of the transition zone (Fig. 3c) [1]. The geometry of the steel part was optimised, the shape of the connectors and the layout of the rebars were changed. Long-hole connectors L-PBL with a diameter of 100 mm and a length of 175 mm were used, which made it possible to pass two reinforcing bars (Fig. 5a). Additional stirrups were also designed to limit the pull out of the structural steel in the vertical direction. The solution was validated by destructive testing and FEM numerical analysis.

a) b)

Figure 5. First generation of transition zone: a) detail of the main rebar with hook passing through the long-hole connectors, b) top and bottom view on the two precast beams.

Testing of two natural scale beams was the main part of the transition zone tests. The transition zones in the zone of positive and negative moment, i.e. respectively in the tension and compression parts of the beam, were checked (Fig. 5b). The steel part of the beams was made by ArcelorMittal in Luxembourg, and the prefabrication of the beam by Betard in Długołęka. All tests were carried out in the Civil Engineering Research Laboratory in Wrocław University of Science and Technology (Fig. 5).

The conducted tests confirmed the high load capacity of the PBL connectors and the effectiveness of the conception of the transition zone of the hybrid beam. Eight connectors were used to transfer approximately 3.7 MN into the formed steel. Both high load capacity and the lack of the characteristic crack pattern which used to occur with pry-out cone failure, show that, although the long-hole connectors are located as close as possible to the lower part of the beam, the steel flange and the constructed reinforcement prevent the concrete from breaking through near the connector. As a result, a complex solution was proposed, in which a hot-rolled section, high-strength steel S460 and concrete class C50/60 were used (Fig. 6a) [1].

a) b)

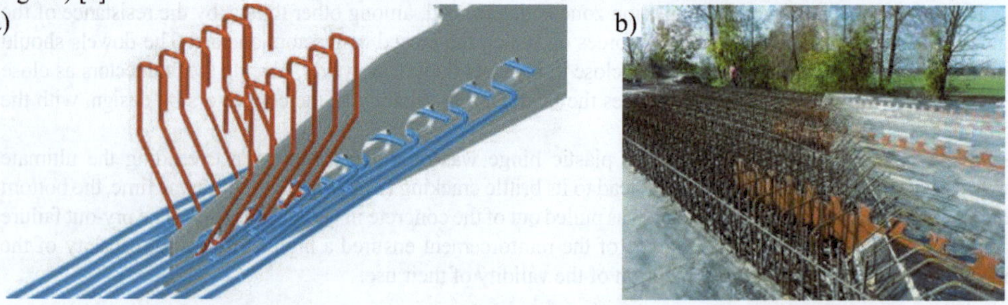

c)

Figure 6. Transition zone concept with hooks applied in bridge in Krzemień Drugi in Poland: a) transition zone conception, b) erection of the bridge, c) ready road bridge.

In 2018, first generation of the transition zone was applied to a road bridge in Krzemień Drugi between Janów Lubelski and Flisy (Fig. 6). The structure is designed by FASYS MOSTY office. The superstructure was made as a two-span continues composite beam with a spans of 2 x 14.9 m. The steel part is a ½ HEA 900 profile with MCL-type continuous connectors at the top part of the flange. In the area of the intermediate support, the thickness of the reinforced concrete slab was increased to match the height of the entire steel section and the section was anchored into concrete. In the transition zone, 100x175 mm long-holes connectors were used and the longitudinal reinforcement bars were anchored in the individual holes with hooks. Consequently, only the reinforced concrete section is used in the negative moment zone.

3.2 Second generation of transition zone

Parallel to the development of the first concept of hybrid beams, work began on their optimisation. The attempt to simplify the execution of the beam required changes to the proposed solution. In particular, the anchoring of the main rebars in each holes was abandoned. Consequently, it was possible to reduce the width of the hybrid beam to the flange width of the steel member. Stirrups were used to reinforce the connectors and the main rebars were laid out parallel to the structural steel. This solution simplified formwork preparation and allowed the use of large diameter bars (Fig. 9a). The height of the steel web was reduced over some of the L-PBL connectors, making it easier to place rebars through the holes (Fig. 2b, 9a). The concept of a hybrid beam is presented in Fig. 7.

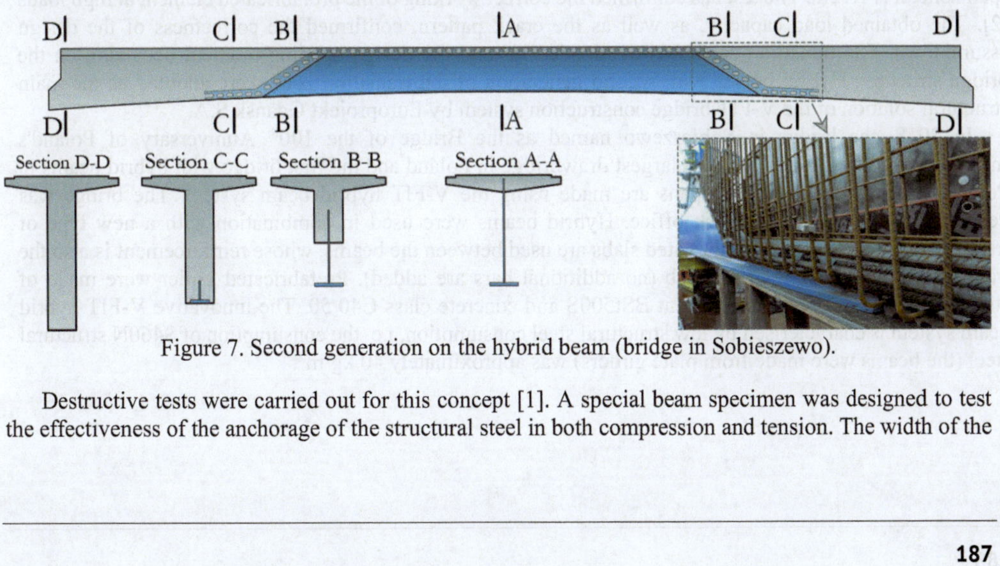

Figure 7. Second generation on the hybrid beam (bridge in Sobieszewo).

Destructive tests were carried out for this concept [1]. A special beam specimen was designed to test the effectiveness of the anchorage of the structural steel in both compression and tension. The width of the beam corresponded to the width of the steel flange and was equal 30 cm. The reinforcing bars were continued to the end of the transition zone. There is a lot of reinforcement in the transition zone, which reduces cracking in this zone and at the same time increases the pry-out cone resistance of the concrete. However, this leads to a higher variation of forces in the individual connectors on the length of the transition zone (compared to first generation). It was also confirmed that the use of sufficiently stiff reinforcement in the area surrounding the connectors has the same function as a wide concrete element, i.e. providing triaxial compression and protection against pry-out failure.

On the basis of the performed destructive tests, a validation of the developed numerical models was performed. The conducted FE numerical analyses showed a convergence with model tests [1]. They confirmed the distribution and magnitude of transferred forces between reinforcing and structural steel in the transition zone at all load levels obtained in experimental tests.

Figure 8. The full scale tests on sull-size bridge girders for Sobieszewo bridge.

The applied research was summarised by destructive tests of two full-size bridge girders with spans of 24.0m (Fig. 8). The height of each beams is variable and in the support zones it is about 1.55m and in the span zone, it is 1.12m. The test has confirmed the correct working of the prefabricated element at high loads [2]. The obtained load capacity, as well as the crack pattern, confirmed the correctness of the design assumptions. The transition zone was overloaded beyond the design bending moment occurring on the bridge structure. Hybrid beams with second generation of the transition zone were adopted as the main structural solution of the V-FIT bridge construction system by Europrojekt Gdańsk S.A.

In 2018, the bridge in Sobieszewo named as the Bridge of the 100th Anniversary of Poland's Independence was opened. It is the largest drawbridge in Poland and the first bridge with hybrid beams in Poland - four external, access spans are made using the V-FIT hybrid beam system. The bridge was designed by Europrojekt Gdańsk office. Hybrid beams were used in combination with a new type of transverse prefabrication. Prefabricated slabs are used between the beams, whose reinforcement is also the transverse reinforcement of the slab (no additional bars are added). Prefabricated girder were made of structural steel S460N, reinforcement BSt500S and concrete class C40/50. The innovative V-FIT hybrid beam system is characterised by low structural steel consumption, i.e. the consumption of S460N structural steel (the beams were made from plate girders) was approximately 40 kg/m^2.

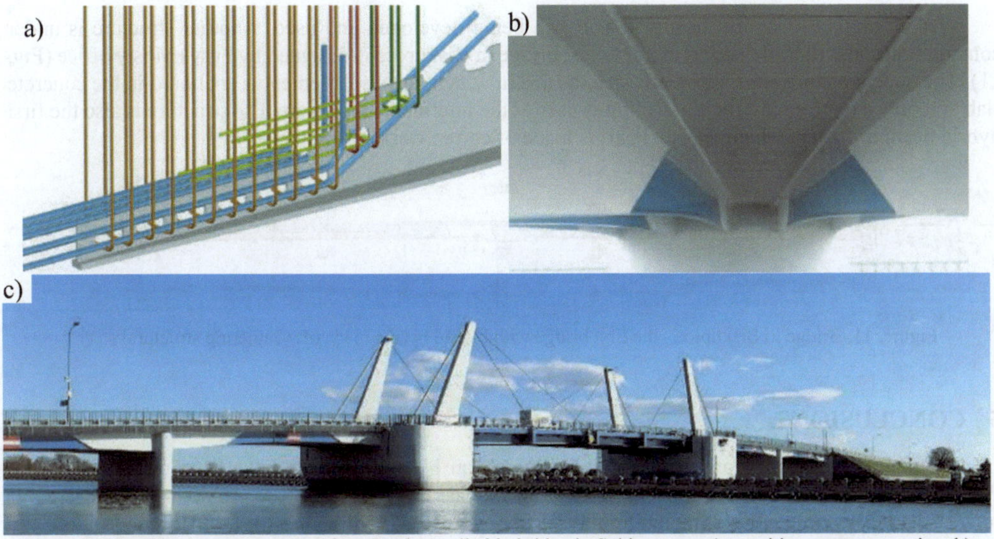

Figure 9. Transition zone concept without hooks applied in bridge in Sobieszewo: a) transition zone conception, b) view to the bottom part of the bridge , c) ready road bridge.

4 FURTHER DEVELOPMENT AND BRIDGES UNDER CONSTRUCTION

Current works carried out by the Wrocław University of Science and Technology are aimed at further developing composite beams, particularly hybrid beams. Optimisation of structural solutions translates into new, previously unobserved failure mechanisms. Particular attention is focused on the failure mechanisms of the concrete part and the development of recommendations for its design and execution. The latest tests concerns the continuous Strong Reinforced Composite Dowels (SRCD) connection, in which the failure mechanism of the pry-out cone type has been structurally eliminated and in addition the thickness of the concrete part has been significantly reduced. Currently, this solution is in the research phase. Destructive tests under static and cyclic load are being carried out. It is expected that the SRCD composite will form the basis for third generation of hybrid beams as an optimisation of second generation.

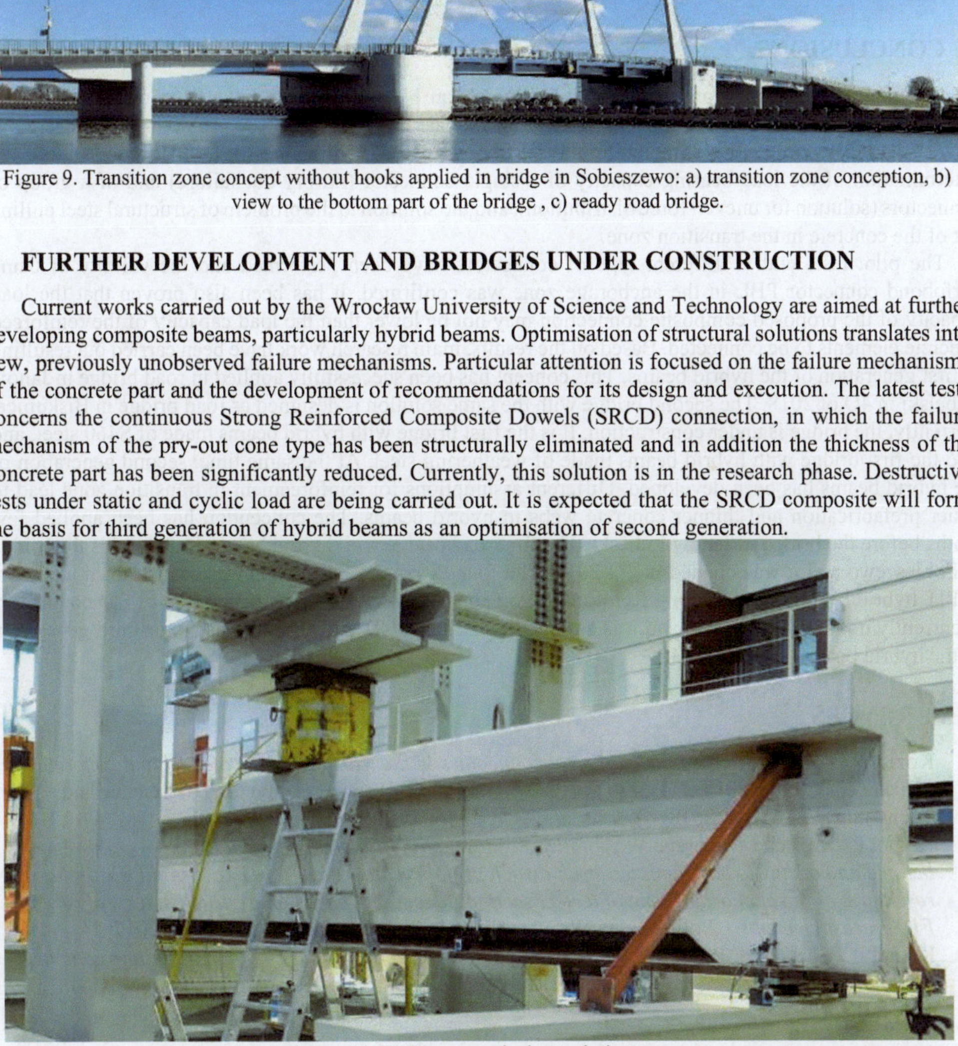

Figure. 10. SRCD composite beam during tests.

First generation of hybrid beams is also still being developed and used. Another structure is under construction and will be delivered in 2021. The bridge in Biskupice is designed by Fasys Mosty office (Fig. 11). The superstructure consists of two-span beams in which, the steel element is anchored in the concrete slab and only a reinforced concrete section is used at the intermediate support. The structure is also the first hybrid beam bridge, in which the steel part is made of corten weathering steel.

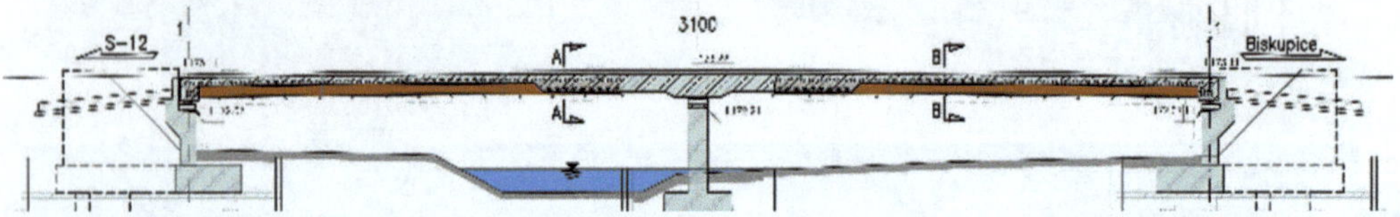

Figure. 11. Bridge in Biskupice - the first bridge with hybrid beams made of weathering structural steel.

5 CONCLUSIONS

Comprehensive R&D works and applied research on hybrid elements have been performed at the Wrocław University of Science and Technology for the last few years [1]. Key problems in the design of such structures were identified, including the selection of the type of connector (L-PBL connectors), the determination of the load-bearing capacity of a single connector (no pry-out failure) and of a group of connectors (solution for uneven force distribution), and the solution to the problem of structural steel pulling out of the concrete in the transition zone.

The pilot tests with a different type of connectors have been performed and the validity of using perfobond connector PBL in the anchorage zone was confirmed. It has been also proven that the load capacity of the proposed composite connection may not be lower than the load capacity of the reinforced concrete elements to be connected. Based on the results, main research work have been carried out resulting in first generation of the hybrid beams. This concept has been successfully applied in road bridge in Janów Lubelski in 4Q of 2018. The second bridge with the same solution is designed as road bridge in Biskupice. Currently, the bridge is under construction. It is the first bridge with hybrid beams made of S460 steel, and also the first bridge with hybrid beams made of weathering steel. At the same time, second generation of the hybrid beams has been developed. Different assumptions for reinforcement in transition zone lead to easier prefabrication and thinner concrete webs in hybrid beams. The conception has been applied few moths before the bridge in Janów Lubeski. The first road bridge with hybrid beams in Poland is the bridge in Sobieszewo and it was opened in 3Q 2018. The four external access spans were constructed using the V-FIT hybrid beam system. Strong Reinforced Composite Dowels as a new generation of CD connectors is currently under development, and it is assumed, that the SRCD will form the basis for third generation of the hybrid beam.

REFERENCES

[1] Kozioł P., *Load-bearing capacity of steel connectors in the transition zone of steel and concrete hybrid beam*, Report PRE No. 3/2018 (PhD dissertation), Wrocław University of Science and Technology, Wrocław, 2018.

[2] Kozioł P., Kożuch M., Lorenc W., *Analiza nośności strefy zakotwienia kształtownika stalowego w betonie na podstawie badań niszczących dwóch prefabrykatów typu V-FIT (EN. Analysis of the resistance of the anchorage zone of a steel section in concrete based on destructive tests of two V-FIT precast elements)*, Politechnika Wrocławska, Wrocław, Raport SPR 21/2017, 2017.

[3] Kozioł P., Kożuch M. P., Lorenc W., Rowiński S., *Connection capacity of the transition zone in steel-concrete hybrid beam*, Civil and Environmental Engineering Reports, vol. 25, no. 2, pp. 137–146, 2017.

[4] Kozioł P., Kożuch M. P., Lorenc W., Rowiński S., *Innovative solution of transition zone in steel-concrete hybrid beam*, presented at the Recent progress in steel and composite structures [Dokument elektroniczny]: proceedings of the XIII International Conference on Metal Structures (ICMS2016), Zielona Góra, Poland, 15-17 June 2016, 2016, pp. 137–142.

[5] Kozioł P., Lorenc W., *Analiza połączenia stali z betonem w strefie przejściowej dźwigara zespolonego stalowo-betonowego*, Materiały Budowlane, no. 5, pp. 97–98, 2016.

[6] Luo G., Mensinger M., Näßl A., Seidl G., Feldmann M., Kopp M., *The influence of the pull-out effect on the load-bearing behaviour of external reinforcement elements*, presented at the 10th Japanese German Bridge Symposium, München, Germany, 2014.

[7] Seidl G., Stambuk M., Lorenc W., Kołakowski T., Petzek E., *Wirtschaftliche Verbundbauweisen im Brückenbau - Bauweisen mit Verbunddübelleisten*, Stahlbau, vol. 82, H. 7, no. Journal Article, pp. 510–521, 2013.

[8] Zheng S., Liu Y., Yoda T., Lin W., *Parametric study on shear capacity of circular-hole and long-hole perfobond shear connector*, Journal of Constructional Steel Research, vol. 117, pp. 64–80, 2016.

[9] *Google Maps, Google Maps*. [Online]. Available: https://www.google.pl/maps/@53.0526863,17.9900759,122m/data=!3m1!1e3. [Accessed: 19-Jul-2022].

9TH INTERNATIONAL CONFERENCE ON COMPOSITE CONSTRUCTION IN STEEL AND CONCRETE
M. Knobloch, U. Kuhlmann, W. Kurz, M. Schäfer. (Eds.)
Online, Germany, July 27-29, 2021

USE OF HIGH- STRENGTH STEEL FOR SLENDER MEDIUM SPAN BRIDGES: TWO RECENT CASE STUDIES IN FRANCE

R. Zanon*

* ArcelorMittal, 66 rue du Luxembourg, L-4221 Esch-sur-Alzette, G.D. of Luxembourg
e-mails: riccardo.zanon@arcelormittal.com

Keywords: Composite bridge, high-strength steel, real project application.

Abstract. *This paper presents two cases of existing bridges with prestressed concrete deck replaced by composite steel-concrete decks. The new decks provide a larger carriage width with increased load carrying capacity according to current design requirements. Furthermore, they have less piers (and therefore larger spans) but still keep the same construction height of the previous solutions. This leads to very slender continuous decks which are an optimal field of application for high-strength steel. The technical solution adopted is an evolution of the traditional technology of composite steel-concrete deck, commonly used in the span range between 20m and 40m.*

1 INTRODUCTION

1.1 Small and medium span bridges – a relevant case

In the area of small and medium span bridges, overpassing the obstacle rarely represents a technical challenge and in most cases several effective solutions are available to achieve it. Ensuring the load-bearing capacity of the structures does not constitute an headache for the designer, but beside this aspect there are various other criteria that the designer shall take into account:

Figure 1 Triangle bridge in Dunkirk (photo: R. Zanon)

- Economic criteria: around 80% of road bridges and 90% of railway bridges have a span of less than 30 m [1], [7], [9]. If the cost of each of them is relatively small, the scale effect leads to a large financial stake for the bridge owners. As a logical consequence the financial expense is a major criterion for this abundant group of small civil works, whose structural and architectural impact is often minor.

- Aesthetics: although the appearance of small and medium bridges rarely marks the landscape, a neglected appearance of the infrastructure leads to an unpleasant perception of its environment (there are several examples known to all of us in urban areas built in ’70 and ’80 in Europe). To optimize the impact of the additional expenses the designer shall focus on the elements which mostly contribute to the bridge aesthetics. The simple observation of the existing heritage shows that the architectural perception of the deck is governed by the choice of colours, the geometry of the cap and the equipment rather than by the shape of the load-bearing structure of the deck. Conversely, deck support elements play a major role in the aesthetics and shall be detailed with care. Example of such elements are piers, pier heads, abutments, embankment wall, rendering of concrete surfaces.

- Duration of the construction site: bridge may be built in several different situations such as replacing an existing structure, a new development in an urban area, creating a new crossing link on an existing network. In all cases the indirect costs of the construction site induced on mobility disturbance or land occupancy are by an order of magnitude higher than the direct cost of bridge construction. Even if these indirect costs do not impact the budget of the bridge owner, they shall be considered and minimized in a reasonable manner.

- Simplicity and structural robustness: centuries of bridge construction experience shows that robust construction techniques are the most effective solution in the long term. A clear force flow without sophisticated details are key to achieve this goal. Structural continuity must be preferred to isostatic scheme because it provides redundancy, an additional safety in the event of accidents or unforeseen load situation. Redundancy may also become a resource to justify load increase requirements during bridge life. In practice, simple structures are also the most suitable to be maintained, reinforced, enlarged or modified for future needs.

- Quality insurance: for standard construction sites quality is compatible with reduced budget if an off-site prefabrication is properly planned. The advantage of quality controls in the factory before dispatch to the site also removes the dangers of construction site compromises and aleas in the execution. This strategy has been implemented since the last 30 years ago, e.g. in [1][2].

- Durability and maintenance costs: a major challenge for building owners is to limit the maintenance costs of existing assets. Indeed this budget tend to grow constantly and bridge owner needs to reach a plateau. For standard bridges which are great in number but without major technical challenges, it is necessary to design durable solutions with maintenance costs as low as possible. Several old and new solutions can contribute: create a monolithically junction between the deck and the piles and / or abutments avoiding bearing devices and pavement joints, use of weathering steel which eliminates the need to regularly repaint the bridge, ...

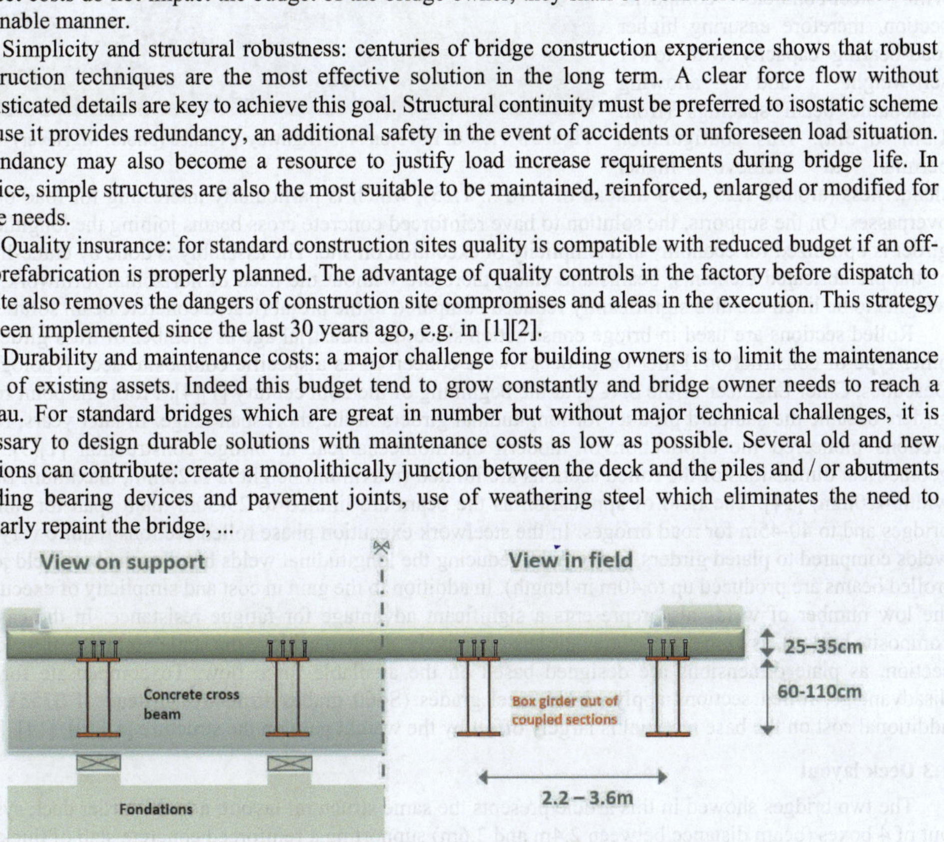

Figure 2. Typical cross-section of the bridge deck.

1.2 The range between 20m and 40m span: the transition zone

The range of spans between 20m and 40m is often considered as the transition zone for road bridges between two construction different construction techniques: prestressed concrete bridges and steel-concrete composite bridges [8]. Precast pre-stressed concrete beams are a solution widely used but they lose the competitiveness at increasing span. For example at 30m about 60% of the load bearing capacity of a road bridge is the self-weight of the concrete beam. When the longitudinal prefabricated concrete elements are closely spaced, it is clear that it is an attempt to push the typology outside its natural field of application.

On the other side the conventional twin-girder steel-concrete composite deck out of welded girders are a very common, economic and robust solution for spans above 30-35m. For smaller spans they may be less competitive from the economic point of view. Configurations using rolled sections can be an effective answer in this span range. They adopt the multi-beam deck configuration of pre-stressed concrete structures, suitable for configurations with a low span / width ratio of the deck. Longitudinal beams are realized with steel-concrete composite section, therefore ensuring higher load-bearing capacity with lower self-weight and allowing reasonable beam spacings (from 1.5m to 3m). This configuration permits to achieve higher slenderness (around 1:25 ... 35 instead of 1:18 ... 1:25), which is particularly interesting for road or rail overpasses. On the supports, the solution to have reinforced concrete cross beams joining the longitudinal girder is optimized for economy and simplicity of execution on site. The assembly is done by crane-lifting of the prefabricated elements, beams and slabs, therefore without the need of horizontal formwork. The weights to be lifted are also significantly reduced compared to the prefabricated concrete beam solutions.

Figure 3 View of PS13 on A85 Highway – France (photo: M. Hever)

Rolled sections are used in bridge construction since the industrial age as member of truss girders or other type of construction. Filler beam decks were conceived as a specific composite deck typology by Descubes, Chief Engineer of the SNCF, at the beginning of the 20th century [1][4]. From this point rolled girders became the standard product for longitudinal girders in the short span range. In later years, rolled sections pioneered the application of modern thermomechanical in bridge construction [1][3]. The geometrical dimensions of the rolled sections are limited (maximum height is 1120mm, maximum flange width 420mm) [14]. The field of application as the beam are limited to 25-30m main span for railway bridges and to 40-45m for road bridges. In the steelwork execution phase rolled sections require very few welds compared to plated girders, not only by reducing the longitudinal welds but also the butt weld joints (rolled beams are produced up to 40m in length). In addition to the gain in cost and simplicity of execution, the low number of welds also represents a significant advantage for fatigue resistance. In the case of composite bridges, asymmetrical plated girders effectively allow to achieve the static optimum of the cross section, as plate dimensions are designed based on the available force flow. To compensate for this disadvantage, rolled sections apply higher steel grades (S460 grades structures instead of S355). The additional cost on the base material is largely offset by the weight gain on the structure [1][10], [14], [16].

1.3 Deck layout

The two bridges showed in this article presents the same structural layout: a multi-girder deck system out of 4 boxes (beam distance between 2.4m and 3.6m) supporting a reinforced concrete slab of thickness between 25cm and 45cm (see Figure 2). The reduced beam distance permits to have prefabricated concrete slab elements in partial or full deck thickness with shear pockets filled on site. The transversal reinforcement ratio remains limited thanks to the reduced span between the beams. No cross beam is foreseen in the field, but the girders are linked at the supports by rigid concrete cross beams. In the span range of 20m and 40m a slenderness ratio between 1:25 and 1:35 can be economically achieved. The fact that both bridges presented were counterproposals on traditional prestressed concrete solutions is the best demonstration. The deck described is very similar to what has become a current solution in recent years ([10], [7], [6], [5], [3], [2]), but in this case there is the specificity of using box girders instead of open I sections. box girders are realized out of standard rolled sections with high strength steel (S460M according EN10025-4 [15]). The choice of joining the two box profiles brings many advantages:

- lateral-torsional buckling phenomena at intermediate supports can be excluded in the final phase. Cross-sections can develop plastic moment resistance with optimal use of HSS.

- the torsional inertia of the box girder contributes to the transversal load repartition and reduces the transversal bending moment in the concrete slab.
- The box girder presents a significant resistance and robustness also in the horizontal plane and can be easily justified in case of impact of abnormal vehicles crossing below the deck or, in the case of bridges on rivers, for extreme flood cases.
- In the two real cases presented in this article, the box girders did not need any stabilizing device in the span, their rotation was blocked only at the supports. They were placed as simply supported beam without propping amongst the supported, loaded with the prefabricated slabs and still could overtake the load due to fresh concrete without need of stabilization devices.
- The upper flange of the box girder is twice wider than the one of single I rolled sections (between 600mm and 850mm), which greatly facilitates placing the shear studs and the interaction with the shear pockets and the reinforcement.
- The boxes are made by single-pass welding carried out with an automatic welding machine. A single pass weld (a = 5 … 7mm) is enough to justify the shear stresses due to torsion, which are quite low. The natural rounded shape of the rolled section flange edge allows this welding to be carried out without any special bevel preparation, which therefore makes this type of finishing economical since the additional assembly-welding cost is offset by the gains in reducing the surface to be painted.
- Box beams are made air-and watertight by the welded end plates so only the external part needs corrosion protection, leading to a dramatic reduction of the surface to paint. To quantify this, the ratio between the exposed metallic surface of the steel structure to the horizontal surface of the deck is compared. For the case of filler beam decks, this value is in a range between 0.4 and 0.7, which is the minimum amongst the common solutions for composite bridges and contributes to explain why this typology is known for outstanding durability. For standard twin girder decks, the ratio is ranging from 1.0 up to 2.5 in the span range between 25 and 45m. The case of box girder deck is in the middle, ranging between 0.6 and 1.2.

This ratio can explain that the impact of corrosion measures is reduced compared to a standard twin girder composite deck by roughly one half. This means strongly decreasing the cost of corrosion protection or reducing the impact of sacrificial thickness for the case of weathering steel. The boxes are small and therefore the inner side is not more accessible after joining together the sections. However, in practice this does not constitute a problem as the beam can be spliced before joining them (necessary also for lengths larger than 40m – which is rarely the case), as well as eventual internal web stiffeners can be added for special cases. The first use of box girders out of rolled sections using watertight welds goes back to 25 years ago for several motorway overpasses. Until now no issues were encountered in terms of maintenance and corrosion protection (PS13 on the A85, OA3 / OA4 / OA5 / OA6 / OA10 on the A13 Luxembourg – Saarland – see Figure 3) [1][5].

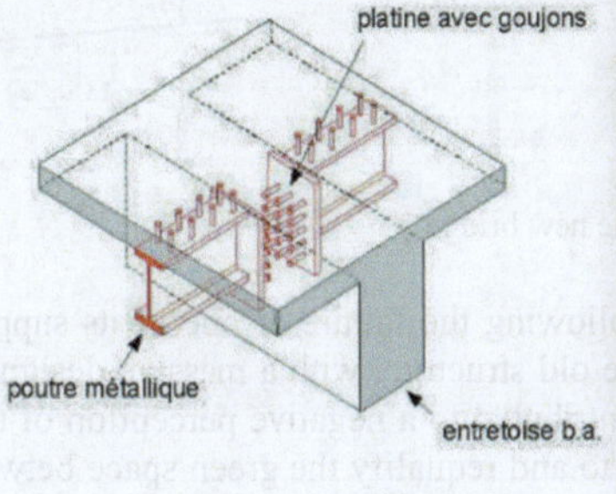

Figure 4. Sketch and seismic test of a concrete crossbeam (photo: R. Zanon)

1.4 Easy and quick erection on the construction site

The prefabrication of all the elements forming the deck (metal frame, prefabricated slabs, sill, cornices) is done in parallel with the preparation of the foundations. On the construction site, only the shear pockets, the joints between the prefabricated elements and the concrete cross beams need to be poured on site. For multi-span structures it is always necessary to have splices for the longitudinal beams. Conventional bolted joint or welded joint solutions are then possible and are used in many cases. However, in order to reduce this splicing work, the technique of reinforced concrete cross beams has been developed to join the main beams onto the intermediate piers. This solution has been the subject of several research programs, both theoretical and experimental, which have made it possible to validate its behaviour under static loads and fatigue. The European-funded project SEQBRI provides some additional guidelines for the use of this technology in the frame of earthquake-prone zones [7], [12], [11], [6]. The static principle is simple and effective (see Fig. 4):

- The bending moment, negative on the support, is decomposed as a couple of opposite forces. Compression in the lower flange is transmitted by contact, whereas traction is transmitted by the reinforcements. In practice the concrete cross beam is the location which is governing the design of the reinforcement in the negative moment zone.
- The shear force is transmitted either by the web of the profiles encased in the concrete beams, or by shear studs welded on the head plate.

2 TRIANGLE BRIDGE IN DUNKERQUE

2.1 Situation

The urban community of Dunkirk (France, Department du Nord 59) refurbished the road from Saint-Pol-sur-Mer to Fort-Mardyck, including the reconstruction of a new bridge as well as other civil works. The project crosses a natural zone used as a public green space, between the urban areas of Saint-Pol-sur-Mer in the East and Fort-Mardyck in the West. The road constitutes one of the main axes of connection in the urban community of Dunkirk to the West.

Figure 5 View of the old bridge being demolished (source: LaVoixduNord)

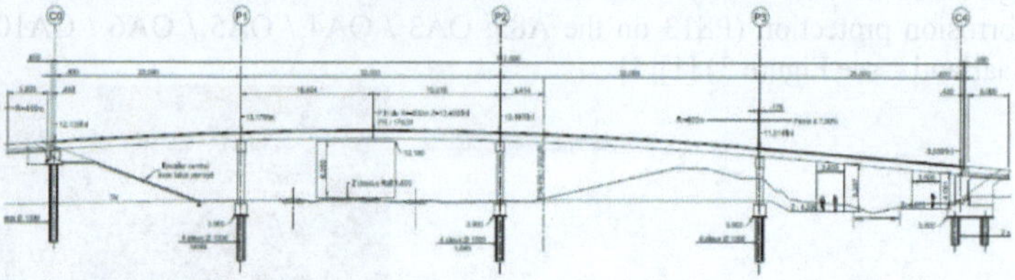

Figure 6 Longitudinal view of the new bridge

The reinforced concrete structure in place was built in 1977, then demolished in March 2011 for safety reasons following the failure of one of its supports. A temporary pedestrian and cyclist overpass was installed. The old structure, with a massive design in pre-stressed reinforced concrete and a neglected appearance, contributed to a negative perception of the area (see Figure 5). The new projects therefore had the ambition to and requalify the green space between the road and the railway line.

2.2 The new project

The bridge crosses a two-track railway line parallel to an artificial waterway used in older times - the Mardyck channel. The ramps are in S-shape both on West and East side so that bridge crosses the obstacles perpendicular to them, without bias for the piers and abutments. The bridge has a total length of 116m with spans of 23m – 32m – 32m – 25m. Piers and abutments are of the classic type with reinforced concrete walls and pile foundations (see Figure 6). The road axis in correspondence of the bridge has a vertical radius of 600m to ensure a free clearance of 6.5m over the railway line and to smoothly connect to the ramps (see Figure 7). The deck has a total width of 13.3m and carries 2 traffic lanes, a sidewalk for pedestrians and a cycleway on both sides of the road. The thickness of the structural deck is 0.9m, thus resulting in a slenderness of the structure of 1:35 on the central span of 32m. The deck is formed by four longitudinal girders (centre distance 3.6m) in the form of a closed box sections. Thanks to the torsional rigidity of these elements, it is possible to avoid any intermediate crossbeam in span, simplifying the structural system. The slab is realized by prefabricated elements with the the same width as the deck and an average thickness of 35cm. In correspondence of the abutments and piers, the beams are connected by reinforced concrete beams.

Figure 7: View of the access ramp on the West side (left) and the central span overpassing the railway line (photo: R. Zanon)

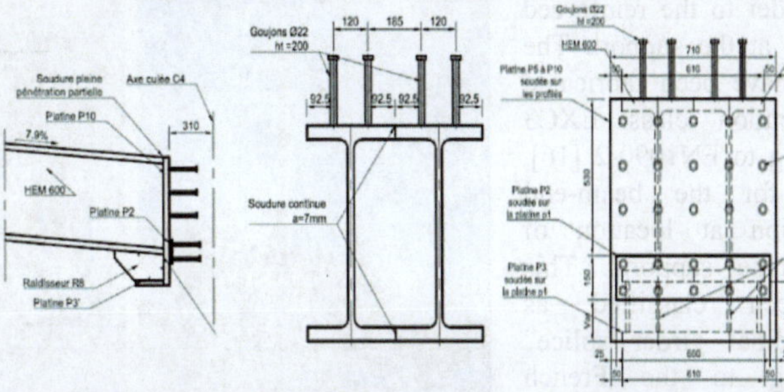

Figure 8: Details about steelwork of the steel girder ends

Figure 9: Steelwork fabrication in the workshop (photo: R. Zanon)

2.3 Execution of the steelwork

The main beams are assembled from two rolled sections HEM600 for the central spans, HEB600 for the edge spans in S460M according to EN10025-4 [15][1][15]. They are produced in the mill directly at the final length to avoid any butt-weld splicing. The sections were bent with a hydraulic press to realize the longitudinal profile of the project: on the main span of 32m the cambering is 410mm, including pre-cambering and permanent curving, and is ensured with a tolerance of +/-5mm, so significantly stricter than the usual fabrication tolerances of L / 1000 (around +/- 32mm) according to standard [16]. The boxes are formed by longitudinal welding of the edges of the flanges with a continuous 7mm fillet weld (see Figure

8, Figure 9). The box is closed by a 40mm thick vertical header plate welded in partial penetration. The plate is doubled in correspondence of the bottom flange by an additional plate 20mm thick to distribute the compression forces in reinforced concrete. Horizontal studs on the vertical plate ensure the transfer of the shear from the steel girder to the reinforced concrete at the support. The girders have been fabricated in execution class EXC3 according to EN1090-2 [16], except for the beam-end preparation at location of intermediate supports. This has been classified as "longitudinal girder splice" according to the French national guidelines for bridges and executed in class EXC4. An anti-corrosion system type ACQPA C4ANV has been applied in the workshop on the whole exposed surface (see Figure 10).

Figure 10: Corrosion protection of the box girder (photo: R. Zanon)

Figure 11: Realization of foundations (left); temporary scaffolding for the pier close to the railway line (right) (photo: R. Zanon)

2.4 Execution on the construction site

Following the demolition of the existing structure, abutments and supports were made on deep foundations already realized during the year 2015 (see Figure 11). Afterwards the steelwork was transported by exceptional rail convoy in Dunkirk and carried to the job site by special truck transport. The girders were placed by crane on the central span during a night interruption of the railroad traffic at the beginning of 2016 (see Figure 12). The weight of the elements is less than 20 tons: the requirements in term of lifting capacities are therefore very reasonable.

Figure 12: Steelwork erection phase (photo: R. Zanon)

The preparation of the end girders was designed to ensure a surface of support also for the temporary phase but still being compatible with the concrete scaffolding system (see Figure 12, Figure 17). Prefabricated slabs corresponding to the entire width of the deck were prepared, including railings fixed on the sides to guarantee the safety of the workers immediately (see Figure 13). It is necessary to guarantee precision in the design and execution precise layout to be able to fit the prefabricated elements on site: on this site the thing was carried out are excessive complications (see Figure 17).

Figure 13: Erection of the prefabricated concrete slab elements (photo: R. Zanon)

Figure 17: Details of the concrete cross beam (photo: R. Zanon)

2.5 Architecture of the structure

Figure 19: Triangle bridge in Dunkerque in its landscape (photo: R. Zanon)

The steel girders are painted in a dark colour, giving the deck an outstanding slender shape. The white metallic cap has the shape of a seagull wing, a typical bird of the Pas de Calais region, giving lightness and ensuring continuity with the ramps (see Figure 19). The traffic lamps are gently inclined towards the inner of the bridge contributing to an overall pleasant feeling. The choice of designer therefore appears as very successful, as it also offers a pleasant view from the new park (see Figure 19). On the 15th July 2016 the majors of the cities of Dunkirk, Saint-Pol-sur-Mer and Mardyck) organized fireworks on the occasion of the National Day in the green space next to the new Triangle Bridge.

3 MOSELLE BRIDGE IN TONNOY

3.1 Situation

In the town of Tonnoy in the Nancy region (France, Department of Meurthe and Moselle, 54) the departmental road D74 crosses the Moselle river in a south-east direction to join the D570 in direction to Flavigny-sur-Moselle. The first modern bridge was built in steel at the beginning of the 20st century, crossing the river with 4 independent spans of around 22m with a total length of 90m. It was realized with arch-shaped steel truss girders positioned laterally to the road with a total deck width of about 9m. This first bridge was destroyed during the first phase of the second world war (see Figure 20). The successive deck was initially rebuilt in 1942, but substantially refurbished in 1953 after the war. It could partially reuse the piers of the previous bridge and similar road layout with four isostatic spans realized by prestressed reinforced concrete beams under the roadway (see Figure 20). This nearly 80 years old bridge was under close surveillance by local authorities. Regular inspections found several disfunctions, raising doubts about the safety level of the structure. In 2012 the weight of the crossing vehicles was limited to 30 tons and the speed to 30 km/h to minimize dynamic effects. The narrow cross section had a 5.10m carriageway and two 0.70m sidewalks which were insufficient to present conditions. Based on all these elements, the demolition of the old bridge and reconstruction with a new bridge were decided. During the construction the connection between the two river shores was disrupted, dividing the village of Tonnoy in two parts and forcing the inhabitants to a long bypass to cross the river. For this reason the construction site had to be as short as possible.

Figure 20: Moselle bridge in Tonnoy: ancient pre-war lateral truss girder deck and post-war prestressed concrete deck

3.2 The new project

The new bridge was planned with only two piers in the river instead of four, decreasing the foundations impact in the Moselle River and ensuring an improved hydraulic flow, relevant. It results a total length of 96.5m divided in 3 continuous spans replacing the previous 90m bridge divided in 4 isostatic spans. The total deck construction height of 1.3m results from the existing road axis and the 100-year flood line of the Moselle River. The central intermediate span is 35m whereas the lateral spans are 30m each (see Figure 21). The new bridge will improve the safety by installing a 2.70 m lane exclusively dedicated to pedestrians and cyclists; the roadway is widened for more comfortable vehicle circulation, increasing from the previous 5.10 m to 6.0 m carriageway. This modification results a new deck width of 10.5m. The longitudinal profile has a constant vertical radius of 2375m. The structure is straight in relation to the abutments and supports. The steel-concrete deck is realized with the typology described in chapter 1 (see Figure 26). The cost of the whole operation is four million euros, financed entirely by the departmental council of Meurthe-et-Moselle.

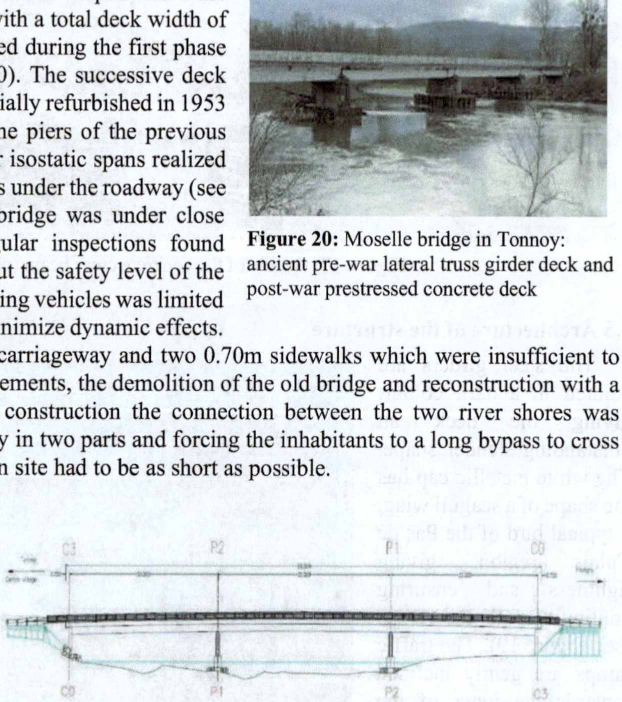

Figure 21: Longitudinal view of the new bridge

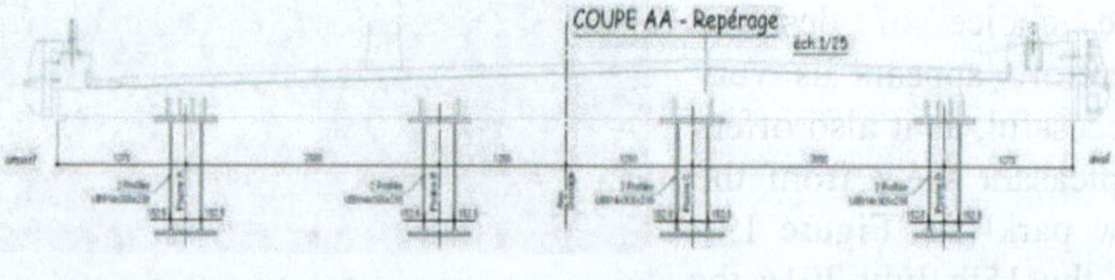

Figure 22: Cross-section of the new bridge

3.3 Steelwork execution

Figure 23: Details about steelwork fabrication at the girder ends (photo: R. Zanon)

The main beams are assembled from two profiles UB914x305x238 according [14] in S460M according to EN10025-4 [15][1][15]. The production and finishing process is very similar to the one presented for the previous bridge. Once beams are rolled the final length to avoid any splicing, they are curved to obtain the needed pre-cambering value to compensate permanent weight deformation and to obtain the final longitudinal axis radius. On the main span of 35m the cambering is 350mm. The boxes are formed by longitudinal welding of the edges of the wings with a continuous 5mm fillet weld.

Figure 24: Realization of the abutments (photo: R. Zanon)

Figure 25: Realization of the intermediate piers and supports (photo: R. Zanon)

The details of the connection of the steelwork with the concrete cross beams were discussed with the Design Office and the Project Manager. In particular, the web of the sections continues up to the concrete cross beam to have a direct transfer of the vertical shear force into the axis of the support. A thick vertical plate is welded at the end of the lower flange to diffuse the horizontal compression forces in the concrete; this plate is fixed with single fillet welds (a = 5mm) due to the fact that it is fully compressed (see Figure 27). Some other plates have been added but only with constructive functions and in particular to ensure the air- and watertightness of the box girder. The shear studs on the upper flange are of 25mm diameter to minimize their number, since the common 22m diameters would have required around 25% more units. This reduced number highly simplifies the fitting of the prefabricated concrete slabs and the reinforcement layout in the shear pockets.

The girders have been fabricated in execution class EXC3 according to EN1090-2 [1][16], except for the beam-end preparation at location of intermediate supports. This has been classified as "longitudinal girder splice" according to the French national guidelines for bridges and executed in class EXC4. An anti-corrosion system type ACQPA C4ANV has been applied in the workshop on the whole exposed surface.

The experience of the previous bridge has permitted to decrease significantly the production efforts and related costs. The size of the welding is really minimal, since their structural function is limited, and the fact to choose higher diameter stud for the upper flange was also very important. In this case no lifting hook was welded on the upper flange (a location where space is really limited) but a lifting bar was bolted on the bottom flange. The details developed for this specific bridge make the typology even more robust and economic.

3.4 Execution on the construction site

A temporary road was built from the south riverbank for the demolition of the existing bridge. This road also gave the access to realization points of intermediate piers. Using sheet pile cofferdams, abutments and piles were made on deep foundations (see Figure 24 and 25).

In September 2019, the steelwork was delivered to the construction site in three consecutive days. Each day the girders corresponding to one span were delivered and directly erected from the truck to the span. The maximum lift weight was 22 tons. The beams were lifted to temporary supports (painted in red, see Figure 25). On these supports were also provided lateral bracing between to beams to block their rotation.

Two types of prefabricated slabs were used: a lateral one placed on edge and internal beam to form the cantilever, and a central element between the two intermediate beams having double length to reduce the longitudinal joints (see Figure 26). At supports the metallic longitudinal girders were linked by reinforced concrete cross beams. The reinforcement of the cross beams was installed before placing the prefabricated slabs (see Figure 26). The pouring of concrete was done in three phases: first the lateral spans with the reinforced concrete beams on the abutments, then the central span and finally the reinforced concrete cross beams with the deck above it. With this simple principle, it can be ensured that the self-weight of the structure is overtaken by the steel girder as simply supported elements and that the reinforcement in the hogging moment region is stressed only for the traffic loads.

The bridge has been opened to traffic early 2020, after less than one year of work on site.

Figure 26: View of the bridge over the Moselle in Tonnoy during achievement of the structural part (photo: R. Zanon)

4 CONCLUSIONS

This article presents a possible evolution of steel-concrete deck to meet the requirements of bridge decks in the medium span range, bringing several advantages compared to the current solutions. These aspects are clearly showed for the case of the two projects presented. It is interesting to conclude reporting the positive feedback of the bridge owner of Moselle bridge in Tonnoy.

"This is an innovative technique which has the advantage of limiting steel construction operations on the construction site, thereby improving the quality and ensuring a better durability of the structure. This technique also reduces the lead time and therefore the overall construction cost. " Department of Meurthe & Moselle, information sheet published for the reconstruction of the Bridge over the Moselle at Tonnoy [17].

5 ACKNOWLEDGMENTS

We are grateful to our customers and our interlocutors for the trust: Benjamin DUPONT, Marc SOMMA (Eiffage TP, Agence Dunkerque), our main contacts for the reconstruction of the Triangle Bridge in Dunkirk. Florent CERCLEY, Adrien LOISY, Sébastien ROTA (Bouygues TP, Agence Nancy), our main partners for the the Moselle Bridge in Tonnoy. Gabriel SINS, Cyril COTTEY (COGECI design office) for good technical collaboration. Damien CHAMPENOY (CEREMA, Metz) for his advice.

REFERENCES

[1] D. Brazillier, G. Fontaine, Un nouveau domaine d'emploi pour les ponts mixtes : Les petits ouvrages, Bulletin Ponts Métalliques N°13, 1989.

[2] J.M. Hottier, J.P. Persy, P. Trouillet, A. Rouby, R. Marcellier - Ponts d'Ehnwihr – Innovation dans le domaine des petits ponts, Préfabrication intégrale du tablier, Bulletin Ponts Métalliques N°14, 1990.

[3] P. Nicodème, P. Meurisse - Des tabliers mixtes industrialisés pour 22 PS sur A16 Amiens – L'Isle Adam, Bulletin Ponts Métalliques N°17, 1994.

[4] Guide SETRA – SNCF, Ponts routes avec tablier à poutrelles enrobées, 1997

[5] S. Barbaux, J-M. Baryla, F. Chevallier, J. Petitjean, A. Piquet, PS Mixte en BHP à connexion différée. Le PS 13 sur A85, Ouvrages d'Art N°36, Decembre 2000.

[6] M. Hever, F. Gluzicki, Poutres en acier avec joints en béton, BTP Grands Travaux N°158, 2002.

[7] G Sedlacek, B Hoffmeister, H Trumpf - Composite bridge design for small and medium spans - Final Report European commission Technical steel report, EUR20583, 2003.

[8] Chabrolin B., Kretz T., Laravoire J. Projet national MIKTI - Ponts Mixtes Acier-béton, 2010.

[9] L. Davaine, Presentation et documents transmis lors du Seminaire Preco+, Etat de l'art des ouvrages métalliques en France, 2013.

[10] R. Zanon, W. Ochojski, O. Hechler, Paweł Klimaszewki, W. Lorenc, Road bridges WD7-WD8 on Route Sucharskiego, Gdansk, Poland - Prefabricated composite beams with high-strength hot rolled sections, Technical Report, IABSE Structural Engineering International (SEI), 2 – 2014.

[11] F. Paolacci, S. Bursi, R. Cremona, S. Stathopoulos, B. Hoffmeister, et al. SEQBRI: Performance based earthquake engineering of Short- and medium-span bridges, Final report, Grant agreement RFSR-CT-2012-00032, 2015.

[12] Bruno Briseghella, Tobia Zordan - An innovative steel-concrete joint for integral abutment bridges, Journal of Traffic and Transportation Engineering (English Edition)Volume 2, Issue 4, August 2015

[13] R. Zanon, D. Zaganelli, Wide Flange Beams: Trends Towards Higher Steel Performance, Current Trends in Civil & Structural Engineering, 2019

[14] EN10365 - Dimensions and masses - Hot rolled steel channels, I and H sections, 2017.

[15] EN10025 - Hot rolled products of structural steels, 2019.

[16] EN1090 - Execution of steel structures and aluminium structures, 2012.

[17] Departement Meurthe & Moselle, Reconstruction du Pont sur la Moselle Tonnoy RD74

1

COMPOSITE COLUMNS

9TH INTERNATIONAL CONFERENCE ON COMPOSITE CONSTRUCTION IN STEEL AND CONCRETE
M. Knobloch, U. Kuhlmann, W. Kurz, M. Schäfer. (Eds.)
Online, Germany, July 27-29, 2021

ASSESSMENT OF GENERAL METHOD FOR COMPOSITE COLUMN DESIGN IN EN 1994-1-1 AND COMPARISON WITH SIMPLIFIED METHOD

Markus Schäfer*, Qingjie Zhang*, Pellumb Zogu*, Marco Bergmann, Ozgun Ergun***

* University of Luxembourg, Department of Engineering, Structural Engineering & Composite Structures

e-mails: markus.schaefer@uni.lu, qingjie.zhang@uni.lu, pellumb.zogu@uni.lu, ozgun.ergun@uni.lu, , , ,
** HRA Ingenieurgesellschaft mbH, Bochum, m.bergmann@hra.de

Keywords: Composite, Columns, General Method, Simplified Method, Non-linear Design Methods

Abstract. *Composite column design in accordance with EN 1994-1-1 can be computed by the general method or by simplified methods. For the general method, EN 1994-1-1 refers to a comprehensive FE-Analysis. However, the recent Eurocode 4 does not provide sufficient guidance for the application of the general method and there is not a harmony in the literature on how to do it. Additional information may be found as non-contradictory complementary information (NCI) in National Annexes (NA). This paper points out the procedure for the general method and refers to the changes and additional information provided by the draft of prEN 1994-1-1:2021. Furthermore, the modelling and computation by FE-Analysis for several software applications are presented by implementing specified initial bow imperfections and predefined residual stresses for the steel section. The results obtained with the general method's approach are compared with that calculated by simplified method by the European buckling curves. In addition, the impact of different stress-strain curves for concrete and strain limits introduced by Codes are investigated.*

1 Introduction

The European design code for composite structures in steel and concrete EN 1994-1-1 [1] provides three different methods for the proof of composite columns and composite compression members:

a) Simplified Method based on the European buckling curves for centric loading. The association to the European buckling curves is based on the cross-section type, buckling axis and degree of reinforcement. Depending on the non-dimensional slenderness the plastic design resistance $N_{pl.Rd}$ of the column is reduced by the reduction factor χ, similar to the design of columns according to EN 1993-1-1.

b) Simplified Method based on the design moment taking into account theory second order effects may be applied for centric and eccentric loading. For this analysis an effective flexural stiffness is used. Column imperfections are considered by equivalent imperfections. These equivalent imperfections depend on the assignment of the cross-section to the European buckling curves. The proof for design will be completed on cross-section level based on the plastic N-M-Interaction relation.

c) General Method represents a more sophisticated approach based on a non-linear analysis. The structural stability analysis shall take account of geometrical and physical non-linear effects including residual stresses, geometrical imperfections, local instability, creep and shrinkage of concrete. The physical non-linearity has to consider cracking of concrete as well as yielding of structural steel and reinforcement steel.

Current EN 1994-1-1 does not provide detailed guidance for the application of the general method. The National Annexes, e.g. DIN EN 1994-1-1/NA [2], and the literature include more information for the application of non-linear design, e.g. [7]. The draft of second generation for Eurocode 4: prEN 1994-1-1 [3] is also filling this gap, detailed guidance have been integrated to fulfil the requirements for ease of use.

Content of this paper is to provide information for application of the general method allowing an assessment of the results based on prEN 1994-1-1. For the example of a rectangular concrete encased composite column (CEC) the analysis according to the general method is discussed. The FE-Analysis (FEA) are provided by various software applications and the results are compared. Furthermore for the cross-section resistance N-M-Interaction curve is computed by plastic and strain-limited resistance. The impacts on the safety factor and the design resistance are investigated.

2 General method according to Eurocode 4

The application of simplified methods requires double symmetrical column cross-section constant over the entire column length. The methods are limited for the design of composite compression members with concrete encased rectangular sections (CEC), partially encased sections (PCEC) and concrete filled rectangular and circular tubes (CFST). Design is based on the full-plastic resistance of the cross-section. Furthermore the contribution of the steel section is limited ($0{,}2 \leq \delta \leq 0{,}9$) and the non-dimensional slenderness shall be smaller than 2,0, while the reinforcement ratio is restricted as well as the c/t relation in regard to local buckling. The concrete compression strength is capped for classes C20/25 until C50/60 and steel grades to S235 until S460. However, many types of composite column do not fulfil those requirements; therefore more elaborated design method becomes necessary.

The general method is characterised by two steps, on the one hand a global analysis of the investigation on column behaviour by an incremental finite element analysis (FEA) and on the other hand, the analysis of the cross-section resistance by the N-M-Interaction curve to determine the overall safety factor. The method can be applied for any type of composite cross-section and any combination of materials covered by Eurocode 4. Thereby the contribution of the structural steel by the δ-factor is not relevant, this factor was introduced for the simplified method to distinguish composite columns from concrete and steel columns [8]. According to the draft of prEN 1994-1-1 [3] concrete compression strength up to C70/85 and reinforcement up to B700 in reference to prEN 1992-1-1and structural steel grade until S690 according to prEN 1993-1-1 can be used if more advanced design methods other than plastic resistance are applied. The FEA shall cover all relevant failure modes. Non-linear effects shall be considered in any direction in which failure might occur, if they affect the structural stability significantly. The design shall ensure that instability does not occur for the most unfavourable combination of actions at the ultimate limit state and that the resistance of individual cross-sections subjected to bending, longitudinal force and shear is not exceeded [3]. Full composite action up to failure may be assumed between steel and concrete components of the member if no impact from slip is attended and adequate shear interaction is provided.

In general, it is possible to apply the equivalent imperfection for the FEA. However, the application of only geometrical bow imperfection and in addition residual stresses results in more economic design. For the geometrical imperfection a bow with a maximum value of $w_o = L/1000$ may be used (Figure 2-1c). This value is also given by prEN 1993-1-1 and approved for the design of steel columns [9]. Reference is given to the discussion in section 3.2 and [17]. The residual stresses are resulting from the rolling and cooling process of the steel profile and welding if applicable. Initial they were introduced by [13], today also taken over in prEN 1993-1-14 [6]. Figure 2-1 shows the residual stresses for hot rolled steel sections.

The general method according to prEN 1994-1-1 is applying an overall safety factor and an additional partial factor for model uncertainties, for resistance and for action effects as given by equation (1).

$$\gamma_{Rd}\ \gamma_{Sd}\ E\left(\gamma_g\, G_k + \gamma_q\, Q_k\right) \leq R\left(\frac{R_m}{\gamma_0}\right) \tag{1}$$

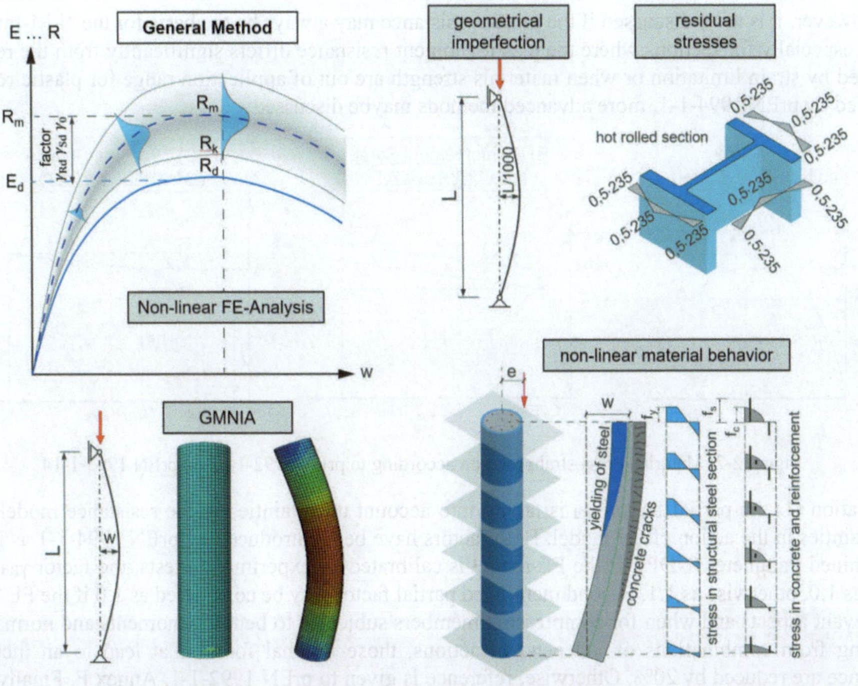

Figure 2-1: Non-linear Analysis of composite columns acc. to prEN 1994-1-1 – General Method [18]

Thereby R_m is the resistance of the compression member obtained from the FEA, that is based on the mean values of the material strength and the non-linear material stress-strain relationship according to prEN 1992-1-1 for concrete and reinforcement and prEN 1993-1-14 for steel, Figure 2-2. Current Eurocode 4 does not provide any information about the material mean values. By the update to prEN 1994-1-1 appropriate references are provided. For the concrete compression strength the mean values of f_{cm} is given by prEN 1992-1-1, Table 5.1 and Table A.1, for reinforcement the value f_{sm} is presented by prEN 1992-1-1, Table A.1, while for structural steel the mean values f_{ym} is to be found by prEN 1993-1-1, Table E.1. However, the tensile strength of concrete shall be neglected for the determination of resistance while the influence of tension stiffening of concrete between cracks on the flexural stiffness may be taken into account. Due to the use of the mean values for the materials, the numerical analysis represents an average of the column load-bearing behavior. The load-deflection path in Figure 2-1 (General Method) shows the variation of the curve when applying characteristic or design values. To determine the design level, the resistance R_m has to be reduced by an overall factor for safety.

According to EC4 the overall partial factor γ_0 can be determined by the full plastic cross-section N-M-Interaction relation obtained by combined bending and compression, Figure 2-3. Thereby the interaction curve should be developed once by using the mean values $f_{i,m}$ and once using the design values $f_{i,d}$ of the material resistances. Thereby the partial safety factors for structural steel (recommended values for γ_{M1} is 1,1, NA may also define other values, e.g. acc. to DIN EN 1993-1-1/NA γ_{M1}=1,1), 1.15 for reinforcement steel and 1.5 for concrete shall be used. For the determination of concrete compression resistance according to prEN 1994-1-1 ($f_{cd} = \eta_{cc} \cdot k_{tc} \cdot f_{ck}/\gamma_c$) the proposed value for k_{tc} is 1,0 for the concrete compression strength at 28-days, unless other values are given by the National Annex to EN 1994-1-1. While the maximum value of η_{cc} is smaller than 0,85 for determination of cross-section resistance, other than elastic resistance and for cross-sections other than CFST. For a defined load combination N_{Ed} and M_{Ed} the overall partial factor γ_0 may be determined by the quotient of the vectors $R_{pl,m}$ and $R_{pl,d}$, Figure 2-3.

However, it is to be discussed if the plastic resistance may always be the basis for the N-M-Interaction curve, especially for sections where the plastic moment resistance differs significantly from the resistance obtained by strain limitation or when materials strength are out of application range for plastic resistance provided by prEN 1994-1-1, more advanced methods maybe discussed.

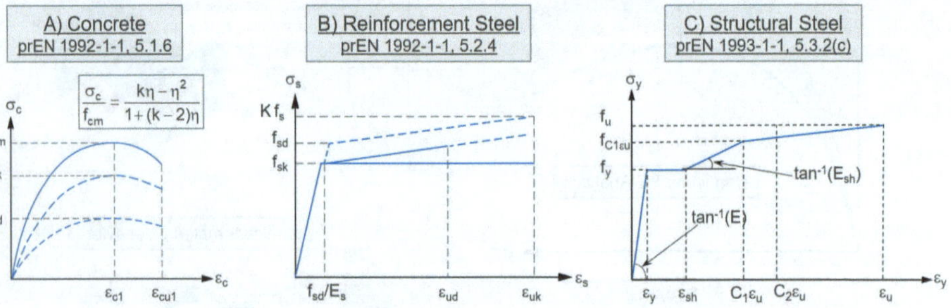

Figure 2-2: Material stress-strain relation according to prEN 1992-1-1 and prEN 1993-1-14

In equation (1) the partial factor γ_{Rd} is taking into account uncertainties in the resistance model and γ_{Sd} uncertainties in the action effect model. Both factors have been introduced by prEN 1994-1-1 as National Determined Parameter (NDP). If the FE-model is calibrated by experimental tests, the factor γ_{Rd} may be taken as 1.0, otherwise as 1.1. Second mentioned partial factor may be considered as 1.0 if the FEA covers all relevant aspects and when for compression members subjected to bending moments and normal forces resulting from combinations of independent actions, those internal forces that lead to an increase of resistance are reduced by 20%. Otherwise, reference is given to prEN 1992-1-1, Annex F. Finally, it is to be proofed that the design values of the action effects increased by the partial factors γ_{Rd} and γ_{Sd} are smaller than the global resistance R according to equation (1). The verification of the column resistance shall be performed for the relevant critical cross-section and decisive load combination [7].

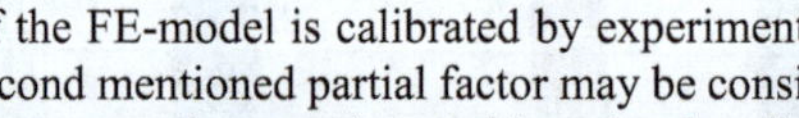

Figure 2-3: Non-linear Analysis – determination of overall factor [3]

3 Application of General Method – Case Study

3.1 Objective and Procedure

In the following case study typical composite encased columns (CEC) are investigated. The rectangular composite cross-section with concrete class C50/60 may be calculated by the simplified methods acc. to EN 1994-1-1 while the application of concrete class C70/85 and reinforcement class B700 are outside of the scope for the simplified methods based on the plastic cross-section resistance. For the FEA commercial software ABAQUS [15] is applied. It can use beam, shell and volume elements as well as combinations of those elements. The outcomes are compared to the results achieved by using software VBST [10], OPENSEES [16] and SL.com [14]. OPENSEES [16] (Open system for earthquake engineering simulation) was developed at the University of California in 2000. The software uses 2D beam elements with fibre sections, so also 3D simulation becomes possible with simplified but fast solver procedure promising efficient investigation in parametric studies. VBST is based on the work of BERGMANN [10] and was used extensively for the development of first German and harmonized European codes for the determination of the initial bow imperfections for the simplified design method [11], [12]. SL.com [14] allows cross-section analysis for composite members, to calculate cross-section resistance, moment-curvature relation, partial-shear diagram as well the M-N interaction curve. In addition, it supports full column nonlinear buckling analysis resulting in load-deflection curve as well as the deformation and strain/stress information of entire section. In SL.com the residual stresses cannot be applied. For the most complex modelling by Abaqus shell elements are applied for steel profile while concrete is modelled by solid elements. The reinforcement and stirrups are modelled by beam elements. The contact between steel and concrete is realised by general contact (explicit), with hard contact, for tangential behaviour a friction coefficient of 0,5 is used. The model is able to consider confinement effects. For OPENSEES confinement can be considered by definition of different concrete areas where the confinement can be computed by predefined analytical approach. For the other applications, it is not possible to take into account those effects.

Table 3.1: Example - Composite Cross-section and material characteristics

Cross-section	**Material Data acc. to drafts of 2nd generation of Eurocodes**		
A1 – C50/60, B500	**Concrete**	**Reinforcement**	**Structural Steel**
A2 – C70/85, B700	prEN 1994-1-1	prEN 1992-1-1	prEN 1993-1-1
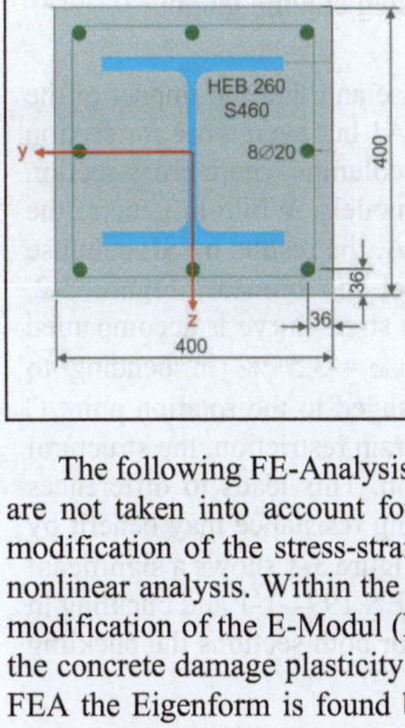 	**C50/60** E_{cm} = 36.773 N/mm^2 f_{ck} = 50 N/mm^2 γ_c = 1,5 f_{cd} = 28,3 N/mm^2 f_{cm} = 58 N/mm^2 **C70/85** E_{cm} = 40.590 N/mm^2 f_{ck} = 70 N/mm^2 f_{cd} = 38,7 N/mm^2 f_{cm} = 78 N/mm^2	**B500** f_{sk} = 500 N/mm^2 f_{sd} = 435 N/mm^2 f_{uk} = 540 N/mm^2 f_{sm} = 538 N/mm^2 f_{um} = 581 N/mm^2 **B700** f_{sk} = 700 N/mm^2 f_{sd} = 625 N/mm^2 f_{uk} = 756 N/mm^2 f_{sm} = 754 N/mm^2 f_{um} = 814 N/mm^2	**S460** E_a = 210000/Nmm^2 f_{yk} = 460 N/mm^2 f_{yd} = 418 N/mm^2 f_{uk} = 540 N/mm^2 f_{ym} = 529 N/mm^2 f_{um} = 594 N/mm^2

The following FE-Analysis are related to short-term loading. Long-term effects as creep and shrinkage are not taken into account for the presentation. However, due to application of shrinkage strains and modification of the stress-strain curve by the creep coefficient also these effects can be included in the nonlinear analysis. Within the simplified design method of Eurocode 4 the creep is taken into account by modification of the E-Modul (E_{cm} => $E_{c,eff}$). In Abaqus also the concrete tension resistance is activated by the concrete damage plasticity model, while it is neglected for the other software applications. Within the FEA the Eigenform is found by a linear buckling analyses for calculation of the assumed imperfection shape, then residual stresses are applied on the steel section, finally the geometrical and physical nonlinear analysis is conducted.

3.2 Impact of slenderness on buckling resistance - comparison of software application

The relative slenderness is determined according to equation (2), based on the plastic resistance of the section $N_{pl,Rk}$. The Eulerian elastic critical axial force $N_{cr,z}$ is computed by equation (4) with the effective flexural stiffness defined by EN 1994-1-1 as given by equation (5). The reduction of the bending stiffness for concrete by factor 0,6 takes into account nonlinear effects as concrete cracking. As mentioned before, the creep effects is not considered for these examples. According to Eurocode 4 creep can be taken into account within the simplified design methods by the effective E-Modul $E_{c,eff}$, equation (6), therefore the distribution of the acting load to permanent and exploitation load have to be defined.

$$\bar{\lambda} = \sqrt{\frac{N_{pl,Rk}}{N_{cr}}} \leq 2.0 \tag{2}$$

$$N_{cr,z} = \frac{E \cdot I_{z,eff} \cdot \pi^2}{L^2} \tag{4}$$

$$EI_{eff} = E_a I_a + E_s I_s + 0.6 E_{cm} I_c \tag{5}$$

$$E_{c,eff} = E_{cm} \cdot \frac{1}{1 + \frac{N_{G,Ed}}{N_{Ed}} \cdot \varphi} \tag{6}$$

In Figure 3-1 the relation of the buckling load $N_{b,k}$, computed by nonlinear FE-Analysis based on the characteristic material resistances for buckling in about the z-axis (in y-y direction) to the characteristic plastic resistance $N_{pl,Rk}$ is presented against the relative slenderness. In addition, the results of the different software applications are compared for the initial geometrical imperfection of L/1000 and application of residual stresses with exception of the investigation SL.com where only geometrical imperfection is considered. The FEA are provided for centric loading. Several benchmarks for the calibration of the model based on available test data for CEC column have been performed before, also a convergence studies have been investigated and the number of elements have been adapted with increasing column length.

It can be concluded, that the results from different software are very close and that the impact of the residual stresses for the presented CEC becomes not significant. For section A1 but even more for section A2 an important difference can be figured out for the results by SL.com for columns where cross-section resistance becomes mandatory. This is related to use of different material models. While in general the nonlinear material for concrete according to Figure 2-2 is applied for the FEA, the results by SL.com use the parabola-rectangle concrete stress-strain curve and for structural steel the bilinear, Figure 3-2. According to EN 1992-1-1 the application of the concrete parabola-rectangle stress curve is accompanied by the limitation of strains under pure compression, it reduces from $\varepsilon_{cu2} = -3{,}5$ ‰ in bending to $\varepsilon_{c2} = -2{,}0$ ‰. To achieve so, the rotation point B (in bending) should be changed to the rotation point C (Figure 3-2) until reaching tension at the bottom concrete fiber. Due to this strain restriction, the structural steel and reinforcement cannot or not completely yield in pure compression. This leads to differences compared to full plastic resistance. While with increasing moment, the bearing resistance may benefit by the contribution of the stress-block ($\varepsilon_{c2} \leq \varepsilon_c \leq \varepsilon_{cu}$) for concrete. Furthermore Figure 3-1 shows a significant difference between the buckling curve c, mandatory for CEC columns acc. to EN 1994-1-1 and buckling in about the weak-axis z, to the computational determined buckling capacity. For both sections the buckling capacity is presenting values around and above buckling curve a.

When comparing the proposed equivalent imperfection of L/150 associated to buckling curve c by Eurocode 4 with the applied imperfection of L/1000 for the general method and in view of the fact that the residual stresses do not play a major role for the here investigated CEC, the reason for this difference can easily be concluded. In [17] the application range for the geometrical imperfection of L/1000 is discussed. This imperfection is originally deviated for steel columns. The question arises, if this approach will be on the safe side for all kind of composite columns independently of the steel contribution and the column shape. On the other hand, EN 1992-1-1 and EN 13670 provide more important imperfections for reinforced concrete compression members. However this approach is too conservative for composite columns with an important contribution from the structural steel section. Therefore an alternative hypothesis by interpolation of the imperfection given by EN 13670 and L/1000 is investigated [17] by about L/500 for a δ-value of 0,5 (A1, for A2 simplified it is similarly applied). The resulting buckling capacity fills the area between buckling curve a and c, Figure 3-1. With respect to this, further investigation and guidance will become necessary but will not be discussed in detail here.

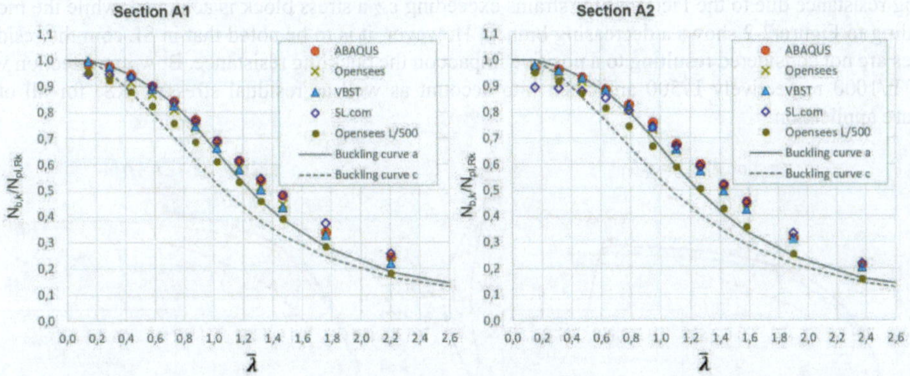

Figure 3-1: Comparison of FE-Analysis for different column length – flexural buckling in y-direction

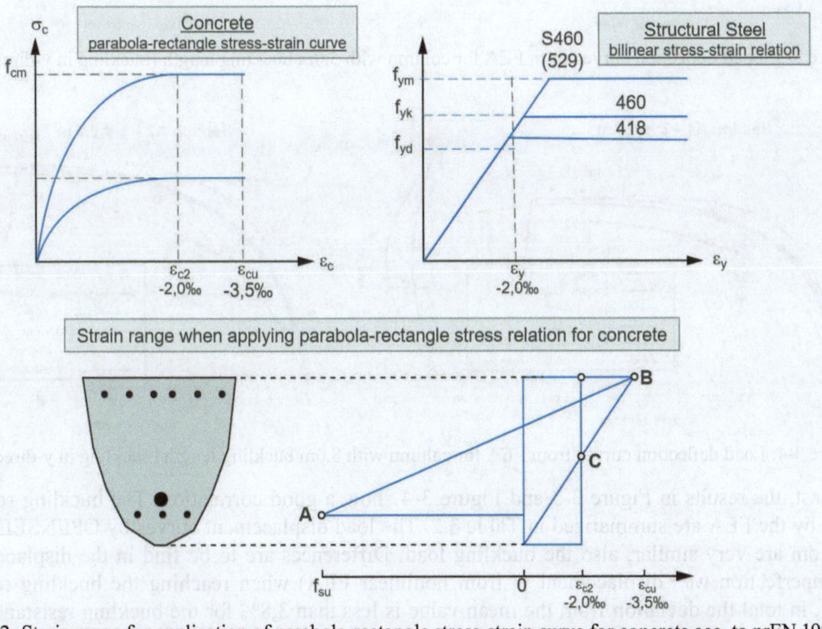

Figure 3-2: Strain range for application of parabola-rectangle stress-strain curve for concrete acc. to prEN 1992-1-1

3.3 FE-Analysis for composite columns

For the column examples of cross-section A1 and A2 (Table 3.1) FE-Analysis are investigated for a column length L equal to 5,0 m (Figure 3-3) and 8,0 m (Figure 3-4). The FEA follows the requirements for the general method in prEN 1994-1-1, thus all material data are based on the mean values, as given by

Table 3.1. Loading is similar to the case studies presented in section 3.2 for centric loading. For the FEA the loading is stepwise increased. Only with Abaqus the post buckling behavior can be investigated. The other software applications represent good result until failure load. For the calculation with the software SL.com both material stress-strain curves for concrete according to Figure 2-2 (SL.com – L/1000 in Figure 3-4) and Figure 3-2 (SL.com – L/1000/concrete PR in Figure 3-4) are investigated for the eight-meter column. Whereby the parabola-rectangle stress strain curve provides with increasing moment higher bearing resistance due to the fact, that for strains exceeding ε_{c2} a stress block is achieved, while the model according to Figure 2-2 shows a decreasing branch. However, it is to be noted that in SL.com the residual stresses are not considered resulting to a positive impact on the buckling resistance. Bow imperfection with w_0 of L/1000 respectively L/500 are taken into account as well as residual stresses (RS) for all other software application.

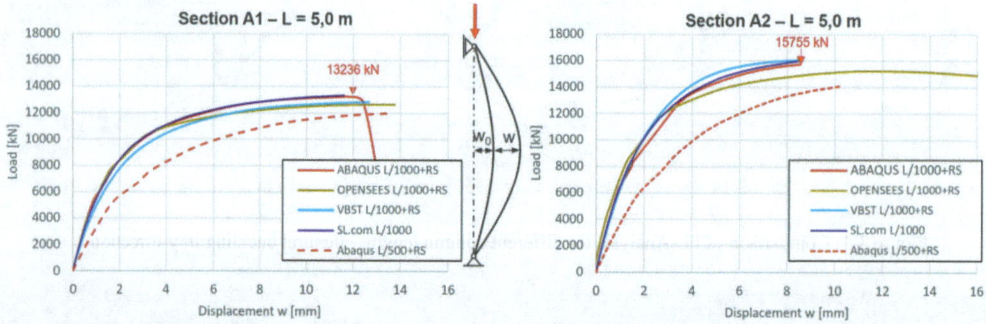

Figure 3-3: Load deflection curves from FEA for column with 5,0m buckling length (buckling in y-direction)

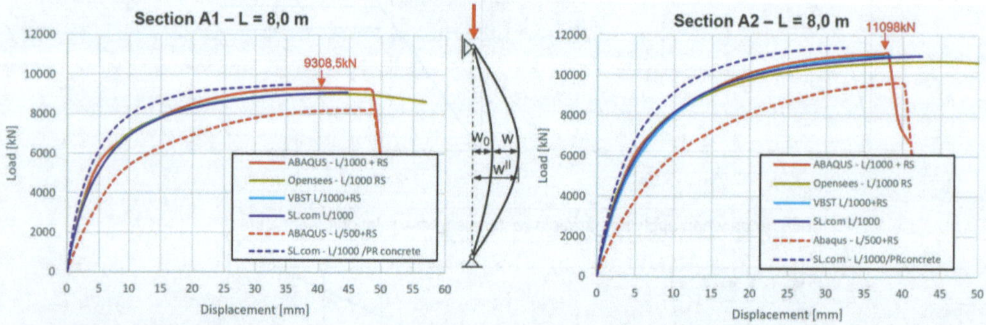

Figure 3-4: Load deflection curves from FEA for column with 8,0m buckling length (buckling in y-direction)

At least, the results in Figure 3-3 and Figure 3-4 show a good correlation. The buckling resistances achieved by the FEA are summarized in Table 3.2. The load displacement curves by OPENSEES, VBST and SL.com are very similar, also the buckling load. Differences are to be find in the displacement w^{II} (initial imperfection w_0 + displacement w from nonlinear FEA) when reaching the buckling resistance. However, in total the deviation from the mean value is less than 3,8% for the buckling resistance $N_{b,mean}$ and less than 5% for the moment M^{II} considering the nonlinear effects. While for applying an initial imperfection of L/500 the bearing capacity is reduced by more than 10% compared to the approach with L/1000. Furthermore it is to be underlined that OPENSEES continues the analysis after the maximum failure load while SL.com and VBST stop the calculation and as mentioned Abaqus is able to simulate the post buckling behavior. Therefore for OPENSEES the max. value of the analysis has to be figured out manually.

Table 3.2: Comparison of results by FEA for 5 m and 8 m columns based on mean material values (buckling in y-y)

	Section A1					Section A2				
	$N_{b,mean}$ kN	Δ %	w^{II} mm	M^{II} kNm	Δ %	$N_{b,mean}$ kN	Δ %	w^{II} mm	M^{II} kNm	Δ %
5,0 m column										
Abaqus – L/1000+RS	13236	+0,6	16,8	222,5	-1,8	15755	+0,1	13,6	213,6	-4,6
OPENSEES L/1000+RS	12648	-3,8	18,3	231,1	+2,0	15222	-3,3	16,3	247,8	+10
VBST-L/1000+RS	13374	+1,7	17,2	230,5	+1,7	16025	+1,8	13,6	218,5	-3,4
SL.com – L/1000	13355	+1,5	16,6	221,8	-2,1	15962	+1,4	13,4	214,7	-4,2
Average values	**13153**	-	**17,2**	**226,6**	-	**15741**	-	**14,2**	**224,0**	-
Abaqus – L/500+RS	12005	-8,7	18,5	222,6	-1,8	14103	-10	15,3	215,4	-3,8
8,0 m column										
Abaqus – L/1000+RS	9308,6	+2,1	49,2	457,7	-1,3	11097	+1,4	46,0	511,0	-2,9
OPENSEES L/1000+RS	9020,5	-1,1	50,8	458,4	+0,1	10715	-2,1	49,6	531,4	+0,9
VBST-L/1000+RS	9050,0	-0,7	48,8	441,3	-3,9	10998	+0,5	46,3	509,2	-3,3
SL.com – L/1000	9087,0	-0,3	50,3	478,3	+4,2	10982	+0,3	50,4	553,5	+5,1
Average values	**9116,5**	-	**42,2**	**459,0**	-	**10948**	-	**48,1**	**526,5**	-
Abaqus – L/500+RS	8201,1	- 10	54,1	443,5	-3,4	9625	-12,1	47,3	455,1	-13,6

3.4 N-M-Interaction relation and verification of design resistance

With the software SL.com and VBST the Normal force – Moment-Interaction curve (N-M-IA) of the cross-sections can be developed by using the material stress-strain curve as presented in Figure 2-2 (concrete NL) and Figure 3-2 (concrete PR). It is to be underlined that by using the parabola-rectangle stress-strain relation (Figure 3-2) with the associated strain limitation of ε_{c2} according to EN 1992-1-1 for cross-section under compression, the reinforcement and steel do not reach plastic strains when the calculation is based on mean values of material resistance under compression. Therefore a significant difference between the N-M-Interaction curves can be seen where compression forces with none or low eccentricity are applied. According to EN 1992-1-1 both stress-strain curves (PR and NL) may be used, whereby for the concrete NL (Figure 2-2) and determination of the curve R_d, the mean values of concrete f_{cm} and the E-Modulus E_{cm} in the equation for the determination of the stress-strain curve (prEN 1992-1-1, eq. (5.6) and (5.7)) are to be replaced by the design values f_{cd} and E_{cd}. When the E-Modulus is taken by prEN 1992-1-1 ($E_{cm} = k_E \cdot f_{cm}^{1/3}$), the design value E_{cd} is defined by division of E_{cm} by a partial safety factor γ_{CE} of 1,5; in current Eurocode this values is defined by 1,2 if no other information is given by a National Annex. If the E-Modulus of concrete is specifically assessed, prEN 1992-1-1 reduces the safety factor to 1,3. It can be stated that for low moments the N-M-IA curve developed by strain-limited resistance and the concrete NL material model is very close to the one based on plastic resistance, but with increasing concrete class the difference becomes also more important. However, with increasing impact by the moment, the difference becomes more important. Figure 3-5 and Figure 3-6 represent the N-M-IA curves and the load-displacement curve for the composite columns resulting by the FEA based on Abaqus. The load vectors R_m/R_d are leading to the overall safety factor γ_0 (Figure 2-3). The functions of this factor over the relation M/N are also given by Figure 3-5 and Figure 3-6 when comparing R_m to R_d by plastic resistance and strain-limited resistance based on the two discussed material models for concrete. The factors γ_{Rd} and γ_{Sd} have been taken as 1,0.

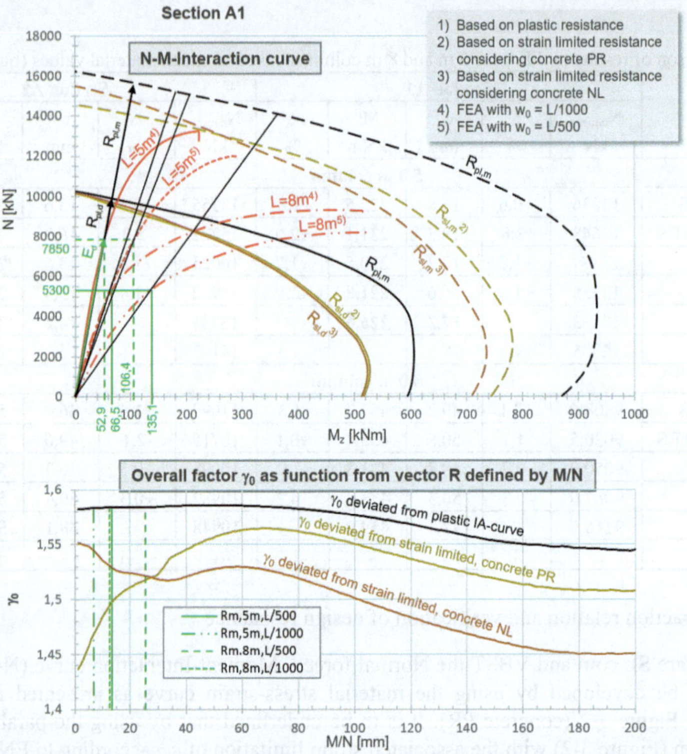

Figure 3-5: Section A1 – N-M-Interaction curve and function of overall factor γ_0

Table 3.3: Comparison of results by general method and based on European buckling curves – Section A1 (A2)

Method (GM: general Method)		w_0 [mm]	L [m]	N_{Ed} [kN]	γ_0 [mm]	$\eta = N_{Ed}/N_{Rd}$	L [m]	N_{Ed} [kN]	γ_0 [mm]	$\eta = N_{Ed}/N_{Rd}$
GM: IA => R_{pl} [1)]	A1	L/1000	5,0	7850	1,583	0,94	8,0	5300	1,614	0,90
GM: IA => $R_{sl,PR}$ [2)]		L/1000	5,0	7850	1,468	0,87	8,0	5300	1,518	0,85
GM: IA => $R_{sl,NL}$ [2)]		L/1000	5,0	7850	1,545	0,91	8,0	5300	1,582	0,87
GM: IA => R_{pl} [1)]		L/500	5,0	7850	1,584	1,04	8,0	5300	1,615	1,02
GM: IA => $R_{sl,PR}$ [2)]		L/500	5,0	7850	1,498	0,98	8,0	5300	1,530	0,98
GM: IA => $R_{sl,NL}$ [2)]		L/500	5,0	7850	1,529	1,00	8,0	5300	1,558	0,98
EN 1994-1-1, 6.7.3.5		L/150	5,0	7850	χ_c=0,70	1,10	8,0	5300	χ_c=0,45	1,16
GM: IA => R_{pl} [1)]	A2	L/1000	8,0	8300	1,614	0,85	8,0	5550	1,614	0,80
GM: IA => $R_{sl,PR}$ [2)]		L/1000	8,0	8300	1,510	0,80	8,0	5550	1,518	0,76
GM: IA => $R_{sl,NL}$ [2)]		L/1000	8,0	8300	1,576	0,83	8,0	5550	1,553	0,78
GM: IA => R_{pl} [1)]		L/500	8,0	8300	1,615	0,95	8,0	5550	1,615	0,93
GM: IA => $R_{sl,PR}$ [2)]		L/500	8,0	8300	1,519	0,89	8,0	5550	1,530	0,88
GM: IA => $R_{sl,NL}$ [2)]		L/500	8,0	8300	1,550	0,91	8,0	5550	1,529	0,88
EN 1994-1-1, 6.7.3.5		L/150	8,0	8300	χ_c=0,67	1,014	8,0	5550	χ_c=0,40	1,13

[1)]Plastic resistance, [2)]Strain limited resistance with concrete PR parabola rectangle, [3)]Strain limited resistance with concrete NL nonlinear

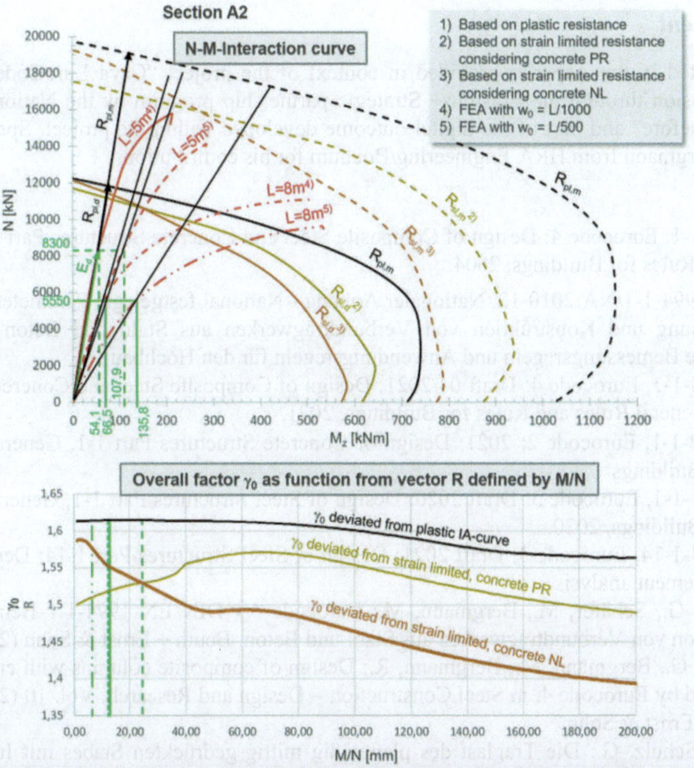

Figure 3-6: Section A2 – N-M-Interaction curve and function of overall factor γ_0

The overall factor γ_0 determined on plastic resistance shows higher safety level compared to the values determined by strain-limited resistance. The differences are up to 10% for the investigated columns (A1, A2). While the difference is more important for low ratio M/N when applying concrete PR material stress-strain relation, it becomes more significant with increasing M/N when using concrete NL relation. According to prEN 1994-1-1 the overall factor γ_0 should be determined based on plastic analysis by the values of $R_{pl,m}$ and $R_{pl,d}$, along the direction of load vector E_d. Further investigation in N-M-Interaction curve based on strain-limited resistance becomes interesting. In addition, the results in Table 3.3 show that the initial imperfection has also an important impact on the buckling resistance. However, all results by the general method can proof higher resistances compared to the simplified method by application of the reduction factor χ determined based on the European buckling curves.

4 CONCLUSION

The paper investigates in the general method for the design of composite columns according to prEN 1994-1-1. Thereby case study on Concrete Encased Composite Columns is presented. The results show an important effect of the initial imperfection on the buckling resistance. Furthermore, the impact on the safety level by using different materials stress-strain curves for the determination of the overall factor based on the N-M-Interaction curves is discussed. The ongoing research will investigate by the next steps in extended parametric studies on different types of composite columns and the safety concept to generate more guidance for application of the general method and numerical design approaches to ensure safe but also economic design.

Acknowledgement

The study presented in this paper is provided in context of the project "Cava EuroCodes" funded by European commission through the Erasmus+ Strategic partnership program by the National Agency in Luxembourg, "Anefore" and part of the digital outcome developed within the project. Special thanks to Dr.-Ing.Marco Bergmann from HRA Engineering/Bochum for his contribution.

References

[1] EN 1994-1-1, Eurocode 4: Design of Composite Steel and Concrete Structures Part 1-1, General Rules and Rules for Buildings, 2004.

[2] DIN EN 1994-1-1/NA:2010-12, Nationaler Anhang - National festgelegte Parameter - Eurocode 4: Bemessung und Konstruktion von Verbundtragwerken aus Stahl und Beton - Teil 1-1: Allgemeine Bemessungsregeln und Anwendungsregeln für den Hochbau

[3] prEN 1994-1-1, Eurocode 4: Draft 04/2021: Design of Composite Steel and Concrete Structures Part 1-1, General Rules and Rules for Buildings, 2021.

[4] prEN 1992-1-1, Eurocode 2: 2021: Design of Concrete Structures Part 1-1, General Rules and Rules for Buildings

[5] prEN 1993-1-1, Eurocode 3: Draft 2020: Design of Steel Structures Part 1-1, General Rules and Rules for Buildings, 2020.

[6] prEN 1993-1-14, Eurocode 3: Draft 2021: Design of Steel Structures Part 1-14: Design assisted by finite element analysis

[7] Hanswille, G., Schäfer, M., Bergmann, M.: Eurocode 4 - DIN EN 1994-1-1 Bemessung und Konstruktion von Verbundtragwerken aus Stahl und Beton, Beuth + Ernst & Sohn (2020)

[8] Hanswille, G., Bergmann, M., Bergmann, R.: Design of composite columns with cross-sections not covered by Eurocode 4, in Steel Construction – Design and Research, Vol. 10 (2017), No. 1, pp. 10-16, Ernst & Sohn

[9] Beer, H., Schulz, G.: Die Traglast des planmäßig mittig gedrückten Stabes mit Imperfektion, Konstruktion und Fertigung KuF, VDI-Z 111, Nr. 21, 1969

[10] Bergmann, R.: Traglastberechnung von Verbundstützen. Technisch-wissenschaftliche Mitteilung Nr. 81-2, Institut für Konstruktiven Ingenieurbau, Ruhr-Universität Bochum, Februar 1981

[11] Lindner, J.: Schlußbericht zum Forschungsvorhaben – Geometrische Ersatzimperfektionen für Verbundstützen, die in die Knickspannungslinien b und c eingestuft sind, Forschungsvorhaben Nr. IV 1-5-787/96 des Deutschen Instituts für Bautechnik (DIBt), Berlin, 1996

[12] Bergmann, R.: Schlußbericht zum Forschungsvorhaben – Geometrische Ersatzimperfektionen für Verbundstützen, die in die Knickspannungslinien a eingestuft sind – zur Bearbeitung der DIN 18800 Teil 5 im Auftrag des IfBt-Berlin, Witten, 1996

[13] European Convention for Constructional Steelwork – Technical Committee 8: Ultimate Limit State Calculation of Sway Frames with Rigid Joints, ECCS-Publication No.33. Brussels, 1984

[14] SL.com, Software for the strain-limited design, Zhang, Q., University of Luxembourg

[15] Abaqus Software 2018, Dassault Systemes, SIMULIA Corp..

[16] Mckenna, F., Fenves, G. L., Scott, M. H. (2000) Open System for Earthquake Engineering Simulation. University of California, Berkeley, http://opensees.berkeley.edu.

[17] Ergun, O., Schäfer, M.: COMPARISON OF GEOMETRICAL IMPERFECTION DEFINITION ON ENCASED COMPOSITE COLUMNS, 9th International Conference on Composite Construction in Steel and Concrete, Germany, July 2021

[18] Schäfer, M., Zahlten, W., Bender, M. Koziol, P., Zogu, P., Zhang, Q., et. al.: CavaEurocode, Computer-based design for structural elements - Analyses and Validation of compliance to EuroCodes, funded by EU Erasmus+ strategic partnerships, 2021

9TH INTERNATIONAL CONFERENCE ON COMPOSITE CONSTRUCTION IN STEEL AND CONCRETE
M. Knobloch, U. Kuhlmann, W. Kurz, M. Schäfer. (Eds.)
Online, Germany, July 27-29, 2021

COMPARISON OF GEOMETRICAL IMPERFECTION DEFINITIONS ON ENCASED COMPOSITE COLUMNS

Ozgun ERGUN*, Markus SCHÄFER*

* University of Luxembourg
e-mails: ozgun.ergun@uni.lu, markus.schaefer@uni.lu

Keywords: Composite, Columns, Imperfection, Abaqus, Finite Element, Construction

Abstract. *The imperfections stemming from manufacturing or construction practices have effects on capacities of structural columns. Bow imperfections or residual stresses are commonly investigated and well known examples of these imperfections. Since the composite construction generally has different stages than reinforced concrete or steel construction, there are other sources of imperfections for this field. Different approaches to consider geometrical imperfections are studied for this paper. These are namely the approach of EN-1992-1-1 [1],EN 1994-1-1 [2] and EN 13670 [3]. Eurocode 2 and Eurocode 4 specify geometric imperfection for columns differently. Former introduces them as load eccentricity and latter introduces them as bow imperfections. EN 13670 giving the construction tolerances for reinforced concrete structures, has different methods to limit geometrical imperfections. For the purpose of comparing different approaches and their effect on capacity calculations, several composite columns were simulated using physical and geometrical non-linear analysis, their load-deflection relationships were monitored and capacities compared. An outlook of the ongoing investigation for the adaptation of geometrical imperfection for composite columns is further discussed.*

1 INTRODUCTION

Geometrical imperfection of structural members are resulting from the production process and the tolerances in state during erection. Those imperfections may be described by eccentricity, bow imperfection or obliquity of the member. While structural imperfections are resulting from the rolling and welding process due to the development of residual stresses and the unequal distributed of the yield strength over the cross-section. Imperfections in the construction and production stages are important factors to consider in any kind of structural design, because they may have a significant impact on the design. Current Eurocode 4 does not provide information for the geometrical imperfections when applying the general design method. At least, guidance is missing for engineering practise. The draft prEN 1994-1-1 is referring to the geometrical imperfection of steel columns. Here, the approaches of Eurocode 2, Eurocode 4 plus prEN 1994-1-1 and EN 13670 are compared in terms of limitation and considered magnitude of the geometrical imperfections for composite columns in braced systems.

2 Imperfection

2.1 Imperfections according to EN 1994-1-1 (Eurocode 4)

Eurocode 4 is the European design code for composite structures in steel and concrete. Its general method for nonlinear design of composite columns in braced frames requires consideration of second-order effects including residual stresses, geometrical imperfections, local instability, cracking of concrete, creep and shrinkage of concrete and yielding of structural steel and of reinforcement [2] for capacity calculations. It also enables the designer to consider both residual stresses and bow imperfections using equivalent bow imperfections. For comparison, both of these methods will be considered within this study. Equivalent bow imperfections are naturally greater than what is typically used as magnitude of only bow imperfections. While the equivalent bow imperfections are given by EN 1994-1-1, no details for the consideration of residual stresses and geometrical imperfection are provided by the code. For the General method (nonlinear design), residual stresses and bow imperfections are modelled explicitly. For the magnitude of geometrical imperfection L/1000 [4] is used as provided by the German National Annex [5]. Current EN 1994-1-1 is not explicitly defining the magnitude of bow imperfections but L/1000 is also considered by the second generation of Eurocodes in reference to prEN 1994-1-1 [6] and prEN 1993-1-14 [7].

In the FE model for the General method, residual stresses on the steel profile will be defined through initial stress field with the distribution adapted from prEN 1993-1-14 [7], Figure 2-1. In the FE Model for the simplified method on the other hand, equivalent bow imperfections are used to represent both geometrical imperfections and residual stresses, For an encased composite column values for curve b are to used for buckling around strong axis and for curve c related to weak axis.

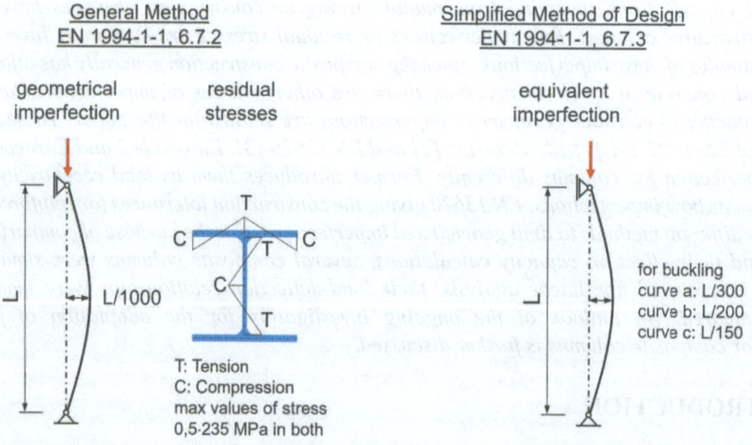

Figure 2-1: Eurocode 4 Imperfection Definition for Concrete Encased and Partially Concrete Encased Section

2.2 Imperfection acc. to EN 1992-1-1 (Eurocode 2)

Eurocode 2 (EC2) is the European design code for designing reinforced concrete structures. In its definition of consideration of imperfections for definition of action effects, for a column with a structure in a braced frame, it requires consideration of eccentricities at column ends for load application in order to take into account geometrical imperfections. The methodology is shown in Figure 2-2. Eurocode 2 also requires considering load eccentricity for planned centrically loaded members' capacity calculations in the ultimate limit state. The amplitude of this eccentricity shall be equal to h/30 but not less than 20 mm where h is the depth of the section. The unfavourable of both is taken as EC 2 approach. Eurocode 2 does not have any specification on residual stresses because it is for concrete structures but to keep consistency, residual stresses on steel profile were taken as in Figure 2-1 for this approach

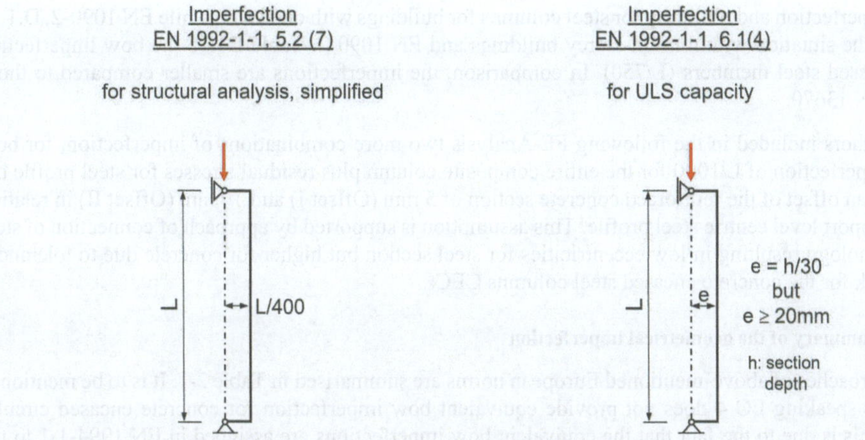

Figure 2-2: Eurocode 2 Geometrical Imperfections

2.3 Imperfection based on the execution standards (EN 13670 and EN 1090-2)

EN 13670 is the European norm for execution of concrete structures. It presents the limits for imperfections of structural members during the construction stage. For concrete columns, two of the norm's limitation approaches are taken here to be compared with previous design codes. These are shown in Figure 2-3. The first case is limiting the bow imperfection and the second case is limiting the end offsets of the axis of the column. The second case will be adapted here by assuming two ends being offset by the same amount to consider an unfavorable situation. It is to be mentioned that EN 13670 considers different tolerances that should respect the limits of a) ultimate limit state, b) serviceability limit state and c) accuracy of fit for the erection process. Within this paper, the tolerances related to ultimate limit state for structural members of class 1 according to EN 13670 are considered. Due to the fact that EN 13670 represent a standard for execution, of course it does not have any clause on residual stresses. For the presented study the residual stresses as presented by Figure 2-1 are considered.

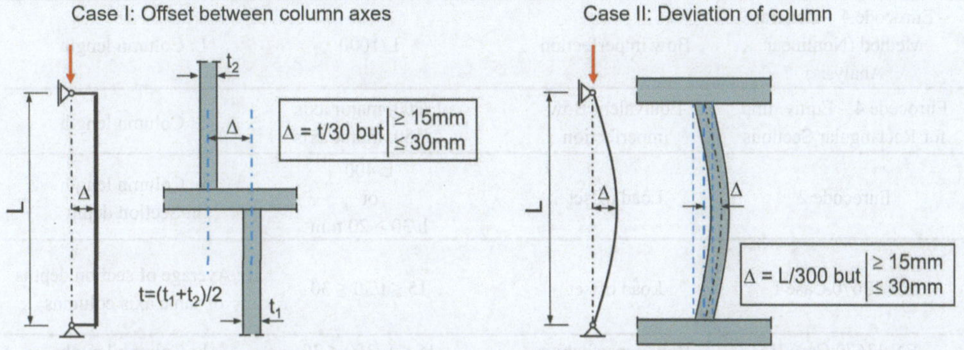

Figure 2-3: Two Cases of Imperfection (tolerances) Limits per EN 13670 10.4, Figure 2

2.4 Other geometrical imperfections

EN 1090-2 represents the execution standard for steel structures and provides tolerances for the bow of the steel column and limiting the end offsets of the axis of the column. EN 1090-2, Table D.1.11 provides the bow imperfection and obliquity for steel columns for buildings with one level, while EN 1090-2, D.1.12 represents the situation within multi-storey buildings and EN 1090-2, Table D1.15 the bow imperfection for compressed steel members (L/750). In comparison, the imperfections are smaller compared to those given by EN 13670.

The authors included in the following FE-Analysis two more combinations of imperfection, for both the bow imperfection of L/1000 for the entire composite column plus residual stresses for steel profile but in addition an offset of the reinforced concrete section of 5 mm (Offset I) and 10 mm (Offset II) in relation to the at support level centric steel profile. This assumption is supported by approach of connection of steel fitting technology resulting in low eccentricities for steel section but higher for concrete due to tolerances of formwork for the concrete encased steel columns CEC.

2.5 Summary of the geometrical imperfection

The approaches of above-mentioned European norms are summarised in Table 2-1. It is to be mentioned that strictly speaking EC 4 does not provide equivalent bow imperfection for concrete encased circular sections. This is due to the fact that the equivalent bow imperfections are assigned in EN 1994-1-1 to the simplified method, that is based on plastic cross-section resistance. The plastic resistance is a simplified approach and can variate form the resistance obtained based on material stress-strain relations. In order to not overestimate the bearing resistance, EN 1994-1-1 is introducing a correction value α_M for the bending resistance determined by the plastic N-M-Interaction diagram. This value was calibrated by comparison of plastic and strain-limited resistance. However, the values was deviated for rectangular composite columns with concrete encased steel section (CEC), rectangular and circular concrete filled steel tube with or without integrated steel profile (CFST) and partial concrete encased composite columns (PCEC), while for circular concrete encased steel section the shape factor is more unfavorable and α_M needs a modification not provided by the code [8]. At least for those sections usually the general method is to be applied. However, the geometrical imperfection of L/200 for the major axis and L/150 for minor axis is also applied for the following analyses.

Table 2-1: Imperfection Definitions

Norm	Geometric Imperfection Type	Size of Imperfection	Definition
Eurocode 4 – General Method (Nonlinear Analysis)	Bow imperfection	L/1000	L: Column length
Eurocode 4 – Equiv. Imp for Rectangular Sections	Equivalent Bow imperfection	L/200, major axis L/150, minor axis	L: Column length
Eurocode 2	Load offset	L/400 or $h/30 \geq 20$ mm	L: Column length h: Section depth
EN 13670 Case I	Load offset	$15 \leq t/30 \leq 30$	t : Average of section depths of same axis columns
EN 13670 Case II	Bow imperfection	$15 \leq L/300 \leq 30$	L: Column length
RC Offset I	Bow imperfection + offset for RC only	L/1000 + offset 5 mm	L: Column length
RC Offset II		L/1000 + offset 10 mm	L: Column length

3 FINITE ELEMENT MODEL

The capacity calculations for different column settings incorporating imperfections were done using Abaqus [9]. The calibration of the proposed model in terms of its accuracy in capacity estimation was assessed through tests of encased composite columns from literature. A criterion for slenderness was set for choosing which tests to incorporate in calibration process. This is due to the fact that stub/short columns are not as sensitive as slender columns to the geometrical imperfections and 2nd order effects. The criterion was that the ratio of effective length to the section size along the bending direction to be greater than 10. The authors could determine several tests with specimens conforming to this criterion. In total 27 specimens from four publications [10], [11], [12], [13] were simulated for calibration and the average of the simulation/test capacity values found to be 1.03 with a standard deviation of 0.08.

3.1 Material, element and modelling

The constitutive relationships for the concrete [1], structural steel [7] and reinforcement [1] were taken from corresponding guidelines. For concrete, nonlinear stress-strain from Eurocode 2 was used. Concrete Damaged plasticity was utilized in Abaqus. For structural steel, multilinear strain hardening steel stress-strain model was used and for the reinforcement, elastic perfectly plastic material model was chosen. All of the material properties were rate independent.

For concrete, first order brick elements were utilized (C3D8R). For structural steel, shell element type with five integration points along thickness was chosen (S4R). Use of shell elements for the profile was quite efficient in terms of computational time compared to solid elements if they were to be used for the flanges and web of the steel profile since the latter would require very small finite elements, which is inefficient for the chosen solver type (explicit). The reinforcements were modelled as wire/beam elements (B31).

Since the sections have three different materials working together, a contact definition was necessary. "General contact" available in the program was used to define the surface interaction between concrete bricks and steel profile shells. Contact definition had friction property in the tangent direction with a friction coefficient of 0.5 and separable and hard property in the normal direction. For the reinforcement, the "embedded region" constraint was used. It essentially ties one part (wire elements) to the surrounding region (concrete brick elements). Figure 3-1 shows the final setting with reinforcements highlighted.

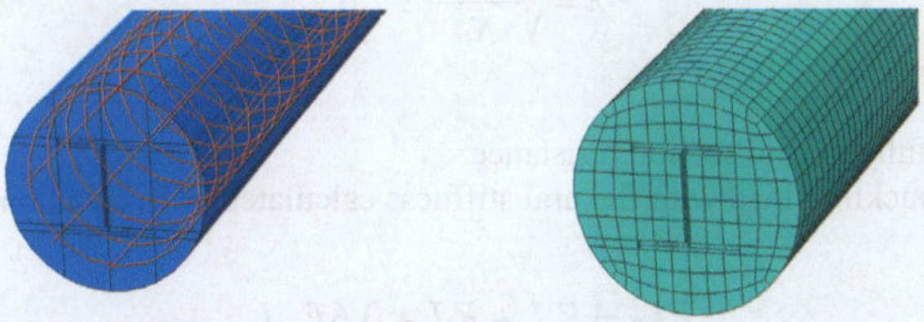

Figure 3-1: View of the parts of specimens and final mesh – circular encased composite column

The boundary condition was defined through a reference point at the loading support, which was then tied with rigid body constraints to the loading surface of the column. The loading was assigned through displacement boundary condition. Half symmetry was used at the column half-length to save computational time. Convergence analysis was done with several different mesh sizes to find the optimal combination of refinement vs computational effort. Global mesh sizes of 10, 20, 40, 100 mm were compared and results did not vary significantly. This can be attributed to the analysis measure being the maximum load at the load path but not any local effect such as buckling or local failure. Finally, 40 mm global mesh size was deemed acceptable (Figure 3-1).

3.2 Solver

The loading considered in the study is monotonic; therefore, the dynamic effects are not of interest. However, after several tries and from the suggestions in the documentation of the software, it was observed that static solver can have convergence problems and even without convergence issues; the concrete stresses might not follow the defined material property especially if there is tension in the concrete. To overcome this, explicit dynamic analysis was used. As is the suggestion by software documentation for highly nonlinear problems. But one must use the dynamic analysis as quasi-static with care. Explicit time integration requires the division of the external loads to very small steps to avoid numerical instability. It does not use any type of Newton-Raphson algorithm but directly integrates through the time. Unlike a static analysis, the mass matrix is in the equation and therefore the kinetic energy of the system has to be monitored over the analysis time verifying the analysis being quasi-static rather than dynamic. For the convergence studies, the ratio of kinetic energy to total internal energy was monitored. For the finally chosen loading rate, the monitored ratio exceeded 5 %, only for the initial 6% of the analysis time.

Use of explicit solver might be tempting since it has no convergence issues but its accuracy is not trivial to assess. Two different convergence studies (mesh convergence and loading rate convergence) have had to be done for a single type of testing (column testing in this case). Then, the calibration was done with test results as mentioned in the first paragraph of this section.

4 COMPARISON

For comparison of different imperfection assumption's effect on capacity calculations, several concrete encased steel columns with varying geometrical properties were modelled and loaded in Abaqus [9]. The variation of section geometry was quantified through the relative slenderness ratio ($\bar{\lambda}$) as defined in Eurocode 4 (EN-1994-1-1 [2]). Use of relative slenderness ratio as the input parameter is reasonable also, because the ultimate limit state resistance under concentric axial load for columns is calculated with so-called European buckling curves in accordance with Eurocode 4 (EN-1994-1-1) adapted from Eurocode 3 (EN 1993-1-1 [14]). There, the capacity depends solely on the slenderness ratio ($\bar{\lambda}$) value. The calculation of lambda per Eurocode 4 is given in equation (1). The geometries of compared columns are given in Table 4-2 and Figure 4-1.

$$\bar{\lambda} = \sqrt{\frac{N_{pl,Rk}}{N_{cr}}} \leq 2.0 \tag{1}$$

Where,

- $N_{pl,Rk}$ is the fully plastic section resistance.
- N_{cr} is ideal buckling load with flexural stiffness calculated as in equation (2) per Eurocode 4.

$$EI_{eff} = E_a I_a + E_s I_s + 0.6 E_{cm} I_c \tag{2}$$

Where,

- E_a, E_s and E_{cm} are modulus elasticity of steel profile, reinforcement and concrete, respectively,
- I_a, I_s, I_c are the second moments of area in the bending axis for steel profile, reinforcement and concrete, respectively.

Four different column types were compared with differing section dimensions but same material properties. Two of the columns had circular sections and the other two had rectangular sections. Their details are given in Figure 4-1 and Table 4-1. In the naming R500, R stands for rectangle and 500 stands for the section edge length in mm. Same goes for C for circular and number for the diameter value.

Table 4-1: Column sections

Column	Steel Profile	Concrete Class	Profile Steel	Rebar Steel	Column Length (m)	Slenderness Ratio ($\bar{\lambda}$)
C350	HEB160	C50	S460	B500	6	1.16
C500	HEB240	C50	S460	B500	6	0.81
R300	HEB160	C50	S460	B500	6	1.17
R500	HEB240	C50	S460	B500	6	0.69

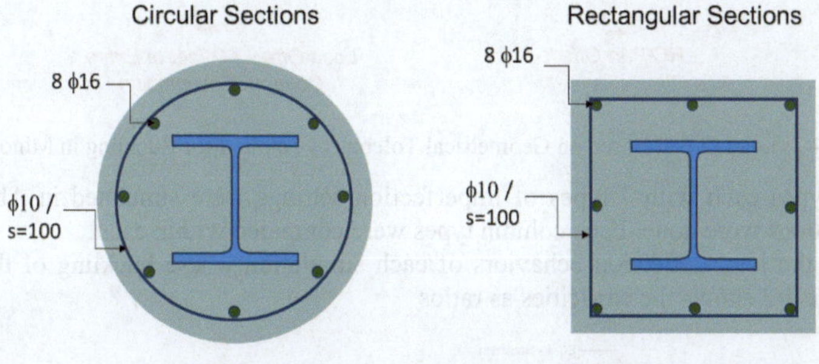

Figure 4-1: Circular and Square Cross-Sections

The overall imperfection values for each approach and geometry are detailed in Table 4-2. For the approaches having two formulas of calculation, the unfavorable values are taken. The approaches of EC 2 and EN 13670 – Case II (Table 2-1) gave the same imperfection values but former introduces them as load offset and latter introduces them as bow imperfections. For all of the approaches except "EC 4 – equiv. Imp." –based on the equivalent bow imperfection, residual stresses on the steel profile are defined.

Table 4-2: Assumption for Imperfection Values

Norm	**EC 4 - Nonlinear**	**EC 4 – Equiv. Imp.**	**EC 2**	**EN 13670 – Case I**	**EN 13670 – Case II**	**Offset I for RC Part + L/1000**	**Offset II for RC Part + L/1000**
Imperfection Type	Bow + Residual Stresses	Equivalent Bow	Load Offset + Residual Stresses	Load Offset + Residual Stresses	Bow + Residual Stresses	Offset RC + Bow + Residual Stresses	Offset RC + Bow + Residual Stresses
	Values (mm)						
C350	6	40	20	15	20	5	10
C500	6	40	20	16.67	20	5	10
R300	6	40	20	15	20	5	10
R500	6	40	20	16.67	20	5	10

The application of RC offset as mentioned on Table 4-2 is illustrated on the left in Figure 4-2. In this type of imperfection, load is applied on the steel centroid. Load offset or offset of entire column is illustrated on the right side of the same figure. Minor axis was taken as the axis of bending and imperfections were defined accordingly.

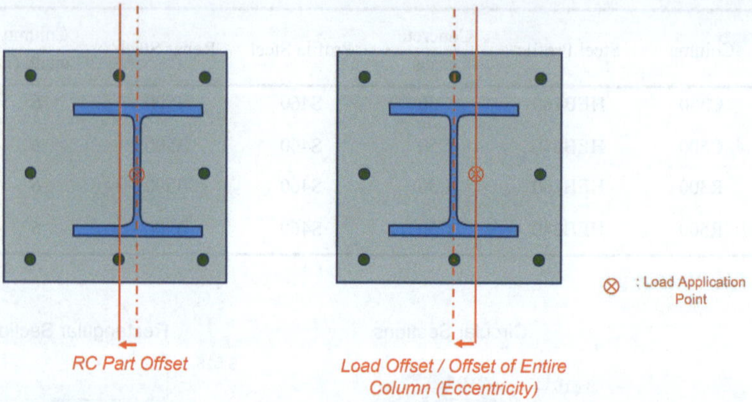

Figure 4-2: Load Offset Based on Geometrical Tolerances Applied for Buckling in Minor Axis

The 4 columns types each with 7 types of imperfection settings were simulated in Abaqus. Therefore in total 28 simulations were done. Each column types were compared within each.
Figure 4-3 shows the load deflection behaviors of each simulation where buckling of the minor axis is investigated. Table 4-3 shows the capacities as ratios.

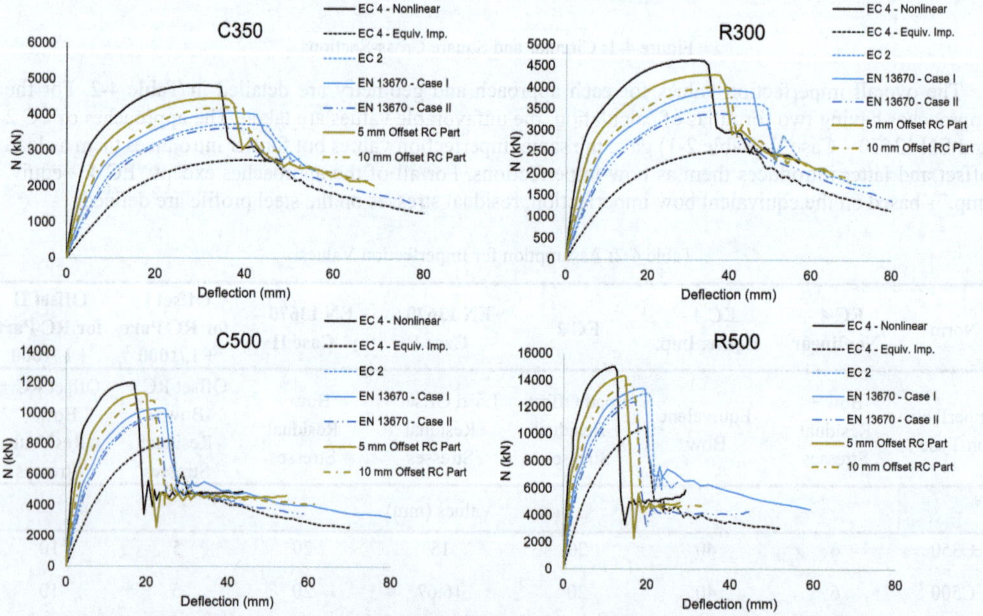

Figure 4-3: Load deflection curves for the investigated CEC with different imperfection setups

In Table 4-3 the resulting capacities are given as ratios to the maximum capacity obtained. The approach of geometrical imperfection of L/1000 is presenting the highest capacity while the EC 4 equivalent imperfection (basis for simplified method) the lowest. Important differences can be detected for the different assumptions for residual stresses. This underlines that further investigation therefore an extended sensitive study is required.

Table 4-3: Capacities with Different Imperfection Settings – Buckling Minor Axis

Imperfection Setup	EC 4 – General Method (Nonlinear)	EC 4 - Equiv. Imp.	EC 2	EN 13670 – Case I	EN 13670 – Case II	5 mm offset RC Part	10 mm offset RC Part
Specimens	Capacities as Ratios to the Highest Capacity						
C350 ($\bar{\lambda}$ = 1.16)	1.0	0.57	0.78	0.84	0.75	0.92	0.87
C500 ($\bar{\lambda}$ = 0.81)	1.0	0.68	0.83	0.86	0.81	0.94	0.90
R300 ($\bar{\lambda}$ = 1.17)	1.0	0.55	0.79	0.85	0.73	0.93	0.87
R500 ($\bar{\lambda}$ = 0.69)	1.0	0.70	0.87	0.89	0.83	0.95	0.92

5 Outlook

For composite columns in steel and concrete no specific execution code is available. Eurocode 4 is referring to EN 1090-2 for execution of steel structures and EN 13670 for execution of concrete structures. However, both norms are presenting different rules for execution tolerances and imperfections. The geometrical imperfection for composite columns are resulting from

- production tolerance of steel profiles leading to bow imperfection,
- fit tolerances for steel profile resulting in axe eccentricity of steel profile,
- inaccuracies for the position of the reinforcement,
- geometrical tolerance for positioning of the formwork resulting in offset of the concrete,
- manufacturing tolerance of the formwork leading to additional bow imperfection.

As shown before, EN 13670 provides higher values of imperfection compared to EN 1090-2. Due to propagation of error, it is highly improbable that the tolerances will overlap in the most unfavourable way. On the contrary, compared to pure concrete structures, it is more likely that a significant higher accuracy can be realised through the steel construction assembly technology and the embedding of the steel profile. Even when the concrete is produced on site, it is aligned with the previously assembled steel profile. Therefore, it is to be expected that the tolerances and thus geometric imperfections are to be classified between those for concrete and steel structures. In addition, it should be noted that the geometric imperfections strongly depend on the column type and the manufacturing process.

For partial encased composite columns (PCEC) it is to be expected that the concrete between the flanges of the steel section follows the shape of the steel profile, also the assembly technology is the same as for steel structures, so that the imperfection for steel may be applied. However, the reinforcement can lead to additional inaccuracies. Furthermore it is essential to ensure that no additional sag is embossed due to inadequate support during the concreting when the composite columns are prefabricated in horizontal position. Also for CFSTs the concrete follows the shape of the steel hollow section so that imperfection may become close to those from steel structures. Nevertheless, increasing degree of reinforcement carries the risk of additional imperfection. When an addition embedded steel profile is provided the centric position of this profile is essential to obtain the maximum bearing capacity. In general, distance plates are to be used to guaranty the position of the embedded core sections. A clearance of 1-2 mm is required on each side between the distance plate and the steel jacket in order to be able to insert the core profile respectively the embedded section into hollow section. Even if this has to be compensated before fixing the position, a certain risk for an offset of the embedded section to the steel jacket may occur. On the other hand, by use of centring pins and steel assembly technology from steel structures the erection follows condition for steel structures. For CEC casted on site, the formwork leads to a further risk of inaccuracies and may result in an offset between steel and concrete section. A sufficient support is also required when prefabricating CFST columns in inclined position.

The discussion illustrates that with all the focus on the design and assessment of tolerances, the construction methods and clear rules of execution are not be disregarded. Therefore suitable guidance for execution of different composite column types is required. Already in the design process, an object-oriented 3D design with a subsequent CNC-controlled production and a corresponding quality control leads to a reduction of the error rate. The steel profile should always be assembled by precise steel assembly technology. Distance keepers should guaranty that the formwork stays with the appropriate distance to the steel profile and is executed true to position. Spacers must also be installed to ensure the position of the reinforcement to the steel profile as well as to the formwork (concrete cover), Figure 5-1.

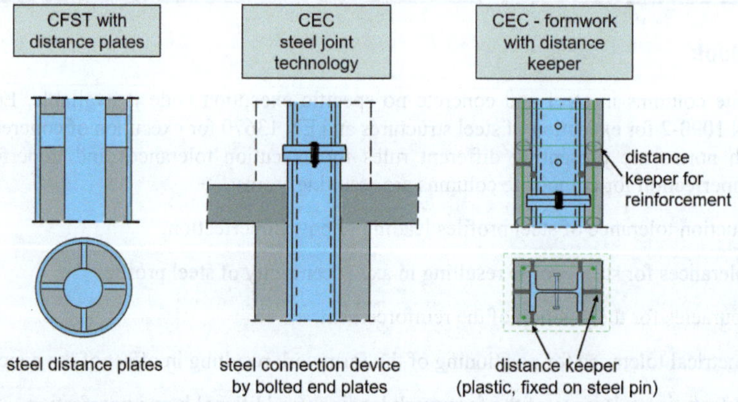

Figure 5-1: Positioning of steel section, concrete and reinforcement for production of composite columns

Imperfections strongly influence the geometrically non-linear calculation of centrically loaded columns. With increasing load eccentricity, the influence of the initial imperfections on the action effects diminishes. The comparison in section 4 underlines the significant impact of the initial geometrical imperfection on the bearing capacity. However, the calibration of the geometric imperfections based on test data from the literature is quite complex. On the one hand, there are not many well-documented data on column tests with centric loading available, and on the other hand, the scatter in the test recalculation often does not result primarily from the imperfection approach, but rather from differ support conditions by the test. The moment rotation capacity of the column support with its restoring moment is often not further investigated and documented. In section 3 the benchmark of the FE-Model by 27 tests from the literature have been mentioned with a very good correlation to the test results. However, four tests with centric loading could be detected in [10] and [11]. For those the difference between FE-analysis and test results have been the most significant while for the columns with eccentric loading high compliance could be confirmed. Based on the application of geometrical imperfection of L/1000 the nonlinear analysis of those four centric loaded tests represents a deviation from 16% in average, where the maximum divergence for the bearing capacity of one test represented an overestimation of more than 30%. By application of an additional offset of 5 mm respectively 10 mm for the reinforced concrete against the steel section the average deviation/max. value could be reduced to 0,95/12% respectively 0,96/11%. However, for engineering practice the application of an offset or different bow imperfection for concrete and steel leads to difficult modelling. Therefore the approach of a simplified bow imperfection for the entire composite column should be investigated.

With increasing amount of reinforcement and increasing bearing capacity of the concrete part the deviation of the tolerances compared to those from steel structures becomes more important. Therefore it seems to be reasonable to define the geometrical imperfection for application within the general method based on the ration δ for composite columns representing the plastic resistance $N_{pl.a.Rd}$ of the steel section in relation to the plastic resistance of the entire composite cross-section $N_{pl.Rd}$. The estimation may determine the geometrical imperfection for columns with a δ-values of zero according to EN 13670 (reinforced concrete columns) with a bow of L/350 while for a value δ equal 0.9 by L/1000 (steel column) with a linear interpolation in-between. However, the target of the here presented investigation is to verify if for the general method under consideration of above mentioned measurements in regard to quality monitoring and strict requirements on the execution, an optimisation of the geometrical imperfection by the definition of an lower limit values L/v_{low} may be achieved, Figure 5-2. For the determination of the upper and lower values of the geometrical imperfections for composite column test data from the literature will be analysed in a larger scale for the next step.

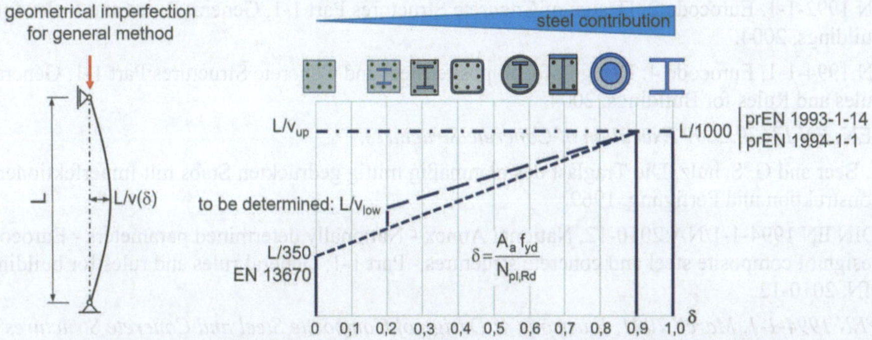

Figure 5-2: Proposal for determination of geometrical bow imperfection based on δ-value

6 CONCLUSION

Current Eurocode 4 separates for the nonlinear design based on the general method the application of residual stresses and geometrical imperfection although EC4 does not provide any information for the geometrical imperfection by chapter 6.7 other than the equivalent imperfection. Furthermore, no specific execution code for composite columns in steel and concrete is available, therefore no specific tolerances and imperfections are defined. The definition of geometrical imperfection for the design in ultimate limit state according to various standard, prEN 1994-1-1, prEN 1993-1-14, EN 1992-1-1, EN 13670 and EN 1090-2, shows very different consideration of the geometrical imperfections by bow imperfection or column eccentricity and different values. The application of those approaches within the nonlinear FE-Analysis for composite columns confirms a significant impact of the initial geometrical imperfection on the bearing resistance of compression members, especially for slender columns with centric loading. The equivalent imperfection defined by EC 4 for the simplified design method gives the lowest capacity from all of the other approaches. However, it should be underlined that those equivalent imperfections have been calibrated for the application of simplified design method where more exact consideration of material non-linearity is not possible compared to the FE-Analysis with volume elements. The most significant result is its difference from the capacity calculated by the geometrical bow imperfection of L/1000 presented by prEN 1994-1-1, it was deviated for steel columns. The approaches taken from reinforced concrete construction norms also result with lower capacity calculations than prEN 1994-1-1 general method but not as low as EC 4 – equivalent imperfection.

The arising question is if the bow imperfection magnitude of L/1000 based on prEN 1994-1-1 and prEN 1993-1-14 may lead to overestimation of the resistance for concrete encased composite columns (CEC) or if the imperfections represented by the tolerances according to EN 13670 are overestimating the imperfection and therefore may be too conservative for composite compression members. It is to be highlighted that with increasing contribution of concrete and reinforcement and also depending on the construction method, e.g. concrete encased composite columns casted on site, the risk for higher tolerances will increase. The authors underlined the importance of quality requirements for the production and erection of the columns, starting with 3D design technology. However, a verification of the assumption of the approach for the geometrical imperfection presented by prEN 1994-1-1 is required. The authors presented the possibility for determination of the geometrical imperfection based on the ratio of steel contribution of the composite column by interpolation between an upper and lower value. A parametric study for the definition of these limit values based on existing test data from literature is in progress.

REFERENCES

[1] EN 1992-1-1, Eurocode 2: Design of Concrete Structures Part 1-1, General Rules and Rules for Buildings, 2004.

[2] EN 1994-1-1, Eurocode 4: Design of Composite Steel and Concrete Structures Part 1-1, General Rules and Rules for Buildings, 2004.

[3] CEN, *EN 13670:2009 Execution of Concrete Structures.*

[4] H. Beer and G. Schulz, Die Traglast des planmäßig mittig gedrückten Stabs mit Imperfektionen, Konstruktion und Fertigung, 1969.

[5] "DIN EN 1994-1-1/NA:2010-12, National Annex - Nationally determined parameters - Eurocode 4: Design of composite steel and concrete structures - Part 1-1: General rules and rules for buildings," DIN, 2010-12.

[6] *prEN 1994-1-1, March 2021, Eurocode 4: Design of Composite Steel and Concrete Structures Part 1-1, General Rules and Rules for Buildings, second draft of EN 1994-1-1, presented by CEN/TC250/SC4.T6, 03/2021.*

[7] CEN/TC250, *prEN 1993-1-14: Design Assisted by Finite Element Analysis.*

[8] G. B. R. Hanswille, "New design methods for composite columns including high strength steel," in *Engineering Foundation Conferences Composite Construction V*, South Africa, 2004.

[9] *Abaqus Software 2018, Dassault Systemes, SIMULIA Corp..*

[10] K. Roik and K. Schwalbenhofer, Experimentelle Untersuchungen zum Plastichen Verhalten von Verbundstuetzen, Instituet fuer Konstruktiven Ingenieurbau, Ruhr-Universitaet Bochum, 1988.

[11] B. Lai, R. Liew and T. Wang, "Buckling Behaviour of High Strength Concrete Encased Steel Composite Columns," *Journal of Constructional Research,* vol. 154, pp. 27-42, 2019.

[12] C. Kim, H. Park, K. Chung and I. Choi, "Eccentric Axial Load Testing for Concrete Encased Steel Columns Using 800 MPa Stee and 100 MPa Concrete," *Journal of Structural Engineering,* vol. 138, pp. 1019-1031, 2012.

[13] C. Kim, H. Park, K. Chung and I. Choi, "Eccentric Axial Load Capacity of High Strength Steel Concrete Composite Columns of Various Sectional Shapes," *Journal of Structural Engineering,* vol. 140, 2014.

[14] EN 1993-1-1, Eurocode 3: Design of Steel Structures Part 1-1, General Rules and Rules for Buildings, 2005.

[15] CEN, *EN 1090-2:2008 Execution of Steel Structures and Aluminium Structures - Part 2: Technical requirements for steel structures.*

9TH INTERNATIONAL CONFERENCE ON COMPOSITE CONSTRUCTION IN STEEL AND CONCRETE
M. Knobloch, U. Kuhlmann, W. Kurz, M. Schäfer. (Eds.)
Online, Germany, July 27-29, 2021

SLENDER COLUMN STRENGTH OF INNOVATIVE CONCRETE FILLED STEEL TUBE COLUMNS WITH HIGH-PERFORMANCE BUILDING MATERIALS

Przemyslaw Schurgacz*, Martin Neuenschwander, Markus Knobloch***

*Chair of Steel, Lightweight and Composite Structures, Ruhr-Universität Bochum, Germany,
przemyslaw.schurgacz@rub.de, markus.knobloch@rub.de
**Structures Pacific Earthquake Eng. Research Center, Univ. of California Berkeley, United States,
m.neuenschwander@berkeley.edu

Keywords: Slender columns, concrete-filled steel tubes, composite columns, buckling behavior, finite-element analysis.

Abstract. *This paper presents a comprehensive experimental and numerical study on the load-carrying behaviour of innovative high performance materials steel-concrete composite columns: Namely concrete-filled steel tube columns with solid steel core (CFSTCSC) and concrete-filled double steel tube columns (CFDST). The experimental results exemplify the basic load-carrying behaviour. In addition to that, the application range of high-strength materials is evaluated in relation to the slenderness. Finally, the influence of geometric imperfections on the load-carrying behavior is investigated based on the manufacturing tolerances of such multi-component cross-sections.*

1 INTRODUCTION

Concrete-filled steel tube columns with solid steel core (CFSTCSC) and concrete-filled double steel tube columns (CFDST) with high-performance building materials are innovative prefabricated composite structural components. They facilitate slender architectural solutions for high load-levels and speedy construction processes. Moreover, they possess outstanding attributes of structural fire performance, which makes them particularly appealing for high-rise buildings, as well as industrial and mega-structures. This paper presents a comprehensive experimental and numerical study on the load-carrying behavior of such composite columns, fabricated with high-strength building materials (nominal yield strength up to 890 MPa and concrete strength up to 100 MPa). The experimental and numerical results form the basis of a future simplified design proposal within the framework of the European design rules.

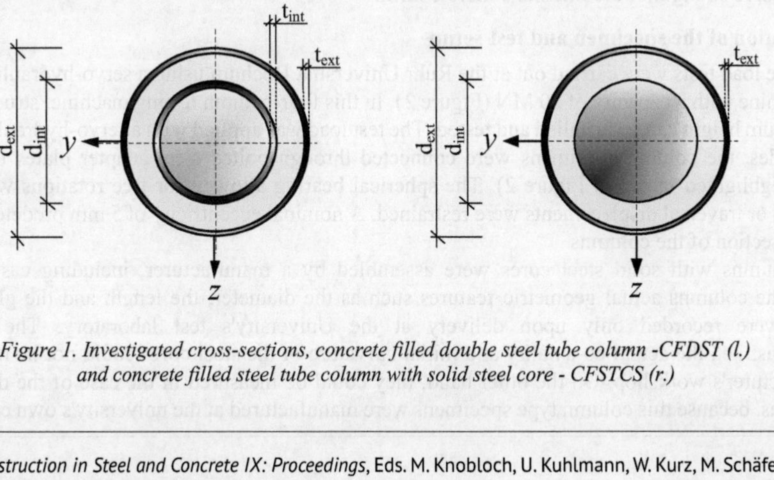

Figure 1. Investigated cross-sections, concrete filled double steel tube column -CFDST (l.) and concrete filled steel tube column with solid steel core - CFSTCS (r.)

2 EXPERIMENTEL STUDIES

2.1 Test Specimen

An experimental program of 24 ultimate load tests with CFSTCSC and CFDST was carried out within a FOSTA national research project, funded by the German Federal Ministry for Economic Affairs and Energy. The test program consisted of 16 slender column (SC) tests and 8 cross-sectional (CS) capacity tests, also referred to as stub-column tests. This paper presents the results of the slender column tests. The test series encompassed slender columns of 4 different cross sections, 5 different steel grades and 2 different concrete strengths. One- and two-storey columns of 4.0 and 8.0 m height respectively were tested to assess the slender buckling strength.

Table 1: Overview of the test specimen properties with SC_II-series for CFDST and SC_I-series for CFSTCS columns.

	Specimen No.	D_{ext} [mm]	t_{ext} [mm]	$f_{y,ext}$ [MPa]	$f_{c,ext}$ [MPa]	D_{int} [mm]	t_{int} [mm]	$f_{y,int}$ [MPa]	$f_{c,int}$ [MPa]	L [mm]	B.C [-]
CFDST	SC_II.01	355.6	8.8	396	68	219.1	10	391	68	4000	P-P
	SC_II.02	355.6	8.8	396	68	219.1	10	391	68	8000	P-P
	SC_II.03	355.6	8.8	396	68	219.1	10	855	94	4000	P-P
	SC_II.04	355.6	8.8	396	68	219.1	10	855	94	8000	P-P
	SC_II.05	355.6	8.8	396	68	219.1	8.8	976	94	4000	P-P
	SC_II.06	355.6	8.8	396	68	219.1	8.8	976	94	8000	P-P
	SC_II.07	273	8.8	383	65	177.8	10	315	65	4000	P-P
	SC_II.08	273	8.8	383	65	177.8	10	315	65	8000	P-P
	SC_II.09	273	8.8	383	65	177.8	8.0	813	97	4000	P-P
	SC_II.10	273	8.8	383	65	177.8	8.0	813	97	8000	P-P
	SC_II.11	273	8.8	383	65	177.8	10	948	97	4000	P-P
	SC_II.12	273	8.8	383	65	177.8	10	948	97	8000	P-P
CFSTCSC	SC_I.01	355.6	5.6	344	61	200	-	301 / 281	-	4000	P-P
	SC_I.02	355.6	5.6	344	61	200	-	301 / 281	-	8000	P-P
	SC_I.03	273	5.0	294	61	150	-	317 / 270	-	4000	P-P
	SC_I.04	273	5.0	294	61	150	-	317 / 270	-	8000	P-P

Table 1 gives the geometric properties and boundary conditions of the tested columns specimens together with the strengths of the steels and concretes employed; where d_{ext}, stands for the outer column diameter; t_{ext}, for thickness of the outer steel tube; d_{int}, for the outer diameter of the inner steel tube; t_{int}, for the thickness of the inner steel tube; L, for the column height; P-P for pinned-pinned boundary conditions; $f_{y,ext}$, for the yield strength of the external steel tube; $f_{c,ext}$ for the compressive strength of the external concrete infill; $f_{y,int}$ for the yield strength of the internal steel tube respectively the steel core; and $f_{c,int}$ for the compressive strength of the internal concrete infill.

2.2 Fabrication of the specimen and test setup

The ultimate load tests were carried out at the Ruhr Universität Bochum using a servo-hydraulic universal testing machine with a capacity of 20 MN (Figure 2). In this four-column testing machine, structures of 10 m in maximum height can be installed and tested. The test load was applied with a servo-hydraulic actuator. On both sides, the column specimens were connected through bolted steel adapter plates to spherical bearings (highlighted in red in Figure 2). The spherical bearing allowed for free rotations whereas any longitudinal or traversal displacements were restrained. A nominal eccentricity of 5 mm predetermined the buckling direction of the columns.

The columns with solid steel cores were assembled by a manufacturer, including casting of the concrete. The columns actual geometric features such as the diameter, the length and the global initial curvature were recorded only upon delivery at the University's test laboratory. The geometric imperfections, i.e., the actual co-axiality and initial curvature of the steel cores could not be measured in the manufacturer's workshop. On the other hand, they could be measured in the case of the double steel tube columns, because this column type specimens were manufactured at the university's own construction laboratory KIBKON. At the time of testing, specimens SC_II.01-06 had a concrete age between 34-55 days, specimens SC_II.07-12 between 142 to 192 days and specimens SC_I.01-04 between 199 to 223 days.

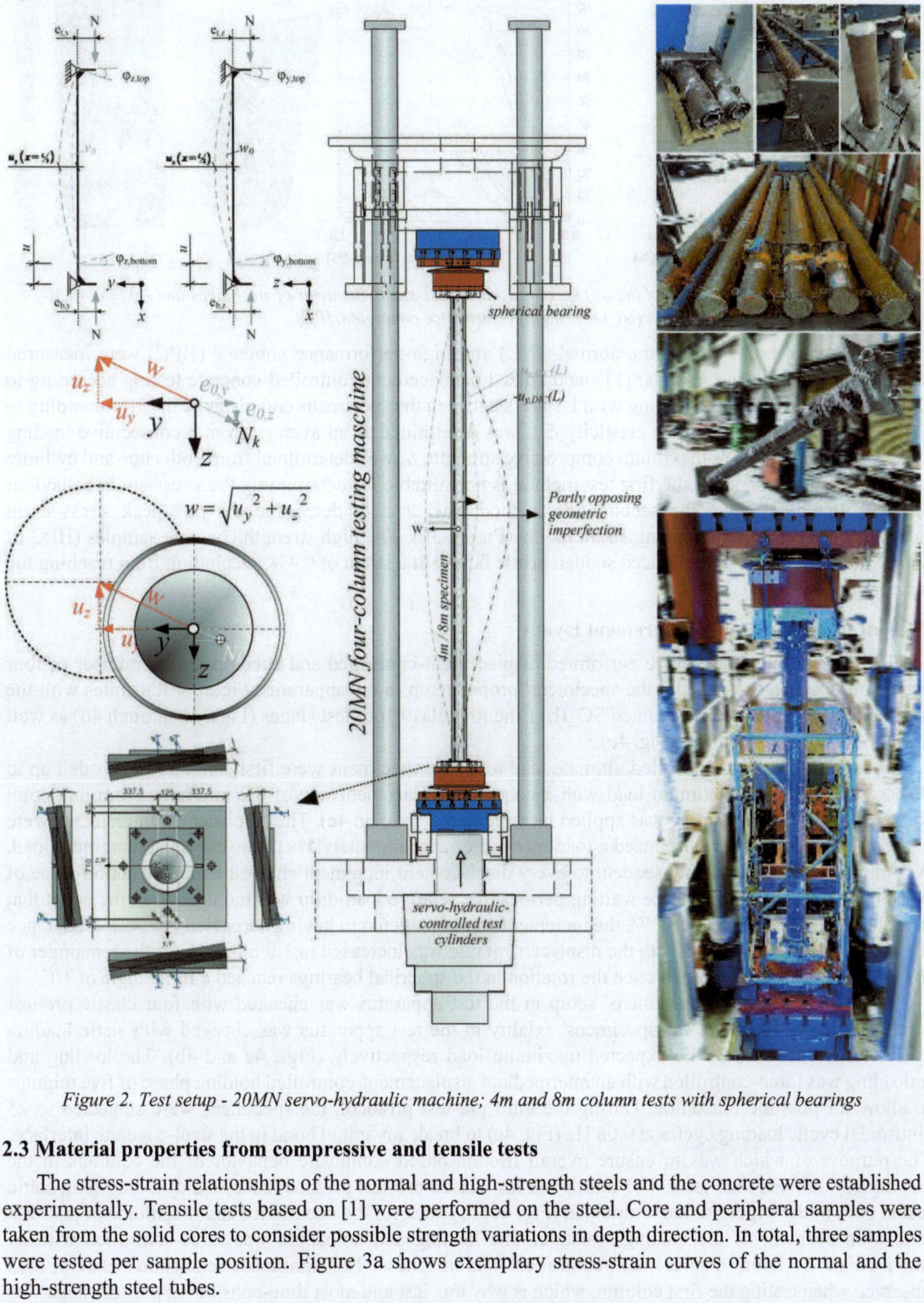

Figure 2. Test setup - 20MN servo-hydraulic machine; 4m and 8m column tests with spherical bearings

2.3 Material properties from compressive and tensile tests

The stress-strain relationships of the normal and high-strength steels and the concrete were established experimentally. Tensile tests based on [1] were performed on the steel. Core and peripheral samples were taken from the solid cores to consider possible strength variations in depth direction. In total, three samples were tested per sample position. Figure 3a shows exemplary stress-strain curves of the normal and the high-strength steel tubes.

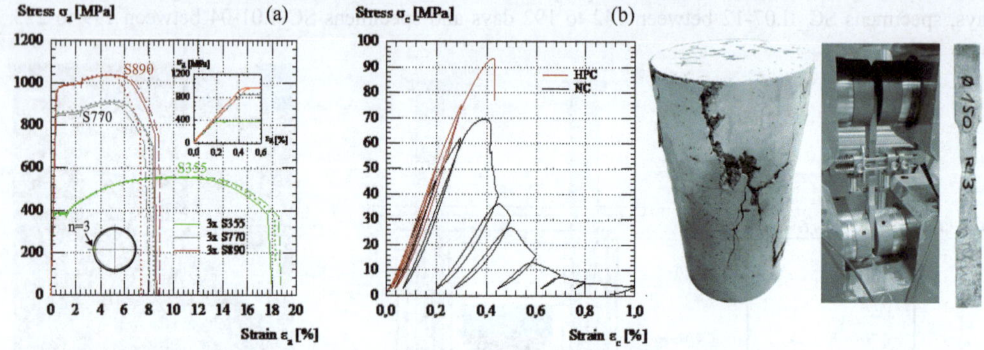

Figure 3. Material properties of the test specimen; (a) Stress-strain behavior of steel tubes and (b) of normal- (NC) and high-performance concrete (HPC)

The material properties of the normal- (NC) and high-performance concrete (HPC) were measured using two different test methods: (1) standardized displacement-controlled concrete testing according to [2] and (2) strain-controlled testing with LVDT´s between the specimens end plates as in [3]. According to the test method, the modulus of elasticity $E_{c,m}$ was determined as an average from 3 consecutive loading cycles; and the concrete's maximum compressive strength, f_c, was determined from both cube and cylinder specimens. However, since the first test method is not suitable for determining the stress-strain behaviour in the softening regime, the second test method was used to determine the post-peak stress-strain relationship up to a total crushing strain of 1% (Figure 3b). The high strength concrete samples (HPC of approximately 96MPa) experienced sudden brittle failure at a strain of 0.4%, precluding from reaching the target softening strain.

2.4 Test procedure and measurement layout

The ultimate load tests were performed displacement-controlled and encompassed a number of four elastic pre-test stages to ensure the specimens' proper setup in the apparatus. Figure 4 illustrates with the axial load time-histories of specimen SC_II.11 the four elastic pre-test stages (Figs. 4a through 4d) as well as the final ultimate load test (Fig. 4e).

In the displacement-controlled ultimate load tests, the specimens were first continuously loaded up to 80% of the expected maximum load with a constant displacement rate of 0.2 mm/min. From this point forward, however, the loading was applied incrementally (see Fig. 4e). The displacement increments were chosen such that they each produced a load increase of approximately 5% of the expected maximum load. A 5-minute waiting period subsequent to every displacement increment ensured complete taking place of possible relaxation. During these waiting periods, the relative load-drop was monitored. In the event that the relative load drop exceeded 5% the increment was considered to having surpassed the peak load. Once the ultimate load was established, the displacement rate was increased to 1.0 mm/min for the remainder of the test. The tests were stopped once the rotation in the spherical bearings reached a limit angle of 10°.

According to [4], the specimens' setup in the test apparatus was checked with four elastic pre-test protocols. In the first two, the specimens' axiality in the test apparatus was checked with static loading cycles to 20 and 40% of the expected maximum load respectively (Figs. 4a and 4b). The loading and unloading was force-controlled with an intermediate displacement-controlled holding phase of five minutes to allow for possible relaxation. During the third pre-test protocol, the specimens were subjected to 25 sinusoidal cyclic loading cycles at 0.08 Hz (Fig. 4c) to break any initial bond in the steel-concrete interface. The purpose of which was to ensure overall friction-based composite behavior of the columns in the subsequent ultimate load tests. The fourth and last elastic pre-test protocol consisted of a long-term static test at 60% of the expected maximum load (Fig. 4d) to check for the occurrence and magnitude of possible relaxation behavior of the loading structure. The loading and unloading was force-controlled with an intermediate 24 hours displacement-controlled holding phase. No significant relaxation behavior was observed when testing the first column, which is why this last and most time-consuming pre-test stage was then omitted in the remainder of the test program.

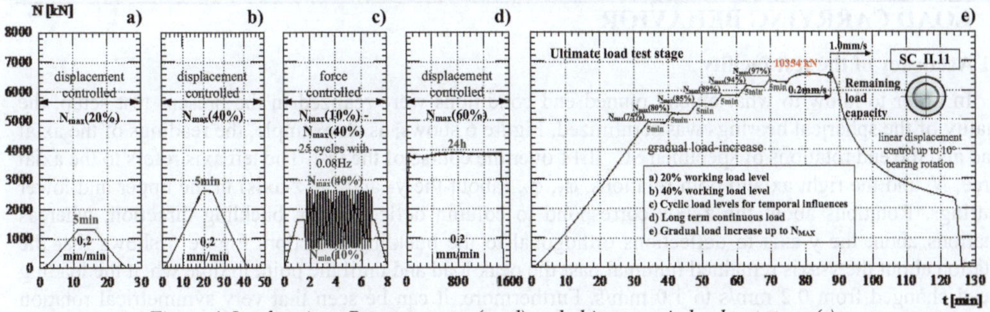

Figure 4. Load regime: Pre-test stages (a - d) and ultimate main load test stage (e)

Two different measurement techniques were used to record the specimen's geometry before any loading took place or track the specimen's deformations during the elastic pre-test stages and the ultimate load tests. Digital image correlation (DIC) was exclusively used to measure geometric pre-test features of the specimens such as the initial curvature or the actual eccentricity in the load application (Figure 5a). On the other hand, LVDTs and DIC were used to track specimen deformations during the elastic pre-test stages and the ultimate load tests. Additionally, the specimens' cross-sections at mid height were equipped with three strain gauges (120° apart) on the inner and outer steel tubes to monitor for deformations of the different cross-section components relative to each other (see Fig. 5b). This measurement layout allowed to construct from the results cross-sectional strain planes and track possible slip between the inner and outer components. As shown in Fig. 5c, LVDT´s at three different heights measured the specimen deflections in y-direction. However, since the specimen mid-height deflections, w, were composed of components in the y- and the z-direction (i.e. $w = \sqrt{u_y^2 + u_z^2}$) the LVDT measurements served more as redundant verification for the DIC system which in turn accurately recorded the deflection components in y- and z-direction during the tests.

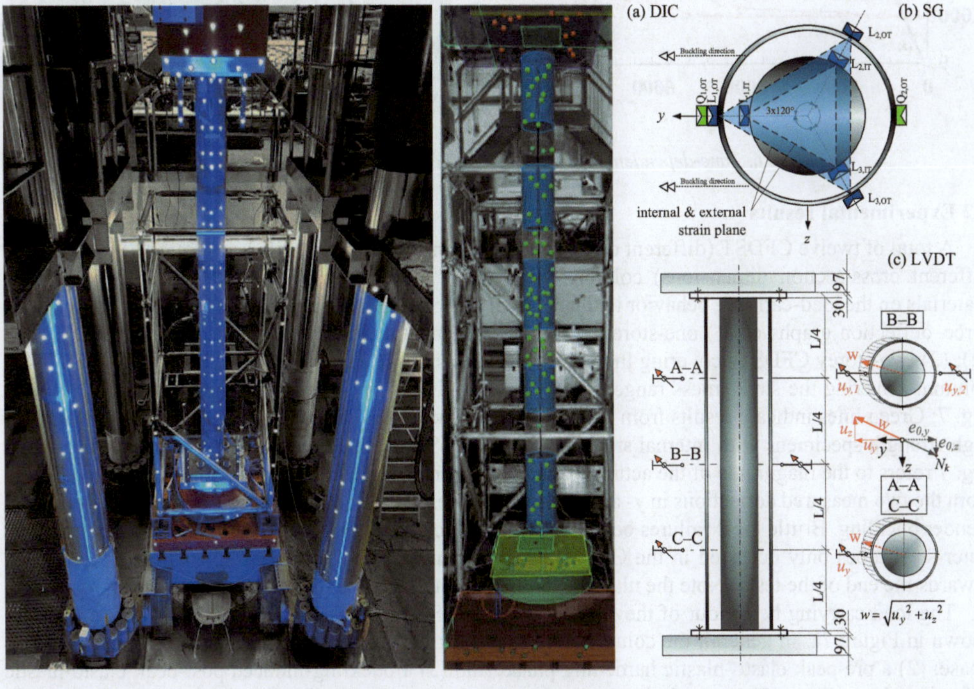

Figure 5. Measurement layout: (a) scanned 3D model using DIC, (b) arrangement of multiple strain gauges (SG) in cross-section and (c) positioning of the LVDT.

3 LOAD CARRYING BEHAVIOR

3.1 Accuracy of the test facility

In order to know to what extend pinned end conditions were realized in the present test setup, the quality of the spherical bearings was scrutinized. Figure 6 shows, as an example, the readings of the axial load and the end rotations of specimen SC_II.11 over the course of the test. The left axis refers to the axial force, *N*, and the right axis to both rotations, φ_y, φ_z, (about the y- and the z-axis) in the upper and lower bearings. Rotations about the z-axis correspond to column deflections in buckling direction, whereas rotations about the y-axis to deflections orthogonal to the buckling direction. Figure 6 shows that the rotation about the y-axis remained minimal past the peak load and until the point in time when the loading speed changed from 0.2 mm/s to 1.0 mm/s. Furthermore, it can be seen that very symmetrical rotation patterns in, φ_z, resulted over the entire course of the test, leading to almost ideal *Euler Case 2* buckling failure modes (see Fig. 6 right).

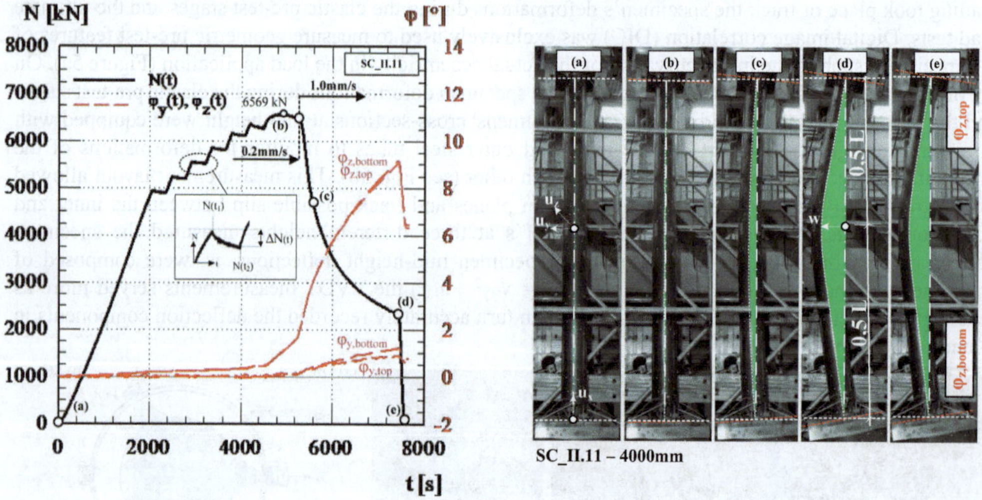

Figure 6. Time-dependent load-deformation behavior of the specimen SC_II.11

3.2 Experimental results

A total of twelve CFDST (different cross-section dimensions and strengths) and four CFSTCSC (only different cross-section dimensions) columns were tested to determine the influence of high-strength materials on the load-carrying behavior in the medium and high slenderness range. Figures 7 plots the axial force–deflection graphs of (1) one-storey CFDSTs covering the slenderness range, $\bar{\lambda}$=0.62–0.93, (Fig. 7a); (2) two-storey CFDSTs covering the slenderness range, $\bar{\lambda}$=1.21–1.86, (Fig. 7b); and (3) CFSTCSC columns covering the slenderness range, $\bar{\lambda}$=0.75–1.89, (Fig. 7c). The following colour-code is used in Fig. 7: Green lines indicate results from normal strength specimens, whereas black and red lines stand for high strength specimens with internal steel tubes of grade S770 and S890 respectively. The deflection in Fig. 7 refers to the magnitude of the actual spatial deflection *w* at the mid-height of the column, computed from the two measured deflections in y- and z-directions, i.e. $w = \sqrt{w_y{}^2 + w_z{}^2}$. All columns failed with global slender buckling. Brittle local failures before the ultimate load did not occur. Local buckling failure of the outer steel tubes only occurred in the CFSTCSC specimens at very large bearing rotation of around 6° towards the end of the tests (Note the ultimate loads were attained in average at ~0.65° bearing rotations).

The load-carrying behaviour of the single-storey CFDST columns of different material strengths are shown in Figure 7a. In general, the columns' behaviour showed three distinct phases: (1) A linear-elastic phase; (2) a pre-peak elasto-plastic hardening phase; and (3) a buckling-induced post-peak elasto-plastic softening phase. The stiffness evolution in the pre-peak regime was similar for all columns, even between different steel and concrete strengths. The initial stiffnesses of specimens II.01 and II.05 were almost identical, while specimen II.03 showed a more flexible pre-peak response. The stiffness differences can be explained by the different initial curvatures of the specimens. Specimens II.01 and II.05 had an almost identical initial curvature opposite to the buckling direction of $-L/1281$ and $-L/1232$ respectively, whereas specimen II.03 had an initial curvature in buckling direction of $+L/891$. The influence of deviating load eccentricities on the stiffness, however, could be ruled out as all three specimens (II.01, II.03 and II.05) exhibited almost identical eccentricities of 7.0mm, 6.3mm, and 7.3mm respectively.

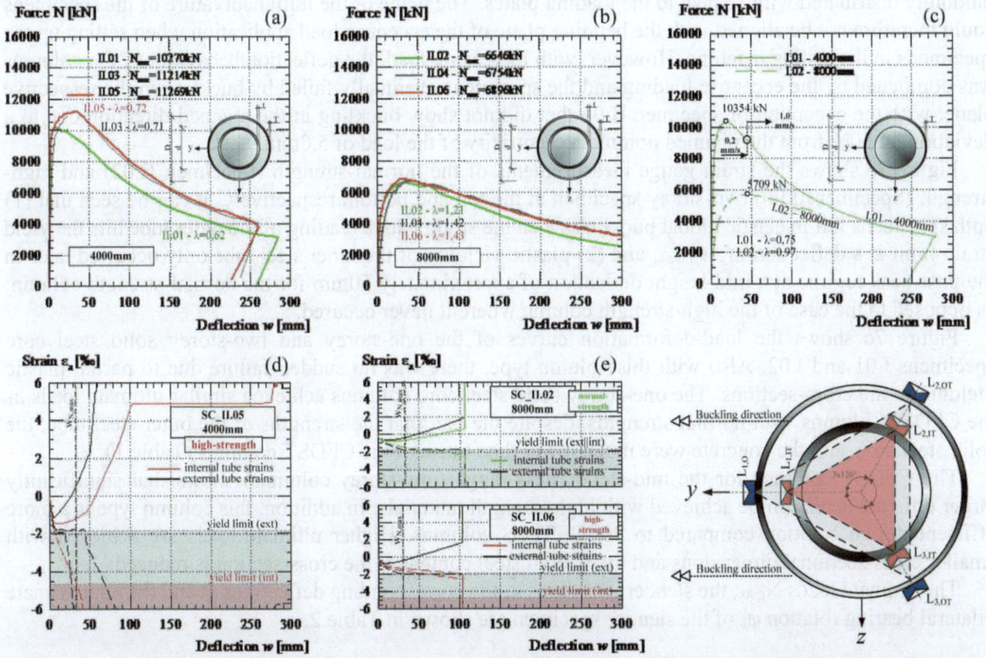

Figure 7. Comparison of the experimental results for one-storey (a) and two-storey (b) CFDST & (c) CFSTCSC column; (d)-(e) strain behavior at mid-span for different material strengths and slenderness ratios

As illustrated in Fig. 7a, high-strength materials clearly increased the load-carrying capacity of the single-storey columns (λ=0.62 – 0.93). Despite different steel grades, test specimens II.03 (t_{int}=10mm) and II.05 (t_{int}=8.8mm) achieved very similar ultimate loads. This can be explained, however, by the different nominal wall thicknesses of the internal high-strength steel tubes. The softening behaviour in the post-peak range of specimen II.03 and II.05 were similar and showed no influence of different tube thicknesses or material-strength influences.

Figure 7d shows the strain gauge measurements of specimen II.05 (high-strength specimen) as a function of the mid-height deflection w. Different colours and line types indicate the strain gauge's location in the cross-section: Black and red lines stand for strain gauges on the external and internal steel tube respectively; and solid and dashed lines for strain gauges on convex (tensile) and concave (compressive) side of the column. Moreover, Fig. 7d shows the experimentally established yield strains of the external (grey shading at 1.83‰) and internal (red shading at 4.65‰) steel tubes. Up to a deflection w of 10mm, there was a uniform increase in compressive strain in the various steel parts of the cross-section. The external steel tube reached its yield point and then temporarily stagnated on the yield-plateau. The increase in deflection w resulted in an ever-greater bending moment and an internal redistribution of strain states (formation of compression- and tension zones) to accommodate the increasing internal forces and moments. It follows from Fig. 7d that the column failed by elasto-plastic global buckling when the inner high-strength tube started to yield at a deflection of 33.3mm. The dashed red curve hits the yield strain limit exactly where the vertical line of the deflection at ultimate load, $w_{N,max}$, intersects with the yield limit strain. This observation also held for all other single-storey specimens.

Figure 7b compares load-carrying behaviour of the two-storey CFDST columns of different material strengths (Specimens II.02, II.04, and II.06). It is apparent from Fig. 7b that no significant increase in load-carrying capacity was achieved through the use of high-strength materials in this high slenderness range.

The columns essentially failed with elastic global buckling and well before a full activation of the strengths of the inner high-strength steel tubes. In the initial linear elastic load phase, the two-storey specimens adjusted their deformation state under a combined influence of initial curvature and eccentricity. At times, the specimens started deflecting in a different direction than expected based on the eccentric load application. This occurred mainly due to the fact that the geometric imperfections of the specimens were randomly distributed with respect to the loading plates. The plane of the initial curvature of the specimens could therefore not be aligned with the bending plane of the eccentric load application when setting up the specimens in the testing machine. However, with increasing load, the deflection behaviour of the columns was dominated by the eccentric loading and the specimens eventually failed by buckling in the respective plane; with the exception of specimen II.02 that did not show buckling in the planned direction due to a deviation by 30% from the planned nominal eccentricity of the load of 5.0mm.

Figure 7e shows the strain gauge measurements of the normal-strength (Specimen II.02) and high-strength (Specimen II.06) two-storey specimen at the top and bottom respectively. It can be seen that (1) both specimens fail in elastic global buckling when the strain gauge reading of the outer tube hits the yield strain limit at a deflection of, $w_{N,max}$; and (2) plastic yielding of the inner steel tube only occurred later in the post-peak regime at a mid-height deflection of approximately 50mm for the normal-strength column; as opposed to the case of the high-strength column where it never occured.

Figure 7c shows the load-deformation curves of the one-storey and two-storey solid steel core specimens I.01 and I.02. Also with this column type, there was no sudden failure due to partial plastic yielding in the cross-sections. The one-storey solid steel core columns achieved similar ultimate loads as the CFDST columns with normal strengths, despite the fact that the strengths of the outer steel tube, the solid steel core, and the concrete were much lower than those of the CFDST columns (Table 1).

The CFDST columns for the mid-slenderness range (one-storey columns) shows that significantly larger ultimate loads can be achieved with high-strength materials. In addition, this column type is a more efficient design solution compared to the CFSTCS columns. Higher ultimate loads are achieved with smaller cross-sectional dimensions and the overall steel content in the cross-section is reduced.

The ultimate loads N_{max}, the slenderness ratios $\bar{\lambda}$, the corresponding deflections w and the approximate bilateral bearing rotation φ_z of the slender specimen are shown in Table 2.

Table 2: Overview of the maximum loads and the corresponding deflections of the slender specimen.

SC_No.	II.01	II.02	II.03	II.04	II.05	II.06	II.07	II.08	II.09	II.10	II.11	II.12	I.01	I.02	I.03	I.04
N_{max} [kN]	10270	6646	11214	6896	11269	6896	5695	2850	6225	3091	6569	3027	10354	5709	5426	2115
$\bar{\lambda}$ [kN]	0.62	1.21	0.71	1.42	0.72	1.43	0.76	1.51	0.89	1.78	0.93	1.86	0.75	1.51	0.96	1.89
w [mm]	9.2	30.7	33.3	44.5	33.3	29.0	13.1	70.0	24.9	58.1	23.1	62.8	18.9	42.6	16.1	41.8
φ_z [°]	0.49	0.70	1.29	0.86	1.20	0.54	0.57	1.66	0.82	1.15	1.32	0.99	0.70	0.68	0.65	0.89

3.3 Parameters influencing the load-carrying behavior

Different influencing variables on the load carrying behavior were investigated. Specifically, the cross-sectional dimensions, the slenderness ratio and the material strength. The results and discussion is presented in two separate sub-sections pertaining to insights from the experimental and the numerical FEM study.

3.3.1 Experimental study

Figure 8a shows fully normalized load–deformation graphs of four CFSTSC column tests (Specimens I.01-04). The axial loads, N, and the mid-height deflections, w, are plotted relative to the actual maximum loads, N_{max}, and deflections at maximum load, w_{max}, respectively. It is apparent from Fig. 8a that this column type exbibits a normalized master load–deformation behavior that is independent of the cross-sectional dimensions and the slenderness; with the relative load-deformation graphs of all four tests virtually coinciding in Fig. 8a.

Within the applied concrete strain-limit ($\varepsilon_c \approx 4.0‰$), the internal high-strength steel tube is still in an elastic stress state with a maximum stress of 565 N/mm² and is not fully utilized, which in turn leads to a significantly lower bending moment. Provided that the high-strength steel tube ($\varepsilon_a \approx 4.5‰$) was fully utilized, partially higher bending moments could be achieved. At the same time, decreasing concrete strengths had to be taken into account in the calculations for larger compressions. The N-M curve of the one-storey specimen II.11 shows that when the maximum concrete compressive strength is reached, no increase in ultimate load is achieved. From this point up to the yield point of the high-strength steel tube, the load stagnates and the bending moment increases continuously. In the case of the two-storey specimen II.12, however, it can be seen that the cross-sectional load-carrying capacity will never be reached due to instability failure.

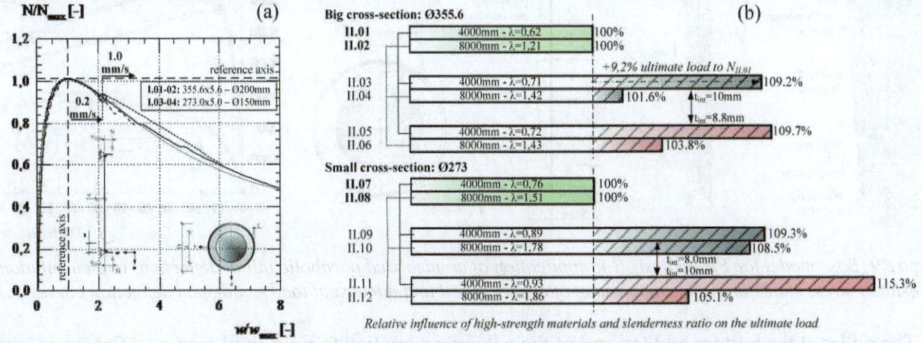

Figure 8. Related load-deformation behavior of CFSTCS (a) *and CFDST* (b)*, Percentage load increase for medium and high slenderness ratios using high-strength materials* (c)

Figure 8b illustrates with bar charts the relative influence of the material strength and the slenderness on the ultimate load capacity of double-tube composite columns (CFDST columns). The bar chart separately gives the relative ultimate load gains for columns of small (⌀273 mm) and large (⌀355.6 mm) total (outer) diameters. For each cross-section type the relative ultimate load gains in percent refer to the ultimate loads of reference columns with normal-strength material (i.e. $f_y = 355$ N/mm^2 and $f_c = 60$ N/mm^2) of both low-slender one-storey columns (4000 mm in height) and high-slender two-storey columns. The respective reference specimen labels are highlighted in bold print in Fig. 8b.

Figure 8b shows that the use of high-strength inner steel tubes (and to some extent also their high-performance concrete infill) effectively increases the column strength of low-slender single-storey columns by 10-15%. This finding is in line with the experimentally observed elasto-plastic global buckling failure of these columns with significant cross-sectional plastic yielding and hence activation of the beneficial effects of high-strength materials. An effect that could, however, not be observed in the case of the high-slender two-storey columns that failed with elastic global buckling. Accordingly, the ultimate load of high-slender columns was not significantly increased by the use of high-strength materials, leading on average to an increase in the ultimate load of 6.8% and 2.7% in the case of the small and the large diameter columns, respectively. Hence, in this case, the use of high-strength materials was of limited benefit.

3.3.2 Finite Element Model – Validation and sensitivity study

A three-dimensional Abaqus FE model was developed and validated against both independent test results from the literature [4, 5] and the program of the present study (Fig. 9). The numerical model shows good agreement with the experimental results even in the post-peak regime of the load-displacement curves presented in Figure 9; and the model is considered suitable for representing the load-carrying and deformation behaviour of the column tests. The DIC measurements of the upper and lower load eccentricities, and the initial global deflections and rotations of the column ends were incorporated into the model as initial geometric imperfections. The measured initial deflection distribution along the column height was idealised by a parabola with a maximum magnitude equal to the measured maximum deflection of the specimen. Some specimens such as II.01 and II.02 had initial deflections opposite to the planned buckling direction, which was taken into account in the numerical model as illustrated in Fig. 9. As in the tests, the column's mid-height deflections were computed from the deflection components in the *y*- and *z*-direction, i.e. $w = \sqrt{u_y^2 + u_z^2}$.

The mesh of the column models was divided into three (CFSTCS) or four (CFDST) separate parts: the concrete core, the concrete ring and the external and internal steel tubes or the internal solid steel core. Linear 8-node continuum elements (C3D8) were used to mesh the parts.

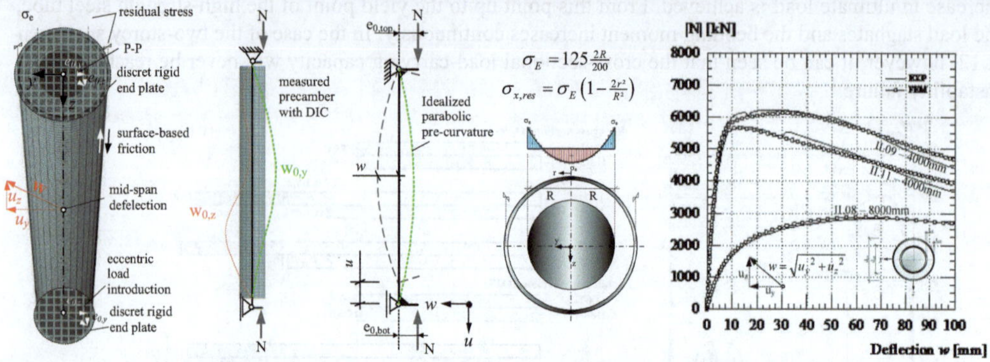

Figure 9. Base model for FE analysis (l.); application of an idealized parabolic initial deflection, implementation of residual stress states (m.) and comparison of experimental and numerical load - mid-span deflection curves (r.).

The nodes at the bottom and top end of the columns were tied to rigid shell-surfaces. Eccentric loading of the columns was realized by applying concentrated loads at the eccentrically defined reference points of the rigid shell-surfaces. Likewise, the pinned-pinned end conditions of the columns were realized by specifying boundary conditions at the reference points.

Contact between the different components of the columns was modelled with a surface-based contact approach with the default penalty method as contact enforcement method. Frictional contact with a friction coefficient of 0.2 was implemented in the tangential direction, while hard contact was chosen for the behavior in normal direction. The contact between the rigid shell-surfaces and the column cross-section was modelled frictionless.

The classic metal plasticity model of Abaqus was calibrated with experimentally determined stress–strain relationships of the steel tubes and the solid steel cores. The concrete-damaged plasticity (CDP) model of Abaqus was calibrated with input data from strain rate controlled cyclic compression tests, i.e. the nonlinear compression hardening/softening behavior (*CONCRETE COMPRESSION HARDENING) and the damage evolution relation in compression (*CONCRETE COMPRESSION DAMAGE).

Residual stresses in the steel tubes and cores were implemented via a *Fortran* scripted user-subroutine (*SIGINI* subroutine). After yielding realistic simulation outcomes in [4], a parabolic rotationally symmetric distribution according to [4] was implemented to represent the residual stresses in the solid steel cores (Fig. 9). The idealized but experimentally calibrated initial deflection distribution was applied through a buckling analysis step, i.e. by calibrating the magnitude of the first eigenmode with the experimentally established maximum magnitude of the mid-height deflection w.

Table 3: Comparison of the experimental and numerical maximum loads of the slender specimen.

SC_No.	II.01	II.02	II.03	II.04	II.05	II.06	II.07	II.08	II.09	II.10	II.11	II.12	I.01	I.02	I.03	I.04
N_{EXP} [kN]	10270	6646	11214	6896	11269	6896	5695	2850	6225	3091	6569	3027	10354	5709	5426	2115
N_{FEM} [kN]	9627	6293	11314	6609	10903	6711	5654	2878	6125	2978	6520	3147	11708	6472	5542	2788
N_{FEM} / N_{FEM} [-]	0.94	0.95	1.01	0.96	0.97	0.97	0.99	1.01	0.98	0.96	0.99	1.04	1.13	1.13	1.02	1.32
$\bar{x}$ [-]												0.981				1.151
σ [-]												0.029				0.123

A comparison of the numerical and experimental ultimate loads is shown in Table 3 and Figure 10a. The numerical loads for CFDST columns are within a tolerance range of 1% to 9% to the experimental results. However, for some cases of the CFSTC columns, the numerically predicted loads exceeded the test results up to 32%. A sensitivity study was carried out to understand why some of the simulations of the columns with steel cores compared relatively poorly against the tests. In this respect, the influence of structural as well as geometrical imperfections was investigated. Among the possible influencing factors, [4] had already shown that the residual stresses in the solid steel core can significantly reduce the load carrying capacity. However, imperfections in the coaxial placing and the initial distribution of the deflections of the steel cores need to be considered as well, which will be presented in this paper.

The load-carrying behaviour of 25 different imperfection models was investigated. However, in this paper, only five particular cases (out of the total of 25 imperfection models) will be discussed. They are illustrated schematically in Fig. 10b and are representative for imperfection combinations as they typically result from current manufacturing practice. Starting from a reference case with geometric imperfections as established experimentally with the DIC system (case I in Fig. 10b), the four remaining cases considered separately and combined the geometric imperfections of (1) offsetting the coaxial placement of the steel core by an equal amount at the bottom and at the top of the column as in case II in Fig. 10b; (2) offsetting the coaxial placement by differing amounts at the bottom and the top as in case III in Fig. 10b; (3) increasing the maximum magnitude of the initial deflection distribution along the column height as in case IV in Fig. 10b; and (4) a worst-case combination of increased initial deflection magnitude and varied offsets in the coaxial placement at the bottom and the top as in case V in Fig. 10b.

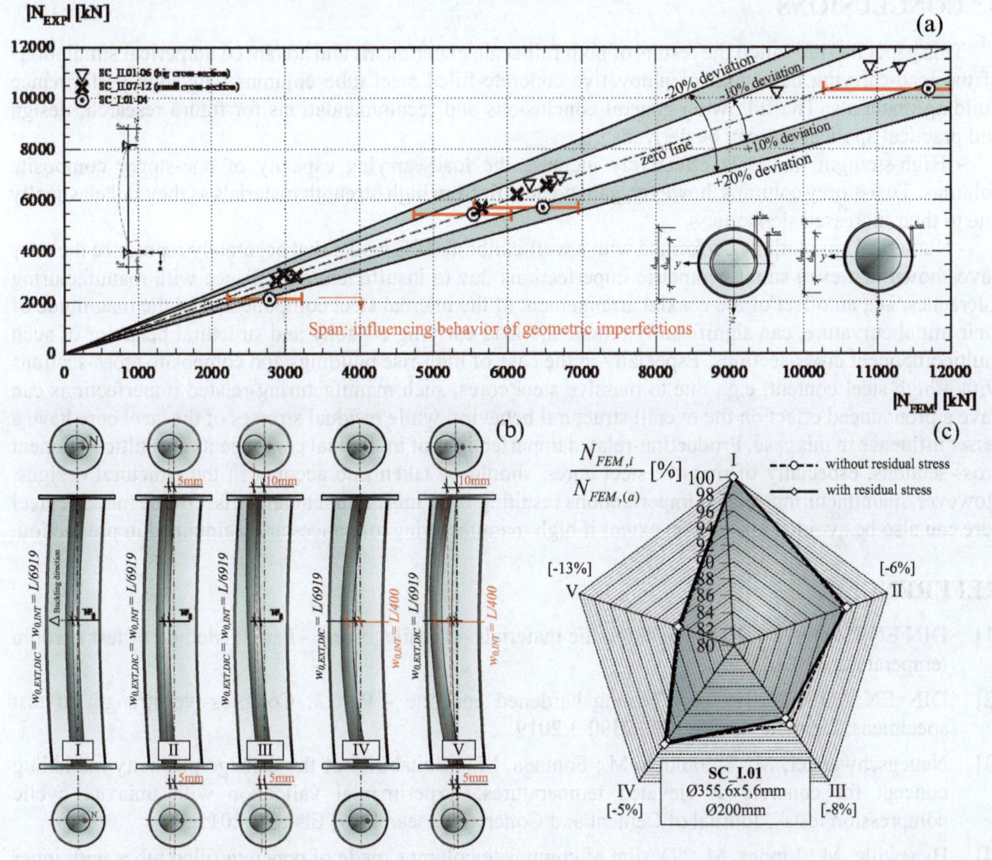

Figure 10. Validation of the numerical models (a), influence of unfavorable geometric imperfections on the load-carrying capacity (b) and corresponding percentage loss of the maximum load compared to the initial situation (c).

Figure 10c illustrates the results of the sensitivity analysis for Specimen I.01 (Ø355.6x5.6mm steel tube with Ø 200 mm steel core) as an example. The diagram shows the relative decreases in the ultimate load when applying the imperfection cases II through IV (relative decreases with respect to the reference imperfection case I). Figure 10c shows that the load-carrying capacity of concrete-filled steel tube columns with solid steel cores is strongly influenced by the geometric imperfections of the solid steel cores; with the ultimate column strength being reduced up to 13% in a worst-case imperfection scenario.

The span of the results from all 25 imperfection models are shown in Figure 10a with red whisker plots. Note that there are favourable as well as unfavourable imperfection combinations when varying over all the 25 different models. Favourable imperfections such as an initial deflection of the solid steel core opposite to the planned buckling direction significantly increased the load capacity, whereas, in turn, unfavourable imperfections such as increased initial imperfections combined with offsets in the axiality of the core significantly reduced the load capacity. When considered alone, even as small offsets of 5 mm in the axiality of the steel core lead to considerably higher bending moments and thus have a significant influence on the load-carrying capacity. It is therefore advisable to prevent such imperfections to the extent possible by enforcing tight manufacturing standards. The second major factor is the steel mill production-related initial deflections of solid steel core bars which, unfortunately, can only be reduced by cost-intensive additional steps during the manufacturing process of the composite columns (heat treatment and forging of the steel core bars).

4 CONCLUSIONS

This paper has presented the results of both full-scale experiments and advanced numerical simulations of the load-carrying behaviour of innovative concrete-filled steel tube columns with high-performance building materials. The following general conclusions and recommendations for future research, design and practical applications can be derived:

- High-strength materials effectively increase the load-carrying capacity of one-storey composite columns. Two-storey columns, however, gain no benefit from high-strength materials as they fail elastically due to their increased slenderness.
- Numerical simulations performed with a well-calibrated FE model that accurately reproduce the tests have shown that even small geometric imperfections due to insufficient compliance with manufacturing tolerances, i.e., an offset of the coaxial arrangement of the internal steel components and the magnitude of their initial curvature, can significantly reduce the load-carrying capacity and structural behavior of such multicomponent cross-sections. Especially in the case of high-rise buildings and composite cross-sections with a high steel content, e.g., due to massive steel cores, such manufacturing-related imperfections can have a pronounced effect on the overall structural behavior, while residual stresses of the steel core have a lesser influence in this case. Production-related imperfections of individual components of multicomponent cross-sections, especially of massive steel cores, should be taken into account in the structural designs. However, manufacturing-related imperfections resulting from misalignment and offset of the massive steel core can also be avoided to a certain extent if high manufacturing tolerances are maintained in production.

REFERENCES

[1] DIN EN ISO 6892-1:2020-06, "Metallic materials - Tensile testing - Part 1: Method of test at room temperature", 2020.

[2] DIN EN 12390-3:2019-10, "Testing hardened concrete - Part 3: Compressive strength of test specimens; German version EN 12390-3:2019.

[3] Neuenschwander, M.; Knobloch, M.; Fontana, M., „Suitability of the damage-plasticity modelling concept for concrete at elevated temperatures: Experimental validation with uniaxial cyclic compression tests", Journal of Cement and Concrete Research 79, Elsevier 2015.

[4] Hanswille, M.; Lippes, M., "Design of composite columns made of concrete filled tubes with inner massive core profiles and high strength materials", Conference: International Conference on Composite Construction in Steel and Concrete, 2008.

[5] Romero, M.L.; Espinos, A.; Portolés, J.M.; Hospitaler, A.; Ibañez, C., "Slender double-tube ultra-high strength concrete-filled tubular columns under ambient temperature and fire", Journal of Structural Engineering, 2015.

[6] ABAQUS/CAE User's Manual, Dassault Systèmes, 2018.

9TH INTERNATIONAL CONFERENCE ON COMPOSITE CONSTRUCTION IN STEEL AND CONCRETE
M. Knobloch, U. Kuhlmann, W. Kurz, M. Schäfer. (Eds.)
Online, Germany, July 27-29, 2021

VARIABLE STIFFNESS REDUCTION FACTOR FOR STABILITY DESIGN OF STEEL-CONCRETE COMPOSITE COLUMNS

Mark D. Denavit*

* Assistant Professor, University of Tennessee, Knoxville
e-mails: mdenavit@utk.edu

Keywords: Composite Construction, Columns, Stability, Design, Stiffness Reduction.

Abstract. *In United States practice, the strength of composite columns is assessed with available strengths that are computed considering member length effects. This approach has advantages; however, developing axial compression-bending moment interaction diagrams that account for member length effects is not a straightforward task. Current methods result in unconservative error for highly slender members with large bending moments. While it is theoretically possible to develop more accurate interaction diagrams, a simpler approach, and one which more directly addresses the cause of the unconservative error, is to improve the accuracy of the stiffness used in the elastic analyses to determine required strengths. In this work, new recommendations are developed for this stiffness for steel-concrete composite columns. The recommended stiffness is a function of required axial compression and bending moment, which increases accuracy and enables improved available strength calculations, resulting in more efficient designs.*

1 INTRODUCTION

The development of design standards for composite columns has often lagged behind that of more traditional structural steel and reinforced concrete columns. As a result, design provisions for composite columns are often developed by adapting provisions for columns of other materials, sometimes discounting the advantages of these efficient, strong, stiff, and ductile structural members. In United States practice, the design of composite columns for buildings is governed by the AISC *Specification for Structural Steel Buildings* [1]. In this standard, design for strength is performed by comparing required strengths, typically computed from a second-order elastic analysis to available strengths computed using the methods and equations defined within the standard. For members with combined axial compression and bending moment, the available interaction strength is computed with consideration of member length effects. Developing appropriate methods or equations for accurate interaction strength is not a straightforward task. Current methods for composite columns involve computing an interaction diagram for a cross section, then applying a transformation that reduces the axial compressive strength of each point on the diagram, however, this produces illogical results in the region of the interaction diagram near the balance point. Other methods involve calibrating a beam-column interaction diagram directly, however, this requires many analyses and results are limited to specific member types and ranges of parameters. Both of these methods result in errors for highly slender members subjected to large bending moments [2,3]. Part of the challenge is that the required strength is based on second-order elastic analyses that employ a stiffness that is a fixed value, even though the level of cracking within the composite member depends on the magnitude of axial compression and bending moment.

This work builds upon previous work [2,3] to develop and validate a stiffness reduction factor that can reduce the unconservative errors observed for highly slender members with large bending moments. The results of this work will lead to more efficient designs and accelerated development of safe and reliable design provisions for newer forms of composite columns, including those with multiple steel shapes and high-strength materials.

Mark D. Denavit

2 METHODOLOGY

The goal of this work is to verify that a variable stiffness reduction factor, when used within the direct analysis method as defined in the AISC *Specification* [1], results in reduced unconservative errors in the strength evaluation of steel-concrete composite columns. Error in the design methodology will be quantified following standard procedures which involve comparing the maximum permitted applied loads per the design methodology to the applied loads which cause failure according to a second-order inelastic analysis. The comparison will be made for a large number of small composite frames that span a wide range of material and geometric properties.

2.1 Parametric Suite

The small composite frames investigated in this work consist of a single composite column as shown in Figure 1. The suite of cross-sectional and frame level parameters used in this work is identical to that used in previous investigations [2,3]. Four categories of cross section were investigated: 1) circular concrete-filled steel tubes (CCFT), 2) rectangular concrete-filled steel tubes (RCFT), 3) steel reinforced concrete (SRC) members subjected to major-axis bending, and 4) SRC subjected to minor-axis bending. Within these groups, sections were selected to span practical ranges of concrete strength, steel ratio, and for the SRC sections, reinforcing ratio. Steel yield strengths were selected as F_y = 345 MPa (50 ksi) for wide flange shapes, F_y = 290 MPa (42 ksi) for round HSS shapes, F_y = 317 MPa (46 ksi) for rectangular HSS shapes, and F_{ysr} = 414 MPa (60 ksi) for reinforcing bars. Three concrete strengths were selected: f'_c = 28, 55, and 110 MPa (4, 8, and 16 ksi). With the selected concrete-filled steel tube (CFT) sections the full range of permitted steel ratios is examined, including those associated with noncompact and slender sections. However, local buckling is neglected in this study, both by not modeling it in the inelastic analyses and by not including the strength reductions in the design strength calculations. Thus, the results of this study are only strictly applicable to compact sections. Both sidesway inhibited and sidesway uninhibited conditions were investigated. The frames are based on and expanded from those used in previous studies on structural steel columns [4,5]. For the sidesway inhibited frames, a range of column lengths (L) and end moment ratios (β) were investigated. For the sidesway uninhibited frames, a range of column lengths (L), leaning column load ratios (γ), and end restraints (rotational spring stiffnesses $k_{\theta,top}$ and $k_{\theta,bot}$) were investigated. Each cross section was run with each frame resulting in 1,200 individual cases for each of the CCFT and RCFT groups and 2,880 individual cases for each of the SRC groups. Full details of the selected benchmark frames are reported by Denavit et al. [2].

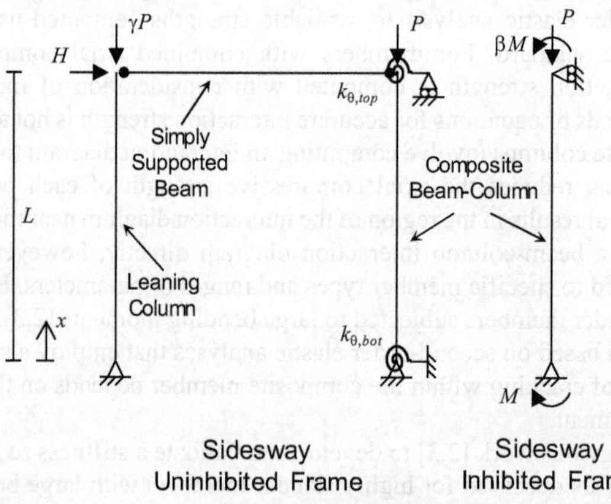

Figure 1. Benchmark frames

2.2 Second-Order Inelastic Analysis

Quantifying error in a design methodology requires a benchmark against which the results of the design methodology can be compared. Physical experiments can provide trusted results, but while thousands of experiments on composite columns have been performed, few have been performed in the range critical to this investigation. Geometric and material nonlinear analysis provides a suitable alternative for which results can be efficiently obtained. Analyses utilizing beam elements (in contrast to continuum and shell elements) were selected to efficiently accommodate the breadth of the parametric suite. The analyses were performed in the OpenSees framework with a mixed beam finite element formulation and fiber cross sections. The uniaxial constitutive relations defined within the fiber representations of the cross sections were calibrated specifically for composite columns. As noted previously, local buckling of the steel tube and other steel components was also neglected. Initial system and member geometric imperfections were directly modeled. Full details of the analyses, including validation against the results of hundreds of physical experiments are reported by Denavit et al. [2] and Denavit and Hajjar [6].

A sample of analysis results is presented in Figure 2 for a CCFT cross section and various lengths of the sidesway inhibited frame with β = 1. The CCFT cross section has an outside diameter, D = 559 mm (22 in), steel tube thickness, t = 9.5 mm (3/8 in), steel yield strength, F_y = 290 MPa (42 ksi), and concrete compressive strength, f'_c = 55 MPa (8 ksi). Note, example results shown later in this paper also use these frame and cross-sectional parameters. Both the applied moments and maximum internal forces at the limit point are shown in Figure 2. The limit point was defined in each analysis as when the lowest eigenvalue of the stiffness matrix was equal to zero. This coincides with the maximum applied moment.

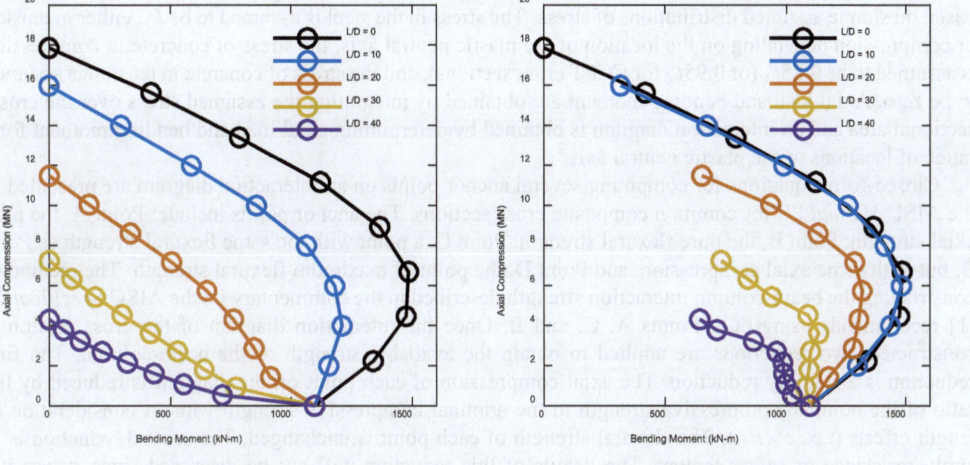

(a) Applied Loads (b) Internal Forces

Figure 2. Second-order inelastic analysis results for an example CCFT cross section

2.3 Design Methodology and Comparison

The direct analysis method defined within the AISC *Specification* [1] consists of a set of rules for calculating required strengths and a set of rules for calculating available strengths. A design is deemed satisfactory if the available strengths meet or exceed the required strengths, often evaluated using an interaction diagram.

For calculation of required strengths, the three key requirements are that 1) required strengths must be computed using a suitably accurate second-order elastic analysis, 2) initial system imperfections (i.e., out-of-plumbness) must be considered, and 3) stiffness reductions must be employed. Full details of the requirements are presented in Chapters C and I of the AISC *Specification* [1]. Details of how these requirements were met in this work are presented here. The elastic analyses in this work were performed by evaluating closed-form solutions to the governing differential equation obtained for the benchmark frames from a computer algebra system. The closed-form solutions were verified against finite-element analyses and are highly accurate within the assumptions made during the derivation (e.g., only flexural deformations are considered). Initial system imperfections were considered for the sidesway uninhibited frames by including a notional lateral load of 0.002 times the vertical load. This notional load was added to the applied lateral load, H (Figure 1). Provisions in the AISC *Specification* permit the notional load to be taken as a minimum lateral load in some circumstances, however, these provisions were not employed in this work. No consideration of initial system imperfections was made for the sidesway inhibited frames. The nominal flexural stiffness of the composite columns is taken as EI_{eff} as defined within the AISC *Specification*. All stiffnesses (i.e., the flexural stiffness of the composite column and the rotational stiffness of the rotational springs in the sideway uninhibited frames) are reduced by 0.8. Additionally, the flexural stiffness of the composite column is reduced further by the factor τ_b. The factor τ_b is defined as a constant 0.8 in Chapter I of the AISC *Specification*. Other versions of the factor τ_b were also investigated as described later.

For calculation of available strengths, Chapter I of the AISC *Specification* [1] provides detailed provisions for the axial compressive strength and the flexural strength. The available strength of members subjected to combined axial compression and bending moment is less well defined. The commentary on the AISC *Specification* [1] describes a method for constructing the interaction strength of a beam column by reducing the interaction diagram of the cross section as calculated using the plastic stress distribution method. The plastic stress distribution method is a method of determining the strength of a cross section based on simple assumed distributions of stress. The stress in the steel is assumed to be F_y, either in tension or compression depending on the location of the plastic neutral axis, the stress of concrete in compression is assumed to be $0.85f'_c$ (or $0.95f'_c$ for CCFT cross sections), and the stress of concrete in tension is assumed to be zero. Axial load and bending moment are obtained by integrating the assumed stress over the cross-sectional area and an interaction diagram is obtained by determining axial load and bending moment for a range of locations of the plastic neutral axis.

Closed-form equations for computing several anchor points on an interaction diagram are presented in the AISC *Manual* [7] for common composite cross sections. The anchor points include: Point A, the pure axial strength; Point B, the pure flexural strength; Point C, a point with the same flexural strength as Point B, but with some axial compression; and Point D, the point of maximum flexural strength. The method of constructing the beam-column interaction strength described in the commentary on the AISC *Specification* [1] recommends using only Points A, C, and B. Once the interaction diagram of the cross section is constructed, two reductions are applied to obtain the available strength of the beam-column. The first reduction is a stability reduction. The axial compression of each point on the diagram is reduced by the ratio of the nominal compressive strength to the nominal compressive strength without consideration of length effects (i.e., P_n/P_{no}). The flexural strength of each point is unchanged. The second reduction is to apply resistance or safety factors. The details of this reduction will not be discussed since it was not performed in this study (i.e., the comparisons were made at the nominal strength level).

An alternative interaction diagram which includes Point D in addition to Points A, C, and B has been recently proposed [3]. Both the interaction diagram constructed in accordance with the recommendations in the commentary on the AISC *Specification* [1] (termed the ACB interaction) and that constructed using Point D (termed the ACDB interaction) are investigated in this work. The use of these interaction diagrams with a constant stiffness reduction as currently defined in the AISC *Specification* form two of the four design methods investigated in this work (Table 1).

Comparing second-order inelastic analysis results and results from the design methodology must be done carefully to ensure a fair comparison. Comparing the maximum internal forces at the limit point of the second-order inelastic analysis to the available strengths from the design methodology would not be a fair or generally useful comparison. The available strengths from the design methodology are intended for comparison against the required strengths computed from an elastic analysis with a specific stiffness. Safe design methodologies can be formulated with stiffnesses that over- or underpredict the true stiffness if the available strength accommodates the resulting differences in geometric nonlinear effect (i.e., P-Δ and P-δ effects). For example, the effective length factor method for structural steel columns utilizes nominal flexural stiffness. This overpredicts the stiffness of the column after the initiation of yielding and thus would generally underpredict the true P-Δ and P-δ effects at the ultimate limit state. However, the use of the effective length factor correspondingly reduces the available strength, resulting in a safe methodology. A better comparison is between the applied loads at the limit point of the second-order inelastic analysis and the maximum permitted applied loads from the design methodology.

Table 1: Design methods

Index	Stiffness Reduction	Interaction Diagram
1	Constant ($\tau_b = 0.8$)	ACB
2	Constant ($\tau_b = 0.8$)	ACDB
3	Variable (Equation 1)	ACB
4	Variable (Equation 1)	ACDB

The applied loads at the limit point of the second-order inelastic analysis are a direct output of the analyses (Figure 2a). The maximum applied loads permitted by the design methodology are obtained from an automated iterative process as the applied loads that result in required strengths that equal the available strength. In other words, the applied loads that produce maximum internal forces from an elastic analysis that lay directly on the interaction diagram (either the ACB or the ACDB interaction). A sample of results is presented in Figure 5 for the example CCFT cross section and the same frames investigated in Figure 2.

The thesis of this work is that the direct analysis method can be improved for steel-concrete composite columns using a stiffness reduction factor τ_b that varies with axial load and bending moment. Such a stiffness reduction factor could more faithfully represent the true behavior of composite members since stiffness does indeed depend on the internal forces. Performing an elastic analysis with a stiffness reduction that varies with internal forces can be cumbersome. However, there is precedent in United States practice. For structural steel members, the factor τ_b varies with axial compression. For reinforced concrete members, the ACI *Code* [8] includes provisions for an effective flexural rigidity that varies with both axial compression and bending moment.

The form of the variable stiffness reduction factor used in this work is shown in Equation 1 based on prior research [6]. The variation of the stiffness reduction factor with axial compression and bending moment is shown in Figure 3.

$$\tau_b = 1.25\frac{M_r}{M_n}\left(1-3\frac{P_r}{P_{no}}\right) \leq 0.8 \tag{1}$$

With a constant stiffness reduction, the maximum permitted applied loads will always result in required strengths that equal available strengths. With a variable stiffness reduction, it is possible that the maximum permitted applied loads are obtained for a case where the required strengths are less than the available strengths. This complicates the determination of maximum permitted applied loads, but they can still be determined in an automated process. To detect cases where the maximum applied load occurs before the maximum internal force reaches the interaction strength, the whole load path is computed, not just the final point. For non-proportional analyses, a loop is made over the internal forces from M_r of zero to the moment strength. At each step of the loop, the reduced stiffness is computed, and the applied load is back calculated in an iterative process. If the applied load for the current step is found to be smaller than the previous step, this indicates a stability-type failure, and that the maximum applied load has been reached.

The effect of the variable stiffness reduction on the maximum permitted applied loads for the example CCFT cross section is shown in Figure 4c for the ACB interaction and Figure 4d for the ACDB interaction. These figures are similar to Figure 4a and Figure 4b where the line with markers represents the maximum permitted applied loads and the line without markers represents the interaction diagram. For the variable stiffness reduction, there are also dashed lines. These represent the internal forces at the maximum permitted applied loads. They are only visible when they are less than the available strengths, which in this case occurs for the more slender members and large applied bending moments. Some attractive features of the variable stiffness reduction can be seen by comparing the results in Figure 4. The variable stiffness reduction has no effect on the pure bending strength, nor does it affect the strength when the axial compression is high. Also, it has a greater effect on more slender members.

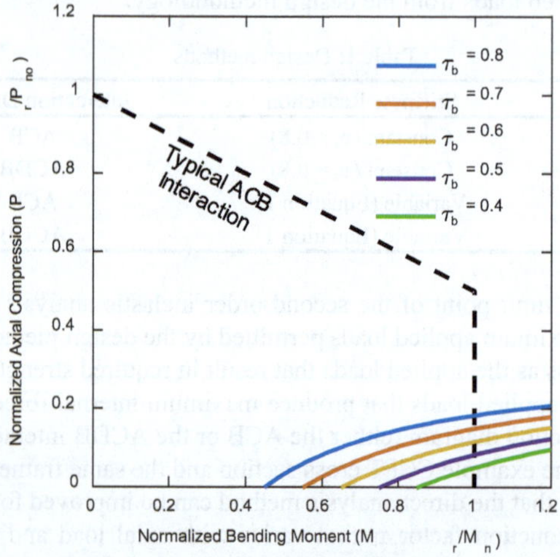

Figure 3. Contour plot showing the variation of the alternative stiffness reduction factor.

A comparison between the maximum permitted applied loads from the design methodology and the maximum applied loads from the second-order inelastic analysis are presented in Figure 5. The cases presented in Figure 5 are drawn from Figure 2 and Figure 4, but only show $L/D = 20$ and $L/D = 40$ and only the ACB interaction equation. The key result from these comparisons is the error measured along a radial line from the origin between the interaction diagrams constructed from the maximum applied loads permitted by the design methodology and the applied loads at which failure occurs according to the second-order inelastic analyses. For higher axial loads the interaction diagram constructed from the inelastic analyses is outside the interaction diagram constructed from the design methodology, indicating conservative error. For higher bending moments and the constant stiffness reduction, the opposite is true, albeit to a lesser degree. The design methodology permits applied loads that the inelastic analysis indicates would result in failure. Use of the variable stiffness reduction pulls the maximum permitted applied loads back within the region that is shown to be safe by the second-order inelastic analyses.

3 RESULTS

To more broadly evaluate the design methodologies, the radial error described in the previous section is computed for each frame and cross-section in the parametric suite, for each design method listed in Table 1, and at many angles within the *M-P* plane. While the selected cross sections and frame parameters can be considered to span the practical range, the distribution of parameters within the selection may not be representative of what is expected in practice. For instance, the selected set contains a far higher proportion of very slender frames than would be expected in typical construction. Accordingly, maximum and minimum error values are more meaningful than median or average error values. Two of the most influential parameters within the set are the steel ratio, ρ_s, and the slenderness. Slenderness is defined by the parameter λ_{oe} (Equation 2) which is proportional to the effective length of the column. An effective length factor is computed and used for determining λ_{oe} despite not being used within the direct analysis method. The frames are separated into bins based on ranges of steel ratio and slenderness to better understand the error. The ranges used to separate the frames based on slenderness are shown in Table 2.

$$\lambda_{oe} = \sqrt{\frac{P_{no}}{P_e}} \tag{2}$$

$$P_e = \frac{\pi^2 EI_{eff}}{(KL)^2} \tag{3}$$

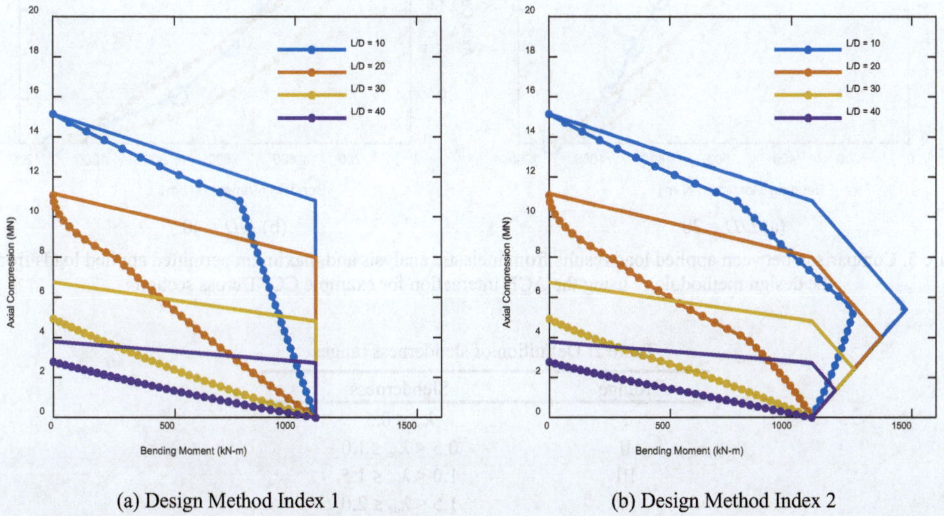

Figure 4. Design methodology results for example CCFT cross section

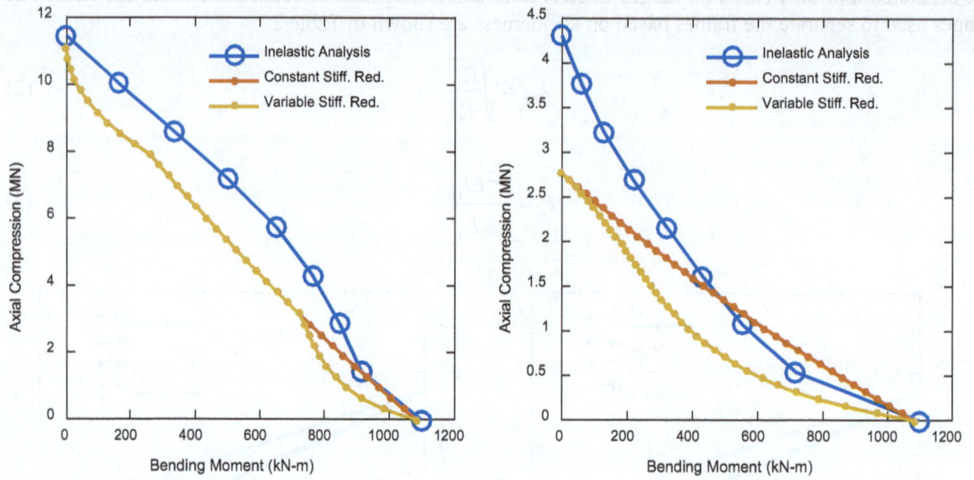

(a) $L/D = 20$ (b) $L/D = 40$

Figure 5. Comparison between applied load results from inelastic analysis and maximum permitted applied loads from the design methodology using the ACB interaction for example CCFT cross section

Table 2: Definition of slenderness ranges

Range	Slenderness
I	$\lambda_{oe} \leq 0.5$
II	$0.5 < \lambda_{oe} \leq 1.0$
III	$1.0 < \lambda_{oe} \leq 1.5$
IV	$1.5 < \lambda_{oe} \leq 2.0$
V	$2.0 < \lambda_{oe} \leq 3.0$
VI	$3.0 < \lambda_{oe}$

The maximum unconservative error for each of the bins for the ACDB interaction is shown in Table 3 for the constant stiffness reduction (i.e., Design Method 2 as defined in Table 1) and in Table 4 for the variable stiffness reduction (i.e., Design Method 4). The maximum unconservative error varies significantly with section type, slenderness, and steel ratio. For the constant stiffness reduction, the greatest unconservative errors are seen for the slenderest and most concrete-dominant cases. There is no specified limit on the level of unconservative error that can be tolerated within a design methodology. One reference identifies 5% unconservative error as a reasonable maximum [9]. Another study identified unconservative errors as large as 16% for structural steel columns designed according to the direct analysis method with direct modeling of member imperfections [10].

Use of the variable stiffness reduction essentially eliminates the unconservative error for the slenderest and most concrete-dominant cases. Some error remains for the SRC members, however, upon review of detailed results it is possible that the errors are an artifact of the discretization of the interaction diagram. In this work, interaction diagrams from second-order inelastic analyses are computed with nine points equally spaced in axial load. Additional points may be necessary to accurately characterize the interaction strength for these cases.

Table 3: Maximum unconservative error based on slenderness and steel ratio for constant stiffness reduction and ACDB interaction (Design Method 2)

	ρ_s	I	II	III	IV	V	VI
CCFT	0.25	6.2%	14.6%	12.4%	13.6%	3.9%	4.1%
	0.18	6.7%	12.4%	14.0%	15.8%	6.6%	7.0%
	0.11	5.2%	9.5%	14.5%	17.9%	8.0%	10.2%
	0.06	6.2%	9.0%	11.7%	12.8%	19.3%	14.2%
	0.02	4.6%	1.0%	7.3%	15.6%	24.9%	30.2%
RCFT	0.28	0.8%	1.7%	1.9%	2.9%	0.0%	0.0%
	0.19	2.2%	3.0%	5.2%	7.0%	0.0%	0.0%
	0.11	4.0%	5.1%	8.3%	11.4%	4.1%	4.8%
	0.06	3.2%	5.0%	8.8%	2.7%	16.3%	11.9%
	0.03	0.0%	2.3%	5.0%	10.5%	18.9%	17.6%
SRC *major-axis*	0.12	5.7%	3.6%	3.6%	4.7%	7.0%	1.4%
	0.09	4.6%	3.5%	3.7%	6.8%	9.4%	2.8%
	0.04	1.2%	0.9%	2.4%	10.6%	14.9%	11.0%
	0.01	2.1%	2.1%	5.3%	7.9%	14.9%	25.3%
SRC *minor-axis*	0.12	14.2%	17.9%	15.8%	14.1%	14.3%	7.4%
	0.09	13.4%	15.0%	10.5%	13.0%	13.4%	6.3%
	0.04	5.4%	6.0%	8.1%	11.5%	14.3%	10.4%
	0.01	2.1%	2.1%	4.6%	8.0%	12.4%	27.4%

Table 4: Maximum unconservative error based on slenderness and steel ratio for variable stiffness reduction and ACDB interaction (Design Method 4)

	ρ_s	I	II	III	IV	V	VI
CCFT	0.25	6.2%	14.6%	2.5%	0.0%	0.0%	0.0%
	0.18	6.7%	12.4%	1.8%	0.0%	0.0%	0.0%
	0.11	5.2%	8.2%	0.6%	0.2%	0.0%	0.0%
	0.06	6.2%	5.3%	2.3%	2.4%	2.9%	0.0%
	0.02	4.6%	1.0%	0.1%	2.0%	3.0%	0.0%
RCFT	0.28	0.8%	1.7%	0.0%	0.0%	0.0%	0.0%
	0.19	2.2%	2.3%	0.0%	0.0%	0.0%	0.0%
	0.11	4.0%	2.6%	0.0%	0.0%	0.0%	0.0%
	0.06	3.2%	0.9%	0.0%	0.0%	0.0%	0.0%
	0.03	0.0%	2.3%	0.0%	0.0%	0.0%	0.0%
SRC *major-axis*	0.12	5.7%	3.6%	1.8%	2.0%	2.4%	1.4%
	0.09	4.6%	3.5%	1.6%	2.4%	3.0%	1.5%
	0.04	1.2%	0.8%	0.8%	2.3%	3.5%	0.7%
	0.01	2.1%	2.1%	2.1%	3.2%	3.5%	3.7%
SRC *minor-axis*	0.12	14.2%	17.9%	15.8%	8.9%	8.9%	7.4%
	0.09	13.4%	15.0%	9.6%	7.8%	7.9%	6.3%
	0.04	5.4%	6.0%	4.0%	4.5%	4.5%	2.6%
	0.01	2.1%	2.1%	2.1%	3.2%	3.6%	4.3%

Other areas of unconservative error remain even with the variable stiffness reduction factor. One case is steel-dominant CCFT members with intermediate slenderness. This error has been noted before [2] and is due to an overprediction of the axial strength by the design equations relative to the second-order inelastic analysis. Another case is steel-dominant SRC members bent about their minor axis. This error has also been noted before [2] and is due to an overprediction of the flexural strength by the plastic stress distribution method relative to the second-order inelastic analysis. Further work has identified the lack of consideration of strain compatibility in the plastic stress distribution method as the cause of the error and proposed correction factors to address the issue [11].

Results for the ACB interaction are similar, although the maximum unconservative errors are not as severe for the constant stiffness reduction. Increased unconservative error is one of the reasons that the commentary on the AISC *Specification* [1] recommends neglecting Point D and using the ACB interaction. The data presented in Table 3 and Table 4 demonstrates that neglecting Point D is not necessary if the variable stiffness reduction presented in this work is employed. It is anticipated that use of the variable stiffness reduction factor presented in this work, or a similar one, can enable other methods of calculating the available strength. For example, it may be possible to base the available strength on that of the composite cross section without any slenderness reductions at all. Developing a safe and efficient design methodology for composite columns that is based on cross-sectional strength would be tremendously beneficial since determining cross-sectional strength is more straightforward and design recommendations for newer forms of composite columns could be developed more quickly.

Evaluation of unconservative error is critical for safety. Maximum conservative errors are a measure of the efficiency of the design method. The difference in maximum conservative error between Design Methods 2 and 4 (as defined in Table 1) is presented in Table 5. Given that the variable stiffness reduction is always less than or equal to the constant stiffness reduction, use of the variable stiffness reduction can only decrease the maximum permitted applied loads and thus increase conservative error. The increase in maximum conservative error is generally limited to the slenderest frames and the concrete-dominant frames. The maximum conservative error for steel-dominant frames with low to moderate slenderness is unaffected by the introduction of the variable stiffness reduction.

Table 5: Difference in maximum conservative error based on slenderness and steel ratio, Design Method 2 vs. Design Method 4

	ρ_s	I	II	III	IV	V	VI
CCFT	0.25	0.0%	0.0%	0.0%	0.0%	3.3%	4.3%
	0.18	0.0%	0.0%	0.0%	0.1%	3.9%	5.2%
	0.11	0.0%	0.0%	0.0%	0.7%	3.1%	5.2%
	0.06	2.6%	11.8%	5.2%	4.1%	1.8%	3.8%
	0.02	20.1%	27.5%	21.7%	14.9%	10.5%	7.0%
RCFT	0.28	0.0%	0.0%	0.0%	0.0%	5.4%	6.2%
	0.19	0.0%	0.0%	0.0%	0.0%	5.5%	6.7%
	0.11	0.0%	0.0%	0.0%	0.4%	3.7%	7.3%
	0.06	2.4%	4.7%	3.7%	2.5%	0.0%	4.0%
	0.03	7.3%	23.0%	10.5%	7.1%	3.3%	5.7%
SRC *major-axis*	0.12	0.0%	0.0%	0.0%	0.0%	0.7%	9.0%
	0.09	0.0%	0.0%	0.0%	0.0%	0.8%	2.1%
	0.04	2.0%	1.9%	0.0%	0.7%	2.0%	1.8%
	0.01	19.3%	24.3%	12.4%	6.4%	5.5%	0.0%
SRC *minor-axis*	0.12	0.0%	0.0%	0.0%	0.0%	0.3%	2.7%
	0.09	0.0%	0.0%	0.0%	0.0%	0.5%	1.6%
	0.04	2.3%	0.0%	0.0%	0.5%	1.6%	1.5%
	0.01	18.9%	24.1%	12.1%	6.1%	5.3%	0.0%

4 SUMMARY AND CONCLUSIONS

This work investigated the potential for a variable stiffness reduction factor to improve stability design of steel-concrete composite columns. Error in the direct analysis method defined within the AISC *Specification* [1] was quantified using established procedures. As with previous work, unconservative errors were noted for concrete-dominant members with high slenderness. Use of the variable stiffness reduction factor given by Equation 1 effectively eliminated these errors, while only minorly increasing conservative error. This observation held even when using a strength interaction equation that includes the balance point. Further development is warranted, for example, the specific factors in Equation 1 could potentially be refined. However, these results have established that a variable stiffness reduction factor can provide a solution to one of the most pressing issues in stability design of steel-concrete composite columns.

REFERENCES

[1] AISC, *Specification for Structural Steel Buildings*. American Institute of Steel Construction, Chicago, Illinois, 2016.

[2] Denavit, M. D., Hajjar, J. F., Perea, T., and Leon, R. T., "Stability Analysis and Design of Composite Structures," *Journal of Structural Engineering*, ASCE, 142(3), 2016.

[3] Denavit, M. D., "Interaction Strength of Steel-Concrete Composite Beam-Columns Including the Balance Point," *Proceedings of the Annual Stability Conference*, Structural Stability Research Council, 2020.

[4] Kanchanalai, T., *The Design and Behavior of Beam-Columns in Unbraced Steel Frames*, Structures Research Laboratory, Department of Civil Engineering, The University of Texas at Austin, Austin, Texas, 1977.

[5] Surovek-Maleck, A. E. and White, D. W., "Alternative Approaches for Elastic Analysis and Design of Steel Frames. II: Verification Studies," *Journal of Structural Engineering*, ASCE, 130(8), 1197–1205, 2004.

[6] Denavit, M. D. and Hajjar, J. F., *Characterization of Behavior of Steel-Concrete Composite Members and Frames with Applications for Design*, University of Illinois at Urbana-Champaign, Urbana, Illinois, 2014.

[7] AISC, *Steel Construction Manual*, 15th Edition, American Institute of Steel Construction, Chicago, Illinois, 2017.

[8] ACI, *Building Code Requirements for Structural Concrete and Commentary*, American Concrete Institute, Farmington Hills, Michigan, 2019.

[9] ASCE, *Effective Length and Notional Load Approaches for Assessing Frame Stability: Implications for American Steel Design*, American Society of Civil Engineers, Reston, Virginia, 1997.

[10] Wang, Y. and Ziemian, R. D., "Design by Advanced Elastic Analysis - An Investigation of Beam-Columns Resisting Minor-Axis Bending," *Proceedings of the Annual Stability Conference*, Structural Stability Research Council, 2019.

[11] Behnam, A. and Denavit, M. D., "Plastic Stress Distribution Method for Predicting Interaction Strength of Steel-Concrete Composite Cross Sections," *Journal of Constructional Steel Research*, 170, 2020.

COMPOSITE DECKS

9TH INTERNATIONAL CONFERENCE ON COMPOSITE CONSTRUCTION IN STEEL AND CONCRETE
M. Knobloch, U. Kuhlmann, W. Kurz, M. Schäfer. (Eds.)
Online, Germany, July 27-29, 2021

CHARACTERIZATION OF THE LOAD-SLIP BEHAVIOUR OF HEADED STUD SHEAR CONNECTIONS IN NARROW PROFILED SHEETING

V. Vigneri*, C. Odenbreit*, M. Schäfer*, S. Hicks, D. Lam*** and F. Hanus******

* University of Luxembourg, Faculty of Science, Technology and Medicine, Department of Engineering
e-mails: vigneri@ibk.baug.ethz.ch, christoph.odenbreit@uni.lu, markus.schaefer@uni.lu

** University of Warwick, School of Engineering
Stephen.J.Hicks@warwick.ac.uk

*** University of Bradford, Faculty of Engineering and Informatics,
Department of Civil and Structural Engineering
d.lam1@bradford.ac.uk

**** Steligence Engineering, ArcelorMittal
Francois.Hanus@arcelormittal.com

Keywords: Composite beams, Shear connectors, Headed studs, Mechanical models, Load-bearing mechanisms, Strut and tie modelling.

Abstract. *Although headed stud shear connections in profiled sheeting represent an efficient solution to transfer the longitudinal shear along composite beams, current EN 1994-1-1 may overestimate their design resistance in case of open trough profile steel sheeting with narrow ribs. Project team CEN/TC250/SC4.T3 proposed new equations based on a "cantilever" model which depends on the concrete tensile strength and the bending resistance of the connector. However, by increasing the slips, the internal forces redistribute, and the system can be represented by a "modified strut and tie" model. At large displacements, a "strut and tie" model was also developed accounting for the influence of the tensile forces in the connector. The respective analytical equations for predicting the load per stud were finally derived and compared with experimental load-slip curves. It was found that the analytical models proposed lead to satisfactory predictions of the load-slip behaviour of the shear connection.*

1 INTRODUCTION

Composite steel-concrete beams represent an optimal way to exploit the compressive strength of the concrete together with the high yield strength and ductility of structural steel. However, to ensure the composite action of steel-concrete beams, the longitudinal shear force between the concrete slab and the steel elements shall be transferred by appropriate shear connectors. In the frame of typical composite downstand beams, the most common shear connector is the headed stud which provides good structural performance in terms of strength and ductility as well as quick and easy installation. Among the different floor solutions available, two main alternatives are shown in Figure 1: the former consist of a solid slab which is typically cast-in-place whereas in the latter an additional profiled steel sheeting is employed.

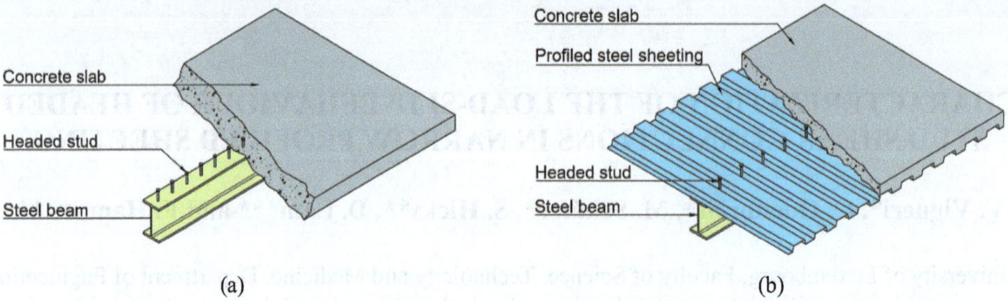

Figure 1. Typical composite steel-concrete beam with (a) solid slab and (b) transverse profiled steel sheeting

In the last decades, the construction industries have favoured the use of headed studs in combination with profiled steel sheeting transverse to the direction of the supporting beam for the use in buildings. This represents a cost-effective solution as the stud can be easily through-deck welded or placed in pre-punched holes. In parallel, the steel sheeting does not only act as formwork but it also increases the bending resistance of the slab and it reduces the concrete volume, thereby reducing the self-weight of the slab. For this purpose, several research studies were carried out alongside the appearance of more modern and sophisticated steel sheeting geometries.

The current work focuses on the analytical characterization of the entire load-slip behaviour of headed stud shear connections in modern, open trough profiled steel sheeting with narrow ribs. The motivation behind this investigation is explained in Section 2, while Section 3 provides the methodology undertaken in this contribution to achieve the objectives. The identification of three separate load bearing phases arising at different displacements is given in Section 4 and the corresponding analytical models and equations proposed are defined in Section 5. Finally, a consistent comparison between analytical predictions and experimental results is provided in Section 6 before summarizing the key outcomes and outlook for further research in Section 7.

2 MOTIVATION

Current European Standard for the design of composite constructions, namely EN 1994-1-1 [1], gives a semi-empirical formulation for determining the design strength of the connector but it does not account for the real mechanical behaviour and failure mode. This design model based on the resistance of headed studs in solid slabs including an additional reduction factor to be used with profiled sheeting was developed in the 1980-90s [2, 3, 4, 5]. It provides a sufficient level of safety for the resistance of headed studs in combination with profiled steel sheeting used at that time, which are still representing an important contribution on the market today. In the last years, the geometry of profile sheeting has indeed partially changed, some open profiled sheeting with particularly narrow ribs are available: in such cases, the distance between the headed stud and the edge of the concrete rib may be less than 60 mm at half height of the concrete rib, while no sufficient embedment depth of the headed stud above the profiled sheeting is provided or no reinforcement layer under the head of the stud is available. Those conditions are not covered by the Eurocode approach. To overcome this issue, project team CEN/TC250/SC4.T3 developed a mechanical-based approach on the basis of the push-out tests carried out within the RFCS project DISCCO [6]. However, the shear connection undergoes different stages involving the gradual damage of the concrete and the bending deformation of the connector, as already pointed out in [7] and further investigated in more recent studies [8, 9]. This work presents the analytical characterization of the load-slip behaviour of headed stud shear connections in profiled steel sheeting with narrow troughs in the frame of the research project ICO-SHEAR [10].

3 METHODOLOGY

The observation of the failure modes after testing does not generally give a clear picture of the mechanical behaviour and the resistance mechanisms activated at low displacements. Therefore, some of the experimental push-out tests carried out within this project [10] have been intentionally stopped at different slips in order to outline the progressive damage pattern of the concrete rib, as well as the bending deformation of the connector. In accordance with the objectives of the project, all the test specimens used the modern profiled steel sheeting ArcelorMittal Cofraplus 60 [11] which has particularly narrow ribs with a bottom and top width of the trough equal to 62 mm and 101 mm, respectively. Headed studs with a nominal length of 100mm and 125mm were considered in this experimental campaign using different position of the reinforcement layer.

In parallel, a validated 3D FE model was developed using the commercial software ABAQUS [12] in order to quantify the compressive stresses in the concrete and to determine the degree of activation of the plastic hinges in the stud [13]. From the evaluation of experimental evidence and the additional results from the numerical model, the mechanical response of the connection was divided into three main phases, where different load bearing mechanisms have been identified, see Figure 2. These load bearing stages and respective analytical models proposed are described in the next sections.

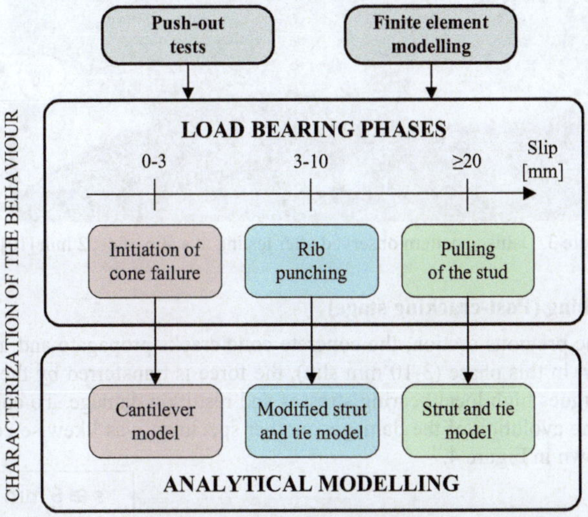

Figure 2. Methodology adopted in the present work for characterizing the load-slip behaviour of headed studs

4 LOAD BEARING PHASES

Based on the aforementioned experimental and numerical tools, three load-bearing phases in headed shear connectors employed in profiled steel sheeting were distinguished:

- Phase 1 (0-3 mm slip): Initiation of concrete cone failure (Pre-cracking stage)
- Phase 2 (3-10 mm slip): Rib punching (Post-cracking stage)
- Phase 3 (ca. 25 mm slip): Pulling of the stud (Large displacements)

4.1 Phase 1: Initiation of concrete cone failure (Pre-cracking stage)

At early stages (0-3 mm slip), the response of the shear connection is characterized majorly by the local bending deformation of the connector and the initiation of the typical concrete cone cracks. This is confirmed by the cut section of a specimen with 100 mm long stud at a slip of ca. 2 mm given in Figure 3 where it can be observed that:

(i) Owing to the high bending stiffness of the whole concrete trough, the crack at the edge of the rib occurs at very low displacements and it starts to propagate across the rib. The crack pattern is not fully developed at this stage.

(ii) In parallel, the high bearing compressive stresses at the shank-concrete interface lead to the first diagonal crack which further propagate in the next phases.

Although no clear bending deformation can be noticed in the connector after testing, based on the supplementary numerical evaluation of the normal stresses in the stud [13], it is thought that it develops a double curvature in proximity of the weld collar.

Figure 3. Damage pattern observed after testing at a slip of ca. 2 mm [10].

4.2 Phase 2: Rib punching (Post-cracking stage)

As mentioned in the previous section, the concrete cone cracks propagate and all the internal forces redistribute accordingly. In this phase (3-10 mm slip), the force is transferred by the concrete around the stud shank which undergoes high local bearing stresses and resulting damage. To check the development of the crack path and the evolution of the damage, another specimen was likewise cut longitudinally at a slip of ca. 6 mm, as shown in Figure 4.

Figure 4. Damage pattern observed after testing at a slip of ca. 6 mm [10].

On that basis, the following considerations can be made:

(i) The concrete cone crack pattern spreads across the rib in a ductile manner allowing for the redistribution of the forces.

(ii) The bearing stresses acting on the stud shank increases while the damage of the concrete in front of the stud is now clearly visible: this corresponds to the so called "rib punching failure".

(iii) As already pointed out in [14], part of the bearing forces is transferred to the steel sheeting via the "concrete wedge" which pushes against the right corner of profile.

(iv) Simultaneously, the tensile forces along the connector are balanced by the compression strut developing on the rear side of the rib, leading to the vertical crack appearing underneath the stud head.

4.3 Phase 3: Pulling of the stud (Large displacements)

By further increasing the slip, see Figure 5, a significant part of the load is carried by the connector through tension forces in this phase. The high shear and tensile stresses acting in the shank may also lead to the fracture of the stud itself.

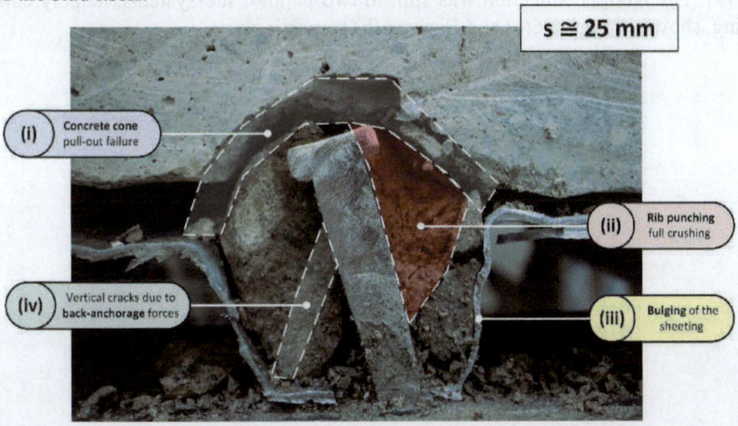

Figure 5. Damage pattern observed after testing at a slip of ca. 25 mm [10].

By analysing the longitudinal cut section of the specimen after testing (at a slip of ca. 25 mm) in Figure 5, the following considerations can be drawn:

(i) Concrete pull-out failure occurs where the cone gradually detaches from the slab due to the increasing pulling forces.

(ii) The front part of the rib is fully damaged and the bearing forces at the stud-concrete interface decrease.

(iii) The "concrete wedge" pushes further against the steel sheeting which bulges. For further displacements, the steel deck may be torn as confirmed by other studies [6].

(iv) The compressive forces acting on the rear side of the concrete rib (back-anchorage) grow further leading to the propagation of the vertical cracks.

Although the 100 mm long stud considered in this case exhibits only one plastic curvature, when longer studs are employed, it is likely that a second plastic curvature is still visible at such high displacements, as confirmed by other experimental tests [10].

5 ANALYTICAL MODELLING

The failure conditions of each model were defined in line with experimental evidences in order to determine the respective value of the load carried by the connection. These analytical values of the resistance were finally collected for the characterization of the load-slip curve of the shear connection.

For each phase identified in the previous section, a respective mechanical model was developed, respectively:

- Phase 1: Cantilever model
- Phase 2: Modified Strut and Tie model (MSTM)
- Phase 3: Strut and Tie model (STM)

Unlike Phase 1, where the concrete was modelled as a cantilever beam element (according to Euler-Bernoulli theory), the other two models considered the rib as an equivalent system of struts in accordance with the direction of the principal compressive stresses identified in the FE model. In all cases, the failure conditions of each mechanical model were consistently defined according to the experimental results and the support of further numerical studies. A detailed description of each model is given in the following subsections.

5.1 Phase 1: Cantilever model (Pre-cracking stage)

The shear connection in Phase 1 was associated to a cantilever beam model which is based on the work of Nellinger [9]. The whole connection was split in two parallel sub-systems, namely concrete cone and stud in bending, shown in Figure 6(a) and Figure 6(b) respectively.

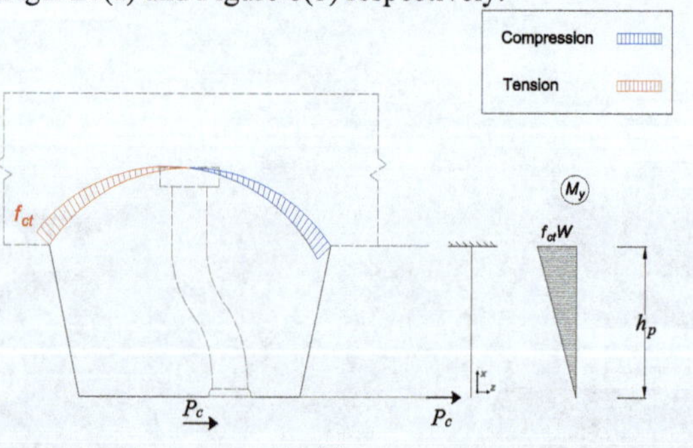

(a)

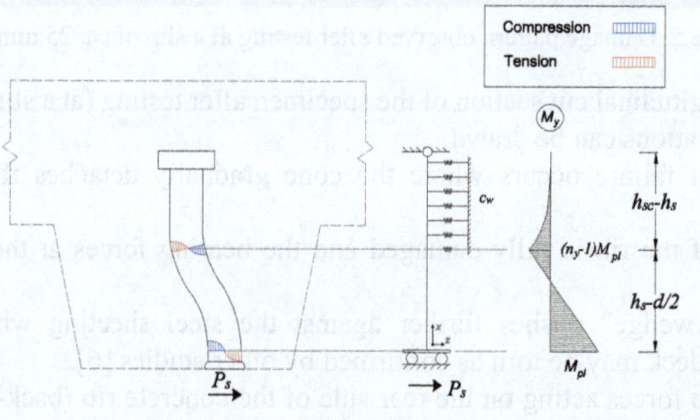

(b)

Figure 6. Cantilever model: (a) concrete cone and (b) stud in bending component [10].

It is worth noticing that the positive action of the transversal load which may apply on the slab [15] is not implicitly considered in this model although it may be implemented as an equivalent stabilizing moment preventing the rotation of the concrete rib. However, in practical applications this positive effect may not always be guaranteed due to possible counteracting lifting forces (e.g. in the zone of web openings, by continuous concrete slabs or loading directly on the steel profile), thus this effect was conservatively neglected. From the stress distribution of a linear elastic cantilever beam shown in Figure 6(a), the concrete cone failure is assumed to initiate at the edge where the stress reaches the tensile strength of the material f_{ct}. From the rotational equilibrium equation, the respective load P_C can be easily computed as follows:

$$P_c = \frac{f_{ct}W}{n_r h_p} \tag{1}$$

And the section modulus of the concrete cone failure surface estimated in [9] is given by:

$$W = [2.4h_{sc} + (n_r - 1)e_t]\frac{b_{max}^3}{6b_{top}} \tag{2}$$

Where f_{ct} is the tensile strength of the concrete, n_r indicates the number of studs placed in each rib and e_t is the transversal spacing between the studs. b_{max} represents the maximum value between the widths b_{top} and b_{bot}. However, in case of open trough profiles b_{max} is always equal to b_{top}.

In parallel, the bending deformation in the connector stud induces high normal stresses. While the bending moment at the base reaches always the plastic bending resistance of the cross section (full plastic hinge), a second bending curvature may develop at a vertical coordinate h_s. As in the case of pile foundations, the degree of activation of the second plastic hinge is strongly related to the embedment provided by the surrounding concrete. Such effect was introduced in the model through a system of horizontal springs with stiffness c_w reproducing the bearing forces of the concrete, as shown in Figure 6(b). Since the value c_w can be hardly quantified analytically, the equivalent parameters n_y and h_s were used for determining the resistance component P_s. Assuming that the degree of activation of the plastic hinges n_y is known, and the bottom plastic hinge develops at a height of $d/2$ [9], the system can be solved through equilibrium equations which leads to the following value of the load P_s:

$$P_s = \frac{n_y M_{pl}}{h_s - d/2} \tag{3}$$

Where h_s and n_y can be simplified as follows:

$$h_s = 0.82h_p \tag{4}$$

$$n_y = \begin{cases} 2 & n_r = 1 \\ 1.92\,\dfrac{h_{sc} - h_p}{2d} - 2.84 \leq 2, \text{but not lower than } 1.0 & n_r = 2 \end{cases} \tag{5}$$

And:

$$M_{pl} = \frac{f_u d^3}{6} \tag{6}$$

Assuming that P_c and P_s are acting in parallel at this stage, the resulting resistance in phase 1 P_1 is given by the following expression:

$$P_1 = C \cdot k_u \left(\frac{f_{ct}W}{n_r h_p} + \frac{n_y M_{pl}}{h_s - d/2}\right) \tag{7}$$

Where the factor C was statistically estimated and defined by Eq. (8) as a function of the average width of the trough b_0. The correction factor k_u accounts for the effect of the eccentric position as well as the welding procedure and the thickness of the sheeting t, see Table 1.

$$C = 1.85\frac{h_p}{b_0}, \quad 1.0 \leq C \leq 1.35 \tag{8}$$

Table 1: Values of the correction factor k_u.

k_u	Profiled sheeting with pre-punched holes	Through-deck welded studs $t < 1$ mm	Through-deck welded studs $t \geq 1$ mm
Centred or staggered position	1.0	1.05	1.25
Favourable position	1.1	1.16	1.38
Unfavourable position	0.8	0.95	1.0

5.2 Phase 2: Modified strut and tie model (Post-cracking stage)

Unlike the strut and tie models for headed studs proposed in the past [14, 16], the "modified" strut and tie model (MSTM) proposed in this paper includes not only concrete compression struts and tension elements, but also the bending resistance component of the stud. In accordance with the numerical evaluation of the stress field in the concrete [8], the rib was replaced by a system of equivalent struts which follows the direction of the principal compressive stresses. The connector was again replaced by beam elements with two local plastic hinges. In consideration of the relatively small flexural stiffness of the profiled steel sheeting, this was assumed to transfer only tension forces, i.e. tie elements. Finally, the assembled structural system of the MSTM is illustrated in Figure 7 indicating all the main geometric variables.

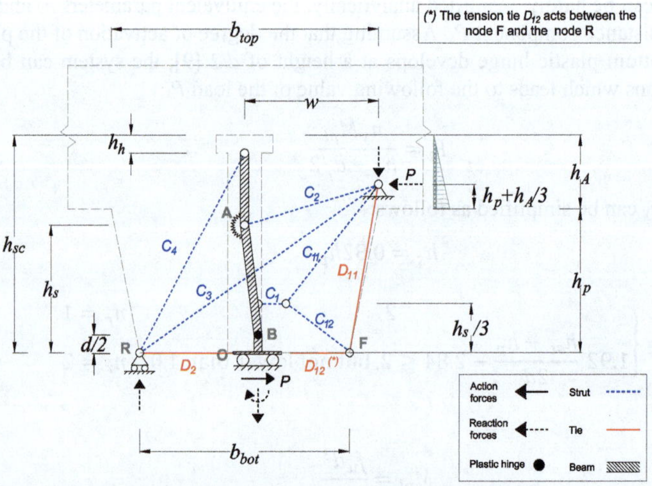

Figure 7. Modified strut and tie model [10].

Recent numerical studies on studs in profiled sheeting [17] showed the bearing forces acting at the shank of the stud are approximately linearly distributed along the segment O-A. As a result of that, the resultant force C_1 applies at a height of $h_s/3$.

As pointed out in the previous subsection, the bending moment of the stud at the base reaches the sectional plastic bending resistance while the upper hinge may partially or fully develop. Additionally, based on experimental observations, it was assumed that the sheeting locally yields and the tie element D_2 reaches its plastic resistance. Under these considerations, the system in Figure 7 is statically determined and the load P_2 can be calculated from equilibrium equations and it is given by:

$$P_2 = C_1 \cdot \zeta + \frac{n_y M_{pl}}{h_s - d/2} + D_2 \tag{9}$$

With:

$$\zeta = \frac{2h_s}{3(h_s - d/2)} \tag{10}$$

The position of the upper hinge is based on the expressions estimated by Lungershausen [7] including the influence of other parameters and it is given in Eq.(11).

$$h_s = 110\alpha_E \cdot (0.4n_r + 0.2)\left[0.8\left(\frac{n_y M_{pl}}{h_s - d/2}\right)^2 + 0.6\right] [\text{mm}] \leq 1.1h_p \tag{11}$$

Assuming that the degree of activation of the plastic hinges n_y is known and the tension tie D_2 reaches its yielding capacity $D_{2,y}$, the system in Figure 7 becomes statically determined. As in the case of h_s, the value n_y was also estimated with the support of numerical simulations [13]:

$$n_y = \begin{cases} 2 & n_r = 1 \\ 1.67\,\dfrac{h_{sc} - h_p}{2d} - 0.17 \leq 2, \text{but not lower than } 1.0 & n_r = 2 \end{cases} \tag{12}$$

While the capacity of the tension tie $D_{2,y}$ is equal to:

$$D_2 = D_{2,y} = k_w \cdot f_{yp} \pi d \tag{13}$$

With:

$$k_w = \begin{cases} 0.7 & \text{Pre} - \text{punched holes} \\ 1.0 & \text{Through} - \text{deck welding} \end{cases} \tag{14}$$

Where f_{yp} is the yield strength of the profiled sheeting material.

To determine the maximum resistance of this modified strut and tie model, the so called "rib punching" failure condition observed in Figure 4 was implemented. Specifically, it was supposed that the strut C_1 capacity is fully exploited considering an effective width of $2d$ and a reduced compressive strength of $0.6f_c$ in cracked concrete, as suggested by EN 1992-1-1 [18]. An analogy with the analytical modelling of RC corbels [19] was finally employed to calculate the effective depth ($k{\cdot}w$) of the strut C_1 , which was equal to the height of the section at the rib-slab interface according to linear-elastic theory applied to an equivalent corbel of depth w. Under these circumstances, the capacity of the strut $C_{1,max}$ can be written as follows:

$$C_1 = C_{1,max} = k_n \cdot (1.2f_c \cdot d \cdot k \cdot w) \tag{15}$$

With:

$$k_n = \begin{cases} 1 & n_r = 1 \\ 0.8 & n_r = 2 \end{cases} \tag{16}$$

$$k = \sqrt{x^2 + 2x} - x \tag{17}$$

Assuming the effective width of the strut equal to $2d$, x is given by:

$$x = \frac{E_s}{E_c}\frac{\pi d}{8w} \tag{18}$$

Where f_c and E_c are the compressive strength and elastic modulus of the concrete respectively while E_s is the elastic modulus of the stud material.

5.3 Phase 3: Strut and tie model (Large displacements)

Unlike the previous stages, the second-order geometric effects become predominant in this phase where the connector can transfer a significant part of the load via tensile stresses [7]. As confirmed by Figure 5, the concrete rib in front of the stud is fully damaged and thus, the corresponding bearing stresses can be neglected. The rotation of the stud head results in the gradual reduction and loss of the upper plastic hinge developed at lower displacements. Although a single plastic hinge is still activated, the analytical model representing this phase was improperly named "strut and tie" because the bending effects become comparably smaller. Since Phase 3 covers the behavior of headed stud shear connections at high displacements, the equilibrium equations may be significantly influenced by the deformation of the system. Therefore, the equilibrium equations take into account the deformed shape of the mechanical model at a slip value *s*, as shown in Figure 8.

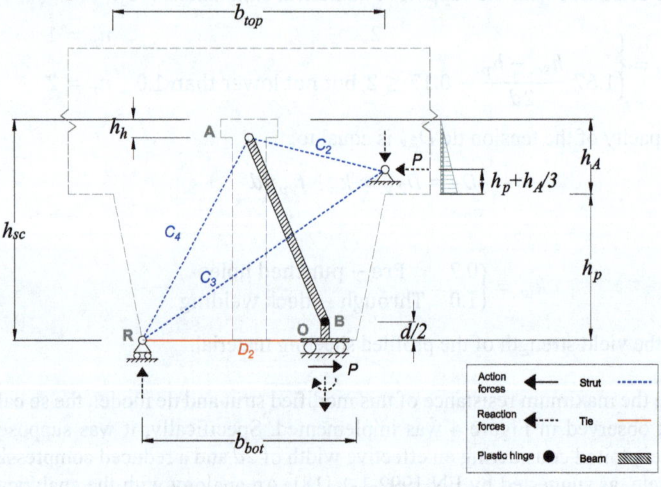

Figure 8. Strut and tie model [10].

Based on these considerations, the load P_3 can be written as:

$$P_3 = \frac{M_B \sin\phi}{(h_{sc} - h_h - d/2)^2 + s^2 - d/2} + N_B \frac{\pi d^2}{4} \cos\phi + D_2 \tag{19}$$

As done in the previous phase, it is assumed that the tension tie D_2 is yielded, i.e. $D_2 = D_{2,y}$. In consideration of the increasing tensile stress in the connector which may lead to the fracture of the shank, the bending resistance of the stud at the node B is influenced by the tensile force N_B acting on the same node. From the evaluation of the axial force at 25 mm slip extracted from FE models [10], it was found that approximately 30% of the axial capacity of the section is exploited. Hence, the maximum value of the bending moment M_B can be taken as the reduced bending resistance accounting for the *M-N* (i.e. Moment-Normal force) interaction, where the axial force N_B=0.3N_{Pl}. By using the parabolic *M-N* relationship in EN 1993-1-1 [20], the value of the bending moment M_B at failure is equal to:

$$M_B = M_{pl,N}(N_B = 0.3N_{pl}) = 0.91M_{pl} \tag{20}$$

6 COMPARISON BETWEEN ANALYTICAL AND EXPERIMENTAL RESULTS

The analytically-computed load values obtained from Equations (7), (9) and (19) for phase 1, 2 and 3 respectively, can be gathered to extract an analytical load-slip relationship which was finally compared with the experimental results in this section. In order to have a consistent comparison between the experimental and analytical prediction of the load-slip curve, the corresponding experimental values P_{e1}, $P_{e,2}$ and $P_{e,3}$ were calculated as the maximum load value within the slip range of each phase, see Figure 9. A good agreement between test and analytical load-slip curves was reached, as confirmed by Figure 10, especially at low displacements, i.e. in phase 1 and 2. On the other hand, the strut and tie model of Phase 3 provides a conservative estimation of the respective load at high displacements which may be due to the fact that the resistance of the concrete was completely neglected at this stage. Owing to the nature of the load-slip behaviour of the shear connector in the first slip displacements, it can be said that both "cantilever model" and modified strut and tie model (MSTM) can be used to determine appropriately the load bearing capacity of these shear connections.

These outcomes confirm the suitability of the "MSTM" and, more importantly, the "cantilever model" for studs in sheeting with narrow troughs. The latter was included in the design proposal of the revised version of EN 1994-1-1 presented by project team CEN/TC250/SC4.PT3 to cover cases with relatively slender rib geometries and it is currently under consideration.

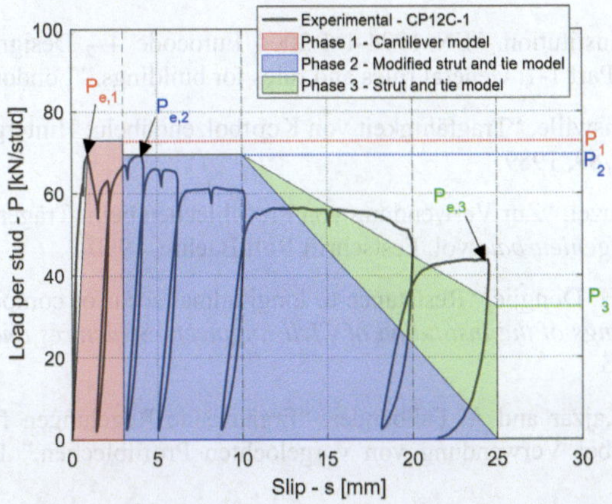

Figure 9. Example of experimental and analytical load-slip curve [10].

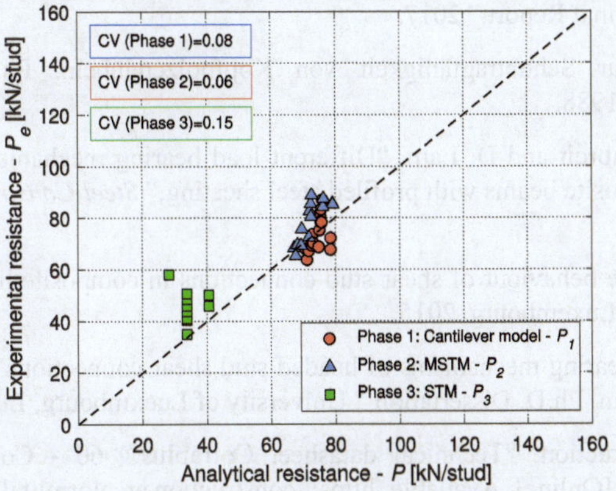

Figure 10. Comparison between experimental and analytically calculated value of the load in each phase for the push-out tests with narrow profiled steel sheeting [10].

7 SUMMARY AND OUTLOOK

The sequence of different load bearing mechanisms in headed stud shear connections employed in profiled steel sheeting with narrow troughs was identified through experimental tests and numerical studies on push-out test specimens. Three load-bearing phases were distinguished at different slip displacements and the corresponding mechanical models were developed for predicting the load-slip behaviour of the connection. The suitability of the models presented was finally confirmed by the good agreement with experimentally obtained load-slip curves. It was also seen that the cantilever model (Phase 1) and modified strut and tie model (Phase 2) are suitable for predicting the maximum load carried by the shear connector.

However, a wider push-test database may be considered in the future to assess the reliability as well as the scope of application of such analytical equations enabling the calibration of design proposals.

REFERENCES

[1] British Standards Institution, "EN 1994-1-1:2004 Eurocode 4 - Design of composite steel and concrete structures Part 1-1: General rules and rules for buildings," London, 2004.

[2] K. Roik and G. Hanswille, "Tragfähigkeit von Kopfbolzendübeln, Hintergrundbericht zu Eurocode 4.," Bericht EC4/12/89, 1989.

[3] H. Bode and R. Künzel, "Zur Verwendung von Profilblechen beim Trägerverbund," *Der Metallbau im Konstruktiven Ingenieurbau,* vol. Festschrift Rolf Baehre, 1990.

[4] R. P. Johnson and H. Dongjie, "Resistance to longitudinal shear of composite beams with profiled sheeting," *Proceedings of the Institution of Civil Engineers-Structures and Buildings,* vol. 110, no. 2, pp. 204-215, 1995.

[5] G. Hanswille, C. Kajzar and A. Faßbender, "Ergänzende Regelungen für die Tragfähigkeit von Kopfbolzendübeln bei Verwendung von vorgelochten Profilblechen," Forschungsbericht 93-01, Wuppertal, 1993.

[6] M. Lawson, E. Aggelopoulos, R. Obiala, F. Hanus, C. Odenbreit, S. Nellinger, U. Kuhlmann, F. Eggert, D. Lam, X. Dai and T. Sheehan, "Development of improved shear connection rules in composite beams - Final Report," 2017.

[7] Lungershausen, "Zur Schubtragfähigkeit von Kopfbolzendübeln. Ph.D. Dissertation," Ruhr universität bochum, 1988.

[8] V. Vigneri, C. Odenbreit and D. Lam, "Different load bearing mechanisms in headed stud shear connectors for composite beams with profiled steel sheeting," *Steel Construction,* vol. 12, no. 3, pp. 184-190, 2019.

[9] S. Nellinger, "On the behaviour of shear stud connections in composite beams with deep decking. Ph.D. Dissertation," Luxembourg, 2015.

[10] V. Vigneri, "Load bearing mechanisms of headed stud shear connections in profiled steel sheeting transverse to the beam. Ph.D. Dissertation," University of Luexmbourg, Luxembourg, 2021.

[11] ArcelorMittal-Construction, "Technical datasheet Cofraplus® 60 - Composite floor decking," ArcelorMittal, 2015. [Online]. Available: https://construction.arcelormittal.com/..

[12] Dassault Systèmes Simulia, "Abaqus Analysis User's Guide v6.14," 2014.

[13] V. Vigneri, C. Odenbreit and M. V. Braun, "Numerical evaluation of the plastic hinges developed in headed stud shear connectors in composite beams with profiled steel sheeting," *Structures,* vol. 21, pp. 103-110, 2019.

[14] S. Ernst, "Factors Affecting the Behaviour of the Shear Connection of Steel-Concrete Composite Beams. Ph.D. Dissertation," University of Western Sydney, Sydney, 2006.

[15] S. Nellinger, C. Odenbreit, R. Obiala and M. Lawson, "Influence of transverse loading onto push-out tests with deep steel decking," *Journal of Constructional Steel Research,* vol. 128, pp. 335-353, 2017.

[16] R. P. Johnson and H. Yuan, "Existing rules and new tests for stud," *Proceedings of the Institution of Civil Engineers - Structures and Buildings,* vol. 128, no. 3, pp. 244-251, 1998.

[17] M. Shen, "Structural behaviour of shear connection in composite structures under complex loading conditions. Ph.D. Dissertation," Hong Kong, 2014.

[18] British Standards Institution, "EN 1992-1-1:2004 Eurocode 2 - Design of concrete structures Part 1-1: General rules and rules for buildings," 2004.

[19] S. Hwang, W. Lu and H. Lee, "Shear Strength Prediction for Reinforced Concrete Corbels," *ACI Structural Journal ,* vol. 97, no. 4, p. 543–552., 2000.

[20] British Standards Institution, "EN 1993-1-1:2004 Eurocode 3 - Design of steel structures Part 1-1: General rules and rules for buildings," 2004.

9TH INTERNATIONAL CONFERENCE ON COMPOSITE CONSTRUCTION IN STEEL AND CONCRETE
M. Knobloch, U. Kuhlmann, W. Kurz, M. Schäfer. (Eds.)
Online, Germany, July 27-29, 2021

SAGGING BENDING RESISTANCE OF COMPOSITE SLABS IN PARTIAL SHEAR CONNECTION

Léopold Sokol*, Anna Palisson*

* SOKOL Consultants, SOKOL PALISSON Consultants
e-mails: leopoldsokol@aol.com, spconsultants@yahoo.fr

Keywords: Slab, Composite, Partial, Sheeting, Plastic, Transition point

Abstract. *This document presents a proposal for improving the partial shear connection method used for the calculation of sagging bending resistance of composite slabs. The improvement is aimed in particular at the equilibrium model of the section of the slab at the ultimate limit state, taking into account the interaction between the tensile force in the sheeting and its resistance moment.*

1 INTRODUCTION

The use of profiled steel sheeting for concrete slabs began in the United States in the 1920s as shuttering. At the end of the 1960s, a first, fully empirical calculation method, called "m-k", was created at Waterloo University (Canada) by R. Schuster and C. Ekberg ([1,2]).

Another, more advanced method, called "partial shear connection", was developed in the late 1970s at TNO-IBBC in the Netherlands by J. Stark ([3,4]) and adopted in EN 1994-1-1: 2005 ([5]).

However, the experience drawn from practical applications as well as the knowledge gained from experimental and theoretical research have provided us with ideas for improving this method, which is the subject of this article.

2 PRINCIPLES OF EN 1994-1-1 FORMULATION

The composite slab section equilibrium is based on the assumption of a rigid plastic behaviour of the sheeting and the concrete, with rectangular stress blocks, as shown in figure 1.

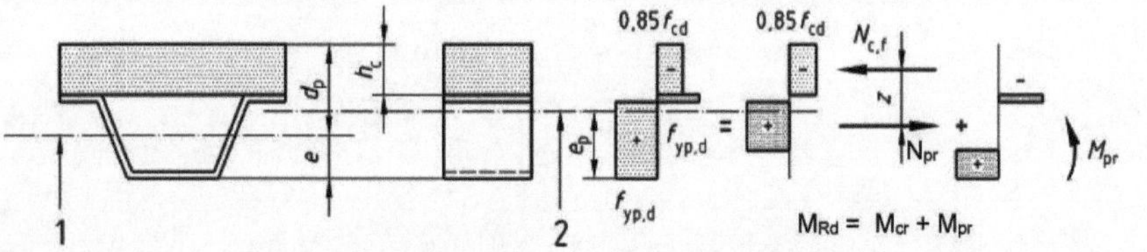

Figure 1. Stress distribution for sagging bending ([5])

By decomposing the stresses as shown in Figure 1, the basic equation for the moment of resistance of the composite section M_{Rd} can be presented as a sum of two components (figure 1):

$$M_{Rd} = M_{cr} + M_{pr} \quad (1)$$

The first term corresponds to the balanced couple of the internal forces $N_{c,f}$ and N_{pr} in the slab section:

$$M_{cr} = N_{c,f}*z \quad (2)$$

$$N_{c,f} = \tau_{u,Rd} b L_x \quad (3)$$

$$N_{pr} = -N_{c,f} \tag{4}$$

where: $\tau_{u,Rd}$ is the design shear strength obtained from testing
b is the calculation width
L_x is the distance of the cross-section being considered to the nearest support
z is the lever arm of the internal forces $N_{c,f}$ and N_{pr},(see figure 1) assumed to be equal to:

$$z = h - x_{pl}/2 - e_p + (e_p - e)\,\eta \tag{5}$$

where: e is the distance of the elastic neutral axis of the sheeting to the lower flange
e_p is the distance of the plastic neutral axis of the sheeting to the lower flange
x_{pl} is the height of the compressive stress block in the concrete:

$$x_{pl} = \frac{N_{c,f}}{0.85bf_{cd}} \tag{6}$$

where: η is the degree of shear connection:

$$\eta = \frac{N_{pr}}{N_{pa}} \tag{7}$$

where: N_{pa} is the tension resistance of the sheeting in the absence of bending moment:

$$N_{pa} = A_{pe}\, f_{y,pd} \tag{8}$$

The above relations lead to the equation:

$$M_{cr} = N_{pa}\eta\left[h - 0.5x_{pl} - e_p + (e_p - e)\eta\right] \tag{9}$$

The second term of equation (1) corresponds to the plastic moment of resistance in the presence of a tensile force and is defined approximately as a bilinear function

$$M_{pr} = 1.25M_{pa}\,(1-\eta) \leq 1 \tag{10}$$

where: M_{pr} is the resistance moment of the sheeting in the presence of tensile force
M_{pa} is the plastic resistance moment of the sheeting in the absence of tensile force

The above relationship is shown in figure 2:

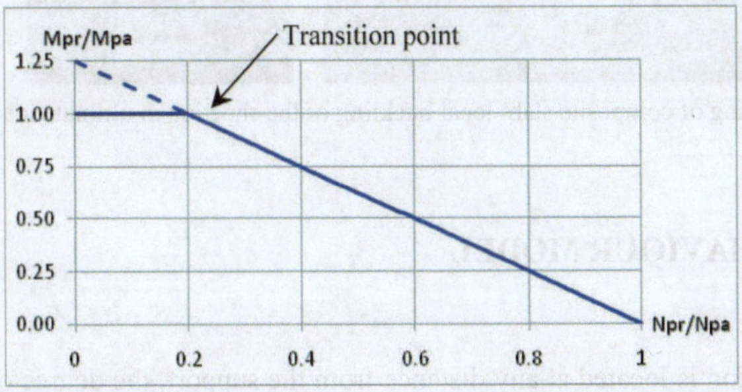

Figure 2. Relationship between tensile force and plastic resistance moment of the sheeting according to [5]

Substituting (9) and (10) in (1), it takes the following form:

$$M_{Rd} = N_{pa}\,\langle \eta\{h - 0.5x_{pl} - [e_p - (e_p - e)]\eta\} + \frac{z_{pa}}{2} * Min[1.25(1-\eta),1)]\rangle \tag{11}$$

3 OBSERVATIONS ON EN 1994-1-1 FORMULATION

The following observations prompted us to look for possible improvements to the EN 1994-1-1 formulation:

1. The first term (M_{cr}) in equation (2) is calculated using the arbitrary approximate simplified expression (5) defining the lever arm z of the internal forces ([4]).

2. The bilinear relationship (10) between tensile force and plastic resistance moment of the sheeting, illustrated in figure 2, is based on a simplified assumption. The transition point (figure 2) is not strictly representative of the mechanical behaviour of structures under normal conditions.

3. The plastic resistance moment M_{pa} of the sheeting is calculated as for section Class 1 according to EN 1993-1-1 ([6]), which implies the full capacity to form a plastic hinge with plastic yielding, without strength reduction due to local buckling. This behaviour is not confirmed be the tests (figure 3), where it is observed that the resistance is governed by the local buckling (figure 4).

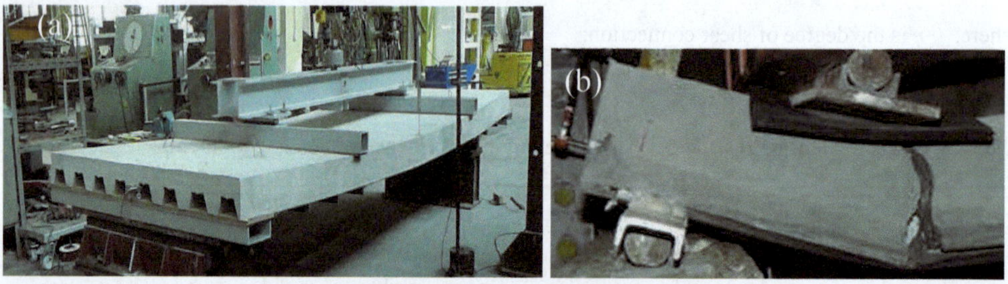

Figure 3. Testing of composite slabs: (a) bending resistance according B.3.2 [5], (b) local resistance near the support under a concentrate load

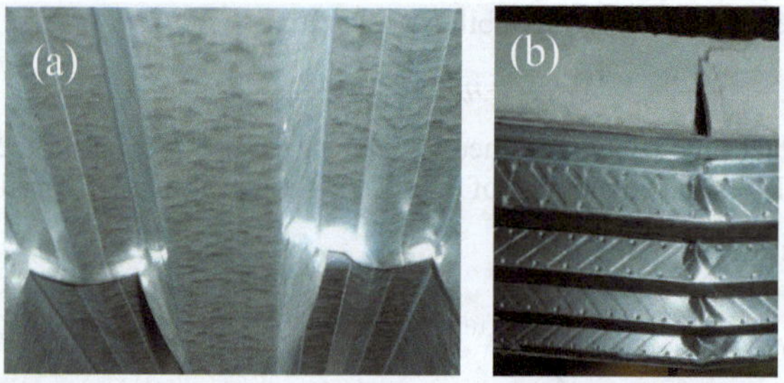

Figure 4. Testing of composite slab: local buckling of the sheeting in ultimate loading phase

4 IMPROVED BEHAVIOUR MODEL

4.1 Assumptions

The given cross-section is located at any distance from the support, the degree of connection is in the range of values $0 \leq \eta \leq 1$.

Although the stability of compressed parts of sheeting section is improved through contact with the concrete, their resistance is still governed by buckling (see figure 4), so it may not be regarded as being Class 1 ([6]).

In calculation of the resistance moment, in order to keep the main line of the EN 1994-1-1 method, we assume that the stress diagram in the components of the section (sheeting and concrete) has still a rectangular block form, but considering the observation no. 3 of Chapter 3, the plastic resistance moment of the sheeting section is reduced by means of a reduction factor $\chi < 1$.

We have verified that depending on various factors such as thickness, yield stress or geometric shape of the cross-section, the loss of plastic resistance of the sheeting remains between 10% to 30 %, compared to a perfect plastic resistance of Class 1 sections. Taking into account the favorable effect of contact with concrete, we have retained for this article the value of the reduction factor $\chi = 0.75$ as reasonably safe. Nevertheless, additional theoretical and experimental studies will be will be desirable to refine the value of this factor.

4.2 Analysis of the behaviour model

We proceed in two stages. First is analyzed the equilibrium of the sheeting section alone. In a similar way as discussed in [7], we decompose the stresses as shown in figure 5.

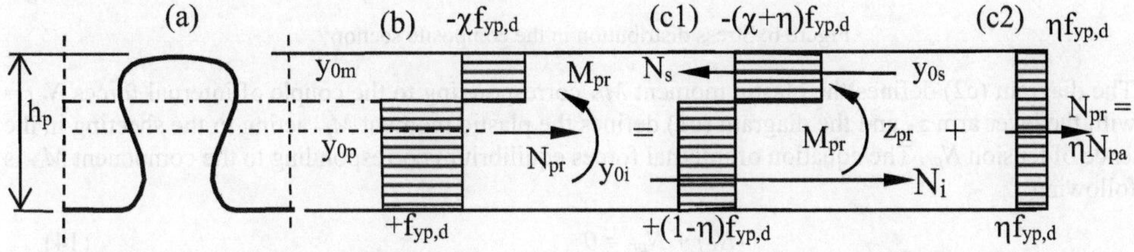

Figure 5. Stress distribution in the sheeting section under the simultaneous action of tension and bending moment

The stress diagram (c1) results from the action of the moment M_{pr} represented by the self-balanced couple of internal forces:

$$N_s = N_i \tag{8}$$

where: $N_s = (\chi + \eta)A_s f_{yp,d}$ (acting in the axis of gravity y_{0s}) (9)

$N_i = (1 - \eta)A_i f_{yp,d}$ (acting in the axis of gravity y_{0i}) (10)

A_s is superior part of the section (under compression)

A_i is inferior part of the section (under traction)

From the equations (8) to (10) we obtain the following relationship defining the areas of compression and tension parts of the section:

$$A_s = A_{pe}\frac{1-\eta}{1+\chi} \tag{11}$$

$$A_i = A_{pe}\frac{\chi+\eta}{1+\chi} \tag{12}$$

where A_{pe} is the total section = $A_s + A_i$

The stress diagram (c2) corresponds to the tension force in the sheeting N_{pr}. The plastic moment M_{pr} in the presence of tension $N_{pr} = \eta N_{pa}$ is:

$$M_{pr} = N_i z_{pr} = A_i(1-\eta)f_{yp,d} z_{pr} = N_{pa} z_{pr}\frac{(1-\eta)(\chi+\eta)}{1+\chi} \tag{13}$$

where: N_{pa} is the tension resistance of the sheeting in the absence of bending moment, given by (2.8)

z_{pr} is the lever arm of the internal forces in the sheeting, to be determined as follows:

- calculate the area of compressed part A_s (11) and tensed part A_i (12) of the section,
- determine the position of the gravity centers y_{0s} and y_{0i} of these parts (figure 4.c1),
- the lever arm z_{pr} is equal to the distance between the gravity centers y_{0s} and y_{0i}.

In the second stage, is analyzed the equilibrium of internal forces in the composite section as a whole, consisting of sheeting and hardened concrete. The stress distribution, shown in the figure 6, provides a model of understandable mechanical behaviour, where there is no need to resort to approximations.

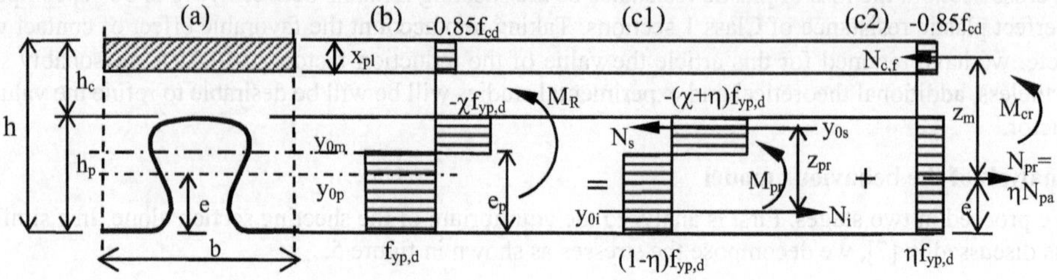

Figure 6. Stress distribution in the composite section

The diagram (c2) defines the plastic moment M_{cr} corresponding to the couple of internal forces $N_{c,f} = N_{pr}$ with the lever arm z_m and the diagram (c1) defines the plastic moment M_{pr} acting in the sheeting in the presence of tension N_{pr}. The equation of internal forces equilibrium corresponding to the component M_{cr} is the following:

$$-N_{c,f} + N_{pr} = 0 \tag{14}$$

where: $N_{c,f} = \tau_{u,Rd} b L_x$ (15)

$\tau_{u,Rd}$ is the design shear strength obtained from tests

L_x is the distance of the cross-section being considered to the nearest support

$N_{pr} = \eta N_{pa}$ (16)

η is the degree of shear connection:

$N_{pa} = A_{pe} f_{y,pd}$ (17)

The force corresponding to the stress block f_{cd} in the concrete is:

$$N_{c,f} = 0.85b\, x_{pl} f_{cd} \tag{18}$$

from which one obtains the height of the compressed portion of the concrete:

$$x_{pl} = \frac{\eta N_{pa}}{0.85bf_{cd}} \tag{19}$$

By decomposing the stresses as shown in Figure 6 (c1) and (c2), the basic equation for the moment of resistance of the composite section can be presented as a sum of two components:

$$M_{Rd} = M_{cr} + M_{pr} \tag{20}$$

where: $M_{cr} = \eta N_{pa} z_m$ (21)

$z_m = h - 0.5x_{pl} - e$ (22)

M_{pr} is defined by the equation (13)

Finally, the basic equation (20) takes the form:

$$M_{Rd} = N_{pa}\left[\eta\left(h - \frac{\eta N_{pa}}{2*0.85bf_{cd}} - e\right) + \frac{(1-\eta)(\chi+\eta)}{1+\chi} z_{pr}\right] \tag{23}$$

We observe that:

- the equation (13) defining the interaction between the tension N_{pr} and the plastic moment resistance M_{pr} results from an analysis of a model of mechanical behaviour, which is not the case of the EN 1994-1-1 method (see observation no. 2 in chapter 3)
- the lever arm z_m according to the equation (22) is determined on the basis of internal forces equilibrium, without using any arbitrary hypotheses (see the observation no. 1 in chapter 3).

5 RESULTS OF COMPARATIVE NUMERICAL APPLICATIONS

To assess the impact of the approximations used in the method of EN 1994-1-1, we carried out a comparative numerical study of the resistance as a function of the degree of connection η, for two geometric shapes of the sheeting (figure 7), with the proposed reduction factor $\chi = 0.75$.

To ensure the representativeness of the comparisons, the following geometric data were taken from existing profiles:

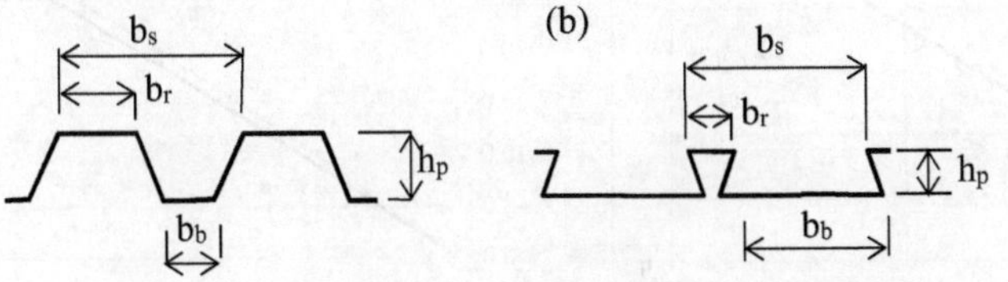

Figure 7. Open trough profile (a) and re-entrant trough profile (b)

(a) open profile	(b) re-entrant profile
h = 140 mm	h = 150 mm
t = 0.71mm	t = 0.71mm
h_p = 77 mm	h_p = 56
b_s = 192 mm	b_s = 150 mm
b_r = 82 mm	b_r = 40 mm
b_b = 54 mm	b_b = 137 mm

The results of the analysis are shown in figures 8 to 11 below, where (a) corresponds to the open trough profile and (b) corresponds to the re-entrant trough profile.

(1) Comparison of the resistance moment M_{pr} of the sheeting

(a) open trough profile (b) re-entrant profile

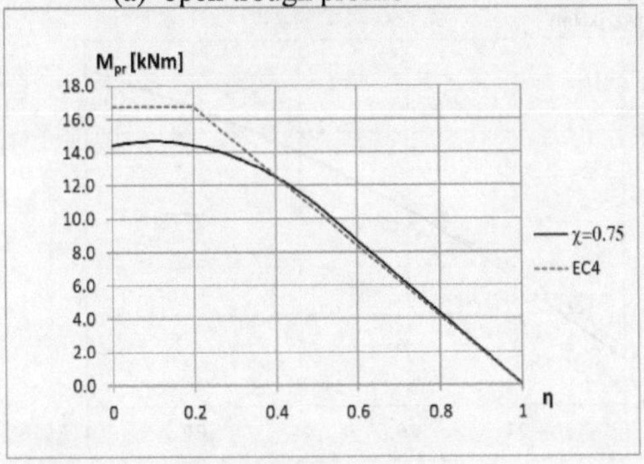

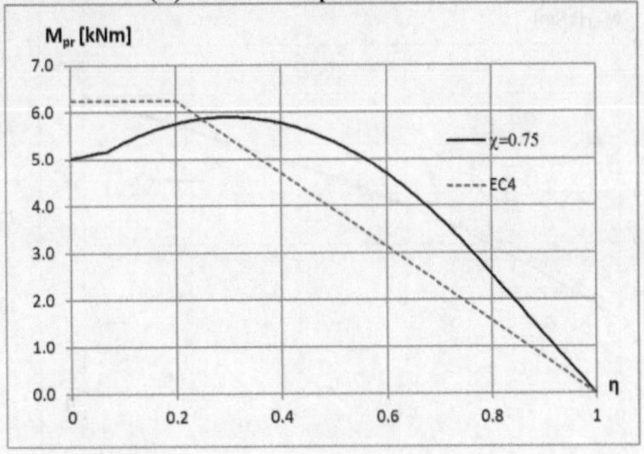

Figure 8. Resistance moments of the sheeting M_{pr} according to EN 1994-1-1, eqn. (10) and according to the proposed equation (13) with $\chi = 0.75$

The results presented in the figure 8 show that:

(a) for the open trough profile, the plastic resistance moment M_{pr} of the sheeting, defined according to EN 1994-1-1, is overestimated for values $\eta < 0.4$, i.e. near the end support.

(b) for the re-entrant trough profile, the plastic resistance moment M_{pr} defined according to EN 1994-1-1 is overestimated for values $\eta < 0.25$ and then it is underestimated.

(2) Comparison of the component resistance moment M_{cr}

(a) open profile (b) re-entrant profile

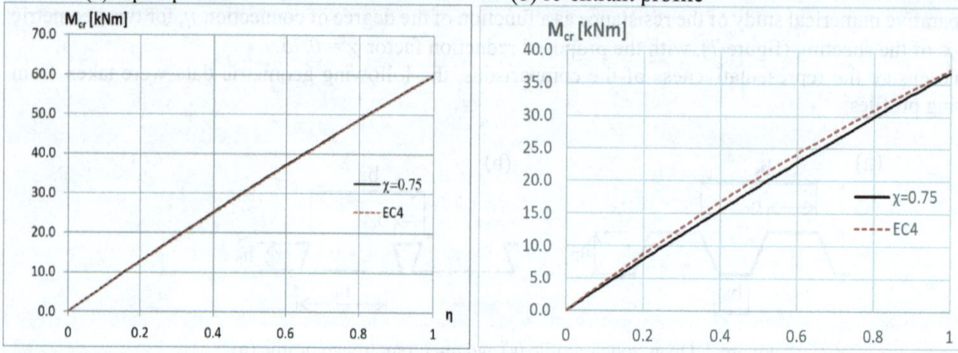

Figure 9. Component resistance moment M_{cr} according to EN 1994-1-1, eqn. (9) and according to the proposed equation (21) with $\chi = 0.75$

The results presented in the figure 9 show that:

(a) for the open trough profile, the values of resistance moment M_{cr}, defined according to EN 1994-1-1 and to the proposed equation (9) are very close,

(b) for the re-entrant trough profile, the resistance moment M_{cr} defined according to EN 1994-1-1 is overestimated, with nearly 10% for $0.35 < \eta < 0.75$.

(3) Comparison of the resistance moment M_{Rd} of the composite slab

The impact of the not accurateness of the sheeting resistance M_{pr} on the plastic resistance M_{Rd} of the slab is shown in the figures 10.a and 10.b

(a) open profile (b) re-entrant profile

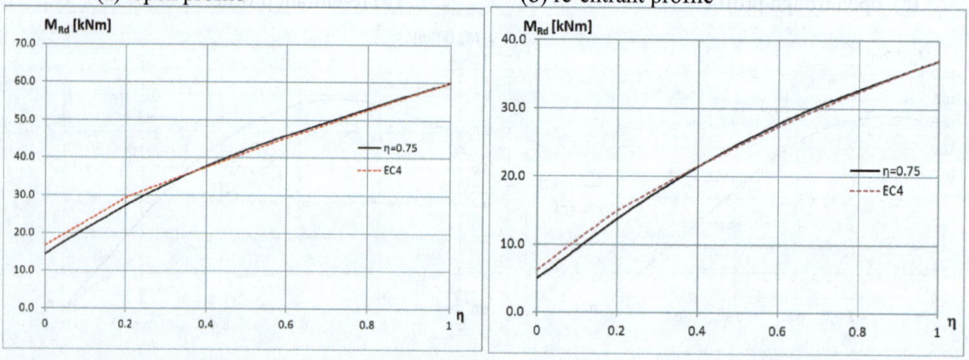

Figure 10. Sagging bending resistance M_{Rd} according to EN 1994-1-1 and according to the proposed equation (23)

The comparison between the results of EN 1994-1-1 and the formulation proposed above, expressed in terms of the ratio of values, are presented in Figure 11.

(a) open profile (b) re-entrant profile

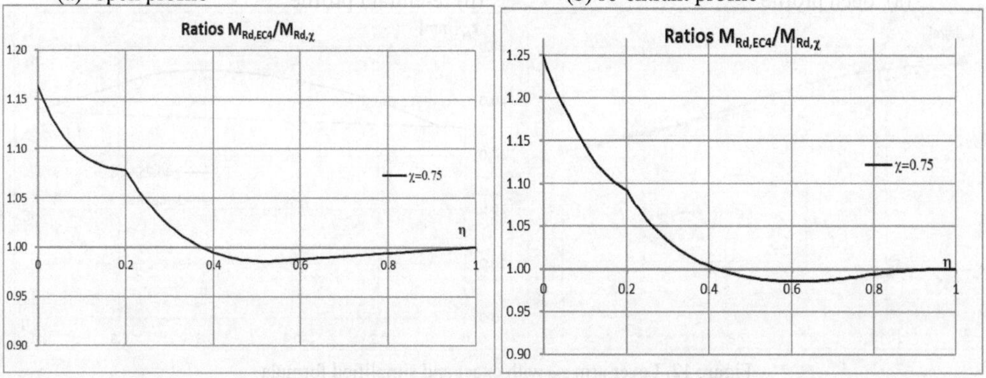

Figure 11. Ratios (M_{Rd} according to EN 1994-1-1)/(M_{Rd} according to proposed equation (23))

For the open profile, the difference on the dangerous side goes up to about 17% and up to 25% for the re-entrant profile. This can seriously affect safety, especially with concentrated loads applied near the support.

6 POSSIBILITY OF A PRACTICAL SIMPLIFICATION

The lever arm z_{pr} used in the equation (13) defining the plastic resistance moment in the presence of tension, is to be determined on the basis of elementary rules of the strength of materials. This calculation being tedious, we thought that it might be interesting to simplify it.

First, we observe that:

- The curves z_{pr} (figure 12) are convex,
- For $\eta=0$, z_{pr} is equal to the lever arm of internal forces z_{pa} in the sheeting corresponding to the plastic resistance moment M_{pr} in the absence of tension, which is easy to calculate. The attention is drawn to the fact, that unlike to the EN 1994-1-1, within the proposed method the resistance moment M_{pr} is calculated considering the reduction of the resistance in compression zone of the sheeting section by means of the factor χ (see figure 6.c1)
- For $\eta=1$, the lever arm z_{pr} is equal to the distance between the upper flange and the gravity center of section: $z_{pr} = h_p - e$.

The idea of safe simplification is to take the value z_{pr} as a linear function passing through the points corresponding to $\eta=0$ and $\eta=1$:

$$z_{pr} = z_{pa} + (h_p - e - z_{pa})\,\eta \qquad (24)$$

where: z_{pa} is the lever arm of internal forces corresponding to the plastic resistance moment M_{pr} of the sheeting, calculated in the absence of tension (see above),
h_p is the height of the sheeting,
e is the position of the elastic neutral axis of the sheeting.

A comparison between simplified and exact z_{pr} value is shown in figure 12.

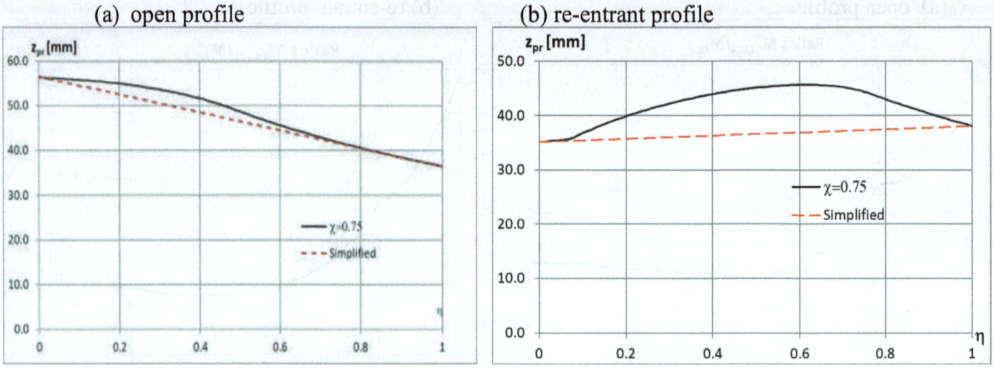

Figure 12. Lever arm z_{pr} with exact and simplified formula

Substituting (24) into (23) we obtain:

$$M_{\mathrm{Rd}} = N_{pa}\left\{\eta\left(h - \frac{\eta N_{\mathrm{pa}}}{2{*}0.85\mathrm{b}f_{\mathrm{cd}}} - e\right) + \frac{(1-\eta)(\chi+\eta)}{1+\chi}\left[z_{\mathrm{pa}} + \left(h_p - e - z_{\mathrm{pa}}\right)\eta\right]\right\} \tag{25}$$

To assess the impact of the proposed simplification, we performed the calculations with the same data as those used previously for comparisons with EN 1994-1-1.
These calculations take into account the reduction of plastic moment due to the formation of the plastic hinge with the reduction factor $\chi = 0.75$.

The results are presented in diagrams in figure 13. As shown, this simplification is close and on the safe side compared to the exact formula (23).

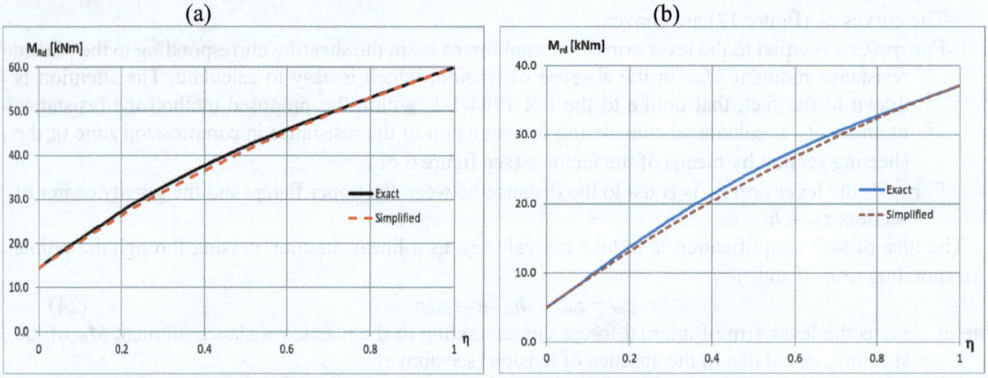

Figure 13. Diagrams of the exact and simplified moments M_{Rd}

7 CONCLUSIONS

In this paper, we first presented the principles of verification of the resistance of composite slab according to the partial connection method of EN 1994-1-1 and we presented our observations which gave us ideas for improvements.

Next, we developed an improved formulation for the sagging bending resistance of composite slabs, which is more precise than that given in EN 1994-1-1 as:
- it takes into account the fact that the thin-walled steel decking does not have the ability to form a perfect plastic hinge without reduction of resistance of compressed section part,

- it provides a solution for the lever arm of internal forces based on a mechanical model.

The presented analysis reveals that the sagging bending resistance of composite slabs defined according to the current version of the EN 1994-1-1 is overestimated for values of the connection degree generally in the range $0 < \eta < 0.40$. The resistance deficit is especially high for $\eta < 0.2$, which can seriously affect safety, especially in the presence of concentrated forces located near the supports.

Two versions of improved method are proposed:
- complete version,
- simplified safe version.

Finally, it should be emphasized that the objective of the present paper is to improve the current formulation of EN 1994-1-1, without calling into question its bases and general principles.

REFERENCES

[1] Ekberg, C. E., Schuster, R. M. "Floor Systems with Composite Form Reinforced Concrete Slabs" *IABSE New York, Final Report* (1968), 385-394

[2] Porter, M. L., Ekberg, C. E. "Investigation of cold-formed steel-deck-reinforced concrete floor slabs", *Proc. 1st. International Specialty Conference on Cold-Formed Steel Structures*, Ed.: W. W. Yu, (1971), 179-185.

[3] Stark, J. "Design of composite floors with profiled steel sheeting", *Proc. 4th. International Specialty Conference on Cold-Formed Steel Structures*, Ed.:W. W. Yu, (1978) 893-922.

[4] Stark, J. W. B., Brekelmans, J. W. P. M. "Plastic design of continuous composite slabs", *Journal of Construction Steel Research*, (1990), 15, 23-47.

[5] EN 1994-1-1 Eurocode 4: Eurocode 4: Design of composite steel and concrete structures - Part 1-1: General rules and rules for buildings, June 2005

[6] EN 1993-1-1 Eurocode 3: Design of steel structures - Part 1-1 : General rules and rules for buildings, 2005

[7] Sokol L. "Sollicitations ultimes dans une section circulaire", *Construction Métallique* N° 2-1988, CTICM

COMPOSITE STRUCTURAL ELEMENTS

9TH INTERNATIONAL CONFERENCE ON COMPOSITE CONSTRUCTION IN STEEL AND CONCRETE
M. Knobloch, U. Kuhlmann, W. Kurz, M. Schäfer. (Eds.)
Online, Germany, July 27-29, 2021

BEHAVIOUR AND DESIGN OF HIGH-PERFORMANCE STEEL AND STEEL-CONCRETE COMPOSITE STRUCTURES

D. Li*, B. Uy*, Z. Huang* and M. Khan*

* School of Civil Engineering, The University of Sydney, Sydney, NSW 2006, Australia
e-mails: dongxu.li@sydney.edu.au, brian.uy@sydney.edu.au, mahbub.khan@sydney.edu.au, zhichao.huang@sydney.edu.au

Keywords: High-strength bolts, High-performance steel, Steel-concrete composite column

Abstract. *Owing to the significant benefits, an increasing popularity to adopt the high-strength materials in the construction industry was witnessed over the recent years. With it comes a series of studies that investigated the mechanical performance of high-strength steel components, which aimed to encourage a more widespread utilisation in engineering practice. Many of the previous research relating to high-strength steel/composite structures have been focused on hot-rolled and cold-formed sections, whereas applications have been focused on thick and fabricated components in mostly high-rise and heavily loaded structures. Therefore, a research program funded by the Australian Research Council was initiated in 2018 and comprehensive experimental and numerical studies were performed over the past three years. This paper summarises the major findings on high-strength bolts, as well as fabricated high- and ultra-high-strength steel/composite structural members.*

1 INTRODUCTION

High-performance materials have found an increasing popularity in the construction industry over the recent years. In particular, high-strength steel (HSS) and ultra-high-strength steel (UHSS) have been attracting attention from the construction industry, where the buildings and bridges are striving to be built taller, lighter and longer. Together with the high-strength concrete, the application of HSS/UHSS and steel-concrete composite structural form can significantly reduce the members' dimension and self-weight, which might serve as an ideal solution for the future building structures. In addition, the relevant savings in production, fabrication and transportation also indicate that the high-performance materials are promising for the construction sector. Pioneering investigations on steel and composite structural components that incorporated high-strength materials have been performed by a number of researchers [1-13]. However, the insufficient of experimental results and design guidance still limits the inclusion of high-strength materials in the current design standards, which potentially hinders their applications in the engineering practice. To encourage a wider application, comprehensive and thorough investigations into the behaviour of fabricated HSS/UHSS and composite structural members are required. A discovery research project funded by the Australian Research Council was therefore initiated, which has produced significant outcomes on HSS/UHSS and composite members. As some studies of the project are still undergoing, this paper only presents parts of the results that were carried out by Uy's research group at the University of Sydney.

2 BEHAVIOUR AND DESIGN OF HIGH-STRENGTH BOLTS

As high-strength materials have found increasing applications in the engineering practice, the structural members can be fabricated with significantly reduced sizes. With it comes the need to utilise higher strength connectors (such as Grade 12.9 bolts) to transfer the load within limited joint regions due to the reduced member dimensions. Over the past decades, most of the studies simply tested the tensile and shear capacities of structural bolts [14,15]. These loading scenarios, however, rarely exist in the building structures as the structural bolts at the connections have higher chances to be subjected to combined loads. Pertinent studies on the behaviour of Grade 12.9 bolts under various loading scenarios are still insufficient, especially for the Grade 12.9 bolts due to the concerns of ductility. The authors herein therefore tested the ultimate strength capacities of higher strength structural bolts (including partially and fully threaded bolts, as well as countersunk bolts) under six loading angles, with which the behaviour of high-strength steel bolts under combined tension and shear can be clearly identified. Comparisons between the Grade 10.9 and Grade 12.9 bolts in terms of the strength and ductility behaviour were performed. It is noteworthy that the current European and American standards only allow the high strength bolts up to Grade 10.9 to be used. However, due to the lack of experimental data and proper design guidance, both Grade 10.9 and Grade 12.9 bolts are not involved in the Australian design codes.

2.1 Test program and facility

The authors tested a total of 180 Grade 10.9 and Grade 12.9 high-strength steel bolts, which included five different types of bolts, as illustrated in Figure 1(a). Each type of bolt was repeated three times under various loading angles to assure the accuracy of the experimental data. Due to the lack of standardised testing rig, Chesson et al. [14] in 1965 developed the first purpose made rig that allowed the bolts to be subjected to combined forces. This testing rig was thereafter adopted by Kawohl and Lange [16] and the authors herein for the investigations of Grade 10.9 and Grade 12.9 bolts. In particular, the authors modified the clamping blocks so that the countersunk bolts can be tested in a similar way [17]. As shown in Figure 1(b), the tested bolts can be subjected to different proportions of tension and shear by changing the loading angles from 0° to 90°.

2.2 Numerical modelling

Over the past few years, an increasing number of studies attempted to analyse the full range behaviour of steel beam-column connections, where, in some cases, the structural bolts were the weak points and fractured under a combination of shear and tensile forces prior to the formation of plastic hinges [18]. The conventional numerical models for structural bolts included the actual bolt threads, which rendered the models not computational efficient and consequently hindered its wider applications. Moreover, some researchers proposed the fracture modelling for structural bolts based on the tensile loading only [19], which might not be suitable for those that are subjected to combined actions.

In addition to the bolt strength tests, the authors herein also tested a series of coupons from the spare specimens with various notch dimensions. Together with the normal coupons results (without notch), the fracture parameters of such high-performance connectors can be obtained by establishing the relationship between the plastic strain at fracture ($\varepsilon_{pl,frac}$) and stress triaxiality (η). Moreover, as shear specimens are

(a) Five types of high-strength bolts (b) Tested bolts at 0°, 30° and 67.5° loading angles

Figure 1. Tests on five types of Grade 12.9 bolts.

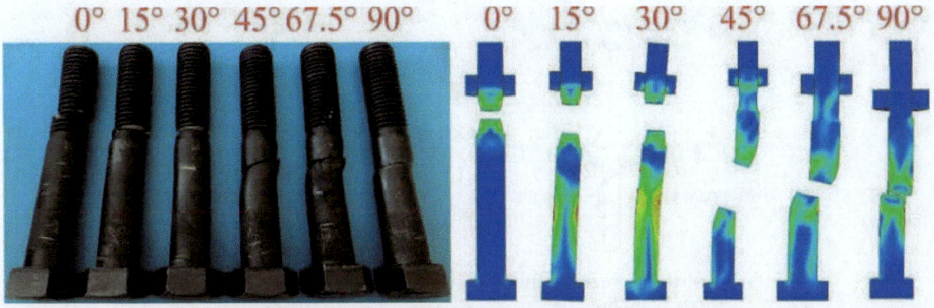

Figure 2. Validation of the failure modes (M24) of Grade 12.9 bolts.

difficult to be extracted from the bolts, the authors simply utilised the tested bolts under 90° to directly calibrate the $\varepsilon_{pl,frac}$-η relationships. As shown in Figure 2, the numerical models equipped with these fracture parameters can capture the failure modes of the tested bolts very well.

2.3 Design solutions

The obtained test results on higher strength bolts were compared with the existing design methods proposed by AS 4100 [20], Eurocode 3 [21] and AISC-360 [22], although none of these codes of practice allowed the use of Grade 12.9 bolts. It should be noted that the design solution stated in Eurocode 3 [21] can safely estimate the strength capacities with a slightly higher conservativeness of 10-20%. On the other hand, AS 4100 [20] and AISC-360 [22] can predict the tensile and shear capacity of the higher strength connectors more accurately. However, the design solutions established by the Australian and American standards generally overestimate the strength capacities of higher strength connectors under loading angles of 45° and 67.5°. The measures of lowering the design predictions can be achieved by either adjusting the interaction curve or modifying the tensile and shear force capacities. According to the present test results, modified design equations for the tensile and shear capacities were proposed for Grade 10.9 and Grade 12.9 bolts. The quadratic interaction curve was still adopted as it can better capture the behaviour of higher strength connectors under different loading scenarios (Figure 3).

3 RESIDUAL STRESS MEASUREMENTS FOR UHSS SECTIONS

Both compressive and tensile residual stresses that are induced from the fabrication process have a series of negative influences on the performance of structural elements. Particularly, the compressive residual stresses within a column section may lead to a premature buckling of the steel sections and consequently lower the ultimate strength capacity of the columns. This phenomenon also exists in the welded UHSS sections. Moreover, as the yield strength is significantly higher than that of mild steels, the

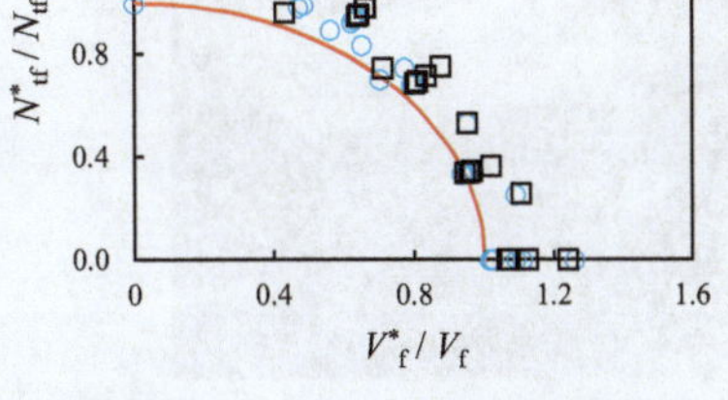

Figure 3. Comparisons of the ultimate strength capacities against the proposed design method

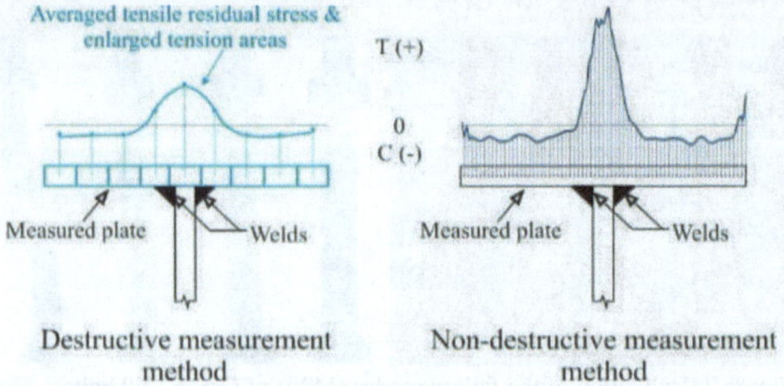

Figure 4. Comparisons between the destructive and non-destructive measurement methods

tensile residual stress regions are more localised and difficult to be detected through the conventional sectioning method. With an aim to provide an accurate estimation of the ultimate strength of the steel sections, advanced measurements (with smaller gauge volumes) of residual stresses for welded UHSS sections are of paramount importance.

3.1 Measurement method

A number of methods have been established for the residual stress measurements, which can be generally classified as destructive and non-destructive ones [23]. The generally used destructive method is the sectioning method with a relatively bigger gauge volume (8-15 mm sectioning slice), which inevitably leads to an averaged and inaccurate tensile residual stress in the weld regions. On the other hand, as one of the non-destructive methods, neutron diffraction method allows the gauge volume to be as small as 1 mm, and neutrons can penetrate deep into the thick specimens. Comparisons between the destructive and non-destructive residual stress measurement methods have been carried out and shown in Figure 4. To obtain an accurate residual stress distribution pattern, the authors thus utilised the non-destructive neutron diffraction method to measure the residual stresses of fabricated UHSS sections.

3.2 Prepared specimens and measurement procedures

Two box sections and two I-sections of various cross-section dimensions were fabricated from 5 mm thick UHSS plates, where the UHSS plates were preheated prior to the fillet welding with ER110S welding consumables. To obtain the undeformed lattice spacing (d_{hkl0}), a small slice (4 mm thick) that is free of longitudinal residual stresses was wire-cut from the original specimen. In addition, the geometric details of the stressed and stress-free specimens were scanned with a laser profile scanner, which was followed by the positioning on top of a measuring table. The accurate positioning of the $2 \times 2 \times 2$ mm^3 gauge volume for neutron penetrations can be achieved, as shown in Figure 5.

3.3 Measurement results and analytical models

The residual stress distribution patterns of the welded UHSS sections are shown in Figure 6. As

Figure 5. Residual stress measurements of UHSS box sections with neutron diffraction method.

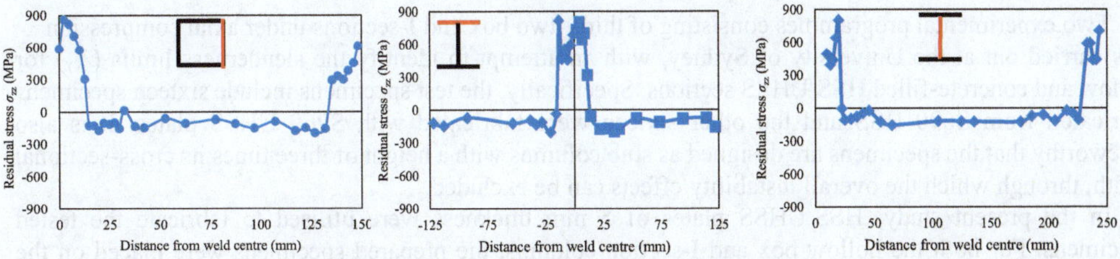

Figure 6. Residual stresses of UHSS welded sections.

illustrated, the general pattern of residual stress distributions is similar to that of conventional mild steel and HSS welded sections. Significantly large tensile residual stresses were recorded near the welds and heat affected zones, whilst the almost flat compressive residual stresses were obtained over the middle proportion of the plates. It is noteworthy that the obtained maximum tensile residual stresses near the welds are higher than those obtained through the destructive methods [24,25]. This difference can be attributed to the unique characteristics of neutron diffraction method, with which the tensile residual stresses within a small region can be accurately identified.

Up to date, only limited analytical models concerning the residual stress patterns of fabricated UHSS sections were developed [24,25]. Most of the previous measurements were carried out based on the sectioning method. Although the obtained tensile residual stress was significantly underestimated, the proposed analytical model simply increased the tensile residual stress to ensure the stress balance can be achieved. The authors herein thus established analytical models for UHSS welded sections based on the accurate neutron diffraction results, which is summarised in Figure 7. Specifically, the authors suggest that the tensile residual stresses and tensile regions of the UHSS welded sections are related to the plate thickness, as more heat input can be produced when multi-pass welds are utilised for the thicker sections. In addition, the compressive residual stresses and compression regions are calculated based on the stress equilibrium theory.

4 LOCAL BUCKLING OF HOLLOW AND COMPOSITE UHSS COLUMNS

As mentioned above, the most significant benefit of using HSS/UHSS plates as the major structural components is the reduced dimension, which can however render the thin-walled plates more susceptible to local buckling. In order to clearly identify the yield slenderness limits for hollow and composite HSS/UHSS columns, the authors herein conducted extensive experimental and numerical analysis [12,13], through which the local buckling of HSS/UHSS welded box and I-sections (with and without concrete infill) can be identified. The obtained yield slenderness limits can be supplemented into the current design standards to promote the wider applications of HSS and UHSS.

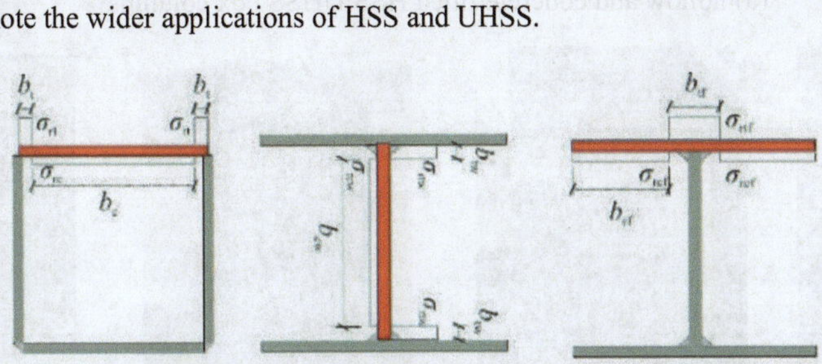

σ_{rt}/f_y=0.3ln(t)+0.3 $\quad$ σ_{rtw}/f_y=0.3ln(t_w)+0.3 $\quad$ σ_{rtf}/f_y=0.3ln(t_f)+0.3

b_t=6.5ln(t)+2 $\quad$ b_{tw}=8ln(t_w)+2 $\quad$ b_{tf}=20ln(t_f)+2

Figure 7. Analytical models of residual stresses for UHSS welded sections.

4.1 Test program and specimen preparation

Two experimental programmes consisting of thirty-two box and I-sections under axial compression was carried out at the University of Sydney, with an attempt to identify the slenderness limits (λ_{ey}) for hollow and concrete-filled HSS/UHSS sections. Specifically, the test specimens include sixteen specimens fabricated from S690 HSS and the other sixteen were fabricated with S960 UHSS plates. It is also noteworthy that the specimens are designed as stub columns with a height of three times its cross-sectional width, through which the overall instability effects can be excluded.

In the present study, HSS/UHSS plates of 5 mm thickness were utilised to fabricate the tested specimens. For both the hollow box and I-section columns, the prepared specimens were placed on the testing machine (with a capacity of 5000 kN) and the axial compression was applied through a thick steel plate to ensure a uniform load can be achieved. The tests were proceeded with displacement control so that the descending part can be better obtained until the load dropped to 80% of the ultimate capacity. A number of strain gauges were attached at the critical regions of the steel sections to facilitate the identification of elastic local buckling. As for the composite box and I-section columns, a 20 mm thick plywood was positioned at the bottom of each steel section, which was followed by the concrete casting. It should be noted that the concrete was poured into the steel sections until another 20 mm recess was left on the top. By implementing this approach, the compressive load was subjected to the steel section only, whilst the filled concrete only restrained the steel plates from inward buckling. This testing method was firstly utilised by Uy [6] and was demonstrated to be particularly effective for the composite sections, as the load distributed on the steel sections is difficult to be extracted if the compression was subjected to both the filled concrete and steel sections. Figure 8 illustrates some failure modes of the hollow and concrete filled UHSS box and I-sections.

4.2 Numerical modelling

The authors herein adopted the shell elements (S4R) and brick elements (C3D8R) to simulate the steel

(a) hollow and concrete filled HSS/UHSS box columns

(b) hollow and concrete-filled HSS/UHSS I-section columns

Figure 8. Failure modes of tested specimens

Table 1. Slenderness limits of HSS/UHSS box and I-sections.

Slenderness limits	Box		I-section	
	Hollow	Composite	Hollow	Composite
λ_{ey}	35	50	14	20

tubes and concrete infill (if any), respectively. In particular, shell elements can better capture the buckling behaviour of the steel tubes. Together with the simplified bilinear stress-strain curves, the residual stresses and geometric imperfections were also included in the developed numerical model, with which the load-strain curves and failure modes of the tested specimens can be accurately mirrored. It is noteworthy that the residual stress distribution model was established by Li et al. [26] and was utilised herein.

4.3 Slenderness limits

The proposed slenderness limits for hollow and composite HSS/UHSS welded sections are compared with the existing design standards and summarised as shown in Table 1. Specifically, the non-dimensional slenderness limit (λ_{ey}) of hollow HSS/UHSS box section is identified as 35. For composite HSS/UHSS box sections, the non-dimensional slenderness limit (λ_{ey}) is found to be around 50, which was improved by 40%, compared with the counterpart without concrete infill. In terms of the HSS/UHSS I-sections, the slenderness limit was improved from 14 (hollow) to 20 (composite).

5 STRENGTH CAPACITIES OF COMPOSITE UHSS BOX COLUMNS

5.1 Test programme and failure modes

Over the past few years, an increasing number of experimental tests were carried out to explore the behaviour of composite columns that incorporated high-strength materials [11,27-30]. It is generally

(a) concentric compression

(b) eccentric compression

(c) four-point bending

Figure 9. Test setup of composite UHSS box columns

recommended that the utilisation of HSS/UHSS should be accompanied by the high-strength or even the ultra-high-strength concrete. Through this kind of combination, the benefits of HSS/UHSS plates can be maximised. It is noteworthy that for the concrete-filled composite columns, concrete is confined within the steel tubes and still exhibits some inelastic strains after the peak stress is reached [31], which is different from that of concrete encased or partially encased composite columns. Set against this background, an experimental programme involving eighteen concrete-filled composite box columns under various loading conditions was performed. The composite box columns were fabricated with four UHSS plates and infilled with two different grades of high-strength concrete (C70 and C100). Moreover, the test specimens were classified into two series. As shown in Figure 9, Series I included 8 specimens that were tested under compression and flexural bending. Series II involved another eight specimens with various member slenderness ratios (L_e/r) that were tested under axial compression. The test setup for specimens in Series II is similar to that shown in Figure 9(a). All the box steel tubes were designed with λ_{ey} smaller than 50 to achieve compact sections.

5.2 Numerical modelling

Similar numerical modelling methods have been utilised by the authors, where the UHSS tubes and concrete infill were simulated with the S4R shell elements and C3D8R brick elements, respectively. Once again, initial geometric imperfections of the steel tubes were measured by the laser scanner and included in the numerical model. Furthermore, residual stresses were included based on the analytical models proposed by [26]. Accuracy of the numerical models was validated with the recorded test data and the failure modes shown in Figure 10.

5.3 *N-M* interaction curve for stub columns

For the stub composite columns with compact sections, existing design standards utilised the plastic stress distribution method and established the similar *N-M* interaction curves. In particular, American design standard includes two methods, with one of them similar to that proposed by the European and

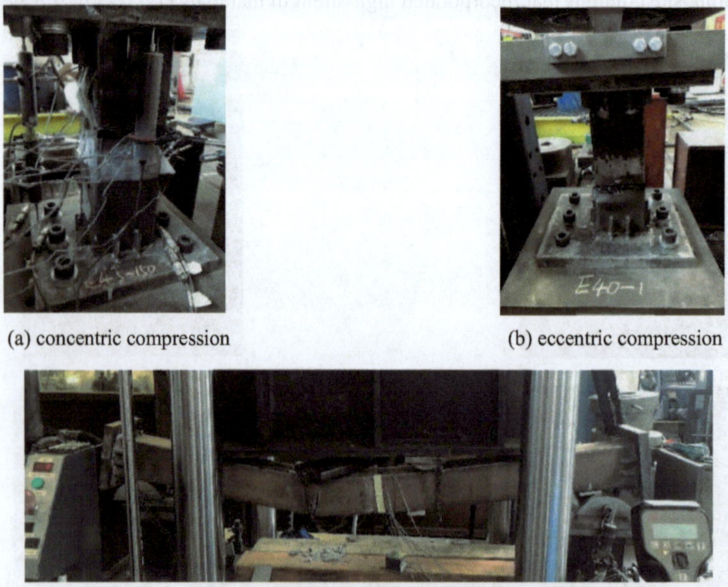

(a) concentric compression (b) eccentric compression

(c) four-point bending

Figure 10. Failure modes of the tested specimens

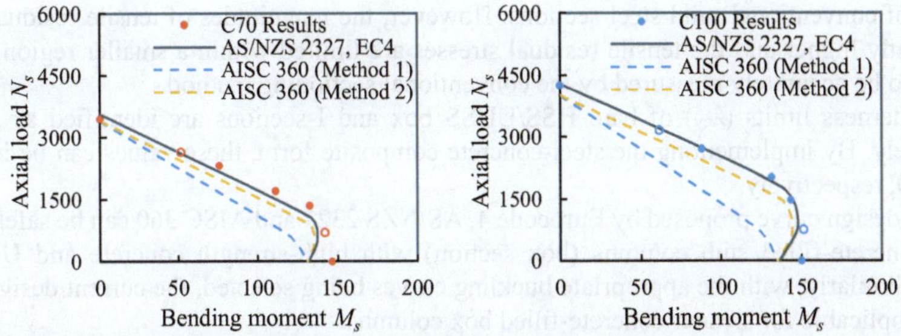

Figure 11. Comparisons of short and slender high-performance composite columns against design standards

Australian design provisions. As shown in Figure 11, another simplified solution is also included in the American code, which can facilitate the use of structural engineers. It should be noted that the method 2 in American standard adopted a reduction factor (0.85) for the strength contribution from infilled concrete, which means the confinement effect was simply ignored and the prediction results are conservative.

5.4 Member buckling curve for slender columns

Existing design standards (AS/NZS 2327, Eurocode 4 and AISC 360) propose different design solutions for predicting the axial compressive capacities of CFST columns. In particular, the European and Australian design standards developed various buckling curves that correspond to different geometric imperfections and residual stresses. However, the American code simply adopts one buckling curve for the design of slender composite columns, which can be readily utilised by the structural engineers. In addition to the experimental tests, a series of numerical analysis were carried out based on the validated model, through which the relationship between normalised axial compression and relative slenderness can be obtained in Figure 12. As illustrated, the current design specifications can be safely extended to the CFST columns incorporating UHSS and high-strength concrete.

6 CONCLUSIONS

Over the past three years, significant research efforts on bare steel and steel-concrete composite structural members that incorporate high-performance materials have been performed by Uy's research group at the University of Sydney. Based on the experimental results and extended numerical analysis, the following conclusions can be drawn:

- The mechanical behaviour in terms of the strength and ductility of high-strength steel bolts of Grade 12.9 is similar to that of Grade 10.9 bolts, which indicates that the Grade 12.9 bolts can be potentially included in the current design standards.
- The general trend of residual stress distribution patterns for HSS/UHSS welded sections are similar

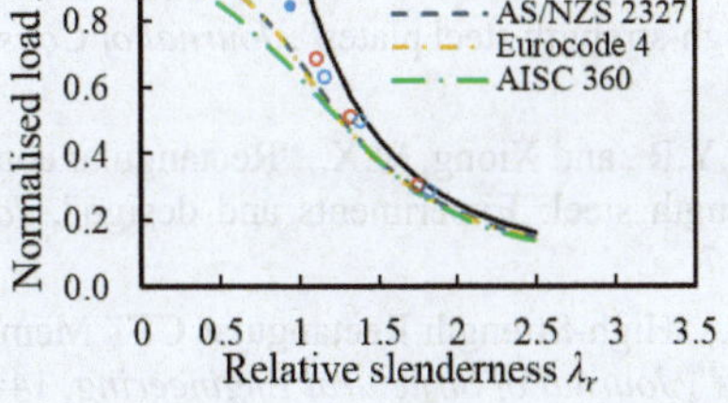

Figure 12. Selection of relative slenderness buckling curves based on different design standards.

to those of conventional mild steel sections. However, the magnitudes of tensile residual stress are significantly higher and the tensile residual stresses are limited within a smaller region, which are difficult to be accurately measured by the conventional sectioning method.
- The slenderness limits (λ_{ey}) of bare HSS/UHSS box and I-sections are identified as 35 and 14, respectively. By implementing the steel-concrete composite form, these values can be increased to 50 and 20, respectively.
- The *N-M* design curve proposed by Eurocode 4, AS/NZS 2327 and AISC 360 can be safely extended to the concrete-filled stub columns (box section) with high-strength concrete and UHSS being utilised. Similarly, with the appropriate buckling curves being selected, the current design solutions are still applicable for slender concrete-filled box columns.

ACKNOWLEDGEMENTS

The present project was financially supported by the Australian Research Council (ARC) under its Discovery scheme (DP170100001). Also, the study of residual stress measurements was supported by Australian Nuclear Science and Technology Organisation (ANSTO) through Proposal 6851.

REFERENCES

[1] Shi, G.; Ban, H., and Bijlaard, F.S.K., "Tests and numerical study of ultra-high strength steel columns with end restraints", *Journal of Constructional Steel Research*, 70, 236-247, 2012.

[2] Ban, H.; Shi, G.; Shi, Y.J., and Bradford, M.A., "Experimental investigation of the overall buckling behaviour of 960 MPa high strength steel columns", *Journal of Constructional Steel Research*, 88, 256-266, 2013.

[3] Wang, Y.B.; Li, G.Q.; Chen, S.W., and Sun, F.F., "Experimental and numerical study on the behaviour of axially compressed high strength steel box-columns", *Engineering Structures*, 58, 79-91, 2014.

[4] Ma, T.Y.; Hu, X.F.; Liu, X.; Li, G.Q, and Chung, K.F., "Experimental investigation into high strength Q690 steel welded H-sections under combined compression and bending", *Journal of Constructional Steel Research*, 138, 449-462, 2017.

[5] Ma, J.L.; Chan, T.M., and Young, B., "Design of cold-formed high-strength steel tubular stub columns", *Journal of Structural Engineering*, 144(6), 04018063, 2018.

[6] Uy, B., "Strength of short concrete filled high strength steel box columns", *Journal of Constructional Steel Research*, 57, 113-134, 2001.

[7] Uy, B., "Axial compressive strength of short steel and composite columns fabricated with high strength steel plate", *Structural Engineering and Mechanics*, 1(2), 1-15, 2001.

[8] Mursi, M., and Uy, B., "Strength of slender concrete filled high strength steel box columns", *Journal of Constructional Steel Research*, 60, 1825-1848, 2004.

[9] Aslani, F.; Uy, B.; Tao, Z., and Mashiri F., "Behaviour and design of composite columns incorporating compact high-strength steel plates", *Journal of Constructional Steel Research*, 107, 94-110,2015.

[10] Du, Y., Chen, Z.; Liew, J.Y.R., and Xiong, M.X., "Rectangular concrete-filled steel tubular beam-columns using high-strength steel: Experiments and design", *Journal of Constructional Steel Research*, 131, 1-18, 2017.

[11] Lai, Z., and Varma, A.H., "High-Strength Rectangular CFT Members: Database, Modeling, and Design of Short Columns", *Journal of Structural Engineering*, 144(5), 04018036, 2018.

[12] Huang, Z.; Li, D.; Uy, B.; Thai, H.T., and Hou, C., "Local and post-local buckling of fabricated high-strength steel and composite columns", *Journal of Constructional Steel Research*, 154, 235-249, 2019.

[13] Li, D.; Huang, Z.; Uy, B.; Thai, H.T., and Hou, C., "Slenderness limits for fabricated S960 ultra-high-strength steel and composite columns", *Journal of Constructional Steel Research*, 159, 109-121, 2019.

[14] Chesson, E.; Faustino, N.L., and Munse, W.H., "High-strength bolts subjected to tension and shear", *Journal of Structural Division. ASCE*, 91(5), 155-180, 1965.

[15] D'Anello, M.; Cassiano, D., and Landolfo, R., "Monotonic and cyclic inelastic tensile response of European preloadable gr10.9 bolt assemblies", *Journal of Constructional Steel Research*, 124, 77-90, 2016.

[16] Kawohl, A.K., and Lange. J., "Tests on 10.9 bolts under combined tension and shear", *Acta Polytechnica*, 56(2), 112-117, 2016.

[17] Li, D.; Uy, B.; Wang, J., and Song, Y., "Behaviour and design of high-strength Grade 12.9 bolts under combined tension and shear", *Journal of Constructional Steel Research*, 174, 106305, 2020.

[18] Coelho, A.M.G., and Bijlaard, F.S.K., "Experimental behaviour of high strength steel end-plate connections", *Journal of Constructional Steel Research*, 63, 1228-1240, 2007.

[19] D'Anello, M.; Cassiano, D., and Landolfo, R., "Simplified criteria for finite element modelling of European preloadable bolts", *Steel and Composite Structures*, 24(6): 643-658, 2017.

[20] Standards Australia, AS 4100-1998: Steel structures, 2016.

[21] Eurocode 3, Design of Steel Structures, Part 1.4: General Rules-Supplementary rules for stainless steels. BS EN 1993-1-4, 2006 (London, UK; 2006).

[22] ANSI/AISC 360-10. Specification for structural steel buildings. Chicago: AISC; 2010.

[23] Withers, P.J., and Bhadeshia, H.K.D.H., "Residual stress Part 1-Measurement techniques", *Materials Science and Technology*, 17, 355-365, 2001.

[24] Somodi, B., and Kovesdi, B., "Residual stress measurements on welded square box sections using steel grades of S235–S960", *Thin-Walled Structures*, 123, 142-154, 2018.

[25] Ban, H.; Shi, G., and Shi, Y., "Experimental study on residual stress in 960MPa high strength steel welded box sections and unified model", *Tumu Gongcheng Xuebao/China Civil Engineering Journal*, 46(11), 63-69, 2013.

[26] Li, D.; Paradowska, A.; Uy, B.; Wang, J., and Khan, M., "Residual stresses of box and I-shaped columns fabricated from S960 ultra-high-strength steel", *Journal of Constructional Steel Research*, 166, 105904, 2020.

[27] Khan, M.; Uy, B.; Tao, Z., and Mashiri, F., "Concentrically loaded slender square hollow and composite columns incorporating high strength properties", *Engineering Structures*, 131, 69-89, 2017.

[28] Huang, Z.; Uy, B.; Li, D., and Wang, J., "Behaviour and design of ultra-high-strength CFST members subjected to compression and bending", *Journal of Constructional Steel Research*, 175, 106351, 2020.

[29] Xiong, M.X.; Xiong, D.X., and Liew, J.Y.R., "Axial performance of short concrete filled steel tubes with high- and ultra-high-strength materials", *Engineering Structures*, 136, 494-510, 2017.

[30] Lai, B., and Liew, J.Y.R., "Axial-moment interaction of high strength concrete encased steel composite columns: Experimental investigation", *Journal of Constructional Steel Research*, 175, 106370, 2020.

[31] Tao, Z.; Wang, Z.B., and Yu, Q., "Finite element modeling of concrete-filled steel stub columns under axial compression", *Journal of Constructional Steel Research.*, 89, 121–131, 2013.

9TH INTERNATIONAL CONFERENCE ON COMPOSITE CONSTRUCTION IN STEEL AND CONCRETE
Q. Zhang*, M. Schäfer
Online, Germany, July 27-29,2021

INNOVATIVE NUMERICAL APPROACHES FOR STRAIN LIMITED DESIGN OF COMPOSITE BEAMS

Q. Zhang*, M. Schäfer*

* University of Luxembourg, Department of Engineering
e-mails: Qingjie.zhang@ext.uni.lu, Markus.schaefer@uni.lu

Keywords: Composite beams, Strain limited design, Numerical approaches.

Abstract. *For composite cross-section in steel and concrete plastic bending resistance is often used. However, also for cross-section of class 1 and 2 rotation capacity may be reduced, then strain-limited or elastic resistance becomes important. The strain-limited resistance refers to the design of concrete sections according to EN 1992-1-1, considering the material nonlinearity. For the numerical approaches, the conventional "finite fibre method" demands high programming and calculation efforts due to the vast amount of fibres to be computed. In this article, three alternative numerical approaches: the finite cell method, the integral strain method and the direct analytical method are introduced. These approaches reduce the number of elements compared to the fibre method. They take different levels of simplification, varies in application ranges, accuracy and calculation speed. For conventional composite beam and slim-floor beam cross-sections, an efficient alternative to the fibre method is presented by the alternates.*

1 INTRODUCTION

Strain limited design is used for the determination of bending resistance of concrete structures according to EN1992-1-1 [5]. It may become important for composite cross-section, e.g. for beam cross-section with a deep neutral axis position, a class 3 cross-section when benefits from plastic strain in tension is assumed, and for other specific cases or when advanced design methods are required [1] - [3]. For strain limited resistance determined according to the procedure in EN 1992-1-1 [5], the strains of the external fibres of the cross-section parts are specified under the consideration of the ultimate strains of materials and the strain distribution is varied until a state of equilibrium of the internal forces is reached by the integral of the stresses over the cross-sectional area. Based on this equilibrium the bending moment resistance is computed. The stress-strain relationship for concrete and reinforcement steel can be found in EN 1992-1-1 [5] and for steel in EN 1993-1-5 [6], Annex C or the new draft prEN 1993-1-14 [8]. When applying strain limited design method, resistance may benefit from strain hardening of steel.

For the numerical approaches, a key procedure is to calculate the inner forces such as bending moment and normal forces of a cross-section from a given strain distribution. Conventionally the finite fibre method is used. According to this method, the cross-section can be divided into a finite number of tiny fibres (Figure 2 A), the average stress and inner forces in each fibre can be computed to further obtain the inner forces of each fibre and afterwards for the entire cross-section. This method is straight forward; however, it requires the cross-section to be divided into vast amounts of small elements (fibres) and may lead to significant computing effort. Furthermore, the cross-section meshing may be difficult to realize. Therefore, alternative calculation methods have been investigated for simple programming and fast cross-section analysis.

2 GENERAL CONCEPTS FOR THE SIMPLIFIED STRAIN LIMITED DESIGN

Compared to the plastic resistance of composite beams, strain limited resistance considers further the nonlinear material stress-strain relationship. Due to both the complex cross-section geometry and nonlinear material stress-strain relationship, it is difficult to get a universal analytical solution. Thus, numerical solutions are often needed. The most used solution is the finite fibre method. To apply it, the cross-section can be horizontally divided into many tiny rectangle fibres with the area (A_i). The average stain (ε_i) and stress (σ_i) values at centre of each fibre are used to further calculate the normal force in the fibre ($N_i = A_i\sigma_i$). The normal force (N) and bending moment (M) of the whole cross-section result by $N = \sum N_i$ and $M = \sum M_i z_i$, where z_i is the distance of centre of each fibre to the reference point. The accuracy of the results is depending on the fibre numbers, usually a few hundreds or thousands of fibres are needed.

If complex geometrical and material non-linearity can be separated and handled separately, the calculation can be simplified. Cross-sections of conventional composite beam normally consist of a rectangle concrete slab and steel section below it. The cross-section can be simplified into a group of rectangles (or lines) by neglecting the fillet of steel profile. This simplification allows to only consider material nonlinearity inside each rectangle in the first step and further combine all the calculation results of each rectangle to obtain the cross-section bending moment and normal force. In this way, the results can be obtained by a few rectangle elements. These simplifications can also apply to slim-floor beams, or composite columns with similar characteristics.

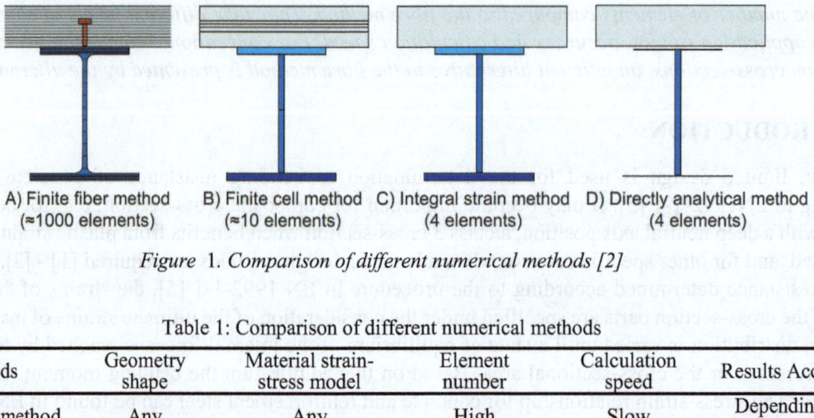

Figure 1. *Comparison of different numerical methods [2]*

Table 1: Comparison of different numerical methods

Methods	Geometry shape	Material strain-stress model	Element number	Calculation speed	Results Accuracy
Finite fibre method	Any	Any	High	Slow	Depending on elements number
Finite cell method	Rectangle simplification	Multi-linear simplification	Medium	Medium	Depending on elements number
Integral strain method	Rectangle simplification	Any	Low	Fast	High
Direct analytical method	Rectangle simplification	Only bilinear and rectangle-parabolic	Low	Very fast	medium

Based on the simplification level on material stress-strain relationship and cross-section geometry, three alternative approaches (the finite cell method, integral strain method and directly analytical method) have been introduced. The "finite cell" method (Figure 1 B) divides the cross-section into a few cells. Inside each cell a linear stress distribution (instead of the average stress value of fibre method) is assumed. The assumption results in a bending moment and normal force pair (N_i, M_i) inside each cell. The "integral strain" method (Figure 1 C) calculates first two newly defined dimensionless parameters by integral based on the material stress-strain relationship. Then the normal force and bending moment of each rectangle can be directly calculated through the newly defined parameters in section 4.1 and the rectangle size and position. The "direct analytical method" (Figure 1 D) is a further simplification. It provides the simplified equations for the bilinear and parabolic-rectangle stress-strain relationship to directly calculate bending moment and normal force inside each rectangle from the assumed neutral axis position. Table 1 provides the comparison of these methods. All the new methods greatly reduce the required elements compared to the fibre method and increase the calculation speed.

3 FINITE CELL METHOD

For the finite fibre method, the cross-section needs to be divided into many small fibres to reach enough accuracy. If the material stress-strain model can be simplified as bi-linear or multi-linear curves, the inner forces can be computed by dividing the cross-section at the turning points of stress curve and geometry in cells – Finite Cell Method – to allow the integration of stresses over the cell area. In Figure 2 this method is visualised where a bi-linear material model is assumed for the rectangle cross-section. We can divide the cross-section of Figure 2 into only two cells, with linear stress distribution inside each. The normal force of each cell is $N_i = bh_i(\sigma_i + \sigma_{i+1})/2$ and the bending moment is $M_i = bh_i^2\,(\sigma_{i+1} - \sigma_i)/12$. The cross-section inner force can be calculated by $N = \sum N_i$ and $M = \sum(M_i + N_i z_i)$. In this way only two cell elements are needed. In general, the number of cells depends on the shape of the section and the stress-strain relation.

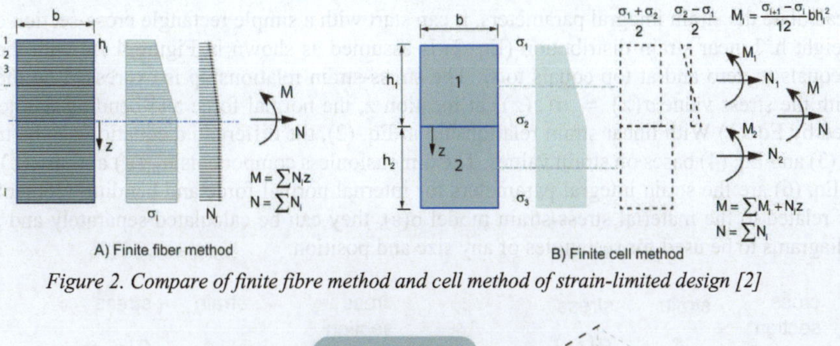

Figure 2. Compare of finite fibre method and cell method of strain-limited design [2]

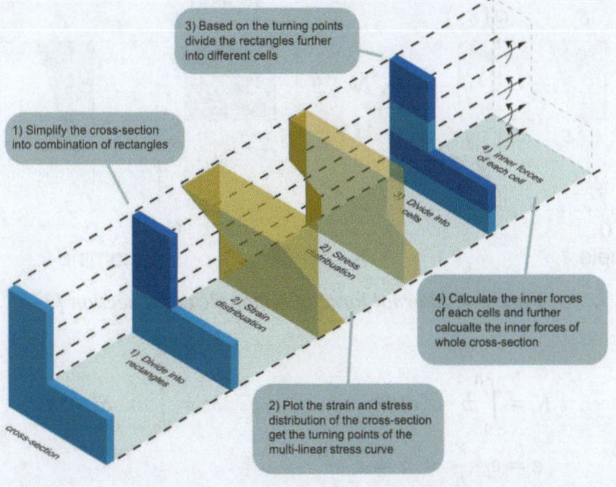

Figure 3. Cell method for strain-limited resistance [2]

For a more complex cross-section as shown in Figure 3, to use the finite cell method, the material model should first be simplified as multilinear curves, and the cross-section should be divided into different rectangles. Based on the material model and strain curve, the rectangles are further divided into separate cells. And afterward, the normal forces and bending moments of each cell can be obtained, and the inner forces of the whole cross-section can be calculated.

The finite cell method can significantly reduce the element number. However, there are still a few limitations. First, the cross-section should be able to be simplified into rectangles easily. Second, the material stress-strain curves should also be easily simplified as a multi-linear curve such as the bilinear or trilinear relationship of structural steel. For concrete, a simplified five-points polyline curve can represent the parabolic-rectangle relationship with acceptable accuracy for the integration. The last, due to that the mesh needs to be continuously updated based on the stress distribution, for each strain input the cross-section needs to be re-meshed, thus reduces the efficiency.

4 INTEGRAL STRAIN METHOD

For a rectangle cross-section, with a given material stress-strain relationship, due to the constant cross-section width, it is possible to first calculate the dimensionless material related parameters and then multiply them with the dimensional values to calculate the inner force. These dimensionless parameters can be obtained by the integral based on the function related to strain and stress relationship. Thus, it can be named as "integral strain parameters".

4.1 General concept of strain integral parameters

To calculate the strain integral parameters, it can start with a simple rectangle cross-section of breadth b and height h. Linear strain distribution (Eq. 2) is assumed as shown in Figure 4 A) with the strain at bottom equals to zero and at top equals to ε_i. The stress-strain relationship is expressed by the function σ(ε). With the stress value $\sigma(z) = \sigma(\varepsilon(z))$ at location z, the normal force and bending moment can be calculated by Eq. (1) With linear strain relationship in Eq. (2), the differential equation can be transferred into Eq. (3) and Eq. (4) bases on strain values. The dimensionless components $n_m(\varepsilon)$ and $m_m(\varepsilon)$ from Eq. (5) and Eq. (6) are the strain integral parameters for internal normal force and bending moment. As they are only related to the material stress-strain model σ(ε), they can be calculated separately and plotted in design diagrams to be used for rectangles of any size and position.

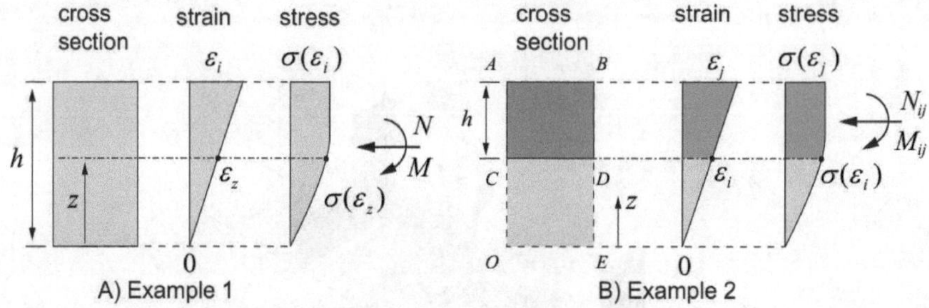

Figure 4. Internal force of rectangle cross-section [2]

$$N = \int_0^h b \cdot \sigma(\varepsilon) dz; \qquad M = \int_0^h b \cdot \sigma(\varepsilon) z dz \tag{1}$$

$$\varepsilon = \varepsilon_i \cdot \frac{z}{h} \Rightarrow z = \varepsilon \cdot \frac{h}{\varepsilon_i} \tag{2}$$

$$N = \int_0^{\varepsilon_i} b \cdot \sigma(\varepsilon) d\left(\varepsilon \cdot \frac{h}{\varepsilon_i}\right) = \frac{bh}{\varepsilon_i} \int_0^{\varepsilon_i} \sigma(\varepsilon) d\varepsilon \tag{3}$$

$$M = b\left(\frac{h}{\varepsilon_i}\right)^2 \int_0^{\varepsilon_i} \sigma(\varepsilon)\varepsilon d\varepsilon \tag{4}$$

Set:

$$n_m(\varepsilon_i) = \int_0^{\varepsilon_i} \sigma(\varepsilon) d\varepsilon; \qquad m_m(\varepsilon_i) = \int_0^{\varepsilon_i} \sigma(\varepsilon)\varepsilon d\varepsilon; \qquad k = \frac{h}{\varepsilon_i} \tag{5}$$

$$N_{0,i}(\varepsilon_i) = bk \cdot n_m(\varepsilon_i) \qquad M_{0,i}(\varepsilon_i) = bk^2 \cdot m_m(\varepsilon_i) \tag{6}$$

However, usually, the strain does not start from zero, as shown in Figure 4 B). If the strain at top and bottom has same sign (the whole cross-section either in compression or tension), the normal forces and the bending moment of rectangle ABCD can simply be calculated as the difference between the rectangle ABOE and CDOE ($N_{ij} = N_i - N_j$; $M_{ij} = M_i - M_j$). For a rectangle with strain value $\varepsilon_i, \varepsilon_j$ at top and bottom, the normal force and bending moment can be calculated by Eq. (8) and Eq. (9). In Eq. (7) the parameter "h" is different from the one in Eq. (5), however both "k" and "k_{ij}" represent curvature of the cross-section.

$$k_{ij} = \frac{h}{\varepsilon_i - \varepsilon_j} \tag{7}$$

$$N_{ij}(\varepsilon_i) = bk_{ij}\left(n_m(\varepsilon_i) - n_m(\varepsilon_j)\right) \tag{8}$$

$$M_{ij}(\varepsilon_i) = bk_{ij}^2\left(m_m(\varepsilon_i) - m_m(\varepsilon_j)\right) \tag{9}$$

When the neutral axis is inside the cross-section, it results in different signs of the strain at the top and bottom. If a point symmetric material model is used which can be expressed as $\sigma(-\varepsilon) = -\sigma(\varepsilon)$, e.g. the structural steel, the resistance can be calculated by Eq. (10) and Eq. (11). These equations are also suitable for the whole cross-section either in compression or tension. If the material stress-strain relationship is different for compression and tension e.g. for concrete, the cross-section needs to be further divided into compression parts and tensile parts to be calculated separately. Alternatively, integral from the ultimate compression strain to tensile strain instead of starting from strain equal to zero can be applied.

$$N_{ij} = bk_{ij}\left(n_m(|\varepsilon_i|) - n_m(|\varepsilon_j|)\right) \tag{10}$$

$$M_{ij} = bk_{ij}^2\left(m_m(|\varepsilon_i|) - \frac{|\varepsilon_j|}{\varepsilon_j} \cdot m_m(|\varepsilon_j|)\right) \tag{11}$$

4.2 Concrete - $n_c(\varepsilon)$ and $m_c(\varepsilon)$ for parabolic-rectangle strain-stress relationship

The parabolic-rectangle concrete model (Eq.12) according to EN1992-1-1 is used here. For concrete class up to C50/60 the ε_{c2} and ε_{cu2} are constants according to the current EN1992-1-1. Thus, it can also be expressed as $\sigma_i = f_{cd} \cdot f(\varepsilon_i)$, where f_{cd} is the design resistance and $f(\varepsilon_i)$ is the basic shape function of stress-strain curve. As $f(\varepsilon_i)$ is same for all concrete classes up to C50/60, we can separate the f_{cd} and get the parameters $n_c(\varepsilon)$ and $m_c(\varepsilon)$ as shown in Eq. (13) and Eq. (14) The subscript of parameters n_m and m_m by Eq. (8) and Eq. (9) changed here from "m" to "c" due to the representation of concrete. A dimensionless diagram is provided in Figure 5 for these two equations.

$$\sigma_c(\varepsilon) = f_{cd}\begin{cases} 1-\left(1-\dfrac{\varepsilon}{\varepsilon_{c2}}\right)^2 & 0 \le \varepsilon_c \le \varepsilon_{c2} \\ 1 & \varepsilon_{c2} \le \varepsilon_c \le \varepsilon_{cu2} \end{cases} \tag{12}$$

$$n_c(\varepsilon_i) = \int_0^{\varepsilon_i} \sigma_c(\varepsilon) d\varepsilon = f_{cd}\begin{cases} \dfrac{\varepsilon^2}{\varepsilon_{c2}} - \dfrac{\varepsilon^3}{3\varepsilon_{c2}^2} & 0 \le \varepsilon_c \le \varepsilon_{c2} \\ \varepsilon - \dfrac{1}{3}\varepsilon_{c2} & \varepsilon_{c2} \le \varepsilon_c \le \varepsilon_{cu2} \end{cases} \tag{13}$$

$$m_c(\varepsilon_i) = \int_0^{\varepsilon_i} \sigma_c(\varepsilon)\varepsilon d\varepsilon = f_{cd}\begin{cases} \dfrac{2\varepsilon^3}{3\varepsilon_{c2}} - \dfrac{\varepsilon^4}{4\varepsilon_{c2}^2} & 0 \le \varepsilon_c \le \varepsilon_{c2} \\ \dfrac{5}{12}\varepsilon_{c2}^2 - \dfrac{1}{2}(\varepsilon^2 - \varepsilon_{c2}^2) & \varepsilon_{c2} \le \varepsilon_c \le \varepsilon_{cu2} \end{cases} \tag{14}$$

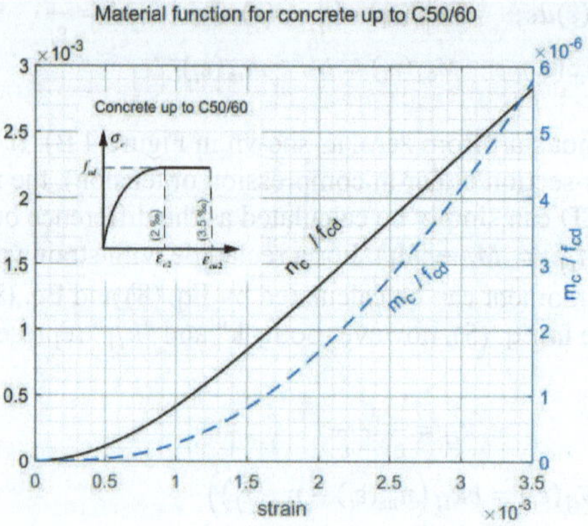

Figure 5. Material functions for Concrete up to C50/60 and parabolic-rectangle model [2]

4.3 Steel - $n_s(\varepsilon)$ and $m_s(\varepsilon)$ for bilinear strain-stress relationship with / without strain-hardening

The bilinear strain-stress relationship (Eq.16) for structural steel according to EN1993-1-5 with strain hardening of E/100 after reaching steel yielding strain ε_y is applied. The resulting strain integral functions are given by Eq. (17) and Eq. (18). In Figure 6 the results are plotted in for different steel grades. When neglecting strain-hardening, the functions Eq. (19) and Eq. (20) are further simplified. For other steel models such as a tri-linear model or non-linear model, similar approach can be followed. The subscript of parameters n_m and m_m by Eq. (8) and Eq. (9) changed from "m" to "s" to represent steel.

$$\varepsilon_y = \frac{f_y}{E_a} \tag{15}$$

$$\sigma_a(\varepsilon) = \begin{cases} E_a\varepsilon & |\varepsilon| \le \varepsilon_y \\ f_y + E_a(\varepsilon - \varepsilon_y)/100 & \varepsilon_y < |\varepsilon| \le \varepsilon_u \end{cases} \tag{16}$$

$$n_{s,sh}(\varepsilon_i) = \int_0^{\varepsilon_i} \sigma_a(\varepsilon)d\varepsilon = f_y \cdot \begin{cases} \dfrac{\varepsilon^2}{2\varepsilon_y} & |\varepsilon| \le \varepsilon_y \\ \left(\varepsilon - \dfrac{\varepsilon_y}{2}\right) + \dfrac{(\varepsilon - \varepsilon_y)^2}{200\varepsilon_y} & \varepsilon_y < |\varepsilon| \le \varepsilon_u \end{cases} \tag{17}$$

$$m_{s,sh}(\varepsilon_i)\& = \int_0^{\varepsilon_i} \sigma_c(\varepsilon)\varepsilon d\varepsilon = f_y \cdot \begin{cases} \dfrac{\varepsilon^3}{3\varepsilon_y} & |\varepsilon| \le \varepsilon_y \\ \dfrac{\varepsilon_y^2}{3} + \dfrac{\varepsilon^2 - \varepsilon_y^2}{2} + \dfrac{(\varepsilon - \varepsilon_y)^2(\varepsilon_y + 2\varepsilon)}{300\varepsilon_y} & \varepsilon_y < |\varepsilon| \le \varepsilon_u \end{cases} \tag{18}$$

$$n_{s,bl}(\varepsilon_i) = f_y \cdot \begin{cases} \dfrac{\varepsilon^2}{2\varepsilon_y} & |\varepsilon| \le \varepsilon_y \\ \varepsilon - \dfrac{\varepsilon_y}{2} & \varepsilon_y < |\varepsilon| \le \varepsilon_u \end{cases} \tag{19}$$

$$m_{s,bl}(\varepsilon_i) = f_y \cdot \begin{cases} \dfrac{\varepsilon^3}{3\varepsilon_y} & |\varepsilon| \le \varepsilon_y \\ \dfrac{\varepsilon_y^2}{3} + \dfrac{\varepsilon^2 - \varepsilon_y^2}{2} & \varepsilon_y < |\varepsilon| \le \varepsilon_u \end{cases} \tag{20}$$

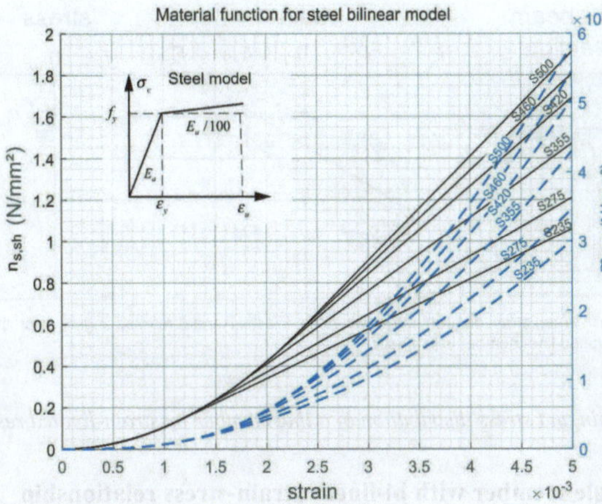

Figure 6. Strain-integral parameters for steel strain-hardening mode [2]

5 DIRECT ANALYTICAL METHOD

With the direct analytical method, the analytical solution of normal force and bending moment of each rectangle or line part related to the strain distribution are provided. However, due to complex mathematical expressions with multi-linear or non-linear material strain-stress relationships the computation may become complex. Simple material stress-strain models are more suitable for this method. Thus, only the steel bi-linear model and concrete parabolic-rectangle model are discussed by this paper. For the analysis the strain distribution is determined by a strain limit point ($z = 0, \varepsilon = \varepsilon_{cu}$) at top of the cross-section and neutral axis position z_i. (Figure 7)

5.1 Solutions of a line member with bi-linear strain-stress relationship

A line member is a rectangle whose height considerably smaller than the total cross-section height. For example, the flange of an IPE steel profile. The moment of inertia based on its own centroid is small enough to be neglected. Considering at the top of the cross-section, the concrete strain limit $(0, \varepsilon_{cu})$ is reached and the centre of the line member is at location z_i. With an ideal elastic-plastic (bi-linear) material model, the two boundary values (z_{ti} and z_{ci}) for the position of neutral axis (z_i) indicate yielding due to tension or compression as shown in Figure 7. Their values are given in Eq. (21). When $z_i \leq z_{ti}$ the considered line member yields due to compression, similarly $z_i \geq z_{ci}$ for tension. In between, the line member is in the elastic stage. The resulting stress fi can be obtained by Eq. (22) and the normal forces N_i and bending moment M_i (calculated based on the neutral axis) are given in Eq. (23).

$$z_{ti} = z_i \cdot \frac{\varepsilon_{cu}}{\varepsilon_{cu} + \varepsilon_y}; \qquad z_{ci} = z_i \cdot \frac{\varepsilon_{cu}}{\varepsilon_{cu} - \varepsilon_y} \tag{21}$$

$$f_i = f_y \cdot \begin{cases} 1 & z_i \leq z_{ti} \\ \left(\frac{z}{z_i} - 1\right)\frac{\varepsilon_{cu}}{\varepsilon_y} & z_{ti} < z_i \leq z_{ci} \\ -1 & x_{ci} < z_i \end{cases} \tag{22}$$

$$N_i = f_i A_i \qquad M_i = N_i \cdot (z - z_i) \tag{23}$$

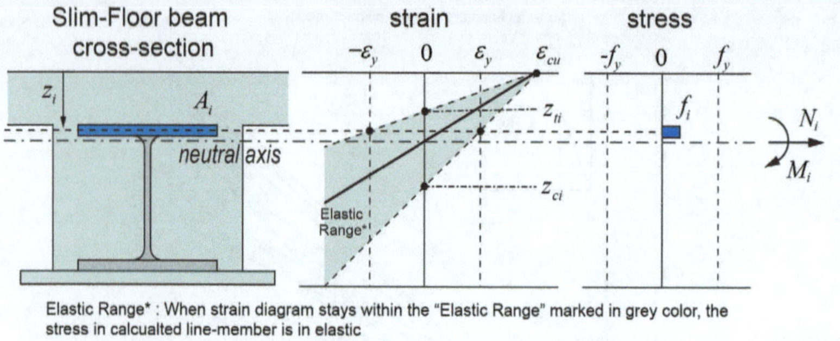

Figure 7. Strain and stress distribution of a line member by strain-limited resistance [2]

5.2 Solutions of a rectangle member with bi-linear strain-stress relationship

For a rectangle cross-section with a bi-linear material model, there are four boundary points of stress related to the position of neutral axis z_i. Two of them ($z_{t,t}$ and $z_{c,t}$) is calculated based on location of the rectangle top surface z_t, the other two ($z_{t,b}$ and $z_{c,b}$) related to the location of bottom z_b. Similar as the line member explained in section 5.1 when $z_i \leq z_{t,t}$ the considered rectangle member yields due to compression, when $z_i \geq z_{c,b}$ the whole rectangle yields due to tension. In between, part of the rectangle is in elastic stage (non-full-plastic range in Figure 8). The resulting normal forces N_i and bending moment M_i of location of neutral axis z_i are given in Eq. (26) and Eq. (27). In order to calculated them, two parameters h_t and h_c related to the plastic zone height, as well as the stress at top and bottom fiber f_t and f_b should be calculated first by Eq. (24) and Eq. (25).

$$z_{t,t} = z_t \cdot \frac{\varepsilon_{cu}}{\varepsilon_{cu} + \varepsilon_y}; \qquad z_{c,t} = z_t \cdot \frac{\varepsilon_{cu}}{\varepsilon_{cu} - \varepsilon_y} \tag{21}$$

$$z_{t,b} = z_b \cdot \frac{\varepsilon_{cu}}{\varepsilon_{cu} + \varepsilon_y}; \qquad z_{c,b} = z_b \cdot \frac{\varepsilon_{cu}}{\varepsilon_{cu} - \varepsilon_y} \tag{22}$$

$$h_t = \left(1 + \frac{\varepsilon_y}{\varepsilon_{cu}}\right) \cdot z_i - z_t; \qquad h_c = z_b - \left(1 - \frac{\varepsilon_y}{\varepsilon_{cu}}\right) \cdot z_i \tag{23}$$

$$f_t = f_y \cdot \begin{cases} 1 & z_i \leq z_{t,t} \\ \left(\frac{z_t}{z_i} - 1\right) \frac{\varepsilon_{cu}}{\varepsilon_y} & z_{t,t} < z_i \leq z_{c,t} \\ -1 & z_{c,t} < z_i \end{cases} \tag{24}$$

$$f_b = f_y \cdot \begin{cases} 1 & z_i \leq z_{t,b} \\ \left(\frac{z_b}{z_i} - 1\right) \frac{\varepsilon_{cu}}{\varepsilon_y} & z_{t,b} < z_i \leq z_{c,b} \\ -1 & z_{c,b} < z_i \end{cases} \tag{25}$$

$$N_i = b \cdot \begin{cases} f_y h - \frac{1}{2}(f_b - f_t) h_t & z_i \leq z_{c,t} \\ f_y (h_c - h_t) & z_{c,t} < z_i < z_{t,b} \\ \frac{1}{2}(f_b - f_t) h & z_{t,b} < z_i < z_{c,t} \\ \frac{1}{2}(f_b - f_t) h_c - f_y h & z_{t,b} \geq z_i \end{cases} \tag{26}$$

$$M_i = N_i \cdot \left(\frac{z_t + z_b}{2} - z\right) + \frac{b(f_b - f_t)}{12} \cdot a_m \tag{27}$$

$$a_m = \begin{cases} 3hh_t - 2h_t^2 & z_i \le z_{c,t} \\ 3h_t(h-h_t) + 3h_c(h-h_c) + (h_c + h_t - h)^2 & z_{c,t} < z_i < z_{t,b} \\ h^2 & z_{t,b} < z_i < z_{c,t} \\ 3hh_c - 2h_c^2 & z_{t,b} \ge z_i \end{cases} \tag{28}$$

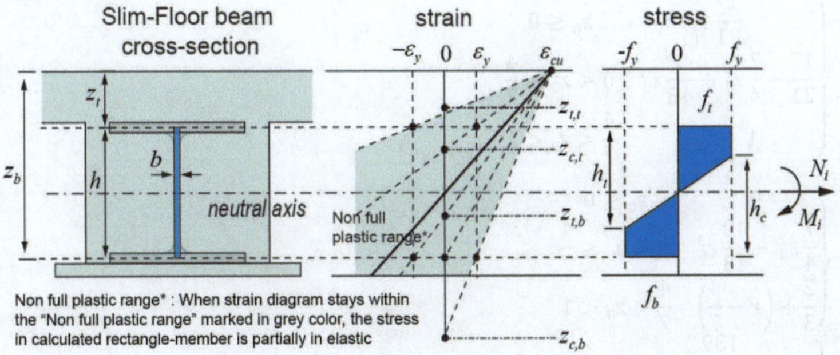

Figure 8. Strain and stress distribution of a rectangle member by strain-limited resistance [2]

5.3 Solution of a concrete T-section member with parabolic-rectangle strain-stress relationship

For a slim-floor beam with concrete flange, the concrete part can be divided into two sections: the one of the concrete flanges and the one in the chambers (direct encasement of the beam web) of the steel sections. For simplification, the area occupied by steel profile and reinforcement embedded into the concrete is not deducted. The parabolic-rectangle stress-strain relationship by Eurocode 2 is used as provided in Eq. (21). For concrete class up to C50/60, the parameters ε_{cu2} and ε_{c2} are fixed to -3.5‰ and -2.0‰ respectively following a parabolic-rectangle stress-strain relationship. For the part above profiled sheeting, the reduction factor for normal forces α_{N1} and for bending moment α_{M1} can be calculated by Eq. (23) and Eq. (25). Similarly, the values α_{N2} and α_{M2} for direct concrete encasement can be obtained by Eq. (24) and Eq. (26). To avoid the complex calculation, a dimensionless chart in Figure 9 allows the values to be directly checked. The normal forces and bending moments of each part can be calculated by Eq. (27). It is to notice that the following equations are only validated when the neutral axis position is inside the concrete T-section and the controlling ultimate strain is at top of the T-section. For a rectangle cross-section, only N_{c1} and M_{c1} are needed.

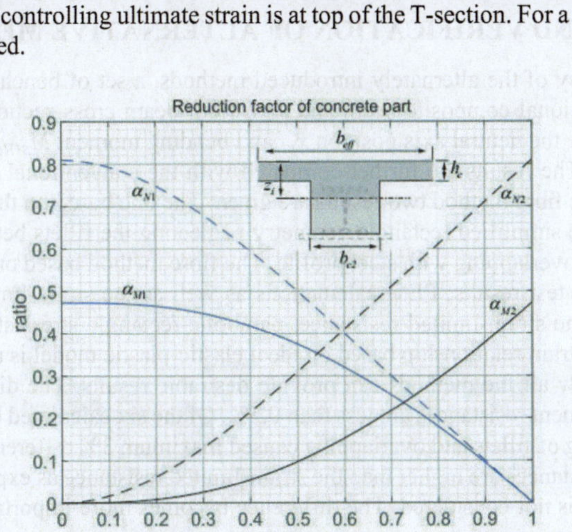

Figure 9 Reduction factors related to concrete up to C50/60 [2]

$$\sigma_c = f_{cd}\begin{cases}\left(1-\left(1-\dfrac{\varepsilon}{\varepsilon_{c2}}\right)^2\right) & 0 \le \varepsilon_c \le \varepsilon_{c2} \\ 1 & \varepsilon_{c2} \le \varepsilon_c \le \varepsilon_{cu2}\end{cases} \tag{21}$$

$$z_0 = 1 - \frac{h_c}{z_i} \tag{22}$$

$$\alpha_{N1} = \begin{cases}\dfrac{17}{21} & z_0 \le 0 \\ \dfrac{17}{21} - \dfrac{7}{4}z_0^2 + \dfrac{49}{48}z_0^3 & 0 < z_0 \le \dfrac{4}{7} \\ 1 - z_0 & \dfrac{4}{7} < z_0 < 1\end{cases} \tag{23}$$

$$\alpha_{N2} = \begin{cases}0 & z_0 \le 0 \\ \dfrac{7}{4}z_0^2 - \dfrac{49}{48}z_0^3 & 0 < z_0 \le \dfrac{4}{7} \\ \dfrac{2}{3} + \left(x - \dfrac{4}{7}\right) & \dfrac{4}{7} < z_0 < 1\end{cases} \tag{24}$$

$$\alpha_{M1} = \begin{cases}\dfrac{139}{294} & z_0 \le 0 \\ \dfrac{139}{294} - \dfrac{7}{6}z_0^3 + \dfrac{49}{64}z_0^4 & 0 < z_0 \le \dfrac{4}{7} \\ \dfrac{1}{2}(1 - z_0^2) & \dfrac{4}{7} < z_0 < 1\end{cases} \tag{25}$$

$$\alpha_{M2} = \begin{cases}0 & z_0 \le 0 \\ \dfrac{7}{6}z_0^3 - \dfrac{49}{64}z_0^4 & 0 < z_0 \le \dfrac{4}{7} \\ \dfrac{20}{147} + \dfrac{1}{2}\left(z_0^2 - \dfrac{16}{49}\right) & \dfrac{4}{7} < z_0 < 1\end{cases} \tag{26}$$

$$\begin{aligned} N_{c1} &= \alpha_{N1} \cdot f_{cd} b_{eff} z_i \\ N_{c2} &= \alpha_{N2} \cdot f_{cd} b_{c2} z_i \\ M_{c1} &= \alpha_{M1} \cdot f_{cd} b_{eff} z_i^2 \\ M_{c2} &= \alpha_{M2} \cdot f_{cd} b_{c2} z_i^2 \end{aligned} \tag{27}$$

6 BENCHMARK AND VERIFICATION OF ALTERNATIVE METHODS

To verify the accuracy of the alternately introduced methods, a set of benchmarks have been carried out based on the conventional composite beam and slim-floor beam cross-sections as shown in Table 2. The benchmarks compare the neutral axis position x_{sl} and bending moment M_{slRd} calculated according to each method (Table 3). The results are further compared with the conventional fibre method and plastic resistance M_{plRd}. For the fibre method two results are given: The one based on the exact geometry as well as the result based on the simplified rectangle geometry neglecting the fillets between web and flange as well as the concrete area overlapping with steel profile. The fibre method based on the exact geometry was benchmarked in [2] with test results, FEM calculations as well as the strain-limited design results from different literature. For the strain-limited resistance, parabolic-rectangle stress-strain relationship is used and bilinear steel stress-strain relationship based on ideal elastic-plastic model is used.

The benchmarks show all the methods can provide desirable results. The differences in neutral axis position and bending moment resistances are less than 0.5%. Of the six calculated beams, the simplification of geometry by neglecting of fillets and overlapping caused maximum 3% difference according to the fibre methods. The plastic resistances are higher than the strain-limited resistance as expected due to fact that the strain-hardening of steel is not considered. The difference becomes more important when relative neutral axis position x_{pl}/h is increasing.

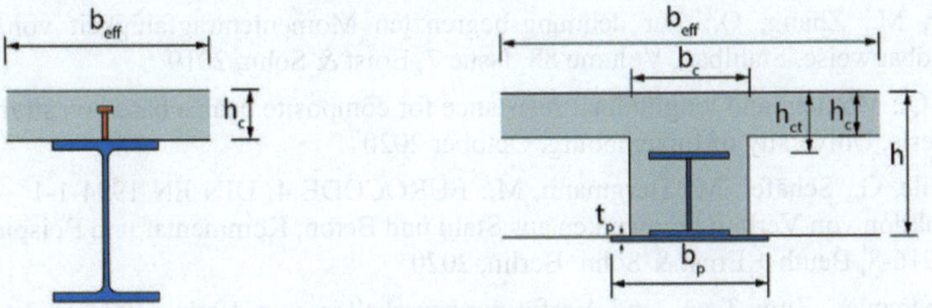

Figure 10. Nomination of cross-section

Table2: Cross-sections for the benchmark (nomination in Figure 10)

ID	h_c	b_{eff}	b_c	Concrete	h_{ct}	Profile	Steel	b_p	t_p
NC1	100	2000	---	C20/25	---	IPE200	S355	---	---
NC2	100	2000	---	C20/25	---	HEB200	S460	---	---
NC3	100	1000	---	C20/25	---	HEB200	S460	---	---
SF1	120	3000	150	C20/25	80	IPE200	S355	200	10
SF2	120	2000	150	C20/25	80	IPE200	S355	200	10
SF3	120	2000	250	C20/25	80	HEB200	S355	300	15

Table 3: Cross-sections resistances according to different method

ID	Plastic		Fiber Method*		Fiber Method		Cell Method		Integral strain Method		Direct Method	
	z_{pl} (mm)	M_{plRd} (kNm)	z_{sl} (mm)	M_{slRd} (kNm)	z_{sl} (mm)	M_{slRd} (kNm)	z_{sl} (mm)	M_{slRd} (kNm)	z_{sl} (mm)	M_{slRd} (kNm)	z_{sl} (mm)	M_{slRd} (kNm)
NC1	44.86	180.87	55.51	174.4	53.02	173.34	53.4139	172.17	52.6	172.38	52.71	172.3
NC2	107.2	468.69	110.82	453.83	108.01	441.62	108.64	440.87	107.97	442.58	108.05	441.12
NC3	113.36	400.27	151.49	377.83	149.03	367.47	149.45	367.08	148.8	367.91	149.08	367.18
SF1	50.6	341.14	60.52	338.33	60.12	333.15	59.9567	332.2	59.381	332.86	59.45	332.54
SF2	75.96	319.34	78.18	313.67	77.09	308.91	77.776	308.48	77.0613	309.23	77.133	308.75
SF3	97.68	661.63	127.99	648.04	127.07	639.61	127.13	639.55	125.97	642.44	126.24	639.65

Fibre Method*: calculated with exact geometry

7 CONCLUSIONS

Strain-limited resistance is important for composite beams and slim-floor beams for the special situations e.g. cross-sections with deep neutral axis where plastic strains cannot be reached and therefore plastic moment resistance may overestimate the bearing capacity. Furthermore strain-limited design can benefit from strain hardening of structural steel and reinforcement, this can also lead to beneficial higher design resistance. Conventional fibre method is widely used however is difficult for programming and often provided by complex calculation sheets. In this article, based on the simplification of cross-section into a group of rectangles, three innovative methods are presented. The alternative methods are faster and easier to perform and proven to able get accurate results. Especially for extended parametric studies this leads in benefit to reduced computational time.

REFERENCES

[1] Schäfer, M., Zhang, Q.: Zur dehnungsbegrenzten Momententragfähigkeit von Flachdecken in Verbundbauweise, Stahlbau, Volume 88, Issue 7, Ernst & Sohn, 2019

[2] Zhang, Q.: Moment and longitudinal resistance for composite beams based on strain limited design, PhD-thesis, University of Luxembourg, October 2020

[3] Hanswille, G., Schäfer, M., Bergmann, M.: EUROCODE 4, DIN EN 1994-1-1 – Bemessung und Konstruktion von Verbundtragwerken aus Stahl und Beton, Kommentar und Beispiele, ISBN 978-3-410-26916-8, Beuth + Ernst & Sohn, Berlin, 2020

[4] S. Heinemeyer. Zum Trag- und Verformungsverhalten von Verbundträgern ausultrahochfestem Beton mit Verbundleisten. Dissertation, Eigenverl. Aachen, 2011.

[5] EN1992-1-1 : Eurocode 2: Design of concrete structures - Part 1-1: General rules and rules for buildings, 2004

[6] EN1993-1-1: Eurocode 3: Design of steel structures - Part 1-1: General rules and rules for buildings, 2005

[7] EN1994-1-1: Eurocode 4 : Design of composite steel and concrete structures -Part 1-1: General rules and rules for buildings, 2004

[8] prEN1993-1-14: Eurocode 3 : Design of steel structures -Part 1-14: Design assisted by finite element analysis, 2020

9TH INTERNATIONAL CONFERENCE ON COMPOSITE CONSTRUCTION IN STEEL AND CONCRETE
M. Knobloch, U. Kuhlmann, W. Kurz, M. Schäfer. (Eds.)
Online, Germany, July 27-29, 2021

NUMERICAL ANALYSIS OF EARLY AGE MOVEMENT IN GROUTED CONNECTIONS

Henneberg, J.* and Schaumann, P.*

* Institute for Steel Construction, Leibniz University Hannover-ForWind
e-mails: henneberg@stahl.uni-hannover.de, schaumann@stahl.uni-hannover.de

Keywords: Early Age Movement, Grouted Connection, Numerical Analysis, Grout Curing, Shear Rates.

Abstract. *Grouted connections (GC) of offshore structures are susceptible to wave loading and current directly after grouting. Early age movement (EAM) during the curing process can affect the material properties and the interface quality between the grout and the steel structure. Such effects were observed in laboratory tests performed at the Leibniz University Hannover. Generally, EAM can be characterised by high shear rates and identified as a damage mechanism. In this contribution, the authors present a numerical approach that transfers the observed damage mechanism into an ultimate strength analyses of the GCs. For this purpose, EAM simulations and laboratory tests with fluid grout material are combined to determine areas of high shear rates, which mainly affect the grout properties. With an element-based adjustment approach, the material law of each element is adjusted according to the level of shear rate. The adjusted finite element model is analysed on its maximum axial load resistance.*

1 INTRODUCTION

Offshore structures, as build in the wind energy sector and the oil and gas industries, are currently experiencing a continuous growth in size. On the one hand, an increasing number of current offshore wind turbines (OWT) forces new installations to greater water depths. Site scarcity is a common phenomenon experienced already in the onshore wind energy. The best sites are always occupied first, or the necessary technologies do not exist for more complex sites yet [1]. On the other hand, turbine sizes are developing rapidly in the offshore wind energy sector, to decrease the levelized costs of energy (LCE) and make offshore wind energy competitive with existing energy sources [2].

A typical offshore bottom fixed support structure in deeper waters is the jacket. Due to increasing loading conditions with deeper waters, the support structures grow accordingly. Jackets are usually anchored to the seabed using foundation piles, which are connected to the support structure via grouted connections (GCs). A GC is a hybrid connection of two infiltrating steel tubes by grouting the annulus between them. For offshore installations, these connections are subjected to special boundary conditions that are very rare in common civil engineering problems. Since the grouting procedure is done under offshore conditions, the substructure and thus the GCs are already exposed to dynamic stresses and movements during the curing process, which can hardly be avoided. Grouted components are usually not to be loaded until the green strength to have an ideal composition of the material as well as a good bonding quality in hybrid structures.

First investigations on movements during the curing of grout (early age cycling *EAC* or early age movement *EAM*) were investigated in the 1980th and 90th in the oil and gas industries [3-6]. After several years the influence of EAM was taken into corresponding guidelines like DNV, ISO19902 and API. Over the years the limit for EAM changed as well as how to behave when it comes to limit overdues. Currently DNV limits the EAM during the first 24h after grouting to 1mm relative displacement [7]. Realistic on-bottom simulations from Henneberg [8] show that this limit is regularly exceeded for typical offshore structures. As there is no approach besides experimental testing to reduce the loading capacity of GCs in current guidelines when exceeding the limit of 1mm [7,9], this paper focuses on a numerical approach to consider a reduction of the compressive strength of the grout material due to shear movements caused by EAM, to reduce the connections capacity.

2 GROUT MATERIAL

In the following section, the state of research in the field of EAM is shown with a special focus on possible damage mechanisms in the grout material. In this context the hardening process of common offshore grouts is described. Furthermore, the material behaviour of cementitious materials in liquid and solid state are described as basis for upcoming numerical simulations: This is important for transferring the damage mechanism from the liquid to the solid grout simulation.

2.1 Influence of early age movements on the grout strength

Grout material that hardens under permanent movement can be damaged by several mechanisms. The kinematic energy applied to the liquid grout material, which is to be avoided for self-compacting concrete [10,11], can cause the aggregates to sediment and the cement paste to agglomerate [12], cf. Fig.1. This can lead to deviating temperature developments during the hydration. An increased concentration of cement paste leads to a stronger temperature development. This increases the risk of temperature-induced and shrinkage cracks. Both have an effect on the durability of the material.

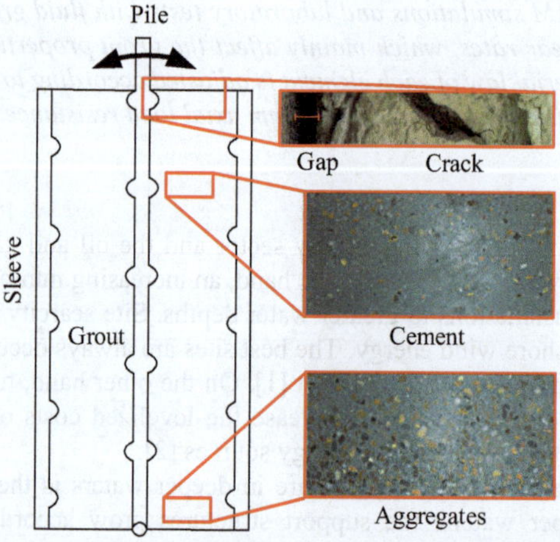

Figure 1. Damage mechanisms caused by massive EAM according to Lohaus et al [12].

Investigations by Lohaus and Cotardo [12,13] show various influences of the EAM on the strength development of the grout. In areas of large movements, there is an increased sedimentation of the aggregates and deaeration. Besides the positive effect on the strength development through deaeration, the segregation/sedimentation leads to an insignificant decrease of the final strength of the grout material. Compressive strength reductions of up to 45% were observed, when applying a relative movement of approximately 1mm. It has to be mentioned, that the movement was applied continuously over a period of 24h. The strengthening effect of the grout was not considered by a force related reduction of the movements, like it would be expected from real offshore conditions [12].

In contrast, investigations by DNV and Wimpey in the 1980th and 90th [3,4] showed severe damage in the shear planes of GCs filled with ordinary portland cement (OPC), depending on the shear key height h to shear key spacing s ratio (h/s ratio); cf. Figure 2. The continuous axial shearing caused significant disturbances in the grout material. As a result, the strength-building CSH fibres were continuously prevented from forming. A porous structure was left along the shear area, cf. Fig. 2 right. Higher h/s ratios lead to a straight shear damage, which leads to a significant reduction of the ULS and FLS capacity of the connection, as the shear keys cannot contribute to the connection strength.

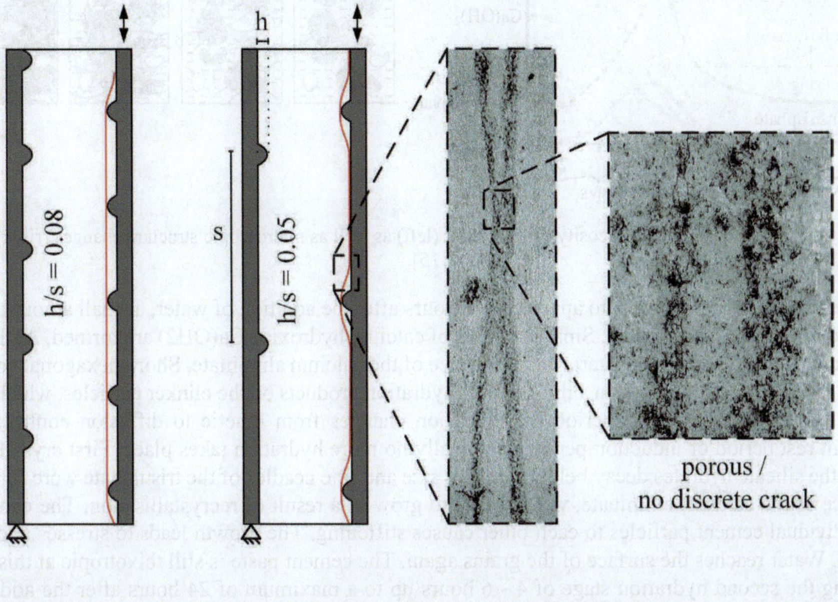

Figure 2. Possible shear planes due to vertical EAM depending on the h/s ratio [4,6] (left) and a qualitative depiction of the damage due to EAM along the shear area according to [4].

Concludingly tests from DNV showed a significant influence of EAM in the form of a shear area damage inside the OPC. Lohaus et al. showed effects of sedimentation, segregation and deaeration in a more generalized way for typically used offshore grouts. Besides the use of different materials, the tests differed in the applied direction of EAM. DNV applied their EAM axially and Lohaus et al. applied their EAM horizontally (pendular movement) fixed at the bottom end.

2.2 Curing process

Grout material is a special kind of concrete. Due to the complex installation situation offshore, the material must be reliably filled and evenly distributed in the grout annulus to ensure ideal shear strength. Due to the location of GCs in jacket structures near the seabed and the accompanying bad inspection conditions, grout material needs to be a kind of self-compacting concrete [10,11]. Another distinguishing feature is the size of the aggregates. In relation to concrete, a typical offshore grout has a maximum aggregate size of 4mm. This reduces the risk of pump blockages and creates a homogeneous structure in the entire grout annulus. Deviations due to the grouting process nevertheless occur [14].

With increasing viscosity, the liquid to plastic cement paste changes continuously through the phases of stiffening, solidification and hardening to a solid cement paste. Fig. 3 shows the hydration process, to clarify the chemical reactions underlying the curing process.

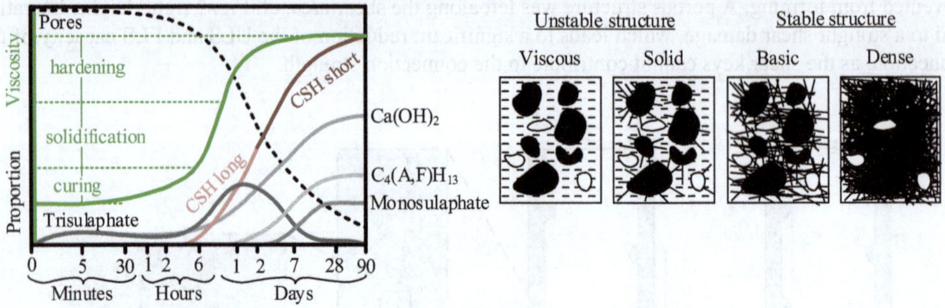

Figure 3. Course of hydration and viscosity development (left) as well as microscopic structure changes (right) acc. to [15].

In the first hydration stage up to approx. 4 - 6 hours after the addition of water, a small amount of lime is released from the clinker phases. Small amounts of calcium hydroxide Ca(OH2) are formed. At the same time, a short, intensive hydration starts on the surface of the calcium aluminate. Short, hexagonal columnar ettringite crystals (trisulphate) form a thin layer of hydration products on the clinker particles, which hinder the entry of water. Within this period the hydration changes from kinetic to diffusion control. In the subsequent rest period or induction period, practically no more hydration takes place. First crystallisation nuclei of the silicate hydrates decay below a critical size and fine needles of the trisulphate were formed on the surface of the calcium aluminate, which starts to grow as a result of recrystallisation. The connection of the individual cement particles to each other causes stiffening. The growth leads to stresses and cracks in the gel. Water reaches the surface of the grains again. The cement paste is still thixotropic at this stage.

During the second hydration stage of 4 - 6 hours up to a maximum of 24 hours after the addition of water, the ettringite crystals continue to grow and cause progressive solidification. In parallel, the crystal growth of the silicate hydrates has become more energetically useful than their decomposition. The resulting long-fibre CSH crystals bridge the water-filled spaces between the cement grains and increasingly solidify the structure (solidification).

Within the third hydration stage that now follows, from day 1 after adding the water, the microstructure is further compacted and the strength of the hardened cement paste increases. The existing pore spaces are increasingly filled by the short-fibre CSH crystals and C4(A,F)H13 crystals. In addition, calcium hydroxide Ca(OH2) with a hexagonal morphology is formed without a significant increase in strength. The primarily formed ettringite shells are broken and slowly transformed into monosulphate.

According to König the strength of the hardened cement paste does not follow the additive law of the individual mineral phases [16]. The strength is limited by the newly formed solid phases and, above all, by the porosity. The kinetics of cement hydration are significantly influenced by the level of the water-cement ratio. If more water is available to the cement than is needed for a complete hydration, the remaining water is deposited in cavities - the so-called capillary pores [17]. As the capillary pore spaces increase, the quality of the hardened cement paste decreases. These effects can be amplified by EAM in a positive way by deaeration [12] and negatively by a disturbed grout hardening in the sheared areas [4,6].

2.3 Fluid grout material

Rheological investigations Cotardo [13] and numerical investigations of liquid concrete and grout material from Roussel et al. [18,19] show good accordance by using computational fluid dynamics (CFD) simulations depicturing the fluid material properties. Due to the thixotropic material behaviour the Herschel-Bulkely material law gives a good approximation for cementitious materials; cf. equation 1.

$$\tau(\dot{\gamma}) = \tau_0 + \eta \cdot \dot{\gamma}^p \tag{1}$$

Where τ is the time and shear rate $\dot{\gamma}$ dependent shear stress, τ_0 the yield stress, η the plastic viscosity and p the unitless Herschel-Bulkely index.

Based on the test results of Cotardo et al. [20] different stages during the curing process are modelled. Therefore, the stiffness-relevant parameters for the Herschel-Bulkley material law for different time stages after grouting are shown in Figure 4. Both parameters, the yield stress and the plastic viscosity, increase over time with a significant increase at approximately 6h. The viscosity development corresponds in a good approximation to the qualitative graph in Fig. 3 (green line). Due to the acting oscillation – representing the EAM - the formation of the strength-increasing hydration products accelerates compared to an undisturbed reaction. Both extrinsic agglomeration and deaeration of the grout material favour the increase of the rheological material properties [21]. For the numerical simulations, following in section 3.2, three points in time (30min, 380min and 430min) after grouting are investigated to see the influence of the stiffening grout; cf. Fig. 3. It should be noted that thermal effects were completely neglected.

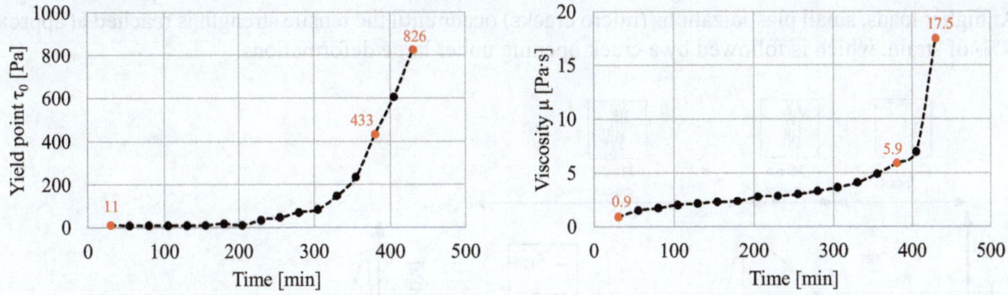

Figure 4. Experimental results of the rotational shear tests according to [20].

2.4 Solid grout material

To model the material behaviour of solidified concrete/grout Zobel [22] recommends to use a material law showing a high compressive strength compared to the tensile strength, very brittle tensile failure and an increased multi-axial compressive strength compared to the uniaxial compressive strength. Todays offshore grout materials are high performance materials designed for special use in harsh environmental conditions. Depending on the area of application, e.g. in platforms or wind turbines, and depending on the designer, different grout materials are used. A depiction of typical grout materials is shown in Fig. 5 Generally, a trend in the offshore wind sector is seen to high performance grouts (HPG) with an increased compressive strength. There is a clear correlation between the compressive strength and the stiffness of the materials, visualized by the dotted trend line. The tested material from Fig. 4 is an HPG with a compressive strength of $f_c = 140$ MPa, a young's modulus of $E = 45$ GPa and a poisson's ratio of $\mu = 0.2$ (marked with a red dot); cf. Fig. 5.

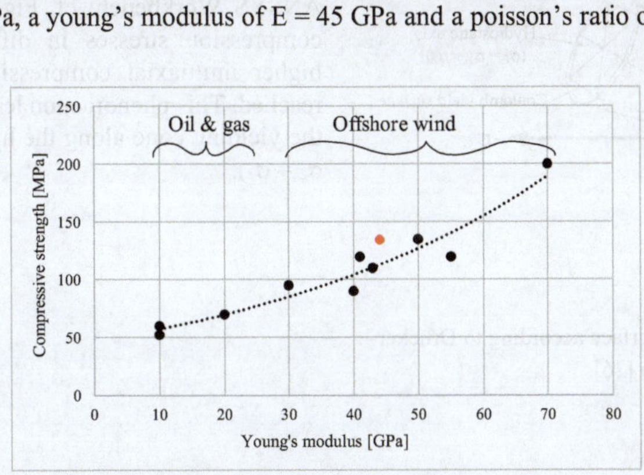

Figure 5. Typical grout materials according to their offshore use with a depiction of compressive strength over young's modulus [23].

The general failure behaviour of grout materials in their solid state is relatively equal to high strength concretes. Under uniaxial compressive loading, a linear elastic material behaviour usually occurs under small utilisations up to approximately 40% of the compressive strength f_{cm}. This range is marked as intact in Fig. 6 a). With further increasing stresses, the first plasticization, in the form of micro cracks, occurs until the compressive strength is reached. By reaching the fracture limit, micro cracks bundle into macro cracks. The post-fracture behaviour begins (marked with a dotted line).

The higher the strength of cementitious materials, the stiffer and more brittle they become. This leads to a lower rearrangement capacity due to plasticization. As a result, the post-fracture behaviour becomes less important and the maximum compressive strength can be selected as the limiting condition [24].

The tensile behaviour can be taken into account via the volume-specific fracture energy $g_{f,int}$. Up to a utilisation of 90% of the tensile strength f_{ctm}, a linear elastic material behaviour can be expected; cf. Fig. 6 b). At higher loads, small plasticizations (micro cracks) occur until the tensile strength is reached at approx. 0.15 ‰ of strain, which is followed by a crack opening under large deformations.

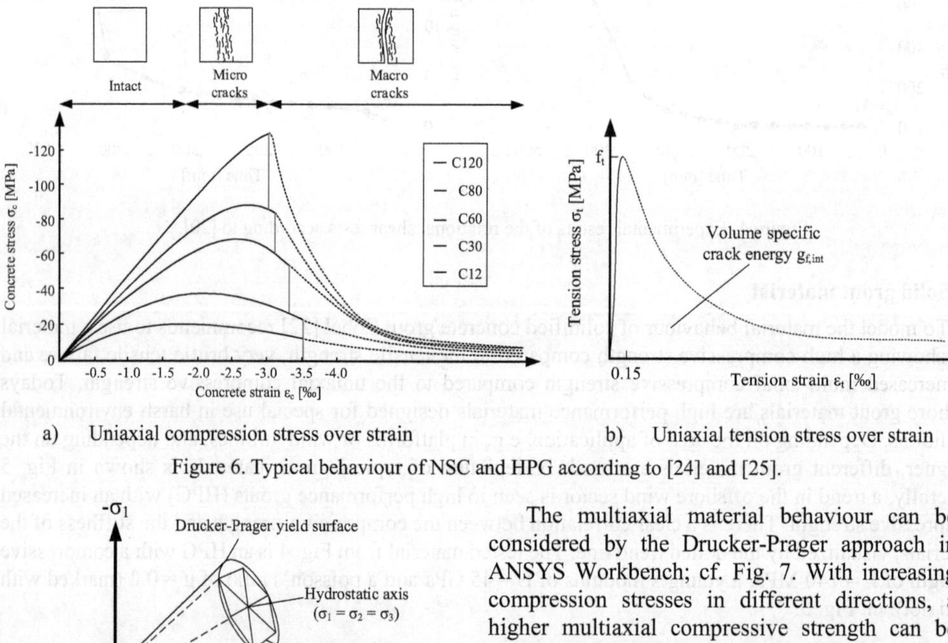

a) Uniaxial compression stress over strain b) Uniaxial tension stress over strain

Figure 6. Typical behaviour of NSC and HPG according to [24] and [25].

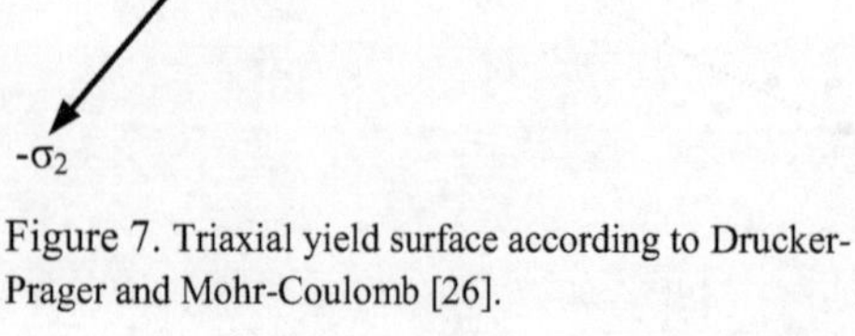

Figure 7. Triaxial yield surface according to Drucker-Prager and Mohr-Coulomb [26].

The multiaxial material behaviour can be considered by the Drucker-Prager approach in ANSYS Workbench; cf. Fig. 7. With increasing compression stresses in different directions, a higher multiaxial compressive strength can be reached. This phenomenon leads to the opening of the yielding cone along the hydrostatic axis ($\sigma_1 = \sigma_2 = \sigma_3$).

3 NUMERICAL SIMULATIONS

3.1 Grouted connection

As an example, a typical grouted connection of a pre-piled jacket support structure is investigated in this paper. Jacket structures are mainly loaded by cyclic axial forces, which, according to [27], are predominantly compression forces. Figure 8 provides an overview of a typical jacket structure of an OWT and the geometric dimensions of the investigated GC.

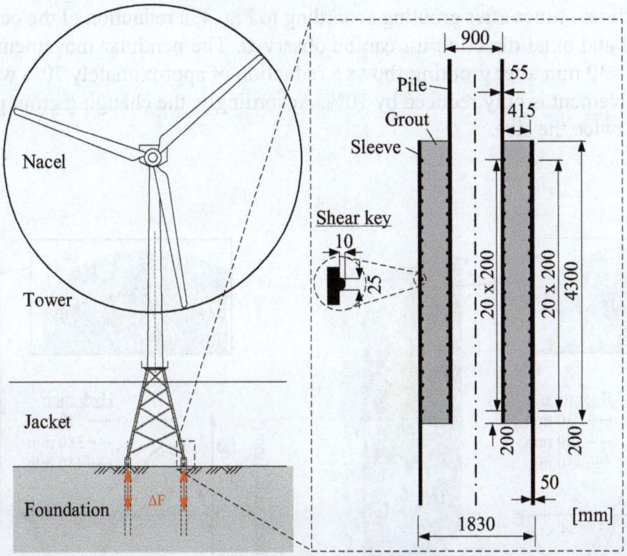

Figure 8. Overview of a typical grouted connection of a jacket support structure.

The typical movement occurring in GCs at the base points of a pre-piled jacket structure is a pendular movement described in [8]. At the upper end of the GC, the infiltrating steel tube (jacket leg) is often fixed to the outer steel tube (foundation pile) by a pile gripper to reduce EAM. The continuous loading by the waves causes the lower end of the jacket leg to oscillate in the foundation pile [8]. Depending on how much time has passed after grouting, reduced movements occur due to the curing of the grout material. To analyse the movements inside the grout material in a first stage a fluid grout simulation (section 3.2) is performed in order to get input data for the transfer function to adjust the solid grout simulation in section 3.3.

3.2 Fluid grout simulation

For the investigation of the EAM, a 2-way fluid structure interaction approach (FSI) is chosen. Such a simulation can be used to model the interaction between solids (e.g. steel tubes) and fluids (e.g. grout material). The meshing is done with brick elements (solids) or with volume elements (fluid). These two calculation methods, which are based on different differential equations, are coupled by means of a system-internal transfer of velocities and forces (2-way FSI). As the sleeve / foundation pile is rammed into the seabed, no movements are expected. Consequently, the sleeve is replaced in the simulation by a no velocity boundary condition ($v_1=v_2=v_3=0$). On the other hand, the pile movement represents the wave induced deformation of the jacket structure.

The investigation of EAM in GCs is carried out in two different damage scenarios. The first scenario involves the horizontal displacement - an approximation of the wave-induced pendular motion - of the pile. The second damage scenario concerns a possible vertical / axial cyclic movement of the pile.

For both models, a cyclic load boundary condition is applied to the top end of the pile, leading to a corresponding EAM. The time derivative of the displacement is applied to the volume elements of the grout by coupling the two models by a Dirichlet boundary condition. Hence the generated velocity field leads to forces in the grout annulus. With increasing viscosity and yield stress, the reaction force becomes larger, which leads to an increasing force transfer to the pile. An increasing resistance of the pile leads to a reduced EAM; cf. Fig. 9. For the first damage scenario (scenario 1) - a horizontal movement of the pile - a moment is applied. For the second (scenario 2) - a vertical movement of the pile – an axial force is applied to the upper edge of the pile.

Investigating different times after grouting according to Fig. 4, a reduction of the occurring movements in radial direction u_r and axial direction u_{ax} can be observed. The pendular movement in radial direction between 30 min and 430 min after grouting shows a reduction of approximately 20% within the first 7h. In contrast the axial movement is only reduced by 10%. Accordingly, the changing grout properties affect the pendular movement twice the size.

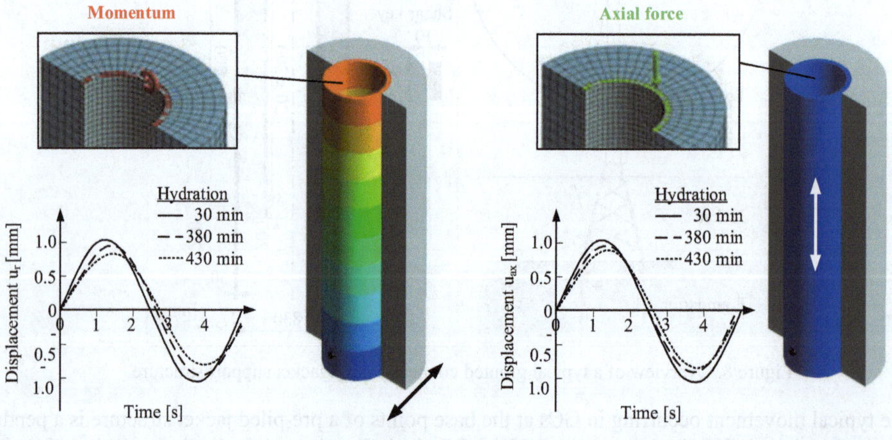

Figure 9. Pendular EAM applied by a momentum (left) and axial EAM applied by an axial load (right).

One reason for the observed differences between pendular and axial EAM is illustrated in Fig. 10. The pendular EAM has a much larger influencing zone, as the pile is pushing the grout material away. The axial EAM is causing shear forces in axial direction by carrying away a few particles of the grout material in a close area near the pile. This phenomenon is equal to the boundary layer effect in a flow through a pipe or along other fixed layers ($v = 0$). The particles in direct contact to the pile move with exactly the same velocity as the pile itself. With increasing distance to the pile, the velocity decreases drastically. This results from the shearing behaviour of the fluid, when exceeding the yield stress.

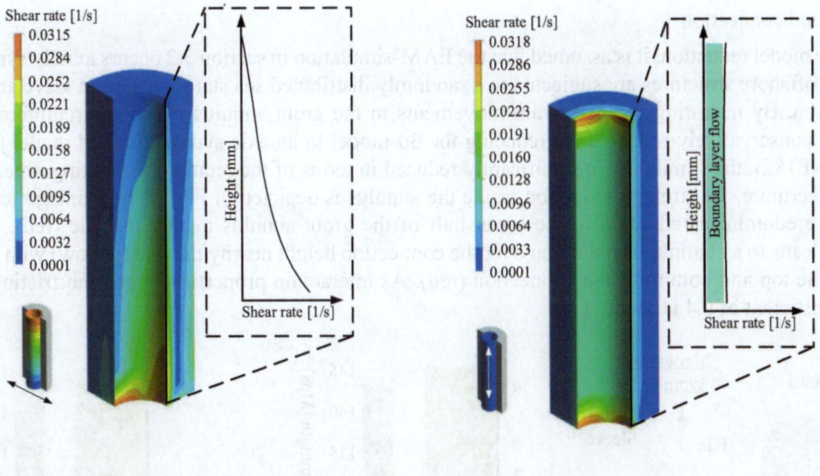

Figure 10. Shear rate analysis in grout annulus according to pendular (left) and axial EAM (right).

3.3 Solid grout simulation

Grouted connections show two main damage mechanisms in ultimate limit state tests. The arrangement of shear keys on the steel tubes result in a concentrated load transfer from one shear key at the pile to the opposing one at the sleeve [27, 23, 28]. Depending on the spacing and size of the shear keys, local grout matrix failure or compression strut failure can occur. Due to the formation of discrete compression struts, an almost one-dimensional stress state exists in this area of the grout material. Failure of the compression strut results from the transverse tension orthogonal to the compression direction. Nearby the shear keys, on the other hand, a three-dimensional compression stress state is formed. Due to the constriction effect caused by the surrounding steel tubes, increasing compression stresses in the grout material can be reached without failure.

3.3.1 Grout specimen

In order to validate the applied material model for the investigated high-strength grout material, a numerical analysis on a reduced model of a grout specimen was carried out to analyse the load deformation behaviour; cf. Fig. 11. In order to create a one-dimensional stress state, the geometry is supported laterally by weak springs. This allows a reduction of constraints with a high degree of numerical stability to depicture the post failure behaviour. The results confirm the implemented material model according to chapter 2.3.

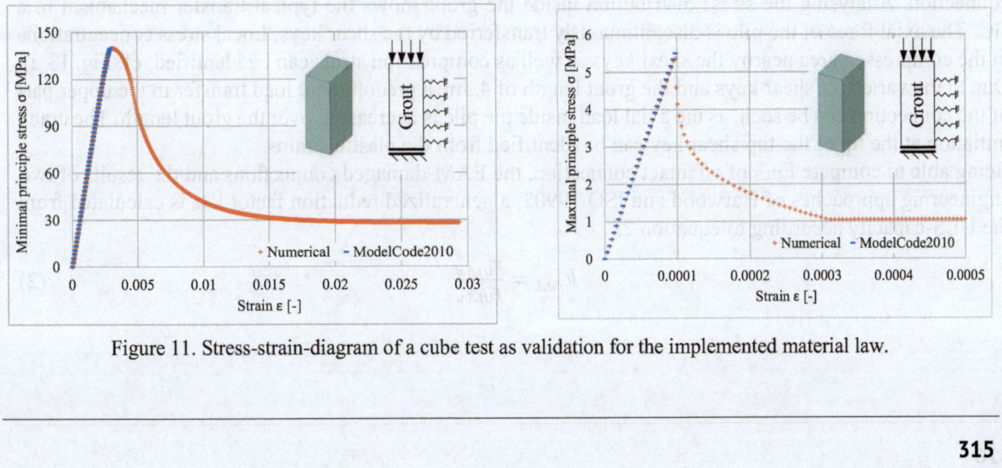

Figure 11. Stress-strain-diagram of a cube test as validation for the implemented material law.

3.3.2 Grouted connection

For the model reduction, it is assumed that the EAM-simulation in section 3.2 occurs axially symmetric. Typically, offshore structures are subjected to a randomly distributed sea state, varying in wave amplitude and direction. By mirroring the maximum movements in the grout annulus over the circumference, the strength is conservatively reduced. By reducing the 3d-model to an axis symmetrical 2d model (element type PLANE182), the simulation is significantly reduced in terms of the required computing time; cf. Fig. 12 a). Furthermore, the strength reduction inside the annulus is depicted in Fig. 12 b). Damage scenario 1 leads to a predominant reduction in the lower half of the grout annulus nearby the pile (red), whereas scenario 2 leads to a continuous reduction over the connection height nearby the pile (yellow) with singular effects at the top and bottom of the connection (red). As interaction properties a coulomb friction with a friction coefficient of 0.4 is used.

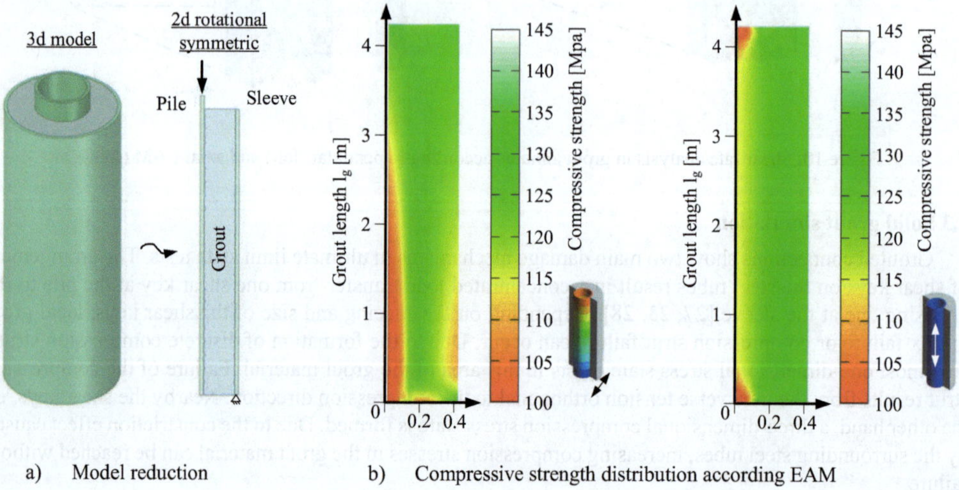

Figure 12. Numerical model and results of the FE-Simulation.

3.3.2 Ultimate limit state results

The nonlinear post-failure behaviour does not show up in the simulation of the entire grouted connection, as the local damage nearby the shear keys leads to a termination of the simulation. This point in time is conservatively assumed to be the limit state for the ultimate loading capacity F_{ULS} of the connection. Analysing the stress distribution inside the grout shows the typical transfer mechanism in a GC. The axial force of the pile is discontinuously transferred by the shear keys. Local stress concentrations in the compressed area nearby the shear keys as well as compression struts can be identified; cf. Fig. 13 a). Due to the variety of shear keys and the grout length of 4.3m, a predominant load transfer in the upper part of the connection can be seen, as the axial load inside the pile is decreasing over the grout length. The crack initiation at the tip of the top shear key can be identified from the plastic strains.

Being able to compare F_{ULS} of an intact connection, the EAM-damaged connections and the results of two engineering approaches of Harwood and ISO 19902, a generalized reduction factor k_{red} is calculated from the ULS-capacity according to equation 2.

$$k_{red} = \frac{F_{ULS,d}}{F_{ULS,u}} \tag{2}$$

With $F_{ULS,d}$ as ULS capacity of a damaged connection and $F_{ULS,u}$ as ULS capacity of an undamaged connection. The numerical results seem to lay in the expected range, as Merxhani showed the unfavourable reduction due to EAM according to ISO 19902 when using high performance grouts; cf. Fig. 13 b). The reduction factor is directly proportional to the compression stress, when exceeding the EAM limit while having $h/s \leq 0.06$, cf. equation 3 [29].

$$k_{red} = 1.0 - 0.1 \cdot \frac{h}{s} \cdot f_{cu} \tag{3}$$

With the shear key height h, the shear ley spacing s and the compression strength f_{cu}. The original reduction factor from Harwood's investigations [5], which were the basis for ISO 19902 looked slightly different; cf equation 4 if $0.06 \geq h/s \geq 0.02$

$$k_{red} = 1.2 - 10 \cdot \frac{h}{s} \tag{4}$$

The simulation results presented in this paper are ranging between the results of Harwood and F_{ULS} of the undamaged connection. Furthermore, it can be concluded, that the axial EAM is less critical than the pendular EAM. This hypothesis has still to be proven by experimental tests.

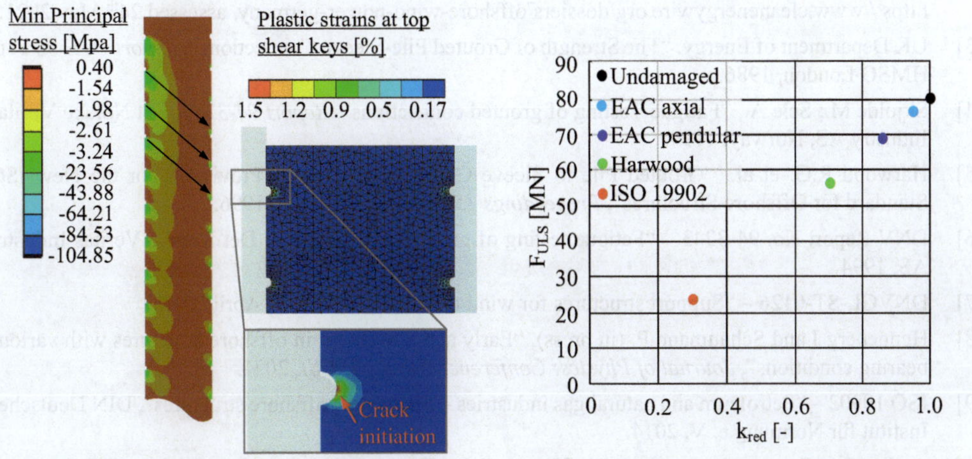

a) Stress and strain distribution in the grout b) Influence of the damage scenarios on the ultimate limit state capacity F_{ULS}

Figure 13. FE-analysis of grouted connection.

4 CONCLUSIONS

Within the presented numerical investigations, a new approach to consider early age movement (EAM) in grouted connections (GCs) is established. By simulating shear movements inside the grout annulus, using computational fluid dynamics (CFD) / fluid structure interaction (FSI) and coupling the simulation results with a transfer function to a FE simulation, the influence of EAM on the curing grout material and hence the ULS capacity of GCs can be calculated. The current approach is based on a simplified transfer function considering a proportional relation between EAM in the form of the shear rate and a reduction of the compressive strength. This transfer function will be validated and extended in future experimental tests.

5 ACKNOWLEDGEMENTS

The presented numerical simulations and results are achieved within the research project 'Grouted Joints – Influence of Early-Age-Movement – GREAM' (BMWi, funding sign: 0324257). The authors thank the German Federal Ministry for Economic Affairs and Energy for funding and all accompanying industrial partners (DNV, Siemens Gamesa, Wilke & Schiele Consulting GmbH and the Federal Institute for Materials Research and Testing) for their expert advice. Moreover, the authors thank the material manufacturers Densit, Holcim and Pagel. Further acknowledgement goes to the Institute for Building Materials Science for the cooperative partnership and Niklas Römgens, who supported the numerical investigations within his master thesis.

REFERENCES

[1] Global Offshore Wind Farms Database - 4C Offshore, [online] available: http://www.4coffshore.com/windfarms/, accessed 24th May 2021.

[2] Clean Energy Wire CLEW, Stiftung Offshore Windenergie, 4th May 2018, [online] available: https://www.cleanenergywire.org/dossiers/offshore-wind-power-germany, assessed 24th May 2021.

[3] UK Department of Energy. "The Strength of Grouted Pile-Sleeve Connections", *Report OTH 86 210*, HMSO London, 1986.

[4] Skjolde M.; Sele A., "Fatigue Testing of grouted connections". *Report 94-3243*, Det Norske Veritas Industry AS, Norway, 1994.

[5] Harwood R.G. et al., "Grouted Pile to Sleeve Connections: Design Provisions for the New ISO Standard for Offshore Structures", *Proceedings 15th OMAE*, Florence, 1996.

[6] DNV Report No. 94-3243 – "Fatigue testing of grouted connections", Det Norske Veritas Industry AS, 1994

[7] DNVGL-ST-0126 – "Support structures for wind turbines", DNV-GL, April 2018

[8] Henneberg J and Schaumann P. (in press), "Early age movement in offshore structures with various bearing conditions", *Journal of Physics: Conference Series (JPCS)*, 2021.

[9] ISO 19902 – "Petroleum and natural gas industries – Fixed steel offshore structures", DIN Deutsches Institut für Normung e. V, 2014.

[10] Anders S., "Betontechnologische Einflüsse auf das Tragverhalten von Grouted Joints", Institute of Building Materials Science, Leibniz Universität Hannover, Hanover, Germany, 2007.

[11] DAfStb Guideline for Self-compacting Concrete (SCC-Guideline), 2003.

[12] Lohaus L., Cotardo D., and Werner M., "A Test System to Simulate the Influence of Early Age Cycling on the Proper-ties of Grout Materials," *Proceedings 24th ISOPE*, Busan, Korea, 2014.

[13] Cotardo D., Haist M. and Lohaus L., "Early-age Movement in Grouted Joints for Offshore Applications - Determination of the Development of Grout-stiffness", *Proc 29th Int ISOPE*, 2019.

[14] Werner M., "Realitätsnahe Materialeigenschaften von hochfesten Feinkornbetonen in Grouted Joints", Dissertation, *Berichte aus dem Institut für Baustoffe*, Heft 15, Hannover, 2017.

[15] Herb A.T., "Indirekte Beobachtung des Erstarrens und Erhärtens von Zementleim, Mörtel und Beton mittels Schallwellenausbreitung", Dissertation, *Institut für Werkstoffe im Bauwesen*, Universität Stuttgart, 2003.

[16] König H., "Untersuchung zur Kinetik grundlegender Reaktionsschritte bei der Hydratation von Portland- und Hochofenzement", Dissertation, *Lehrstuhl für Bauchemie*, TU München, März 2010.

[17] Schneider M., "Hydratation des Zements und Gefüge des Zementsteins", *Zement Taschenbuch*, Verein Deutscher Zementwerke, Düsseldorf, 2002.

[18] Roussel N. and Gram A., "Simulation of Fresh Concrete Flow – State of the Art Report of the RILEM", *Springer Verlag*, Netherland, 2014.

[19] Roussel N., "A thixotropy model for fresh fluid concretes - theory, validation and applications", *Cement and Concrete Research. 36*, 2006.

[20] Cotardo D. et al., "Rotatorische und oszillatorische Scherversuche zur Ermittlung steifigkeits-relevanter Kenngrößen von Offshore-Vergussmörteln unter dem Einfluss des Early-age Movement", *Proceedings of the 28th Conference and Laboratory Workshops*, Regensburg, 2019.

[21] Cotardo D., Haist M. and Lohaus L., "Zum Einfluss des Early-age Movement auf die Rheologie von Offshore-Vergussmörteln", *Proceedings of the 60th DAfStb-Forschungskolloquium*, Hannover, 2019.

[22] Zobel R., "Verbundmodellierung von Beton- und Spannstahl unter Querzug", Dissertation, *Institut für Massivbau*, TU Dresden, 2017.

[23] Bechtel A., "Fatigue behaviour of axially loaded grouted connections in jacket structures". Dissertation, *Institute for Steel Construction*, Leibniz Universität Hannover, 2016.

[24] fib Bulletin 42, "Constitutive modelling for high strength/high performance concrete", *International Federation for Structural Concrete*, 2008.

[25] Schlegel R., „Nichtlineare Berechnung von Beton und Stahlbetonstrukturen nach DIN 1045-1 mit ANSYS", *23rd CADFEM Users' Meeting*, Bonn, 2005.

[26] Kathibinia M. et al., "Shape optimization of concrete gravity dams considering dam-water-foundation interaction and nonlinear effects", *Journal of Optimization in Civil Engineering*, University of Birjand, 2016.

[27] Raba A., "Fatigue Behaviour of Submerged Axially Loaded Grouted Connections", Dissertation. *Institute for Steel Construction*, Leibniz Universität Hannover, 2018.

[28] Schaumann P., Henneberg J. and Raba A., "Axially loaded grouted connections in offshore conditions using ordinary portland cement", *Proceedings 12th ASCCS*, Valencia, 2018.

[29] Merxhani A. et al., "Critical Review of Early Age Cycling Effects on the Capacity of pile to Sleeve Grouted Connections as Treated in ISO 19902", *Proceedings of the ASME 2019 38th OMAE*, Glasgow, 2019.

9TH INTERNATIONAL CONFERENCE ON COMPOSITE CONSTRUCTION IN STEEL AND CONCRETE
M. Knobloch, U. Kuhlmann, W. Kurz, M. Schäfer. (Eds.)
Online, Germany, July 27-29, 2021

PERFORMANCE AND DESIGN OF STAINLESS-STEEL COMPOSITE STRUCTURES – BEAMS, COLUMNS AND JOINTS

Sina Kazemzadeh Azad*, Brian Uy*, Yifan Zhou*, Yuchen Song*, and Jia Wang*

* School of Civil Engineering, The University of Sydney, Sydney, Australia
e-mails: sina.kazemzadehazad@sydney.edu.au, brian.uy@sydney.edu.au,
yifan.zhou@sydney.edu.au, yuchen.song@sydney.edu.au, jia.wang@sydney.edu.au

Keywords: Stainless steel, Composite beam, Composite column, Beam-to-column joint, Experiments.

Abstract. *The paper summarises a series of investigations conducted in recent years at the University of Sydney on the behaviour and design of stainless steel-concrete composite structures, with a focus on the mechanical performance of composite beams, columns, and joints. Three stainless steel grades were considered in the experiments, namely austenitic, duplex, and lean duplex. Details of the experiments as well as some of the experimental observations are presented herein, along with the comparison of the test results with the results of simulations. The test results are also compared with the requirements and predictions of international design codes, which are generally developed for carbon steel composite members, in order to investigate their applicability for stainless steel composite members. A number of design recommendations are also developed and presented. The studies are expected to help engineers achieve a deeper understanding of the behaviour and design of stainless steel-concrete composite members.*

1 INTRODUCTION

The idea of using stainless steel in structural members was first investigated in the 1950s [1, 2]. These pioneering studies and many other following investigations all concluded that stainless steel can be a viable alternative to carbon steel in structural members. When the long-term cost of maintenance of a structure located in a corrosive environment is taken into account, stainless steel typically becomes the logical choice over carbon steel. In line with this, in recent decades, stainless steel has been used in many structural applications [3], from the Sölvesborg bridge in Sweden to the recently-completed first stainless steel bridge of the United Kingdom over River Eamont. Nevertheless, the initial material cost of a stainless steel structure is generally considered to be several times that of its carbon steel counterpart. A method to reduce the material cost would be to efficiently use stainless steel in a composite form with concrete. Roufegarinejad et al. [4] proposed the idea of using stainless steel tubes as concrete-filled columns. A similar concept was used in the towers of the Stonecutters bridge in Hong Kong. As noted in the review study of Baddoo [5], there are many other cases where structural stainless steel members can be used in a composite form with concrete to achieve an economic, corrosion-resistant, safe, and architecturally appealing outcome.

In line with this, a comprehensive research programme was initiated in 2018 at the University of Sydney with the main aim of improving the understanding of the structural response and design of stainless steel-concrete composite members and structures. The programme investigated the behaviour and design of stainless steel-concrete composite beams, columns, and beam-to-column joints. The investigation on joints also consisted of a comprehensive study on stainless steel bolts. In total, over 110 specimens and 200 bolts were tested as a part of the programme. Although stainless steel sections are mainly intended for large infrastructure projects, the studies on large fabricated (welded) stainless steel sections are rather limited in the literature. Consequently, all the specimens tested in the experiments were fabricated from austenitic, duplex, and lean duplex stainless steel plates. The experiments were typically followed by comprehensive finite element (FE) analyses. An important objective was to provide reliable and convenient design recommendations for practicing engineers. While the research programme is still ongoing, a general outline of the tests conducted as well as some of the experimental observations, numerical simulation results, and developed design recommendations are presented in this paper.

2 STAINLESS STEEL-CONCRETE COMPOSITE BEAMS

2.1 Stainless steel I-beams under bending and/or shear

While a major part of the research programme was focused on the behaviour and design of stainless steel-concrete composite I-beams, in the early stages of the project, it was deemed fruitful to first conduct a series of experiments on bare stainless steel I-beams to better understand their behaviour. A review of the literature revealed that only a few studies investigated the shear response of stainless steel I-beams, and those which did, focused mainly on plate girders [6, 7]. The studies on the moment-shear interaction of stainless steel I-beams were even more scarce [8]. It was also noted that, similar to the case of carbon steel beams, most design codes ignored the contribution of strain hardening in the design of stainless steel I-section beams – which can be significant.

Consequently, a series of experiments were designed and carried out on compact stainless steel I-sections. In total, eleven specimens were fabricated from austenitic 304, duplex 2205, and lean duplex LDX2101 stainless steel plates. Four-point loading tests were conducted to investigate the flexural behaviour of the specimens whereas three-point loading tests considering different beam length were carried out to study the shear and moment-shear interaction responses of the stainless steel I-beams. Details of the test setups are schematically illustrated in Figure 1.

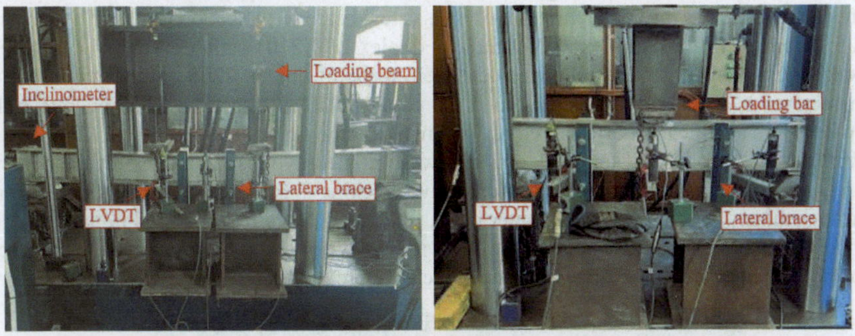

Figure 1. Four-point and three-point loading test setups for stainless steel I-beams.

Based on the specimen length and the applied loading scenario, flexural failure triggered by flange local buckling, shear failure due to web local buckling, or a combination of these two failure modes were observed. Some of the typical failure modes are depicted in Figure 2. The experimental results were also used to validate FE models which were then used for a parametric study.

Figure 2. Typical flexural and shear failure modes observed for the tested stainless steel I-beams.

While the investigation is still ongoing [9], a preliminary comparison of the experimental and numerical results with the predictions of the Australian [10], American [11], and European [12] design standards/guides revealed that the flexural and shear strengths of the stainless steel I-beams were significantly (20 to 50%) higher than the codified predictions. This was mainly attributed to the strain hardening effect which was particularly high for the austenitic specimens. Accordingly, the moment and shear capacities of a compact stainless steel I-beam could be determined as follows:

$$M_u = \alpha M_{pl} = \alpha Z f_y \quad (1)$$

$$V_u = \beta V_{pl} = \beta(0.6 f_y A_w) \quad (2)$$

where M_{pl} and V_{pl} are the plastic moment and shear capacities, Z is the plastic section modulus, f_y is the yield stress, A_w is the web area, and $\alpha,\beta > 1.0$ are factors which shall be calibrated based on the test and FE results to reflect the effect of strain hardening on the strengths. Preliminary investigations suggested that α and β are dependent upon the stainless steel grades and the section slenderness. The results also indicated that the accuracy of the shear-moment interaction curves stipulated in international design standards could be significantly increased if the above expressions are used instead of the codified pure moment and shear capacities.

2.2 Compact stainless steel-concrete composite beams under bending and/or shear

In the next phase of the investigation, a new set of experiments was carried out on compact stainless steel-concrete composite I-beams. This was the first comprehensive experimental study on such specimens and included four-point and three-point loading tests conducted on composite beams of different lengths in order to investigate their flexural, shear, and moment-shear interaction responses.

As shown in Figure 3, a total of eighteen composite specimens were tested using experimental setups similar to those used previously for testing the bare stainless steel beams (Figure 1). Both bolted and welded shear connectors were used for connecting the stainless steel beam to the concrete slab. The I-beams were fabricated from austenitic 304, duplex 2205, and lean duplex LDX2101 stainless steel plates, whereas the stainless steel shear connectors and reinforcement were made of duplex 2205 or austenitic 316L grades. The beam sections had similar dimensions to those of a 180UB18.1 section. The concrete slab had a thickness of 100 mm and the cylindrical strength of concrete at the start of testing was 42 MPa. The flexural failure dominated the response of longer beams and was initiated by tensile cracking of the concrete slab and yielding of the bottom flange, followed by extensive cracking in the slab and finally concrete crushing. For the shear-dominant specimens, on the other hand, fine cracks were first detected in the early stages of loading which extended gradually. The shear failure was however marked by significant diagonal cracking in the slab and web bucking in the I-beam. These typical failure modes are depicted in Figure 3.

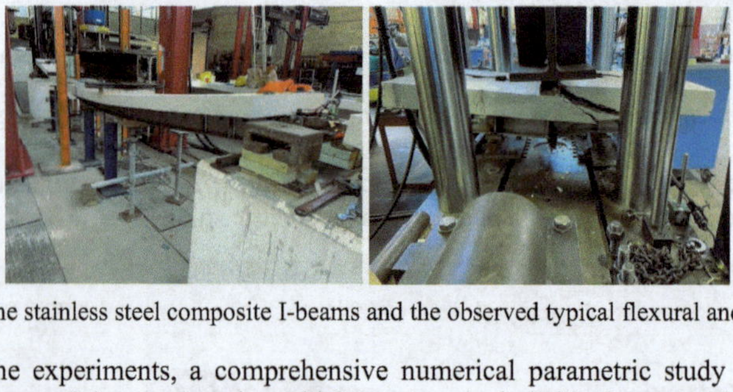

Figure 3. Details of the stainless steel composite I-beams and the observed typical flexural and shear failure modes.

In addition to the experiments, a comprehensive numerical parametric study is also underway to investigate other degrees of shear connection which were not included in the tests. The investigation of the test and FE data is still ongoing [13], however, the initial results strongly suggested that the flexural capacity of a compact stainless steel composite I-beam could be underestimated by a *Plastic Stress Distribution Method* (PSDM), which is the typical approach used in most international design standards developed for carbon steel members. The reasons for this underestimation (which, based on the test data, could be in excess of 20%) is that the significant strain hardening effect, which is an important characteristic of a stainless steel beam, is ignored in PSDM. The preliminary results suggested that, for a more efficient design, a modified approach needs to be developed which would incorporate the strain hardening in the stress distribution of the stainless steel I-beams [13].

Regarding the shear response, the initial results revealed that the method used by some international design standards (such as Eurocode 4 [14]) in which the shear strength of a composite I-beam is limited to that of its steel web only (i.e. $V_u = 0.6 f_y A_w$) can drastically underestimate the shear capacity of a stainless steel composite I-beam. An approach to alleviate this underestimation is to add the contribution of the concrete slab (V_{slab}) to the total shear strength:

$$V_u = 0.6 f_y A_w + V_{slab} \tag{3}$$

For instance, AS/NZS 2327 [15] allows for such an approach to be followed for the design of carbon steel-concrete composite I-beams. The preliminary results revealed that, while the AS/NZS 2327 [15] approach is much more accurate (compared to ignoring the slab contribution altogether), when applied to stainless steel composite I-beams, it could sometimes overestimate the total shear strength. The reason for this appears to be rooted in the occurrence of concrete failure prior to the development of the full shear resistance in the stainless steel I-beam. A revised design formula is currently being developed [13] to provide safe and accurate predictions for the shear capacity of stainless steel-concrete composite I-beams.

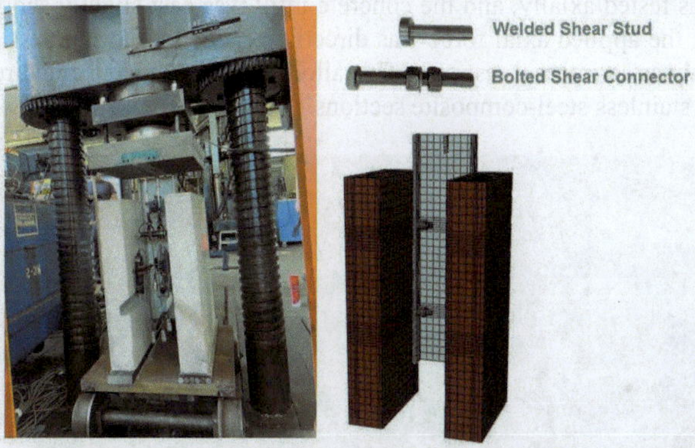

Figure 4. Testing and simulation of stainless steel welded and bolted shear connectors.

2.3 Stainless steel welded and bolted shear connectors

In parallel to the experiments on bare and composite stainless steel I-beams, a series of pushout tests were also conducted for the first time at the University of Sydney. Both welded and bolted stainless steel shear connectors of different diameters, made from austenitic 316L and duplex 2205 grades were included in the testing programme. Furthermore, both normal and high strength concrete (42 and 70 MPa) were considered. The utilised test setup and the general details of the shear connectors are illustrated in Figure 4. During the tests, depending on the details of each specimen, failure initiated either in the shear connector or the concrete slab.

The initial results suggest that if the current limit of 500 MPa, stipulated in AS/NZS 2327 [15] and Eurocode 4 [14] on the ultimate strength (f_u) of shear connectors, is used instead of the actual f_u of high strength stainless steel shear connectors, the codified predictions could be overly conservative. A solution could be to ignore this limit in the design process; however, the initial test results revealed that doing so could this time lead to overestimation of the strength of the stainless steel shear connectors. The results highlighted the need for revised design expressions to be developed for both the limit states of shear connector fracture and concrete crushing. As shown in Figure 4, detailed FE analyses are underway to complement the experimental data in order to propose such design expressions for stainless steel shear connectors [16].

3 STAINLESS STEEL-CONCRETE COMPOSITE COLUMNS

3.1 Slenderness limits for stainless steel-concrete composite columns

Stainless steel-concrete composite columns have a high potential to be used in infrastructure projects. For instance, concrete-filled stainless steel tubular (CFSST) columns can be used in bridges as members with high strength and stiffness, as well as high corrosion resistance. A review of the literature revealed that, while the behaviour of stainless steel composite columns was investigated previously [4, 17-19], there were no studies dedicated to the local and postlocal buckling behaviour of these members. As a result, the *slenderness limits* required for the design of stainless steel composite columns were not developed. These limits are traditionally used to distinguish compact sections from slender section for which the local buckling typically limits the strength. Consequently, as a part of the research programme at the University of Sydney, a comprehensive series of tests were conducted to develop such slenderness limits.

As shown in Figure 5, the testing programme comprised of concrete-filled box and circular sections as well as partially-encased I-section columns. Specimens were fabricated from three stainless steel grades, namely austenitic 304, duplex 2205, and lean duplex LDX2101. Based on a preliminary study [20], for each column type and material grade, a wide range of section sizes were considered; from very compact sections to highly slender ones. Some of the tested specimens were among the largest CFSSTs tested to date. Each specimen was tested axially, and the concrete infill was cast slightly shorter than the stainless steel section. Therefore, the applied axial force was directly resisted by the stainless steel section and the concrete infill served only as an internal restraint. This allowed for an isolated and direct investigation into the local stability of the stainless steel-composite sections.

Box CFSST Circular CFSST Partially-encased Stainless Steel I-Section

Figure 5. Slenderness tests on stainless steel-concrete composite columns.

In order to determine the axial slenderness limits, the ultimate load (P_u) resisted by each specimen was recorded and used to calculate the effective width (or diameter) ratio, b_e/b (or d_e/d). For a compact section this ratio is 1.0 or more, and for slender section the ratio is below 1.0 and indicates in relative terms a reduction in the section capacity below its yield strength due to local buckling. As a representative, the results for box CFSSTs is shown in Figure 6. The plot includes the test data as well as the results of a numerical parametric study. As shown in the figure, a limit of $\lambda_e = 50$ can be considered as an appropriate axial slenderness limit for box CFSST; which is lower than those recommended in AISC 360 [21] and AS/NZS 2327 [15] for carbon steel concrete-filled tubular (CFST) sections. The test and FE results were also used to develop formulae for predicting the effective width/diameter ratios of the studied stainless steel composite sections. For instance, as shown in Figure 6, the following calibrated Winter's formula could be used to estimate b_e/b for box CFSSTs:

$$\frac{b_e}{b} = \left(1.49\frac{50}{\lambda_e}\right)\left[1-0.22\left(1.49\frac{50}{\lambda_e}\right)\right] \le 1.0 \tag{4}$$

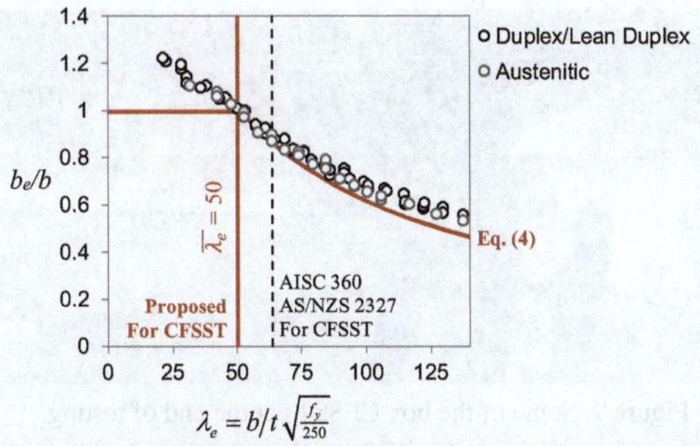

Figure 6. Determination of axial slenderness limit for box CFSSTs.

A similar approach was followed to determine the axial slenderness limits for the other studied section types as well. The proposed limits are summarised in Table 1. Further details can be found in [22, 23].

Table 1: Proposed slenderness limits for stainless steel-concrete composite sections.

Section Type[+]	Under axial compression, the section is slender if	Under flexural bending, the section is slender if
Box CFSST	$\lambda_e = b/t \times (f_y/250)^{1/2}$ is greater than **50**	$\lambda_e^f = b/t \times (f_y/250)^{1/2}$ is greater than **50**
Circular CFSST[++]	$\lambda_e^* = d/t \times (f_y/250)^{1/3}$ is greater than **148**	-
Partially-Encased I-Section	$\lambda_e = b/t \times (f_y/250)^{1/2}$ is greater than **18**	-

[+] For box sections b is the clear plate width; for I-sections b is the outstand width; t is thickness
[++] Note that for circular CFSSTs, the power used in the definition of the slenderness parameter is 1/3

A separate investigation was also carried out to recommend slenderness limits for box CFSSTs under flexural bending. To this end, box CFSST beams were tested under four-point bending (see Section 3.2) and a special technique was utilised to extract the moment carried only by the stainless steel section (M_u^{st}) based on strain gauge readings. The test data was then complemented by a comprehensive numerical study. The M_u^{st} values were then normalised by M_{pl}^{st} in order to investigate whether the stainless steel section was able to attain its full capacity or not; where M_{pl}^{st} is the share of the stainless steel section when the entire CFSST beam reaches to its plastic bending strength. The obtained results suggested that, in order to have a compact box CFSST beam, the flange slenderness, $\lambda_e^f = b/t\ (f_y/250)^{1/2}$, should be limited to 50 (as noted in Table 1). It was also observed that the web slenderness ($\lambda_e^w = h/t\ (f_y/250)^{1/2}$) did not alter this conclusion as long as it was kept below 160 (which is extremely slender) [24]. It should be noted that the flexural slenderness limits for circular CFSSTs and partially-encased I-sections were not included in the investigation and remain potential topics for future investigations.

3.2 Behaviour and design of compact and slender box CFSSTs under compression and/or bending

Another phase of the research programme was dedicated to investigating the behaviour of compact and slender box CFSSTs under compression, bending, and combined compression & bending, with the final goal of recommending complete design procedures. To this end, box CFSSTs with different section sizes, fabricated from austenitic 304 and lean duplex LDX2101 stainless steel plates were tested. The strength of concrete was 37 MPa at the start of the testing. A special post-processing routine was developed in order to estimate the share of the stainless steel section and the concrete infill in the total resistance of the CFSST section. This was deemed necessary for recommending accurate and reliable design expressions.

Based on the specimen size and the applied loading scenario, different failure modes including yielding, local buckling, global buckling, weld fracture, and concrete crushing were observed during the tests. Some of the failed specimens are depicted in Figure 7. In addition to the experiments, a comprehensive parametric study was also conducted. All the test and FE results were then synthesised to recommend design procedures for box CFSSTs. For instance, the following was proposed for design under compression:

Figure 7. Some of the box CFSSTs at the end of testing.

$$N_{prop} = \begin{cases} A_s f_y + \gamma_u A_c f_c & if \quad \lambda_e \le 50 \\ A_e f_y + \gamma_u A_c f_c & if \quad 50 < \lambda_e \le 140 \end{cases} \tag{5}$$

where A_s and A_c are the areas of the stainless steel section and the concrete infill, respectively, f_c is the concrete cylindrical strength, γ_u is a reduction factor for the *size effect*, and A_e is the *effective area* of the stainless steel section determined using Eq. (4). Similarly, expressions could also be developed for compact and slender box CFSSTs under bending, and combined compression & bending [24].

4 STAINLESS STEEL-CONCRETE COMPOSITE BEAM-TO-COLUMN JOINTS

4.1 Stainless steel bolts under tension and/or shear

Stainless steel bolts could serve as fastening components in beam-to-column joints and as demountable shear connectors in composite beams (Section 2.2). An individual bolt could therefore undergo complex states of stress due to tension, shear, or a combination of these two. A detailed review of the relevant design guides [11, 14] revealed that the present design principles for stainless steel bolts are essentially same as those developed for carbon steel bolts. This is due to a lack of sufficient experimental evidence to support a more rational design method. In line with this, a comprehensive experimental programme incorporating more than 200 stainless steel bolts was conducted. A wide range of stainless steel bolt grades including the most commonly used austenitic grades (i.e. A4-70 and A4-80) [25], as well as high strength austenitic and duplex grades with strength classes up to 1200 MPa [26] were tested. Figure 8 illustrates the purpose-built test setup allowing the bolt to be loaded under pure tension, pure shear, and combined actions. Typical failure modes observed in the tests are also shown in the figure. In general, it was observed that the tested stainless steel bolts exhibited a much more ductile behaviour compared to their carbon steel counterparts.

Figure 8. Test setup for the bolt tests under combined tension and shear, along with fractured specimens.

The obtained experimental results were compared with the predictions of AS 4100 [10], Eurocode 3 [27], and AISC 360 [21]. Overly conservative predictions were found from Eurocode 3 [27] whereas the predictions of the Australian [10] and American [21] specifications were more reasonable, though slightly unconservative in some cases. Consequently, revised design expressions were proposed which were specifically calibrated to provide safe and accurate predictions for stainless steel bolts [26]:

$$N_{design} = f_u \hat{A} \qquad \textit{pure tension}$$

$$V_{design} = \begin{cases} 0.65 f_u \hat{A} & \textit{pure shear – shear plane passing through threads} \\ 0.59 f_u A & \textit{pure shear – shear plane passsing through unthreaded shank} \end{cases} \tag{6}$$

$$\left(\frac{N}{N_{design}}\right)^2 + \left(\frac{V}{V_{design}}\right)^2 \leq 1.0 \qquad \textit{combined actions}$$

where N and V are the applied tension and shear force, whereas $\hat{A}$ and A are the tensile stress area and gross area of the stainless steel bolt respectively.

In addition to proposing design equations, the experimental results were also utilised for calibrating FE models which could simulate the fracture behaviour of stainless steel bolts. A complete framework was proposed to calibrate the constitutive and fracture models of stainless steel bolts with a minimum number of required material tests [25]. The framework could readily be applied to FE models in order to predict the fracture of a bolt individually or in a beam-to-column joint.

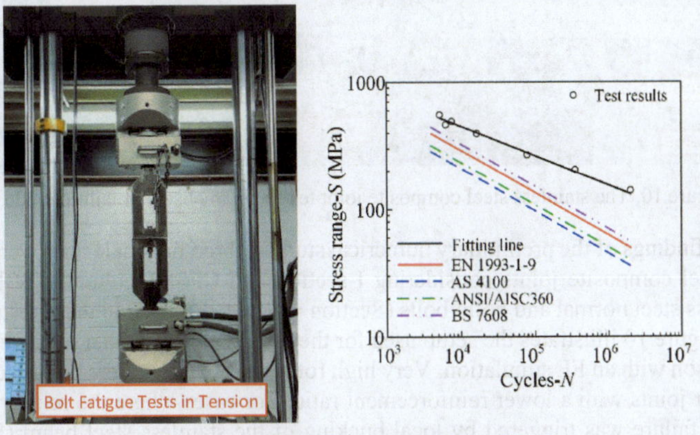

Figure 9. Stainless steel bolt fatigue tests.

Parallel to the investigations on stainless steel bolts under static loads, experiments were also conducted under high-cycle fatigue in tension and shear [28]. In addition to normal stainless steel bolts (hexagon bolts), stainless steel hollo bolts were also tested in these series. The utilised test setup is shown in Figure 9. As shown in the figure for a representative case, "Stress Range vs Cycles to Failure" (S-N) curves were plotted based on the test results and compared with the prediction of international design standards which are mainly intended for carbon steel bolts. In most cases, the predictions of the design standards were reasonable, although they could be overly conservative for the low tensile stress range. The predictions of BS 7608 were found to provide the closest predictions compared to the test data.

4.2 Stainless steel-concrete composite beam-to-column joints

The final phase of the research programme focused on the behaviour and design of stainless steel-concrete composite beam-to-column joints. To this end, first a series of numerical simulations were conducted in order to have a better understanding of the response of these joints. In these numerical studies, a wide range of joint configurations were investigated including stainless steel composite joints with I-section columns [29] and concrete-filled circular columns [30]. It was concluded that the unique mechanical properties of stainless steel, particularly the nonlinear stress-strain and significant strain-hardening, play important roles in the overall joint behaviour. Therefore, modifications to the existing design methods (developed mainly for carbon steel composite joints) were deemed necessary.

It is worth noting that the numerical studies were also extended to investigate the progressive collapse of stainless steel composite beam-to-column joint sub-models and moment resisting frames under column removal scenarios [31]. The analysis results demonstrated the importance of the catenary action, connection design, and stainless steel strain hardening in the progressive collapse performance. Based on the simulations, a progressive collapse design process was recommended along with a revised and safe formula for determining the dynamic increase factor (DIF).

Figure 10. The stainless steel composite joint tests and an observed failure mode.

In line with the findings of the preliminary numerical studies, three full-scale tests were recently carried out on stainless steel composite joints considering I-section and CFSST columns [32]. The previously investigated stainless steel normal and hollo bolts (Section 4.1) were utilised in these tests as fasteners and shear connectors. Figure 10 illustrates the setup used for these tests along with an observed failure mode as well as the comparison with an FE simulation. Very high rotation capacities were recorded for all the tested joints, especially for joints with a lower reinforcement ratio. For cases where the reinforcement ratio was relatively high, the failure was triggered by local bucking of the stainless steel beam (Figure 10) which limited the development of the full connection ductility.

Based on the test results and numerical analyses, an updated design method was developed for stainless steel composite joints [32], which was modified based on the component method in Eurocode 4 [14] and AS/NZS 2327 [15]. The effect of strain-hardening on the moment resistance, as well as the reduced rotational stiffness due to the nonlinear stress-strain response of stainless steels were considered in the updated method.

5 SUMMARY

The experimental and numerical studies conducted in recent years at the University of Sydney on the behaviour and design of stainless steel-concrete composite members, fabricated from austenitic, duplex, and lean duplex materials, were summarised in the present paper. Over 110 specimens and 200 bolts were tested as a part of the programme, with the main aim of developing convenient and safe procedures for the design of stainless steel-concrete composite systems.

The investigations on bare and composite stainless steel beams revealed the need for revised processes for the design of these members under bending and/or shear in order to take into consideration their significant strain hardening characteristics. Pushout test results conducted on welded and bolted stainless steel shear connectors also revealed that the current codified approaches which are intended for carbon steel shear connectors are not directly applicable to the stainless steel counterparts.

A comprehensive set of slenderness tests and simulations was conducted to recommend slenderness limits for box and circular CFSSTs as well as partially-encased stainless steel I-sections. The study was then extended to investigate the strength response of compact and slender box CFSSTs under compression, bending, and combined compression & bending, with an aim of recommending full design procedures.

A special setup was built to test stainless steel bolts under tension and/or shear. Based on the test results the need for revised design formulae was highlighted and new design expressions were developed accordingly. The test data was also utilised to develop a framework for calibrating FE models which could predict the fracture of stainless steel bolts under tension, shear, and combined actions. The study on bolts was also extended to investigate the high-cycle fatigue response of conventional and hollo stainless steel bolts under tension and shear.

In the last phase of the project, the behaviour of stainless steel-concrete composite beam-to-column joints was investigated. The initial numerical results suggested that the overall joint behaviour is influenced notably by the unique properties of stainless steel, and therefore, modifications in the codified design methods, which are mainly developed for carbon steel joints, were deemed necessary. As a part of the numerical investigation, the progressive collapse of stainless steel-concrete composite systems was also studied under column removal scenarios. Finally, full-scale experiments were conducted on stainless steel composite joints where the effect of reinforcement ratio on the observed failure mode was highlighted. On the basis of the test and FE results a revised method for the design of stainless steel composite joints was also developed.

6 ACKNOWLEDGMENTS

The research programme was supported by the Australian Research Council (ARC) under its Discovery Scheme (Project No.: DP180100418). The support of the following manufacturers/suppliers in the form of partial or full material donations is gratefully acknowledged: Outokumpu, Stirlings, Bumax, Lindapter, Ancon, and Hobson. Special thanks to Dr Dongxu Li, Dr Xinpei Liu, Dr Zhichao Huang, and Mr Jun Mo as well as Dr Mohanad Mursi and all the technical staff of the J. W. Roderick Materials and Structures Laboratory for their assistance during the concrete castings and experiments. Some of the authors were also supported during the research programme by the University of Sydney International Scholarship (USydIS), University of Sydney Postgraduate Award (UPA), Engineering and Information Technologies Research Scholarship (EITRS), and Postgraduate Research Support Scheme (PRSS). The authors would also like to acknowledge the support provided by the Sydney Informatics Hub (SIH) in facilitating access to Artemis, the high performance computing cluster of the University of Sydney.

REFERENCES

[1] Hammer E. W. and R. E. Petersen, "Column curves for type 301 stainless steel," *Journal of the Franklin Institute,* 261(2), 257-260, 1956

[2] Dubuc J., V. Krivobok, and G. Welter, "Studies of Type 301 Stainless Steel Columns," *Proceedings of the ASTM 2nd Pacific Area National Meeting*, Los Angeles, CA, 1956.

[3] International Stainless Steel Forum, *Stainless Steel in Infrastructure: Bridges*. Available Online: https://www.worldstainless.org, 2020

[4] Roufegarinejad A., B. Uy, and M. A. Bradford, "Behaviour and design of concrete filled steel columns utilizing stainless steel cross sections under combined actions," *Proceedings of the 18th Australasian conference on the mechanics of structures and materials*, Perth, Australia, 2004, 159-165.

[5] Baddoo N. R., "Stainless steel in construction: A review of research, applications, challenges and opportunities," *Journal of Constructional Steel Research,* 64(11), 1199-1206, 2008

[6] Saliba N. and L. Gardner, "Experimental study of the shear response of lean duplex stainless steel plate girders," *Engineering Structures,* 46, 375-391, 2013

[7] Saliba N., E. Real, and L. Gardner, "Shear design recommendations for stainless steel plate girders," *Engineering Structures,* 59, 220-228, 2014

[8] Chen X. W., H. X. Yuan, E. Real, X. X. Du, and B. W. Schafer, "Experimental behaviour of stainless steel plate girders under combined bending and shear," *Journal of Constructional Steel Research,* 166, 105900, 2020

[9] Zhou Y., B. Uy, J. Wang, and D. Li, "Behaviour and design of welded stainless steel beams with compact sections under flexure and shear," *Journal of Constructional Steel Research*, 2021 (In preparation)

[10] AS 4100, *Steel Structures*. Standards Australia, Sydney, Australia, 2020

[11] Baddoo N. R., *AISC Steel Design Guide 27: Structural Stainless Steel.* American Institute of Steel Construction, Chicago, IL, 2013

[12] Eurocode 3, *Design of steel structures - Part 1-4: General rules - Supplementary rules for stainless steels, EN 1993-1-4:2006 +A1:2015*. European Standard, Comité Européen de Normalisation, Brussels, Belgium, 2006 (+A1:2015)

[13] Zhou Y., B. Uy, J. Wang, D. Li, Z. Huang, and X. Liu, "Behaviour and design of stainless steel-concrete composite beam," *Journal of Constructional Steel Research,* 185, 106863, 2021

[14] Eurocode 4, *Design of composite steel and concrete structures – Part 1-1: General rules and rules for buildings, EN 1994-1-1:2004*. European Standard, Comité Européen de Normalisation, Brussels, Belgium, 2004

[15] AS/NZS 2327:2017 (+Amd 1:2020), *Composite structures — Composite steel-concrete construction in buildings*. Standards Australia/Standards New Zealand, Sydney, Australia/Wellington, New Zealand, 2020

[16] Zhou Y., B. Uy, J. Wang, D. Li, and X. Liu, "Behaviour and design of stainless steel shear connectors in composite beam," *Steel and Composite Structures*, 2021 (In preparation)

[17] Young B. and E. Ellobody, "Experimental investigation of concrete-filled cold-formed high strength stainless steel tube columns," *Journal of Constructional Steel Research,* 62(5), 484-492, 2006

[18] Lam D. and L. Gardner, "Structural design of stainless steel concrete filled columns," *Journal of Constructional Steel Research,* 64(11), 1275-1282, 2008

[19] Uy B., Z. Tao, and L. H. Han, "Behaviour of short and slender concrete-filled stainless steel tubular columns," *Journal of Constructional Steel Research,* 67(3), 360-378, 2011

[20] Kazemzadeh Azad S., B. Uy, D. Li, and C. Hou, "A Numerical Study on Local and Post-Local Buckling of Stainless Steel-Concrete Composite Columns," *Proceedings of the 9th International Conference on Advances in Steel Structures (ICASS'2018)*, Hong Kong, China, 2018.

[21] AISC, *Specification for Structural Steel Buildings, ANSI/AISC 360-16*. American Institute of Steel Construction, Chicago, IL, 2016

[22] Kazemzadeh Azad S., D. Li, and B. Uy, "Axial Slenderness Limits for Austenitic Stainless Steel-Concrete Composite Columns," *Journal of Constructional Steel Research,* 166, 2020

[23] Kazemzadeh Azad S., D. Li, and B. Uy, "Axial Slenderness Limits for Duplex and Lean Duplex Stainless Steel-Concrete Composite Columns," *Journal of Constructional Steel Research,* 172, 2020

[24] Kazemzadeh Azad S., D. Li, and B. Uy, "Compact and Slender Box Concrete-Filled Stainless Steel Tubes under Compression, Bending, and Combined Loading," *Journal of Constructional Steel Research,* 184, 106813, 2021

[25] Song Y., J. Wang, B. Uy, and D. Li, "Experimental behaviour and fracture prediction of austenitic stainless steel bolts under combined tension and shear," *Journal of Constructional Steel Research,* 166, 105916, 2020

[26] Song Y., J. Wang, B. Uy, and D. Li, "Stainless steel bolts subjected to combined tension and shear: Behaviour and design," *Journal of Constructional Steel Research,* 170, 106122, 2020

[27] Eurocode 3, *Design of steel structures – Part 1-8: Design of Joints, EN 1993-1-8:2005*. European Standard, Comité Européen de Normalisation, Brussels, Belgium, 2005

[28] Wang J., B. Uy, D. Li, and Y. Song, "Fatigue behaviour of stainless steel bolts in tension and shear under constant-amplitude loading," *International Journal of Fatigue,* 133, 105401, 2020

[29] Song Y., B. Uy, and J. Wang, "Numerical analysis of stainless steel-concrete composite beam-to-column joints with bolted flush endplates," *Steel and Composite Structures,* 33(1), 143-162, 2019

[30] Wang J., B. Uy, and D. Li, "Initial stiffness and moment capacity assessment of stainless steel composite bolted joints with concrete-filled circular tubular columns," *Steel and Composite Structures,* 33(5), 681-697, 2019

[31] Wang J., B. Uy, D. Li, and Y. Song, "Progressive collapse analysis of stainless steel composite frames with beam-to-column endplate connections " *Steel and Composite Structures,* 36(4), 427-446, 2020

[32] Song Y., J. Wang, B. Uy, and D. Li, "Behaviour and design of stainless steel-concrete composite beam-to-column joints," *Journal of Constructional Steel Research,* 184, 106800, 2021

FATIGUE AND FRACTURE

9TH INTERNATIONAL CONFERENCE ON COMPOSITE CONSTRUCTION IN STEEL AND CONCRETE
M. Knobloch, U. Kuhlmann, W. Kurz, M. Schäfer. (Eds.)
Online, Germany, July 27-29, 2021

COMPOSITE BRIDGES WITH CRACKED CONCRETE DECK SPANNING BETWEEN TRANSVERSE BEAMS UNDER FATIGUE SHEAR LOADING

Lena Stempniewski*, Ulrike Kuhlmann*

* Institute of Structural Design, University of Stuttgart
e-mails: lena.stempniewski@ke.uni-stuttgart.de, sekretariat@ke.uni-stuttgart.de

Keywords: composite bridges, cracked concrete deck, fatigue, shear loading.

Abstract. *For concrete deck of large steel-concrete composite road bridges cracking occurs due to tension because of negative bending moments at the support area, at the same time traffic load with high wheel loads passes the deck spanning between transverse beams. They induce cyclic shear loading in the cracked concrete deck. In composite bridges the application of prefabricated concrete elements has led to economical designs with a benefit in construction time. In this case, the prefabricated concrete elements are supported by the transverse beams when the on-site concrete is poured. For this construction, the concrete is subjected to tension forces resulting from the global load-carrying effect, which needs to be superimposed with local effects as wheel loads acting as shear fatigue loading. In this paper, the fatigue strength of cracked concrete deck under tension and shear fatigue loading is discussed.*

1 INTRODUCTION

In coming years numerous large road bridges with spans of 50 - 60 m and more need replacement in Germany. For this span range, steel composite structures proved to be an economical concept for the superstructures. Recently, pre-fabricated concrete elements have increasingly been used for the erection of the concrete deck leading to a more economic and faster manufacturing of the decks. In this case a transversely oriented steel structure is required to support the prefabricated concrete elements. One development in recent years, which is the focus of the research described here, is the arrangement of transverse beams also as cantilever for supporting the prefabricated elements and thus predominantly concrete decks spanning longitudinally. As an example of this construction *Heidingsfeld viaduct* is shown in Figure 1.

Figure 1. Heidingsfeld viaduct (A3) in construction stage

As a result of the global structural behaviour, for the design of composite bridges with transverse beams, the concrete deck is subjected to tension due to the negative bending moment in the support area. Due to the tensile stress in the concrete, cracks often appear over the entire height of the concrete deck even in the serviceability state. These cracks are passed by cyclic wheel loads inducing transverse forces in the concrete deck. In this case, the global tensile effects in the deck need to be superimposed with local cyclic shear loading transferred between the cross beams. The results of a current research project funded by the Ministry of Transport and Digital Infrastructure [1] addressing this situation are presented in the paper.

2 REINFORCED CONCRETE SLABS UNDER TENSILE STRESSES WITHOUT SHEAR REINFORCEMENT UNDER FATIGUE LOADING

Whilst the static behaviour of cracked reinforced concrete decks without shear reinforcement under direct shear loading has already been investigated by an experimental test programme [2] and the corresponding rules were included in Eurocode 4 (EC4-2, 6.2.2.5 (3)) [3], the fatigue behaviour still needs clarification. The influence of vertical fatigue shear loading on the behaviour of cracked concrete decks has not yet been investigated sufficiently. Separation cracks are induced by tensioning of the concrete. As a result of repeated shear stress the state of the concrete may change, e.g. a failure starting from the tension cracks might be initiated. In addition, the cracks may reduce the dowel effect of the longitudinal reinforcing bars to a greater extent, and the bond zone between concrete and reinforcing steel may also fail.

To determine the load-bearing behaviour of longitudinally reinforced concrete decks under tensile stress without shear reinforcement, a test matrix has been developed to determine S-N diagrams under vertical shear fatigue loading, see chapter 4. First, for each test series a static test is realised in order to determine the maximum load capacity and to compare the behaviour under static and fatigue loading. Then, fatigue tests are conducted in order to determine a S-N-diagram. The outcome of these tests will be further discussed in this paper.

3 SHEAR STRENGTH OF REINFORCED CONCRETE MEMBERS WITHOUT TRANSVERSE REINFORCEMENT

3.1 Mechanisms of the shear force transfer

The shear force transfer in reinforced concrete members without shear reinforcement has several load-bearing components that influence and superimpose each other. A number of publications and dissertations [2], [4], [5], [6] point out the following load-bearing effects, see Figure 2 and Figure 3:

- Shear force transfer in the non-cracked concrete of the compression zone due to bending $V_{c,c}$
- Dowel action of the longitudinal reinforcement $V_{c,d}$
- Aggregate interlock forces $V_{c,r}$ leading to crack interlocking
- Residual tensile strength of concrete in the fracture process zone
- Direct compression strut in the area of supports

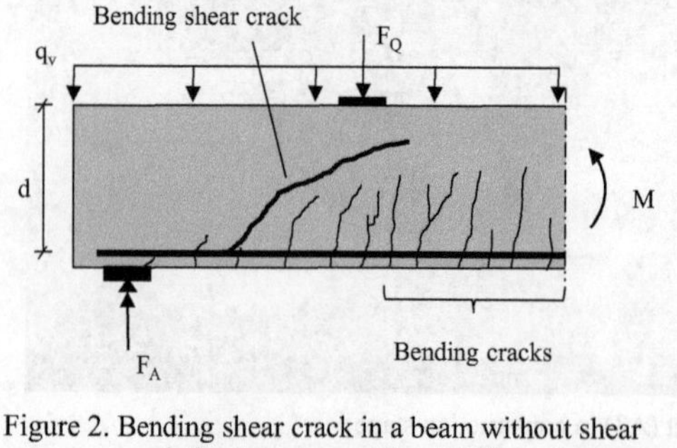

Figure 2. Bending shear crack in a beam without shear reinforcement according to [7]

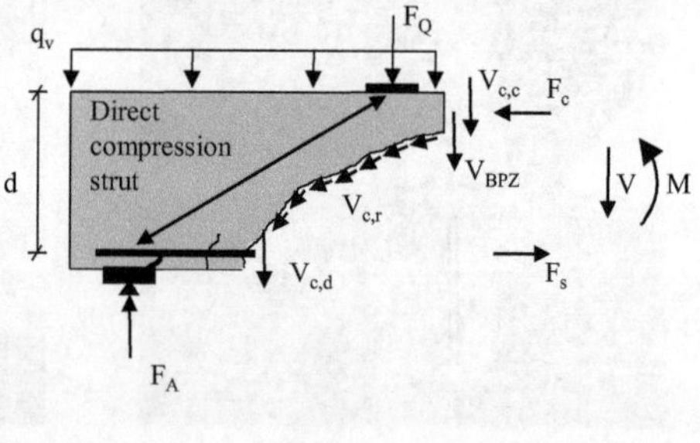

Figure 3. Mechanisms of shear force transfer according to [7]

3.2 Normative rules according to EN 1992 and 1994

The verification of the fatigue strength of structural components of reinforced concrete without shear force reinforcement of EN 1992 [8], [9] is based on the CEB Bulletin 188 [10] and the fib Model Code 1990 [11] and 2010 [12]. The rules are based on the assumption that the cracking behaviour under fatigue loading is similar to that under static loading. There is a relationship between static and alternating stress capacity, i.e. the verification under fatigue is carried out by a reduction of the static load-bearing capacity [6].

The static vertical shear force bearing capacity of a structural component without shear force reinforcement is calculated according to EN 1992-2 [8], [9] as given by equation (1). The influence of a normal force is considered in this equation by the stress σ_{cp}.

$$V_{Rd,c} = \left[C_{Rd,c} \cdot k \cdot \left(100 \cdot \rho_l \cdot f_{ck}\right)^{1/3} + k_1 \cdot \sigma_{cp}\right] \cdot b_w \cdot d \quad (1)$$

With a minimum of:

$$V_{Rd,c} \geq \left(v_{min} + k_1 \cdot \sigma_{cp}\right) \cdot b_w \cdot d \quad (2)$$

where:

- f_{ck}: Characteristic compressive cylinder strength of concrete at 28 days [N/mm²]
- k: $k = 1 + \sqrt{200/d}$; d in [mm]
- d: effective depth of a cross-section [mm]
- ρ_l: $\rho_l = A_{sl} / (b_w \cdot d) \leq 0{,}02$ [-]
- A_{sl}: is the area of the tensile reinforcement, which extends (l_{bd} + d) beyond the section considered
- b_w: is the smallest width of the cross-section in the tensile area [mm]
- σ_{cp}: $\sigma_{cp} = N_{Ed} / A_c < 0{,}2 \cdot f_{cd}$
  for tensile stress in composite members acc. to EN 1994-2 [3] (tensile stress $\sigma_{cp} < 0$):
  $\sigma_{cp} = N_{Ed} / A_c > -1{,}85$ N/mm²
- N_{Ed}: is the axial force in the cross-section due to loading or prestressing [N] ($N_{Ed} > 0$ for compression). The influence of imposed deformations on N_{Ed} may be ignored.
- A_c: Is the area of concrete cross section [mm²]

The values of $C_{Rd,c}$, v_{min} and k_1 are given in the corresponding National Annex. The recommended value for $C_{Rd,c}$ is $0{,}18/\gamma_c$ that for v_{min} is given by eq. (3) and that for k_1 is 0,15.

$$v_{min} = 0{,}035 \cdot k^{3/2} \cdot f_{ck}^{1/2} \quad (3)$$

According to the German National Annex DIN EN 1992-1-1 NA [8] and EN 1994-2 [3]: The recommended value for $C_{Rd,c}$ is $0{,}15/\gamma_c$ that for v_{min} is given by eq. (4) and (5) and that for k_1 is 0,12.

$$v_{min} = 0{,}525/\gamma_c \cdot k^{3/2} \cdot f_{ck}^{1/2} \quad \text{for } d \leq 600 \text{ mm} \quad (4)$$

$$v_{min} = 0{,}0375/\gamma_c \cdot k^{3/2} \cdot f_{ck}^{1/2} \quad \text{for } d > 800 \text{ mm} \quad (5)$$

For 600 mm < d ≤ 800 mm interpolation is allowed.

In EN 1992-2 [8], [9] therefore, the minimum and maximum shear forces occurring are related to the static vertical shear force bearing capacity. For structural elements without a shear force reinforcement a sufficient resistance to fatigue may be assumed, if the conditions according to equation (6) or (7) are met. By graphing the equations, a Godman diagram [8] is obtained that clearly displays the permissible ranges, as shown in Figure 4.

For $V_{Ed,min}/V_{Ed,max} \geq 0$: $\frac{|V_{Ed,max}|}{|V_{Rd,c}|} \leq 0{,}5 + 0{,}45 \cdot \frac{|V_{Ed,min}|}{|V_{Rd,c}|}$ $\begin{array}{l} \leq 0{,}9 \text{ up to C50/60} \\ \leq 0{,}8 \text{ greater than C55/67} \end{array}$ (6)

For $V_{Ed,min}/V_{Ed,max} < 0$: $\frac{|V_{Ed,max}|}{|V_{Rd,c}|} \leq 0{,}5 - \frac{|V_{Ed,min}|}{|V_{Rd,c}|}$ (7)

where:

$V_{Ed,max}$: is the design value of the maximum applied vertical shear force under frequent load Combination

$V_{Ed,min}$: is the design value of the minimum applied shear force under frequent load combination in the cross-section where $V_{Ed,max}$ occurs

$V_{Rd,c}$: is the design value for shear-resistance according to eq. (1)

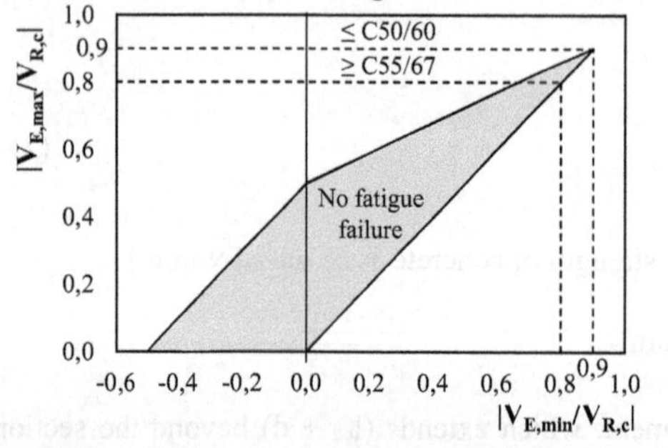

Figure 4. Godman-diagram according to EN1992 [8]

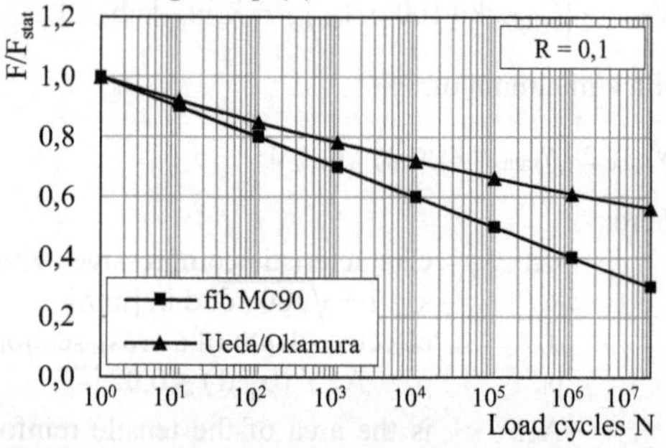

Figure 5. S-N diagram according to the fib model code [11], [12] and Ueda/Okamura [15]

The influence of tensile axial stresses (i.e. with cracked cross-sections) has not yet been considered in the verification under fatigue. Since neither experimental data nor a mechanical model are available, a more stringent verification of the crack width is currently required (see [13] and DIN EN 1994-2/NA, NDP to 7.4.1 (4) [3]).

3.3 S-N diagram according fib model code 1990 [11] and 2010 [12] and Ueda/Okamura [15]

The fatigue check of reinforced concrete components without shear force reinforcement is considered to have been carried out if the number of load cycles during the service life n is less than or equal to the maximum tolerable number of cycles to failure N.

$$n \leq N \tag{8}$$

In Model Code 1990 [11] and 2010 [12], a linear relationship is assumed between the ratio of the maximum load to the static shear force bearing capacity and the number of failure cycles N, see Figure 5. The number of cycles to failure are determined as follows:

$$\log N = 10 \cdot (1 - V_{max}/V_{ref}) \tag{9}$$

where:

V_{max}: is the maximum shear force under the relevant representative values of permanent loads including prestress and maximum cyclic loading

V_{ref}: Vertical shear capacity of a component without shear reinforcement; $V_{ref} = V_{Rd,c}$
Model Code 1990: $V_{ref} = 0{,}12 \cdot \xi \cdot (100 \cdot \rho \cdot f_{ck})^{1/3} \cdot b_{red} \cdot d$
Model Code 2010: $V_{Rd,c} = k_v \cdot \sqrt{f_{ck}}/\gamma_c \cdot z \cdot b_w$

ξ: $\xi = 1 + \sqrt{200/d}$; d in [mm]

ρ: Degree of longitudinal reinforcement

f_{ck}: Characteristic concrete compressive strength [N/mm²]
(Model Code 1990: $f_{ck} \leq 50$ N/mm²; Model Code 2010: $\sqrt{f_{ck}} \leq 8$ N/mm²)
b_{red}: Width of the cross-section minus clamping channels [mm]
d: effective depth of a cross-section [mm]
k_v: Components without significant normal force stress with $f_{yk} \leq 600$ N/mm², $f_{ck} \leq 70$ N/mm² and $d_g \geq 10$ mm (Level I):

$$k_v = \frac{180}{1000 + 1{,}25\, z}$$

Level II:

$$k_v = \frac{0{,}4}{1 + 1500\varepsilon_x} \cdot \frac{1300}{1000 + k_{dg} \cdot z}$$

k_{dg}: $k_{dg} = \frac{32}{16+d_g} \geq 0{,}75$
d_g: Largest grain size of the aggregate [mm]
z: lever arm z = 0,9 d
ε_x: mid-depth strain at the control section;

$$\varepsilon_x = \frac{1}{2 \cdot E_S \cdot A_S} \cdot \left[\frac{M_{Ed}}{z} + V_{Ed} + N_{Ed}\left(\frac{1}{2} \pm \frac{\Delta e}{z}\right)\right] \leq 3{,}0 \cdot 10^{-3}$$

N_{Ed}: axial (normal) force (positive for tension)

This approach was extended by Ueda and Okamura [15] on the basis of their own experiments to consider influences from the ratio R of the minimum to maximum load on the total number of cycles, see eq. (10). The relationship is shown graphically in Figure 5.

$$\log(V_{max}/V_{ref}) = -0{,}036 \cdot (1 - R^2) \cdot \log N \qquad (10)$$

where:

R:	$R = F_{min}/F_{max}$
F_{min}:	minimum applied force under cyclic load
F_{max}:	maximum applied force under cyclic load

4 EXPERIMENTAL TEST PROGRAM AND PROCEDURE

4.1 Overview of the experimental tests

For a more detailed investigation of the load-bearing behaviour of the cracked concrete slabs under vertical shear force loading, fatigue tests are conducted. 9 tests with identical geometry are tested under fatigue loading with variable stress ratio to define a basic S-N diagram (series S), see Table 1. In addition, a static test is also carried out allowing a direct link to previous research findings (see [2]). Based on the static test the applied load in the fatigue tests may be determined, as the fatigue strength is dependent on the static strength. All fatigue tests are conducted with a frequency of the transverse load of 3,5 Hz and a ratio of minimum to maximum applied load ($R = F_{min}/F_{max}$) of 0,1.

Table 1: Test series S: static and cyclic tests with varied parameters.

Series number	Axial stress	$R = F_{min}/F_{max}$	Maximum transverse load F_{max}	Frequency f [Hz]
S-01	$\sigma_N = f_{ctm} A_c$	-	F_{ult}	Static loading
S-02	$\sigma_N = f_{ctm} A_c$	0,1	0,6* 0,7	3,5
S-03	$\sigma_N = f_{ctm} A_c$	0,1	0,7	3,5
S-04	$\sigma_N = f_{ctm} A_c$	0,1	0,7	3,5
S-05	$\sigma_N = f_{ctm} A_c$	0,1	0,65	3,5
S-06	$\sigma_N = f_{ctm} A_c$	0,1	0,6	3,5
S-07	$\sigma_N = f_{ctm} A_c$	0,1	0,6	3,5
S-08	$\sigma_N = f_{ctm} A_c$	0,1	0,65	3,5
S-09	$\sigma_N = f_{ctm} A_c$	0,1	0,42* 0,5* 0,7	3,5
S-10	$\sigma_N = f_{ctm} A_c$	0,1	0,55* 0,65	3,5

* This load level did not cause any signs of failure of the specimen. Thus, the loads were increased in order to initiate failure of the specimen.

4.2 Geometry of the test specimens

The aim of the tests is to cause shear failure of the test specimen. However, the shear and bending capacity depend on the same parameters such as concrete compressive strength, longitudinal reinforcement ratio and the geometry of the specimen [16]. Therefore, the load capacity of shear and bending failure are directly related. As a result, it is necessary to determine these parameters very carefully to really achieve shear failure. For this purpose, the geometry especially the shear slenderness a/d must be chosen in harmony with the longitudinal reinforcement ratio. In particular, the shear slenderness a/d should be in a range leading neither to a bending failure (a/d < 6) nor to a shear failure as a result of a direct compression strut into the support (a/d > 2,0 (EN 1992 [8], [9] or 3,0 [17]).

Kani [17] compared the loads in case of shear failure and of bending failure. The quotient of flexural to shear failure can be given as a function of the longitudinal reinforcement ratio and the shear slenderness. The range, in which shear force failure becomes decisive, is referred to by Kani as the "valley of diagonal failure". In this case, the bending tensile reinforcement usually are not fully utilised.

A similar procedure was carried out as a parametric study to determine the failure mode of the test specimens depending on the shear slenderness and the longitudinal reinforcement ratio. Thus, the failure mode may be determined for different levels of shear slenderness and ratios of longitudinal reinforcement. The utilisation may be calculated as a function of those, see Figure 6. In order to achieve the widest possible parameter range, a/d must be chosen as small as possible, but at the same time failure due to a direct compression strut into the support should be avoided. Therefore, a/d has to be chosen as large as possible: a shear slenderness of a/d = 4 has been selected.

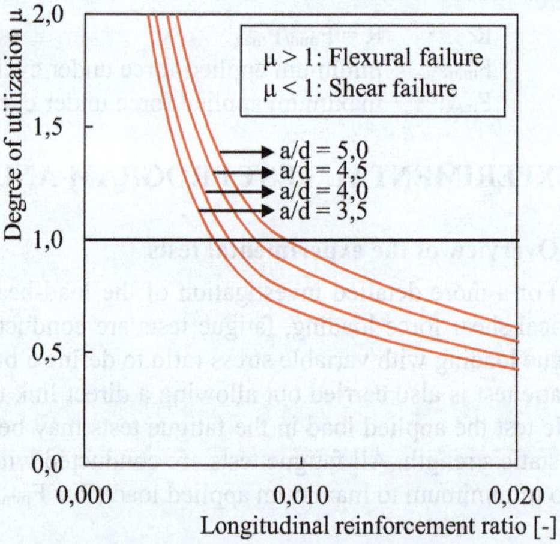

Figure 6. Shear valley for the selected test specimen geometry as a function of the shear slenderness a/d and the longitudinal reinforcement ratio. [1]

Finally, the reinforcement for the standard specimen is selected with 5 Ø 20 mm at 8 cm spacing according to ZTV-ING [18]. With the selected degree of longitudinal reinforcement, a shear failure can be assumed with a shear slenderness of 4. The final geometry of the test specimen (standard specimen, series S) is given in Figure 7.

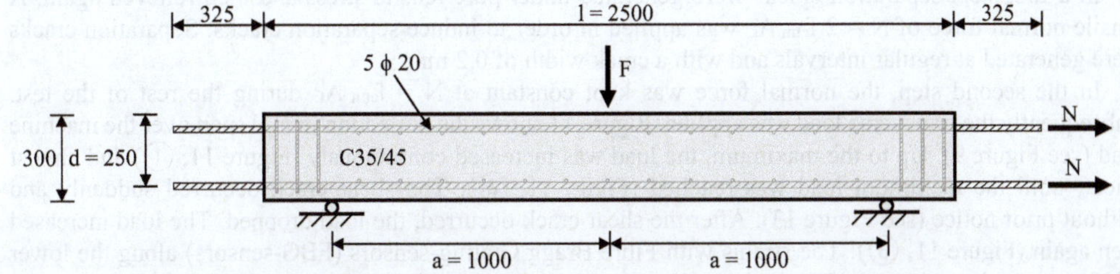

Figure 7. Geometry of the test specimen (standard specimen, test series S).

4.3 Test procedure

Prior to the start of the beam tests the compressive and tensile strengths were determined. Based on the tensile concrete strength the axial force was defined as $N = 2\ f_{ctm}\ A_C$. In the first step of the beam tests, the tensile axial force is introduced to the test specimen leading to separation cracks. Therefore, a completed crack pattern is achieved. Afterwards the test specimen is unloaded. In the next step, the normal force is increased to the planned axial force (for test series S: $N = f_{ctm}\ A_C$) and kept constant during the rest of the test. Finally, the transverse load is increased stepwise up to the upper load limit during fatigue tests or until failure in static tests. In fatigue tests, the cyclic transverse loading is started afterwards. The upper load limit in the fatigue tests is determined based on the results of the static reference test.

4.4 Experimental test setup

Figure 8 shows the test setup. The axial tensile force as well as the shear load are applied to the test specimen by a frame construction. For the application of the axial tensile force, reinforcement rods with a rolled-up thread are used, that are guided through the test specimen and connected to the frame. The vertical shear load is applied to the centre of the test specimen via a portal frame with hydraulic cylinder, resulting in a symmetrical system with equal shear slenderness on both support sides.

Figure 8. Test set-up

5 RESULTS OF THE BEAM TESTS OF SERIES S

5.1 Static beam test S-01

In a first step, separation cracks were generated under pure tensile stress and then relieved again. A tensile normal force of $N = 2\, f_{ctm}\, A_C$ was applied in order to induce separation cracks. Separation cracks were generated at regular intervals and with a crack width of 0,2 mm.

In the second step, the normal force was kept constant at $N = f_{ctm}\, A_C$ during the rest of the test. Subsequently the transverse load was applied. Figure 11 shows the deflection in mid span over the machine load (see Figure 9). Up to the maximum, the load was increased continuously (Figure 11, ①) and almost linear, until the maximum load was reached (Figure 11, ②). The shear crack occurred suddenly and without prior notice (see Figure 13). After the shear crack occurred, the load dropped. The load increased then again (Figure 11, ③). The strains with Fibre Bragg Grating-sensors (FBG-sensors) along the lower reinforcement layer increased strongly in the area of the shear crack (see Figure 10 and Figure 12, sensor D10). The loads were still transferred via the dowel action of the longitudinal reinforcement. Subsequently, a transverse crack occurred on the bottom of the specimen (see Figure 14), which led to the failure of the bond between the concrete and reinforcement and finally the failure of the specimen (Figure 11, ④).

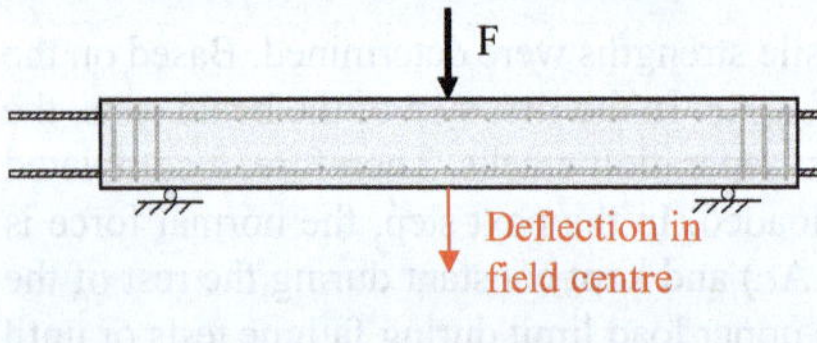

Figure 9. Measurement of the deflection in field centre.

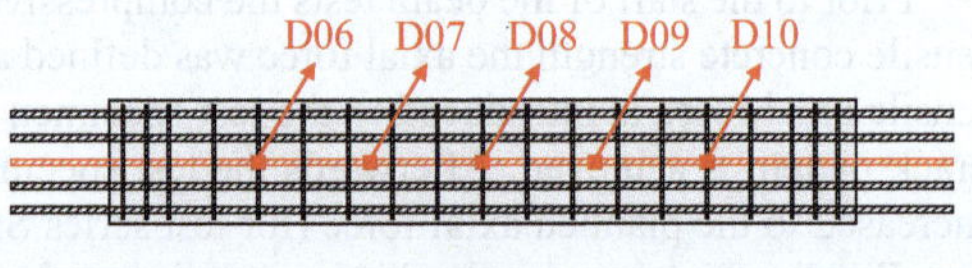

Figure 10. Measurement of the strain in the lower reinforcement.

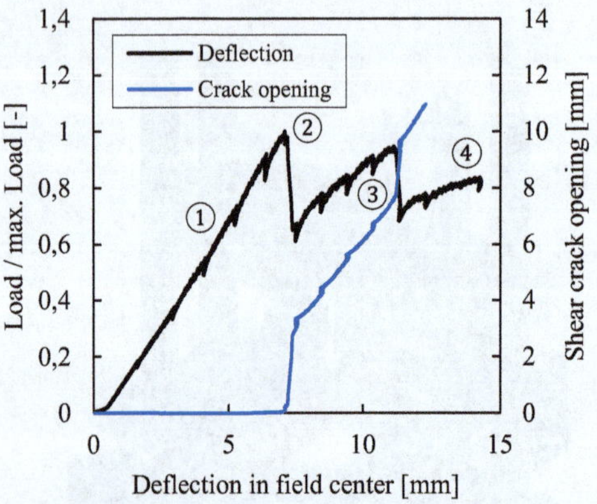

Figure 11. Load-deflection diagram (test S-01).

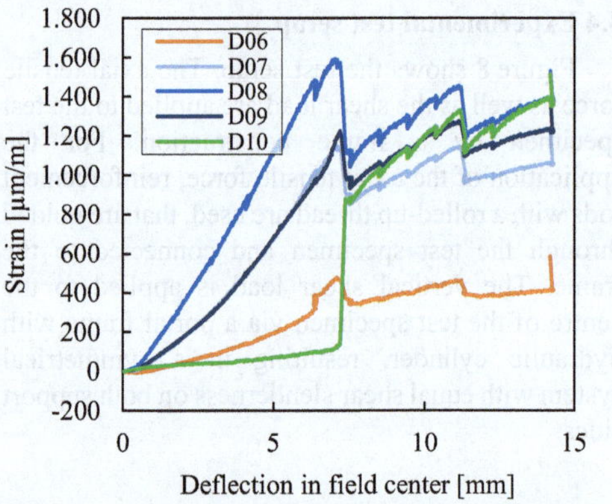

Figure 12. Strain due to transverse loading in the lower reinforcement vs. deflection in field centre (test S-01).

Figure 13. Failure of test specimen S-01 (static test).

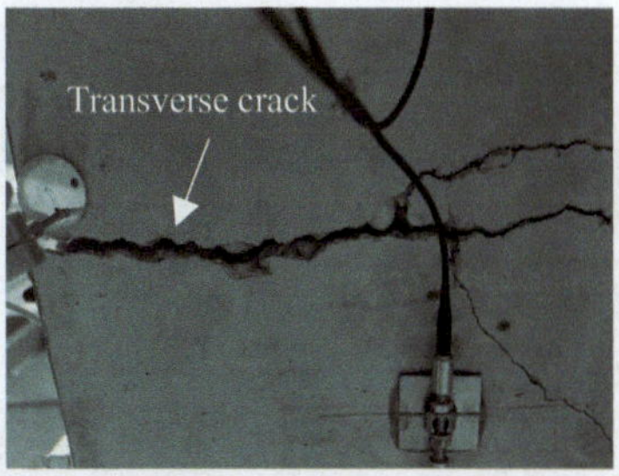

Figure 14. View from below (test S-01).

5.2 Fatigue beam tests S-02 to S-10

In the fatigue tests, also separation cracks were generated in the first step by purely axial loading. A tensile axial force of $N = 2\ f_{ctm}\ A_C$ was applied. Separation cracks were generated at regular intervals and with an average crack width of 0,2 mm. In the next step, the axial force was kept constant at $N = f_{ctm}\ A_C$ for the rest of the test. Subsequently the transverse load was applied in a cyclic constant amplitude mode. In the following, the course of test S-08 is presented exemplarily. Figure 15 shows the deflection in the centre span over number of load cycles. Before the first shear crack occurred at approximately 83.200 load cycles on side III, the deflection remained constant. When the shear crack occurred (Figure 15, ①), the deflection increased. The strain in the area of the shear crack increased strongly (sensor D06 in Figure 16). In the further course of the test, the deflection remained almost constant (Figure 15, ②), but the shear crack opening increased (see Figure 15). At approximately 666.455 load cycles, a second shear crack appeared on side IV, cf. Figure 17, leading strongly increasing deflections (Figure 15, ③) as well as the strains in sensor D10 (see Figure 16). Finally, also a transverse crack occurred at the bottom of the specimens leading to the failure of the bond between the reinforcement and concrete and thus the failure of the specimen (see Figure 18).

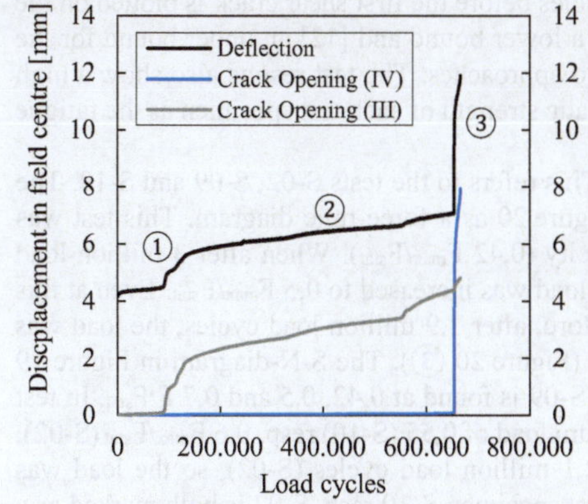

Figure 15. Deflection in field centre vs. load cycles (test S-08).

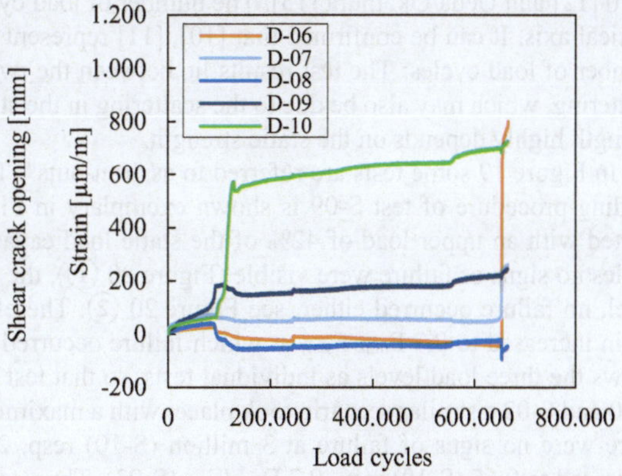

Figure 16. Strain in the lower reinforcement due to transverse loading vs. load cycles (test S-08).

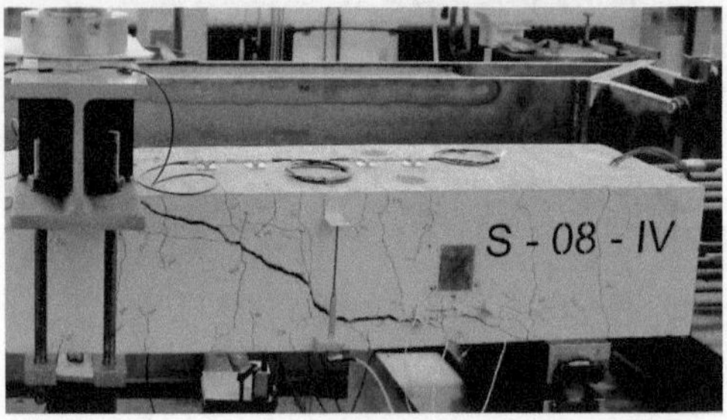

Figure 17. Failure of test specimen S-08 (fatigue test).

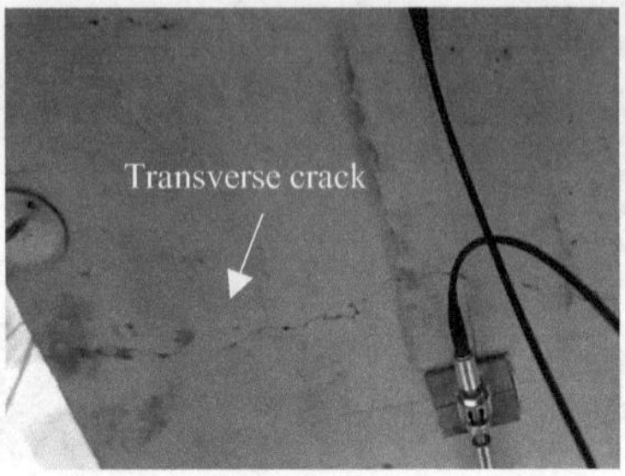

Figure 18. View from below (test S-08).

6 ANALYSIS AND EVALUATION OF THE TEST RESULTS

6.1 Failure mode and influence of separation cracks

In the static test, the shear crack developed independently of the separation cracks. The failure mode corresponds to the results from the static tests of Ehmann [2]. Here the separation cracks are crossing the shear crack and are not influencing the path of the shear crack. For the corresponding conducted static test S-01 (see Figure 13), the same behaviour was observed. In the fatigue tests, the shear crack has also developed independently of the separation cracks (see e.g. Figure 17). A comparison shows, that the failure mode of the static and fatigue tests is similar. As a result, it may be assumed that the behaviour under fatigue loading is directly linked to static load capacity. Therefore, static and fatigue tests are evaluated together in the S-N-diagram.

6.2 S-N diagram

Figure 19 compares the test results in the S-N-diagram according to the fib model code 1990 [11] and 2010 [12] and Ueda/Okamura [15]. The number of load cycles before the first shear crack is plotted on the vertical axis. It can be confirmed that [10], [11] represent a lower bound and [12] an upper bound for the number of load cycles: The test results lie between the two approaches. The test results also show a high scattering, which may also be due to the scattering in the static strength of each test specimen as the fatigue strength highly depends on the static strength.

In Figure 19 some tests are referred to as "run-outs". This refers to the tests S-02, S-09 and S-10. The loading procedure of test S-09 is shown exemplary in Figure 20 as a force-time diagram. This test was started with an upper load of 42% of the static load capacity (0,42 F_{max}/F_{stat}). When after 3 million load cycles no signs of failure were visible (Figure 20 ①), the load was increased to 0,5 F_{max}/F_{stat}. Even at this level, no failure occurred either, see Figure 20 ②. Therefore, after 1,9 million load cycles, the load was again increased to 0,7 F_{max}/F_{stat}, at which failure occurred (Figure 20 ③). The S-N-diagram in Figure 19 shows the three load levels as individual tests, so that test S-09 is found at 0,42, 0,5 and 0,7 F/F_{stat}. In test S-10 and S-02 a similar scenario took place: with a maximum load of 0,55 (S-10) resp. 0,6 F_{max}/F_{stat} (S-02), there were no signs of failure at 3 million (S-10) resp. 2,1 million load cycles (S-02), so the load was increased to 0,65 (S-10) resp. 0,7 F_{max}/F_{stat} (S-02). Thus, test specimen S-10 resp. S-02 is both marked as a run-out at 0,55 resp. 0,6 F/F_{stat} and shown at 0,65 resp. 0,7 F/F_{stat} in Figure 19.

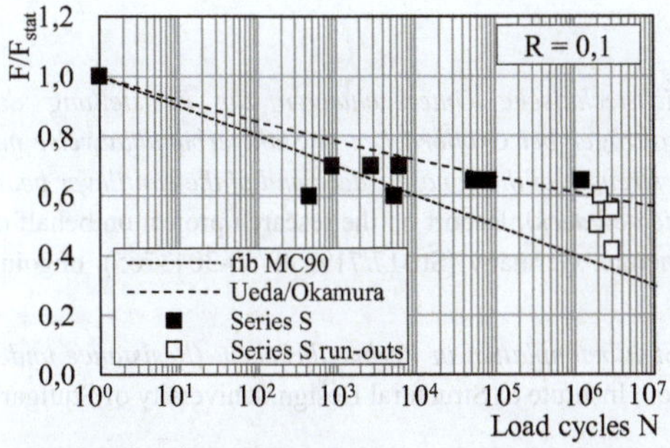

Figure 19. S-N diagram according to the fib model code [11], [12] and Ueda/Okamura [15] with test results of series S from [1]

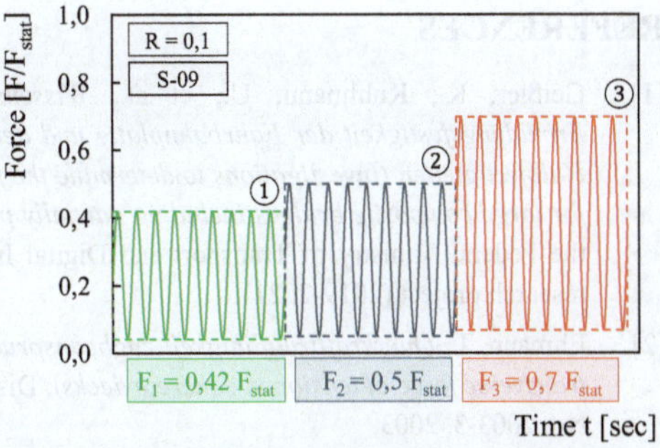

Figure 20. Force-time diagram for test specimen S-09

7 SUMMARY AND OUTLOOK

In this paper, tests of reinforced concrete slabs under tensile stresses without shear reinforcement under shear loading were discussed. Fatigue tests as well as a static tests as reference tests were conducted. With the results of the static test, the fatigue loading has been specified, as it is directly connected to the static strength. The test results showed that the failure mechanisms under static and fatigue loading are similar: the separation cracks did not open during fatigue loading. The shear crack was crossing the separation cracks and had developed independently. Further, the test results were evaluated in S-N-diagrams according to the fib model code [11], [12] and Ueda/Okamura [15]. The comparison showed that the results lie in between both approaches. As a conclusion, the fib model code [11], [12] represents a lower limit function for the fatigue strength and Ueda/Okamura an upper one.

Further tests with varying axial forces and longitudinal reinforcement ratios are underway. Furthermore, an additional test series with prefabricated concrete elements is going to be conducted soon. Also, additional tests are planned in order to analyse the effect of realistic varying loading conditions in bridges to check the applicability of the Palmgren-Miner-rule.

ACKNOWLEDGEMENT

The work presented is carried out as part of the research project [1] of the Federal Ministry of Transport and Digital Infrastructure. The authors wish to thank for that funding. The project [1] is a joint project with colleagues from the Technical University of Berlin, Ingenieurbüro Prof. Dr.-Ing. Ulrike Kuhlmann, GMG Ingenieurgesellschaft and the Technical University of Munich. The authors would like to thank all project partners for the good and fruitful cooperation. Moreover, we thank the Material Testing Institute of the University of Stuttgart for their great support in conducting the tests.

REFERENCES

[1] Geißler, K.; Kuhlmann, U.; et al., *Wissenschaftlich-technische Untersuchungen zur Feststellung der Ermüdungsfestigkeit der Fahrbahnplatte und der Konsolträger bei Großbrücken in Stahlverbundbauweise mit Halbfertigteilen (Investigations to determine the fatigue strength of the concrete deck and of the cantilever beam for large composite bridges including partially prefabricated deck)*, Report on the research project on behalf of the Federal Ministry of Transport and Digital Infrastructure, Germany (StB17/7192.40/80-3043765), ongoing research project (2019-2021).

[2] Ehmann, J., *Querkrafttragfähigkeit zugbeanspruchter Stahlbetonplatten in Verbundbrücken, (Resistance under transverse force of tensioned concrete decks)*, Dissertation, Institute of Structural Design, University of Stuttgart, No. 2003-3, 2003.

[3] EN 1994-2, *Eurocode 4: Design of composite steel and concrete structures. Part 2: General rules and rules for bridges*, Brussels, Belgium: Comité Européen de Normalisation, 2005.

[4] ACI-ASCE Committee 445, *Recent Approaches to Shear Design of Structural Concrete: State-of-the-Art-Report by ASCE-ACI Commitee 445 on Shear and Torsion*, ASCE-Journal of Structural Engineering 124, p. 1375-1417.

[5] Hegger, J.; Görtz, S., *Querkrafttragfähigkeit von Stahlbeton- und Spannbetonbalken aus Normal- und Hochleistungsbeton*, Deutscher Ausschuss für Stahlbeton im DIN Deutsches Institut für Normung e.V., Beuth Verlag, Berlin, 2007.

[6] Kohl, M., *Tragverhalten von Stahlbetontragwerken ohne Querkraftbewehrung unter Ermüdungsbeanspruchungen,* Dissertation, Hamburg University of Technology, 2014.

[7] Latte, S., *Zur Tragfähigkeit von Stahlbeton-Fahrbahnplatten ohne Querkraftbewehrung*, Dissertation, Hamburg University of Technology, 2010.

[8] EN 1992-1-1: *Eurocode 2: Design of concrete structures - Part 1-1 : General rules and rules for buildings*, 2011, and German National Annex: DIN EN 1992-1-1/NA: Nationaler Anhang zu Eurocode 2: Bemessung und Konstruktion von Stahlbeton- und Spannbetontragwerken – Teil 1-1: Allgemeine Bemessungsregeln und Regeln für den Hochbau, Beuth Verlag, 2013.

[9] EN 1992-2: *Eurocode 2 - Design of concrete structures - Concrete bridges - Design and detailing rules*, 2010, and German National Annex: DIN EN 1992-2/NA: Nationaler Anhang zu Eurocode 2: Bemessung und Konstruktion von Stahlbeton- und Spannbetontragwerken – Teil 2: Betonbrücken, Beuth Verlag, 2013.

[10] Comité Euro-International du Béton (CEB): Fatigue of Concrete structures, CEB Bulletin d'information, 1988.

[11] Comité Euro-International du Béton (CEB); Fédération Internationale de la Précontrainte (FIP): CEB-FIP Model Code 1990, 1993.

[12] Fédération internationale du béton (fib): fib Model Code for Concrete Structures 2010, 2012.

[13] Federal Ministry of Transport and Digital Infrastructure, *Konstruktions- und Bemessungshinweise für Stahl- und Stahlverbundbrückenkonstruktionen*, Berlin, 2018.

[14] Mager, M.; Geißler, K., *Ersatzneubau der Rader Hochbrücke - Gutachterliche Stellungnahme zur Bauweise mit Konsolträgern und längsgespannter Fahrbahnplatte (1, Teilbericht)*, Report, GMG Ingenieurgesellschaft, Berlin, 2018.

[15] Ueda, T., Okamura, H., *Behavior in shear of reinforced concrete beams under fatigue loading*, Journal of the Faculty of Engineering, Vol. 37, No. 1, The University of Tokyo, 1983.

[16] Zilch, K., Zehetmaier, G., *Bemessung im konstruktiven Betonbau - Nach DIN 1045-1 (Fassung 2008) und EN 1992-1-1 (Eurocode 2)*, 2. Edition, Springer-Verlag, Berlin, 2010.

[17] Kani, G. N. J,. *Basic Facts Concerning Shear Failure*, Journal of the American Concrete Institute, No. 6, p. 675-692, 1966.

[18] *Zusätzliche Technische Vertragsbedingungen und Richtlinien für Ingenieurbauten (ZTV-ING) - Teil 4: Stahlbau, Stahlverbundbau*, Federal Highway Research Institute, 2012.

9TH INTERNATIONAL CONFERENCE ON COMPOSITE CONSTRUCTION IN STEEL AND CONCRETE
M. Knobloch, U. Kuhlmann, W. Kurz, M. Schäfer. (Eds.)
Online, Germany, July 27-29, 2021

FATIGUE ANALYSIS OF COMPOSITE BEAM WITH BOLTED SHEAR CONNECTORS

S.M. Hosseini *, O. Mirza, F. Mashiri *****

* School of Engineering, Design and Built Environment , Western Sydney University, Sydney, Australia

emails:f.mashiri@westernsydney.edu.au,o.mirza@westernsydney.edu.au,
m.hosseini@westernsydney.edu.au,

Keywords: Composite beam; Bolted shear connector , Fatigue life, FE-Safe, FEM

Abstract. *This paper describes the simulation technique developed to predict the fatigue life of bolted shear connector. As existing composite structures experience complex loading conditions over the entire life of the structure, therefore, the fatigue behaviour of the shear connectors is always critical. Among different types of bolted shear connectors, the installation procedure of the blind bolts is less complex and more rapid than that of conventional systems. Therefore, the concept of fatigue analysis of the blind bolt shear connector has been presented, in this paper here in, under constant amplitudes cyclic loading, using ABAQUS/explicit and FE-SAFE programs. All fatigue simulations were conducted under load-control mode based on constant amplitude conditions and a constant mean force. The results showed that the number of cycles to failure decreases with the increasing stress level while every stress amplitude has a certain number of cycles until failure.*

1 INTRODUCTION

Steel-concrete composite bridges are often subjected to fatigue loads caused by usual traffic loading. Therefore, high cycle fatigue damage is generally experienced by shear connectors as an essential aspect in the design of the composite beam. However , demolition of conventional composite systems (Figure1(a)) is associated with waste and uses significant energy, hence, to replace conventional headed shear studs between the steel beam and concrete slab of a composite beam, demountable shear connector are proposed (Ataei et al. 2014; Hitch 2009). Due to the lack of detailed research and design rules regarding fatigue behaviour of bolted shear connector, these connections are rarely used in composite structures (Pavlović 2013). A relatively new initiative in the composite construction industry is the use of the blind bolts (BB) as a shear connector (Figure 2 (b)), which provides a sustainable and practical solution to the primary disadvantage of using welded shear studs in composite structures.

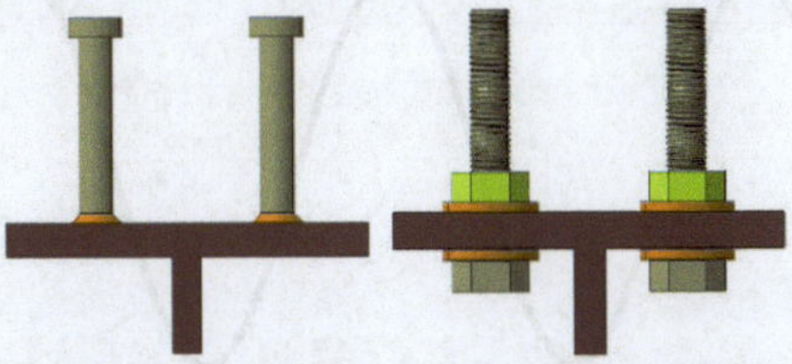

(a) Welded Stud (WS)   (b) Blind Bolt (BB)

Figure 1. Different types of shear connector

Accordingly, in this study, ABAQUS/explicit and FE-SAFE programs were employed to quantify the fatigue failure of the blind bolt (BB) shear connector. As a result, the logarithmic life distributions of the bolted shear connector were calculated using different constant amplitudes. These investigations showed that the fatigue strength of the blind bolt shear connectors is considerably higher than welded stud shear connector.

2 HIGH CYCLE FATIGUE ANALYSIS

Initially, loads were applied at the top side of the steel plate in the form of pressure as a cyclic loading, with sinusoidal constant amplitude (Figure 2). Afterward, the stress datasets obtained from dynamic explicit analysis for all cycles were then read by the Fe-Safe program for high cycle fatigue (HCF) assessment using FE-SAFE software.

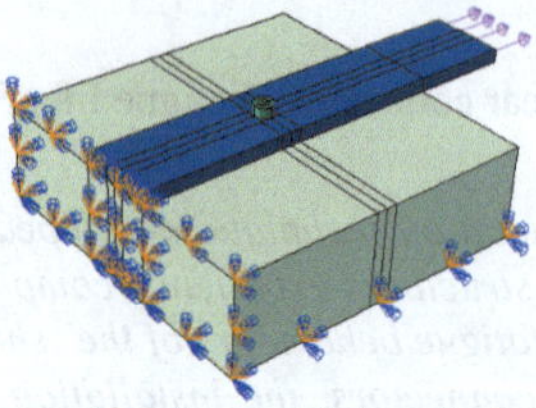

Figure 2. Boundary and loading conditions for BB-DT-FEM model

2.1 Loading history

In this analysis, three specimens were modelled at different stress ranges based on the ultimate strength values obtained from the static tests to estimate the HCF behaviour as shown in Table 1. As shown in Figure 3, the forces were applied at the top side of the steel plate for constant amplitude loading (CAL) as sinusoidal function with magnitudes varying from 5% to 50% of its maximum capacity at a rate of 3Hz.

Table 1: HCF program for FEM blind bolt shear connector

Specimen	fc(MPa)	Stress range (MPa)	Cyclic load (kN)
BB-195HFT-FEM	32	195	5 to 40% (f_u)*
BB-220HFT-FEM	32	223	5 to 45% (f_u)*
BB-250HFT-FEM	32	251	5 to 50% (f_u)*

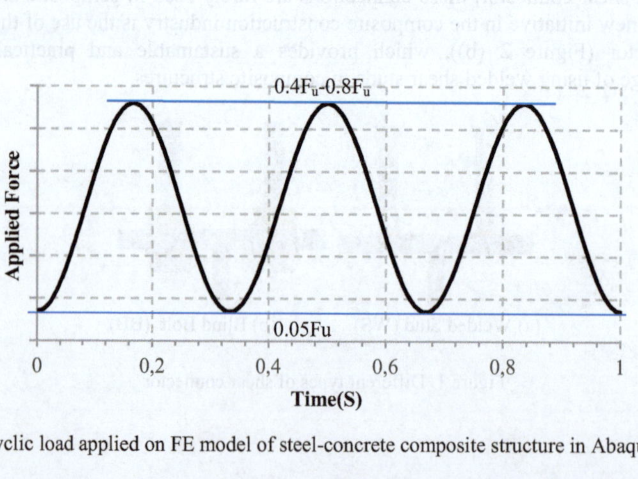

Figure 3. Cyclic load applied on FE model of steel-concrete composite structure in Abaqus/explicit

2.2 Material properties

Material properties used for the fatigue analysis of Blind Bolt are shown in Table 2 based on the material testing report provided in the study conducted by Hosseini et al. (2020), and material database supplied with Fe-Safe.

Table 2: Material properties for blind bolt shear connector

Mechanical Properties	**Cyclic Elastic properties**		
	Parameters	Symbols	Values
	Young's Modulus	E	213870 MPa
	Poisson's Ratio	υ	0.33
	Cyclic Plastic properties		
	Yield stress (MPa)	855.3	888.53
	Plastic strain	0	0.00373
Fatigue Properties (Fe-Safe 2017)	Parameters	Symbols	Values
	Ultimate Tensile Strength	UTS	889 MPa
	Strength at 10^4 cycles	-	452 MPa
	Strength at 10^6 cycles (Fatigue Limit)	-	313.3 MPa
	Strain Hardening Coefficient	K'	1186 MPa
	Strain Hardening Exponent	n'	0.15
	Fatigue Strength Coefficient	σ'_f	1000 MPa
	Fatigue Strength Exponent	b	-0.08
	Fatigue Ductility Coefficient	ε'_f	0.32
	Fatigue Ductility Exponent	c	-0.58
	Surface Finish Factor	K_t	16 <Ra<=40μm

2.3 Fatigue algorithm

In this analysis, the Brown-Miller criterion with morrow's mean stress correlation was selected for the fatigue analysis as shown in Equation 1:

$$\frac{\Delta\gamma_{max}}{2} + \frac{\Delta\varepsilon_n}{2} = C_1 \frac{\sigma'_f}{E}(2N_f)^b + C_2\varepsilon'_f\,(2N_f)^c \qquad (1)$$

Where:

$\frac{\Delta\gamma_{max}}{2}$ is shear strain amplitude
$\frac{\Delta\varepsilon_n}{2}$ is normal strain amplitude
$2Nf$ is endurance in reversals
σ'_f is fatigue strength coefficient
ε'_f is fatigue ductility coefficient

b is fatigue strength exponent
c is fatigue ductility exponent
C_1 is 1.65 (constant)
C_2 is 1.75 (constant)

2.4 Modelling with ABAQUS

Hard contacts used in the normal direction interface, and in the tangential direction, penalty frictional formulation contact method (μ=0.5) was employed between the hole surface and bolt. A finer mesh was employed to model the bolted shear connectors to obtain a detail stress distribution. The stress distribution at 3 different load levels for BB specimens are shown in Figure 4 for high cycle fatigue (HCF). The finite element analysis using ABAQUS/explicit predicted that the severe stress concentrations were observed around the blind bolt shank at the steel concrete interface (Figures 4).

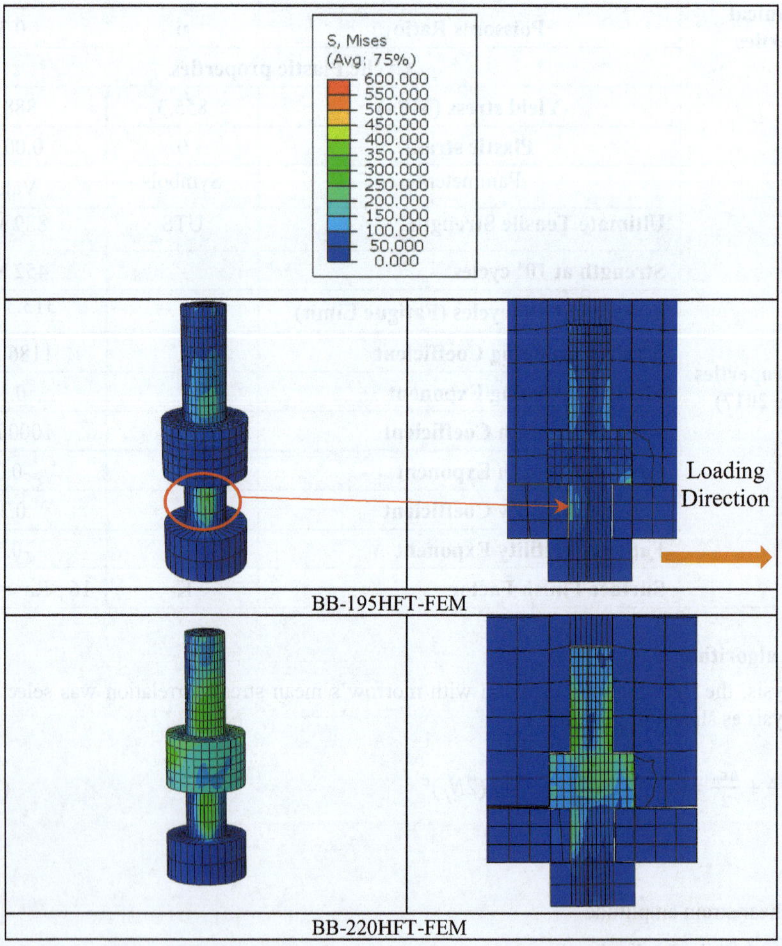

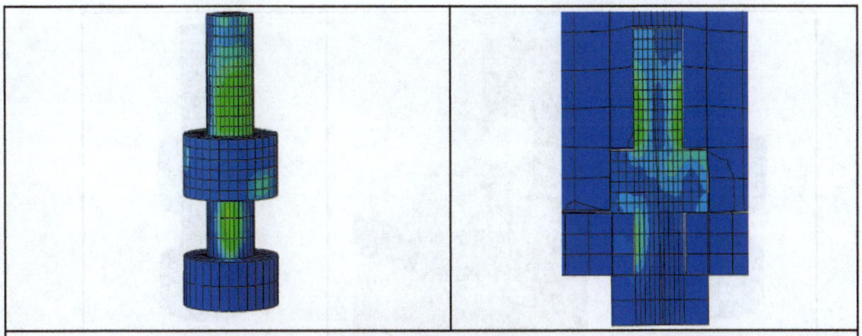

BB-250HFT-FEM

Figure 4. Misses stress distribution at the different loading levels (HCF)

3 HIGH CYCLE FATIGUE FINITE ELEMENT RESULTS

The results of the fatigue analysis were presented in Table 3, whereas all stresses are within elastic range. In order BB-250HFT-FEM test was used to verify and validate the fatigue test results. Consequently, fatigue life estimation of blind bolt shear connector in term of cycles to failure and the factor of strength (FOS), under constant amplitude loading conditions can be observed in Table 3. The factor of strength (FOS) in Table 3 shows how much the blind bolt is over- or under-strength in terms of FEA stresses. As shown in Figure 5 , the results obtained from FE analysis showed failure mode characterization of blind bolt were in good agreement with the experimental results conducted by Hosseini et al. (2020).

Table 3: HCF Finite element results for Blind-Bolt specimens.

Specimen	Stress range (MPa)	Mean load (kN)	Cycles to Failure(FEM)	Cycles to Failure (EXP)	FOS
BB-195HFT-FEM	195	24.5	7690216	1500000*	0.981
BB-220HFT-FEM	220	28.0	6117275	2000000*	0.962
BB-250HFT-FEM	250	31.5	378742	369164	0.781

*run out tests

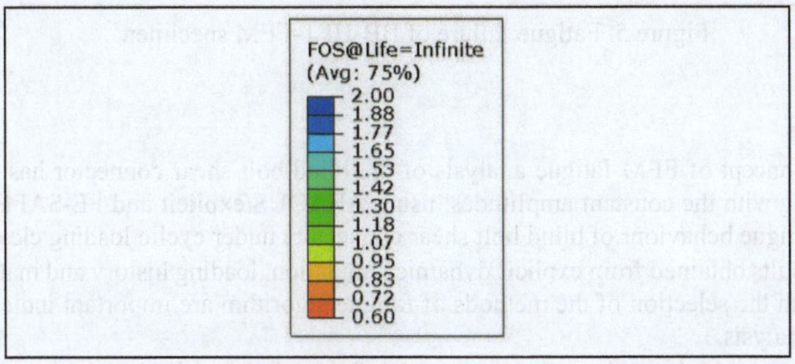

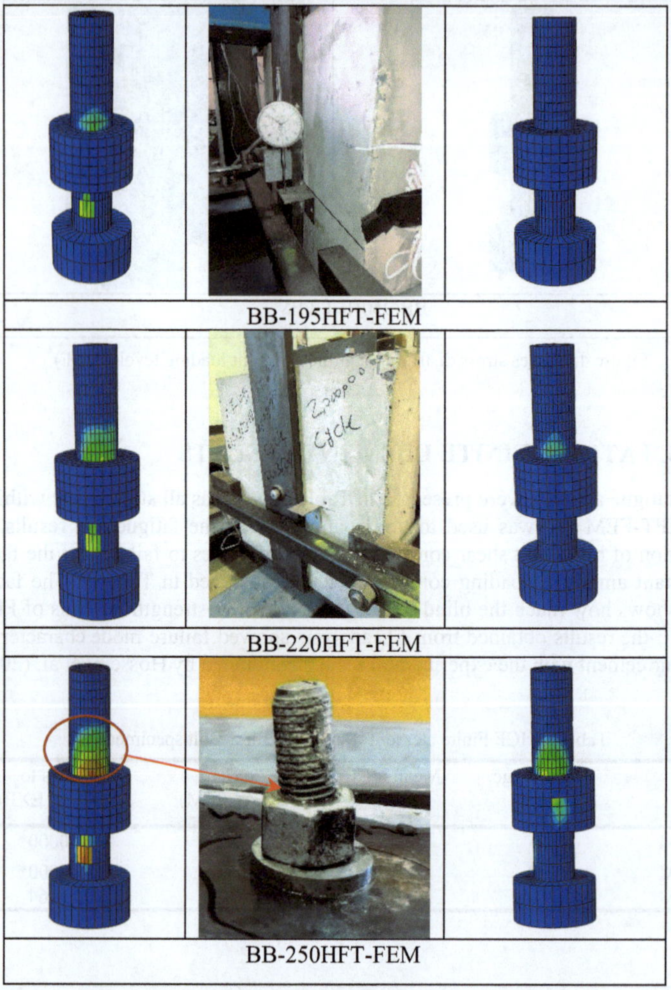

Figure 5. Fatigue failure of BB-HFT-FEM specimen

4. CONCLUSION

In this study, the concept of FEM fatigue analysis of the blind bolt shear connector has been presented, under cyclic loading with the constant amplitudes, using ABAQUS/explicit and FE-SAFE programs. The FEM analysis of fatigue behaviour of blind bolt shear connectors under cyclic loading clearly shows:

- Stress results obtained from explicit dynamic simulation, loading history and material properties, along with the selection of the methods of fatigue algorithm are important indicators during the fatigue analysis.
- HCF requires a high number of cycles to reach fatigue failure where stress is mainly elastic, and the stress levels are also lower than the yield strength of a section.
- The factor of safety of 0.981 in HCF analysis shows that the stress range is almost at the allowable limit, accordingly, the BB shear connector life can be considered as infinite.

- The simulation analysis shows that failure modes of blind bolt were in good agreement with the experimental results.
- More experimental tests are required to precise prediction of the fatigue life of blind bolt shear connectors.

REFERENCES

[1] Ataei, A, Bradford, MA & Liu, X 2014, 'Sustainable composite beams and joints with deconstructable table bolted shear connectors', vol. II.

[2] Fe-Safe 2017, *fe-safe 2017*, Dassault Systemes Simulia Corp.

[3] Hitch, R 2009, 'Structures Assessment', in *Engineering Manual Structures*, pp. 1-83.

[4] Hosseini, M, Mamun, MS, Mirza, O & Mashiri, F 2020, 'Behaviour of blind bolt shear connectors subjected to static and fatigue loading', *Engineering Structures*, vol. 214, p. 110584, DOI https://doi.org/10.1016/j.engstruct.2020.110584, <http://www.sciencedirect.com/science/article/pii/S0141029619334480>.

[5] Pavlović, M 2013, *Resistance of Bolted Shear Connectors in Prefabricated Steel-Concrete Composite Decks*, Belgrade, University of Belgrade.

9TH INTERNATIONAL CONFERENCE ON COMPOSITE CONSTRUCTION IN STEEL AND CONCRETE
M. Knobloch, U. Kuhlmann, W. Kurz, M. Schäfer. (Eds.)
Online, Germany, July 27-29, 2021

INFLUENCE OF HOT DIP GALVANIZING ON THE FATIGUE BEHAVIOUR OF T-STUDS FOR A GROUTED JOINT OF INTEGRAL COMPOSITE FRAME BRIDGES

Martin Mensinger*, Florian Oberhaidinger* and Lukas Stimmelmayr**

* Chair of Metal Structures, Technical University of Munich, Germany
** Chair of Concrete Structures, Technical University of Munich, Germany
e-mails: mensinger@tum.de, f.oberhaidinger@tum.de, lukas.stimmelmayr@tum.de

Keywords: grouted joint, hot dip galvanizing, bridge, fatigue

Abstract. *Recent investigations paved the way for hot dip galvanizing in the field of bridge constructions [1]. Due to technical limitations of hot dip galvanizing, steel girders longer than 16 meters have to be divided and connected afterwards. The Chair of Metal Structures works in cooperation with the Chair of Concrete Structures of the Technical University of Munich (TUM) and the Bundesanstalt für Materialforschung und -prüfung (BAM), to develop a grouted joint connection, which is suitable for hot dip galvanizing. In contrast to welding the segments on site or using prestressed bolt connections, the hot dip galvanization of the grouted joint remains undamaged for the whole service life time of the structure. As part of this research, the main component of the grouted joint - a hot dip galvanized T-stud - is investigated regarding its fatigue behaviour. The present paper outlines the idea and the structure of the grouted joint and classifies the hot dip galvanized T-stud according to a fatigue class given in EN 1993-1-9 [2]. Furthermore, the fatigue assessment for the grouted joint and an optimization of the geometry of the T-Studs is carried out using finite element analysis.*

1 INTRODUCTION

As a result of the increasing traffic and the rapid growth in heavy goods, many highways are currently being widened to six lanes and will be in the near future. Considering the large number of bridge structures along highways, it is obvious that a correspondingly number of new bridges will have to be built in the course of this expansion. For the required span of 35 m to 45 m and up to 60 m, the integral frame bridge in composite construction (see figure 1) has proven itself in recent years [3]. The whole span can be built without the need of a center bearing. This results in short assembly times as well as a minimal impact on traffic. It also provides low maintenance costs, as a transition construction or elastomeric bearings are not required as well.

Figure 1. integral composite frame bridge

Except for the organic corrosion protection system of the steel girder, all main components of this bridge type can be built maintenance free over the whole service life time of 100 years. As the share of integral composite frame bridges in the road and highway system has been growing over the last decades, it is a serious approach to further reduce maintenance and therefore face the service life time of the corrosion protection system.

As one of the most sustainable options to protect steel from environmental influences, hot dip galvanizing (HDG) is able to last more than 100 years even in heavily polluted environments. During the process of HDG, the steel parts are immersed into tanks with approximately 450°C hot liquid zinc. Due to limitations in the process, the length of common zinc tanks is up to 16 m and in rare cases up to 19 m [4]. Hence, longer girders have to be divided into segments and connected afterwards. Welding is possible but comes along with disadvantages regarding maintenance. Since the welding can only take place after the process of HDG, the area close to the seam has to be protected by zinc thermal spraying, which has a service lifetime of 25 to 35 years and therefore does not offer an overall solution. While I-girders of composite sections can be connected using prestressed bolt connections, there is no suitable solution for box girders so far. A research project concerning the highway systems of South-Germany showed that box girders are favoured over I-Sections, due to several advantages. Box girders have a 25 to 30% smaller surface exposed to chloride contaminated spray, which is accompanied by lower ablation rates of the zinc coating. Furthermore, box girders offer a higher torsional stiffness, which make them suitable for curved layouts in top view.

For this reason, the Chair of Metal Structures works in cooperation with the Chair of Concrete Structures at the Technical University of Munich (TUM) and the Bundesanstalt für Materialforschung und -prüfung (BAM), to develop a joint for box-girders, which meets the technical, structural and architectural requirements. After preliminary studies, a grouted joint connection turned out as the most suitable.

2 GROUTED JOINT

Considering the length of the zinc tank to be 16 m, the required bridge span of 35 to 45 m results in three segments with lengths from 12 to 16 m (figure 2).

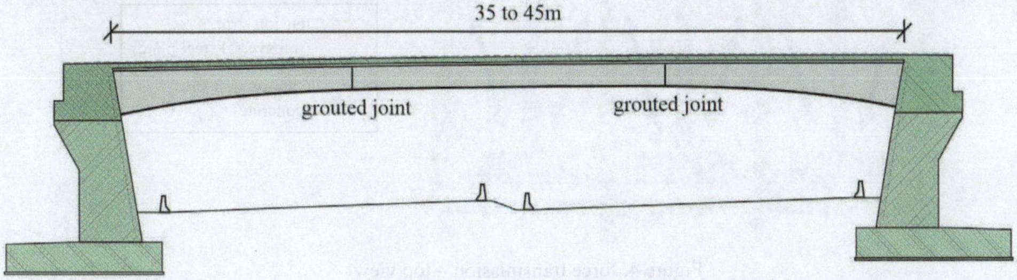

Figure 2. positioning of grouted joints

To avoid temporary propping towers at the joint position and therefore ensure a short assembly period without affecting traffic, the grouting of the joint and the pouring of the prefabricated slab should take place before the girder is lifted into place. To reduce the bending moments acting on the joint, the position of the joint should be chosen as close as possible to location where the global bending moment is zero. For this reasin, the middle segment has to be designed as long as possible. Considering the global position of the joint, mainly positive bending moments occur during assembly and over service lifetime. Nevertheless, the joint is also capable of transmitting negative bending moments. The design of the grouted connection consists of interlocking steel T-studs, diaphragms and longitudinal plates embedded in a box girder. To ensure a form- and force-locking connection, the hollow space surrounding the T-studs is grouted with Ultra-High-Performance Fiber-Reinforced Concrete (UHPFRC). Figure 3 shows the grouted joint before and after connecting the steel segments. The diaphragms, which are closer to the T-studs, enclose the space for grouting.

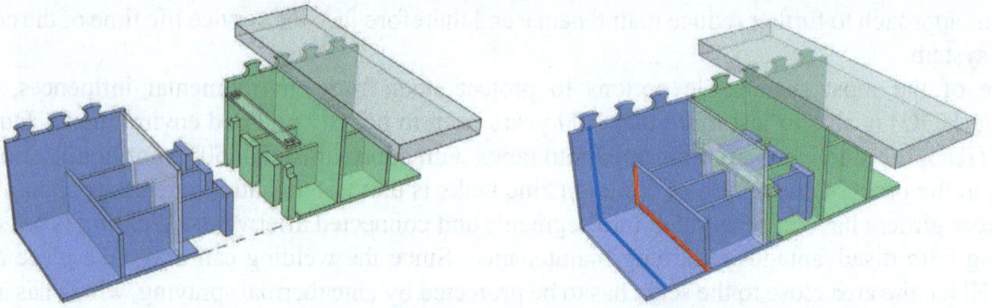

Figure 3. grouted joint before (left) and after assembly (right)

In the cross section outside the joint (figure 3; blue line), the tension force resulting from the bending moment can be assigned to the steel box section, while the compression force acts on the concrete slab, which consists of a prefabricated slab and an in-situ-concrete slab. At the transition area to the joint (figure 3; red line), the tension force is transmitted to the webs of the T-studs through the extended webs, welded to the lower steel chord of the box girder. At the end of the T-studs, the flanges of the T-studs forward the tension force to the UHPFRC, where it is transmitted to the flanges of the opposite segment by compression struts. Since two compression struts are acting on the T-stud flanges symmetrically from both sides, there is no transverse bending. At the special position of the outer flange, where the half-sized flange is welded to the web of the steel box, the compression strut is acting asymmetrically on the flange. By using a transversal steel plate (see figure 4), which is connected to the upper ends of the flanges, the outer compression struts are short-cutted.

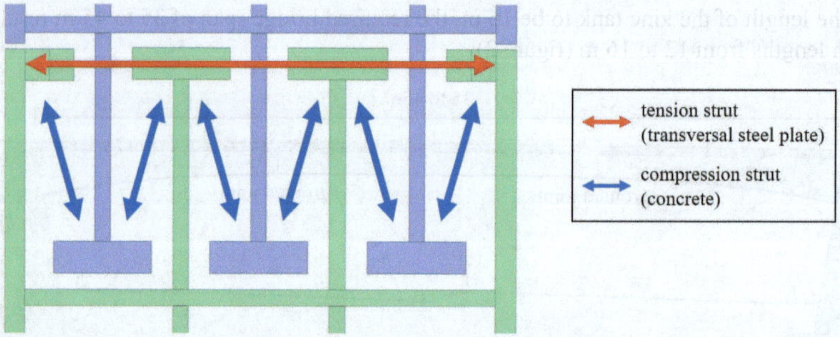

Figure 4. force transmission – top view

The bottom surfaces of the T-stud flanges are placed on top of the lower box steel chord of the opposite segment (see figure 3). In this way, the shear force can be transmitted to the diaphragms of the opposite segment.

The main load carrying component - the T-stud - has to transmit internal forces due to static and dynamic actions according to EN 1991-2 [5]. The resulting stresses can be determined by using simplified calculation models. These models are currently developed based on known mechanical relations and verified by practical experiments and finite element studies.

When determining the resistance of the T-studs according to EN 1993, a distinction between static and fatigue resistance is required. Since hot-dip galvanizing does not affect the static resistance of steel components, the static verifications can be carried out conventionally. For the fatigue assessment as per EN 1993-2 [6] however, the unfavourable influence of hot-dip galvanizing has to be taken into account.

3 RECENT INVESTIGATIONS ON THE FATIGUE BEHAVIOUR OF HOT DIP GALVANIZED DETAIL CATEGORIES AS PER EN 1993-1-9

During the process of hot dip galvanizing, the pre-treated steel component is immersed in a bath of zinc with approximately 450 °C. The liquid zinc reacts with the steel and forms a coating on the surface. During the subsequent cooling process, micro defects can form in the zinc coating due to the different shrinkage and cooling rates of steel and zinc (see figure 5). If the component is subjected to fatigue stresses during its service life, these micro defects can propagate into the base material and initiate fatigue failure.

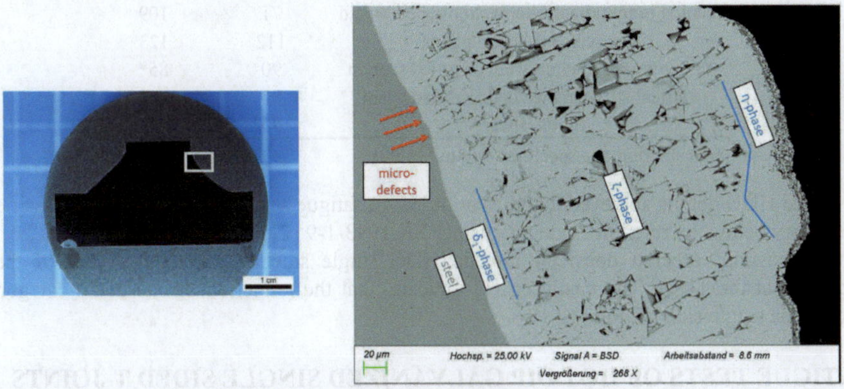

Figure 5. phases of the hot dip galvanize coating

For a long time, it was assumed that hot-dip galvanizing was generally unsuitable as a corrosion protection system for steel components which are exposed to fatigue loads. This was one of the reasons why the influence of hot-dip galvanizing on the fatigue strength was not scientifically investigated. Nevertheless, recent investigations have disproved this assumption. According to [1] the negative impact of hot dip galvanization on the fatigue resistance can be considered conservatively by the reduction of the reference detail category as per EN 1993-1-9 [2] by one category. Figure 6 shows the results from the tests carried out in [1] for a gas-cutted plate, which is assigned to fatigue class 125 (FC 125).

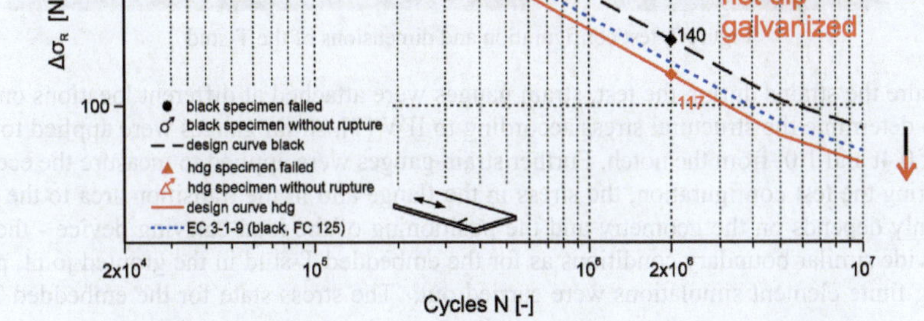

Figure 6. fatigue tests of gas-cutted plates (FC 125), black and hot dip galvanized [1]

Although a reduction of the fatigue strength is evident, the hot dip galvanized specimens achieved a reference fatigue strength $\Delta\sigma_{C,HDG}$ of 117 N/mm² at design level and are thus only 8 N/mm² below the reference detail category 125. Further detail categories investigated in [1] are shown in Table 1.

Table 1: test results for hot dip galvanized fatigue classes according to [1]

detail description acc. EC3 [2]	$\Delta\sigma_{C,EC3}$	$\Delta\sigma_{C,HDG}$
manual longitudinal fillet weld (small scale)	100	147
manual longitudinal fillet weld (big scale)	100	134
manual longitudinal fillet weld over butt weld	71	109
transverse splice in plates	112	123
full cross section butt weld of rolled section	90	85*
vertical stiffeners welded to rolled beam	80	82
Studs (shear)	90	80

*considering one specimen without rupture

All tested HDG details achieved a higher or at least a fatigue strength, which can be classified as one detail category below the reference category as per EN 1993-1-9 [2].

As the existing research does not include HDG single sided T-joints, it is part of the current investigations at the TUM. The results in [1] indicate, that the reduction of one detail category is also suitable for the single sided T-joint.

4 FATIGUE TESTS OF HOT DIP GALVANIZED SINGLE SIDED T-JOINTS

27 specimens of single sided T-joints were tested for their fatigue strength. All specimens are of the same steel material (S355) and dimensions. In order to compare to black steel, 16 specimens were hot dip galvanized while 11 remained without corrosion protection. The dimensions of the T-studs and the test setup are shown in figure 7.

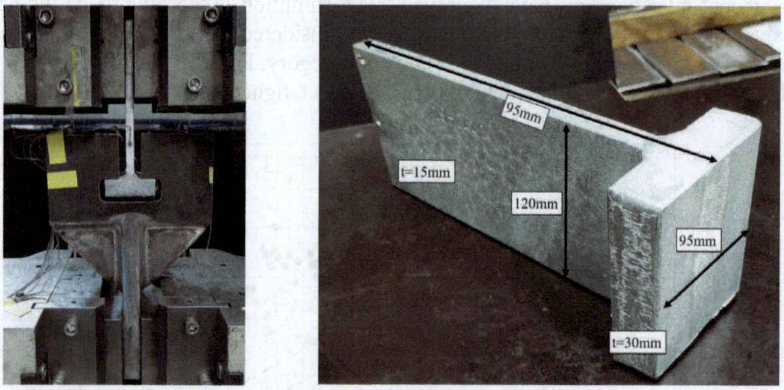

Figure 7. test configuration and dimensions of the T-stud

To measure the strains during the test, strain gauges were attached at different locations on the flange and web. To determine the structural stress according to IIW [7], strain gauges were applied to the web at a distance of 0.4t and 1.0t from the notch. Further strain gauges were applied to measure the eccentricities.

Considering the test configuration, the stress in the flange and in the transition area to the web of the T-joint, mainly depends on the geometry and the positioning of the load-carrying device - the clamp. In order to provide similar boundary conditions as for the embedded T-stud in the grouted joint, preliminary studies using finite element simulations were carried out. The stress state for the embedded T-stud was evaluated and the geometry of the clamp was iteratively adjusted until an almost identical stress state was obtained.

The specimens are organized in groups according to different stress ranges $\Delta\sigma_{R,Test}$. The estimated load cycles were calculated by using the SN curve for fatigue class 80 for black and fatigue class 71 for hot dipped galvanized specimens according to EN 1993-1-9 [2]. The SN curves as per code are given for a probability of survival of 95%. To calculate the estimated load cycles for each stress range, the SN curves was transformed to a probability of survival of 50% according to [8]. All tests were performed under sinusoidal load cycles with a stress ratio of R = 0,1 and a frequency of 2 Hz. Table 2 shows the specimen groups and estimated load cycles.

Table 2: fatigue tests of single sided T-joints with and without zinc coating

	Quantity of specimen	$\Delta\sigma_{R,Test}$ in N/mm²	Estimated load cycles (PS=50%)
	2	282	100.000
black steel	2	223	200.000
FC 80	4	165	500.000
	3	131	1.000.000
	3	282	70.000
hot dip galv.	3	223	141.000
	4	165	347.000
FC 71	5	131	694.000

The tests were evaluated on the basis of the nominal stress concept according to EN 1993-1-9 [2]. The *pearl string method* according to DIN 50100 [9] was applied to calculate the regression curve and the design curves were determined for a survival probability of 95%. Figure 8 shows the results for the specimens with and without hot dip galvanizing.

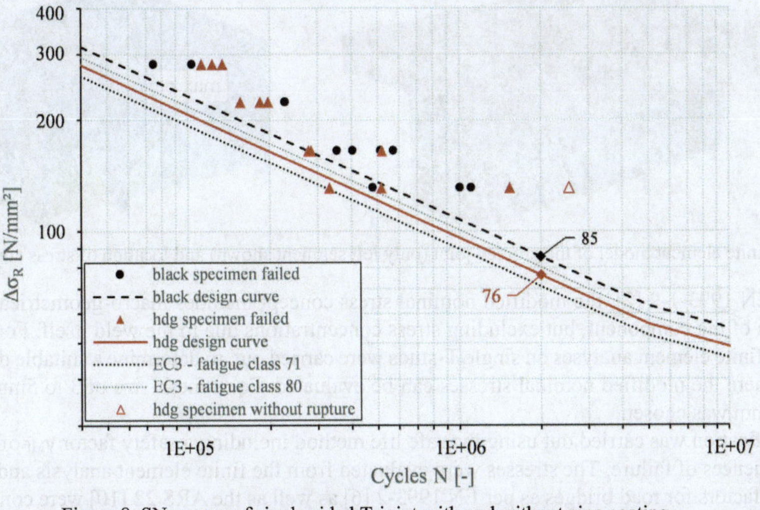

Figure 8. SN-curves of single sided T-joint with and without zinc coating

According to expectation, all specimen failed at the toe of the notch and the initial crack appeared at the edge of the web plate. The design curve of the blank steel specimens reaches a resistant stress range of 85 N/mm² for 2x10⁶ cycles and therefore is on the safe side compared to the reference detail category as per EN 1993-1-9 [2]. The design curve for HDG specimens reaches a resistant stress range of 76 N/mm² and are on the safe side as well, considering the reduction of one fatigue class according to [1].

For the stress ranges of 165 to 282 N/mm², the test results for both the black and the hot-dip galvanized specimens were close to the expected number of load cycles. For the stress range of 131 N/mm², a significantly larger scatter was observed for both groups. This, in combination with the relatively small number of specimens, has a significant effect on the design curves. It can be assumed that the test results for a higher number of specimens would be further on the safe side.

5 FATIGUE ASSESSMENT OF THE GROUTED JOINT USING THE MODIFIED NOMINAL STRESS CONCEPT

The experimental fatigue tests were performed under tension force only, which results in a constant stress distribution over the height of the T-stud. Considering the T-stud in the real joint geometry, the external forces result in an uneven stress distribution and stress concentrations, especially at the bottom of the T-stud near the transition area between web/flange and web/diaphragm as shown in figure 9. These stress concentrations have to be critically assessed regarding fatigue failure.

For the verification of the bridge joint, the modified nominal stress concept as per EN 1993-1-9 [2] is applied. This concept provides the advantage, that the stress concentration resulting from macro-geometrical effects due to the detail geometry and from load effects are considered. Furthermore, the fatigue class of the HDG T-stud is known from the experimental tests. To quantify the stresses due to macro-geometrical effects, EN 1993-1-9 [2] gives stress concentrations factors for specific details. Alternatively, a finite element analysis can be performed. Since there is no detail matching with the specific geometry of the joint, the joint was modeled as shown in figure 9. The finite element model was loaded with the fatigue load model 3 according to EN 1991-2 [5]. According to results of the experimental tests, the detail category was determined to 71.

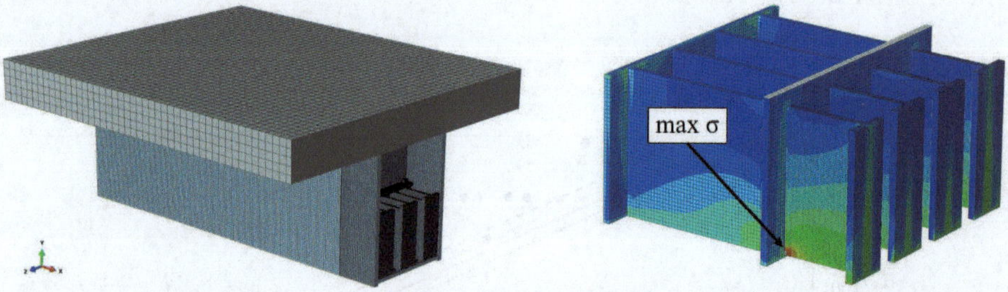

Figure 9. finite element model of the grouted joint (only left segment shown) and location of stress concentration

As per EN 1993-1-9 [2], the modified nominal stress concept includes macro-geometrical effects due to the design of the component, but excluding stress concentrations due to the weld itself. For this reason, preliminary finite element analyses on single T-studs were carried out, to determine a suitable distance from the notch where the modified nominal stresses can be evaluated. For a mesh size of 3 to 5mm, a distance of 10 to 15 mm was chosen.

The verification was carried out using the safe life method including a safety factor γ_{Mf} of 1,35, due to high consequences of failure. The stresses were evaluated from the finite element analysis and the damage equivalence factors for road bridges as per EN 1993-2 [6] as well as the ARS 22 [10] were considered. The utilization for the fatigue assessment at the guiding location – the bottom of the T-stud close to the diaphragm - is up to 95%.

A comparison of all required verifications for the bridge joint showed that the fatigue assessment is guiding in all cases, due to distinct stress concentrations and high loads. On the one hand, the design of the joint has to be suitable for HDG which requires a geometry where stress concentrations are unavoidable. On the other hand, the load assumption is very conservative regarding the specifications given in the ARS

22. The λ_2 coefficient of the damage equivalent factors as per EN 1993-2 [6], which allows the load assumption to be adapted to the traffic volume, is prescribed with a value of 1,1 according to ARS 22 [10]. This corresponds to a crossing of $2x10^6$ vehicles with an average weight of 400 kN per year and is usually chosen for roads and highways with two or more lanes per direction and high flow rates of lorries according to EN 1991-2 [5]. The overall damage equivalent factor λ results in 2,3 with λ_{max} guiding.

Since the present type of composite bridge including the grouted joint is designed for country roads overpassing a highway, an approach as per EN 1991-2 [5] considering medium flow rates of lorries and an average weight of 300 kN per vehicle would be sufficient. The overall λ factor would result in 1,59 which corresponds to a 30% reduction of the assumed load.

6 OPTIMIZATION OF THE FATIGUE BEHAVIOUR OF THE T-STUD

Due to the high fatigue stresses, it is of particular interest to reduce the stress concentrations and therefore improve the fatigue behavior of the T-stud. For this reason, the geometry of the T-stud was optimized at the critical locations by providing cut-outs.

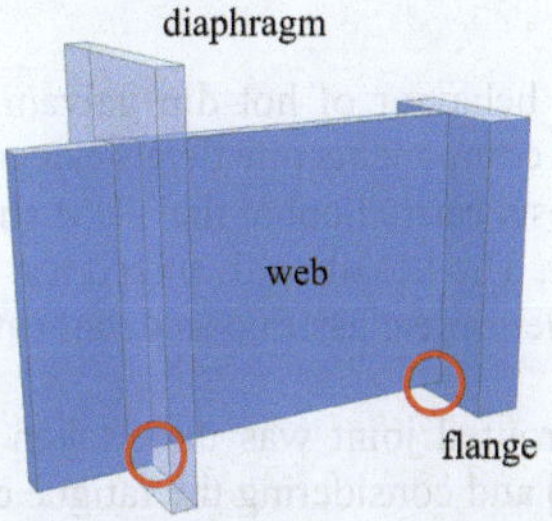

(a) critical locations with stress concentration in the web

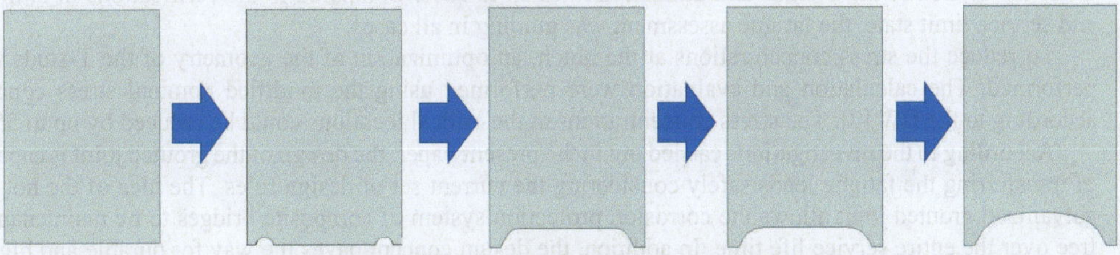

(b) geometrical optimization process of the web

Figure 10. optimization of the T-stud geometry at the location of the stress concentrations

To quantify the reduction of the fatigue relevant stresses due to the cut-out and to consider the more complex geometry, a finite element analysis using the modified nominal stress method was performed. Two models with and without cut-out were meshed and evaluated according to the corresponding concept in the IIW [7]. Figure 11 shows the principal stresses occurring at the locations of stress concentrations for the model with and without cut-out. The stresses were evaluated at a distance of 10 mm from the notch to consider the stress concentration due to the geometry but neglect the stress concentration due to the weld. The model was loaded by the fatigue load according to the fatigue load model 3 as per EN 1991-2 [5].

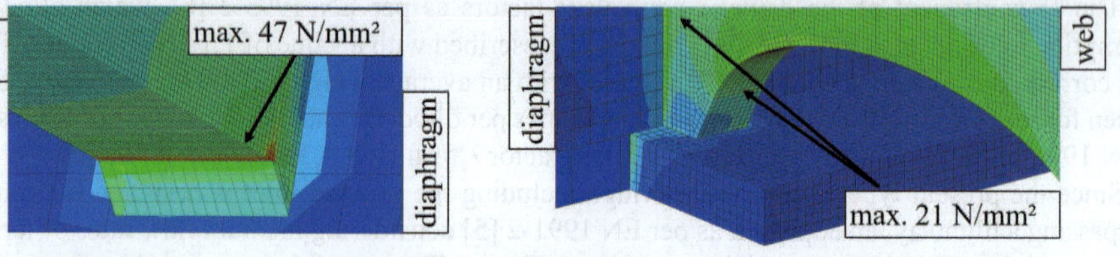

Figure 11. principal stresses at lower transition area web/diaphragm for models without (left) and with cut-out (right)

The stress concentrations in the model without cut-out are localized at one single area at the weld seam and reach a maximum of 47 N/mm² at a distance of 10 mm from the notch. In the model with cut-out, the stress concentrations are localized at the transition area to the web and to the lower cord. The maximum principal stresses are 21 N/mm². Due to the cut-out, the stresses can be distributed over a greater height at the diaphragm. Comparing the maximum stresses of both models, the reduction results in ca. 55%.

6 CONCLUSIONS

Previous research on the fatigue behavior of hot dip galvanized details recommend to classify the fatigue strength of hot dip galvanized components one detail category below the reference category as per EN 1993-1-9 [1]. The experimental tests carried out at the TUM showed, that this recommendation is also applicable for the single sided T-joint. The hot dip galvanized test specimens reaches a stress range of 76 N/mm² at $2x10^6$ load cycles and are therefore classified as detail category 71 for the specific application of the grouted joint.

The fatigue assessment of the grouted joint was undertaken applying the modified nominal stress concept according to EN 1993-1-9 [2] and considering the fatigue class of 71 for the T-studs. The relevant stresses were evaluated using a finite element analysis. The analysis showed distinct stress concentrations at the lower area of the T-studs and utilization rates up to 95%. Compared to the verifications in ultimate and service limit state, the fatigue assessment was guiding in all cases.

To reduce the stress concentrations at the notch, an optimization of the geometry of the T-studs was performed. The calculation and evaluation were performed using the modified nominal stress concept according to the IIW [7]. The stress concentrations at the critical locations could be reduced by up to 55%.

According to the investigations carried out in the present paper, the design of the grouted joint is capable of transferring the fatigue loads safely considering the current set of design rules. The idea of the hot dip galvanized grouted joint allows the corrosion protection system of composite bridges to be maintenance-free over the entire service life time. In addition, the design concept paves the way for durable and highly loadable connections in other fields of composite construction.

REFERENCES

[1] D. Ungermann e. al., „Forschungsvorhaben P 835 / IGF-Nr. 351 ZBG - Feuerverzinken im Stahl- und Verbundbrückenbau" Verlag und Vetriebsgesellschaft mbH, Düsseldorf, 2014.

[2] DIN EN 1993-1-9:2010-12, Eurocode 3: Bemessung und Konstruktion von Stahlbauten – Teil 1-9: Ermüdung, 2010

[3] U. Kuhlmann e. al., Forschungsvorhaben P 843 / IGF-Nr. 353 ZN - Ganzheitliche Bewertung von Stahl- und Verbundbrücken nach Kriterien der Nachhaltigkeit, Düsseldorf: Verlag und Vertriebsgesellschaft mbH, 2016.

[4] D. Ungermann e. al. "Entwurfshilfe zum Einsatz von feuerverzinkten Bauteilen im Stahl- und Verbundbrückenbau" bauforumstahl e.V, Düsseldorf, 2016

[5] DIN EN 1991-2:2010-12, Eurocode 1: Einwirkungen auf Tragwerke – Teil 2: Verkehrslasten auf Brücken, 2010

[6] DIN EN 1993-2:2010-12, Eurocode 3: Bemessung und Konstruktion von Stahlbauten – Teil 2: Stahlbrücken, 2010

[7] A. F. Hobbacher „Recommendations for Fatigue Design of Welded Joints and Components, Second Edition, IIW Collection", Basel: Springer International Publishing AG, 2016 UND Genoa: International Institute of Welding, 2016

[8] Dipl.-Ing. Heise, F.-J., "Forschungsvorhaben P 778 - Bemessung von ermüdungsbeanspruchten Bauteilen aus hoch- und ultrahochfesten Feinkornbaustählen im Kran- und Anlagenbau; Rechnerische Nachweise - Kerbdetails- Lastkollektive", KIT – RWTH Aachen – Liebherr, 2013

[9] DIN 50100, Schwingfestigkeitsversuch – Durchführung und Auswertung von zyklischen Versuchen mit konstanter Lastamplitude für metallische Werkstoffproben und Bauteile, 2016

[10] Bundesanstalt für Straßenwesen, „Allgemeines Rundschreiben Straßenbau Nr. 22/2012", Bonn: Bundesministerium für Verkehr, Bau und Stadtentwicklung, 2012

9TH INTERNATIONAL CONFERENCE ON COMPOSITE CONSTRUCTION IN STEEL AND CONCRETE
M. Knobloch, U. Kuhlmann, W. Kurz, M. Schäfer. (Eds.)
Online, Germany, July 27-29, 2021

STUDIES OF THE FATIGUE STRESS AND STRENGTH OF LARGE COMPOSITE BRIDGES WITH CANTILEVERS AND PRECAST CONCRETE ELEMENTS

Josef Karl Kraus* and Karsten Geißler*

* Technische Universität Berlin, Fachgebiet Entwerfen und Konstruieren - Stahlbau
e-mails: josef.kraus@tu-berlin.de, karsten.geissler@tu-berlin.de

Keywords: composite bridges, cantilever, partial-depth precast elements, fatigue, hot spot stress method.

Abstract. *When building large composite bridges with steel cantilevers as supports for partial-depth precast deck-elements, currently there are only recommendations available, which based on experiences with regard to the fatigue stressed connections of the cantilevers. In a research project currently being worked on for the Federal Ministry of Transport and Digital Infrastructure (BMVI) appropriate recommendations are developed based on experimental tests, accompanied by numerical analyses and structural monitoring. The article provides an insight into the ongoing research work at the Technical University of Berlin.*

1 INTRODUCTION

In the area of federal roads, numerous large bridges with spans of 50 – 70 m and more will have to be replaced in the next few years. Composite constructions have established themselves as an economical construction method for the superstructures. Recently, partial-depth precast elements have increasingly been used for the production of the deck, which allow a more cost-effective and even faster execution. The focus of the research described here, is the possibility to arrange steel cantilevers in the transverse frame spacing to support the precast elements, see Fig. 1. These designs were used for example at the Heidingsfeld viaduct (A3) or the Langenfelder bridge Hamburg (A7). Other such structures are currently being planned by DEGES or are already under construction.

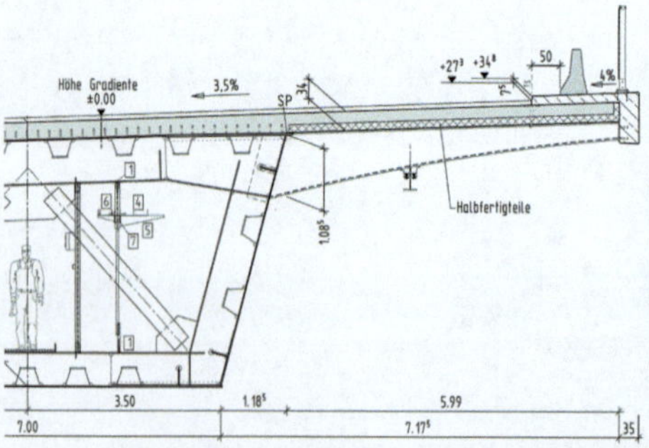

Figure 1: Cross section of the Heidingsfeld viaduct, A3 (left); Langenfelder bridge Hamburg, A7 (right)

One key aspect of the research project [1] currently being carried out on behalf of the BMVI, is the development of fatigue strength parameters and design recommendations for the connection details of the cantilevers. Regarding the cantilever's web and lower flange (with the aim of avoiding free cuts in the main girder's web [2]) a possible offset to the internal transverse frame of approximately 5 - 10 mm cannot be ruled out and which is currently not covered by the fatigue details of the nominal stress concept in this order of magnitude. Additionally, the supporting effect of the intermediate main girder's web on the misalignment moment has not yet been systematically investigated by means of tests. Regarding the connection of the cantilever's upper flange, the current recommendation with rounding (R $\geq$ 150 mm) and additional thickness compensation (1:4) means a very high manufacturing effort. The relatively low stresses due to the close centre of gravity have a positive effect here, which are realistically determined by numerical investigations and structural monitoring. For the mentioned fatigue-critical points, experimental tests are carried out, including current databases of fatigue tests, which will lead to reliable recommendations. The article is intended to provide an insight into this ongoing research work.

Beyond that, in the same research project [1] the Institute for Construction and Design at the University of Stuttgart is dealing with the fatigue strength of the reinforced concrete deck, which is simultaneously subjected to tensile and shear forces. However, this aspect is not covered in this article.

2 NUMERICAL INVESTIGATIONS

FE models, which are developed and calibrated in parallel with a structural monitoring, are used for the realistic calculation of the local stress level, see Fig.2. The modelling is entirely based on shell elements and easily scalable by using parameters. Only components out of focus, that lie outside the examined transverse frame or cantilever section are modelled as slab elements (cantilever, transverse frame, transverse bracing and longitudinal ribs). The cracking of the concrete is considered in a simplified way via an effective E-module. The evaluation focuses on the stress concentration on the cantilever's lower flange, the low stress level at the cantilever's upper flange and the distribution of axle loads in the area of the cantilevers.

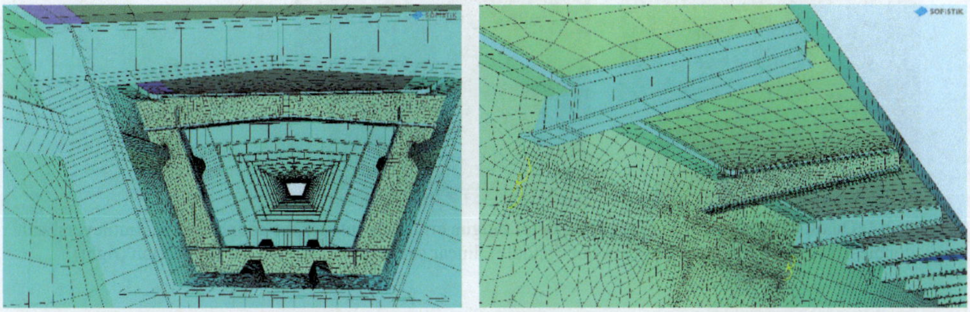

Figure 2: Large-scale FE model of the Langenfelder bridge Hamburg, A7 - transverse frame (left), cantilevers (right)

Additionally, detailed models for the connection of the cantilevers are analysed before and after the experimental tests. Those are mainly used to identify critical points and after calibration for the determination of stress concentration factors. The modelling is carried out with both tetrahedral and hexahedral elements, the former being used only for subordinate components and the latter at the critical points for the hot spot stress method. Regarding the modelling of welds, the goal is to depict the worst case in the model. Elevated parts of butt welds (additional fillet welds) are therefore not considered. On the other hand, for the hot spot stress method relevant fillet welds are considered, such as at the free cuts in the cantilever's web at the connection of the upper flange. A selection of the most critical points can be seen in Fig. 3 and Fig. 4. This includes for example, at the connection of the upper flange, the lower plate edge at the notch-free rounding (section 3001). When planning a cantilever with hollow section, the overall most critical point can be identified at the lower plate edge of the web (section 6001). However, in order to verify the numerical results, real component tests are required.

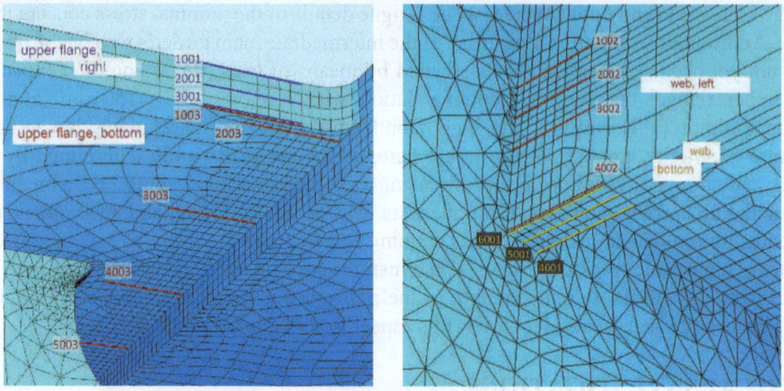

Figure 3: Small-scale FE models for the connection of the cantilevers with sections for the hot spot stress method, double-T section's upper flange (left), hollow section's lower flange (right) [3]

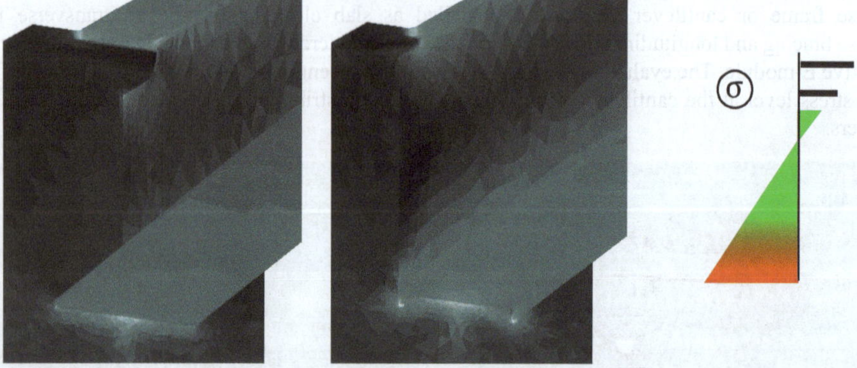

Figure 4: Qualitative comparison of the location of fatigue critical points for the connection of cantilevers with double-T section (left) and hollow section (middle); linear stress distribution at the connection (right)

3 COMPONENT TESTS ON A SCALE OF 1:1

In a first test program the fatigue safety is tested on specimens on a scale of 1:1. The aim of these tests is, on the one hand, to measure the actual stress levels for further calibration of the FE models and, on the basis of this, to determine the stress concentration factors at the connection of the cantilever's upper and lower flange. On the other hand, they serve to take into account realistic residual stresses and to identify the actual position of cracking. In total, the test program consists of three variants, including the connection of cantilevers as double-T section with increased and usual offsets as well as the connection of a cantilever as hollow section. The former is shown in Fig. 5 to Fig. 7. Each specimen consists of a part of the transverse frame and an attached cantilever. The applied load at the edge of the cantilever is a pulsating compressive force.

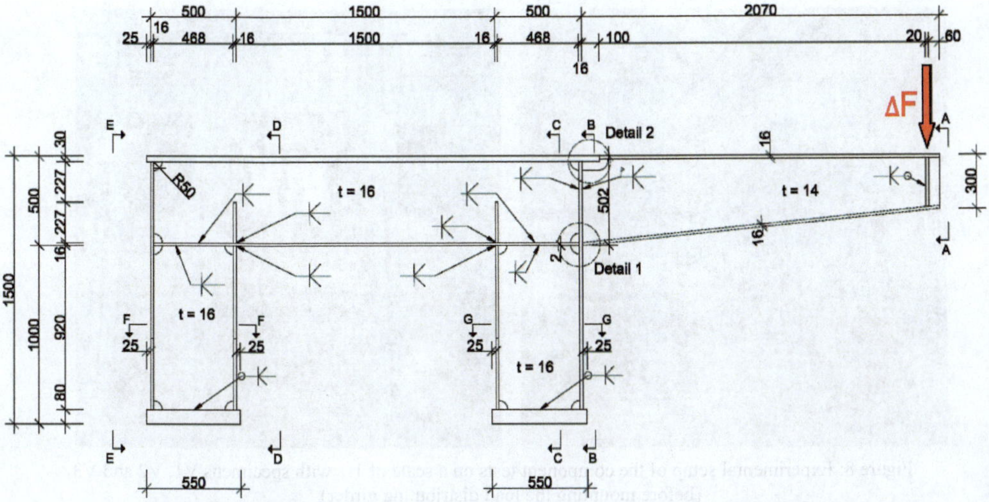

Figure 5: Specimens on a scale of 1:1 for S-N tests and the analysis of hot spot stresses

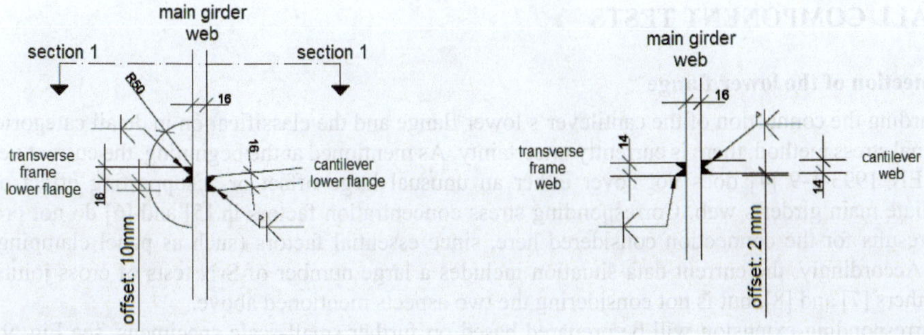

Figure 6: Welding detail on the cantilever's lower flange (left) and web (right), considering a possible offset

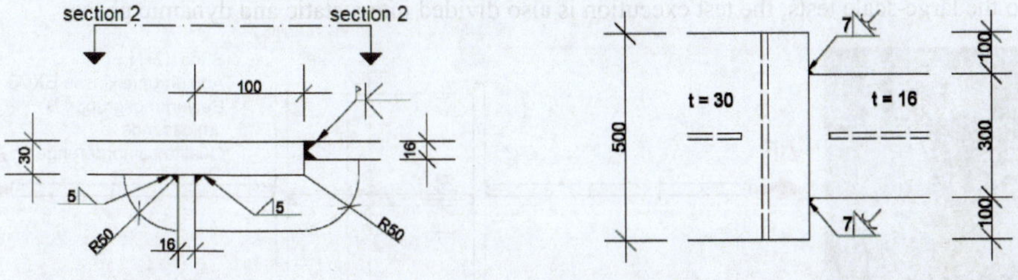

Figure 7: Welding detail on the upper flange, aiming at low manufacturing effort and adequate fatigue resistance

The first test phase is carried out under a static load and is used to determine the stress concentration factors at the connection. In the second test phase the actual S-N test takes place. The test specimens V1, V2 and V3, manufactured by Stahlbau Magdeburg GmbH, can be seen in Fig. 8 in the test station of the TU Berlin. The experimental setup allows a simultaneous loading of the three specimens. The load distributing girder used for this purpose is screwed to the hydraulic cylinder and has contact with the cantilever via roller bearings. After cracking of a specimen in phase 2, the test continues by removing individual roller bearings on either one or two of the remaining specimens.

Figure 8: Experimental setup of the component tests on a scale of 1:1 with specimens V1, V2 and V3 (before mounting the load distributing girder)

4 SMALL COMPONENT TESTS

4.1 Connection of the lower flange

Regarding the connection of the cantilever's lower flange and the classification in detail categories for the nominal stress method, there is currently uncertainty. As mentioned at the beginning, the current version of DIN EN 1993-1-9 [4] does not cover either an unusual large offset or a supporting effect of the intermediate main girder's web. Corresponding stress concentration factors in [5] and [6] do not provide reliable results for the connection considered here, since essential factors (such as panel clamping) are unclear. Accordingly, the current data situation includes a large number of S-N tests of cross joints, see among others [7] and [8], but is not considering the two aspects mentioned above.

A corresponding extension will be prepared based on further small-scale specimens, see Fig. 9. The geometry of the specimens is designed for double use, that means after cracking and repairing the first side, the S-N test can be continued until the fracture of the second side, based on the previous stress cycles. Similar to the large-scale tests, the test execution is also divided into a static and dynamic phase.

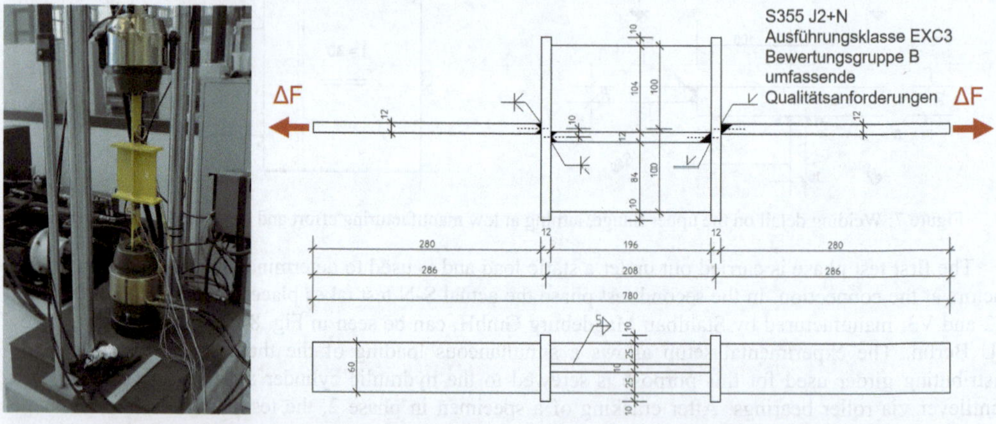

Figure 9: S-N tests on small-scale specimens for the connection of the cantilever's lower flange

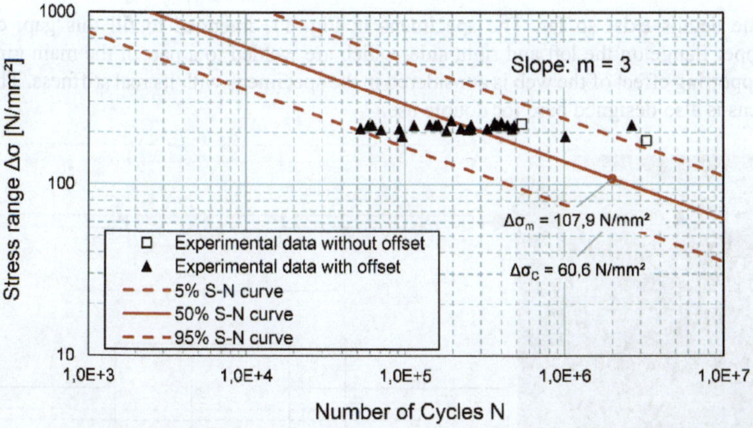

Figure 10: Results of the S-N tests for the connection of the cantilever's lower flange

In total, the test program consists of 27 S-N tests at a constant stress ratio of R = 0.1, but different variants regarding the offset, plate thickness and weld design (DHV and HV), see Tab. 1. Results are shown in Fig. 10.

Table 1: Overview of the specimens for the connection of the cantilever's lower flange

	U1	U2	U3	U4	U5
Cantilever's lower flange (1)	12 mm	12 mm	12 mm	20 mm	20 mm
Transverse frame's lower flange (2)	12 mm	12 mm	12 mm	20 mm	20 mm
Transverse frame's web (3)	12 mm	12 mm	12 mm	12 mm	12 mm
Main girder's web (4)	12 mm	12 mm	12 mm	12 mm	20 mm
Offset	0 mm	5 mm	10 mm	10 mm	10 mm
Weld design	50% DHV weld / 50% HV weld				
No. of S-N tests	3-4	6-8	6-8	6-8	6-8

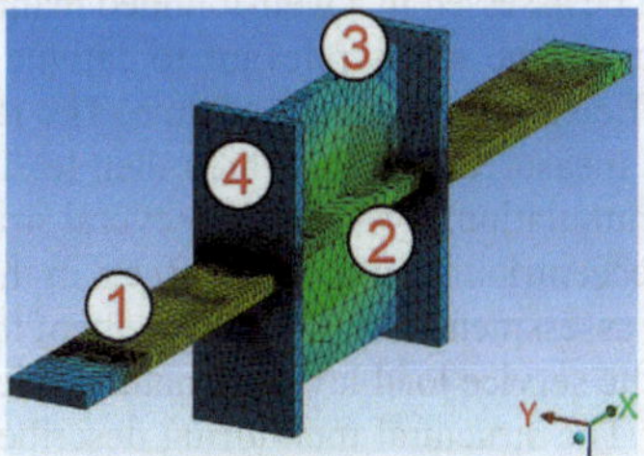

4.2 Connection of the upper flange

For the connection of the cantilever's upper flange according to the current recommendation, as mentioned at the beginning with rounding (R ≥ 150 mm) and thickness compensation (1:4), a well clarified fatigue detail can be assumed, see among others [9], [10] and [11]. However, this variant means a very high manufacturing effort in contrast to the low local stress level. Therefore, it must be examined to what extent this rather high requirement can be reduced without limiting the durability. All variants considered here avoid the thickness compensation and differ regarding the rounding:

1) Radius with R = 20 mm (build-up welding, grinding to notch-free rounding), see Fig. 11
2) Radius with R = 7 mm (grinding of the existing fillet welds to notch-free rounding)
3) Without radius

All variants have in common to be a significantly more economical construction, for which however, due to the high difference in plate thickness and the interaction with the lateral rounding, no experimentally clarified fatigue details exist so far. The specimens in Fig. 11, intended to fill this gap, consist of a cantilever's upper flange on the left and right side, which are welded to a part of the main girder's upper flange. The supporting effect of the web is considered in the specimens with its real stiffness. The geometry of the specimens is also designed here for double use.

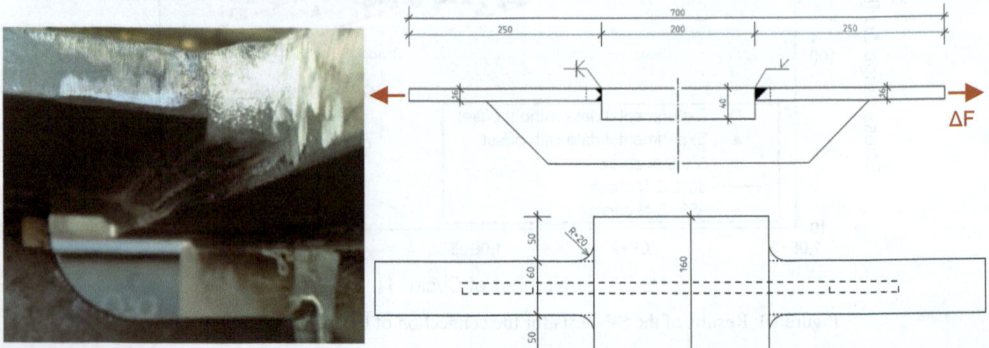

Figure 11: S-N tests on small-scale specimens for the connection of the cantilever's upper flange

5 STRUCTURAL MONITORING

By using structural monitoring, the suitability of this new construction method with its structural details can be assessed in a well-founded manner, especially regarding questions of fatigue. With the help of strain gauges, the stress relevant to fatigue as a result of traffic loads at critical points is measured over a sufficiently long period of time. The rainflow counting method is used to determine load collectives from the measured signals, which then lead directly to fatigue damage in the measurement period via damage accumulation. By arranging several strain gauges in the fatigue-critical details, local additional stresses due to eccentricities and notch effects in the vicinity of weld seams can be recorded very precisely. Regarding the assessment of the load capacity of the structure or its components in the limit state, in-situ measurements at the service load level are naturally less suitable but can still provide valuable impulses.

The structural monitoring described here focusses on the Heidingsfeld viaduct (A3). The 635 m long superstructures are curved in plan and consist of single-cell steel box girders with a composite deck made of in-situ concrete and partial-depth precast elements, which are supported outside the box girder by cantilevers on both sides, see Fig. 1. The deck is approximately 22.3 m (northern superstructure) and 20.3 m (southern superstructure) wide. The bridges span the valley at a maximum height of 50 m with seven openings and spans of 50 m, 80 m, 90 m, 100 m, 110 m, 120 m and 80 m. The construction height of the main girder is 4.60 m.

The measurement system was installed in July 2019, initially inside the box girder and, in a second step, outside in the cantilever area. Additionally, a camera was installed which monitors the traffic situation. The monitoring system permanently records the signals from 32 channels (strain gauges, crack width sensors) at a sampling rate of 100 Hz, which are regularly transmitted via LTE to an external server.

Results of the measurement so far show, especially in comparison with older motorway bridges, the extremely low stress level. The maximum range of strain measured so far on the main girder's lower flange was approximately 143 µm/m, which corresponds to 30 N/mm². Responsible for this was a heavy transport with a total mass of approximately 140 t in the direction of Nuremberg. The stresses in traffic jams are significantly lower compared to the stresses from heavy transports. Under normal truck traffic, the measured strains are only approximately 50-60 µm/m, i.e. 10-13 N/mm². The stresses measured at the connection of the cantilever are even lower.

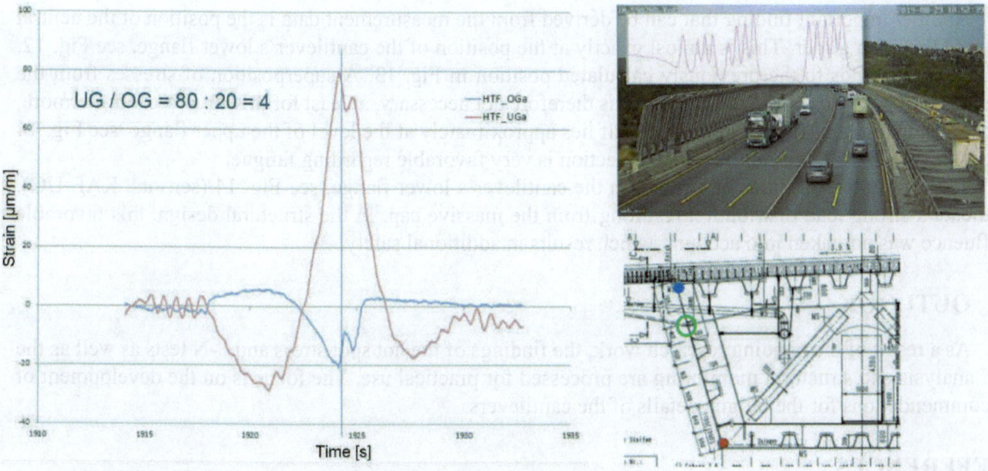

Figure 12: Determination of the position of the neutral axis in the main girder's web from the signal curve at the upper and lower edge, Heidingsfeld viaduct, A3

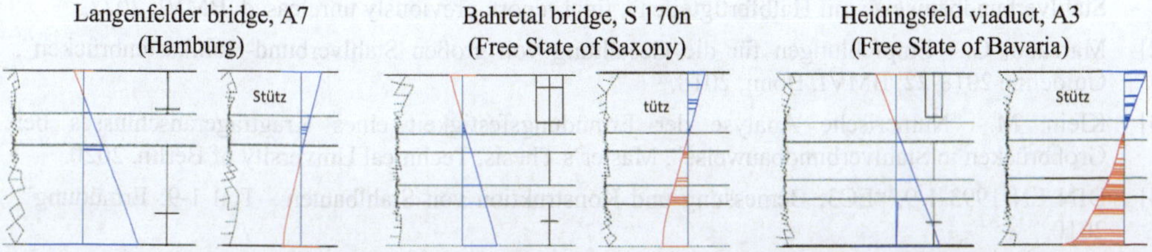

Figure 13: Calculated longitudinal stress in the main girder's web with neutral axis close to the connection of the cantilever's lower flange

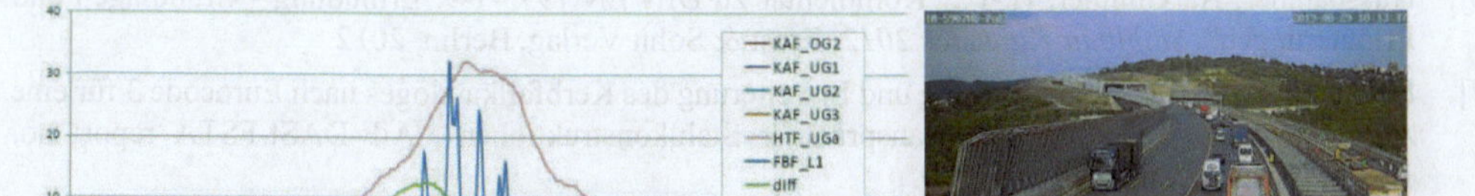

Figure 14: Stresses in the cantilever with different lengths of influence and low stress level in the upper flange (sensor: KAF-OG2) due to the close centre of gravity, Heidingsfeld viaduct, A3

Another important finding that can be derived from the measurement data is the position of the neutral axis of the main girder. This is almost exactly at the position of the cantilever's lower flange, see Fig. 12, which corresponds to the previously calculated position in Fig. 13. A superposition of stresses from the main girder with those from the cantilever is therefore not necessary, at least for this structure. Furthermore, since the neutral axis of the cantilever itself lies approximately at the level of the upper flange, see Fig. 14 (sensors: KAF-OG), the cantilever's connection is very favorable regarding fatigue.

The long influence lines of stresses in the cantilever's lower flange, see Fig. 14 (sensors: KAF-UG), indicate a strong load distribution resulting from the massive cap. In the structural design, this favorable influence was not taken into account, which results in additional safety.

6 OUTLOOK

As a result of the ongoing research work, the findings of the hot spot stress and S-N tests as well as the FE analysis and structural monitoring are processed for practical use. The focus is on the development of recommendations for the fatigue details of the cantilevers.

REFERENCES

[1] Geißler, K. et al, "Wissenschaftlich-technische Untersuchungen zur Festlegung der Ermüdungsfestigkeit der Fahrbahnplatte und der Konsolträger bei Großbrücken in Stahlverbundbauweise mit Halbfertigteilen", final report, previously unreleased, BMVI, 2022.

[2] Marzahn, G., "Empfehlungen für die Gestaltung von großen Stahlverbund-Hohlkastenbrücken", Guideline 2018-22, BMVI, Bonn, 2018.

[3] Klein, M., "Numerische Analyse der Ermüdungsfestigkeit eines Kragträgeranschlusses bei Großbrücken in Stahlverbundbauweise", Master's Thesis, Technical University of Berlin, 2020.

[4] DIN EN 1993-1-9, "EC3: Bemessung und Konstruktion von Stahlbauten - Teil 1-9: Ermüdung", 2010.

[5] Hobbacher, A., "Recommendations for fatigue design of welded joints and components", IIW document IIW-1823-07 ex XIII-2151r4-07/XV-1254r4-07, International Institute of Welding, Paris, 2008.

[6] Nussbaumer, A., Günther, H.-P., "Kommentar zu DIN EN 1993-1-9: Ermüdung - Grundlagen und Erläuterungen", *Stahlbau-Kalender 2012*, Ernst & Sohn Verlag, Berlin, 2012.

[7] Feldmann, M. et al, "Neubewertung und Erweiterung des Kerbfallkataloges nach Eurocode 3 für eine zukunftsfähige Auslegung hochbeanspruchter Stahlkonstruktionen", AiF-DASt-FSTA report no. 19178, previously unreleased, 2019.

[8] Booth, G.S., "Constant amplitude fatigue tests on welded steel joints performed in air", *European Offshore Steel Research Seminar*, Vol. 1, The Welding Institute, Abington, 1980.

[9] Hirt, M. A., Crisinel, M., "La résistance à la fatigue des poutres en âme pleine composées-soudées: effet des plaquettes et goussets soudés à l'aile", Institut de la construction métallique, EPFL, Lausanne, 1975.

[10] Berger, P., "Investigation on the Fatigue Strength of Welded Joints by Full Scale Tests", Doc. XIII-983-80, Zentralinstitut für Schweißtechnik der DDR, Halle, 1980.

[11] Ram, P. et al, "Beurteilung des Ermüdungsverhaltens von Krankonstruktionen bei Einsatz hoch- und ultrahochfester Stähle", *Forschung für die Praxis*, FOSTA, Düsseldorf, 2006.

FIRE BEHAVIOUR

9TH INTERNATIONAL CONFERENCE ON COMPOSITE CONSTRUCTION IN STEEL AND CONCRETE
M. Knobloch, U. Kuhlmann, W. Kurz, M. Schäfer. (Eds.)
Online, Germany, July 27-29, 2021

A NOVEL STRATEGY FOR SLIM-FLOOR FIRE PROTECTION

M.L. Romero*, V. Albero, A. Espinós*, E. Serra*, A. Hospitaler*and D.Pons***

* ICITECH, Universitat Politècnica de València, Spain
e-mails:mromero@mes.upv.es, aespinos@mes.upv.es,ensermer@upv.es,ahospitaler@cst.upv.es

** Department of Mechanical Engineering and Construction, Universitat Jaume I, Castellón, Spain
e-mail: valbero@uji.es

Keywords: Steel-concrete composite beam; slim-floor beam; fire resistance; thermal behaviour.

Abstract. *This paper presents a novel slim-floor configuration which consist of an I-section steel profile welded to a wider bottom steel plate with an insulation layer of 5mm thick placed in between the two. The insulation layer is made up of a flexible and non-combustible material, which protects the I-section profile from the temperature rise through its bottom surface. Besides, this insulation layer remains protected from external agents with no need of maintenance because it is placed in the cavity created on purpose between the bottom plate and the lower flange of the steel profile thank to the weld between them. This new slim-floor configuration is currently right-protected by the Spanish Office of Patents and Trademarks with the reference number P201930438 and has been tested in the Universitat Politècnica de Valencia testing facilities to assess its thermal behaviour in fire. Specifically, the specimen was tested using an electric furnace where the slim-floor is placed at the top surface. Therefore, the composite beam is exposed to the heat source only from the bottom surface, as it is the case in real fire scenarios. Moreover, the experimental campaign includes also a common slim-floor specimen, without the insulation layer, in order to compare their thermal behaviour. The steel parts used in all the test specimens were grade S355, having a nominal yield strength of 355 MPa, the nominal compressive strength (cylindrical) of the cast in-situ concrete was 30 MPa and the precast concrete hollow core slabs were manufactured with 45 MPa nominal compressive strength (cylindrical). In turn, the embedded reinforcing bars were grade S500, with 500 MPa nominal yield strength. The first experimental results have shown that the temperature gap between the bottom steel plate and the lower flange of the steel profile can reach up to 200ºC. This impressive thermal gap delays the temperature rise along the I-section steel profile which produces a significant improvement on the composite section fire resistant time, up to 90 or 120 minutes. Additionally, it should be highlighted that the observed improvement on the cross-section thermal behaviour has been achieved using a novel configuration which no need maintenance in contrast with the common external protection systems using intumescent coating or other fireproofing materials.*

1 INTRODUCTION

The fire action is one of the most important hazards that may affect any structural element. Many researchers have been working on this topic during the second half of the twentieth century and significant advances have been done conductive to improve the fire safety in buildings.

Regarding this subject, it is worth mentioning that composite concrete-steel structural members have demonstrated a good behaviour against fire as compared with bare steel members, which are highly affected by the elevated temperatures produced during the fire event. For instance, composite steel-concrete columns like concrete-filled steel tubular columns have showed a very good fire behaviour due to the synergy of the steel tube and the concrete core working together [1-3]. Additionally, recent studies [4-12] have confirmed a promising behaviour at elevated temperatures of the slim-floor beam. This slim-floor beam is a type of composite beam made up by an I-section welded to a bottom steel plate, which was developed by the most important steel manufacturers as *ArcelorMittal*, *Peikko*, *Tata Steel*, etc.

Concerning the fire performance of this system, the slim-floor beam presents a very good behaviour during fire, since the steel beam is totally embedded into the concrete floor and therefore it results only exposed to fire from its lower surface. Owing to this configuration, the temperature within the steel members of the section increases slower than in other steel or composite concrete-steel beams, which are most frequently exposed to fire from the three sides. Additionally, recent research studies [4], [11] have proved that the Shallow Floor Beam (SFB) configuration, which is formed by a double steel sheet at the bottom part -lower flange of the steel profile and bottom plate- shows a very good thermal behaviour during fire. The interesting fire behaviour observed by the SFB configuration can be explained by the action of a thermal gap which appears at the contact interface between the bottom plate and the I-section lower flange due to its thermal bow, that was experimentally confirmed in previous research works [5], [11].

Usually, the thermal behaviour of steel beams in fire can be enhanced by using fireproof materials, that require an accurate application and an appropriate maintenance during their whole life cycle to ensure a proper fire protection. However, the described benefits of the thermal gap on the SFB beams suggested the authors to improve the thermal behaviour of this slim-floor beam by increasing this gap using any insulation material.

Related to this last idea, the main objective of this paper is to present a new slim-floor beam configuration designed to increase the insulation of this composite system against fire by using a new strategy that, among other benefits, does not needed for maintenance. This paper also presents the results of an experimental campaign carried out at the *Universitat Politècnica de Valencia* (Spain) testing facilities, focused on the thermal behavior of this new slim-floor configuration.

2 A NOVEL SLIM-FLOOR BEAM PROTECTION IN FIRE

2.1 Specimen description

The novel fire protection presented in this paper for SFB consists of an I-section steel profile welded to a wider bottom steel plate with an insulation layer of 5mm thick placed in between the two, see Figure 1. This insulation layer is made up of a flexible and non-combustible material, which protects the I-section profile from the temperature rise through its bottom surface.

Figure 1. Slim-Floor beam with the protection layer.

This insulation layer remains placed in a cavity created on purpose between the bottom plate and the lower flange of the steel profile thank to the weld between them. Therefore, the insulation material is totally protected from any aggressive ambient conditions and no maintenance is required. This insulation layer is made of a material which shows a density between 200-400 kg/m^3, a thermal conductivity lower than 0.08 W/m·K at 600ºC and a specific heat value around 1000 J/Kg·K. Besides, this material is flexible enough to follow any deflection of the bottom plate without showing any crack.

2.2 Thermal experiments

The experimental campaign presented in this section was carried out at the fire testing facilities of the *Universitat Politècnica de Valencia, Spain* and was focused on the thermal behaviour of the previously described fire protected slim-floor [13].

Specifically, three specimens were tested in order to assess the thermal behaviour of the new slim-floor configuration, see Table 1. The first specimen (S1) follows the common SFB configuration and was used as a reference experiment for comparison. This specimen was made up with an HEB200 steel profile welded to a 15x360 mm bottom plate. Hollow core slabs were used as the secondary element for the floor system and an in-situ concrete topping of 5 cm thick was added to achieve a total floor height of 250 mm. Moreover, two reinforcing bars of 20 mm diameter were embedded within the concrete mass with 30 mm cover measured from the top face of the lower steel flange and placed at a distance of 40 mm to the profile web.

Table 1: Experimental campaign [13].

Specimen	Description
S1	SFB: HEB200 + Plate 15
S2	IFP-SFB: HEB200 + Plate 15 + **Insulation Layer**
S3	IFP-SFB: HEB200 + Plate 15 (**Stainless steel**) + **Insulation Layer**

The second specimen (S2) was identical to the first one, with the only difference of adding the insulation layer described previously in-between the two steel plates generating the novel fire protection. Finally, the third specimen (S3) was equal to S2 but in this case with a bottom plate made up of austenitic stainless steel grade 1.4301 (AISI 340).

The steel parts used in all the test specimens were grade S355, having a nominal yield strength of 355 MPa, the nominal compressive strength (cylindrical) of the cast in-situ concrete was 30 MPa and the precast concrete hollow core slabs were manufactured with 45 MPa nominal compressive strength (cylindrical). In turn, the embedded reinforcing bars were grade S500, with 500 MPa nominal yield strength.

In turn, the thermal tests were carried out using an electric furnace made of four radiative panels of 3kW each manufactured with electric wires embedded in ceramic bricks. The radiative panels were placed creating a rectangular furnace with dimensions 1020 mm wide, 1800 mm long and 510 mm high. The configuration of panels permits that the specimens are placed at the open top surface, see Figure 2. Therefore, the slim-floor beam remains exposed to the heat source from its lower surface, as it is the case of real fire conditions.

Figure 2. Electric furnace configuration for slim-floor beams.

The power and configuration of the testing furnace allows temperatures up to 800ºC inside the chamber. In order to measure the temperature evolution along the slim-floor cross-section, four type-K thermocouples were placed within the specimen, see Figure 3. Specifically, thermocouple TC1 registers the temperature evolution at the bottom steel plate; TC2 measures the lower flange temperature; TC3 is placed at the web of the steel profile and TC4 measures the evolution of the reinforcing bar temperature.

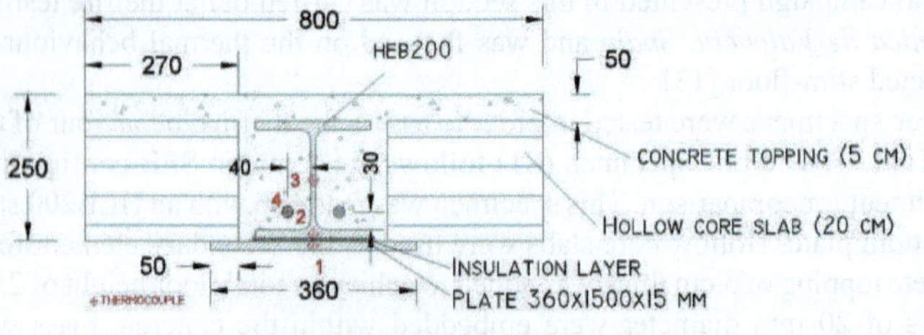

Figure 3. Thermocouple's layout.

The control device was set up to reach automatically a target temperature of 800ºC at the maximum allowed heating rate. The specimens were kept inside the furnace during 240 minutes. The temperature registered in thermocouples TC1 to TC4 is shown Figure 4.

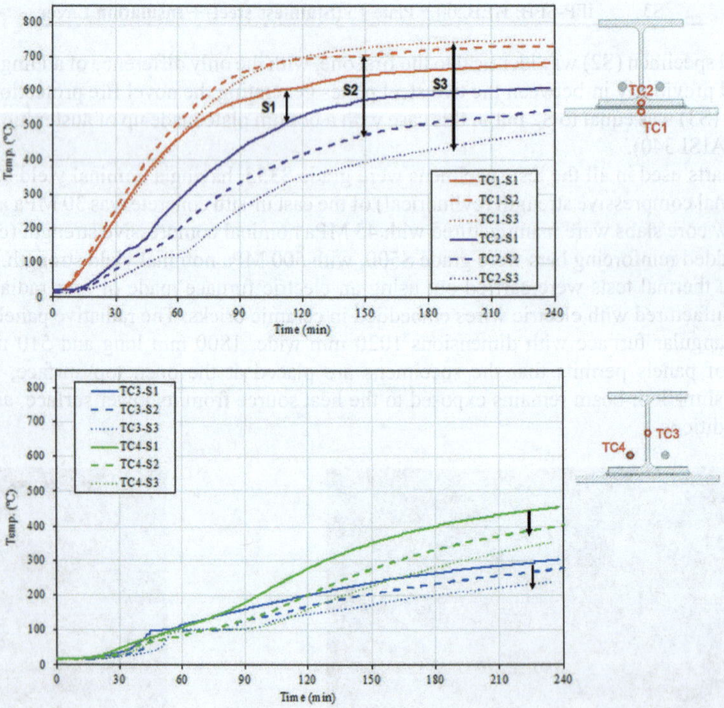

Figure 4. Thermocouple's results.

It can be observed that the difference between TC1-TC2, which shows the temperature gap between bottom plate and lower steel flange, increases from S1 to S2 and also from S2 to S3, see black arrows in

Figure 4. Additionally, it is worth reminding here that specimen S1 was configured as the common slim-floor SFB while specimen S2 is the novel fire protected slim-floor beam. Therefore, the effect of the insulation layer placed into the cavity between bottom plate and lower steel flange results enough to increase the temperature gap up to 200ºC. Specimen S3 which includes stainless steel at the bottom steel plate increases the temperature gap up to 250ºC. It can be concluded that the insulation layer creates a heating delay that permits a slower temperature increment of the lower steel flange.

About the temperature evolution of the steel profile web (TC3) and reinforcing bars (TC4), both show a slower increase of the temperature in case of S2 and S3 specimens, which are fire protected.

3 ASSESSMENT UNDER STANDARD CONDITIONS

3.1 Finite element thermal model

It is worthy to notice that the experimental tests described in the previous section were not developed under standard fire conditions, due to electric furnace limitations. Therefore, the subsequent necessary steps to confirm the enhanced fire behaviour of this novel slim-floor configuration consists of the development of a finite element (FE) thermal model in order to evaluate the fire performance of this new configuration when it is exposed to a standard ISO834 time-temperature curve, as well as the assessment of the mechanical response of the cross-section at elevated temperatures.

The main objective of the FE model consists of solving the temperature *(θ)* evolution along the slim-floor cross-section through the *Fourier's* heat transfer differential equation:

$$\nabla(\boldsymbol{k} \cdot \nabla\theta) = \rho c_p \frac{\partial\theta}{\partial t} \tag{1}$$

The thermal material properties implicit in this equation are: Thermal conductivity (**k**); specific heat (c_p) and material density (ρ). These material properties are temperature dependent and were obtained for concrete and steel in composite structures from EN 1994-1-2 [14]. In particular, for the concrete specific heat, a 4% of moisture content for the peak value was established and the upper conductivity level was assumed. In case of stainless steel, for specimen S3, the thermal material properties from EN1993-1-2 Annex C [15] were used.

The insulation layer used in specimens S2 and S3 was defined by a conductivity value of 0.08 W/m·K at temperatures lower than 600ºC and 0.17 W/m·K at temperatures higher than 1000ºC. Between 600 and 100ºC, a linear interpolation function is defined. The specific heat of this insulation material was defined with a value of 1000 J/kg·K and the density was 300 kg/m^3.

The slim-floor cross-section was meshed by using 2D four-noded quadrilateral elements. The maximum mesh size was 5 and 2.5 mm for the concrete and steel parts respectively, see Figure 5.

Regarding the heat convection and radiation, this heat transfer mechanisms were taken into account as boundary conditions in both the bottom 'exposed' and the upper 'unexposed' surfaces. These conditions are defined as follows:

$$-\boldsymbol{n} \cdot \boldsymbol{k} \cdot \nabla\theta = h\,(\theta - \theta_\infty) + F \cdot \varepsilon \cdot \sigma\,(\theta^4 - \theta_\infty^4) \tag{2}$$

In this equation, the emissivity -ε- value for the heat radiation was taken, following from EN1994-1-2 [14], as 0.7 for the concrete and carbon steel surfaces, while for the stainless steel surface (specimen S3) a value of 0.4 was used, following from EN 1993-1-2 Clause 2.2 [15]. In turn, the convective coefficient -h- was defined as 25 W/m^{2}K at the exposed surface and 4 W/m^{2}K at the unexposed one, following from EN 1991-1-2 [16].

Additionally, the contact properties at the interface between different materials should be defined in the thermal FE model. Specifically, according to previous works from the authors [4], [11] a thermal conductance value of 250 W/m^{2}K was assumed for the concrete-steel interaction, while a value of 100 W/m^{2}K was established for the steel-steel contact.

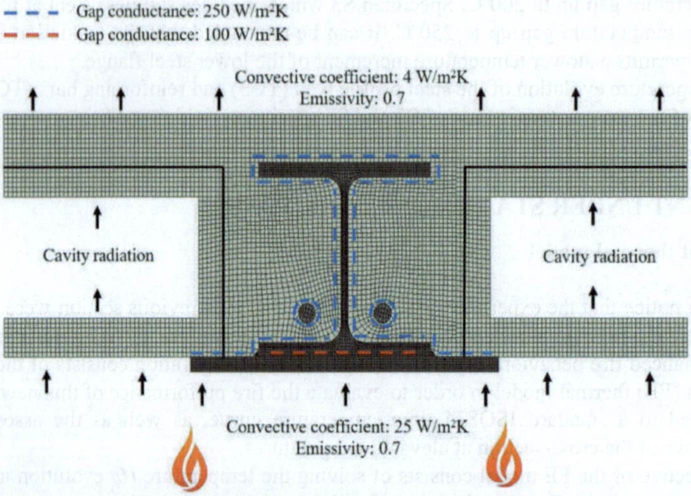

Figure 5. FE model. Boundary conditions.

The FE model previously described was validated against the experiments carried out (section 2) obtaining a good agreement. Figure 6, shows the FE validation on the specimen S2.

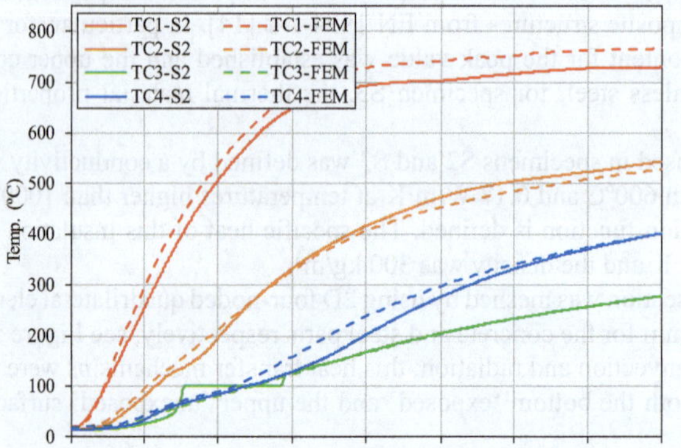

Figure 6. FE validation. Specimen S2.

3.1 Mechanical assessment

Once the thermal FE model was validated against the previously described experiments, the three configurations listed in Table 1 were computed numerically under a standard ISO-834 time-temperature curve.

Next, the bending moment plastic resistance at elevated temperatures ($M_{pl,Rd,\theta}$) was obtained by an iterative incremental analysis of the equilibrium equation:

$$\begin{bmatrix} \delta N_{Rd,\theta} \\ \delta M_{pl,Rd,\theta} \end{bmatrix} = \begin{bmatrix} E_{11} & E_{12} \\ E_{21} & E_{22} \end{bmatrix} \cdot \begin{bmatrix} \delta \varepsilon_0 \\ \delta \chi_y \end{bmatrix} \tag{3}$$

In this equilibrium equation, the non-linear behaviour of the materials is taken into account through (E_θ) and the specimens are considered as axially unrestrained. The external net axial force is fixed as $\delta N_{Rd,\theta} = 0$.

The secant stiffness matrix for each increment can be obtained by dividing the slim-floor beam cross-section into cells matching with the FE elements, where the temperature evolution (θ_i) was obtained from the previous thermal analysis:

$$E_{11} = \sum E_{\theta,i} \, A_i; \; E_{12} = E_{21} = \sum E_{\theta,i} \, y \, A_i; \; E_{22} = \sum E_{\theta,i} \, y_i^2 \, A_i \tag{4}$$

Additionally, the thermal strains (ε_{th}) were considered by substracting them from the total strain before solving the equilibrium equation.

The plastic bending moment resistance of specimens S1, S2 and S3 under different standard ISO-834 times is displayed in Figure 7.

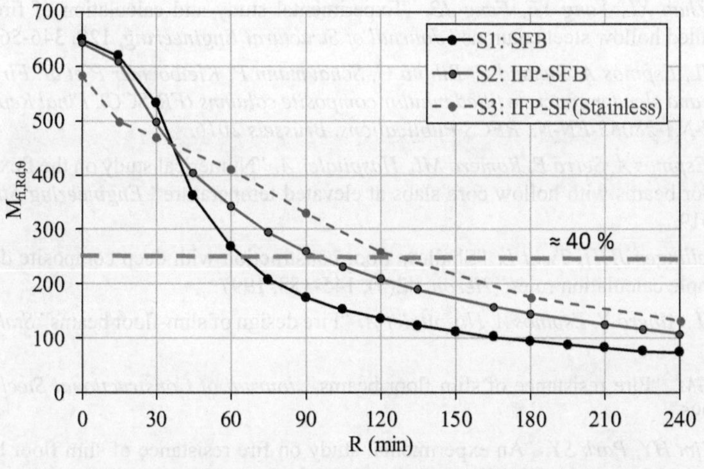

Figure 7. Plastic bending moment resistance under ISO-834 standard exposure.

It is observed that, in general, under the same bending moment applied, S2 and S3 configurations show higher fire resistance times. Specifically, under approximately a 40% of the bending moment resistance at room temperature, the SFB configuration shows 60 minutes of fire resistance, while the novel fire protected slim-floor (S2) increases this fire endurance up to 90 minutes and even up to 120 minutes with bottom plate made of stainless steel (S3). Therefore, under a 40% of load level - which is a frequent value in fire design - the new type of beam proposed in this work increases 30 minutes the fire resistance time and even 60 additional minutes when the bottom plate is made of stainless steel.

4 CONCLUSIONS

This paper presented a novel fire protected slim-floor beam configuration which consist of a modified SFB with an insulation layer placed between the bottom steel plate and the lower flange. This new configuration produces a heating delay of the beam profile in fire. Specifically, this delay generates a temperature gap up to 200-250ºC between the bottom steel plate and the beam lower flange.

In turn, under standard fire conditions, the new SFB configuration has shown an increase of 30 minutes of fire resistance (from 60 to 90) at 40% load level as compared with the common SFB type. Additionally, the use of stainless steel at the bottom plate can enhance even more this beneficial fire behaviour, increasing the fire resistance of the system up to 120 minutes at the referred 40% load level.

In conclusion, this novel fire protection may become a promising strategy for manufactures in order to increase the fire resistance of slim-floor beams without the required maintenance of traditional fire protection using fireproof materials.

This new slim-floor configuration is currently right-protected by the Spanish Office of Patents and Trademarks with the reference number P201930438.

REFERENCES

[1] *Kodur VKR, Mackinnon DH.* "Design of concrete-filled hollow structural steel columns for fire endurance". *Engineering Journal,* 37, 13-24, 2000.

[2] *Han LH, Zhao XL, Yang YF, Feng JB.* "Experimental study and calculation of fire resistance of concrete-filled hollow steel columns" *Journal of Structural Engineering,* 129, 346-56, 2003.

[3] *Romero ML, Espinós A, Renaud C, Bihina G, Schaumann P, Kleiboemer P et al. Fire resistance of innovative and slender concrete filled tubular composite columns (FRISCC). Final Report, Catalogue number KI-NA-28082-EN-N. RFCS Publications. Brussels 2016.*

[4] *Albero V, Espinós A, Serra E, Romero ML, Hospitaler A.* "Numerical study on the flexural behaviour of slim-floor beams with hollow core slabs at elevated temperature" *Engineering Structures,* 180, 561-73, 2019.

[5] *Both C, Fellinger JHH, Twilt L.* "Shallow floor construction with deep composite deck: From fire tests to simple calculation rules" *Heron,* 42(3), 145-158, 1997.

[6] *Romero ML, Albero V, Espinós A, Hospitaler A.* "Fire design of slim-floor beams" *Stahlbau,* 88, 665-74, 2019.

[7] *Newman GM.* "Fire resistance of slim floor beams" *Journal of Constructional Steel Research,* 33, 87-100, 1995.

[8] *Kim HJ, Kim HY, Park SY.* "An experimental study on fire resistance of slim floor beam" *Applied Mechanics and Materials,* 752-757, 2011.

[9] *Ma A, Mäkeläinen P.* "Structural behaviour of composite slim floor frames in fire conditions" *Journal of Constructional Steel Research,* 62(12), 1282-1289, 2006.

[10] *Zaharia R, Franssen JM.* "Simple equations for the calculation of the temperature within the cross-section of slim floor beams under ISO Fire" *Steel and Composite Structures,* 13, 171-85, 2012.

[11] *Albero V, Serra E, Espinós A, Romero ML, Hospitaler A.* "Innovative solutions for enhancing the fire resistance of slim-floor beams: Thermal experiments" *Journal of Constructional Steel Research,* 165, 105897, 2020.

[12] *Hanus F, Zaganelli D, Cajot L-G, Braun M.* "Analytical methods for the prediction of fire resistance of "reinforced" slim floor beams" *In: Sohn E, editor. EuroSteel 2017. Copenhagen, Denmark 2017.*

[13] *Albero V, Serra E, Espinós A, Romero ML, Hospitaler A.* "Internally fire protected composite steel-concrete slim-floor beam" *Engineering Structures,* 227, 111447, 2021.

[14] *CEN, EN 1994-1-2, Eurocode 4: Design of composite steel and concrete structures. Part 1-2: General rules - Structural fire design., Comité Européen de Normalisation, Brussels, Belgium, 2005.*

[15] *CEN, EN 1993-1-2, Eurocode 3: Design steel structures. Part 1-2: General rules - Structural fire design., Comité Européen de Normalisation, Brussels, Belgium, 2005.*

[16] *CEN, EN 1991-1-2, Eurocode 1: Actions on structures. Part 1-2. General actions - actions on structures exposed to fire, Comité Européen de Normalisation, Brussels, Belgium, 2002.*

9TH INTERNATIONAL CONFERENCE ON COMPOSITE CONSTRUCTION IN STEEL AND CONCRETE
M. Knobloch, U. Kuhlmann, W. Kurz, M. Schäfer. (Eds.)
Online, Germany, July 27-29, 2021

ADVANCED NUMERICAL FIRE DESIGN OF INDUSTRIAL COMPOSITE SLABS WITH UNPROTECTED STEEL BEAMS – A CASE STUDY

Pascal Lequime

Gruner AG, Basel
pascal.lequime@gruner.ch

Keywords: Composite slab, structural steel, structural fire design, numerical analysis.

Abstract. *In this paper, the development and the application of a framework for advanced numerical thermal and mechanical analysis of a composite slab with unprotected steel beams which is subjected to fire is described. A short overview over existing calculation methods is given to describe the state of the art in fire design of composite slabs. For the numerical model in the software framework SOFiSTiK, the main advantages, features and characteristics are described. To analyse the reliability of the model, the process of validation with existing fire tests is presented. Finally, the model is applied in an engineering project by means of a case study.*

1 INTRODUCTION

In engineering practice, the importance and spread of numerical methods are generally increasing. The application of finite element methods for the design of structures corresponds to the current state of the art and is anchored in practice. However, for the design of structures in case of fire, simplified design methods are still mostly used in practice. The use of advanced finite element design methods for fire design is already established in research projects, but not yet in daily engineering practice.

In this paper, the computer-oriented structural fire design using the finite element method is applied to the example of a reinforced concrete composite floor with unprotected steel girders. The investigations have several objectives which are described as follows.

By applying an advanced design procedure for the structural design in case of fire, the load-bearing reserves existing in the structure can be utilised, which remain partly unused by applying only simplified design methods. If possible, the design should make it possible to dispense with fire protection coatings on the steel girders. Dispensing with fire protection measures has a direct influence on the economic efficiency of the steel structures, represents a significant simplification of the approval and construction processes, and contributes to protection of the environment and the sustainable use of resources. With these factors, the attractivity of steel as a building material increases compared to reinforced concrete.

The establishment and acceptance of advanced design methods in engineering practice should be promoted. In the sense of a practical applicability of the design method, the software framework SOFiSTiK, which is strongly anchored in engineering practice and yet powerful, is used for the finite element analyses. However, software-specific limitations and simplifications must be accepted.

A reinforced concrete composite floor with steel girders is used as a practical example. With the aim of dispensing with fire protection coatings, the focus is on a fire resistance of the supporting structure of REI30, as it is set in Switzerland, for example, in connection with a sprinkler system with an overall requirement of REI60. For the fire effects, a limitation is made to the application of the standard temperature-time curve as given in EN 1991-1-2 [1].

To achieve the formulated objectives, the current state of the art regarding available analytical and numerical calculation methods and their application is shortly presented and potential improvements for implementation in an advanced calculation model are identified from this. The development of an own completely parametric calculation model and its validation with selected fire tests is described more in detail. The developed advanced model is finally used and tested in the context of a real engineering project of an industrial composite slab with steel beams.

2 EXISTING FIRE DESIGN CALCULATION METHODS FOR SLABS

For composite or fully concrete slabs, several analytical methods for calculation in case of fire are disponible.

The analytical method according to Bailey [2] concentrates on the verification in the ultimate limit state. The method is based on a mechanism of linear plastic yield lines in the slab with rigid parts and the enhancement of the resistance by a membrane effect in the slab in case of elevated temperatures and high deflections. It is however not designed to predict deflections in case of fire. Furthermore, it is only permissible for unprotected edge beams up to a fire resistance of 30 minutes and is therefore no longer universally applicable for higher fire resistance durations of the edge beams. The method is strongly anchored and established in practice - it is implemented in the software MACS+ of ArcelorMittal [3] and in tabled design values [4] for chosen cases. Furthermore, it will be presented in an annex of the new generation of EN 1994-1-2 [5]. The calculation is described completely by analytical formulae and can be carried out completely by hand with reasonable effort. The contribution of the steel beams within the slab can be included. These and the edge beams are to be designed with the help of analytical formulae.

The analytical calculation method according to Zhang and Li [6] is an enhancement of the Bailey method. It is assumed that for small deformations the load bearing system of the slab is functioning via bending. With an increasing deflection the membrane effect starts, yield lines appear in the same pattern as assumed by the Bailey method. In the ultimate limit state, however, the two approaches differ; Zhang and Li propose an approach with an elliptical field in the middle of the slab (as shown in Figure 1). It is implicated that the calculation method is suitable for predicting deflections in addition to design.

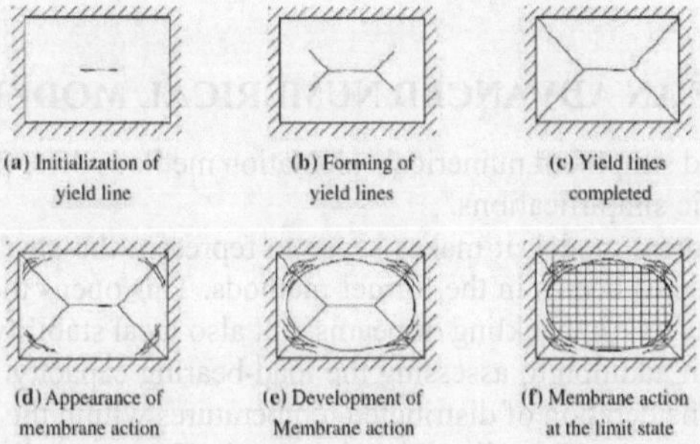

Figure 1. Development of membrane action in a concrete slab [6].

Although the calculation is done using analytical formulae, solving the equations by hand is not convenient. An iterative procedure with the help of Excel spreadsheets or a computer aided calculation is necessary. The method has not yet been implemented in a known software. The contribution of the steel girders within the bay is not considered, the edge beams must be designed separately.

The Burgess method (as described in [7] and developed in [8]) is a very recent analytical calculation method. It is also a development of the Bailey method. It uses a kinematically consistent mechanism for the membrane action. A main aspect of the method is the fact that the orientation of the yield lines is adjusted on the slab geometry and its reinforcement ratio in both directions. The yield lines are not limited to the arrangement parallel to the longer slab side as it is in the former methods, they appear already at small deformations. In contrast to the methods by Bailey or Zhang and Li, not the entire cross section of the slab is assumed to be stressed only in tension or compression, but numerous different stress states are considered. An important input variable for the method with a considerable impact on the result is the crack width of the concrete when the load-bearing capacity of the reinforcement is reached. This critical crack width can practically only be determined by tests. Although the calculation is carried out completely with the help of analytical formulae, it is practically impossible to manage by hand due to a mandatory iterative procedure and numerous cross dependencies of the calculation variables among each other. With the help of an Excel spreadsheet, this can be solved, but still with a considerable effort. It would be favourable to implement this in a calculation software, but this has not yet been done. The contribution of the steel girders within the slab is considered in the method via their contributing tensile force.

A simplified numerical calculation with the software framework SOFiSTiK is part of the work done by Stadler [9]. In this simplified approach, the temperature dependent parameters of the material are simulated by equivalent temperature loads in form of strains. The calculation itself is done with a linear elastic material law. The finite element model considers the steel girders as beam elements. In contrast to the Bailey method, the Stadler method is also suitable for calculating deflections. In the calculation according to Stadler, the unprotected internal beams are completely neglected due to their small contribution to the load-bearing effect of the slab and are therefore not represented in the model. Although no nonlinear material laws are used in the calculation, they are indirectly considered in the determination of the thermal equivalent loads. In a recent development of the method [10], the reinforced concrete slab is calculated materially nonlinear, so that no thermal equivalent loads have to be determined by hand. Despite finite element modelling, the verification of the steel girders is carried out with the help of analytical formulae at the level of the internal forces, which are not implemented directly in the finite element calculation. So, there are still a few analytical calculations to be done for practical application of the method.

In all these methods the slab is considered as a slab and the steel girders as beams. By considering the girders as beams, they are assessed at the level of internal forces and their load-bearing capacity in case of fire is described with moments and forces. All methods have in common that constrained internal forces from prevented thermal expansion and their influence on steel profiles and connections are not taken into consideration. In none of the presented methods the connections are explicitly investigated in thermal or mechanical terms.

3 DEVELOPMENT OF AN ADVANCED NUMERICAL MODEL

The existing analytical and simplified numerical calculation methods offer potential for extensions. All are provided with their specific simplifications.

In an advanced finite element model, it makes sense to represent the steel girders by means of shell elements, which are considered as beams in the former methods. This opens the advantage of considering not only global stability effects as e.g. buckling of beams, but also local stability effects as e.g. buckling of plates directly in the model, in addition to assessing the load-bearing capacity. The shell elements for the steel girders also allow the consideration of distributed temperatures within the cross sections which is not always possible with beam elements. This allows local thermal effects as well as the thermal strains to be considered more realistically.

For the development of an advanced numerical model, the software framework SOFiSTiK is chosen. This software is an established high-performance engineering tool. One of the main characteristics of the model is the fact that the steel girders are not modelled by beam elements, but by shell elements. The fact that the complete structure is modelled by shell elements leads to the obligation to analyse and evaluate stresses and strains in the steelwork and not forces and moments as it is done in classical approaches. This manner of modelling combined with a geometrically and materially nonlinear calculation considers local and global instability effects. The thermal analysis of the steel and concrete sections is directly linked to the main mechanical model of the structure. There is a full implementation of nonlinear material properties for the structural steel and the reinforced concrete including plastic analysis as a standard in the SOFiSTiK software (described in [11] and [12]). This implicates the consideration of plastic redistribution of stresses in the steelwork and of cracking in the concrete. The temperature dependent stress-strain properties of the materials are also included in the mechanical analysis (example of structural steel in Figure 2).

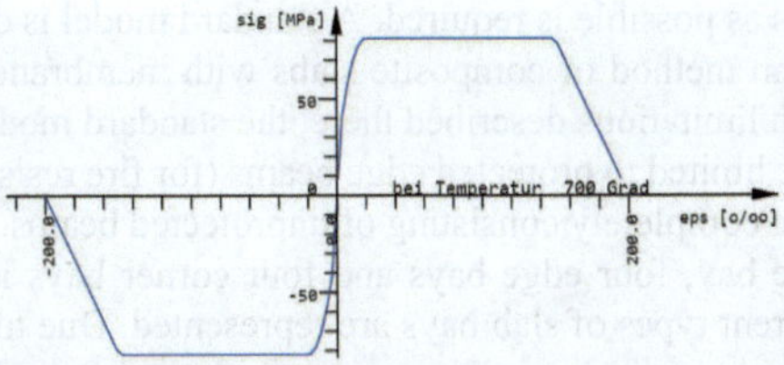

Figure 2. Temperature dependent material law for structural steel in SOFiSTiK (example S355 at 700°C).

Steel connections are modelled by shell elements for the plates and by nonlinear springs with temperature dependent material laws for the bolts. As specific material laws for bolts are not available in EN 1993-1-2 [13], an approach based on the component method of EN 1993-1-8 [14] combined with the temperature dependant factors of EN 1993-1-2 is chosen (Figure 3). For the shear deformation capacity, the results of experimental investigations for bolts submitted to shear forces is used [15].

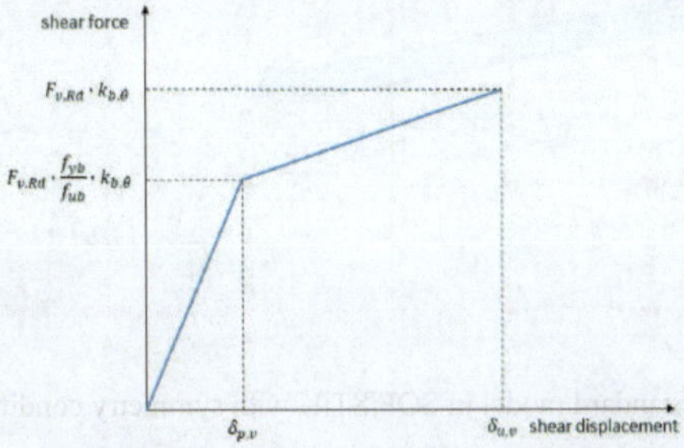

Figure 3. Implemented material law for bolts (example shear force).

In particular, the step of switching from a beam formulation to the use of shell elements in the modelling of steel beams results in a significant increase in complexity and effort in system generation. In the commercial application of such models, the effort for the manual generation of the geometric model plays a central role. For economic use, it is therefore expedient to parameterise and automate the modelling as far as possible (see an example of parametrical input in Figure 4).

```
$
$ Allgemeine Optionen
$
STO#verbund      'LXPMX'  $ voller Verbund: 'LPPMX', ohne Verbund: 'LXPMX'
STO#bolts        4        $ 0: Modellierung ohne Schrauben
$                            1: Modellierung mit Schrauben und konstantem Bettungswert
$                            2: Modellierung mit Schrauben und linearer Arbeitslinie für Zug und Druck
$                            3: Modellierung mit Schrauben und linearer Arbeitslinie nur für Zug
$                            4: Modellierung mit Schrauben und plastischen temperaturabhängigen Arbeitslinien mit Ansatz nach Henriques
$                            5: Modellierung mit Schrauben und plastischen temperaturabhängigen Arbeitslinien mit Ansatz nach Wald
$
$ Lasten
$
STO#lastg1       1        $ Ständige Deckenlast Mittelfeld
STO#lastg2       1        $ Ständige Deckenlast Randfeld oben
STO#lastg3       1        $ Ständige Deckenlast Randfeld links
STO#lastg4       1        $ Ständige Deckenlast Eckfeld
STO#lastq1       20       $ Veränderliche Deckenlast Mittelfeld
STO#lastq2       20       $ Veränderliche Randfeld oben
STO#lastq3       20       $ Veränderliche Randfeld links
STO#lastq4       20       $ Veränderliche Eckfeld
STO#PSINUTZ      0.50     $ PSI2-Faktor für den Ansatz der Nutzlast im Brandfall
$
$ Struktur
$
STO#feldxm       10.000   $ Spannweite Nebenträger Mittelfeld
STO#feldxr       10.000   $ Spannweite Nebenträger Randfeld
STO#feldym       8.750    $ Spannweite Hauptträger Mittelfeld
STO#feldyr       8.750    $ Spannweite Hauptträger Randfeld
STO#nsecm        2        $ Anzahl Nebenträger Mittelfeld
STO#nsecr        2        $ Anzahl Nebenträger Randfeld
$
```

Figure 4. Full parametrical pre-processing of the SOFiSTiK model (short excerpt of the input file).

This eliminates a great part of modelling effort as the time requirement is almost limited to the pure calculation time of the model as well as its post-processing for evaluation and assessment. An approach that can cover as many use cases as possible is required. A standard model is developed, which is based on the model used in the calculation method of composite slabs with membrane bearing effect described in [4]. In contrast to the model with limitations described there, the standard model must be as generally valid as possible. Its use should not be limited to protected edge beams (for fire resistance above R30) but should allow the calculation of a system completely consisting of unprotected beams. A slab model with a total of 9 bays, which includes a centre bay, four edge bays and four corner bays is chosen. Thus, all possible continuity effects between different types of slab bays are represented. Due to the complexity of the finite element model, the calculation times are no longer in a range of a few minutes, as is the case in a formulation with steel girders as beam elements with linear calculation. To minimise the numerical effort, a model reduced to a quarter of the total system is used. The system is divided along two symmetry axes, and symmetry boundary conditions are used at the cut lines. The transition from the whole system to the subsystem using symmetry conditions is shown in Figure 5.

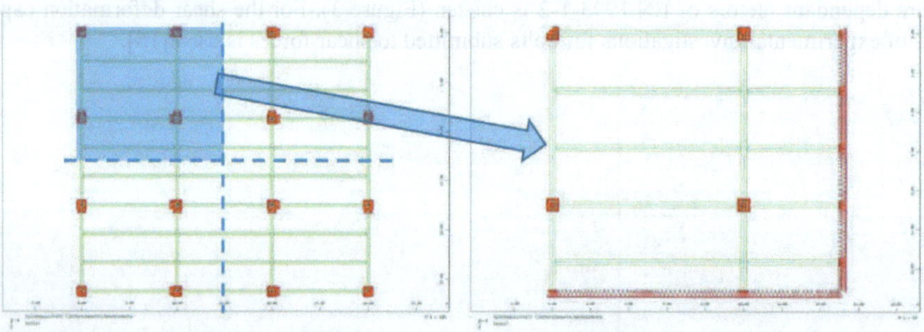

Figure 5. Standard model in SOFiSTiK with symmetry conditions.

The calculation time of such a reduced system for a selected example with a fire duration of 30 minutes with a time step size of 5 minutes on a recent computer system (in the year 2020) is approximately 45 minutes in total.

4 VALIDATION OF THE MODEL

To verify the finite element model and its assumptions, three existing fire tests are analysed with the finite element model. For the FRACOF test [3], which is part of a European research campaign, and two fire tests performed at TU Munich [9], the calculated results of the mechanical analyses are compared to the test results.

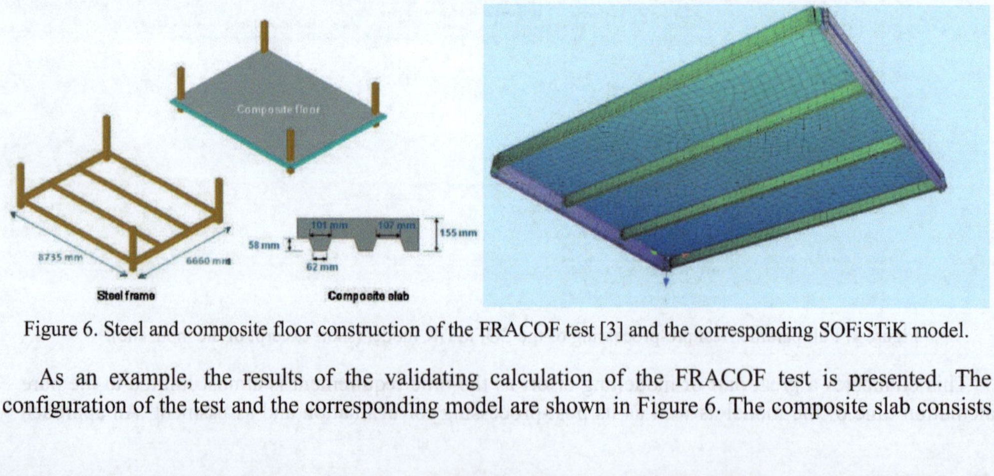

Figure 6. Steel and composite floor construction of the FRACOF test [3] and the corresponding SOFiSTiK model.

As an example, the results of the validating calculation of the FRACOF test is presented. The configuration of the test and the corresponding model are shown in Figure 6. The composite slab consists of a COFRAPLUS60 sheet with concrete cast in site. The slab is supported by fire protected edge steel beams and two intermediate secondary beams within the slab which remain unprotected. The composite slab is submitted to a fire with a duration of 120 minutes. The fire almost follows the standard temperature-time curve and achieves almost about 1100°C at the end of the period.

The load-bearing effect of the system or its change over the fire duration can be read very well via the main stresses in the concrete, especially via the compressive stresses on the upper side of the slab. In the case of a classic flexural load-bearing effect of a slab, there are compressive stresses in the centre of the field at the top. The concrete stresses at the top of the panel at 120 minutes at the end of the fire duration are shown in Figure 7. The compressive stresses are marked in red and the tensile stresses in blue. Cracked areas due to exceeding the tensile strength of the concrete are not coloured.

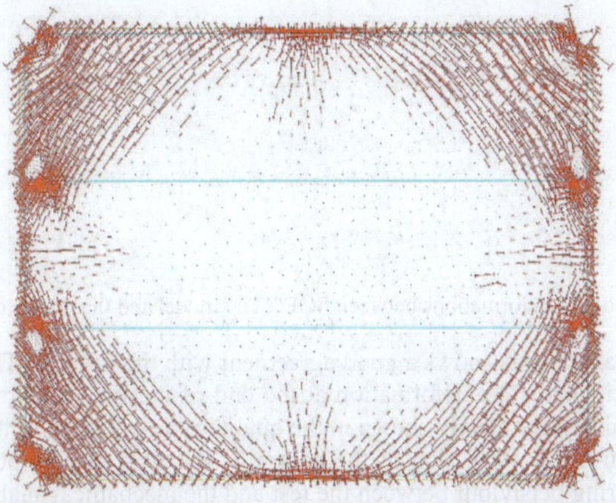

Figure 7. Membrane stresses in the SOFiSTiK model of the FRACOF test.

A compression ring with compressive principal stresses is clearly visible around the slab. In the centre of the slab there are no more stresses. Here the concrete is cracked at the top and no longer absorbs compressive stresses. This would not be possible with a flexural load-bearing effect. Based on the stress state shown, it can be deduced that the slab functions primarily via membrane action with tensile forces in the reinforcement.

The equivalent stresses in the structural steel also change significantly over the duration of the fire. At the beginning the stresses remain below 200 N/mm² and thus in the linear-elastic range. During the fire duration, the stresses clearly redistribute. The unprotected secondary beams in the centre of the field practically no longer absorb any stresses as they reach temperatures of more than 1000°C and thus only have about 2% to 4% of their initial strength. Due to their fire protection, the edge beams remain in temperature ranges below 300°C and thus practically do not change their strength properties. The strains reach local peak values of approx. 46% at the time of 120 minutes. The peaks occur in the connections of the secondary beams. Here, individual finite elements can be considered as stretched beyond the ultimate strain and thus as having failed. Nevertheless, a failure of the general structure does not occur. The unprotected secondary beams practically no longer contribute to the load bearing. The strains in the structure otherwise remain below 15%.

Figure 8 shows the vertical deformations and the corresponding deformation velocities in comparison to the limit values $L^2/400d$ for the deformation and $L^2/9000d$ for the deformation velocity. These limit values are used during standard fire tests on single steel girders [16]. In the test and the mechanical analysis, the deformation limit value are slightly exceeded. The limit value for the deformation velocity only takes effect from minute 10, from then on, the velocities remain clearly below it. High velocities occur up to the 10th minute. The limit values for deformation are slightly exceeded, nevertheless no failure can be observed in the test.

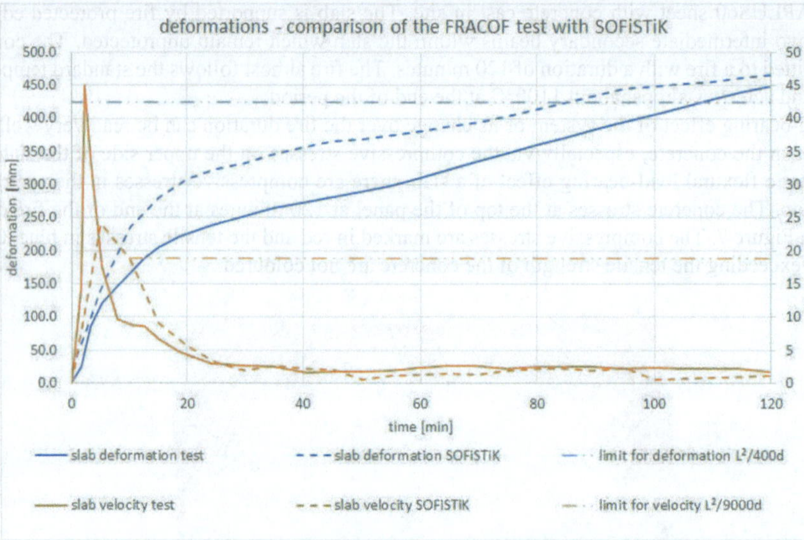

Figure 8. Comparison of the deformations between SOFiSTiK model and the results of the FRACOF test.

Overall, the mechanical analysis shows a good agreement with the fire test. The vertical deformations are relatively well approximated. The deformation at the end of the fire test agrees almost exactly. The membrane effect of the slab is very clearly shown in the qualitative evaluation of the stresses and the crack widths. In the test, a clear crack formation in the centre of the field can be observed. The qualitative agreement of the load-bearing behaviour between the test and the mechanical analysis is given.

The membrane effect can be observed clearly in the mechanical analyses of all three fire tests basing on the stress distribution in the slab. A good qualitative agreement of the load-bearing behaviour of models and tests can be observed. The characteristic compression ring running around the slab as well as the change from compressive stresses in the centre on the upper side of the slab to cracked areas without stresses can be recognized in all three models.

For the unprotected steel girders in the centre of the field, it can be observed in each case that practically no more stresses can be absorbed there due to the high temperatures and that these are no longer significantly carried. The slab compensates for this by developing its membrane action. Primarily the areas around the connections are submitted to the maximum strains in the structural steel. These are also heavily stressed due to the thermal constraints. The shell formulation makes it quite simple to check and assess locally overstretched areas. This would not be possible with a beam formulation for the steel girders.

In the FRACOF test, a good agreement between calculation and test is achieved with an overestimation of the deformations by 4%. In the 2nd test at the TU Munich, the deflections are overestimated by about 30%. To be able to evaluate the influence of the composite sheet not used in the calculation, the first test at the TU Munich without composite sheet is also calculated. The agreement there is better with an overestimation of the deformations by 14% but does not come as near as it is in the FRACOF test. It seems that the Holorib composite sheet with the undercut profile contributes more to the load bearing of the slab at TU Munich than it is with the open profile of the FRACOF slab. Another aspect of the Munich tests is that the system extends over 2 bays, in contrast to the single-bay system of the FRACOF test. This makes the load-bearing system significantly more complex due to the continuity of the slab and the cracking in the reinforced concrete above the dividing protected beam acting as a support. This also explains the only satisfactory agreement of the results in the 1st fire test without composite sheet.

In all three fire tests and associated calculations, the recommended limit values for deflections and deflection velocities are partially exceeded. In each case, no failure of the load-bearing structure or an announcement of this can be seen in the test and mechanical analysis. This underlines the thesis that the limit values resulting of standard fire tests on single steel girders are to be understood only as guideline values and that the structural safety must be assessed in each individual case.

5 APPLICATION OF THE MODEL TO A CASE STUDY

The behaviour of a composite slab structure in case of fire is to be assessed for a building structure for production in the pharmaceutical industry. For this purpose, the parametrical finite element model in SOFiSTiK is to be applied.

The aim of the design in the case study is to ensure the fire resistance without the use of a fire protection coating on the steel girders in the form of a sprayed fire protection or an intumescent fire protection coating.

Assuming a fire load density of maximum 1000 MJ/m² according to the fire protection concept of the building, the structural system must have a fire resistance of R30 to be in-line with the requests by the local fire authorities. The building will be equipped with a sprinkler system, which enables the R30 requirement - without a sprinkler system, the load-bearing system would have to be designed for R60. At the client's request, the design in case of fire is to be carried out using the standard temperature-time curve to be totally independent of the usage of the building and to ensure absolute flexibility in use.

The production building is being constructed as a reinforced concrete composite structure. A typical storey of the building (122.5 m x 85 m) is shown in Figure 9. The slab structure consists of main girders HEA900 and secondary girders HEA700 in structural steel S355. Above the steel structure there is a reinforced concrete composite floor with Holorib sheet SHR51 with t = 1.25 mm and in-situ concrete C30/37 with 20 cm in total. The building is braced by reinforced concrete walls arranged in core zones. The vertical load transfer takes place via walls and reinforced concrete composite columns. Due to the insufficient load-bearing capacity of the ground for a flat foundation, the building is founded with piles under the core zones and under the columns.

Figure 9. Overview of a standard storey with composite slab.

The construction method with steel girders and composite slabs is explicitly requested by the client to ensure a high degree of flexibility. It should be possible to extend or convert the production areas located in the north and south wings of the building with the subsequent installation of reinforced concrete composite slabs or grating levels. As far as possible, any arrangement of openings in the slabs should be possible. For this purpose, the steel girder structure must be adaptable. In this context of ensuring a high flexibility, a design without a bond between steel girders and composite slab is also requested by the client.

According to the client's specifications, the total height of the initially designed slab structure should not be noticeably changed by the design in case of fire. This means that the slab thickness should remain at 20 cm and the beam heights of approx. 70 cm and approx. 90 cm should not be increased. In fact, only the modification of the HEA profiles to other series such as HEB or HEM or the use of welded profiles with a similar construction height are conceivable alternatives.

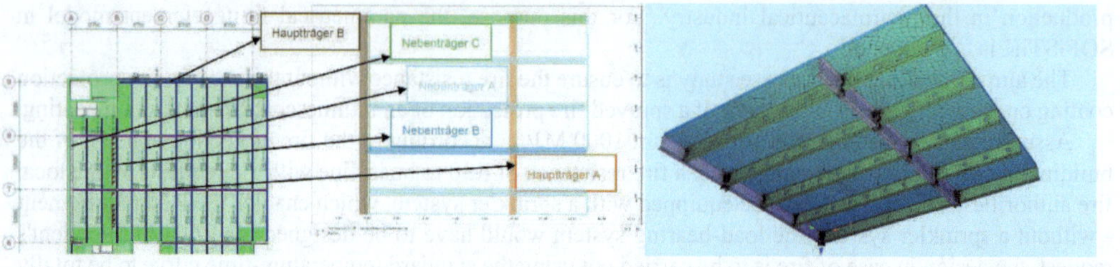

Figure 10. Assignment of the beam types and corresponding numerical model of the composite slab in SOFiSTiK.

For the application of the parametrical finite element model for the present project, a corner area of the standard storey is chosen (Figure 10). This allows the consideration of all existing bay types: inner bay, edge bay with main girder as termination and edge bay with secondary girder as termination. In the project, there is a reinforced concrete core in each corner bay. Therefore, although this bay is present in the calculation model, it does not exist in this form in the real structure. The resulting finite element model shown in Figure 10 is therefore not 100% congruent with reality, but it represents all relevant components in a sufficiently accurate way.

The most representative results of the calculation are the deflections of the slab and the strains in the structural steel, both are presented in detail as follows.

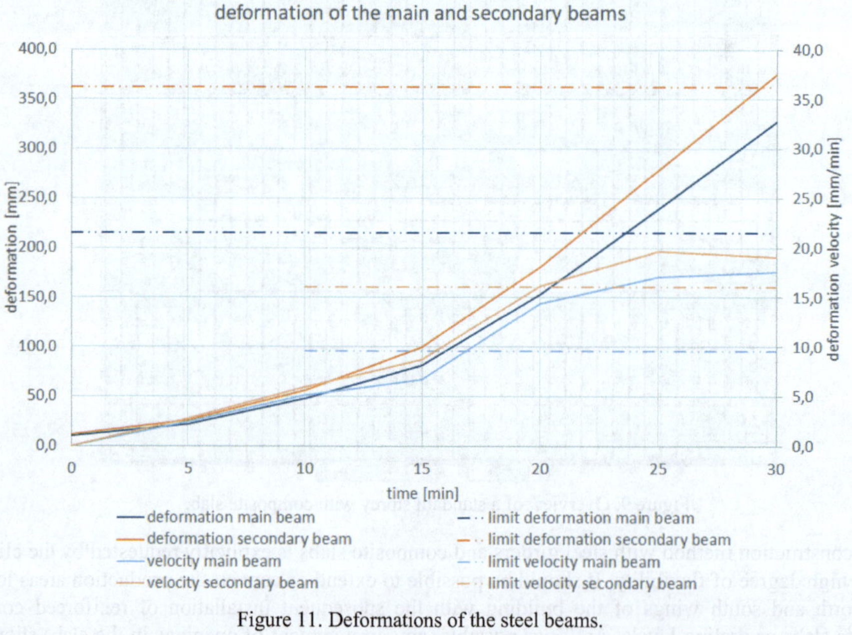

Figure 11. Deformations of the steel beams.

The deflections of primary and secondary steel beams are shown in Figure 11. For the secondary girder, the limit values for deflection and deflection velocity are slightly exceeded. For the main girder, the exceedances are more pronounced. On the one hand, this clear exceeding results from the fact that the span of the main girder with 8.75 m is shorter than that of the secondary girder with 10 m - and thus the limit value $L^2/400d$ is smaller in comparison. On the other hand, the limit value also includes the girder height, which in the case of the main girder leads to a further reduction of the limit value. With 327 mm and 375 mm, the deflections of the two beams are of a similar order of magnitude, but the limit values differ considerably. In this context, the question arises to what extent these limit values resulting of standard tests on single girders are meaningful for the practical application in a model consisting of several parts in a comprehensive structural system. They do not represent an actual serviceability or load-bearing capacity criterion. In the present case, as in the fire tests used for the validation of the model, no loss of load-bearing safety of the static system is announced despite the limit values for deflections being exceeded. With increasing fire duration, the deflection velocity does increase up to the 20th minute, but from this point onwards it only increases moderately or even decreases. This and the observations from the real fire tests lead to the evaluation that the structural safety is given from the point of view of deflection behaviour despite of the theoretically exceeded limit values.

The maximum strains in the structural steel occur for all steel girders around the connecting end plates. In the secondary beams the strains have a maximum of 12.4%, in the main beams 19.3%. All are below the limit value of 20%. As an example, the stress-strain behaviour of the main girder HEA900 as shown in Figure 12 is examined more in detail. The maximum strain of 19.3% occurs in this main girder type. The maximum strains occur in the web around the connection to the column.

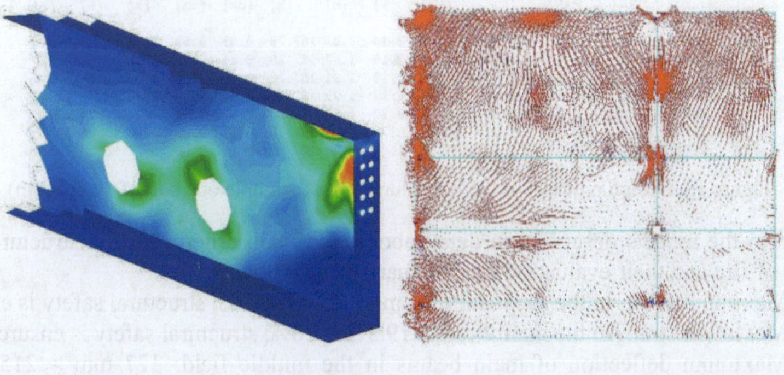

Figure 12. Strains in the main steel beams and membrane forces in the concrete slab (upper side) at end of fire.

In the horizontal direction of the main beam, both compressive and tensile stresses (and strains) occur in the beam web. Tension is evident in the upper connection area, compression in the lower connection area - corresponding to the beam restraint around the bolted end plate connection. In the vertical direction, there is a concentrated compressed zone in the upper web area at the girder end. Due to the already strongly reduced bending stiffness of the girder in the last time step and the quasi unchanged remaining stiffness of the support, a concentrated load introduction from the concrete slab into the steel structure occurs at this point. In the area of the bolted end plate connection, the maximum shear stresses occur corresponding to the introduction of the shear force from the beam into the end plate. The maximum value of the strain in the last time step of 19.3% occurs locally in one finite element, in the directly adjacent elements the strains drop to values of approx. 17% and in elements that are no longer directly adjacent they quickly drop to values of approx. 10% or below. At the point stressed by maximum strains, compressive stresses prevail in both the horizontal and vertical directions.

In the centre bay as well as in the edge bay, there is pronounced cracking at the top of the slab. This indicates tension in the upper reinforcement. In the centre bay, a membrane effect is also visible due to the stress trajectories (centre bay in the down right corner in Figure 12). The top of the slab is subject to tensile stresses in the centre bay. The slab top is no longer subject to compressive stresses in the centre of the field. In the edge field at the top, there are still pronounced compressive stresses in the direction transversal to the secondary beams even in the last time step - but not in the direction parallel to the secondary beams, matching to the crack formation. There, the slab carries the load with flexion between the secondary beams. In the direction of the secondary beams, tension is activated in the upper reinforcement. Thus, during the fire duration, the effect of membrane action in this case study only develops clearly in the middle part of the slab. In other parts, the load bearing of the slab is still realised via flexural action or via mixed forms between flexion and membrane action.

Based on the bolt forces (example of bolt forces in Figure 13), several facts can be deduced regarding the load-bearing behaviour at the connections. The change in the resulting vertical shear forces over time indicates load redistribution in the load-bearing system. Changes in the tensile forces in the bolts over time indicate a change in the load-bearing effect of the connection itself. At the beginning of the fire duration, tensile forces are already present in parts of the bolts, resulting from a restraining effect of the connections. Afterwards, these forces tend to decrease due to the thermal expansion in the system combined with the increase of compressive forces. Towards the end of the fire duration, tensile forces build up again when the deformations in the system increase and the constraining forces in the connections decrease again. Unequal vertical shear forces on both sides of the beam and sign changes of the upper to the lower horizontal shear forces are also noticeable. This indicates a torsion of the beam, combined with corresponding shear forces on the bolts.

Schraube 590024

LF	t [min]	Fv,Ed [kN]	Fv,Rd [kN]	Ft,Ed [kN]	Ft,Rd [kN]	n_int [-]	gam [%]	gam_u [%]	uz [mm]	uz_u [mm]	eps [%]	eps_v [%]	eps_u [%]
300	0.00	22.50	339.36	40.19	403.92	0.14	1.04	187.20	0.03	2.50	0.21	0.73	120.00
301	5.00	41.30	325.16	20.49	387.02	0.16	1.97	187.20	0.06	2.50	0.11	1.32	120.00
302	10.00	50.56	307.51	-0.01	366.01	0.16	2.91	187.20	0.09	2.50	0.00	1.94	120.00
303	15.00	45.14	235.91	-0.01	280.79	0.19	3.23	187.20	0.09	2.50	0.00	2.15	120.00
304	20.00	42.35	138.14	20.09	164.42	0.39	4.13	187.20	0.12	2.50	0.22	2.76	120.00
305	25.00	35.02	68.27	21.81	81.26	0.70	5.75	187.20	0.17	2.50	0.41	3.85	120.00
306	30.00	24.09	55.22	52.91	65.73	1.01	4.96	187.20	0.15	2.50	4.24	5.37	120.00

Figure 13. Design protocol of the bolts during fire (example one bolt of main beam).

In addition to the aspects described in detail above, the main criteria for the structural safety of the system in case of fire and their evaluation are summarised as follows:

- Maximum strain in the secondary beams: 12.4% < 20%, structural safety is ensured.
- Maximum strain in the main beams: 19.3% < 20%, structural safety is ensured.
- Maximum deflection of main beams in the middle field: 327 mm > 215 mm, criterion exceeded - however, the structural safety is evaluated as ensured.
- Maximum deflection of secondary beams between columns: 375 mm > 362 mm, criterion exceeded - however, the structural safety is evaluated as ensured.
- Maximum concrete compression strain: in general up to approx. 3%, structural safety is given.
- Maximum average crack width 0.76 mm, structural safety is ensured.
- Maximum reinforcing steel strain: in general up to approx. 5%, structural safety is ensured.
- All bolt verifications regarding interaction, elongation and shear deformation are complied, the structural safety is ensured.

Thus, for the REI30 requirement of the structure in case of fire, the component "R" for structural safety is ensured. Due to the crack behaviour with the calculated moderate crack widths and no recognisable tendency for cracks which run through the entire cross section, the criterion "E" for the integrity is also, assessed as fulfilled. Since the upper surface of the slab remains quasi at its initial temperature until the end of the fire duration, the criterion "I" for insulation is also fulfilled.

6 CONCLUSION

As a result of the case study on the chosen example of an industrial composite slab, the steel structure can be designed for the fire safety requirement of REI30 without the use of an insulating coating. The developed finite element model can be used for the design of the concrete slab, the steel girders and the bolts. During the validation of the model based on different fire tests and as well during the calculation of the industrial slab, the assumed limit values for deflections and their velocity are partly exceeded without any indication of a failure of the structure. It appears that these limits must be evaluated as indicative limits. The assessment of the structural safety in case of fire must be realised with the aid of qualitative evaluations as well as with limits for strains in the structural and reinforcement steel and in the concrete, crack widths in the concrete and bolt forces with their corresponding deformations.

REFERENCES

[1] SN EN 1991-1-2:2002, *Eurocode 1, Einwirkungen auf Tragwerke - Teil 1-2: Allgemeine Einwirkungen - Brandeinwirkungen auf Tragwerke,* Schweizerischer Ingenieur- und Architektenverein, Zürich, 2002.

[2] Bailey C.G., *Membrane action of slab/beam composite floor systems in fire,* Engineering Structures 26, p. 1691-1703, Elsevier, Amsterdam, 2004.

[3] Vassart O., Zhao B., *MACS+ Membrane Action of Composite Structures in Case of Fire, Brandverhalten von Stahl- und Verbunddeckensystemen, Background Document,* URL: http://www.macsfire.eu/de-start.html (download 27.11.2018), 2012.

[4] SZS *C1/12 steelwork, Verbundbau Bemessungstafeln,* Stahlbau Zentrum Schweiz, Zürich 2012.

[5] prEN 1994-1-2:2021, *Eurocode 4, Design of composite steel and concrete structures – Part 1-2: General – Structural fire design,* CEN/TC250/SC 4, 2021.

[6] Zhang N., Li G., *A New Method to Analyze the Membrane Action of Composite Floor Slabs in Fire Condition,* Fire Technology, 46, p. 3-18, Springer Science+Business Media, New York, 2010.

[7] Burgess I., *Yield-line plasticity and tensile membrane action in lightly-reinforced rectangular concrete slabs,* Engineering Structures 138, p. 195-214, Elsevier, Amsterdam, 2017.

[8] Burgess I., Sahin M., *Tensile membrane action of lightly-reinforced rectangular composite slabs in fire,* Structures 16, p. 176-197, Elsevier, Amsterdam, 2018.

[9] Stadler M., *Design of Composite Slab Systems in Case of Fire Using Simplified Finite Element Analyses,* PhD thesis, Technische Universität München, 2012.

[10] Müller A., Lange N., Gaigl C., Mensinger M., *Methode zur effizienten Modellierung von Verbunddeckensystemen im Brandfall,* Bauingenieur Bd. 95 (2020) Nr. 2, p. 48-54, VDI Fachmedien. Düsseldorf, 2020.

[11] SOFiSTiK, *ASE – Allgemeine Statik Finiter Element Strukturen, ASE-Manual Version 2018-8,* SOFiSTiK AG, Oberschleißheim, 2019.

[12] SOFiSTiK, *AQUA – Materialien und Querschnitte, AQUA-Manual Version 2018-8,* SOFiSTiK AG, Oberschleißheim, 2019.

[13] SN EN 1993-1-2:2005, *Eurocode 3, Bemessung und Konstruktion von Stahlbauten - Teil 1-2: Allgemeine Regeln - Tragwerksbemessung für den Brandfall,* Schweizerischer Ingenieur- und Architektenverein, Zürich, 2005.

[14] SN EN 1993-1-8:2005, *Eurocode 3, Bemessung und Konstruktion von Stahlbauten - Teil 1-8: Bemessung von Anschlüssen,* Schweizerischer Ingenieur- und Architektenverein, Zürich, 2005.

[15] Henriques J., Jaspart J.-P., Simões da Silva, L., *Ductility requirements for the design of bolted lap shear connections in bearing,* Advanced Steel Construction, Volume 10, Number 1, p. 33-52, The Hong Kong Institute of Steel Construction, Hong Kong, 2014.

[16] SN EN 13501-2:2016, *Klassifizierung von Bauprodukten und Bauarten zu ihrem Brandverhalten - Teil 2: Klassifizierung mit den Ergebnissen aus den Feuerwiderstandsprüfungen, mit Ausnahme von Lüftungsanlagen*, Schweizerischer Ingenieur- und Architektenverein, Zürich, 2016.

9TH INTERNATIONAL CONFERENCE ON COMPOSITE CONSTRUCTION IN STEEL AND CONCRETE
M. Knobloch, U. Kuhlmann, W. Kurz, M. Schäfer. (Eds.)
Online, Germany, July 27-29, 2021

CONCRETE CONE FAILURE OF POST INSTALLED FASTENERS DURING FIRE

Hitesh Lakhani* and Jan Hofmann*

* Institute of Construction Materials, University of Stuttgart
e-mails: hitesh_lakhani06@rediffmail.com, jan.hofmann@iwb.uni-stuttgart.de

Keywords: Concrete cone failure, fire resistance, fasteners, expansion anchors.

Abstract. *The fire resistance of fasteners needs to be verified for all failure modes applicable at ambient conditions. Generally, in most cases, for unprotected fasteners loaded in tension, steel failure mode is decisive. But for fasteners made of stainless steel and/or larger (bolt) sizes, concrete cone failure may be the decisive failure mode for small anchorage depths. Due to practical difficulties associated with loading an anchor during a fire test, very limited experimental data is available in literature and that to for relatively small embedment depths. The paper presents the results (6 nos) of the fire tests conducted on expansion anchors (made of stainless steel) with sizes M12 (h_{ef} = 70 mm) and M20 (h_{ef} = 100 mm). Moreover, the paper also compares the reduction in the concrete cone capacity with exposure duration, predicted as per the current design guidelines and the new set of experimental data obtained in this study.*

1 INTRODUCTION

The use of anchors in concrete structures to connect elements to beams, walls, etc., has increased in recent years, due to the flexibility it offers in planning and construction [1]. Since, designing a structure against fire is an integral part of the standard design procedure. Any anchor used for structural applications should also, at least, have the same fire rating as the elements being connected. Hence, the fire resistance of fasteners has gained attention of increasing number of researchers [1,2].

As per the current design guidelines [3], the fire resistance of fasteners is checked for all failure modes as for ambient conditions. The failure modes which are required to be verified at ambient conditions under tension loads are namely, steel failure, concrete cone failure, pull-out failure, combined pull-out & concrete failure (for bonded anchors), concrete splitting failure (non-cracked concrete) and concrete blow-out failure. In an event of fire, steep thermal gradients are induced in reinforced concrete members due to its low thermal conductivity and high specific heat. Due to these thermal gradients and degradation of mechanical properties of concrete with increase in temperature, the capacities of failure modes of fasteners associated with concrete is also reduced significantly. In general, for most of the cases, with unprotected fasteners made of carbon steel and loaded in tension, steel failure is found to be decisive. It should be noted that the previous statement is not applicable to bonded fasteners, for which pull-out failure mode would be decisive [4,5]. But for protected anchors or for example expansion anchors made of stainless steel, concrete cone failure may be the decisive failure mode.

The current prescriptive design for steel failure is based on the evaluation of a large experimental data set for steel failure by Reick (2001) [6]. These characteristic steel stresses for different fire resistance durations are dependent on the diameter of the fastener and steel viz., carbon steel and stainless steel.

In contrast to steel failure, the characteristic pull-out resistance of mechanical fasteners is based on very limited experimental results [1] and is given as 25% & 20% of the value at cold state, for 90 minutes & 120 minutes of standard fire exposure, respectively.

In case of concrete cone failure, the reduction factor is a linear function of embedment depth. For fire exposure up to 90 minutes the reduction factor is given by $h_{ef}/200$, which is further reduced in a step to

$0.8 \times h_{ef}/200$ for fire exposure between 90 to 120 minutes. The influence of fire on concrete cone capacity decreases with increasing embedment depth, because the cracking starts at the deepest point of the anchorage depth. For fasteners with h_{ef}= 200 mm which are far away from edge, the capacity is unaffected up to 90 minutes of fire exposure. The basis for these guidelines is the extensive numerical study by Periskic (2009) [7] on headed studs. The numerical model used by Periskic was validated against very limited experiments on headed studs (with embedment depth of 40, 50 & 60 mm) [8]. It should be acknowledged that the existing guideline for concrete cone failure lacks an extensive experimental backing for post installed mechanical fasteners. Moreover, no attempts were made in the past to expand the (very limited) experimental database to embedment depths commonly used in practice. This is mainly because of two main reasons:

1. As mentioned previously, in general, for fasteners made of carbon steel with embedment depths commonly used in practice, mostly steel failure is the decisive failure mode.
2. Performing fire tests is very demanding both economically and technically. Moreover, the high levels of load that need to be applied on stainless-steel fasteners during the fire tests imposes a serious practical limitation and challenges.

Recently, a new medium scale fire testing facility for anchors was developed by Department for Fastening and strengthening, Institute of Construction materials, at MPA, University of Stuttgart, to overcome the practical limitation mentioned above in point 2. This new test setup made it possible to generate new experimental data for post installed mechanical fasteners with load levels from 8 – 28 kN, presented in this paper.

2 EXPERIMENTAL INVESTIGATION

Experiments were conducted on torque-controlled expansion anchors with (bolt) thread sizes M12 and M20, made of stainless steel. The effective embedment (h_{ef}) was 70 mm & 100 mm, for sizes M12 & M20, respectively. The concrete slabs in which the anchors were installed had an age of more than 90 days and were made of C20/25 grade concrete. The setting position of the anchors was defined with respect to the edge distance (c_1 / c_2) from the two nearest edges of the slab. Table 1 gives the details of the concrete slab and the anchors tested.

Table 1: Details: setting position and concrete slabs.

	Anchor size M12	Anchor size M20
1	2	3
h_{ef} [mm]	70	100
Setting position c_1 / c_2 [mm]	100 / 115	140 / 200
Load levels [kN]	18.75 / 12.25 / 8.00	28.40 / 26.00 / 22.00
Slab thickness [mm]	150	200
Cube compressive strength @ 28 days [MPa]	29.75	28.5
Age of concrete slab on the day of testing [days]	107 / 177	131 / 133

In case of fire tests performed in bigger furnaces, the concrete slabs, with anchors installed in them, are placed on top of the furnace as covering slab. But in contrast to these traditional fire tests, in the present experimental setup, the concrete slab was placed on the floor of the furnace. This was possible due to the appropriate medium scale of the furnace. Such a placement of specimen in furnace allows us to use a loading setup very similar to that used for ambient tests. The furnace being used is a single burner small test furnace as shown in figure 1.

The concentric tension load was applied on the anchors through a loading fixture which was connected to a loading rod of size M30. The loading fixtures were designed as per the guidelines provided in EOTA TR020 [9]. The loading rod extended all the way through the furnace covering blocks. The loading rod was then connected to the extractor rod outside the furnace, passing through the hydraulic cylinder and the load cell. The required loads on the anchors were applied using the hydraulic hand pumps. The loads were applied approx. 5 minutes before the fire test and were maintained constant during the test duration. The load history was continuously monitored and recorded using the data acquisition system. The loading histories for 6 reported tests are shown in figure 2.

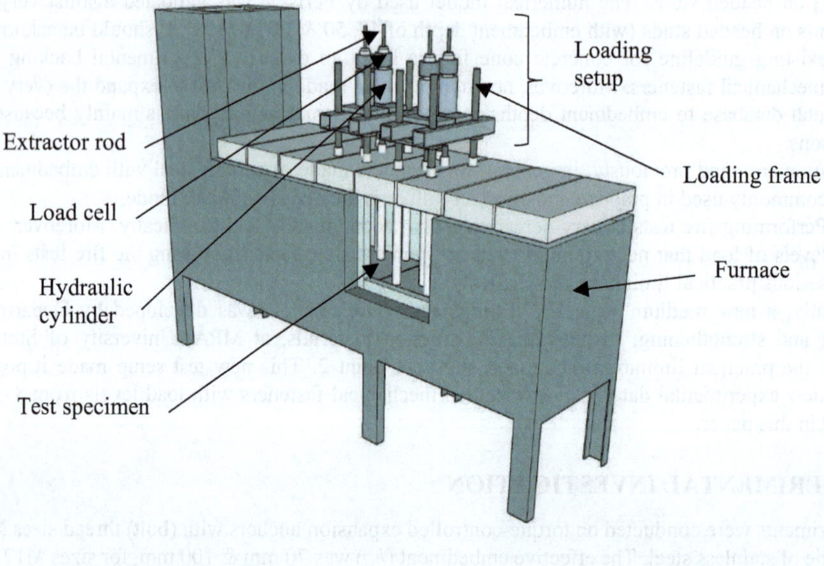

Figure 1. Schematic representation of the test facility and loading setup.

The loaded anchors were then exposed to fire defined by standard temperature time curve [10]. The slab was exposed to fire only from the side on which the anchors were installed and all other sides were insulated. The furnace temperature was measured using two plate type thermocouples. Figure 3 shows the average furnace temperature along with the target standard fire exposure.

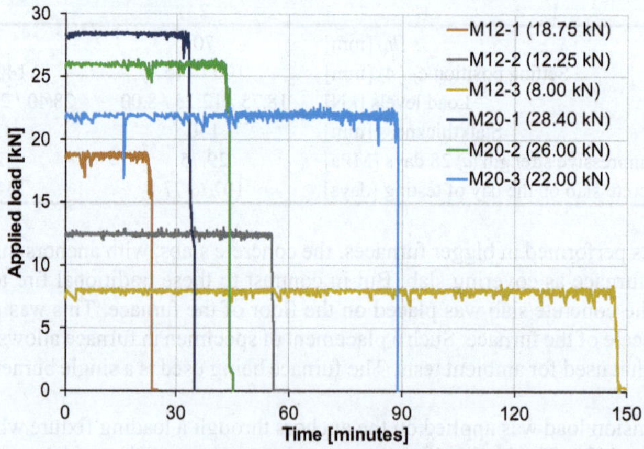

Figure 2. Applied loading histories.

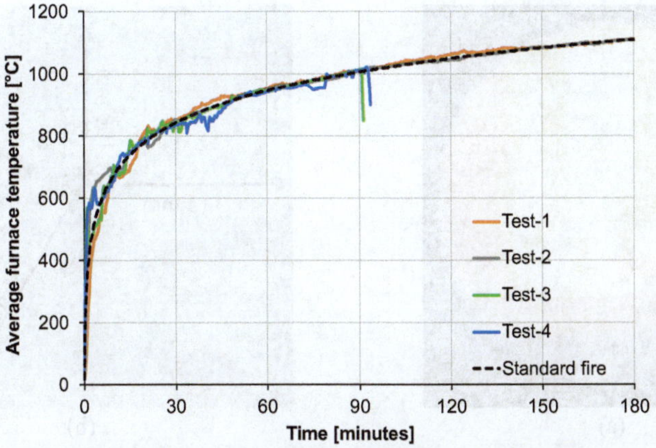

Figure 3. Average furnace temperature during different fire tests.

3 EXPERIMENTAL RESULTS

The time at which the applied load could no longer be maintained on the anchor, was noted as the time to failure. The time to failure of the anchor can be easily identified and read from the applied load histories shown in figure 2. The experimentally obtained variation of concrete cone capacity with exposure time is shown in figure 4. It was observed that for both the embedment depths investigated the reduction in concrete cone capacity is steep for short fire duration as compared to longer fire durations.

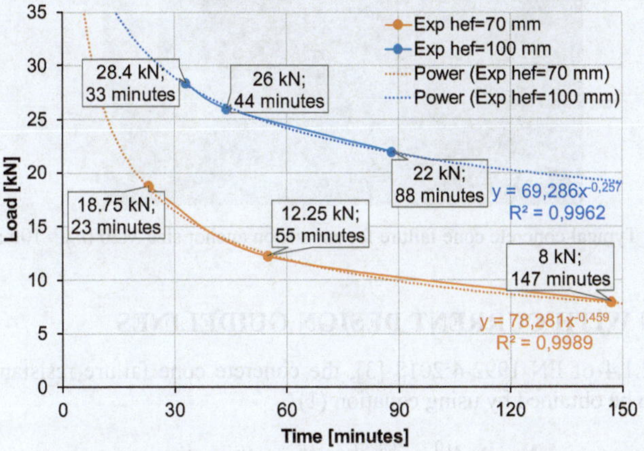

Figure 4. Experimental results for concrete cone failure.

The typical concrete cone failure observed in the experiments is shown in figure 5 & 6, for sizes M12 and M20, respectively. It should be noted that the formation of a complete concrete cone breakout was prevented due to the loading frame support points, shown by red dotted circles in figure 5 & 6.

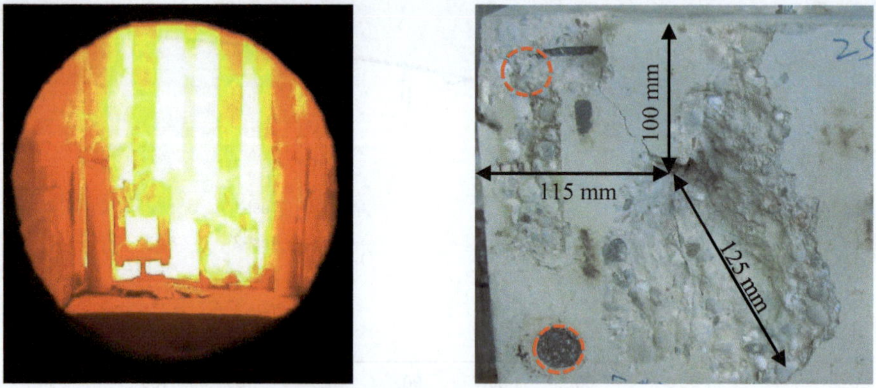

(a) (b)

Figure 5. Typical failure mode for expansion anchor size M12 (h_{ef} = 70 mm) (a) image during the fire test and (b) after the fire test.

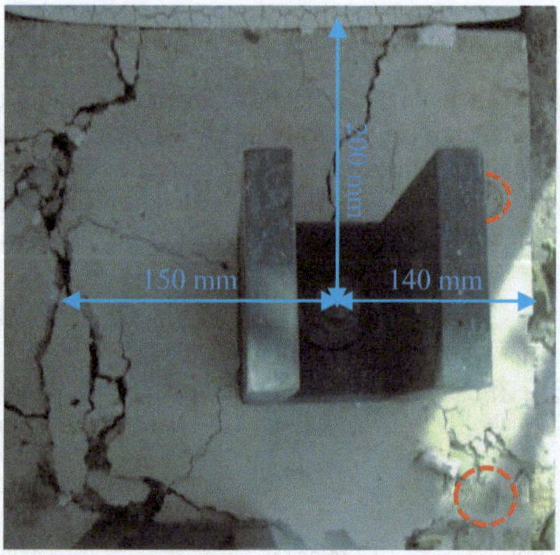

Figure 6. Typical concrete cone failure for expansion anchor size M20 (h_{ef} = 100 mm).

4 COMPARISON WITH CURRENT DESIGN GUIDELINES

As per section 7.2.1.4 of EN 1992-4:2018 [3], the concrete cone failure resistance of a fastener at ambient conditions can be obtained by using equation (1).

$$N_{u,C} = N_{u,C}^{0} \frac{A_{Cn}}{A_{CN}^{o}} \psi_{s,N} \, \psi_{re,N} \, \psi_{ec,N} \, \psi_{M,N} \tag{1}$$

Where,

A_{cN} / A_{cN}^{o} accounts for the geometric effects like spacing and edge distances. In case of fire exposure, the reference projected area (A_{cN}^{o}) is calculated as *16* times the square of effective embedment depth (h_{ef}) as compared to 9 times at ambient conditions. This is due to increased requirements on characteristic spacing ($s_{cr,N} = 4 \times h_{ef}$) and edge distance ($c_{cr,N} = 2 \times h_{ef}$),

$\psi_{s,N}$ is a factor that account for the change in stress distrbution in concrete due to the proximity of an edge of the concrete member. It is a function of smallest edge distance *(c)* and characteristic edge distance ($c_{cr,N} = 2 \times h_{ef}$*; in case of fire*) (equation (2)),

$$\psi_{s,N} = 0.7 + 0.3\,\frac{c}{c_{cr,N}} \leq 1 \qquad (2)$$

The factors to account for shell spalling ($\psi_{re,N}$) and eccentric loading ($\psi_{ec,N}$) are not applicable in the present case. Hence, are taken as 1.0. Moreover, there are no investigations available until now if this factor is applicable under fire conditions at all. Since spalling above the reinforcement is very likely under fire perhaps this factor may also be valid for a lower amount of reinforcement.

The factor $\psi_{M,N}$, which accounts for the positive effect of presence of compression force between fixture and concrete is conservatively taken as 1.0. Since the stress state in the concrete under fire does not only depend on the applied forces but also on the thermal stress distribution. Hence, this factor should not be accounted for during fire. The beneficial stress state may be changed by superposing the thermal stresses.

$N^0_{u,c}$ is the resistance of a single fastener not influenced by adjacent fasteners or edge of the concrete member. In case of fire exposure, this value is to be replaced with the values given by equations (3) & (4) for standard fire exposure up to 90 minutes and between 90 & 120 minutes, respectively.

$$N^0_{u,fi(90)} = \frac{h_{ef}}{200} N^0_{u,c,20°C} \leq N^0_{u,c,20°C} \qquad (3)$$

$$N^0_{u,fi(120)} = 0.8 \times \frac{h_{ef}}{200} N^0_{u,c,20°C} \leq N^0_{u,c,20°C} \qquad (4)$$

The concrete cone failure capacity at ambient conditions, $N^0_{u,c,20°C}$, is calculated using equation (5). Due to the presence of thermal cracks, the calculation assumes that the concrete is cracked during fire. The value of k_{cr} is typically taken as 0.7 times the coefficient (k_{ucr}) for uncracked concrete [11].

$$N^0_{uc,20°C} = k_{cr}\sqrt{f_c}\,h^{1.5}_{ef} \qquad (5)$$

To make comparisons at mean level, the value of k_{ucr} proposed in Eurocode cannot be used because those values are applicable at characteristic level. Hence, the value proposed by Eligehausen et al., (2012) at mean level are used, i.e., k_{ucr} = *14.6* when fc is cylinder compressive strength. Thus, k_{cr} is equal to 0.7 × 14.6 = *10.22*.

The cube strengths for the concrete were measured only at age of 28 days but tests were performed at age between 107-177 days. Hence, the concrete compressive strength on the day of testing was calculated using equation (6), taken from fib Model Code 2010 [12]. Table 2 summarises the compressive strength calculations for concrete on different days of testing.

$$f_c(t) = exp\left\{s\left[1-\sqrt{\frac{28}{t}}\right]\right\} \times f_{cm} \qquad (6)$$

Table 2: Calculation of cylinder compressive strength on the test day as per fib Model Code 2010.

Specimen	$f_{c,cube}$ *@28 days*	$f_{c,cyl}$ *@28 days = 0.8 ×* $f_{c,cube}$	Age of concrete on testing day	Coefficient *"s" (for 32.5 N cement type)*	$f_{c,cyl}$ on test day *(as per equation 6)*
1	2	3	4	5	6
	[MPa]	[MPa]	[days]	[-]	[MPa]
M12 – 1 (18.75 kN)	29.75	23.8	107	0.25	26.89
M12 – 2 (12.25 kN)	29.75	23.8	177	0.25	27.67
M12 – 3 (8.00 kN)	29.75	23.8	177	0.25	27.67
M20 – 1 (28.40 kN)	28.5	22.8	131	0.25	26.08
M20 – 2 (26.00 kN)	28.5	22.8	133	0.25	26.10
M20 – 3 (22.00 kN)	28.5	22.8	133	0.25	26.10

The computed cylinder compressive strengths are then used for computing $N^0_{uc,20°C}$ using equation (5). The computed concrete cone capacity of fasteners not affected by concrete edges at ambient conditions and under fire (90 & 120 minutes), in cracked concrete are summarised in Table 3. The concrete cone capacity of fasteners accounting for the edge distance are given in Table 4.

The comparison between the average concrete cone capacity under fire as per EN 1992-4:2018 and the experimental results obtained in this study are shown in figure 7.

Table 3: Computation of concrete cone capacity (for cracked concrete).

Age [Days]	h_{ef} [mm]	k_{cr} [-]	f_c [MPa]	$N^0_{uc,20°C}$ [kN]	$N^0_{u,fi(90)}$ [kN]	$N^0_{u,fi(120)}$ [kN]
1	2	3	4	5	6	7
107	70	10.22	26.89	31.04	10.86	8.69
177	70	10.22	27.67	31.48	11.02	8.82
131	100	10.22	26.08	52.19	26.10	20.88
133	100	10.22	26.10	52.22	26.11	20.89

Table 4: Computation of concrete cone capacity accounting for edge influence.

h_{ef} [mm]	c_1/c_2 [mm]	Age [Days]	$N^0_{u,fi(90)}$ / $N^0_{u,fi(120)}$ [kN]	Acno [mm²]	Acn [mm²]	$\frac{Acn}{Acn0}$ [-]	$\Psi_{s,N}$ [-]	$\Psi_{re,N}$; $\Psi_{ec,N}$; $\Psi_{M,N}$ [-]	$N_{u,fi(90)}$ / $N_{u,fi(120)}$ [kN]
1	2	3	4	5	6	7	8	9	10
70	115 / 100	107	10.86 / 8.69	78400	61200	0.78	0.914	1.00	7.75 / 6.20
70	115 / 100	177	11.02 / 8.82	78400	61200	0.78	0.914	1.00	7.86 / 6.29
100	140 / 200	131	26.10 / 20.88	160000	136000	0.85	0.910	1.00	20.19 / 16.15
100	140 / 200	133	26.11 / 20.89	160000	136000	0.85	0.910	1.00	20.19 / 16.16

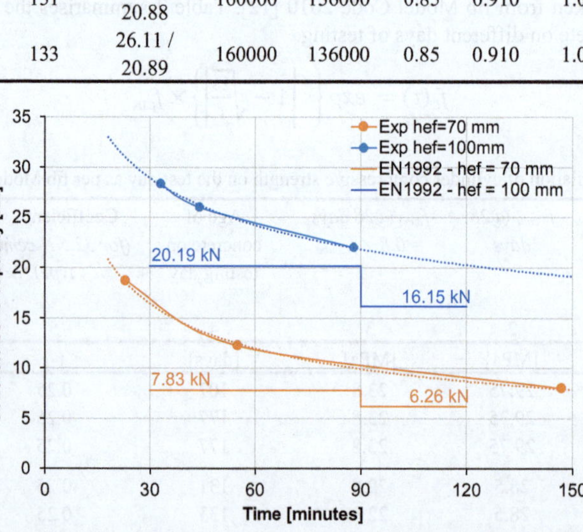

Figure 7. Comparison between the experimental results and current design guidelines.

5 CONCLUDING REMARKS

The paper presented the first set of experimental results for concrete cone capacity of expansion anchors under fire. The paper also briefly described the new test facility developed by Department for Fastening and strengthening, Institute of Construction materials, at MPA, University of Stuttgart, which now makes it possible to test anchors under high loads and fire exposure.

The results validated the EN 1992-4:2018 design guidelines for concrete cone resistance of post installed fasteners under fire. It was found that the current design guidelines are on the conservative side and the safety margins are high for short fire durations, but they reduce with increase in fire duration and embedment depth.

Moreover, there is a significant lack of knowledge regarding the validity of the reduction and increasing factors given in equation (1) in case of fire exposure. In addition, the first results show that a step function related to the fire durations of 30, 60 ,90 and 120 minutes could be more accurate.

REFERENCES

[1] Eligehausen, R. and Fuchs, W., "Recent developments and open problems in fastening technique", *Connections between Steel and Concrete - Proceedings of the 2nd International symposium*, Stuttgart, Germany, 2007.

[2] Fuchs, W. and Silva, J., "EOTA Approach to Qualification and Design of Post-Installed Adhesive Anchors for Fire Exposure", *ACI SP-283: Understanding Adhesive Anchors: Behavior, Materials, Installation*, 2012. https://doi.org/10.14359/51683774.

[3] EN1992-4:2018. Eurocode 2 - Design of concrete structures - Part 4: Design of fastenings for use in concrete, European Committee for standardization, Brussels, 2018.

[4] Hofmann, J.; Lakhani, H. and Aggarwal, J., "Post installed rebars - Pull-out capacity during fire", *Otto-Graf-Journal*, 18, 2019.

[5] Lakhani, H. and Hofmann, J., "A numerical method to evaluate the pull-out strength of bonded anchors under fire", *Connections between Steel and Concrete - Proceedings of 3rd International Symposium*, Stuttgart, Germany, 2017.

[6] Reick, M., "Brandverhalten von Befestigungen mit großem Randabstand in Beton bei zentrischer Zugbeanspruchung", Phd Thesis, Institute of Construction Materials, University of Stuttgart, 2001. (In German)

[7] Periskic, G., "Entwicklung eines 3D thermo-hygro-mechanischen Modells für Beton unter Brandbeanspruchung und Anwendung auf Befestigungen unter Zuglasten", Phd Thesis, Institute of Construction Materials, University of Stuttgart, 2009. (In German)

[8] Hofmann, J.; Lakhani, H. and Periškić, G., "Design concept for the behaviour of fasteners under fire", *Commentary to EN1992-4: Design of fastenings for use in concrete*, DAfStb, Berlin, Report No. 615(en), 2019

[9] EOTA TR 020. Evaluation of Anchorages in Concrete concerning Resistance to Fire, European Organisation for Technical Approvals, Brussels, 2004.

[10] ISO 834-1:1999 (en), Fire-resistance tests - Elements of building construction - Part 1: General requirements. International Organization for Standardization, 1999.

[11] Eligehausen. R.; Mallee, R. and Silva, J.F., *Anchorage in Concrete Construction*, John Wiley & Sons, Ltd, 2012.

[12] fib. *Model Code for Concrete Structures 2010*. John Wiley & Sons, Ltd; 2013.

9TH INTERNATIONAL CONFERENCE ON COMPOSITE CONSTRUCTION IN STEEL AND CONCRETE
M. Knobloch, U. Kuhlmann, W. Kurz, M. Schäfer. (Eds.)
Online, Germany, July 27-29, 2021

LOAD TRANSFER FROM HOLLOWCORE SLABS TO DELTABEAM® SLIM FLOOR BEAMS IN CASE OF A FIRE

Oliver Beckmann*, Michael Cyllok**

* Peikko Group Corp, R&D, Product development composite structures
e-mail: oliver.beckmann@peikko.com
** LGA Landesgewerbeanstalt Bayern, Research & Development
e-mail: michael.cyllok@lga.de

Keywords: Composite, beam, prestressed, hollowcore, slab, fire

Abstract. *This paper presents the background, results and conclusions of a series of fire tests with DELTABEAM® composite beams supporting prestressed hollowcore slabs (HCS) concerning the shifted direct support in fire situation. When designing a slab system consisting of DELTABEAM® and HCS in fire situation, the current European design codes and product approvals for both products do not cover the special issues "indirect / shifted direct" and "flexible" support that arise from combining beams and HCS. A test program of single HCS unit shear tests and system fire tests has been carried out to assess these issues. The presented paper will give an overview of the test program and a detailed look at the shifted direct support situation for load transfer from the slab to the beam.*

1 INTRODUCTION

In modern slim-floor slab structures as shown in Figure 1, HCS are supported on the steel bottom plate ledge during the construction phase. After filling the composite beam and the slab joints with concrete, the slab is acting as a composite member. While the load transfer mechanisms at ambient temperature are widely accepted, the interaction of HCS and supporting beams in fire case has been challenged in several European countries due to the lack of specific code regulations and approvals.

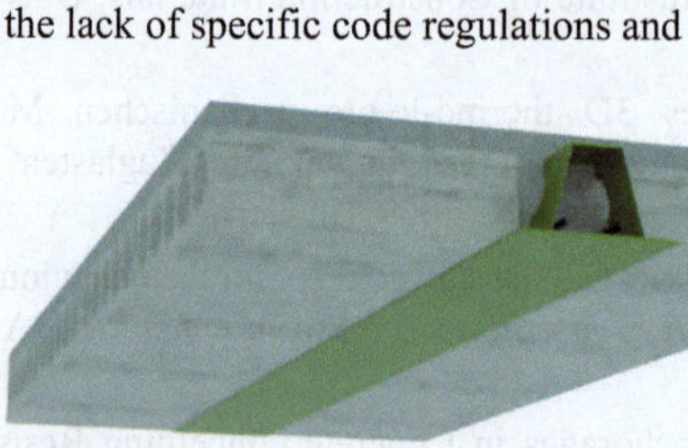

Figure 1 – Deltabeam cross section with hollow-core slabs

The following two issues have been the main concerns in fire situation.

1.1 Shifted direct support in fire case

The direct support of HCS provided by the steel bottom plate ledge changes when the steel bottom plate of steel or composite beams is exposed to fire and therefore heated up to ~1000°C after 90 minutes of ISO fire exposure. The strength and stiffness of the unprotected steel material is reduced and the location of the resultant support force underneath the ends of the slab elements shifts towards the steel web (see Figure 2a).

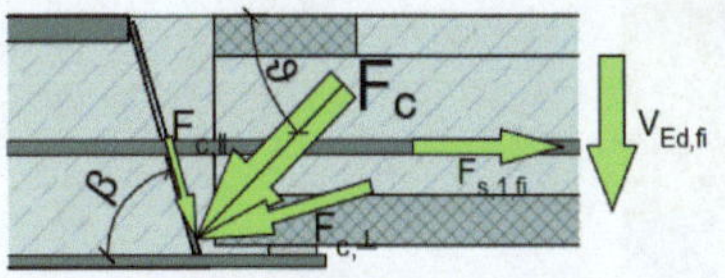

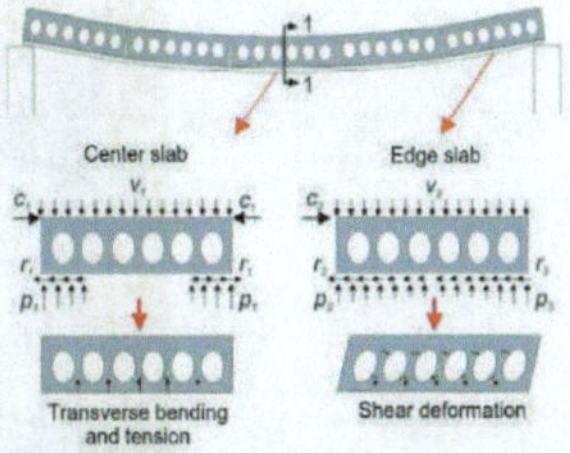

a) Simplified load transfer model for the shifted direct support

b) Imposed distortion and crack pattern in flexible support situation of HCS (from Roggendorf [1])

Figure 2 - Models of shifted direct and flexible support of HCS

1.2 Flexible support in fire case

When HCS are supported on flexible supports (beams bending due to vertical loads from the slab) in ambient temperature, their vertical shear resistance at ambient temperature is reduced due to additional stress in the HCS webs (cf. Roggendorf [1], Borgogno [2] et. al.) as shown in Figure 2b. It has been discussed in the past years if this effect persists in fire situation. To be on the safe side for application in practice, limitations for the allowable beam deflection or a requirement for additional fire-protection of the bottom plate have been imposed in some countries. The German national technical approvals for HCS (e.g. [3]) have required a deflection limit for the supporting beam of L/100 including thermal deflection – a demand that can hardly be met by shallow beams in practice.

The issue of flexible support is not covered by this paper but will be analyzed in subsequent publications.

2 EXISTING RESEARCH BACKGROUND AND PREVIOUS TESTS

2.1 Load transfer tests without a supporting ledge

In 2009, a series of three ambient temperature tests have been carried out at RWTH Aachen and assessed by H+P engineers, Aachen [4]. The test setup consisted of a Deltabeam that was rigidly supported, with HCS types VMM VSD and L-EPD at both sides, cf. Figure 3. The height of the slabs and Deltabeam was varied in the three tests. The slabs were connected to the beam with straight reinforcement passing through the web holes and anchored in the slab joints and opened voids. Table 1 gives an overview of the cross sections of HCS and Deltabeam and the amount of connecting reinforcement. The thermal softening of the bottom plate ledge in the event of a fire was simulated on the safe side by completely removing the ledge at the outside of the web weld. Thus, the slab elements did not have any direct support from underneath. The concrete grout between the slab ends and the Deltabeam together with the transverse reinforcement established the connection between HCS and Deltabeam. The influence of thermal effects (thermal expansion of the heated HCS soffit with resulting compressive strains, reduction of concrete strength etc.) could not be analysed in these ambient temperature tests.

Table 1 - Test dimensions of load transfer tests at RWTH Aachen

No.	Deltabeam	HCS	Reinforcement
1	D26-300	1x VMM VSD 25 (60cm)	5Ø12mm, every 0,3m
		2x VMM VSD 25 (45cm)	a_s=3,77cm²/m
2	D37-400	1x VMM L-EPD 35 (60cm)	2Ø14mm, every 0,6m
		2x VMM L-EPD 35 (45cm)	a_s=2,05cm²/m
3	D26-300	1x VMM L-EPD 27 (60cm)	2Ø12mm, every 0,6m
		2x VMM L-EPD 27 (45cm)	a_s=1,51cm²/m

Figure 3 - Test setup of indirect support tests

Table 2 – Test results from Aachen load transfer tests – $V_{u,test}$ is the test result without a direct support; $V_{u,ref,test}$ is the result of a directly supported reference test

No.	HCS	$V_{u,test}$ [kN/m]	$V_{u,ref,test}$ [kN/m]
1	VMM VSD 25	242,5	240,5
2	VMM L-EPD 35	315,2	458,8
3	VMM L-EPD 27	169,0	273,7

As shown in Table 2, the directly supported reference tests have equal or higher resistances than the indirectly supported slabs connected to the Deltabeam. Nevertheless, both the indirectly supported slabs and the directly supported reference tests have reached significantly higher values than the calculated characteristic cracked resistance $V_{Rk,ct,II}$ and even more than the uncracked resistance $V_{Rk,ct,I}$ listed in Table 3, calculated according to EN 1992-1-1 [5] and EN 1168 [6]. It is noted that for all tested slabs, the uncracked design resistance $V_{Rd,ct,I}$ was lower than the cracked resistance $V_{Rd,ct,II}$ and was therefore raised to the level of the cracked resistance. This is caused by the high degree of prestress which results in a higher cracked resistance.

Table 3 - Design and characteristic shear resistance of HCS with material values from Eurocode 2 or material tests

	Design resistance, EC2 values			Char. Resistance, EC2 values			Char. Resistance, test values		
No.	$V_{Rd,ct,I}$ [kN/m]	$V_{Rd,ct,II}$ [kN/m]	$V_{Rd,fi}$ [kN/m]	$V_{Rk,ct,I}$ [kN/m]	$V_{Rk,ct,II}$ [kN/m]	$V_{Rk,c,fi}$ [kN/m]	$V_{Rk,ct,I}$ [kN/m]	$V_{Rk,ct,II}$ [kN/m]	$V_{Rk,fi}$ [kN/m]
1	114	114	93	148	144	93	164	150	99
2	115	115	85	173	147	85	193	154	91
3	103	103	70	136	131	70	155	139	76

$V_{Rd/k,ct,I}$ are the characteristic uncracked resistances acc. EN 1168, 4.3.3.2.2.2 [6]
$V_{Rd/k,ct,II}$ are the characteristic cracked resistances acc. formula 6.2a, DIN EN 1992-1-1 [5];
$V_{Rd/k,fi}$ are the resistances after 90 minutes fire exposure acc. EN 1168, Annex G [6]

Note: There were some uncertainties in H+P test report [4] with respect to the prestress level of the slab elements. For the evaluation given in Table 3 it was assumed that the slab elements in the tests were types VSD 25 VD4X, L-EPD 27 ZP5D and L-EPD 35 ZP5D, which are the slab elements with the highest prestress level according to the manufacturers specification [7] [8] but still less than described in the test report [4].

This test series proves that, even in the extreme case of a complete loss of a direct support underneath the slab ends, the load transfer from slab to Deltabeam can be maintained with a sufficient safety level. The inclined Deltabeam webs together with the stiff welded corner at the bottom end of the webs and the tie reinforcement connecting the slab to the Deltabeam can provide a sufficiently strong connection. The vertical shear resistance of HCS connected in this manner is reduced when compared with a direct support, but the resistance still exceeds the calculated characteristic resistances.

2.2 System fire tests at SP in Sweden

In 2009, Peikko carried out a set of four loaded system fire tests with HCS combined with Deltabeam at the Technical Research Institute of Sweden SP (now RI.SE) in Boras [9] [10] [11] [12]. The slab systems consisted of one main Deltabeam D26-400, two edge Deltabeam DR26-230 and Consolis Xtrumax 27 HCS located between the beams and supported directly on the fire-exposed steel bottom plate ledge, see Figure 4. The beam span was 3,92 m and the length of the HCS elements at both sides of the main beam was 2,43 m. A connecting reinforcement of 1∅12 mm was placed in each joint of the slab elements, at a distance of 1200 mm. The test setup and geometry have been equal for all four tests. The only differences were the load level, the duration of fire exposure and the axis distance of the prestressing tendons (35/50 mm, depending on the planned fire duration). The bottom plate supporting the slab elements had a thickness of 8 mm. An ISO fire was applied to the unprotected bottom side of the specimen after the mechanical loads had been imposed with two line loads on the slabs at both sides of the main Deltabeam.

c) Loading equipment on HCS

d) Aerial view of test setup

Figure 4 – Slab system fire test setup

The slabs were loaded with a line load at a distance from the slab end of around 2,5-times the slab height prior to starting the fire exposure.

The imposed vertical shear force in the slabs, the vertical shear resistance at the end of the fire exposure times and the utilization ratio $\eta_{fi}=V_{Ed,fi}/V_{Rd,fi}$ of the four tests are listed in Table 4. In addition, the relative deflection of the Deltabeam L/w_{max} at midspan at the end of the fire exposure is shown.

Table 4 - Test results of SP loaded slab tests

No.	T_{fi} [min]	$V_{Ed,fi}$ [kN/m]	$V_{Rd,ct,II}$ [kN/m]	$V_{Rd,fi}$ [kN/m]	η_{fi} [-]	L/w_{max} [mm]
A	60	38,0	75,3	47,1	0,81	54
B	60	44,8	75,3	47,1	0,95	53
C	120	38,0	72,5	44,2	0,86	42
D	180	25,0	72,5	38,1	0,66	40

During the fire test, it could be observed that the steel bottom plate ledges supporting the HCS elements followed the deflection of the slab. The ledges of the main Deltabeam remained in contact with the soffit of the slabs throughout the whole duration of the fire exposure, without any visible gaps. From this behaviour, it can be derived that the position of the support reaction must have shifted towards the webs of Deltabeam during the fire exposure. Otherwise, the deflection of the ledge would have increased from plastic bending and a lift-off from the soffit of the slab would have occurred at the outer end of the ledge. It can furthermore be assumed that a contact pressure between steel ledge and slab persisted throughout the fire test.

No failure of the load-bearing resistance of the HCS has occurred in any of the four tests although significant cracking occurred. As a result of the tests, it can be stated that the load transfer on a level of up to almost 100% of the vertical shear resistance of the slabs could be maintained. The applied straight horizontal tie reinforcement of $a_s = 1{,}13\ cm^2 / 1{,}2\ m = 0{,}94\ cm^2/m$, together with the remaining strength of the supporting bottom plate ledge, were able to ensure the load transfer mechanism described later in this document. It is noted that the applied amount of tie reinforcement is clearly a lower boundary of the reinforcement applied in practice.

3 TEST PROGRAM OF THE CURRENT RESEARCH PROJECT

3.1 Single unit shear tests

A set of five different HCS units have been tested for their behaviour under combined loading with a vertical shear force and a horizontal forced distortion under elevated temperatures. The aim of these tests was to assess the ductility, influence of longitudinal cracking and elevated temperatures to vertical shear resistance and to calibrate FEA models. Figure 5 shows the schematic test setup. To reflect an overview on the HCS sections available on the market, four different heights of slabs from two different manufacturers with a varying amount and shape of voids was tested.

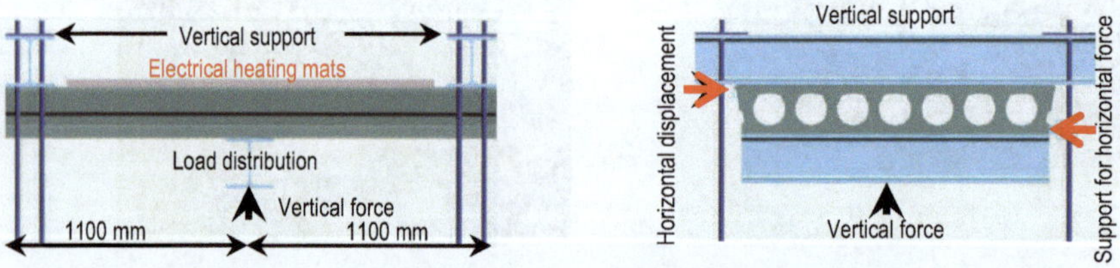

Figure 5 – Setup of single unit loaded shear tests (from [13])

All slab units have shown a ductile behaviour with a horizontal distortion between 5 and 12 mm before failure at a vertical load of more than 100% of $V_{Rd,c,fi}$ acc. EN 1168. The temperature level in the HCS webs exceeded the value corresponding to a 90-minute ISO fire exposure.

The single unit tests were carried out to analyse the flexible shear issue only and are therefore not further exploited in this paper.

3.2 System fire tests

Two full-scale, loaded system fire tests with a height of 200 and 320 mm have been carried out to verify the design models for the shifted direct and flexible support. The test setup and geometry are shown in Figure 6 and Figure 7. A straight horizontal tie reinforcement of ∅12 mm every 60 cm has been applied.

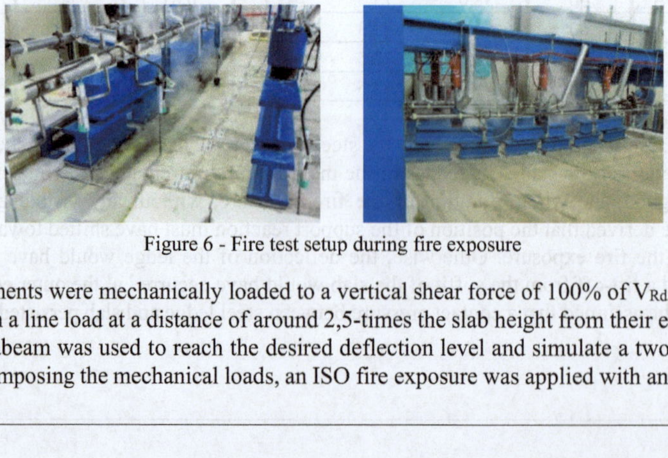

Figure 6 - Fire test setup during fire exposure

The slab elements were mechanically loaded to a vertical shear force of 100% of $V_{Rd,c,fi}$ acc. EN 1168, Annex G [6] with a line load at a distance of around 2,5-times the slab height from their end. An additional load on the Deltabeam was used to reach the desired deflection level and simulate a two-sided loading of the beam. After imposing the mechanical loads, an ISO fire exposure was applied with an exposure time of

90 and 120 minutes. Both tests were stopped without a failure. The beam deflections at the end of the fire exposure time were as high as L/32.

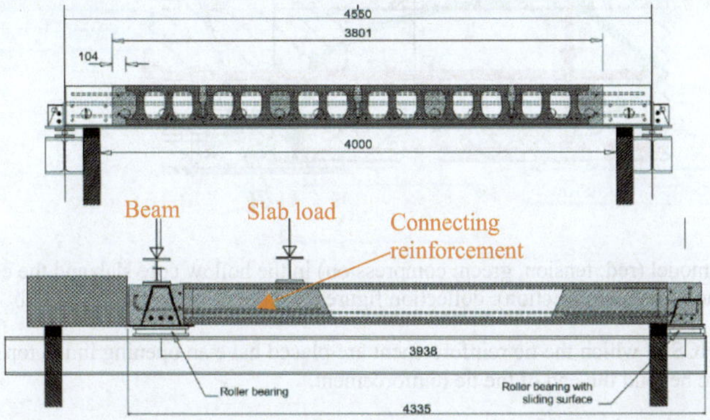

Figure 7 - Sections of system test setup (Deltabeam D32-300 with A32V slabs)

As observed in the previous SP tests, the bottom plate ledge followed the deflection of the slabs. No separation of the ledge and the soffit of the slabs was visible. The load transfer mechanism provided by the bottom plate ledge, the inclined web and the horizontal tie reinforcement was able to safely resist the applied loads. As described in section 2.2 of this paper, these observations indicate a gradual shift of the position of the support reaction towards the beam web during the fire exposure.

The strain gages applied to the tie reinforcement showed the onset of yielding of most reinforcement bars in the load transfer area after less than 20 minutes of fire exposure, which was likely caused by thermal expansion of the fire-exposed soffit of the specimen. The horizontal tie force indicates an inclination of the support reaction towards the beam web, as proposed in Figure 2-a.

The load transfer mechanism consisting of the heated bottom plate ledge, the concrete grout, the inclined Deltabeam web and the horizontal tie reinforcement anchored in the HCS elements was able to safely resist the vertical shear force on a level of more than 100% of the design vertical shear resistance of the HCS based on EN 1168, Annex G [6]. It is noted that this loading level is in most cases significantly higher than an additional limitation of the allowed vertical shear resistance of HCS in fire case of 0,6 x $V_{Rd,ct,II}$ imposed in Germany.

4 DESIGN FOR SHIFTED DIRECT SUPPORT

In a slim floor system consisting of Deltabeam and HCS, the Deltabeam bottom plate ledge supports the HCS at ambient temperatures. In fire situation an unprotected bottom plate of the Deltabeam quickly loses its bearing capacity (cf. Borgogno [2]). Hence, the HCS's direct support shifts from the Deltabeam bottom plate ledge to the inclined web of the Deltabeam: the vertical support force of the HCS equals the HCS's shear force $V_{Ed,fi}$ at the support and is modelled as a compression strut. The strut tilts and its horizontal component has to be anchored with a horizontal tie (see Figure 8). Due to compatibility a crack opens between Deltabeam top plate and the concrete grout.

The elements of the bearing mechanism including the strut-and-tie model at the support will be described in this section. The boundary conditions of these elements will be discussed.

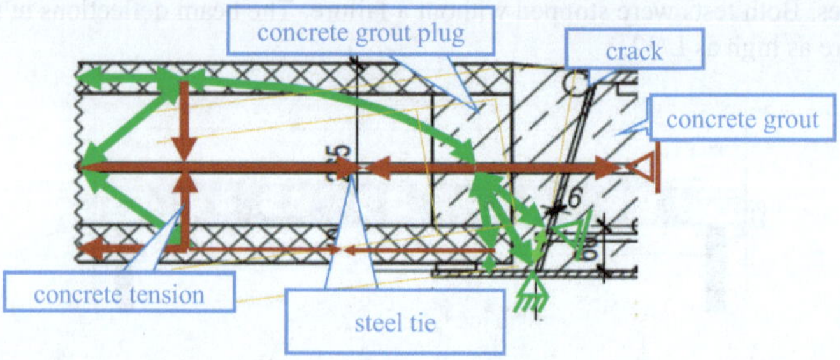

Figure 8 – Strut-and-tie model (red: tension, green: compression) in the hollow core slab and the composite structure of the Deltabeam (section); deflection figure (orange) of the hollow core slab

Note: The voids of the HCS in which the tie reinforcement are placed have an opening in the top hull for grouting and are grouted with concrete beyond the end of the tie reinforcement.

4.1 HCS's vertical shear resistance

Away from the support a linear strain distribution exists in the HCS. Hence, Borgogno [2] assumes shear force resistance $V_{Rd,c}$ resp. $V_{Rd,c,fi}$ according to EN 1992-1-1 [5] resp. EN 1992-1-2 [14] for cracked members without shear reinforcement in the fire situation. Here, the vertical shear force is in the non-cracked compression zone and in the webs of the HCS including the tendons running there.

Looking deeper into vertical shear resistance $V_{Rd,c}$ resp. $V_{Rd,c,fi}$ for cracked members without shear reinforcement, there are models for individual load-bearing mechanisms available, such as the aggregate interlock, the shear resistance of the compression zone and the dowel force, see Figure 9, and also the arch and tie mechanism (cf. Zilch et al. [15]). Nevertheless, so far [15] no consensus could be found on the realistic combination of the individual models for the determination of the failure load, which is why partly empirical solutions are resorted to.

For the **arch and tie mechanism**, the anchorage of the bottom reinforcement at the support is decisive. However, the shifting of the HCS's direct support has no influence on the anchorage of the pretensioned bonded tendons in the hollow slab. According to EN 1992-1-1 [5], the anchorage length of the pretensioned bonded tendons l_{bqd} does not depend on the support – in contrast to the anchorage length of reinforcing steel l_{bd}. Furthermore, when verifying the anchoring of the HCS's pretensioned bonded tendons according to Annex G of EN 1168 [6], the governing temperature of the tendon is calculated without the protective effect of a support. The essential boundary conditions of the arch and tie mechanism are thus complied.

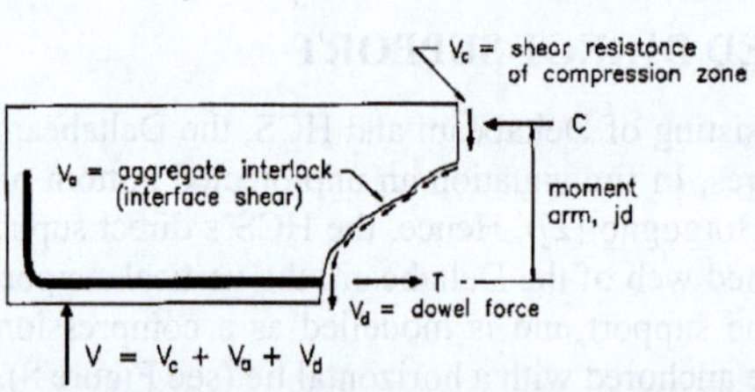

Figure 9 - Shear transfer mechanisms (cf. Kim et al. [16])

For the **shear resistance of the compression zone** plus aggregate interlock (interface shear) and dowel force, the vertical shear force is distributed evenly over the depth of the cross-section, see Figure 9. The compression zone exists mainly in the HCS's top flange but near the support it reaches down to the bottom flange – this is true also for the fire situation. This does not contradict the vertical shear resistance $V_{Rd,c,fi}$ in Annex G of EN 1168 [6].

This finding is validated by hybrid small-scale tests on the effects of an indirect support of HCS [4]. These tests show that vertical shear failure takes place in the investigated HCS itself and not at the concrete grout including the plug, see Figure 8. These tests furthermore show that the full vertical shear resistance (there: $V_{Rd,c}$ for uncracked cross-sections) is achieved even with indirect support and no reduction of the vertical shear resistance is required.

Hence, vertical shear resistance in fire situation $V_{Rd,c,fi}$ may be calculated according to EN 1168 [6].

4.2 Strut-and-tie model for shifted direct support

At the support of the HCS a non-linear strain distribution exists. Hence, a strut-and-tie-model is introduced there, see Figure 8.

1) At the support of the HCS, the shear force $V_{Ed,fi}$ creates struts in the bottom flange of the slab and via the flanks of the web also in the concrete grout plug (see Figure 8). A further strut builds up in the concrete grout via the face of the webs. The strut in the bottom flange is supported by the residual bearing capacity of the bottom plate ledge of the Deltabeam and thus by the Deltabeam cross-section.
2) The inclined strut in the concrete grout is supported by the web of the Deltabeam (see Figure 3). Since the strut is not perpendicular to the Deltabeam web in every case, there is usually a force component parallel to the web. This parallel force component is supported by friction between concrete and steel or by the "pocket" formed between the web and the load-bearing bottom plate ledge.
3) Due to the inclined compression strut, the HCS pushes away from the Deltabeam cross-section. To establish the force equilibrium, a tie F_s is needed. The tie is provided by horizontal reinforcing steel – taking into account its temperature in the fire case – and is anchored in the HCS by grouting concrete in the joint or the central hollow chamber. For the use of the Deltabeam as an edge beam, the reinforcement must also be anchored in the Deltabeam composite cross-section.
4) The Deltabeam is continuously “loaded” by the compression struts, but only "supported" every 60 cm by the joint reinforcement or the reinforcement in the central hollow chamber. Therefore, a small transverse bending moment acts in the Deltabeam.
5) A unilateral loading (e.g. edge beam) would lead to torsion in the Deltabeam. Bearing without torsion can be established by building up an equilibrium force at the top flange of the HCS.

The issues of the residual bearing capacity of the Deltabeam bottom plate ledge, the compression strut’s way into the concrete grout and the compression strut’s way into the Deltabeam cross section including its effect on the tie are discussed in detail in the following paragraphs.

4.2.1 Residual bearing capacity of the bottom plate ledge of the Deltabeam

The HCS on the Deltabeam bottom plate ledge leave a gap of 20 mm between the HCS’s face and the Deltabeam web, see Figure 10.

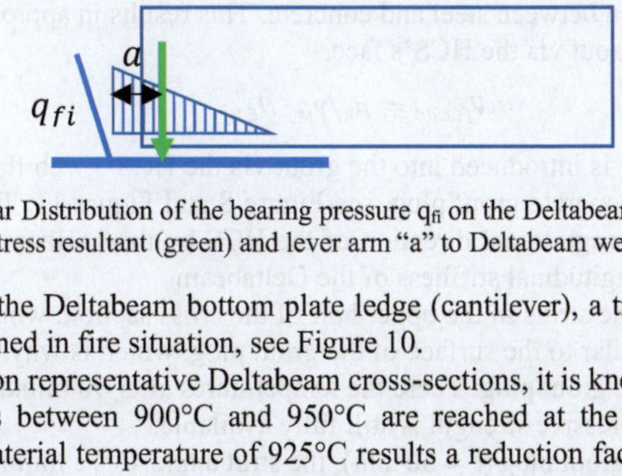

Figure 10 – Triangular Distribution of the bearing pressure q_{fi} on the Deltabeam bottom plate ledge; stress resultant (green) and lever arm “a” to Deltabeam web

Due to the stiffness of the Deltabeam bottom plate ledge (cantilever), a triangular distribution of the bearing pressure q_{fi} is assumed in fire situation, see Figure 10.

From thermal analysis on representative Deltabeam cross-sections, it is known that after 90 minutes of standard fire, temperatures between 900°C and 950°C are reached at the bottom plate. Considering EN 1993-1-2 [17] and a material temperature of 925°C results a reduction factor $k_{y,\vartheta}$=0.055 for the yield strength f_y at normal temperature (ϑ=20°C) and thus for the material S355 a temperature dependent yield strength $f_{y,\vartheta}$=0.055·35.5 kN/cm².

Given the usual bottom plate thickness of t=8 mm, S355 and ϑ=925°C, the plate’s resistance equals $M_{fi,,d}$=20.5 kNcm/m and $V_{fi,t,Rd}$=90.2 kN/m, using eq. (1) and (8) with symbols acc. to EN 1993-1-1 and EN 1993-1-2 [17].

$$M_{fi,t,Rd} = W \cdot f_y \cdot k_{y,\vartheta} \quad (1)$$

$$V_{fi,t,Rd} = A_v \cdot f_y \cdot k_{y,\vartheta}/\sqrt{3} \quad (2)$$

For a calculated lever arm a=10 mm, in the fire situation the bottom plate ledge resists a HCS's force of 20.5 kN/m. This results in a bearing pressure under the HCS with a resultant of 1/9·20.5 kN/m = 2.27 kN/m. The support provided by the residual load-bearing capacity of the Deltabeam bottom plate ledge is used as a "hidden reserve" and as a safety element. Additionally, a favourable bearing pressure for the HCS's support is generated.

4.2.2 Compression strut into the concrete grout

Borgogno states [2] that with the low residual bearing capacity of a beam's support flange in the fire situation the HCS's support force $V_{Ed,fi}$ must be introduced into the beam via interlock between concrete and grouting and a horizontal steel reinforcement. As validation laboratory tests with defined HCS and Deltabeams including system fire tests were undertaken.

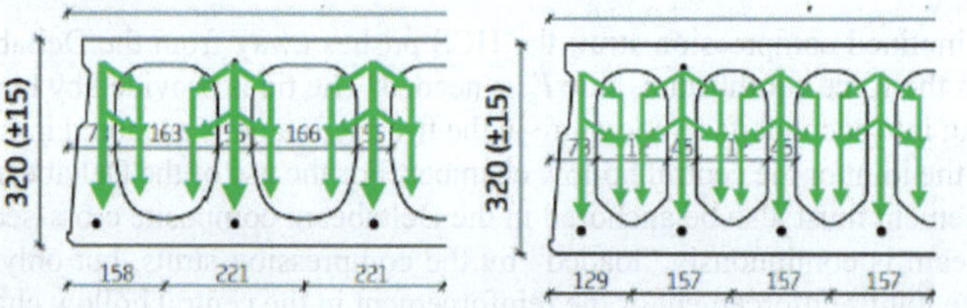

Figure 11 – HCS's section with introduction of the vertical shear force (green) into the concrete grout plug in the hollow core; resulting tensile forces (not shown) are borne by the Deltabeam

In the following, the load introduction will be quantified to be able to transfer the results found in the laboratory tests to other HCS types on Deltabeams. The introduction of $V_{Ed,fi}$ into the grout takes place continuously over the entire cross-section depth: via the **HCS's face** in the longitudinal direction, see Figure 8, and via the **HCS's web flank** in the transverse direction into the grout plug, see Figure 11.

To quantify the introduction of $V_{Ed,fi}$ coming from the HCS's face into the concrete grout, rules following the code DIN V 4141 [18] are used. Thus, the interface resistance $V_{fi,t,Rd}$ is calculated with eq. (3), with $N_{Ed,fi} = F_s$, the tie-force and $\gamma_M = 1{,}0$. The HCS's face is relatively smooth from being cut by a diamond saw in the production process. Hence, friction factor $\mu_k = 0{,}6$, following DIN V 4141 [18], in which 0,6 is used for friction between steel and concrete. This results in approximately $0{,}6 \cdot V_{Ed,fi}$ being introduced into the concrete grout via the HCS's face.

$$V_{fi,t,Rd} = \mu_k/\gamma_M \cdot N_{Ed,fi} \quad (3)$$

The remaining $0{,}4 \cdot V_{Ed,fi}$ is introduced into the grout via the HCS's web flank. At first inclined struts take the $0{,}4 \cdot V_{Ed,fi}$ into the concrete grout plug, see Figure 8 and Figure 11. The tensile forces resulting from the struts are borne in longitudinal direction of the HCS by the horizontal reinforcing steel and in transverse direction by the longitudinal stiffness of the Deltabeam.

According to Figure 11, the struts in the upper half of the cross section, which are inclined at an angle of approx. 45°, are perpendicular to the surface of the grout plug, which is why no shear forces form there between the web flank and the grout plug. There the temperatures after 90 minutes of standard fire are still so low that the concrete compressive strength is still fully available.

With the geometry of the grout plug (t = 50 mm), the strut angle 45°, a minimum compressive strength of the grouted concrete in the fire situation of $f_{ck,fi} \geq 20$ N/mm^2, a remaining $0{,}4 \cdot V_{Ed,fi}$ for the web flank and approximately 25% of the breadth b of the HCS applicable for the force introduction via the web flank, eq. (8) follows for the upper limit of $V_{Rd,c,fi}$ of a standard HCS with b = 1200 mm.

$$V_{Rd,c,fi} \leq \frac{1}{0,4} \cdot \sin 45° \cdot 0,25 \cdot 1200\,\text{mm} \cdot 50\,\text{mm} \cdot 20\,\text{N/mm}^2 = 530\,\text{kN} \tag{4}$$

A shear resistance of 530 kN is normally not reached by HCS and is thus normally not governing in the design process. This presented model for the save shear load introduction into the concrete grout is furthermore supported by the laboratory tests that show that shear failure does not take place in the concrete grout or grout plug.

4.2.3 Compression strut into the Deltabeam cross section and effect on tie

According to the German Deltabeam approval [19] the web inclination β of the Deltabeam is between 72° and 78° to its bottom plate. Therefore, the compression strut F_c is not perpendicular to the Deltabeam web and generates a force component $F_{c||}$ parallel to the web, see Figure 12.

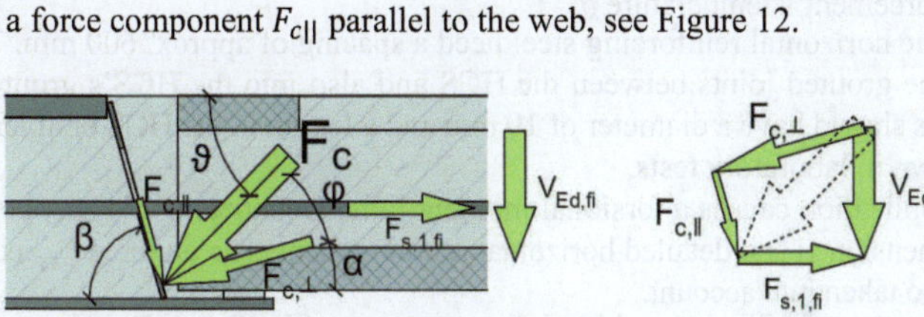

Figure 12 – Applied shear force $V_{Ed,fi}$ generates inclined compression strut F_c and tension strut F_s. F_c at $\vartheta = 42°$ generates force components $F_{c||}$ parallel to the inclined web and $F_{c\perp}$.

The forces (components) $F_{c||}$ and $F_{c\perp}$ are calculated according to eq. (5) and (8) with the angles from Figure 12.

$$F_{c||} = \sin(\varphi = 30°) \cdot F_c = 0,50 \cdot F_c \tag{5}$$

$$F_{c\perp} = \cos(\varphi = 30°) \cdot F_c = 0,866 \cdot F_c \tag{6}$$

Using the friction between concrete and steel web with $\mu_k = 0,6$ according to eq. (3) the resistance R exceeds the action E, see eq. (8).

$$R = 0,6 \cdot F_{c\perp} = 0,52 \cdot F_c > E = F_{c||} = 0,50 \cdot F_c \tag{7}$$

With the compression strut's angle $\vartheta = 42°$ according to Figure 12, the tie force F_s is calculated according to eq. (8) and must be accounted for with a steel reinforcement taking into account its temperature due to the standard fire.

$$F_s = \cot(\vartheta = 42°) \cdot V_{Ed,fi} = 1,11 \cdot V_{Ed,fi_c} \tag{8}$$

Taking into account, that the vertical shear resistance via a compression strut (see $V_{Rd,max}$ according to EN 1992-1-1 [5]) always exceeds the vertical shear resistance of a cracked concrete cross-section without shear reinforcement (see $V_{Rd,c}$ according to EN 1992-1-1 [5]) the introduction of the shear force $V_{Ed,fi} \leq V_{Rd,c,fi}$ from the HCS into the Deltabeam is validated.

4.3 Proposed design for shifted direct support

A design for the fire situation where the HCS's direct support shifts from the Deltabeam bottom plate ledge to the inclined web of the Deltabeam can be deduced from the presented strut-and-tie model and the laboratory tests. It provides rules for design and construction.

The design concept is valid for HCS according to DIN EN 1168 [6] with depth h and a resistance $V_{Rd,c,fi}$ in the fire situation and covers only the shifted direct support, but not the effect of a flexible support. Latter is subject to further investigations. The following requirements apply:

1) For HCS supported on Deltabeams, fire resistance classes REI 30, REI 60 and REI 90 may be verified.

2) When calculating the shear resistance $V_{Rd,c,fi}$ of the HCS in the fire situation according to DIN EN 1168 [6], it is allowed to assume a direct support and a support length $b_{AL} = 0$ mm for for calculating the anchorage length x. This could be shown in laboratory tests.
3) The HCS's depth h is 200 mm $\leq h \leq$ 500 mm. The reduction of the shear resistance by 10% according to DIN EN 1168 4.3.3.2.2.1 (3) [6] for HCS with $h \geq 450$ mm must also be accounted for in the fire situation.
4) The Deltabeam bottom plate $t_{fb} \geq 8$ mm and its web's inclination $\beta \leq 78°$.
5) The shifted direct support of the HCS with an acting shear force $V_{Ed,fi}$ at the support results in a tie force $F_s = 1{,}11 \cdot V_{Ed,fi} \leq A_s \cdot f_{sy,\vartheta}$. This tie force has to be accounted for with horizontal reinforcing steel, with steel's cross section A_s, the temperature-dependent steel yield strength $f_{sy,\vartheta}$ and the reinforcement's temperature ϑ.
6) The bars of the horizontal reinforcing steel need a spacing of approx. 600 mm. Thus, they should be put into the grouted joints between the HCS and also into the HCS's grouted central hollow core. The bars should have a diameter of 10 mm and a length in the HCS of at least 700 mm. This could be shown in laboratory tests.
7) If the load application causes a torsional moment inthe Deltabeam, this may be compensated by a suitably dimensioned and detailed horizontal reinforcement. The tie force F_{ss}according to clause 5) shall be also taken into account.

If these requirements are fulfilled, the shifted direct support of HCS on a Deltabeam is verified for the fire situation.

5 SUMMARY AND CONCLUSIONS

Based on extensive fire test references, a design model for the fire situation where the HCS's direct support shifts from the Deltabeam bottom plate ledge to the inclined web of the Deltabeam was developed, based on a strut-and-tie model. It provides rules for design and construction of the load transfer area between HCS and supporting Deltabeam and is based upon EN 1168 [6] with additional requirements.

The investigated issues are the basis for an ongoing DIBt construction method approval application. The proposed design model does not cover the effect of a flexible support, the latter being subject to further, ongoing investigations. For further investigation, it is suggested to enlarge the scope of application to other beam types and geometries since the current tests and analyses have been focusing on the distinct cross section of Deltabeam composite beams. The application of the design specifications described in this paper is limited to the boundary conditions defined by the special Deltabeam geometry.

It is suggested to carry out additional FEA calculations to obtain a deeper understanding of the load transfer mechanism and verify the load path suggested in this paper.

REFERENCES

[1] T. Roggendorf, „Zum Tragverhalten von Spannbeton-Fertigdecken bei biegeweicher Lagerung," Rheinisch-Westfälische Technische Hochschule Aachen, Fakultät für Bauingenieurwesen, Aachen, 2010.

[2] W. Borgogno und M. Fontana, „Tragverhalten von Slim Floor Decken mit Betonhohlplatten bei Raumtemperatur und Brandeinwirkungen," Eidgenössische Technische Hochschule, Institut für Baustatik und Konstruktion, Zürich, 1997.

[3] DIBt - Deutsches Institut für Bautechnik, „Z-15.10-279 - Allgemeine bauaufsichtliche Zulassung DW Systembau," DIBt, Berlin, 2014.

[4] J. Hegger, N. Kerkeni und H.-P. Doser, „Zur Tragfähigkeit von Spannbeton-Hohlplatten System VMM bei indirekter Auflagerung auf Peikko Delta-Träger," unveröffentlicht, Aachen, 2009.

[5] DIN - Deutsches Institut für Normung, NABau, „DIN EN 1992-1-1 +NA - Eurocode 2: Bemessung und Konstruktion von Stahlbeton- und Spannbetontragwerken - Teil 1-1: Allgemeine

Bemessungsregeln und Regeln für den Hochbau; Deutsche Fassung EN 1992-1-1:2004+AC2010,“ Beuth Verlag, Berlin, 2011-01.

[6] DIN Deutsches Institut für Normung, NA Bau, „DIN EN 1168 - Betonfertigteile - Hohlplatten,“ Beuth Verlag, Berlin, 2011.

[7] Thüringer Landesverwaltungsamt - Prüfamt für Standsicherheit, „Prüfbericht Nr. 4117.30-2774/2010-01 - Verlängerung und Änderung der statischen Typenprüfung Nr. 4117.20-012/01/06 und Nr. 4117.20-1192/2008, Spannbeton-Fertigdecken Typ VMM-L EPD,“ Weimar, 2010.

[8] Thüringer Landesverwaltungsamt - Prüfamt für Standsicherheit, „Prüfbericht Nr. 4117.30-2774/2010-02 - Verlängerung und Änderung der statischen Typenprüfung Nr. 4117.20-012/02/06, Spannbeton-Fertigdecken Typ VMM-VSD,“ Weimar, 2010.

[9] SP Technical Research Institute of Sweden; Hilling, Rolf, „Fire test of a loadbearing floor construction, P802216A,“ Boras, Sweden, 2010.

[10] SP Technical Research Institute of Sweden; Hilling, Rolf, „Fire test of a loadbearing floor construction, P802216B,“ Boras, Sweden, 2010.

[11] SP Technical Research Institute of Sweden; Hilling, Rolf, „Fire test of a loadbearing floor construction, P802216C,“ Boras, Sweden, 2010.

[12] SP Technical Research Institute of Sweden; Hilling, Rolf, „Fire test of a loadbearing floor construction, P802216D,“ Boras, Sweden, 2010.

[13] M. Cyllok, S. Pessel, S. Hothan und D. Häßler, „Gutachtliche Stellungnahme "Zur biegeweichen und indirekten Auflagerung von Spannbetonhohlplatten auf Slim-Floor-Trägern "DELTABEAM" im Brandfall - Abstimmung Validierungsversuche",“ 2018.

[14] DIN - Deutsches Institut für Normung, NABau, „DIN EN 1992-1-2 +NA - Eurocode 2: Bemessung und Konstruktion von Stahlbeton- und Spannbetontragwerken - Teil 1-2: Allgemeine Regeln - Tragwerksbemessung für den Brandfall; Deutsche Fassung EN 1992-1-2:2004 + AC:2008,“ Beuth Verlag, Berlin, 2010-12.

[15] K. Zilch und G. Zehetmaier, „Bemessung im konstruktiven Betonbau nach DIN 1045-1 und DIN EN 1992-1-1,“ Springer, Berlin, Heidelberg, 2006.

[16] J. Kim und Y. Park, „Prediction of shear strength of reinforced concrete beams without web reinforcement,“ *Materials,* Bd. 93, pp. 213-222, 1996.

[17] DIN - Deutsches Institut für Normung, NABau, „DIN EN 1993-1-2 + NA - Eurocode 3: Bemessung und Konstruktion von Stahlbauten - Teil 1-2: Allgemeine Regeln - Tragwerksbemessung für den Brandfall; Deutsche Fassung EN 1993-1-2:2005+AC:2009,“ Beuth Verlag, Berlin, 2010-12.

[18] DIN - Deutsches Institut für Normung, NABau, „DIN V 4141-1 - Lager im Bauwesen - Teil 1: Allgemeine Regelungen,“ Beuth Verlag, Berlin, 2003-05.

[19] DIBt - Deutsches Institut für Bautechnik, „Bauaufsichtliche Zulassung Z-26.2-49, DELTA Verbundträger,“ DIBt, Berlin, 2020.

9TH INTERNATIONAL CONFERENCE ON COMPOSITE CONSTRUCTION IN STEEL AND CONCRETE
M. Knobloch, U. Kuhlmann, W. Kurz, M. Schäfer. (Eds.)
Online, Germany, July 27-29, 2021

FINITE ELEMENT ANALYSES FOR EUROPEAN MODEL OF THE STEEL AND FIBRE REINFORCED CONCRETE CIRCULAR HOLLOW SECTION COLUMN IN FIRE

A. Tretyakov*, I. Tkalenko *, K. Cábová* and F. Wald*

*Czech Technical University in Prague, Czechia
e-mails: alexey.tretyakov@fsv.cvut.cz, illia.tkalenko@fsv.cvut.cz, kamila.cabova@fsv.cvut.cz, frantisek.wald@fsv.cvut.cz

Keywords: Steel and concrete columns, fibre reinforced concrete, fire design, Eurocode 4.

Abstract. *This paper is focuses on advanced numerical FEM model of the global mechanical behaviour of CHS columns with SFRC infill at elevated temperature description. The proposed model was validated on experimental data and examples from literature and it is a foundation for numerous of numerical simulations generation in aim to develop simplified design tool, which will follow the coming changes in European standard for composite member design in fire in EN1994-1-2:2020.*

1 INTRODUCTION

The recent development in technology of production and transport of the steel fibre-reinforced concrete (SFRC) enables its utilization in composite steel-concrete structures. Presence of steel fibres in mixture increases ductility, compressive and tensile strength and post cracking plastic capacity of concrete. Therefore, using of SFRC infill for composite columns is very beneficial for the member resistance at ambient temperature and mainly in fire situation. This work is a part of project which focuses on description of mechanical behaviour, advanced finite element method (FEM) models and design models of circular hollow section (CHS) composite steel and SFRC columns at elevated temperature.

In an experimental and numerical study conducted by Kodur [1] it was found that the addition of steel fibre reinforcement to the concrete core significantly increases the fire resistance of the element compared to plain concrete use and may be more economical than bar-reinforced concrete [2] and [3]. Recent experimental studies on axially loaded columns in a vertical furnace were made by Romero et al [4]. The European team of authors managed to prepare a European design approach for composite columns with plain and bar-reinforced concrete. Notwithstanding the advantages of using a SFRC core at elevated temperatures, the European analytical fire design model for CHS columns with SFRC infill has not been developed yet.

With regard to these questions, the GAČR 15-19073S project was prepared. Project focuses on the behaviour of steel fibre reinforced concrete filled circular columns at fire. The project carried out an extensive experimental study which included experiments on concrete cubes in compression and split tensile tests at elevated temperature [6], experiments on heat transfer in column cross-section [7] and experiments in four-point bending [8], axial and eccentric loading [9] on real size specimens. The experimental study was performed with the aim of developing an advanced numerical model, a simplified numerical model and an analytical design model for this type of columns.

This paper is focussed on the description of the development of an advanced numerical model of SFRC filled CHS columns in fire. The advanced model was developed in the FEM software ATENA Červenka. The calculation of temperature models was performed using the ATENA Transport module and calculation of the mechanical models in the ATENA Static module. All models were made in accordance with the recommendations of software approved procedures [10]. The preparation of an advanced model presents a complex question which involves the determination of thermal properties, mechanical properties and the way of interaction of steel and fibre concrete at ambient and elevated temperature.

2 FINITE ELEMENT MODEL

2.1 Thermal properties of materials

Entering the thermal properties of materials in the model is limited to two parameters, thermal conductivity and thermal capacity. The thermal material properties of the steel were modelled nonlinear in accordance with the relevant standard EN 1993-1-2 [11]. The thermal capacity and conductivity of concrete are greatly affected by its humidity. These parameters have been deeply studied accepted in accordance with EN 1992-1-2 [12]. The thermal capacity was entered for the lower limit and the thermal conductivity for 3 % concrete moisture. The thermal fire load was entered according to the standard fire curve EN 13501-2. The thermal model was validated on experiments performed in a vertical furnace. A comparison of the experimental results with the results of numerical model is shown in figure 1.

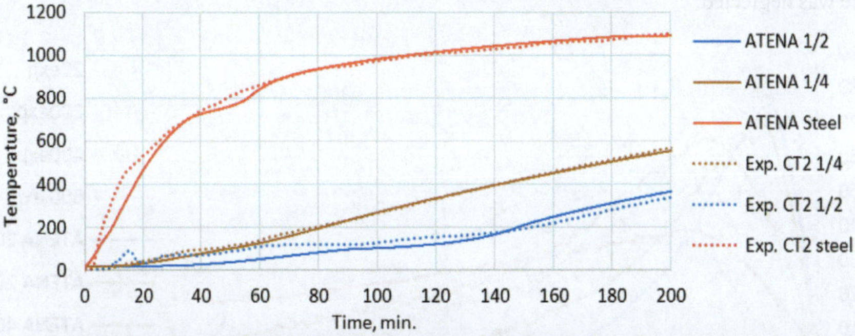

Figure 1. Comparison between thermal FEM model and experiment CT2 from [9].

2.2 Mechanical properties of materials at elevated temperature

The Mises condition is used to calculate steel structures in the ATENA software. The mechanical properties of the steel including modulus of elasticity, yield strength and thermal expansion. The mechanical properties of the steel and its degradation at elevated temperature were specified in accordance with the relevant standard EN 1993-1-2 [11] and verified on an extra tensile test model. The results of the calculation are shown on figure 2.

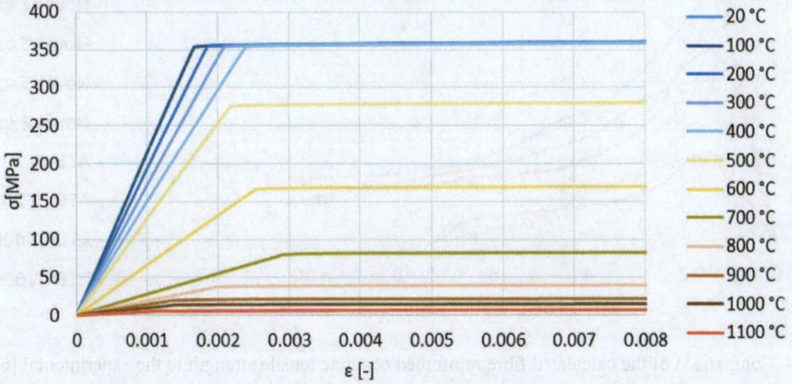

Figure 2. Calculated stress – strain relationship for S355 steel under [11]

The mechanical material model of SFRC was developed on the basis of an experimental study performed within the project GACR 15-19073S, Models of steel and fibre composite columns exposed to fire [7]. The study included compressive and split tensile experiments that were performed on concrete cubes at elevated temperature [6]. For the material model of concrete, it was necessary to determine the compressive strength, tensile strength, modulus of elasticity and fracture energy at normal and elevated temperature. Based on compressive test and tensile test models were developed to calibrate the mechanical properties. The NonLinCementitious2 model with the influence of temperature was used as the basic material. Compressive strength and modulus of elasticity settings were decisive for the compressive test. By adjusting the fracture energy and tensile strength, the correct plastic tensile reserve was achieved in the material model of concrete. Consistent with the experiment, the calculation was performed for four temperatures: 20, 200, 400 and 600 °C. Reduction factors for the degradation of mechanical properties at elevated temperatures were derived from the values for plain concrete recommended in EN1992-1-2 [12]. Comparison of results of FEM calculation and experiments from [6] see on figures 3 and 4. Derived reduction coefficients of degradation due to temperature are summarised in table 1. The thermal expansion of concrete was neglected.

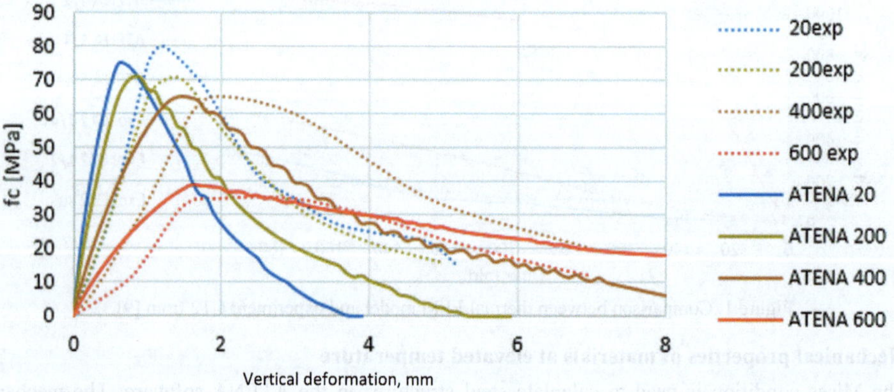

Figure 3. Comparison of the calculated fibre reinforced concrete compression strength to the experimental [6] one.

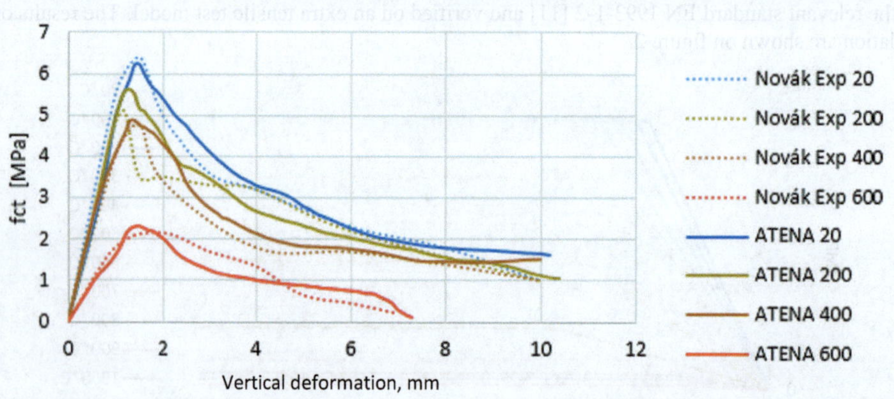

Figure 4. Comparison of the calculated fibre reinforced concrete tensile strength to the experimental [6] one.

Table 1: Reduction factors of fibre reinforced concrete at elevated temperature.

Temperature °C	20	200	400	600	800	1 000	1 100
$f_{c,fi}/f_c$	1.00	0.9	0.75	0.50	0.15	0.04	0.01
$f_{ct,fi}/f_{ct}$	1.00	0.84	0.74	0.13	0.01	0.01	0.01
E_{fi}/E	1.00	0.410	0.190	0.180	0.015	0.004	0.001
$G_{f,fi}/G_f$	1.00	0.88	0.68	0.33	0.25	0.23	0.20

2.3 Global model of column in fire

The global advanced numerical model consists of a thermal model performed in ATENA Transport module, and a mechanical model performed in ATENA Static module. In the thermal model, was calculated heat transfer in section under fire load. Subsequently, the cross-sectional temperature data are imported into a mechanical model, where the mechanical boundary conditions were specified and the corresponding materials were used, considering the degradation of material properties due to elevated temperature. Both models are created in a 3D environment. The rounded surfaces were idealized for a segmental shape. Shapes of the individual volumes were chosen in the form of quadrilaterals, so that it was easier to enter areas and volumes and more simple process of mesh setup was reached. The mesh was made up of hexahedra elements, which reduces the number of elements compared to using tetrahedra elements. With an assumed symmetrical deflection along the vertical axis the model was made as one half. View on the model cross-section and the geometry of the individual elements of which the steel hollow section and the column infill are composed so as to avoid inappropriately small angles in the volumetric elements, is shown on figure 5.

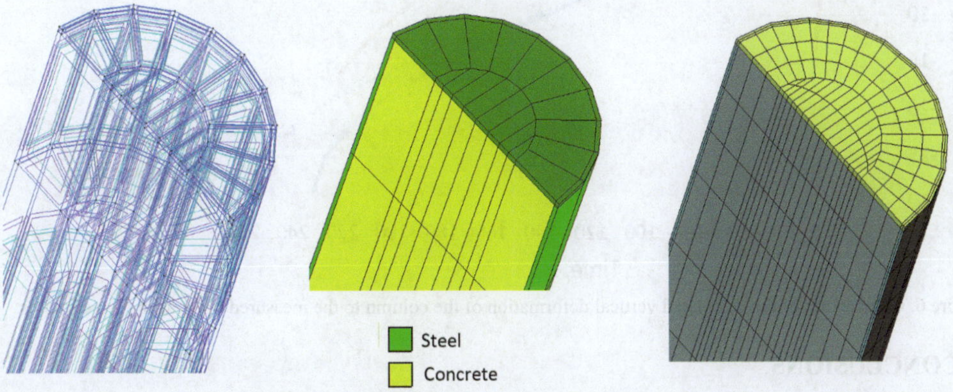

a) b) c)

Figure 5. Geometry of the model, top end view, a) surfaces and volumes, b) materials, c) mesh.

The calculation in the thermal model takes place within one load interval. In this step, the fire load is applied to the outer surfaces according to the standard fire curve, Fire Boundary for surface - Fire curve ETK. The monitoring points in the model are selected to record the calculated values of temperature. Dimensions of the elements, application of materials and the monitored points are chosen in accordance with the experiments.

The mechanical model repeats the geometry of the temperature model but with different materials and boundary conditions. A temperature-resistant elastic rigid material was used to specify the boundary conditions at the ends of the column. For the correct interaction of steel and concrete, contact volumes with interface material were settled between. The contact volume setting is specified according to the recommendations in the software manual. The calculation of the global column was performed in three loading intervals. In the first interval imperfections are applied as 1/1000 h, as recommendation according to [5]. To avoid effect of residual stresses soft elastic materials were used in first interval. In the second interval change of materials was applied and the vertical load on the column was entered. In the third interval the load from the thermal model was imported. The boundary conditions at the ends of the column are different for each interval. When performing validation on experiments from [13], it was found that definition of the supports on the ends of column plays a major role. Despite the fact that in both experiments specimens were supported with a hinges in the numerical model CT1 case corresponds to hinged support of both ends and for CT2 more to the fixed - fixed supports. In order to eliminate the effect of the error of supports the model was validated to experiments performed by Kodur [13], in which specimens were fixed in the furnace on both ends as fixed – fixed. Results of the calculations in comparison with results of the experiments CT1, CT2 and C69 [1] are shown on figure 6.

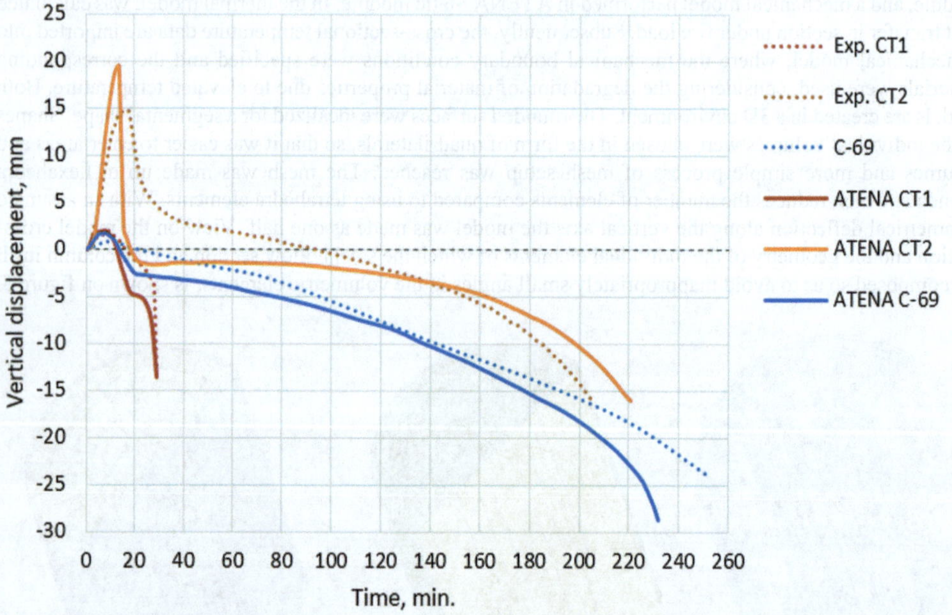

Figure 6. Comparison of the calculated vertical deformation of the column to the measured ones during experiments.

3 CONCLUSIONS

An advanced numerical model of the CHS columns with SFRC infill in fire was developed on the basis of an extensive experimental study. The mechanical properties of materials were approved on extra numerical models. A material model of SFRC was created in the interface of the ATENA software. The FEM results of global model analyses were validated on experiments performed in a vertical furnace. According to the validation performed, the developed model provides accurate results, see [14]. The columns with SFRC infill show higher fire resistance. The difference depends on slenderness of columns. Slender columns behave closer to steel ones.

The functionality of the model was validated on a larger selection of experiments. Based on an advanced numerical model, a series of numerical experiments was performed with the aim of developing an analytical design approach for the fire design of columns.

The paper was prepared under grant SGS22/141/OHK1/3T/11working on results of the project GAČR 15-19073S Models of steel and fibre concrete composite columns exposed to fire.

REFERENCES

[1] Kodur V. K. R. and Latour J. C., „*Experimental studies on the fire resistance of hollow steel columns filled with high-strength concrete,*" Institute for Research in Construction, National Research Council of Canada (NRCC) Ottawa, Canada., 2005. NRC-IRC-17882.

[2] Kodur V. K. R. and Lie T. T., „*Experimental Studies on the Fire Resistance of Circular Hollow Steel Columns Filled with Steel-Fibre-Reinforced Concrete*" NRC CNRC Internal Report, No. 691, 1995.

[3] Kodur V. K. R. and Mackinnon D. H., „Design of concrete filled hollow structural steel columns for fire endurance" *Engineering Journal, 13-24, 2000.*

[4] Romero M. L., Moliner V., Espinos A., Ibañez C. and Hospitaler A., „Fire behaviour of axially loaded slender high strength concrete-filled tubular columns," *Journal of Constructional Steel Research*, 12, 67, 1953-1965, 2011.

[5] Espinos A., Romero M. L. and Hospitaler A., „Advanced model for predicting the fire response of concrete filled tubular columns", *Journal of Constructional Steel Research,* 66, 1030-1046, 2010.

[6] Novak J. and Kohoutkova A., „Mechanical properties of concrete composites subject to elevated temperature", *Fire Safety Journal*, 95, 66-76, 2018.

[7] Štefan R., Sura J., Procházka J., Kohoutková A., Wald F., "Numerical investigation of slender reinforced concrete and steel-concrete composite columns at normal and high temperatures using sectional analysis and moment-curvature approach", *Engineering Structures*. 190, 285-305, 2019.

[8] Tretyakov A., „*Flexural stiffness of composite steel and fibre reinforced concrete columns at elevated temperature",* in Proceedings of the Seminar of doctoral students of the Department of Steel and Timber Structures, Praha, 2019.

[9] Tkalenko I., Tretyakov A., Wald F., Novak J., Stefan R. and Kohoutková A., „*The steel and fibre-reinforced concrete circular hollow section composite column exposed to fire",* in EUROSTEEL, Copenhagen, Denmark, 2017.

[10] Červenka V. and Červenka J., „*ATENA Program Documentation Part 2-2 User's Manual for ATENA 3D*", Červenka consulting s.r.o., Praha, 2017.

[11] *Eurocode 3: Design of steel structures - Part 1-2: General rules - Structural fire design,* Brussels, CEN, 2005.

[12] *Eurocode 2: Design of concrete structures - Part 1-2: General rules - Structural fire design,* Brussels, CEN, 2004.

[13] Lie T. T. and Kodur V. K. R., "Thermal and mechanical properties of steel fibre-reinforced concrete at elevated temperatures", *Canadian Journal of Civil Engineering,* 23, 1996.

[14] Tretyakov, A.; Tkalenko, I.; Wald, F. "Fire response model of the steel fibre reinforced concrete filled tubular column", *Journal of Constructional Steel Research*, 186, 2021. 10.1016/j.jcsr.2021.106884

9TH INTERNATIONAL CONFERENCE ON COMPOSITE CONSTRUCTION IN STEEL AND CONCRETE
M. Knobloch, U. Kuhlmann, W. Kurz, M. Schäfer. (Eds.)
Online, Germany, July 27-29, 2021

MEMBRANE ACTION OF REINFORCED CONCRETE SLABS IN FIRE

T. Hirashima*, F. Ozaki, T. Yoshida***, K. Kimura**** and J. Suzuki*******

* Chiba University, e-mail: hirashima@faculty.chiba-u.jp
** Nagoya University, e-mail: ozaki@nuac.nagoya-u.ac.jp
*** JFE Steel, e-mail: tor-yoshida@jfe-steel.co.jp
**** Nippon Steel, e-mail: kimura.ze8.kei@jp.nipponsteel.com
***** Building Research Institute, e-mail: junichi-szk@kenken.go.jp

Keywords: RC slabs, Membrane action, Composite flooring system, Load-bearing fire tests.

Abstract. *For RC slabs supported in 2 ways, when deflection increases, membrane stress occurs within the cross section of the slab, and as a result the load-bearing capacity of the slab is improved due to this membrane action. In this paper, the results of load-bearing fire test are introduced that targets a composite flooring system composed of a flat RC slab in which two-tier reinforcements are arranged and an unprotected beam. The RC slab to which the bending collapse load based on the yield-line theory was given sustained load-bearing capacity even after being subjected to 3 hours and more of standard fire heating.*

1 INTRODUCTION

In case of fire, the wall and floor slab applied as compartment elements are required to possess thermal insulation and fire integrity, and further the column, beam, floor slab and bearing wall are required to possess appropriate load-bearing capacity. Accordingly, it is general practice in steel-structure construction to provide fire protection with columns and beams and to ensure fire resistance. This also applies to the secondary beams that support the 1-way slabs. Meanwhile, it may be feasible to eliminate the application of fire protection for secondary beams located in the middle of the 2-way slabs.

When the unprotected steel beam applied as a secondary beam is exposed to fire, it loses strength at an early stage of fire. On the other hand, for reinforce concrete (RC) slabs supported in 2 ways, when deflection increases, membrane stress occurs within the cross section of the slab, and as a result the load bearing capacity of the slab is improved due to this membrane action. In RC slabs subjected to heating from the lower side during fire, because of large deflection caused by reduction in the rigidity of the material and thermal deflection due to the temperature difference between the upper and lower sides of slab, the membrane action becomes more striking at that stage of fire than at the room temperature stage, and as a result the slab can withstand fire heating for a long time [1 - 4]. The load-carrying capacity of an RC slab during fire surpasses by a great margin the bending collapse load based on yield-line theory, and the ratio of the load-bearing capacity to the bending collapse load depends on the amount of deflection [5]. Further, a characteristic feature of the membrane action of RC slab lies in that compression stress occurs to the circumferential direction of the peripheral section of the RC slab and this compression stress balances the tensile stress widely occurring at the centre section of slab, and as a result it is unnecessary to restrict the horizontal-direction displacement at the four peripheral sections of the slab [6].

In flat RC slabs in which two-tier reinforcements are arranged, when subjected to fire heating, the temperature at the upper-side reinforcement becomes lower than that at the lower-side reinforcement, and thus the tensile strength at the upper-side reinforcement is kept to a high level for a long time. As a result, it is forecast that the retention of strength by the slab due to the membrane action during fire will become more conspicuous. In order to understand these behaviours, a load-bearing fire test was carried out by means of standard fire heating in accordance with ISO 834-1, targeting a composite flooring system composed of a two-tier reinforcement-arranged flat RC slab and an unprotected steel beam. In this paper, the results of load-bearing fire test are introduced that targets a composite flooring system composed of a flat RC slab and an unprotected beam.

2 OUTLINE OF TESTING

2.1 Test Conditions

Table 1 shows the conditions of the load-bearing fire test. The first aim was to understand the effect of the headed stud arranged at the periphery of the slab on the deflection behaviour of the RC slab during fire by comparing specimens No. 1 and No. 2. The second aim was to understand the effect of applying unprotected secondary beams on the fire resistance of the composite flooring system by comparing specimens No. 2 and No. 3.

In the test, a load of 11.23 kN/m², including the dead load of the floor slab, was given as the target value. The live load was the collapse load based on yield-line theory in the case when the RC slab was simply supported at the four peripheral sections of the slab. The collapse load was calculated using the following Equation (1) paying no attention to upper-side reinforcement.

$$W_{p,AT} = \frac{24 \times 0.9\, d_L\, A_{s,L}\, \sigma_{y,AT}}{l^2} \left(\frac{n}{\sqrt{1+3n^2}-1} \right)^2 \qquad (1)$$

where,

$W_{p,AT}$: Collapse load of the slab at ambient temperature based on yield-line theory (=11.23 kN/m²)

l : Short-side span (= 4.2 m)

n : Aspect ratio, i.e. ratio of long-side span to short-side span (= 1.38)

d_L : Effective depth of lower-side reinforcement (= 0.09 m)

$A_{s,L}$: Cross-sectional Area of lower-side reinforcement per unit width (= 366.6 mm²/m)

$\sigma_{y,AT}$: Yield stress of lower-side reinforcement shown in mill sheet (= 380 x 10^{-3} kN/mm²)

The live load was set at 8.35 kN/m², assuming that a unit weight of concrete applied to the floor slab is 24 kN/m³. The test specimen was heated based on the standard time/temperature curve in accordance with ISO 834-1.

2.2 Test Specimens

Fig. 1 shows an outline of specimen No. 1. Fig. 2 shows the specimen before the placement of concrete. The span of the flooring system specimen was set at 5.8 x 4.2 m in a distance between primary beam cores. The thickness of the flat RC slab was set at 120 mm, and a 1 mm-thick flat deck was applied as a concrete permanent form.

Regarding specimen No. 1, the headed stud was arranged in a primary steel beam that supported the four peripheral sections of the RC slab so as to restrict rotational- and horizontal-direction displacement at the slab edge under conditions roughly similar to those in a practical steel structure. The unprotected steel secondary beam was structured to a complete composite beam together with the RC slab. The cross section of the secondary beam was H-300 x 150 x 6.5 x 9, and the steel grade adopted was SN400B (JIS G 3136). The lower-limit yield point of the SN400B was 235 N/mm², and the lower-limit tensile strength was 400 N/mm². Three high-strength bolts (diameter: 20 mm) were arranged at the end of the secondary beam web to form a pin joint as shown in Fig. 3. The cross section of the primary steel beam applied for the peripheral section of the slab was H-350 x 175 x 7 x11. The orthogonal primary steel beams were joined by means of full-penetration flange welding.

Regarding specimen No. 2, the same specifications as those of specimen No. 1 were adopted except for the headed stud arranged in the primary steel beam that supported the four peripheral sections of the floor slab.

Regarding specimen No. 3, the same specifications as those of specimen No. 2 were adopted except for the secondary beam arranged in the centre section of the floor slab. That is, specimen No. 3 had a structure in which the RC slab was only arranged on the primary steel beam.

Deformed steel bars (nominal diameter: 10 mm) were applied for RC slab reinforcement and arranged at 200 mm-intervals between reinforcing bars for both the upper- and lower-side reinforcements and with a covering thickness of 20 mm. According to the mill sheet for the deformed steel bar applied, its yield stress was 380 N/mm^2. Table 2 shows mix proportion of the concrete. The water to cement ratio of the concrete was set at 51.9%, and polypropylene fibre was mixed with the concrete to prevent explosions in the case of fire. The strength of the concrete after 28-days aging was 36.1 N/mm^2. A ceramic-wool blanket with a thickness of 100 mm for use as fire protection was provided to the heating side of the peripheral primary steel beams, and the same fire-protection blanket with a thickness of 25 mm was provided to the high-strength bolt joints at the end of the secondary beam.

2.3 Test Methods

The test was conducted using a heating furnace for horizontal elements operated by the Building Research Institute of Japan. The heating area at the lower surface of the RC slab was set at 5.6 x 4.0 m. A total of 24 weights (6 x 4 weights) were placed at equal intervals on the upper surface of the RC slab, and the slab was then heated from its lower surface. The weight per weight was set at 8.62 kN. Fig. 4 shows the specimen on which the weight was placed.

The vertical displacement of the slab was measured at 18 positions at the centre section of the short and long sides and on the diagonal line. The temperature of the unprotected secondary beam was measured at the centre section and at the end of the bolt. The temperature within the cross section of the RC slab was measured at the centre section and at a position at a distance 1/4 of the diagonal line, as shown in Fig. 1. The temperature of reinforcement within the RC slab was measured at positions where the long- and short-side reinforcement bars crossed or contacted each other.

2.4 Mechanical Properties of Steel Products Applied at Elevated Temperatures

The high-temperature tension test based on JIS G 0567 was conducted for reinforcing bars and secondary beams used for the specimens in the current load bearing fire test. Figs. 5 (a) and (b) show the relationship between the 0.2% proof stress, stress at 1% strain and tensile strength and the temperature, which was obtained from the test results. The effective yield stress shown in the figure is the value for the stress at 1% strain.

As shown in Fig. 5 (a), the strength of the reinforcing steel bars was similar to that at ambient temperature before the temperature reached 400°C, but the strength rapidly lowered when the temperature surpassed 400°C, and when the temperature reached 700°C, the strength lowered to a level of 1/4 to 1/5 that at ambient temperature. The trend of strength reduction was nearly close to the high-temperature strength retention rate of hot-rolled reinforcing bars in Eurocode 2 [7].

As shown in Fig. 5 (b), the strength of the steel member used as the unprotected secondary beam lowered to a level of nearly 1/10 that at ambient temperature when the temperature surpassed 800°C. It is forecast from these test results that unprotected secondary beam subjected to standard fire heating would lose its strength in an earlier stage of a fire.

Table 1: Parameters in load bearing fire tests

Specimen	Headed stud at the ends of the slab	Unprotected secondary steel beam	Load including dead load of the slab	Heating
No.1	diameter: 15.4 mm, distance: 150mm	H-300 x 150 x 6.5 x 9 (SN400)	11.23 kN/m^2	Standard fire curve in accordance with ISO 834-1
No.2	None			
No.3		None		

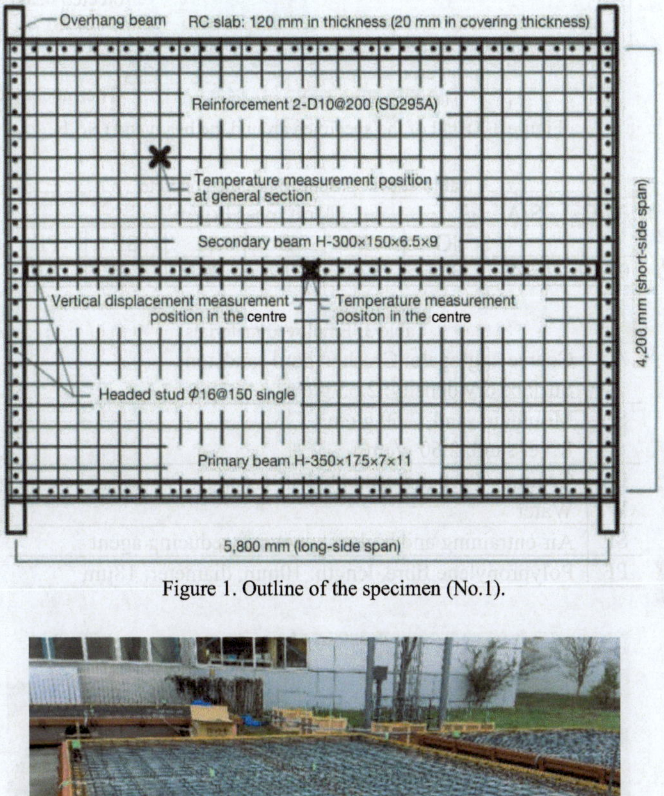

Figure 1. Outline of the specimen (No.1).

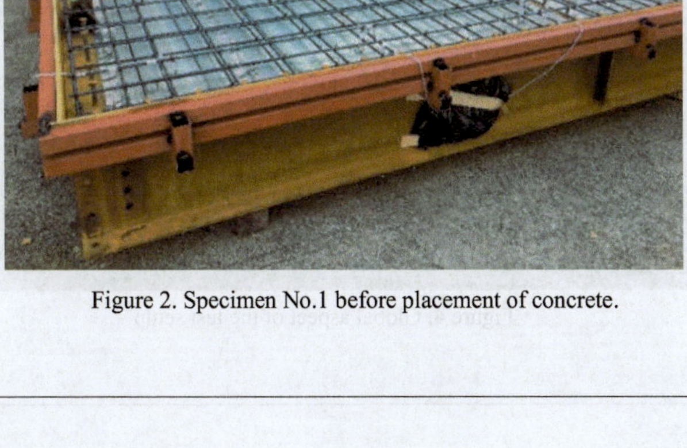

Figure 2. Specimen No.1 before placement of concrete.

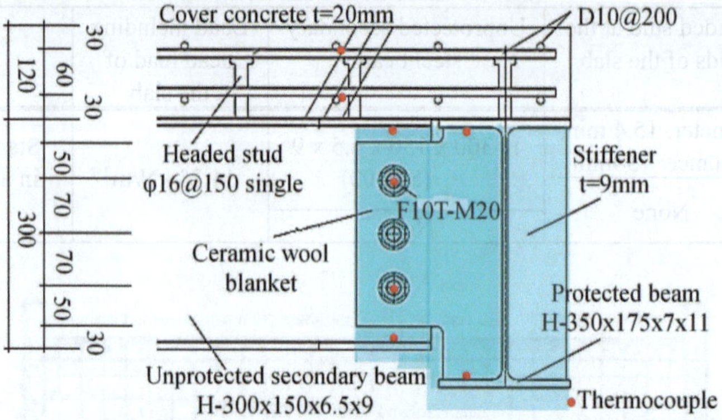

Figure 3. Detail of the specimen around the beam end (No.1).

Table 2: Mix proportion of the concrete

W/C	S/A	Unit mass [kg/m³]					
		C	W	S	A	SP	PF
0.519	0.443	345	179	767	991	3.45	1.0

Table 3: Employed materials

A	Coarse aggregate: Crushed hard sandstone Surface dry density: 2.65 g/cm³
S	Mountain sand, 2.59 g/cm³ River sand, 2.60 g/cm³
C	Ordinary Portland cement, 3.15 g/cm³
W	Water
SP	Air entraining and high-range water reducing agent
PF	Polypropylene fibre, length: 10mm, diameter: 18μm

Figure 4. Global aspect of the test setup

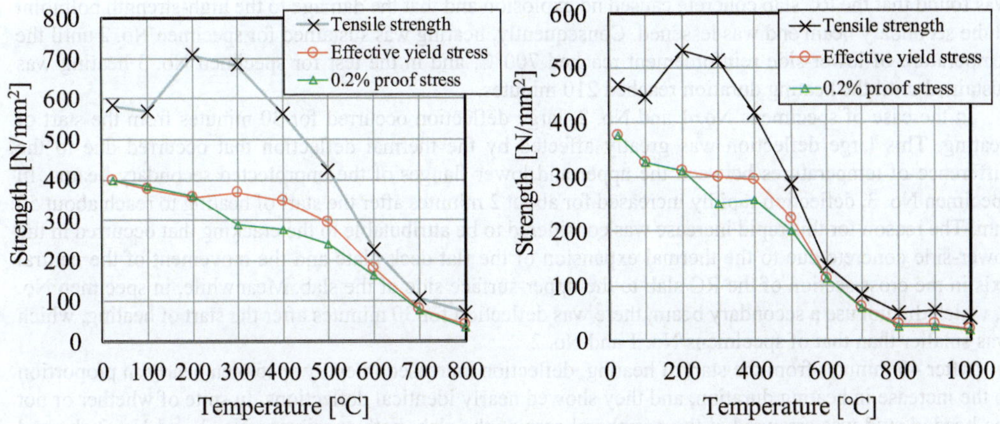

(a) Reinforcement (SD295) (b) Secondary steel beam (SN400B)

Figure 5. High-temperature strength of steel products used for specimens.

3 TEST RESULTS

3.1 Temperature Measurement Results

Fig. 6 shows the temperature measurement results for specimen No. 2 as an example of the temperature measurements results for the composite RC flooring system. In the test of specimen No. 2, heating stopped at the stage when the temperature of the lower-side reinforcement in the slab reached 700°C. The heating duration was 202 minutes. The temperature inside the heating furnace was controlled following the standard fire curve in accordance with ISO 834-1.

The temperature of the lower flange of the unprotected beam surpassed 700°C after 30 minutes of heating, and 1,000°C after 120 minutes of heating. The temperature of the web reached a level similar to that of the upper flange. The temperature of the upper flange came closer to that of the lower flange as the heating duration increased, and it surpassed 950°C after 120 minutes of heating. The temperature of the RC slab lower surface was low in comparison with the temperature inside the furnace, and it reached about 800°C after 120 minutes of heating. For the test specimens in this test, a flat deck was applied as a permanent form. Therefore, the lower surface of the RC slab was not exposed directly to flames, and a gap occurred between the flat deck and the RC slab lower surface during heating, and as a result it was considered that the temperature of the RC slab lower surface was lower than that inside the furnace.

While the temperature of the lower-side reinforcement in the RC slab was low, or about 250°C after 60 minutes of heating, due to stagnation from the temperature rising in the vicinity of 100°C accompanied by the evaporation of water contained in the concrete, and the temperature then increased as the heating duration lapsed, reaching about 650°C after 180 minutes of heating. On the other hand, the temperature of the upper-side reinforcement in the RC slab showed a longer stagnation time from the rising temperature at 100°C compared to the lower-side reinforcement, and it was 200°C even after 180 minutes of heating. The temperature of the RC slab upper surface was about 120°C even after 180 minutes of heating, and the difference in temperatures between the upper and lower-surfaces after 180 minutes of heating was about 700°C.

3.2 Deflection at the Centre Section of RC Slabs

Fig. 7 shows changes in deflection occurring in the centre section of the RC slab. The heating duration was 132 minutes for specimen No. 1, 202 minutes for specimen No. 2 and 217 minutes for specimen No. 3. Because specimen No. 1 was first subjected to the test, heating was stopped when the temperature of the lower-side reinforcement reached 500°C. When the damage condition of specimen No. 1 was confirmed, it was found that the RC slab concrete caused no explosion and that the damage to the high-strength bolt joint at the secondary beam end was lessened. Consequently, heating was sustained for specimen No. 2 until the temperature of lower-side reinforcement reached 700°C, and in the test for specimen No. 3 heating was sustained until the heating duration reached 210 minutes.

In the case of specimens No. 1 and No. 2, large deflection occurred for 30 minutes from the start of heating. This large deflection was greatly affected by the thermal deflection that occurred due to the difference of temperatures between the upper and lower flanges of the unprotected secondary beams. In specimen No. 3, deflection rapidly increased for about 2 minutes after the start of heating to reach about 50 mm. The reason for this rapid increase was considered to be attributable to the cracking that occurred in the lower-side concrete due to the thermal expansion of the flat deck plate and the movement of the neutral axis in the cross section of the RC slab to the upper-surface side of the slab. Meanwhile, in specimen No. 3, which did not use a secondary beam, there was deflection for 30 minutes after the start of heating, which was smaller than that of specimens No. 1 and No. 2.

After 40 minutes from the start of heating, deflection increased in every specimen nearly in proportion to the increase in heating duration, and they showed nearly identical deflections. In spite of whether or not the headed stud was arranged at the peripheral area of the slab, both specimens No. 1 and No. 2 showed nearly identical deflection behaviours, and therefore it was found that the effect of the restraint of rotational and horizontal displacement at the slab end on the deflection behaviour of composite flooring systems during fire was small. Further, in spite of whether or not the unprotected secondary beam was applied, both specimens No. 2 and No. 3 showed nearly identical deflection behaviours after 40 minutes from the start of heating, and therefore it was considered that the live load of the composite flooring system was supported only by the RC slab after 40 minutes from the start of heating. Further, these test results can clearly be understood from that the lower flange temperature surpassed 800°C after 40 minutes from the start of heating in the temperature measurement results in Fig .6 and that the strength of the secondary beam at 800°C lowered to nearly 1/10 that at ambient temperature as shown in Fig. 5 (b).

The maximum deflection of specimens No.2 and No. 3 after stopping heating was 1/15 the deflection of the short-side span of the RC slab. Because large deflection occurred in the slab, the membrane stress occurred within the slab, and it was considered that the slab could sustain the load due to this membrane action. In addition, because the temperature of the upper-side reinforcement was low, it was considered that the tensile resistance of the upper-side reinforcement worked effectively in terms of membrane action. Further, these test results can be understood by comparing the temperature of the upper and lower-side reinforcements after 3 hours from the start of heating shown in Fig. 7 and the high-temperature strength of reinforcing steel bars shown in Fig. 5 (a).

While heating was stopped for specimen No. 1 after 132 minutes, when taking into account the test results for specimens No. 2 and No. 3, it was considered that specimen No. 1 sustained the load-bearing capacity for 3 hours or more as with the case of specimens No. 2 and No. 3.

3.3 Crack Occurrence Condition at Upper Surface of RC Slab

Figs. 8 (a) and (b) show the occurrence of cracking on the upper surface of the RC slab. In specimen No. 1 shown in Fig. 8 (a), because the heating duration was short, the cracking width in the centre section of the RC slab was narrow, but because the end of the RC slab was restricted, it showed remarkable cracking in the vicinity of the end. In specimen No. 2 shown in Fig. 8 (b), large cracks vertical to the long-side direction occurred in the centre section of the RC slab. In the centre section of the RC slab, it was considered that the full cross section experienced a tension region in the long-side direction area. Meanwhile, no cracking vertical to the short-side direction was observed. Because the curvature in the short-side direction area was large, it was considered that the compression due to bending and the tension due to membrane stress were offset at the upper surface of the slab. In specimen No. 3, the cracking condition was similar to that of specimen No. 2.

In specimen No. 1, oval-shaped cracking was observed in the vicinity of the peripheral section. In specimens No. 2 and No. 3, diagonal cracking like diamond-shaped cracking was observed. The reason for such cracking was considered to be the suppression of floating of the RC slab at the corner section by the weight arranged in the corner section. At the stage where the membrane action of the RC slab worked, the tension force occurred to the radial direction in the centre section of the RC slab, and the compression force occurred to the circumferential direction in the peripheral section of the RC slab.

The trends in cracking in the RC slabs under the condition of the membrane action were confirmed from the cracking conditions in these three test specimens used in this load-bearing fire test for the composite flooring system.

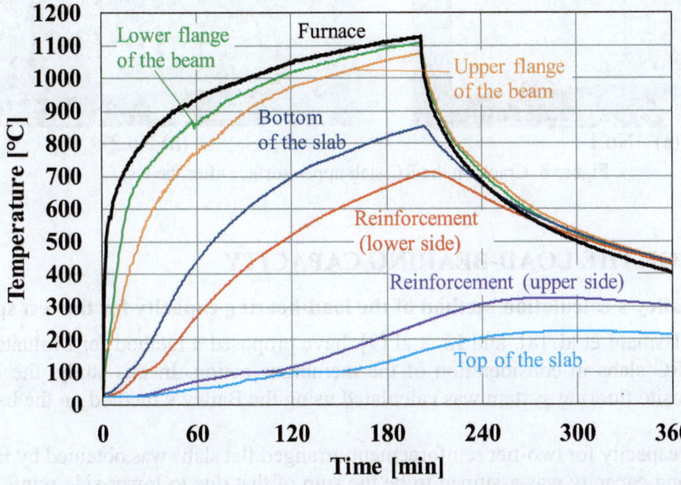

Figure 6. Temperature measurement results (specimen No. 2).

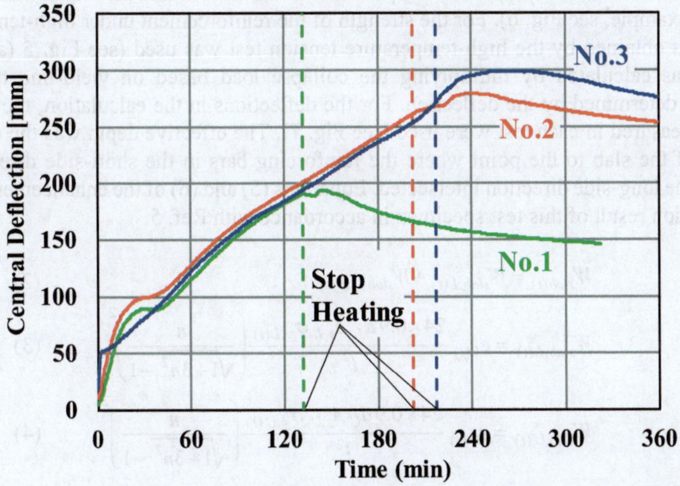

Figure 7. Temperature measurement results (specimen No. 2).

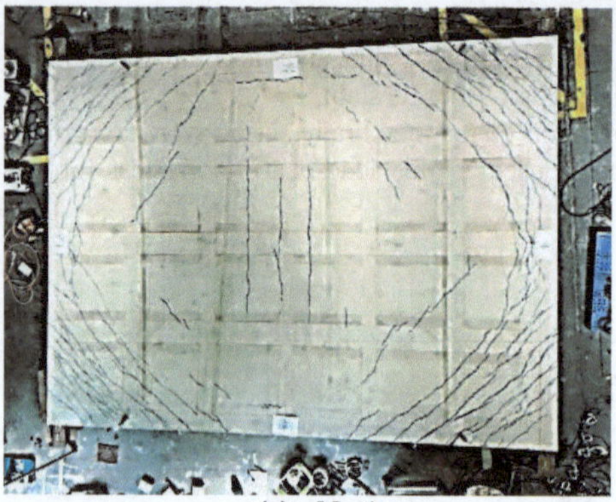

(a) No.1

(b) No.2

Figure 8. Crack on the RC slab upper-surface after the test.

4 DISCUSSION OF THE LOAD-BEARING CAPACITY

4.1 Application of Bailey's calculation method of the load-bearing capacity for the test specimens.

Bailey et al [5]., Usmani et al. [8], and Li et al [9]. have proposed a method for evaluating the load-bearing capacity of RC slabs in consideration of the membrane action. In this study, the load-bearing capacity of the composite flooring system was calculated using the Bailey's method on the basis of yield-line theory model [5].

The load-bearing capacity for two-tier reinforcement-arranged flat slabs was obtained by Equations (2) to (6). The load-bearing capacity was assumed to be the sum of that due to lower-side reinforcement and that of due to upper-side reinforcement. Crushing of the concrete was not considered. Each load bearing capacity depends on the high-temperature strength of the reinforcing steel bars, which changes with the passage of heating time. For the reinforcement temperatures in the calculation, the values measured in each test were used (for example, see Fig. 6). For the strength of the reinforcement under high-temperature, the effective yield stress obtained by the high-temperature tension test was used (see Fig. 5 (a)). The load-bearing capacity was calculated by multiplying the collapse load based on yield-line theory by the enhancement factor determined by the deflection. For the deflections in the calculation, the values of the central deflection measured in each test were used (see Fig. 7). The effective depth was the distance from the upper surface of the slab to the point where the reinforcing bars in the short-side direction and the reinforcing bars in the long-side direction intersected. Equations (5) and (6) of the enhancement factor were obtained by calculation result of this test specimen in accordance with Ref. 5.

$$W_{slab(t)} = W_{slab,L(t)} + W_{slab,U(t)} \tag{2}$$

$$W_{slab,L(t)} = e_{L(t)} \frac{24 \times 0.9 d_L A_{s,L} \sigma_{y,L(t)}}{l^2} \left(\frac{n}{\sqrt{1+3n^2}-1} \right)^2 \tag{3}$$

$$W_{slab,U(t)} = e_{U(t)} \frac{24 \times 0.9 d_U A_{s,U} \sigma_{y,U(t)}}{l^2} \left(\frac{n}{\sqrt{1+3n^2}-1} \right)^2 \tag{4}$$

$$e_{L(t)} = 0.252 \frac{v_{(t)}}{d_L} + 0.93 \tag{5}$$

$$e_{U(t)} = 0.252 \frac{v_{(t)}}{d_U} + 0.93 \tag{6}$$

where,

$W_{slab(t)}$: Load-bearing capacity of the slab at time t under high-temperature (kN/m²)

$W_{slab,L(t)}$: Load-bearing capacity of the slab due to lower-side reinforcement at time t (kN/m²)

$W_{slab,U(t)}$: Load-bearing capacity of the slab due to upper-side reinforcement at time t (kN/m²)

$e_{L(t)}$ ($e_{U(t)}$): Enhancement factor of the slab due to lower-side (upper-side) reinforcement at time t

l : Short-side span (= 4.2 m), n : Aspect ratio (= 1.38)

d_L (d_U): Effective depth of lower-side (upper-side) reinforcement (d_L = 0.090 m, d_U = 0.030 m)

$A_{s,L}$ ($A_{s,U}$): Cross-sectional Area of lower-side (upper-side) reinforcement per unit width (mm²/m)

$\sigma_{y,L(t)}$ ($\sigma_{y,U(t)}$): Effective yield stress of lower-side (upper-side) reinforcement at time t under high-temperature (kN/mm²)

$v_{(t)}$: Central deflection of the slab at time t (m)

The load-bearing capacity for composite flooring systems was obtained by Equations (7) and (8). For the temperature of unprotected secondary steel beam in the calculation, the average value of each steel temperatures at the central section of the beam measured in the test was used (see Fig. 6). For the strength of the steel beam under high-temperature, the effective yield stress obtained by the high-temperature tension test was used (see Fig. 5 (b)).

$$W_{total(t)} = W_{slab(t)} + W_{beam(t)} \quad (7)$$

$$W_{beam(t)} = \frac{8}{(nl)^2 (l/2)} Z_p \, \sigma_{y,b(t)} \quad (8)$$

where,

$W_{total(t)}$: Load-bearing capacity of the flooring system at time t under higt-temperature (kN/m²)

$W_{beam(t)}$: Load-bearing capacity due to the unprotected beam at time t under higt-temperature (kN/m²)

Z_p : Plastic section modulus of the beam (= 5.42 x 10^{-4} m³)

$\sigma_{y,b(t)}$: Effective yield stress of the unprotected beam at time t under high-temperature (kN/m²)

4.2 Calculation results of the load-bearing capacity

Fig. 9 shows the calculation result of the load-bearing capacity for the two-tier reinforcement-arranged flat slab (regarding specimen No.3). The initial value of the load-bearing capacity of the slab due to lower-side reinforcement $W_{slab,\,L(t)}$ was the same as the applied constant load in the test determined by Equation (1). $W_{slab,\,L(t)}$ decreased from about 120 minutes when the temperature at the lower-side reinforcement surpassed 500°C, and became lower than the load-bearing capacity of the slab due to upper-side reinforcement $W_{slab,\,U(t)}$ after 160 minutes of heating. $W_{slab,\,U(t)}$ continued to increase as the deflection increased because the temperature at the upper-side reinforcement was below 300°C during the heating. Fig. 9 indicates that the upper-side reinforcement contributed considerably to supporting the applied live load. It seemed that specimen No.3 had not reached the limit state although the load-bearing capacity was close to the applied load when the heating was stopped.

Fig. 10 shows the calculation result of the load-bearing capacity for the composite flooring system (regarding specimen No.2). Load-bearing capacity due to the unprotected beam $W_{beam(t)}$ decreased significantly after the start of heating and was lost after 60 minutes of heating. The change in the load-bearing capacity of the slab $W_{slab(t)}$ of test No.2 was similar to that of test No.3. Fig. 10 indicates that the applied live load was supported by the slab and the unprotected secondary beam did not contribute to supporting the load.

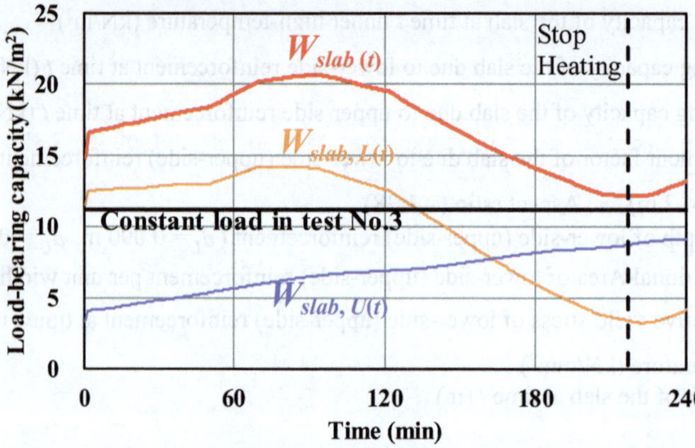

Figure 9. Calculation result of the load-bearing capacity of the slab (test No.3).

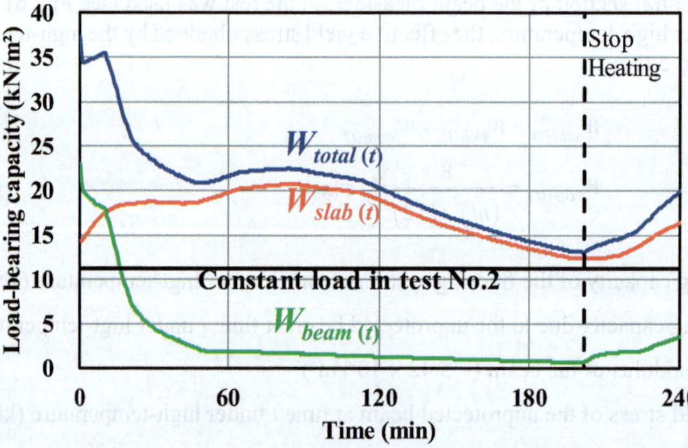

Figure 10. Calculation result of the load-bearing capacity of the flooring system (test No.2).

5 CONCLUSIONS

The following will summarize the results of the load-bearing fire test that was conducted for the composite flooring system employing the two-tier reinforcement-arranged flat RC slabs and unprotected steel beams:

- The RC slab to which the bending collapse load based on the yield-line theory was given sustained load-bearing capacity even after being subjected to 3.5-hours of standard fire heating while involving deflection surpassing 1/20 that of the short-side span.
- In the RC slab in which two-tier reinforcements were arranged, even when the temperature of the lower-side reinforcement reached 700°C, the temperature of the upper-side reinforcement was about 300°C, thereby allowing for the slab to retain its load-bearing capacity. The reason for this successful retention was considered to be that the tension resistance of the upper-side reinforcement worked effectively due to the membrane action of the RC slab.
- In the composite flooring system used in this test, the effect of the application of an unprotected steel beam on fire resistance was small.
- The effect of the headed studs arranged in the four peripheral sections of the RC slab on the deflection behavior of the RC slab during fire was small. Due to the small effect, even when the rotational and

horizontal displacement at the end of the slab was not restricted, the membrane action of the RC slab worked effectively, and it was thus shown that the load-bearing capacity of the composite flooring system was retained during fire.

- It was shown from these test results that the application of an unprotected secondary beam in the composite flooring system is made possible by means of the membrane action of the RC slab.
- The calculation results of the load-bearing capacity on the basis of Bailey's method showed the influence of membrane action of the RC slab and indicated the significant effect of the upper-side reinforcement.

ACKNOWLEDGEMENT

This work was funded by The Japan Iron and Steel Federation. These tests were conducted as a joint research with the Building Research Institute. The deck plate used for the specimen was provided by JFE Metal Products & Engineering Inc. And we got a lot of cooperation from staffs of TOA-RIKA Co., Ltd. and graduate students of Chiba University. We gratefully thank the participants.

REFERENCES

[1] British Steel plc, Swinden Technology Centre, A European Joint Research Programme, The Behaviour of Multi-Storey Steel Framed Buildings in Fire, 1999

[2] Linus C. S. Lim, Membrane Action in Fire Exposed Concrete Floor Systems, Ph D thesis, University of Canterbury, 2003

[3] Bin Zhao, Mohsen Roosefid, Olivier Vassart, Full Scale Test of a Steel and Concrete Composite Floor Exposed to ISO Fire, *Proceedings of the Fifth International Conference on Structures in Fire*, 539-550, 2008

[4] Na-Si Zhang, Guo-Qiang Li, Guo-Biao Lou, Shou-Chao Jiang, Kun-Chao Hao, Experimental Study on Full Scale Composite Floor Slabs under Fire Condition, *Proceedings of Application of Structural Fire Engineering, Prague*, Czech Republic, 502-511, 2009

[5] C. G. Bailey, D. B. Moore: The structural behaviour of steel frames with composite floor slabs subject to fire: Part 1: Theory, *The Structural Engineer* 78, No.11, 19-27, 2000

[6] Zhaohui Huang, Ian W. Burgess, Roger J. Plank: Modeling Membrane Action of Concrete Slabs in Composite Buildings in Fire. II: Validations, *Journal of Structural Engineering*, ASCE, 129 (8), 1103-1112, 2003

[7] EN 1992-1-2, Eurocode 2: Design of concrete structures – Part 1-2 General rules – Structural fire design, 2004

[8] A.S. Usmani, N. J. K. Cameron, Limit capacity of laterally restrained reinforced concrete floor slabs, *Cement and Concrete Composites* 26 (2), 127-140, 2004

[9] Guo-Qiang Li, Shi-Xiong Guo, Hao-Sheng Zhou, Modeling of membrane action in floor slabs subjected to fire. *Engineering Structures* 29 (6), 880-887, 2007

9TH INTERNATIONAL CONFERENCE ON COMPOSITE CONSTRUCTION IN STEEL AND CONCRETE
M. Knobloch, U. Kuhlmann, W. Kurz, M. Schäfer. (Eds.)
Online, Germany, July 27-29, 2021

NUMERICAL ANALYSIS OF LOAD-BEARING FIRE TEST FOR REINFORCED CONCRETE SLABS

K. Kimura*, T. Hirashima, F. Ozaki***, T. Yoshida**** and J. Suzuki*******

* Nippon Steel Corporation, e-mail: kimura.ze8.kei@jp.nipponsteel.com

** Chiba University, e-mail: hirashima@faculty.chiba-u.jp

*** Nagoya University, e-mail: ozaki@nuac.nagoya-u.ac.jp

**** JFE Steel Corporation, e-mail: tor-yoshida@jfe-steel.co.jp

***** Building Research Institute, e-mail: junichi-szk@kenken.go.jp

Keywords: RC slabs, Membrane action, Composite flooring system, Numerical analysis.

Abstract. *A numerical analysis is conducted for the load-bearing fire test of composite flooring systems composed of flat RC slabs and unprotected beams (beams with no fire protection covering), and the test results and numerical analysis results are compared. The main aim of the numerical analysis is to verify the accuracy of the numerical analysis approach as well as to analyse the stress distribution within the composite floor when the membrane functions as intended, the data for which are difficult to obtain from the load-bearing fire test.*

1 INTRODUCTION

Composite structural members comprising steel and concrete exhibit excellent structural and fire-resistance performances; therefore, they are expected to reduce the use of fireproofing materials. Owing to their favourable characteristics, various composite structural members, including concrete-filled steel tube columns and steel-framed reinforced concrete beams, have been developed and implemented. Recently, research activities pertaining to the tensile membrane action of a composite flooring system under increasing temperatures have been conducted in Europe [1 – 3]. Design guidelines for tensile membrane action have been proposed such that secondary beams can remain unprotected against fire [4 – 5]. Both tension and compression zones were formed in the slab as slab deflection increased, thereby providing a stable tensile membrane action to gain resistance against the vertical load. To apply these design guidelines to structural buildings more extensively, the behaviour of slabs with unprotected secondary beams must be simulated numerically.

In this study, a numerical analysis was conducted for the load-bearing fire test of composite flooring systems composed of flat RC slabs and unprotected secondary beams. First, a thermal analysis was conducted to evaluate the temperature of the member elements. Subsequently, a structural analysis was conducted to evaluate the deformation of the structural members. The results from the numerical analysis were compared with those from the fire test, and the accuracy of the numerical approach was verified. In addition, a detailed observation was conducted to study the stress distribution within the slab section when the membrane functioned as intended, the data for which were difficult to obtain from the load-bearing fire test.

2 ANALYSIS TARGETS AND NUMERICAL APPROACH

The numerical analysis was performed for the load-bearing fire test of a composite flooring system, as shown in Figure 1. The plane size of the floor slab in the composite flooring system measured 5,800 mm × 4,200 mm with a thickness of 120 mm, and 10 mm diameter reinforcing steel bars were arranged in the slab in two layers at a pitch of 200 mm. Four floor perimeters were supported with protected steel beams, and an unprotected secondary beam was arranged so that it was connected to the centre of the short sides of the slab. Two specimens were used in this study, i.e. No. 1 and No. 2. The No. 1 specimen comprised headed studs at the ends of the slab with a diameter of 15.4 mm and a pitch of 150 mm, whereas the No. 2 specimen did not have any headed studs.

A uniformly distributed load of 11.23 kN/m^2 (including the dead load) was loaded on the floor slab. The slab was heated from its lower surface using a standard fire curve in accordance with ISO 834.

Subsequently, a numerical analysis was conducted on test specimen No. 1, where headed studs were arranged at the perimeter beams.

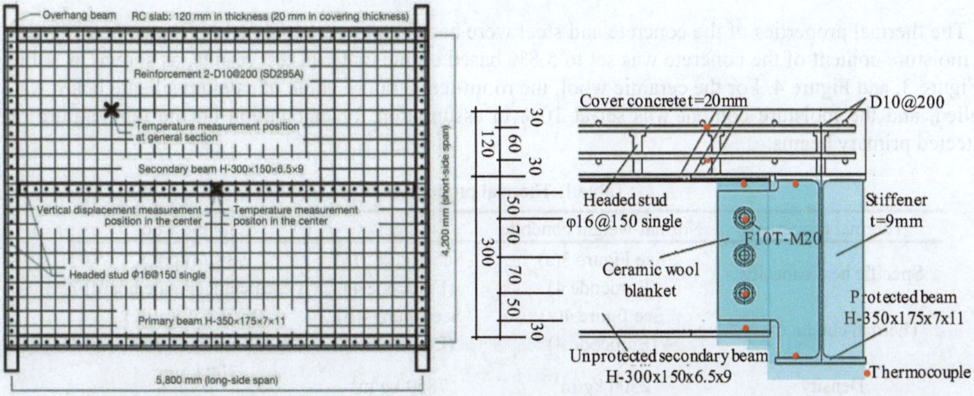

(a) Outline of specimen No.1

(b) Details of specimen around beam end (No. 1)

Figure 1. Illustration of analysis target.

For the numerical analysis, the finite element analysis software "SAFIR" developed at the University of Liege in Belgium was adopted. This software comprises a thermal analysis program and a structural analysis program.

First, the distribution and hysteresis of the temperature in the cross section of the structural members were calculated using the thermal analysis program, and the stress and deformation conditions of these members were calculated using the structural analysis program, which employs the results obtained from the calculation above. For more details regarding SAFIR, please refer to the reference [6].

3 THERMAL ANALYSIS

3.1 Analysis models

Figure 2 shows the thermal analysis model, in which (a) shows the RC slab, (b) the unprotected secondary beam, and (c) the protected primary beam. For the thermal analysis model shown in Figure 2, the thickness of the RC slab was 120 mm, the cross section of the secondary beam was H-300 × 150 × 6.5 × 9, and the cross section of the primary beam was H-350 × 175 × 7 × 11 (unit : mm). As shown in Figure 2 (c), ceramic wool was coated on three surfaces of the H beam, excluding its upper flange, and the heat transfer caused by convection and radiation in the cavity was considered.

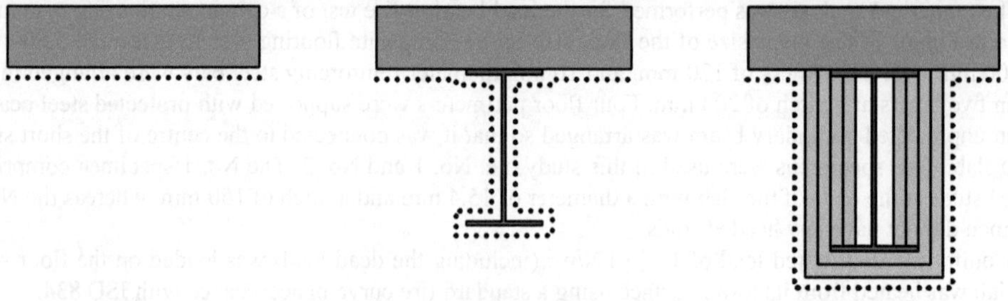

(a) RC slab (b) Unprotected secondary beam (c) Protected primary beam

Figure 2. Thermal analysis models.

The thermal properties of the concrete and steel were based on those stipulated in Eurocode 4 [7], and the moisture content of the concrete was set to 5.8% based on the material test results, as shown in Table 1, Figure 3, and Figure 4. For the ceramic wool, the room-temperature value of calcium silicate board was applied, and the moisture content was set at 10% in assumption for calculation on the temperature of protected primary beams.

Table 1: Thermal properties.

Thermal property	Normal-weight concrete	Steel	Ceramic wool
Specific heat capacity	See Figure 3(a) (Eurocode 4)	See Figure 3(b) (Eurocode 4)	938 J/(kg·K) (= calcium silicate board)
Thermal conductivity	See figure 4(a) (Eurocode 4)	See figure 4(b) (Eurocode 4)	0.0748 W/(m·K) (= calcium silicate board)
Density	2300 kg/m^3	7850 kg/m^3	300 kg/m^3 (= calcium silicate board)
Water content	5.8% (Material test result)	-	10% (Assumption)

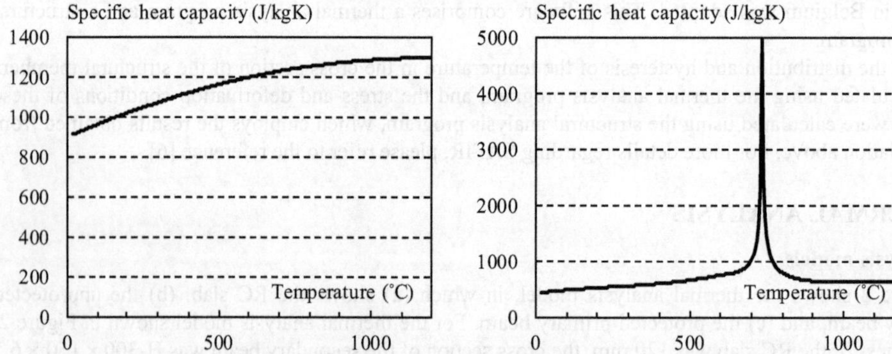

(a) Normal-weight concrete (b) Steel

Figure 3. Specific heat capacity.

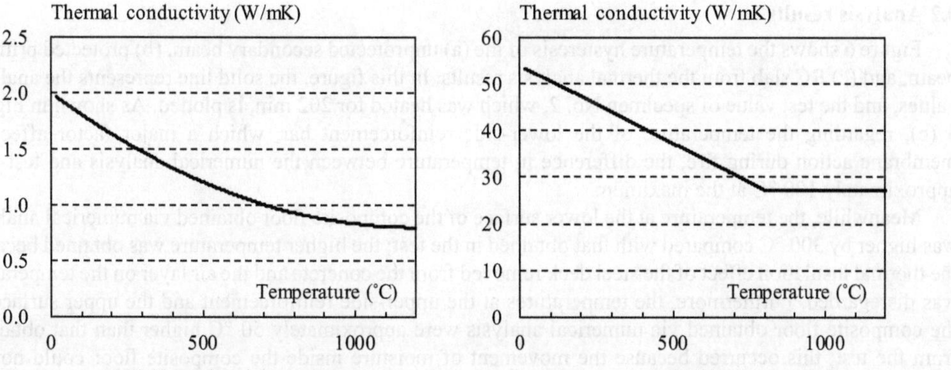

(a) Normal-weight concrete (b) Steel

Figure 4. Thermal conductivity.

To simulate the condition in which heating at the flashover stage of fire is applied to the respective structural members from their lower side, the atmospheric temperature at the heated surface was set based on the standard heating temperature stipulated in ISO 834, as shown in Figure 5, and the atmospheric temperature at the non-heated surface was set at 20 °C. As shown in Table 2, the convection heat transfer coefficient at the heated surface was set at 23 W/(m^2·K), and that at the non-heated surface at 4 W/(m^2·K). The reinforcing steel bars and headed studs in the RC slab and steel decks were not modelled, and the effects of these materials on the member temperature were disregarded.

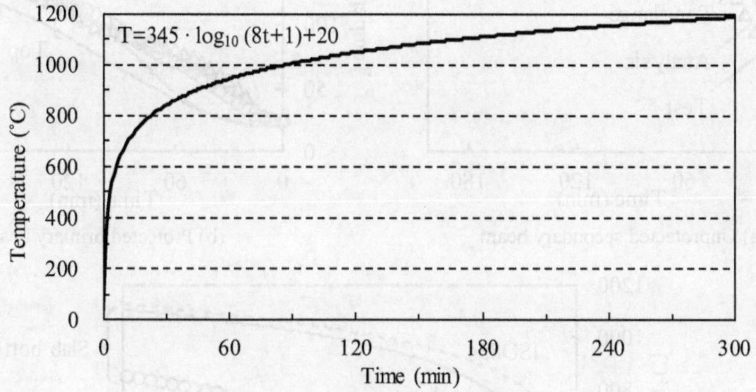

Figure 5. Standard temperature–time curve.

Table 2: Boundary conditions.

Type	Value
Convection heat transfer coefficient at heated surface	23 W/(m^2·K)
Convection heat transfer coefficient at non-heated surface	4 W/(m^2·K)
Convection heat transfer coefficient for cavity	4 W/(m^2·K)
Emissivity of material	0.7
Emissivity of flame	1.0

3.2 Analysis results

Figure 6 shows the temperature hysteresis of the (a) unprotected secondary beam, (b) protected primary beam, and (c) RC slab from the thermal analysis results. In this figure, the solid line represents the analysis values, and the test value of specimen No. 2, which was heated for 202 min, is plotted. As shown in Figure 3 (c), regarding the temperature of the lower-side reinforcement bar, which a major factor affecting membrane action during fire, the difference in temperature between the numerical analysis and test was approximately 100 °C at the maximum.

Meanwhile, the temperature at the lower surface of the composite floor obtained via numerical analysis was higher by 300 °C compared with that obtained in the test; the higher temperature was obtained because the thermal insulation effect of the steel deck removed from the concrete and the air layer on the temperature was disregarded. Furthermore, the temperatures at the upper-side reinforcement and the upper surface of the composite floor obtained via numerical analysis were approximately 50 °C higher than that obtained from the test; this occurred because the movement of moisture inside the composite floor could not be considered.

Although the temperatures of the structural members obtained via numerical analysis were higher than those obtained from the test, it was assumed that these temperatures can be assessed qualitatively. Hence, these numerical analysis results were adopted as the input data in the structural analysis.

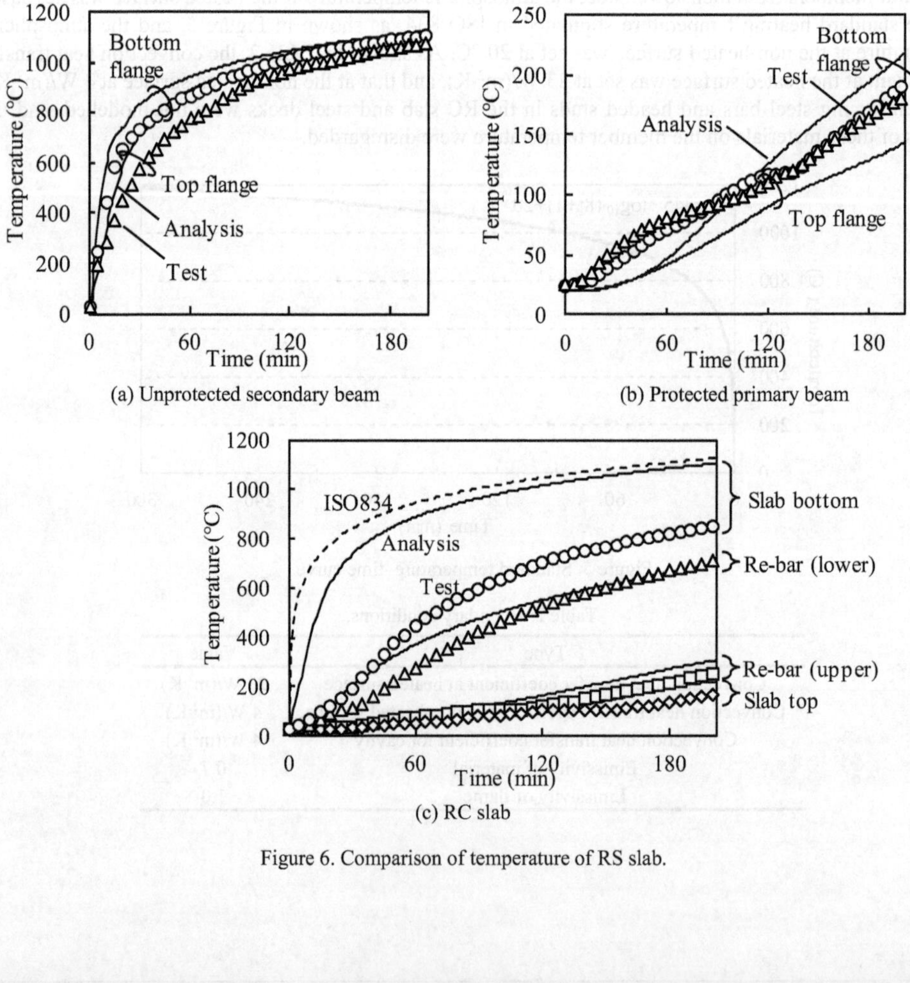

Figure 6. Comparison of temperature of RS slab.

4 STRUCTURAL ANALYSIS

4.1 Analysis models

Figure 7 shows the structural analysis model. The RC slab was modelled using the shell element, and both the unprotected secondary beam and protected primary beam were modelled using the beam element. The vertical displacement at the floor edge was restricted, and a vertical load of 11.2 kN/m^2 was uniformly applied to the entire surface of the composite floor.

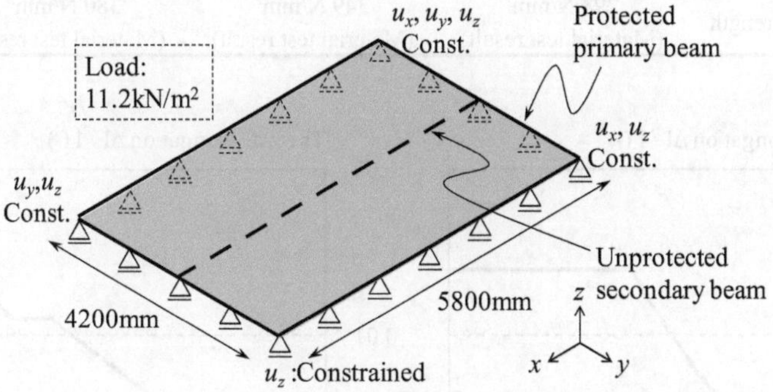

Figure 7. Structural analysis model.

The stress–strain relationships of concrete and steel at high temperature reflected those published in Eurocode 4, as shown in Figure 8, and the strength at room temperature was set based on the material test results. Specifically, the compression strength of concrete was set to 36.1 N/mm^2, and its tensile strength to 0 N/mm^2. Meanwhile, the yield point of the reinforcing bars was set to 380 N/mm^2, whereas those of the secondary beam and primary beam were set to 349 and 298 N/mm^2, respectively (see Tables 3 and 4).

The thermal elongation behaviours of concrete and steel at high temperatures reflected those published in Eurocode 4, as shown in Figure 9.

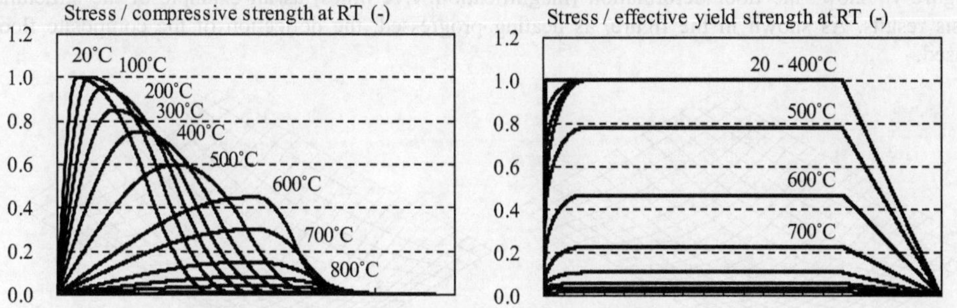

(a) Normal-weight concrete

(b) Steel

Figure 8. Stress–strain relationship.

Table 3: Strength of normal weight concrete.

Type	Value
Compressive strength	36.1 N/mm^2 (Material test result)
Tensile strength	0 N/mm^2 (design)

Table 4: Strength of steel.

Type	Perimeter beam	Secondary beam	Reinforcing bar
Yield strength	298 N/mm^2 (Material test result)	349 N/mm^2 (Material test result)	380 N/mm^2 (Material test result)

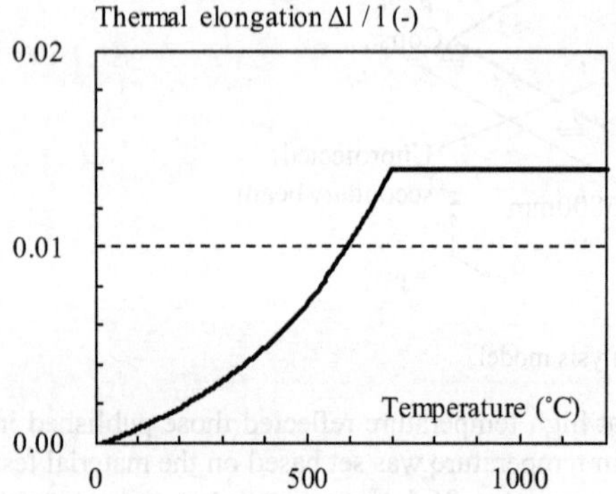

(a) Normal-weight concrete

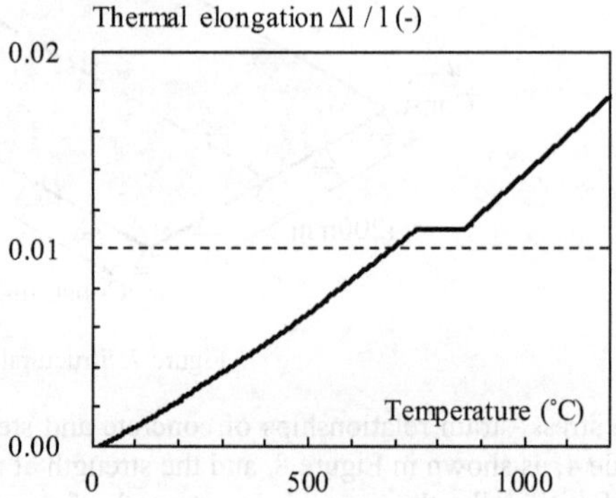

(b) Steel

Figure 9. Thermal elongation.

4.2 Analysis results

Figure 10 shows the floor deformation (magnification: five times) as an example of the structural analysis results. As shown in the figure, as heating progressed, the deflection of the composite floor increased.

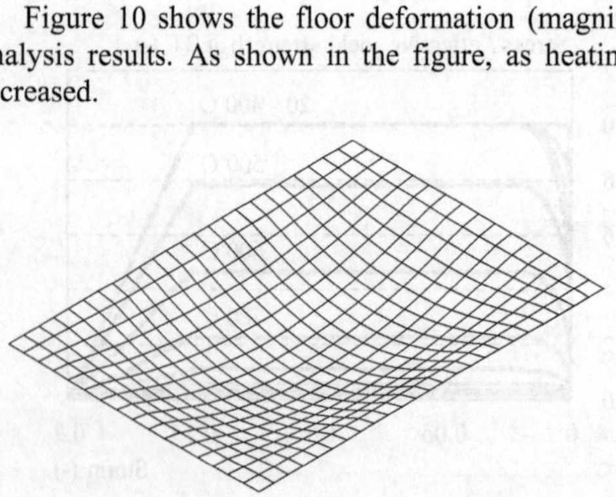

(a) 60 min

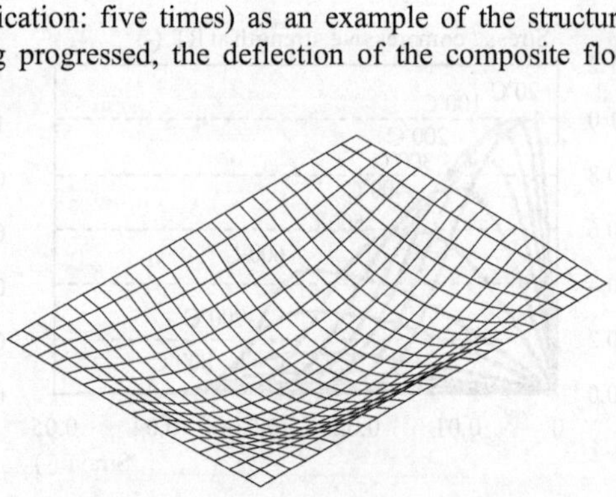

(b) 120 min

Figure 10. Floor deformation.

Figure 11 shows the deflection hysteresis at the centre of the composite floor. As shown in the figure, whereas the analysed value was smaller than the test value up to 45 min of heating, the increase trends in deflection after 45 min of heating were similar for the analysis and test values. The deflection at 132 min (stop heating) was 191 mm in the test and 197 mm in the analysis, demonstrating consistency between them.

Figure 12 shows a comparison of the floor deflection curve in the diagonal direction between the numerical analysis and the fire test. As shown in the figure, the deflection obtained from the analysis and test were similar at both 60 and 120 min. In addition, based on the analytical results, a negative curvature was observed around the slab ends.

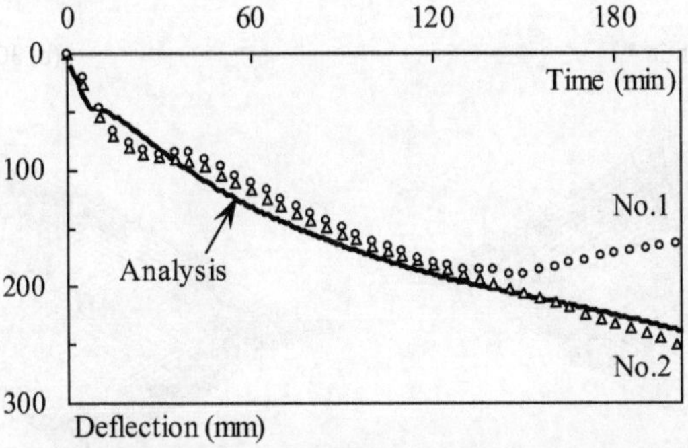

Figure 11. Comparison of floor deflection at centre of slab.

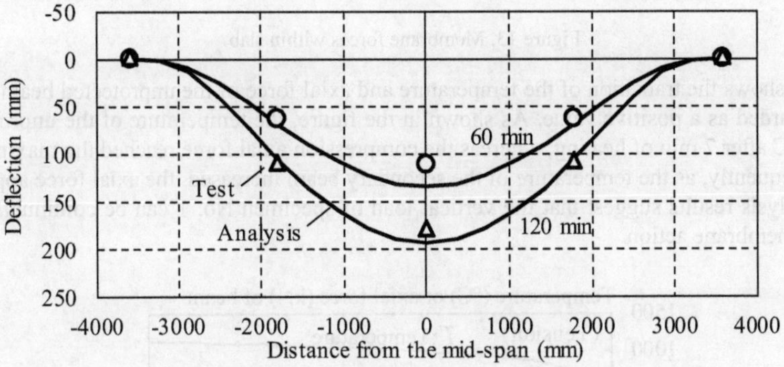

Figure 12. Comparison of deflection curve (diagonal).

Figure 13 shows the distribution of the membrane forces within the slab, i.e. the compression (blue) and tension (red) forces. Based on the figure, it was confirmed that the floor long-side-direction tensile force due to the thermal expansion of the unprotected beam occurred after 5 min of heating; additionally, the compressive force occurred in the circumferential direction in the peripheral area of the composite floor after 30 min of heating. Simultaneously, a tension sphere was formed in the centre of the composite floor. Subsequently, as heating continued, the temperature of the lower-side reinforcement increased, causing a reduction in the strength of the composite floor; consequently, the tension sphere expanded.

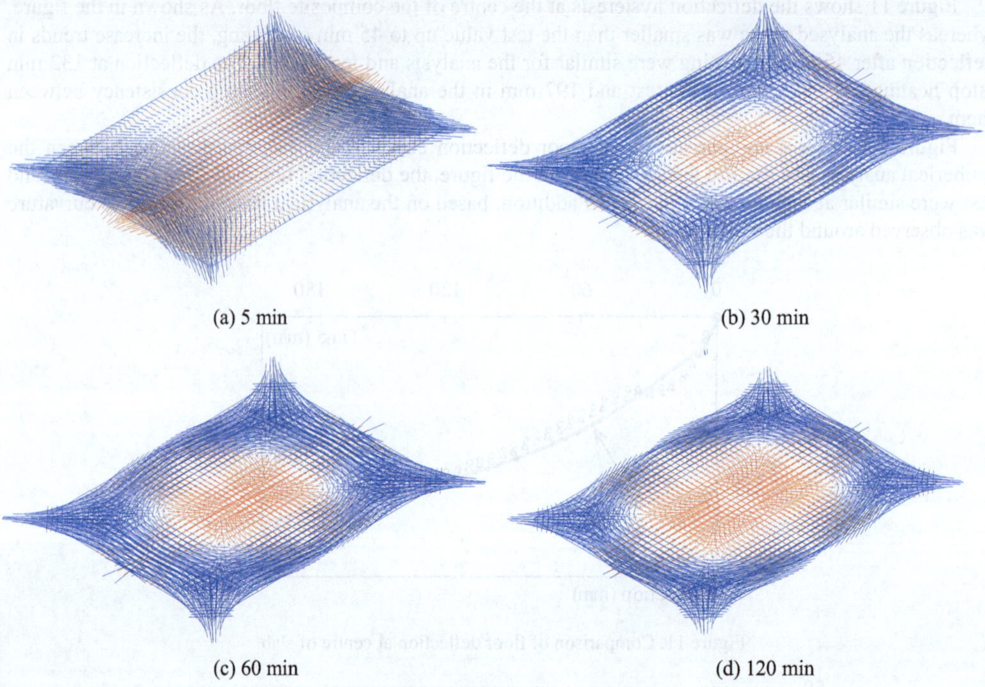

(a) 5 min (b) 30 min

(c) 60 min (d) 120 min

Figure 13. Membrane forces within slab.

Figure 14 shows the transition of the temperature and axial force of the unprotected beam. The tensile force was regarded as a positive value. As shown in the figure, the temperature of the unprotected beam reached 450 °C after 7 min of heating, whereas the compression axial force reached the maximum level at 930 kN. Subsequently, as the temperature of the secondary beam increased, the axial force approached 0.

These analysis results suggest that the vertical load of specimen No. 1 can be continually supported owing to the membrane action.

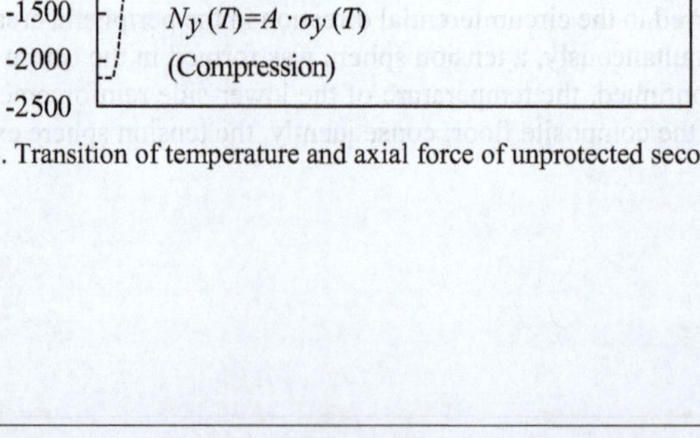

Figure 14. Transition of temperature and axial force of unprotected secondary beam.

Figure 15 shows the output positions of the sectional data. In other words, the temperature, stress, and strain of the top concrete and upper or lower reinforcement bars at positions A to D were verified. Table 5 shows these values at 60, 120, 180, and 211 min when the numerical analysis was terminated. Positive values indicate tension, whereas negative values indicate compression. In addition, Figure 16 shows the stress and strain at positions A and B, respectively. In Figures 16 (a) and (b), the dashed line indicates the hystereses of the stress and strain.

These data show that stress tended to decrease, whereas strain tended to increase as heating progressed. Based on the upper reinforcement bar, it was discovered that the stress at the slab end was higher than that at the centre, and the strain at the slab end was larger than that at the centre in both the X- and Y-directions. Therefore, it can be concluded that the vertical load was supported by the tensile force of the upper reinforcement bar. In addition, based on the top concrete, it was discovered that the compression stress at the floor centre became much smaller at 211 min and could have indicated the beginning of transition to tension stress. In this analysis, the tensile strength of concrete was assumed to be zero; therefore, the analysis was terminated.

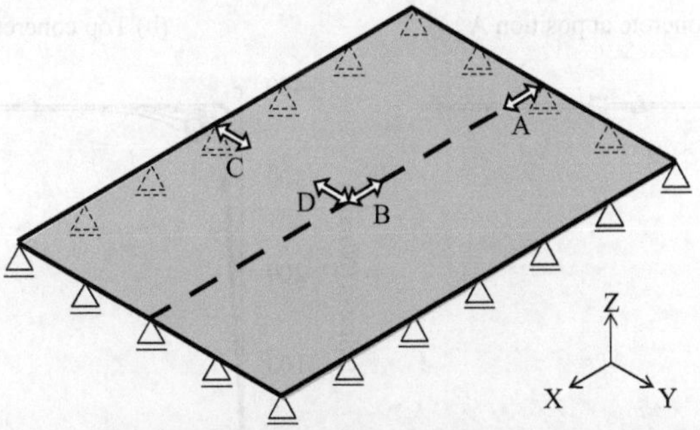

Figure 15. Output position of sectional data.

Table 5: Temperature, stress, and strain in slab section.

		60 min			120 min			180 min			211 min		
		T (°C)	σ (N/mm²)	ε (%)	T	σ	ε	T	σ	ε	T	σ	ε
A	Top concrete	58	-14.7	-0.63	107	-8.2	-1.30	195	-2.6	-1.29	229	-0.8	-1.14
	Upper re-bar	78	380.0	1.98	164	380.0	3.06	272	379.2	3.58	319	375.4	3.79
	Lower re-bar	370	261.0	0.39	551	177.5	0.65	662	78.8	0.88	710	42.3	0.96
B	Top concrete	58	-10.1	-0.91	107	-7.3	-1.64	195	-2.7	-1.71	229	-0.9	-1.57
	Upper re-bar	78	372.1	0.25	164	369.0	0.50	272	357.8	0.62	319	348.4	0.67
	Lower re-bar	370	85.5	0.06	551	112.8	0.16	662	63.8	0.27	710	35.2	0.32
C	Top concrete	58	-19.7	-1.10	107	-13.3	-2.08	195	-6.7	-2.34	229	-4.3	-2.30
	Upper re-bar	78	380.0	2.83	164	380.0	4.29	272	380.0	4.95	319	375.4	5.16
	Lower re-bar	370	82.4	0.06	551	65.4	0.09	662	30.5	0.10	710	14.7	0.08
D	Top concrete	58	-8.0	-0.85	107	-7.1	-1.63	195	-2.8	-1.73	229	-1.0	-1.60
	Upper re-bar	78	266.7	0.12	164	340.6	0.17	272	329.6	0.23	319	313.9	0.26
	Lower re-bar	370	199.1	0.17	551	123.7	0.20	662	64.2	0.28	710	35.1	0.32

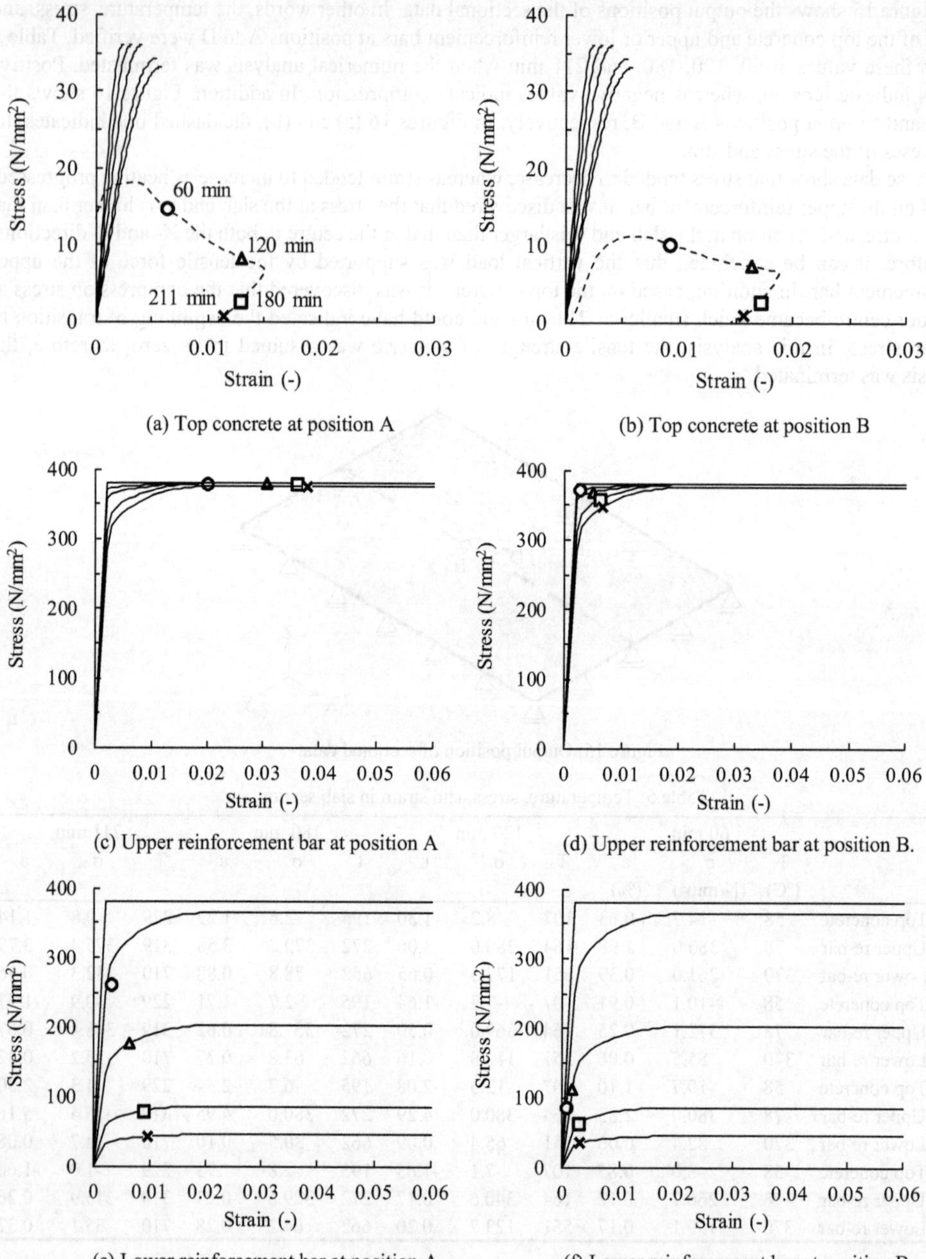

(a) Top concrete at position A

(b) Top concrete at position B

(c) Upper reinforcement bar at position A

(d) Upper reinforcement bar at position B.

(e) Lower reinforcement bar at position A

(f) Lower reinforcement bar at position B

Figure 16. Stress and strain within slab section.

5 CONCLUSION

The outline and results of a numerical analysis for the load-bearing fire test of a composite flooring system composed of a flat RC slab and an unprotected secondary beam were presented herein. The temperature of the member elements obtained from the thermal analysis were consistent with the fire test results. The deflection hysteresis and curve of the slab obtained from the structural analysis were consistent with the fire test results. In addition, the membrane forces within the slab and the stress in the slab section were evaluated via structural analysis. Because these data are difficult to obtain from the fire test, structural analysis is useful for evaluating them and analysing the behaviour of composite flooring systems with unprotected secondary beams. The numerical approach explained herein is applicable to the design of structural buildings and obviates the use of fire-proofing materials in secondary beams.

ACKNOWLEDGEMENT

This work was funded by The Japan Iron and Steel Federation. These tests were conducted as a joint research with the Building Research Institute. We gratefully thank the participants.

REFERENCES

[1] Wang, Y.C., "Tensile membrane action in slabs and its application to the Cardington fire tests", *Proceeding of the second Cardington conference*, 55-67, 1996.

[2] Bailey, C.G. and Moore, D.B., "The structural behavior of steel frames with composite floor slabs subject to fire, Part1: Theory & Part2: Design", *The Structural Engineer*, 78(11), 19-33, 2000.

[3] Nadjai, A.; Bailey, C.G.; Han, S.H.; Vassart, O.; Zhao, B.; Hawes, M.; Franssen, J.M. and Simms, I., "Full scale fire test on a composite floor slab incorporating long span cellular steel beams", *The Structural Engineer*, 89(21), 18-25, 2011.

[4] Steel Construction Institute (SCI)., *Fire safe design: A new approach to multi-storey steel framed buildings (P288)*, 2006.

[5] Vassart, O. and Zhao, B., *Membrane action of composite slab in case of fire: Background document & Design guide (Edition 2012-1)*, 2012.

[6] Gerna, T. and Franssen, J.M., "Modeling structures in fire with SAFIR®: Theoretical background and capabilities", *Journal of Structural Fire Engineering*, Vol. 8, issue 3, 300-323, 2017.

[7] EN 1994-1-2, Eurocode 4: Design of composite steel and concrete structures, Part 1.2: General rules - Structural fire design, 2005.

INNOVATIVE STRUCTURES

9TH INTERNATIONAL CONFERENCE ON COMPOSITE CONSTRUCTION IN STEEL AND CONCRETE
M. Knobloch, U. Kuhlmann, W. Kurz, M. Schäfer. (Eds.)
Online, Germany, July 27-29, 2021

ADVANCED ANALYSIS OF STEEL-CONCRETE COMPOSITE BUILDINGS

H. Tran*, H.T. Thai*, T. Ngo*, B. Uy, D. Li** and J. Mo****

* Department of Infrastructure Engineering, The University of Melbourne, Parkville, VIC 3010, Australia
e-mails: haut@student.unimelb.edu.au, tai.thai@unimelb.edu.au, dtngo@unimelb.edu.au

** School of Civil Engineering, The University of Sydney, Sydney, NSW 2006, Australia
e-mail: brian.uy@sydney.edu.au, dongxu.li@sydney.edu.au, jumo4858@uni.sydney.edu.au

Keywords: Composite frame, Fiber section, Nonlinear simulation, OpenSees, Semi-rigid connection.

Abstract. *This paper presents a nonlinear simulation method for composite framing systems constituted from concrete-filled steel tubular columns (CFST) and composite beam systems. A force-based fibre beam-column element in OpenSees was adopted. This element was capable of accurately capturing the local buckling of steel and the confining effect of concrete using the modified stress-strain relationships of the steel and concrete fibres. A source code for the connection element was also developed in OpenSees to capture the semi-rigid behaviour of the beam-to-column connections of the composite buildings. Through the verification with numerous experiments, the model has shown its capability of accurately simulating composite frames with simplicity and less computational cost. An extensive parametric study was conducted to examine the effect of the bracing systems and the rigidity of the connections on the behaviour and instability of the whole composite buildings.*

1 INTRODUCTION

Nowadays, the application of composite components in the construction of mid-rise buildings and high-rise buildings has become prevalent. These composite buildings can comprise of concrete-encased steel columns or steel columns linked to composite beams by welded connections or blind bolted connections. Recently, a new approach using concrete-filled steel tubular columns (CFST) and concrete-filled steel plates shear walls has been increasingly utilized in the construction. In this approach, concrete that is filled inside steel tubes/plates will refrain steel tubes/plates from buckling and the compressive strength of concrete is also improved due to the circumferential stress provided by steel tubes/plates, which contributes to the higher strength and better ductility of this kind of composite component [1, 2]. In addition, these composite components can also be pre-fabricated in the factories to make the construction process faster with higher precision. Due to numerous advantages of this approach, a lot of studies have been conducted [2-7]. However, these studies only focused on the behaviour of the composite components such as CFST columns or composite beams, the study on the simulation procedure and behaviour of the composite systems that comprise of these composite components have not been conducted elaborately.

In this paper, the advanced analysis for composite structures including both composite components (CFST columns and composite beams) and composite systems will be conducted in OpenSees using fibre beam-column element with appropriate material models. Moreover, the source code in which the moment-rotation relationship is integrated is developed in OpenSees to simulate the behavior of semi-rigid connections that have great influence on the stiffness and the loading capacity of the composite structures. After being validated with numerous experimental data, the presented model is utilized to conduct the case study of a 10-story composite frame and a plenty of parameters, especially the semi-rigid connection and bracing systems, are investigated.

2 MATERIAL MODELS AND FINITE ELEMENT MODEL IN OPENSEES

2.1 Material models

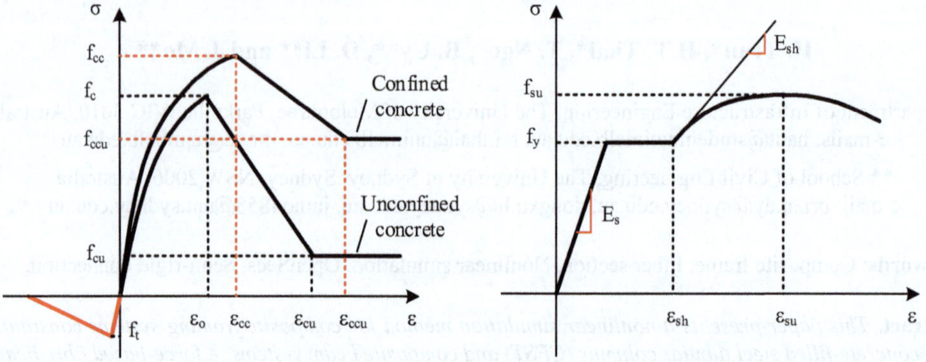

a) Concrete model b) Steel model

Figure 1. Material models.

For normal concrete, a simple model proposed by Kent and Park [8] is applied. This model considers the behaviour of concrete at pre-peak stage as nonlinear and linear at post-peak stage (Figure 1a). For concrete in the steel tube, the concrete compressive strength will increase due to the confined effect caused by circumferential stress from steel skin, and it is also taken into account in this paper. The compressive strength of confined concrete is given by Susantha et al. [9] as following:

$$f_{cc} = f_c + 4.0 f_{rp} \tag{1}$$

where f_{cc} is the compressive strength of confined concrete (MPa); f_c is the compressive strength of unconfined concrete (MPa); f_{rp} is the confining stress from steel tubes (MPa) and it is determined by the below equations for square tubes:

$$f_{rp} = -6.5R\frac{f_c^{1.46}}{f_y} + 0.22 f_c^{1.03} \tag{2}$$

$$R = \frac{b}{t}\sqrt{\frac{12\left(1-\upsilon_s^2\right)}{4\pi^2}}\sqrt{\frac{f_y}{E_s}} \tag{3}$$

For circular CFST columns, the value of f_{rp} can be calculated as following [10]:

$$f_{rp} = \begin{cases} 0.7(\upsilon_e - \upsilon_s)\dfrac{2t}{D-2t} f_y & for \ \dfrac{D}{t} \le 47 \\ \left(0.006241 - 0.0000357\dfrac{D}{t}\right) f_y & for \ 47 < \dfrac{D}{t} \le 150 \end{cases} \tag{4}$$

where D and t are the diameter and thickness of the steel tube, respectively; υ_e is Poisson's ratio of the steel tube with in-filled concrete defined by Tang et al. [11] as:

$$\upsilon_e = 0.2312 + 0.3582\upsilon_e' - 0.1524\left(\frac{f_c}{f_y}\right) + 4.843\upsilon_e'\left(\frac{f_c}{f_y}\right) - 9.169\left(\frac{f_c}{f_y}\right)^2 \tag{5}$$

$$\upsilon_e' = 0.881\times10^{-6}\left(\frac{D}{t}\right)^3 - 2.58\times10^{-4}\left(\frac{D}{t}\right)^2 + 1.953\times10^{-2}\left(\frac{D}{t}\right) + 0.4011 \quad (6)$$

The model for concrete in tension is also presented and it is considered bilinear as depicted in Figure 1a. The tensile strength of concrete is determined by CEB-FIB 1990 [12] as $f_t=1.4(f_c/10)^{2/3}$. For steel, the available material namely *ReinforcingSteel* in OpenSees is applied. This material allows the hardening of steel in tension and the buckling of steel in compression to be taken into consideration. In this material, some initial parameters to characterize the behaviour of steel need to be input. They are the steel yield strength f_y, the steel ultimate strength f_{su}, the initial elastic modulus E_s, the modulus at the initial strain hardening E_{sh}, the strain corresponding to the initial strain hardening ε_{sh} and the strain corresponding to the ultimate strength ε_{su}. The stress-strain relationship of steel is illustrated in Figure 1b.

2.2 Finite element model in OpenSees

In this paper, only fibre force-based beam-column element in OpenSees is utilized to simulate the behaviour of structures that comprise of CFST columns and composite beams. Therefore, the interface between concrete-to-steel of CFST columns and composite beams are assumed to be rigid and this assumption does not affect the accuracy of numerical results as observed in the verification part. Using beam-column element with fibre section, the presented constitutive laws for concrete and steel are assigned explicitly to each concrete/steel fibre (Figure 2a), which allows the material nonlinearity to be captured precisely. In addition, the geometric nonlinearity due to large deformation and rotation is also simulated well by the co-rotational formulation in the fibre beam-column element. In Figure 2b, the mesh sensitivity on the results of a composite beam using fibre force-based beam-column is illustrated, it shows that the results are not affected by the number of elements. Therefore, only one element per structural component is required to achieve the necessary accuracy and it leads to the significant reduction in the modelling computational cost.

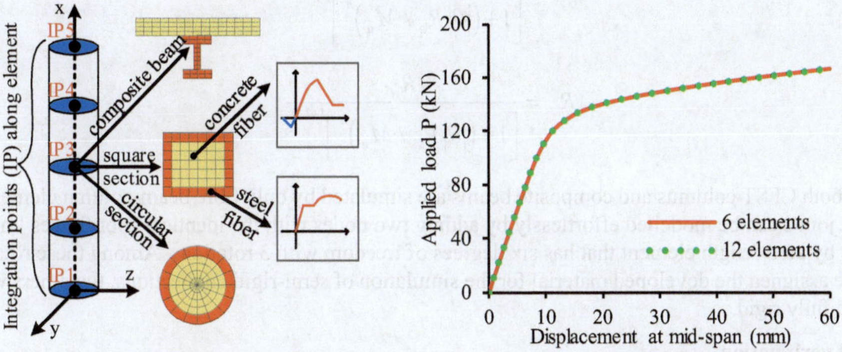

a) Fibre section of some components b) Mesh sensitivity analysis

Figure 2. Fibre sections and mesh sensitivity of force-based beam-column element.

Traditionally, the joints between CFST columns and composite beams are considered as either pin or rigid connections, but this conventional approach is not able to accurately capture the behaviour of these joints since they are normally bolted connections. Hence, the proper simulation of the semi-rigid behaviour of such connections is critical.

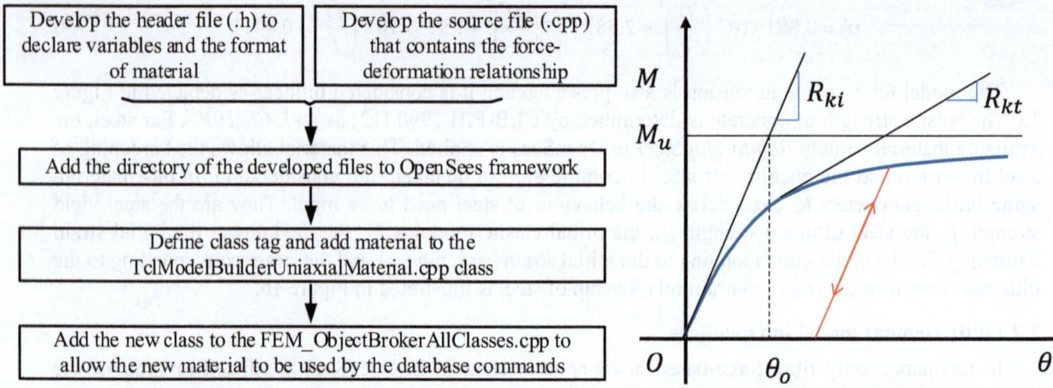

a) Procedure to add new material to OpenSees b) Moment-rotation relationship

Figure 3. Adding a new material to simulate semi-rigid connections based on Kishi-Chen's model [13] into OpenSees.

In this paper, a source code for a new material is developed and added into OpenSees whilst it is lacking appropriate material to model rigorously the behaviour of semi-rigid connections. The procedure to add a new material to OpenSees is depicted in Figure 3a. In the developed code of the new material, the moment-rotation curve based on the work of Kishi-Chen [13] is integrated (Figure 3b), and this material can be used by inputting the shape factor, the initial stiffness and the ultimate moment of the connections. The moment-rotation relationship is:

$$M = \frac{R_{ki}\theta}{\left[1+\left(|\theta| R_{ki} / M_u\right)^n\right]^{1/n}} \tag{7}$$

$$R_{kt} = \frac{R_{ki}}{\left[1+\left(|\theta| R_{ki} / M_u\right)^n\right]^{1+1/n}} \tag{8}$$

Since both CFST columns and composite beams are simulated by only fibre beam-column element, the semi-rigid joints can be modelled effortlessly by adding two nodes with the identical coordinates linked to each other by zero-length element that has six degrees of freedom with 3 rotations. Among these rotations, one will be assigned the developed material for the simulation of semi-rigid connections, the others will be considered fully rigid.

2.3 Model verification

To verify the presented advanced analysis procedure, the simulation for numerous test data has been conducted and the results are compared with those from the experiments.

2.3.1 Composite beams

The simply supported composite beam subjected to gravity loads namely SCB-1 is analysed [14]. The concrete compressive strength and the steel yield strength of this beam are 27.7 MPa and 310 MPa, respectively. Another continuous composite beam tested by Slutter and Driscoll [15] is taken into consideration. The concrete compressive strength and the steel yield strength of this beam are 16 MPa and 252.4 MPa, respectively. More details about the loading schemes and geometry of two composite beams are illustrated in Figure 4.

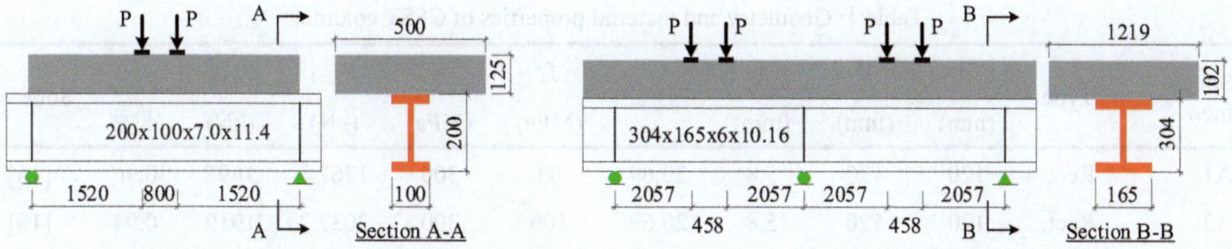

Figure 4. Geometry and loading schemes of simply supported and continuous composite beam (Unit: mm).

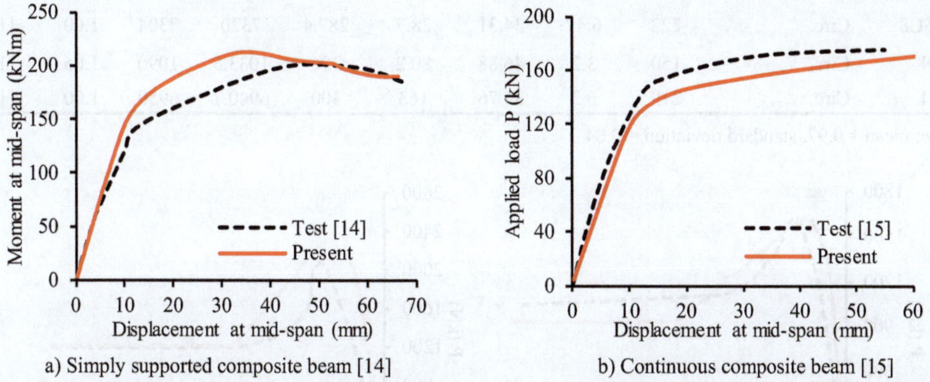

Figure 5. Comparison of the load-displacement curve between simulation and experiment.

Figure 5 shows the comparison between the experimental results and numerical results. Generally, good agreement between experiment and simulation has been observed, the small discrepancy can be explained by the errors due to experimental procedure.

2.3.2 CFST columns

A lot of test of CFST columns subjected to axial loads has been collected and analysed to verify the presented nonlinear model. The geometry of these CFST columns is presented in table 1 in which *"Rect."* and *"Circ."* is rectangular and circular CFST columns, respectively; *B* is the width of rectangular columns, *D* is the height of rectangular columns and the diameter of circular columns, *t* is the thickness of CFST columns.

The results and the comparison between simulation and test are illustrated in table 1 and Figure 6. The mean and deviation of numerical ultimate load-to-experimental ultimate load are respectively 0.97 and 0.04 for CFST columns with various slenderness, which indicates that the model can predict well the loading capacity of CFST columns. Figure 6 also shows good correlation of load-displacement curves between test and simulation.

Table 1: Geometry and material properties of CFST columns

Speci-men	Type	B (mm)	D (mm)	t (mm)	D/t	f_c (MPa)	f_y (MPa)	$P_{u,cal}$ (kN)	$P_{u,exp}$ (kN)	$P_{u,exp}/P_{u,cal}$	Source
A1	Rect.	120	120	5.8	20.69	83	300	1767.8	1697	0.96	[16]
A2	Rect.	120	120	5.8	20.69	106	300	2037.2	1919	0.94	[16]
A4-1	Rect.	100	130	5.8	22.41	83	300	1628.7	1601	0.98	[16]
A7-1	Rect.	100	180	5.8	31.03	83	300	2169.8	2059	0.95	[16]
CC4D41	Circ.	-	450	2.96	152.0	77	843	7484.9	6870	0.92	[17]
CC8C8	Circ.	-	222	6.47	34.31	28.7	287.4	7320	7304	1.00	[17]
3HN	Circ.	-	150	3.2	46.88	80.2	306	1033.5	1090	1.05	[18]
CS-1	Circ.	-	219	6.3	34.76	163	300	6940.7	6950	1.00	[19]

Note: mean = 0.97, standard deviation = 0.04

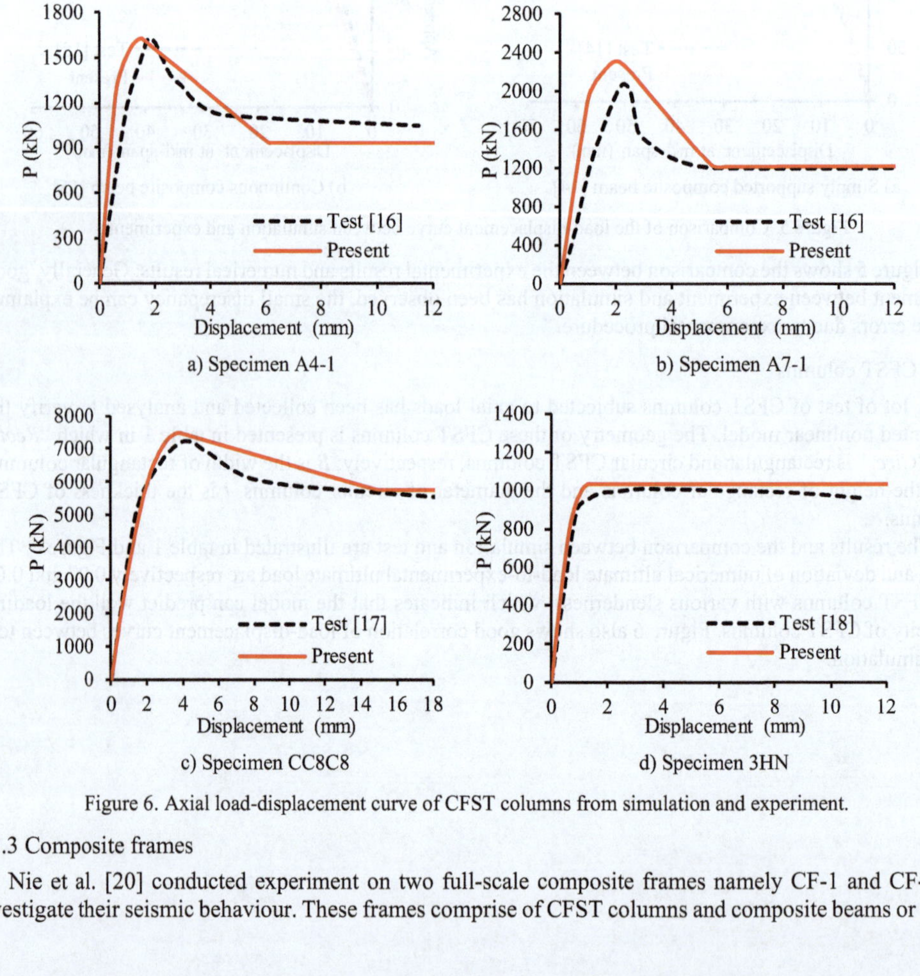

a) Specimen A4-1

b) Specimen A7-1

c) Specimen CC8C8

d) Specimen 3HN

Figure 6. Axial load-displacement curve of CFST columns from simulation and experiment.

2.3.3 Composite frames

Nie et al. [20] conducted experiment on two full-scale composite frames namely CF-1 and CF-2 to investigate their seismic behaviour. These frames comprise of CFST columns and composite beams or steel beams subjected to lateral cyclic loads. The concrete compressive strength of concrete slab is 30.4 MPa and that of CFST column is 36.6 MPa, the steel yield strength is presented in table 2. The details about the geometry of these frames are illustrated in Figure 7.

Table 2: The yield strength of steel of frame CF-1 and CF-2

	CF-1			CF-2		
	8 mm	10 mm	12 mm	8 mm	10 mm	12 mm
Yield strength (MPa)	428	337	354	435	332	354
Ultimate strength (MPa)	586	494	535	591	494	533

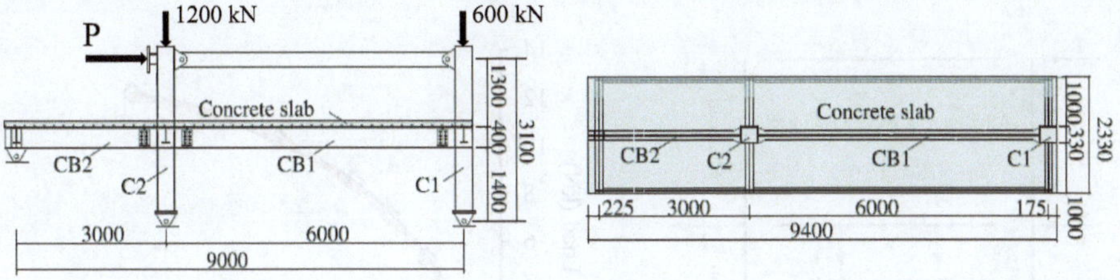

a) Frame CF-1

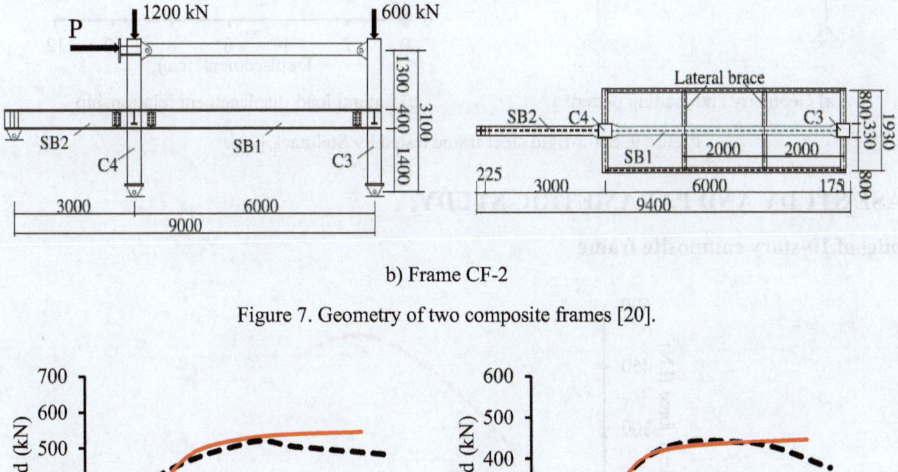

b) Frame CF-2

Figure 7. Geometry of two composite frames [20].

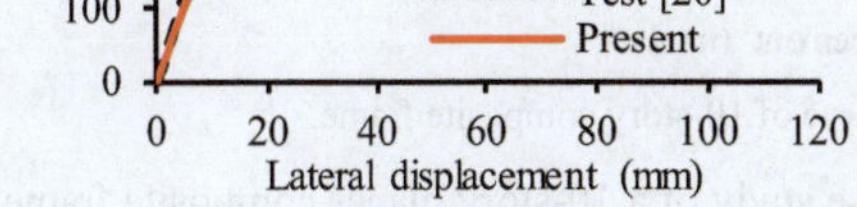

a) Frame CF-1

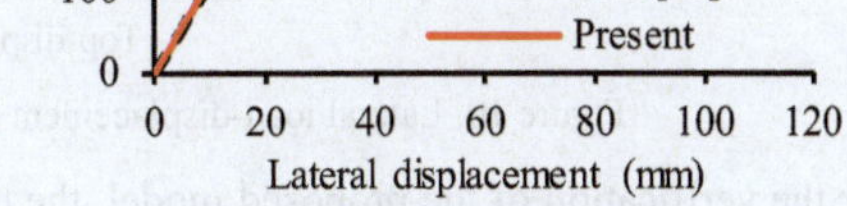

b) Frame CF-2

Figure 8. Lateral load-displacement curve of CFST columns from simulation and experiment.

The back-bone curve of the lateral load-displacement are shown in Figure 8. As can be seen in the figure, good agreement between simulation and experiment has been observed, which demonstrates that the presented model can capture well the nonlinear behaviour of these composite frames.

2.3.4 Frame with semi-rigid connections

To verify the developed source code for the simulation of semi-rigid connections in OpenSees, the 2-story planar steel frame tested by Stelmack et al. [21] was analysed. All beams of the frame are linked to columns by bolted top and seat angles connections that have the semi-rigid behaviour when subjected to external loads. The frame is assembled from members with the section of W5x16 made from A36 steel. The geometry and loading pattern of the frame are presented in Figure 9a. In Figure 9b, the comparison of the result from this study and that from the experiment is depicted. A good correlation between two studies has been observed and it indicates that the added source code shows high reliability.

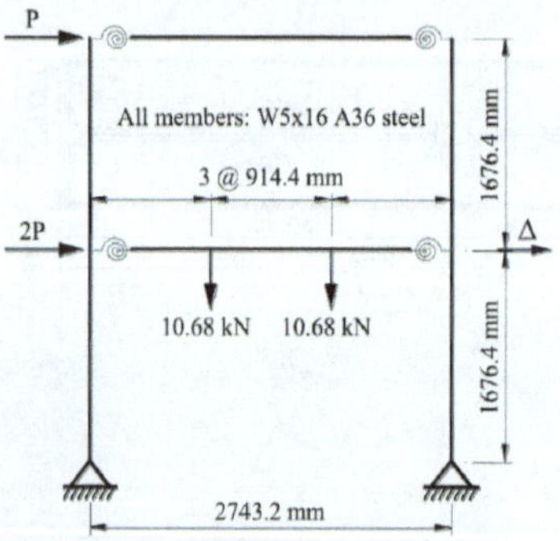

a) Geometry and loading pattern

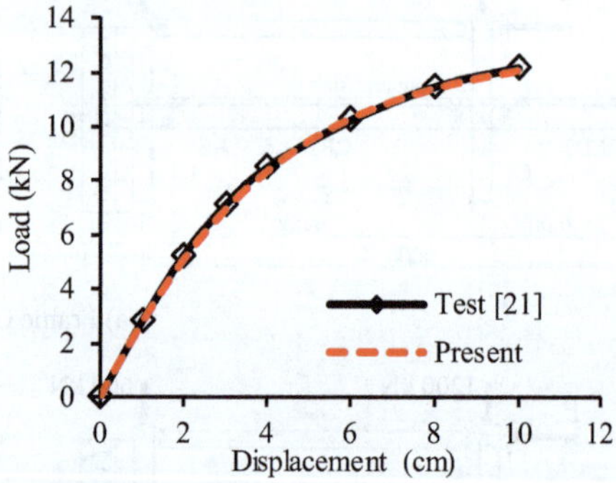

b) Lateral load-displacement relationship

Figure 9. Semi-rigid steel frame tested by Stelmack [21].

3 CASE STUDY AND PARAMETRIC STUDY

3.1 Model of 10-story composite frame

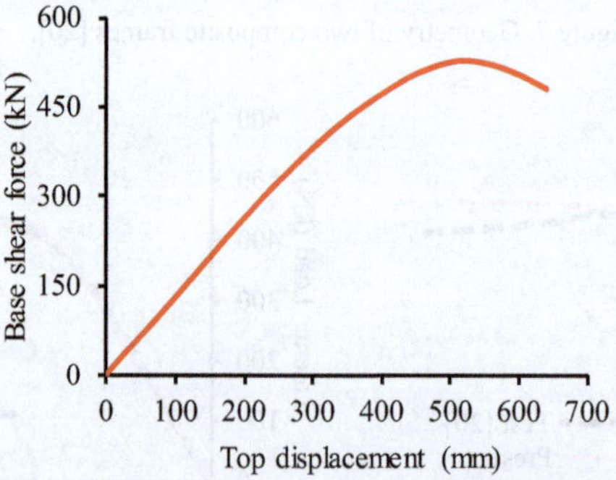

Figure 10. Lateral load-displacement curve of 10-story composite frame.

After the verification of the proposed model, the case study of a 10-story planar composite frame that comprises of CFST columns and composite beams is conducted. This frame is subjected to both gravity load and wind load and is designed based the strength limit state given in AS/NZS 2327:2017. The distributed gravity load is 36 kN/m, the wind load is calculated based on AS/NZS 1170.2:2011 with the wind speed of 36.274 m/s taken at Melbourne. The concrete compressive strength is 35 MPa and the steel yield strength is 450 MPa. The steel tubes of all square CFST columns have the section of 250x250x8 mm, the section of the composite beams is H250x150x10x15 for steel beam with the dimension of concrete slab of 1200x125 mm, and the section of braces is H80x80x3x3. More details about the composite frame are presented in Figure 11a. The load-displacement curve of this frame is shown in Figure 10.

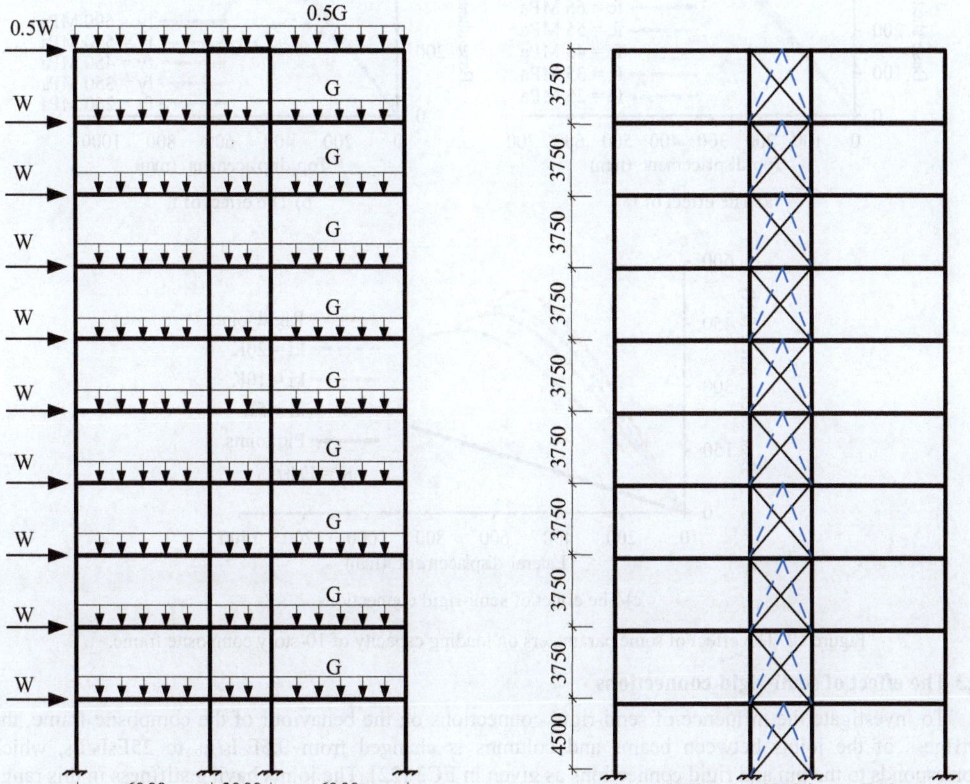

a) Geometry of 10-story composite frame b) Bracing systems (X braces and V braces)

Figure 11. Geometry of 10-story composite frame and bracing systems.

3.2 The effect of material properties

The effect of material properties on the load-displacement curve of the composite frame is illustrated in Figure 12a and b. As can be seen in this figure, the yield strength of steel has significant influence whilst the concrete compressive strength has negligible effect on the lateral loading capacity of the composite building. The compressive strength of concrete is fluctuated from 25 MPa to 65 MPa but the corresponding loading capacity only increases slightly from 512 kN to 565 kN. In contrast, the fluctuation of the steel yield strength from 250 MPa to 690 MPa has led to great growth in the loading capacity which increases from 300 kN to 782 kN, respectively.

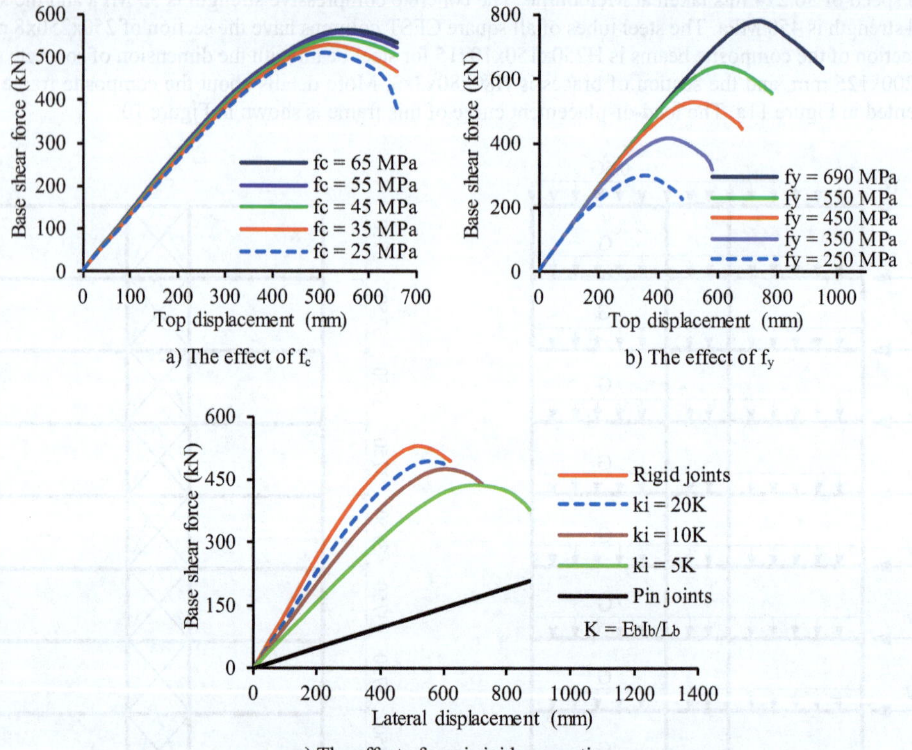

c) The effect of semi-rigid connections

Figure 12. The effect of some parameters on loading capacity of 10-story composite frame.

3.3 The effect of semi-rigid connections

To investigate the influence of semi-rigid connections on the behaviour of the composite frame, the stiffness of the joints between beams and columns is changed from $0.5E_bI_b/L_b$ to $25E_bI_b/L_b$, which corresponds to the pin and rigid connections as given in EC3 [22]. The joints having stiffness in this range are considered as semi-rigid connections. In Figure 12c, the load-displacement curves of the composite frame with semi-rigid connections are depicted. It can be seen that when the stiffness (k_i) of the joints is adjusted from $0.5E_bI_b/L_b$ (pin connection) to $10E_bI_b/L_b$ the stiffness and the peak of the load-displacement curve increase significantly, but when k_i is larger than $10E_bI_b/L_b$, the load-displacement curve does not change much.

3.4 The effect of bracing system

In this study, the bracing systems with two configurations: X-shape and V-shape are considered (Figure 11b). These bracing systems are assigned to the rigid composite frames and semi-rigid composite frames to investigate their influence on the loading capacity of the composite structure. The results of the simulation are shown in Figure 13. It is obvious that when bracing systems are implemented, the loading capacity of the composite frame increases dramatically. When utilizing X-shape bracing systems, the ultimate base shear force of the rigid composite frame and the semi-rigid composite frame surges from 513 MPa to 844 MPa and 437 MPa to 772 MPa, respectively. In case of V-shape bracing system, a remarkable growth in the ultimate base shear force of the rigid composite frame and the semi-rigid composite frame has been observed, which rises from 513 MPa to 714 MPa and 437 MPa to 640 MPa. Also from the data, it can be concluded that the X-shape bracing systems are more efficient than the V-shape bracing systems.

Another feature that should be noted is the bracing systems can minimize the influence of semi-rigid connections since the ultimate base shear force of the rigid composite frame and semi-rigid composite frame are nearly identical after the bracing systems are added (844 MPa and 772 MPa with X braces, 714 MPa and 640 MPa with V braces).

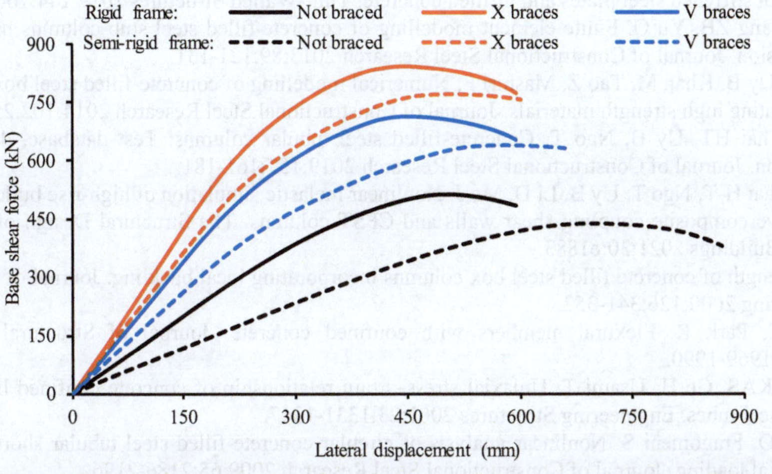

Figure 13. The effect of bracing systems on loading capacity of 10-story composite frame.

4 CONCLUSIONS

This study has proposed a nonlinear model for the analysis of composite frames with semi-rigid connections. These frames comprise of CFST columns and composite beams that are 100% composite without slip at the interface between concrete and steel. The model has been verified by numerous experimental data before it is used to conduct the case study and parametric study of the 10-story composite frame. Based on the results of the study, some conclusions can be drawn following:

(1) The proposed model with the developed source code for the simulation of semi-rigid connections shows high reliability since the results from the model agree well with those from experiment.

(2) The compressive strength of concrete has minor influence on the lateral loading capacity of the composite structures since the increase of concrete strength from 25 MPa to 65 MPa only leads to the growth of the loading capacity from 512 kN to 565 kN. In contrast, when the steel yield rises from 250 MPa to 690 MPa, the loading capacity surges from 300 kN to 782 kN, which shows great effect of the steel yield strength on the response of composite structures.

(3) The semi-rigid connections affect the behaviour of composite structures significantly. However, the loading capacity of this kind of structure does not change much when the stiffness of the connections exceeds $10E_bI_b/L_b$.

(4) The bracing system can contribute greatly to reduce the effect of semi-rigid connections on the behaviour of the composite structures since the loading capacity of the rigid composite frame and the semi-rigid composite frame are nearly identical when bracing system are considered (844 MPa and 772 MPa with X braces, 714 MPa and 640 MPa with V braces). The X-shape bracing systems show superior performance in comparison to the V-shape bracing system.

ACKNOWLEDGEMENTS

The present study was supported by the Australian Research Council (ARC) under its Discovery scheme (DP200100112). The financial support is gratefully acknowledged.

REFERENCES

[1] Elremaily A, Azizinamini A. Behavior and strength of circular concrete-filled tube columns. Journal of Constructional Steel Research 2002;58:1567-1591.

[2] Zhang W, Wang K, Chen Y, Ding Y. Experimental study on the seismic behaviour of composite shear walls with stiffened steel plates and infilled concrete. Thin-Walled Structures 2019;144:106279.

[3] Tao Z, Wang ZB, Yu Q. Finite element modelling of concrete-filled steel stub columns under axial compression. Journal of Constructional Steel Research 2013;89:121-131.

[4] Thai HT, Uy B, Khan M, Tao Z, Mashiri F. Numerical modelling of concrete-filled steel box columns incorporating high strength materials. Journal of Constructional Steel Research 2014;102:256-265.

[5] Thai S, Thai HT, Uy B, Ngo T. Concrete-filled steel tubular columns: Test database, design and calibration. Journal of Constructional Steel Research 2019;157:161-181.

[6] Tran H, Thai H-T, Ngo T, Uy B, Li D, Mo J. Nonlinear inelastic simulation of high-rise buildings with innovative composite coupling shear walls and CFST columns. The Structural Design of Tall and Special Buildings 2021;30:e1883.

[7] Uy B. Strength of concrete filled steel box columns incorporating local buckling. Journal of Structural Engineering 2000;126:341-352.

[8] Kent DC, Park R. Flexural members with confined concrete. Journal of Structural Division 1971;97:1969-1990.

[9] Susantha KAS, Ge H, Usami T. Uniaxial stress–strain relationship of concrete confined by various shaped steel tubes. Engineering Structures 2001;23:1331-1347.

[10] Liang QQ, Fragomeni S. Nonlinear analysis of circular concrete-filled steel tubular short columns under axial loading. Journal of Constructional Steel Research 2009;65:2186-2196.

[11] Tang J, Hino S, Kuroda I, Ohta T. Modeling of stress - strain relationships for steel and concrete in concrete filled circular steel tubular columns. . Steel Construction Engineering 1996;3:35-46.

[12] Eom TS, Park HG, Lee CH, Kim JH, Chang IH. Behavior of double skin composite wall subjected to in-plane cyclic loading. Journal of Structural Engineering 2009;135:1239-1249.

[13] Kishi N, Chen WF. Moment-rotation relations of semirigid connections with angles. Journal of Structural Engineering 1990;116:1813-1834.

[14] Nie J, Cai CS. Steel–concrete composite beams considering shear slip effects. Journal of Structural Engineering 2003;129:495-506.

[15] Slutter RG, Driscoll Jr GC. Flexural strength of steel-concrete composite beams. Journal of the Structural Division 1965;91:71-99.

[16] Liu D, Gho WM. Axial load behaviour of high-strength rectangular concrete-filled steel tubular stub columns. Thin-Walled Structures 2005;43:1131-1142.

[17] Fujimoto T, Mukai A, Nishiyama I, Sakino K. Behavior of eccentrically loaded concrete-filled steel tubular columns. Journal of Structural Engineering 2004;130:203-212.

[18] Tomii M. Experimental studies on concrete filled steel tubular stub columns under concentric loading. Paper presented at: Proceedings of International Colloquium on Stability of Structures Under Static and Dynamic Loads, SSRC/ASCE/Washington, DC, 1977.

[19] Liew JYR, Xiong DX, Zhang MH. Experimental studies on concrete filled tubes with ultra-high strength materials. Proceedings of the 6th Internaltional Symposium on Steel Structures, Seoul, Korea 2011:377-384.

[20] Jianguo N, Yuan H, Weijian Y, Jiansheng F. Seismic behavior of CFRSTC composite frames considering slab effects. Journal of Constructional Steel Research 2012;68:165-175.

[21] Stelmack TW, Marley MJ, Gerstle KH. Analysis and tests of flexibly connected steel frames. Journal of Structural Engineering 1986;112:1573-1588.

[22] Eurocode 3. Design of steel structures-Part 1-8: Design of joints. Design of steel structures-Part 1-8: Design of joints; 2005.

9TH INTERNATIONAL CONFERENCE ON COMPOSITE CONSTRUCTION IN STEEL AND CONCRETE
M. Knobloch, U. Kuhlmann, W. Kurz, M. Schäfer. (Eds.)
Online, Germany, July 27-29, 2021

EXPERIMENTAL AND NUMMERICAL INVESTIGATION OF THE FRICTION BASED DEMOUNTABLE SHEAR CONNECTOR

Jovan FODOR*, Markus SCHÄFER*

* University of Luxembourg, Department of Engineering, Structural engineering and composite structures
e-mails: jovan.fodor@uni.lu, markus.schaefer@uni.lu

Keywords: Steel-concrete composite beams, Demountable shear connector, Friction behaviour, Modularity, Reusability, Sustainability

Abstract. *In order to provide sustainable structural solution for the steel-concrete downstand composite beams in buildings by providing modularity and reusability, new demountable shear connector device based on frictional behaviour is devised. The mechanical behaviour of the newly developed connector is investigated both experimentally and numerically and its applicability in aforementioned application is assessed.*

1 INTRODUCTION

The connector device presented in this paper is developed in an effort to improve the sustainability of the steel-concrete composite floors (beams) by providing modular, demountable, and reusable structural solution while allowing for required execution tolerances in the construction stage regarding modularity.

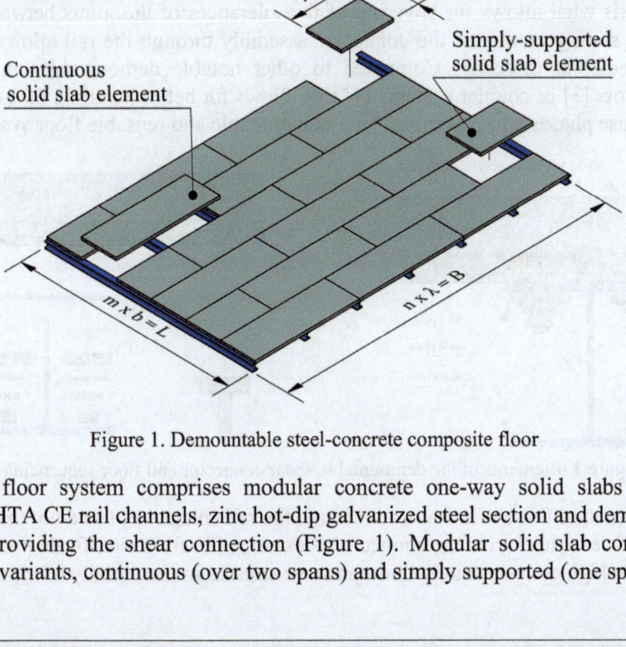

Figure 1. Demountable steel-concrete composite floor

The demountable floor system comprises modular concrete one-way solid slabs with equidistantly embedded Halfen HTA CE rail channels, zinc hot-dip galvanized steel section and demountable connector device assembly providing the shear connection (Figure 1). Modular solid slab concrete elements are represented in two variants, continuous (over two spans) and simply supported (one span). This allows for improvement of the inplane behaviour of the floor as the slab elements are placed in offsetted horizontal layout. In order to decrease the resource demand during the execution and reduce the load on the shear connection, the floor is unpropped in the construction phase.

2 SLAB ELEMENT, FLOOR AND CONNECTOR LAYOUT

The modularity of the slab element is dictated by the floor grid, by the span (L) and the spacing (λ) of the composite beams (Figure 1). In order to have a modular system whose elements may be reused, the spacing (e) between the concrete-embedded rail channels is uniform along the width of the slab element (b), (Figure 2). This implies that the shear connectors are equidistantly spaced along the composite beam and that preferable mechanical behaviour of the shear connector after reaching ultimate capacity is ductile. In this way the full potential in terms of capacity of the shear connection is exploited as connectors may not be concentrated at the supports if the elastic distribution of the longitudinal shear is assumed [1],[2].

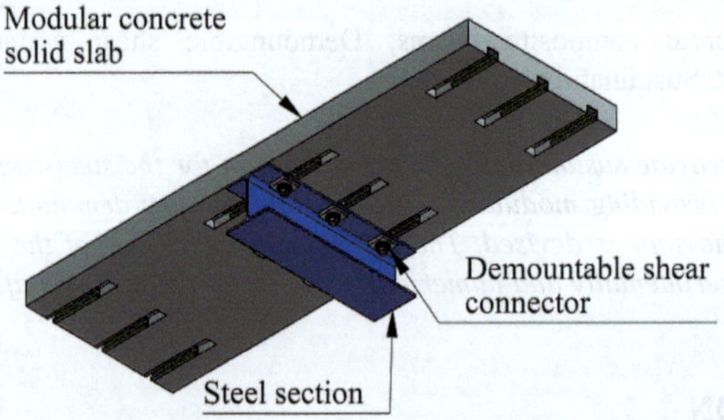

Figure 2. Main elements of the demountable composite floor

The connector device comprises rail plate fitting the socket of the rail channel, HV pretension bolt assembly, zinc hot-dip galvanized back plate, enlarged diameter washer and two (or one) disc springs (Figure 3). The bolt hole of the steel section is long-slotted while the bolt holes of the rail plate and the back plate are normal or fitted. The long-slotted bolt holes allow for large tolerance in execution in the direction of beam axis what allows for covering of the tolerances of the joints between the consecutive slabs. Similarly, the sliding motion of the connector assembly through the rail allows for managing the tolerances in perpendicular direction. Compared to other notable demountable connector devices as friction-grip connectors [3] or coupler systems [4] this allows for better handling of the tolerances during the execution and reuse phases what is crucial for a demountable and reusable floor system.

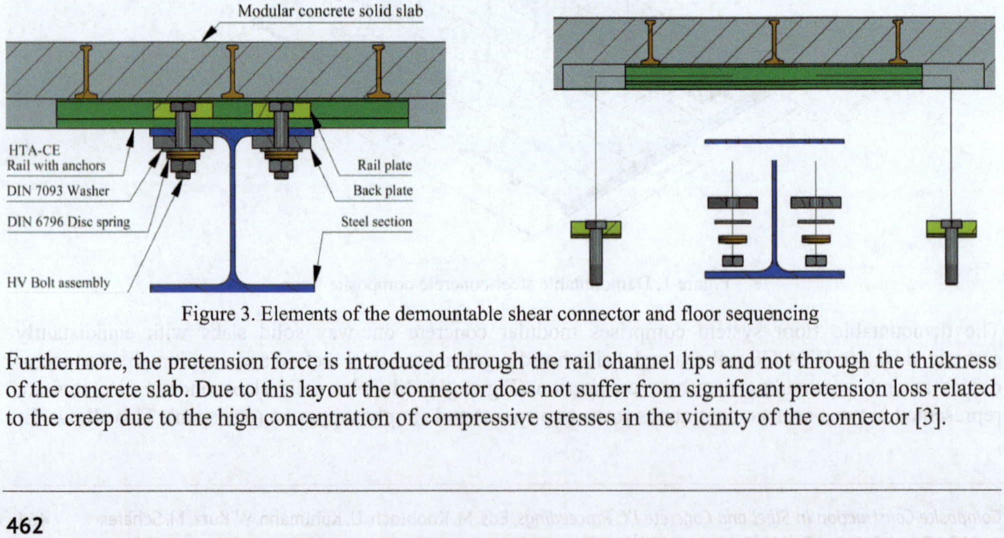

Figure 3. Elements of the demountable shear connector and floor sequencing

Furthermore, the pretension force is introduced through the rail channel lips and not through the thickness of the concrete slab. Due to this layout the connector does not suffer from significant pretension loss related to the creep due to the high concentration of compressive stresses in the vicinity of the connector [3].

The composite floor is unpropped in the execution stage. The steel beam is precambered regarding self-weight deflection that arises from the weight of the steel beam and the weight of the concrete slabs. After the placement of the slabs, that are provided with rail plates and bolts, due to the precamber the floor is in levelled state. Afterwards, the high strength HV pretension bolts are preloaded to the design preload force F_{pc} and the shear connection is therefore activated.
The pretension force of the bolts exerts equal clamping force between the concrete slab and the steel section and between the back plate and the bottom face of the steel section top flange. Obtained mechanical connection is frictional and it is composed out of two aforementioned faying surfaces. Mechanically, the system is equivalent to the lap-joint friction dampers for bracing systems in seismic design [5]. The application of the DIN 6796 disc springs allows for redundancy of the pretension force regarding contraction of the pretension assembly due to slipping or creep of the surface material [6].

3 EXPERIMENTAL CAMPAIGN

3.1 Push test layout and procedure

In accordance with Annex B of EN 1994-1-1 [7], three push tests were performed on three identical test specimens. The concrete slabs (Figure 4) were provided with the recess at the bottom side. The reinforcement was denser ($\emptyset 10/75$) compared to the code provisions ($\emptyset 10/150$) in order to mitigate the concrete related failure modes. The transversal reinforcement (T) might be reduced in the direction of slab module span for more efficient design as its importance lies within the connector load introduction zone.

The first specimen (PO-B-1) was tested only in displacement control mode with displacement speed of $0.25 mm/min$ reproducing quasi-static response of the system. Specimens PO-B-2 and PO-B-3 were firstly loaded in force control mode with 25 cycles of harmonic loading with amplitudes respectively 5% and 40% of the failure load defined by the results of the test specimen PO-B-1 and load rate of $360 kN/min$.

Table 1. PO-B push test series schedule

Specimen	Bolt assembly	Target pretension [kN]	Cycling step
PO-B-1	M24 HV	175	NO
PO-B-2	M24 HV	175	YES
PO-B-3	M24 HV	175	YES

After the placement of the assembly the bolts were preloaded to the target pretension level $F_{pc} = 175 kN$. The applied pretension force was 70% of the designated pretension force for the M24 HV bolt ($247 kN$). The substantial elastic reserve was designated in order to ensure elastic response of the bolt avoiding pretension loss during the slipping due to the bending of the bolt shank [6]. The right value of the pretension torque M_t was obtained by performing torque calibration test in accordance with Annex H of EN 1090-2 [8]. The experiment procedure is explained in the chapter *Pretension tests*.

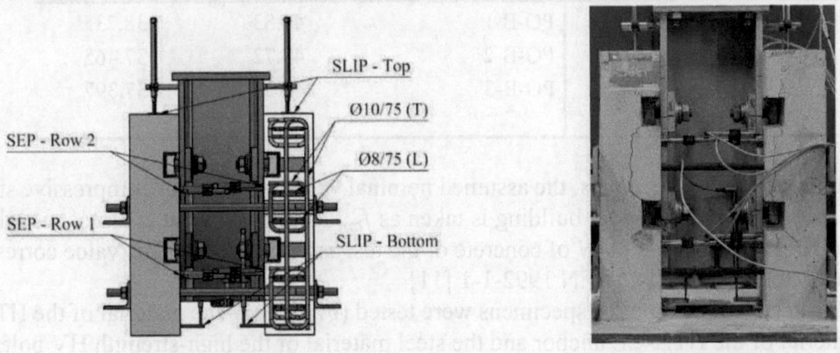

Figure 4. PO-B push test series layout

In total 14 (8 + 6) LVDT sensors were placed to measure displacements during the tests. Eight sensors were measuring the separation displacement in the vicinity of each shear connector (SEP) while six sensors were measuring the slip displacement (SLIP), four on the top of the assembly and two at the bottom (Figure 4). Especially important were the slip readings from the bottom two sensors as they were not affected by the deformation of the grout at the base of the assembly. Besides the displacement measurements the total push force was measured as the force of the hydraulic jack.

3.2 Material tests

In order to assess the material properties of the main assembly parts that will be used later for the building of the corresponding material models for the finite element analysis, the standard material tests were performed on the relevant concrete and steel materials.

The evaluation of the modulus of elasticity and the compressive strength tests of the hardened concrete were done in accordance with EN 12390-13 [9] and 12390-3 [10] respectively (Figure 5). All the test specimens were casted at the same time of the casting of the test slabs. The test specimens (FC 1-4) were kept in climactic chamber and tested after 28 days while the test specimens (PO-B 1-3) were cured alongside the slabs and tested in the period of the execution of the push test series PO-B.

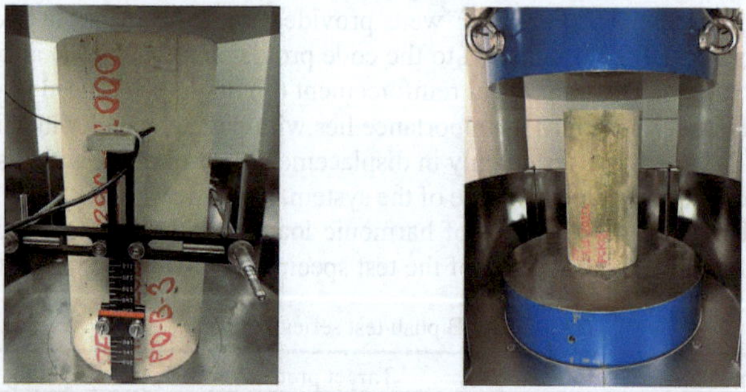

Figure 5. Test specimens for the evaluation of material properties of the hardened concrete

The mean concrete compressive strength in the first case is $f_{cm,28} = 51.9\ kN$ while in second $f_{cm,B} = 47.9\ kN$. In both cases coefficient of variation is less than 2%. The higher strength obtained in the first case despite the fact that the concrete material was tested earlier comes from the better curing conditions. In the second case the measured secant modulus E_{cm} is slightly higher than the one predicted in accordance with the Table 3.1 of EN 1992-1-1 [11] based on the mean compressive strength $f_{cm,B}$ ($E_{cm} = 36.056\ GPa$).

Table 2. Results of the compressive tests of hardened concrete

Specimen	$f_c\ [MPa]$	Specimen	$f_c\ [MPa]$	$E_{cm}\ [GPa]$
FC-1	51.4	PO-B-1	47.83	38.735
FC-2	53.04	PO-B-2	47.72	37.465
FC-3	51.85	PO-B-3	48.25	37.397
FC-4	51.41			

Due to the limited number of test specimens, the assumed nominal value of the mean compressive strength of the concrete for the later material model building is taken as $f_{cm} = 48MPa$ what is almost equal to the mean compressive strength $f_{cm,B} = 47.9kN$ of concrete of the test assembly slabs. This value corresponds to the concrete class C 40/50 according to EN 1992-1-1 [11].

Regarding the steel material, three types of specimens were tested (Figure 6). The material of the HTA-CE rail channel, the material of the HTA-CE anchor and the steel material of the high-strength HV bolt.

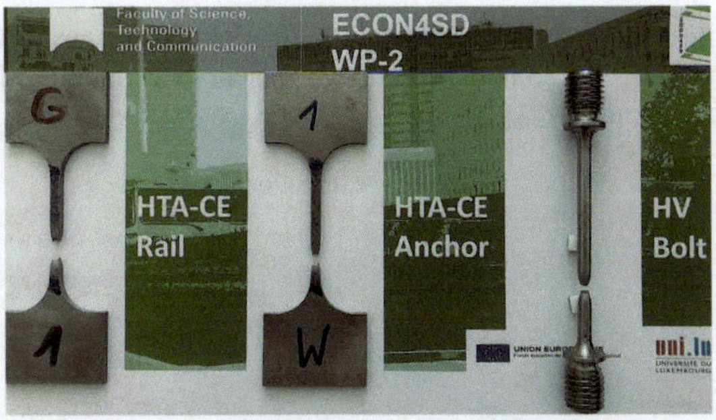

Figure 6. Test specimens after the tensile tests

For the rail and anchor parts, flat specimens with original cross-sectional area of $S_0 = 12\,mm^2$ were prepared while for the bolt cylindric specimen with cross-sectional area of $S_0 = 19.635\,mm^2$ ($D_0 = 5mm$). The tests were performed in accordance with EN 6892-1 [12].

Table 3. Results of the tensile tests for rail and anchor steel material

Specimen	$S_0\,[mm^2]$	$A\,[\%]$	$R_{eH}\,[MPa]$	$R_{eL}\,[MPa]$	$R_m\,[MPa]$	$E\,[GPa]$
HTA-CE Rail	12.14	28.0	380	364	443	211.2
HTA-CE Anchor	12.18	41.0	299	277	374	212.8

Table 4. Results of the tensile tests for the bolt steel material

Specimen	$S_0\,[mm^2]$	$A\,[\%]$	$R_{p0.2}\,[MPa]$	$R_m\,[MPa]$	$E\,[GPa]$
HV M24 Bolt	19.78	17.5	966	1055	210.7

The results obtained for the rail and anchor correspond to properties of mild steels while the results obtained for the bolt specimen correspond well to the properties of 10.9 bolt material class (HV).

3.3 Pretension tests

In order to assess the torque M_t that corresponds to the target pretension force F_{pc}, the pretension test in accordance with the EN 1090-2 Annex H [8] was performed. The test assembly corresponds as close as possible to the actual assembly in the push test (Figure 7). In order to uniformize the torque results and to decrease the target pretension torque M_t, Molykote G-Rapid Plus lubricant was used to lubricate the threads and the contact surface between the nut and the disc spring.

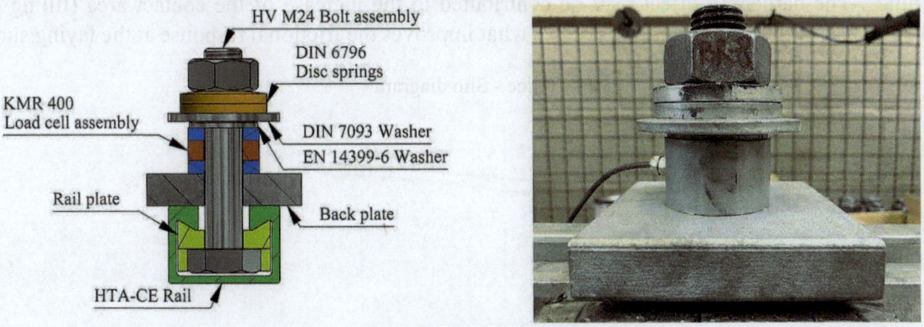

Figure 7. Pretension calibration test layout

Obtained results were quite consistent (Figure 8) and for the target pretension force $F_{pc} = 175kN$, the derived target pretension was $M_t = 525\ Nm$ with the coefficient of variation of $V_M = 5\%$ respecting the required precision for 8 specimens ($V_{M,max} = 6\%$).

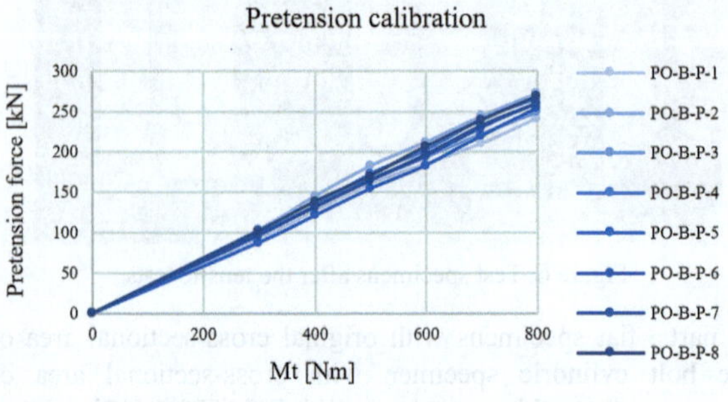

Figure 8. Pretension calibration test results

The pretension force in the push test was applied in two steps as per EN 1090-2 [8]. In the first step the torque of $M_{t,1} = 0.75M_t \cong 350Nm$ is applied to all bolts and consecutively the torque $M_{t,2} = 1.10M_t \cong 575Nm$. The additional 10% increase of the torque covers closely the short-term pretension loss that happens between the torqueing of the bolts and the execution of the experiment.

4 PUSH TEST RESULTS

The force-slip diagrams per connector are presented bellow in the Figure 9. The initial response of the assembly is almost rigid like. This is attributed to the slip capacity at the slab-steel section interface. After the aforementioned slip capacity is reached, secondary faying surface is activated (back plate-top flange) in addition. The friction resistance generated at the secondary faying surface includes the transfer of the force from the second faying surface to the concrete slab via the bending and shearing of the bolt. This process is observable as the second part of the force-slip diagram characterized by the finite stiffness response. The stiffness (slope) of the curve in this region depends on the bending stiffness of the bolt and its boundary conditions within the connector assembly (rail lips). The outlying of the specimen PO-B-1 in this region is attributed to the geometrical imperfection of the long-slotted hole where one of the bolts was bearing on the bolt hole what resulted in stiffer response.

After the additional capacity at the secondary faying surface is exhausted as well, the system reaches third and last part of the force-slip diagram that is characterized by the sliding behaviour followed by limited "hardening". The hardening effect may be contributed to the increase of the contact area (filling of the slotted hole) that decreases the contact stresses what improves the frictional response at the faying surfaces.

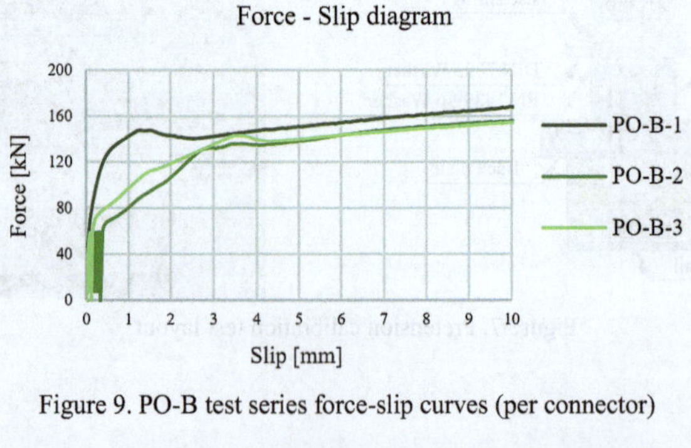

Figure 9. PO-B test series force-slip curves (per connector)

The force-slip behaviour qualitatively fulfils the requirements for the application of ductile shear connection. Initial rigid-like response will allow for complete interaction response of the beam when fulfilling serviceability limit states while high ductility of the shear connector would allow for the uniform redistribution of longitudinal shear force hence justify the application of equidistant rail spacing (e) within the modular solid slab element.
One of the issues may be the flexible behaviour in the second part of the force-slip response what may lead to deviation from the complete interaction composite beam behaviour model at serviceability load levels. Having in mind that the system is tailored as full degree of shear connection system ($\eta \geq 1.00$) it was demonstrated that the influence of the softer response in the second force-slip region on the beam response is negligible [13]. Based on the explained mechanics of the connector assembly characterized by three force-slip regions, appropriate tri-linear capacity model (TB) may be derived.

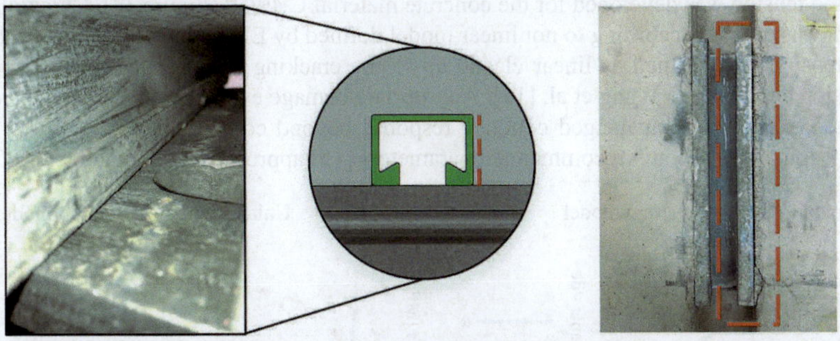

Figure 10. The aftermath of the push test

After the experiments no damage was observed in the critical parts (Figure 10). The geometry of the rail remained intact and the connector assembly was easily disassembled proving the demountability of the system. The only visible damage was observed at the rail wall-slab interface opposite to the load introduction face where the rail wall separated from the concrete matrix. Possible anchoring of the wall would increase the overall stiffness of the connector in the second force-slip region.

5 NUMERICAL ANALYSIS

The numerical analysis of the behaviour of the proposed shear connector device was performed using Abaqus FEA software suite. Advance FE model was developed and processed using Abaqus/Explicit dynamic solver.
Exploiting the planes of symmetry localized FEA assembly was developed (Figure 11). The support face of the concrete slab was coupled to the support reference point RP-1 that was fixed. The load face of the steel section was coupled to the reference point RP-2 where the force and displacement load were assigned.

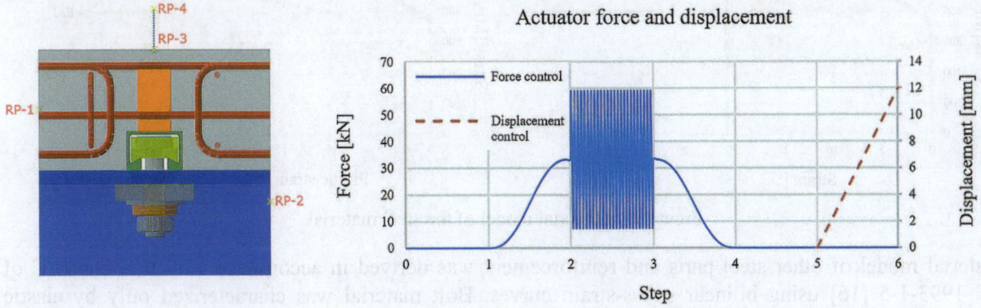

Figure 11. FEA assembly model and the load sequencing

The top face of the concrete slab was coupled with the reference point RP-3 that was connected with the spring element to the fixed support reference point RP-4. The spring has the stiffness of half-length of the threaded rod of the lateral restraining assembly.
Two sequencing procedures were developed (Figure 11). The first considering only the monotonic push displacement load (PO-B-1) and the one considering the load cycling (PO-B 2-3) as well. In the first case after the pretension step (0-1) where the pretension force is introduced as a temperature difference $\Delta T < 0$, the steel section is pushed monotonically executing the displacement load step (5-6). In the second case after the pretension step, the mean value of the cycling load is applied (1-2) and then throughout the cycling step (2-3) the load is cycled 25 times as in the experiment. In the next step the assembly is deloaded (3-4), rested (4-5), and as in the experiment pushed monotonically in the displacement mode (5-6).
The concrete part was meshed with refined C3D8R brick solid element mesh. The uniaxial behaviour in compression and tension was developed for the concrete material C 40/50 (Figure 12). Uniaxial behaviour in compression was defined according to nonlinear model defined by EN 1992-1-1 [11] while the uniaxial behaviour in tension was defined as linear elastic up to the cracking strain followed by hyperbolically decaying function according to Wang et al. [14]. Appropriate damage evolution functions were defined by comparing the damaged and undamaged concrete response beyond compressive strength f_{cm} or tensile strength f_{ctm}. Taking into account recommended parameters [15] appropriate CDP model was defined.

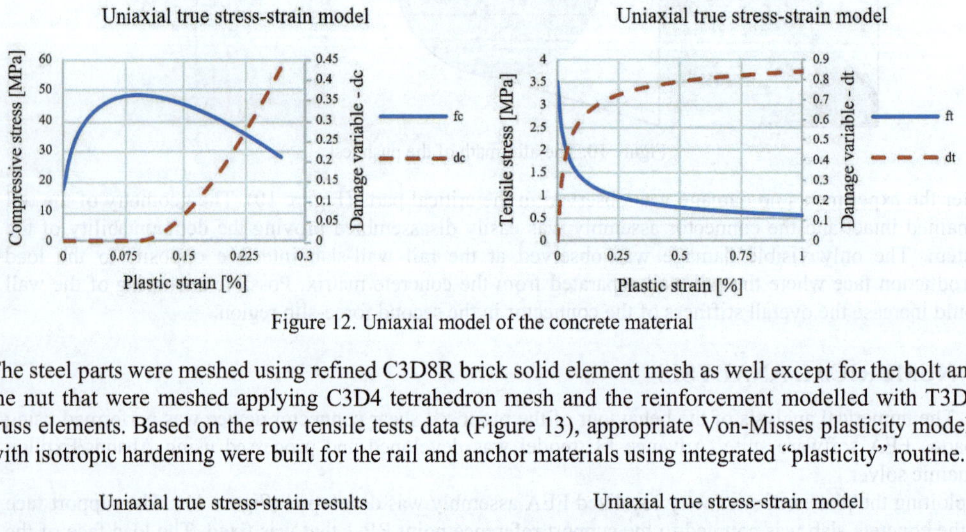

Figure 12. Uniaxial model of the concrete material

The steel parts were meshed using refined C3D8R brick solid element mesh as well except for the bolt and the nut that were meshed applying C3D4 tetrahedron mesh and the reinforcement modelled with T3D2 truss elements. Based on the row tensile tests data (Figure 13), appropriate Von-Misses plasticity models with isotropic hardening were built for the rail and anchor materials using integrated “plasticity” routine.

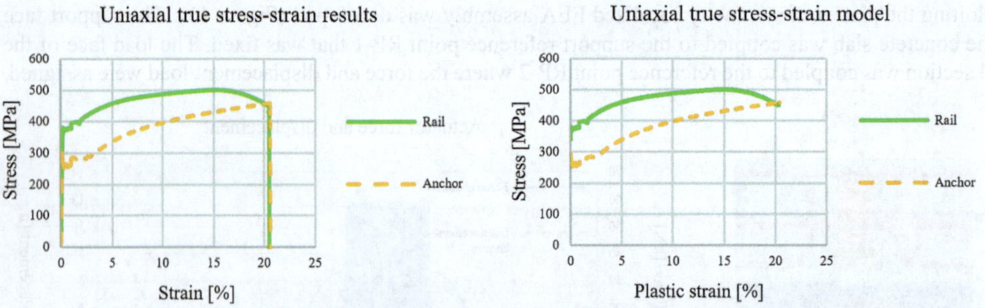

Figure 13. Uniaxial model of the steel material

Material model of other steel parts and reinforcement was derived in accordance with the Annex C of EN 1993-1-5 [16] using bilinear stress-strain curves. Bolt material was characterized only by elastic behaviour as due to the reduced pretension it remains completely elastic during the course of the analysis.

The contacts between the parts were modelled as surface-to-surface contact pairs. The interaction property in normal direction was modelled as "hard" while in the tangential direction "penalty" formulated tangential behaviour was modelled. The friction coefficient of the concrete-steel interface at the top faying surface was assigned as $\mu_1 = 0.45$ while at the steel-steel brushed galvanized interface the value of $\mu_2 = 0.3$ was assigned.

In order to rationalize the required time for the execution of the simulation while obtaining quasi-static response of the system in explicit analysis, semi-automatic mass scaling was applied. The validity of the results was justified by keeping the ratio of kinetic and artificial strain energy over internal energy well under 5% throughout the whole analysis.

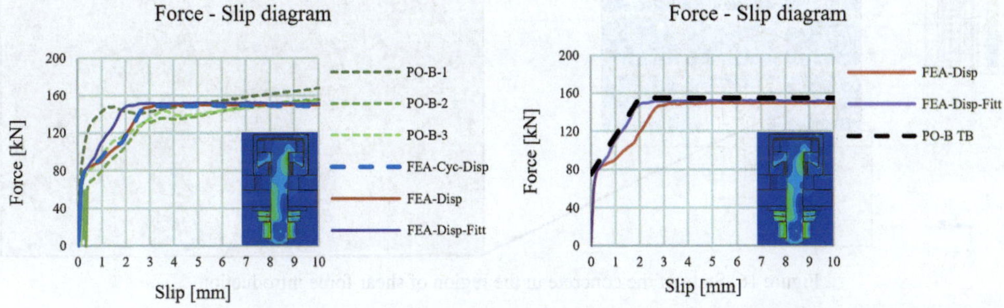

Figure 14. Experimental vs. numerical force-slip data (per connector)

Numerically obtained results fit well the experimental force-slip data (Figure 14). The three-stage force-slip behaviour is clearly visible as in the experiment and may be modelled by a trilinear capacity curve (PO-B TB). Besides assembly that corresponds to the actual push test (FEA), one additional assembly without bolt shank tolerance (FEA-Fitt) was simulated. This model assumes that the rail plate and back plate bolt holes are fitted. If this adjustment is applied to the assembly it significantly improves the system response in terms of stiffness. Switching to the fitted bolt holes regarding these parts is completely feasible as it does not impair overall tolerance flexibility of the connector and the floor assembly.

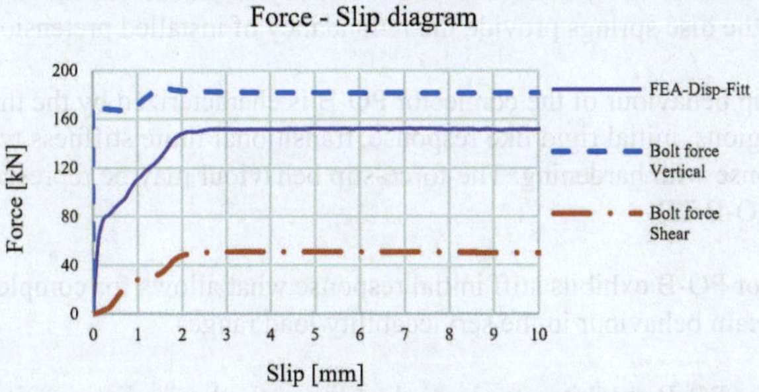

Figure 15. Development of the connector shear capacity (per connector)

The three-stage force-slip behaviour is also observable by following the development of the shear capacity of the connector, vertical bolt force and the bolt shear force (Figure 15). Initially, while the slip is very small the capacity is provided only by the friction at the primary faying surface (slab-steel section). As the slip displacement increases it activates the second faying surface (bolt shear). Due to the restraint of the bolt and its elastic behaviour the vertical bolt force slightly increases during the activation of the secondary faying surface. After exhaustion of the capacity at the both faying surfaces, the sliding phase under constant vertical bolt force (pretension) occurs.

The results of the numerical analysis show that only limited concrete damage related to the tensile stresses is concentrated in the corner of the rail on the side of the shear force introduction up to the first reinforcement bar. The same was observed during the experimental stage after the cutting of the slab element through the bolt axis (Figure 16). This means that even at the ultimate load level the concrete slab remains intact and that the concrete related failure modes are omitted. Hence, the modular concrete solid slab element is robust and has great reuse potential due to the introduction of the shear force through the walls of the rail channel.

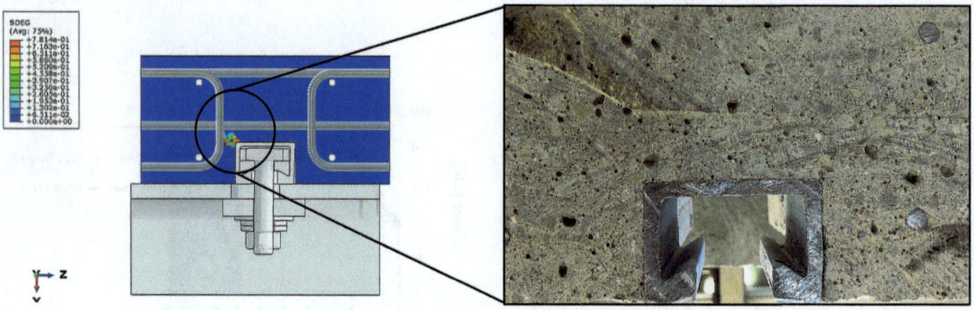

Figure 16. State of the concrete in the region of shear force introduction

6 CONCLUSIONS

- The proposed floor system is modular, demountable and reusable with great flexibility regarding the construction tolerances.
- The steel-concrete composite floor is unpropped and tailored for the full degree of shear connection layout ($\eta \geq 1.00$).
- The pretension force is introduced through the body of the rail channel avoiding the concrete creep while the disc springs provide the redundancy of installed pretension force.
- The force-slip behaviour of the connector PO-B is characterized by the three distinctive force-slip regions, initial rigid like response, transitional finite stiffness response and ideal plastic response with hardening. The force-slip behaviour may be represented by trilinear force-slip model PO-B TB.
- The connector PO-B exhibits stiff initial response what allows for complete interaction composite beam behaviour in the serviceability load ranges.
- The connector PO-B exhibits very large ductility with plastic slip capacity $\delta_u \geq 10mm$ what allows for uniform redistribution of longitudinal shear force and justifies the equidistant arrangement of connectors (rails).
- The stiffness of the connector in the second force-slip behaviour region may be improved by anchoring of the rail walls into the concrete matrix and by providing the rail plate and back plate with fitted bolt holes.
- The modular concrete slab element with its elements remains undamaged even after reaching the ultimate shear capacity of the connector rendering it suitable for further reuse.

7 AKNOWLEDGEMENT

The presented research work is the result of research of Work package 2 within the parent research project Eco-Construction for the Sustainable Development (ECON4SD) [17] performed at the University of Luxembourg and supported by FEDER (Fonds Européen de Développement Régional). During the course of the research work invaluable assistance in terms of technical expertise and provision of material required for the execution of the experimental work was provided by:

- Research support, Department of Engineering, University of Luxembourg
- Spannverbund GmbH, Waldems, Germany/Spannverbund Luxembourg S.A., Luxembourg
- Halfen GmbH, Langenfeld, Germany
- Astron Buildings S.A., Diekirch, Luxembourg

REFERENCES

[1] G. Hanswille, M. Schäfer, and M. Bergmann, *Eurocode 4 DIN EN 1994-1-1 Bemessung und Konstruktion von Verbundtragwerken aus Stahl und Beton*. Ernst, Wilhelm & Sohn, 2020.

[2] R.P. JOHNSON, *Composite Structures of Steel and Concrete*, 3rd ed. Blackwell Publishing Ltd, 2004.

[3] K. Roik and G. Hanswille, "Beitrag zur Ermittlung der Tragfähigkeit von Reib-Abscherverdüblungen bei Stahlverbundträgerkonstruktionen," *Stahlbau*, vol. 2, pp. 41–46, 1984.

[4] A. S. Kozma, "Demountable Composite Beams: Analytical calculation approaches for shear connections with multilinear load-slip behaviour," 2020.

[5] T. Sano, K. Shirai, Y. Suzui, and Y. Utsumi, "Loading tests of a brace-type multi-unit friction damper using coned disc springs and numerical assessment of its seismic response control effects," *Bull. Earthq. Eng.*, vol. 17, no. 9, pp. 5365–5391, 2019.

[6] S. Ramhormozian, G. C. Clifton, G. A. MacRae, and G. P. Davet, "Stiffness-based approach for Belleville springs use in friction sliding structural connections," *J. Constr. Steel Res.*, vol. 138, pp. 340–356, 2017.

[7] CEN, "EN 1994-1-1: Design of composite steel and concrete structures - Part 1-1: General rules and rules for buildings." 2004.

[8] CEN, "EN 1090-2: Execution of steel structures and aluminium structures - Part 2: Technical requirements for steel structures." 2008.

[9] CEN, "EN 12390-13: Determination of the modulus of elasticity under pressure (secant module)." 2014.

[10] CEN, "EN 12390-3: Testing hardened concrete: Compressive strength of test specimens." 2002.

[11] CEN, "EN 1992-1-1: Design of concrete structures - Part 1-1 : General rules and rules for buildings Eurocode." 2004.

[12] CEN, "EN ISO 6892-1: Metallic materials - Tensile testing - Part 1: Method of test at room temperature." 2009.

[13] J. Fodor and M. Schäfer, "Behavior of downstand simply supported steel-con- crete composite beam applying friction based demountable shear connection," in *Eurosteel 2021*, 2021.

[14] T. Wang and T. T. C. Hsu, "Nonlinear finite element analysis of concrete structures using new constitutive models," *Comput. Struct.*, vol. 79, no. 32, pp. 2781–2791, 2001.

[15] P. KMIECIK and M. KAMIŃSKI, "Modelling of reinforced concrete structures and composite structures with concrete strength degradation taken into consideration," *Arch. Civ. Mech. Eng.*, vol. 11, no. 3, pp. 623–636, 2011.

[16] CEN, "EN 1993-1-5: Design of steel structures - Part 1-5: Plated structural elements." 2006.

[17] "Eco-Construction for Sustainable Development (ECON4SD) – Just another UL Multisite site." [Online]. Available: https://econ4sd.uni.lu/. [Accessed: 13-Sep-2020].

9TH INTERNATIONAL CONFERENCE ON COMPOSITE CONSTRUCTION IN STEEL AND CONCRETE
M. Knobloch, U. Kuhlmann, W. Kurz, M. Schäfer. (Eds.)
Online, Germany, July 27-29, 2021

INNOVATIVE COMPOSITE STRUCTURAL SYSTEMS FOR MODULAR TALL BUILDINGS

H.T. Thai*

*Department of Infrastructure Engineering, The University of Melbourne, Parkville, VIC 3010, Australia
E-mails: tai.thai@unimelb.edu.au

Keywords: Composite Structural System, Modular Building, Modular Connection.

Abstract. *Modular or offsite construction is believed to shape the future of the construction industry as it possesses significant benefits over traditional onsite construction methods. However, most of its application are limited to steel or concrete buildings. Although steel-concrete composite structural system has many merits over the steel and concrete systems, its application in modular buildings is very limited. This paper explores recent developments of composite systems for modular high-rise buildings. They include modular units for resisting vertical gravity loads and lateral structural systems for resisting horizontal forces from wind and earthquake loadings and progressive collapse due to accidental loads such as fire, explosions and impact. Various inter-module joining methods developed in the literature will also be reviewed. Finally, a case study of the most efficient connection is presented to explore its applicability to high-rise modular buildings.*

1 INTRODUCTION

Modular construction, also known as prefabricated or offside construction, refers to any construction methods in which building components (up to 90%) are prefabricated in a quality controlled manufacturing environment before being transported onsite for assembly. Therefore, modular construction is believed to shape the future of the construction industry because of offering greener (reducing onsite construction waste up to 90%), faster (saving construction time up to 50%) and thus cheaper (up to 20% equivalent to 22 billion dollars globally) benefits compared to traditional onsite methods [1]. Although the concept of modular construction is not a new, it has gained more attention because the industry starts to realize the potential of modular building in resolving the affordable housing shortage. Recently, the use of modular construction has also been increased in Australia, Singapore, UK and US where housing shortages and labour costs are of significant concern [1].

In the last decade, a large number of studies on modular construction have been carried out. From a structural engineering perspective, it is feasible to apply the modular construction method to multi-storey buildings [2], and thus a number of structural systems for modular tall buildings have been developed. They can be found in the review articles published recently on connection performance [3, 4] structural response [5, 6], structural stability [7], seismic performance [8], and structural systems [1]. It is revealed from the literature review that most of the structural systems developed for modular buildings are limited to steel or concrete systems [1]. The development of composite structural systems for modular buildings is very limited.

This paper will highlight the potential application of steel-concrete composite structural systems for modular multi-storey buildings. In addition, the numerical study of an innovative connection system carried out recently by the author will also be introduced. Modular construction can be classified into three different forms based on the degree of prefabrication: 1D single member, 2D panelised system and 3D volumetric system. However, this paper only focuses on the 3D volumetric system for tall buildings because it is considered as the most effective system that can allow up to 95% of a building to be prefabricated in a factory [1]. An example of a typical 3D volumetric modular tall building composed of modular unit systems for resisting gravity loads as shown in Figure 1a, core systems for resisting lateral loads as shown in Figure 1b, and joining systems for ensuring the structural integrity as shown in Figure 1c.

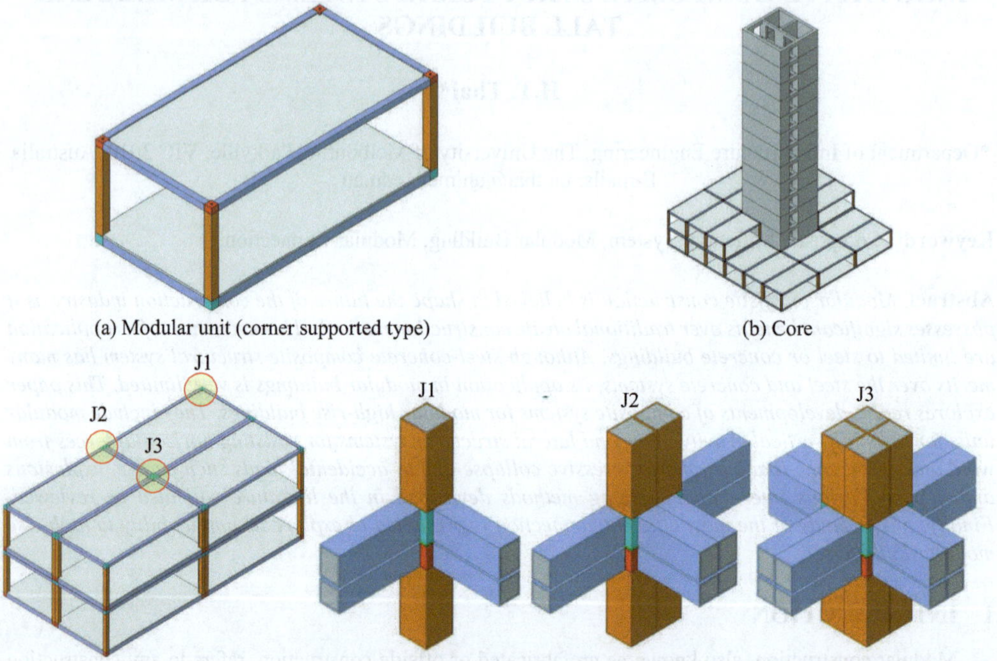

(a) Modular unit (corner supported type) (b) Core

(c) Inter-module connection

Figure 1. Example of a modular tall building.

2 BENEFITS OF USING COMPOSITE SYSTEMS FOR TALL BUILDINGS

Steel-concrete composite structures in the forms composite slabs, composite beams, composite walls and composite columns have significant benefits compared with the steel and concrete counterparts because it employs the merits of two constituent materials (i.e., offering the performance of reinforced concrete (RC) structures with the speed benefits of steel construction). In particular, the composite assemblies in the forms of concrete filled steel tubular (CFST) columns [9] and steel-plate composite (SC) walls [10] as shown in Figure 2 are the most effective and efficient composite systems thanks to their high load-carrying capacity. This is because the steel tube/plate confine the infilled concrete core, and simultaneously the infilled concrete core acts as a restraint to prevent inward buckling of the steel tube. In addition, the steel tube/plate can be used as reinforcement and permanent formwork for concrete casting, and thus saving construction time and costs [11]. Therefore, they have been increasingly used in civil and structural applications especially for tall buildings [12] as shown in Figure 2 where the structural members at the base floors usually experience heavy gravity and lateral loadings.

The most well-known example of using CFST columns and SC walls for high-rise buildings is the Rainier Square Tower project, a 58-storey building in Seattle, Washington, US. In this project, both CFST columns and SC walls were respectively used in a framing system and a SpeedCore system as illustrated in Figure 2. SpeedCore is a steel-concrete composite core system composed of SC walls and coupling beams as shown in Figure 3. The success of this real-world SpeedCore project revealed that the erection of SpeedCore took over 40% less time compared with the construction of traditional RC core thanks to the benefits of modularity and prefabrication of steel construction.

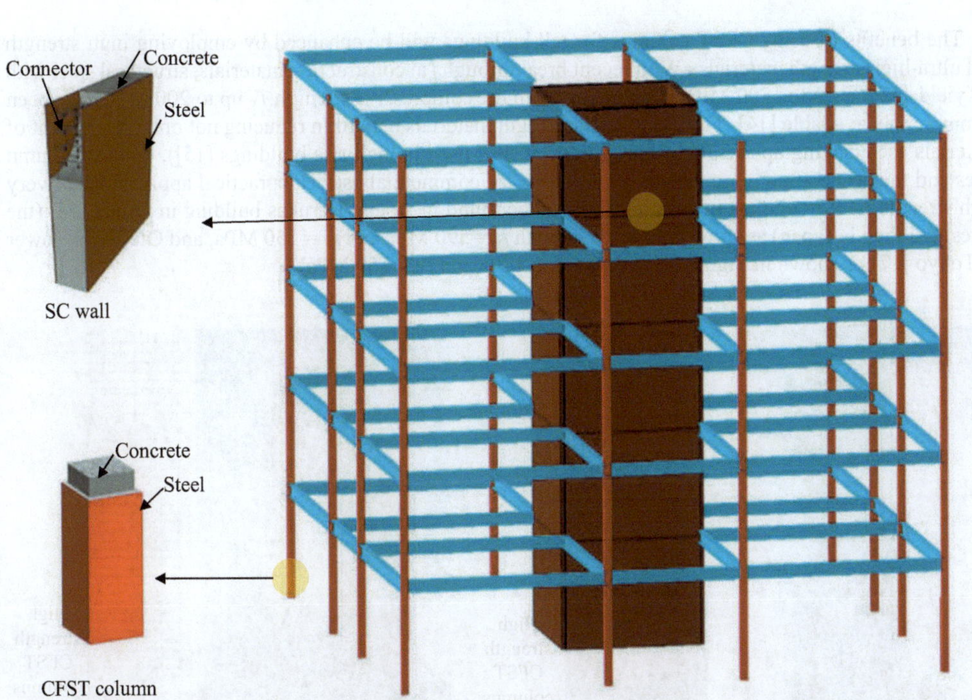

Figure 2. Example of a tall building using CFST columns and SC walls.

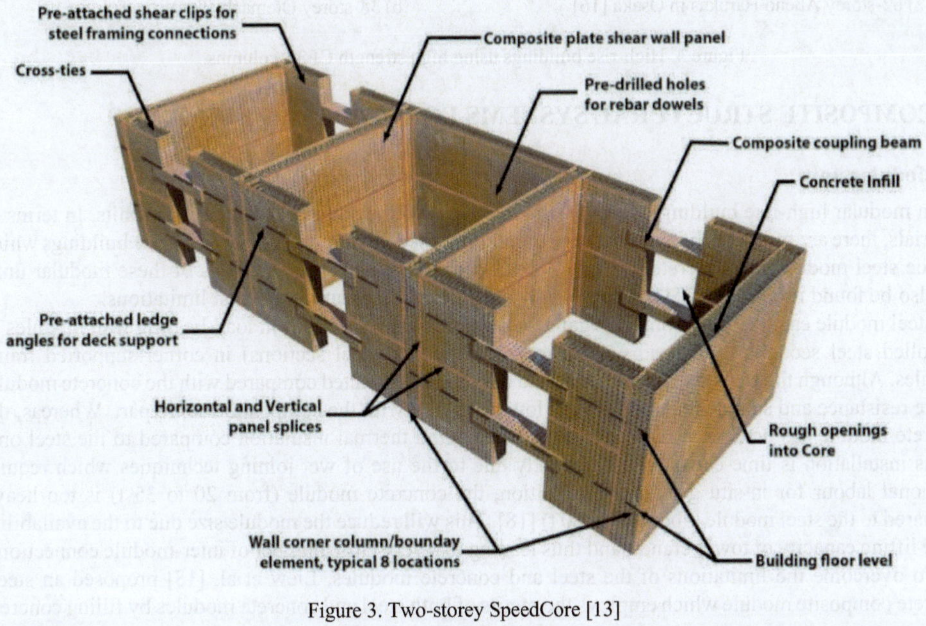

Figure 3. Two-storey SpeedCore [13]

The benefits of using CFST columns for tall buildings will be enhanced by employing high strength and ultra-high strength materials. With recent breakthroughs in construction materials, structural steel with the yield stress f_y up to 1300 MPa and concrete with the compressive strength f'_c up to 200 MPa have been commercially available [14]. The use of high strength materials benefit in reducing not only the amount of materials used (saving up to 50% of the amount of steel used in high-rise buildings [15]), but also column sizes and thus generating more valuable workspace for commercial use. The practical applications of very high strength CFST columns in tall buildings can be found in Abeno Harukas building in Osaka [16] (the tallest building in Japan) as shown in Figure 4a with f_y = 590 MPa and f'_c = 150 MPa, and Otemachi Tower in Tokyo [17] as shown in Figure 4b with f_y = 780 MPa and f'_c = 150 MPa).

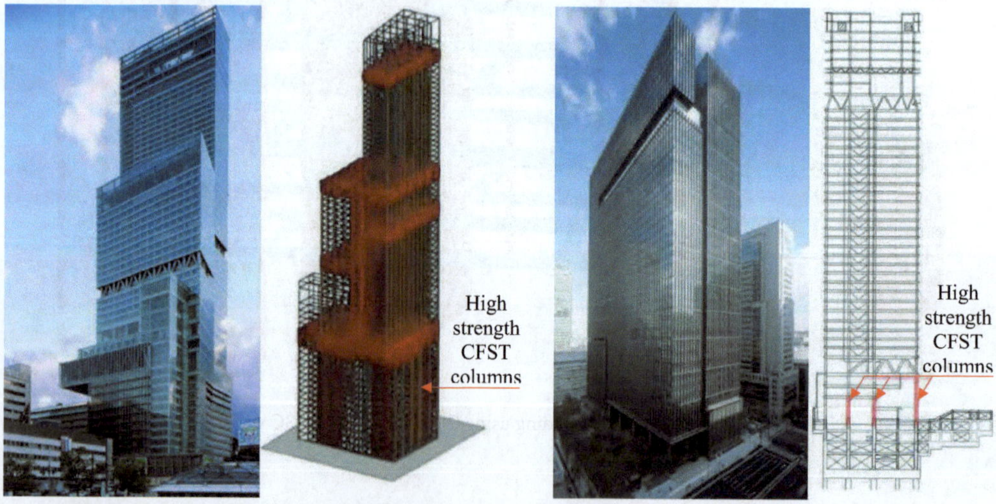

(a) 62-storey Abeno Harukas in Osaka [16]

(b) 38-storey Otemachi Tower in Tokyo [17]

Figure 4. High-rise buildings using high strength CFST columns

3 COMPOSITE STRUCTURAL SYSTEMS FOR MODULAR BUILDINGS

3.1 Modular unit

In modular high-rise buildings, vertical gravity loads will be supported via modular units. In terms of materials, there are two types of modular units commonly used in current modular high-rise buildings which include steel module and concrete module. A review of the practical applications of these modular units can also be found in Thai et al. [1]. All steel and concrete modules currently exist limitations.

Steel module can be made from light gauge cold-formed steel sections in load-bearing wall modules or hot-rolled steel sections (or hybrid cold-form and hot-rolled steel sections) in corner-supported frame modules. Although the steel module is lighter and easy to be connected compared with the concrete module, its fire resistance and sound insulation is very low compared with the composite counterpart. Whereas, the concrete module has better fire resistance and acoustic and thermal insulation compared to the steel one, but its installation is time-consuming and costly due to the use of wet joining techniques which require additional labour for in-situ grouting. In addition, the concrete module (from 20 to 35 t) is too heavy compared to the steel module (about 15 to 20 t) [18]. This will reduce the module size due to the availability of the lifting capacity of tower cranes, and thus leading to a very large number of inter-module connections.

To overcome the limitations of the steel and concrete modules, Liew et al. [18] proposed an steel-concrete composite module which employs the merits of both steel and concrete modules by filling concrete into the hollow sections of the steel posts and steel beams as shown Figure 5. This idea came from Mills et al. [19] as an solution for modular high-rise buildings. The use of composite sections can significantly enhance the load-carrying capacity of structural members as explained in Section 2. To reduce the self-weight of the composite module and improve the performance of acoustic and thermal insulation, Liew et al. [18] also suggested the use of lightweight structural concrete for the slab. This can potentially reduce the weight of a module up to 40% as indicated in a preliminary study [18]. In addition, the use of slim floor composite beams [20] instead of conventional composite beams can also be beneficial for the composite modular unit with span from 9 to 12 m since they can increase the headroom of the module up to 25% [18].

One of the disadvantages of using steel modules for high-rise buildings is that the column sizes of the steel module will vary along the height of the building due to the variation of the gravity load. This makes the connection between the upper and lower modules become more difficult. However, this can be resolved by using composite modules with high strength materials as shown in Figure 5.

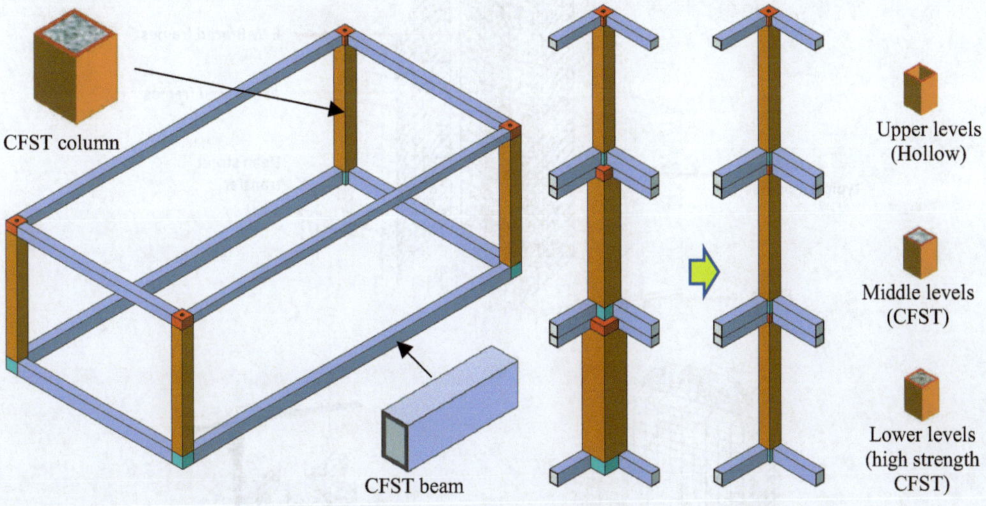

Figure 5. Composite modular unit.

3.2 Lateral structural system for modular high-rise buildings

Although modular construction method has been successfully used in low and medium-rise buildings, its practical application in high-rise buildings is very limited due to the concern about (i) the resistance to horizontal forces such as seismic and wind loadings, and (ii) the structural robustness against progressive collapse due to accidental loads such as fire, explosions and impact. Unlike conventional high-rise buildings where many lateral structural systems have been well developed (see Ali and Moon [21] for a critical review of these systems), the development of the lateral systems for resisting horizontal forces and withstanding progressive collapse of modular high-rise buildings is very limited. In modular multi-storey buildings, there are three main lateral systems including: (i) diaphragm system, (ii) braced frame system and (iii) shear wall core system [22]. A review on the performance of the lateral systems for modular buildings can be found in Srisangeerthanan et al. [3].

The diaphragm system is usually integrated into the walls and/or the floor and ceiling of the modular unit. The stiffness and strength of this system are strongly dependent on the performance of inter-module connections [3], and thus this system is only suitable for modular low-rise buildings from four to six storeys [22]. For modular high-rise buildings with high lateral forces, additional systems like the braced frame and shear wall core ones should be used. The braced frame system which is usually made from hot-rolled steel members can be applied to modular buildings up to 32 stories as in the case of B2 Tower in New York (see

Figure 6). The shear wall core system can be made from either steel or reinforced concrete (RC). Unlike the braced frame system, the shear wall core system can accommodate stairs, lifts and service risers. Therefore, it is commonly used in modular high-rise buildings especially the RC core. It is evidenced by the fact that four out of five of the world's tallest modular buildings including the 44-storey Croydon Tower in London were constructed using the RC core system [1]. In current practice, the RC shear wall core system of modular high-rise buildings can be built using slipform and jumpform approaches. However, these approaches are time-consuming and costly due to the nature of RC construction (e.g., requiring the installation of formwork and steel reinforcement).

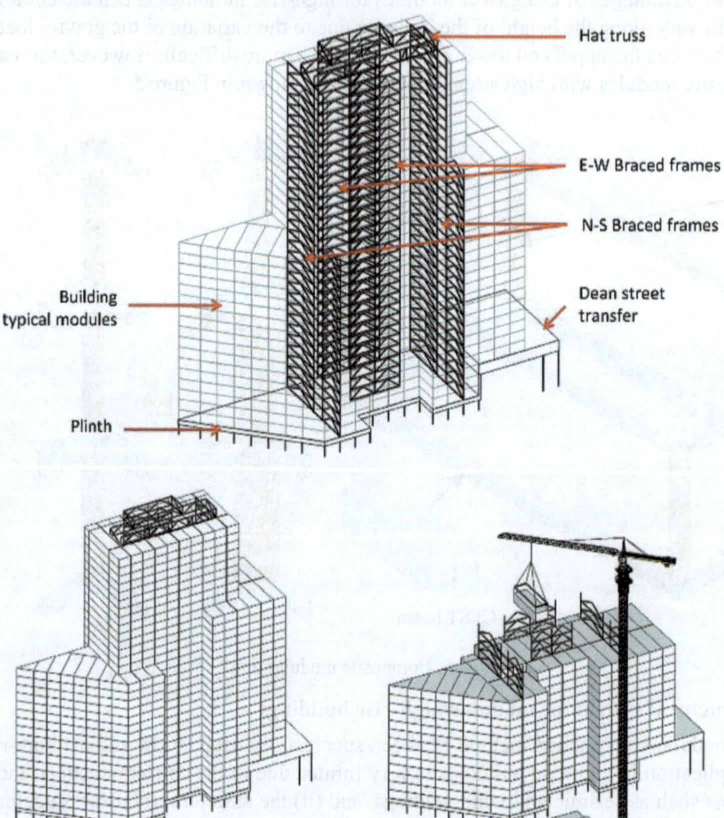

Figure 6. Typical modular high-rise buildings using steel braced frame system [23].

The successful use of the innovative SpeedCore system (see Figure 3) in a 58-storey composite building in the Rainier Square Tower project revealed that it can save construction time up to 52% compared with the traditional RC core thanks to the benefits of modularity and prefabrication of steel construction. SpeedCore is therefore considered as a promising solution to maximise the efficiency and productivity of modular construction of high-rise buildings. Currently, the SpeedCore system is erected by onsite welding the SC walls using the vertical and horizontal splice plates as shown in Figure 3. The welding method is only suitable for the countries with more relaxed WorkSafe practices like US. For the countries with stricter WorkSafe practices like Australia, onsite welding is discouraged due to safety hazards and time-consuming inspection after welding. More importantly, the welding method will incur more labour costs compared with the bolting method. Therefore, the bolting method is suggested for the erection of the SpeedCore system in Australia. Figure 7 illustrates the installation process of the SpeedCore system using blind bolting technique which can install from the outside of the steel plates.

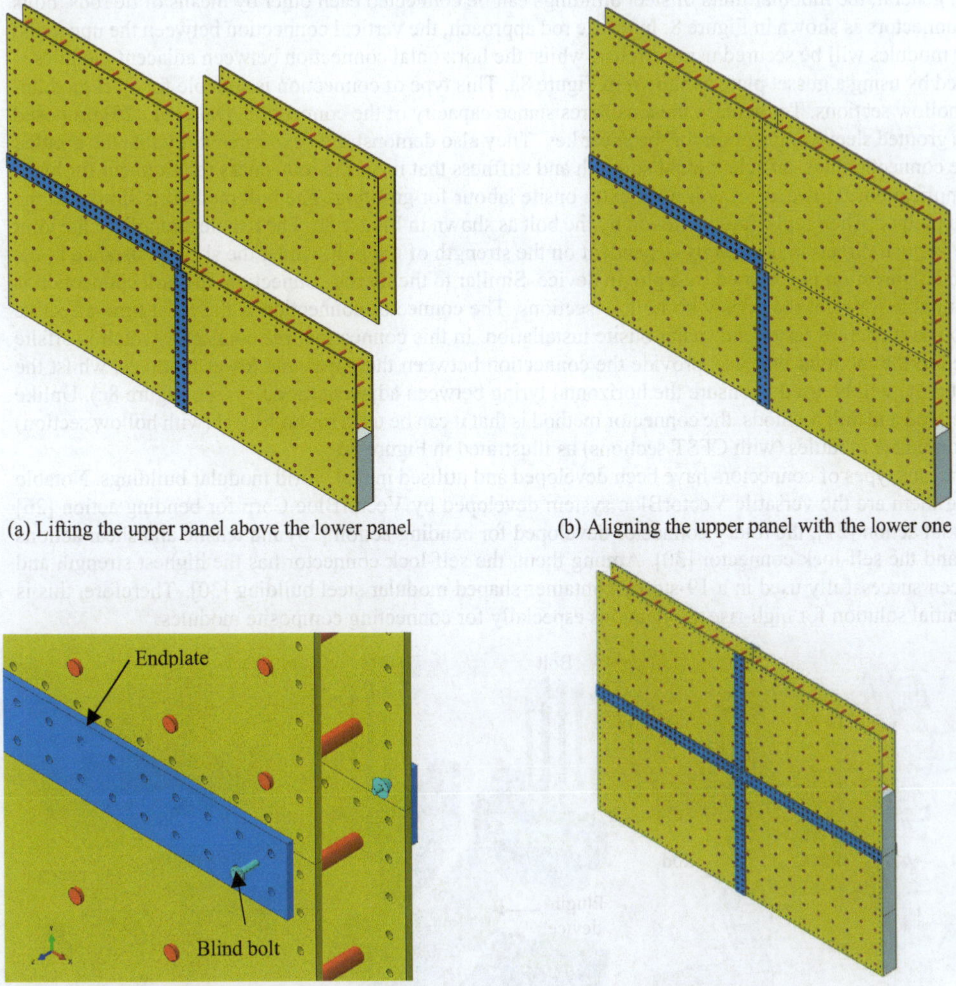

(a) Lifting the upper panel above the lower panel (b) Aligning the upper panel with the lower one

(c) Installing endplate using blind bolt (d) Filling concrete core

Figure 7. Illustration of the erection of the SpeedCore system using the blind bolting technique.

3.3 Inter-module connection

The connections between modules (see Figure 1c) play a critical role in ensuring the structural integrity and overall stability of the modular tall building. In addition, to meet the robustness requirements against disproportionate or progressive collapse of modular buildings, the inter-module connection should also have sufficient tensile strength or minimum level of tying forces to enable the development of alternative load paths for redistributing loads from damaged areas to intact members [24]. A number of joining techniques for modular buildings have been developed in the literature, and a critical review on such connections can also be found in Refs. [1, 3, 4]. It should be noted that the welding connection (in steel modular buildings) and wet connection (in concrete modular buildings) will not be considered in this study due to its time-consuming and costly.

In general, the modular units of steel buildings can be connected each other by means of tie rods, bolts and connectors as shown in Figure 8. In the tie rod approach, the vertical connection between the upper and lower modules will be secured using a rebar, whilst the horizontal connection between adjacent modules is ensured by using a gusset plate as shown in Figure 8a. This type of connection is suitable for steel modules with hollow sections. To increase the tensile resistance capacity of the connection, Dai et al. [25] proposed to use grouted sleeve connection for the shear key. They also demonstrated experimentally that the grouted sleeve connection has sufficient axial strength and stiffness that meet the robustness requirement for high-rise applications. However, it will need extra onsite labour for grouting. The bolt method is similar to the tie rod method when replacing the tie rod by the bolt as shown in Figure 8b. The tensile strength of the joint in the vertical direction is strongly dependent on the strength of the bolt, whilst the shear resistance in the horizontal direction is provided by a plugin device. Similar to the tie rod connection, the bolt connection is also suitable for steel modules with hollow sections. The connector connection is the most time-effective method as it enables faster and easier onsite installation. In this connection, the connector which is offsite welded to the modular unit will provide the connection between the upper and lower modules, whilst the gusset plate will be used to ensure the horizontal tyring between adjacent modules (see Figure 8c). Unlike the tied rod and bolt methods, the connector method is that it can be used for both steel (with hollow section) and composite modules (with CFST sections) as illustrated in Figure 8c.

Various types of connectors have been developed and utilised in real world modular buildings. Notable among them are the versatile VectorBloc system developed by VectorBloc Corp for bending action [26] and axial action [27], the rotary connector developed for bending action [28] and tensile and shear actions [29], and the self-lock connector [30]. Among them, the self-lock connector has the highest strength and has been successfully used in a 19-storey container-shaped modular steel building [30]. Therefore, this is a potential solution for high-rise applications especially for connecting composite modules.

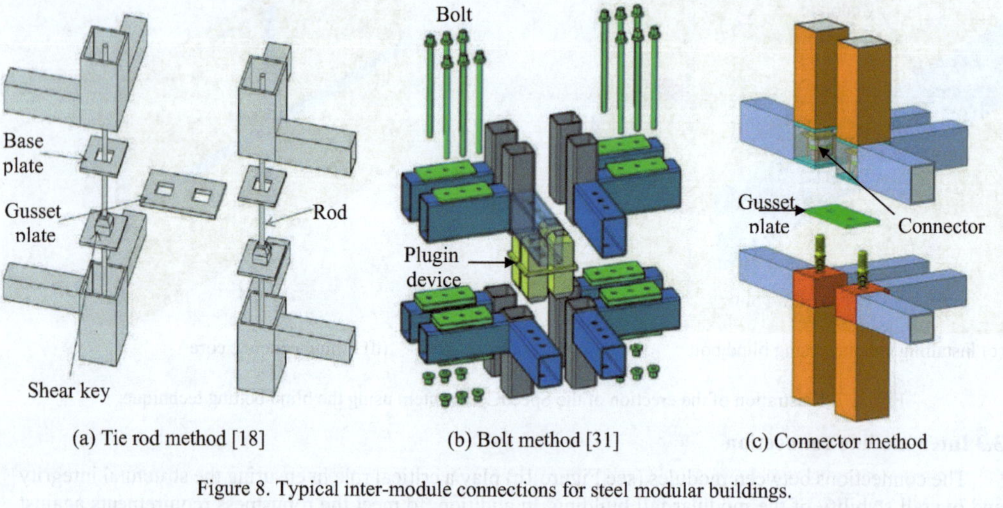

(a) Tie rod method [18] (b) Bolt method [31] (c) Connector method

Figure 8. Typical inter-module connections for steel modular buildings.

4 CASE STUDY OF AN INTER-MODULE CONNECTION

As discussed in the previous Section, the self-lock connector is the most effective jointing method for composite modular buildings due to its fast and easy installation as well as the applicability with composite sections. In addition, this joining technique was successfully used in a 19-storey steel modular building.

Therefore, this study will present a finite element (FE) study on this connection to explore its applicability to high-rise buildings.

Figure 9 shows the detail and installation process of the connector. In this self-lock connector method, the plug-in connector is located inside the bottom joint box of the upper module via four screws, whilst the stud is fixed to the top surface of the top joint box of the lower module by thread (see Figure 9a). On onsite erection, the upper module is lifted above the lower module, and then laid down once it aligns with the lower module to plug the stud into the connector (see Figure 9c). During the connecting process, the self-lock mechanism will be activated by means of an action spring when the top surface of the stud is in touch with the trigger plate (see Figure 9d). The trigger plate is pushed out of the cone-shaped latches, the pull-out force from the action spring is transferred to the cone-shaped latches through the sleeve, and thus friction locking the stud as shown in Figure 9e.

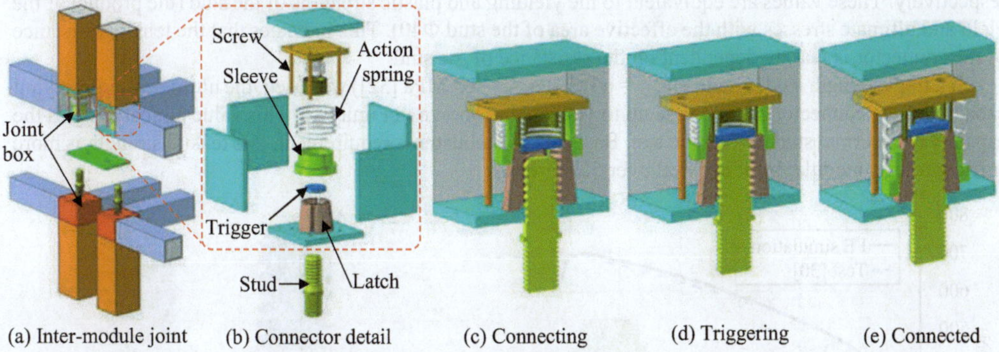

Figure 9. Detail and installation of the self-lock connector.

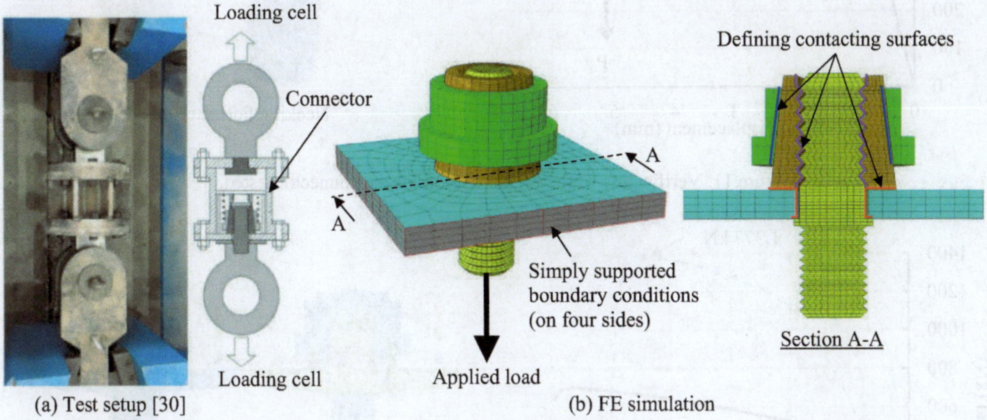

Figure 10. Test setup and FE modelling

The tensile strength of the connector was tested by Dai et al. [30] under pull-out force from two loading cells as shown in Figure 10a. To simplify the simulation, the joint box is not considered in this study by assuming that its stiffness is rigid in comparison with the stiffness of the connector. This assumption is reasonable based on the observation from the failure of the tested specimen. In the FE modelling, simply supported boundary conditions are applied to the bottom plate of the joint box and the pull-out force is applied to the end section of the stud as shown in Figure 10b. All components of the connector including the cone-shaped latches, sleeve, stud and the bottom plate of the join box are modelled by solid elements (C3D8R) in ABAQUS. Mesh convergence is also performed to obtain the optimal mesh size as shown in Figure 10b. The surface-to-surface contact interaction between the connector components is also included. Ductile damage model in ABAQUS is used to simulate the necking failure of the stud observed in the test. The stud with a diameter of 50 mm (effective diameter of 40 mm) is made from normal steel with a yield stress f_y of 280 MPa and ultimate stress f_u of 550 MPa. Other geometric and material properties of the connector can be found in Dai et al. [30]. Explicit dynamic analysis is used to avoid the convergence issues which are commonly encountered in the problem with large deformations and multiple contact interactions.

Figure 11 shows the comparisons between the experimental results and FE simulation. A very good agreement is obtained for both the load-deflection behaviour and necking failure mode of the stud. As shown in Figure 12, the elastic and the ultimate resistances of the connector are 350 kN and 690 kN, respectively. These values are equivalent to the yielding and plastic strengths of the stud (the product of the yield and ultimate stresses with the effective area of the stud Φ40). This indicates that the tensile resistance of the connector is mainly dependent on the resistance of the stud.

If a high strength stud (Grade 8.8, f_y = 668 MPa, f_u =865 MPa [32]) was used, the ultimate load-carrying capacity of the connector can increase up to 1,377 kN as shown in Figure 12. This value almost doubles the capacity of a normal stud used in the test. Based on the robustness requirement, this tensile strength is more sufficient for a modular building well over 50 storeys.

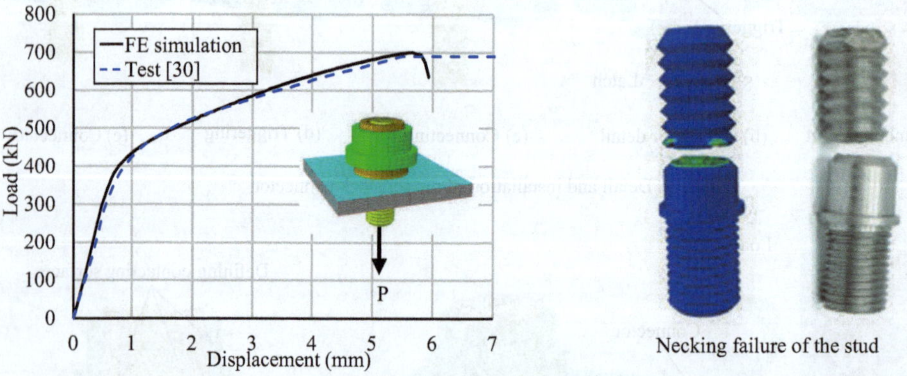

Figure 11. Verification of the FE simulation for the connector tested.

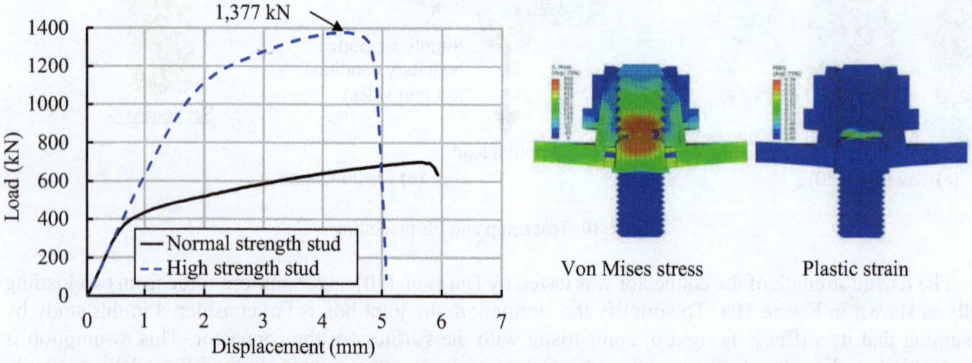

Figure 12. FE simulation of the connector with a stud of 50mm diameter and Grade 500 steel.

5 CONCLUSIONS

A review on the recent developments of structural systems for modular buildings has been presented in this paper with an emphasis on steel-concrete composite systems for high-rise applications. By combining the merits of both steel and concrete materials, the composite systems have a great potential in improving the efficiency and productivity of modular construction. The findings obtained from this study are:

(1) By using high strength and lightweight materials, the composite modular unit will be lighter and stronger. This will enable modular buildings to be built taller to maximise the benefits of modular construction.
(2) SpeedCore has shown significant benefits in terms structural performance and economy compared with the RC core. Therefore, it can be used for modular high-rise buildings to replace the RC core. However, the use of onsite welding method for erection in current practices in US is not suitable for the countries with stricter WorkSafe practices like Australia. Therefore, further work on SpeedCore needs to be done to develop alternative erection approaches such as blind bolting techniques to promote its application.
(3) Among inter-module joining methods developed in the literature, the self-lock connector method shows great potential for modular construction due to its fast and easy installation. This will enable modular buildings to be built faster and thus cheaper than current practices allow. In addition, it can provide sufficient tensile strength which is required to ensure the robustness of modular high-rise buildings (well over 50 storeys as indicated in the numerical simulation). Therefore, this is the most potential joining method for modular high-rise buildings.

ACKNOWLEDGEMENTS

This is research was supported by the Australian Research Council (ARC) under its Future Fellowship Scheme (Project No: FT200100024). The financial support is gratefully acknowledged.

REFERENCES

[1] Thai, H.T.; Ngo, T. and Uy, B., "A review on modular construction for high-rise buildings", *Structures* 28, 1265-1290, 2020.

[2] Shan, S.; Looi, D.; Cai, Y.; Ma, P.; Chen, M.-T.; Su, R.; Young, B. and Pan, W., "Engineering modular integrated construction for high-rise building: a case study in Hong Kong", *Proceedings of the Institution of Civil Engineers - Civil Engineering* 172, 51-57, 2019.

[3] Srisangeerthanan, S.; Hashemi, M.J.; Rajeev, P.; Gad, E. and Fernando, S., "Review of performance requirements for inter-module connections in multi-story modular buildings", *Journal of Building Engineering* 28, 101087, 2020.

[4] Lacey, A.W.; Chen, W.; Hao, H. and Bi, K., "Review of bolted inter-module connections in modular steel buildings", *Journal of Building Engineering* 23, 207-219, 2019.

[5] Lacey, A.W.; Chen, W.; Hao, H. and Bi, K., "Structural response of modular buildings – An overview", *Journal of Building Engineering* 16, 45-56, 2018.

[6] Ferdous, W.; Bai, Y.; Ngo, T.D.; Manalo, A. and Mendis, P., "New advancements, challenges and opportunities of multi-storey modular buildings – A state-of-the-art review", *Engineering Structures* 183, 883-893, 2019.

[7] Ye, Z.; Giriunas, K.; Sezen, H.; Wu, G. and Feng, D.-C., "State-of-the-art review and investigation of structural stability in multi-story modular buildings", *Journal of Building Engineering* 33, 101844, 2021.

[8] Deng, E.F.; Zong, L.; Ding, Y.; Zhang, Z.; Zhang, J.F.; Shi, F.W.; Cai, L.M. and Gao, S.C., "Seismic performance of mid-to-high rise modular steel construction - A critical review", *Thin-Walled Structures* 155, 106924, 2020.

[9] Han, L.-H.; Li, W. and Bjorhovde, R., "Developments and advanced applications of concrete-filled steel tubular (CFST) structures: Members", *Journal of Constructional Steel Research* 100, 211-228, 2014.

[10] Mo, J.; Uy, B.; Li, D.; Thai, H.T. and Tran, H., "A review of the behaviour and design of steel-concrete composite shear walls", *Structures* 31, 1230-1253, 2021.

[11] Thai, S.; Thai, H.-T.; Uy, B. and Ngo, T., "Concrete-filled steel tubular columns: Test database, design and calibration", *Journal of Constructional Steel Research* 157, 161-181, 2019.

[12] Zhong, S.-T. and Zhang, S., "Application and development of concrete-filled steel tubes (CFST) in high rise buildings", *Advances in Structural Engineering* 2, 149-159, 1999.

[13] Traut-Todaro, J., "SpeedCore - Lateral system innovation for today's construction challanges ", *Structure Magazine* November, 12-15, 2019.

[14] Thai, H.-T.; Thai, S.; Ngo, T.; Uy, B.; Kang, W.-H. and Hicks, S.J., "Reliability considerations of modern design codes for CFST columns", *Journal of Constructional Steel Research* 177, 106482, 2021.

[15] Liew, J.Y.R.; Xiong, M. and Xiong, D., "Design of Concrete Filled Tubular Beam-columns with High Strength Steel and Concrete", *Structures* 8, 213-226, 2016.

[16] Hirakawa, K.; Saburi, K.; Kushima, S. and Kojima, K., "Performance-based design of 300 m vertical city Abeno Harukas", *International Journal of High-Rise Buildings* 3, 35-48, 2014.

[17] Matsumoto, S.; Hosozawa, O.; Narihara, H.; Komuro, T. and Kawamoto, S.-i., "Structural design of an ultra high-rise building using concrete filled tubular column", *International journal of High-rise buildings* 3, 73-79, 2014.

[18] Liew, J.Y.R.; Chua, Y.S. and Dai, Z., "Steel concrete composite systems for modular construction of high-rise buildings", *Structures* 21, 135-149, 2019.

[19] Mills, S.; Grove, D. and Egan, M., "Breaking the pre-fabricated ceiling: Challenging the limits for modular high-rise", *Proceedings of CTBUH International Conference* 416-425, 2015.

[20] Ahmed, I.M. and Tsavdaridis, K.D., "The evolution of composite flooring systems: applications, testing, modelling and eurocode design approaches", *Journal of Constructional Steel Research* 155, 286-300, 2019.

[21] Ali, M.M. and Moon, K.S., "Advances in structural systems for tall buildings: emerging developments for contemporary urban giants", *Buildings* 8, 104, 2018.

[22] Lawson, R.M.; Ogden, R.G. and Bergin, R., "Application of modular construction in high-rise buildings", *Journal of Architectural Engineering* 18, 148-154, 2012.

[23] Pasquale, J.D.; Innella, F. and Bai, Y., "Structural concept and solution for hybrid modular buildings with removable modules", *Journal of Architectural Engineering* 26, 04020032, 2020.

[24] Thai, H.-T.; Ho, Q.V.; Li, W. and Ngo, T., "Progressive collapse and robustness of modular high-rise buildings", *Structure and Infrastructure Engineering* 1-13, 2021.

[25] Dai, Z.; Pang, S.D. and Liew, J.Y.R., "Axial load resistance of grouted sleeve connection for modular construction", *Thin-Walled Structures* 154, 106883, 2020.

[26] Dhanapal, J.; Ghaednia, H.; Das, S. and Velocci, J., "Behavior of thin-walled beam-column modular connection subject to bending load", *Thin-Walled Structures* 149, 106536, 2020.

[27] Dhanapal, J.; Ghaednia, H.; Das, S. and Velocci, J., "Structural performance of state-of-the-art VectorBloc modular connector under axial loads", *Engineering Structures* 183, 496-509, 2019.

[28] Chen, Z.; Liu, Y.; Zhong, X. and Liu, J., "Rotational stiffness of inter-module connection in mid-rise modular steel buildings", *Engineering Structures* 196, 109273, 2019.

[29] Chen, Z.; Wang, J.; Liu, J. and Cong, Z., "Tensile and shear performance of rotary inter-module connection for modular steel buildings", *Journal of Constructional Steel Research* 175, 106367, 2020.

[30] Dai, X.-M.; Zong, L.; Ding, Y. and Li, Z.-X., "Experimental study on seismic behavior of a novel plug-in self-lock joint for modular steel construction", *Engineering Structures* 181, 143-164, 2019.

[31] Chen, Z.; Liu, J. and Yu, Y., "Experimental study on interior connections in modular steel buildings", *Engineering Structures* 147, 625-638, 2017.

[32] Pang, X.-P.; Hu, Y.; Tang, S.-L.; Xiang, Z.; Wu, G.; Xu, T. and Wang, X.-Q., "Physical properties of high-strength bolt materials at elevated temperatures", *Results in Physics* 13, 102156, 2019.

9TH INTERNATIONAL CONFERENCE ON COMPOSITE CONSTRUCTION IN STEEL AND CONCRETE
M. Knobloch, U. Kuhlmann, W. Kurz, M. Schäfer. (Eds.)
Online, Germany, July 27-29, 2021

INTEGRAL BRIDGE WITH RS-OVERPASS TECHNOLOGY – STEP INTO THE FUTURE OF STANDARD HIGHWAY OVERPASSES

R. Zanon*, D. Rademacher*, G. Seidl, D. Pak*****

* ArcelorMittal, 66 rue du Luxembourg, L-4221 Esch-sur-Alzette, G.D. of Luxembourg
e-mails: riccardo.zanon@arcelormittal.com, dennis.rademacher@arcelormittal.com

** FH Potsdam, University of Applied Sciences, Kiepenheuerallee 5, 14469 Potsdam, Germany
e-mail: seidl@fh-potsdam.de

*** University of Siegen, Paul-Bonatz-Str. 9-11, 57076 Siegen
e-mail: pak@bau.uni-siegen.de

Keywords: Composite bridge, integral bridges, highway overpass, weathering steel, galvanized steel.

Abstract: *Starting from Germany, the VFT bridge girder technology (prefabricated composite beam) has spread in several European countries in the last decades as further development of the standard steel-concrete composite solution. Its configuration fits perfectly to integral frame bridges built over existing infrastructure with the scope to minimize the traffic disturbance. In this paper a further development with the focus on cost efficiency in span ranges of 40-55 m is proposed: the RS-Overpass technology. The change consists in using standard rolled sections by adapting them into the integral frame bridge configuration, further enhancing the structural efficiency with the benefits of high-steel strength. The constructive details to achieve the polygonal form have been studied to allow easy fabrication and a maximal clearance height over highway traffic lanes. The RS-Overpass solution is conceived as the economic option of standard highway overpasses, with the aim of a overall construction cost reduction of 5%. The possibility of weathering steel or hot-dip galvanization create the unique offer for a maintenance-free solution.*

1 INTRODUCTION

Roads overpassing highways or expressways are a rather common case, as such bridges occur on average every 2 km on highways in densely populated areas. Due to their high number and similar boundary conditions, the design and realization tends to be standardized and repetitive. The total bridge length ranges mostly between 40 and 55 m, which are achieved either by a double span with a pile between the lanes (solution very common in the past decades) or by a single span (solution which is preferred in later years). The construction height is often a limiting factor since minimum clearance heights must be guaranteed both for safety as well as for driving comfort (see Figure 1). The choice of construction material for such bridges strongly depends on the boundary conditions [1]. Reinforced concrete and prestressed concrete frame bridges are built on falsework and are therefore unsuitable for bridges built over running traffic. Precast prestressed concrete bridges are usually limited to spans of 30 - 35 m due to their piece weight; they come to application only when the bridge is composed by two spans. For the case of single span highway overpasses, steel-framed composite decks have proven to be the most suitable solution [5][7][9].

A specific construction type for this kind of highway overpasses – the so-called "VFT" (Verbundfertigteil-Träger = prefabricated composite girder) – was developed in Germany by the design office SSF Ingenieure in the 1990s under the lead of V. Schmitt and G. Seidl (Figure 1) [7]. This solution showed to be effective and economical and is now widespread in several countries across Europe as common construction practice.

Figure 1: Typical VFT-frame bridge as highway overpass ©SSF

The VFT-method is based on the successful construction type of prefabricated prestressed concrete elements, combining it with the advantages of composite construction. A steel girder is completed on top of its upper flange with a partial-precast reinforced concrete slab of 2-3m width (Figure 2). The first composite stage is already achieved in the precast factory, achieving therefore the composite action for the self-weight stage. This slab stabilizes the steel girder in horizontal direction and LTB during transport and erection phase. Further, the stiffness is already increased in the construction stage due to a composite effect.

On the construction site, the girders can be placed by mobile cranes (Figure 2) next to each other on temporary support frames (arranged in front of the abutments and pile shafts), directly on the support benches or in case of integral frame structures on the frame walls. Assembly of the girders is realized with a clear joint distance of 2 cm between the concrete upper chords. In addition to structural aspects, the wide concrete upper chords serve as formwork for the cast in-situ concrete supplement. The main advantage is that the impact on the traffic under the new constructed bridge is minimized. Closure times are usually needed for the short lifting phases only, casting of the supplementary concrete slab can be executed without any traffic disturbance due to elimination of all temporary constructions such as falsework under the bridge.

The cast-in-place concrete on top connects furthermore the composite girders to the cast in-situ cross girders in transversal direction in the area of the bearing axes to form a monolithic bridge deck. From a structural point of view, the concrete topping creates a second composite action. Thanks to the sufficient stabilizing effect of the precast concrete plate, no specific devices against lateral torsional buckling are required. After setting the upper chord's concrete, only the loads from bridge equipment and traffic act on the full composite cross section. Considerable material savings can be made for the structural steel compared to traditional composite construction, in particular taking into consideration that span cross-beams and bracings are not more necessary. The VFT girder shows analogies to common prestressed concrete girder construction but has the major advantage of lower piece allowing elements up to 60 m.

Figure 2: Schematic cross-section of a VFT girder ©SSF (left), Prefabricated composite girder ©ArcelorMittal (right)

2 DEVELOPMENT OF THE RS-OVERPASS CONCEPT

2.1 Starting point

Over the past decades, VFT construction method has proven to be a very effective solution for single span frame bridges serving as highway overpasses. The girders are designed with a pronounced variable inertia, with important construction height at the abutments and slender construction height in span. The frame corner moments increase due to the resulting high stiffness and reduce the span moment. As a result, large slenderness's are possible in the span area. In most cases, the steel girders are produced as a welded steel construction, as the geometry of the cross sections can thereby be optimized as a function of the loads and the construction phases. Steel sections are uniaxially symmetrical with narrower and thinner upper flanges (in the region of the composite sections centre line) and wider thicker lower flanges. In this way, the amount of steel used is minimal.

From the technical point of view the VFT Frame system is therefore already optimized. For those projects that do not raise any difficulties from a statical point of view and feature current spans, the only limiting factor for the choice of VFT construction method may be the cost face to prestressed concrete decks. To develop the competitiveness of the solution further in this range, a study was carried out using rolled sections instead of the commonly used welded sections: the RS-Overpass technology [22].

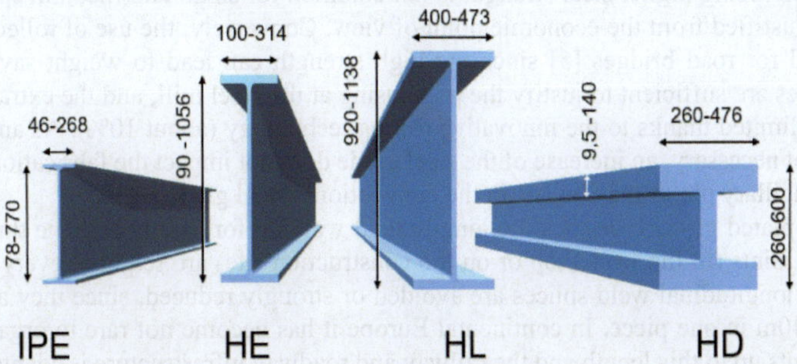

Figure 3: Overview of available H structural shapes according [17][13]

2.2 Availability of rolled sections

Rolled profiles (L, I, H, U) were developed at the end of the 19th century, answering the need for simplification of plated sections assembled by rivets. The advantages in terms of weight savings, fabrication simplification and cost reduction were integral to the acceptance of rolled sections in every field. Today, the geometric range of available H structural shapes is extensive with beam heights of 80-1150mm, flange widths of 50-480mm and flange thicknesses of 4-140mm (Figure 3) with a well-established presence of production sites around the world, making rolled sections a standard product well-known by engineers and contractors. In addition to the extension of geometric properties, the continuous development of optimized rolling procedures proposed more advanced steel grades over the decades. Since the 1990s, thermo-mechanical rolling has become a standard for the most advanced plants. To enhance the benefits of thermomechanical rolling, the quenching and self-tempering process (QST) was developed specifically for hot-rolled sections with thick flanges [3]. Implementing this innovative procedure made it possible to economically obtain high steel strengths (up to S500 with 500MPa yield strength) for heavy sections without the costly addition of alloying elements. At today status, the last commercially available step achieved is Grade 80 (550 MPa) [19].

Beside advantages from the structural point of view, such top-end products ensure also significant benefits from the sustainability point of view. In this case it is not only recycled steel, but truly up-cycled since from basic steel can be obtained high-strength steel thanks to the thermo-mechanical rolling [12]. This permits to substantially reduce the amount of steel needed for a given construction.

Figure 4: WD7 and WD8, Gdansk, Poland with VFT technology with rolled sections – maximum span 41m [13]

2.3 Considerations about steel grade and fabrication

Plated girders for small and medium span bridges are usually constructed in steel grade S355. This grade is quite common – therefore easier to purchase for the steel contractor – and implies well known welding procedures. Using higher steel strength is not common for small and medium span bridges – and its utilization not justified from the economic point of view. Conversely, the use of rolled girders in S460 is well established for road bridges [5] since the high strength can lead to weight savings in the steel structure. Quantities are sufficient to justify the purchasing at the steel mill, and the extra costs of the raw material are quite limited thanks to the innovative rolling technology (about 10%). As an assembly of the rolled section is not necessary, an increase of the steel grade does not impact the fabrication. Stiffeners, end plates and other ancillary plates may be left in the conventional steel grade S355.

In the case of plated girders, besides the longitudinal welding for assembling the three plates to an I shape, plate/beam joints (in the workshop or on the construction site) are required every 12m-18m. With rolled profiles the longitudinal weld splices are avoided or strongly reduced, since they are manufactured up to a length of 40m in one piece. In continental Europe it has become not rare to organize exceptional transport of elements up to this length and the railway and roadway infrastructure is adapted; in most cases the last km to access the construction site may be the most complicate, but in the case of highway overpasses is not a constraint. In this way, the manufacturing effort is strongly reduced, decreasing the costs, and improving the delivery time. These advantages can usually compensate for the higher transportation costs, in the cases where the construction site is accessible by heavy trucks. Because of the benefits of weight savings and simplified manufacturing, rolled sections in high strength steel grade (HSS) are being used more and more frequently in different innovative bridge deck types as the more economical alternative to welded built-up sections [2].

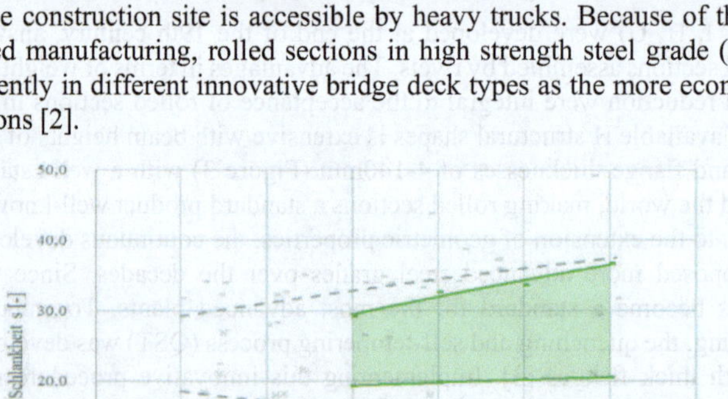

Figure 5: Slenderness ratio: VFT frame bridges – realized projects (grey); RS-Overpass frame bridge (green). Upper line indicates the maximal slenderness in span, bottom line indicates the minimal slenderness at support.

2.4 Consideration about the structural layout in longitudinal view

Standard highway overpasses are considered with spans between 40m and 55m. Using rolled sections the construction height in midspan is fixed at around 1.5m, formed by a 1.1m deep beam, 12cm of prefabricated concrete slab and 25-30cm concrete supplement, poured on site. This results in a slenderness range of 1 / 27 – 1 / 37, which enters in the common range for VFT frame bridges (Figure 5).

The situation is more complex at the support, where usual VFT frames have a slenderness ratio in the range of 1 / 15 - 1 / 18. This would require achieving a construction height up to 2.2 - 3.1m (Table 1). For the case of the welded steel girders this is easily achieved by increasing the web height. In the case of standard rolled sections other solutions have to be considered.

Table 1 VFT-Integral frame overpass: Span, slenderness and support sections construction heights

Span [-]	Slenderness [-]	Construction height at support [m]	Haunch at support [mm]
30	20	1.5	-
35	19.5	1.8	300-400
40	19	2.1	600-700
45	18.5	2.4	800-1000
50	18	2.8	1300-1400
55	17.5	3.1	1600-1700

2.5 Options to achieve the haunch at support

First option considered is to apply a triangular haunch cut and rewelded out of a rolled section, how it is usual done for industrial buildings (see Figure 6). This allows a section height increase up to 900mm. Nevertheless, for the case of bridges an accurate detailing has to be provided since the structure is subjected to cyclic loading and fatigue. This option is simple to understand, allows for easy fabrication of the steel structure and does not require any change of the precast concrete element. Nevertheless, it has some significant limits in terms of aesthetics as well as in the increase of steel consumption due to the double bottom flange.

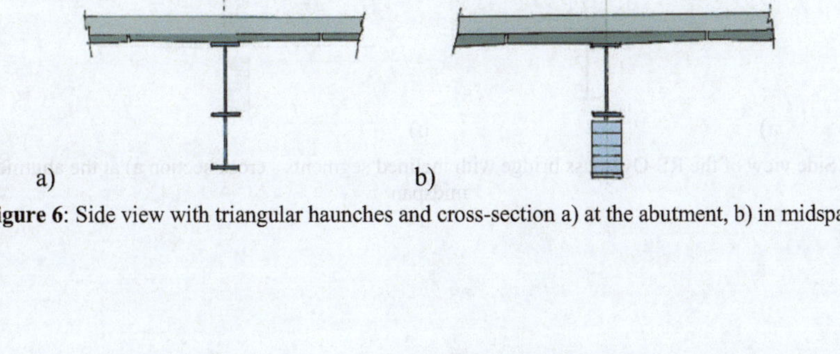

a) b)

Figure 6: Side view with triangular haunches and cross-section a) at the abutment, b) in midspan

A second option is realized by bending of rolled steel sections, obtaining two segments to be welded to the ends of the straight central part (see Figure 7). This solution clearly provides an interesting aesthetic aspect but requires more work both on the structural steel side as well as on the side of the prefabrication concrete elements.

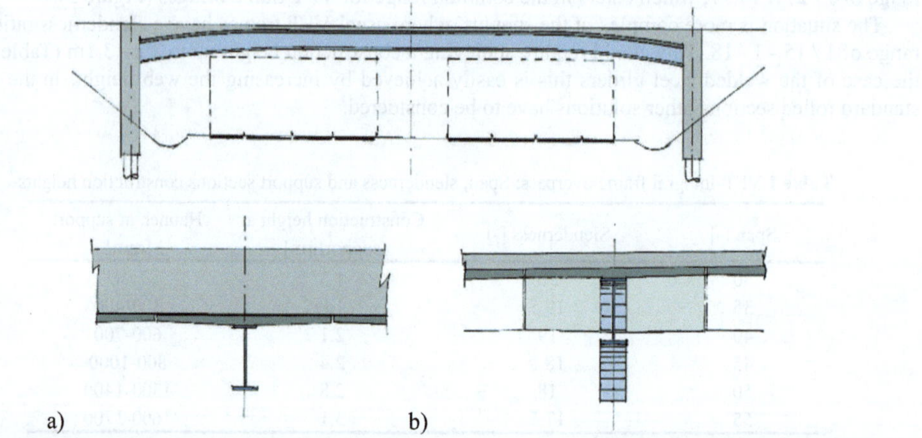

Figure 7: Side view of the RS-Overpass bridge with arched segments - Cross-section a) at the abutment, b) in midspan

A third options has been considered: straight segments welded to the ends of the straight middle section under a defined angle. Differently from previous concept, the steelwork fabrication as well as the concrete prefabrication becomes easier, since no curved beams are used (Figure 8). The haunches are realized with two butt weld joints in the kinks. Considering that for beams longer than 40m at least one joint is required anyway, only one additional joint must be executed.

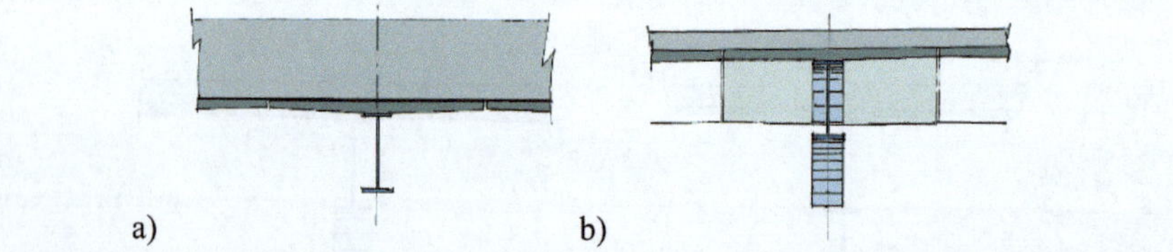

Figure 8: Side view of the RS-Overpass bridge with inclined segments - cross-section a) at the abutment, b) in midspan

3 DEVELOPMENT OF RS-OVERPASS SOLUTIONS FOR STANDARD CASES

The third option described in the previous section has been considered as the best option for RS-Overpass technology. It allows for a central span of constant construction height to maximize the clearance height above the traffic lanes, marking a distinctive feature from existing solutions on the market. The length of the inclined part has been chosen based on the typical distance between the traffic lane border and the implantation of the abutment wall.

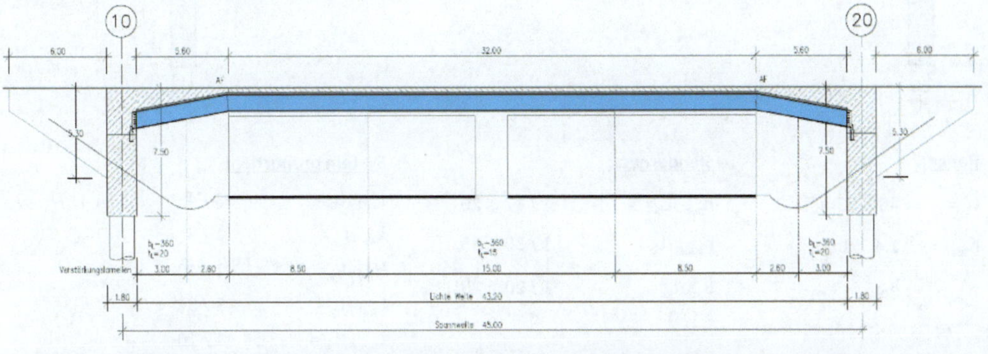

Figure 9: Example of longitudinal view for a 45m span RS-Overpass frame - Units in [m].

In order to assess the possible benefits of this technology, various standard cases have been investigated to obtain bill of materials to compare with the existing VFT frame solution. Design has used the commercial software SOFISTIK [23] and have been benchmarked with calculations with AdvanceBridge Expert, a commercial software with a dedicated module for calculation of composite bridges which is made freely available [24]. Following parameters have been varied:

- 4 different spans: 40m / 45m / 50m / 55m
- 3 different cross-sections corresponding to the standard roadway cross-sections: local street (width 5.80m), national street (width 9.90m) and highway (total deck width 16.70)

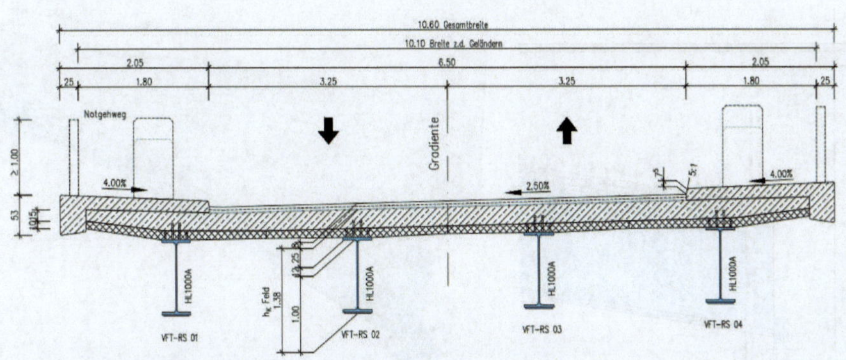

Figure 10: Example of deck cross-section for the national road- Units in [m].

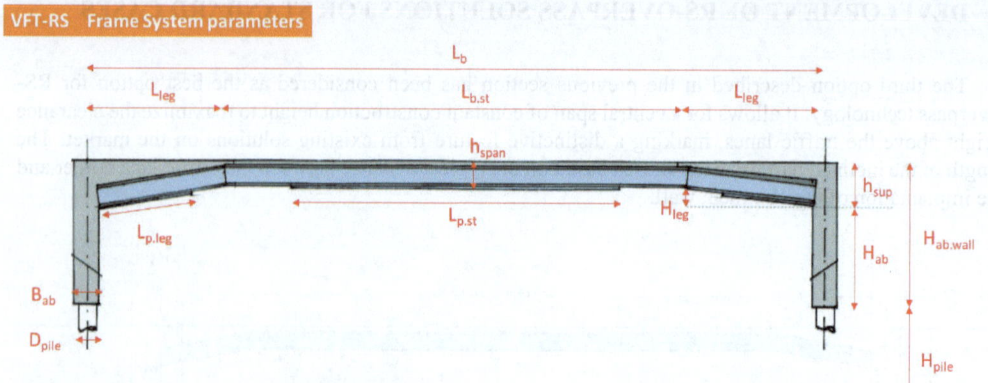

Range:		Slenderness:		System proportion:	
L_b =	40 ... 55m	h_{sup} / L_b =	1 / 17 ... 20	$L_{b.st} / L_b$ =	3 / 4
h_{span} =	1.4 ... 1.6m	h_{span} / L_b =	1 / 30 ... 35	L_{leg} / L_b =	1 / 8
D_{pile} =	0.9 ... 1.5m	B_{ab} / L_b =	1 / 20 ... 25	H_{leg} / L_b =	1 / 40
				$L_{p.st} / L_b$ =	3 / 5
				$L_{p.leg} / L_b$ =	1 / 10

Figure 11: Geometric parameters of the RS-Overpass frame system (based on steel S460 and concrete C35/45)

Design loads according to the Eurocode and German national annexes have been assumed. The foundations are realized by a single pile row with a proportion of one pile per beam line. A sensibility study has showed that for normal soil conditions the soil parameters do not excessively impact the results in term of steelwork consumption for this kind of frames with a slender deck. In the same way the impact of the skewness has been analysed, concluding that its impact was minor up to 70 gons. Therefore, results for the most unfavourable cases are showed considering that they are valid in a skewness range between 70 and 100 gons.

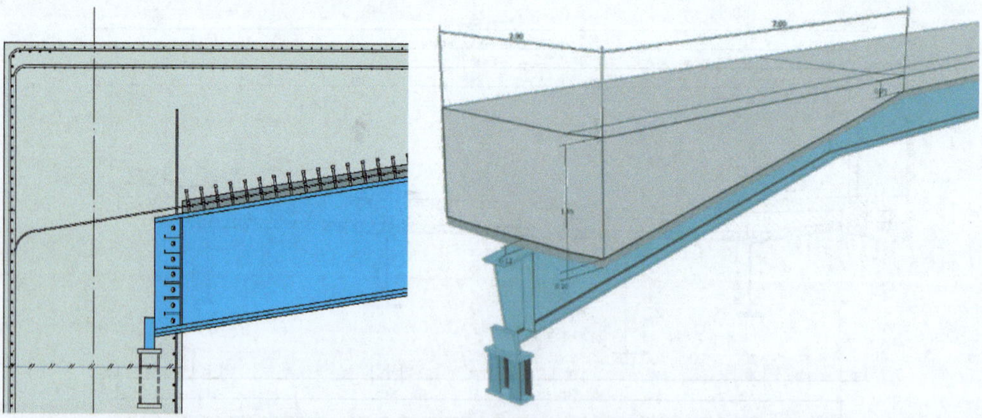

Figure 12: Proposed execution of the frame angle, with the additional concrete to be poured on site

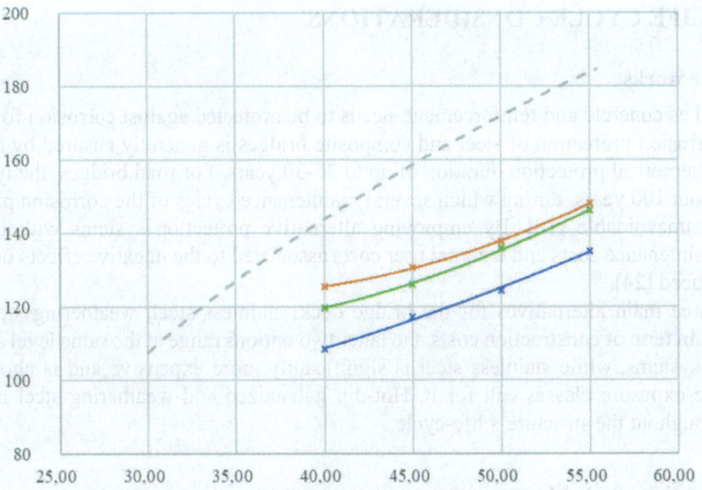

Figure 13: Steel consumption vs bridge span for RS-Overpass solution (orange, green, blue line in function of the road category) compared to VFT solution (grey line)

The results compared with the numerical simulations have been compared with the optimal steel consumption of VFT integral frames, for which a wide experience exists. From the steel consumption point of view, it can be clearly seen that RS-Overpass technology achieves a significant improvement. Nevertheless on the other hand it also requires an higher concrete consumption and effort for the execution of the prefabricated elements.

An economic study considering the different variables (steelwork supply and fabrication, concrete prefabrication and in-situ, erection, painting,…) has showed that in the considered range RS-Overpass could secure an average economic weight of 140 € / m2 compared to a standard VFT integral frame [19]. This difference is relevant since it can constitute something like 5% of the total cost of the structure.

4 BRIDGE LIFE CYCLE CONSIDERATIONS

4.1 Maintenance works

Steel, as well as concrete and reinforcement, needs to be protected against corrosion for their intended lifetime. The corrosion protection of steel and composite bridges is generally ensured by organic coating systems with a theoretical protection duration of up to 25-30 years. For road bridges, the typical expected service life is about 100 years, during which several maintenance cycles of the corrosion protection of the bridge deck are unavoidable [18]. By employing alternative protection systems without any need of renewal, both maintenance costs and external user costs associated to the negative effects on the traffic are significantly reduced [24].

There are three main alternatives for the bridge deck: stainless steel, weathering steel and hot-dip galvanized steel. In term of construction costs, the latter two options range at the same level as conventional organic coating systems, while stainless steel is significantly more expensive and is chosen only when extremely severe exposure classes call for it. Hot-dip galvanized and weathering steel may require no maintenance throughout the structure's life-cycle.

4.2 Hot-dip galvanizing in bridges

Experience shows that structures designed with protection by hot-dip galvanizing allow for maintenance intervals of 80 years and more [12]. Comparing this protection duration with common organic coating systems in an atmosphere of corrosivity class C4 according to ISO 9223, a minimum of two maintenance intervals can be avoided for the steel members over a service life of 100 years. However, most bridges are in environments classified into corrosivity category C2-C3, although recent studies show that micro-climatic effects could prevail over atmospheric exposure. For instance, exposure tests carried out on a service bridge over the German motorway A4 found that the local application of de-icing salts leads to the high corrosivity category C4-C5. Based on these exposure tests and the maximum measured corrosion loss of 4.52 μm during the first year, it was concluded that a > 200 μm zinc coating thickness yields a theoretical protection duration of at least 80 years [18].

While more than 1,000 bridges in Japan are constructed with hot-dip galvanization as corrosion protection, in many European countries it has only been used for predominantly statically loaded structures, as knowledge about the effect of hot-dip galvanisation on the fatigue resistance of the steel members is not available. Therefore, research has developed fatigue design specifications for hot-dip galvanised structures subjected to fatigue loading, closing the gap in current knowledge [18].

In combination with hot-dip galvanizing, hot-rolled sections offer additional advantages that further enhance the benefits. In terms of process handling, the homogeneous hot-rolled profiles fit very well for the galvanizing process, where they are completely and uniformly coated with zinc by being immersed into the liquid-hot galvanizing bath. Furthermore, hot-rolled sections offer a better fatigue behaviour than built-up sections, whereby a required reduction by one detail category for galvanized steel has less effect on the design [18]. The worldwide first PreCoBeam bridge with a hot-dip galvanized steel structure from rolled HD320x300 sections was built near Halle in Germany in 2017.

4.3 Weathering steel in bridges

The use of weathering steel offers similar advantages for design, as it has similar mechanical properties to non-alloyed structural steel (according to EN 10025 parts 2 to 4), but significantly better durability properties without any need for additional coating [4][15]. Currently rarely used in some countries, weathering steel is growing in popularity in many countries, such as the UK. Being part of the design rules in Eurocode 3, weathering steels are included in the harmonized European standard series EN 10025-5 [21], which was recently published in a revised version. Having the additional letter "W" in the designation, e.g. S460J2W, weathering steel grades S355W and S460W are now available for bridge design, for rolled sections up to 63mm flange thickness. National and international guidelines (e.g. DASt guideline 007 [25],

ECCS guide [4] or SCI guidance note [26]) provide additional information on design and construction. Slightly higher procurement costs for weathering steel, due to the additional alloying elements and the consideration of material loss over the lifetime, are in many cases not exceeding the savings for the elimination of protective coating systems. Considering the 100-year life-cycle of a bridge, the advantage of weathering steel is even greater, especially in places, where the repair of corrosion protection is difficult, time-consuming and expensive, such as for the here presented motorway overpasses [11].

Figure 14: 3D rendering of the bridge with rolled girders with a span of 55.00 m overpass on a 8-traffic lanes highway

5 CONCLUSIONS

This paper introduces a proposal for a new configuration of a steel-concrete composite section adapted for highway overpasses called RS-Overpass. These configurations are compared in terms of their economic efficiency with the ones already established on the market. The new RS-Overpass technology is still based on existing methods but offers further economic advantages and innovative features thanks to the use of high-strength steel in combination with maintenance-free solutions, such as hot-dip galvanized or weathering steel.

The use of RS-Overpass frame bridges is the best fit for a standardized highway overpass. Even if the central deck is very slender, the flow of forces is clearly visible ensuring a simple structural form. This solution guarantees the maximal clearance height above the traffic lanes. The prefabricated element between the VFT-concrete plate and the cap ensures a qualitative pleasant aspect of the wing-wall giving a lighter aspect to it (see Figure 9). The structure thus appears harmonious and uniform design even for different spans and higher wingwalls.

REFERENCES

[1] Braun A., Seidl G., Weizenegger M.: Rahmentragwerke im Brückenbau, Beton- und Stahlbetonbau 101 (2006), Heft 3, S. 187-197

[2] Rademacher D., Ochojski W., Lorenc W., Kożuch, M.: Advanced solutions with hot-rolled sections for economical and durable bridges, Steel Construction, Vol 11, No. 3, 2018 (pp. 196-204)

[3] Bouchard S., Axmann G., The Steel of Choice, Structural Engineer, 2000.

[4] ECCS - European Convention for Constructional Steelwork (2001): The Use of Weathering Steel in Bridges, Brussels: ECCS.

[5] G Sedlacek, B Hoffmeister, H Trumpf - Composite bridge design for small and medium spans - Final Report European commission Technical steel report, EUR20583, 2003.

[6] SCHMITT V. et al., Untersuchungen zum verstärkten Einsatz von Stahlverbundkonstruktionen bei Brücken kleiner und mittlerer Stützweiten. final report, FOSTA P629, 2006

[7] Pak D. (2012), Zu Stahl-Verbundbrücken mit integralen Widerlagern, PhD thesis, RWTH Aachen, Schaker Verlag.

[8] KUHLMANN, U., et al., Sustainable steel-composite bridges in built environment (SBRI). Brussels : RFCS Publications, 2013. RFCS final report, RFSR-CT-2009-00020

[9] UNGERMANN, D., et al., Hot-dip galvanizing in steel- and composite bridge construction, FOSTA P835, final report, 2014

[10] R. Zanon, W. Ochojski, O. Hechler, Paweł Klimaszewki, W. Lorenc, Road bridges WD7-WD8 on Route Sucharskiego, Gdansk, Poland - Prefabricated composite beams with high-strength hot rolled sections, Technical Report, IABSE Structural Engineering International (SEI), 2 – 2014.

[11] UNGERMANN, D., et al., Application of weathering steel in bridges. Boston, 2016. Connections VIII

[12] IBU (Institut Bauen und Umwelt e.V.). EPD-ARM-20170033-IBD1-EN: Structural steel sections in HISTAR grades. Berlin, 21.02.2017

[13] EN10365 - Dimensions et masses - Hot rolled steel channels, I and H sections, 2017.

[14] Rademacher, D. (2017): Zur sicheren Anwendung feuerverzinkter Bauteile im Stahl- und Verbundbrückenbau, Dissertation.

[15] FRIEDRICH, H., Current challenges and future developments for bridge constructions in the federal highway network. N° 6, 2018, Stahlbau, p. 87.

[16] R. Zanon, D. Zaganelli, Wide Flange Beams: Trends Towards Higher Steel Performance, Current Trends in Civil & Structural Engineering, 2019

[17] EN 10025-5 (2019): Hot rolled products of structural steels - Part 5: Technical delivery conditions for structural steels with improved atmospheric corrosion resistance.

[18] Pak D., Seidl G. (2020) Eine kurze Geschichte der Rahmenbrücken in Deutschland - Vom stählernen Rahmen zur integralen Verbundbrücke, Stahlbau 89/3.

[19] Seidl G., Rademacher D., Zanon R., Yu C., (2021), Weitgespannte Rahmenbrücken mit Walzträgern in Verbundbauweise: VFT-RS, 21. Symposium Brückenbau Leipzig, Wiederspahn Verlag.

[20] KUHLMANN, U., et al., SBRI+ Design Manual II - Advanced Applications. 2018

[21] DASt - Deutscher Ausschuss für Stahlbau (1993): Guideline 007: Lieferung, Verarbeitung und Anwendung wetterfester Baustähle. Düsseldorf.

[22] SCI - Steel Construction Institute (2015): P185 - Use of weather resistant steel. Ascot: 11-2015.

[23] SOFiSTiK, www.sofistik.com, 2021.

[24] AdvanceBridge Expert, www.graitec.de, 2021.

JOINTS

9TH INTERNATIONAL CONFERENCE ON COMPOSITE CONSTRUCTION IN STEEL AND CONCRETE
M. Knobloch, U. Kuhlmann, W. Kurz, M. Schäfer. (Eds.)
Online, Germany, July 27-29, 2021

INFLUENCE ON THE LOAD-DISPLACEMENT BEHAVIOUR OF STEEL-TO-CONCRETE CONNECTIONS WITH POST-INSTALLED ADHESIVE ANCHORS

Nilde Maçi*, Jan Hofmann*

* Institute of Construction Materials, University of Stuttgart
e-mails: nilde.maci@outlook.com; jan.hofmann@ iwb.uni-stuttgart.de

Keywords: Adhesive, Anchors, Cyclic, Incremental, Post-Installed.

Abstract. *Designing frame structures requires knowing the behaviour of each member of the frame (beam, column, joint, etc.) regarding the axial forces, shear forces, and bending moments. Furthermore, the ductility of the structure under seismic loading is essential in earthquake regions. Nowadays, structural frames are built from different materials such as concrete and steel, to achieve better performance. Therefore, the behaviour of the connections between steel and concrete is essential. The steel-to-concrete joints were the focus of the INFASO project, where joint solutions with easy fabrication, quick assembly, applicability in old structures, sufficient ductility, and high loading capacity were developed. They proposed the use of anchor plates with welded studs or post-installed fasteners such as adhesive anchors to connect the steel and concrete members. This paper focuses on the performance of post-installed adhesive anchors. During their service life, post-installed anchors are subjected to monotonic, constant, and seismic loading. Each of these loading approaches is described in the current standards. Seismic and constant loading tests are of importance to the long-term behaviour of the anchors. Seismic loading tests are carried out using a predefined cycle pattern. According to TR049, seismic tests of category C1 (tension and shear) are performed with 140 load cycles, where the load amplitude decreases after 10, 30, and 100 cycles. On the contrary, C2 category tests increase the amplitude within 75 or 59 cycles, depending on the test. Constant loading tests apply when the anchors are installed in cracked concrete. In contrast to the mentioned patterns, this study observes the behaviour of the adhesive anchors when the amplitude of each cycle is increased by 5 % until the anchor fails. Standard short-term pull-out tests are carried out to determine the load increments. Various parameters such as the embedment depth, bond line thickness, hole cleaning, wet concrete, and elevated temperatures are studied. Confined and unconfined tests are performed. Overall, the reference short-term failure loads are higher than those in the tests with incremental and cyclic loading. The anchors installed with reduced hole cleaning, in wet concrete or subjected to elevated temperatures have a lower failure load compared to the reference series. Two additional bond line thicknesses are used for comparison. The increase in the thickness influenced the failure load differently for confined and unconfined test setups.*

1 INTRODUCTION

Knowing the behaviour of each member of a frame structure is important especially in earthquake regions [1]. One of these members is the connection between steel and concrete. In recent years, there has been a significant advance in the research of these type of joints. The project INFASO focused on the joint solution with easy fabrication and applicability using anchor plates with headed studs or post-installed fasteners such as adhesive anchors [2].

Adhesive anchors are classified as capsule systems and injection systems [8]. This classification is based on their method of placement in concrete. They consist of two parts, the resin and the steel component which is either a threaded rod or a reinforcing bar. The resins can be vinyl ester, epoxy or unsaturated polyester resin. The load is transferred through micro-keying and adhesion (chemical interlock) [9]. The behaviour of adhesive anchors is influence by different factors such as embedment depth, concrete condition, hole cleaning, temperature (installation and when testing), bond line thickness, anchor diameter, support diameter, long-term loading, etc. [10].

During their service life, adhesive anchors are subjected to different types of loading (monotonic, constant, and seismic). Different standards describe the testing procedure to evaluate the performance of adhesive anchors [3,4,5,6]. Figure 1 illustrates the different types of loadings. Anchors subjected to short-term loading follow mostly the monotonic loading pattern. Constant and seismic loading are used to investigate the long-term performance of anchors [6,7]. According to TR049, there are two categories to test the performance of post-installed anchors in concrete, namely C1 category and C2 category. Both categories consist of pulsating or alternating tension and shear loads using the sinusoidal pattern. The main difference between them is that the cyclic loading in the C1 category decreases within 140 cycles, whereas in the C2 category, the load increases within 75 or 59 cycles. The applied load is decreased or increased in a predefined number of cycles which are specified in the standard. Within each cycle group, the load is kept constant [6].

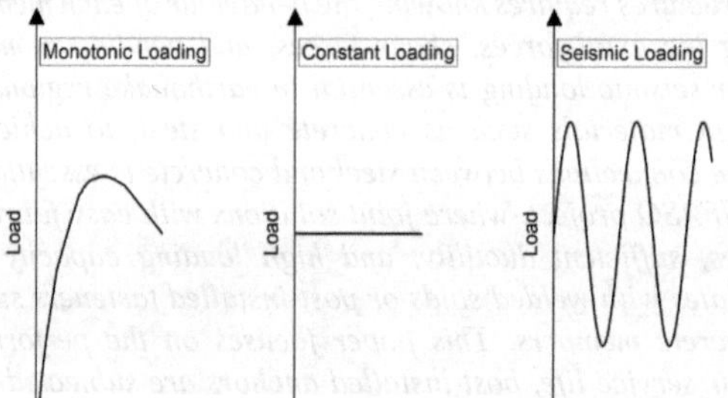

Figure 1. Types of anchor loading.

This paper is part of a research project at the Institute of Construction Materials [11]. The objective of this paper is to investigate the behaviour of adhesive anchors subjected to incremental loading in non-cracked low-strength concrete using confined and unconfined test setups. The loading pattern investigated is described in section 2.2.

2 TESTING PROGRAM

Concrete of the class C20/25 was used as anchorage ground. The cubic compressive strength of the specimen was 32.31 MPa. The compressive strength was tested on three concrete cubes with dimensions 150 x 150 x 150 mm. The adhesive anchor tested was a two-component epoxy resin. The components were mixed through a mixing nozzle with the help of an application gun. Threaded rod M12 with steel strength 12.9 served as a fastener.

Table 1 summarizes the range of the parameters used in this test program. Short-term and incremental loading tests using confined and unconfined setup were performed. The first column describes the test series name, followed by the type of test carried out (R - reference test, ShT - short-term test, IL - incremental loading test). The third and fourth columns show the embedment depth and the drilling diameter. The fifth column describes the concrete condition, whereas the sixth column the hole cleaning (HP - hand pumping, MB - machine brushing). The last two columns display the temperature when testing and the test support used to test the anchors.

Table 1: Testing program.

Test Series	Type of Test	Embedment depth	Drilling diameter	Concrete Condition	Hole Cleaning	Testing temperature	Support
[-]	[-]	[mm]	[mm]	[-]	[-]	[C]	[-]
D028	R-ShT	60	14	dry	HPxMBxHP	20	confined
D029	R-IL	60	14	dry	HPxMBxHP	20	confined
D030	R-ShT	60	14	dry	HPxMBxHP	20	unconfined
D031-D032	R-IL	60	14	dry	HPxMBxHP	20	unconfined
D033	IL	60	14	dry	1xHP	20	confined
D034	ShT	60	14	dry	1xHP	20	confined
D035	IL	60	14	dry	1xHP	20	unconfined
D036	ShT	60	14	dry	1xHP	20	unconfined
D037	IL	60	16	dry	HPxMBxHP	20	confined
D038	IL	60	16	dry	HPxMBxHP	20	unconfined
D039	IL	60	20	dry	HPxMBxHP	20	confined
D040	IL	60	20	dry	HPxMBxHP	20	unconfined
D041-D042	R-ShT	80	14	dry	HPxMBxHP	20	unconfined
D043	R-IL	80	14	dry	HPxMBxHP	20	unconfined
D044	IL	80	14	dry	1xHP	20	unconfined
D045	IL	80	16	dry	HPxMBxHP	20	unconfined
D046	IL	80	20	dry	HPxMBxHP	20	unconfined
D047	IL	60	14	wet	HPxMBxHP	20	unconfined
D048	IL	60	14	wet	HPxMBxHP	20	confined
D049	IL	60	14	dry	HPxMBxHP	43	confined
D050	IL	60	14	dry	HPxMBxHP	43	unconfined
D051	IL	80	14	dry	HPxMBxHP	43	unconfined
D065	IL	60	14	dry	1xHP	43	confined
D066	IL	60	20	dry	HPxMBxHP	43	confined
D067	IL	80	14	dry	HPxMBxHP	43	unconfined
D068	IL	60	20	dry	HPxMBxHP	43	unconfined
D069	IL	80	20	dry	HPxMBxHP	43	unconfined
D070	R-ShT	80	14	dry	HPxMBxHP	20	confined
D071	R-ShT	105	14	dry	HPxMBxHP	20	unconfined
D072	R-IL	105	14	dry	HPxMBxHP	20	unconfined
D073	IL	105	14	dry	HPxMBxHP	20	unconfined
D074	R-IL	80	14	dry	HPxMBxHP	20	confined
D075	IL	80	14	dry	1xHP	20	confined
D076	IL	80	20	dry	HPxMBxHP	20	confined

2.1 Specimen preparation

The holes were drilled using a hammer drilling system mounted on a drill stand to ensure perpendicularity on the concrete slab. Three drill bit diameters d_{cut} were used: 14 mm (standard drill bit diameter according to the manufacturers' printed instructions MPII), 16 mm, and 20 mm. After drilling the holes were cleaned as described in table 1. The fasteners were then installed at room temperature. The mortar was pressed in the hole and the steel component was inserted. The anchors were cured at room temperature for 24 to 48 hours.

The anchors in wet concrete were installed similarly. However, before drilling, the concrete specimens were stored underwater for a minimum of 28 days.

2.2 Testing

A total of 36 series (D028-D051 and D065-D076) were tested. Six parameters were changed throughout the program: the embedment depth, the bond line thickness, hole cleaning, hole saturation, temperatures when tested, and the test support width. The reference tests were drilled with a drill bit diameter of 14 mm and cleaned according to the MPII cleaning procedure: hand pumping to remove dust, machine brushing, and hand pumping.

The anchors were installed at room temperature at three embedment depths, 60 mm, 80 mm, and 105 mm. The bond line thickness investigated were 1 mm (d_{cut}= 14 mm), 2 mm (d_{cut}= 16 mm) and 4 mm (d_{cut}= 20 mm). Besides the standard cleaning procedure, a reduced cleaning effort was investigated (only hand pumping). The anchors were tested at 20°C and 43°C using a confined and unconfined test setup.

The anchors in the short-term tests were loaded to failure with a constant loading rate within 3 minutes. The anchors subjected to incremental loading were loaded and unloaded with 5 % load increments within 5 minutes. The increments were calculated using the reference short-term failure loads. Figure 2 illustrates a typical loading curve for short-term and for incremental loading tests.

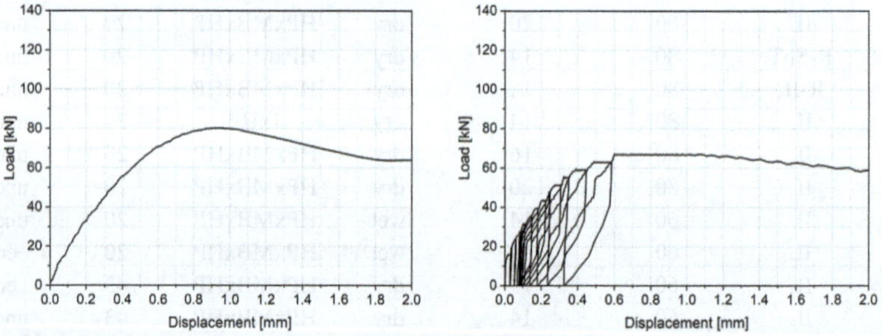

Figure 2. Illustration of load vs displacement curves for short-term and incremental loading tests.

The fastener was axially loaded with the help of a hydraulic cylinder. The load was transmitted from the cylinder to the fixture using a steel rod. Figure 3 illustrates the test setups used. The clearance hole of the fixture for confined tests had a diameter of 24 mm. The diameter of the support for the unconfined test was chosen such that its diameter was more than four times the embedment depth. Thus, for an embedment depth of 60 mm a 240 mm support was used, and for 80 mm and 105 mm, a diameter of 350 mm. The load was measured using a load cell, whereas the displacement using a displacement transducer. The measurements were recorded using commercial software.

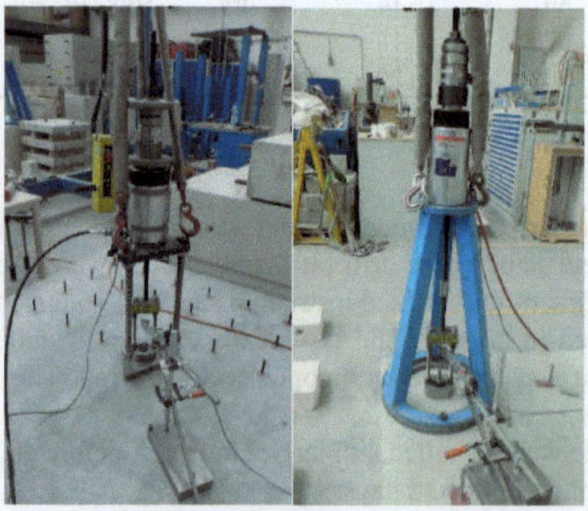

Figure 3. Confined and unconfined test setup.

3 RESULTS

The results of the tests are listed in table 2. The first three columns are the same as in table 1. The fourth and the fifth columns give the mean failure load and standard deviation for each test series.

The anchors exhibited the following failure modes: 12 % pull-out failure, 25 % pull-out with mortar failure, 11 % combined pull-out and pull-out with mortar failure, 17 % concrete failure, and 35 % mixed concrete and pull-out failure.

Table 2: Test results.

Test Series [-]	Type of Test [-]	Embedment depth [mm]	Mean Load $N_{u,m}$ [kN]	Standard deviation [kN]	Support [-]
D028	R-ShT	60	78.62	2.34	confined
D029	R-IL	60	68.71	2.27	confined
D030	R-ShT	60	39.22	3.30	unconfined
D031-D032	R-IL	60	33.94	2.61	unconfined
D033	IL	60	42.13	2.36	confined
D034	ShT	60	47.00	4.62	confined
D035	IL	60	30.86	2.28	unconfined
D036	ShT	60	35.86	2.19	unconfined
D037	IL	60	80.16	8.24	unconfined
D038	IL	60	39.38	0.07	confined
D039	IL	60	77.62	2.42	unconfined
D040	IL	60	39.02	4.45	confined
D041-D042	R-ShT	80	57.87	3.25	unconfined
D043	R-IL	80	46.99	3.17	unconfined
D044	IL	80	41.53	4.50	unconfined
D045	IL	80	58.26	0.41	unconfined
D046	IL	80	57.08	3.42	unconfined
D047	IL	60	32.83	1.20	unconfined
D048	IL	60	65.06	5.15	confined
D049	IL	60	56.68	2.48	confined
D050	IL	60	31.42	2.19	unconfined
D051	IL	80	45.63	3.27	unconfined
D065	IL	60	45.92	8.23	confined
D066	IL	60	65.8	2.32	confined
D067	IL	80	81.56	8.69	confined
D068	IL	60	33.98	1.75	unconfined
D069	IL	80	49.44	2.97	unconfined
D070	R-ShT	80	118.23	2.35	confined
D071	R-ShT	105	89.99	12.92	unconfined
D072	R-IL	105	81.58	4.64	unconfined
D073	IL	105	90.22	4.41	unconfined
D074	R-IL	80	100.77	5.93	confined
D075	IL	80	61.88	2.01	confined
D076	IL	80	110.95	3.47	confined

3.1 Influence of incremental loading

The influence of incremental loading was observed on a total of ten test series. Table 3 lists the reference test series and the series with incremental loading. Figure 4 illustrates the results for the confined setup tests (left figure) and unconfined tests (right figure). The difference between short-term tests and incremental loading tests was more apparent in the confined tests. The anchors installed at 60 mm embedment depth had a 12 % lower failure load because of incremental loading. Similarly, with an increase in the embedment depth to 80 mm, the failure load reduced by 17 %. The test series with reduced cleaning effort (D034 and D033) had a 10 % drop in failure load, due to incremental loading.

Table 3: Influence of incremental loading.

Confined		Unconfined	
reference	incremental	reference	incremental
D028	D029	D030	D031/D032
D034	D033	D036	D035
D070	D074	D041/D042	D043
		D071	D072

The series tested with an unconfined setup and incremental loading had a drop in the loads between 10 % and 19 % for the three embedment depths and the reduced cleaning effort tests.

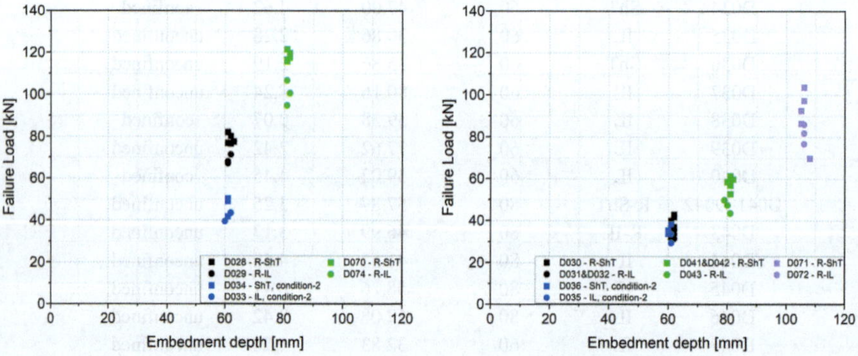

Figure 4. Failure load vs embedment depth curves - influence of the type of loading.

3.2 Influence of hole cleaning

The influence of hole cleaning was observed on two embedment depths for both support widths. Table 4 summarizes the series which were considered.

Table 4: Influence of hole cleaning.

Confined		Unconfined	
reference	hole cleaning	reference	hole cleaning
D028	D034	D030	D036
D029	D033	D031/D032	D035
D049	D065	D043	D044
D074	D075		

The comparison was carried not only on the incremental loading tests but also in the short-term tests. As it can be seen in figure 5, the short-term tests with reduced cleaning effort had 40 % (confined setup) and 9 % (unconfined setup) lower failure load compared to the reference short-term tests with the standard cleaning procedure.

The incremental loading tests under confined setup at room temperature displayed a drop in the carrying capacity of 39 % at both embedment depths, whereas the tests at elevated temperature (43°C) dropped only 19 % in the load.

The unconfined tests revealed the same pattern, for the embedment depth 60 mm, the load reduced by approximately 9 % and for the 80 mm depth tests 12 %.

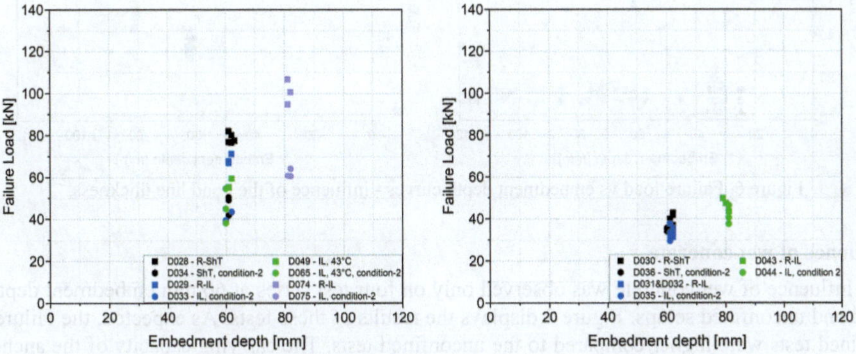

Figure 5. Failure load vs embedment depth curves - influence of the hole cleaning.

3.3 Influence of bond line thickness

The bond line thicknesses of 1 mm (standard), 2 mm (RS2), and 4 mm (RS4) were investigated. All the anchors were tested at both temperatures. A summary of the test series is given in table 5.

Table 5: Influence of bond line thickness.

Confined		Unconfined	
reference	bond line thickness	reference	bond line thickness
D029	D037, D039	D031/D032	D038, D040
D049	D066	D043	D045, D046
D074	D076	D050	D068
		D051	D069
		D072	D073

Figure 6 illustrates the failure loads for confined and unconfined setups. As it can be seen, the tests with embedment depth 60 mm at 20°C had an increase in the failure load of 17 % for bond line thickness RS2, however only 13 % for RS4. At a higher embedment depth, 80 mm, the load of the tests with 4 mm bond line thickness was 10 % higher compared to its reference series. Furthermore, the anchors at elevated temperature, also revealed an increase in the failure load when drilling with a larger drill bit (16 % rise in load).

A similar behaviour pattern exhibited in the unconfined tests. The results of the tests at room temperature with 2 mm bond line thickness at the embedment depths 60 mm and 80 mm, failed at 16 % and 24 % higher loads compared to the reference tests with 1 mm bond line thickness. A further increase to 4 mm bond line thickness, also showed an increase of the load compared to the reference tests, however a slight decrease compared to the RS2 tests. The tests at 43°C and RS4 bond line thickness revealed 16 % and 8 % higher loads (60 mm and 80 mm depth respectively) compared to the tests with 1 mm bond line thickness. Lastly, the anchors installed at 105 mm and RS4 thickness had a rise of 11 % in the failure load.

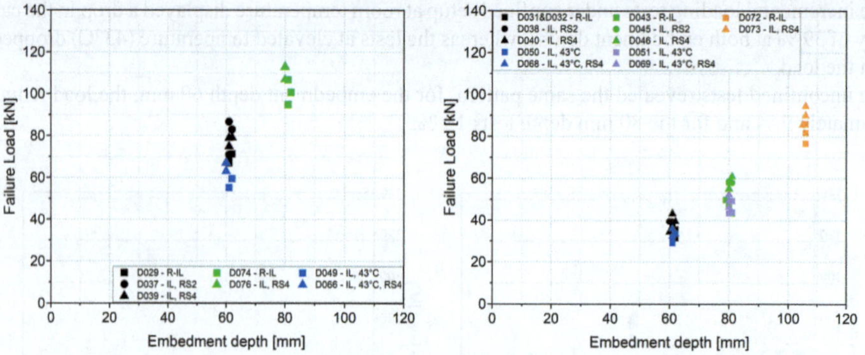

Figure 6. Failure load vs embedment depth curves - influence of the bond line thickness.

3.4 Influence of wet concrete

The influence of wet concrete was observed only on four test series at 60 mm embedment depth with confined and unconfined setups. Figure 7 displays the results of these tests. As expected, the failure loads for confined tests were higher compared to the unconfined tests. The carrying capacity of the anchor was influenced minimally by the condition of the concrete. For the confined tests, the anchors installed in wet concrete had a 5 % lower failure load, whereas the unconfined tests only 3 % lower load compared to the reference tests in dry concrete.

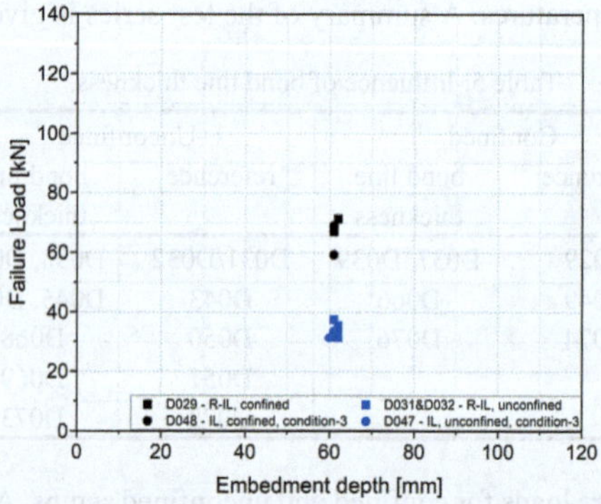

Figure 7. Failure load vs embedment depth curves - influence of the wet concrete.

3.4 Influence of temperature

Table 6 lists the test series where the influence of the temperature when testing was observed. For each support width, four test series were considered. Aside from the temperature, the embedment depth, hole cleaning procedure, and bond line thickness were varied.

Figure 8 illustrates the results of the tests. The results showed an influence of the temperature on the carrying capacity of the anchors. At the embedment depth of 60 mm, the load decreased approximately 18 % for confined tests and 7 % for unconfined tests with an increase in the testing temperature to 43°C. At a higher embedment depth (80 mm), the same trend was observed, 19 % and 3 % lower failure loads for confined and unconfined tests.

Table 6: Influence of temperature when testing.

Confined		Unconfined	
reference	temperature	reference	temperature
D029	D049	D031/D032	D050
D033	D065	D040	D068
D039	D066	D043	D051
D067	D074	D046	D069

The next parameter taken into consideration was the bond line thickness of 4 mm or RS4. The confined tests at 60 mm had 15 % lower residual capacity at increased temperature compared to the room temperature tests. Similarly, for unconfined tests, the loads had a 13 % drop for both embedment depths.

The last parameter tested was the hole cleaning procedure, however only for confined setup at 60 mm embedment depth. The high testing temperature and reduced cleaning effort influenced positively the residual capacity of the anchors; the mean failure load increased from 42.13 kN for reference tests to 45.92 kN (9 % load increase).

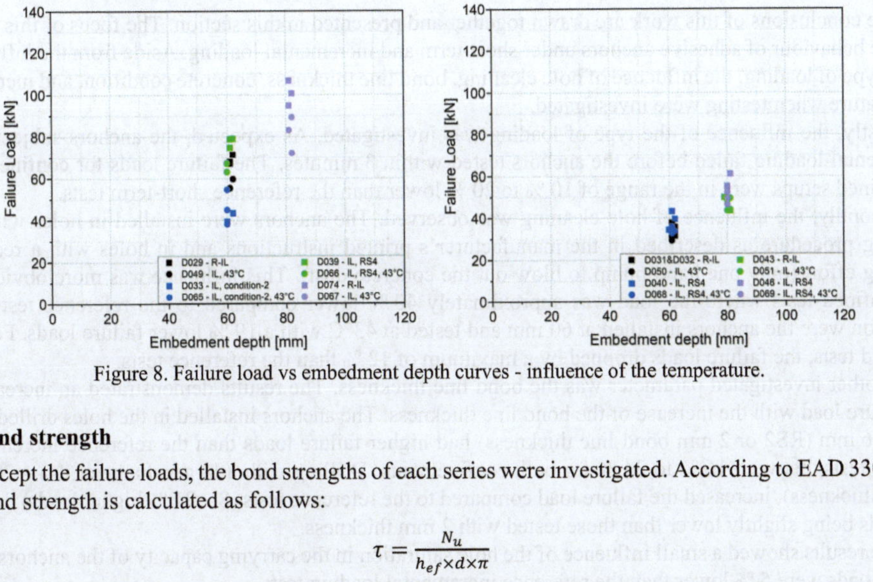

Figure 8. Failure load vs embedment depth curves - influence of the temperature.

3.5 Bond strength

Except the failure loads, the bond strengths of each series were investigated. According to EAD 330499, the bond strength is calculated as follows:

$$\tau = \frac{N_u}{h_{ef} \times d \times \pi} \tag{1}$$

where: τ is the bond strength in N/mm^2, N_u is the failure load in kN, d is the diameter of the fastener in mm and h_{ef} is the embedment depth [4].

Figure 9 summarizes the bond strength for each test. In this figure, the values are grouped according to the embedment depth and the support width. The mean bond strengths and their standard deviation are plotted. The bond strength of the anchors installed at 60 mm depth and tested with confined setup varied from 18.21 N/mm^2 (the tests with reduced cleaning effort) to 34.66 N/mm^2 (the tests with bond line thickness 2 mm). The bond strength increased by approximately 10 % with an increase in the embedment depth to 80mm.

Similarly, the unconfined tests, exhibited lower bond strengths at 60 mm (13.42 - 17.27 N/mm^2) compared to 80 mm (13.60 - 19.24 N/mm^2) and 105 mm (20.42 - 22.65 N/mm^2).

The trend of the results illustrated that the bond strength of the anchors was the lowest for the anchors installed with reduced cleaning effort and the highest for the increased bond line thickness tests (both cases).

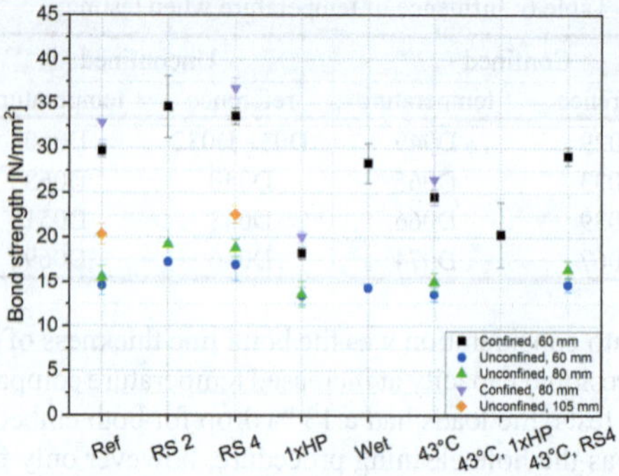

Figure 9. Summary of the mean bond strength for each embedment depth and parameter.

4 CONCLUSIONS

The conclusions of this work are drawn together and presented in this section. The focus of this paper was the behaviour of adhesive anchors under short-term and incremental loading. Aside from the influence of the type of loading, the influence of hole cleaning, bond line thickness, concrete condition, and increased temperature when testing were investigated.

Firstly, the influence of the type of loading was investigated. As expected, the anchors subjected to incremental loading failed before the anchors tested within 3 minutes. The failure loads for confined and unconfined setups were in the range of 10 % to 20 % lower than the reference short-term tests.

Secondly, the influence of hole cleaning was observed. The anchors were installed in holes with the cleaning procedure as described in the manufacturer's printed instructions and in holes with a reduced cleaning effort (only one hand pump to blow out the concrete dust). This influence was more obvious in the confined tests where the load was approximately 40 % lower compared to the reference tests. An exception were the anchors installed at 60 mm and tested at 43°C with a 19 % lower failure loads. For the confined tests, the failure loads dropped by a maximum of 12 % than the reference tests.

Another investigated parameter was the bond line thickness. The results demonstrated an increase of the failure load with the increase of the bond line thickness. The anchors installed in the holes drilled with d_{cut} = 16 mm (RS2 or 2 mm bond line thickness) had higher failure loads than the reference incremental loading tests, between 16 % to 24 % for both confinements. Drilling with a larger diameter d_{cut} = 20 mm (4 mm thickness), increased the failure load compared to the reference tests (8 – 21 % higher loads) despite the loads being slightly lower than those tested with 2 mm thickness.

The results showed a small influence of the hole saturation in the carrying capacity of the anchors. The failure loads were 5 % lower than the reference incremental loading tests.

Lastly, the influence of increased testing temperature was observed. Overall, the increase of the temperature lowered the residual capacity of the anchors. The failure loads for confined tests were 15 % to 19 % lower and for unconfined tests between 3 % and 13 % lower than their reference loads. An exception was the confined tests at 60 mm with a reduced hole cleaning, where the load increased with 9 % with an increase of the testing temperature.

REFERENCES

[1] Park, R.; Paulay, T., "Reinforced Concrete Structures", John Wiley & Sons Ltd, 1975.

[2] Wald, F.; Hofmann, J.; Kuhlman, U.; et al, "Design of Steel-to-Concrete Joints, Design Manual I", *European Convention for Constructional Steelwork*, 2014.

[3] ACI 355.4-11, “Qualification of Post-Installed Adhesive Anchors in Concrete”, *American Concrete Institute*, 2011.

[4] EAD 330499-00-0601, “Bonded Fasteners for Use in Concrete”, *European Organisation for Technical Assessment,* 2017.

[5] E488/E488M-15, “Standard Test Methods for Strength of Anchors in Concrete Elements”, *ASTM International*, 2015.

[6] TR049, “Post-Installed Fasteners in Concrete under Seismic Action”, *European Organisation for Technical Assessment,* 2016.

[7] DIBt, “Leitfaden für Dübelbefestigungen in Kernkraftwerken und anderen kerntechnischen Anlagen“, *Deutsches Institut für Bautechnik*, 2010.

[8] fib Bulletin 58, “Design of anchorages in concrete”, *federation internationale du béton*, 2011.

[9] Eligehausen, R.; Mallée, R.; Silva, J.F., “Anchorage in concrete construction”, Ernst & Sohn, 2006.

[10] Cook, R.A.; Konz, R.C., “Factors influencing bond strength of adhesive anchors”, *ACI Structural Journal*, 2001.

[11] Maçi, N., “Influences on the bond behaviour of anchorages under short-term and long-term loading”, Dissertation, in preparation.

9TH INTERNATIONAL CONFERENCE ON COMPOSITE CONSTRUCTION IN STEEL AND CONCRETE
M. Knobloch, U. Kuhlmann, W. Kurz, M. Schäfer. (Eds.)
Online, Germany, July 27-29, 2021

SFRC SLAB IN COMPRESSION IN STEEL-CONCRETE COMPOSITE FRAME JOINTS

Petr Cervenka*, Jakub Dolejs*

* Faculty of Civil Engineering, Czech Technical University in Prague; Thakurova 7; 166 29, Prague; CZ
e-mails: petr.cervenka.1@fsv.cvut.cz, dolejs@fsv.cvut.cz

Keywords: Composite, Frame, Joint, FRC, Slab, Compression.

Abstract. *The objective of the research is a steel fiber-reinforced concrete slab in compression in a composite frame joint. EN 1998-1 relates only to composite joints with common concrete slab and it prescribes two mechanisms. By these mechanisms, it is prescribed inaccurate formula for the effective width calculation of a common concrete slab in compression in the nodal zone, as the standard procedure does not consider the contribution of the confinement effect to the resistance of mechanism 1. Moreover, some publications describe different inclination of the concrete struts of mechanism 2. The aim of this work is to derive a more accurate formula for calculation the effective width of the steel fiber-reinforced concrete slab in compression in the nodal zone. Experiment with real-sized isolated composite joint, validation of numerical models and parametric study of composite joints with fiber-reinforced concrete slab are the main steps of this research.*

1 INTRODUCTION

The concrete slab in compression as the component of a steel-concrete composite joint occurs mainly in following situations: 1) asymmetric joint loading by hogging moments; 2) exceptional internal column loss; 3) horizontal seismic loading.

Current code formula [1] for effective width calculation of this component is inaccurate, as is stated in many publications [2], [3], [4], [5], [6], [7]. Until now, no research has dealt with the bearing capacity and the stiffness of steel fibre-reinforced concrete (SFRC) slab in compression in a nodal zone of steel-concrete composite frame.

2 STATE OF ART

A formula for calculating of the effective width b_{eff} of the common concrete slab is prescribed in an Appendix C of EN 1998-1 [1]:

$$b_{eff} = 0{,}7h_c + b_c \tag{1}$$

where h_c and b_c are cross-sectional dimensions of the column depicted on Figure 1. This standard formula is based on two mechanisms: mechanism 1 acts in the concrete slab with a concentrated compressive force on the column flange; mechanism 2 is created by struts and a tie model, where concrete struts compress concave parts of the column cross-section.

Prescribed formula for an effective width calculation of a common concrete slab in compression in the nodal zone is inaccurate, because it does not consider the contribution of the confinement effect to the resistance of the mechanism 1. There is a concrete slab in compression confined by means of the transverse reinforcement, continuity of the slab around the column flange and friction on the column flange. This positive state of stress leads to a higher compression strength of the concrete slab in this area [2], [3], [4], [5]. Moreover, some authors described different inclination of the concrete struts of mechanism 2 [3], [6],

[7]. Standard formula took the angle 45° into account. The resistance of mechanism 2 significantly depends on this inclination.

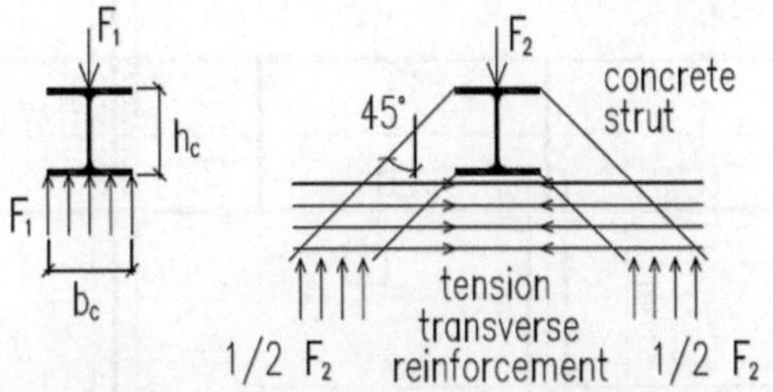

Figure 1. Transfer of forces – F_1 is developed by the mechanism 1, F_2 is developed by the mechanism 2; h_c and b_c are cross-sectional dimensions of the column [8].

3 EXPERIMENTAL STUDY

3.1 Steel-SFRC composite joint

An experiment [8] on inner steel-concrete composite frame joint was performed. The aims of the experiment were to measure distribution of normal compression deformation in concrete slab in the nodal zone and to measure the inclination of minimal principle deformation of mechanism 2. The joint was subjected to a hogging moment in one side of the joint and to the sagging moment in the other until the joint collapsed. Basic test configuration is given in Figure 2: specimen of composite steel-concrete joint (1), test frame girder (2), a member inserted between test frame girders (3), hydraulic jacks applying compressive force (4) and tensile force (5), column of main test frame (6), beam of minor test frame (7), position of lateral restrains of steel-concrete cantilever (8), pin connection (9).

A scheme of designed joint is shown in Figure 3 [9]. A steel-concrete composite joint design used in the experiment ensured that a sagging moment resistance of the joint was smaller than a hogging moment resistance. Components of the steel part of the joint were designed to be ductile to enable plastic distribution of forces. The column web panel in shear is usually a critical part of joints exposed to significantly unequal moments. A column web panel was equipped with supplementary web plates on both sides. As a result, a bearing capacity of the component web panel in shear increased.

A solid slab was designed with a thickness 100 mm and a concrete class C25/30 reinforced by steel fibres (the amount of the fibres is 80 kg/m3, TriTreg 50-1). Material properties of SFRC were determined by coupon tests (tab. 1). Average cubic strength in compression was 61.8 MPa by coupon tests. Cylinder strength in compression was 45.6 MPa and that value was determined from cubic strength by multiplying coefficient 0.738 used for SFRC [10].

Table 1: Results of coupon tests - SFRC.

Average material properties	[MPa]
Modulus of elasticity	35 800
Cubic compression strength	61.8
Transverse tension strength	5.8
Flexural tension strength (100x100x400 mm)	8.2
Flexural tension strength (150x150x700 mm)	5.4

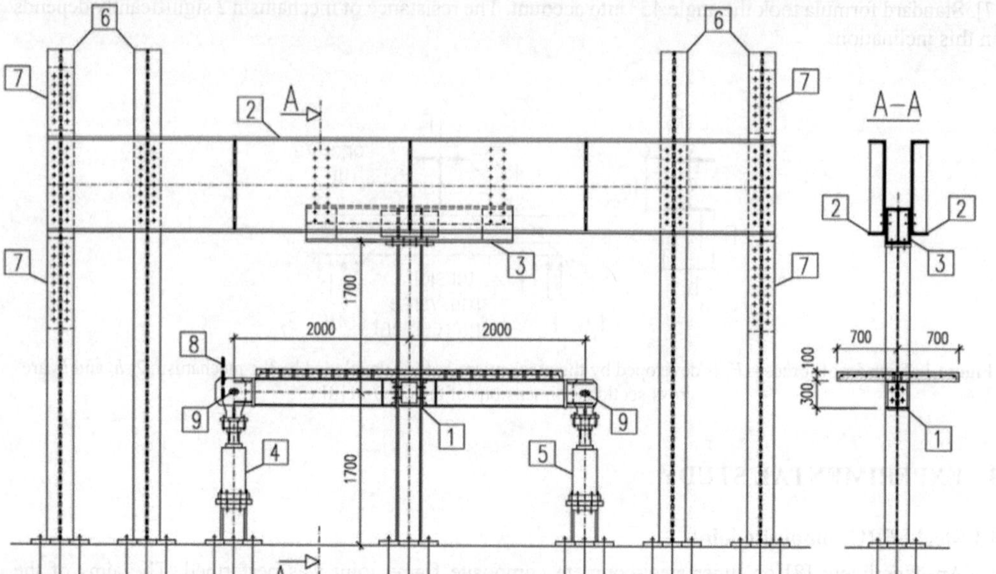

Figure 2. Test configuration [9].

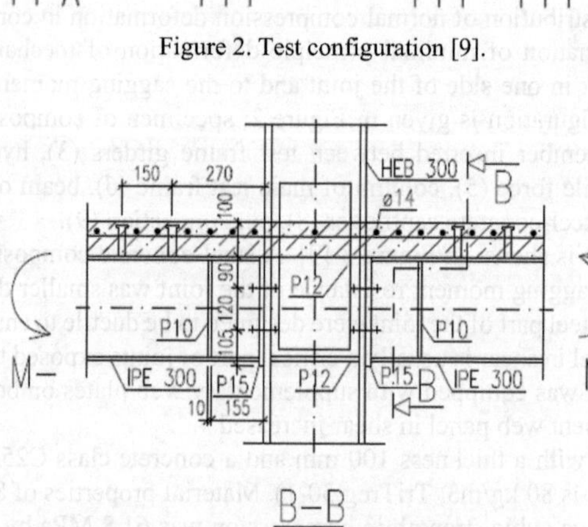

Figure 3. Scheme of designed joint [9].

3.2 Instrumentation for the measurement

The strain gauges were located on top of the concrete slab in the sagging moment area and their location is described in Figure 4. Along with the strain gauge measurement, the distribution of deformation on the concrete slab was measured by the digital image correlation (DIC) method on an area of 300x300 mm next to the column to obtain principal direction and minimal principal deformation value by mechanism 2, Figure 4.

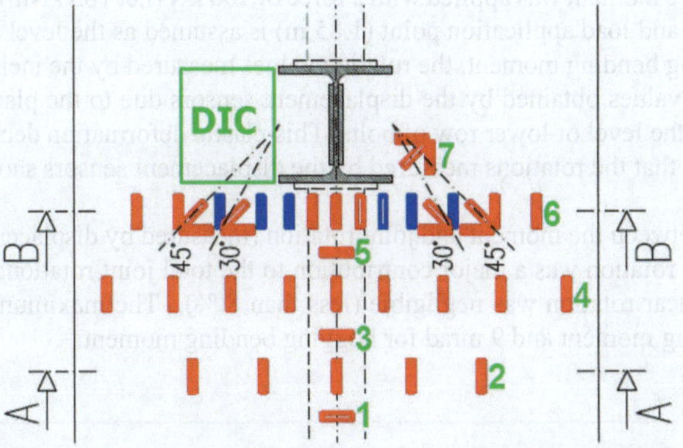

Figure 4. Location of strain gauges on the concrete slab and location of area measured by DIC method [8].

The displacement sensors along with the inclinometers were used to obtain the relationship between the moment and rotation of the connection, the overall joint and the column web panel, Figure 5. For the evaluation of column web panel rotation there were five displacement sensors connected to an independent structure. The relationship between moment and rotation obtained by inclinometers and displacement sensors was compared according to [11]. Further two sensors were located on the top and bottom of the column to get a horizontal movement on its supports. Other two displacement sensors were placed in the ends of steel-concrete composite cantilevers to get a slip between the steel beams and the concrete slab. The strain gauges were placed on following components: column flange in bending, end plate in bending, beam web in tension, beam flange in compression and the strain gauge rosette is situated in the column web panel in shear. The strain gauges were located also on longitudinal reinforcement in two cross-sections in hogging moment area.

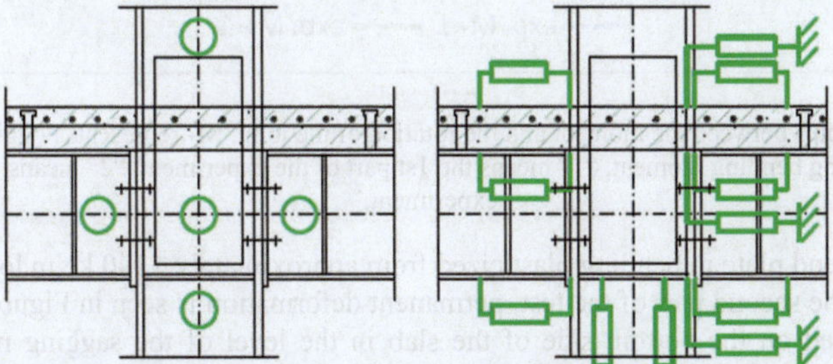

Figure 5. Location of inclinometers (left) and displacement sensors (right) [8].

3.3 Results of experiment

The monotonic loading test was performed by hydraulic jacks, which were controlled by deformation. The test consisted of two parts: 1) applied the displacement gradually by hydraulic jacks up to 60 mm and unloaded; 2) applied the displacement gradually by hydraulic jacks up to 105 mm and end of the test. In the first part, the loading speed was 1 mm/min for positive moment side and 1,25 mm/min for negative moment side. In the second part, the loading speed was 10 mm/min for both sides. The hydraulic jack limitation on the negative moment was applied with a force of 100 kN (i.e. 185 kNm). The distance between the front column flange and load application point (1.85 m) is assumed as the level arm.

In the side of sagging bending moment, the rotation values measured by the inclinometers are less than about 20% the rotation values obtained by the displacement sensors due to the plastic deformation of the beam web in tension at the level of lower row of bolts. This plastic deformation decrease rotation obtained by the inclinometers, so that the rotations measured by the displacement sensors should give more relevant values.

The relationships between the moment and joint rotation (measured by displacement sensors) is shown in Figure 6. Connection rotation was a major contribution to the total joint rotation; on the other hand, the contribution of panel shear rotation was negligible (less than 8 %). The maximum joint rotation was 50 mrad for sagging bending moment and 9 mrad for hogging bending moment.

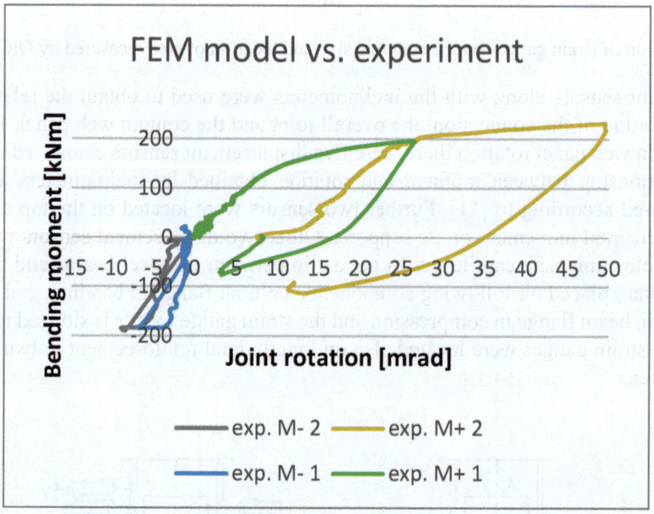

Figure 6. The relationships between the moment and the rotation of the joint: M- represents hogging bending moment, M+ represents sagging bending moment, “1” means the 1st part of the experiment, “2” means the 2nd part of the experiment.

As expected, the end plate in bending plasticized from approximately +140 kNm load level (Figure 7). At the beginning of the second part of the test, permanent deformation is seen in Figure 6. Tension cracks were visually observed on the bottom side of the slab in the level of the sagging moment connection approximately from loading +60/-130 kNm, Figure 7. These cracks were caused by rotation of semi-rigid connection, position of neutral axis in level of SFRC slab and anchored longitudinal reinforcement tensioned by hogging moment. After the test, further local damage to the top of the slab in front of the column flange was observed due to rotation of semi-rigid connection on side of sagging moment, Figure 8. This damaged part of slab was approximately 3 cm wide along the column flange and in average 1 cm deep.

Figure 7. Plasticization of end plate in bending and crack on the bottom side of SFRC slab in the level of the connection.

Figure 8. Local damage of slab in front of the column flange.

In Figures 9, 10 and 11, the distribution of normal deformation on the top surface of the slab is shown in different cross-sections and in different parts of the test (parts of the test are described in Figure 6). In the distance of 500 mm from sagging bending moment connection (Figure 9), a shear lag can be observed. In the distance of 270 mm (Figure 10), concrete trusses of the mechanism 2 are at the cross-sectional edges. In the distance of 75 mm (Figure 11), these concrete trusses are closer to the slab axis.

In the Figure 11a, deformation on left side of the cross-section is higher than deformation in the right side because of imperfect shape of the column flanges. This manufacturing imperfection led to horizontal curvature of composite cantilevers, which were connected to imperfect the column flanges (end of cantilever had different position by 24 mm in sagging section and by 3 mm in hogging section). These imperfections should be assumed in validated numerical model.

Local damage of the concrete slab in front of the column flange (Figure 8) leads to a decrease of normal deformation, Figure 11. After a drop of normal deformation in front of the column flange, normal deformation of concrete trusses rose. The deformation in front of the column flange exceeded the compressive strain in the concrete at the peak stress in the first part of the test (2,4‰ according to an analytic calculation for common concrete). The deformation in front of the column flange also exceeded the ultimate compressive strain in the concrete in the second part of the test (3,5‰ for common concrete), but there was no permanent deformation at the place of the strain gauges after the test by visual observation.

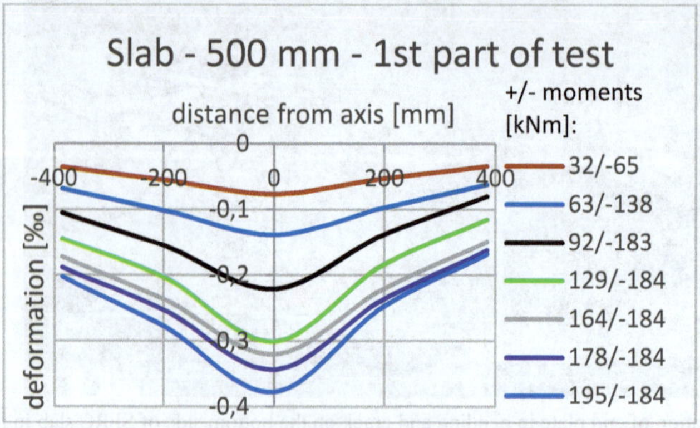

Figure 9. Distribution of the normal deformation on the top surface of the slab in the distance of 500 mm from the connection for the first part of experiment.

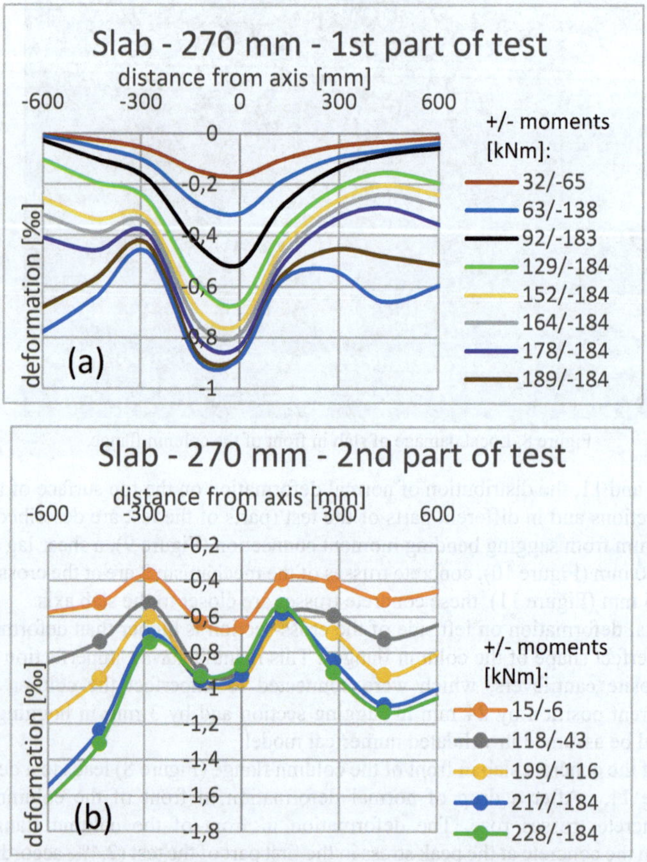

Figure 10. Distribution of the normal deformation on the top surface of the slab in the distance of 270 mm from the connection for the first part of experiment (a), for second part of experiment (b).

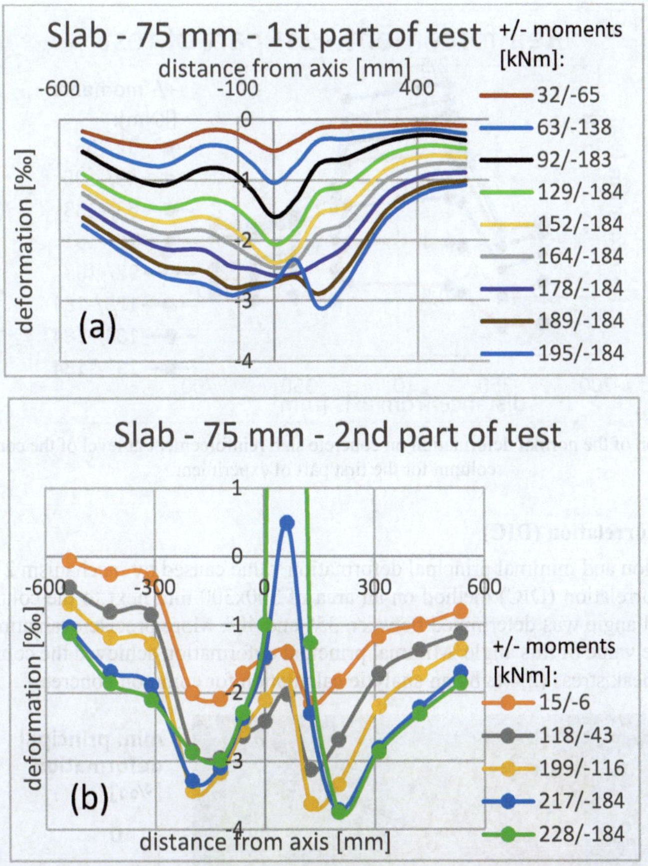

Figure 11. Distribution of the normal deformation on the top surface of the slab in the distance of 75 mm from the connection for the first part of experiment (a), for second part of experiment (b).

The distribution of the normal deformation on the concrete slab reinforcement at the level of hogging bending moment connection of the beam to the column is shown in Figure 12. The effect of the imperfect column shape is also evident here (there is higher normal deformation on the left side of the slab than on the right side), but that influence is less than in case of SFRC slab.

The hydraulic jack limitation on the negative moment was applied with a moment of -184 kNm. At a constant hogging moment of -184 kNm, the normal reinforcement deformation near the slab axis increased slightly (Figure 12). Compression stress in the slab from the sagging moment area were probably carried partly by the longitudinal reinforcement back in front of the column flanges to the side of sagging moment, so that additional truss model could be created. This additional truss model is described in [12].

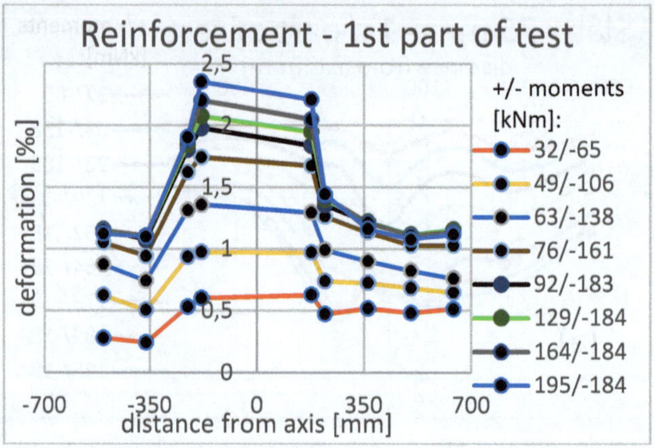

Figure 12. Distribution of the normal deformation on concrete slab reinforcement at level of the connection of beam to column for the first part of experiment.

3.4 Digital image correlation (DIC)

Principal direction and minimal principal deformation value caused by mechanism 2 was measured by the digital image correlation (DIC) method on an area of 300x300 mm next to the column, Figure 4 and Figure 13. Principal angle was determined between 35° and 40°. More precise evaluation should be done to get more accurate value of this angle. Minimal principal deformation achieved the compressive strain in the concrete at the peak stress (2,4‰ by an analytic calculation for common concrete).

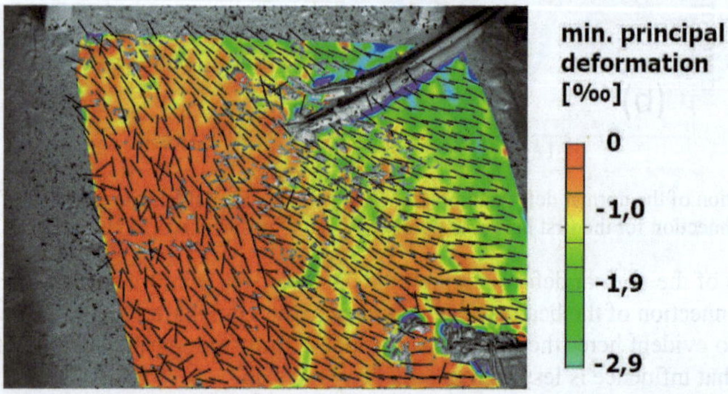

Figure 13. DIC: minimal principal deformation and minimal principal direction.

4 NUMERICAL MODEL IN ATENA SCIENCE

Numerical material nonlinear model was created in Atena Science program [13], Figure 14. Measured material properties of steel members, SFRC slab and reinforcement were assumed. Model is validated through comparisons to experimental results, Figure 15. Numerical model shows higher bending stiffness because of preliminary modelled contact of some composite joint components (end plate in bending, concrete slab in compression, shear studs). In addition, the model does not yet take into account mentioned imperfections of construction that greatly influenced the joint behaviour during the experiment. The model is not validated completely yet, but some preliminary results can be observed, e.g. confinement effect is present and inclination of concrete struts could be 35°-40°, Figure 16.

Moreover, the result of numerical model of steel-SFRC composite joint (material properties of SFRC slab were determined by coupon tests, tab. 1) were compared to numerical model of composite joint with common concrete slab (class C45/55), Figure 17.

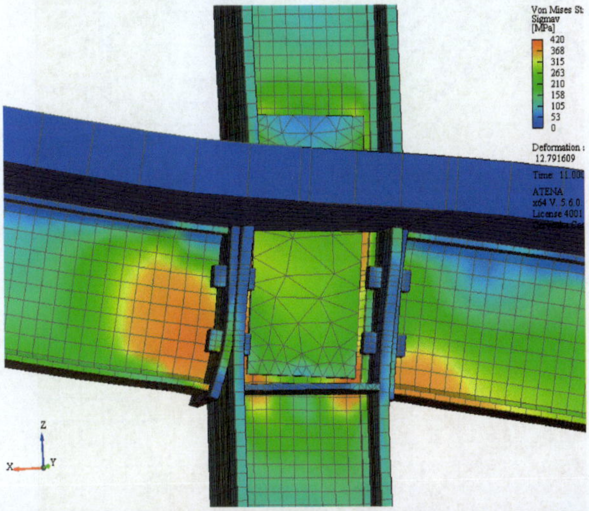

Figure 14. Numerical material nonlinear model in program Atena Science.

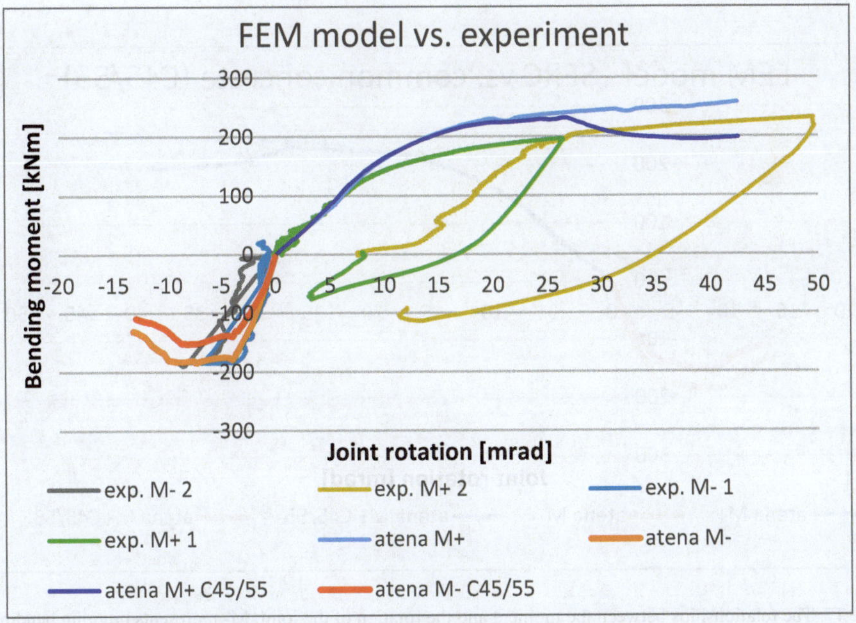

Figure 15. The relationships between the moment and the rotation of the joint: M- represents hogging bending moment, M+ represents sagging bending moment, "1" means the 1st part of the experiment, "2" means the 2nd part of the experiment.

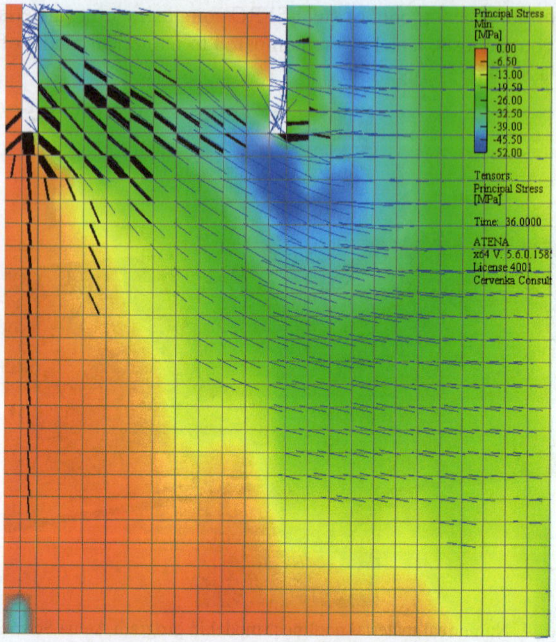

Figure 16. Minimal principal stress, minimal principal direction and distribution of cracks on the top surface of slab on sagging moment area at column.

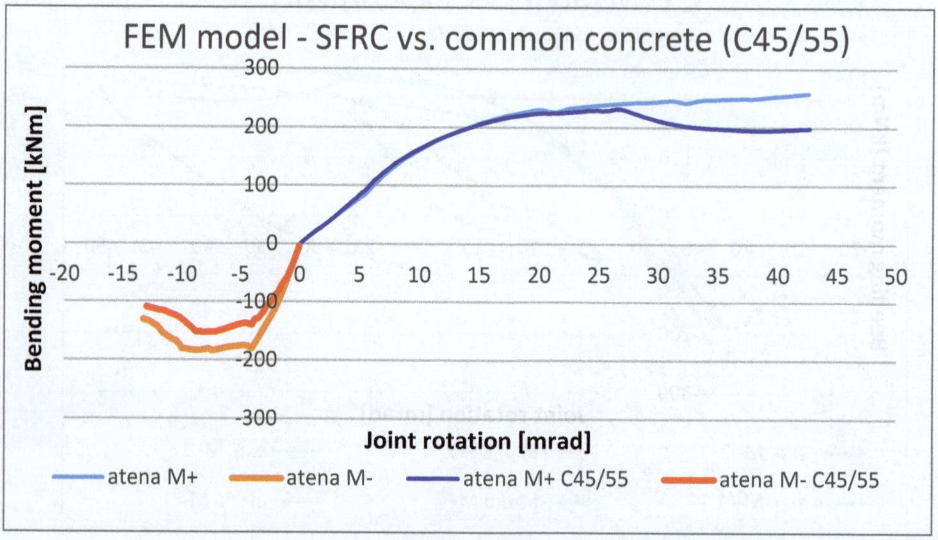

Figure 17. The relationships between the moment and the rotation of the joint: M- represents hogging bending moment, M+ represents sagging bending moment, "1" means the 1st part of the experiment, "2" means the 2nd part of the experiment.

5 CONCLUSION

The subject of this work are results of the experimental test on the steel-concrete composite joint with the SFRC slab and basic information about numerical model. The results of the experiment showed that: 1) the inclination of the concrete struts is between 35° and 40°; 2) apparently the confinement effect is present. A larger parametric study will be performed out with validated and verified numerical models to obtain an analytical procedure.

REFERENCES

[1] EN 1998-1: Eurocode 8: Design of structures for earthquake resistance - Part 1: General rules, seismic actions and rules for buildings. European committee for standardization, 2004.

[2] Plumier, A. & Doneux, C. 2001. Seismic behaviour and design of composite steel concrete structures. LNEC Edition.

[3] Bennacer, M.A., Beroual, A., Kriker, A. & Demonceau, J.-F. 2015. Analytical model for composite joints under sagging moment. *Engineering Structures* 101: 399-411.

[4] du Plessis, D.P. & Daniels, J.H. 1973. Strength of Composite Beam-to-Column Connections. Fritz Engineering Lab. Lehigh University, Bethlehem, Pennsylvania.

[5] Kato, B. & Tagawa, Y. 1985. Strength of composite beams under seismic loading. *Composite and Mixed Construction*, ASCE.

[6] Somja, H. & Marie, F. 2017. Concrete Structures Reinforced by Steel Profiles. SMARTCOCO Project. RFCS GRANT AGREEMENT N° RFSR-CT-2012-00031. Deliverable D8.3 Strut-and-tie models for shear keys. Contribution to the calibration report. INSA Rennes, Internal Report.

[7] Civjan, S.A., Engelhardt, M.D. & Gross, J.L. 2001. Slab Effects in SMRF Retrofit Connection Test. *Journal of Structural Engineering* 127(3).

[8] Cervenka, P. & Dolejs, J. Behaviour of a concrete slab in compression in composite steel-concrete frame joints. In: WALD, F. and M. JANDERA, eds. Proceedings of the International Colloquia on Stability and Ductility of Steel Structures (SDSS 2019). *Stability and Ductility of Steel Structures 2019*, Praha, 2019-09-11/2019-09-13. Leiden: CRC Press/Balkema, 2019. p. 229-234. 1. vol. 1. ISBN 9781000752243.

[9] Cervenka, P. Experimental test on composite steel-concrete frame joint. 2018. In: *Sbornik seminare doktorandu katedry ocelovych a drevenych konstrukci 13.2. a 20.9. 2018*. Prague: CVUT. pp. 23-26.

[10] Zhu, H., Li, Ch., Gao, D. & Cheng, S.. Study on mechanical properties and strength relation between cube and cylinder specimens of steel fiber reinforced concrete. 2019. *Advances in Mechanical Engineering* 11(4).

[11] Jaspart, J.P. & Marquoi, R. 1992. Investigation by testing of the structural response of semi-rigid joints. *Testing of metals for structures*; edited by F. M. Mazzolani.

[12] Macrae G.A. et al. Slab Effects on Beam-Column Subassemblies—Beam Strength and Elongation Issues. *International Conference on Composite Construction in Steel and Concrete.* 2013.

[13] Cervenka Consulting s.r.o. Atena Science – Atena Studio v5. Downloaded 7/2019.

9TH INTERNATIONAL CONFERENCE ON COMPOSITE CONSTRUCTION IN STEEL AND CONCRETE
M. Knobloch, U. Kuhlmann, W. Kurz, M. Schäfer. (Eds.)
Online, Germany, July 27-29, 2021

STRENGTHENING OF ANCHOR CHANNELS ON THE CONCRETE SURFACE

Michael C. Petrasch*, Jan Hofmann**

* fischerwerke GmbH & Co. KG, Klaus-Fischer-Straße 1, 72178 Waldachtal
e-mail: michael.petrasch@fischer.de

** Institute of Construction Materials, University of Stuttgart, 70569 Stuttgart
e-mail: jan.hofmann@iwb.uni-stuttgart.de

Keywords: Anchor channels, FEM, supplementary reinforcement, shear loads, concrete edge failure.

Abstract. *For the installation of curtain wall systems, the positions of anchor channels gradually move closer to the edge of the concrete slabs. But close to the edge of the slab, conventional supplementary reinforcements have problems to redistribute the loads to the reinforcement due to the short bond length that is available after concrete edge failure occurred. Therefore, tests have been conducted to investigate the behaviour of anchors channels, close to the edge. To achieve this, a strengthening system on the surface of the concrete with post installed fasteners was developed. First results indicate a high potential for strengthening the channels close to the edge under shear load. With these test results numerical simulations with Atena Science were performed and, after its calibration, the model was used for further investigation.*

1 INTRODUCTION

1.1 General

For anchor channels close to the edge concrete edge failure is often the decisive failure mode when loaded in shear. Therefore, the presence of supplementary reinforcement is essential for the load bearing capacity of the anchor channel. According to national codes [1] a supplementary reinforcement needs a minimum bond length of $l_{1,min} = 4 \cdot d_s$, where d_s is the diameter of the reinforcement. As an example, a reinforcement with a diameter of 8 mm would need at least 32 mm bond length within the concrete cone. For an anchor channel casted with a minimum edge distance of 50 mm and assuming a minimum concrete cover of 20 mm, the minimal possible bond length is barely obtained. A reinforcement with a diameter of 12 mm would not be able to reach this minimal bond length at all.

Therefore, a reinforcement respectively strengthening system which is not dependent of the bond length is necessary to effectively handle the loads at such small edge distances. In this study a system, preventing deformations of the anchor channel itself, was tested. The system consists of post-installed fasteners and metal lugs which are connected to the anchor channel.

The results of the tests are used to calibrate an FEM model which has been modelled with Atena Science. With the calibrated model a test program shall be conducted where the test parameters are created according to a Design of Experiments (DoE) approach. This paper is part of a research project at the Institute of Construction Materials [2]. The goal of this project is to develop a new approach of strengthening anchor channels close to the edge.

1.2 Codes and guidelines

According to DIN EN 1992-4 [1] and EOTA TR047 [3] the calculation of the characteristic concrete edge breakout strength $V_{Rk,c,re}$ for supplementary reinforced anchor channels, loaded perpendicular to the edge shall be carried out to the following equation:

$$V_{Rk,c,re} = \psi_{cr} \cdot (V_{Rk,c,hook} + V_{Rk,c,bond}) \tag{1}$$

where:

ψ_{cr} is the modification factor for cracked or uncracked concrete. $V_{Rk,c,hook}$ and $V_{Rk,c,bond}$ are calculated according to the following equations:

$$V_{Rk,c,hook} = \psi_1 \cdot \psi_2 \cdot \psi_3 \cdot A_{S,i} \cdot f_{yk} \cdot \left(\frac{f_{ck}}{25}\right)^{0,1} \tag{2}$$

$$V_{Rk,c,bond} = \pi \cdot d_s \cdot l'_{1,i} \cdot f_{bk} \tag{3}$$

The hook and bond capacity of one stirrup depends on the factor l_1 which in case of the hook capacity is defined within ψ_2 and for the bond capacity is defined by $l'_{1,i}$. It is calculated according to the following equation for stirrups crossing the assumed failure crack with an angle < 90°:

$$l_1 = c_1 - c_c - 0.7 \cdot (e_i - b_{ch}) \geq 4 \cdot d_s \tag{4}$$

And for stirrups crossing the assumed failure crack with an angle of 90°:

$$l_1 = c_1 - c_c \geq 4 \cdot d_s \tag{5}$$

With l_1 calculated according to equation 4 or 5, $l'_{1,i}$ can be calculated through the following equation:

$$l'_1 = l_1 - 3 \cdot d_s \tag{6}$$

For stirrups crossing the assumed failure crack with an angle of 90° where:

c_1 is the edge distance of the anchor channel,
c_c is the concrete cover over the stirrup,
e_i is the distance from the loading point to the middle axis of the stirrup,
b_{ch} is the width of the anchor channel,
d_s is the diameter of the stirrup.

Equation 4 shows, that for stirrups that are not positioned underneath the anchor channel, the bond length, depending on the point of load introduction, will be further decreased. Therefore, the installation of anchor channels at a minimum edge distance of 50 mm is not possible, since according to equations 4 to 6 the supplementary reinforcement does not increase the resistance at all because the minimal bond length is not reached.

1.3 Purpose of the investigation

Since there are no test results available for strengthening anchor channels from the surface, it is necessary to determine the effectivity of such a system. Therefore, the aim of the described investigation is to get

- a first impression of the load increasing potential from surface "reinforcements"
- an understanding of the influence of the anchor channels profile
- an understanding of the crack propagation when the concrete does not need to be cracked to introduce the loads to the strengthening module
- an understanding of the influence of the positioning of the strengthening module to the loading point

2 EXPERIMENTAL INVESTIGATIONS

2.1 Introduction of the investigated strengthening module

The investigated strengthening module used to increase shear resistance of the anchor channels from the top through a metal lug which engages into the channel is shown in figure 1.

Shear loads introduced to the special screws of the anchor channel will be transferred by bending of the channel to the strengthening module. The module engages into the channel by using a hook, a bolt screw, or a comparable mechanism. Thus, the metal lug of the strengthening module will be loaded in tension.

The idea behind this concept is, that the hook engages into the anchor channel before the bending of the channel will reach its next anchor. Therefore, loads will first activate the strengthening module. After the activation of the module the lug will reach its yielding stress, and then the loads will be transferred to the anchor of the channel. This mechanism should delay the cracking of the surrounding concrete, redirecting the loads thorough the metal lug to the fastener away from the edge of the concrete member.

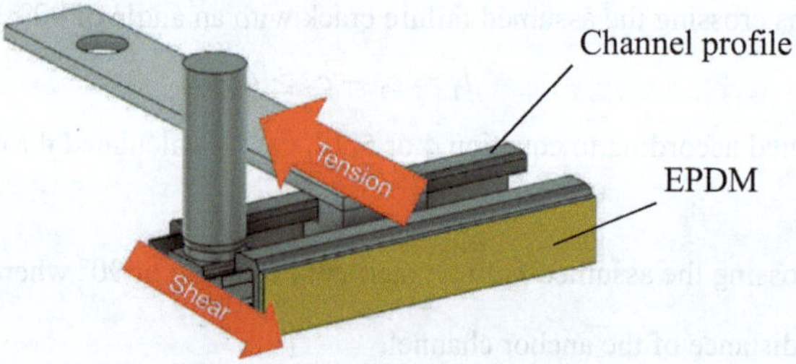

Figure 1. Illustration of the post-installed strengthening module engaging in the anchor channel profile by a hook.

The EPDM tape on the front of the anchor channel acts as a clearance of 2 mm between the concrete and the anchor channel loaded in shear. Due to this provided clearance, the strengthening module is activated before the shear loads are introduced to the surrounding concrete in front of the channel. A similar method is already used by anchor plates, where slotted holes are provided for the front anchors so that the shear loads must be transferred by the anchors far away from the edge.

The anchor channel between two strengthening modules can be idealized as a beam on two supports, as it is shown by Krause [4] for tension loading.

With this type of system, the cracking of the concrete is not a prerequisite to activate the lug, like for strengthening systems based on supplementary reinforcement. Also, the problem of providing the minimal bond length of 4 times d_s with this system is no longer a limit. An open question is, if the hook or bolt screw can be activated before loads are introduced into the concrete cover before the channel.

2.2 Test phase 1 – Investigation to determine the mechanism and effectivity of strengthening modules engaging from the top of the anchor channel

To investigate the effectivity of the strengthening module explained above, shear tests according to Table 1 were performed. Tests with and without strengthened anchor channels in unreinforced concrete are used to show the mechanism of strengthening.

The tests have been conducted in C- and V-shaped anchor channels, which are standard profile types. The used V-profile was comparable to fischer's FES 40/22 profile, because this profile type can be cast in with an edge distance of 50 mm according to its European Technical Assessment (ETA) [5].

For loading, a bolt screw with a diameter of 16 mm and a length of 80 mm, was used. The bolt screws were made of 8.8 grade galvanized steel to avoid the possibility of steel failure. In the tests, no other failure mode than concrete edge failure was aimed.

All tests were performed under the same conditions and the

- C- and V- shaped profiles
- embedment depth h_{ef}
- and anchor bolts

were kept constant for all tests. The concrete class was C20/25 with a constant edge distance c_1 of 50 mm starting from the anchors of the anchor channel. The distance of the strengthening modules to the loading point *s* was 47,5 mm to both sides of the loading point. The anchor channel length was 150 mm with a spacing of the anchors of 100 mm.

Table 1: Overview of the test program to determine the effectivity of surface elements to strengthen anchor channels close to the edge.

Test	Test series description	Support spacing	Number of tests
[-]	[-]	[mm]	[-]
Reference C	Reference tests in unreinforced, uncracked concrete, C-Profile	550	4
Reference V	Reference tests in unreinforced, uncracked concrete, V-Profile	550	4
V1	Strengthened anchor channel C-Profile	550	4
V2	Strengthened anchor channel V-Profile	550	4
V3	Strengthened anchor through rebar module	550	4

In figure 2 the spacing between the supports can be seen. The distance was fixed to 550 mm acc. to [3], so that the formation of a full concrete cone at the edge was possible. The cracks on both sides of the anchor channel and the displacement of the loading plate were monitored through three LVDTs. For the strengthened tests, also the displacement of the fasteners was monitored by two additional LVDTs.

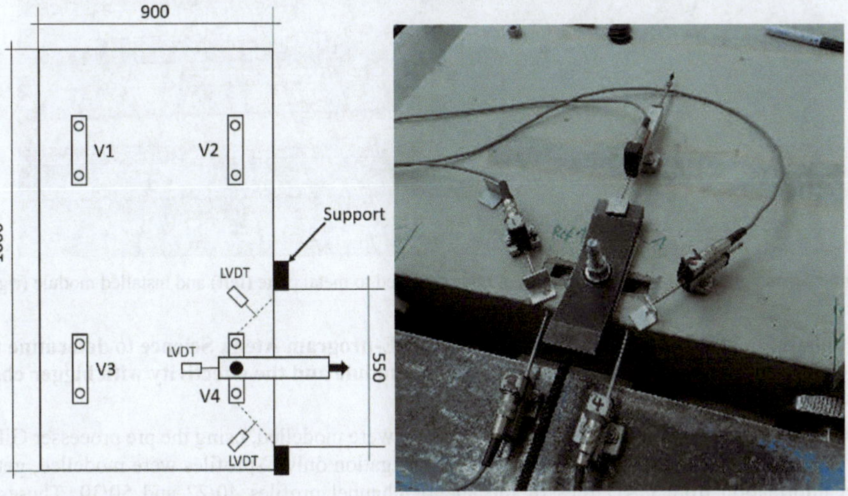

Figure 2. Schematic of the test setup of the reference tests (left) and the conducted tests (right).

The strengthened tests were performed with metal lugs having a length of 200 mm, a hole distance of 150 mm and a width of 30 mm (see figure 3). The lugs had a thickness of 4 mm and consisted of standard steel (S235). For an easy installation of the strengthening module, the same bolt screw was used as for the loading point.

Because for the tests with the C-profile, the annular gap between the fastener and the metal lug was not filled, a slipping during the loading process was possible. The annular gap of the V-profiles was filled with fischer FIS V to prevent this slipping.

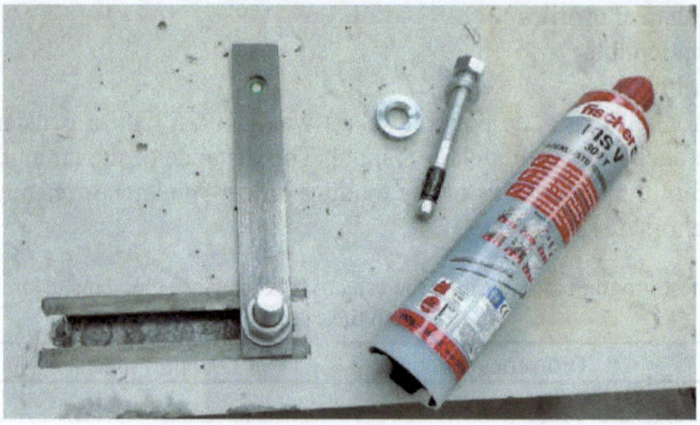

Figure 3. Installation process for the strengthening module with the fastener fischer FAZ II and fischer FIS V to fill the annular gap between the fastener and the metal lug.

To be able to compare the strengthening effect of the post installed module and a strengthening module using reinforcement according to figure 4. Additional tests with the V-profile using the module according to figure 4 were performed. The module consists of a metal plate with a width of 200 mm and a height of 50 mm. The metal plate was attached to the formwork through screws. Four rebars with a diameter of 12 mm and a spacing of 50 mm were welded to the steel plate. This module is activated when cracks in the concrete developed and a cone is restrained and held back by the plate. Therefore, the rebars can only be fully activated for larger crack widths (normally much larger than the limit of serviceability of 0,3 mm).

Figure 4. Strengthening module consisting of rebars welded to metal plate (left) and installed module (right).

2.3 Test phase 2 –Numerical investigation with the FE-program Atena Science to determine the influence of variable positions of the strengthening module and the effectivity with bigger channel profiles.

In the first step, the anchor channel and the concrete were modelled, using the pre processer GiD which is provided by Atena Science. For the numerical investigation only C-profiles were modelled, getting the required information from CAD files of the anchor channel profiles 40/22 and 50/30. Theses anchor channels have been embedded in the concrete as shown in figure 5. For all contact surfaces between the anchor channel and concrete, interface elements have been used. Those elements can only transfer compression stresses.

The gap of 2 mm between the front of the anchor channel and the concrete was also modelled. The connection between the anchor channel and the concrete was realized through separated interface elements which react the same way as interface elements when getting in contact after a displacement of 2 mm.

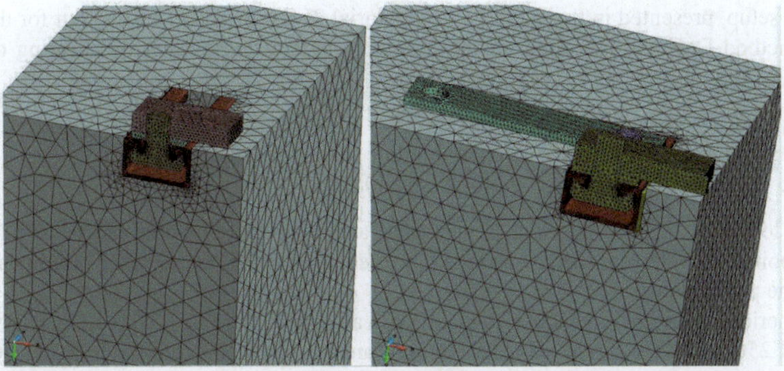

Figure 5. Half FE-model of the reference tests (left) and the strengthened anchor channel (right) with loading plate.

As given in figure 5 (right side), the fastener of the metal lug was not modelled because the tests did not show any damage of the fastener. Therefore, the clearance hole of the lug was defined as fixed in the loading direction. Using its boundary conditions, yielding or bending of the lug is still possible. For the accurate estimation of the load-deformation of the lugs some calibration tests were conducted (see figure 6).

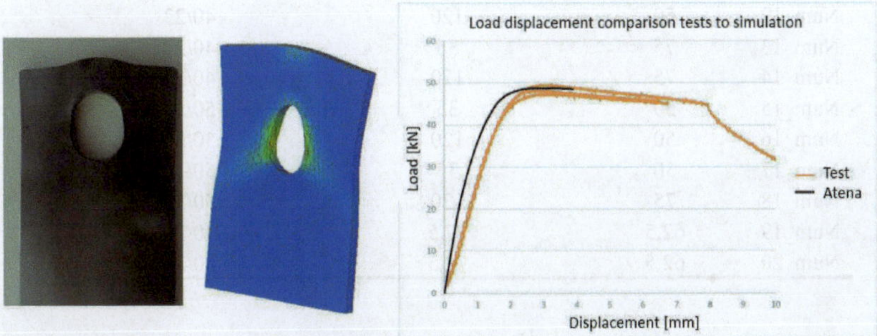

Figure 6. Calibration steps from the tested to the modelled lug with the corresponding load-displacement behaviour (from left to right).

Table 2: Overview of the numerical test program to determine the effectivity of surface elements to strengthen anchor channels close to the edge for a channel length of 150 mm.

Test	Edge distance c_1	Distance to loading plate s	Anchor channel profile p
[-]	[mm]	[mm]	[-]
Num_1	50	35	40/22
Num_2	50	50	40/22
Num_3	75	35	40/22
Num_4	75	50	40/22
Num_5	50	35	50/30
Num_6	50	50	50/30
Num_7	50	35	50/30
Num_8	75	50	50/30
Num_9	62,5	42,5	40/20
Num_10	62,5	42,5	50/30

For the test setup, presented in this paper, a full factorial 2^3 test plan was carried out for the numerical part with the method DoE and the effect of the following parameters on the load bearing capacity was investigated:

- c_1: Edge distance [mm]
- s: Distance of the strengthening module to the loading plate [mm]
- p: Anchor channel profile (40/22 and 50/30) [-]

The factor levels of the full-factorial 2^3 test plan and their combination can be seen in Table 2. Additionally, points in-between where designed, to see a possible linear resp. nonlinear behaviour correlation of the parameters.

For the numerical investigation anchor channels with a length *l* of 150 and 300 mm and with a spacing of 100 mm and 250 mm were modelled. The 300 mm anchor channels were modelled to determine the maximum distance of the strengthening module to the possible loading point. The test program for the anchor channels with a length of 300 mm can be seen in Table 3.

Table 3: Overview of the numerical test program to determine the effectivity of surface elements to strengthen anchor channels close to the edge for a channel length of 300 mm.

Test	Edge distance c_1	Distance to loading plate s	Anchor channel profile p
[-]	[mm]	[mm]	[-]
Num_11	50	35	40/22
Num_12	50	120	40/22
Num_13	75	35	40/22
Num_14	75	120	40/22
Num_15	50	35	50/30
Num_16	50	120	50/30
Num_17	50	35	50/30
Num_18	75	120	50/30
Num_19	62,5	77,5	40/20
Num_20	62,5	77,5	50/30

3 INVESTIGATIONS TO DETERMINE THE MECHANISM AND EFFECTIVITY OF STRENGTHENING MODULES ENGAGING FROM THE TOP OF THE ANCHOR CHANNEL

3.1 Evaluation of the test results at failure load levels

The comparison of the reference tests shows, that this close to the edge, the difference of the mean failure load of the C- or V-profile was negligible.

Table 4: Failure loads of the reference tests.

Test	$V_{u,test}$	$V_{um,test}$	$f_{c,test}$	$V_{um,test,25}$	Variation coefficient
[-]	[kN]	[kN]	[N/mm²]	[kN]	[%]
Reference c-profile	12.84 12.58 13.50 13.41	13.06	28.62	12,.21	3.48
Reference v-profile	12.71 12.72 13.62 12.42	12.87	28.62	12.03	4.05

While the C-profile had a mean failure load $V_{um,test,25}$ of 12.21 kN, the failure load of the V-profile was at 12.03 kN which corresponds to a difference of 1.5 % with a comparable coefficient of variation under 5 % (Table 4).

A noticeable difference in the reference tests with respect to the profile type was the starting point of the crack propagation towards the edge. While the crack for the C-profiles starts from the middle to the back of the anchor channel, the cracks of the V shaped profiles start at the front of the anchor channel (figure 7).

When comparing the bending behaviour of the anchor channel profiles from the reference tests, there was no visible deformation of the channel itself, as the channel moved together with the concrete cone.

Figure 7. Comparison of the starting point of the crack propagation for the V-profile (left) and the C-profile (right).

The strengthened anchor channels showed a comparable behaviour compared to the reference tests regarding the small difference of the mean failure loads. While the mean failure load of the strengthened C-profile was at 37.58 kN, the strengthened V-profile failed at 38.89 kN. This equals a difference of 3.4% which is within the expected scattering of the concrete of 15%.

With this small difference between the failure loads, no effect of the filled or not filled annular gap could be evaluated.

In Table 5 the failure loads of each test are shown. The failure loads of anchor channels having a strengthening module, using a supplementary reinforcement are 33% higher and reach a level of 51.66 kN. This failure load is higher than the comparable strengthened tests of the V-profiles with the post-installed strengthening module. Therefore, a detailed analysis of the load behaviour considering the crack widths was performed in the following evaluation.

Table 5: Failure loads of the strengthened tests.

Test [-]	$V_{u,test}$ [kN]	$V_{um,test}$ [kN]	$f_{c,test}$ [N/mm²]	$V_{um,test,25}$ [kN]	Variation coefficient [%]
Strengthened c-profile	39.87 36.23 36.35 37.88	37.58	28.62	35.12	4.52
Strengthened v-profile	37.26 31.74 43.70 42.85	38.89	28.62	36.35	14.29
Strengthened V- Profile with rebars	50.51 54.01 56.15 60.40	55.27	28.62	51.66	7.48

Compared to the reference tests, the post-installed strengthened anchor channels showed a noticeable increase of their mean failure load of appr. 2.88 to 3 times of the reference loads for the C- and V-profiles.

This strengthening effect is comparable to the maximal possible strengthening effect through supplementary reinforcement $V_{Rk,re,max}$, by Schmid [6].

$$V_{Rk,re,max} = 4{,}2 \cdot c^{-0{,}12} \cdot V_{Rk,c} \quad (7)$$

For an edge distance of 50 mm equation 7 states, that the maximal effect of the reinforcement would be 2.63 times $V_{Rk,c}$, where $V_{Rk,c}$ is the possible load of the unreinforced anchor channel.

In figure 8, a deformation of the anchor channel through bending could be observed, while the failed anchor channels of the reference tests moved along the concrete cone with no bending. This indicates that the desired load transferring effect to the strengthening module by bending was achieved.

Figure 8. Failed strengthened anchor channels v-profile (left) and c-profile (right)

3.2 Evaluation of the test results at serviceability load levels

When comparing the results for defined load levels, limited by serviceability, the crack width over the loads must be considered. The test results in figure 9 show the comparison of the loads for a crack width of 0.2, 0.3 and 0.4 mm. The crack widths are chosen according to [1] as recommended maximal crack widths for prestressed concrete, limit of serviceability in normal environments, and limit of serviceability in dry environments.

C-profiles, where the annular gap between the strengthening module and the fastener was not filled, show a strengthening effect for an increasing crack width of appr. 40 % to 60 % compared to the reference values (see figure 9 left). V-profiles, where the annular gap was filled with fischer FIS V, showed a load increasing effect of 80 % to 104 % of the reference loads (see figure 9, grey scale on the right).

This difference of the load increasing effect is assumed to be caused by uncontrolled slipping of the strengthening module through the loading process. Through this slipping the C-Profile gets into contact with the surrounding concrete before the module could be fully activated, causing an earlier cracking of the concrete.

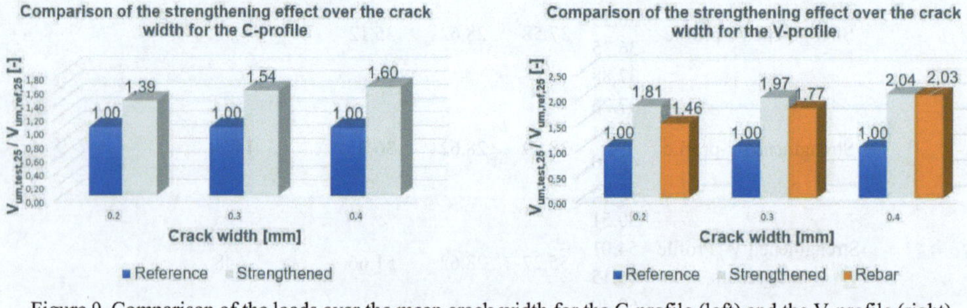

Figure 9. Comparison of the loads over the mean crack width for the C-profile (left) and the V-profile (right).

Compared to the strengthening module using reinforcement, presented in figure 4, the load increasing effect of the post-installed strengthening module for the crack widths of 0.2 and 0.3 mm is 20 % to appr. 40 % higher. At a mean crack width of 0.4 mm the loads are on a comparable level.

This indicates that the tested post-installed strengthening module prevents cracking up until yielding of the module. As expected, the failure load of the module using reinforcement is higher at larger crack widths after leaving the serviceability load level, while the post-installed strengthening module, is most effective prior to the crack opening respectively at smaller crack widths.

4 NUMERICAL INVESTIGATION TO DETERMINE THE INFLUENCE OF VARIABLE POSITIONS OF THE STRENGTHENING MODULE AND THE EFFECTIVITY OF BIGGER CHANNEL PROFILES.

4.1 Evaluation of the numerical investigation

For the numerical investigation conducted for this paper, only symmetrical cases were modelled. Therefore, the position of the strengthening module was placed on both sides of the loading point within the distance *s* according to Table 2. Therefore, only half of the system could be used in the numerical simulations to decrease the time needed for the FEM analysis.

The results of the numerical investigation were compared with the shear load resistances calculated according to equation 8 as suggested by Schmid [6].

$$V_{um,c} = V^0_{um,c} \cdot \alpha_{s,V} \cdot \alpha_{c,V} \cdot \alpha_{h,V} \tag{8}$$

The mean failure loads $V_{um,c}$ for all conducted tests of the numerical investigation given in Table 2 and 3 have been calculated with equation 8. For the following evaluation of the obtained strengthening effects those calculated loads are called V_{calc} while the results of the numerical investigation will be called V_{Sim}.

In figure 10 The influence of the anchor channel profile on the strengthening effect of the module is shown. The results of the FE investigation show, that the mean strengthening effect for the modelled profiles 40/22 and 50/30 are both by appr. 2.3 times the load of the unreinforced channels. Therefore, it can be stated that the influence of the tested anchor profiles on the effectivity of the strengthening module is neglectable.

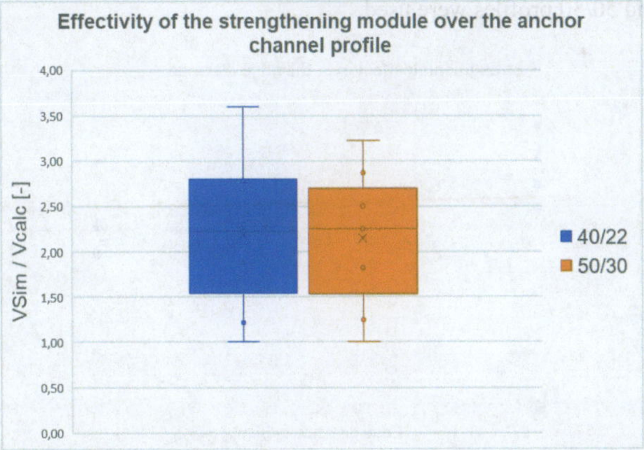

Figure 10. Influence of the anchor channel profile on the strengthening effect.

The authors assumed, that with an increasing distance *s* of the post-installed strengthening module to the position of loading, the effectivity of the system should decrease. The results of the numerical investigation confirmed this assumption. For both anchor channel profiles, the strengthening effect close to the loading plate, 35 mm to the bolt screw, was appr. 2.5 times. In contrast to that at 120 mm, the strengthening effect was neglectable. Figure 11 shows the trend of the strengthening effect depending on the distance of the module from the loading point between 35 and 120 mm.

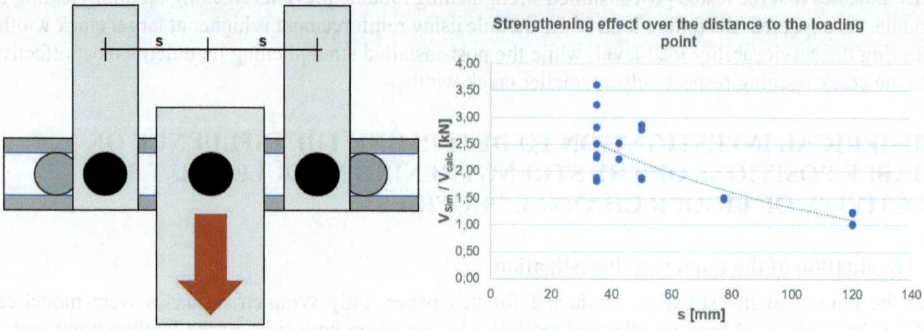

Figure 11. Influence of the distance s between the loading point and the strengthening module on the strengthening effect.

The last part of the numerical investigation was to determine the effectivity of the strengthening module with an increasing edge distance. As it can be seen in equation 7, the effectivity of normal supplementary reinforcement decreases with an exponent of $c^{-0.12}$ up to the point, where the concrete edge failure won't be the decisive failure mode anymore.

Since the post-installed strengthening module was designed for the area, where normal supplementary reinforcement would not reach the minimal bond length, the edge distance was only variated from 50 mm to 75 mm. In figure 12 the strengthening effect as a function of the edge distance, compared to the maximal possible value according to equation 7, can be seen. For this comparison, it must be considered, that the minimal bond length for normal reinforcement would most likely not be reached. Therefore, the load increasing effect according to the standards could not be used for design.

Like normal reinforcement, the post-installed strengthening module has a decreasing effect with an increasing edge distance. For this evaluation, only the strengthening modules close to the loading point, at 35 mm, for the 40/22 and 50/30 profiles were used.

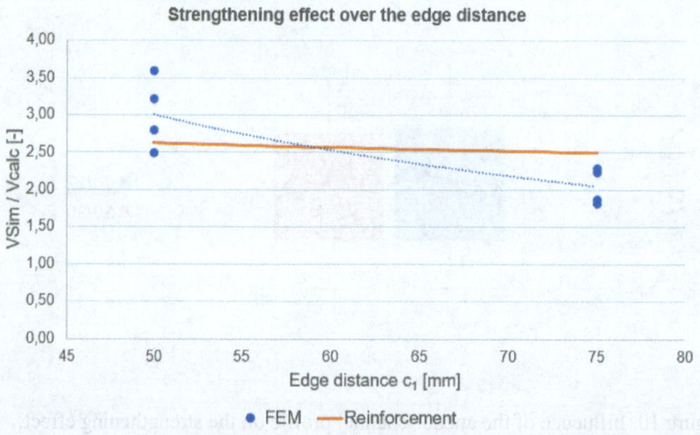

Figure 12. Comparison between the strengthening effect over the edge distance c_1 for the post-installed module (dotted line) to the maximal effect of a reinforcement (full line).

As it can be seen, the strengthening effect of the post-installed module has an effectivity of appr. 3 times at 50 mm edge distance while the normal reinforcement is at 2.63 times. Which is a difference of appr. 15 % towards the post-installed module. At a larger edge distance of 75 mm the post-installed module has a strengthening effect of 2.1 while the reinforcement would be at 2.5, which is a difference of 18 % towards the maximal effect that can be provided by reinforcement according to [5].

5 CONCLUSIONS

Experimental investigations were performed to determine the effectivity of strengthening modules for anchor channels that interact with the anchor channel. The numerical and experimental study was performed with C- and V-shaped strengthened anchor channels with an edge distance c_1 of 50 mm. The tests have been evaluated for the failure load and the cracking behaviour. For this reason, the load levels at 0.2, 0.3 and 0.4 mm crack width were determined. Also, a comparison with the strengthening method using a steel plate with welded reinforcement has been performed, to determine the effectivity of the new system compared to already existing and used systems.

In a second part, the results of the experimental tests were used to calibrate a numerical model for further investigation on the influence of the position, edge distances and larger anchor channel profiles.

Based on the results of the study, following conclusions can be drawn:

- The failure loads of anchor channels in unreinforced concrete using the post-installed strengthening module are increased by a factor of 2.6 to 3.0 times. This strengthening effect is comparable with the calculation using equation 7 that limits the maximal strengthening effect of channels using supplementary reinforcement.
- The annular gap between the fastener and the strengthening module had no significant influence on the failure load, but on the load limited by the serviceability level.
- The load evaluated over the crack width of anchor channels with the strengthening module and filled annular gap showed an increase of 20 to 40 % compared to the module using reinforcement.
- The assumption, that the load is transferred by bending of the channel could be confirmed.
- The numerical investigation indicates that the size of the profile between 40/22 and 50/30 will have no influence on the strengthening effect.
- The effectivity of the strengthening module decreases with an increasing distance to the load point at the channel. As of now it seems for $s \geq 120$ mm the strengthening effect is nearly zero.
- The effectivity of the strengthening module decreases with an increasing edge distance c_1.

REFERENCES

[1] "DIN EN 1992-4, Eurocode 2 - design of concrete structures. Part 4, Design of fastenings for use in concrete." *Deutsche Fassung EN 1992-4:2018. Berlin: Beuth Verlag GmbH, April 2019.*

[2] Petrasch, M; „Bewehrungselemente für Ankerschienen", Dissertation, in preparation.

[3] EOTA.: "Technical Report 047 (EOTA TR 047). Calculation method for the performance of anchor channels.", September 2015, amended March 2018.

[4] Kraus, J.: "Tragverhalten und Bemessung von Ankerschienen unter zentrischer Zugbelastung", *Dissertation, University of Stuttgart*, 2002.

[5] DIBt: „ETA 18/0862 fischer Anchor Channel FES with fischer Channel Bolts FBC", https://www.dibt.de/pdf_storage/2020/ETA-18%210862%288.06.01-15%2120%29e.pdf, 16.06.2020.

[6] Schmid, K.: „Tragverhalten und Bemessung von Befestigungen am Bauteilrand mit Rückhängebewehrung unter Querlasten rechtwinklig zum Rand", *Dissertation, University of Stuttgart, 2010*

9TH INTERNATIONAL CONFERENCE ON COMPOSITE CONSTRUCTION IN STEEL AND CONCRETE
M. Knobloch, U. Kuhlmann, W. Kurz, M. Schäfer. (Eds.)
Online, Germany, July 27-29, 2021

STUDIES ON STEEL-TO-CONCRETE JOINTS WITH CONCENTRATED LOADING CONDITIONS

Maximilian Ziwes*, Jakob Ruopp* and Ulrike Kuhlmann*

* Institute of Structural Design, University of Stuttgart
e-mails: maximilian.ziwes@ke.uni-stuttgart.de, sekretariat@ke.uni-stuttgart.de

Keywords: Steel-to-concrete joints, headed studs, reinforcement, anchor plates, edge influence

Abstract. *In mixed buildings of steel and concrete an economical design depends on intelligent solutions for steel-to-concrete joints. The design and verification of steel-to-concrete joints is challenging, as the structural behaviour and the properties of different materials have to be considered. Large forces often have to be transferred in localized areas, e.g. bridge bearings, and various failure modes of steel and concrete may occur in the joints due to load concentrations. An imposed load preventing uplift and a well-placed reinforcement can strengthen these joints decisively. In this contribution, experimental and numerical investigations and recent approaches for the design of steel-to-concrete joints are presented in order to extend and harmonise current rules leading to new chances of optimisation.*

1 INTRODUCTION

Mixed buildings of steel and concrete allow to utilize the best performance of the structural materials and therefore an economical design. However the design of the joints between the different materials is often a point of discussion. Different failure modes may occur in these joints and have to be considered within the design for steel and concrete components. In case of column base plates in strip foundations, bridge bearings or on top of columns forces have to be transferred in a localized concentrated area and concrete failure mechanisms such a concrete edge failure may become decisive. Conservative assumptions for the edge influence or the contribution of supplementary reinforcement in current rules only allow for small load carrying capacities or require high edge distances. Recent research has shown that a holistic approach for the design of these joints is possible, if the failure mechanisms of the anchorage according to the fastening technology are integrated in the design concept of the component method for steel and composite structures. An economic and in comparison to concrete structures competitive design of steel and composite structures is possible, as the load-carrying behaviour in the concrete part can be captured. In the frame of this paper current research results [1], [2] are presented, where tests were conducted to develop an analytical model for steel-to-concrete- joints under concentrated loading conditions. An increase of the load carrying capacity can be reached, if the concrete components are strengthened, by considering load distributions due to edge influences and the contribution of reinforcement. A first step for standardization and implementation of the fastening technique in the design rules for these joints has partly been realized in prEN 1993-1-8 [3]. This is be presented in the contribution.

2 STATE OF THE ART

Steel-to-concrete joints may be designed according to current guidelines for steel, composite or concrete structures (see Figure 1). For steel and composite structures the design of these joints is based on the component method [4], [5] according to EN 1993-1-8 [6] and EN 1994-1-1 [7]. Within this approach the joint is subdivided into its different components, e.g. the T -stub model for the anchor plate (Figure 1 a)) or the web panel and flange in tension/shear for the steel column (Figure 1 b)), and the load bearing capacity of the joints is defined by the component with the smallest load bearing capacity. The component method may only be used for steel-to-concrete joints, if a failure of the concrete components can be excluded and the steel components become decisive. This is not the case for most base plates with anchors of low embedment depth or small edge distances. In addition, the influence of a grout layer between anchor plate and concrete is not addressed in the design rules [6], [7].

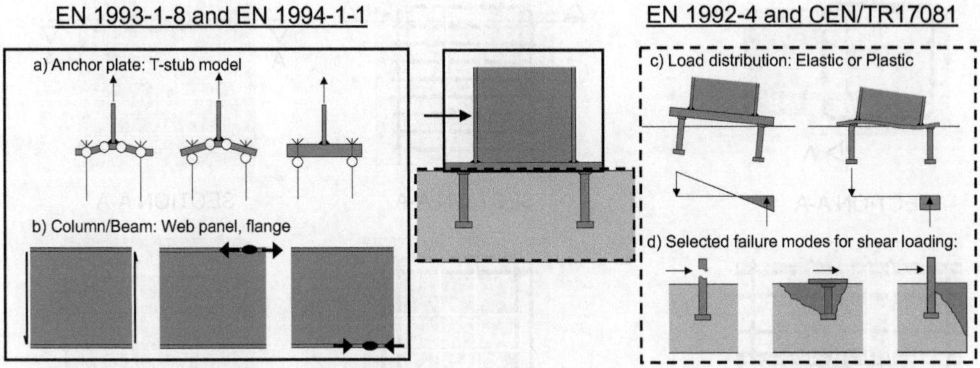

Figure 1: Steel and concrete components for column base plates acc. to current design rules

The concrete components may be designed according to EN 1992-4 [8] based on the CC-method [9]. In these guidelines different failure modes for anchors in concrete under tension and shear are defined (see Figure 1 d)). Different parameters such as edge influence, influence of a grout layer, cracked and non-cracked concrete and different kinds of fasteners are considered. The use of supplementary reinforcement is covered in a conservative way as the interaction of steel and concrete components regarding the load transfer is not considered. In addition, assumptions have to be made regarding the load distribution among the headed studs. Axial forces in the anchors may be distributed between the anchor rows according to an elastic design concept assuming that all anchors have the same stiffness and that the anchor plate is sufficiently rigid or in case of flexible anchor plates according to a plastic design concept [10], see Figure 1 c).

For anchor plates under shear close to the edge assumptions for the load distribution between the anchor rows and for the crack propagation have to be made. In current rules [8] it is assumed that the crack failure starts from the anchors closest to the edge and that the embedment length of the supplementary reinforcement is defined by the assumed concrete breakout body. In this case, it is assumed that a load redistribution to the anchors remote to the edge is not possible and the bearing capacity of the joint is determined in a conservative way by the rules.

3 INVESTIGATIONS ON STEEL-TO-CONCRETE JOINTS WITH CONCENTRATED LOADING CONDITIONS

3.1 Experimental investigations

Fifty tests were conducted on anchor plates under shear loading with 2- and 4-sided edge influence (Figure 2) at the MPA (Materials Testing Institute) of University of Stuttgart to gather information about the load bearing behaviour of steel-to-concrete joints under concentrated loading conditions and to apply these results as a validation for further numerical investigations. In a second step a design model for steel-to-concrete joints was developed based on the experimental and numerical studies.

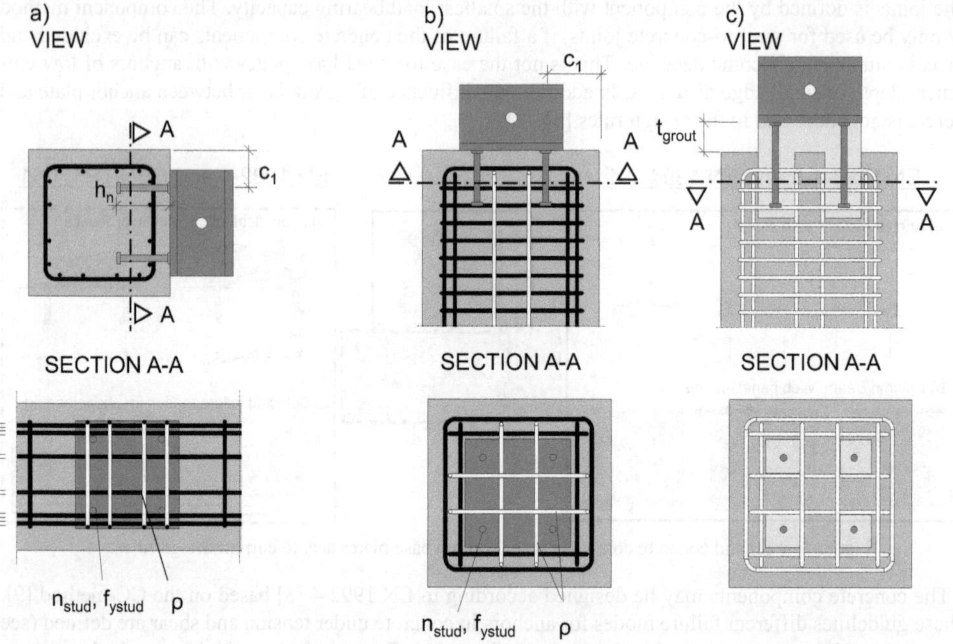

Figure 2: Test specimen for test on anchor plates with 2- sided edge influence (a) and with 4-sided edge influence without (b) and with grout layer (c)

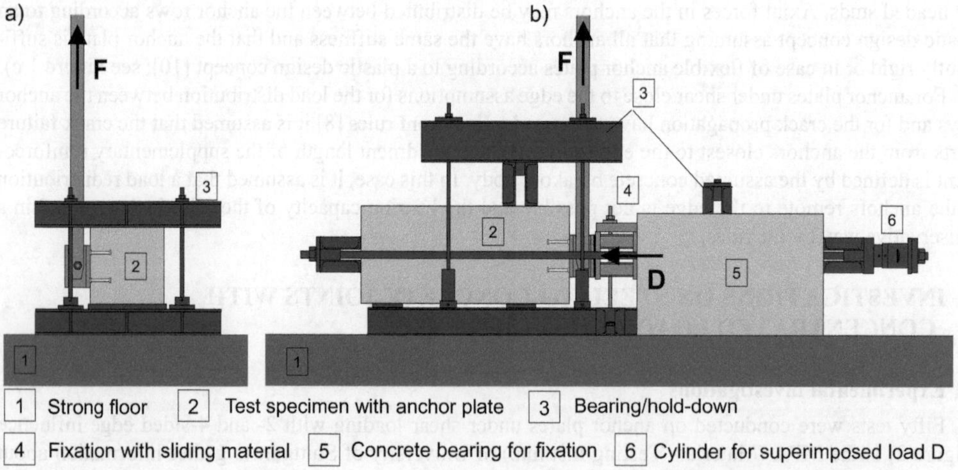

Figure 3: Test setup for tests on anchor plates with 2- sided (a) and 4-sided (b) edge influence under shear loading

Table 1: Testing scheme of shear tests on anchor plates with 2- and 4-sided edge influence

Series	Constraint/ superimposed load D[1)]	c_1 [mm]/ Edge influence [sides]	ρ [mm]	h_n [mm]	n_{stud}	t_{grout} [mm]	C_c	$f_{y,stud}$	n_{tests}
B1	B1-1/B1-2: no B1-3: fixed	100/ B1-1: 2 B1-2/B1-3: 4	1 ø12	150	4	-	C20/25	S235J2+C450	3
R2	R2-1/R2-2: no R2-3: fixed	100/ R2-1: 2 R2-2/R2-3: 4	2 ø12	150	4	-	C20/25	K800	3
R3	no	100/ R3-1(1)/R3-1(2): 2 R3-1(1)/R3-2(2):4	2 ø12	150	4	-	C20/25	K800	4
R4	R4-1/ R4-2: no R4-1-3/R4-2-3: fixed	100/ R4-1: 2 R4-2/R4-1-3/ R4-2-3:4	2 ø12	100	4	-	C20/25	R4-1/ R4-2/R4-1- 3: K800 R4-2-3: S235J2+C450	4
R5	R5-1/R5-2: no R5-3: fixed	100/ R5-1:2 R5-2/R5-3: 4	2 ø12	200	9	-	C20/25	K800	3
R6	R6-1-1/R6-1-2/R6-2- 1/R6-2-2: no; R6-1-3/R6-2-3/R6-3-3/ R6-4-3: fixed	75-200/ R6-1-1/R6-2-1: 2	2 ø12	150- 200	4	-	C20/25 - C50/60	K800	8
Z1	Z1-1: fixed; Z1-2: D[1)] = 50% Z1-3: D[1)] = 100% Z1-4: D[1)] = 150%	100/ 4	1 ø12	150	4	-	C20/25	S235J2+C450	4
Z2	Z2-1: fixed; Z2-2: D[1)] = 50% Z2-3: D[1)] = 100% Z2-4: D[1)] = 150%	100/ 4	2 ø12	150	4	-	C20/25	S235J2+C450	4
Z3	Z3-1: fixed; Z3-2: D[1)] = 50% Z3-3: D[1)] = 100% Z3-4: D[1)] = 150%	150/ 4	2 ø12	150	4	-	C20/25	S235J2+C450	4
Z4	Z4-1: fixed; Z4-2: D[1)] = 50% Z4-3: D[1)] = 100% Z4-4: D[1)] = 150%	250/ 4	2 ø12	150	4	-	C20/25	S235J2+C450	4
Z5	Z5-1: fixed; Z5-2: D[1)] = 100% Z5-3: D[1)] = 150%	150/ 4	2 ø12	150	4	20	C20/25	S235J2+C450	3
Z6	Z6-1: fixed; Z6-2: D[1)] = 100% Z6-3: D[1)] = 150%	150/ 4	2 ø12	150	4	20	C20/25	S235J2+C450	3
Z7	Z7-1: fixed; Z7-2: D[1)] = 100% Z7-3: D[1)] = 150%	150/ 4	2 ø12	150	4	50	C20/25	S235J2+C450	3

[1)] [%] load bearing capacity of the respective reference test of each series

Aim of the tests was to initiate different failure modes and to investigate the load distribution between the anchor rows considering the edge influence. Therefore, the following parameters (see also Figure 2) were varied within the specified limits according to Table 1: Edge distance c_1 (75, 100, 150,200 and 250 mm), reinforcement ratio ρ between the two outer anchor rows parallel/ perpendicular to the load direction ρ (1x ø12 mm and 2x ø12 mm), grout layer between anchor plate and concrete t_{grout} (20 and 50 mm), embedment length of headed studs h_n (100, 150 and 200 mm), number of headed studs n_{stud} (2x2 and 3x3), material of the headed studs $f_{y,stud}$ (S235J2 + C450 and K800), concrete strength C_c (C20/25 and C50/60).

In all tests rigid anchor plates with a thickness of 30 mm were loaded by a shear force with an eccentricity of 100 mm. Therefore, the anchor plate was loaded in tension and shear. In several tests the anchor plate was additionally constraint by a steel block and fixed perpendicular to the loading direction (see Table 1 “fixed”) or additionally loaded by a superimposed load D that prevents the uplift of the anchor plate (see Figure 3). As a result, the headed studs were loaded by shear only. A sliding material was used to ensure a frictionless contact between steel block and anchor plate. This setup is adapted to investigate the load bearing behaviour of an anchor plate, e.g. in a bridge bearing or a loaded column. In order to identify the influence of the different parameters, the testing scheme in Table 1 was chosen in a way that reference tests had been defined from which the other tests differed by only one single parameter. Thus, the influence of the various single parameters could clearly be identified.

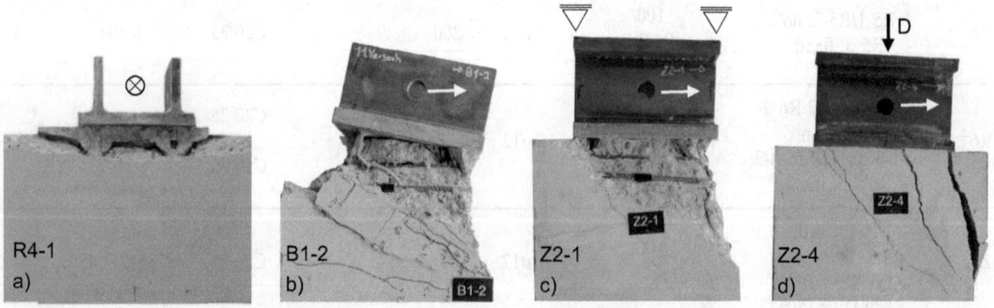

Figure 4: Different concrete failure modes of shear tests with anchor plates under edge influence:
a) Concrete cone failure, b) Concrete edge/Pry-Out failure, c) Concrete edge failure, d) Concrete edge failure with restrained crack propagation due to a superimposed loading

Different failure modes could be observed, mainly influenced by the edge distance, constraint of the anchor plate or the superimposed load and the material of the headed studs (Figure 4). The observed failure modes matched with the concrete failure modes according to the design rules of fastenings [8]. In most of the tests a concrete failure towards the loaded edge could be observed. The cracks propagated from the front as well as from the back row. The crack propagation of the tests indicated that the load can be distributed from the front to the back row in case of a concrete edge failure. The tests showed, that the edge distance had a strong influence on the load distribution and the failure mechanism of the joint. If a free rotation of the anchor plate was possible, a pry-out mechanism could be observed in the tests, due to the eccentric loading of the anchor plate, see Figure 4 a), b). In this case, the load bearing capacity decreased in tests with shorter embedment depth. If the uplift of the anchor plate was prevented by a fixation or a superimposed load, higher load bearing capacities could be achieved. In all tests with a fixed anchor plate and a S 235J2+C450 steel for the headed studs, a ductile failure mechanism developed by forming a plastic hinge in the headed studs at the base of the shank. In the tests with a fixed anchor plate and a high strength material of the headed studs, a shearing-off of all headed studs occurred. In the tests with an additional superimposed load, higher load bearing capacities could be achieved compared with tests with a fixation of the anchor plate. It could be observed, that the crack development happened at higher levels of the applied shear force. It may be assumed that the superimposed load compresses the concrete and prevents the cracking of the concrete, see Figure 4 c), d).

A higher load bearing capacity could be achieved in the tests with higher reinforcement ratio (Series R2, Z2) compared with tests with a lower ratio (Series B1, Z1). The crack propagation was restrained by the reinforcement and a ductile load bearing behaviour of the anchor plate could be observed. In the tests with a cracked concrete (Series R3), no significant influence of the concrete condition on the load bearing behaviour could be observed. It may be assumed that the crack propagation of the cracks induced before the test was restrained by the reinforcement.

An increase of the load bearing capacity was also reached in all tests with an increased concrete strength (Series R6). The load increase was in accordance with the estimations according to the current rules [8]. In the tests with 3x3 headed studs (Series R5) the assumption of a rigid anchor plate could be verified by monitoring the strains in the headed studs with strain gauges (SGS). An elastic load distribution of the axial forces could be observed.

For the tests with a grout layer (Series Z5, Z6, Z7), the grouting mortar PAGEL® V1®/50 was used. At the day of testing, the average strength of the mortar was $f_{cm,cyl}$ = 61 N/mm². The tests showed that by increasing the grout layer, it was more likely that the steel failure of the headed studs governed the load bearing behaviour. The load bearing capacity decreased compared with the tests without a grout layer (Series Z3). It may be assumed that the grout layer cracks and deforms at a lower load level than the enclosed concrete when increasing the load. This results in an increased loading of the headed studs by shear and bending.

3.2 Numerical investigations

The FE program MASA [11] was used for the numerical studies. The numerical model was validated based on the test results by comparing the maximum loads and the failure modes. The numerical recalculations of the tests suggested that the stirrups placed next to the headed studs are the most effective, whereas the stirrups remote from the studs are only partially loaded.

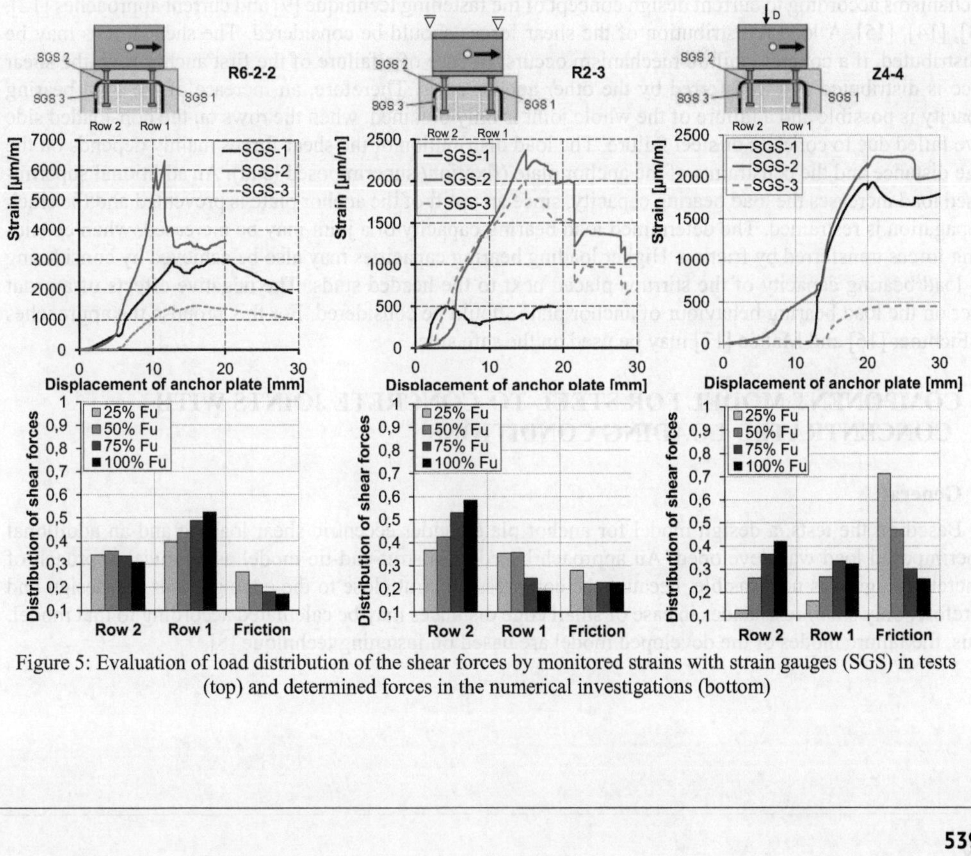

Figure 5: Evaluation of load distribution of the shear forces by monitored strains with strain gauges (SGS) in tests (top) and determined forces in the numerical investigations (bottom)

The load bearing behaviour of the joint regarding the distribution of the shear forces considering edge influences and friction forces was investigated in further numerical studies. In the tests, the load distribution was investigated by monitoring the strains of the strain gauges placed on the stirrups next to the different rows of the headed studs (Figure 5, top). In the numerical studies the shear forces transferred by friction and by the section at the base of the headed studs in the different rows were determined for different load levels and compared to the strains monitored in the tests, see Figure 5, bottom.

In the tests with a free rotation of the anchor plate and sufficient edge distance, the loads were distributed to the loaded side and mainly transferred by row 1, due to a pry-out failure mechanism (R6-2-2). In tests with short edge distances and a fixed anchor plate, the stirrups next to row 2 was highly activated (R2-3). When increasing the load level, the shear forces were distributed from row 1 to row 2, due to concrete edge failure and the shear forces were mainly transferred by row 2.

In the tests with an additional superimposed load and sufficient edge distance the shear forces were distributed equally between the anchor rows (Z4-4). The superimposed load prevented a pry-out failure as well as spalling of the concrete edge, and a steel failure of the headed studs governed the load bearing behaviour of the anchor plate. When decreasing the superimposed load or the edge distance, concrete failure mechanisms were more likely to govern the load bearing behaviour of the anchor plate. In all tests, a significant amount of shear forces was transferred by friction. Especially in the tests with a superimposed load a huge amount of shear was transferred by friction. At higher load levels the relative amount of shear transferred by friction decreased (Figure 5, bottom), however, the total amount increased, due to an increase of the resulting compressing force.

3.3 Conclusions

In the following, the main suggestions for a design model according to the presented experimental and numerical investigations are given. The failure mechanisms observed in the tests match with the failure mechanisms according to current design concept of the fastening technique [9] and current approaches [12], [13], [14], [15]. A load redistribution of the shear forces should be considered. The shear forces may be redistributed, if a concrete failure mechanism occurs. In case of a failure of the first anchor row, the shear force is distributed and transferred by the other anchor rows. Therefore, an increase of the load bearing capacity is possible and a failure of the whole joint is only obtained, when the rows on the non-loaded side have failed due to concrete or steel failure. The load distribution of the shear forces mainly depends on the edge distance and the constraints of the anchor plate (fixation/ superimposed load). An additional superimposed load increases the load bearing capacity, since an uplift of the anchor plate is prevented and the crack propagation is restrained. The determined load bearing capacity of a joint may be increased, when considering forces transferred by friction. Higher loading bearing capacities may also be achieved by considering the load bearing capacity of the stirrups placed next to the headed studs. The negative effects of a grout layer on the load bearing behaviour of anchor plate should be considered. For this propose the approaches by Fichtner [16] and Mallée [17] may be used on the safe side.

4 COMPONENT MODEL FOR STEEL-TO-CONCRETE JOINTS WITH CONCENTRATED LOADING CONDITIONS

4.1 General

Based on the tests, a design model for anchor plates under eccentric shear loading and an additional superimposed load was developed. An approach based on a strut-and-tie model as a typical approach of concrete design was not feasible, because the compression strut close to the edge governs the design and therefore, only small resistances in case of small edge distances may be calculated according to that model. Thus, the failure modes of the developed model are based on fastening technique [8].

4.2 Single components under tension und shear

The component model for steel-to-concrete joints under concentrated loading is based on existing design concepts for the failure modes of single components. The components, concrete cone, concrete edge failure, steel failure of the headed studs with and without consideration of a grout layer, pull-out and pry-out failure according to the CC-method [8], [9] are expended by approaches [12], [13], [14], [15], [16], [17] to consider a grout layer, supplementary reinforcement and additional failure modes such as concrete strut failure.

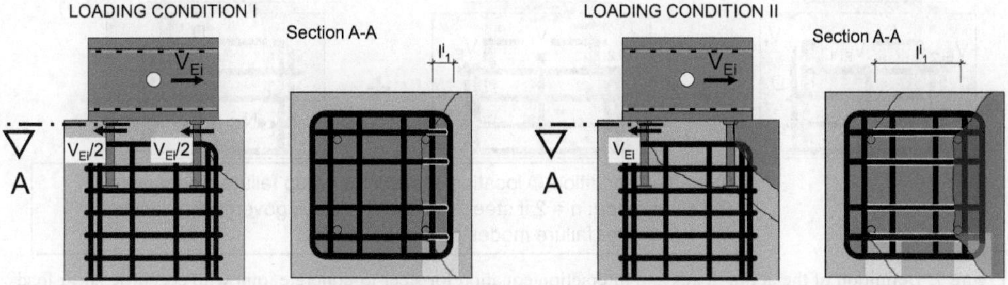

Figure 6: Distribution of shear forces and embedment lengths for loading condition I and loafing condition II in case of a concrete edge failure

The load bearing capacity of the joint can be increased, if the supplementary reinforcement placed next to the headed studs prevents a spalling of the concrete edge and the failure of the reinforcement governs the load bearing behaviour. For this propose a sufficient embedment length of the stirrups in the assumed breakout failure body of the concrete is a necessary condition, otherwise a concrete failure occurs due to an anchorage failure of the reinforcement. However, even with a sufficient embedment length of the stirrups the increase of the load bearing capacity due to supplementary reinforcement is limited by a concrete cone failure in between the stirrups and a failure of the concrete strut. This may be prevented by a sufficient embedment length of the headed studs. If the failure of the stirrups can be excluded, a common load transfer of concrete and reinforcement is considered according to [12] for the verification of the anchor rows in tension. For the verification of the rows in shear, the common load transfer is neglected on the safe side, because of an early crack propagation of the concrete, due to small edge distances.

The influence of an additional grout layer is considered by a reduction of the shear strength of the headed studs. In case of a small grout layer ($0{,}5\ d_{stud} < t_{grout} < 5d_{stud}$) the shear strength is reduced as a function of the grout thickness according to Fichtner [16]. For thicker grout layers ($t_{grout} > 5d_{stud}$) the studs in row 1 are assumed to be loaded in bending and the shear strength of row 1 depends on the moment resistance of the headed studs according to Mallée [17].

The load distribution between the different rows is considered by different loading conditions. Depending on the loading condition, forces are redistributed between the different rows and different concrete breakouts and appropriate embedment lengths can be derived, see Figure 6. In loading condition I, it is assumed that cracks propagate from the front row to the loaded edge (Figure 6, left). The shear forces are distributed equally between the different rows. In loading condition II this crack pattern is finalised by a developed breakout failure from the front row to the loaded edge (Figure 6, left). Therefore, the forces are redistributed to the back row assuming cracks, which propagate from the back row to the loaded edge. In case of a steel failure of the headed studs or a pry-out failure in loading condition I, loading condition II does not apply and a redistribution of the shear forces is not possible. If an anchor row is compressed by a superimposed load, it is assumed that a verification for pry-out failure is not necessary.

4.3 Component model

In the following, an analytical design model is proposed for steel-to-concrete joints under concentrated shear loading, which considers the load distribution and the resistances depending on the different failure mechanisms and loading conditions. For the definition of the shear components the loading of the joint is split into a bending moment for the definition of the axial forces and into pure shear forces, see Figure 7.

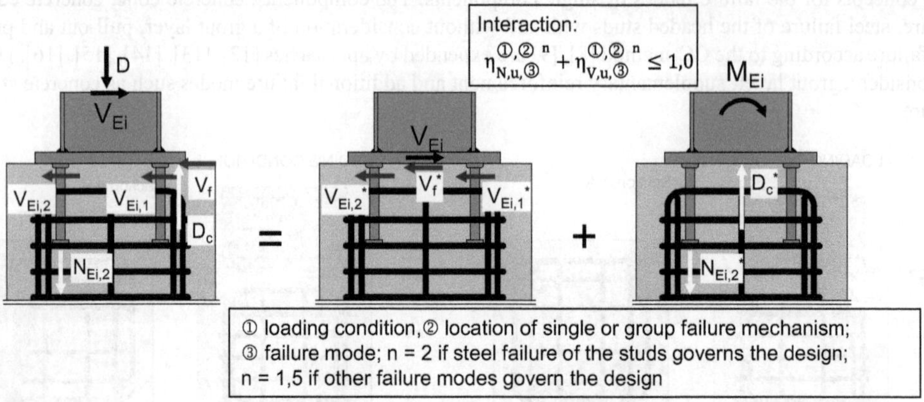

Figure 7: Definition of the acting loads and interaction equation for steel-to-concrete joint with eccentric shear loading

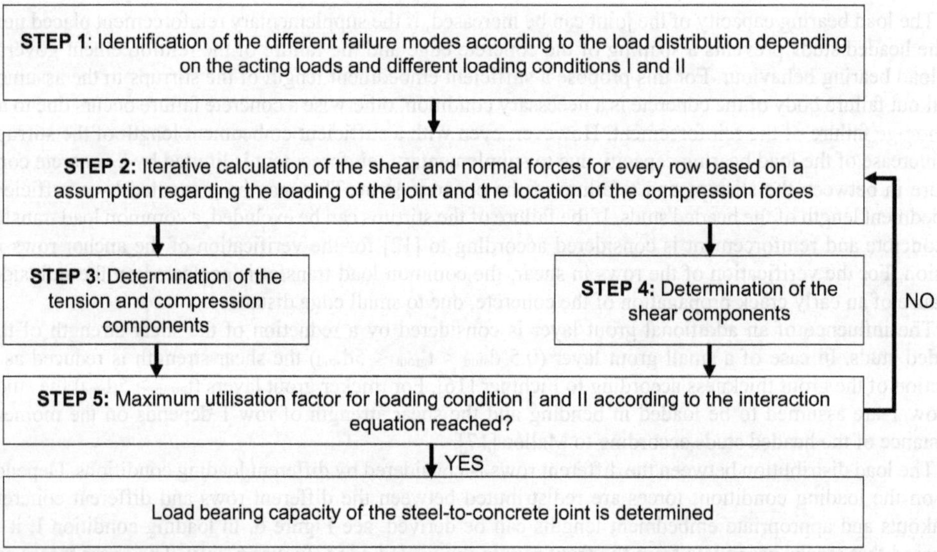

Figure 8: Flow chart for the design approach for steel-to-concrete joints [18]

This definition applies to all failure modes given in section 4.2. The shear forces transferred by friction are considered by the equilibrium of the axial forces and a friction coefficient. The axial forces are distributed between the different rows according to an elastic design concept. In case, the joint is loaded by an additional superimposed load, the superimposed load is distributed proportionally to the tensions forces in each row. The resulting axial, shear and friction forces under consideration of the superimposed load (Index "*", see Figure 7) are used for the verification. The verification for every component has to be done iteratively considering the different loading conditions and every row individually. The load bearing capacity of the joints is determined by the maximum utilisation factor according to the interaction equation given in Figure 7. For the different interactions, it has to be distinguished between different failure modes, their location (row1, 2 or 3) and single failure or group failure mechanism. Therefore, the interaction for single failure mechanisms only apply, if failure mechanisms become decisive at the same location and in the same loading condition. A simplified flow chart for the determination of the load bearing capacity of steel-to-concrete joints is given in Figure 8.

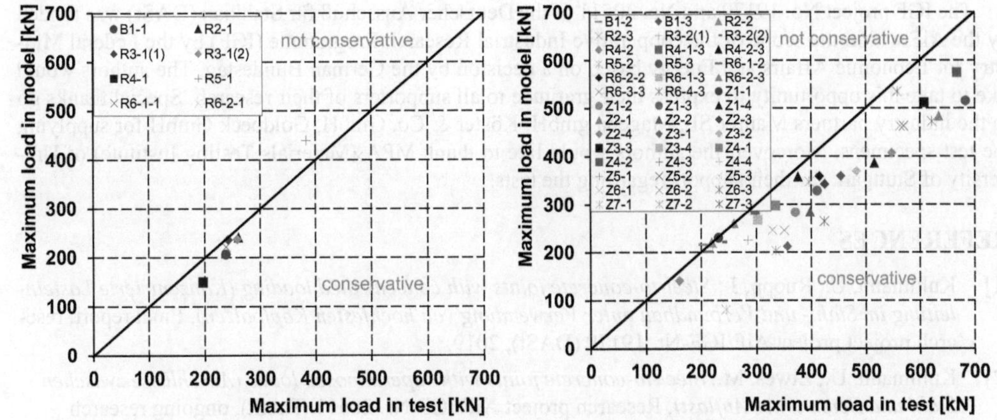

Figure 9: Comparison of maximum loads of the analytical model and the experimental results for joints with 2-sided edge influence (left) and with 4-sided edge influence (right) [2], [18]

The results from the analytical design model for steel-to-concrete joints with 2- and 4-sided edge influence for the different failure modes and load bearing capacities agree sufficiently well with the test results, see Figure 9. A derivation of design values for the component model for steel-to-concrete joints under concentrated loading is given in [18].

4.4 Implementation in current design concepts and remaining issues

The presented design model considers the contribution of the reinforcement, friction forces, a superimposed loading and the influences of a redistribution of the shear forces on the load bearing behaviour of steel-to-concrete joints in concentrated loading conditions and allows an economic and feasible design. The design model is based on the design concept of the fastening technique and its failure modes. Therefore, an easy implementation in the current design rule for fastenings in concrete [8] is possible. A first step for implementation in the design concept for joints according to the component method has partly been realized in prEN 1993-1-8 [3]. The new given opening clauses in Annex A (A. 17, A. 19) and Annex D (D. 1.1) point out, that for the design of a steel-to-concrete joints a verification of the concrete components according to the fastening technique [8] is necessary and allow a consideration of an increased resistance due to supplementary reinforcement. As a result, a holistic design concept for steel-to-concrete joints with compatible design rules considering steel as well as concrete components is possible.

However, there are some remaining issues, as described in the following. A possible reduced resistance of the anchorage, due to a grout layer is not considered in [3] and an additional verification for a steel failure of the anchorage in the grout layer - as suggested within this contribution - is not required. This may lead to an unsafe design in some cases. The presented design concept was only verified for a maximum of 3x3 anchorages. For configurations with larger anchor plates the plastic design concept according to CEN/TR17081 [10] seems a promising approach, which may be extended for joints in concentrated loading conditions. In this case, the possible flexible behaviour of the anchor plate may be considered according to the T-Stub model [3], [18]. The CC-method of the fastening technique [8] is a force-based concept, in contrast to the component method [6] considering also the stiffness and deformation of a joint. In terms of stiffness and deformation the concrete components in steel-to-concrete joints in tension without edge influences may be calculated according to [12]. An approach adjusted for joints loaded in shear was introduced in [13]. These approaches are not verified for steel-to-concreted joints in concentrated loading conditions, because of the complex load bearing behaviour due to edge influences and load redistributions.

ACKNOWLEDGEMENTS

The IGF project No. 19179 and No. 20511 of the Deutscher Ausschuß für Stahlbau (DASt) was funded by the AiF within the scope of the cooperative Industrial Research Programme (IGF) by the Federal Ministry for Economic Affairs and Energy based on a decision by the German Bundestag. The authors would like to take this opportunity to express their gratitude to all supporters of their research. Special thanks go to the industry partners Maurer SE, Mageba gmbH, Köster & Co. GmbH, Goldbeck GmbH for supplying the test specimens. Moreover, the authors would like to thank MPA (Materials Testing Institute) of University of Stuttgart, for their support regarding the tests.

REFERENCES

[1] Kuhlmann, U., Ruopp, J.: *Steel-to-concrete joints with concentrated loading (Konzentrierte Lasteinleitung im Stahl- und Verbundbau unter Verwendung von hochfesten Kopfbolzen),* Final report, research project project AiF/IGF-Nr. 19179 (DASt), 2019.

[2] Kuhlmann, U., Ziwes, M.: *Steel-to-concrete joints with superimposed loads (Anschlüsse zwischen Stahl und Beton unter Auflast)*, Research project AiF/IGF-Nr. 20511 (DASt), ongoing research 2019 – 2021.

[3] prEN 1993-1-8 (Eurocode 3): *Design of steel structures – Part 1-8: Design of joints*, Draft 2020.

[4] Tschemmernegg, F.; Huber, G.; Huter, M.; Rubin, D.: *Component method and tests for composite constructions (Komponentenmethode und Komponentenversuche zur Entwicklung von Baukonstruktionen in Mischbauweise)*, Stahlbau 66, pp. 624 – 639, Ernst & Sohn Verlag, Berlin,1997.

[5] Jaspart, J.-P., Weynand, K.: *Design of Joints in Steel and Composite Structures: Eurocode 3: Design of Steel Structure, Part 1-8 - Design of Joints; Eurocode 4: Design of Composite Steel and Concrete Structures, Part 1-1 - General Rules and Rules for Building*, ECCS Design Manual, ECCS, Brussels, 2016.

[6] EN 1993-1-8: (Eurocode 3): *Design of steel structures – Part 1-8: Design of joints*, 2005.

[7] EN 1994-1-1: (Eurocode 4): *Design of composite steel and concrete structures – Part 1-1: General rules and rules for buildings*, 2004.

[8] EN 1992-4:(Eurocode 2): *Design of concrete structures –Part 4: Design of fastenings for use in concrete*, 2018.

[9] Fuchs, W.; Eligehausen, R.: *The CC-method to determine the concrete break out load of fastenings (Das CC-Verfahren für die Berechnung der Betonausbruchlast von Verankerungen)*, Beton- und Stahlbetonbau 90, Vol. 1, pp. 6-9, Vol. 2, pp. 38-44, vol. 3, pp.73-76, 1995.

[10] CEN/TR 17081: *Design of fastenings for use in concrete – Plastic design of fastenings with headed and post-installed fasteners*, 2018.

[11] Ožbolt, J.: *MASA – Microplane Analysis Program*, Institute of Construction Materials, University of Stuttgart, 2014.

[12] Kuhlmann, U.; Hofmann, J.; Wald, F.; da Silva, L.; Krimpmann, M.; Sauerborn, N. et al.: *New Market Chances for Steel Structures by Innovative Fastening Solutions between Steel and Concrete (INFASO).* Final report, Report EUR 25100 EN, European Commission, 2012.

[13] Kuhlmann, U., Ruopp, R. Hofmann, J.: *Steel-to-concrete joints (Anschlüsse zwischen Stahl und Beton)*, Stahlbaukalender 2018, pp. 743-804, editor: Ulrike Kuhlmann, Ernst & Sohn Verlag, Berlin, 2018.

[14] Schmid, K.: *Behaviour and design of fastenings at the edge with anchor reinforcement under shear loads towards the edge (*Tragverhalten und Bemessung von Befestigungen am Bauteilrand mit Rückhängebewehrung unter Querlasten rechtwinklig zum Rand*)*, PhD thesis, Institute of Construction Materials, University of Stuttgart, 2010.

[15] Kuhlmann, U.; Ožbolt, A.: *Improvement of the load carrying capacity of anchor plates with welded headed studs in narrow reinforced concrete members (Verbesserung der Tragfähigkeit von Ankerplatten mit angeschweißten Kopfbolzen in stabförmigen Stahlbetonbauteilen),* Final report (in German), research project AiF/IGF-Nr. 17028 (DASt), 2013.

[16] Fichtner, S.: *Investigations on the load behaviour of group fastenings considering the baseplate thickness and a grout layer (Untersuchungen zum Tragverhalten von Gruppenbefestigungen unter Berücksichtigung der Ankerplattendicke und einer Mörtelschicht)*, PhD thesis, Institute of Construction Materials, University of Stuttgart, 2011.

[17] Malleé, R.: *Influence of Mortar Layer on the Shear Capacity of post-installed anchors (Einfluß einer Mörtelschicht auf die Quertragfähigkeit von Dübelbefestigungen)*, Beton-und Stahlbetonbau 99 (10), pp. 798-805, 2004.

[18] Ruopp, J.: *Studies on the Load Carrying Behaviour of Anchor Plates with Headed Studs for Normal and Shear Loads (Untersuchungen zum Tragverhalten von Ankerplatten mit Kopfbolzen zur Abtragung von Zug- und Querkräften)*, PhD thesis, Institute of Structural Design, University of Stuttgart, 2020.

PRACTICAL APPLICATIONS

9TH INTERNATIONAL CONFERENCE ON COMPOSITE CONSTRUCTION IN STEEL AND CONCRETE
M. Knobloch, U. Kuhlmann, W. Kurz, M. Schäfer. (Eds.)
Online, Germany, July 27-29, 2021

APPLICATION OF ROLLED SECTIONS IN COMPOSITE BRIDGES WITH SPAN OVER 50 METERS

M.Kożuch*, W.Lorenc*, B.Bartoszek*, A.Stempniewicz, H.Windorpski***, M.Struczyński***, R.Sęk**** and W.Ochojski*******

* Wrocław University of Science and Technology
e-mails: maciej.kozuch@pwr.edu.pl, wojciech.lorenc@pwr.edu.pl, blazej.bartoszek@pwr.edu.pl

** Fasys Mosty Sp. z o.o.
e-mails: stempniewicz-adam@o2.pl

*** Pracownia Inżynierska CREATOR Sp. z o.o. Sp. K
e-mails: h.windorpski@pi-creator.pl, m.struczynski@pi-creator.pl

**** RS Projekt
e-mails: radoslaw.sek@gmail.com

***** ArcelorMittal
e-mails: wojciech.ochojski@arcelormittal.com

Keywords: Composite, Bridges, Hot rolled sections, Composite dowels.

Abstract. *Solutions using rolled sections instead of plated beams are strongly developed in Poland [1-7]. One of the ideas is to use rolled sections HL1100 out of S460 steel for span region and variable height of girder at the piers (using different composite solutions) what has become a standard bridge type for continuous min. 3-span bridges with internal spans range 50-60 m. It took less than 10 years to come from the idea to standard market product. Appropriate R&D works have been carried out. The idea of using rolled section leaded to implementation of the double-composite action and composite dowels and this way new structural forms of bridges appeared in Poland. The driving force for innovation in composite bridges in Poland is the design-and-build formula, international collaboration, open-minded (not so conservative) young engineers and good economy for bridge market in Poland nowadays. In the paper applications of the considered solutions are presented.*

1 DEVELOPEMENT

In the year 2012 the engineers from Wrocław (Poland) suggested that it was possible to implement rolled sections HL1100 out of S460 steel into bridge engineering for spans range about 50 meters and more, creating competitive solutions to plated girders, that usually adopt S355 or S420 steel grade. Promost Wrocław design office launched R&D project with City of Wrocław [1] and designed the first bridge with span over 40 m (intermediate span of 42 m) based entirely on rolled sections (bridge in Olsztyn). In order to increase the resistance for hogging moments, double composite action was assumed in support regions (and constant construction height) – see Figure 1. That was the starting point for further investigations.

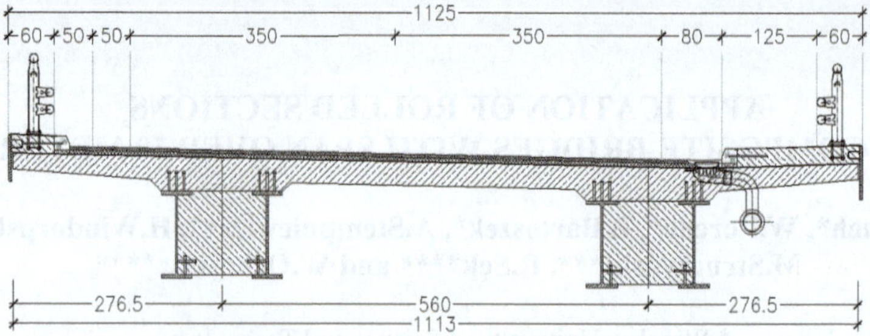

Figure 1. Cross section over the intermediate supports – bridge in Olsztyn.

It is well known that simple application of rolled sections without any adjustments is limited to certain span range (approx. 40 m) which is an effect of existing assortment of HL sections (up to HL1100). Exact limits of span ranges depend on the applied analysis method (plastic or elastic bending resistance), but in Poland old standard's requirements forced the designers to use elastic methods. If one compares bending resistance governed by structural steel yielding of an exemplary steel section alone (HL1100B), a composite steel – concrete section with slab above (assumed width of 2,00 m, thickness of 0,25 – 0,35 m), and a composite steel – concrete section in negative (hogging) bending moment, where concrete is cracked and only reinforcement 2#20/100 (2 layers of 20 mm diameter bars spaced at 100 mm between centres) cooperates with steel girder, then it can be proved that bending resistance of uncracked composite section is about 150% of that of steel alone, and cracked composite section is about 120% of aforementioned, as presented in Figure 2 (section modulus corresponding to bottom flange is calculated for each considered section).

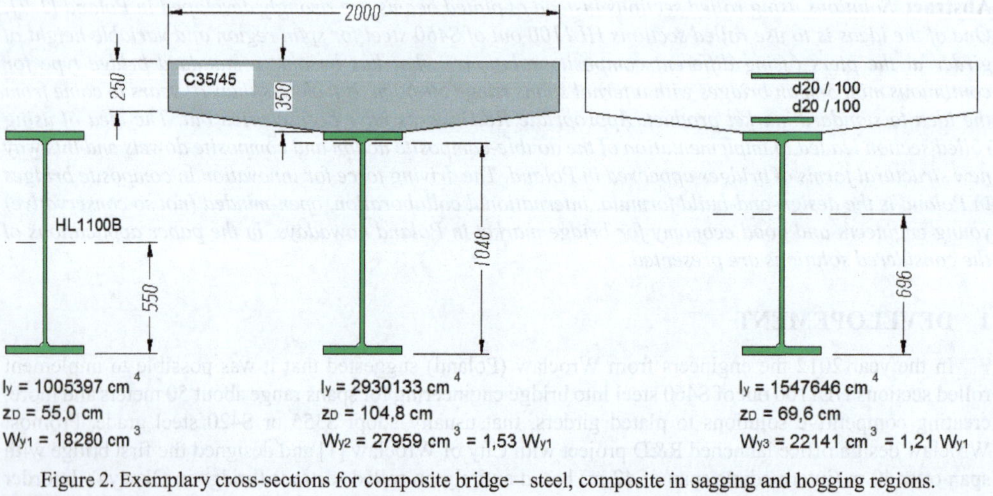

Figure 2. Exemplary cross-sections for composite bridge – steel, composite in sagging and hogging regions.

Difference between resistance of cracked and uncracked composite sections is about 20% (various for different reinforcement ratio). Assuming that 50% of loads are dead loads, which are carried only by steel beams itself in construction stages, one can state that composite beam is designed optimally through all its length when hogging bending moment is about 10% smaller then sagging one (approximately equal). In both continuous beams and frames with rigid corners hogging moments are significantly bigger then sagging ones if uniformly distributed load is considered (the longer span, the greater influence of distributed loads over concentrated ones in total bending moment, so in the first approach uniform loads may be considered) – see Figure 3.

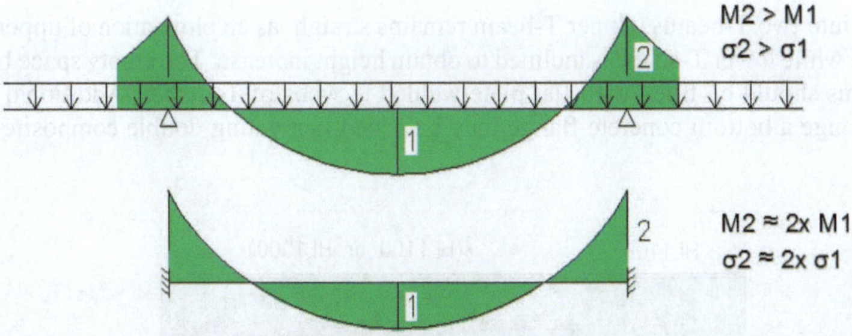

Figure 3. Bending moments and stresses comparison in mid-span and support regions for continuous beams and spans with restrained end joints (e.g. frames with rigid corners).

Thus supports of the beam should be characterized with particular stiffness, be not rigid, to obtain equal utilization of support and mid-span sections. It can be achieved by shifting supports to increase span length over 40 m and shaping support segments with much higher stiffness, what is a key aspect if rolled girders are used. Sufficient support stiffness may be achieved by:

1. double composite action in segments close to intermediate supports without changing construction height (only for small span range increase, see Figure 1, applied in bridge in Olsztyn – not considered anymore in this paper),
2. variable construction height of beam's segments close to intermediate supports,
3. both above-mentioned applied together.

The idea of this is presented in Figure 4. Length of segments with variable height and increased height itself are parameters for individual investigations in a particular project, but recent experiences proof that approx. 0,25 L is a good assumption.

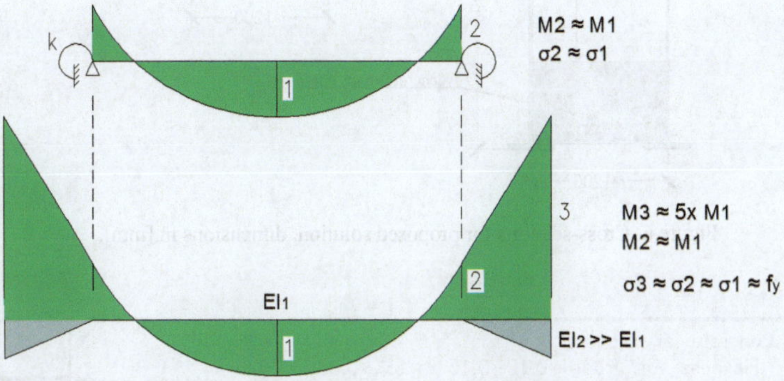

Figure 4. Bending moments and stresses comparison for beams with flexural end supports.

The last 10 years of experiences has shown that detailing of intermediate support zone can be done on a variety of ways reaching assumed goals and that a final steel consumption is not always the only determinant of the solution's quality (depending on the bridge contractor many aspects influence the financial profitability).

2 R&D AND OPOLE BRIDGE (MAIN SPAN OF 54.0 METERS)

In 2015 ArcelorMittal from Luxembourg and Pontes design office (Wrocław, Poland) started cooperation to develop an idea of application hot rolled sections in bridges with spans over 50 m [2] (report freely available). Among many possible solutions of increasing the construction height of the span in region of intermediate supports, one solution was chosen for further investigation. This solution is based on the cut along I-beam into two T-beams. Upper T-beam remains straight as an elongation of upper flange of the mid-span section, while lower T-shape is inclined to obtain height increase. The empty space between upper and lower T-beams should be filled with flat plate welded to webs of T-shapes. Additional to strengthen the lower steel flange a bottom concrete flange may be added (activating double composite action) – see Figure 5, 6 and 7.

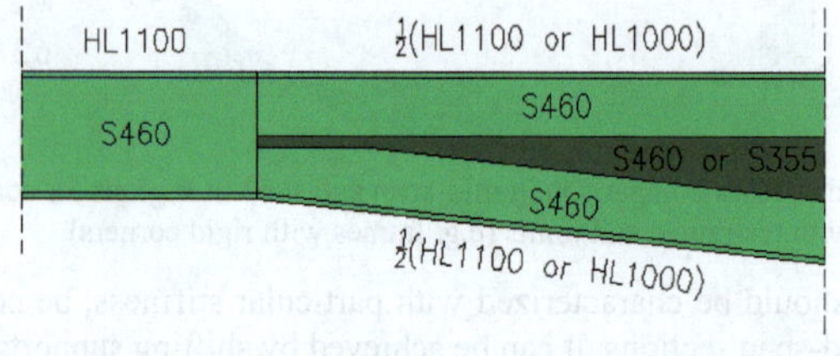

Figure 5. Side view for proposed solution, steel grade S460.

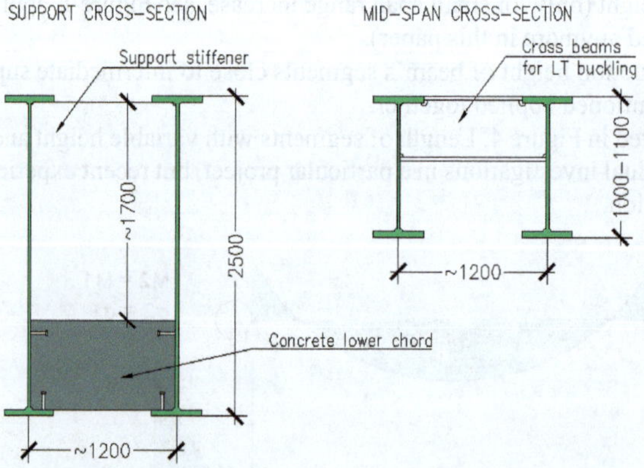

Figure 6. Cross-sections for proposed solution, dimensions in [mm].

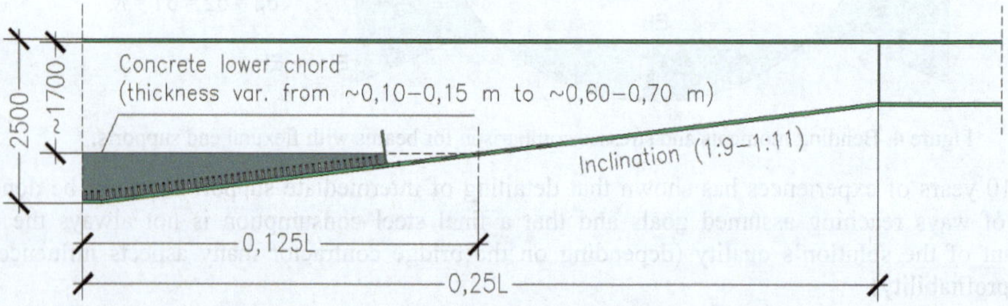

Figure 7. Recommended support zone parameters, dimensions in [mm].

Application of additional bottom concrete flange seems to be very reasonable, because:

1. increases significantly resistance and stiffness of the section,
2. ensures (or at least improves) local stability of the compressed web,
3. ensures global stability of the bottom, compressed flange of the I-beam in hogging bending moment regions.

Moreover, this is a quite cheap and easy solution. Formworks for concrete casting are not complex, and depending on the inclination and length of strengthening concrete can be casted with almost horizontal upper surface what generates its various thickness along the girder (bigger in the axis of pier where internal forces are the highest). If the bottom concrete flange is casted before top slab casting, this helps a lot in optimization of steel consumption but causes a big compressive stresses in a concrete flange itself. Then its yielding should be allowed, what brings nonlinearity into FE model. A simplification in a global analysis was proposed taking into account reduction of concrete stiffness as an effect of both its creeping and yielding – please refer to Figure 8 and equ. (1).

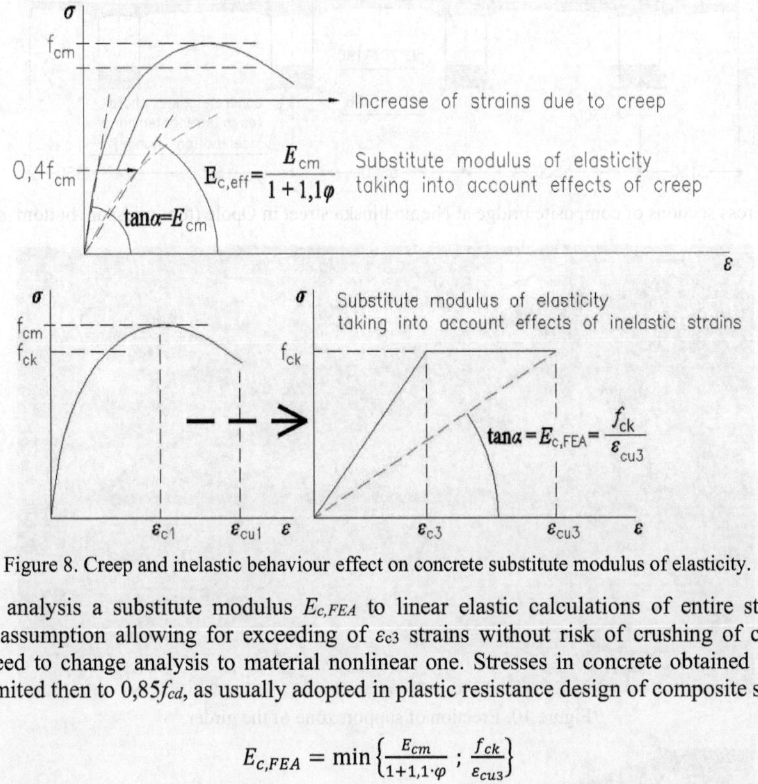

Figure 8. Creep and inelastic behaviour effect on concrete substitute modulus of elasticity.

Taking into analysis a substitute modulus $E_{c,FEA}$ to linear elastic calculations of entire structure is a simplifying assumption allowing for exceeding of ε_{c3} strains without risk of crushing of concrete and without a need to change analysis to material nonlinear one. Stresses in concrete obtained from design should be limited then to $0{,}85f_{cd}$, as usually adopted in plastic resistance design of composite sections.

$$E_{c,FEA} = \min\left\{\frac{E_{cm}}{1+1{,}1\cdot\varphi}\ ;\ \frac{f_{ck}}{\varepsilon_{cu3}}\right\} \tag{1}$$

This simplification allows to substitute a material non-linear analysis by a standard linear analysis with an assumption that is safe-sided for compressed structural steel and ensures also resistance of concrete flange under compression within its nonlinear limits. However, due to reduction of concrete stiffness it can lead to underestimation of shearing forces acting on the composite connection. Thus, it is needed to design shear connection of concrete bottom flange for full composite action. Of course, as an alternative approach, non-linear analysis can always be performed, or standard modular-ratio method acc. to EC 1994-1-1 – if final superposition of stresses in concrete flange will not exceed f_{cd}. For the shrinkage effects the same reduced $E_{c,FEA}$ modulus is used by the authors and appropriate temperature load case is solved (but other approaches can also be considered).

Proposed solutions and assumptions [2] were applied directly during design of the bridge in a city of Opole (Niemodlińska street) with spans of 42.0 + 54.0 + 42.0 m [3]. The bridge was erected in 2017, see Figures 9 – 12.

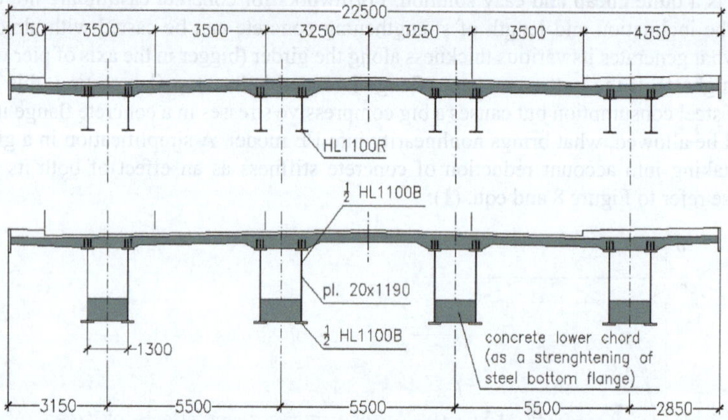

Figure 9. Cross sections of composite bridge at Niemodlińska street in Opole (top: midspan, bottom: supports).

Figure 10. Erection of support zone of the girder.

Figure 11. Girders during assembling on the site.

Figure 12. Ready bridge in Opole (Niemodlińska street)

3 RZUCHÓW BRIDGE (MAIN SPAN OF 68.0 METERS)

In the same year of 2017, a construction of a second bridge with a similar solution started next to Rzuchów (Poland) [4]. The span lengths were 24.5 + 39.7 + 68.0 + 39.6 + 25.0 + 24.8 m. Proposed solutions were slight modification of those subjected to Opole bridge, because of much longer span. Modification assumed that HL sections will be cut along asymmetrically (both for mid-span and support zone) and welded together again longitudinally into one asymmetric I-section (its web thickness differs along the web height as a result of connection 2 different rolled sections). The main span (68.0 m) was constructed using this new system (see Figures 13-14), for the side spans standard rolled sections were used, steel grade was S460.

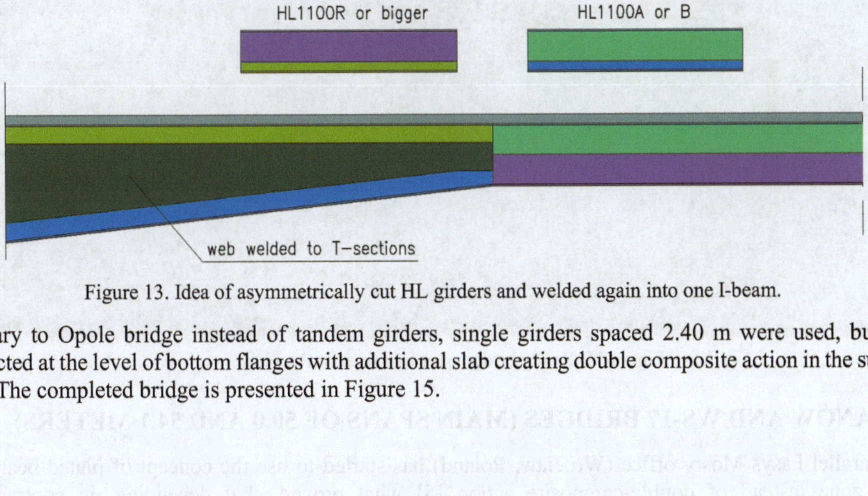

Figure 13. Idea of asymmetrically cut HL girders and welded again into one I-beam.

Contrary to Opole bridge instead of tandem girders, single girders spaced 2.40 m were used, but also connected at the level of bottom flanges with additional slab creating double composite action in the support zone. The completed bridge is presented in Figure 15.

Figure 14. Application of described idea in a bridge over San river in Rzuchów

Figure 15. Ready bridge over San river in Rzuchów

4 ULANÓW AND WS-17 BRIDGES (MAIN SPANS OF 50.0 AND 54.1 METERS)

In parallel Fasys Mosty office (Wrocław, Poland) has started to use the concept of plated beams in support zone instead of double-composite action [5] what proved, that depending on contractor's specification different solutions for negative bending zone are economically comparable. The idea was to combine hot-rolled sections for mid-span regions with ordinary welded plated girder (with transversal and longitudinal stiffeners) in the intermediate support zone, all elements made out of S460 steel. This was implemented for bridge over Tanew river in Ulanów (30.0 + 50.0 + 30.0 m), see Figure 16-17.

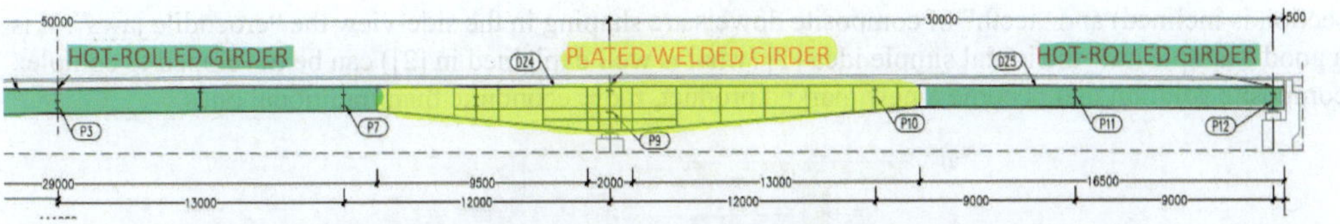

Figure 16. Idea of implementation of both rolled girders and welded ones into one bridge

Figure 17. Bridge in Ulanów

In 2020 PI Creator office (Gdańsk, Poland) designed WS-17 bridge using a similar concept. Mid-span regions were designed with rolled girders, while intermediate support zones similar to Rzuchów bridge (asymmetrical cut-along HL section to obtain 2 T-sections and weld the web in-between). Contrary to the Rzuchów bridge instead of double composite action an ordinary additional bottom flange plate was added to ensure both resistance and lateral stability of the girders in hogging moments regions, see Figure 18. Again, S460 steel grade was chosen for all main elements.

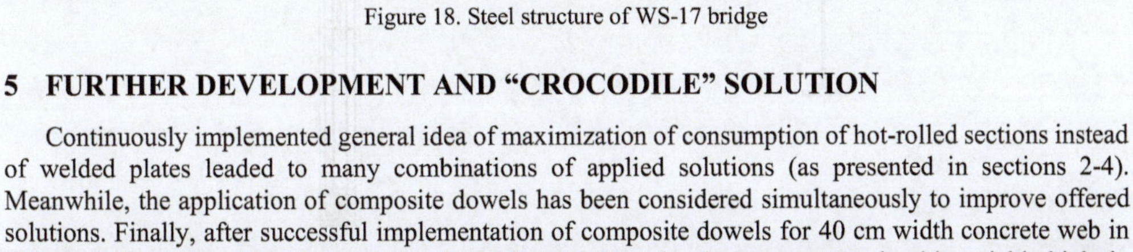

Figure 18. Steel structure of WS-17 bridge

5 FURTHER DEVELOPMENT AND "CROCODILE" SOLUTION

Continuously implemented general idea of maximization of consumption of hot-rolled sections instead of welded plates leaded to many combinations of applied solutions (as presented in sections 2-4). Meanwhile, the application of composite dowels has been considered simultaneously to improve offered solutions. Finally, after successful implementation of composite dowels for 40 cm width concrete web in railway bridge [6] and implementation of steel bottom T-section in a concrete web with variable hight in Elbląg bridge [7], the new concept, so called "crocodile", was proposed and it is being developed in Poland right now (first bridges are already offered with this solution). "Crocodile" girder consists of longitudinally cut HL section with composite dowel line (obtaining two steel T-sections with composite dowels) and the concrete web between this T-sections (see Figure 19-21, the idea of this solution have origins in an upside-down cross-section of railway bridge, as shown in Figure 19a). Variable height of the girder (bottom T-section is inclined) and "teeth" of composite dowels are shaping in the side-view the "crocodile jaws". It is a good example how the initial simple idea [1] (what is well explained in [2]) can be developed to complex composite solution that become a new marked product, more economic than traditional ones.

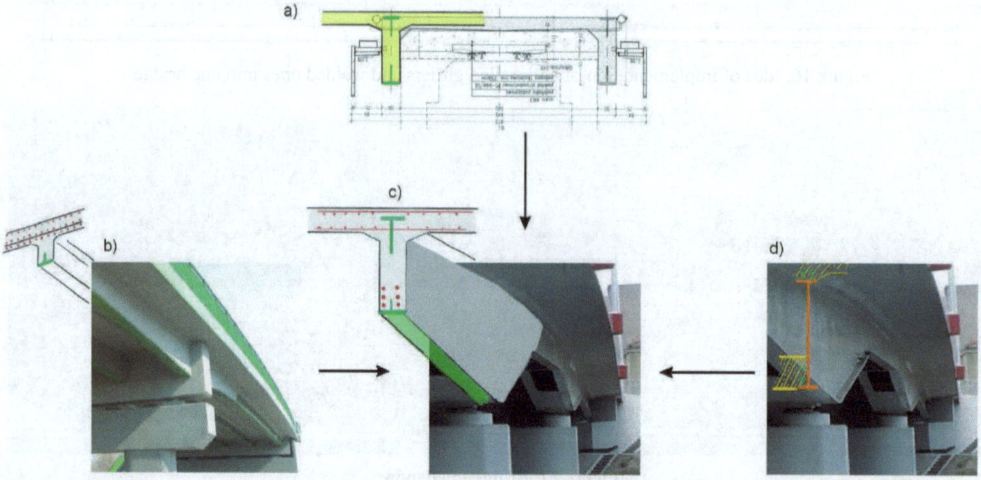

Figure 19. Chosen ideas for intermediate support zones in bridges with spans out of HL1100 sections leading to development of a "crocodile" solution: a) upside-down view of cross section of Dąbrowa bridge [6], b) Elbląg bridge [7], c) "crocodile" solution, d) Opole or Rzuchów bridge [3]

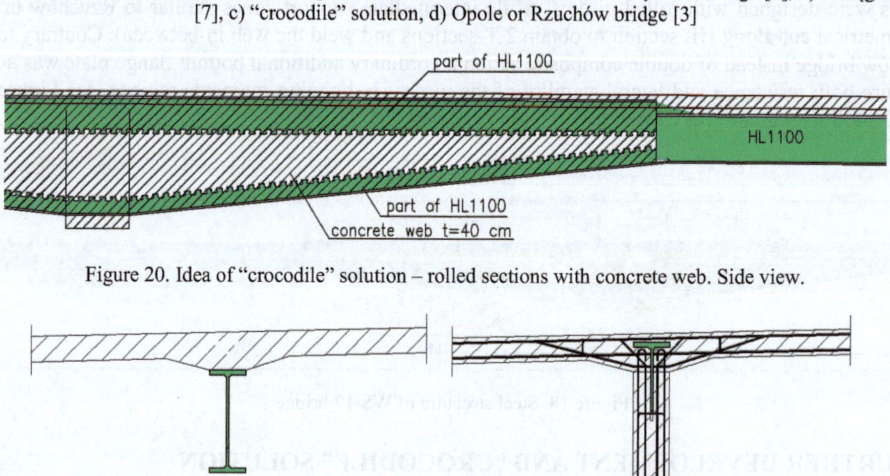

Figure 20. Idea of "crocodile" solution – rolled sections with concrete web. Side view.

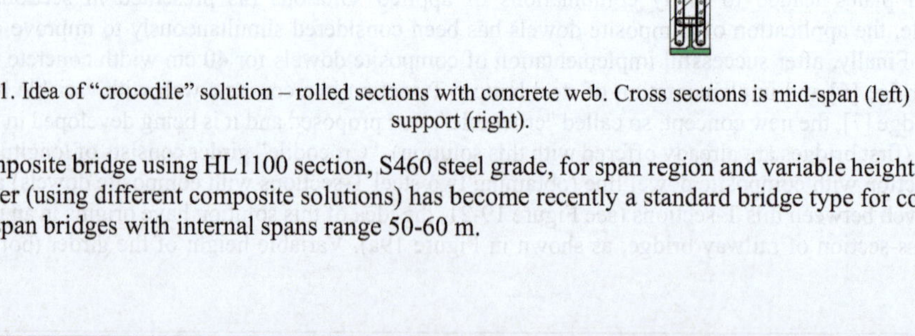

Figure 21. Idea of "crocodile" solution – rolled sections with concrete web. Cross sections is mid-span (left) and at the support (right).

The composite bridge using HL1100 section, S460 steel grade, for span region and variable height of girder at the pier (using different composite solutions) has become recently a standard bridge type for continuous min. 3-span bridges with internal spans range 50-60 m.

6 CONCLUSIONS & ECONOMIC ANALYSIS

All analysed solutions have variable construction height over a distance of approx. $0.25L$ (L – middle span range). Assuming that increase of the girder web's height is made by welded plate between cut-along HL beam and that the bridge has spans in the $(0.75 + 1.00 + 0.75)\cdot L$ system, one can calculate the additional web surface needed to construct the support zones as $(4\cdot 0.25/2)/(0.75+1+0.75) = 0.2$ of the web surface in HL beams (for 1 girder 4 support zones over 2 intermediate supports are needed, each with length of $0.25L$ and triangularly shaped). Assuming that in the case of a rolled I-section, half of the steel is in the web and half in flanges, it is possible to determine an additional relative steel consumption per welded plate of the web at 10%.

In the case of the Opole bridge, the steel consumption was approx. 483 tons for rolled beams, 40 tons for triangular sheets forming variable height of webs and 83 tons for other elements (cross members, transversal and longitudinal stiffeners, diaphragms etc.). Steel consumption for triangular sheets in relation to rolled sections is approx. 40 / 483 = 0.08, so about 10% of rolled sections. Steel consumption for the remaining elements was relatively high in this bridge and amounted to 83 / 483 = 0.17. This was due to the use of tandems. It seems reasonable to assume a 20% allowance for all additional steel elements in the bridge like in Opole but designed from single beams and using good optimization. The consumption of welded studs d22 for the upper flange of the girders was 3.6 pcs/m^2 and for the lower flange 0.8 pcs/m^2. Consumption of rolled steel of the Histar460 grade is 154 kg/m^2 for this bridge and total steel consumption 193 kg/m^2. In the Rzuchów bridge, an asymmetric method of cutting rolled I-sections was used, what resulted in the trapezoidal shaped additional web sheets. Due to the assembly and the larger span, cross members were also used in the bridge. In the further quantitative analysis, the shape of the sheets in the form of a rectangle, i.e. gross, is assumed (overestimating its contribution into steel consumption). The consumption of rolled steel was approx. 316 tons, 50 tons for web plates and 57 tons for other elements (stiffeners, cross beams etc.). 16% of additional weight for web plates (without ribs) and 18% for other elements was obtained in relation to the rolled girders. Consumption of rolled steel of the Histar460 grade was 155 kg/m^2 for this bridge and total steel consumption of 207 kg/m^2. In the Ulanów Bridge (30 + 50 + 30), the weight of the welded elements (entire support zone + cross beams, stiffeners etc.) exceeds the weight of the rolled elements almost twice, and this solution is a significant departure from the idea of the system: rolled sections complement the welded structure. Consumption of rolled steel of the Histar460 grade was only 72 kg/m^2, while total steel consumption amounted to 209 kg/m^2. The most recent example is the WS-17 bridge. In this object concrete crossbeams were used. There are trapezoidal sheets of webs and bottom flange cover plates (used instead of double composite action). In this facility, additional sheets of webs, ribs and the bottom flange amounted to 23% (10% for cover plates and 13% for ribbed web plates) of rolled steel. Consumption of rolled steel of the Histar460 grade was 139 kg/m^2, while total steel consumption reached 172 kg/m^2. The same bridge calculated as a "crocodile" solution had a total steel consumption of approx. 139 kg/m^2 (this solution was not chosen due to formal issues). It is emphasized that all the above structures were calculated in accordance with the old Polish standard, i.e. adopting elastic resistance. If plastic design would be adopted (according to Eurocode) steel consumption factors would be smaller, but the relative ratio of additional plates to rolled girders will not significantly change, therefore the conclusions are general. Steel consumption factors for considered brides are presented in Table 1.

Table 1: Steel consumption factors.

Bridge	Load class*	Steel consumption** in tons for					
		rolled sections	stiffeners, cross-beams	web plates	lower flange cover plates	total in t	total in kg / m^2
Opole	B	482,6	82,6 (17,1%)	39,7 (8,2%)	0 (0,0%)	604,9	192,6
Rzuchów	A	315,9	57,5 (18,2%)	49,8 (15,8%)	0 (0,0%)	423,2	207,1
Ulanów	A	117,2	223,3 (190,4%)		0 (0,0%)	340,5	209,4
WS-17	A	308,1	16,3 (5,3%)	25,3 (8,2%)	29,7 (9,6%)	379,4	171,6
WS-17 (crocodile)	**A**	**308,1**	**0 (0,0%)**	**0 (0,0%)**	**0 (0,0%)**	**308,1**	**139,4**

* according to PN-85/S-10030 (Polish standard)

** in brackets percentage in relation to rolled steel is indicated

The latest bridge (WS-17, now under construction) was designed based on the experience of previous constructions. According to the authors, the indicators from this bridge are a good point of reference for further optimization of this type of structure. The analysis of the above shows that the steel consumption for ribbed steel webs can be estimated as approx. 15% of rolled steel, and the use of welded cover plates instead of bottom concrete results in an additional steel consumption of 10%. The decision to design cover plates instead of a double composite action was dictated by the owner – the authors consider the first solution to be more justified.

As a general conclusion, it may also be stated that the application of steel sections with composite dowels create a wide range of possible solutions for shaping the bridge girders. The WS-17 "crocodile" bridge is just an example of possible optimization.

REFERENCES

[1] Lorenc W., Mozart Project (online: *https://wca.wroc.pl/dr-inz-wojciech-lorenc-promost-wroclaw-sp-z-oo-mosty-zespolone-o-rozpietosci-przesel-ponad-50-metrow-z-wykorzystaniem-ksztaltownikow-walcowanych-o-duzych-wymiarach)*

[2] Lorenc W., Kożuch M., Innovative design of composite bridges for medium spans based on rolled sections (online: *https://sections.arcelormittal.com/repository2/Sections/Steel%20Research%20 Reports/5_Innovative_design_Tandem_solution_AM.pdf)*

[3] Kożuch M., Lorenc W., Rabiega J., Ochojski W., „Technologia budowy mostu drogowego o rozpiętości przęsła powyżej 50 m z dźwigarów walcowanych", *Mosty: przemiany w projektowaniu i technologiach budowy: Seminarium Wrocławskie Dni Mostowe 2017*, 229-237, 2017.

[4] Sęk R., Wyrwas P., Dzida Ł., Tokarczyk D., Ochojski W., Lorenc W., Kożuch M., Przęsło z podwójnym zespoleniem o rozpiętości 68 m z zastosowaniem rozcinanych belek walcowanych. *Mosty hybrydowe: Seminarium Wrocławskie Dni Mostowe 2018*, 127-133, 2018

[5] Lorenc W., Nowe technologie budowy mostów zespolonych. *Mosty hybrydowe: Seminarium Wrocławskie Dni Mostowe 2018,* 101-118, 2018.

[6] Bartoszek B., Stempniewicz A., Ochojski W., Adamczyk A., Natonek G., Lorenc W., Railway bridge in Dąbrowa Górnicza using composite dowels: new system development of composite railway bridges. *Mosty a środowisko: Seminarium Wrocławskie Dni Mostowe 2019,* 447-464, 2019.

[7] Rademacher D., Ochojski W., Lorenc W., Kożuch M., Advanced solutions with hot-rolled sections for economical and durable bridges. *Steel Construction*, vol. 11, no. 3, 196-204, 2018.

9TH INTERNATIONAL CONFERENCE ON COMPOSITE CONSTRUCTION IN STEEL AND CONCRETE
M. Knobloch, U. Kuhlmann, W. Kurz, M. Schäfer. (Eds.)
Online, Germany, July 27-29, 2021

DESIGN, CALCULATION AND CONSTRUCTION WORK OF A PRESTRESSED COMPOSITE CONSTRUCTION TO SUPPORT THE FAÇADE COLUMNS OF A HIGH-RISE BUILDING

U. Breuninger*, J. Landsberger*

* Weiske und Partner Beratende Ingenieure VBI GmbH
e-mails: ulrich.breuninger@weiske-partner.de, jonas.landsberger@weiske-partner.de

Keywords: Composite Structural Elements, Practical Application, Codification, Innovative Structures, High-Rise Building.

Abstract. *The publication describes the design, the structural analysis and the execution of two similar, high-rise residential buildings in Frankfurt on the Main. The high-rise buildings have 15 respectively 17 floors above ground. The socle of the high-rise buildings contain the entrance (ground floor) and the building technology (1st and 2nd floor). The other 12 respectively 14 floors serve only residential purposes. The residential floors of the buildings protrude over the three-storey high socle. The façade columns of the high-rise buildings are existing only in the residential floors. Their load must be transferred to the socle. This load transfer is made possible by the all-around cantilevered, prestressed composite construction above the socle.*

1 INTRODUCTION

In 2016 B&L Real Estate, the owner of the project \`Hafenpark Quartier Süd´ in Frankfurt on the Main, launched an architectural competition for hotel- and residential utilisation. This competition was won by the architect Hadi Teherani. The competition contribution includes one hotel building and 11 residential buildings. They form a block of buildings and are completely underlaid by a two-levelled underground garage (Figure 1).

Figure 1. Project \`Hafenpark Quartier Süd´, hotel and residential buildings (B&L Gruppe/Panoptikon)

The essential aspect of the competition contribution are two high-rise residential buildings (component 7+8, Figure 2) on the south side of the block. Component 7 has 15 overground floors, component 8 has 17. Both buildings have 12 respectively 14 residential floors, in their layout considerably reduced, above the socles (ground floor, 1st and 2nd floor). The residential floors form therefore an all-around protruding, elevated structure above the socle. The two high-rise buildings are, apart from their number of floors, very similar. This publication refers generally to the higher component 8.

Figure 2. Project ´Hafenpark Quartier Süd´, component 7 und 8 (B&L Gruppe/XOIO)

2 SCHEMATIC DESIGN OF THE BUILDING

2.1 Structure of the regular floors

The raised residential floors are designed as reinforced concrete skeleton constructions. The reinforced concrete slabs relieve their load through façade columns, internal columns and the staircase and lift core. The staircase and lift core also are bracing the building.

In the basement levels as well as from the ground floor up to the 2nd floor, the socle is being arranged as the load-bearing element of the building. Inside the socle is the staircase and the lift core. All vertical interior columns of the residential floors are in their layout placed in a way where they can release the vertical load directly onto the socle of the building. Figure 3 depicts the simplified layout plan of the residential floor and the outline of the building socle of component 8.

The reinforced concrete skeleton construction was chosen to allow the flexibility of the building regarding the flat outlines. The internal flat walls as well as the partition walls don´t hold any load. Their position is therefore adjustable. Fixed spots do however result from the ducts for supplies and disposal. They establish the positions of sanitary and kitchen areas in the layout.

The ceilings of the standard storeys are built with hollow bodies, by the Cobiax-System, to minimize their weight. Integrated into the flat slab are additional air supply and exhaust pipes, sprinkler pipelines and electrical lines. Areas near columns and the core walls must be kept clear of hollow bodies and pipes to control the punching shear forces of the slab. This requires a deepened alignment of HVACR engineering and structural engineering.

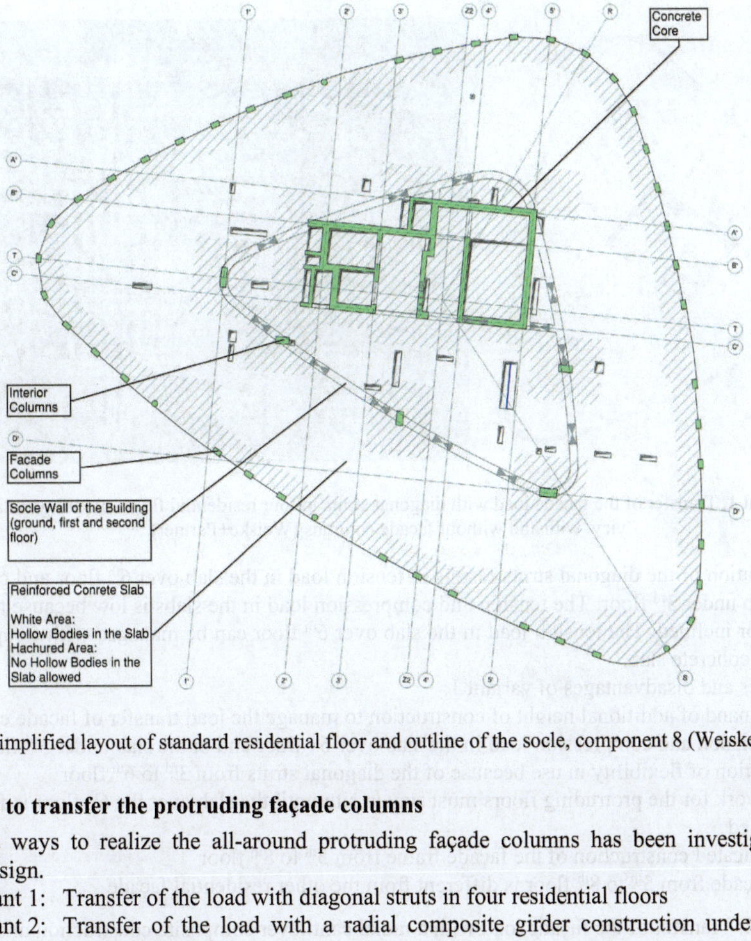

Figure 3. simplified layout of standard residential floor and outline of the socle, component 8 (Weiske+Partner)

2.2 Variants to transfer the protruding façade columns

Different ways to realize the all-around protruding façade columns has been investigated in the schematic design.

- Variant 1: Transfer of the load with diagonal struts in four residential floors
- Variant 2: Transfer of the load with a radial composite girder construction under the lowest residential floor

2.2.1 Variant 1: Transfer of the façade load with diagonal struts in four residential floors

11 diagonal struts are arranged in the lowest four residential floors (3rd to 6th floor) of the building. The façade columns must form a frame with the façade girder from the 5th to 8th floor. This frame concentrates the façade load to the 11 diagonal struts. The diagonal struts transfer the façade load to the socle of the building. The façade columns from the 3rd to the 6th floor are tension columns. Figure 4 shows the main principle of variant 1 in a ground view and in isometric views.

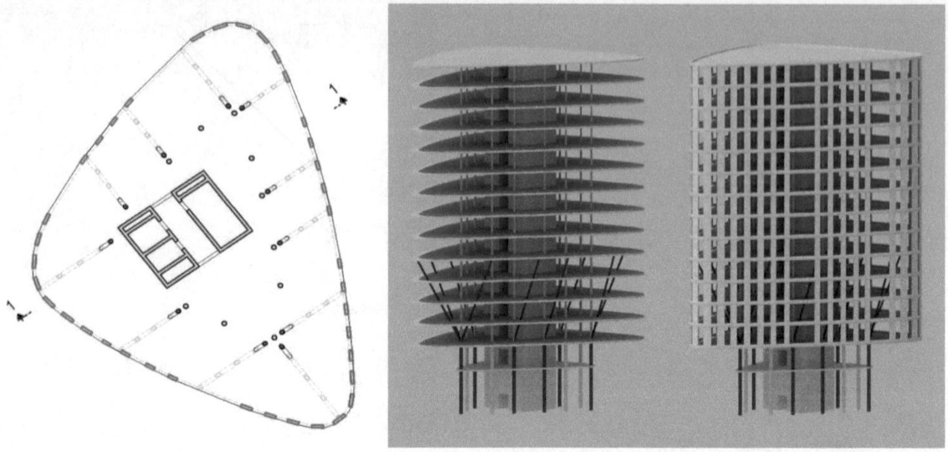

Figure 4. Variant 1, Transfer of the façade load with diagonal struts in four residential floors, floor layout and isometric view with and without façade columns (Weiske+Partner)

The inclination of the diagonal struts results in tension load in the slab over 6th floor and compression load in the slab under 3rd floor. The tension and compression load in the slabs is low because the diagonal struts are minor inclined. The tension load in the slab over 6th floor can be managed with a supplement in the reinforced concrete slab.

Advantages and disadvantages of variant 1:

\+ No demand of additional height of construction to manage the load transfer of façade columns
\+ Low tension and compression load in the slab over 6th floor and in the slab under 3rd floor
– Limitation of flexibility in use because of the diagonal struts from 3rd to 6th floor
– Formwork for the protruding floors must stay in situ until the slab over the 6th floor is finished and hardened
– Complicated construction of the façade frame from 5th to 8th floor
– The façade from 5th to 8th floor is different from the other residential façade

2.2.2 Variant 2: Transfer of the façade load with a radial cantilever composite construction under the residential floors

An all-around cantilevered and radial support grid is arranged above the socle of the building. The radial, haunched steel girders are connected to the concrete slab under the 3rd floor to a composite construction. The maximum static height of the grid is 1.5 m. An edge beam is arranged under the façade columns. It must transfer the load to the façade columns to the radial support grid. Figure 5 shows the main principle of variant 2 in isometric views.

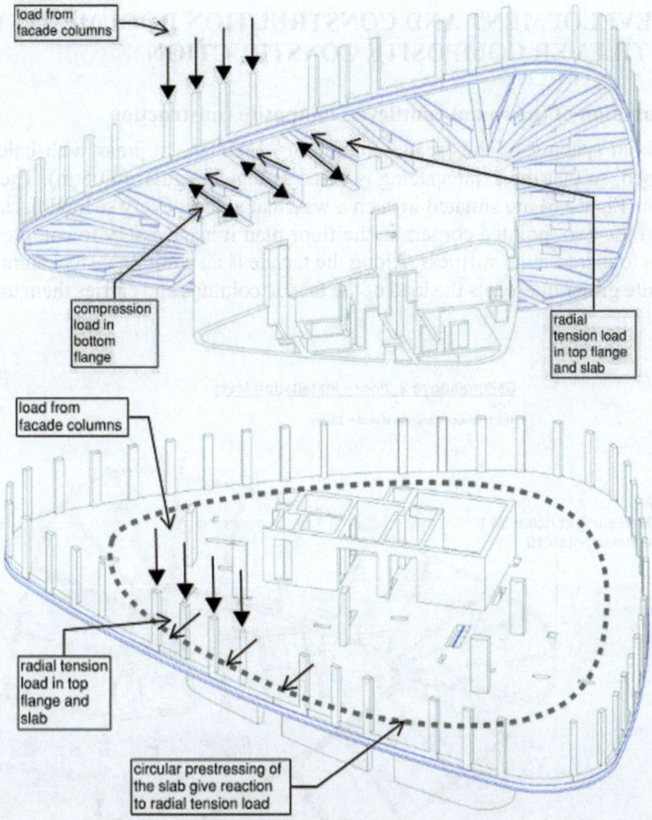

Figure 5. Variant 2, Transfer of the façade load with a radial cantilever composite construction, isometric view from below (upper figure) and from above (lower figure, Weiske+Partner)

The slab under the 3rd floor is used as upper flange of the all-around cantilevered support grid. It gets radial tension load because of the cantilever function. Circular prestressing cables of the concrete slab give the reaction to the radial tension load caused by the radial support grid.

The compression load in the bottom flange of steel girders is shorted with a concrete slab in the socle of the building.

Advantages and disadvantages of variant 2:

- \+ No limitation of flexibility in use in the residential floors
- \+ Formwork for the protruding floors must stay in situ only until the slab under the 3rd floor is finished and hardened
- \+ Easy construction of the façade columns
- – Additional construction height of 1.5 m required to realize the protruding façade
- – radial construction steel grid required
- – circular prestressing cables required
- – The cost for the shell construction is higher than in variant 1 because of construction steel and prestressing cables

The owner decided to realize variant 2. It was decisive that the flexibility in use is not limited in the residential floors and that the formwork can be removed earlier than in variant 1.

3 DESIGN DEVELOPMENT AND CONSTRUCTION DOCUMENTS OF THE RADIAL CANTILEVER COMPOSITE CONSTRUCTION

3.1 Design and function of the radial cantilever composite construction

The regular axial spacing of the facade columns is 2,025 m. In areas with balconies some facade columns are skipped, so that the axial spacing is twice as big as regular (4,05 m). The radial steel girders above the 2nd floor (Figure 6) are situated in such a way that at least every second facade column is placed on a cantilever. At the two, pointed corners in the floor plan it is necessary to compress the axial spacing of the steel girders to increase the stiffness. Along the facade is an additional steel composite girder placed. This steel composite girder intercepts the load of the facade columns and carries them to the steel composite cantilevers.

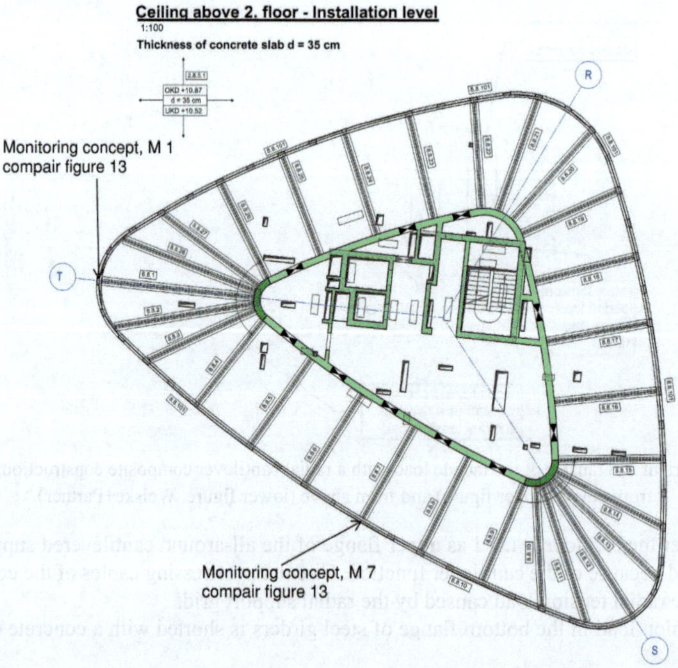

Figure 6. Slab above 2nd floor, positioning of the radial steel composite cantilevers (Weiske+Partner)

Figure 7 shows two different radial steel composite cantilevers in context to the concrete slab above and the compressed concrete slab below which short circuits the compressing forces. The lower- and compression chord of the steel composite cantilever is designed for the compression due to the cantilever function. By a load sharing construction, the compression is transferred into the concrete slab above the 1st floor. All steel components are made in grade S355. The concrete slab is designed with a thickness of 50 cm in C50/60. In this concrete slab all compressive forces from the radial steel composite cantilevers are short circuited. Several openings (especially elevator shaft and stairwell) should be noted.

The upper flange of the steel cantilever has a minor structural part and is mainly used to fix the headed stuts. To compound Steel and concrete there will be used headed stuts in three rows. The traction of the steel girder is transformed straight and continuously into the concrete slab above the 2nd floor due to the headed studs. The web of the steel girder is equipped with a regular opening for the implementation of various, different house-technical trades.

Due to the choice of the structural system, there are circumferentially, radial tractions in the concrete slab above the second floor (thickness 35 cm, C30/37). Summing up, the result of this radial tractions is circular tensile stress. With circular prestressed reinforcement with mono strands, this tensile stress can be absorbed.

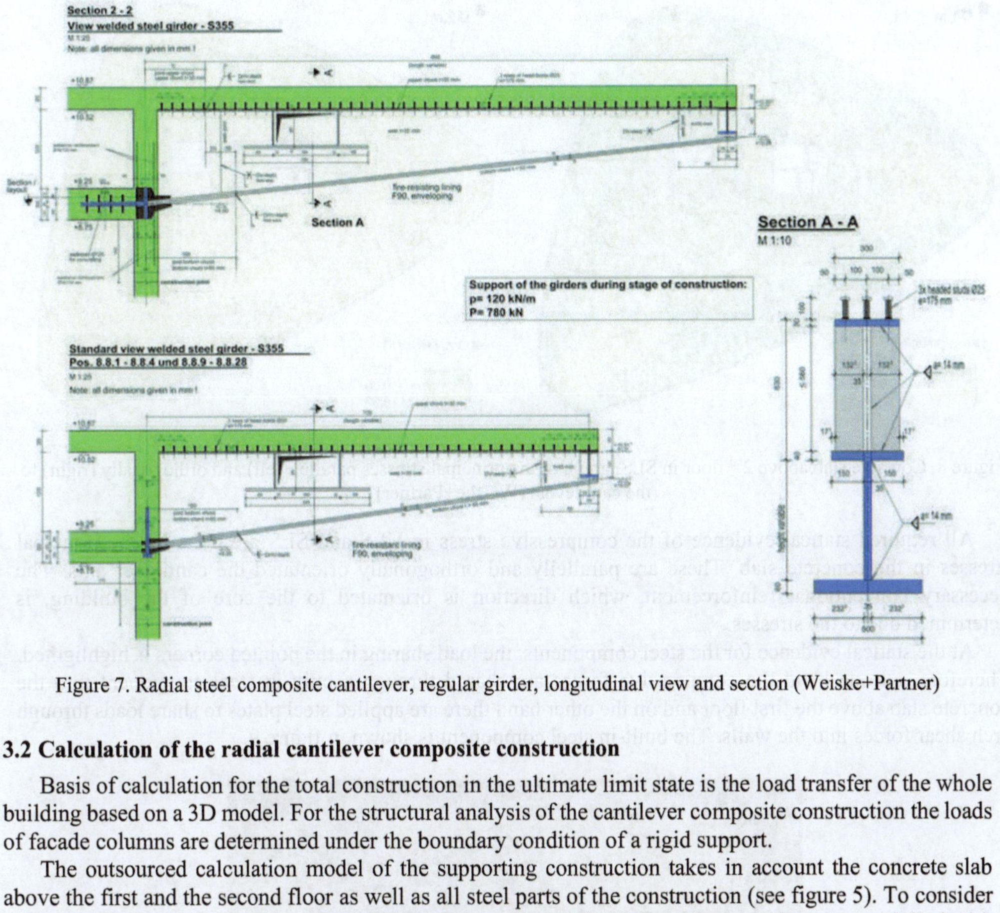

Figure 7. Radial steel composite cantilever, regular girder, longitudinal view and section (Weiske+Partner)

3.2 Calculation of the radial cantilever composite construction

Basis of calculation for the total construction in the ultimate limit state is the load transfer of the whole building based on a 3D model. For the structural analysis of the cantilever composite construction the loads of facade columns are determined under the boundary condition of a rigid support.

The outsourced calculation model of the supporting construction takes in account the concrete slab above the first and the second floor as well as all steel parts of the construction (see figure 5). To consider the bond between steel cantilever and prestressed concrete slab, the coupling members are modelled with linear elastic behaviour in the ULS. The stiffness of the connectors between steel and concrete is chosen according to the secant stiffness of the headed studs.

The prestressing cables are cumulated in groups of five to eight pieces. The resulting force of each group is applied on one tendon in their middle position. Also, the prestressing losses can be calculated in sum for each group. The main part of the prestressing losses are results of friction due to deflection. Departing from standard tendons, in this case a ring post-tensioning system is used, which short circuited the forces in a recess (see figure 11 right). Thereby the prestressing losses due to planned deflection can be halved from 360° to 180°. Additional to the planed prestressing losses, the losses of unplanned deflection according to the technical approval should be noted. Furthermore, there are time depending, losses through creeping and shrinking of the concrete and relaxation of the prestressing tendons. At the time of the last prestressing process the concrete slab already has the age of 140d. So, in case of creeping, only the difference between t = 140d and t = ∞ has to be considered.

The prestressing force is chosen with a value which ensures that the concrete slab above the second floor is compressed or just under low tensile stress level smaller than f_{ctm} in the SLS. Figure 8 shows the principal stress parallelly and orthogonally to the cantilevers subjected to total service load. In building 8 are 288 mono strands of type St 1570/1770 applied.

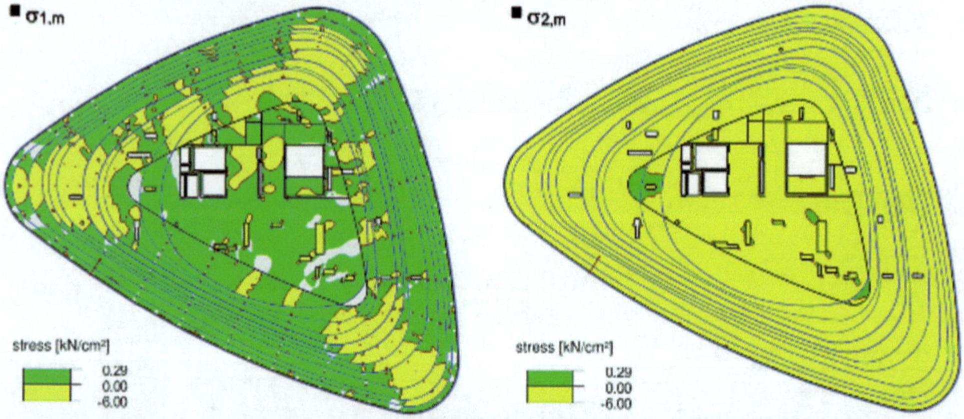

Figure 8. Concrete slab above 2nd floor in SLS, graph of the principal stresses parallel (left) and orthogonally (right) to the cantilevers (Weiske+Partner)

All required statical evidence of the compressive stress in ULS and SLS are done by the principal stresses in the concrete slab. These are parallelly and orthogonally orientated the cantilever axis. The necessary conventional reinforcement, which direction is orientated to the core of the building, is determined due to the stresses.

At the statical evidence for the steel components, the load sharing in the pointed corners is highlighted. Therefore, on one hand the compressive forces are shared through a built-in steel component into the concrete slab above the first floor and on the other hand there are applied steel plates to share loads through arch shear forces into the walls. The built-in steel component is shown in figure 9.

Figure 9. Built-in steel component in pointed corners (Weiske+Partner)

3.3 Deformation and prestressing steps of the radial cantilever composite construction

The radial steel composite cantilever construction is a system that adsorbs the facade loads through to bending (pressure in the lower flange and tension in the concrete slab). The loads are consisting circa 61% of the self-weight of the building, circa 24% additional dead loads of the building, circa 4% facade loads and circa 11% service loads of the building.

The tensile stress of the concrete slab above the 2nd floor will be assigned to the radial fixed prestressing tendons. The deformations of the system, especially the tensile extensions of the concrete slab, will be prevented this way. During the design development of the construction, it becomes clear that the prestressing of the concrete slab above the 2nd floor, had to be increased continuously as the loads of the raising building shell. There are three prestressing steps determined.

The deformation of the system depends on diverse influences:

- Load of facade columns (time dependent, building shell, facade and additional dead loads)
- Elastic deformation behaviour of the steel cantilever
- Shrinkage behaviour of the compression concrete slab (slab above 1st floor)
- Elastic deformation behaviour of the compression concrete slab (slab above 1st floor)
- Delayed elastic deformation behaviour (creep) of the compression concrete slab (slab above 1st floor)
- Shrinkage behaviour of the tensile stressed concrete slab (slab above 2nd floor)
- Elastic deformation behaviour of the compression concrete slab (slab above 2nd floor)
- Delayed elastic deformation behaviour (creep) of the tensile stressed concrete slab (slab above 2nd floor)
- Prestressing of the tensile stressed concrete slab (slab above 2nd floor)
- Time dependent prestressing losses (slab above 2nd floor)

For the deformation calculation (SLS) a second 3D-Model is defined. In this model for the service limit state the shear connection is modelled as a rigid connection due to the low level of the loads. On the one hand the deformations will be calculated for the final state, considering the different load cases, and on the other hand for the different states of constructions.

Based on the deformation calculations the required superelevation is set. It has been considered that the calculated results should be evaluated critically due to the fuzzy boundaries. The Prestressing is able, due to the gradually rise of the force, to influence the deformations of the system within limits. Therefor a monitoring-program (see section 4.2) has been developed to monitor the deformations of the steel composite construction during the increasing of the loads and the degree of pre-tensioning.

4 CONSTRUCTION WORK OF THE RADIAL CANTILEVER COMPOSITE CONSTRUCTION

4.1 Installation of the radial cantilever composite construction and the prestressing cables

Basis of the implementation planning is the developed construction process, which is a result of all boundary conditions in the static calculations. Especially it is about the introducing of the prestressing forces in relation to construction progress.

After the 1st floor walls are completed, the temporary supporting construction for the steel girders can be built up which are carrying the load down to the ground plate of the building. The steel cantilevers are placed on the 1st floor walls and the temporary supporting construction. In this situation all parts of the steel structure can be adjust and the superelevation of the cantilever can be fixed. Figure 10 shows this situation of component 7 in the background. Most of the steel girders can be produced completely in the factory. Only the cantilevers in both pointed corners have to be welded with the built-in component in the concrete plate above the first floor at the construction side.

Figure 10. Work on the prestressed concrete slab of component 8 (slab above 2th floor, in front) and installed steel cantilevers of component 7 (background) (Weiske+Partner)

In the next step the compression concrete slab (concrete slab above 1st floor) and the rising concrete components of the 2nd floor will be built. After the formwork for the concrete slab above the 2nd floor is installed, the regular reinforcement and the prestressing tendons can be installed (see Figure 10, component 8 in front of the picture).

Figure 11. Installed prestressing cables (left) and recess for prestressing (right, Weiske+Partner)

All of the 36 prestressing tendons consisting of 8 cables which are placed by the use of a crane lifted transport roll. Start- and ending point of the cables in each case, is the recess (figure 11) which will be filled with concrete after the last step of prestressing. After complete positioning of the tendons and the following installing of the regular reinforcement and the concreting, all works at the supporting construction are done.

4.2 Stepwise prestressing of the cantilever composite construction and monitoring of building deformations

As mentioned above the prestressing of the concrete slab above the second floor will be prestressed in three steps. The timing of the prestressing steps in context to structural work is displayed in figure 12. The different states of construction can be described as follows:

- Construction step 1: 28 days after the concreting of the slab above the 2nd floor and after building up the first two regular floors, the first step of prestressing follows. At this point the slab above the 4th floor is built. Due to the application of 1/3 of the maximum prestressing force the supporting structure is stress free and can be rebuild.
- Construction step 2: Slab above 7th floor is completed. The prestressing force is still 1/3 of the maximum prestressing force.
- Construction step 3: The prestressing force will be raised to 2/3 of the maximum Prestressing force.
- Construction step 4: Slab above 12th floor is completed. The prestressing force is still 2/3 of the maximum prestressing force.
- Construction step 5: The prestressing force will be raised to the maximum prestressing force.
- Construction step 6: Slab above 16th floor is completed (shell of the building is finished). The maximum of the prestressing force is applied without time-depending losses.
- Construction step 7: Additional dead loads and service loads are applied. The maximum of the prestressing force is applied without time-depending losses.
- Final state: Additional dead loads and service loads are applied. The maximum of the prestressing force is applied with time-depending losses. ($t = \infty$).

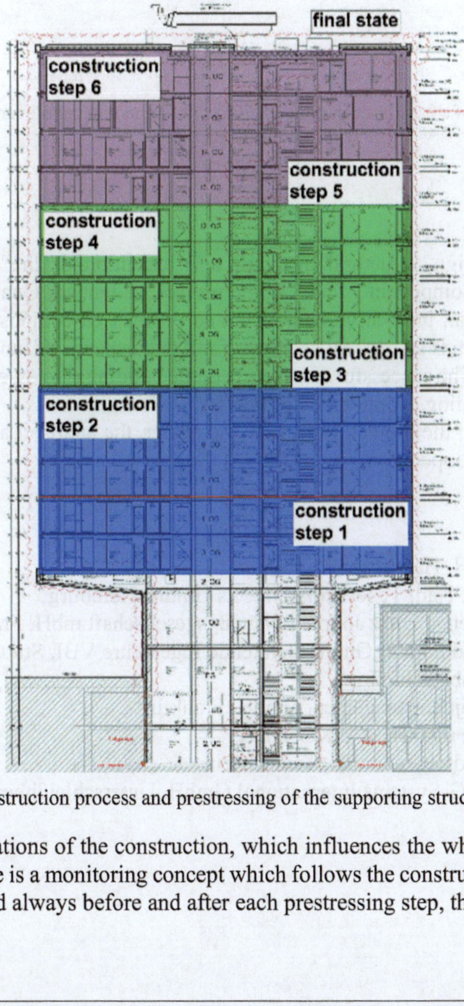

Figure 12. Construction process and prestressing of the supporting structure (Weiske+Partner)

To control the deformations of the construction, which influences the whole building, and to compare the theoretical results, there is a monitoring concept which follows the construction works. After concreting the slab above 2nd floor and always before and after each prestressing step, the vertical deformations at the cantilevers ends along the facade is measured up. This enables the possibility to react on deviations during the following prestressing steps and to specify the final prestressing force if necessary.

Currently the 10th floor of component 8 is in progress. Bevor the last step of prestressing is started, it is required to evaluate the results of the monitoring and to draw the right conclusions.

Figure 13 shows the curve of vertical deformations of a cantilever in the pointed corner (left side) as well as deformations of a cantilever in the area with less curvature (right side). These cantilevers are marked in figure 6. The solid line represents the calculated and the dashed line represents the measured results of deformation. The graphs are almost parallel. Apart from differences at the beginning, due to installation tolerances, there is a very good alignment between calculated and measured deformations.

Conclusively it is fixed that also the third and last prestressing step will be executed as planned.

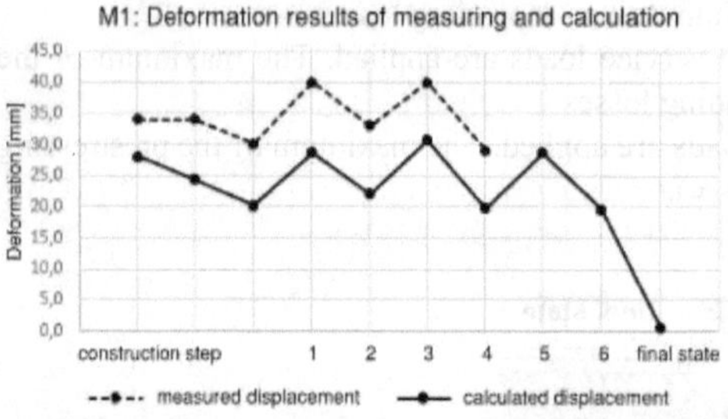

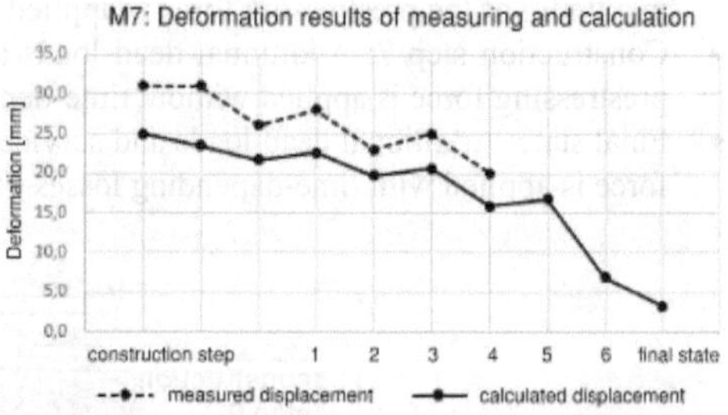

Figure 13. Vertical deformation of a cantilever in pointed corner (left) and of a cantilever in area with less curvature (right, Weiske+Partner)

5 CONCLUSIONS

This report illustrates design, structural analysis, and execution of the support construction for protruding façade columns of component 7 and 8 of the project `Hafenpark Quartier Süd´ in Frankfurt on the Main. To realize the task, an innovative, effective, and prestressed composite construction has been developed. The selected solution is an optimal choice to fulfil the demands of functionality and appearance of the owner and the architect. The successful completion of the building is to owe the intensive cooperation between all planning and executing people in the project.

We thank all participants in the project, especially the owner, the architect and the construction firms for the good and constructive cooperation.

6 PARTICIPANTS

Owner: B&L Real Estate GmbH, Hamburg
Architect Design Development: Hadi Teherani Architects GmbH, Hamburg
Architect Construction Documents: Dietz Joppien Planungsgesellschaft mbH, Frankfurt
Structural Engineer: Weiske und Partner GmbH Beratende Ingenieure VBI, Stuttgart
HVACR Engineer: IGTech GmbH, Hamburg
Project Management: Witte Projektmanagement GmbH, Frankfurt
Shell Construction: Adolf Lupp GmbH + Co KG, Nidda
Steel Construction: AL Verbundträger GmbH, Taunusstein
Prestressing Cables: DYWIDAG-Systems International GmbH, Unterschleißheim

9TH INTERNATIONAL CONFERENCE ON COMPOSITE CONSTRUCTION IN STEEL AND CONCRETE
M. Knobloch, U. Kuhlmann, W. Kurz, M. Schäfer. (Eds.)
Online, Germany, July 27-29, 2021

KUNMING TOWER: COMPOSITE SYSTEMS IN SUPERTALL DESIGN

Xuemei Li*, Ashpica Chhabra, Han Ding*****

* Skidmore Owings and Merrill
e-mails: xuemei.li@som.com, ashpica.chhabra@som.com

** Skidmore Owings and Merrill
e-mail: han.ding@som.com

Keywords: Supertall, tied-braced frame, composite mega columns, FVD, CFT, ductility, damping, creep and shrinkage, shake table testing, finite element analysis.

Abstract. *For the past 20 years, composite systems have dominated supertall construction. This paper will present composite construction solutions at system and component level in the design of supertall buildings using the example of the 407m tall Kunming Tower. The tower utilizes an innovative tied-braced frame structural system at the perimeter and strategically placed energy dissipating mechanisms to meet the Chinese code requirements and to achieve seismic performance objectives. Given the complexity of the non-prescriptive structure, many advanced analyses were performed to verify building behavior and optimize the structural system.*

1 INTRODUCTION

Kunming Eyes of the Spring Tower aspires to be a landmark for the city of Kunming that has seen unprecedented growth over the last decade. In design of supertall structures the most critical first step is the selection of an appropriate structural system of the project. The hierarchy of optimization for a tall building starts with the selection of the overall structural system, followed by the optimization of the sub system and finally the optimization of the members themselves. Wind and seismic load effects, creep and shrinkage and soil structure interaction are to be considered in the selection of the right structural system such that it minimizes the impact on design from these effects.

Kunming is located in a low wind environment but a high seismic zone with a design intensity of 8 according to the National Seismic Code of China. The structural design is therefore primarily controlled by seismic forces. At over 400m tall, the shear walls around the central vertical circulation and mechanical service zones alone can no longer provide sufficient stiffness and strength. A trussed tube with corner mega columns would most efficiently achieve the target stiffness for a building at this height. However, since the region is highly seismic, the system also needs to be sufficiently ductile to meet the required performance targets under moderate and rare earthquake events. The links that tie the braced frames are carefully proportioned such that the structure at the perimeter of Kunming tower achieves building drift limits and remains elastic under ultimate wind loads and frequent earthquake events. Under larger earthquakes, the steel links are designed to yield and dissipate energy thereby limiting damage to the structure.

The structural system selection is also in response to an integrated approach to architectural design and is celebrated in the architectural expression of the building. The system effectively controls the total seismic mass and its distribution over the height of the building, achieves appropriately tuned stiffness that manages structural deformations under minor and rare earthquakes while limiting the resulting seismic forces and allows to strategically design energy dissipating mechanisms to limit damage to primary structural components.

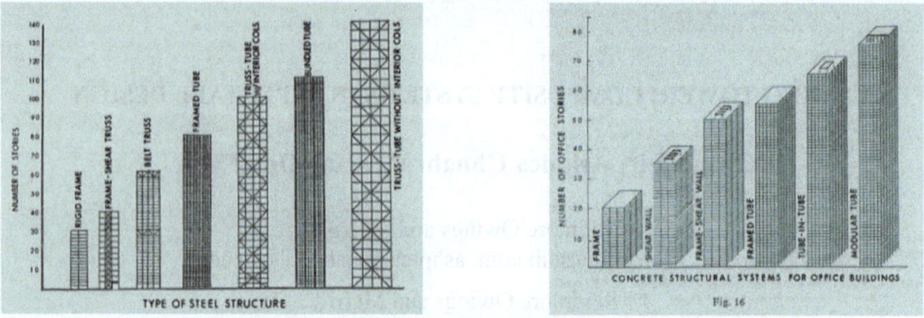

Figure 1. Appropriate structural systems with variation in building height

2 ARCHITECTURAL DESIGN

The project has a mixed-use program consisting of retail, office, service apartment, hotel and observation decks in that order over the height of the building. There are a total of 81 stories above grade and 4 stories below grade, with the total area of the superstructure at about 197,000 sqm. Aspiring to be a landmark building that defines the skyline of the city, the project would have a significant impact on the future urban texture of Kunming. The architectural design along with considering the main visual landscape, dominant wind direction and sustainable approaches, organically integrates its structural form.

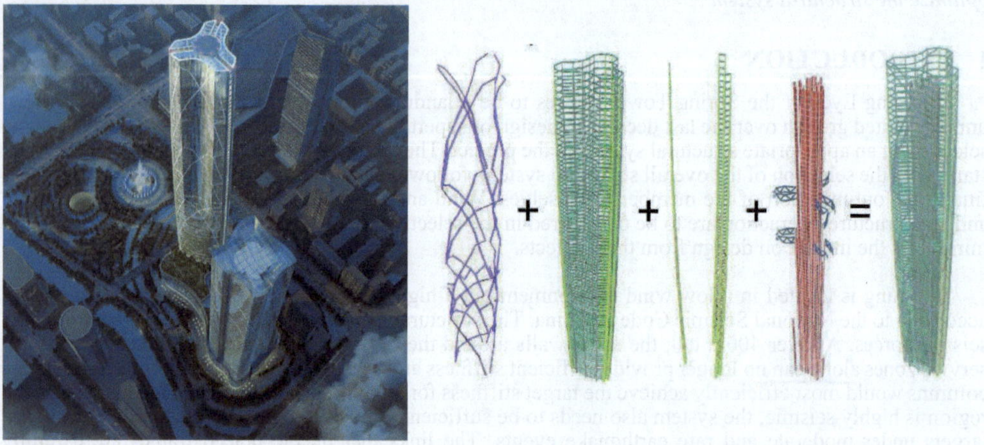

Figure 2. Architectural rendering and structural system diagram

The mixed-use tower has a triangular form with each side defined by convex curves for the lower two thirds of the tower that gradually transforms into concave curves as the tower tapers towards the top into a Y-shape. The proportioning is derived to meet the functional requirement of larger lease spans and floor area for the lower office portion, and reduced spans and area near the top for hotel and residential floors. The three faces of the proportioning are arranged by introducing a “gap” between any two faces, visually dividing the volume into three parts. The three parts are then woven back into a whole by the mega braces that are architecturally expressed on the exterior wall. An ecological chimney is created in the "gap", which functions like the "lungs" of the building, providing natural ventilation for the entire building, and at the same time corresponds to the energy-dissipating tie beams provided between the structural diagonally braced faces.

3 STRUCTURAL SYSTEM AND MATERIAL SELECTION

Steel floor framing with composite metal deck slab was selected as the gravity floor system to reduce the total structural weight. This minimizes the loads on the foundation as well as the seismic mass which is critical in high seismic zones. Due to the undulating exterior form, the in-plane forces generated at the perimeter braced frame need to be transferred by the structural diaphragm to the building core. By using structural steel floor framing it is ensured that there's a direct load path for these forces after the concrete slab cracks.

The lateral system consists of a perimeter braced frame and an interior shear wall core structure. The core walls are arranged in a "Y" configuration with a hexagonal central portion, providing high torsional stiffness to the tower. While the upper half of the core is reinforced concrete, the lower half is composite. The composite portion of the core has embedded steel column shapes at the boundary zones and steel plates along its length. Such an arrangement has the advantage of strategically using higher cost structural steel only where it's most efficient to do so by taking advantage of the high stiffness of reinforced concrete shear walls throughout the tower height and high strength and ductility of the composite structure in the lower portion only. Energy dissipating core link beams are selected to be either reinforced concrete or composite with embedded steel profiles based on the strength and ductility demands at a given location.

At the perimeter of the tower, a mega-braced frame acts to maximize the overall 'footprint' of the structure. The high stiffness of the perimeter mega-braced frame system is augmented with ductile tie beams between the mega-columns creating a Tied Braced Frame (TBF). The diagonals are typically structural steel box sections to meet the higher tension demands. The top end bay of each bracing module incorporates buckling restrained braces (BRBs). The BRBs and the steel ties act together as fuses during rare earthquake events protecting the primary braces and columns from damage by reducing the stiffness of the braced frame.

Concrete filled tube (CFT) sections are used for all the frame columns with high compression loads. At the upper levels where compression loads are smaller, structural steel columns are used to reduce the building mass. Circular CFT column shape maximizes concrete confinement resulting in higher strength and ductility thus used for all face columns. However, it is more difficult to procure such columns for the large diameters required in the mega-columns. Therefore, a polygon shape subdivided by steel plates for better confinement was selected for the mega columns to maximize the usable floor area at the perimeter of the building.

4 ELASTIC ANALYSIS

ETABS was used as the elastic analysis software, with modeling assumptions and loadings defined according the requirements of the Chinese codes. The first three mode shapes of the tower are presented in Figure 3: Mode 1 represents the translation along the narrow face of the building (Y-direction), Mode 2 the translation along the broad face of the building (X-direction) and Mode 3 the rotational modal shape of the building.

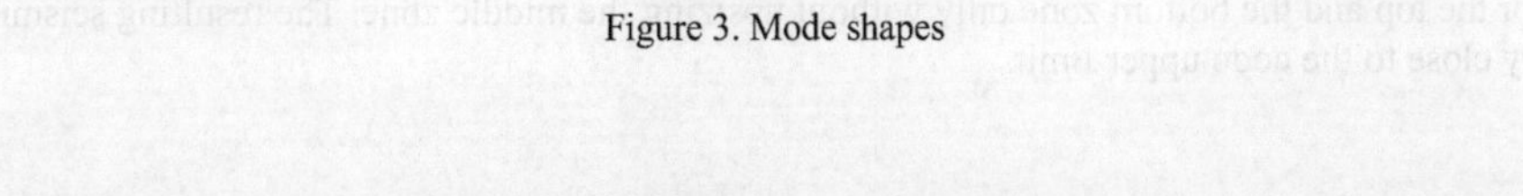

Mode-1: T1=5.85 sec Mode-2: T2=5.79 sec Mode-3: T3=3.10 sec

Figure 3. Mode shapes

The shear force distribution between the perimeter TBF system and central core is in between that of a 'Core + Moment Frame' building and a 'Core + Concentric Braced Frame' building. Perimeter mega frame resists a significant portion of storey shear over the height of the building. Similar behavior is observed in the distribution of seismic overturning moment (OTM) with the perimeter resisting 63% of the total OTM at base and a range of 60-80% over the height of the building. In Chinese and other major international building codes, for a dual system, the perimeter moment frame is required to be capable of resisting up to 20-25% of the design seismic forces. However, most 'Core + Moment Frame' structures need to be artificially oversized to meet this requirement, thereby leaving a gap between actual demand and capacity. To the contrary, Kunming's mega-frame naturally forms a competitive lateral system at the perimeter that works in parallel with the core to increase overall material efficiency.

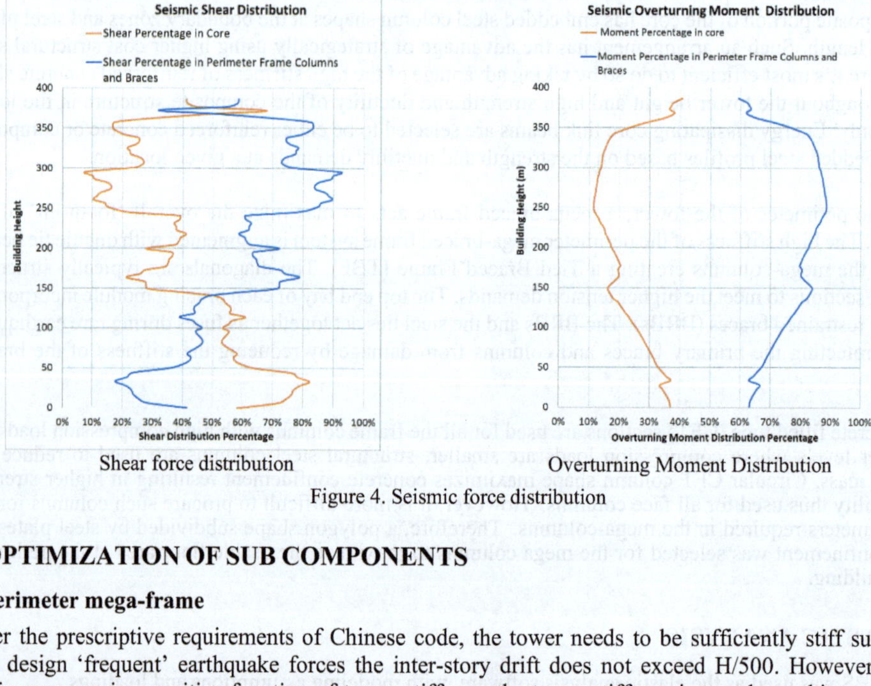

Shear force distribution Overturning Moment Distribution

Figure 4. Seismic force distribution

5 OPTIMIZATION OF SUB COMPONENTS

5.1 Perimeter mega-frame

Per the prescriptive requirements of Chinese code, the tower needs to be sufficiently stiff such that under design 'frequent' earthquake forces the inter-story drift does not exceed H/500. However, since seismic response is a positive function of tower stiffness, the tower stiffness is tuned to achieve optimal member sizing to meet the drift requirements. This was done using an in-house optimization program based on the paper by W. Baker [1]. The theory is based on the Lagrange multiplier method for finding extremum of multivariate.

For axial members, the local extremum is given by:

$$\frac{\partial}{\partial A_i}\left(\sum \frac{nFL}{EA}\right) + \lambda \frac{\partial}{\partial A_i}\left(\sum AL - V\right) = 0 \qquad (1)$$

For flexural members, the local extremum is given by:

$$\frac{\partial \Delta}{\partial A_i} = \frac{-L_i M_i}{E_i I_i^2}\frac{d\Delta I_i}{dA_i} + \lambda L_i = 0 \qquad (2)$$

The perimeter mega-frame members were initially sized to meet strength requirements and were input as minimum member sizes for the first iteration of the optimizer. At this stage, the tower drift exceeded the code criteria in the upper third of the building. The optimizer revealed a strategy to increase stiffness for the top and the bottom zone only without upsizing the middle zone. The resulting seismic drift is very close to the code upper limit.

5.2 Central Core

The final configuration of perimeter braces results in denser braces in some zones and more sparse bracing in others (highlighted in blue in Figure 5). In these zones, the core experiences a higher distribution of seismic shear, resulting in high shear and moments at the link beams. Maintaining a consistent link beam depth over the entire height of the building is no longer feasible without adding significant steel shapes at approximately 50% of all link beams. A strategy was employed to tune the shear force distribution by providing shallow link beams in the dense bracing zones and deeper or double link beams in the sparse bracing zone. After employing this strategy, the number of links in the beams that exceed the requirement to fulfil the capacity ratios as reinforced concrete sections was reduced by one-third. The remaining link beams were designed as composite sections.

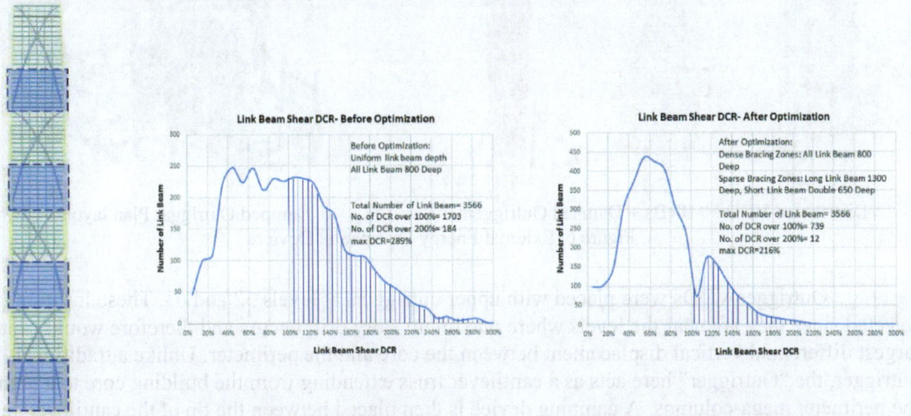

Sparse Bracing Zones | Uniform link beam depth | Variable link beam depth

Figure 5. Optimization of core link beams

5.3 Energy Dissipation Devices

The building as designed meets the drift and strength requirements set by Chinese national codes. However, damping devices were introduced to achieve the enhanced performance required by the Kunming code requirements. Located in a high seismic zone, Kunming City Bureau requires all newly designed supertall towers to be equipped with external energy dissipation measures, in addition to the tower's inherent damping. The goal is to increase the tower's energy dissipation capability in a rare earthquake event and reduce damage to structural and non-structural components. The provincial code sets a prescriptive requirement to further reduce the inter-story drift to H/550 (H/500 without dampers) under frequent earthquake and H/100 (H/130 without dampers) under rare earthquake by utilizing damping devices.

The tower utilized Buckling Restrained Braces (BRB) in the perimeter frame and Viscous Damper Devices(VDD) with outriggers and between corner mega columns as the external energy dissipation mechanisms to meet these requirements. The BRBs were located at Levels 21, 42 and 63 in the perimeter frame and replaced the original single level braces. The brace remains elastic in frequent earthquake but is allowed to yield in the rare earthquake. The location of the BRBs allows for a redundant load path for the gravity loads. The columns and beam tie members were designed for a scenario where the BRBs yield in a large earthquake event.

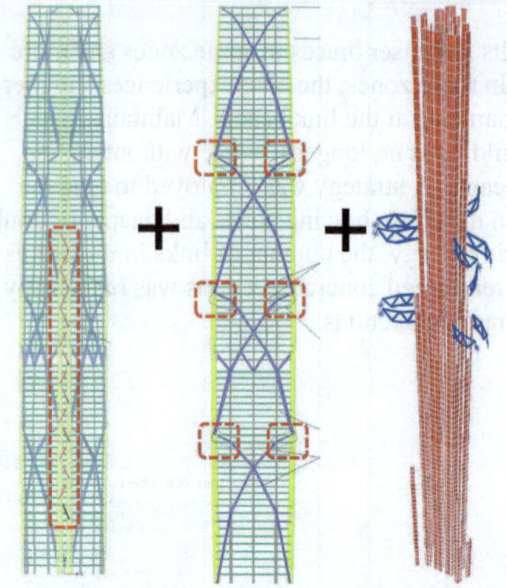

Distributed VDDs + BRBs + Damped Outriggers

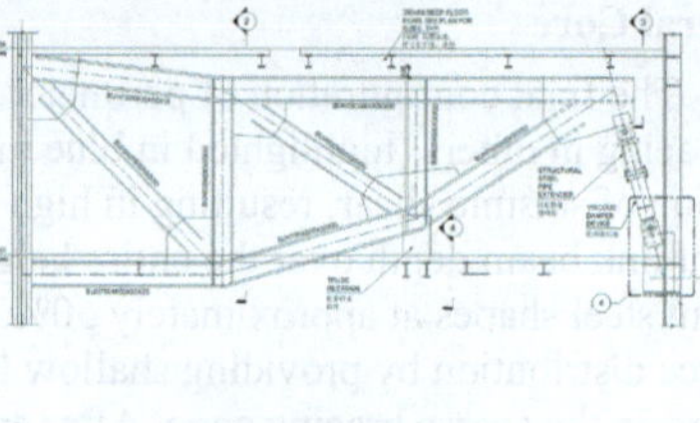

Damped Outrigger Section

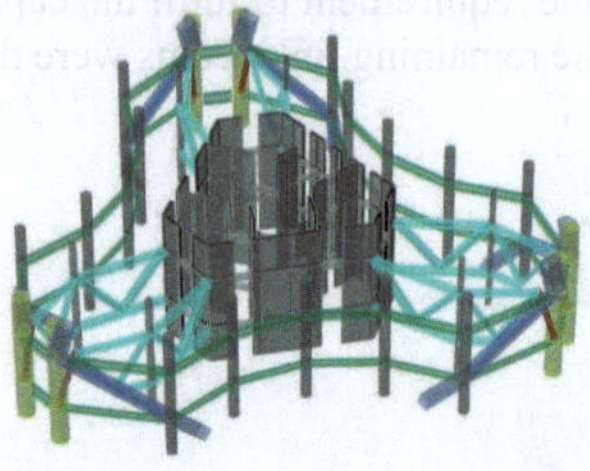

Damped Outrigger Plan layout

Figure 6. External Energy Dissipation Devices

Outrigger VDDs were placed with upper outriggers at levels 52 and 63. These locations were selected since they are near the levels where maximum story drift occurs and therefore would capture the largest differential vertical displacement between the core and the perimeter. Unlike a traditional outrigger, the "Outrigger" here acts as a cantilever truss extending from the building core to a point near the perimeter mega-columns. A damping device is then placed between the tip of the cantilever and the mega-column at the level below. VDDs were also placed between corner mega columns up to the mid-height of the tower taking advantage of the large racking between the mega-columns pairs. The extent of these dampers over the height of the building was based on their relative contribution to the building behavior and impact on architectural coordination.

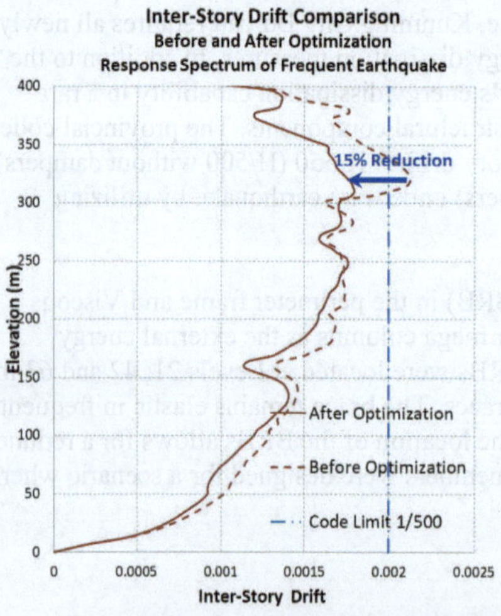

15% Reduction due to Perimeter Frame Optimization

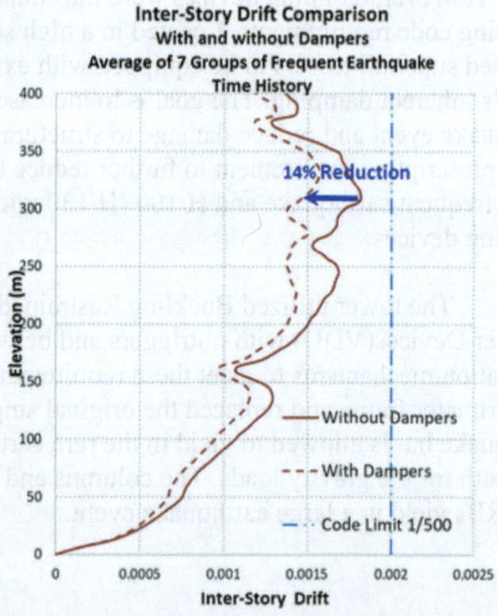

14% Reduction due to adoption of Dampers

Figure 7. Comparison of Elastic Inter-story Drift

6 SPECIAL STUDIES

6.1 Creep and Shrinkage Analysis

Construction sequence and creep & shrinkage analysis of the tower was conducted using ETABS structural analysis software. Two separate non-linear static staged load cases were analyzed. The first load case analyses the impact of anticipated constructions sequence by sequentially adding structural elements and loads. The second load case provides analytical outputs at specific points in time after the completion of tower construction and occupancy to capture the long-term time-dependent material effects.

It was found that the anticipated construction sequence with delayed construction of braces pushes more gravity load into the perimeter columns and pulls load out of the mega-columns. This effect is slightly mitigated by the long-term creep and shrinkage effects which pull more load into the mega-columns and out of the side columns. However, the net result is still a reduction of mega-column forces and increase of perimeter column forces when compared to the linear elastic analysis. This effect can also be seen in the plots of column and brace forces over the building height. The maximum difference between the core and perimeter was calculated to be 70mm. While this is not a small value, such differential displacement can normally be accommodated with a construction compensation program, and tuned to provide floor levelness at a desired point in time.

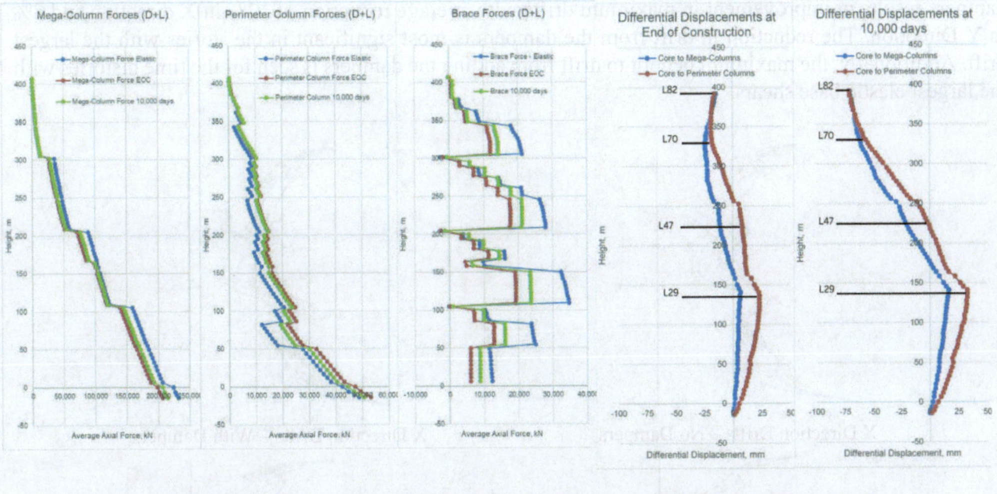

Figure 8. Impact of construction sequence/ creep and shrinkage on building behavior

6.2 Elasto-Plastic Analysis

PERFORM 3D was used to perform nonlinear time history analysis to evaluate the inelastic response of the building and assess its performance under rare earthquakes that have a probability of 2% exceedance in 50 years. Seven sets of site specific time histories were provided by the local seismic consultant and applied bi-directionally using code compliant combinations. Element inelastic properties are defined per FEMA 356. Inelastic energy dissipation is achieved by yielding of core link beams at the core and yielding of buckling restrained braces and steel ties at the perimeter.

Figure 9 shows the energy distribution of the building at each time step under one of the rare earthquake of different time histories. The average amount of inelastic energy dissipation varies between 20% - 40% of the total energy for the seven time history groups. If fluid viscous dampers are provided, the energy in fluid dampers varies from 6-8% of the total energy.

Providing tie beams between mega columns, adding dampers and BRBs in the perimeter has a huge impact on the distribution of inelastic dissipation leading to larger energy dissipation in the perimeter and thereby protecting elements in the core. In a typical perimeter frame + core + outrigger building at this height, the majority of energy dissipation and OTM is concentrated within the core putting a huge demand on the link beams and net tension in the core elements. With 64% of dissipated energy in the core and 36% at the perimeter, the structural system for this building results in an balanced distribution of elements dissipating the inelastic energy.

Figure 9. Distribution of inelastic energy distribution

Figure 10 below presents a comparison of the elasto-plastic inter-story drifts in the building resulting from the bi-directional time histories with and without the addition of non-linear dampers. The addition of dampers results in improvement in maximum drifts with average reduction of 12% in X direction and 8% in Y Direction. The reduction in drift from the dampers is most significant in the stories with the largest drift. Additionally, the maximum benefit to drift from adding the dampers is seen for the time histories with the largest elastic base shears.

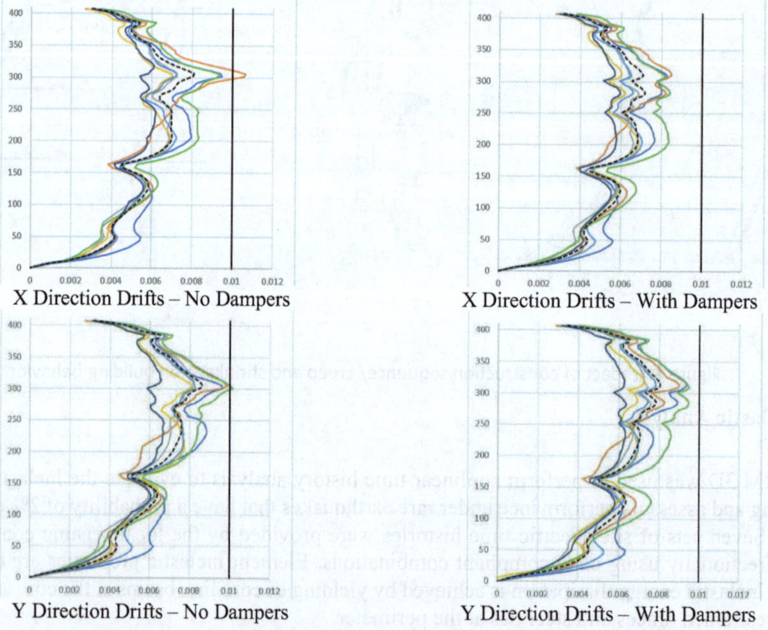

Figure 10. Impact of Dampers on Inelastic Drift

The sequence of hinging observed in EP analysis provides further insight into the building behavior. The shorter link beams at the core yield first, followed by the longer link beams. This is followed by yielding of the steel tie beams in shear. The plots shown in Figure 11 are for bidirectional time histories with primary time history in Y, hence it's seen that tie beams parallel to Y yield before tie beams parallel to X, BRBs start to yield after about 2/3rds of the tie beams yield with the top and bottom BRBs yielding before the middle BRBs. Similar hinging sequence was observed during shake table testing.

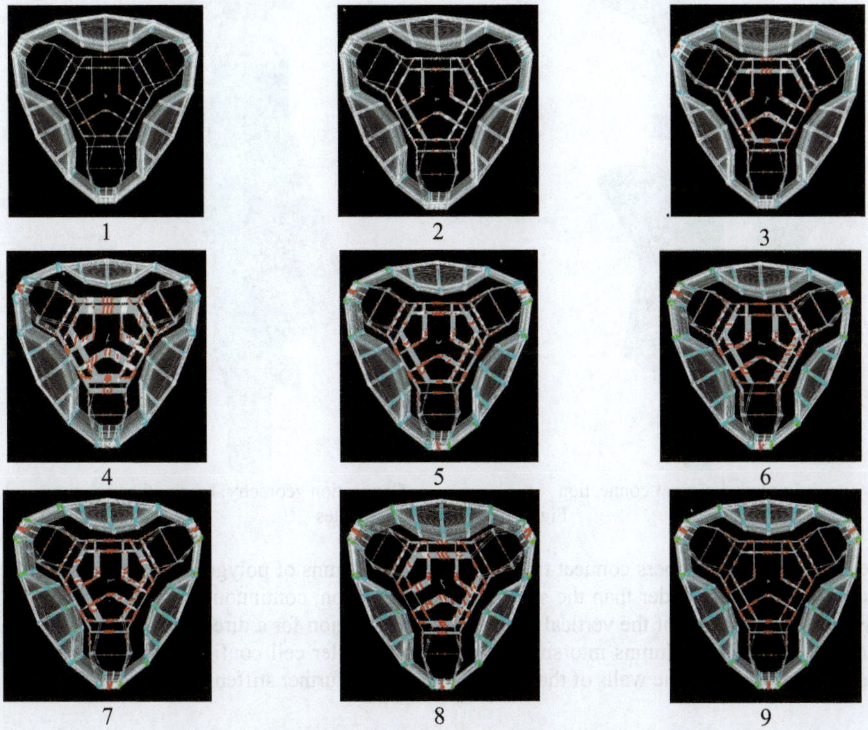

Figure 11. Sequence of hinging

6.3 Connection Configurations and Analysis

With the concave and convex nature of the tower massing, the bracing members that lace around the surface change angles and orientations along the height. This typically results in connections of steel plates that are not coplanar. The connections between the bracing members of boxed steel sections and columns of circular CFT sections add an extra issue of complexity. Since both bracing members and CFT columns are primary load resisting members in the system, it becomes very important to maintain continuity of the cross-section of both members. After comparison of a few different connection concepts, the following approach was used to ensure the continuity of load transfer while maximizing the number of coplanar plates:

1. Change the circular section to a rectilinear one above and below the connection, so as to simplify the geometry at the connection. This is realized by including a partial cone surface, which is a developable surface within the transition zone.
2. Reduce the number of non-coplanar plates by taking the advance of the symmetry in the center connections, and generate plate surface using three points and reduce it down to the required profile.
3. Typically maintain the size of the circular column to be smaller than the size of the boxed section in the bracing so that the majority of the sections in both members continue through the connection without only part of the load transferred via the gusset plates.

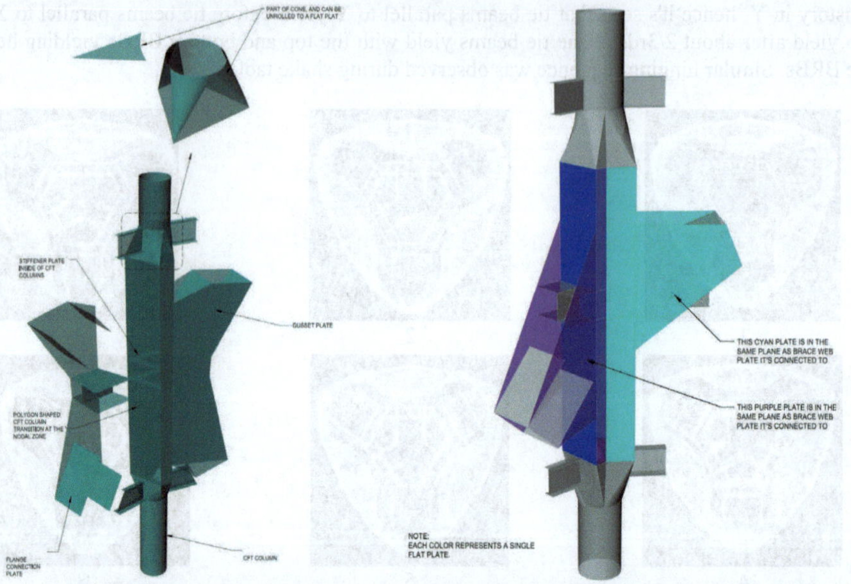

Exploded view at connection Connection geometry: brace - face column

Figure 12. Connectivity Rules

4. Where the bracing members connect to the CFT mega columns of polygon shapes, because the mega columns are typically wider than the steel bracings in section, continuous vertical stiffener plates are provided at the location of the vertical plates of the box section for a direct load transfer, as well as to divide the large CFT columns into smaller cells. The smaller cell configuration helps to reduce the hydrostatic pressure on the walls of the mega columns, and further stiffen the section.

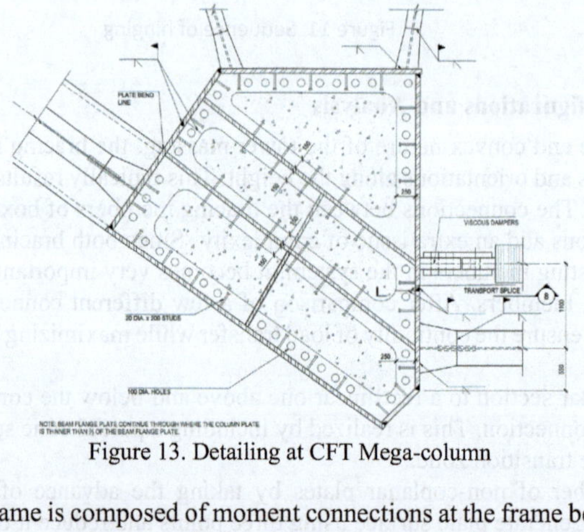

Figure 13. Detailing at CFT Mega-column

5. Since the exterior braced frame is composed of moment connections at the frame beam to column joints and frame beam to bracing connections, continuity ring plates are provided within the CFT sections to ensure the load transfer while allowing the concrete pass through. Considering the large size of the bracing members, ring plates are provided inside of the section to reduce the protrusion of structural members into exterior wall connections. As the width of frame beams are typically about one third of the bracing sections, ring plates are used instead of full width continuity plates.

With these connectivity rules the connection geometries were simplified while maintaining the continuity of the load transfer, although the tower geometry varies along the building height.

6.4 Foundation Settlement Analysis

The project site and adjacent region is composed of soft soil stratum of silty sand and silty clay. A pile supported mat foundation system was recommended based on the geotechnical conditions at the site. Piles were typically required to be 75m long to achieve the required capacity of 15000kN for a 1m diameter pile, from a combination of skin friction and end bearing. Tower mat was designed to be 5000mm thick, to distribute the load out to more piles outside of core and column footprint. Due to the low compressive modulus of the soil stratum at the base of the piles, initial analysis showed large settlement of approximately 150-175mm, which is consistent with measured settlement of a few tall buildings at nearby sites. Considering the sensitivity of a super tall tower to differential settlement, CABR Foundation Engineering Co. Ltd performed foundation and soil interaction finite element analysis. Detailed analyses indicated that in the worst case the settlement would be 196 mm, with the differential settlement of 0.11%, just over 0.1% of the targeted code limit. To control the differential settlement below 0.1%, additional piles were added along the east side, as shown below with the asymmetric layout of the piles.

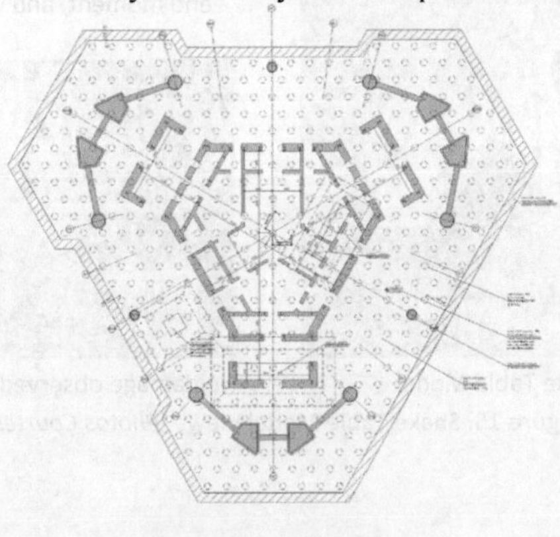

Figure 14. Tower Foundation Plan

6.5 Shake Table Analysis

A 1/35 scaled shake table test was performed at China Academy of Building Research (CABR). The goals for the test was to evaluate the structural performance under frequent earthquake (63% exceedance probability), moderate earthquake (10% exceedance probability), and rare earthquake (10% exceedance probability). Structural deformations and responses of individual members were monitored to verify whether performance targets for the project can be achieved. The test is also intended to find any weak local areas in the structure that may not have been captured by the analysis models. The testing model was built using the same material as the actual structure where possible. Where materials were replaced with other similar materials, material testing was performed to achieve a consistent relationship between materials. The shake table simulates motion in three directions and six degrees of freedom. Testing was performed using seven time history curves for frequent earthquake, three for moderate earthquakes, and one for rare earthquakes. These time histories matched those used in structural analysis and therefore allowed for the testing results to be compared with the corresponding analysis results. Acceleration and strain sensors were placed at various locations over the height of the model, including critical locations at mega columns, dampers and outriggers.

During testing, no apparent damage was observed at critical elements including transfer belt trusses, exterior mega braces and CFT columns which had been designed to remain elastic for rare earthquake events. As expected, under rare earthquake time history, cracking was observed at link beams ends and portions of the core walls due to tensile stress. The damage was found to be distributed, not concentrated, and the extent of the damages was not severe, which meant that the integrity of the structure was not compromised. The maximum interstory drifts observed were at 1/103 and 1/90 for X and Y direction correspondingly, close to drifts predicted by the elasto-plastic analysis and therefore considered satisfactory.

Typical link beam cracking due to shear and end moment, and wall tension cracking.

Shake Table Model

No damage observed at transfer belt trusses

Figure 15. Shake Table Model Views *(Photos Courtesy of CABR)*

7 CONCLUSIONS

This paper has presented an overview of the design innovations and processes that were implemented in the design of the Kunming Tower. The tower is currently under construction with anticipated completion in 2022. Extensive linear and non-linear analyses were performed to demonstrate that the structure met the prescriptive requirements of the Chinese code and achieved the targets for performance based design.

ACKNOWLEDGEMENTS

The authors would like to thank their colleagues Ronald Johnson and Zhaofan Li for their insight, expertise and individual contributions that resulted in successful design of the project. Additionally, the authors would like to thank the local design institutes East Architectural Design Institute and China Academy of Building Research for their collaboration during the design process.

REFERENCES

[1] Baker, W. F., "Energy Based Design of Lateral Systems", *Structural Engineering International Journal*, 2(2):99-102, 1992.

[2] Besjak, C., Biswas, P., Petrov, G., Streeter, M. and Devin A., "Effects of Perimeter to Core Connectivity on Tall Building Behavior", *International Journal of High-Rise Buildings* , 6(1):1-9, 2017

[3] Chen, J., Ding, H., Rankin, D., Baker, W., "Influence of The Mega Columns Selection For 528m Super Tall Tower Considering Creep and Shrinkage Construction Analysis.", *Building Structure*, 48(17):52–59, 2018

9TH INTERNATIONAL CONFERENCE ON COMPOSITE CONSTRUCTION IN STEEL AND CONCRETE
M. Knobloch, U. Kuhlmann, W. Kurz, M. Schäfer. (Eds.)
Online, Germany, July 27-29, 2021

NPS COMPOSITE BEAMS AND COLUMNS USED FOR THE ODENSE UNIVERSITY HOSPITAL

Stefano China* and Aroldo Tegon**

* Tecnostrutture S.r.l.
e-mail: schina@tecnostrutture.eu

** Odense Hospital Project Team Joint Venture I/S
e-mail: aroldo.tegon@cmbcarpi.it

Keywords: Composite Beams, Composite Columns, NPS, Composite Structural Elements, Practical application

Abstract. *This document presents the case study of the Odense University Hospital, built with the NPS steel-concrete composite frame. This is one of the most challenging projects currently in Europe: 250,000 square meters to be built in just over three years. The client has taken up the challenge of respecting timeframe and budget by focusing on offsite construction and digitalization of processes. Time constrains, safety and sustainability are the three main drivers of the project. Among these, the strict delivery time represents the main critical aspect. The expected turnkey delivery in 2022 assumed the choice of an extremely simple and fast construction method. Further advantages of using steel-concrete composite structures with reference to the Odense University Hospital Case are explained.*

1 CHARACTERISTICS OF THE PROJECT

The new Odense University Hospital is one of the most challenging building projects currently in Europe in terms of size and schedule objectives. About 250,000 square meters that must be built in just over three years. Almost 500 hospital bedrooms, 71 operating rooms, 68 intensive care stations, more than 450 laboratories, 57 diagnostic imaging rooms with the most advanced solutions in the sector, modern and innovative services for patients and visitors, systems for efficient and widespread logistics.

Hospital buildings at the OUH are closely clustered together in a dense circle, which is surrounded by a pedestrian ring and walkways that shoot off in spokes from the centre.

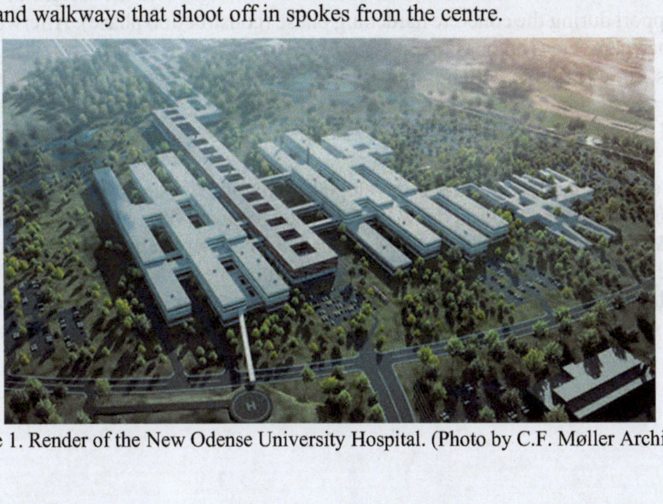

Figure 1. Render of the New Odense University Hospital. (Photo by C.F. Møller Architects)

Visitors and patients are encouraged to stroll around the property and take in the landscape, which features dense old forests, extensive fields, hedges, ponds and channels. The buildings are all designed to maximize daylighting. As Henning Larsen Architects says, "Numerous research projects have demonstrated that nature has a positive influence on patients' recovery. The objective of this project is to create a truly healing hospital environment merging all accessible knowledge, space, landscape and technology. "

The Odense hospital consists of three main bodies divided into seven buildings of four or five floors. The total area is 250,000 m². All of them have the same structural grid: main beams with 7.2 m span; hollow-core slab length is 7.7 or 4.5m. More than 1800 NPS PDTI columns and more than 5000 NPS beams are used for the construction. Approximately 40,000 linear meters of NPS beams and 25,000 meters NPS columns.

2 REQUIREMENTS OF THE CLIENT

From the beginning, the Odense Hospital Project Team Joint Venture I/S intended to manage this immense project by focusing on offsite construction and digitalization of processes. They have chosen Tecnostrutture as a partner since the company matches this working method and expectations. Time constrains, safety and sustainability are the three main drivers of the project.

Time constrains require to a productivity up to 15,000 m² per month of slabs cast. For the design, BIM collaboration tools with model sharing are required. Compact sections are welcome to allow optimization of interior spaces. "Quality of manufacture, onsite safety and the environmental issues are three fundamental processes for managing the construction site, especially given the big size. Building in a simple way makes control tasks easier", refers Elisa Galli. QHSE Manager of the Project.

3 REASONS FOR CHOOSING A STEEL-CONCRETE COMPOSITE FRAME

The expected turnkey delivery for 2022 means to build 250,000 m² in three years. This requires an extremely simple and rapid construction system. After careful technical and economic evaluations, the NPS composite steel-concrete solution was chosen.

3.1 New Performance System (NPS): description and research work

3.1.1 Description of the NPS construction system

Composite steel-concrete trussed beams were invented by Salvatore Leone, an engineer, in the 1960s. This prefabricated beam (a steel work truss) can bear its own weight and the weight of the slabs without any provisional support during the concrete hardening phase (construction phase). Afterword it collaborates with the cast in place concrete (end phase).

Over the years, the profitable collaborations between universities and companies led to the design, construction, and verification of beam-column joints for possible applications in earthquake-resistant frames (Scotta and Tesser [1], Amadio et al. [2], Tesser [3], Amato et al. [4], Mazzotti et al. [5], Scotta and Tesser [6]).

Starting from one beam, Tecnostrutture developed a complete system composed by self-supporting beams and columns. Finally, in 2011, the New Performance System NPS by Tecnostrutture was launched.

The columns (known as NPS PDTI columns) are made of steel with a metal reinforcement inside. They are produced in the factory and then filled with concrete on site. The same applies to the NPS beams: a steel truss structure, welded in the factory and completed with concrete after the placement. Beams and columns match any type of slabs, even with large spans. It ends up thus in a complete construction system.

Figure 2. NPS frame system under construction.

3.1.2 Research works

To develop the NPS column-beam construction system, several research works have been carried out over the years. Some are mentioned here as examples.

Tesser [3] has conducted a numerical analysis and experimental verifications about the behaviour of composite steel truss and concrete beams. Specific steel damage indexes, and a new concrete compressive damage evolution law have been developed to better evaluate the effect of confining reinforcement in RC structure. The developed method has been used to predict and analyse the experimental tests carried out at the University of Padua (Italy). The experimental results have demonstrated the efficiency of the proposed design method and the interesting features of the studied structural type such as its strength and ductility properties. An innovative composite beam-column joint has been studied for applications in medium-low seismicity regions. The results confirmed the efficiency of the proposed composite beam-column joint.

Later, Scotta and Argentoni [7] continued this research of University of Padua through cyclic loading tests on NPS composite structures. The goal was to evaluate the behaviour of NPS structure subjected to seismic equivalent static actions. The project proves the NPS response is comparable to the performance of traditional RC structures.

Figure 3. Experimental setup at the University of Padua (Italy).

Bellotti, Bolognini and Nascimbene of EUCENTRE [8] developed simplified parametric procedures for evaluating the equivalent stiffness of a composite steel truss and concrete beam, to properly calibrate structural analysis models using beam elements.

Research activities continue constantly. Currently there are ongoing independent research projects at the University of Washington, EUCENTRE of Pavia, ETH Zurich, University College of London, University of Camerino and University of Padua.

3.2. Benefits of NPS offsite construction system

3.2.1 Structural performance

Once the concrete casting on site has been completed, the structure has the typical characteristics of monolithic reinforced concrete frames. This guarantees the expected robustness required by the Danish legislation.

The high ductility of NPS composite structures ensures the strength and sturdiness of the work in the event of exceptional events such as earthquakes, wind, and explosions/subsidence, compliant with the method prescribed by the Danish "key element" standard.

The self-supporting NPS frame is simply completed with a slab with the same features: the hollow-core slab. The metal structure of beams and columns can ensure resistance and stability during the assembly phases even in the presence of an extremely heavy type of floor, in this case hollow-core.

3.2.2 Construction speed

NPS enables an instant load bearing ceiling, which permits to work underneath immediately after placement. There is no need to wait 28 days until the concrete cures completely. This allows an overlap of building phases: during the placement of NPS beams at level 2, the installer can work on level 0 and so on until the top of the building. The average time saved compared to traditional systems is 40%, as demonstrated by the comparative GANTT.

The offsite system eliminates the risk of delays and other external variables. With NPS, weather conditions are not as crucial as with the traditional way of building. The final cost meets budget estimates because there are very few unknown factors to manage.

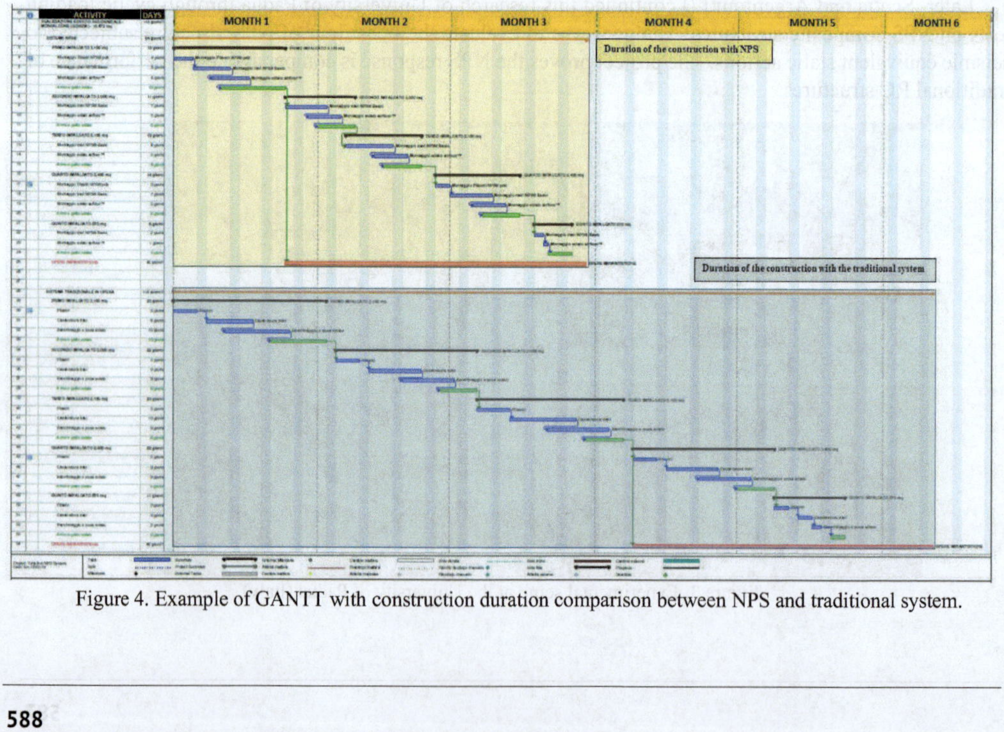

Figure 4. Example of GANTT with construction duration comparison between NPS and traditional system.

3.2.3 Safety on site

The Covid-19 health emergency has changed both the way of living and that of working. To achieve the primary objective of ensuring health and safety of employees on the construction site, physical distancing plays an important role. To address this topic, one solution consists in the application of offsite construction [9].

Critical construction phases move from the construction site to a controlled industrial environment. Only the assembly phase remains on site, thus drastically reducing the number of employees involved and ensuring safer working methods. The following are some among the most positive aspects: construction process speed-up, fewer but more skilled workers, protected and controlled working environment, design with precision, reducing errors as well as the production of waste materials, transferring the quality and performance of industrial processes into the building sector.

NPS beams and columns are installed by a team of three people. 80% less manpower at the construction site compared to traditional systems reduces the risks of accidents. Safety devices integrated in the beams allow to safely work at height.

Figure 5. An integrated setup accommodates safety parapets, allowing installation on the ground.

3.2.4 Low environmental impact

The integrated use of steel and concrete leads to greater structural efficiency. NPS structures use less raw materials and have compacter sections compared to traditional structures, for example in reinforced concrete.

The comparative Life Cycle Assessment carried out by the University of Genoa [10] notes that in the analysed cases a building made with NPS BASIC beams and NPS PDTI columns has a lower environmental impact of over 20% compared to steel or reinforced concrete structures, regarding primary energy consumption and CO2 equivalent emissions.

Structural efficiency, high recycled content, reuse of materials after deconstruction make the system compliant with environmental certification (e.g. LEED) and protocols of sustainability. The limited environmental impact of NPS is proven by an Environmental Product Declaration for each element.

The shipment was organised in order to minimize the environmental footprint. Structures' transportation took place for 1400 km by rail. Only the first and last kilometres are covered by road.

4 STRUCTURAL MODELLING AND BIM

From the beginning, the digitalization of processes was selected as a key factor for the success the project.

4.1 Structural modelling

Tecnostrutture introduced the concept of a continuous resistant frame connected to the shear walls. A concrete topping on the hollow-core enabled the creation of rigid floors.

The modelling of NPS composite beams took in consideration both construction and end phase.

During the assembly phase on site, NPS beams are steel elements with reduced stiffness in an isostatic scheme. Thus, the acting loads are only the own weights of the raw structures.

In the end phase, after the casting of concrete on site, NPS structures are steel-concrete composite elements, with stiffness equal to the composite section, in a hyperstatic pattern. Therefore, the loads acting are the permanent and variable design overloads; in addition to actions caused by events such as wind and earthquakes.

To correctly model the structural behaviour described above, beam elements of controlled stiffness were inserted according to the modelling phases. For the construction phase, the stiffness is reduced to simulate a "hinge" type scheme. For the end phase the stiffness is set equal to that of the beam and the continuity is correctly simulated.

The overlap of the effects allows the linear summation of the schemes described above, allowing Tecnostrutture´s designer to correctly define the design stresses.

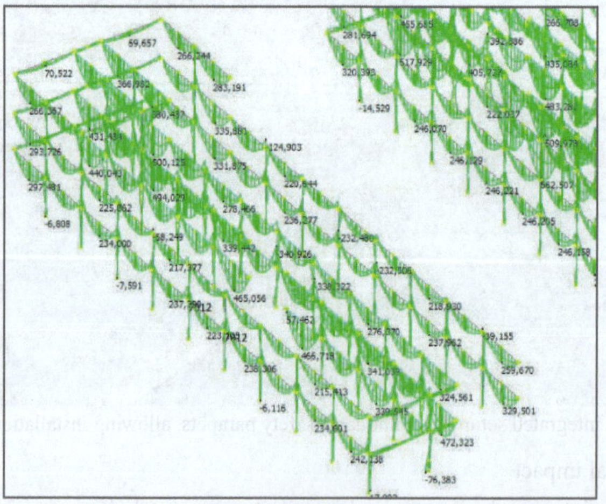

Figure 6. Detail of bending moments of NPS beams and PDTI columns designed with Straus FEM model.

The design of the structure was completed by creating a BIM model, using Tekla Structures software, with a high level of detail (LOD400).

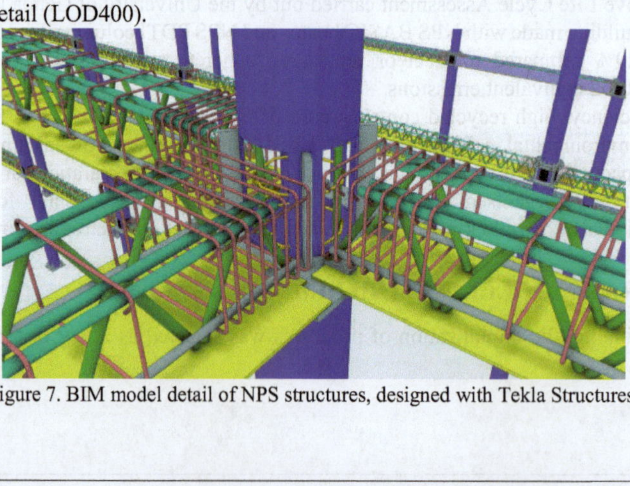

Figure 7. BIM model detail of NPS structures, designed with Tekla Structures.

4.2 The BIM managed design phase

The solid starting point for all Tecnostrutture's designs is the use of Tekla Structures. BIM is used for the modelling of the works, for the creation of shop drawings and for the management of the assembly phases and related traceability.

Even more due to the Covid-19 health emergency, the processes of digitalization and the adoption of agile work have recently seen a strong acceleration. In this period, BIM collaboration tools, such as Tekla Model Sharing, have become essential also for all the stakeholders of the Odense University Hospital team.

Designing with Tekla Model Sharing software made it possible to:

- Manage multiple revisions of the model, uploaded by different users, thus also viewing each change made;
- Retrieve previous versions in case of need;
- Work locally and upload the model update at a later stage.

5 THE PLACEMENT OF NPS FRAME

5.1 The process of installation

The NPS system, consisting of beam and column, allows to build with a team of only three members, respecting physical distancing requirements.

The installation of each NPS column or beam requires one crane operator and two specialized workers, and it takes about five minutes. Such a team mounts up to 400 m2 per day consisting of beams, columns, and slabs.

It starts with the installation of the multi-storey columns: bolt the base and unhook the top. Inside the column rebar adherence connections will create the continuity between the composite structure and the other elements.

It continues with the placement of the NPS beams. They will be laid with the safety parapets already installed, leaning the end of the beam on the column. There is no need to bolt them, simple support is sufficient. Then we proceed with the laying of the slab.

Figure 8. Unload of NPS PDTI column.

Figure 9. Lifting of NPS PDTI column.

Figure 10. Placement of the NPS PDTI column.

Figure 11. NPS PDTI column base is bolt.

Figure 12. Unhook of the NPS PDTI column top.

Figure 13. Laying of NPS beam.

Figure 14. Laying of hollow-core slab.

Only at this point the top casting of the slabs, beams, and columns occurs, all in the same day.

Figure 15. Cast of concrete.

As beams, columns and slabs are installed precast façades have been fitted. Considering the big size of the project, this operation requires maximum precision. NPS structures met these needs. Façades were easily hanged on the column.

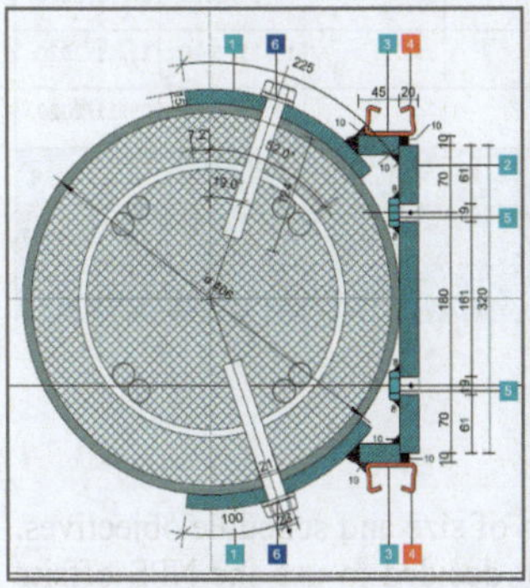

Figure 16. Drawing of setup for façades.

Figure 17. Set up on site.

Figure 18. Façades hanged on the columns.

In ten months, 120,000 m^2 of beams, columns and slabs have been installed, despite the winter snow and the lockdown due to the pandemic. For the client, having a unique supplier for the structural frame has simplified the project management, making the work safer and faster.

5.2 The planning

The client defines the installation schedule that has to be respected. Taking Building 34 as an example, the program for assembling the NPS frame includes:

- 3 days for the assembly of 50 pillars, including plumbing
- 3 days for the assembly of 62 beams, including alignment and perimeter shoring, where required
- 5/6 days for the assembly of 2200 m2 of hollow core slab
- 6 days for the casting of the floor.

OUH.DP05.B34.L00.ST.ZA4	ZA4	119d	26/05/2020	06/11/2020
OUH.DP05.B34.L00.ST.ZA4.15	Element Pillar PRC - 00 - ZA4	3d	05/06/2020	09/06/2020
OUH.DP05.B34.L00.ST.ZA4.16	Element beam PRC - 00 - ZA4	3d	16/06/2020	18/06/2020
OUH.DP05.B34.L00.ST.ZA4.19	Concrete hollow deck - 00 - ZA4	7d	19/06/2020	29/06/2020
OUH.DP05.B34.L01	LEVEL 01	487d	16/07/2020	27/05/2022
OUH.DP05.B34.L01.ST	STRUCTURE	84d	16/07/2020	10/11/2020
OUH.DP05.B34.L01.ST.ZA4	ZA4	84d	16/07/2020	10/11/2020
OUH.DP05.B34.L01.ST.ZA4.15	Element Pillar PRC - 01 - ZA4	3d	30/07/2020	03/08/2020
OUH.DP05.B34.L01.ST.ZA4.16	Element beam PRC - 01 - ZA4	3d	04/08/2020	06/08/2020
OUH.DP05.B34.L01.ST.ZA4.19	Concrete hollow deck - 01 - ZA4	7d	07/08/2020	17/08/2020
OUH.DP05.B34.L02	LEVEL 02	375d	02/09/2020	08/02/2022
OUH.DP05.B34.L02.ST	STRUCTURE	52d	02/09/2020	12/11/2020
OUH.DP05.B34.L02.ST.ZA4	ZA4	52d	02/09/2020	12/11/2020
OUH.DP05.B34.L02.ST.ZA4.15	Element Pillar PRC - 02 - ZA4	3d	17/09/2020	21/09/2020
OUH.DP05.B34.L02.ST.ZA4.16	Element beam PRC - 02 - ZA4	3d	22/09/2020	24/09/2020
OUH.DP05.B34.L02.ST.ZA4.19	Concrete hollow deck - 02 - ZA4	7d	25/09/2020	05/10/2020
OUH.DP05.B34.L03	LEVEL 03	419d	20/10/2020	27/05/2022
OUH.DP05.B34.L03.ST	STRUCTURE	25d	20/10/2020	23/11/2020
OUH.DP05.B34.L03.ST.ZA4	ZA4	25d	20/10/2020	23/11/2020
OUH.DP05.B34.L03.ST.ZA4.15	Element Pillar PRC - 03 - ZA4	3d	05/11/2020	09/11/2020
OUH.DP05.B34.L03.ST.ZA4.16	Element beam PRC - 03 - ZA4	3d	10/11/2020	12/11/2020
OUH.DP05.B34.L03.ST.ZA4.19	Concrete hollow deck - 03 - ZA4	7d	13/11/2020	23/11/2020
OUH.DP05.B36	BUILDING 36	533d	07/04/2020	21/04/2022
OUH.DP05.B36.L00	LEVEL 00	497d	05/05/2020	30/03/2022
OUH.DP05.B36.L00.ST	STRUCTURE	173d	05/05/2020	31/12/2020

Figure 19. Project schedule for Building 34 relevant to the structural frame.

6 CONCLUSIONS

The new Odense University Hospital is a challenging project in terms of size and schedule objectives. About 250,000 m^2 that must be built in just over three years. The client decided to use the NPS offsite construction solution, composed of self-supporting steel-concrete beams and columns, offering the expected robustness required by the Danish legislation.

Roberto Pettinella, Construction Manager, reports "We have decided to adopt this construction system for its impressive installation speed, and for the reduced size of the elements, which allowed us to optimise all the interior spaces. A further key factor was teamwork, together with the project's Joint Venture members." All stakeholders are used to interacting through advanced BIM three-dimensional models.

This offsite system made it possible to achieve the expected productivity levels up to 15,000 square meters per month of slabs cast. This happened respecting workplace safety and pandemic prescriptions of physical distancing.

Structural efficiency, high recycled content, reuse of materials after deconstruction and intermodal transport are compliant with the environmental requirements set for the project.

Finally, The NPS offsite system has met the efficiency requirements given by an evolved and challenging public client such as the Danish Region Syddanmark, that appreciated the versatility and simplicity of the structural elements.

Figure 20. Odense University Hospital construction site in August 2020

REFERENCES

[1] Scotta R, Tesser L. *Preliminary experiences and basic concepts on the structural performance of hybrid trussed beams*. 4th Spec. Conference on The Conceptual Approach to Structural Design, Venice, 2007.

[2] Amadio C., Macorini L., Sorgon S., Suraci G, *Analisi di un sistema ibrido sismo-resistente costituito da elementi tralicciati in acciaio inglobati nel calcestruzzo*, 17° Congresso C.T.E. Rome: 801-810, 2008.

[3] Tesser, L., *Composite Steel Truss and Concrete Beams and Beam-Column Joints for Seismic Resistant Frames: Modelling, numerical analysis and experimental verifications*, Ph.D. Thesis, University of Padua, 2009.

[4] Amato G., Badalamenti V., Colajanni P., La Mendola L, *Comportamento ciclico delle connessioni tra travi prefabbricate reticolari miste e pilastri in c.a.*, 18° Congresso C.T.E. Brescia: pp. 3-12, 2010.

[5] Mazzotti C., Vincenzi L., Savoia M., Ferrari M., *Comportamento di nodi trave-colonna di strutture parzialmente prefabbricate*, 18° Congresso C.T.E. Brescia: pp. 255-265, 2010.

[6] Scotta R., Tesser L., *Comportamento di nodi trave-pilastro sismo-resistenti in struttura mista di tipo tralicciato soggetti ad azioni cicliche*, Progettazione Sismica 3, pp. 47-62, 2011.

[7] Scotta R., Argentoni A., *Prove cicliche su nodi in struttura mista*, Università degli Studi di Padova, 2011.

[8] Bellotti D, Bolognini D, Nascimbene R, *Sviluppo di modelli per la definizione della rigidezza equivalente di differenti configurazioni di travi tralicciate*, memoria XIV convegno ANIDIS, Bari 18-22 Settembre 2011.

[9] Calderini C., Piccardo C., *Comparative LCA of traditional structures and structures made with the NPS® System*, Tecnostrutture ACADEMY https://academy.tecnostrutture.eu/en/wp-content/uploads/sites/2/2019/09/Comparative-LCA-of-traditional-structures-and-the-NPS%C2%AE-System.pdf, 2019.

[10] Daniele G., Covid-19 health emergency: what changes in the way you work on site?, Tecnostrutture ACADEMY, https://academy.tecnostrutture.eu/en/publications/covid-19-health-emergency-what-changes-in-the-way-you-work-on-site/, 2020.

SEISMIC BEHAVIOUR

9TH INTERNATIONAL CONFERENCE ON COMPOSITE CONSTRUCTION IN STEEL AND CONCRETE
M. Knobloch, U. Kuhlmann, W. Kurz, M. Schäfer. (Eds.)
Online, Germany, July 27-29, 2021

CYCLIC BEHAVIOR OF COMPOSITE CONNECTIONS IN COMPOSITE FLOOR DIAPHRAGMS

Nicholas E. Briggs*, Raul E. Avellaneda-Ramirez, Kyle Coleman*, Benjamin W. Schafer***, Matthew R. Eatherton**, W. Samuel Easterling**** and Jerome F. Hajjar***

* Department of Civil and Environmental Engineering, Northeastern University
e-mails: briggs.n@northeastern.edu, ky.coleman@northeastern.edu, jf.hajjar@northeastern.edu

** Department of Civil and Environmental Engineering, Virginia Polytechnic Institute and State University
e-mail: rear93@vt.edu, meather@vt.edu

*** Department of Civil Engineering, Johns Hopkins University
e-mail: schafer@jhu.edu

**** College of Engineering, Iowa State University
e-mail: wse@iastate.edu

Keywords: Composite, Steel, Deck, Diaphragm, Seismic, Earthquake.

Abstract. *This research explores the cyclic behavior and strength of the composite concrete-filled steel deck diaphragm system through experimental pushout and cantilever diaphragm tests. This work is part of the Steel Diaphragm Innovation Initiative (SDII), a multi-institutional project that is investigating the behavior of composite steel deck diaphragms in steel buildings structures in high seismic zones. Sixteen unique pushout tests were conducted at Northeastern University as part of SDII. These specimens had variations in the orientation of the steel deck, concrete density, effective slab width, shear stud position and grouping. Additionally, seven of these tests were cyclic. Eight cantilever diaphragm tests were conducted at Virginia Tech as part of SDII. The parameters included deck height, slab thickness, concrete density, layout of shear studs, and reinforcement. These experimental tests provide valuable knowledge that aids in the characterization of the behavior and strength prediction of this structural system.*

1 INTRODUCTION

During seismic events, structural floor and roof diaphragms are responsible for translating the structure's inertial forces horizontally to the lateral force resisting systems. Multiple structures experienced inelastic diaphragm failures during the 1994 Northridge Earthquake when the diaphragms exhibited insufficient strength. One particular structure was the California State University (CSU) at Northridge Parking Garage. The CSU parking garage was a precast concrete structure that had interior gravity columns and exterior perimeter columns designed as a ductile concrete lateral force resisting frame. Figure 1 [1] shows the collapsed structure, in which the exterior lateral force resisting columns exhibited extreme ductility. The diaphragm, in this case, had insufficient strength to transfer the inertial loads to the exterior lateral force resisting system. The 1994 Northridge Earthquake resulted in eight parking structure failures (precast and cast-in-place), 40,000 heavily damaged structures, and 58 deaths. These led to changes to the Universal Building Code (UBC), which governed seismic design in this portion of the United States at the time. The changes sought to address potential inadequacies of collector elements and diaphragm systems as well as required deformation compatibility for structural members that are not components of the lateral force resisting system.

Figure 1. California State University parking garage collapse [1].

This type of failure presented concerns to designers of other structural diaphragm systems. Five steel industry organizations involved in the manufacture, design, and construction processes for composite steel deck diaphragms sponsored an initiative to investigate the seismic stability and resulting safety of this commonly used structural floor and roof system. The goal of the SDII project is to develop an understanding of composite concrete-filled steel deck diaphragms as an integrated system within steel building structures, integrate that understanding into accurate building models and design procedures, and utilize those advances to explore and prototype novel steel deck diaphragm solutions to create optimized structural response and increase building structural resilience. This paper discusses two particular research activities that are part of SDII's research efforts. A set of sixteen experimental pushout test specimens are presented. These specimens utilize a redesigned testing setup to permit cyclic testing of the headed shear stud connector in the composite concrete-filled steel deck diaphragm system. Additionally, a set of eight cantilever diaphragm specimens are presented, which characterize the strength, stiffness, ductility, residual strength, and failure modes of composite concrete-filled steel deck diaphragms.

2 LITERATURE REVIEW

Previous experimental research into the structural stability of composite concrete-filled steel deck diaphragms has centered mainly around the use of three types of tests: pushout tests, beam tests, and cantilever diaphragm tests. This research investigates the performance of this structural system in high seismic regions and is thus primarily interested in the in-plane strength of the diaphragm system. Pushout and cantilever diaphragm tests best demonstrate the in-plane strength of composite steel deck diaphragms. A literature review of previous experimental specimens of this type is presented.

2.1 Pushout tests

Prior to the use of steel deck, pushout tests were conducted on solid slab diaphragm specimens [2-5]. As the use of corrugated steel deck in composite diaphragm construction became increasingly common, this spurred additional research and experimental testing. Robinson conducted 17 pushout specimens investigating the effect that steel deck orientation had on stud strength [6]. Jayas and Hosain [7-8] expanded understanding of the effect of the steel deck on performance by investigating the effect of the rib geometry and longitudinal spacing of the shear connectors on strength. Forty-two pushout tests were performed by Lloyd and Wright [9], which further examined the strength performance of composite steel deck diaphragm pushout specimens while varying basic parameters, such as steel deck geometry, stud position and slab size. Sublett [10] performed a set of 36 pushout tests, which demonstrated that for a range of parameters the AISC strength equation was unconservative. This finding was reinforced by Easterling et al. [11] who emphasized the findings of Sublett [10], pointing to how the strength prediction was unconservative for particular configurations of the tested parameters, which include depth of steel deck, slab reinforcement, shear stud height, concrete density, shear stud position, and number of shear studs in a rib. Lyons [12] also performed 48 pushout tests to determine how stud tensile strength and concrete compressive strength affect this connection. Rambo-Roddenberry [13-15] utilized strength results from many of these experimental pushout tests to demonstrate that the AISC shear strength equation for headed shear connectors was overestimating strength and successfully proposed and implemented adjustments accordingly.

Few composite steel deck pushout tests have investigated cyclic performance of headed shear connectors and even fewer have examined low-cycle failure of the studs. Hawkins and Mitchell [16] tested 10 reverse cyclic pushout specimens and 13 companion monotonic specimens. These specimens explored how the presence, geometry and orientation of steel deck affects the strength and behavior of these specimens. This research analyzed the failure mechanisms exhibited by the specimens and suggested that this affects the ductility, strength, and cyclic loading response of the specimens. Additional cyclic composite steel deck pushout tests were conducted by Bursi and Gramola [17] to investigate the seismic performance of this structural system and the influence that boundary conditions and loading protocol had on the system. Seven of the eleven pushout tests conducted were cyclic and were tested with one of the four cyclic protocols employed in this work. These specimens were larger, having eight headed stud connectors per slab instead of two used in the slabs tested by Hawkins and Mitchell [16].

2.2 Cantilever diaphragm tests

To document the ultimate strength, ductility, stiffness, and failure mechanisms exhibited by composite steel deck diaphragms, cantilever diaphragm tests have been conducted in prior work. Luttrell [18] conducted 11 monotonic cantilever diaphragm tests to determine the effect that lightweight insulating concrete, purlin spacing, and steel deck profile had on the strength and behavior of diaphragms that used welding washers to adhere the steel deck to the underlying steel members. This was followed by additional research by Davies and Fisher [19], who tested four monotonic cantilever composite steel deck diaphragm specimens. These specimens were designed to study failure modes and to develop strength and flexibility prediction models for this structural system. Porter and Greimann [20] and Porter and Easterling [21-22] worked on a two phase research project testing 32 experimental cantilever diaphragm specimens. Investigated parameters of this research include, the effect that slab size and thickness, concrete compressive strength, steel deck type, depth, thickness, strength, and connector quantity and type had on the behavior and strength performance of the specimens. The specimens utilized either arc spot welds or headed shear connectors and were all tested cyclically. Altoubat et al. [23] conducted twelve monotonic cantilever diaphragm specimens with two types of steel deck profiles to identify the effect that fiber reinforcement of the concrete slab has on the strength and behavior of the structural system. Three cantilever diaphragm tests were conducted by Rezeian et al. [24], who provided more realistic boundary conditions that emulate the support provided by the seismic resisting system. These specimens used headed shear stud connectors and were monotonically loaded.

3 PUSHOUT TESTING PROGRAM

A set of sixteen composite concrete-filled steel deck pushout specimens were tested to explore the cyclic behavior and strength of the headed steel shear stud connectors in this structural system. The specimens are presented in Table 1 and are designed to test the effect that various parameters have on performance. The parameters of interest include steel deck orientation, concrete density, edge effects, shear stud grouping, shear stud position, and loading profile. All specimens used a 76 mm, 20 gauge steel deck and 19 mm diameter shear studs with a post welded height of 114 mm. Welded wire reinforcement for the lightweight and normal weight concrete specimens was 6x6 – W1.4/W1.4 and 6x6 – D2.1/D2.1

respectively. Specimens with the steel deck oriented perpendicular to the steel beam have additional reinforcement. A #4 longitudinal reinforcing bar with 19 mm of clear cover was positioned at the top outer edges of these slabs to provide tensile strength in the slab to prevent upward buckling of the concrete, an undesirable failure mechanism. Additionally, a reinforcement cage consisting of two transverse #4 reinforcing bars spaced 76 mm apart vertically with #3 stirrups spaced at 127 mm on center was constructed in the last down rib at each of the two bearing ends of these concrete slabs. There are no shear studs present in these down rib and thus the reinforcement does not affect composite action and instead aids in strengthening the concrete slab in bearing at this boundary condition. Center specimens have slab dimensions of 1524 mm long by 1219 mm wide, edge specimens maintain the same 1524 mm length, but are 813 mm wide. The lightweight and normal weight concrete mixes were designed to achieve a 28-day compressive strength of 32.4 MPa with a nominal strength of 27.6 MPa, but reached strengths of 28.3 MPa and 35.1 MPa respectively as an average of three 152 mm diameter concrete cylinder tests which were cured while immersed in water. Test day concrete strengths are presented in Table 3 and are an average of three 152 mm diameter concrete cylinders cured in a comparable manner as the specimens that they represent.

Table 1: Pushout test matrix.

Test Number	Steel Section	Steel Deck	Concrete Weight	Edge or Center Specimen	Studs	Stud Location	Loading Protocol
Pe-LW-C-1-S-M	W10x39	Perpendicular	Lightweight	Center	1 @ 305 mm	All Strong	Monotonic
Pe-LW-C-1-W-M	W10x39	Perpendicular	Lightweight	Center	1 @ 305 mm	All Weak	Monotonic
Pe-LW-C-1-W-C	W10x39	Perpendicular	Lightweight	Center	1 @ 305 mm	All Weak	Cyclic
Pe-LW-C-1-B-M	W10x39	Perpendicular	Lightweight	Center	1 @ 305 mm	Balanced	Monotonic
Pe-LW-C-1-B-C	W10x39	Perpendicular	Lightweight	Center	1 @ 305 mm	Balanced	Cyclic
Pe-LW-C-2-S-M	W10x39	Perpendicular	Lightweight	Center	2 @ 305 mm	All Strong	Monotonic
Pe-LW-C-2-W-M	W10x39	Perpendicular	Lightweight	Center	2 @ 305 mm	All Weak	Monotonic
Pe-LW-C-2-W-C	W10x39	Perpendicular	Lightweight	Center	2 @ 305 mm	All Weak	Cyclic
Pe-LW-C-2-B-M	W10x39	Perpendicular	Lightweight	Center	2 @ 305 mm	Balanced	Monotonic
Pe-LW-C-2-B-C	W10x39	Perpendicular	Lightweight	Center	2 @ 305 mm	Balanced	Cyclic
Pa-LW-C-1-A-M	W10x39	Parallel	Lightweight	Center	1 @ 305 mm	Alternate Sides	Monotonic
Pa-LW-C-1-A-C	W10x39	Parallel	Lightweight	Center	1 @ 305 mm	Alternate Sides	Cyclic
Pa-NW-C-1-A-M	W10x39	Parallel	Normal	Center	1 @ 305 mm	Alternate Sides	Monotonic
Pa-NW-C-1-A-C	W10x39	Parallel	Normal	Center	1 @ 305 mm	Alternate Sides	Cyclic
Pa-LW-E-1-A-M	W10x39	Parallel	Lightweight	Edge	1 @ 305 mm	Alternate Sides	Monotonic
Pa-LW-E-1-A-C	W10x39	Parallel	Lightweight	Edge	1 @ 305 mm	Alternate Sides	Cyclic

The test setup is presented in Figure 2 and depicts a deck perpendicular pushout specimen in the testing rig. Key features of this test setup relative to other common setups for pushout tests are that it has four groups of studs along the length, rather than the usual one or two studs in most pushout tests [9-15]. This enables a more detailed investigation of the distribution of shear transfer along the length due to cyclic loading as may be seen in a collector in a diaphragm. In addition, gravity loading is applied to the specimen with a high degree of consistency. The specimen beam, a W10x39, is supported below by three Hilman rollers and the concrete slab is restrained laterally as demonstrated in the diagram. An actuator frames horizontally into each end of the specimen beam. These actuators are centered vertically and aligned with the interface at the top of the specimen beam at the composite interface to reduce eccentricities. A third actuator applied downward load to the specimen through a spreader beam to simulate gravity loading and prevent an undesirable failure mechanism. This actuator applies a constant load throughout the test equal to 10% of the expected specimen cumulative stud shear strength. The two horizontal actuators act in antisymmetry and either apply the monotonic or cyclic loading protocol. The monotonic protocol is presented in Table 2. The cyclic loading protocol is adapted from the beam-to-column cyclic protocol outlined in the AISC Seismic Provisions [25]. The protocol used in all seven of the cyclic pushout tests is presented in Table 3.

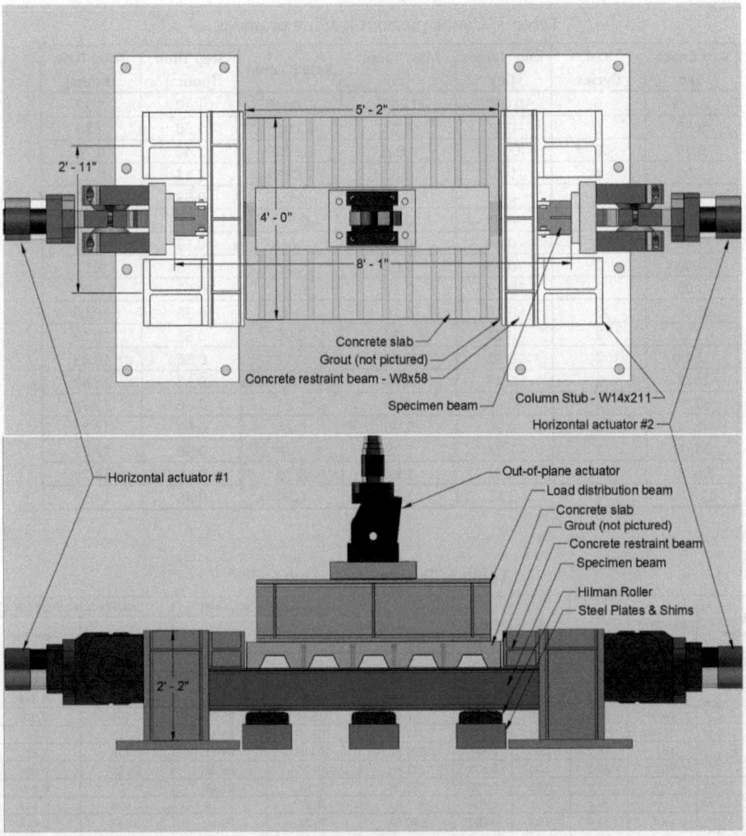

Figure 2. Pushout test setup.

The ultimate strength of the pushout test specimens and the expected strength specified by AISC are presented in Table 4 [26]. On average, considering measured properties, the test to predicted strength ratio was 0.807 with a standard deviation of 0.135. However, the results are more consistent, and have a lower standard deviation, if they are separated into the following subgroups. Deck perpendicular specimens with 1 stud every 305 mm have an average test to predicted ratio of 0.802 with a standard deviation of 0.0462. Deck perpendicular specimens with 2 studs every 305 mm have an average test to predicted ratio of 0.663 with a standard deviation of 0.0677. Lastly, deck parallel specimens have an average test to predicted ratio of 0.957 with a standard deviation of 0.0345. The peak strength reached in each loaded direction of the cyclic specimens is reported, however the East direction is the initial direction of loading and is used as the governing specimen strength.

Table 2: Monotonic pushout loading protocol.

Beginning Location ["]	Ending Location ["]	Rate ["/min]	Step Time [min]	Start Time [min]	End Time [min]
0	0.06	0.0025	24.00	0.00	24.00
0.06	0.24	0.015	12.00	24.00	36.00
0.24	0.96	0.06	12.00	36.00	48.00
0.96	2.56	0.24	6.67	48.00	54.67

Table 3: Cyclic pushout loading protocol.

% of Critical Slip	# of Cycles	Min. Target [in]	Max. Target [in]	Rate ["/min]	Step Time [hours]	Total Time [hours]
0.375	6	-0.0075	0.0075	0.0025	1.20	1.20
0.50	6	-0.01	0.01	0.0025	1.60	2.80
0.75	6	-0.015	0.015	0.0025	2.40	5.20
1.00	4	-0.02	0.02	0.0025	2.13	7.33
1.50	2	-0.03	0.03	0.0075	0.53	7.87
2.00	2	-0.04	0.04	0.015	0.36	8.22
3.00	2	-0.06	0.06	0.015	0.53	8.76
4.00	2	-0.08	0.08	0.03	0.36	9.11
6.00	2	-0.12	0.12	0.03	0.53	9.64
8.00	2	-0.16	0.16	0.06	0.36	10.00
12.00	2	-0.24	0.24	0.06	0.53	10.53
16.00	2	-0.32	0.32	0.12	0.36	10.89
24.00	2	-0.48	0.48	0.12	0.53	11.42
32.00	2	-0.64	0.64	0.24	0.36	11.78
48.00	2	-0.96	0.96	0.24	0.53	12.31
64.00	2	-1.28	1.28	0.48	0.36	12.67
96.00	2	-1.92	1.92	0.48	0.53	13.20
128.00	2	-2.56	2.56	0.96	0.36	13.56

Table 4: Pushout test results.

Test Number	Direction	f'_c [MPa]	R_g	R_p	Specimen Strength [kN]	AISC Strength [kN/stud]		Test/Predicted Strength	
						Measured	Nominal	Measured	Nominal
Pe-LW-C-1-S-M	N/A	36.1	1.00	0.750	390.24	115.10	95.80	0.848	1.018
Pe-LW-C-1-W-M	N/A	36.1	1.00	0.600	295.72	92.08	76.64	0.803	0.965
Pe-LW-C-1-W-C	East	36.1	1.00	0.600	267.96	92.08	76.64	0.728	0.874
	West	36.1	1.00	0.750	352.61	115.10	95.80	0.766	0.920
Pe-LW-C-1-B-M	N/A	36.1	1.00	0.675	312.93	103.59	86.22	0.755	0.907
Pe-LW-C-1-B-C	East	36.1	1.00	0.675	303.81	103.59	86.22	0.733	0.881
	West	36.1	1.00	0.675	311.69	103.59	86.22	0.752	0.904
Pe-LW-C-2-S-M	N/A	39.4	0.85	0.750	553.04	97.83	81.43	0.707	0.849
Pe-LW-C-2-W-M	N/A	39.4	0.85	0.600	435.97	78.27	65.14	0.696	0.837
Pe-LW-C-2-W-C	East	39.4	0.85	0.600	340.69	78.27	65.14	0.544	0.654
	West	39.4	0.85	0.750	440.02	97.83	81.43	0.562	0.675
Pe-LW-C-2-B-M	N/A	39.4	0.85	0.675	411.77	88.05	73.29	0.585	0.702
Pe-LW-C-2-B-C	East	39.4	0.85	0.675	424.18	88.05	73.29	0.602	0.723
	West	39.4	0.85	0.675	429.79	88.05	73.29	0.610	0.733
Pa-LW-C-1-A-M	N/A	39.1	1.00	0.750	424.94	115.10	95.80	0.923	1.109
Pa-LW-C-1-A-C	East	39.1	1.00	0.750	399.18	115.10	95.80	0.867	1.042
	West	39.1	1.00	0.750	414.33	115.10	95.80	0.900	1.081
Pa-NW-C-1-A-M	N/A	45.2	1.00	0.750	456.70	115.10	95.80	0.992	1.192
Pa-NW-C-1-A-C	East	45.0	1.00	0.750	465.10	115.10	95.80	1.010	1.214
	West	45.0	1.00	0.750	481.52	115.10	95.80	1.046	1.257
Pa-LW-E-1-A-M	N/A	39.7	1.00	0.750	440.19	115.10	95.80	0.956	1.149
Pa-LW-E-1-A-C	East	39.7	1.00	0.750	412.57	115.10	95.80	0.896	1.077
	West	39.7	1.00	0.750	408.97	115.10	95.80	0.888	1.067

Through analysis of the data collected during these tests, the load transferred through each shear stud or shear stud group was calculated. It was shown that up until the occurrence of significant initial damage, such as preliminary concrete cracking, the load transferred was approximately equal between the studs or stud groups. Due to the large specimen size, each specimen has four shear studs or four groups of two shear studs. This finding validates a key quality assumed composite connection.

The strength versus slip results from the monotonic deck perpendicular and deck parallel pushout tests are presented in Figures 3a and 3b respectively. The shear stud position in the deck perpendicular specimens influence the specimen behavior, consistent with findings in the literature [13-16]. Specimens that had all of their shear studs in the strong position were initially very stiff, rising approximately linearly to their ultimate strength. This was then followed by a drop in strength due to concrete cracking and rib failure. Conversely, specimens that had all of their shear studs in weak position were initially just as stiff as those with studs in the strong position, but progressively softened until the ultimate strength was reached at a much larger slip value than the strong position specimen counterparts. This behavior is due to the limited concrete material supporting the shear stud, which allows the shear stud to slowly break through this concrete and peal through the steel deck. Following ultimate strength, these weak position specimens slowly lost their strength as the shear studs could gradually break free of the concrete slab. The deck perpendicular specimens however required significant concrete damage to release the shear studs, which resulted in a shaper decline in post peak strength. Specimens with 50% of their shear studs in strong and 50% of their shear studs in weak positions exhibited a mixture of this behavior, falling between the strong and weak specimens. The deck parallel pushout specimens were initially even stiffer than the deck perpendicular specimens. The force supported by these specimens quickly rose, until some initial concrete cracking occurred, which caused a small drop in strength. After additional slip, the load carried by the specimens increased up until ultimate strength was reach. This was followed by an initially gentle decline in strength, characterized by concrete cracking, followed by a significant drop in strength as the shear studs fail in shear.

The test to predicted strength ratios, as demonstrated in Table 4, are generally less than 1.0. This is not surprising when put in context of other pushout tests from the literature. As part of this research, a pushout test database has been compiled and analysed. This work confirmed that historical tests are often being over predicted by the current AISC shear strength equation for headed shear connectors [6-10, 12-15]. The effect that an edge condition had on specimens Pa-LW-E-1-A-M and Pa-LW-E-1-A-C was negligible. This is because these edge cases had the steel deck oriented parallel to the steel beam. Deck parallel specimens did not exhibit any need for large effective slab widths to provide strength; this is because the concrete failure exhibited in these specimens was localized compared to the deck perpendicular specimens. The effect that concrete density had on specimens Pa-NW-C-1-A-M and Pa-NW-C-1-A-C was also small. The behavior of these two specimens was similar to that of their companion lightweight concrete specimens Pa-LW-C-1-A-M and Pa-LW-C-1-A-C, and the strength was 7% to 12% larger. More testing would need to be conducted to determine any significant effects that concrete density exhibits on composite steel deck diaphragms.

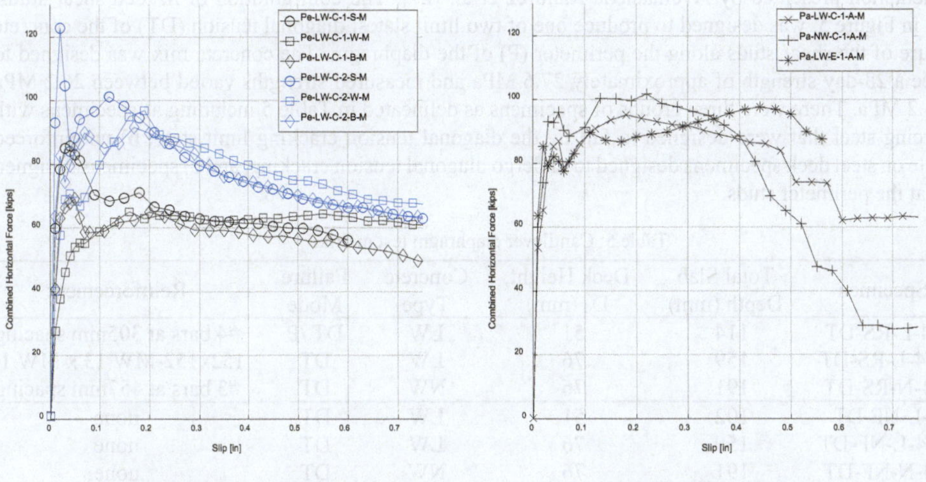

a) Deck perpendicular pushout tests b) Deck parallel pushout tests

Figure 3. Monotonic pushout strength vs. slip results.

The applied load versus displacement plot for cyclic specimen Pe-LW-C-1-W-C is presented in Figure 4 as representative of typical cyclic results. Also plotted are the results of monotonic specimens Pe-LW-C-1-S-M and Pe-LW-C-1-W-M, each of which represent the monotonic counterpart of specimen Pe-LW-C-1-W-C in their respective direction of loading. Comparing the cyclic and monotonic results demonstrates similar initial stiffness values as well as similar ultimate strengths. Between all seven cyclic tests the ultimate strength was only reduced by 5.1% on average, compared to their monotonic counter parts. This reduction is small compared to the 25% strength reduction required by AISC for headed shear studs used in intermediate or special seismic force resisting systems [25].

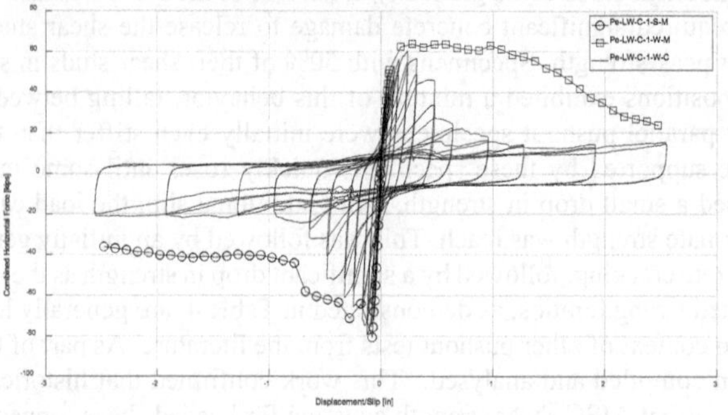

Figure 4. Cyclic pushout strength vs. displacement plot.

4 CANTILEVER DIAPHRAGM TESTING PROGRAM

A cantilever diaphragm experimental program was conducted to characterize the cyclic behavior of composite concrete-filled steel deck diaphragms including their strength, stiffness, ductility, residual strength, and failure modes. The test matrix, shown in Table 5 consisted of eight specimens with varying slab depth, deck height, concrete type, headed shear stud configuration, and reinforcing steel. These experimental specimens are part of a comprehensive study of diaphragm behavior with further documentation presented by Avellaaneda-Ramirez et al. [27]. The configuration of headed shear studs, shown in Figure 5, was designed to produce one of two limit states: diagonal tension (DT) of the concrete or failure of the shear studs along the perimeter (P) of the diaphragm. The concrete mix was designed to produce a 28-day strength of approximately 27.6 MPa and measured strengths varied between 26.2 MPa and 33.2 MPa. There were three groups of specimens as delineated in Table 5 including a) specimens with reinforcing steel that were designed to fail by the diagonal tension cracking limit state, b) unreinforced concrete on steel deck specimens designed to undergo diagonal tension cracking, and c) specimens designed to fail at the perimeter studs.

Table 5: Cantilever diaphragm test matrix.

Test Specimen	Total Slab Depth (mm)	Deck Height, D_d (mm)	Concrete Type	Failure Mode	Reinforcement
2/4.5-4-L-RS-DT	114	51	LW	DT /P	#4 bars at 305mm spacing
3/6.25-4-L-RS-DT	159	76	LW	DT	152x152-MW 13 x MW 13
3/7.5-4-N-RS-DT	191	76	NW	DT	#3 bars at 457mm spacing
2/4-4-L-NF-DT	102	51	LW	DT	none
3/6.25-4-L-NF-DT	159	76	LW	DT	none
3/7.5-4-N-NF-DT	191	76	NW	DT	none
3/6.25-4-L-NF-P	159	76	LW	P	none
3/7.5-4-N-NF-P	191	76	NW	P	none

The test setup is shown in Figure 5 which consists of a fixed beam along one edge and a loading beam along the opposing edge that is subjected to the FEMA 461 cyclic displacement protocol [28]. Loading is applied with two MTS servo-controlled hydraulic actuators working in tandem. All beams in the test setup were W610x230x125 with double angle connections between beams. ASTM A29 19 mm headed shear studs were installed in a symmetric pattern along the beams at the spacings described in Figure 5. ASTM A615/A615M reinforcing steel was centered in the concrete cover, while the ASTM A1064/A1064M welded wire mesh was placed directly on the steel deck with no chairs (although the wire mesh was pushed upward during concrete placement).

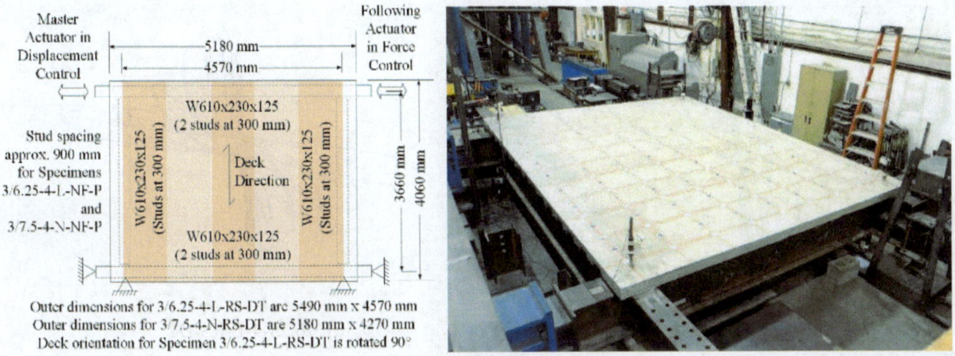

Figure 5. Cantilever diaphragm experimental setup.

Experimental results are summarized in Table 6 and examples of the hysteretic behavior are shown in Figure 6. Shear angle was measured using displacement sensors attached to the upper part of the steel beams and is defined as the lateral displacement of the loading beam divided by the distance between the loading beam and fixed beam. Shear force was calculated as the applied force divided by the 4570 mm width between beams. The residual shear strength is the remaining strength after displacement cycles up to 0.025 rad shear angle and the shear angle associated with a point when the strength deteriorated to 80% of the ultimate strength is reported as a measure of deformation capacity.

Table 6: Summary of experimental cantilever diaphragm results.

Specimen	Ultimate Experimental Shear Strength (kN/m)	Residual Shear Strength (kN/m)	Shear Angle at Ultimate Strength (rad)	Shear Angle at 80% of Ultimate Strength (rad)
2/4.5-4-L-RS-DT	257	47	0.0041	0.0059
3/6.25-4-L-RS-DT	194	61	0.0024	0.0055
3/7.5-4-N-RS-DT	315	61	0.0012	0.0030
2/4-4-L-NF-DT	130	50	0.0031	0.0069
3/6.25-4-L-NF-DT	140	50	0.0015	0.0049
3/7.5-4-N-NF-DT	225	63	0.0013	0.0023
3/6.25-4-L-NF-P	58	19	0.0011	0.0013
3/7.5-4-N-NF-P	82	13	0.0037	0.0039

Some of the trends in the results include an increase in deformation capacity with the addition of reinforcing steel which is observed in Table 6 by comparing the shear angle at 80% of ultimate strength for specimens with identical configuration, but adding reinforcing steel (i.e. 3/6.25-4-L-RS-DT and 3/7.5-4-N-RS-DT compared to 3/6.25-4-L-NF-DT and 3/7.5-4-N-NF-DT). The difference in behavior with the addition of reinforcing steel is also demonstrated in Figure 6 wherein the specimen with reinforcing steel (3/6.25-4-L-RS-DT) experiences a more gradual decrease in strength after reaching peak strength as compared to a sharper drop in strength after diagonal tension cracking for the unreinforced specimen (3/6.25-4-L-NF-DT). The residual strength of concrete-filled steel deck diaphragms undergoing the diagonal tension cracking limit state appears to be related to the deck type as specimens with 51 mm tall deck have similar residual strengths and the same is true of specimens with 76 mm tall deck with the exception of Specimen 3/6.25-4-L-NF-DT. Finally, the deformation capacity of composite concrete-filled steel deck diaphragms as measured by the shear angle at 80% of ultimate strength is generally decreased as the total thickness of the assembly is increased as demonstrated in Table 5. More discussion of the test details and results can be found in Avellaneda-Ramirez et al. [29].

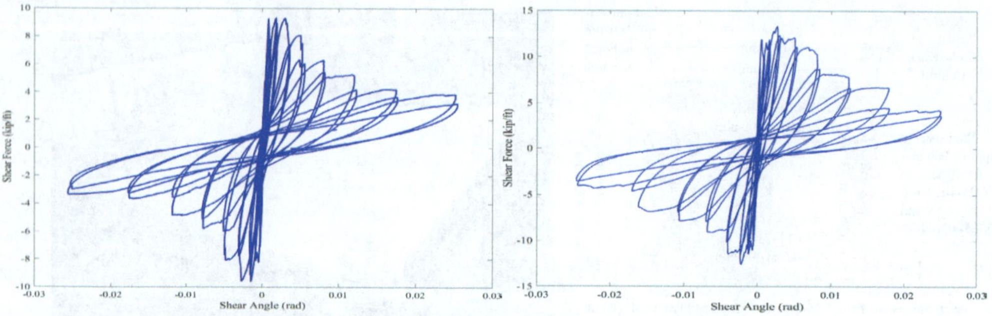

a) Specimen 3/6.25-4-L-NF-DT b) 3/6.25-4-L-RS-DT

Figure 6. Load-deformation behavior for cantilever diaphragm specimens with and without reinforcing steel.

5 CONCLUSIONS

Experimental research has been conducted to characterize the cyclic behavior of composite concrete-filled steel deck diaphragms including their strength, stiffness, ductility, residual strength, and failure modes. Specimens in the form of pushout tests and cantilever diaphragm tests have been utilized to identify the effect that the steel deck height and orientation, concrete density, headed shear stud configuration, and concrete reinforcement has on these structural systems. The pushout tests demonstrated an overprediction of monotonic strength and a conservative prediction of cyclic strength for headed shear studs provided by the AISC provisions [25-26]. Additionally, these specimens demonstrated a uniform load distribution between shear studs until initial failure occurred. Specifically, the lightweight concrete pushout tests exhibited a comparable, but slightly lower ultimate strength in relation to their normal weight counterparts. The deck parallel pushout specimens with a closer edge condition on one side demonstrate a negligible difference in ultimate strength compared to corresponding specimens with the full effective width. The cantilever diaphragm tests demonstrate how reinforcement leads to an increase in deformation capacity and a more gradual decrease in strength. Additionally, the residual strength of the structural system was identified to be a function of the steel deck type used. Lastly, the deformation capacity of the specimens generally decreased with the increase in total thickness of the assembly. Research on this project is ongoing to fully classify and document the ductility of composite steel deck diaphragms.

6 ACKNOWLEDGEMENTS

This work was supported through the Steel Diaphragm Innovation Initiative which is funded by the American Institute of Steel Construction (AISC), the American Iron and Steel Institute, the Steel Deck Institute (SDI), the Metal Building Manufacturers Association (MBMA), the Steel Joist Institute (SJI)., and the National Science Foundation under Grant No. 1562669. In-kind funding was provided by Capone Iron Corporation, Canam, S&F Concrete and Nelson Stud Welding. The contributions to this research of R. B. Bond are gratefully acknowledged. Any opinions, findings and conclusions expressed in this material are those of the authors and do not necessarily reflect the views of the National Science Foundation or the other sponsors.

REFERENCES

[1] Gould, N. C., Kallros, M. K., and Dowty, S. M. (2019). Concrete Parking Structures and the Northridge Earthquake: Performance and Resulting Building Code Changes. *Structure Magazine*.

[2] Slutter, R. G. and Driscoll, G. C. (1961). Research on Composite Design at Lehigh University. *Thirteenth National Engineering Conference of AISC*, Minneapolis, MN. American Institute of Steel Construction.

[3] Slutter, R. G. and Driscoll, G. C. (1963). Flexural Strength of Steel and Concrete Composite Beams. *Fritz Laboratory Reports*. Paper 1806, Lehigh University, Bethlehem, PA.

[4] McGarraugh, J. B. and Baldwin, J. W. (1971). Lightweight Concrete-on-Steel Composite Beams. *Engineering Journal*, American Institute of Steel Construction, Vol. 8, pp. 90-98.

[5] Ollgaard, J. G., Slutter, R. G. and Fisher, J. W. (1971). Shear Strength of Stud Connectors in Lightweight and Normal-Weight Concrete. *Engineering Journal*, American Institute of Steel Construction, Vol. 8, pp. 55-64.

[6] Robinson, H. (1988). Multiple Stud Shear Connections in Deep Ribbed Metal Deck. *Canadian Journal of Civil Engineering*, 15(4), 553-569.

[7] Jayas, B. S., and Hosain, M. U. (1988). Behaviour of Headed Studs in Composite Beams: Push-Out Tests. *Canadian Journal of Civil Engineering*, 15(2), 240-253.

[8] Jayas, B. S., and Hosain, M. U. (1989). Behaviour of Headed Studs in Composite Beams: Full-Size Tests. *Canadian Journal of Civil Engineering*, 16(5), 712-724.

[9] Lloyd, R. M., and Wright, H. D. (1990). Shear Connection Between Composite Slabs and Steel Beams. *Journal of Constructional Steel Research*, 15(4), 255-285.

[10] Sublett, C. N. (1991). *Strength of Welded Headed Studs in Ribbed Metal Deck on Composite Joists*. Master's., Thesis, Virginia Polytechnic Institute and State University, Blacksburg, Virginia.

[11] Easterling, W. S., Gibbings, D. R., and Murray, T. M. (1993). Strength of Shear Studs in Steel Deck on Composite Beams and Joists. *Engineering Journal*, American Institute of Steel Construction, 30(2), 44-54.

[12] Lyons, J. C. (1994). *Strength of Welded Shear Studs* (Publication No. 19683), Master's, Thesis, Virginia Polytechnic Institute and State University, Blacksburg, Virginia.

[13] Roddenberry, M. R., Easterling, W. S. and Murray, T. M. (2002). *Behavior and Strength of Selded Stud Shear Connectors*, Report No. CE/VPI-ST 02/04. Virginia Polytechnic Institute and State University, Blacksburg, VA, 266 pages.

[14] Roddenberry, M. R., Easterling, W. S. and Murray, T. M. (2002). *Behavior and Strength of Welded Stud Shear Connectors-Data Report,* Report No. CE/VPI-ST 02/05. Virginia Polytechnic Institute and State University, Blacksburg, VA, 513 pages.

[15] Rambo-Roddenberry, M. D. (2002). *Behavior and Strength of Welded Stud Shear Connectors* (Publication No. 14958), Ph.D., Dissertation, Virginia Polytechnic Institute and State University, Blacksburg, Virginia.

[16] Hawkins, N. M., and Mitchell, D. (1984). Seismic Response of Composite Shear Connections. *Journal of Structural Engineering*, 110(9), 2120-2136.

[17] Bursi, O. S., and Gramola, G. (1999). Behaviour of Headed Stud Shear Connectors under Lowcycle High Amplitude Displacements. *Materials and Structures*, 32(4), 290-297.

[18] Luttrell, L. D. (1971). Shear Diaphragms with Lightweight Concrete Fill. *Proc., 1st Specialty Conf. on Cold-Formed Steel Structures*, Univ. of Missouri-Rolla, MO, 111-117.

[19] Davies, J. M., and Fisher, J. (1979). The Diaphragm Action of Composite Slabs. *Proceedings of the Institution of Civil Engineers*, 67(4), 891-906.

[20] Porter, M. L., and Greimann, L. F. (1980). Seismic Resistance of Composite Floor Diaphragms. *ISU-ERI-Ames-80133,* Iowa State University, Ames, Iowa.

[21] Easterling, W. S. and Porter, M. L. (1994). Steel-Deck-Reinforced Concrete Diaphragms – Part I. *Journal of Structural Engineering*, ASCE, 120(2), 560-576.

[22] Easterling, W. S. and Porter, M. L. (1994). Steel-Deck-Reinforced Concrete Diaphragms – Part II. *Journal of Structural Engineering*, ASCE, 120(2), 577-596.

[23] Altoubat, S., Ousmane, H., and Barakat, S. (2016). Experimental Study of In-Plane Shear Behavior of Fiber-Reinforced Concrete Composite Slabs. *Journal of Structural Engineering*, 142(3), 04015156.

[24] Rezaeian, H., Clifton, G. C., MacRae, G. A., Hogan, L., and Lim, J. B. (2019). In-plane Behaviour of Composite Floor Slab Diaphragm Interfaces under High Shear Demand. *Journal of Constructional Steel Research*, 162, 105715.

[25] AISC (2016). *Seismic Provisions for Structural Steel Buildings*. Chicago, IL: American Institute of Steel Construction.

[26] AISC (2016). *Specification for Structural Steel Buildings*. Chicago, IL: American Institute of Steel Construction.

[27] Avellaneda-Ramirez, R. E., Eatherton, M. R., Easterling, W. S., Schafer, B. W., and Hajjar, J. F. (2021). Strength of Concrete-Filled Steel Deck Composite Diaphragms with Reinforcing Steel. In M. Knobloch, U. Kuhlmann, W. Kurz, and M. Schafer. (Eds.), *Composite Construction in Steel and Concrete IX – Proceedings of the 9th International Conference on Composite Construction in Steel and Concrete*, Stromberg, Germany, July 26-30, 2021.

[28] FEMA. (2007). *Interim Testing Protocols for Determining the Seismic Performance Characteristics of Structural and Nonstructural Components*, (FEMA 461). Applied Technology Council, Federal Emergency Management Agency.

[29] Avellaneda-Ramirez R. E., Eatherton, M. R., Easterling, W. S., Schafer, B., and Hajjar, J. F. (2021). *Cantilever Testing of Concrete-Filled Steel Deck Composite Diaphragms Using Various Types of Steel Reinforcing*, Cold-Formed Steel Research Consortium Report Series, Report No. CFSRC R-2021-04, Johns Hopkins University, Baltimore, MD.

9TH INTERNATIONAL CONFERENCE ON COMPOSITE CONSTRUCTION IN STEEL AND CONCRETE
M. Knobloch, U. Kuhlmann, W. Kurz, M. Schäfer. (Eds.)
Online, Germany, July 27-29, 2021

EVALUATION OF THE PLASTIC DEFORMATION CAPACITY OF COMPOSITE BEAMS THROUGH THE CONNECTION COEFFICIENT

Yuko Shimada*

* Chiba University
e-mail: yshimada@faculty.chiba-u.jp

Keywords: Composite beam, Connection coefficient, Plastic deformation capacity, Ductility factor.

Abstract. *For safer structural design, a simple index that describes the plastic deformation capacity of composite beams is needed, similar to the connection coefficient in the case of strength. In this study, the relationships between the plastic deformation capacity and strength under varying material properties and geometric shapes of concrete and steel members were considered through finite element analyses. Based on the obtained results, the plastic deformation factor was defined as the ratio of two rotation angles. The minimum value of the connection coefficient in composite beams corresponding to the ductility factor under severe earthquakes was determined because the plastic deformation factor can be used similar to the ductility factor. In addition, the minimum value of the connection coefficient in bare steel beams that will be safe even if a concrete slab is added was determined; therefore, the design of bare steel beams and composite beams are connected by the connection coefficient.*

1 INTRODUCTION

Composite beams consisting of steel beams and concrete slabs have higher strength and stiffness than bare steel beams. According to the composite effect of concrete slabs, the neutral axis of composite beams moves from the centre of the beams closer to the concrete slabs. The concentration of deformation in the lower flange of steel beams resulting from this movement of the neutral axis was one of the causes of the fracturing of steel beams during the Northridge earthquake [1] and the 1995 Southern Hyogo Prefecture Earthquake [2]. After the damage due to these earthquakes, steel beam ends were designed to maintain significant plastic deformation and prevent early fracturing under ground motions. The steel material properties, welding conditions including the details of the weld access hole, and effect of dynamic loading were studied to modify the energy absorption capacity of steel buildings [3-5]. The current Japanese seismic design manual [6] includes a unique index for the seismic safety of connections according to the strain hardening and the variation in the yield stress of bare steel based on the above previous studies. This index is called the connection coefficient (α), as shown in Eq. (1).

$$ {}_{j}M_{u} \geq \alpha \cdot {}_{b}M_{p}, \qquad (1)$$

where ${}_{j}M_{u}$ is the maximum strength of the connection, and ${}_{b}M_{p}$ is the fully plastic moment of bare steel members.

The connection coefficient is generally recommended to be greater than 1.3; however, this value can only be used for bare steel members. The connection coefficient for composite beams has not yet been determined, although it is useful for designing buildings with high seismic performance. The main reason for this is that the plastic deformation data for composite beams is significantly less abundant than that for bare steel beams. The previous data for composite beams are not sufficient to clarify the plastic deformation capacity of composite beams under various conditions and material properties, as mentioned above. To determine the connection coefficient of composite beams, a series of experiments on composite beams with a wide range of conditions are necessary, and these are quite difficult to conduct. This paper presents an analytical consideration of the issue through an analysis model calibrated using a composite beam specimen in a previous study. The analysis model consists of a combination of bare steel beams and a rectangular hollow section (RHS) column, which has a connection coefficient greater than 1.3. When a concrete slab is added to the bare steel beam-to-column connections with sufficient energy absorption capacity, the plastic deformation capacities and strengths of the composite beams are changed by the concrete slab height, concrete strength, and stress–strain relationship of steel. Based on the evaluation of each parameter, the recommended value of the connection coefficient for composite beams can be clarified. In addition, the ductility factor of the composite beams is evaluated corresponding to the plastic deformation capacity, similar to bare steel beams with sufficient energy absorption capacity.

2 FINITE ELEMENT ANALYSIS

To understand the relationship between the strength and plastic deformation capacity of composite beams under various conditions, a series of finite element analyses were conducted.

2.1 Analysis model

The analysis model used in this study is shown in Figure 1. The model reproduced a partial steel frame specimen, including a beam-to-column connection and a concrete slab [7]. H-shaped beams, an RHS column, and a concrete slab were all modelled using shell elements based on the analysis model developed by Yamada et al. [8]. The rigid beams were tied from the top and bottom of the column to points set as pin supports, similar to the boundary condition in the experiment [7]. The beam web had weld access holes, the shape of which was a combination of two circles to reflect that the welding area between the diaphragm and column had less deformation than the column itself, and shell elements were added to the outer surfaces of the column at locations of +20 mm and -20 mm from the diaphragm. The thickness of the extra shell elements was 0.5 times the thickness of the column.

The headed studs that transfer shear stress from the concrete slab to the steel beam flange were modelled as rigid beam elements. The rigid beam elements had two parts: the shaft area and the pin area; all of the edges of the rigid beam elements were fixed. The length of the shaft area was 1 mm shorter than the length of the headed stud in the experiment [7]. Otherwise, the pin area was located between the beam flanges and the shaft area. To correspond to the pinned joint at the base of the headed studs, the length of the pin area was 1 mm, and the diameter of the pin area was 10% of the diameter of the shaft area.

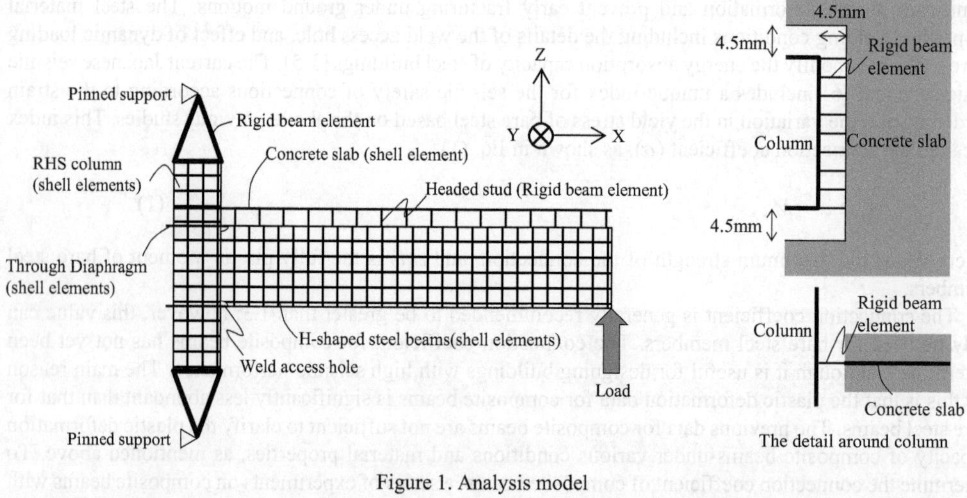

Figure 1. Analysis model

To reproduce the pushing and transfer of stress from the concrete slab to the column, another series of rigid beam elements were defined. The gap between the edges of the concrete slab and the outer surface of the column was 4.5 mm to insert the rigid beam elements. All of the edges of the rigid beam elements were also fixed, and the diameter of the rigid beam elements was quite large. A shell element with a size of 209 mm × 10 mm was attached to the top of the concrete slab to prevent the collapse of the concrete near the column during loading. The thickness of the extra shell element on the concrete slab was 10 times the thickness of the concrete slab.

2.2 Material properties

The stress–strain relationship of steel used the results of a coupon test in the case of verification of the analysis model. In the parametric study, the stress–strain relationship of steel was based on the prediction equation proposed by Ono and Sato [9]. The prediction equation can calculate the stress–strain relationships based on the yield stress and tensile stress for each steel classification. The input data for the steel material properties in the analysis were the real stresses and logarithmic strains calculated from the above stress–strain relationships, assuming a constant volume of the steel member. The input data for the beam flange and beam web were the same as those of H-shaped steel with a grade of 400 N/mm^2, and the input data for the column were those of RHS steel with a grade of 400 N/mm^2.

The material properties of concrete used the results of the compression test of a cylindrical specimen in the case of verification of the analysis model. In the parametric study, the material properties of concrete were a combination of the elasticity and concrete damaged plasticity. The Young's modulus was calculated using Equation (2) [10].

$$E_c = 3.35 \times 10^4 \times \left(\frac{\gamma}{24}\right)^2 \times \left(\frac{F_c}{60}\right)^{1/3}, \quad (2)$$

where γ is the unit volume weight, and F_c is the standard strength of concrete in the design.
The stress–plastic strain relationship in compression was based on the method reported by Umemura [11] using the compressive strength of concrete, as shown in Figure 2. The input data in tension were a stress of 10% of the compressive strength of concrete and zero plastic strain.

The rigid beams were defined as elastic with a large Young's modulus, assuming a steel material.

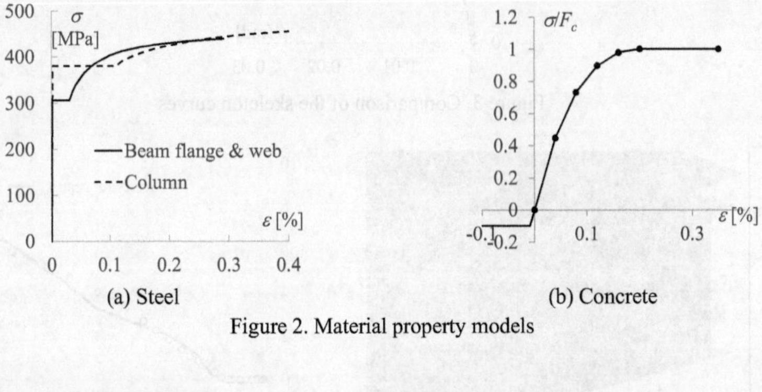

Figure 2. Material property models

2.3 Loading and boundary conditions

The aim of this study is to establish an index that describes the strength and plastic deformation capacity of composite beams. Only the positive bending direction of composite beams was considered. The load in the analysis was applied at the free end of the beam in the +Z direction, as shown in Figure 1. To concentrate the loading point in a specific direction, the H-shaped section of the free end of the beam was constructed and fixed in the X, Y, and RZ directions.

As mentioned above, the top and bottom of the column were pinned supports through the rigid beam elements, and these pinned supports were fixed in the X, Y, and RZ directions.

2.4 Verification of the analysis model

Before the parametric analysis, the reproducibility of the test results using the analysis model should be verified. The size of the member, boundary conditions, and material properties were defined as the same as the test specimen (A-09-200) of a composite beam reported by Shimada et al. [12]. Figure 3 shows a comparison of the relationship between the bending moment and rotation angle of specimen A-09-200 at the end of the beam. The test result in Figure 3 is the skeleton curve under positive bending obtained from the cyclic loading data. Both elastic stiffnesses corresponded well, as shown in Figure 3. After the composite beam yielded, the bending moment in the analysis results was 92% of the bending moment of the test results. The maximum rotation angle in the analysis results was almost the same as the maximum rotation angle in the test results.

According to the von Mises stress distribution at the final step of the analysis, as shown in Figure 4, the force could be transferred from the beams to the column surface through the concrete slab. The stress was high on the surface of the column facing the concrete slab and the lower beam flange at the beam end. Figure 5 shows a comparison of the strains of the beam flange in the analysis and test results. The strains shown in Figure 5 are the average of those on the lower beam flange located a position 70 mm from the column surface. Both strains corresponded from yielding to a large plastic range, and the rotation angles when the strains reached their maximum values were almost the same. Therefore, the analysis model has sufficient reproducibility of the test results and is reasonable to use in the following study.

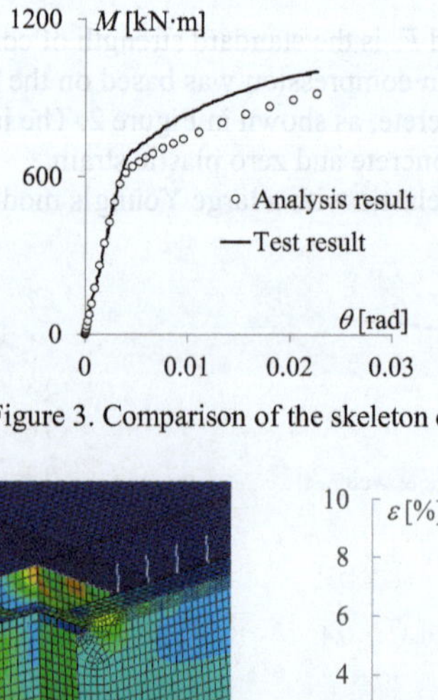

Figure 3. Comparison of the skeleton curves

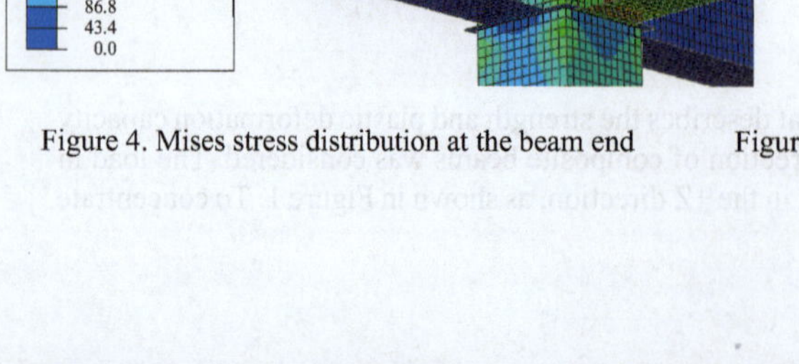

Figure 4. Mises stress distribution at the beam end

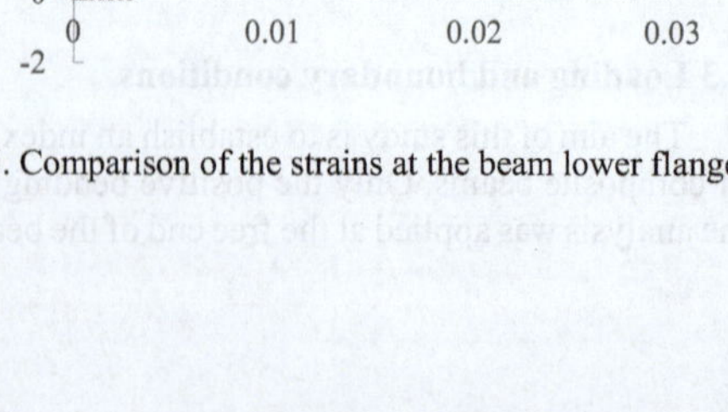

Figure 5. Comparison of the strains at the beam lower flange

3 RELATIONSHIP BETWEEN THE CONNECTION COEFFICIENT OF THE COMPOSITE BEAM AND PLASTIC DEFORMATION

3.1 Parameter details

To evaluate the effects on the connection coefficient of the composite beam and plastic deformation, five parameters were considered in this analysis: (1) the width-to-thickness ratio of the column surface facing the concrete slab, (2) the height of the steel beam, (3) the stress–strain relationship of steel, (4) the compression strength of concrete, and (5) the thickness of the concrete slab.

Parameters (1) and (2) are affected by the strength and elastic stiffness of the beams. Table 1 lists the combinations of the width-to-thickness ratio of the column surface facing the concrete slab (parameter (1)) and the height of the steel beam (parameter (2)) in this study. The sizes of the columns and beams were based on the composite beam test specimen [12], which was assumed to represent the steel members in medium- and low-rise buildings. The columns and beams were selected to have a connection coefficient of bare steel of approximately 1.3, as indicated in Table 1.

Table 1: Combinations of columns and beams

		Column width: 300			
Thickness of column surface		9	12	16	19
Width-to-thickness ratio of column		33	25	19	16
Beams	H-600×200×11×17	1.21	1.24	1.27	1.30
	H-500×200×10×16	1.29	1.32	1.35	1.38
	H-400 ×200×8×13	1.38	1.41	1.43	1.43

Parameter (3) was based on the conditions for determining the connection coefficient of the bare steel beam-to-column connection [13]. The connection coefficients in [10] were determined by comparing the variations in the yield stress against the real value and the design value. The stress–strain relationship of steel in this study was also determined by considering variations in both the yield stress and tensile stress. Based on the averages and standard deviations of the yield stress and the tensile stress obtained in the statistical study by Fujisawa [14], three stress–strain relationships were defined: a case using the average values (Case μ), a case using the sum of the average and standard deviation (Case μ+σ), and a case using the difference between the average and standard deviation (Case μ−σ).

The stress distribution on the compression side of the composite beams differs depending on parameters (4) and (5). The compressive strength of concrete (F_c) in composite beams is 18–24 N/mm^2 based on the general design strength. The real value of the compressive strength is often greater than the design strength; however, a difference of 5–10 N/mm^2 in the compressive strength will not significantly change the composite effect [15]. In this study, the compressive strengths were set as $F_c = 21$ N/mm^2 and $F_c = 42$ N/mm^2. The thickness of the concrete slab (t_c) in medium- to low-rise buildings is generally 150–200 mm. Based on the composite beam test specimen [12], the thicknesses of the concrete slabs in this study were $t_c = 150$ mm, $t_c = 175$ mm, and $t_c = 200$ mm.

Table 2: Parameter combinations

		Case μ																Case μ+σ				Case μ−σ			
		F_c=21				F_c=21				F_c=42				F_c=21				F_c=21				F_c=21			
		t_c=150				t_c=175				t_c=175				t_c=200				t_c=175				t_c=175			
Width-to-thickness ratio of column		33	25	19	16	33	25	19	16	33	25	19	16	33	25	19	16	33	25	19	16	33	25	19	16
Beams	H-600×200×11×17																								
	H-500×200×10×16																								
	H-400×200×8×13																								

		Case μ				Case μ+σ				Case μ−σ			
		non slab				non slab				non slab			
Width-to-thickness ratio of column		33	25	19	16	33	25	19	16	33	25	19	16
Beams	H-600×200×11×17												
	H-500×200×10×16												
	H-400×200×8×13												

Table 2 summarises the models reflecting the combinations of parameters, including the analysis model without a concrete slab for comparison with the composite beam model. A total of 104 models were analysed.

3.2 Effect of parameters on the connection coefficient

The connection coefficient in this study is defined as the ratio of the maximum bending moment (M_{max}) on the positive side to the plastic bending moment (M_{pla}). In the composite beams, the maximum bending moment was calculated using the load in the last step. The maximum bending moment of non-slab beams was calculated as the moment when the average strain of the lower beam flange located 70 mm from the column surface was greater than the maximum strain of the lower beam flange end in the test specimen of the composite beam [12].

Here, the plastic bending moment shown in Figure 6 is not the full plastic moment (M_p) in the previous recommendation [15]. Based on a database of test results for composite beams [16], the plastic bending moment was shown to match the bending moment at which the beam began to exhibit plasticity. Therefore, the denominator of the connection coefficient is defined as the plastic bending moment in both the composite and non-slab beams. According to the above definition, the connection coefficient of the composite beams (${}_c\alpha$) and the connection coefficient of the non-slab beams (${}_s\alpha$) were obtained.
Figure 7 shows the relationship between the ratio of connection coefficient and the width-to-thickness ratio of the column. The ratio of the connection coefficient is the ratio of the connection coefficient of the composite beams to the connection coefficient of non-slab beams with the same material properties. The ratio of the connection coefficient in the case of beam height of 500 mm and 600 mm were approximately 0.8-1.0, similar to that mentioned in [12]. On the other hand, the ratio of the connection coefficient in the case of beam height of 400 mm was 0.9-1.1. The difference was based on the fracture location of composite beams. In the case of beam height of 500 mm and 600 mm, the maximum mises stress were shown at the lower beam flange, while, in the case of beam height of 400 mm, the maximum mises stress was shown at the lower end of beam web. In a previous study, the ratio of the connection coefficient increased as the width-to-thickness ratio of the column decreased [12]; however, the ratio of the connection coefficient in this study did not depend on the width-to-thickness ratio of the column. The transmission of force in the beam web was confirmed to contribute to the bending moment of the composite beams and the connection coefficient because the previous trend was obtained by numerical analysis ignoring the transmission of stress at the centre of the beam web. In this study, the stress–strain relationship of steel had the greatest effect on the ratio of the connection coefficient. When the compressive strength of concrete was 42 N/mm^2, the ratio of the connection coefficient was approximately 0.1 higher than when the compressive strength was 21 N/mm^2, regardless of the width-to-thickness ratio of the column. If the beam height is large, the composite effect of the concrete slab is often small because the ratio of the slab thickness to the whole beam height is small. In this study, the ratio of the connection coefficient do not exhibit a clear trend with respect to the beam height and thickness of the concrete slab.

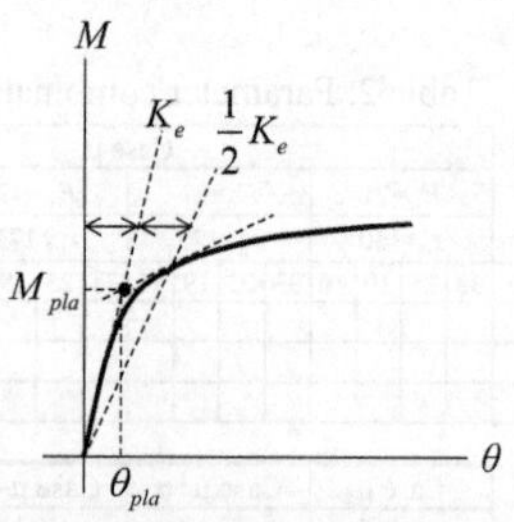

Figure 6. Definition of the plastic bending moment

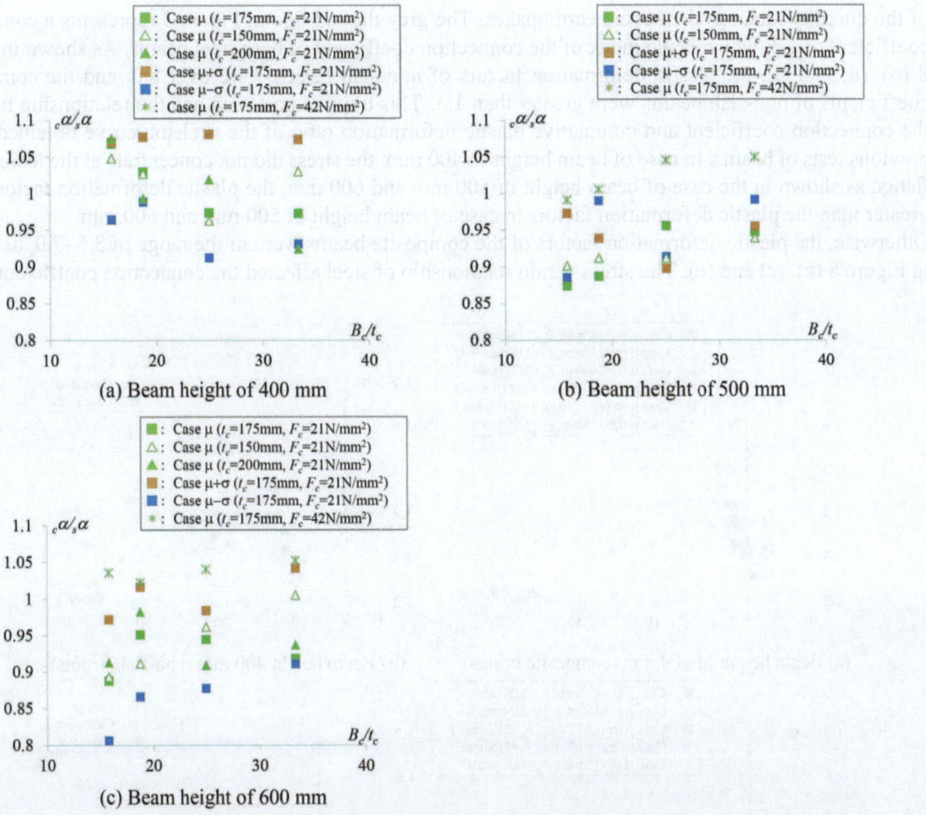

Figure 7. Relationships between the width-to-thickness ratio and connection coefficient

3.3 Effect of parameters on the plastic deformation capacity

To consider the effect of the parameters on the plastic deformation capacity of the composite beams, the maximum rotation angle (θ_{max}) and plastic rotation angle (θ_{pla}) were considered. The maximum rotation angle of the non-slab beams was obtained as the rotation angle when the maximum bending moment occurred. Here, the plastic rotation angle is defined as the quotient of the plastic bending moment (M_{pla}) and the elastic stiffness of the beams, as shown in Figure 6. Generally, the elastic limit rotation angle calculated as the quotient of the fully plastic bending moment and the elastic stiffness is used as the denominator to show the plastic deformation capacity. In the case of non-slab beams, the plastic rotation angle was matched to the elastic limit rotation angle. As mentioned in the previous section, the plastic bending moment at which the beam began to exhibit plasticity was the denominator of the connection coefficient. Corresponding to this definition, the plastic rotation angle is defined as the denominator of the plastic deformation capacity.

In this study, the plastic deformation factor was defined as the ratio of the maximum rotation angle to the plastic rotation angle to describe the ultimate deformation capacity of the beams, similar to the ductility factor. Compared with the ductility factor, the plastic deformation factor was safer for verification based on the definition.

Figure 8 shows the relationships between the connection coefficient and plastic deformation factor. The black dashed line in Figure 8 shows a plastic deformation factor of 4.0, based on the minimum value of the ductility factor under severe earthquakes. The grey dashed line in Figure 8 represents a connection coefficient of 1.3, which is the index of the connection coefficient of bare steel beams. As shown in Figure 8 (b), (d) and (f), the plastic deformation factors of non-slab beams were over 8.0, and the connection coefficients of non-slab beams were greater than 1.3. This trend almost matched the relationship between the connection coefficient and cumulative plastic deformation ratio of the skeleton curve obtained in the previous tests of beams. In case of beam height of 400 mm, the stress did not concentrate at the lower beam flange as shown in the case of beam height of 500 mm and 600 mm, the plastic deformation factors were greater than the plastic deformation factors in case of beam height of 500 mm and 600 mm.

Otherwise, the plastic deformation factors of the composite beams were in the range of 3.5–7.0, as shown in Figure 8 (a), (c) and (e). The stress–strain relationship of steel affected the connection coefficient;

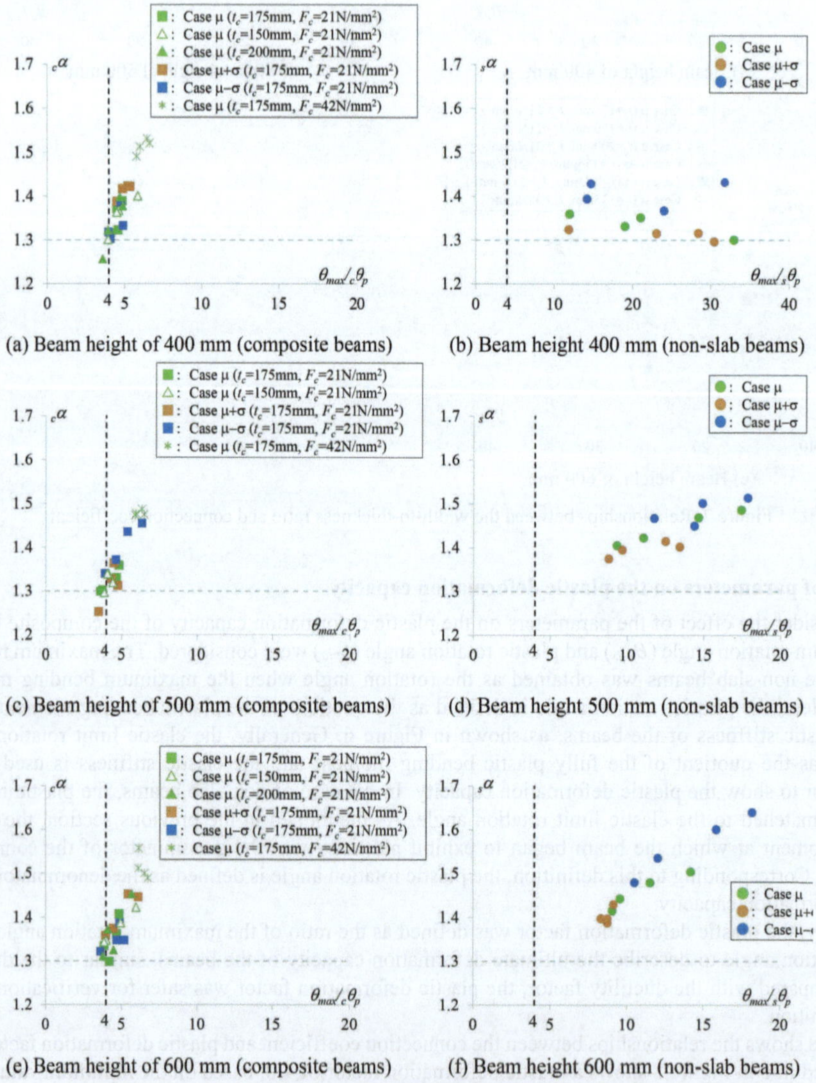

Figure 8. Relationships between the plastic deformation factor and connection coefficient

however, it did not affect the plastic deformation factor. This means that a design focusing only on the connection coefficient is at high risk of having an insufficient plastic deformation capacity. Focusing on the models with a normal compressive strength of concrete, the relationships between the connection coefficient and plastic deformation factor were linear. The plastic deformation factor with beam heights of 400 mm and 500 mm was smaller than the plastic deformation factor with a beam height of 600 mm, as in the trend of the connection coefficient. According to these trends, the connection coefficient of composite beams should be greater than 1.3 to maintain sufficient plastic deformation capacity under severe earthquakes. According to the trend mentioned in the previous section, the connection coefficient of bare steel beams should be secure at 1.45 (=1.3/0.9). For enhanced safety, a structural design with a larger beam height is better. The two models with beam heights of 500 mm and four models with beam heights of 600 mm had relatively high plastic deformation factors. The width-to-thickness ratio of the column and slab thickness in these models varied; however, even the models with large width-to-thickness ratios of the columns had large plastic deformation factors depending on the conditions.

In the models with a high compressive strength of concrete in both beam height case, the plastic deformation factors were almost 6.0 or greater. This is because the elastic stiffness is smaller than that of the other models, and the maximum rotation angle is larger than in the other models when the compressive strength is large. The effect of the width-to-thickness of the column on the plastic deformation factor is not shown in the models with a high compressive strength of concrete. Note that the plastic deformation capacity of the composite beam should not rely excessively on the compressive strength of concrete because few concrete slabs with high-strength concrete are used in typical medium- and low-rise buildings.

4 CONCLUSIONS

The connection coefficient is useful for easily designing with a focus on safety. In the current design manual in Japan, the connection coefficient of bare steel beams is defined considering that the plastic deformation capacity corresponds to the strain hardening and variation in the material properties of bare steel. By expanding the connection coefficient method to composite beams, the relationships between the plastic deformation capacity and the strength of the composite were considered through finite element analyses. The main parameters were the geometric and material properties. From the obtained results, the ratio of the connection coefficient in this study varied regardless of the width-to-thickness ratio of the column. The relationships between the connection coefficient and plastic deformation factor are linear; hence, the connection coefficient of composite beams should be greater than 1.3 to maintain sufficient plastic deformation capacity under severe earthquakes. As a result, the recommended value of the connection coefficient of bare steel beams is 1.45 to ensure safety even if a concrete slab is added to the bare steel beams with sufficient plastic deformation capacity.

The plastic deformation capacity of beams in the skeleton curves considered in this study is the basis of seismic design; however, the cumulative plastic deformation of beams under cyclic loading is also important for seismic safety. For the general use of the connection coefficient of composite beams in seismic design, further development of this study under cyclic loading is needed.

REFERENCES

[1] Earthquake Engineering Research Institute, *Notrhridge Earthquake January 17, 1994, preliminary Reconnaissance Report*, John Hall (ed.), EERI, Oakland, 1994.

[2] The committee of steel structures in the Kinki Branch of Architectural Institute of Japan, *the 1995 Southern Hyogo Prefecture Earthquake Report of damage in Steel structures*, AIJ Publications, Osaka, 1995.

[3] Inoue, K.; Kohzu, I.; Tada, M.; Tateyama, E.; Tabuchi, M.; Toyoda, M.; Nakashima, M.; Maruoka, Y.; Morisako, K.; and Morita, K., "Full-Scale Test on Plastic Rotation Capacity of Steel Wide-Flange Beams Connected with Square Tube Steel Columns (Part 1 - Part 5) ", *Journal of Steel Structures*, 4(16), 27-104, 1997.

[4] Okada, K.; Yamada, S. and Oh, S., "Experimental Study on Deformation Capacity of Composite Beams with Improved Type Beam-to-column Connections". *Journal of Structural and Construction Engineering*, Architectural Institute of Japan, 554, 123-130, 2002.

[5] Suita, K., "Flexural Strength Demand for Beam-to-column Connections in Earthquake Resistant Design of Steel Moment Frames". *Journal of Structural and Construction Engineering*, Architectural Institute of Japan, 567, 165-171, 2003.

[6] Editorial committee of technical criteria references on structure of buildings, "Technical criteria references on structure of buildings" Ver. 2020., 2020.

[7] Shimada, Y.; Yamada, S.; and Yasuda, B., "Calculation Method of Connection Coefficient of Composite Beam". *Journal of Structural and Construction Engineering*, Architectural Institute of Japan, 724, 1005-1014, 2016.

[8] Yamada, S.; Satsukawa, K.; Kishiki, S.; Shimada, Y.; Matsuoka, Y.; and Suita, K., "Elasto-plastic Behavior of Panel Zone in Beam to External Column Connection with Concrete Slab". *Journal of Structural and Construction Engineering*, Architectural Institute of Japan, 644, 1841-1849, 2009.

[9] Ono, T.; and Sato, A., "Modeling of Stress-strain Relationships on Metallic Materials". *Journal of Structural and Construction Engineering*, Architectural Institute of Japan, 532, 177-184, 2000.

[10] Architectural Institute of Japan, *AIJ Standard for Structural Calculation of Reinforced Concrete Structures*, 2018.

[11] Umemura, H., "Plastic Deformation and Ultimate Strength of Reinforced Concrete Beames". *Journal of Architectural Institute of Japan*, Architectural Institute of Japan, 42, 59-70, 1951

[12] Shimada, Y.; Yamada, S.; and Yasuda, B., "Calculation Method of Connection Coefficient of Composite Beam". *Journal of Structural and Construction Engineering*, Architectural Institute of Japan, 724, 1005-1014, 2016

[13] Architectural Institute of Japan, *AIJ Recommendations for Design of Connections in Steel Structures*, 2021.

[14] Fujisawa, K.; Inoue, Y.; Sugimoto, M.; and Sonoda, M., "Statistical Study on Mechanical Properties and Chemical Compositions of SN Steels". *Summaries of Technical Papers of Annual Meeting Architectural Institute of Japan, Structure III*, 699-700, 2013.

[15] Architectural Institute of Japan, *Recommendations for Limit State Design of Steel Structures*, 2010.

[16] Yamada, S.; Kishiki, S.; Shimada, Y., "Evaluation on Strength of Full Composite Beam under Positive Bending". *Proceeding of Constructional Steel*, Japanese Society of Steel Construction, 18, 93-98, 2010.

9TH INTERNATIONAL CONFERENCE ON COMPOSITE CONSTRUCTION IN STEEL AND CONCRETE
M. Knobloch, U. Kuhlmann, W. Kurz, M. Schäfer. (Eds.)
Online, Germany, July 27-29, 2021

SEISMIC PERFORMANCE OF PREFABRICATED CORE STEEL TUBE REINFORCED CONCRETE COLUMNS

Y.X. Zhang*, Y. Liu*, J. Ruan* and P.Z. Wang*

* College of Civil Engineering, Huaqiao University, Xiamen, China
e-mails: zhangyixin@hqu.edu.cn, lyliuyang@hqu.edu.cn, 240511062@qq.com, 835936359@qq.com

Keywords: Prefabrication, Core steel tube reinforced concrete column, Seismic performance, Pseudo-static test.

Abstract. *Prefabricated assembly accelerates the building construction process. A primary concern related to prefabricated assembly is the increased fragility of the connection between precast members compared to cast-in-situ members. On the other hand, core steel tube reinforced concrete (CSTRC) columns are increasingly being used in high-rise structures to provide effective axial and lateral force resistance. This study proposes a prefabricated core steel tube reinforced concrete (PCSTRC) column which capitalizes on both the strength of the CSTRC column and the advantages of precast manufacturing, as shown in Fig. 1. The connection is designed to guarantee the integrity of the whole column. Two full-scale PCSTRC columns with varying axial loads and two controlled specimens, namely a monolithic CSTRC column and a conventional reinforced concrete (RC) column, were fabricated and tested under quasi-static cyclic loading in order to investigate their seismic performance. The test results demonstrate that all the columns experienced typical flexure-controlled failure, while the PCSTRC columns suffered less visual damage than the monolithic controlled specimens. The PCSTRC column performed satisfactorily in terms of strength, stiffness, ductility, and general hysteretic behavior and had significantly greater load-carrying and energy dissipation capacities than the conventional RC column. In addition, there were no apparent differences between the capacities of the CSTRC and PCSTRC columns. As axial load increased, the deformation capacity of the PCSTRC column decreased. Overall, the proposed prefabricated composite column demonstrates good seismic performance and can be considered for use in building construction.*

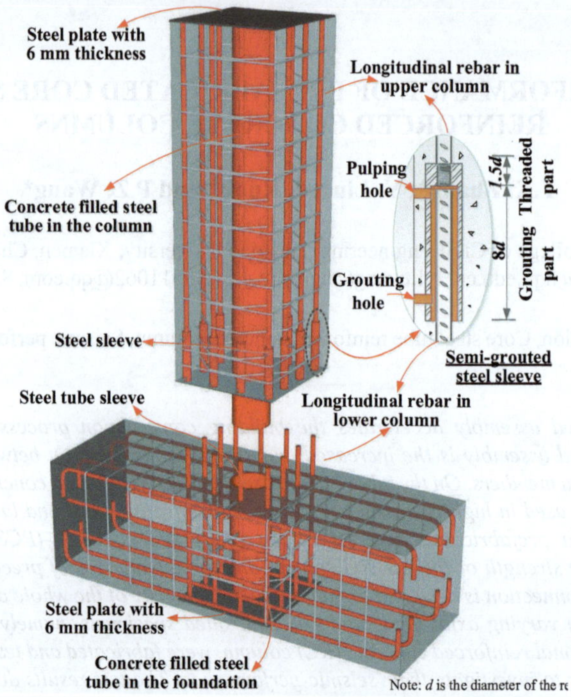

Figure 1. Schematic configuration of PCSTRC columns.

1 INTRODUCTIONS

Use of precast structures simplifies the construction process, enables rapid on-site assembly, and offers greater control over construction quality because industrialization replaces manual labor on the construction site. Therefore, precast components and systems have been widely adopted in many countries. The major concern regarding the adoption of precast concrete segments is the performance of precast connections during earthquakes [1–2].

The seismic behavior of precast structures is highly dependent on the mechanism characteristics of the connections between the precast members as well as between the precast members and the foundation. Special attention is required to ensure good performance. Chong et al. [3] proposed an enhanced horizontal joint for superimposed reinforced concrete (RC) shear walls. Haber et al. [4] employed high-strength reinforcing bars for the sleeve region to increase the rigidity of the joint. Xue et al. [5] combined precast and cast-in-place parts for steel RC.

Core steel tube reinforced concrete (CSTRC) columns are cost-effective composite components with good performance. The column contains a small-section steel tube at the core of the section, which can be filled with concrete of different strengths according to the structural needs. The core steel tube can provide external confinement on the concrete infill [6–8]. In addition, it can improve the axial stability of the column [9-10]. Although the structural performance of CSTRC columns has been studied extensively [11–13], research on the prefabrication approach for CSTRC columns remains scarce.

Therefore, this paper proposes a prefabricated CSTRC column (PCSTRC), as shown in Fig. 1. The precast connection for the columns uses both semi-grouted steel sleeves and steel tube sleeves, which is referred to as a “hybrid connection” in this study. Fig. 1 presents the details of the semi-grouted steel sleeve, which comprises a threaded connection between the reinforcing bar and sleeve at the top end and a grouted connection at the bottom end. The core steel tube uses a tenon connection to transmit the bending moment that is generated during the construction process. Four column specimens, including monolithic and precast columns, were designed and tested to study the seismic performance of the proposed precast connections. The seismic performance of the test columns in terms of failure mode, hysteretic curves, ductility, and energy dissipation are discussed.

2 EXPERIMENTAL PROGRAM

2.1 Specimen design

Four full-scale specimens with the same geometry and different structural details were designed. These included two PCSTRC columns with different axial loads, a monolithic CSTRC column as a counterpart, and a conventional RC column. Table 1 presents the design parameters for the specimens. The total column length was 1,350 mm and the lateral load was applied 150 mm from the top face of the test columns. The columns had a shear span-to-depth ratio of 3. Fig. 2 depicts the cross sections and reinforcement details of the test columns and footings. A circular steel tube with a thickness of 6 mm was adopted for both the core steel tube and its sleeve in the foundation. The upper and lower ends of the PCSTRC test column were equipped with 6 mm-thick steel plates for the positioning of the longitudinal bars and the steel tube.

The PCSTRC test columns were prefabricated in a factory and assembled in the lab using a four-step procedure: (1) the longitudinal bars and the core steel tube were positioned by welding them to the steel plate in a manufacturing factory, where the longitudinal bars were mechanically connected to the sleeves at the threaded part; (2) the columns and the footings were cast separately; (3) the columns were inserted into the column footings after the concrete had hardened; and (4) the rebar sleeves were filled with grout. The construction procedure is illustrated in Fig. 3.

Table 1: Design and test parameters for specimens.

Specimen No.	Construction method	D/t	Axial load ratio	Axial load (kN)
RC-1	in-situ	–	0.3	624
CSTRC-1	in-situ	23.3	0.3	624
PCSTRC-1	precast	23.3	0.3	624
PCSTRC-2	precast	23.3	0.5	1040

Note: *D/t is the diameter-to-thickness (D/t) ratio for core steel tube.

2.2 Material properties

The mechanical properties of steel were tested. The results are summarized in Table 2. All specimens, as well as pedestals and footings, were made from commercial concrete. To determine the concrete strength, standard cubic specimens with a side length of 150 mm were fabricated using the same batch of concrete for the test columns. The compressive concrete strength value was 43.60 MPa.

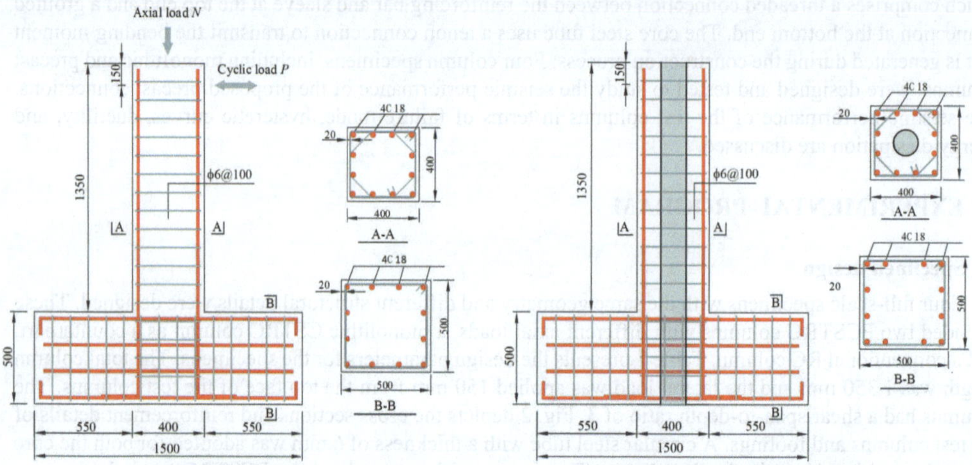

(a) RC column (b) CSTRC and PCSTRC columns

Figure 2. Specimen dimensions and cross section details.

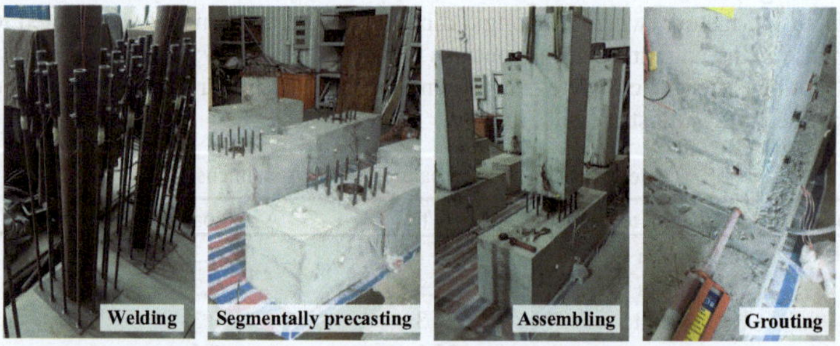

Figure 3. Construction procedure of PCSTRC columns.

Table 2: Mechanical properties of steel.

Reinforcement	Specification	Diameter (mm)	f_y (MPa)	f_u (MPa)	E_s (MPa)
Stirrup	HPB300	8	336.0	545.0	1.85×10^5
Longitudinal rebar	HRB400	18	454.9	595.0	2.07×10^5
Steel tube	Q235	6*	325.2	485.0	2.06×10^5

Note: *The thickness of the steel tube is 6 mm.

2.3 Test setup and instrumentation

Two hydraulic actuators were used to apply the constant axial load and lateral cyclic load, respectively. The lateral load application was displacement controlled and each load cycle was repeated three times at the same displacement magnitude after the column yielding. The test setup and the lateral load protocol are shown in Fig. 4.

A laser displacement sensor was installed to record the lateral displacement of the loading point. Linear variable differential transformers (LVDTs) were installed on the footings to monitor the axial and lateral displacement of the foundation. Strain gauges were attached to the selected points including those on the longitudinal bars, transverse bars, and steel tubes within the plastic hinge of test columns.

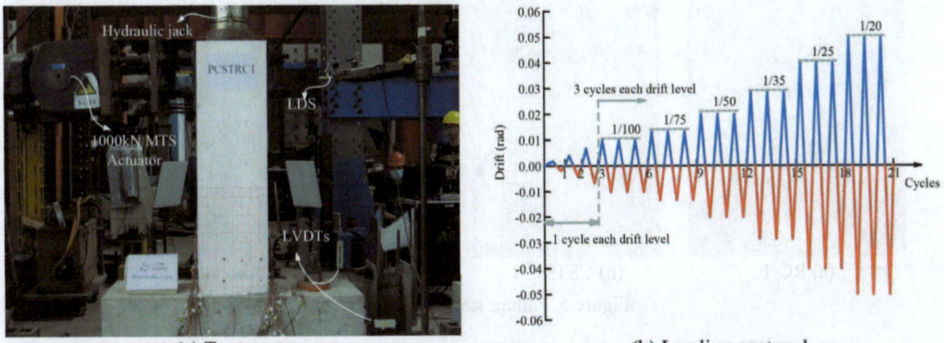

(a) Test setup (b) Loading protocol

Figure 4. Test setup and loading protocol.

3 EXPERIMENTAL RESULTS

3.1 Crack pattern and failure modes

At the end of the experiments, all test columns exhibited typical flexure failure after a plastic hinge was formed at the base of each column as shown in Fig. 5. All specimens underwent a similar damage process. Taking the conventional specimen RC-1 as an example, horizontal cracks began forming at the bottom of the column when loaded to a drift ratio (defined based on the effective column length), $\theta = 1/250$. As the drift level increased, so did the number and length of the horizontal cracks along the column's length. Subsequently, the longitudinal bar yielded, and shortly thereafter, horizontal cracks developed obliquely with an angle of 45°. The applied lateral load reached its maximum when $\theta = 1/75$. As the loading displacement increased, the concrete in the compression zone at the corners of the column spalled, and the broken area gradually expanded. When $\theta = 1/25$, buckling of the longitudinal bars appeared at the bottom of the column, and vertical cracks appeared in the concrete on the compression side and gradually extended upwards. When $\theta = 1/20$, general buckling of the longitudinal bars at the bottom of the specimen, including the bars at the core section, was observed, and most of the corner concrete was crushed, demonstrating failure of the specimen. For specimen CSTRC-1, the peak load appeared later, at $\theta = 1/50$. Local buckling was also initiated at $\theta = 1/25$, while only the longitudinal bars at the edges ultimately buckled.

Despite the similarity of the failure process across the precast and cast-in-situ specimens, some discrepancies still appeared. Fig. 6 compares the failure progression of a CSTRC specimen (CSTRC-1) and a precast specimen (PCSTRC-1). PCSTRC-1 showed fewer visible horizontal cracks compared to its monolithic counterpart, as the contact surface of the precast column cracks easily during loading, resulting in insufficient development of tensile cracks along the column height. Overall, the two specimens showed an analogous failure progression, especially before $\theta = 1/25$. At the failure state, in contrast to CSTRC-1, PCSTRC-1 did not buckle and displayed only slightly crushed concrete at the corners of the column. The cover concrete spalled at the assembled connection. This lack of damage at the steel coupler area (191 mm from the top of the base) indicates that the use of hybrid steel sleeves and a continuous steel tube to connect the lower and upper columns created a strong connection element.

In PCSTRC-2, the precast specimen with a larger axial load, the first cracks appeared at $\theta = 1/200$, and yielding occurred at $\theta = 1/100$. These values are larger than those of PCSTRC-1. Compared to PCSTRC-1, horizontal cracks developed more slowly and were shorter in PCSTRC-2, indicating that axial pressure can delay the appearance of cracks in the specimen and slow the development of cracks.

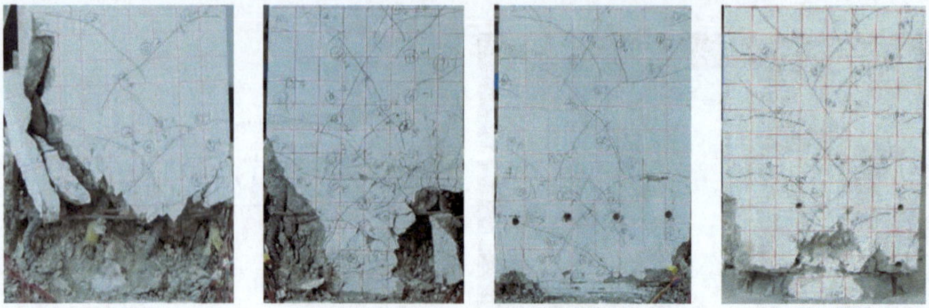

(a) RC-1 (b) CSTRC-1 (c) PCSTRC-1 (d) PCSTRC-2

Figure 5. Failure status of specimens.

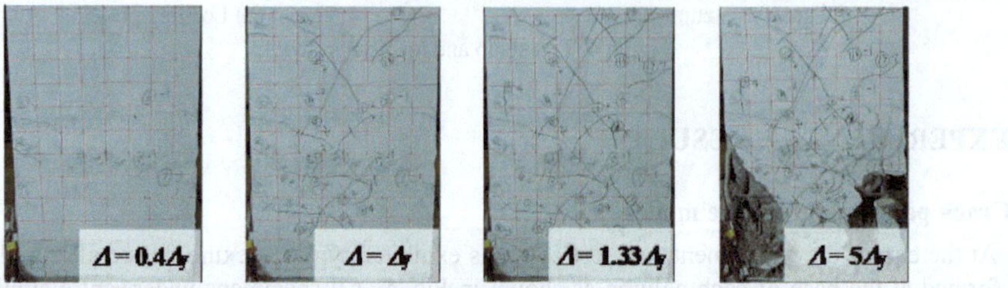

(a) CSTRC-1

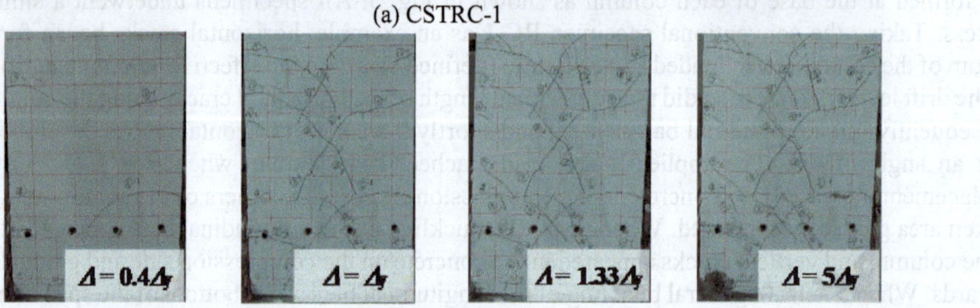

(b) PSTRC-1

Figure 6. Comparisons of failure progress.

3.2 Hysteretic behavior

Fig. 7 depicts the relationship between the lateral load, *P*, and the lateral displacement, *Δ*, measured at the loading point. In the figure, the events of initial cracking, yielding, and concrete spalling are indicated separately using symbols.

The loops in Fig. 7 illustrate a generally symmetric and stable hysteretic behavior for each specimen. It can be observed that the load–displacement relationship formed an almost-straight line before rebar yielding, and no stiffness degradation appeared at this stage. Subsequently, both the reloading stiffness and unloading stiffness of the specimens decreased gradually, and the area of hysteretic loops expanded. As soon as the peak value was reached, the degradation of both the reloading and unloading stiffness became more evident, and the residual deformation after unloading increased. The hysteretic loops were still shuttle-like, which indicates that the specimens still possessed good energy dissipation capacity.

Generally, the cast-in-place specimens (RC-1 and CSTRC-1) exhibited a fuller hysteretic curve than the precast specimens (PCSTRC-1 and PCSTRC-2), indicating that the cast-in-place specimens had better energy dissipation capacity, as shown in Fig. 7. Nonetheless, the hysteric curves of the precast specimens resembled that of the monolithic CSTRC counterpart (comparing the curves in Fig. 7b–c), implying that the proposed precast columns had a global cyclic response akin to that of the monolithic column.

Evidently, the hysteretic curves of the precast specimens were influenced by the axial compression ratio (comparing the curves in Fig. 7c–d). In the case of the higher axial load (PCSTRC-2), the column displayed more rapid deterioration of strength, especially in the push direction, compared to the test column PCSTRC-1 that had a lower applied axial load. Furthermore, the disparity in the post-peak behavior became increasingly evident as the lateral displacement increased.

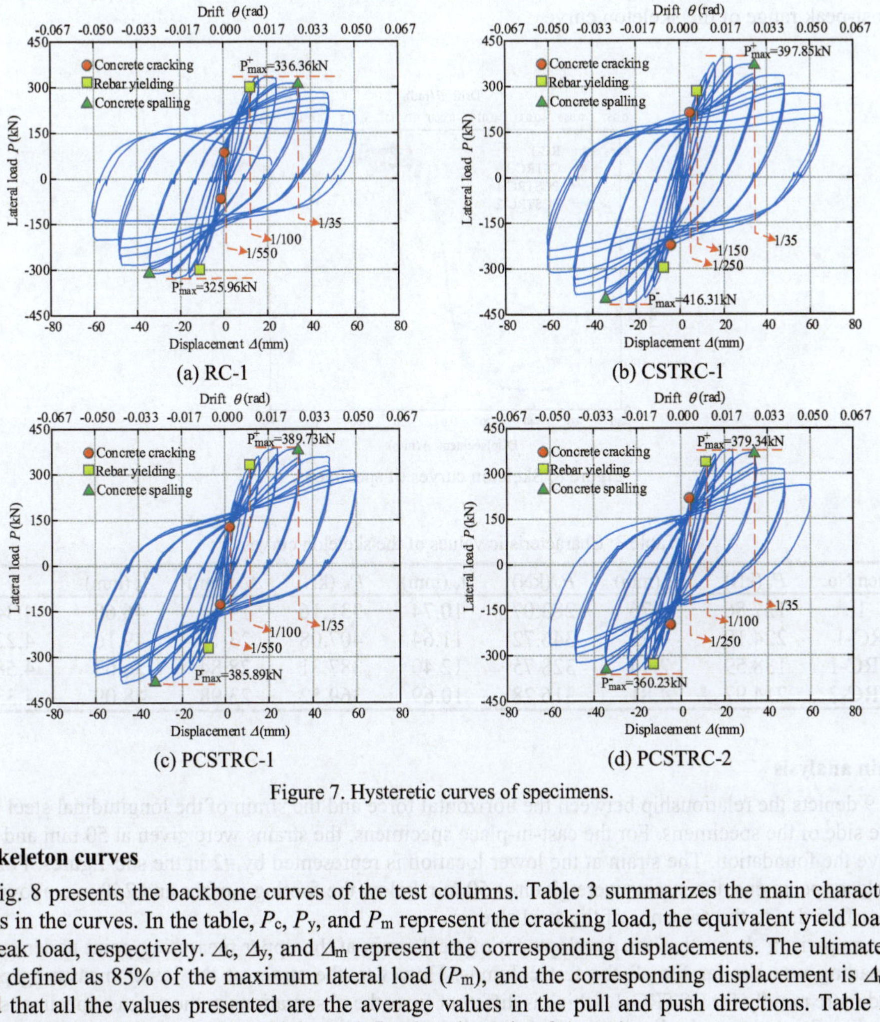

Figure 7. Hysteretic curves of specimens.

3.3 Skeleton curves

Fig. 8 presents the backbone curves of the test columns. Table 3 summarizes the main characteristic values in the curves. In the table, P_c, P_y, and P_m represent the cracking load, the equivalent yield load, and the peak load, respectively. Δ_c, Δ_y, and Δ_m represent the corresponding displacements. The ultimate load, P_u, is defined as 85% of the maximum lateral load (P_m), and the corresponding displacement is Δ_u. It is noted that all the values presented are the average values in the pull and push directions. Table 3 also presents the ductility values (μ), which are defined as the ratio of Δ_u to Δ_y.

Table 3 shows that compared to the conventional RC-1 column, the CSTRC-1 column had greater forces and deformations at the critical levels (i.e., the measured crack, yield, and maximum).

The comparison of the overall responses of specimens CSTRC-1 and PCSTRC-1 clarifies the performance of the proposed precast connection. The load corresponding to the initial crack of PCSTRC-1 was significantly lower than that of CSTRC-1 because, as mentioned in Section 3.1, the connection section cracks easily. The average values of P_y and P_m of PCSTRC-1 were approximately 5.18% and 4.73% lower than those of CSTRC-1, indicating that the precast column has comparable bearing capacity to the in- situ column. The average measured ductility of PCSTRC-1 was slightly larger than that of CSTRC-1.

The comparison of the overall responses of specimens PCSTRC-1 and PCSTRC-2 demonstrates the influence of axial load. The average values of P_c, P_y, and P_m of PCSTRC-1 were approximately 1.05%, 2.57%, and 4% greater than those of PCSTRC-2, respectively. The load-carrying capacity and lateral stiffness remained almost unchanged with the increase of axial load. The deformation capacity decreased in the post-peak range of the skeleton curve.

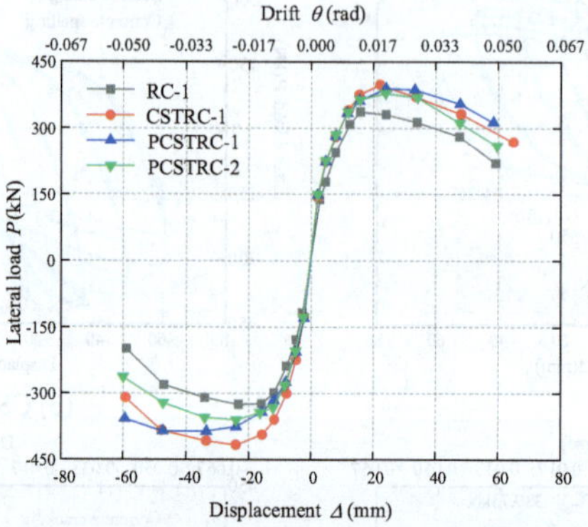

Figure 8. Skeleton curves of specimens.

Table 3: Characteristic values of the skeleton curves.

Specimen No.	P_c (kN)	Δ_c (mm)	P_y (kN)	Δ_y (mm)	P_m (kN)	Δ_m (mm)	Δ_u (mm)	μ
RC-1	127.86	2.76	283.07	10.74	331.16	19.54	46.60	4.34
CSTRC-1	224.10	4.72	346.72	11.64	407.08	23.13	49.16	4.22
PCSTRC-1	138.59	2.14	328.75	12.40	387.81	28.84	56.63	4.58
PCSTRC-2	214.97	4.81	316.28	10.69	369.52	23.98	48.09	4.33

3.4 Strain analysis

Fig. 9 depicts the relationship between the horizontal force and the strain of the longitudinal steel bars along the side of the specimens. For the cast-in-place specimens, the strains were given at 50 mm and 350 mm above the foundation. The strain at the lower location is represented by #2 in the sub-figures. For the cast-in-place specimens, the strains were given at 50 mm below the footing surface and 240 mm above the footing surface due to the presence of the steel sleeve.

The comparison shows that the development of steel strain at the upper measuring point under cyclic lateral loading was similar for all the test columns. However, the strain at the lower measuring point increased more rapidly for CSTRC-1, which is different from the observations in the precast specimens. As shown in Fig. 8c–d, the strain distributions of the reinforcement bar connected by the sleeve were similar, although the strain value of the lower reinforcement bar was slightly greater than that of the upper reinforcement bar after yielding. The phenomenon is related to the cohesive force between the reinforcement bar and the grouting material in the sleeve. This shows that stress could be transferred effectively between the longitudinal bars.

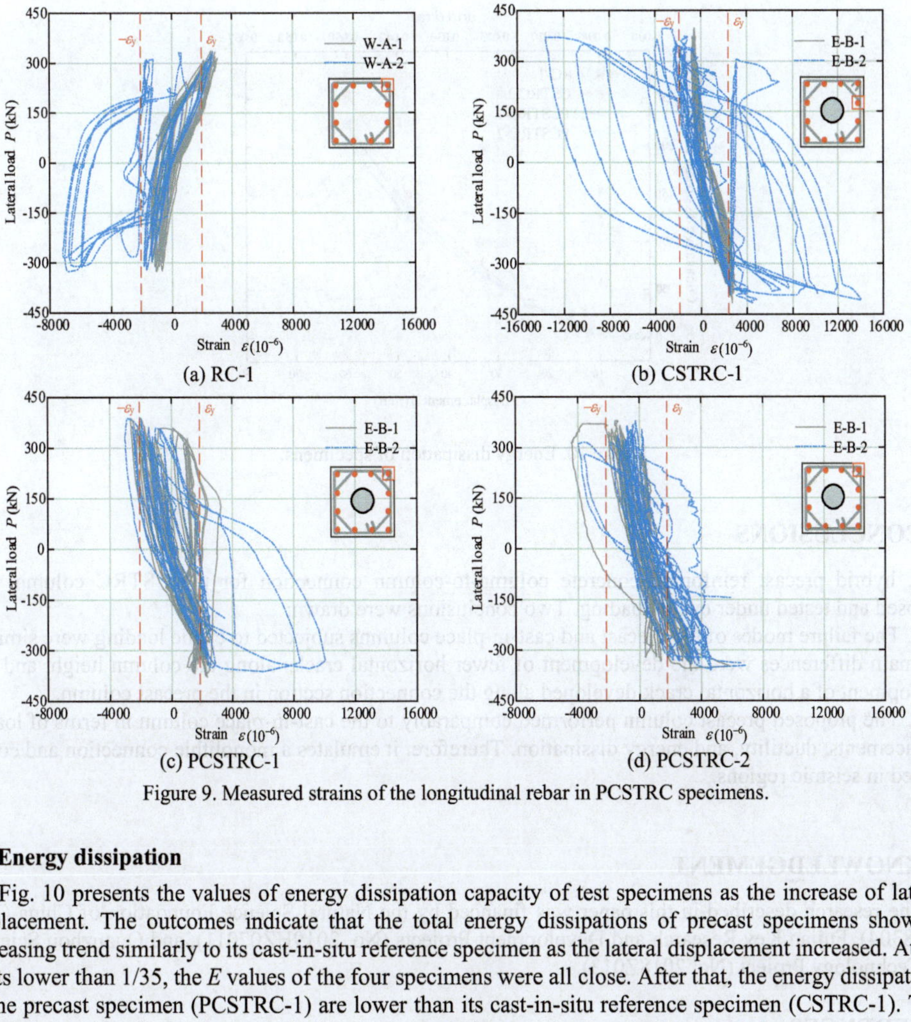

(a) RC-1 (b) CSTRC-1 (c) PCSTRC-1 (d) PCSTRC-2

Figure 9. Measured strains of the longitudinal rebar in PCSTRC specimens.

3.5 Energy dissipation

Fig. 10 presents the values of energy dissipation capacity of test specimens as the increase of lateral displacement. The outcomes indicate that the total energy dissipations of precast specimens show an increasing trend similarly to its cast-in-situ reference specimen as the lateral displacement increased. At the drifts lower than 1/35, the E values of the four specimens were all close. After that, the energy dissipations of the precast specimen (PCSTRC-1) are lower than its cast-in-situ reference specimen (CSTRC-1). This is mainly because the total energy dissipations are sensitive to the shape of hysteresis loops.

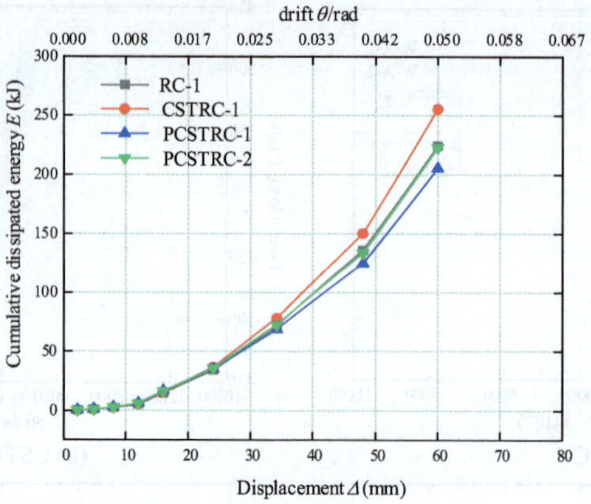

Figure 10. Energy dissipation of specimens.

4 CONCLUSIONS

A hybrid precast reinforced concrete column-to-column connection for the CSTRC column was proposed and tested under cyclic loading. Two conclusions were drawn:

1. The failure modes of the precast and cast-in-place columns subjected to cyclic loading were similar. The main differences were the development of fewer horizontal cracks along the column height and the development of a horizontal crack developed along the connection section in the precast column.

2. The proposed precast column performed comparably to the cast-in-place column in terms of loads, displacements, ductility, and energy dissipation. Therefore, it emulates a monolithic connection and could be used in seismic regions.

ACKNOWLEDGEMENT

The research described in this paper was financed by the Natural Science Foundation of China (No. 51878304), Fujian Key Research and Development Projects (No. 2019HZ07011), and Quanzhou Science and Technology Project (No. 2018Z013).

REFERENCES

[1] Kurama, Y. C.; Sritharan, S.; Fleischman, R. B., et al., "Seismic-resistant precast concrete structures: state of the art", Journal of Structural Engineering, 144(4): 03118001, 2018.

[2] Ye, Z.; Giriunas, K.; Sezen, H., et al., "State-of-the-Art Review and Investigation of Structural Stability in Multi-Story Modular Buildings", Journal of Building Engineering, 101844, 2020.

[3] Chong, X.; Xie, L.; Ye, X., et al., "Experimental study on the seismic performance of superimposed RC shear walls with enhanced horizontal joints", Journal of Earthquake Engineering, 23(1): 1-17, 2019.

[4] Haber, Z. B.; Mackie, K. R.; Al-Jelawy, H. M., "Testing and analysis of precast columns with grouted sleeve connections and shifted plastic hinging", Journal of Bridge Engineering, 22(10): 04017078, 2017.

[5] Xue, Y.; Yang, Y.; Yu, Y., "Cyclic behavior of partially precast steel reinforced concrete short columns: Experiment and theoretical analysis", Engineering Structures, 199: 109658, 2019.

[6] Xu, W.; Han, L. H.; Li, W., "Seismic performance of concrete-encased column base for hexagonal concrete-filled steel tube: experimental study" Journal of Constructional Steel Research, 121: 352-369, 2016.

[7] Hu, H. S.; Wang, H. Z.; Guo, Z. X., et al., "Axial compressive behavior of square spiral-confined high-strength concrete-filled steel-tube columns" Journal of Structural Engineering, 146(7): 04020136, 2020.

[8] Han, L. H.; Li, W.; Bjorhovde, R., "Developments and advanced applications of concrete-filled steel tubular (CFST) structures: Members" Journal of Constructional Steel Research, 100: 211-228, 2014.

[9] Liu, Y.; Guo, Z. X.; Ouyang, W. J., et al., "Experimental study of the seismic behaviour and axial compression ratio limit of CSRC column", China Civil Engineering Journal, 6787(53): 4, 2010.

[10] Hu, H. S.; Liu, Y.; Zhuo, B. T., et al., "Axial compressive behavior of square CFST columns through direct measurement of load components" Journal of Structural Engineering-ASCE, 144(11): 04018201, 2018.

[11] Liu, Y.; Guo, Z. X.; Jia, L. P.; Chen, Q. M., " Experimental study on axial compression performance and design method of core steel tube reinforced concrete short columns", Journal of Building Structures, 36(12): 135-142, 2015.

[12] CECS 188: 2005. Technical specification for steel tube reinforced concrete column structure. Beijing: China Planning Press, 2005: 18-21. (in Chinese)

9TH INTERNATIONAL CONFERENCE ON COMPOSITE CONSTRUCTION IN STEEL AND CONCRETE
M. Knobloch, U. Kuhlmann, W. Kurz, M. Schäfer. (Eds.)
Online, Germany, July 27-29, 2021

EXPERIMENTAL INVESTIGATION OF INTERIOR AND EXTERIOR STEEL-CONCRETE COMPOSITE NPS® BEAM-COLUMN JOINTS

J. Ann Albright*, Alessio Argentoni and Paolo M. Calvi***

* University of Washington
e-mails: pmc85@uw.edu, ann1ea@uw.edu

** Tecnostrutture S.r.l.
e-mail: AArgentoni@tecnostrutture.eu

Keywords: Internal and External Beam-column joints, Reinforced concrete, Seismic performance, Cyclic experiments, NPS® System.

Abstract *This paper presents the results of an experimental program involving a series of five full scale NPS® (three interior and two exterior) and two moderately ductile reinforced concrete (one interior and one exterior) beam-column joints, designed per Eurocode requirements. The NPS® System utilizes steel wrapped reinforced concrete columns and composite steel truss concrete (CSTC) beams reinforced with steel truss elements made with bottom plates to double as partial formwork. The specimens were tested under quasi-static reversed cyclic horizontal loading, with a constant axial load applied to the column. The variables considered were the level of shear reinforcement provided within the joint panel zone and the presence/absence of additional shear reinforcement in the beam end-regions.*

1 INTRODUCTION

The NPS® is a steel-concrete composite system that aims at providing an advanced and efficient technological solution to attain ductile moment resisting frames (MRFs) capable of high seismic performance. The NPS® frame consists of a self-supporting system made of hollow steel columns, with square, circular or rectangular cross section and internal pre-installed reinforcing cage made of longitudinal and transverse steel bars, and steel truss beams with steel plate bottom cord, completed with concrete cast on situ. The steel HSS and beam truss elements support all gravity loads during construction, while also serving as formwork. A view of the NPS® system under construction is shown in Figure 1.

Moment continuity at the beam-column connection is achieved through the placement of integrative steel bars that are inserted through the joint openings in the multilevel column after the beam trusses have been set in position and before concrete is cast, as shown in Figure 1. The HSS corners are reinforced with stiffeners around the joint opening locations to prevent buckling and avoid local failure of the tubes during transportation.

In this configuration, the NPS® system is weaker in flexure at the beam-column interface, where strength is provided solely by the continuity reinforcement. Under seismic loads, flexural yielding is expected to first occur at this location and, provided that the joint region is adequately protected against the high shear loads that arise in the panel zone following yielding of the continuity elements, a localized flexural plastic hinge is expected to occur in the form of a single major crack at the beam-column interface. This is expected to serve as an efficient flexural fuse, limiting the demand on all other structural elements and allowing the prevention of undesired failure modes (e.g. shear failures).

To properly work, the system needs to possess the following attributes: (i) adequate anchorage of the continuity elements; and (ii) adequate strength and integrity of the joint panel zone. Furthermore, the localized plastic hinge needs to have sufficient rotation capacity.

These aspects were investigated experimentally by testing a total of five full-scale NPS® beam-column joint specimens (three interior and two exterior). Two traditional Eurocode compliant reinforced concrete specimens were also designed and tested to provide a tangible target performance that was considered in the assessment and classification of the NPS® specimen response.

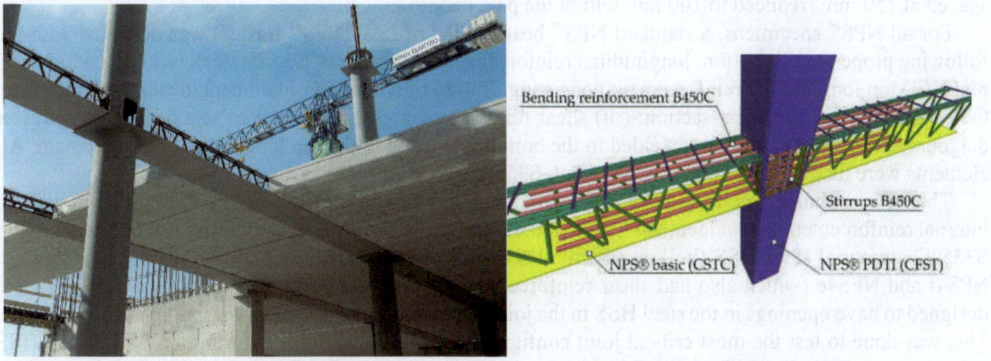

(a) NPS® frame system under construction (b) View of NPS® joint

Figure 1. View of NPS® frame system.

The specimens were tested under quasi-static reversed cyclic horizontal loading, with a constant axial load applied to the column, and the variables considered included the level of shear protection of the joint, and the presence/absence of additional shear reinforcement in the beam end-regions (i.e. the "critical" or "plastic hinge" length).

In the sections that follow, the results of the experiments are discussed and used to draw some preliminary conclusions pertaining to the seismic performance of MRFs realized using NPS® technology.

2 EXPERIMENTAL PROGRAM

2.1 Specimen Design and Fundamental Properties

A total of seven full-scale beam-column joint specimens were tested in the Eucentre Laboratory (Pavia, Italy). The specimens were given the designations shown in Table 1 and were grouped into two Test Series, namely MDI (moderately ductile interior) and MDE (moderately ductile exterior). Within each Test Series, a traditional reinforced concrete joint intended to serve as reference was also designed and tested. All specimens included a 3.5 m long column and a 5 m long beam, both with 400 x 400 mm cross sections. The specimens were designed to represent realistic beam-column joints of a hypothetical three-bay four-story moderately ductile MRF (behaviour factor q = 3.9), subjected to earthquake-induced lateral loads.

All beam-column assemblies were designed with Eurocode (4 and 8) recommendations, following capacity design principles and considering the same demand (i.e. flexural demand at beam ends, M_b = 275 kNm, and axial load demand in the column, N_c = -300 kN) [1, 2]. In design, assumed material properties were typical of construction practice, with a concrete compressive strength, f_c', of 40 MPa and reinforcement yield strength, f_y, of 450MPa. The key objective targeted in this phase was to design a set of capacity protected specimens that could approximately achieve the same strength and overall performance.

The MDI specimens were designed first and the MDE specimens were subsequently assigned the same reinforcement layouts, adequately modified to meet the necessary reinforcement anchorage requirements. It should be noted that the demand on the column and the joint region would be lower for an exterior joint than for its interior companion, and the amount of reinforcement utilized in those regions of the structure could theoretically be reduced. However, identical steel reinforcement was used for MDI and MDE specimens in the interest of exploring the response of systems which had the benefits of a higher level of protection in the joint and the overall response is governed by the flexural response of the beam.

The reference specimens (REFi and REFe) included a beam reinforced longitudinally with a total of 16-18 mm diameter deformed steel bars. The beam was reinforced in shear with 4 leg-10 mm transverse steel reinforcement, spaced at 100 mm (within the critical region) and 200 mm. The column longitudinal reinforcement consisted of 12-28 mm steel bars, and shear reinforcement of 4 leg-12 mm diameter stirrups spaced at 150 mm (reduced to 100 mm within the panel zone).

For all NPS® specimens, a standard NPS® beam (referred to a "NPS® Basic") was designed with the following properties: (i) bottom longitudinal reinforcement consisting of a 5 mm thick (400 mm wide) steel plate; (ii) top longitudinal reinforcement consisting of 4-26 mm diameter plain bars located 360 mm from the bottom fibre of the cross section; (iii) shear reinforcement consisting of two sets of 20 mm diameter diagonal bars spaced at 400 mm, welded to the bottom plate and to the top longitudinal reinforcement. All elements were made of EN 10025-2 compliant, S355-J0 structural steel.

The NPS® column (referred to as "PDTI Column") consisted of a reinforced concrete column with an internal reinforcement layout identical to that of the reference specimens, encased in an EN 1090 compliant, S355-J2 structural steel, HSS (hollow structural section) tube (400 x 400 x 8 mm). All specimens except NPS4i and NPS4e (which also had shear reinforcement spaced at 150 mm throughout the column) were designed to have openings in the steel HSS in the joint panel region, on the sides without connecting beams. This was done to test the most critical joint configuration and to gauge the contribution of the steel HSS column encasement to the integrity of the joint region. It should be noted that under normal circumstances, the steel HSS serves the purpose of temporarily supporting all gravity loads during construction and to enhance strength, ductility and energy dissipation under service conditions as a composite system. However, this work focused on the response of the joint panel zone, the beam-column connection and the beam-end regions. In this context, the column was a capacity protected element, intended to respond without experiencing any damage, thus the HSS beneficial features were not actualized. Nevertheless, the HSS was included in the interest of testing NPS® systems as realistically as possible.

Table 1: Fundamental properties of the specimens and key experimental results.

Specimen ID	Continuity bars	Joint shear reinforcement	Beam-end shear reinforcement	F_n^+[kN] F_n^-[kN]	F_{peak}^+[kN] F_{peak}^-[kN]	$F_{5\%}^+$[kN] $F_{5\%}^-$[kN]
REFi	8 ϕ18 (straight)	Yes (4 ϕ12 @ 100)	Yes (4 ϕ10 @ 100)	199.0 -199.0	195.9 -191.5	133.0 -141.0
NPS2Ai	3 ϕ30 (straight)	Yes (4 ϕ12 @ 100)	No	236.5 -231.0	230.8 -219.7	167.9 -147.6
NPS2Bi	3 ϕ30 (straight)	Yes (4 ϕ12 @ 100)	Yes (2 ϕ14 @ 100)	236.5 -231.0	239.6 -226.5	173.4 -159.6
NPS4i	3 ϕ30 (straight)	Yes (4 ϕ12 @ 150) HSS not cut-out	No	236.5 -231.0	238.8 -234.3	198.7 -192.5
REFe	8 ϕ18 (hooked)	Yes (4 ϕ12 @ 100)	Yes (4 ϕ10 @ 100)	99.5 -99.5	97.8 -102.8	NA NA
NPS2Ae	3 ϕ30 (hooked)	Yes (4 ϕ12 @ 100)	No	118.3 -115.5	104.6 -114.0	79.0 NA
NPS4e	3 ϕ30 (hooked)	Yes (4 ϕ12 @ 150) HSS not cut-out	No	118.3 -115.5	94.9 -115.4	65.0 NA

Moment continuity in the NPS® specimens was achieved via six (equally split between top and bottom) 30 mm diameter bars. Large diameter bars were selected in this program because they represent a desirable solution to address constructability and reinforcement congestion issues in NPS® frame systems. Finally, to investigate different beam-end confinement solutions, specimen NPS2Bi had additional shear reinforcement consisting of 2 leg-14 mm diameter stirrups spaced at 100 mm within the beam critical region (650 mm), while all other NPS® specimens had no integrative stirrups.

The fundamental specimen properties are summarized in Table 1, alongside the estimated nominal strength F_n, and the peak and residual (at 5% drift ratio) experimental strengths (referred to as F_{peak} and $F_{5\%}$, respectively). The final design of all MDE specimens and the main material properties measured experimentally are summarized in Figure 2.

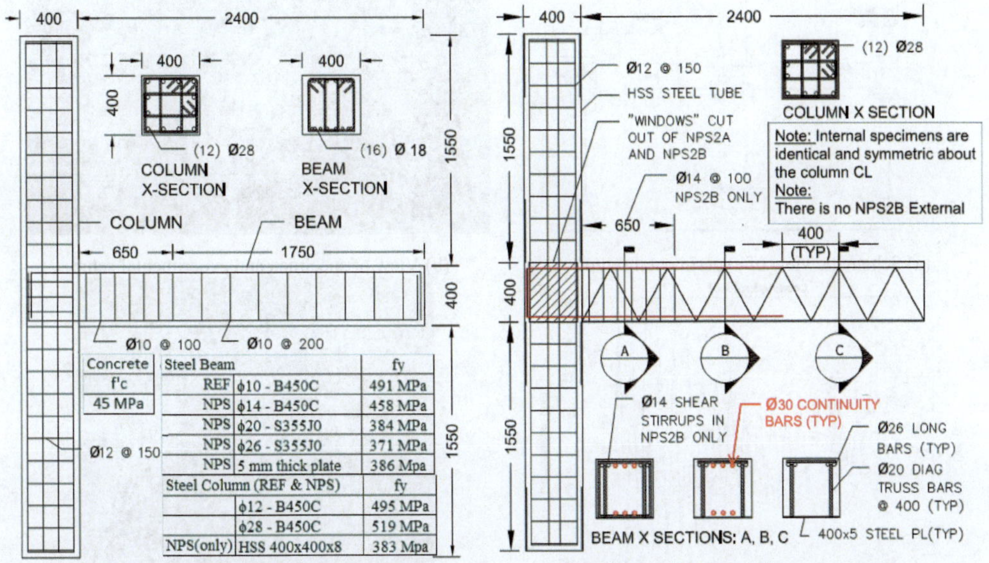

(a) REF Specimens reinforcement layout (b) NPS Specimens reinforcement layout

Figure 2. Specimen reinforcement layout.

2.2 Experimental Setup and Loading Protocol

The experimental program described in this report was conducted in the Shake Lab at Eucentre, Italy. The experimental setup adopted in this program is schematically shown in Figure 3a and in the photograph in Figure 3b. In the configuration the specimen is connected to the strong floor through two steel rods pinned to the extremity of each beam and through a cylindrical hinge located at the base of the column. The top of the column is rigidly connected to two servo-controlled hydraulic actuators intended to simultaneously apply the desired loads. All specimens were tested under quasi-static reversed cyclic horizontal loads and constant vertical loads (300 kN, applied at the top of the column). The loads were applied through two ±500 kN, ±500 mm stroke, servo-controlled actuators. The tests were run in displacement control, following the loading protocol summarized in Figure 3c.

Primary real-time measurements of the global displacements and average strain state of all specimens were provided by up to 49 linear potentiometers (42 for exterior specimens) mounted on the front and back surfaces of the beam-column joints around the panel region (see Figure 3d) or arranged to monitor the horizontal displacements of the systems. Each specimen was also instrumented with at least 10 strain gauges (6.25 mm gauge length) attached directly to selected reinforcing bars such as the continuity reinforcement, the joint shear reinforcement, the column longitudinal bars and some of the beam longitudinal and shear reinforcement.

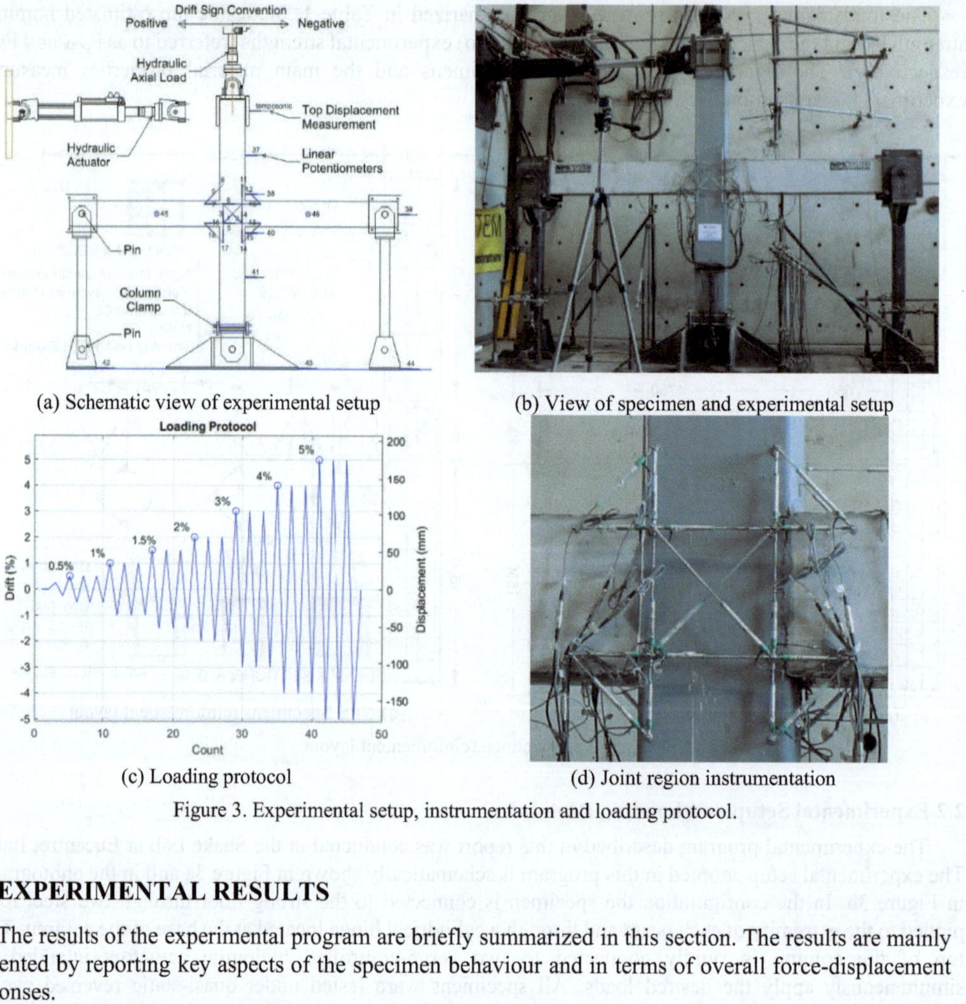

(a) Schematic view of experimental setup

(b) View of specimen and experimental setup

(c) Loading protocol

(d) Joint region instrumentation

Figure 3. Experimental setup, instrumentation and loading protocol.

3 EXPERIMENTAL RESULTS

The results of the experimental program are briefly summarized in this section. The results are mainly presented by reporting key aspects of the specimen behaviour and in terms of overall force-displacement responses.

3.1 MDI Specimens

Peak and residual strength results obtained from the MDI Test Series are summarized in Table 1. Specimens NPS2Ai, NPS2Bi and NPS4i displayed similar strengths (peak loads within 3% of each other) and on average were about 20% stronger than the reference specimen, REFi. These strengths are reflective of the nominal strength of the various systems, which was reached during testing in all cases. Table 1 also lists residual strengths, taken as the peak loads recorded on positive and negative side, during the first loading cycle to 5% drift. Specimens REFi, NPS2Ai and NPS2Bi show similar levels of residual capacity, around 70% of peak on average, while NPS4i displays residual strength ratio of about 85%.

All systems reached flexural beam yielding as per the design objective, developing full nominal strengths. To this end, it is interesting to note that the strength of all specimens is predicted accurately using the Eurocode 8 [2] approach, with average test-to-predicted strength ratio of 0.99 and COV of only 2.3%.

Specimens REFi to NPS2Bi exhibited the same failure mode that involved beam continuity reinforcement yielding (which generally occurred between 1.5% and 2.0% drift ratio) followed by progressive joint deterioration. The damage sequence observed during these tests involved some minor beam cracking (see Figure 5a) followed by severe joint deterioration (see Figure 4), with no notable response differences to report. Similar extent of damage can be observed for the three joints that include extensive diagonal cracking, concrete spalling, and fully exposed reinforcing bars. The joint region of specimen NPS4i benefitted from higher protection, thanks to the intact steel HSS surrounding part of the joint core. This resulted in a different failure mode of the system that involved flexural beam yielding, with the formation of a major vertical crack at the beam-column interface and the beam exhibiting minor cracking. The failure of specimen NPS4i was classified as beam yielding accompanied by nontrivial bond-slip deterioration of the continuity bars around the beam-column interface region (identified as the main source of the pinching effects observed in the force-displacement response).

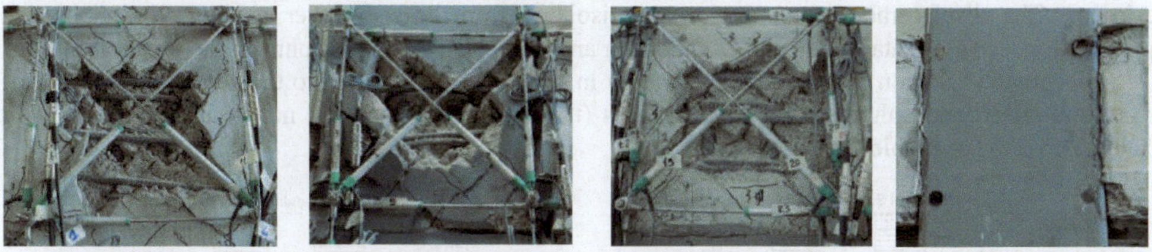

(a) REFi at 5% drift (b) NPS2Ai at 5% drift (c) NPS2Bi at 5% drift (d) NPS4i at 5% drift

Figure 4. Joint region of MDI specimens at 5% drift.

(a) REFi at 5% drift (b) NPS2Ai at 5% drift

Figure 5. Beams of representative MDI specimens at 5% drift.

The presence/absence of added shear reinforcement in the beam-end regions represents the first variable of interest in this experimental program. The effects of this variable were studied by analysing the response of specimens NPS2Ai and NPS2Bi. The force-displacement response of the two specimens is shown in Figure 6a. NPS2Ai reached a peak load of 231 kN, while NPS2Bi was about 4% stronger, showing a maximum resistance of 240 kN. Virtually identical performance was also observed in terms of residual strength and stiffness degradation.

The second aspect of interest was the potential beneficial effects of the column HSS with concurrent reduction of internal shear reinforcement, on the integrity of the joint region. This was studied through the response of specimens NPS2Ai and NPS4i. The force-displacement response of the two specimens is shown in Figure 6b. Specimen NPS2Ai reached a peak load of 231 kN, while NPS4i was about 3% stronger, showing a maximum resistance of 239 kN. The two systems exhibited qualitatively similar behaviour.

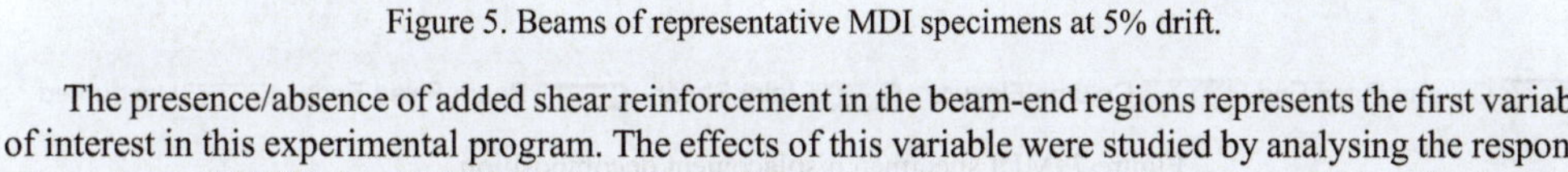

(a) NPS2Ai vs NPS2Bi (b) NPS2Ai vs NPS24i (c) NPS2Ai vs REFi

Figure 6. Force-displacement response of MDI specimens.

NPS4i exhibited superior ability to maintain peak strength over multiple loading cycles, as the displacement demand increased. However, NPS4i also displayed a more severely pinched response, to the point that the re-loading stiffness assumed negative values, at times.
The comparison between the response of specimens REFi and NPS2Ai (chosen as a representative NPS® "baseline" specimen) also provides interesting behavioural insights. The force-displacement response of both specimens is shown in Figure 6c. NPS2Ai displayed an experimental peak strength of about 231 kN, while REFi reached an approximate peak load of 196 kN and, pre-yielding, REFi was characterized by a softer response than its NPS® companion. Despite their different strengths and initial stiffness, the specimens exhibited nearly identical performance in terms of residual strength.

Figure 7 shows the comparison of displacement decomposition at drift ratios 3% and 5% for all MDI specimens. The displacement components were isolated from potentiometer readings, based on the approach proposed by Ashtani et al. and Gombosurer and Maki [3, 4], and on Mohr's circle transformations. The "unaccounted" portion of the total drift shown in Figure 7, is attributed to the combination of beam flexure and shear, and column shear deformations (these components could not be broken down with certainty, from the available measurements).

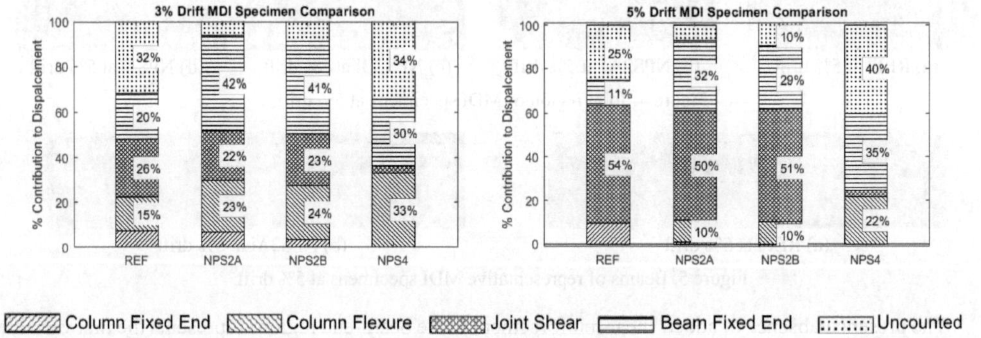

Figure 7. MDI specimen displacement decomposition.

Consistently with what discussed earlier, the major sources of story drift were beam flexure and joint-panel shear for specimens REFi, NPS2Ai and NPS2Bi, with joint-panel shear deformations becoming predominant with increasing drift demands (54%, 50% and 51% of the total drift at 5% ratio for REFi, NPS2Ai and NPS2Bi, respectively). The total drift of specimen NPS4i was mainly due to beam flexure (64% and 75% of the total at 3% and 5% drift ratio, respectively), with joint-panel shear deformations playing a minor role (about 3% of the total drift at 5% drift).

3.2 MDE Specimens

Experimental strength results pertaining to the MDE Test Series are summarized in Table 1. The MDE specimens showed somewhat non-symmetrical response, displaying higher resistance in the negative loading direction in all cases. Specimens NPS2Ae and NPS4e exhibited identical negative strength (about 115 kN), while NPS2Ae was approximately 10% stronger under positive loading. The strength of REFe was slightly lower, with peak load recorded in the negative loading direction of 102.8 kN. As discussed for MDI Test Series, these strengths are reflective of the nominal strength of the various systems, which was reached during testing in all cases.

Overall, the strength of all specimens is predicted accurately using the simplified Eurocode-based approach, with average test-to-predicted strength ratio of 0.95 and COV of 9%. However, the positive strength of the specimens was somewhat lower than expected, and some unconservative strength predictions can be seen in Table 1.

Both NPS® specimens exhibited the same flexural failure mode that involved beam continuity reinforcement yielding. Minor beam cracking was observed (Figure 9) and the joint region exhibited minor to no (NPS4e) diagonal cracking (Figure 8). Instead, a major vertical crack formed at the beam-column interface that worked as an effective localized "plastic hinge" (visible in Figure 8). REFe suffered a premature out-of-plane failure at the base of the column due to a problem with the experimental setup that occurred during the second loading cycle to 3% drift. Up to that point, the specimen was displaying an analogous behaviour to that of its NPS® companions. Photographs of the specimens at failure are shown in Figure 8.

It is worth pointing out that, unlike what seen for MDI specimens, the panel zone of MDE specimens remained virtually undamaged. This outcome is consistent with the fact that the panel zone of MDI and MDE specimens had identical strength but the panel region of MDI specimens was subject to forces that were roughly twice as large on account of the presence of two beams.

Potential beneficial effects of the column HSS on the integrity of the joint region (with reduced amount of shear reinforcement within the join core) were examined analysing the response of specimens NPS2Ae and NPS4e.The force-displacement response is shown in Figure 10a. Because NPS2e did not experience any joint failure, enhancing the strength of the joint panel via steel HSS in NPS4e did not produce any notable performance improvement. The response of NPS2e and NPS4e was virtually identical, with NPS2Ae (no HSS around joint core) showing slightly superior ability to sustain the applied peak as the displacement demand increased.

Preliminary performance of exterior NPS® beam-column joints was conducted by comparing the response of specimens REFe and NPS2Ae. The force-displacement response of both specimens is shown in Figure 10b. The available data (only up to 3% drift for REFe), show that the response of NPS2e and REFe was virtually identical. Specimen NPS2Ae was initially slightly stiffer and somewhat stronger in the positive loading direction. REFe showed peak strength of 103 kN and was about 10% weaker than NPS2Ae, which peaked at 114.5 kN. The available data show that the specimens exhibited nearly identical performance in terms of residual strength. Overall, it is reasonable to believe that REFe would have continued to experience similar response to NPS2Ae, had it not suffered a premature twist-induced collapse at the base of the column.

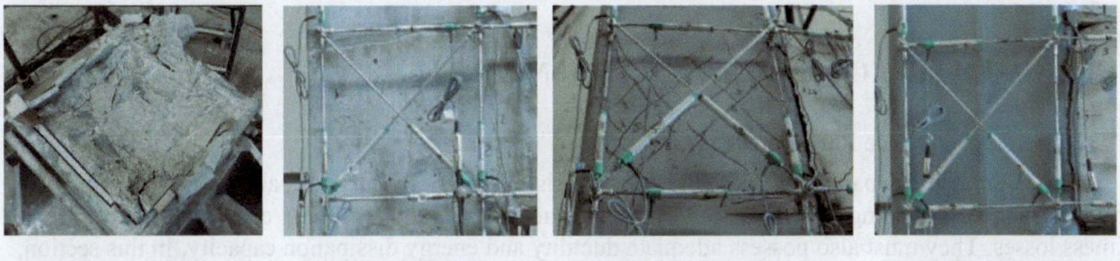

(a) REFe column base (b) REFe (3% drift) (c) NPS2Ae (5% drift) (d) NPS4e (5% drift)

Figure 8. Joint region of MDE specimens.

(a) Beam REFe (3%) (b) Beam NPS2Ae (5%)

Figure 9. Beams of representative MDE specimens.

Figure 11 shows the comparison of displacement decomposition at drift ratios 3% and 5% for all MDE specimens. It is evident that the major source of story drift was beam flexure, with joint-panel shear deformations and column-related displacements playing a minor role.

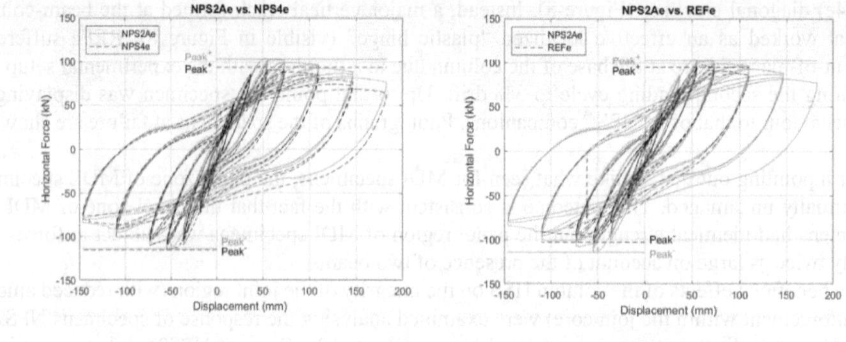

(a) NPS2Ae vs NPS4e (b) NPS2Ae vs REFe

Figure 10. Force-displacement response of MDE specimens.

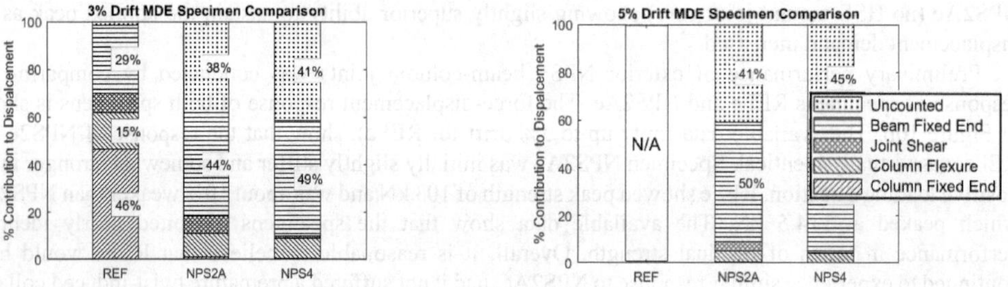

Figure 11. MDE specimen displacement decomposition.

4 SPECIMEN PERFORMANCE ASSESSMENT

4.1 Selected Performance Parameters

Beam-column joints part of a ductile moment resisting frame need to be able to sustain multiple earthquake-induced loading cycles at large displacements, without suffering excessive strength and stiffness losses. They must also possess adequate ductility and energy dissipation capacity. In this section, the performance of the specimens tested in this experimental program is assessed as a function of those parameters.

Specific performance metrics considered include: (i) the ratio between the peak force recorded during each loading cycle, $P_{max,i}$, and the maximum strength of the system in the same loading direction, P_{peak} (computed as a measure of the specimen's ability to sustain its strength); (ii) the ratio between the residual secant stiffness in each cycle, K_0, and the initial stiffness, K_i (computed to quantify the progressive stiffness degradation); (iii) the normalized energy dissipated per cycle $\bar{E}_i$, computed as the ratio between the area encompassed by the force-displacement curve, $E_{n,i}$, and $2\pi P_{max,i}\Delta_{max,i}$; and (ii) the specimen ductility, μ, computed as the ratio of the ultimate displacement, Δ_u, and the yield displacement, Δ_y.

The yield displacement Δ_y was calculated from Equation 1, as recommended by Park [5]. The ultimate displacement Δ_u, was taken as the displacement corresponding to 85% of peak-strength [6, 7].

$$\Delta_y = P * \Delta_{.75P} / P_{75} \tag{1}$$

The initial stiffness of each specimen was calculated based on the yield displacement (see Equation 2). The residual secant stiffness per cycle, K_0, was computed per recommendations of ACI Committee 374 [8], summarized in Equation 3.

$$K_i = \Delta_y / P \tag{2}$$

$$K_0 = \frac{1}{2}\left[\left(\frac{P_{+10\%} - P_{-10\%}}{\Delta_{+10\%} - \Delta_{-10\%}}\right)_{top\ of\ loop} + \left(\frac{P_{+10\%} - P_{-10\%}}{\Delta_{+10\%} - \Delta_{-10\%}}\right)_{bottom\ of\ loop}\right] \tag{3}$$

A graphic representation of how the various parameters were computed from the force-displacement hysteretic response is provided in Figure 12.

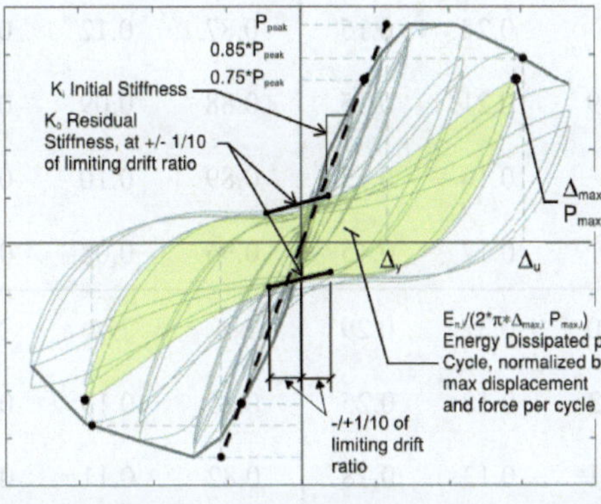

Figure 12. Specimen performance parameter calculation.

4.2 Specimen Response Classification

The outcome of the analysis described in the previous section is summarized in Table 2, for drift ratios of 3% and 4%. The performance results show good levels of compatibility between the different NPS® specimens and between the NPS® specimens and the reference RC specimens.

The results pertaining to the MDI NPS® specimens can be summarized as follows: (i) NPS2Ai and NPS2Bi achieved virtually identical performance with respect to all metrics analysed, which further confirms that adding ductility stirrups at the beam-ends did not enhance the system's performance. The only notable differences are that NPS2Ai appears somewhat less prone to stiffness degradation at 3% drift, while NPS2Bi possesses slightly higher ductility; (ii) NPS4i and NPS2Ai (and NPS2Bi) exhibited analogous performance, but NPS4i was less affected by strength degradation and displayed higher ductility (about 25% higher than its companions).

The results pertaining to the MDE NPS® specimens show that NPS2Ae and NPS4e achieved virtually identical performance with respect to all metrics considered, including ductility. This is in line with the notion that providing additional protection to the joint panel region is not beneficial if the joint itself contains an adequate amount of shear reinforcement.

The interior NPS® and reference specimens also displayed comparable performance with respect to all metrics, with the only notable differences being that the NPS® specimens were generally more ductile while REFi exhibited lower extent of stiffness loss. The same outcome can be observed pertaining to the MDE specimens at 3% drift ratio.

However, both the NPS® and the RC specimens displayed analogous re-loading stiffness values as the loading cycles progressed, so the lower residual stiffness-to-initial stiffness ratios computed for the NPS® specimens, is consistent with their higher initial stiffness. It should be noted that stiffness degradation phenomena are to be partly attributed to the observed bond-slip deterioration at the beam-column interface, that was somewhat more pronounced for the NPS® specimens because of the large diameter continuity bars that were intentionally selected. Thus, the fact that the performance of the NPS® systems was nearly identical to that of the RC reference systems is remarkable, and indicates that capacity protected NPS®

beam-column joints crossed by large diameter bars can exhibit comparable performance to Eurocode 8 compliant moderately ductile RC joints.

Table 2: Performance assessment results.

Specimen	Failure Mode	Performance evaluation at 3% drift			Performance evaluation at 4% drift			Δ_y^+ Δ_y^- (mm)	μ^+ μ^{+-}
		P_0/P_{peak}	K_0/K_i	$\bar{E}$	P_0/P_{peak}	K_0/K_i	$\bar{E}$		
REFi	BJ[1]	1	0.24	0.15	0.87	0.12	0.18	66.0 -61.5	2.23 2.55
NPS2Ai	BJ[1]	0.99	0.21	0.15	0.88	0.09	0.16	58.0 -54.4	2.62 2.60
NPS2Bi	BJ[1]	1	0.16	0.15	0.89	0.10	0.17	57.1 -50.7	2.71 2.74
NPS4i	BY(a)[2]	1	0.17	0.15	0.96	0.08	0.15	51.8 -48.1	3.42 3.33
REFe	BY-O[3]	0.90	0.32	0.29	NA	NA	NA	50.6 -44.9	1.78* 2.04*
NPS2Ae	BY[4]	0.93	0.18	0.25	0.85	0.13	0.31	43.2 -38.1	2.40 2.11
NPS4e	BY[4]	0.91	0.13	0.28	0.82	0.11	0.36	38.8 -32.9	2.12 2.19

[1]Beam yielding followed by joint deterioration; [2]Beam yielding followed by continuity bars debonding; [3]Beam yielding followed by out-of-plane failure; [4]Beam yielding

5 CONCLUSIONS

The major conclusions that can be drawn from the experimental results presented and analysed in this paper can be summarized as follows:

With respect to the design variables studied:

- The response of beams in capacity-designed NPS® beam-column joints is not influenced by the presence/absence of integrative shear reinforcement (i.e. ductility stirrups) in the beam end regions. If properly designed, the NPS® beam truss alone can resist the full shear actions, while most of the displacement demand is absorbed in the form of localized rotation at the beam-column interface (over-reinforced joint cores) or as a combination of localized rotation at the beam-column interface and shear deformation of the joint panel zone (under-reinforced joint cores).
- Shear protection of the joint core is necessary to ensure that flexural plastic mechanisms can effectively develop at the beam-column interface. The amount of necessary joint reinforcement depends on the "ductility class" that is targeted in design and on the level of confinement of the joint. Using the column HSS as a substitute for traditional stirrups represent a very effective way of achieving full protection of the joint panel zone. It proved particularly effective with respect to the response of the planar interior joints tested in this program, for which a failure mode shift was observed. A similar effect would be provided by transverse beams in a 3D joint.

With respect to the NPS® joint performance:

- All systems reached their targeted design strength, displaying a good level of ductility. For the most part, interior joints exhibited joint failure following beam yielding at the beam-column interface, while exterior joints did not show major signs of distress in the joint and experienced beam yield failure. In both cases, the demand on the system was controlled by the yield strength provided by the continuity elements at the beam-column interface.
- All interior joints were characterized by a pinched force-displacement response, due to bond-slip deterioration caused by the high bond demand on the continuity bars. Despite the level of

pinching observed, all systems displayed sufficiently high levels of strength, ductility, stiffness and energy dissipation. This is seen as a very positive and promising outcome, particularly considering that large continuity bars with a joint depth-to-bar diameter ratio of 13 were purposely selected in this program. Evidently, bond-slip deterioration phenomena can be mitigated via smaller diameter continuity bars.

- All exterior beam-column joints exhibited excellent performance with no deterioration of the joint panel zone and trivial bond-slip degradation. Their force-displacement response was characterized by fat hysteretic loops denoting high levels of energy dissipation and ductility.

With respect to design implications:

- The results of this experimental program demonstrated that NPS® beam-column joints can be successfully designed to achieve the same performance of moderately ductile (per Eurocode 8 definition) RC systems. This supports the notion that a behaviour factor of up to 3.9 can be considered in design to estimate the seismic demand.
- The capacity design approach adopted in this study worked well, leading to the design of a series of full-scale NPS specimens that behaved as intended, without displaying undesired mechanisms in beams and columns. This design approach is thus recommended moving forward.
- To address practical/constructability issues, the use of large diameter continuity bars is often required. The specimens tested in this program were characterized by a joint depth-to-bar diameter ratio of 13 and exhibited good performance. For the time being, this value may be used as a reference, knowing that higher ratios would lead to better performance, all else being equal.

REFERENCES

[1] *E. 1994-1-1:2004, "Eurocode 4: Design of composite steel and concrete structures - Part 1-1: General rules and rules for buildings".*

[2] *EN 1998-1:2013, "Eurocode 8: Design of structures for earthquake resistance - Part 1: General rules, seismic actions and rules for buildings".*

[3] *M. S. Ashtani, R. P. Dhakal and A. N. Scott, "Seismic Performance of High-Strength Self-Compacting Concrete in Reinforced Concrete Beam-Column Joints," ASCE Journal of Structural Engineering, vol. 140, no. 5, p. 04014002, 2014.*

[4] *D. Gombosurer and T. Maki, "Prediction of Joint Shear Deformation Index of RC Beam-Column Joints," MDPI Buildings, vol. 10, no. 176, p. 10100176, 2020.*

[5] *R. Park, "Ductility Evaluation from laboratory and Analytical Testing," in Proceedings of Ninth World Conference on Earthquake Engineering, Vol VIII, Tokyo-Kyoto, Japan, 1988.*

[6] *W. Liu and J. Jia, "Experimental Study on the Seismic Behavior of Steel-Reinforced Ultra-High-Strength Concrete Frame Joints with Cyclic Loads," Advances in Structural Engineering, vol. 21, no. 2, pp. 270-286, 2018.*

[7] *J. Wang, L. Cheng and Y. Yang, "Experimental Study of Seismic Behavior of High-strength RC Columns Strengthened with CFRP Under Cyclic Loading," Journal of Structural Engineering, vol. 145, no. 2, 2019.*

[8] *ACI Committee 374, Acceptance criteria for moment frames based on structural testing and commentary (ACI 374.1-05)., Farmington Hills, MI: American Concrete Institute, 2005.*

9TH INTERNATIONAL CONFERENCE ON COMPOSITE CONSTRUCTION IN STEEL AND CONCRETE
M. Knobloch, U. Kuhlmann, W. Kurz, M. Schäfer. (Eds.)
Online, Germany, July 27-29, 2021

RCS MOMENT FRAMES IN HIGH SEISMIC ZONES IN THE UNITED STATES

Luis B. Fargier-Gabaldon[1], Paul Cordova[2], Gustavo Parra-Montesinos[3] and Gregory Deierlein[4]

[1] University of Notre Dame
luis.fargier@nd.edu

[2]Simpson Gumpertz & Heger Inc
PPCordova@sgh.com

[3]University of Wisconsin-Madison
gustavo.parra@wisc.edu

[4] Stanford University
ggd@stanford.edu

Keywords: Composite Structures, Beam-to-Column Connections, Seismic Design, RCS structures, Joint Deformation.

Abstract

The research on composite moment connections for seismic applications is examined to propose an update of the Guidelines for Design of Joints Between Steel Beams and Reinforced Concrete Columns, published by the ASCE in 1994. These guidelines were developed based on limited experimental data and restrict composite moment frames to zones of low to moderate seismicity. Recent experimental data, however, support their use in high seismic regions. A database of full-scale connection tests is included, accompanied by a comparison between strength measured experimentally vs. the calculated joint strength. Results from this comparison show good agreement.

1 INTRODUCTION

In composite moment joints between Reinforced Concrete (RC) columns and Steel (S) Beams, the beam runs continuous through the reinforced concrete column (Fig. 1). This composite structural system has been referred to as RCS frames and is particularly attractive for contractors because it allows simultaneous construction tasks along the height of the building that results in a fast construction process. In one construction approach, light steel columns are used to erect a steel skeleton in the upper stories while few stories below the light steel column is encased by a reinforced concrete column (Fig. 2). At the same time, a construction crew in the lower stories places the steel deck and casts concrete. It is also attractive to designers because of its superior performance under reversed cyclic loading, characterized by full hysteresis loops (Fig. 3). Research programs conducted in the United States focused on typical construction practices preferred by the industry in North America, starting in the mid-80s at the University of Texas at Austin and followed by experimental programs at Cornell University, The University of Michigan, Texas A&M University and at Stanford University. A summary of these research programs can be found in [1].

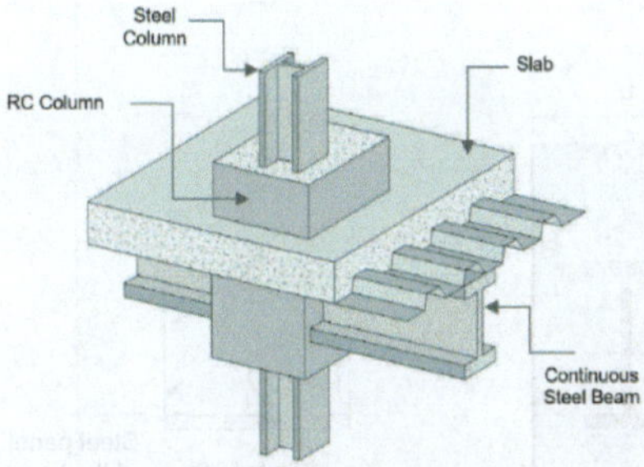

Figure 1. RCS joint [2]

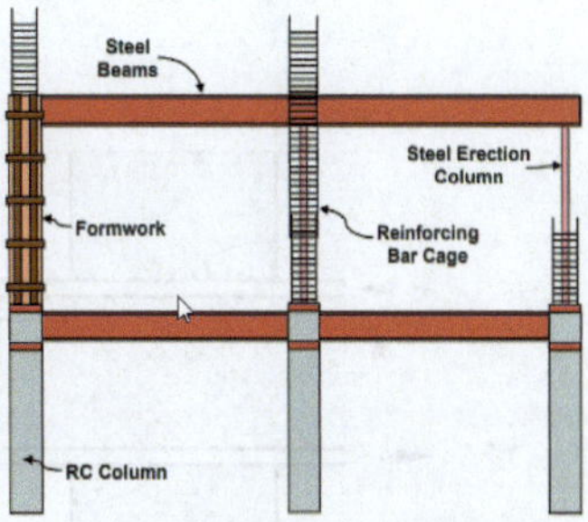

Figure 2. RCS frame [3]

Over 50 full-scale beam-to-column RCS connections have been tested in the United States, including interior, exterior and roof-T connections and a full-scale RCS frame was tested in Taiwan, as part of a collaborative effort. Test results suggest that the joint strength can be idealized as the summation of three components, (a) the strength of the web panel of the steel beam, (b) the strength of a concrete strut that forms between the flanges of the steel beam and (c) the strength of a concrete strut that forms outside the flanges of the steel beam (Fig. 4). In the current design Guidelines [5], these three mechanisms are assigned to two panels. An interior panel (includes the contribution of the steel web panel and concrete strut that forms between flanges of the steel beam) and the exterior panel (contribution of the concrete strut that forms outside the width of the beam flanges). More information can be found in [1].

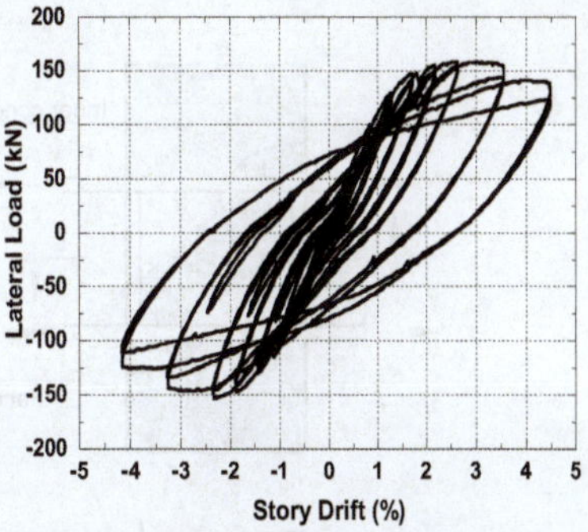

Figure 3. Load vs. Drift.
RCS interior connection [4]

2 STRENGTH OF THE CONNECTION

The connection strength is given by the contribution of the inner and outer panel,

$$V_n = k\,(\phi V_{in} + \phi_s V_{on}) \quad (1)$$

V_{in} and V_{on} are the horizontal shear strength of the inner panel and outer panel, respectively and k is a factor to limit connection deformations (k = 1.0 where large connection deformations are expected and k = 0.85 where the intent is to limit joint deformations), $\phi_s = 0.85$ and the ϕ factor accompanying the term V_{in} is equal to 0.85 for shear failures or 0.75 for bearing failures.

2.1 Strength of inner panel,

The strength of the inner panel (V_{in}) is controlled by the lesser of: (a) the summation of the nominal horizonal shear strength of the steel beam web panel (V_{spn}) and the inner concrete strut (V_{icn}, as shown in Fig 4a, and 4b) and (b) by the bearing strength of the connection, which can be expressed in terms of horizontal shear as $(M_{vb} - V_b\,h)/d_j$ (as shown in Fig. 5).

$$\phi V_{in} = \phi_s\left(V_{spn} + V_{icn}\right) \le \phi_b \frac{(M_{vb} - V_b\,h)}{d_j} \qquad (2) \qquad \phi_s = 0.85\,,\ \ \phi_b = 0.75$$

In Eq. 2, d_j is the joint depth taken as the distance between flange centerlines of the steel beam (Fig. 5), h is the column depth, V_b is the beam shear and M_{vb} is the moment produced by vertical bearing of the concrete column against the flanges of the steel beam (Fig. 5).

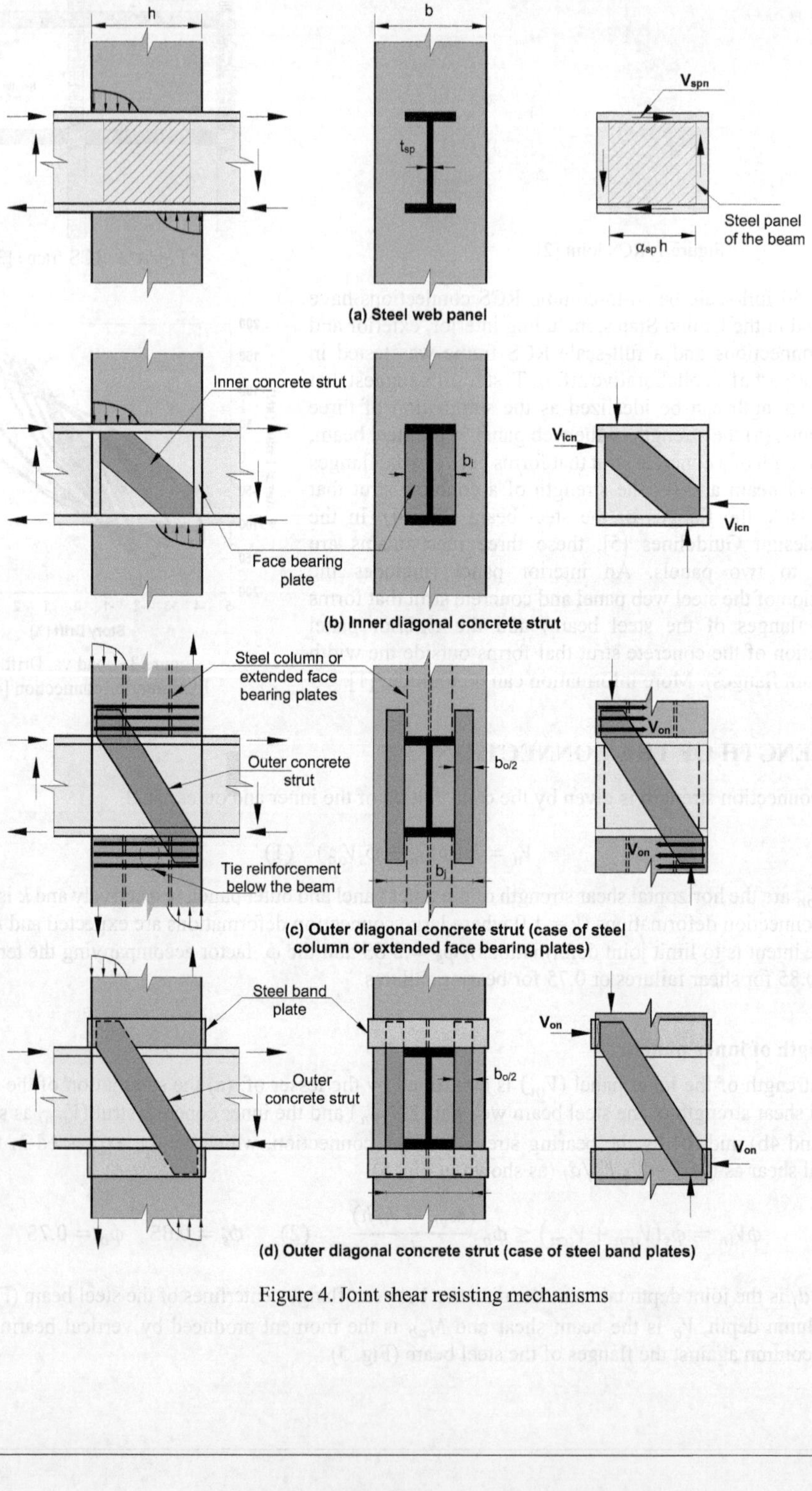

Figure 4. Joint shear resisting mechanisms

2.1.1 Strength of the web panel

The horizontal shear strength of the steel beam web panel (V_{spn}) is given by Eq. 3 (see Fig 4a),

$$V_{spn} = 0.6\, F_{ysp}\, t_{sp}\, \alpha_{sp}\, h \qquad (3)$$

In Eq. 3, F_{ysp} and t_{sp} are the specified yield strength and nominal thickness of the steel beam web panel, respectively. The term 0.6 F_{ysp} is the average yield shear strength. Results from research [6, 7, 8] indicate that a shorter steel web panel length is mobilized in exterior joints compared with interior joints and even shorter web panel lengths are mobilized in joints with T configuration (roof level). This reduction is accounted for with the term α_{sp} (Table 1).

2.1.2 Strength of the inner concrete strut

The horizontal shear strength of the inner concrete strut is given by Eq. 4 (Fig. 4b),

$$V_{icn} = \alpha_{cs}\, 1.7\, \sqrt{f_c'}\, b_i\, h \leq 0.5\, f_c'\, b_i\, d_j \qquad (4)$$

In Eq 4, f_c' is the specified concrete compressive strength in MPa, b_i is the width of the inner panel (Fig. 4b) and d_j is the joint depth. Recommended values for the coefficient α_{cs} are shown in Table 1. The strength of interior joints is assumed equal to that specified in [9] (ACI 318-19), for connections confined by four faces. The upper limit on Ec. 4 is set to prevent concrete crushing at the end of the strut.

Table 1: Recommended α_{sp} and α_{cs} factors

Connection type	α_{sp}	α_{cs}
interior	0.9	1.0
exterior	0.8	0.6
other	0.3	0.3

2.1.3 Bearing strength of the inner panel

In Fig. 5, a fraction of the column moments framing into the joint is replaced with vertical bearing stresses acting on the top and bottom flanges of the steel beam. The resulting bearing force is equal to,

$$C_{cn} = \alpha_{cn}\, f_c'\, b_f\, (0.3\, h) \qquad (5)$$

Forces acting on joint reinforcement, T_{vrn} and C_{vrn} are also shown in Fig. 5. For joints confined with steel band plates, $\alpha_{cn} = 2.5$ and for other cases $\alpha_{cn} = 2.0$. The length of the bearing area is set equal to 0.3 h, [10, 11]. The factor α_{cn} considers the amount of confinement [4, 6, 7, 10, 12, 13]. Summing moments above or below the joint,

$$M_{vb} = C_{cn}\, (0.7\, h) + h_{vr}(T_{vrn} + C_{vrn}) \qquad (6)$$

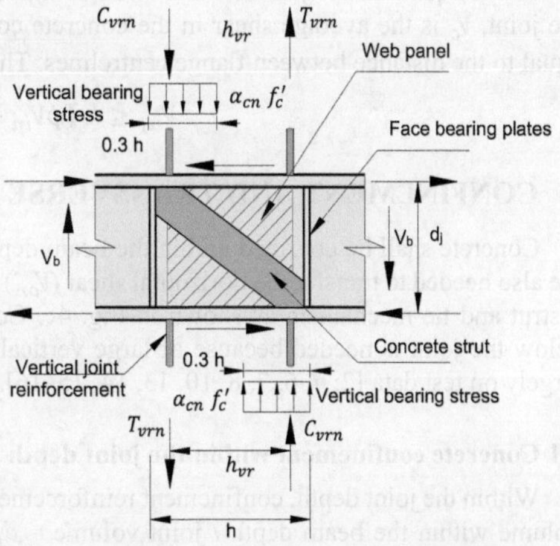

Figure 5. Bearing forces acting on the inner panel

where M_{vb} is the moment produced by bearing of the column against the flange of the steel beam. The contribution of the term $(T_{vrn} + C_{vrn})$ is limited to $C_{cn}/2$ to avoid excessive bearing stresses.

2.2 Strength of outer panel,

The horizontal shear strength of the outer panel is equal to the horizontal shear strength of the outer concrete strut, V_{on},

$$V_{on} = \alpha_{cs} 1.25 \sqrt{f_c'} b_o h \qquad (7)$$

Test data [2, 6, 7] suggest that concrete regions over a width equal to b_o outside the width of the beam flanges are mobilized and contribute to the joint strength (Fig 4c and 4d). Values of α_{cs} are shown in Table 1. When shear keys in the form of extended FBPs or steel columns are provided, the outer panel width (b_o) is calculated with Ec. 8a. For connections with steel band plates, Ec. 8b shall be used,

$$b_o = (d_{scol} + h)/2 + b_{scol} - b_f \leq b_{core} - b_f \quad (8a) \qquad\qquad b_o = 12t_{bp} \leq b_{core} - b_f \quad (8b)$$

In Eq. 8a and 8b, d_{scol} and b_{scol} are the depth and width of the steel column welded to the flanges of the steel beam, respectively, and b_{core} is the width of the core of the concrete column. For the case of extended face bearing plates, $d_{scol} = h$ and b_{scol} is equal to the width of the extended face bearing plate. When more than one shear key is provided, the width of the outer panel can be determined based on the shear key that leads to the largest width. When tension forces are probable in the column the term V_{on} shall be taken equal to zero.

3 REQUIRED STRENGTH

The required connection shear strength, V_{uj},

$$V_{uj} = \frac{\Sigma M_b}{d_j} - V_c \quad (9)$$

In this equation, ΣM_b is the summation of expected moment strength of the steel beam at either side of the joint, V_c is the average shear in the concrete column concurrent with ΣM_b, and d_j is the joint depth, equal to the distance between flange centrelines. The design of the joint shall satisfy,

$$V_{uj} \leq k(\phi V_{in} + \phi_s V_{on}) \qquad (10)$$

4 CONFINEMENT AND TRANSVERSE REINFORCEMENT

Concrete shall be confined within the beam depth as shown in Fig. 6. Ties above and below the beam are also needed to transfer the horizontal shear (V_{on}) from the steel beam to the outer concrete panel through a strut and tie mechanism as shown in Fig. 4c. Confinement of the concrete regions located above and below the joint is needed because of large vertical bearing stresses (Fig. 5). These provisions are based largely on test data [2, 4, 6, 7, 8, 10, 13, 14, 15, 16].

4.1 Concrete confinement within the joint depth

Within the joint depth, confinement reinforcement in a volumetric ratio (ρ) equal to 0.01 is needed (tie volume within the beam depth / joint volume = $d_j\ h\ b$) at a spacing less than 0.25 d and 0.25 h in the absence of steel band plates, as shown in Fig 6a. Steel band plates shall be used when transverse beams frame into the connection (Fig. 6b). In this case, ties are needed to prevent buckling of the column longitudinal bars passing through the joint, as shown in Fig 6b.

4.2 Concrete confinement above and below the steel beam

Concrete regions located above and below the steel beam shall be confined with ties (Fig. 6a) or band plates (Fig. 6b). In both cases, the minimum total cross-sectional area A_t is equal to V_{on}/F_{yt}, were A_t is the cross-sectional area of or ties over a depth of $0.4d$ (Fig. 6a) or band plates (Fig. 6b) and F_{yt} is the specified yield strength of tie reinforcement or band plate. In addition, the required minimum volumetric ratio of each band plate is 0.03 (band plate volume / $d_j\ b\ h$) and its minimum depth and thickness is equal to 0.25 d and 13 mm, respectively.

5 DETAILING CONSIDERATIONS

Table 2: Detailing considerations

Applies to	Equation	Definition
Column-to-beam strength ratio. Reinforced concrete columns shall satisfy the requirements of Chapter 18 of the ACI-318 (2019) Building Code [9]. Flexural strength shall be assumed such that the column moments oppose to the beam moments. Nominal flexural strength of the beams shall be calculated in accordance with the AISC-360 (2016) Specification [19].	$\Sigma M_{nc} \geq \frac{6}{5} M_{nb}$ The contribution of the concrete slab shall be considered when composite beam action is expected.	ΣM_{nc} is the sum of nominal flexural strength of columns framing into the joint, evaluated at the face of the joint and ΣM_{nb} is the sum of nominal flexural strength of the beams framing into the joint, evaluated at the lateral faces of the joint.
Diameter of column longitudinal (d_b). Test results [6, 7] showed that the proposed equation provides an upper limit to the bar diameter to prevent excessive slippage.	$d_b \leq \frac{d_j}{20}\frac{420}{F_{yr}} \leq \frac{d_j}{20}$ d_b may be increased by 25% when steel band plates are used (except connection at the top level).	F_{yr} is the specified yield strength of column longitudinal bars in MPa. For top joints, the use of headed bars or mechanical anchors for column longitudinal reinforcement is recommended [8]. d_j is the distance between centreline of the beam flanges.
Flanges of the steel beam shall resist transverse bending caused by bearing of the inner concrete strut [14, 15].	$t_f \geq 0.30\sqrt{\frac{b_f\, t_{sp}\, d\, F_{ysp}}{h\, F_{yf}}}$	t_{sp}, t_f = nominal thickness of steel beam web panel and flanges, F_{ysp} and F_{yf} specified yield strength of web panel and flanges. d =steel beam depth.
Thickness of the Face Bearing Plates (t_p). FBP shall resist a force equal to the nominal horizontal shear strength of the inner concrete strut [14, 15].	$t_p \geq \frac{\sqrt{3}\, V_{icn}}{2\, b_f\, F_{up}}$	F_{yp} and F_{up} = the specified yield and tensile strength of the FBP, respectively and d_j is the distance between beam centreline.
	$t_p \geq 0.20\sqrt{\frac{V_{icn}\, b_i}{F_{yp}\, d_j}}$	b_i =width of the inner panel (Fig. 4b).
	$t_p \geq b_f/22$	
	$t_p \geq 10\ m$	

6 CALCULATED VS. MEASURED JOINT STRENGTH

A summary of the research programs conducted in the United States relevant to this research can be found in [1].

Seven interior joints were tested at the University of Texas at Austin under monotonic loading and eight joints under reversed cyclic loading [14, 15, 16] in the late 80s. All specimens featured a 510 mm square column and a continuous built-up steel beam (depth ranging from 300 mm to 450 mm). Joint details included face bearing plates, steel erection columns and extended face bearing plates. Five years later, in the mid-90s, nineteen interior joints were tested under reversed cyclic loading at Cornell University [10]. A 406 mm square column was used in all specimens with a continuous steel beam (depths ranging from 210 mm to 350 mm). Joint details included cover plates around the column faces within the beam depth, steel band plates, transverse steel beams, vertical joint reinforcement and headed studs welded to the beam flanges. Three specimens featured high strength concrete (100 MPa) while an axial load corresponding to an average compression stress of approximately $0.3f_c'$ was applied to five specimens. Joint reinforcement ranged from 0.4 to 1.2%. in volume.

Research at the University of Michigan in the late 90s included nine tests of exterior joints under reversed cyclic loading [6, 7]. Joint details included transverse steel beams, face bearing plates, steel band plates, steel columns and horizontal joint reinforcement in volumetric ratios ranging from 0.5 to 0.9%. The use of fibre reinforcement in the joint region was also explored. A 400 mm square column was used in all specimens while the depth of the continuous steel beam ranged from 240 mm to 400 mm. Two joints with a T-configuration were also tested under reversed cyclic loading in the early 2000s [8]. A 400 mm square column was used, and the joints featured face bearing plates, steel band plates and steel columns. The depth of the continuous beam in one specimen was equal to 210 mm and in the other specimen equal to 360 mm. In both cases the column longitudinal reinforcement was terminated at the top of the joint with mechanical anchors to reduce reinforcement congestion. As part of the same research program, two exterior connections were tested under reversed cyclic loading [2]. A rectangular column was used (b=600 mm, h=400 mm) with a 360 mm deep steel beam continuous through the joint. Both specimens featured face bearing plates and an embedded steel erection column. Details in one specimen included horizontal joint reinforcement in a volumetric ratio of 0.7%, while the other specimen featured steel band plates and transverse steel beams.

Research was also conducted at Texas A & M University in the early 2000s [17, 18]. Five interior and one exterior RCS beam-column-slab subassemblies were subjected to reversed cyclic loading in the two principal directions. Joint detailing included face bearing plates, band plates and cover plates encasing the entire joint. The specimens featured composite slab and were design for beam hinging. Test results indicated that the slab width assumed to be effective in transferring force to the joint correlates well to the effective slab width given in [19]. Experimental evidence confirmed the feasibility of RCS connections for use in high seismic zones.

A database of specimens that feature details comparable with those that are anticipated to be included in the updated Guidelines is shown in Table 3. On the other hand, Fig. 7 shows the measured strength (V_{uj},) vs. calculated joint strength (V_n) for the specimens included in the database. In the calculations, $k = 1.0$ and all reduction factors (ϕ) are set equal to one.

An examination of the data shown in Fig. 7 indicates that the proposed equations provide good estimates of joint strength when compared with experimental results, and thus can be used in practice when designing joints with comparable geometries.

The behaviour of the of RCS frames under earthquake-induced loads was studied at the National Center for Earthquake Engineering, Taipei, Taiwan, in collaboration with Stanford University [20]. The frame was designed to meet the code requirements of a structure built in California and Taiwan. Composite action of the continuous steel beams with the concrete slab was achieved with the use of shear studs. Joint were detailed with face bearing plates, steel band plates and transverse steel beams. The structure was subjected to three pseudo-dynamic tests scaled to represent seismic hazard levels with 50%, 10%, and 2% chance of exceedance in 50 years. Results from these tests showed good correlation with limited structural damage, repairable damage and collapse prevention, respectively. It should be noted that slab-composite action was maintained in all tests without sights of shear stud failure. A final push-over test was conducted up to an inter-story drift of approximately 10% which led to a mechanism. Results from this test confirmed that properly designed RCS frame are suitable for high seismic zones.

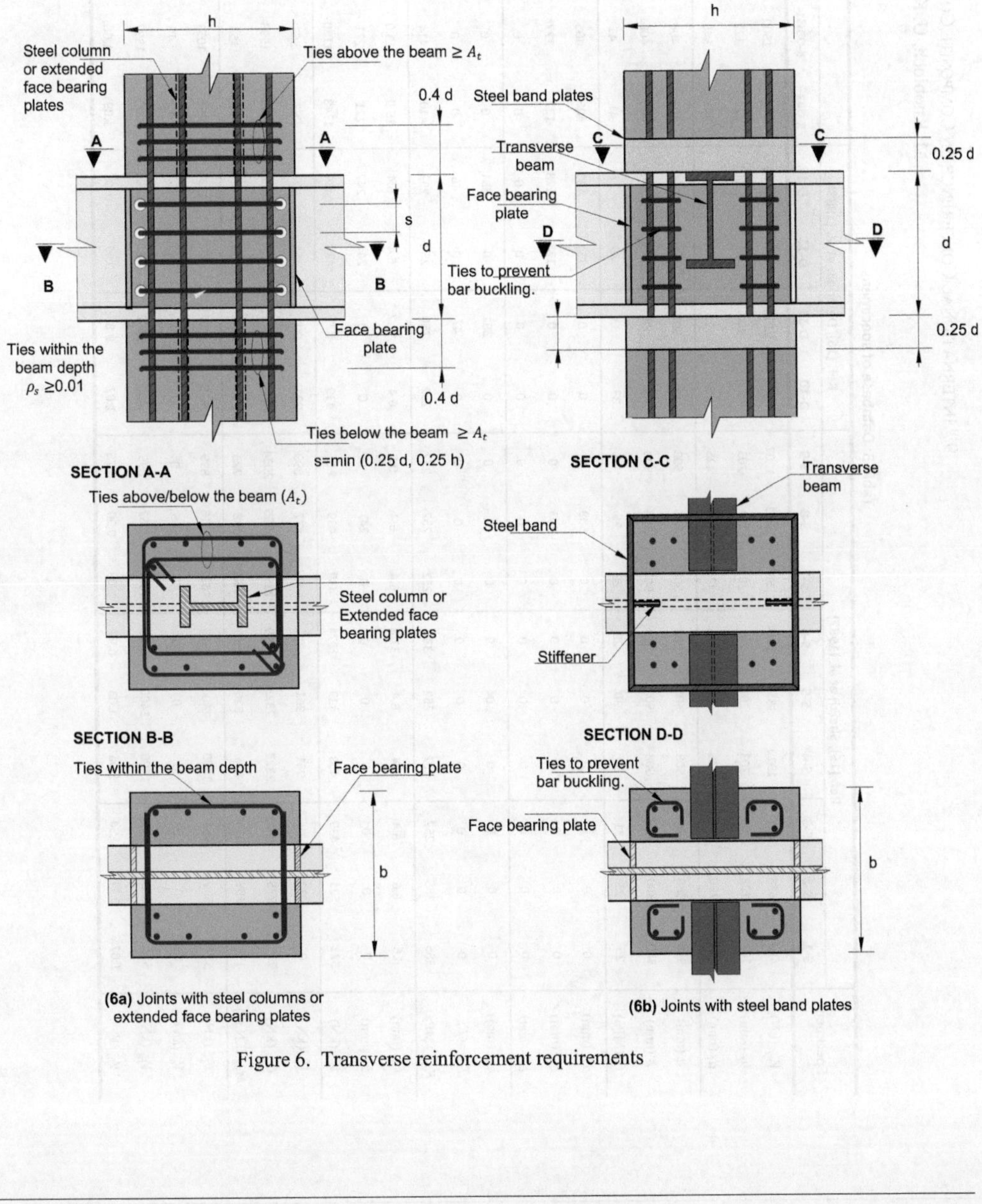

Figure 6. Transverse reinforcement requirements

9TH INTERNATIONAL CONFERENCE ON COMPOSITE CONSTRUCTION IN STEEL AND CONCRETE
M. Knobloch, U. Kuhlmann, W. Kurz, M. Schäfer. (Eds.)
Online, Germany, July 27-29, 2021

Table 3: Database of specimens

Specimen	Ref. [15], Sheikh et al. (1987)									Ref. [16], Deierlein et al. (1988)				Ref. [10], Kanno (1993)					
	S-1	S-2	S-3	S-4	S-5	S-6	S-7	S-8	S-9	D-10	D-11	D-15	D-17	K-OB1-	K-OBJS1-	K-OBJS2-	K-OJS1-	OJS2-0	OJS3-0
V_{uj} (kN)	489	688	807	1319	1359	1664	1580	2061	1207	1397	2152	1764	1679	1473	1500	1294	1341	1653	2181
b_f (mm)	102	102	203	203	203	203	203	203	203	200	200	200	200	127	127	127	127	102	153
d_j (mm)	290	290	425	425	425	425	425	425	425	423	423	423	423	345	345	345	345	210	330
h (mm)	380	380	508	508	508	508	508	508	508	508	508	508	508	406	406	406	406	406	406
b (mm)	380	380	508	508	508	508	508	508	508	508	508	508	508	406	406	406	406	406	406
f_c' (MPa)	25	25	31	30	30	28	28	25	26	33	33	28	27	41	41	45	41	43	43
d_{sc} (mm)	0	0	0	0	0	0	0	508	0	0	0	131	131	406	406	0	0	127	406
b_{sc} (mm)	0	0	0	0	0	0	0	203	0	0	0	128	128	127	127	0	0	102	153
t_{bp} *(mm)*	0	0	0	0	0	0	0	0	0	0	0	0	0	0	0	0	0	0	0
h_{vr} (mm)	0	0	0	0	0	0	0	0	0	0	381	0	381	0	0	305	0	0	305
$T_{vr} = C_{vr}$	0	0	0	0	0	0	0	0	0	0	47	0	46	0	0	23	0	0	23
F_{ysp} (MPa)	385	385	252	252	252	330	252	252	252	252	250	252	252	410	414	338	414	407	407
t_{sp} (mm)	6.6	6.6	6.4	6.4	6.4	31.8	6.4	6.4	0.0	6.4	12.7	6.4	6.4	15.0	13.0	6.0	6.0	6.0	6.0
b_o (mm)	0	0	0	0	0	0	0	247	0	0	0	247	247	221	221	76	76	246	195
V_{sp} (kN)	521	521	439	439	439	2874	439	439	0	439	871	439	439	1348	1180	445	545	535	535
V_{ic} (kN)	0	332	0	961	961	0	1403	877	899	991	991	914	897	563	563	591	563	459	689
C_{cn} (kN)	735	735	2400	2323	2323	2199	2199	1935	2034	2507	2507	2134	2057	1593	1593	1756	1593	1323	1985
M_{vb} (kN.m)	196	196	853	826	826	782	782	688	723	891	1249	759	1082	453	453	638	453	376	703
V_{in} (kN)	521	529	439	1400	1400	1615	1627	1316	899	1430	1862	1353	1336	1056	1051	1035	1079	995	1225
V_{on} *(kN)*	0	0	0	0	0	0	0	546	0	0	0	360	360	546	546	0	0	546	546
V_{jn} (kN)	521	529	439	1400	1400	1615	1627	1862	899	1430	1862	1713	1696	1602	1597	1035	1079	1541	1771
V_{jn}/V_u	1.07	0.77	0.54	1.06	1.03	0.97	1.03	0.90	0.75	1.02	0.86	0.97	1.01	1.09	1.06	0.80	0.80	0.93	0.81

Table 3: Database of specimens (cont.)

Specimen	Ref. [10], Kanno (1993)													Ref [6], Parra-Montesinos			Ref. [21] Fargier-Gabaldón (2005)			
	OJS4-1	OJS5-	OJS6-	OJS7-	HJS1-	HJS2-	HJS3-	OJB-1	OJB2-0	OJB3-0	OJB4-0	OJB5-0	OJB6-1	P- 1	P-6	P-9	1W	2W	1T	2T
V_{uj} (kN)	2274	2394	1943	2512	2529	2608	2991	2044	2440	2947	2223	2233	2040	1711	1114	1293	1639	1785	501	653
b_f (mm)	153	153	153	153	153	153	153	153	153	153	153	153	153	203	172	172	172	172	166	171
d_j (mm)	330	330	330	330	330	330	330	339	339	339	339	339	339	210	360	360	360	360	203	351
h (mm)	406	406	406	406	406	406	406	406	406	406	406	406	406	400	400	400	400	400	400	400
b (mm)	406	406	406	406	406	406	406	406	406	406	406	406	406	400	400	400	600	600	400	400
f'_c (MPa)	43	48	48	48	102	102	102	49	49	46	49	46	46	43	30	29	32	33	30	36
d_{sc} (mm)	406	406	406	406	406	406	406	127	406	406	127	127	127	162	162	162	162	162	162	162
b_{sc} (mm)	153	153	153	153	153	153	153	127	153	153	127	127	127	154	154	154	154	154	154	154
t_{bp} *(mm)*	0	13	0	0	0	0	0	0	0	13	0	0	0	0	0	13	0	13	13	13
h_{vrn} (mm)	305	305	305	305	0	0	0	0	0	0	0	305	0	0	0	0	0	0	0	0
$T_{vrn} = C_{vrn}$	23	23	23	23	0	0	0	0	0	0	0	16	0	0	0	0	0	0	0	0
F_{ysp} (MPa)	407	393	393	393	407	407	407	345	345	345	345	345	272	279	381	320	393	441	333	374
t_{sp} (mm)	6.0	6.0	6.0	6.0	6.0	6.0	6.0	25.0	25.0	25.0	25.0	25.0	25.0	13.0	7.9	7.9	7.9	7.9	7.2	6.9
b_o (mm)	195	195	195	195	195	195	195	195	195	195	195	195	195	137	168	168	253	253	174	169
V_{sp} (kN)	535	517	517	517	535	535	535	1891	1891	1891	1891	1891	1491	696	576	484	594	667	174	185
V_{ic} (kN)	689	734	734	734	1067	1067	1067	737	735	716	735	716	716	543	384	378	397	403	186	209
C_{cn} (kN)	1985	2813	2250	2250	4752	4752	4752	2269	2260	2679	2260	2143	2143	2619	1548	1871	1651	2129	1495	1846
M_{vb} (kN.m)	703	938	779	779	1351	1351	1351	645	642	761	642	707	609	733	433	524	462	596	419	517
V_{in} (kN)	1225	1251	1251	1251	1602	1602	1602	1547	1470	1734	1508	1697	1442	1239	960	862	991	1070	359	394
V_{on}*(kN)*	546	546	546	546	546	546	546	546	546	546	546	546	546	270	276	271	446	453	143	152
V_{jn} (kN)	1771	1797	1797	1797	2148	2148	2148	2093	2016	2280	2054	2243	1988	1510	1236	1113	1437	1523	502	546
V_{jn}/V_u	0.78	0.75	0.92	0.72	0.85	0.82	0.72	1.02	0.83	0.77	0.92	1.00	0.97	0.88	1.11	0.88	0.88	0.85	1.00	0.84

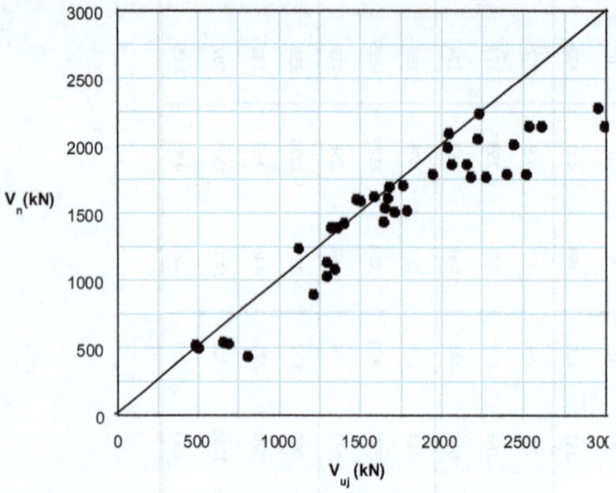

Figure 7. Measured (V_{uj}) vs. Calculated (V_n) horizontal joint shear

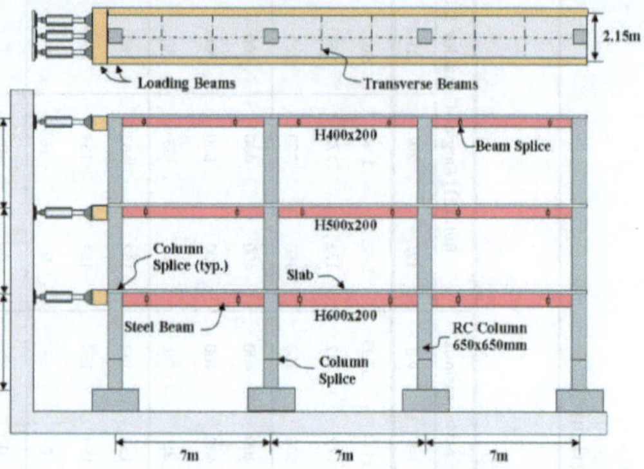

Figure 8. RCS Frame tested at the National Center for Earthquake Engineering, Taiwan, Taipei [3]

7 CONCLUSIONS

Experimental evidence accumulated over the past 20 years provide sufficient information to update the *Guidelines for the Design of Joints Between Steel Beams and Reinforced Concrete Columns* published by the ASCE, [5]. Test results of full-scale connections indicate that properly designed and detailed joints have adequate stiffness and toughness to resist large earthquake-induced demands, while a comparison between the strength measured in 39 full-scale speciments and the calculated strength shows good agreement. A full-scale pseudo-dynamic test of a three story-three bay structure showed that the composite frame exceeded the requirements implied by modern codes sustaining drift demands up to 10%. Thus, it is concluded that properly designed RCS composite frames with reinforced concrete columns and continuous steel beams are suitable as lateral load resisting systems in high seismic regions.

REFERENCES

[1] Fargier-Gabaldon, L.; Cordova P.; Parra-Montesinos G.; and Deierlein G. "Update of Design Provisions for Moment-Resisting Joints between Steel Beams and Reinforced Concrete Columns" *8th International Conference on Composite Construction in Steel and Concrete*. Jackson, Wyoming, 2017.

[2] Fargier-Gabaldón L.; Parra-Montesinos G. and Wight J.K. "Seismic Behavior of RC wide-Column to Steel Beam Joints". *ACI Structural Journal,* 25 (3), 117-128, 2020.

[3] Cordova P. "Validation of the Seismic Performance of Composite RCS Frames: Full-scale Testing Analysis and Design". *PhD Dissertation, Civil and Environmental Engineering, Stanford University*, 2005.

[4] Liang, X. and Parra-Montesinos, G. "Seismic behavior of RCS beam-column-slab subassemblies and frame systems." J. of Struct. Engrg., 130 (2), 310-319, 2004.

[5] ASCE Task Committee on Design Criteria for Composite Structures in Steel and Concrete. "Guidelines for Design of Joints Between Steel Beams and Reinforced Concrete Columns". *J. Struc. Engrg*., 120(8), 2330-2357,1994.

[6] Parra-Montesinos, G., and Wight, J.K. "Seismic behavior, strength and retrofit of exterior RC column-to-steel beam connections." *Report UMCEE 00-09, Dept. of Civil and Envir. Engrg., Univ. of Michigan, Ann Arbor*, 2000a.

[7] Parra-Montesinos, G., and Wight, J.K. "Seismic response of exterior RC column-to-steel beam connections," *J. Struc. Engrg*., ASCE 126(10), 1113-1121, 2000b.

[8] Fargier-Gabaldón, L. and Parra-Montesinos G. "Behavior of Reinforced Concrete Column–Steel Beam Roof Level T-Connections under Displacement Reversals," *J. Struc. Engrg.*, ASCE 132 (7), 1041-1051, 2006.

[9] ACI Committee 318-19. "Building code requirements for reinforced concrete." Report No. ACI 318-05. *American Concrete Institute*, Farmington Hills, MI, 2019.

[10] Kanno, R. "Strength, deformation, and seismic Resistance of joints between steel beams and reinforced concrete columns." *Ph.D. thesis, Cornell University*, Ithaca, NY, 1993.

[11] Kanno, R. and Deierlein, G.G. "Design model of joints for RCS frames." *Composite Construction in Steel and Concrete IV*, Alberta, Canada, 947-958, 2000.

[12] Deierlein, G. G.; Yura, J. A. and Jirsa, J. O. "Design of moment connections for composite framed structures." *PFMSEL Report No. 88-I, Univ. of Texas at Austin*, Austin, Texas, 1988

[13] Minami, K. "Beam to column stress transfer in composite structures." *Composite and Mixed Construction*. C. Roeder, ed., ASCE, New York, N. Y., 215-226, 1985.

[14] Sheikh, T. M.; Yura, J. A. and Jirsa, J. O. "Moment connections between steel beams and concrete columns." *PMFSEL Report No. 87-4. University of Texas, Austin*, 1987.

[15] Sheikh, T.M., Deierlein, G.G.; Yura, J.A. and Jirsa, J.O. "Beam-column moment connections for composite frames: Part 1." *J. Struct. Engrg.*, ASCE, 115(11), 2858-2876, 1989.

[16] Deierlein, G. G.; Sheikh, T. M.; Yura, J. A. and Jirsa, J. O. "Beam-column moment connections for composite frames: Part 2." J. *Struct. Engrg.*, ASCE, 115(11), 2877-2896, 1989.

[17] Bracci, J.M.; Moore, W.P.Jr.; and Bugeja, M.N. "Seismic design and constructability of RCS special moment frames." J. *Struc. Engrg.*, ASCE 125(4), 385-392, 1999.

[18] Bugeja, M.N., Bracci, J.M., and Moore, W.P.Jr. "Seismic behavior of composite RCS frame systems." *J. Struc. Engrg.*, ASCE 126(4), 429-436, 2000.

[19] ANSI/AISC 360. (2016). "Specification for Structural Steel Building." *American Institute of Steel Construction*, Chicago, IL.

[20] Deierlein, G. G.; Cordova P.; Tsai K. C.; Chen, C. H. and Lai, W. C. "Composite moment frames". *Concrete International*, 27(05), 39-44, 2005.

[21] Fargier-Gabaldón L. "Seismic Behavior of Reinforced Concrete Column-To-Steel Beam (RCS) Connections with Special Configuration." *PhD Dissertation, Department of Civil and Environmental Engineering, University of Michigan*, August 2005.

9TH INTERNATIONAL CONFERENCE ON COMPOSITE CONSTRUCTION IN STEEL AND CONCRETE
M. Knobloch, U. Kuhlmann, W. Kurz, M. Schäfer. (Eds.)
Online, Germany, July 27-29, 2021

SLIM-FLOOR BEAM-TO-COLUMN JOINTS FOR SEISMIC-RESISTANT STRUCTURES: JOINT PERFORMANCE AND CASE STUDY ON MOMENT-RESISTING FRAMES

Rafaela F. Don*, Adrian L. Ciutina, Aurel Stratan* and Cristian M. Vulcu*,*****

* Politehnica University of Timisoara / Department of Steel Structures & Structural Mechanics
e-mails: rafaela.don@student.upt.ro, aurel.stratan@upt.ro

** Politehnica University of Timisoara / Department of Overland Communication Ways, Foundation & Cadastral Survey
e-mail: adrian.ciutina@upt.ro

*** RWTH Aachen University / Institute of Steel Construction
e-mail: c.vulcu@stb.rwth-aachen.de

Keywords: Slim-Floor System, Moment-Resisting Frames, Seismic Performance, Experimental Test, Structural Analysis.

Abstract. *The current seismic design of steel and composite frames relies on lateral frame stiffness and on the controlled failure mechanism of dissipative members or connections. The implementation of the slim-floor solution in practice results in several advantages. But due to the reduced height of the slim-floor beam – generally considered to result in lower flexural stiffness – the current practice disregards the usage of moment-resisting slim-floor beam-to-column joints in seismic applications. However, a previous study of the authors has experimentally and numerically confirmed the possibility of expanding the application of composite beams by developing inelastic behaviour in slim-floor beams. The test results allowed the definition of the slim-floor beam's mechanical characteristics and the evaluation of the structure's seismic performance through nonlinear static and dynamic analyses. The usage of the slim-floor system resulted in a vertical space gain - due to its reduced height - and demonstrated the solution's effectiveness when implemented in seismic-resistant structures.*

1 INTRODUCTION

An increasingly interesting form of composite construction is represented by the slim-floor system. Its implementation in civil engineering structures has been shown to lead to a noticeable series of advantages [1]. The solution is feasible, and it exemplifies one possibility of improving the current approach to civil engineering structures. However, this solution is currently put into practice as a mere vertical load-bearing system in the form of simply-supported beams, although several studies [2] [3] [4] [5] have demonstrated the potential of continuous or semi-continuous slim-floor joints. In contrast to the existing practice, continuous and semi-continuous slim-floor joints, able of developing the flexural resistance of the connected beam, could be used in lateral load-resisting systems as dissipative components.

The current seismic design methodology of moment steel and composite frame systems relies on the lateral stiffness of frames and on failure mechanisms that are designed to dissipate seismic energy through plastic deformations in dissipative members or connections. The seismic design [6] [7] of Moment-Resisting Frames (i.e., MRFs) implies the dissipation of energy in plastic hinges - which can develop either in beams or in their connections - and at the base of the first storey's columns. Therefore, the application of slim-floor beams to seismic design requires the usage of either semi-continuous or continuous joints that can develop significant moment resistance.

2 EXPERIMENTAL AND NUMERICAL RESULTS

The experimental tests [2] validated the possibility of applying moment-resisting slim-floor joints (see Figure 1a-c) to seismic applications and proved an adequate performance of the entire assembly. This conclusion was supported by the following test results: ▪ adequate rotation capacity of the joints; ▪ high concentration of plastic deformations in the dissipative zone of the beam; ▪ elastic response and high stiffness of the connection; ▪ symmetrical response of the joint assembly in sagging and hogging bending.

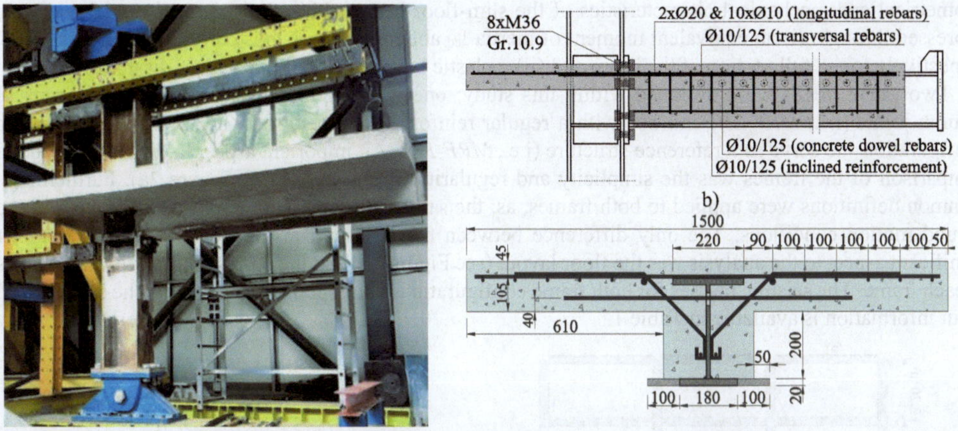

Figure 1. Joint specimen: a) test layout; b) longitudinal and c) transverse section with details.

The components of the joint assemblies had the following material properties: S355 structural steel (f_y= 390 N/mm^2 for the ½ IPE-600 steel profile; f_y= 406 N/mm^2 for the lower flange of the slim-floor beam); C30/37 reinforced concrete (f_c= 38.9 N/mm^2), B500B reinforcing bars (f_y= 515 N/mm^2). The experimental tests were performed in monotonic and cyclic loading conditions, followed by a numerical validation of the results in terms of load transfer and failure mechanism. The tests revealed the development of a plastic hinge in the dissipative zone of the beam, demonstrating that large plastic deformations in slim-floor beams are possible, but exclusively within the dissipative zone. In this case, the ductile member was obtained by implementing a *Reduced Beam Section* (i.e., *RBS*) in the wider lower flange of the beam. The *RBS* led to an overall desired behaviour of the assembly by concentrating large plastic deformations in the dissipative zone, balancing the hogging and sagging responses and maintaining an elastic response of the connection. At the same time, the role of the beam-to-column connection was as important as that of the *RBS* in obtaining this overall response, and to this end, a capacity design was performed. Another important aspect was the behaviour of the concrete slab. As the plastic hinge developed, the concrete slab located in the dissipative zone's proximity sustained significant cracks without dislocations in both loading scenarios. Although considered possible, the presence of the concrete slab did not prevent the development of the plastic hinge (yielding in tension / local buckling in compression) in the dissipative zone - a finding also supported by the symmetric response obtained through cyclic testing.

According to AISC 341-10 [8], beam-to-column joints for seismic design should develop a rotation capacity of at least ± 40mrad; in the current version of EN 1998-1, ± 35mrad [9] are required. The tested joint developed a rotation capacity of ± 45.4mrad at the Significant Damage Limit State, thus exceeding the required values by both standards. As recorded by the local instrumentation and later confirmed by the numerical model, most of the rotation was sustained by the dissipative zone of the beam - underlining the dissipative capability of the slim-floor beam, as well as the development of the plastic hinge within it. The study presented within this paper aimed to integrate experimental test results to structural analyses and then to evaluate the seismic performance of a frame with moment-resisting slim-floor beam-to-column joints.

Currently addressed in EN 1994-1-1 [10], the application of slim-floor beams to seismic design would imply an extension of the current practice to the requirements of EN 1998-1-1 [6] and EN 1998-1-2 [7].

3 DESIGN OF MOMENT-RESISTING FRAMES

The subsequent step in proving the seismic performance of the joint solution was through nonlinear structural analyses. This implied the usage of the results from the previous study in order to define the geometrical and mechanical characteristics of the slim-floor beam within the structural model and were expressed through: (i) an equivalent moment of inertia I_{eq} about the strong axis corresponding to the *RBS*, respectively to the full section of the beam and (ii) a plastic hinge model of the beam's dissipative zone.

Two plane MRFs were designed within this study: one with moment-resisting slim-floor beam-to-column joints (i.e., *MRF-SF*) and one with a regular reinforced concrete slab (i.e., RC slab) over beam configuration - to serve as a reference structure (i.e., *MRF-RF*). An important aspect to the analysis and to comparison of the frames was the simplicity and regularity of the layout (see Figure 2a). Furthermore, common definitions were applied to both frames, as: the same geometry, structural system, materials and ground motion conditions. The only difference between them resided in the flooring system. Another significant aspect to the analysis was the floor layout (see Figure 2b) and the corresponding seismic masses to each frame. The seismic masses for both frame configurations corresponded to one bay. The summarised input information is available in Table 1.

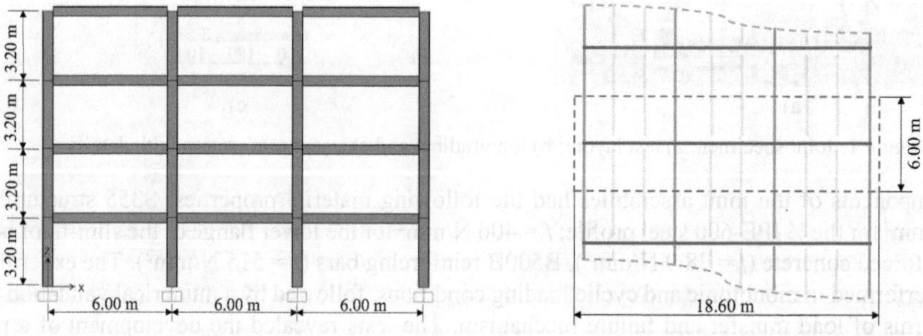

Figure 2. Structural configuration of *MRF-SF* / *MRF-RF*: a) elevation; b) floor layout.

Table 1: Input data for the analysed frames.

Parameters	*MRF-SF*	*MRF-RF*
Spans	3 x 6.00 m	3 x 6.00 m
Storeys	4 x 3.20 m	4 x 3.20 m
Slab type	Slim-floor system	RC slab [h=12 cm]
Site category	Type C	Type C
PGA	$a_g = 0.30 \cdot g$	$a_g = 0.30 \cdot g$
Ductility class	DC3	DC3
Behaviour factor	q = 6.5	q = 6.5
Consequence class & factor [11]	CC2 (normal), KF = 1.0	CC2 (normal), KF = 1.0

The main dissimilarity between the frames resided in the flooring system. Slim-floor beams were considered for *MRF-SF*, while regular I-profile beams, as part of the current standard solution, were used in *MRF-RF*. The columns were modelled with a fixed base. A rigid diaphragm constraint was assigned to each level, tying all nodes of one storey. Loads corresponding to one bay were imposed on each frame: permanent (5 kN/m^2), live (3.8 kN/m^2) and seismic action ($a_g = 0.3 \cdot g$ and soil type C).

Due to the fixed end supports of the slim-floor beams, several design and modelling particularities surfaced. First, as Figure 1a-c shows, the cross-section of the slim-floor beam incorporated the following components: a partially-encased asymmetrical steel beam with *RBS*, several reinforcement types and concrete dowels. Consequently, an accurate model of the beam required the accounting of all constituent parts of the cross-section. Secondly, due to the complexity of the cross-section, automatic software definitions were inadequate, so an elastic calculus of the full-section and the *RBS* in both sagging and hogging bending was necessary to estimate an equivalent moment of inertia I_{eq}. The combination method of the flexural stiffness in sagging and in hogging bending [9] is described by Equation 1. The mechanical properties of the slim-floor beam, respectively of the bare steel beam [12], are presented in Table 2.

$$I_{eq} = 0.6 \cdot I_1 + 0.4 \cdot I_2 \tag{1}$$

Table 2: Mechanical characteristics of the slim-floor beam and of the bare steel beam.

Beam	Cross-section	b_{eff} [mm] M^+/M^-	I_1 [mm^4] M^+	I_2 [mm^4] M^-	I_{eq} [mm^4]
SF beam	Full section	450 / 600	4.388 x 10^8	3.139 x 10^8	3.888 x 10^8
	RBS	450 / 600	2.991 x 10^8	2.514 x 10^8	2.800 x 10^8
Bare steel	IPE-450	-	3.374 x 10^8	3.374 x 10^8	3.374 x 10^8

Notations:

b_{eff}: the elastic effective width of the slim-floor beam.

I_1: moment of inertia about the strong axis in sagging bending (i.e., positive bending).

I_2: moment of inertia about the strong axis in hogging bending (i.e., negative bending).

I_{eq}: equivalent moment of inertia about the strong axis.

In the case of *MRF-RF*, the database of the structural analysis software [13] provided the necessary definitions for regular I cross-sections. Because the slab was positioned over the beam, the structural model was adjusted to account for the lateral-torsional restraining effect and the rigid diaphragm effect on the steel beams. The recommended design guidelines and the verifications of EN 1998-1-1 [6] and EN 1998-1-2 [7] were applied. Following the capacity design philosophy, the beams of both *MRF-SF* and *MRF-RF* were designed as dissipative components, while the columns and connections as non-dissipative members. The importance of the second order effect was evaluated through the value of the interstorey drift sensitivity coefficient θ [14] at each storey. As the value of θ was under 0.1, the choice to disregard the effect of the second order effects was taken. A summary of the preliminary results is presented in Table 3.

Table 3: Preliminary results structural analyses - *MRF-SF* and *MRF-RF*.

MRF-SF	*MRF-RF*
▪ Structural elements - S355 structural steel [storeys 1 ÷ 4]:	
Beams: SF beams (½ IPE-600 + Pl. 380×20 mm)	Beams: IPE-450
Columns: HEB-500	Columns: HEB-500
▪ Fundamental period: $T_{1,SF} = 0.658$ s	$T_{1,RF} = 0.656$ s
▪ Maximum interstorey drifts at Damage Limitation / Significant Damage:	
$d_{r,max,DL} = 0.0072 < 0.0075$ rad	$d_{r,max,DL} = 0.0071 < 0.0075$ rad
$d_{r,max,SD} = 0.015 < 0.020$ rad	$d_{r,max,SD} = 0.015 < 0.020$ rad

4 SEISMIC PERFORMANCE EVALUATION

The nonlinear behaviour of the slim-floor beam was engaged through the definition of a plastic hinge model (see Table 4), which was based on the results of the previously conducted tests [2]. The value of the strain hardening factor was $\gamma_{sh} = 1.2$ [14]. Regarding the plastic behaviour of the columns - a plastic hinge type for members subjected to axial loading with flexure was adopted [15] [16]. The following criteria at Damage Limitation (i.e., DL), Significant Damage (i.e., SD) and Near Collapse (i.e., NC) were defined:

$1 \cdot \theta_y$, $6 \cdot \theta_y$ and $8 \cdot \theta_y$ – also according to [15] [16]. To test the accuracy of the beam model, a Pushover analysis on a single slim-floor joint assembly was performed considering the static scheme shown in Figure 3a. As displayed in Figure 3b, the joint model followed the response of the cyclic-tested specimen in terms of initial stiffness, bending resistance and rotation capacity. This result proved the reliability of the developed joint model.

Similarly to *MRF-SF*, possible plastic regions were defined in the beams and the columns of *MRF-RF*. A plastic hinge type for members subjected to flexure was selected for the beams, while another for elements subjected to axial load and bending was chosen for the columns. The following acceptance criteria were considered: $1 \cdot \theta_y$ at DL, $6 \cdot \theta_y$ at SD, $8 \cdot \theta_y$ at NC [14] [15] [16].

Table 4: Slim-floor beam plastic hinge model and corresponding acceptance criteria.

Plastic hinge properties		
Point	M / M_{pl}	θ / θ_{pl}
A	0	0
B	1	0
C	1.2	4.75
D	0.6	4.75
E	0.6	5.8

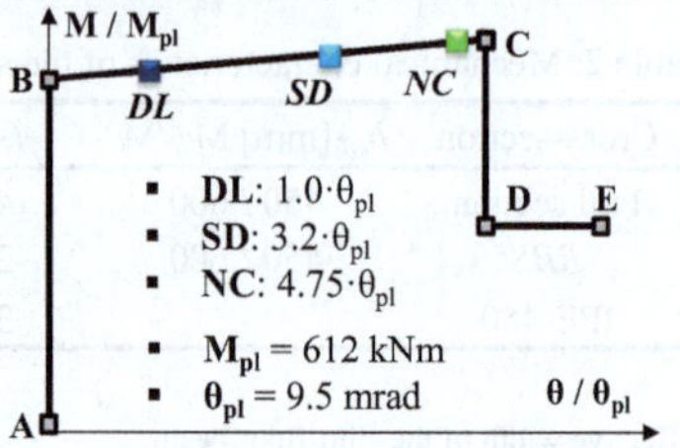

Notations:

θ_{pl}: yield rotation of the beam member.

M_{pl}: plastic bending strength of the beam member.

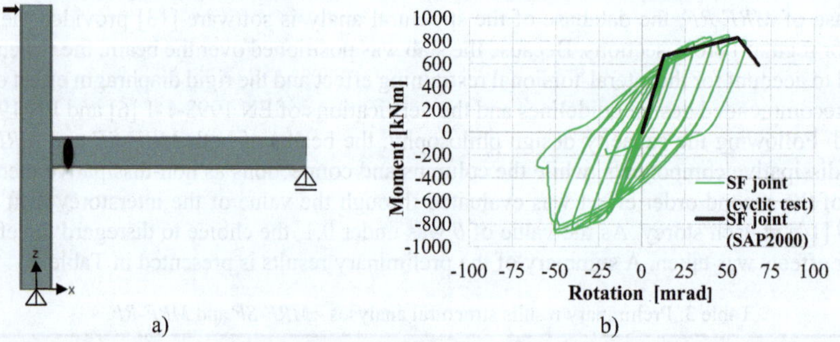

Figure 3. Modelling and response of the slim-floor joint model: a) static scheme; b) comparison of the moment-rotation curves (i.e., cyclic test vs. joint model).

4.1 Nonlinear static analysis

The seismic performance evaluation with the N2 method [17] offered a response in terms of plastic rotation at each individual Limit State and allowed the assessment of the global failure mechanism. To perform this analysis, target displacements at each Limit State were computed. The parameters of Consequence Class 2 (i.e., CC2) are shown in Table 5, along with the resulted target displacements.

Table 5: Seismic action parameters for CC2 and target displacements of *MRF-SF* and *MRF-RF*.

Limit States (LS)	Return period of action $T_{LS,CC}$	Performance factors $\gamma_{LS,CC}$	Target displacements *MRF-SF*	Target displacements *MRF-RF*
Damage Limitation (DL)	60 years	0.5	0.08 m	0.08 m
Significant Damage (SD)	475 years	1.0	0.17 m	0.16 m
Near Collapse (NC)	1600 years	1.5	0.25 m	0.24 m

The results of the nonlinear static analyses are shown in Figure 4a-h. The capacity curves (see Figure 4a-b) evidenced a similar response of the analysed frames in terms of initial stiffness, resistance and rotation. The force-displacement curves slightly diverged after reaching DL (i.e., decrease of 5.8% in resistance in the case of *MRF-SF*).

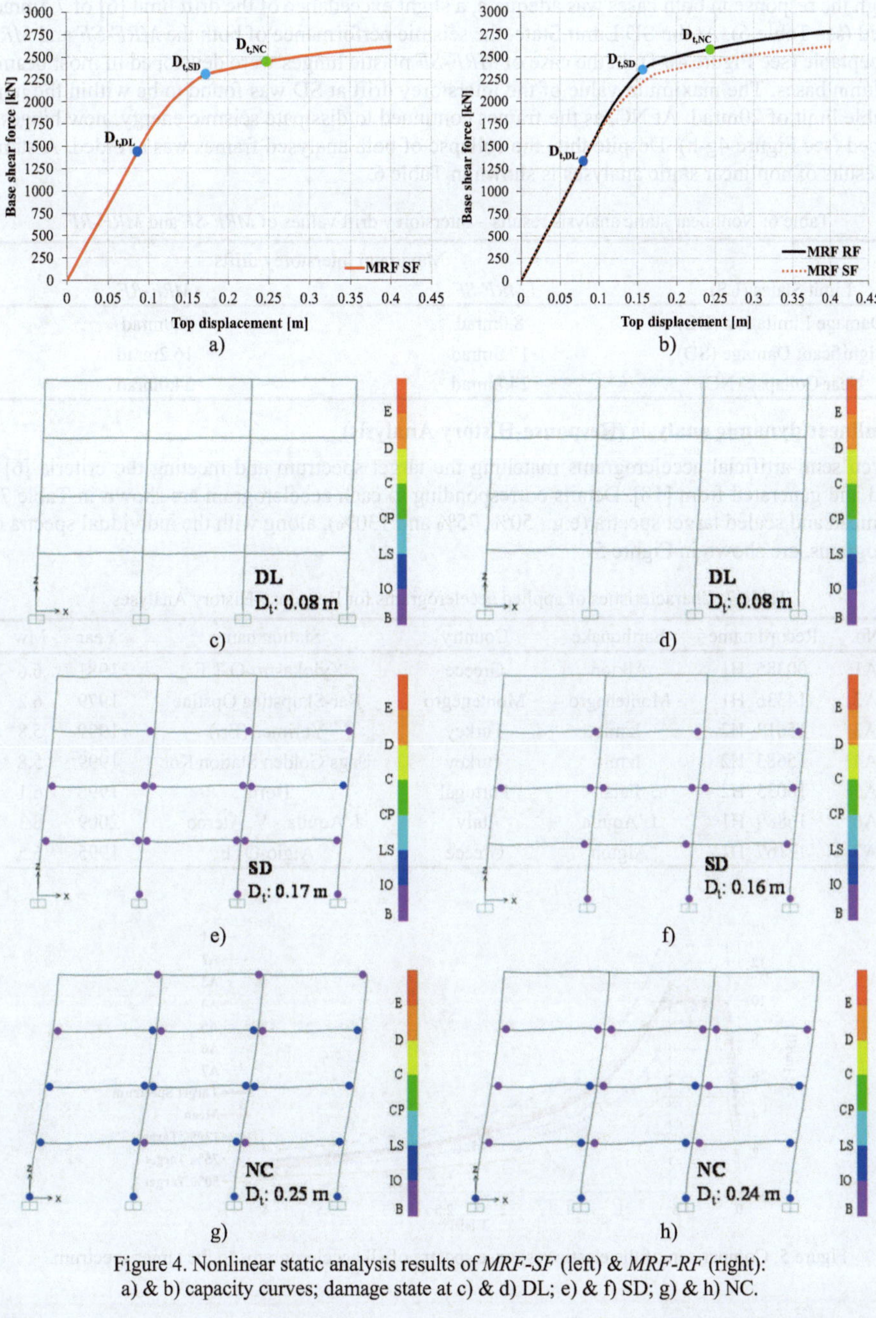

Figure 4. Nonlinear static analysis results of *MRF-SF* (left) & *MRF-RF* (right):
a) & b) capacity curves; damage state at c) & d) DL; e) & f) SD; g) & h) NC.

The legend from Figure 4c-h correlates the level of deformations in the plastic hinges to the Limit States (i.e., IO, LS, CP). It could be accepted that IO (i.e., Immediate Occupancy) corresponds to DL, whereas LS (i.e., Life Safety) to SD and CP (i.e., Collapse Prevention) to NC.

As shown in Figure 4c-d, both analysed frames responded in the elastic range at the DL Limit State. Although the response in both cases was adequate, a slight exceedance of the drift limit [6] of 7.5mrad was observed (see Table 6). At the SD Limit State, the seismic performance of both the *MRF-SF* and *MRF-RF* was acceptable (see Figure 4e-f); in the case of *MRF-SF* plastic hinges were developed in most beam ends and column bases. The maximum value of the interstorey drift at SD was found to be within the imposed acceptable limit of 20mrad. At NC, as the frames continued to dissipate seismic energy, new hinges were developed (see Figure 4g-h). Despite this, the collapse of both analysed frames was avoided. A summary of the results of nonlinear static analyses is shown in Table 6.

Table 6: Nonlinear static analysis results – interstorey drift values of *MRF-SF* and *MRF-RF*.

	Maximum interstorey drifts	
Limit States (LS)	*MRF-SF*	*MRF-RF*
Damage Limitation (DL)	8.0mrad	8.0mrad
Significant Damage (SD)	17.3mrad	16.2mrad
Near Collapse (NC)	24.8mrad	24.0mrad

4.2 Nonlinear dynamic analysis (Response-History Analysis)

Seven semi-artificial accelerograms matching the target spectrum and meeting the criteria [6] were selected and generated from [18]. Details corresponding to each accelerogram are shown in Table 7. The target, mean and scaled target spectra (e.g., 50%, 75% and 130%), along with the individual spectra of the accelerograms, are shown in Figure 5.

Table 7: Characteristics of applied accelerograms for Response-History Analyses.

Nb.	Record name	Earthquake	Country	Station name	Year	Mw
A1	00385_H1	Alkion	Greece	Xylokastro-O.T.E.	1981	6.6
A2	14336_H1	Montenegro	Montenegro	Bar-Skupstina Opstine	1979	6.2
A3	15613_H2	Izmit	Turkey	Yarimca (Eri)	1999	5.8
A4	15683_H2	Izmit	Turkey	Usgs Golden Station Kor	1999	5.8
A5	16035_H2	Faial	Portugal	Horta	1998	6.1
A6	16889_H1	L'Aquila	Italy	L'Aquila - V. Aterno -	2009	6.3
A7	17167_H1	Aigion	Greece	Aigio-OTE	1995	6.5

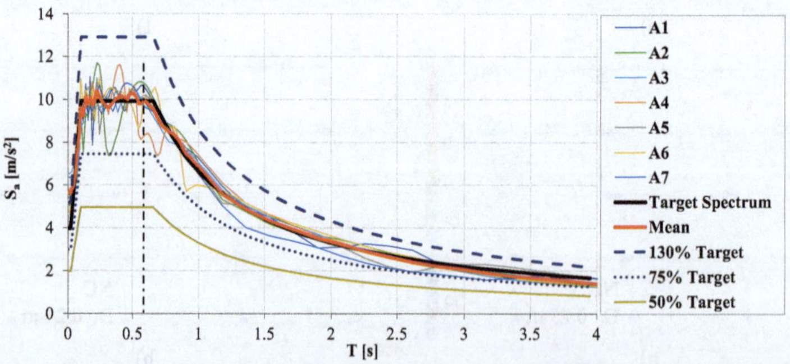

Figure 5. Comparison of the elastic response spectra of all accelerograms to the target spectrum.

Results of Response-History Analyses generated by accelerogram *#A1* are shown in Figure 6a-i. From the set of seven applied accelerograms, accelerogram *#A1* produced the most structural damage. The analyses of the frames at the DL intensity level underlined the similarities of the two structures: responses in the elastic range, minor residual deformations, similar average values of the interstorey drifts (i.e., 6.5mrad for *MRF-SF* and 6.4mrad for *MRF-RF*), similar top displacements (i.e., 0.057 m in *MRF-SF* and 0.054 m in *MRF-RF*) (see Figure 6a). Evident similarities in response were also recorded at the SD Limit State. As shown in Figure 6d, the maximum recorded top displacement of *MRF-SF* was 0.139 m – 1.17 % more than value of the top displacement produced in *MRF-RF*. Although plastic hinges were developed in both frames (Figure 6e-f), the imposed drift limit of 20mrad at SD was not exceeded, as the average interstorey drift value was 13.2mrad in the case of *MRF-SF* and 13.0mrad in that of *MRF-RF*. Unlike the damage state of the frames at SD obtained with the N2 method, the Response-History Analyses evidenced no plastic hinge development at the base of the columns. Finally, at the NC Limit State, in both *MRF-SF* and *MRF-RF*, a maximum top displacement of 0.21 m was recorded (see Figure 6g). Although the number of plastic hinges at the ends of the beams and at the base of the columns (see Figure 6h-i) increased, as higher rotations were attained, the structural collapse of both *MRF-SF* and *MRF-RF* was avoided.

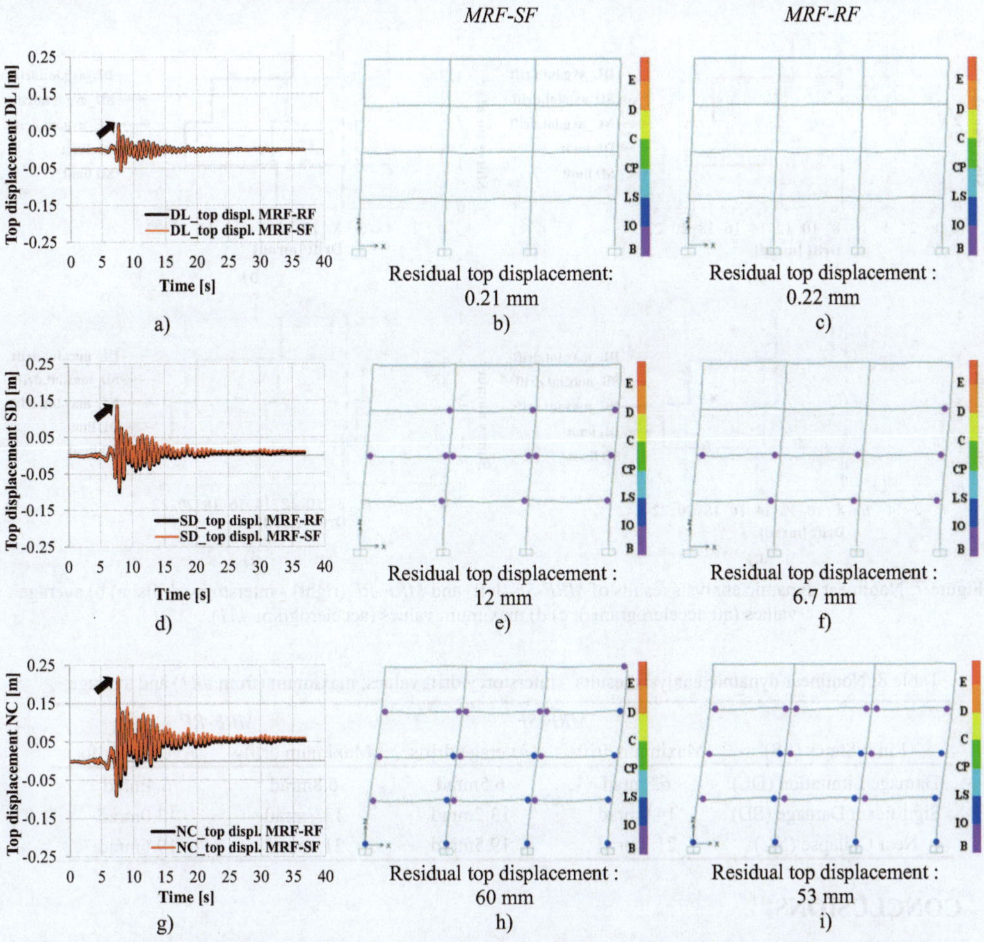

Figure 6. Nonlinear dynamic analysis (with accelerogram *#A1*) results of *MRF-SF* (centre) and *MRF-RF* (right): a) d) g) top displacement comparison at DL, SD, NC; damage state at b) & c) DL; e) & f) SD; h) & i) NC.

The interstorey drifts per Limit State that were produced in each of the analysed frames are shown in Figure 7a-d. These drifts are presented in two distinct situations: (i) average values obtained from all accelerograms and (ii) maximum values generated by the most detrimental accelerogram (i.e., *#A1*). The vertical dashed lines from Figure 7a-d represent the maximum acceptable values of the interstorey drifts at DL (i.e., 7.5mrad) and SD Limit States (i.e., 20.0mrad) according to the European seismic norm [6] [7]. The average interstorey drift values of *MRF-SF* (see Figure 7a) respected the DL and SD criteria at all storeys. Consequently, the seismic performance of this frame was regarded as adequate. The maximum interstorey drifts were produced at the 2nd storey (i.e., 6.5mrad at DL, 13.2mrad at SD). In the case of *MRF-RF*, the average interstorey drift values were 6.4mrad at DL and 13.0mrad at SD (see Figure 7b).

The analysis of *MRF-SF* with accelerogram *#A1* also revealed a good seismic performance, as the interstorey drift values remained within the previously mentioned allowable limits. In the case of *MRF-SF* the maximum obtained drift values were 6.8mrad at DL and 14.0mrad at SD (see Figure 7c). Similar values of 6.8mrad at DL and 13.7mrad at SD were produced in *MRF-RF* (see Figure 7d). A summary of the results obtained with Response-History Analyses is shown in Table 8.

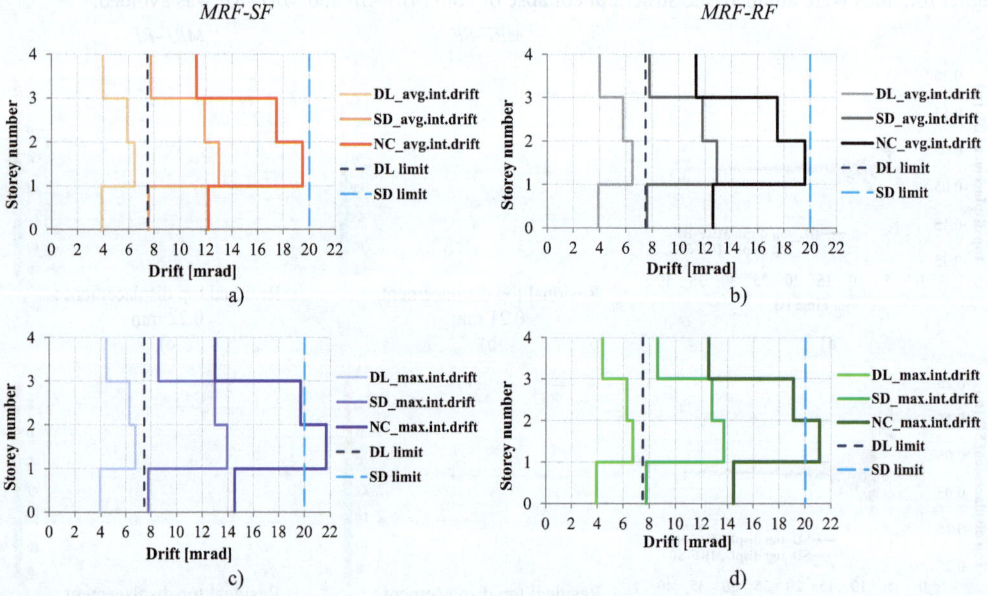

Figure 7. Nonlinear dynamic analysis results of *MRF-SF* (left) and *MRF-RF* (right) - interstorey drifts: a) b) average values (all accelerograms); c) d) maximum values (accelerogram *#A1*).

Table 8: Nonlinear dynamic analysis results – interstorey drift values: maximum (from *#A1*) and average.

	MRF-SF		*MRF-RF*	
Limit States (LS)	Maximum drifts	Average drifts	Maximum drifts	Average drifts
Damage Limitation (DL)	6.8mrad	6.5mrad	6.8mrad	6.4mrad
Significant Damage (SD)	14.0mrad	13.2mrad	13.7mrad	13.0mrad
Near Collapse (NC)	21.7mrad	19.5mrad	21.1mrad	19.6mrad

5 CONCLUSIONS

This paper engages in a parallel discussion about the seismic performance of a moment-resisting frame with slim-floor beam-to-column joints (i.e., *MRF-SF*) and a regular reference frame (i.e., *MRF-RF*) evaluated through nonlinear static (i.e., Pushover, N2 method) and dynamic (i.e., Response-History) analyses. The particularity of the *MRF-SF* resided in the flooring system, which incorporated a slim-floor decking configuration with moment-resisting beam-to-column joints. In contrast, the reference frame *MRF-RF* relied on a regular reinforced concrete slab over beam configuration.

An in-depth description of the experimental and numerical investigations of the slim-floor beam-to-column joints was published [2] and demonstrated the influence of various parameters (e.g., *RBS*, concrete slab, reinforcement type and percentage, concrete dowels and others). Based on the experimental results, the current paper focused on evaluating the seismic performance of the solution through structural analyses (i.e., on the *MRF-SF* structural configuration), while both the design procedure and the detailing of the proposed joint solution will make the subject of an upcoming scientific paper. As for the aim of this paper, the rotation demand of the slim-floor beam-to-column joints - resulted from both nonlinear static and dynamic analyses - was of main interest.

Because during the performed experimental tests, the slim-floor beam-to-column joint assemblies performed adequately in the elastic and plastic domains by responding symmetrically in both bending directions and by concentrating plastic deformations in the dissipative zone of the beam, while maintaining an elastic response of the connection, the adopted slim-floor beam model with elastic and plastic characteristics was expected to: (i) develop plastic deformations after reaching the SD Limit State, while remaining within the imposed interstorey drift limits and (ii) to lead to a desired global mechanism.

The results of the nonlinear static and dynamic analyses revealed an overall acceptable seismic performance of *MRF-SF*, as the frame exhibited a desirable behaviour at each Limit State: responding elastic at DL, sustaining deformations within the imposed limits and leading to a global plastic mechanism at SD and NC. Considering that the tested specimens had developed a rotation capacity of ± 45.4mrad at SD, and that the structural analyses evidenced a demand of ± 13.2mrad, it can be concluded that the rotation capacity of the proposed joint solution is much higher than the resulted demand, even in comparison to the demand at NC intensity level (i.e., ± 19.5mrad). As the results of all nonlinear analyses have shown, if compared, the response of the analysed frames was very similar. This aspect could be also linked to the similarity between the values of the equivalent moment of inertia of the slim-floor beam (i.e., 3.888×10^8 mm^4 in SF beam's full section, 2.800×10^8 mm^4 in the dissipative zone with *RBS*) and the value of the moment of inertia about the strong axis of the IPE-450 bare steel beam (i.e., 3.374×10^8 mm^4). The comparison of the mechanical characteristics of slim-floor beam to those of the bare steel beam led to interesting conclusions. For example, the value of the equivalent moment of inertia of the slim-floor beam's full section was 15.2 % higher than the moment of inertia about the strong axis of the IPE-450. On the other hand, the value of the equivalent moment of inertia corresponding the slim-floor beam's *RBS* was 17.0 % smaller that the moment of inertia about the strong axis of the IPE-450. In the end, the overall flexural stiffness of the slim-floor beam, considering both the full section and the *RBS*, was very similar to the moment of inertia about the strong axis of the IPE-450. Therefore, it could be concluded that the fixed slim-floor beam (i.e., ½ IPE-600 + Pl. 380×20 mm with the corresponding concrete slab) was analogous to the purely steel IPE-450 beam – a hypothesis also supported by similar periods of vibration (see Table 3), respectively by the almost identical initial stiffnesses (see the comparison of the capacity curves in Figure 4b). However, the total height of the slim-floor system was $h_{t,SFbeam} = 0.37$ m, while the total height of the regular package (i.e., 12 cm of concrete slab positioned on top of the IPE-450 beam) was $h_{t,IPE450} = 0.57$ m. Hence, although the height of the current slim-floor beam was significantly reduced, its flexural stiffness was comparable to that of an IPE-450 steel beam.

A simplified evaluation of the steel consumption of each frame was performed. Apart from the main beams and columns, secondary beams on the longitudinal direction corresponding to one bay of 6.0 m were also accounted for: (i) beams connected to columns, plus beams supporting the reinforced concrete slab in the situation of *MRF-RF* and (ii) beams connecting the columns – corresponding to *MRF-SF*. In the case of *MRF-SF* the resulted steel consumption was 22.3 tons (i.e., 51.5 kg/m^2), while in the case of *MRF-RF* the total was 23.2 tons (i.e., 53.7 kg/m^2). Therefore, the steel consumption was 4.0 % lower in the case of *MRF-SF*. However, because the evaluation of the material consumption was performed on 2D frames, excluding components as the reinforcement, the concrete itself and the steel decking, a more realistic approach will be investigated on 3D structural configurations as part of future research activities.

The results of the analyses have demonstrated that the fixed slim-floor beam could be incorporated in a frame's seismic design in a similar manner to a regular steel beam. Moreover, the slim-floor beam could also perform adequately in the plastic domain and lead to a desirable global plastic mechanism. In comparison to a regular frame, for the same structural performance, the current moment-resisting frame with slim-floor (i.e., *MRF-SF*) granted a vertical space gain of 0.20 m per storey. Therefore, the slim-floor joint was not only efficient, but also applicable to seismic design in a similar way to a purely steel solution.

ACKNOWLEDGEMENTS

The last author was supported by the Alexander von Humboldt Foundation through a Research Fellowship for Experienced Researchers. The support of the Alexander von Humboldt Foundation is hereby gratefully acknowledged.

REFERENCES

[1] ArcelorMittal: Slim-floor – an innovative concept for floors. Arcelor Mittal Europe – Long Products Sections and Merchant Bars (2017).

[2] Don, R., Ciutina, A., Vulcu, C., Stratan, A.: Seismic-resistant slim-floor beam-to-column joints: experimental and numerical investigations. Journal of Steel and Composite Structures, 37(3), pp. 307-321 (2020); DOI: http://dx.doi.org/10.12989/scs.2020.37.3.307.

[3] Malaska, M.: Behaviour of a Semi-Continuous Beam-Column Connection for Composite Floors, (PhD Thesis), Helsinki University of Technology, Finland (2000).

[4] Duarte Simões, R.A., da Silva, L.S., Cruz, P.J.S.: Cyclic behaviour of end-plate beam-to-column composite joints. International Journal of Steel and Composite Structures, 1(3), pp. 355-376 (2001).

[5] Ahmed, I.M., Tsavdaridis, K.D.: The evolution of composite flooring systems: applications, testing, modelling and Eurocode design approaches. J. Constr. Steel Resc., 155, pp. 286-300 (2019).

[6] CEN European Committee for Standardization: Eurocode 8: Design of structures for earthquake resistance-Part 1-1: General rules and seismic action (EN 1998-1-1:2020, draft), Brussels, Belgium.

[7] CEN European Committee for Standardization: Eurocode 8: Design of structures for earthquake resistance - Part 1-2: Design of structures for earthquake resistance (EN 1998-1-2:2020, draft), Brussels, Belgium.

[8] AISC: Seismic provisions for structural steel buildings. ANSI/AISC 341, Chicago, USA (2010b).

[9] CEN European Committee for Standardization: Eurocode 8: Design of structures for earthquake resistance - Part 1: General rules, seismic actions and rules for buildings (EN 1998-1:2004), Brussels, Belgium.

[10] CEN European Committee for Standardization: Eurocode 4: Design of composite steel and concrete structures - Part 1-1: General rules and rules for buildings (EN 1994-1-1:2004), Brussels, Belgium.

[11] CEN European Committee for Standardization: Eurocode 0: Basis of structural and geotechnical design (EN 1990:2019, draft), Brussels, Belgium.

[12] ArcelorMittal: Sections and Merchant Bars – ArcelorMittal Long Products (2020).

[13] CSI Berkley: SAP2000 v21. Copyright Computers and Structures (2019).

[14] Landolfo, R., Mazzolani, F., Dubina, D., da Silva, L.S., D'Aniello, M.: Design of Steel Structures for Buildings in Seismic Areas. 1st Edition. ECCS – European Convention for Constructional Steelwork (2017), ISBN (ECCS): 978-92-9147-138-6, ISBN (Ernst & Sohn): 978-3-433-03010-3.

[15] FEMA 356: Prestandard and commentary for the seismic rehabilitation of buildings. Federal Emergency Management Agency, Washington, D.C. (2000).

[16] CEN European Committee for Standardization: Eurocode 8: Design of structures for earthquake resistance - Part 3: Assessment and retrofitting of buildings (EN 1998-3:2005), Brussels, Belgium.

[17] Fajfar, P.: A nonlinear analysis method for performance-based seismic design. Earthquake Spectra, 16(3), pp. 573-92 (2000).

[18] Akkar, S., Sandıkkaya, M.A., Şenyurt, M., Azari Sisi, A., Ay, B.Ö., Traversa, P., Douglas, J., Cotton, F., Luzi, L., Hernandez, B., Godey, S.: Reference database for seismic ground-motion in Europe (RESORCE). Bulletin of Earthquake Engineering, 12(1), pp. 311-339 (2014).

9TH INTERNATIONAL CONFERENCE ON COMPOSITE CONSTRUCTION IN STEEL AND CONCRETE
M. Knobloch, U. Kuhlmann, W. Kurz, M. Schäfer. (Eds.)
Online, Germany, July 27-29, 2021

STRENGTH OF CONCRETE-FILLED STEEL DECK COMPOSITE DIAPHRAGMS WITH REINFORCING STEEL

Raul E. Avellaneda-Ramirez*, Matthew R. Eatherton*, W. Samuel Easterling, Benjamin W. Schafer***, Jerome F. Hajjar******

* Civil and Environmental Engineering Department, Virginia Tech
e-mails: rear93@vt.edu, meather@vt.edu

** College of Engineering, Iowa State University
e-mail: wse@iastate.edu

*** Department of Civil and Systems Engineering, John Hopkins University
e-mail: schafer@jhu.edu

**** Department of Civil and Environmental Engineering, Northeastern University
e-mail: jf.hajjar@northeastern.edu

Keywords: Composite Diaphragms, Large-Scale Testing, Seismic Performance, Steel Reinforcement

Abstract. *Concrete-filled steel deck floors are the most commonly used diaphragm system in steel buildings. While a considerable amount of testing has been performed on bare steel deck diaphragms, a limited amount of data exists on concrete-filled steel deck floor diaphragms subjected to cyclic loads. As part of a testing program at Virginia Tech, a total of eight cantilever diaphragm tests were performed on diaphragm specimens that were 5.2 m by 4.1 m with varied parameters such as deck height, total concrete thickness, concrete type (lightweight and normal weight), layout of headed shear studs, and reinforcement. Three of the specimens included reinforcing steel: two with reinforcing bars and one with welded wire mesh. The goal of this paper is to investigate strength predictions for these concrete-filled steel deck diaphragms with varying types of steel reinforcement. A prediction model for the shear strength of concrete-filled steel deck diaphragms was proposed and validated using the results of this experimental program, as well as the results of a testing program performed at Iowa State University.*

1 OVERVIEW

1.1 Introduction

During extreme lateral loading events such as earthquakes, diaphragms can be subjected to large cycles of shear deformation. There have been few experimental programs in the past investigating the behavior of concrete-filled steel deck diaphragms subjected to cyclic loading and almost none that included reinforcing steel. To understand how concrete-filled steel deck diaphragms will behave during large lateral loading, cyclic full-scale testing is necessary, and analytical models need to be developed to predict the strength of such diaphragms. Furthermore, this evaluation should be conducted for concrete-filled steel deck assemblies that are representative of typical modern construction such as the use of headed studs as the primary perimeter fastener and typical slab thicknesses used for minimum fire rating.

To address these knowledge gaps, eight cantilever diaphragm specimens were tested with varying concrete cover, deck depth, concrete type, and types of reinforcement. The results of the testing program were used to validate proposed strength prediction equations for diagonal tension cracking which account for the presence of steel reinforcement. These equations compute the shear strength of a diaphragm as the combination of a concrete component of the strength and a component for reinforcing steel bars. This approach is similar in concept to ACI 318-19 Section 12.5.3.3 [1], except the concrete component of the strength accounts for the contribution of the deck and concrete in the flutes. The shear strength of the diaphragm is then taken as the minimum of this predicted diagonal tension cracking strength and the perimeter fastener strength which is also evaluated against test results.

1.2 Test Setup and Test Matrix

Cantilever diaphragm tests were conducted at the Thomas M. Murray Structural Engineering Laboratory at Virginia Tech with test setup shown in Figure 1 wherein one edge was restrained against translation in all directions by reaction steel connected to the strong floor and the opposing edge was subjected to a displacement history using a pair of servo-controlled hydraulic actuators with a combined capacity of 2800 kN. The master actuator was displacement-controlled while the force in the following actuator was constrained to be in the opposite direction with a magnitude equal to the force in the master actuator multiplied by the ratio of control to following force capacities.

Composite concrete-filled steel deck diaphragm specimens were 5.18 m by 4.06 m with 4.57 m and 3.66 m between the centerlines of W610x230x125 beams. The deck was 0.91 mm thick Verco W2-36 FormlokTM or W3-36 FormlokTM steel deck (ASTM A653) and the deck direction was as shown in Figure 1, with the exception of specimen 3/6.25-4-L-RS-DT, for which this direction was parallel to the loading. The perimeter fasteners were ASTM A29, 19 mm diameter headed shear studs with nominal strength, Fu=448 MPa and spacing as shown in Figure 1. A typical specimen is shown in Figure 2.

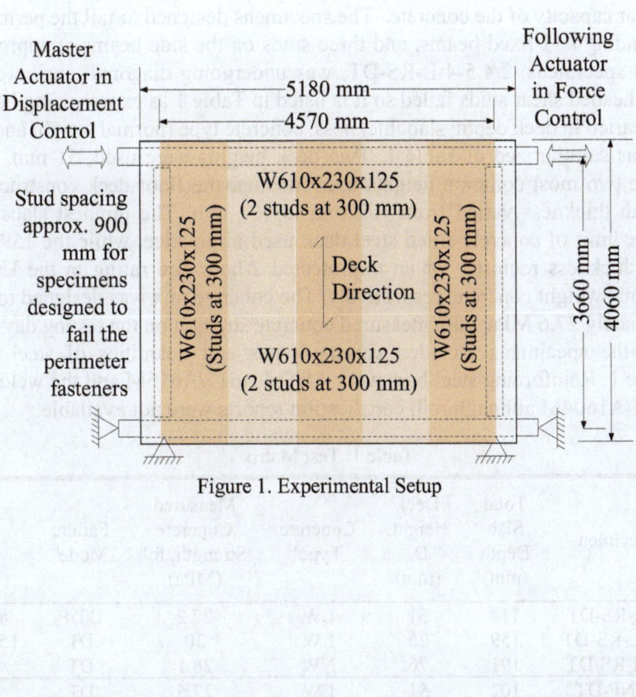

Figure 1. Experimental Setup

Figure 2. Typical Specimen

As given in Table 1, the testing program consisted of eight specimens, of which six were designed to fail in diagonal tension cracking, and two were designed to fail the perimeter shear stud anchors without reaching the full shear capacity of the concrete. The specimens designed to fail the perimeter fasteners had four studs on the loading and fixed beams, and three studs on the side beams at approximately 900 mm spacing. One of the specimens, 2/4.5-4-L-RS-DT, was undergoing diagonal tension cracking when the shear transfer at the headed shear studs failed so it is listed in Table 1 as experiencing both failure modes.

The specimens varied in deck depth, slab thickness, concrete type (normal weight and lightweight), and reinforcement type as summarized in Table 1. Two deck heights were used, 51 mm. and 76 mm deck, because they are the two most common heights used incomposite floor deck construction in the United States. The total slab thickness varied from 102 mm to 191 mm. The thinnest slabs were intended to investigate the lower limit of concrete-filled steel deck used in practice, while the 159 mm and 191 mm slabs represent the thickness required for an unprotected 2-hour fire rating in the United States using lightweight and normal weight concrete, respectively. The concrete mix was designed to produce a 28-day strength of approximately 27.6 MPa, with measured concrete strength on the testing day (28 days) given in Table 1. Three of the specimens included varying types and quantities of steel reinforcement, as summarized in Table 1. Reinforcing steel bars were ASTM A615/A615M and the welded wire mesh was likely ASTM A1064/A1604M although mill certification reports were not available.

Table 1: Test Matrix

Specimen Type	Test Specimen	Total Slab Depth (mm)	Deck Height, D_d (mm)	Concrete Type[a]	Measured Concrete Strength, fc' (MPa)	Failure Mode[b]	Reinforcement
Reinforcing Bars	2/4.5-4-L-RS-DT	114	51	LW	27.2	DT/P	#4 bars at 305 mm C-C
	3/6.25-4-L-RS-DT	159	76	LW	30	DT	152x152 MW 13xMW 13
	3/7.5-4-N-RS-DT	191	76	NW	28.1	DT	#3 bars at 457 mm C-C
Unreinforced	2/4-4-L-NF-DT	102	51	LW	27.5	DT	none
	3/6.25-4-L-NF-DT	159	76	LW	27.2	DT	none
	3/7.5-4-N-NF-DT	191	76	NW	26.2	DT	none
	3/6.25-4-L-NF-P	159	76	LW	31	P	none
	3/7.5-4-N-NF-P	191	76	NW	33.2	P	none

[a]LW = lightweight concrete, NW = normal weight concrete
[b]DT = Diagonal Tension, P = Perimeter Fastener

1.3 Loading Protocol

A cyclic loading protocol was adopted based on FEMA 461 [2], Section 2.9.1, and included at least six cycles before reaching the elastic limit. The elastic limit was calculated as a ratio of the predicted strength for the expected limit state, to the predicted stiffness. It was verified after the tests that this approach resulted in more than six cycles of loading before reaching the elastic limit for all specimens. The loading protocol included two cycles for every displacement step with a 40% increase in amplitude between displacement steps. An example of a loading protocol is provided in Figure 3.

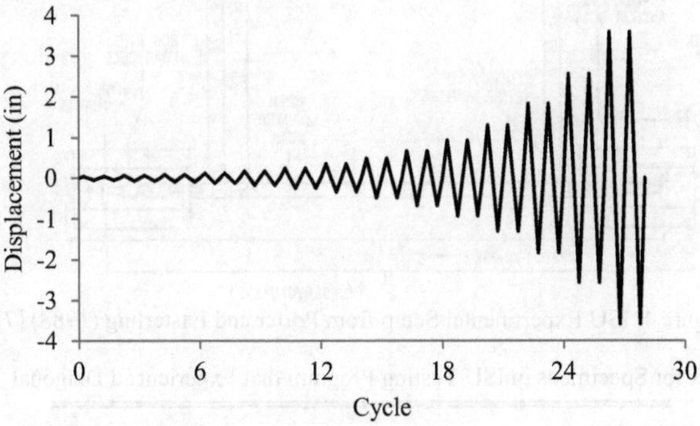

Figure 3. Typical Loading Protocol

1.4. Iowa State University Testing Program

An extensive testing program investigating the performance of concrete-filled steel deck diaphragms was performed at Iowa State University. A total of 32 cantilever diaphragm tests were performed, as reported in Porter and Easterling (1988) [7], 15 of which had diagonal tension cracking listed as the sole reported failure mode. The results of these specimens are used in conjunction with the results of the experimental program performed at Virginia Tech to validate the proposed prediction model. A summary of the specimens used in this validation is provided in Table 2. Specimens 1 through 19 had in-plane dimensions of 15 ft. by 15 ft, while specimens 22 through 29 had in plane dimensions of 12 ft. by 15 ft, similar to the specimens tested at Virginia Tech. Specimens 1, 2, 25, 26, and 29 included headed studs as perimeter fasteners, while the remaining specimens included welds as the perimeter fasteners. All specimens were loaded using a cyclic loading protocol. The experimental setup for this testing program is illustrated in Figure 4. Different deck profiles were used.

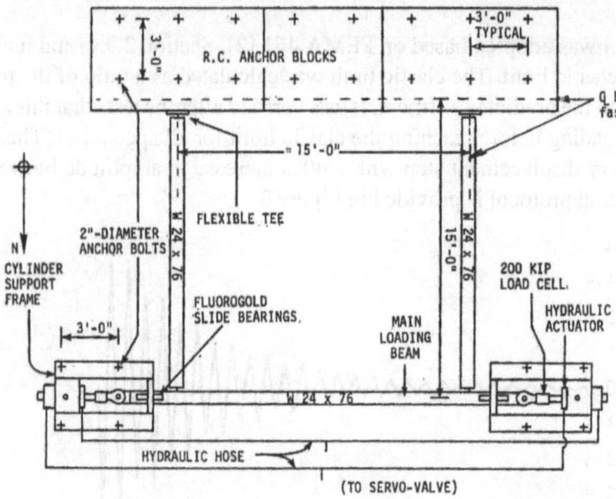

Figure 4. ISU Experimental Setup from Porter and Easterling (1988) [7]

Table 2: Test Matrix for Specimens in ISU Testing Program that Experienced Diagonal Tension Cracking

Test Specimen	Total Slab Depth (mm)	Deck Height, D_d (mm)	Concrete Type[a]	Measured Concrete Strength, fc' (MPa)
1	137	76	NW	38.8
2	140	76	NW	36.2
5	90	38	NW	20.4
9	139	76	NW	37.3
10	140	84	NW	22.8
12	142	84	NW	23.5
13	140	76	NW	42.7
16	106	38	NW	20.4
18	141	84	NW	21.0
19	146	76	NW	18.5
22	144	84	NW	22.8
24	143	76	NW	27.9
25	145	84	NW	32.2
26	120	47	LW	23.9
29	141	76	NW	19.9

2 STRENGTH PREDICTION OF REINFORCED CONCRETE-FILLED STEEL DECK COMPOSITE DIAPHRAGM

This section presents prediction models for the strength of concrete-filled steel deck diaphragms that will be evaluated against the test results. First, existing prediction models are described, and then a proposed approach is presented.

2.1 AISI S310-16

Equation (1) gives the shear strength, S_n, for concrete-filled steel deck diaphragms associated with the diagonal tension cracking limit state as given by AISI S310-16 [5]. The shear strength is calculated as a combination of a contribution of the deck fasteners and the contribution of the shear strength of the concrete above the top of the deck flutes.

$$S_n = \frac{\beta \cdot P_{nf}}{L} + k \cdot b \cdot d_c \cdot \sqrt{f_c'} \qquad \left[\frac{kN}{m}\left(\frac{lbf}{in}\right)\right] \tag{1}$$

Where,
β = factor defining connection interaction contribution to diaphragm shear strength per unit length (see AISI S310-16 [5] for calculation)
P_{nf} = nominal shear strength of a support connection per fastener kN (kip)
L = steel deck panel length, m (ft)
k = factor for structural concrete shear strength
$= \frac{w^{1.5}}{585(10^3)}$, for Imperial
$= \frac{w^{1.5}}{452(10^6)}$, for SI units
w = structural concrete density, kg/m³ (pcf)
b = unit width of diaphragm (1000 mm for SI, 12 inch for Imperial)
d_c = concrete thickness above the top of the deck flute, mm (in.)
f_c' = specified concrete compressive strength at 28 days, MPa (psi)

2.2 AISI S310-20

Equation (2) is an updated version of the equation included in AISI S310-20 [8] (originally proposed by Easterling and Porter [10]) which recognizes that the diagonal tension cracking limit state does not include failure of the deck fasteners (i.e. there is no failure of sidelap or support connections when diaphragms reach their diagonal tension cracking strength), and as such the term associated with fasteners is inappropriate. This prediction equation also includes the concrete in the deck flutes, considers the contribution of the steel deck using an equivalent transformed concrete thickness, uses an increased concrete shear stress calibrated based on previous testing [5],[10], and adds a factor to account for lightweight concrete.

$$S_n = \lambda \cdot k_c \cdot t_e \cdot b \cdot \sqrt{f_c'} \qquad \left[\frac{kN}{m}\left(\frac{lbf}{in}\right)\right] \tag{2}$$

Where,
λ = lightweight factor (1 for NW, 0.75 for LW)
k_c = concrete strength factor, (2.66 for SI, 0.0032 for Imperial)
$t_e = t_c + t_{dt}$, equivalent total transformed concrete thickness, mm (in.)
t_c = effective concrete thickness, mm (in.)
$t_{dt} = n_{sc} * t_{cd} * \frac{d}{s}$, transformed thickness of composite deck, mm (in.) with d and s as defined in Figure 5
$n_{sc} = \frac{E_s}{E_c}$, shear modular ratio of steel deck to concrete
E_s = elastic modulus of composite steel deck
E_c = elastic modulus of concrete fill
t_{cd} = thickness of corrugated deck, mm (in.)

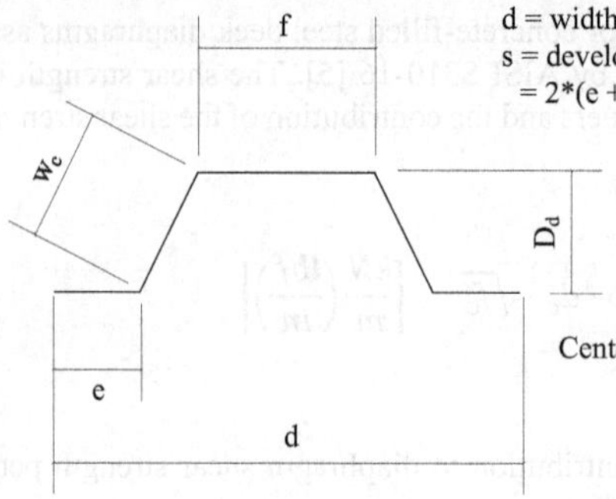

Figure 5. Deck Corrugation Geometry

2.3 ACI 318-19

AISC 341 [9], Section D1.5b requires that the nominal in-plane shear strength of concrete-filled steel deck diaphragms that are part of the seismic force resisting system in steel buildings shall be taken as the nominal shear strength of the reinforced concrete above the top of the steel deck ribs in accordance with ACI 318. ACI 318-19 [1], Section 12.5.3 provides (3) as a method for the prediction of the in-plane shear strength of a diaphragm with steel reinforcement. The shear strength is calculated as a combination of the contribution of the shear strength of the concrete and the tensile strength of the reinforcement oriented parallel to the in-plane shear.

$$S_n = d_c \cdot b \cdot \left(2\lambda\sqrt{f_c'} + \rho_t f_y\right) \quad \left[\frac{kN}{m}\left(\frac{lbf}{in}\right)\right] \tag{3}$$

Where,

λ = modification factor to reflect the reduced mechanical properties of lightweight concrete relative to normal weight concrete of the same compressive strength per section 19.2.4.1 of ACI 318 [1] (0.75 for all-lightweight aggregate, 0.85 for sand-lightweight aggregate, 1 for normal weight aggregate)

ρ_t = ratio of cross-sectional area per unit width of reinforcing steel bars parallel to the direction of diaphragm shear to the unit area of concrete above the deck flutes

f_y = specified minimum yield stress of reinforcement, MPa (ksi)

2.4 Perimeter Fastener Shear Transfer Capacity (AISC 360)

AISC 360 [11], Section I8.2a provides (4) to calculate the nominal shear strength of a headed stud anchor embedded in a composite slab with steel deck. This model was used for the strength prediction of the specimens for which the intended limit state was perimeter fastener failure.

$$Q_n = 0.5 \cdot A_s\sqrt{f_c' \cdot E_c} \leq R_p \cdot R_g \cdot A_{sa} \cdot F_u \ [kN\ (kip)] \tag{4}$$

Where,

R_p = 0.75 for strong position studs with deck oriented perpendicular to the direction of shear, or for studs with deck oriented parallel to the direction of shear

= 0.6 for weak position stud with deck oriented perpendicular to the direction of shear

= 0.52 for staggered position studs

R_g = 1.0 for one stud per rib with deck oriented perpendicular to the direction of shear or for studs with deck oriented parallel to the direction of shear

= 0.85 for two studs per rib

= 0.7 for three or more steel headed stud anchors welded in a steel deck rib with the deck oriented perpendicular to the direction of shear
A_s = Cross sectional area of the headed stud shank, mm² (in²)
F_u = Ultimate tensile strength of headed stud material, MPa (ksi)

2.5 Proposed Prediction Model

This paper introduces a prediction model for the diagonal tension cracking shear strength of concrete-filled steel deck diaphragms with reinforcing bars building upon existing prediction methods. The prediction model was validated using the results of the testing program.

Similar to the prediction model provided by ACI 318 [1], the proposed prediction model (5) calculates the shear strength of the diaphragm as a linear combination of the contribution of the shear strength of the concrete and a contribution from the reinforcement oriented parallel to the in-plane shear. The contribution of the concrete is calculated using the same expression provided by AISI S310-20 [8], which includes the concrete in the deck flutes as well as a contribution from the steel deck.

$$S_n = \lambda \cdot k_c \cdot t_e \cdot b \cdot \sqrt{f_c'} + A_v \cdot f_y \qquad [kN\ (kip)] \tag{5}$$

Where,
A_v = Area of reinforcement oriented parallel to the direction of shear
f_y = yield stress of steel reinforcement

3 EXPERIMENTAL RESULTS

3.1 Diagonal Tension

A comparison of the experimental results to applicable prediction methods is provided in Table 3. It can be observed for that the ACI 318 [1] approach, which only includes the concrete above the flutes and smaller concrete shear stress, tends to underpredict the shear strength of unreinforced the specimens, with experimental strengths being up to almost 4 times larger than predicted. The prediction method provided in AISI S310-20 [8], which accounts for the concrete within the steel deck flutes as well as a contribution from the steel deck itself, more accurately predicted the strength of the unreinforced specimens which underwent diagonal tension cracking, with an average experiment to prediction ratio of 1.09 with a standard deviation of 0.14.

The experimental results for the reinforced specimens showed an increase in strength due to the reinforcement when compared to the expected strength of the unreinforced specimen. The proposed prediction method for specimens with reinforcing bars more accurately predicted the strength of the specimens, with an average experiment to prediction ratio of 0.97 and standard deviation of 0.12, as compared to the ACI 318 [1] method, which exhibited an average experiment to prediction ratio of 1.57 with a standard deviation of 0.35. The proposed model predicts diaphragm shear strengths that are approximately 70% larger than the ACI model because the proposed prediction model considers the concrete in the flutes, includes the contribution of the deck, and uses a larger concrete shear stress compared to ACI 318 [1]. The proposed model is more accurate than the ACI model because the concrete component of the proposed model is based on AISI S310-20 [8] which has been calibrated to concrete-filled steel deck [6],[10]. A graphical comparison of the experimental results to the proposed prediction method is presented in Figure 6.

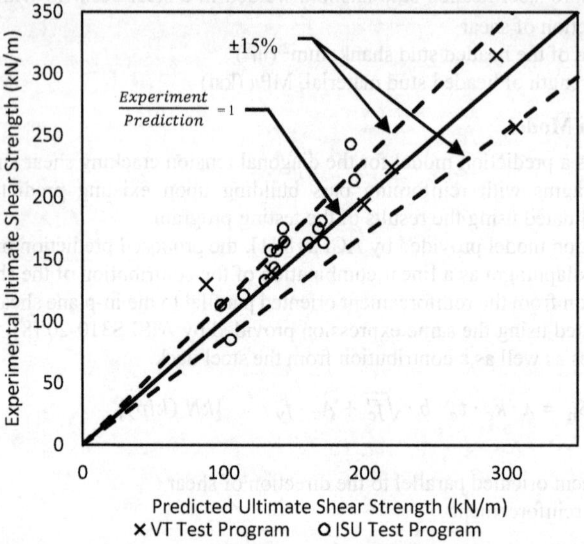

Figure 6. Experimental vs. Predicted Shear Strength of Unreinforced Specimens and Specimens with Reinforcing Bars Using Proposed Prediction Model

Table 3: Comparison of Results to Prediction Methods for Specimens that Experienced Diagonal Tension Cracking

Specimen Type	Test Specimen	Failure Mode	Experimental Shear Strength (kN/m)	Predicted Shear Strength, ACI 318 (kN/m)	Experiment / Prediction	Predicted Shear Strength, AISI S310-20 (kN/m)	Experiment / Prediction	Predicted Shear Strength, Proposed Eq. (5) (kN/m)	Experiment / Prediction
Reinforcing Bars	2/4.5-4-L-RS-DT	DT/P	257	220	1.17	102	2.51	304	0.84
	3/6.25-4-L-RS-DT	DT	194	115	1.69	140	1.39	199	0.98
	3/7.5-4-N-RS-DT	DT	315	172	1.84	222	1.42	293	1.08
Unreinforced	2/4-4-L-NF-DT	DT	130	34	3.87	88	1.48	88	1.48
	3/6.25-4-L-NF-DT	DT	140	54	2.59	134	1.04	134	1.04
	3/7.5-4-N-NF-DT	DT	225	98	2.30	219	1.03	219	1.03
ISU Testing Program	1	DT	163	63	2.61	171	0.96	171	0.96
	2	DT	181	63	2.85	169	1.07	169	1.07
	5	DT	113	39	2.91	98	1.15	98	1.15
	9	DT	214	64	3.35	193	1.11	193	1.11
	10	DT	157	45	3.49	138	1.14	138	1.14
	12	DT	175	47	3.74	142	1.24	142	1.24
	13	DT	243	70	3.49	190	1.28	190	1.28
	16	DT	121	51	2.37	114	1.05	114	1.05
	18	DT	157	44	3.60	133	1.17	133	1.17
	19	DT	143	50	2.87	136	1.05	136	1.05
	22	DT	164	48	3.43	143	1.15	143	1.15
	24	DT	163	59	2.79	160	1.02	160	1.02
	25	DT	175	57	3.06	167	1.05	167	1.05
	26	DT	85	44	1.91	105	0.81	105	0.81
	29	DT	133	48	2.78	129	1.03	129	1.03

DT = Diagonal Tension, P = Perimeter Fastener

3.2 Perimeter Fastener Failure

The shear strength of a diaphragm should be calculated as the lesser strength of the diaphragm for the limit states of diagonal tension cracking and perimeter fastener failure. Perimeter fastener failure was observed as the controlling limit state in the two specimens specifically designed to fail the perimeter fasteners. Perimeter fastener failure was also observed in Specimen 2/4-4-L-NF-DT near peak load after exhibiting extensive diagonal tension cracking. In all three cases, the perimeter failure occurred along the loading and fixed beams. The results for the three specimens are summarized in Table 4. The AISC method described in (4) is used to predict the capacity of the specimens based on the configuration of the perimeter fasteners on the loading and fixed beams for each specimen. The AISC prediction model overpredicted the strength of all three specimens. The specimens had a 200 mm (8 in.) edge distance from the centerline of the fasteners to the closest edge in the direction perpendicular to the shear force and the failure mode was typically due to rib shear (horizontal cracks in the rib across the top of the deck flute), a typical failure mode for small edge distances. This failure mode can result in low strengths compared to AISC predictions and likely contributed to the low experiment to prediction ratios shown in Table 4. These results suggest that the AISC equations to predict shear stud failure may require additional review when the edge distances are small, there are no pour stops, or the stud shear forces are not all parallel to the free edge.

Table 4: Comparison of Results to Prediction Methods for Specimens that Experienced Perimeter Fastener Failure

Test Specimen	Experimental Shear Strength (kN/m)	Predicted Shear Strength, Perimeter Fastener (kN/m)	*Experiment* / *Prediction*
2/4.5-4-L-RS-DT	257.0	557.8	0.46
3/6.25-4-L-NF-P	58.4	84.9	0.69
3/7.5-4-N-NF-P	81.7	87.5	0.93

4 CONCLUSIONS

This paper investigated the strength of concrete filled steel deck composite diaphragms with different types of steel reinforcement through an experimental program consisting of eight cantilever diaphragm tests. A prediction method was proposed and validated using experimental results. The prediction model was derived using mechanics and existing literature and proved to accurately predict the shear strength of concrete-filled steel deck diaphragms for the limit state of diagonal tension cracking. For specimens which failed the perimeter fastener, the existing prediction model given by AISC overpredicted the shear strength. More investigation on shear transfer strength of headed shear studs with small edge distances and shear forces not parallel to the edge is recommended.

5 ACKNOWLEDGEMENTS

This work was supported through the Steel Diaphragm Innovation Initiative which is funded by AISC, AISI, SDI, SJI, and MBMA, and the National Science Foundation under grant No. 1562669. In-kind funding was provided by Verco Decking Inc., Nelson Fastener Systems, AISC, and Banker Steel Company. Any opinions, findings, and conclusions or recommendations expressed in this material are those of the authors and do not necessarily reflect the views of the National Science Foundation or the other sponsors.

REFERENCES

[1] ACI (2019). Building Code Requirements for Structural Concrete, (ACI 318-19). American Concrete Institute

[2] FEMA. (2007). Interim Testing Protocols for Determining the Seismic Performance Characteristics of Structural and Nonstructural Components, (FEMA 461). Applied Technology Council, Federal Emergency Management Agency.

[3] AISI. (2013). Test standard for cantilever test method for cold-formed steel diaphragms, (AISI S907-13). American Iron and Steel Institute

[4] ICC. (2016) Acceptance Criteria for Steel Deck Roof and Floor Systems (AC43-16). ICC Evaluation Service.

[5] AISI. (2016). North American Standard for the Design of Profiled Steel Diaphragms, (AISI S310-16). American Iron and Steel Institute.

[6] O'Brien, P. (2017). "Characterizing the Load-Deformation Behavior of Steel Deck Diaphragms Using Past Test Data." Virginia Tech, Blacksburg, VA, USA

[7] Porter, M. and Easterling, W. (1988) Behavior, Analysis, and Design of Steel-Deck-Reinforced Concrete Diaphragms. Report No. ISU-ERI-Ames-88305 Project 1636, Iowa State University.

[8] AISI. (2020). North American Standard for the Design of Profiled Steel Diaphragms, (AISI S310-20). American Iron and Steel Institute.

[9] AISC (2016). Seismic Provisions for Structural Steel Buildings. (ANSI/AISC 341-16). American Institute of Steel Construction.

[10] Easterling, W. S. (1987). Analysis and design of steel-deck-reinforced concrete diaphragms. Dissertation, Iowa State University.

[11] AISC (2016). Specification for Structural Steel Buildings. (ANSI/AISC 360-16). American Institute of Steel Construction.

SHEAR CONNECTIONS

9TH INTERNATIONAL CONFERENCE ON COMPOSITE CONSTRUCTION IN STEEL AND CONCRETE
M. Knobloch, U. Kuhlmann, W. Kurz, M. Schäfer. (Eds.)
Online, Germany, July 27-29, 2021

DETERMINATION OF SLIP-FACTOR BETWEEN FRICTION SHIMS AND SHOT-BLASTED STEEL SURFACES

Taygun F. Yolaçan*, Markus Schäfer

University of Luxembourg, Department of Engineering – Structural Engineering and Composite Structures
firat.yolacan@uni.lu, markus.schaefer@uni.lu

Keywords: Steel-concrete Composite Joints, Slip Factor, Friction Coefficient, Bolted Connections, Friction Shims.

Abstract. *This paper presents the results of an experimental testing campaign performed to determine slip-factor between the faying surfaces of shot-blasted steel plates and state-of-the-art friction shims. In total five experimental tests were performed as consistent with the provisions of EN1090-2 Annex G to determine the slip-factor between the faying surfaces. The results of the experimental test campaign showed that the surface roughness of the steel plates needs to be suitable for the selected type of the friction shims to enrich the slip-resistance of the bolted-connections. The test procedure was simulated with a commercial finite-element analysis software, Abaqus, to compare the empirically determined slip-factor and the penalty friction formulation of the software. It is shown that the slip-factor definition of EN1090 – 2 Annex G fits in a very good agreement to the penalty friction formulation of the software to estimate the slip-resistance of the bolted connections. The differences between the test results and the outputs of the finite element analysis are discussed in detail with a novel representation of non-linear stick-slip transition path for the slip-resistant bolted connections. A further parametric study was performed to understand the impact of the geometric parameters on the slip-resistance and the load-slip behaviour of the faying surfaces.*

1 INTRODUCTION

By the advancements in computational techniques for the design of civil engineering structures, slip-resistant bolted connections gained importance in the last decades as they offer possibility to optimize dimensions of the bolted connections [1, 2]. Furthermore, slip-resistant interfaces allow development of dismountable and reusable structural components by means of steel and steel-concrete composite beams for building and bridge structures [3-5]. In addition, the recent studies showed that the slip-resistant connections have superior mechanical characteristics and high-energy dissipation capacity without damage incurring under dynamic loading conditions [6-8]. Therefore, it could be stated that the slip-resistant bolted connections may lead reduced construction, maintenance and inspection costs, and improvements for the structural safety at the ultimate limit state.

Slip resistance of the bolted connections mainly depends on the level of the bolt pre-tension load and dry/static friction coefficient between the faying surfaces. The level of the minimum required bolt pre-tension loads for the design of the slip-resistant bolted connections are defined in detail by the structural design codes [9, 10]. However, dry/static friction coefficient between faying surfaces of the bolted connections is an intrinsic variable for the determination of the design capacity of the slip-resistant connections. Although the design codes of civil engineering structures define slip-factors for various steel surface treatment conditions [10], they stipulate experimental test results for the utilization of new technologies such as friction shims and surface coatings. The recent advancements in material science have paved the way towards the production of innovative friction shims that offer reliable and high dry friction coefficient along a slip-interface [11]. To increase the reliability and guarantee the durability of the slip-resistant bolted connections, these state-of-the-art solutions have been started to be used for the design of civil engineering structures. On the other hand, there is lack of knowledge, practical experience and the most importantly standardized design regulations for the use of these innovative solutions in civil engineering practice for a wide range of applications.

Cavallaro *et. al.* [6] performed experimental test campaign similar to the one suggested in EN1090 – 2 Annex G [10] to determine friction coefficient between friction shim enriched faying surfaces of stainless steel plates under cyclic loading conditions, and they confirmed that the shims provided relatively high friction coefficient and high-energy dissipation capacity. However, their loading procedure, number of tests and the evaluation criteria were different from the ones suggested by EN1090 – 2 Annex G. In the literature, to the best of the authors knowledge, there is no other investigation that focuses on the determination of the slip-factor to utilize the friction shims with respect to European code of practice for the execution of steel and aluminium structures [10]. Overall, there is lack of knowledge and data for the use of friction shims as consistent with the code of practice. Therefore, this paper presents experimental and numerical investigations to introduce slip-factor between the friction shim enriched faying surfaces of shot-blasted steel plates to create a basis for the development of optimized, dismountable, reusable and damage-free bolted connections of civil engineering structures.

2 SLIP – FACTOR DETERMINATION TEST CAMPAIGN

2.1 Overall details of the test campaign and test assemblies

The slip-factor determination test campaign was performed as consistent with EN1090-2 Annex G [10]. In total five identical test assembly were tested in the Structural Testing Laboratory of Luxembourg University. The first four tests were performed with normal test speed [10] while the fifth assembly was tested under creep test conditions [10]. The overall geometry of the test specimens and test assembly, configurations of the displacement, LVDTs, and bolt-force, LCs, measurement devices are illustrated in Fig.1 together with the position and the dimensions of the friction shims.

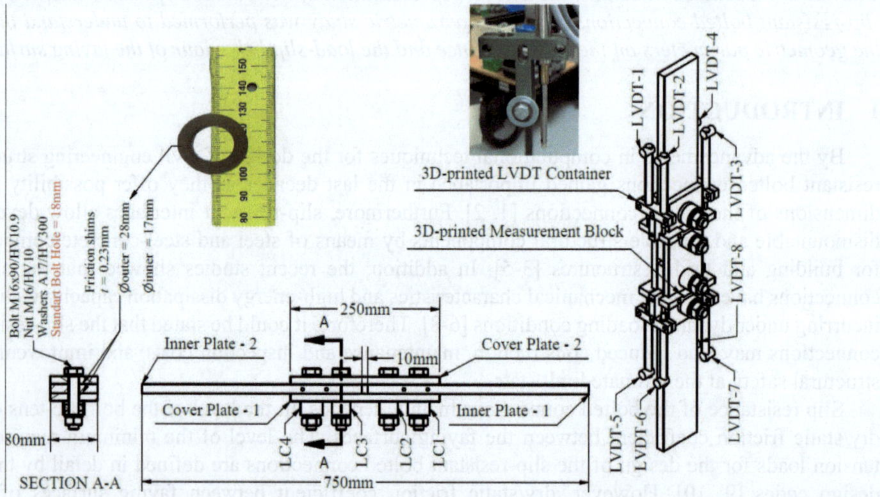

Figure 1. Overall geometry of the test assembly and the friction shims together with the configuration and labels of the measurement devices.

EN1090 – 2 Annex G [10] suggests two different set of dimensions for the test specimens based on the selected bolt size, M16 or M20. For the aforementioned test campaign, M16 bolt size was selected as it was unknown that if the capacity of the universal testing machine, 400kN, would be sufficient to overcome the slip-resistance of the test assembly, thus by selecting the lower bolt size this risks was minimized.

Table 1: Details of the measurement devices

Type / Label of the Measurement Devices	Product ID	Measurement Range
Linear Displacement Transducer / LVDT1 – LVDT8	HBM/WA [14]	10mm[1]
Load Cells / LC1 – LC3	HBM/KMR [15]	200kN
Load Cells / LC2 – LC4	HBM/KMR [15]	400kN

1-A new set of LVDTs with 2mm measurement range were installed for the 5th test.

In addition, it is stated in EN1090 – 2 Annex G [10] that the results of a slip-factor determination test campaign performed with 10.9 grade bolts could be used for the design of a slip-resistant bolted connection of 8.8 grade bolts. Therefore, 10.9 grade bolts were selected also considering that the impact of the bolt size and bolt grade on the slip-factor of the bolted connection is insignificant [12, 13]. S355 material grade was selected for the steel plates to eliminate the risks regarding to the possible plastic deformation on the inner plates between the centre of the bolt groups and the clamps of the testing machine in case of a relatively high slip-resistance. Due to the available sizes of the steel plates for the selected material grade and the dimensional requirements to position the LVDTs as suggested by the code of practice [10], the nominal thicknesses of the inner plates, -1mm, and the cover plates, +4mm, were different than the dimensions suggested by EN1090 – 2 Annex G. However, the impact of such small differences on the plate thicknesses was indicated as insignificant [12, 13], and the impact of these differences were investigated with computer aided finite element analyses presented in Section 3 of this study.

EN1090 – 2 Annex G [10] defines the centre of the bolt groups as the measurement position to record the relative displacement, i.e. slip, between the cover and inner plates. On the other hand, in the past the slip was recorded at different positions such as at the upper and lower edges of the cover plates [1, 6, 16]. However, Stranghöger *et. al.* [12] noted that recording the slip at a position different than the centre of the bolt groups would result in lower slip-factor for the slip criterion of EN1090 – 2 Annex G. This difference is mainly due to the elastic deformation, i.e. elongation, of the inner plates between the clamps of a testing machine and the centre of the bolt groups. Therefore, to eliminate this effect, the measurements were recorded at the suggested positions by utilizing 3D-printed LVDT containers and measurement blocks as shown in Fig.1. The variation in the bolt pre-tension loads were recorded during the entire test duration for all of the tests. The details of the measurement devices are presented in Table 1.

Surface asperities of the inner plates were investigated by means of the maximum surface roughness profile (R_z) to be able to understand the impact of the surface conditions on the test results. Three different measurement were taken from the each faying surfaces of the inner plates, and the measurements were evaluated according to ISO4087 [17]. The mean R_z values are presented in Table 2.

To determine the slip-factor, initially four identical tests were performed with a quasi-static loading condition. Thereafter, the fifth test was performed under a loading condition that initiates with a quasi-static loading period which was followed by a static loading period for three hours at a load level equal to the 90% of the mean slip-resistance of the first four tests, and finally followed by again the quasi-static loading period until an excessive relative displacement [10]. A displacement-controlled loading condition with 0.01mm/s loading speed was applied for the quasi-static loading period of all the five tests.

Table 2 summarize the loading procedure of the tests, material properties, dimensions and the design norms of the test specimens, mean surface roughness of the inner plates.

2.2 Assembly of the test specimens and application of the bolt pre-tension loads

The test specimens and the measurement devices were assembled towards the beginning of the each test to minimize the preload losses on the bolts triggered by the creep and the environmental effects. The shims were placed between the inner plates and cover plates for the all of the faying surfaces. The central axis of the shims were coincided with the central axis of the bolts. Hence, in total eight shims were used for the each test. The bolts were pre-tensioned with a calibrated torque-wrench and the pre-tension forces were recorded and monitored during torqueing of the all four bolts in addition to the recordings during the tests. Once the pre-defined bolt pre-tension level for the selected bolt type, i.e. 110kN for HV 10.9 M16 [10], was achieved for all of the four bolts, the entire assembly was carried carefully to the testing machine.

Table 2: Summary of the test campaign and test specimens

Test ID	Loading Procedure: Load Type	Loading Procedure: Load Speed	Plates: Grade & Norm	Plates: Surface Condition Mean R_z [µm]	Plates: Mean Thickness [mm] Inner	Plates: Mean Thickness [mm] Cover	Bolts/Nuts: Grade & Norm	Bolts/Nuts: Size [mm]	Bolts/Nuts: Pre-load	Washers: Grade & Norm	Washers: Size [mm]
FS-01	qs[1]	0.01 mm/s*	S355J2 EN10025-2 [18]	36.91	15.17	12.17	HV10.9/10 EN14399-4 [19]	M16x90/M16	110kN	HV300 EN14399-6 [20]	A17
FS-02	qs			34.12	15.23	12.19		M16x90/M16			A17
FS-03	qs			38.18	15.27	12.29		M16x90/M16			A17
FS-04	qs			34.33	15.27	12.25		M16x90/M16			A17
FS-05	qs-s*-qs			32.06 (Mean = 35.12)	15.25	12.19		M16x90/M16			A17

1) The abbreviations *qs* and *s* correspond quasi-static and static loading conditions respectively.
*) The loading speed at the static (*s*) step of the fifth test, FS – 05, is zero.

Thereafter, 3D-printed LVDT containers were screwed to the inner plate at the centre of the bolt group and the LVDTs were positioned into the containers. The entire procedure was repeated for the each test. Fig.2 shows the assembly procedure of the test specimens together with the final test-setup including the universal testing machine of 400kN loading capacity.

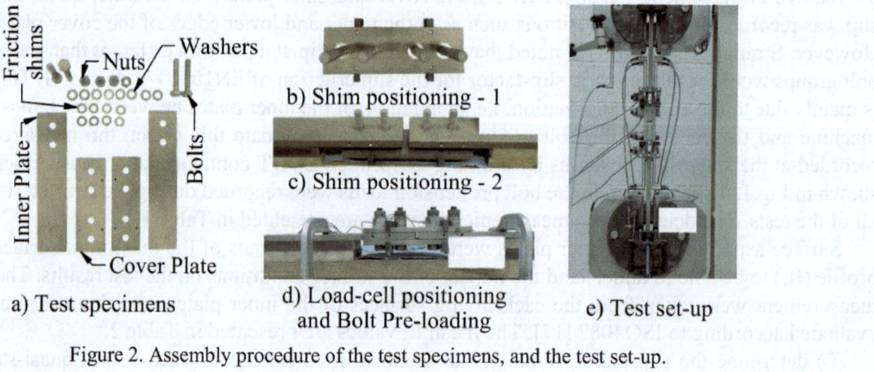

Figure 2. Assembly procedure of the test specimens, and the test set-up.

2.3 Results and discussions

EN1090 – 2 Annex G [10] defines the following equations for the determination of the individual slip-factor for the each faying surfaces of the each test assembly (Eq.1), mean value of the slip loads (Eq.2), mean value of the individual slip-factors (Eq.3) and the standard deviation of the slip-factors (Eq.4):

$$\mu_i = \frac{F_{si}}{4F_{p,C}} \tag{1}$$

$$F_{sm} = \frac{\sum F_{si}}{n} \tag{2}$$

$$\mu_m = \frac{\sum \mu_i}{n} \tag{3}$$

$$s_\mu = \sqrt{\frac{(\mu_i - \mu_m)^2}{n-1}} \tag{4}$$

where; n is the total number of the measurement, F_{si} is the individual slip load for the each couple of the faying surfaces, and $F_{p,C}$ is the nominal bolt pre-tension force.

According to EN1090 – 2 Annex G [10] the nominal bolt pre-tension load shall be used for the calculation of the individual slip-factor (Eq.1). However, during the test the pre-tension loads on the bolts reduce mainly due to the lateral contraction of the inner plate [21]. Therefore, in this study, the actual slip-factor [1, 12, 13] for the each of the faying surfaces are established (Eq.5) based on the bolt pre-tension loads at the pre-defined slip instant of EN1090 – 2 Annex G.

$$\mu_{actual} = \frac{F_{si}}{4F_{p(sllip-instant)}} \tag{5}$$

Fig.3a shows the load-slip curves for the upper and the lower faying surfaces of the first four tests together with the pre-defined slip-criterion, 0.15mm, of EN1090 – 2 Annex G [10] and the average load-slip curve derived from the eight data sets. It could be deduced from Fig.3a that the load-slip behaviours are similar for all of the first four tests. However, the slip-resistance of the third test, FS – 03, is slightly higher (~10%) than the average behaviour. This difference could be explained with the higher mean surface roughness (R_z [17]) for the inner plates of the third test assembly as presented in Table 2. Furthermore, according to the load-slip curve presented for the upper faying surfaces of the second test, FS – 02, it could be seen that the pre-defined slip-criterion was not reached by the end of the test. This is due to the deliberate termination of the test not to damage the load cells by the bolt bearing as an excessive slippage occurred at the lower part of the assembly. The variation of the bolt pre-tension forces with respect to the slippage of the corresponding (upper or lower) faying surfaces are presented in Fig.3b to Fig.3e for the tests FS – 01 to FS – 04, respectively. The test results for the first four tests are summarized in Table 3. It is important to note that the results from the upper faying surface of the second test, FS – 02, are excluded from the calculation of the mean slip-resistance and the slip-factors presented in Table 3. The conditions of the specimens are presented for the test, FS – 04, in Fig.4 as an example. It could be concluded from the figure that the normal and shear stresses, thus the slip were concentrated along the contact between the shims and the inner plates.

The slip rates for the upper and lower faying surfaces of the creep test, FS – 05, are presented in Fig.5a together with the total slippage at the 5th minute and 3rd hour of the static loading period. Fig.5b shows the load-slip curves for the upper and lower faying surfaces of the fifth test for the entire test duration, and Fig.5c shows the variation of the bolt pre-tension loads with respect to the slippage of the corresponding faying surfaces. Table 4 summarize the result of the creep test, FS – 05. According to EN1090 – 2 Annex G [10] if the slip between the faying surfaces for the creep test are more than 2μm by the end of the static loading period, the results shall not be included in the calculation of the mean slip-factor (Eq.2), and extended creep tests are required for the determination of the design values. However, in this study, instead of the establishment of a design value for the slip-factor, it is aimed to investigate the overall load-slip behaviour of the friction shim enriched faying surfaces of the shot-blasted steel plates. Therefore, the extended creep tests were not performed, and only the results from the first four tests, except the ones from the upper faying surfaces of test FS – 02, were considered as suitable for the calculation of the mean slip-factor. Furthermore, it could be concluded from Fig.5b that the upper faying surfaces of the test assembly did not reach the pre-defined slip condition of EN1090 – 2 Annex G [10], this was due to the deliberate termination of the test not to damage the load-cells due to the excessive slip of the lower faying surfaces.

Although the upper and lower faying surfaces of the creep test, FS – 05, did not pass the pre-defined creep criterion, it could be stated from Table 4 that the slip load, nominal and actual slip-factors of the creep test are consistent with the mean values presented in Table 3. In addition, the theoretical minimum slip-factors presented for the upper faying surfaces of the second, FS – 02, and the fifth tests, FS – 05, in which the pre-defined slip-criterion of EN1090 – 2 Annex G [10] was not reached, are consistent with the mean values. According to EN1090 – 2 Annex G, if the coefficient of variation for the mean slip load determined from the five tests, i.e. ten data sets, is higher than 8%, additional specimens shall be tested to establish a design value of the slip-factor. Although the coefficient of variation presented in Table 3 is 6.24%, as earlier indicated that the statistical parameters presented in Table 3 were calculated based on the seven data sets instead of ten as the creep test, FS – 05, did not satisfy the creep-criterion and the upper faying surface of the second test, FS – 02, did not reach the pre-defined slip-criterion.

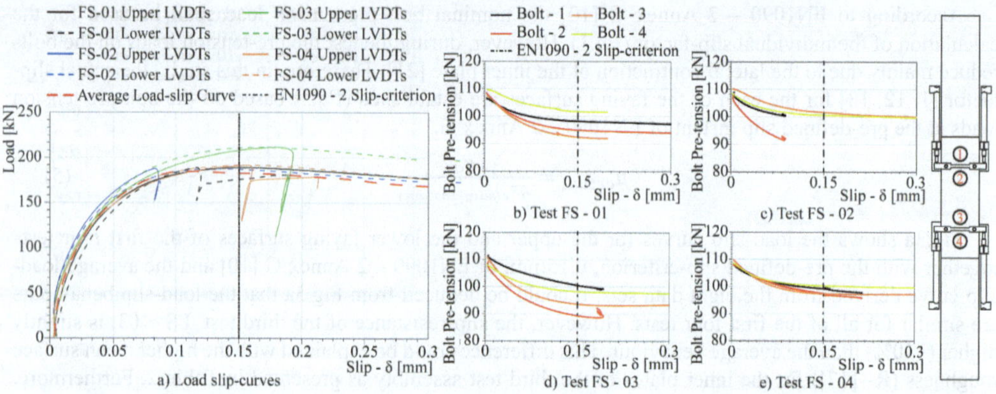

Figure 3. Load-slip curves and the bolt pre-tension load variation for the first four tests, FS – 01 to FS – 04.

Table 3: Summary of the test results for the first four tests, FS – 01 to FS – 04

Test ID	LVDTs Position	Slip Criterion [mm]	Slip-load [kN]	Bolt Pre-tension loads[kN] Initial Nominal	At the slip-instant/criterion Bolt – 1	Bolt – 2	Bolt – 3	Bolt – 4	Slip-factors $\mu_{nominal}$	μ_{actual}
FS-01	Upper	0.15	189.38	110	99.86	95.89	-	-	0.43	0.48
	Lower	0.15	175.75		-	-	95.91	101.98	0.40	0.44
FS-02	Upper	0.10[1]	189.45[1]		101.71	92.40	-	-	> 0.43	> 0.49
	Lower	0.15	188.63		-	-	100.77	103.92	0.43	0.46
FS-03	Upper	0.15	207.98		100.88	90.46	-	-	0.47	0.54
	Lower	0.15	210.88		-	-	94.23	101.69	0.49	0.54
FS-04	Upper	0.15	191.51		99.59	97.43	-	-	0.44	0.49
	Lower	0.15	191.38		-	-	97.93	103.69	0.44	0.48
Mean Slip-load[2]			193.65			Mean Slip-factor[2]			0.44	0.49
Standard Deviation			12.08			Standard Deviation			0.029	0.038
Coefficient of Variation (%)			6.24			Coefficient of Variation (%)			6.61	7.73

1) The maximum slip and corresponding slip-resistance at the end of the test are presented.

2) The results from the upper faying surfaces of the second test, FS – 02, excluded from the statistical evaluations.

On the other hand, considering the congruency between the mean values presented in Table 3 and the results from the fifth test, FS – 05, and the upper faying surfaces of the second test, FS – 02, it could be stated that the selected shims did not provide any enrichment for the slip resistance between the faying surfaces of the shot-blasted steel plates. Because the friction surface class of shot-blasted steel plates is considered as Class A with a characteristic slip-factor larger than 0.5. This result could be attributed to the similarity between the mean surface roughness of the inner plates, $R_{z\text{-innerplates}}$= 35.12µm, and the selected shims, $R_{z\text{-shims}}$= 35µm, as it was noted by the producer of the shims that to have proper performance the mean surface roughness of the base material must be at most the half of the mean surface roughness of the shims.

To sum up, the results showed that further test campaigns are required with different shim types and surface treatments of the base materials to be able to define reliable design values for the slip-factor between the faying surfaces of the bolted connections enriched with friction shims. Furthermore, the surface treatment, thus the surface asperities of the base materials should be carefully investigated during the execution of the civil engineering structures if the structural joints are designed with friction shims.

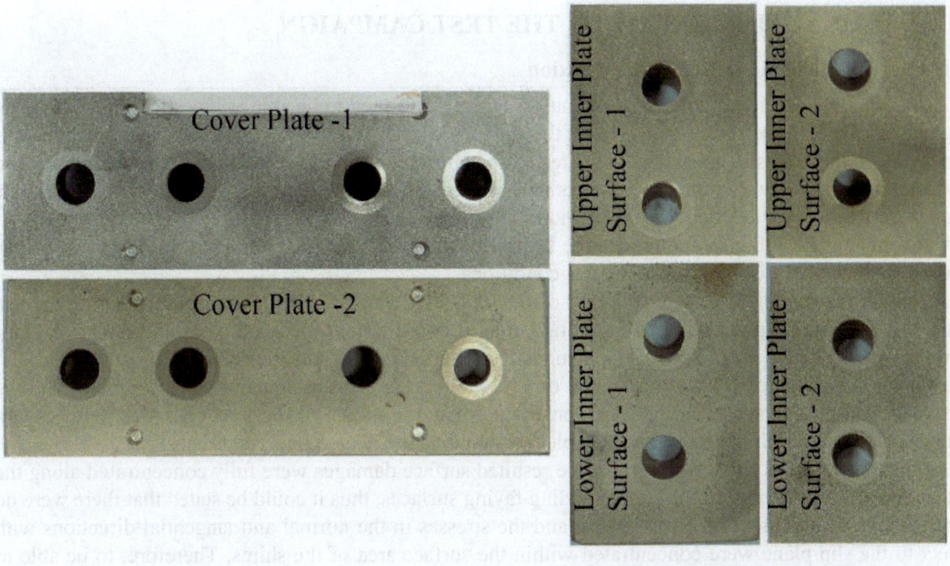

Figure 4. After test condition of the test specimens for the fourth test, FS – 04

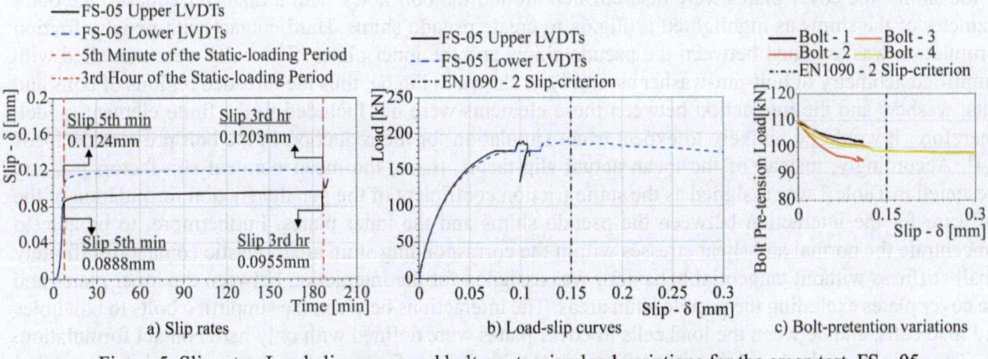

Figure 5. Slip rates, Load-slip curves and bolt pre-tension load variations for the creep test, FS – 05.

Table 4: Summary of the test results for the creep test, FS – 05

Test ID	LVDTs Pos.	Creep Crit. Verification [mm]: Creep Crit. [mm]	5th min	3rd hour	Delay. Slip²	Slip Crit. [mm]	Nom. Slip Load [kN]	Bolt Nominal Pre-tension Loads[kN]: Init.	At the slip-instant: Bolt 1	Bolt 2	Bolt 3	Bolt 4	Slip Factors: μ_{nom}	μ_{act}
FS-05	Upper	0.002	0.0888	0.0955	0.0067	0.010¹	190¹	110	102	95	-	-	>0.43	>0.46
	Lower	0.002	0.1124	0.1203	0.0079	0.015	186		-	-	100	100	0.42	0.46

1) Maximum slip, corresponding slip-resistance and bolt pre-tension loads are presented for the end of the test.

2) Delayed slip (Delay. Slip) corresponds to the difference between the slip measured at the 5th minute and 3rd hour of the static loading period.

3 NUMERICAL SIMULATION OF THE TEST CAMPAIGN

3.1 Numerical model definition and validation

A set of computer aided finite element analyses were performed to be able to investigate the impact of the geometrical parameters such as the size of the shims, presence of the load-cells and the clamping lengths of the bolts on the load-slip behaviour of the tested specimens. Furthermore, the relation between the mean slip-factor and static/dry friction coefficient is investigated. The geometry of the test specimens, except the shim, were explicitly modelled with their nominal dimensions within a finite element analysis (FEA) software, Abaqus [22]. Fig.6a summarize the finite element model geometry and the boundary conditions imposed to the model to approximate the test conditions.

Although the suggested set-up geometry of EN1090 – 2 Annex G [10] is quarter symmetric, the test set-up was only half-symmetric due to the utilization of two different load-cell types and their positions. Therefore, the half of the test assembly was modelled with the related symmetric boundary conditions.

Bilinear strain-hardening material model of EN1993 – 5 [23] was assigned to the steel plates and the bolts with Von-mises yield criterion. The constitutive material model of the load cells was defined with linear-elastic material properties of steel. Table 5 summarize the related material properties.

As demonstrated in Fig.5, the slip and the resulted surface damages were fully concentrated along the contact between the shims and the corresponding faying surfaces, thus it could be stated that there were no contact between the inner and cover plates, and the stresses in the normal and tangential directions with respect to the slip plane were concentrated within the surface area of the shims. Therefore, to be able to define the interaction between the friction shims and the related faying surfaces without explicit geometry of the shims, the cover plates were discretizated around the bolt holes with a diameter equal to the outer diameter of the shims as highlighted in Fig.6a to create pseudo shims. Hard contact with penalty friction formulation was assigned between the pseudo shims and the inner plates. The bolts were modelled with simplified geometry of bolt-nut-washer assembly as shown in Fig.6a, thus the threaded regions of bolts and nuts, washers and the interaction between these elements were not included in the finite element model. Therefore, it would be unlikely to expect exact simulation for the reduction in the bolt pre-tension loads [24]. Accordingly, instead of the mean actual slip-factor, μ_{actual}, the mean nominal slip-factor, $\mu_{nominal}$, presented in Table 3 was assigned as the static friction coefficient of the penalty friction formulation of the software for the interaction between the pseudo shims and the inner plates. Furthermore, to be able to concentrate the normal and shear stresses within the corresponding shim areas, elastic contact of infinitely small stiffness without tangential behaviour was assigned for the interaction between the inner plates and the cover plates excluding the pseudo shim areas. The interactions between the simplified bolts to bolt holes and load cells, and between the load cells to cover plates were defined with only hard contact formulation.

The vicinity of the bolt holes were discretised with 0.5mm finite element mesh size while the global mesh size for the inner and cover plates were selected as 3mm and 2mm respectively. The bolts and the load cells were discretised with 2mm global mesh size. Three-dimensional eight node reduced integration linear brick elements (C3D8R) and three-dimensional six nodes linear triangular prism elements (C3D6) were used for the finite element discretization of the model.

The analyses were performed with Abaqus/Standard solver with two successive solution steps. In the first step, bolt pre-tension loads were applied by the bolt load option of the software, and in the second step the solution was derived by the boundary conditions given in Fig.6a.

Fig.6b shows the comparison between the average load-slip curve of the first four tests and the load-slip curve obtained from the FEA. As consisted with the experimental tests, the load-slip curve of the FEA was determined as the mean relative displacement between the inner and cover plates of the upper and lower bolt groups. The comparison showed that there is a very good agreement for the slip-resistance of the FEA and the mean test result.

The differences between the load-slip curves of the FEA and the average test results could be explained with elastic-slip phenomenon [22]. Fig.7a illustrates stick-slip status and the related shear stress – slip state of a faying surface for a slip-resistant environment, where $\tau_{critical}$ corresponds to a critical shear stress to be exceeded for the faying surface to switch from the sticking state to the slipping state. Obviously, the critical shear stress, $\tau_{critical}$, along the slip plane is a function of friction-coefficient, i.e. slip-factor, and the normal stress that is in the direction of the slip plane normal as illustrated in Fig.7a. Therefore, by the reduction of the normal stress due to the relaxation in the bolt pre-tension loads it is expected that the critical shear stress, $\tau_{critical}$, to reduce for each successive steps of the quasi-static loading period while the required amount of elastic-slip, e.g. pre-defined slip-criterion of EN1090 – 2 Annex G [10], to switch from the sticking state to slipping state does not change. Thus, this leads to highly nonlinear load-elastic slip behaviour for the tests as illustrated in Fig.7b.

On the other hand, as earlier indicated that the finite element model was fed by the nominal slip-factor instead of the actual one because the bolt pre-tension load relaxation was less than the tests for the FEA due to the simplified bolt/nut/washer geometry. As a result, while the mean slip-resistance of the tests was in a very good agreement with FEA results, the amount of the elastic-slip and the nonlinearity of the load-slip behaviour of the FEA were less with respect to the average load-slip behaviour of the test results as illustrated in Fig.7b.

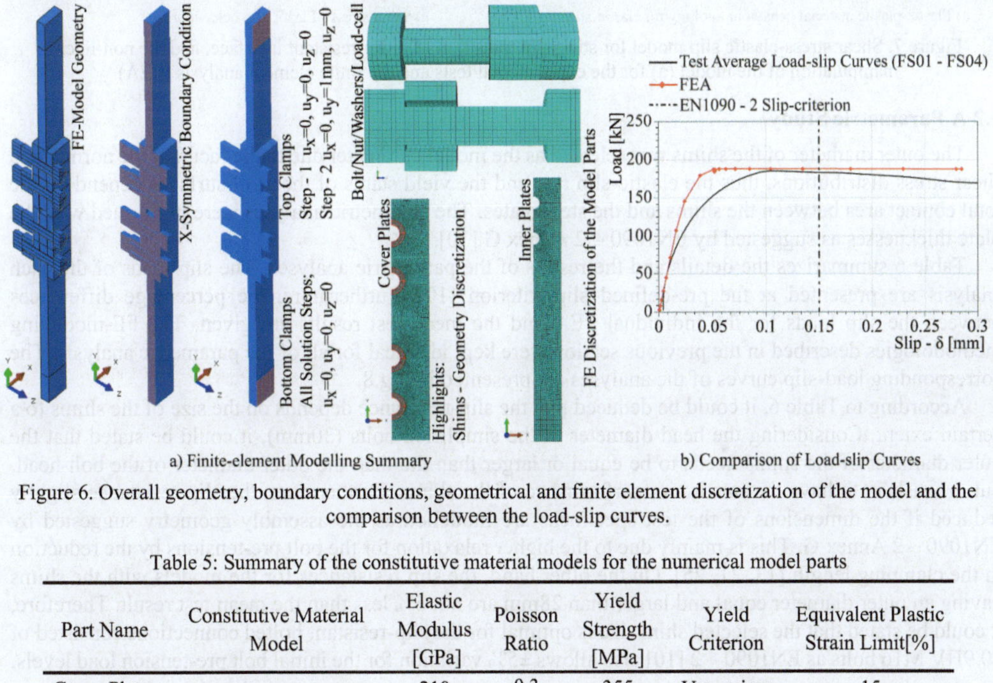

a) Finite-element Modelling Summary   b) Comparison of Load-slip Curves

Figure 6. Overall geometry, boundary conditions, geometrical and finite element discretization of the model and the comparison between the load-slip curves.

Table 5: Summary of the constitutive material models for the numerical model parts

Part Name	Constitutive Material Model	Elastic Modulus [GPa]	Poisson Ratio	Yield Strength [MPa]	Yield Criterion	Equivalent Plastic Strain Limit[%]
Cover Plates	Linear Elastic – Plastic with Strain Hardening	210	0.3	355	Von-mises	15
Inner Plates		210	0.3	355	Von-mises	15
Bolts		210	0.3	900	Von-mises	5
Load-cells	Linear Elastic	210	0.3	-	-	-

It is important to note that the FE-model was not calibrated by means of friction coefficient or elastic-slip stiffness, default elastic-slip threshold of the software was used and the mean nominal slip-factor, $\mu_{nominal}$, presented in Table 3 was inputted as the friction coefficient for the penalty friction formulation of the software. Therefore, considering the simplicity of the modelled geometry of the bolt-nut-washer assemblies, the pseudo shims and the representation of the novel non-linear stick-slip transition path presented in Fig.7b, it is concluded that the presented modelling technique is sufficiently robust for the numerical investigation and the design of slip-resistant bolted connections with friction shims.

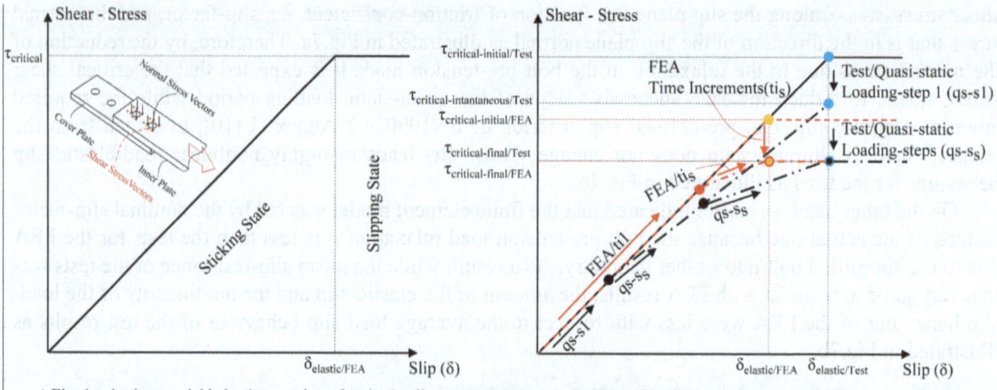

a) Elastic-plastic material behaviour anology for elastic-slip b) Shear stress - Elastic slip behaviours

Figure 7. Shear stress-elastic slip model for stick-slip transition of a slip-resistant interface, and the non-linear manipulation of the model (a) for the experimental tests and the finite element analysis (FEA).

3.2 A Parametric Study

The outer diameter of the shims was selected as the model parameter due to the fact that the normal and shear stress distributions, thus the elastic-slip rate and the yield status of the steel surfaces depend on the total contact area between the shims and the steel plates. The parametric analyses were performed with the plate thicknesses as suggested by EN1090 – 2 Annex G [10].

Table 6 summarizes the details and the results of the parametric analyses. The slip loads of the each analysis are presented at the pre-defined slip-criterion [10]. Furthermore, the percentage differences between the slip loads for the individual FEA and the mean test results are given. The FE-modelling methodologies described in the previous section were kept identical for all of the parametric analysis. The corresponding load-slip curves of the analyses are presented in Fig.8.

According to Table 6, it could be deduced that the slip-resistance depends on the size of the shims to a certain extent. Considering the head diameter of the simplified bolts (30mm), it could be stated that the outer diameter of the shims needs to be equal or larger than the than the outer diameter of the bolt-head, nut or washers to have the maximum performance of the shims. Furthermore, the slip-resistance slightly reduced if the dimensions of the test specimens are modelled as the assembly geometry suggested by EN1090 – 2 Annex G. This is mainly due to the higher relaxation for the bolt pre-tensions by the reduction in the clamping length [13, 21, 25]. On the other hand, the slip resistances for the models with the shims having an outer diameter equal and larger than 28mm are not 5% less than the mean test result. Therefore, it could be stated that the selected shim size is optimal for the slip-resistant bolted connections consisted of 10.9HV M16 bolts as EN1090 – 2 [10] also allows ±5% variation for the initial bolt pre-tension load levels.

Table 6: Details of the parametric study and the summary of the results.

Labels	Clamping Length [mm]	Plate Thickness [mm] Inner	Plate Thickness [mm] Cover	Shim Outer Diameter [mm]	Slip-load [kN]	Difference[%]
Tests	69	15	12	28	193.65	n/a
FS – FE01	69	15	12	28	187.86	-2.99%
FS – FE02	40[2]	16	8	20	70.32	-63.69%
FS – FE03	40	16	8	24	148.75	-23.19%
FS – FE04	40	16	8	28	184.47	-4.74%
FS – FE05	40	16	8	36	185.18	-4.37%
FS – FE06	40	16	8	44	184.57	-4.69%

1)The first model corresponds to the simulation of the test configuration presented in Section 3.1.

2)The clamping length 40mm corresponds to the suggested clamping length for 10.9 Grade M16 bolts [10]

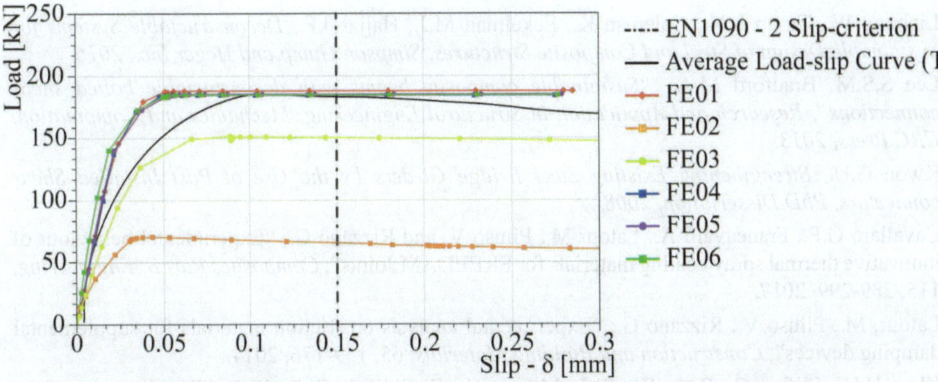

Figure 8. Load-slip curves of the parametric analyses and the average test results

4 CONCLUSION

In this paper, we presented nominal and actual slip-factors between friction shim enriched faying surfaces of shot-blasted steel plates to introduce experimental data set and numerical modelling techniques to create a basis for the utilization of the friction shims on the slip-resistant bolted connections of civil engineering structures. It is shown that the distinctive parameter for the performance of the friction shims is the surface condition of the steel components by means of their mean surface roughness (R_z [17]) which is not determined for the field applications of the steel components among its other mechanical properties. An efficient finite element modelling methodology is presented to be able to directly use the test results obtained from EN1090 – 2 Annex G for the numerical investigation and computer aided design of the slip-resistant bolted connections. A novel representation of non-linear stick-slip transition path is presented for the slip-resistant bolted connections of highly slip-critical engineering structures such as nuclear power plants. It is important to highlight that the presented finite element modelling methodology should be further validated with existing testing campaigns presented in the literature. The parametric numerical study showed that the size of the friction shims could be kept relatively small but not less than the outer diameter of bolt-head, nut or washers to optimize the construction economy of engineering structures where thousands of bolts and shims might be required for the entire structure. This study does not involve impact of bolt hole size on the critical slip-load. Therefore, further investigations with additional testing campaigns of various surface treatments and bolt hole sizes are required to establish design values for the wide range application of the friction shims on the slip-resistant bolted connections of civil engineering structures. Further research and experimental test campaigns may also focus on the variation of the contact area and its roughness.

Acknowledgements

This research is funded by Luxembourg National Research Fund (FNR), grant reference: 2018-1-12651094 and co-founded by ArcelorMittal Belval Differdange S.A.

REFERENCES

[1] Pavlović M., Heistermann C., Veljković M., Pak D., Feldmann M., Rebelo C. and Silva L. S., "Friction connection vs ring flange connection in steel towers for wind converters", *Engineering Structures, 98, 151-162, 2015.*

[2] Pavlović M., Heistermann C., Veljković M., Pak D., Feldmann M., Rebelo C. and Silva L. S., "Connection in towers for wind converters, Part II: The friction connection behaviour", *Journal of Constructional Steel Research, 115, 458-466, 2015.*

[3] Lizhong W., Troup L.N., Coleman K., Eckelman M.J., Hajjar J.F., *Deconstructable Systems for Sustainable Design of Steel and Composite Structures, Simpson Gump and Heger Inc., 2016.*

[4] Lee S.S.M, Bradford M.A., *"Sustainable composite beams with deconstuctable bolted shear connections", Research and Application in Structural Engineering, Mechanics and Computation, CRC Press, 2013.*

[5] Kwon G.U., *Strengthening Existing Steel Bridge Girders by the Use of Post Installed Shear connectors, PhD Dissertation, 2008.*

[6] Cavallaro G.F., Francavalli A., Latour M., Piluso V. and Rizzano G., "Experimental behaviour of innovative thermal spray coating materials for FREEDAM Joints", *Composites Part B: Engineering*, 115, 289-299, 2017.

[7] Latour, M., Piluso V., Rizzano G., "Experimental analysis on friction materials for supplemental damping devices", *Construction and Building Materials*, 65, 159-176, 2014.

[8] Khoo H.H., Clifton C., ButterWorth J., Macrae G., Gledhill S., Sidwell G., "Development of the self-cantering Sliding Hinge Joint with friction ring springs", *Journal of Constructional Steel Research*, 78, 201-212,2012.

[9] *Specification for Structural Joints Using High-Strength Bolts,* Research Council on Structural Connections (RCSC), Chicago/Illinois, 2009

[10] *EN1090-2:2008/A1:2011: Execution of steel structures and aluminium structures – Part 2: Technical requirements for steel structures*, European Committee for Standardization (CEN), Brussels, 2011.

[11] Wohlfeil F., "Radical Technological Innovation: Case Study of the Friction Disc by SKF", Institute for Entrepreneurship; Karlsruhe Institute of Technology (KIT), 2015.

[12] Stranghöner N., Afzali N., Vries de P., Schedin E., Pilhagen J., Cardwell S., "Slip-resistant bolted connections of stainless steel", *Steel Construction*, 10, 4, 333 – 343, 2017.

[13] Stranghöner N., et.al, "Execution and reliability of slip resistant connections for steel structures using CS and SS – SIROCCO", Research Fund for Coal and Steel (RFCS), 2018.

[14] *Data Sheet B00553_11_E00_00 HBM: public - Inductive Standard Displacement Transducers*, Hottinger Baldwin Messtechnik GmbH.

[15] *Mounting Instructions, Force Washer KMR A2114-1.1,* Hottinger Baldwin Messtechnik GmbH.

[16] Cruz A., Simões R., Alves R., "Slip factor in slip resistant joint with high strength steel", *Journal of Constructional Steel Research*, 70, 280-288, 2012.

[17] *ISO 4287:1997, Geometrical Product Specifications (GPS) – Surface texture: Profile method – Terms, definitions and surface texture parameters*, International Organization for Standardization (ISO), Switzerland, 1997.

[18] *DIN EN 10025-2 Hot rolled products of structural steels - Part 2: Technical delivery conditions for non-alloy structural steels*, Deutsches Institut für Normung (DIN), Berlin, 2019.

[19] *DIN EN 14399-4 High-strength structural bolting assemblies for preloading –Part 4: System HV – Hexagon bolt and nut assemblies,* Deutsches Institut für Normung (DIN), Berlin, 2006.

[20] *DIN EN 14399-6 High-strength structural bolting assemblies for preloading –Part 6: Plain chamfered washers,* Deutsches Institut für Normung (DIN), Berlin, 2006.

[21] Otter d.C., Maljaars J., "Preload loss of stainless steel in aluminium plated slip resistant connections", *Thin-Walled Structures*, 157, 2020.

[22] *Abaqus documentation,* Simulia User Assistance, Dassault Systemes, Providence, RI, USA, 2017.

[23] *EN 1993-1-5. Eurocode 3: Design of steel structures - Part-1-5: Plated structural elements,* European Committee for Standardization (CEN), Brussels, 2005.

[24] P. Hradil, A. Chen, N. Baddoo, "Numerical modelling of stainless steel preloaded bolted connections" *Steel Construction*, 10, 344–353, 2017

9TH INTERNATIONAL CONFERENCE ON COMPOSITE CONSTRUCTION IN STEEL AND CONCRETE
M. Knobloch, U. Kuhlmann, W. Kurz, M. Schäfer. (Eds.)
Online, Germany, July 27-29, 2021

DEVELOPMENT OF A CONSISTENT DESIGN CONCEPT FOR COMPOSITE DOWELS

Yannick Broschart*, Wolfgang Kurz, Kevin Wolters***, Georgios Christou***, Markus Feldmann***, Josef Hegger*** and Martin Claßen*****

* BORAPA Ingenieurgesellschaft mbH
e-mail: broschart@borapa.de

** University of Kaiserslautern, Germany
e-mail: wolfgang.kurz@bauing.uni-kl.de

*** RWTH Aachen University, Germany
e-mails: k.wolters@stb.rwth-aachen.de, gchristou@imb.rwth-aachen.de,
feldmann@stb.rwth-aachen.de, jhegger@imb.rwth-aachen.de,
mclassen@imb.rwth-aachen.de

Keywords: Composite Dowels, Design Concept, Technical Specification.

Abstract. *Composite dowels are an economically and technically reasonable alternative to the conventional shear connectors, the headed studs. However, their competitiveness is limited by the fact that they are only approved nationally and for a limited period of time, whereas headed studs are firmly anchored in European regulations. Numerous investigations, performed on a national and international level for decades, were necessary in order to enable an integration of composite dowels into the European standards. This paper summarizes the extensive investigations on the influence of cracked concrete, of tensile and combined shear-tensile loads as well as of composite dowel positions close to the surface on the load-bearing behaviour of composite dowels. Based on these and former investigations, a consistent design concept for composite dowels was developed, which will also be presented within this paper and shall form the basis for the introduction into European standardization.*

1 INTRODUCTION AND STATE OF THE ART

Composite dowels for the transfer of longitudinal shear forces between steel cross sections and concrete slabs have several advantages compared to conventional shear connectors, i.e. headed studs. In addition to increased material efficiency [1-3] and corresponding lower CO_2 emissions, structural components with composite dowels are characterized by a simple manufacturing process. Both the flame cutting process and the welding of the composite dowel strips can be automated [4]. In addition, no special welding qualification or equipment is required in contrast to the welding of headed studs according to [5], which also enables small and medium-sized companies to manufacture such steel beams. The use of composite dowels also offers the advantage of high flexibility, as the dowel geometries can be adapted to the specific requirements of the component. This way, high resistance, stiffness and deformability can be achieved, even when using high-strength materials [6].

Composite dowels have been used more and more frequently in recent years, mostly in bridge construction. Nevertheless, in building construction, the market shares of these innovative shear connectors have been extremely low. In addition to the lack of experience of the engineering offices, their competitiveness is also limited by the fact that they are only approved nationally and for a limited period of time [7], whereas headed studs are firmly anchored in European regulations [8]. In Germany the load-bearing capacity per dowel is determined on the basis of a national technical approval via the minimum of the failure modes steel failure [9], shearing of the concrete dowel [6,7] and pry-out of a concrete cone [10-12]. Under cyclic loading, steel resistance is calculated with the help of the structural stress concept [7,13] and the concrete resistance by means of a stress limitation [7,14]. In this case, further tests are continuously being performed to increase the efficiency of composite dowels [13,15-17].

In order to improve their competitiveness, extensive experimental and numerical investigations were carried out within a joint research project of the RWTH Aachen University and the University of Kaiserslautern, which focused on the corresponding load cases and relevant installation positions in building construction and are shown schematically in figure 1. The aim of the research project was to develop a consistent design concept for composite dowels. This should serve as a basis for the preparation of a CEN Technical Specification. This article summarizes the test program of the small-size specimens within the research project as well as the resulting design model including the underlying engineering models. A more detailed description of individual topics has already been given in [18-20]. Furthermore, large-size component tests were carried out to verify the transferability of the design equations which are presented within the final report [21].

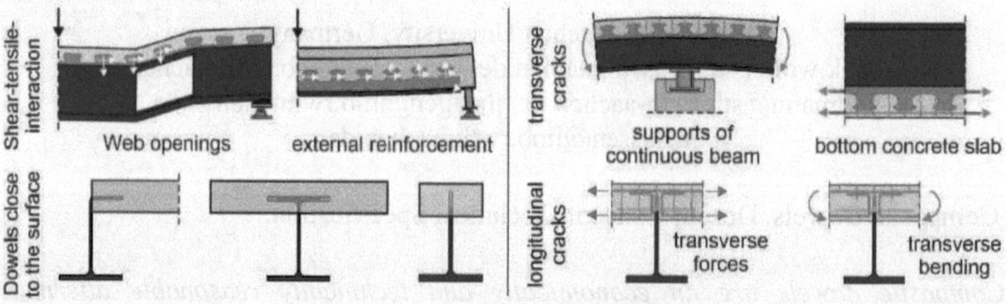

Figure 1. Schematic overview of the examined load situations and installation positions.

2 SHEAR-, TENSION- AND SHEAR-TENSION-RESISTANCE

2.1 General

While the longitudinal shear resistance of composite dowels in non-cracked concrete has already been researched extensively and is regulated within the current German approval, neither their tension- nor their combined shear-tension-resistance is covered in regulations. However, there are conceivable applications where composite dowels are stressed by corresponding loads, for example at the edge of large web openings in steel girders. Within the joint research project, numerous shear-, tension- and shear-tension-tests were performed on single-push-out tests in order to derive an approach for the interaction of both resistances. Furthermore, the calculation of the longitudinal shear-resistance in the current German approval does not consider the influence of concrete cracks. In past research projects [20,22-24], however, a significant reduction of the load-bearing capacity was observed as a consequence of transversal cracks in the concrete slab. In building construction, transversal cracks above central supports of continuous beams, longitudinal cracks above the composite dowels from transversal bending as well as crack formation due to constraining loads in statically indeterminate systems occur. In the research project the resistance of individual dowels in concrete with transversal and longitudinal cracks under shear-, tension- and shear-tension-loads was tested.

2.2 Tension-resistance of composite dowels

The existing approach of the tensile-resistance of composite dowels according to [25-28] was calibrated on small-size tests with puzzle-shaped dowels (PZ) and a uniform embedment depth of 30 mm. In addition, different test series with various dowel geometries (figure 1), embedment depths and positions of the reinforcement bars were performed. The experimental investigations were accompanied by numerical parameter studies.

Figure 2 summarizes the test setup and selected results of the tensile tests based on the *DV* test series (variation of dowel geometry). The specimens were placed centrally under the test cylinder and back-anchored via linear supports parallel to the steel plate (figure 2a). Despite different dowel geometries, the test results were similar with regard to the force-slip curve and the maximum load-bearing capacity (figure 2b). The failure mechanism was usually a large-area concrete pry-out, which settled from the outer edges of the dowel and spread along the stirrup reinforcement (figure 2c+d). In tests with PZ geometry and an arrangement of the reinforcement outside of the steel tooth undercut, the dowels were pulled out without such a large-area failure. On the contrary, the concrete was sheared only in the small undercut areas. This failure at a low load level has to be avoided by a tightly fitting dowel reinforcement in the case of composite dowels under tensile loads.

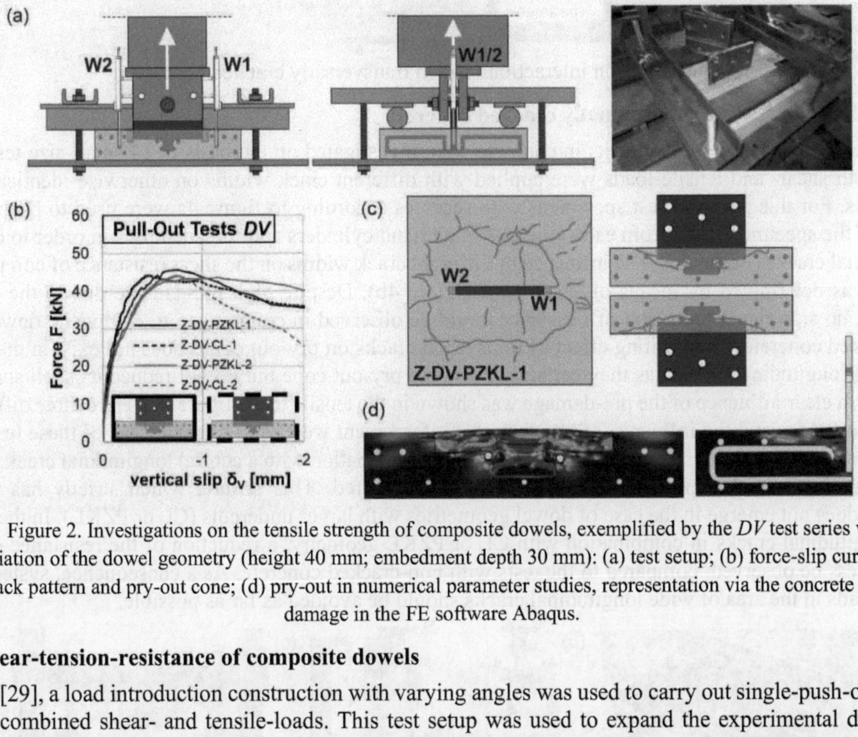

Figure 2. Investigations on the tensile strength of composite dowels, exemplified by the *DV* test series with variation of the dowel geometry (height 40 mm; embedment depth 30 mm): (a) test setup; (b) force-slip curves; (c) crack pattern and pry-out cone; (d) pry-out in numerical parameter studies, representation via the concrete tensile damage in the FE software Abaqus.

2.3 Shear-tension-resistance of composite dowels

In [29], a load introduction construction with varying angles was used to carry out single-push-out tests under combined shear- and tensile-loads. This test setup was used to expand the experimental database within the research project (figure 3, [20]). The influence of the reinforcement arrangement, which had previously been identified as decisive, was systematically investigated. Furthermore, different dowel geometries under five different load angles were considered. The resistance of the examined composite dowels decreased both with increasing distance of the transversal reinforcement from the steel tooth and with increasing angle of load application (increasing tensile component). The position of the transversal reinforcement was only negligible under predominantly shear loads (angle of application <22°). Furthermore, the influence of transversal cracks was systematically investigated in pure tension- and interacting shear-tension-tests. Both the materials used and the measuring techniques are documented in detail in [20].

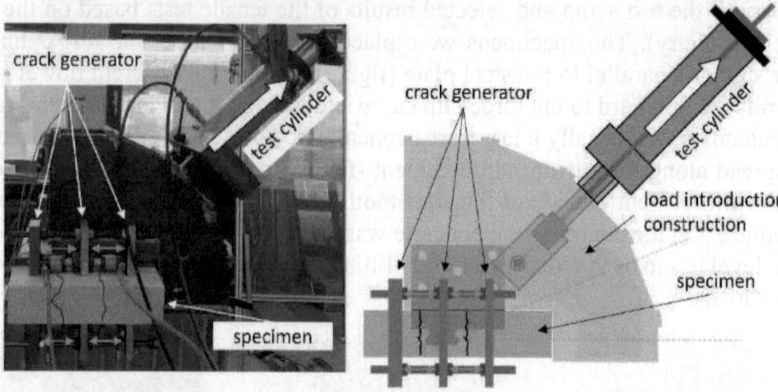

Figure 3. Test setup for shear-tension interaction tests in transversally cracked concrete.

2.4 Small-size tests with longitudinally cracked concrete

The effects of longitudinal cracks in concrete were investigated on the basis of 14 small-size tests, in which both shear- and tensile-loads were applied with different crack widths on otherwise identical test specimens. For this purpose, test specimens with recesses according to figure 4a were used to push both halves of the specimen away from each other by small hand cylinders and threaded rods in order to create longitudinal cracks (figure 4d). The influence of different crack widths on the shear resistance of composite dowels was determined by means of seven tests (figure 4b). Despite high pre-damage due to the crack insertion, no significant reduction of resistance could be observed in comparison to composite dowels in non-cracked concrete. The limiting effect of transversal cracks on pry-out cones does not exist in the case of central longitudinal cracks, as they certainly cross the pry-out cone but do not reduce its shell surface. However, a clear influence of the pre-damage was shown in the tensile tests (figure 4c). Here three different dowel geometries and the influence of transversal reinforcement were tested. In the case of these tests on composite dowels with PZ shape under tensile stress in combination with a central longitudinal crack (with or without transversal reinforcement) pull-out failure occurred. This failure, which strictly has to be avoided, does not emerge in the case of dowel geometries with larger undercuts (CL or PZKL). In the tests with longitudinal cracks in combination with CL or PZKL geometry, a reduction of the resistance could nevertheless be observed, compared to the tests with non-cracked concrete. As a consequence, systematic tensile loads in the area of wide longitudinal cracks should be avoided as far as possible.

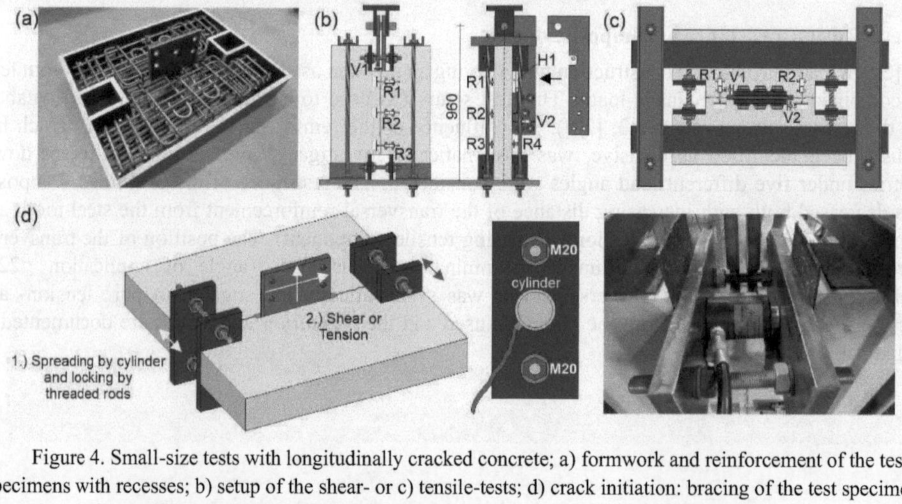

Figure 4. Small-size tests with longitudinally cracked concrete; a) formwork and reinforcement of the test specimens with recesses; b) setup of the shear- or c) tensile-tests; d) crack initiation: bracing of the test specimen halves by spreading plates with the help of hand cylinders and threaded rods.

3 COMPOSITE DOWELS POSITIONED CLOSE TO THE SURFACE

3.1 General

Specifically, in building construction, modern composite structures are being realized increasingly slimmer. To achieve such low slab thicknesses, the composite dowels can be positioned horizontally within the concrete slab (rotated by 90° around their longitudinal axis). In this case, as well as with installation positions close to the free edge of the concrete slab, forms of failure can occur which are not explicitly considered in current regulations, but which are often decisive for the design of slim composite structures. The positions described are referred to as 'close to the surface'. The typical failure mode of composite dowels positioned close to the surface occurs in the form of spalling of the concrete cover and is called edge failure. A horizontal arrangement of the composite dowels allows concrete slab thicknesses far below the limit of $b_c \geq 250$ mm, which is the current restriction of the German technical approval and causes economic disadvantages of the construction method in the field of building construction.

3.2 Small-size tests

In total, more than 40 small-size tests were carried out and accompanied by numerical simulations within the research project to investigate the influence of individual parameters on the load-bearing and deformation behaviour of composite dowels positioned close to the surface. Correspondingly, the influence of the lateral concrete cover $c_{D,s}$, the dowel geometry (PZ, CL or PZKL), the scaling factor e_x, the steel tooth axial spacing, the steel plate thickness t_w, the concrete compressive strength f_c and the arrangement of the transversal reinforcement were inspected. The tests were mainly manufactured in the form of push-out tests with four composite dowels each (figure 5a+b). The force-slip behaviour of specimens with a concrete slab thicknesses b_c of 100 mm, 150 mm and 200 mm is exemplarily shown in figure 5c, while two test specimens of identical parameter configuration were tested in each case. The lateral concrete cover is equal to $c_{D,s} = (b_c - t_w)/2$. Although the maximum load-bearing capacity increased with growing slab thickness, a reduction in stiffness was recorded in all tests at a load level of approximately 265 kN (figure 5c, Detail A). This observation is caused by two overlapping effects.

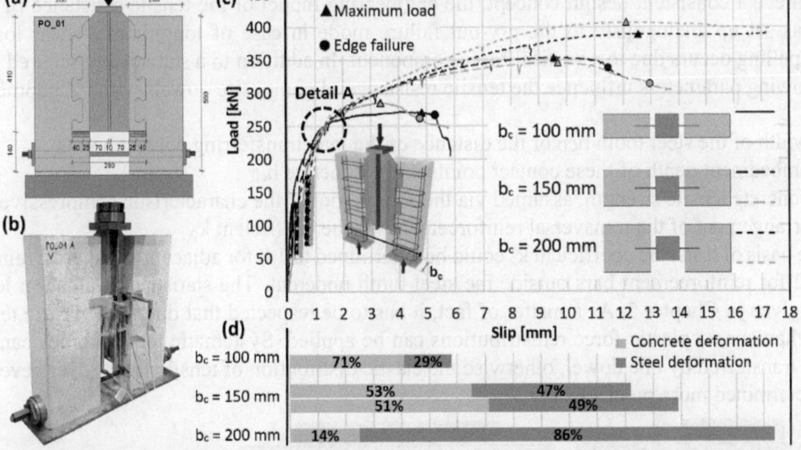

Figure 5. Small-size tests with composite dowels positioned close to the surface: (a) push-out test specimen; (b) test setup; (c) force-slip behaviour with different concrete slab thicknesses b_c; (d) slip composition.

Firstly, it is due to the pulverization of the concrete in the quasi-hydrostatic compressive stress zone directly below the steel teeth, which is also reflected in the predominant concrete deformation in the slip composition of the 100 mm thick specimens (figure 5d) and was described more detailed in [18,19]. Secondly, the 4 mm thick steel teeth reached their yield strength at the said load level, resulting in the large total slip of the 200 mm thick specimens, their load-bearing capacity was ultimately limited by the steel resistance and not by edge failure. Although concrete pulverization occurred there below the steel teeth, too, the greater lateral concrete cover was sufficient to dissipate the radiating compressive forces. Overall, the ratio of concrete deformations decreased with higher concrete slab thicknesses, while steel tooth deformations increased.

4 ENGINEERING MODELS

4.1 General

Based on the results of the small-size tests, engineering models were developed which were then statistically calibrated on own and former small-size tests according to Annex D.8 of Eurocode 0 [30]. Subsequently, the approaches for tensile- and combined shear-tensile-resistance, transversal cracks and composite dowels positioned close to the surface will be described. The statistically validated equations will be compiled in chapter 5 within a consistent design concept for composite dowels with geometries as shown in figure 6. The longitudinal shear resistance which is currently regulated within the German approval [7] is also proposed to be modified by a new statistical evaluation.

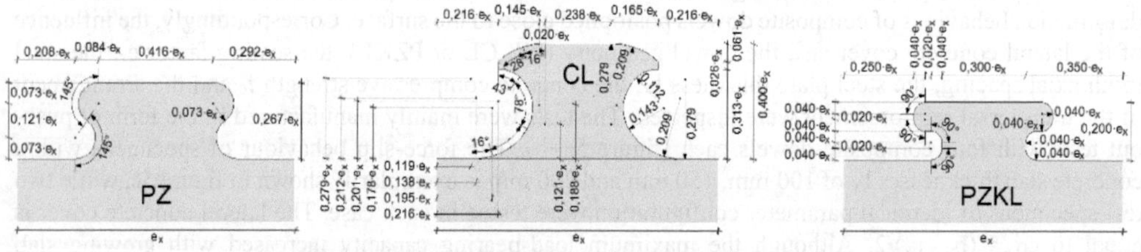

Figure 6. Composite dowel geometries PZ, CL und PZKL.

4.2 Approach to consider the tensile-resistance

To achieve a consistent design concept, the engineering model of the tensile resistance of composite dowels was set up analogously to the pry-out failure mode in case of longitudinal shear loads, where concrete spalling occurs due to a vertical force component. In addition to a statistically secured coefficient k, the following parameters influence the tensile resistance of composite dowels with PZ geometry (figure 7):

- the width of the steel tooth b_{CD} or the distance of the load transferring contact points
- the embedment depth of these contact points in the concrete h_{eff}
- the concrete tensile strength, assumed via the square root of the characteristic compressive strength
- the arrangement of the transversal reinforcement via the coefficient k_ρ

On the basis of tests, the coefficient k_ρ could be determined to 1.8 for adjacent transversal reinforcement bars or 1.2 for reinforcement bars outside the steel tooth undercut. The statistical evaluation leads to the equations given in Chapter 5. As a matter of fact, it has to be respected that due to the brittle tensile load-bearing behaviour no plastic force redistributions can be applied. Systematic tensile forces can therefore, at best, be transferred by one dowel, otherwise the elastic distribution of tensile forces over several dowels must be determined more precisely.

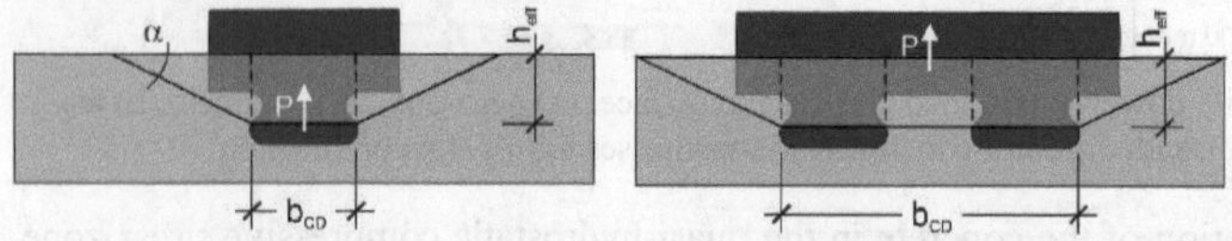

$$P_Z = k \cdot \left(\frac{\pi}{\tan\alpha \sin\alpha} + \frac{2 \cdot b_{CD}}{h_{eff} \cdot \sin\alpha}\right) \cdot h_{eff}^{1,5} \cdot \sqrt{f_{ck}} \cdot k_\rho$$

Figure 7. Determination of the pry-out cone geometry in case of one dowel or block failure and design approach.

4.3 Approach to consider the combined shear-tensile-resistance

In [29] an interaction relationship was proposed (figure 8) which divides both of the effecting shear and tensile forces by the corresponding resistances while each quotient was calibrated with an exponent k to consider the interaction relationship. On the basis of test results, the exponent k could be determined for different dowel geometries and reinforcement arrangements.

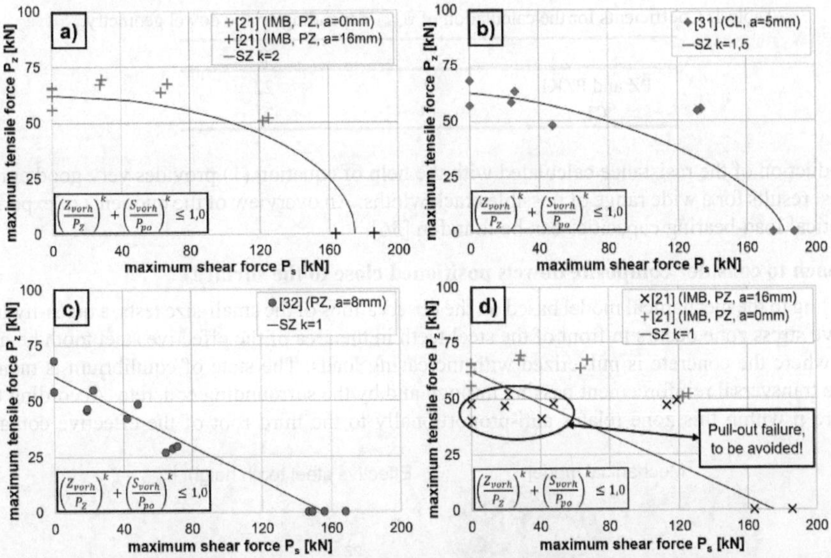

Figure 8. Shear-tensile-interaction relationships for reinforcement bars at the steel dowel (a), within the steel tooth undercut (b+c) and outside the steel tooth undercut (d).

Depending on the position of the transversal reinforcement, three cases can be distinguished. The maximum resistance under a combined shear-tensile load can be achieved if the transversal reinforcement is placed directly at the steel tooth within the undercut (a = 0 mm). In this case, the resulting shear-tension interaction curve can be seen in figure 8a and the exponent k can be set to 2.0. The diagram contains test results from [21]. If the transversal reinforcement does not contact the steel tooth but still is in the undercut area (figure 8b, a = 5 mm) with tests according to [31], the exponent k can be selected as 1.5 for the dowel geometries PZKL and CL, whereas in the case of the PZ geometry (figure 8c, a = 8 mm) with tests according to [32] an interaction coefficient of k = 1.0 must be used. If no explicit attention is paid to the arrangement of the transversal reinforcement, no tensile stress may be applied in the case of the PZ geometry, while with the CL and PZKL geometries still an interaction coefficient of k = 1.0 (figure 8d) with tests according to [21] can be applied. If systematic tensile forces occur, is important to ensure that the transversal reinforcement is positioned as close as possible to the steel tooth to increase the tensile resistance. The proposed interaction relationships have been further developed on the basis of the proposed models according to [33].

4.4 Approach to consider transversally cracked concrete slabs

In [34] an equation for the mechanical derivation of a reduction factor ψ_{crack} was developed which considers the concrete strength, the geometric conditions and the crack pattern. The influence of the crack pattern was very complex to determine due to the mechanical consistency and was simplified in [35] with the help of conservative geometric assumptions. On the basis of a statistical evaluation of the extended database, the reduction coefficient was further developed and statistically verified. In the elaborated proposal according to [1] only the geometric value h_{po} (height of the pry-out cone), the crack width w_r and the corresponding crack spacing s_r were considered. The coefficients α and β can be determined for the different dowel geometries on the basis of Table 1.

$$\psi_{crack} = \frac{s_r}{6 \cdot h_{po} + s_r} + \frac{\sqrt{h_{po}}}{\alpha + \beta \cdot w_r} \leq 1.0 \tag{1}$$

Table 1: Coefficients for the calculation of ψ_{crack} depending on the dowel geometry.

	α	β
PZ and PZKL	9	22
CL	13	31

The reduction of the resistance calculated with the help of equation (1) provides very good agreement with the test results for a wide range of possible crack widths. An overview of the quotients of experimental and theoretical load-bearing capacities can be found in [36].

4.5 Approach to consider composite dowels positioned close to the surface

According to the mechanical model based on the observations of the small-size tests, a quasi-hydrostatic compressive stress zone occurs in front of the steel teeth in the area of the effective steel tooth height $h_{d,eff}$ (figure 9) where the concrete is pulverized with increasing loads. The state of equilibrium is maintained both by the transversal reinforcement passing through and by the surrounding concrete. According to [37], the pressure p within this zone relates anti-proportionally to the third root of the effective contact area $t \cdot h_{d,eff}$.

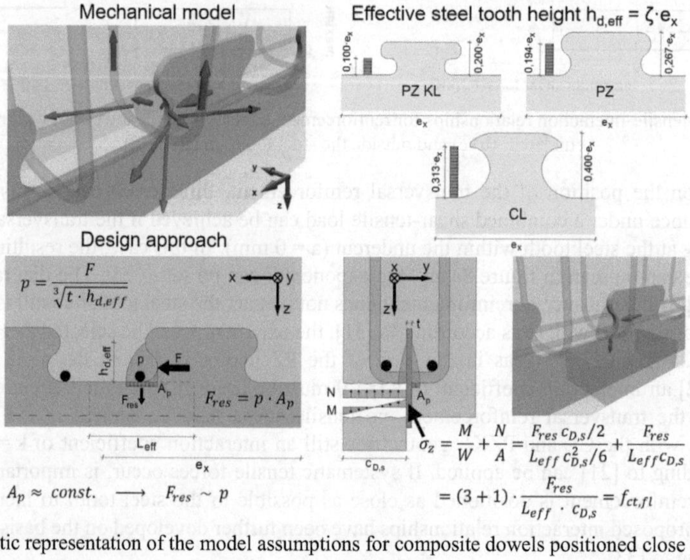

Figure 9. Schematic representation of the model assumptions for composite dowels positioned close to the surface.

The compressive stress zone causes, among other effects, a vertical force component F_{res}, which acts on an approximately constant and parameter-independent area A_p. Because of its eccentricity, it induces bending stresses within the concrete cover around the x-axis. If the concrete tensile stress σ_z locally reaches the flexural tensile strength $f_{ct,fl}$, the concrete edge spalls, the previously encased concrete powder escapes and the loss of load-bearing capacity is initiated. As a consequence, the resistance of composite dowels positioned close to the surface is significantly limited by maintaining the state of equilibrium of the quasi-hydrostatic compressive stress zone by the concrete cover. The resistance of this local zone can be approximated by applying static model assumptions as to be seen in figure 9.

4.6 Neglection of the failure mode 'shearing of the concrete dowel' in thin slabs

The current German approval additionally requires the calculation of the resistance against the failure mode shearing of the concrete dowel. As a matter of fact, when the required transversal reinforcement with an area of $A \geq 0.5 \cdot P/f_{sd}$ per dowel according to [8] is used, the failure mode has not been observed in tests with open dowel geometries and small embedment depth any more with regard to [12,34].

In addition, calculations for parameter configurations with PZ geometry showed that the failure mode shearing of the concrete dowel is not decisive in comparison to the failure mode pry-out for usual dimensions in building construction. Therefore, a design equation for determining the shear strength could be omitted for concrete covers below 100 mm in future regulations [21].

5 CONSISTENT DESIGN CONCEPT FOR COMPOSITE DOWELS

On the basis of the previously illustrated theoretical considerations, a consistent design model for composite dowels was developed which extends the scope of application of the current German approval [8]. For pure longitudinal shear loads, the characteristic value of the resistance per dowel $P_{S,Rk}$, considering the potential failure modes steel failure P_{pl}, pry-out of a concrete cone in non-cracked or transversely cracked concrete P_{po} and edge failure P_{poe}, is derived from the calculative minimum of the equations given in figure 10.

Shear resistance

$$P_{S,Rk} = \min \begin{cases} P_{pl,k} = \lambda_{geo} \cdot e_x \cdot t_w \cdot f_y & \rightarrow \text{Steel failure} \\ P_{po,k} = k_1 \cdot \chi_x \cdot \chi_y \cdot h_{po}^{1,5} \cdot \sqrt{f_{ck}} \cdot (1 + \rho_{D,l}) \cdot \psi_{crack} & \rightarrow \text{Pry-out} \\ P_{poe,k} = 2,16 \cdot \sqrt{f_{ck}} \cdot k \cdot c_{D,s} \cdot \sqrt[3]{t_w \cdot h_{d,eff}} \cdot \sqrt[3]{e_x} & \rightarrow \text{Edge failure} \end{cases}$$

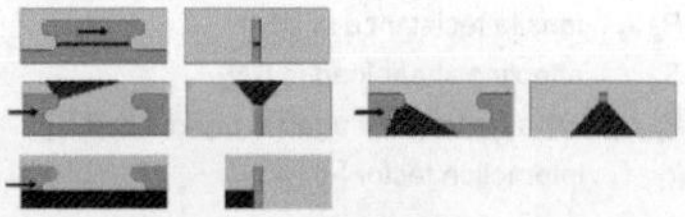

λ_{geo} specific geometry factor:
$\lambda_{geo,PZ} = 0,286$; $\lambda_{geo,PZKL} = 0,186$; $\lambda_{geo,CL} = 0,259$

e_x steel tooth distance in [mm];
$150\ mm \leq e_x \leq 500\ mm$

t_w thickness in [mm]; $4\ mm < t_w \leq 60\ mm$;
maximum $t_{w,max} = 40\ mm$ is used in the calculation

f_y nominal value of the steel yield strength in [N/mm²]

f_{ck} characteristic concrete compression strength in [N/mm²]

k_1 factor for failure mode pry-out:
71 bei PZ- und PZKL-Geometrie
95 bei CL-Geomerie

h_{po} height of the fictitious pry-out cone in [mm]

$$h_{po,PZ} = \min \begin{cases} c_{D,o} + 0,07 \cdot e_x \\ c_{D,u} + 0,08 \cdot e_x \end{cases}$$

$$h_{po,PZKL} = \min \begin{cases} c_{D,o} + 0,05 \cdot e_x \\ c_{D,u} + 0,06 \cdot e_x \end{cases}$$

$$h_{po,CL} = \min \begin{cases} c_{D,o} + 0,07 \cdot e_x \\ c_{D,u} + 0,13 \cdot e_x \end{cases}$$

$\rho_{D,l}$ reinforcement degree for failure mode pry-out : $\rho_{D,l} = \frac{E_s \cdot A_{sf}}{E_{cm} \cdot A_{D,l}}$

A_{sf} transversal reinforcement area in the concrete slab in [mm²]:
$A_{sf} = A_b + A_t$

$A_{D,l}$ participating concrete area in [mm²]: $A_{D,l} = h_c \cdot e_x$

h_c height of the concrete slab in [mm]

E_s modulus of elasticity of the steel in [N/mm²]

E_{cm} modulus of elasticity of the concrete in [N/mm²]

$c_{D,o}$ upper concrete cover to the steel tooth in [mm]; $c_{D,o} \geq 30\ mm$

$c_{D,u}$ lower concrete cover to the steel tooth in [mm]; $c_{D,u} \geq 30\ mm$

$c_{D,s}$ lateral concrete cover to the steel tooth in [mm]; $c_{D,s} \geq 45\ mm$

χ_x reduction factor with regard to e_x
$\chi_x = 1,0$ for $e_x \geq 4,5 \cdot h_{po}$
$\chi_x = \frac{e_x}{4,5 \cdot h_{po}} \leq 1,0$ for $e_x < 4,5 \cdot h_{po}$

χ_y reduction factor with regard to the number of composite dowel stripes, to e_x and to the transversal distance of the composite dowels e_y:
in case of one composite dowel stripe:
$\chi_y = 1,0$
in case of two composite dowel stripes:
$\chi_y = 1,0$ for $e_y \geq 9 \cdot h_{po}$
$\chi_y = \frac{1}{2} \cdot \left(\frac{e_y}{9 \cdot h_{po}} + 1\right) \leq 1,0$ for $120\ mm < e_y < 9 \cdot h_{po}$

ψ_{crack} reduction factor to consider transversal cracks:
$\psi_{crack,PZ/PZKL} = \frac{s_r}{6 \cdot h_{po} + s_r} + \frac{\sqrt{h_{po}}}{9 + 22 \cdot w_r} \leq 1,0$
$\psi_{crack,CL} = \frac{s_r}{6 \cdot h_{po} + s_r} + \frac{\sqrt{h_{po}}}{13 + 31 \cdot w_r} \leq 1,0$

s_r distance between cracks in [mm]

w_r crack width in [mm]

k size factor: $k = 1 + \sqrt{200/c_{D,s}}$

Figure 10. Design equations for longitudinal shear resistance.

In the case of pure tensile loads, the characteristic value of the resistance per dowel $P_{Z,Rk}$ can be determined in a simplified way with regard to the pry-out failure mode, using the equations shown in figure 11 depending on the arrangement of the transversal reinforcement. If the transversal reinforcement contacts the steel tooth within its undercut, an increased tensile resistance can be applied. First of all, it has to be checked whether the effecting tensile force can be transferred by a single dowel. If this is not the case, the elastic distribution of the tensile force over several dowels must be determined more precisely. Then both the tensile resistance of each individual dowel and the block failure, which can be calculated by using b_{CD} as the distance between the outer load transferring areas of several dowels under tension, have to be considered. A plastic distribution of the tensile forces must not be assumed due to the low ductility.

Tensile resistance

$$P_{Z,Rk} = (7 + 2 \cdot \frac{b_{CD}}{h_{eff}}) \cdot h_{eff}^{1,5} \cdot \sqrt{f_{ck}}$$ → reinforcement in border position

$$P_{Z,Rk} = (10,5 + 3,1 \cdot \frac{b_{CD}}{h_{eff}}) \cdot h_{eff}^{1,5} \cdot \sqrt{f_{ck}}$$ → reinforcement in steel undercut

b_{CD} opening width between the steel teeth in [mm]
h_{eff} effective height of the pry-out cone in [mm]
f_{ck} characteristic concrete strength in [N/mm²]

Combined shear-tension resistance

$$\left(\frac{Z_e}{P_{Z,Rd}}\right)^k + \left(\frac{S_e}{P_{po,Rd}}\right)^k \leq 1,0$$

Z_e effecting tensile load in [kN]
$P_{Z,Rd}$ tensile resistance in [kN]
S_e effecting shear load in [kN]
$P_{po,Rd}$ shear resistance against pry-out in [kN]
k interaction factor [-]

reinforcement positions	k-factor		
	PZ	PZKL	CL
	invalid	1,0	1,0
	1,0	1,5	1,5
	2,0	2,0	2,0

Figure 11. Design equations for tensile- and shear-tensile-resistance.

Applications with composite dowels under tensile loads in cracked concrete shall be avoided as far as possible. Otherwise the transversal reinforcement must be placed closely to the steel tooth in the undercut and the crack width must be limited to $w_r = 0.1$ mm according to the current technical regulations. In addition, dowel geometries with a larger undercut (CL or PZKL) should be selected preferably. For composite dowels with PZ geometry, the design value of the calculated tensile resistance shall be reduced to 50% of the tensile resistance according to figure 11, for CL and PZKL geometry to 70%.

The equations to determine the resistance of composite dowels under combined shear-tensile loads are also given in figure 11, where the position of the transversal reinforcement has a decisive influence, too.

6 SUMMARY AND OUTLOOK

On the basis of numerous experimental investigations and corresponding numerical simulations, which were partly presented within this paper, and of former test results, a consistent design concept was developed to consider the tensile- as well as the shear-tensile-resistance of composite dowels, the influence of cracks and of positions close to the surface. This model approach shall form the basis for the preparation of a CEN Technical Specification, which can be introduced into the European standardization process and thus increase the competitiveness of composite dowels.

ACKNOWLEDGEMENT

The presented test results were achieved during a research project which was carried out in the framework of the industrial collective research program (IGF no. 19438 N). It was thankfully supported by the Federal Ministry for Economic Affairs and Energy (BMWi) through the AiF (German Federation of Industrial Research Associations e.V.) based on a decision taken by the German Bundestag.

REFERENCES

[1] Claßen M., Gallwoszus J., Hegger J. et al., "Nachhaltigkeitsbewertung von Deckensystemen mit großen Spannweiten", *Bauingenieur,* 89(3), pp. 125-133, 2014.

[2] Heinemeyer S., Gallwoszus J., Hegger J., "Verbundträger mit Puzzleleisten und hochfesten Werkstoffen", *Stahlbau*, 81(8), pp. 595-603, 2012.

[3] Kopp M., Christou G., Stark A. et al, "Integriertes Deckensystem für den Stahl- und Verbundbau", *Stahlbau*, 87(2), pp. 136-148, 2018.

[4] Hegger J., Claßen M., Schaumann P. et al., "Entwicklung einer integrierten Verbunddecke für nachhaltige Stahlbauten", *Stahlbau*, 82(1), pp. 11-17, 2013.

[5] European Committee for Standardization, *Schweißen - Lichtbogenschweißen von metallischen Werkstoffen*, DIN EN ISO 14555:2010, Brussels, 2010.

[6] Heinemeyer S., *Zum Trag- und Verformungsverhalten von Verbundträgern aus ultrahochfestem Beton mit Verbundleisten,* Phd thesis, RWTH Aachen University, 2011.

[7] DIBt Deutsches Institut für Bautechnik, *Allgemeine Bauartgenehmigung: Stahlverbundträger mit Verbunddübelleisten in Klothoiden- und Puzzleform*, German technical approval Z-26.1-23, Berlin, 2018.

[8] European Committee for Standardization, *Eurocode 4: Design of composite steel and concrete structures - Part 1-1: General rules and rules for buildings*, EN 1994-1-1:2004 + AC:2009, Brussels, 2004.

[9] Claßen M., Herbrand M., Stark A., "Dübelkennlinien von Verbunddübelleisten mit Stahlversagen", *Bauingenieur*, 92(6), pp. 237-244, 2017.

[10] Claßen M., Bielak J., Hürth D., "Verbunddübelleisten mit Betonausbruchversagen unter Zug- bzw. Druckbeanspruchung - Teil 1: Finite-Elemente-Analyse", *Beton- und Stahlbetonbau*, 113(5), pp. 348-359, 2018.

[11] Claßen M., Schmidt P., Wieneke K., "Verbunddübelleisten mit Betonausbruchversagen unter Zug- bzw. Druckbeanspruchung – Teil 2: Datenbank und Bemessungsansätze", *Beton- und Stahlbetonbau*, 113(6), pp. 443-452, 2018.

[12] Kopp M., Wolters K., Claßen M., Hegger J., Gündel M., Gallwoszus J., Heinemeyer S. and Feldmann M., "Composite dowels as shear connectors for composite beams – Background to the design concept for static loading", *Journal of Constructional Steel Research*, 147, pp. 488-503, 2018.

[13] Wolters K., Christou G., Feldmann M., "Untersuchungen zum Risswachstum in Verbunddübelleisten", *Bauingenieur*, 94(6), pp. 216-227, 2016.

[14] Gallwoszus J., *Zur Ermüdung von Verbundkonstruktionen mit Verbunddübelleisten,* Phd thesis, RWTH Aachen University, 2014.

[15] Christou G., Hegger J., Claßen M., "Fatigue of clothoid shaped rib shear connectors", *Journal of Constructional Steel Research*, 171, pp. 106-133, 2020.

[16] Christou G., Ungermann J., Wolters K. et al., "Ermüdung von Verbunddübelleisten", *Beton- und Stahlbetonbau*, 115(5), pp. 355-363, 2020.

[17] Christou G., Ungermann J., Wolters K. et al., "Ermüdung von Verbunddübelleisten – experimentelle Untersuchungen", *Stahlbau,* 89(4), pp. 388-398, 2020.

[18] Broschart Y. and Kurz W., "Tragverhalten von Verbundträgern mit randnahen Verbunddübelleisten", *Stahlbau*, 88(9), 2019.

[19] Broschart Y., Kurz W., Wolters K. et al., "Einflussfaktoren auf das Tragverhalten randnaher Verbunddübelleisten", *Bauingenieur*, 94(6), pp. 199-205, 2019.

[20] Christou G., Wolters K. and Broschart Y., "Verbunddübelleisten unter kombinierter Schub-Zug-Beanspruchung im gerissenen Beton", *Bauingenieur*, 94(6), pp. 228-236, 2019.

[21] Feldmann M., Wolters K., Hegger J., Claßen M, Christou G., Kurz W. and Broschart Y., "Konsistentes Bemessungsmodell für fertigungsoptimierte Verbunddübelleisten", Final report P1208, 2021.

[22] Claßen M., Hegger J., "Ausstanzmodell mit Rissreibung für Verbunddübelleisten im gerissenen Beton", *Beton- und Stahlbetonbau*, 112(3), pp. 155-166, (2017).

[23] Claßen M., Hegger J., "Ausstanzen von Verbunddübelleisten im gerissenen Beton", *Stahlbau*, 86(3), pp. 256-268, 2017.

[24] Claßen M., Gallwoszus J., Hegger J., "Einfluss von Querrissen auf das Schubtragverhalten von Verbunddübelleisten in schlanken Betongurten", *Beton- und Stahlbetonbau*, 109(12), pp. 882-894, 2014.

[25] Claßen M., Hegger J., "Anchorage of composite dowels", *Steel Construction*, 9(2), pp. 138-150, 2016.

[26] Claßen M., Hegger J., "Verankerungsverhalten von Verbunddübelleisten in schlanken Betongurten", *Bautechnik*, 91(12), pp. 869-883, 2014.

[27] Claßen M., Kurz W., Schäfer M. et al., "A mechanical design model for steel and concrete composite members with web openings", *Engineering Structures*, 197, 2019.

[28] Claßen M., Bielak J., Spelter A., "Verbundträger und Verbunddeckensysteme mit Stegöffnungen", *Stahlbau*, 88(7), pp. 685-697, 2019.

[29] Claßen M., Adam V., Kueres D. et al., "Zug-Schub-Interaktion von puzzleförmigen Verbunddübelleisten", *Beton- und Stahlbetonbau*, 112(6), pp. 334-345, 2017.

[30] European Committee for Standardization, *Eurocode 0: Grundlagen der Tragwerksplanung*, DIN EN 1990:2010-12, Brussels, 2010.

[31] Feldmann M., Kopp M., Hegger J. et al., *Endbericht P 1097 - Praxisreifes, integriertes Deckensystem für den Stahl- und Verbundbau*, Düsseldorf, 2017.

[32] Claßen M., *Zum Trag- und Verformungsverhalten von Verbundträgern mit Verbunddübelleisten und großen Stegöffnungen*, Phd thesis, RWTH Aachen University, 2016.

[33] Claßen M., Herbrand M., Adam V. et al., "Puzzle-shaped rib shear connectors subjected to combined shear and tension", *Journal of Constructional Steel Research*, 145, pp. 232-243, 2018.

[34] Claßen M., Hegger J., "Shear Tests on Composite Dowel Rib Connectors in Cracked Concrete", *ACI Structural Journal*, 115 (3), pp. 661-671, 2018.

[35] Claßen M.: "Vorschlag zur Erweiterung der Norm – Proposal for pry-out of composite dowels in cracked concrete", 2017.

[36] Wolters K., Christou G., Broschart Y., Claßen M., Feldmann M., Hegger J., Kurz W., "Entwicklung eines konsistenten Bemessungsmodells für Verbunddübelleisten", *Bauingenieur*, 2021.

[37] Eligehausen R., Mallée R., *Befestigungstechnik im Beton- und Mauerwerkbau*, Ernst & Sohn, Berlin, 2000.

9TH INTERNATIONAL CONFERENCE ON COMPOSITE CONSTRUCTION IN STEEL AND CONCRETE
M. Knobloch, U. Kuhlmann, W. Kurz, M. Schäfer. (Eds.)
Online, Germany, July 27-29, 2021

EXPERIMENTAL AND NUMERICAL INVESTIGATION OF DOWEL STRIPS FOR LONGITUDINAL AND TRANSVERSE LOADING

C. Karge*, R. Stroetmann*, J. Wisnewski*

* TU Dresden, Institut für Stahl- und Holzbau, Dresden, Germany
e-mail: caecilia.karge1@tu-dresden.de, richard.stroetmann@tu-dresden.de, joachim.wisnewski@tu-dresden.de

Keywords: Orthotropic composite slabs, composite dowel strips, experimental tests, structural analysis.

Abstract. *A new field of application for dowel strips are orthotropic composite roadway slabs. In contrast to conventional composite steel bridge construction, the dowel strips create a two-dimensional bond. As a result, they are not only exposed to longitudinal but also to transverse shear forces. Up to now, neither the load-bearing capacity nor the load-deformation behavior under transverse shear forces have been investigated. Furthermore, there are design related deviations from the current approval in form of a lower dowel distance. This paper presents two different types of push-out tests, which provide new insights into small composite dowel strips under longitudinal and transverse loads. Based on these tests and numerical investigations, the effects of smaller dowel strips are analyzed and special features of the load transfer mechanism under transverse shear forces are highlighted. The investigations presented here were carried out as part of the AiF-FOSTA research project P 1265 [1].*

1 INTRODUCTION

Notch sharpness in details and a strongly increased traffic load led to extensive fatigue damage to existing orthotropic steel roadway slabs of bridges. Reinforced concrete bridges and conventional composite bridges are less susceptible to fatigue but considerably heavier. With the so-called ortho-composite-slab (OC-slab) a lightweight and robust construction is currently being developed as an alternative. The construction consists of a modified orthotropic stiffened steel plate and a thin, single-layer reinforced in-situ concrete (Figure 1). Similar roadway slab types were first used by Leonhardt [2] and Robinson [3] in the 1940s and 1950s. However, for most bridges, the concrete was always structurally arranged and not statically involved [4, 5].

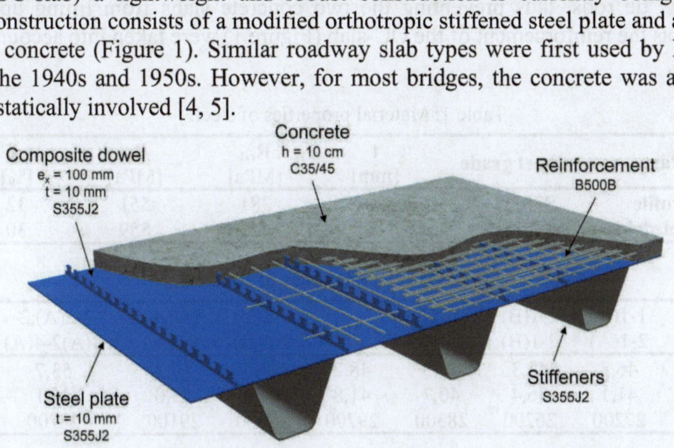

Figure 1. Principal construction of the orthotropic composite slab.

Composite dowel strips (CDS) with a clothoid shape (CL) are provided as shear connectors in the new construction. They are used to transfer the shear forces between the steel plate and the concrete layer. The dowel strips are manufactured from heavy plates with recesses open at the top; they are connected to the cover plate with fillet welds. In Germany, they are currently regulated by the general design approval (GDA) Z-26.4-56 [6]. This includes the determination of the longitudinal shear resistance and the proof of fatigue for dowels in puzzle or clothoid form. The design value of the longitudinal shear load bearing capacity can be determined for the failure modes shear and punching out of concrete as well as for steel failure. According to GDA, dowels with a spacing e_x of 150 to 500 mm are regulated. The dowel height depends on this spacing. Due to the thin top layer of concrete it is necessary to arrange a strip with low height and correspondingly small dowel spacing. With a concrete layer of 10 cm, a dowel spacing e_x of 100 mm results, if the required concrete cover is observed. In addition, unlike in conventional steel composite bridge constructions, the dowel strips create a two-dimensional bond. They are not only exposed to longitudinal shear forces but also to transverse shear forces. So far, neither the load-bearing capacity nor the load transfer mechanism in transverse direction has been investigated.

2 EXPERIMENTAL INVESTIGATIONS ON COMPOSITE DOWEL STRIPS

2.1 Test specimens, production and materials

Two types of shear tests are presented in this paper which are used to investigate the influence of the small tooth spacing and the transverse shear capacity. A total of four identical tests of each type were performed. The results and findings from the experimental tests are shown below.

The test specimens were developed based on the standard push-out test according to Eurocode 4 Annex B [7] for headed stud connectors and those from the research report P 804 [8] with composite dowel strips. Figure 2 shows the individual test specimen. The push-out test in Figure 2a examines the shear resistance in longitudinal direction, the one in Figure 2b the transverse shear resistance of dowel strips. They are welded to the flanges of the steel sections in each case. Due to the relative displacement of the steel section against the concrete slabs, the dowel strips are exposed to shear forces only. For the determination of the longitudinal shear strength, the strips were aligned in the direction of the cylinder compressive force. In order to obtain the transverse shear resistance, the dowel strips were turned by 90° and were thus perpendicular to the cylinder pressure force.

The composite dowel strips were manufactured by autogenous flame cutting. Their material properties and those of the welded profiles can be taken from Table 1. The test setup with a centric steel profile and two adjacent concrete slabs allows the load to be eccentrically transferred. The resulting horizontal forces are absorbed by the tie rods, thus preventing the two concrete slabs from lifting up (Figure 2). The dimensions as well as the reinforcement of the OC-slab (Figure 1) were taken into account in the design of both test specimens.

Table 1: Material properties of steel.

Part	Steel grade	t [mm]	R_{eH} [MPa]	R_m [MPa]	A [%]
Profile	S355J2+N	20	381	551	32
Dowel strips	S355J2+N	10	425	559	30

Table 2: Material properties of concrete.

Series-Specimen (Concrete-Slab)	1-1(A) 2-1(A)	1-1(B) 2-1(B)	1-3(B)	1-3(A)	1-2(B) 1-4(B)	1-2(A) 1-4(A)	2-2(A),2-3(A)2-4(A)	2-2(B),2-3(B) 2-4(B)
$f_{cm,cu}$ [MPa]	46,8	43,3	47,9	48,2	48,6	50,3	53,7	52,1
$f_{cm,cyl}$ [MPa]	41,1	36,4	40,7	41,8	40,0	43,0	46,1	45,8
E_{cm} [MPa]	27200	26200	28300	29700	28600	29100	28700	27800

The push-out test specimens were concreted in horizontal position according to the conditions in the OC-slab. Accordingly, one concrete slab (A) per specimen was concreted first. After two to five days, the specimens were turned over so that the second concrete slab (B) could be produced. Table 2 shows the material properties of the concrete. In addition to the cube and cylinder compressive strength $f_{cm,cu}$ and $f_{cm,cyl}$, the E-modulus E_{cm} was determined for each concrete slab at the time of testing.

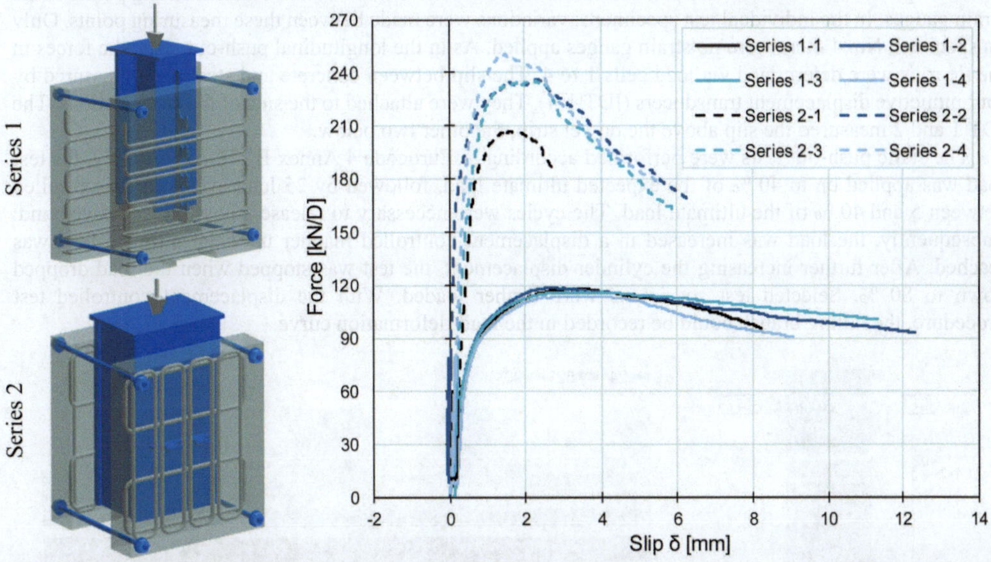

Figure 2. Experimental setup of push-out tests in longitudinal (a) and transverse direction (b) of dowel strips; Force-slip curve of the push-out tests (c).

2.2 Measurement and testing procedure

In the push-out tests, strain gauges (SG) were applied to the dowel strips before concreting. After concreting, shortly before the test was carried out, inductive displacement transducers (IDT) were installed on the test specimen and load cells (LC) were attached to the tie rods.

Figure 3 shows the measurement concept of the push-out tests in longitudinal direction. The application of the strain gauges is shown on the right. They measure the local strains on the composite dowel strips. On the left, the arrangements of the IDT and LC are shown. Each dowel strip is equipped with a total of five strain gauges. The positioning is described by an angle measured in the recess from the origin of the clothoid. Pre-calculations showed that, if steel failure occurs, the crack is to be expected at an angle of 52°. According to this, a strain gauge is located on each dowel at exactly this position. On the first dowel there are two further strain gauges, one at 25°, the other at 90°. From [8] and [9] it can be seen that the greatest stresses in the composite dowels in a beam test occur significantly below 52°. The reason for this is that the dowel strip is stressed by bending of the beam in addition to the shear stress. In order to be able to differentiate the shear stress from the bending stress in the dowel strips in the component tests, a strain gauge was applied at 25°. The strain gauge at 90° was used for comparison purposes with the strain gauge of the push-out tests in transverse direction. In these tests a strain gauge was applied at the same position. All strain gauges were applied at a distance of 5 mm tangential to the outer edge of the clothoid of the dowels. Thus, there was sufficient space to cover the strain gauges in all directions with a protective silicone rubber. The forces in the tie rods were_determined using load cells 1 to 4. The slip between concrete and steel was measured with two inductive displacement transducers at the upper edges of the concrete slabs (IDT 1+2).

The measurement concept of the push-out tests in transverse direction is shown in Figure 4. In contrast to the longitudinal tests, the composite dowels are equipped with strain gauges on both sides. Numerical pre-calculations showed that stresses due to bending and normal force occur in the composite dowels, which should be detected by double-sided applications. Furthermore, the pre-calculation showed that by anchoring the dowels in the concrete, the direction of the bending over the dowel height is reversed. This should be reflected in the strains by a change of sign. Therefore, strain gauges were applied both at the base of the dowel strips and at the level of the smallest width of the dowels. Figure 4 shows all measuring points of the strain gauges. In the individual test specimens, variations were made between these measuring points. Only on specimen No. 4 there were no strain gauges applied. As in the longitudinal push-out tests, the forces in the tie rods were determined via load cells 1 to 4. The slip between concrete and steel were measured by four inductive displacement transducers (IDT 1-4). They were attached to the side of the steel sections. The IDT 1 and 2 measured the slip above the dowel strip, the other two below.

The static push-out tests were performed according to Eurocode 4 Annex B [7]. Accordingly, the test load was applied up to 40 % of the expected ultimate load, followed by 25 load cycles force-controlled between 5 and 40 % of the ultimate load. The cycles were necessary to release a possible adhesive bond. Subsequently, the load was increased in a displacement controlled manner until the ultimate load was reached. After further increasing the cylinder displacement, the test was stopped when the load dropped down to 80 %. Selected test specimens were further loaded. With the displacement controlled test procedure, the falling branch could be recorded in the load-deformation curve.

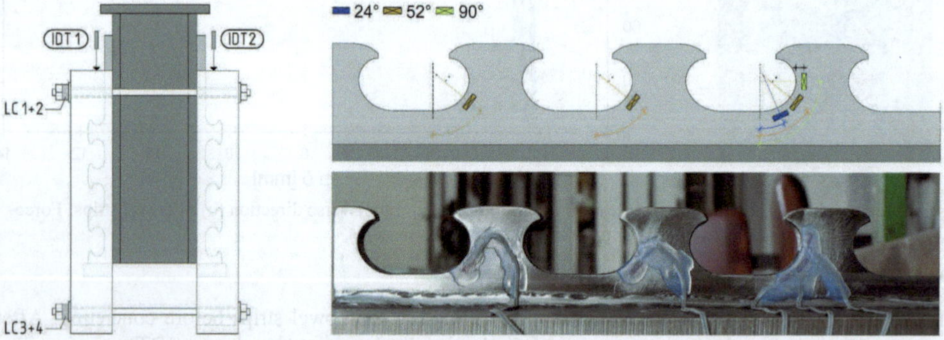

Figure 3. Measurement concept of push-out tests in longitudinal direction a) Arrangement of inductive displacement transducers (IDT) and load cells (LC), b) Arrangement of strain gauges (SG).

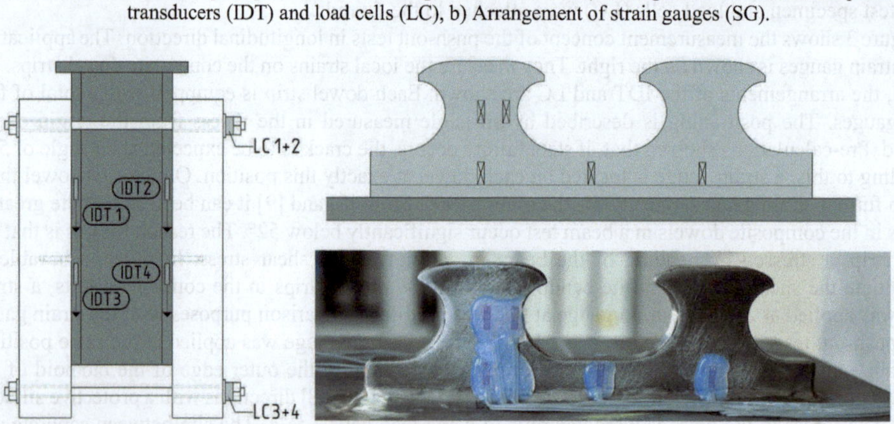

Figure 4. Measurement concept of transverse push-out tests a) Arrangement of IDT and LC, b) Arrangement of SG.

2.3 Evaluation of the shear force-slip curves

Figure 2 shows the shear force-slip curves of all push-out tests. The evaluation was performed according to EN1994-1-1, Annex B [7]. The characteristic resistance P_{Rk}, ductility and stiffness C of the composite dowels were determined by the shear force-slip curves. The characteristic resistance was calculated as 90 % of the ultimate load. The smallest value from all tests becomes decisive [7]. In longitudinal direction, a characteristic shear resistance of 103 kN/D (force per dowel) could be determined. The force-slip curves of the push-out tests in transverse direction show a significant difference between test specimen 1 and the specimens 2, 3 and 4. This is due to the different concrete compressive strengths. While series 2-1 has an average cube compressive strength of 45 MPa, this is 53 MPa for the other tests. The results of series 2 are thus evaluated separately by concrete batch. Under transverse shear force, the dowel strip of the first test specimen reaches a characteristic resistance of P_{Rk} = 187,4 kN/D. The other three test specimens achieved a higher value of 213,35 kN/D, corresponding to the higher concrete compressive strength. If the longitudinal and transverse shear resistance are compared, it becomes clear that the transverse shear capacity is 1,82 times higher than the longitudinal shear capacity.

To evaluate the ductility, the slip is read off when the characteristic shear capacity P_{Rk} is reached again. The characteristic value of the slip δ_{uk} is given by $0.9 \cdot \delta_u$. According to EN1994-1-1 a composite dowel can be classified as ductile if δ_{uk} is at least 6 mm. In the longitudinal direction, a δ_{uk} of 5,45 mm could be determined. The deformation capacity of the dowel strips under transverse shear stresses is significantly lower with a δ_{uk} of 2,64 mm. While the dowel strip with small dowel spacings shows a pronounced deformation capacity in the longitudinal direction, it can still not be classified as ductile either in longitudinal or transverse direction. However, in view of the deformation capacity requirements for OC-slabs, there is no need to comply with the limit derived for composite beams in building construction.

The initial stiffnesses in the longitudinal direction are between 3400 and 4200 kN/cm, while the initial stiffnesses under transverse shear load are 6100 to 8800 kN/cm. Thus, in the transverse direction the strips behave much more rigidly. The results from the shear force-slip curves are summarized in Table 3.

Table 3. Results of the shear force-slip curves.

Series	1-1	1-2	1-3	1-4	2-1	2-2	2-3	2-4
P_{max} [kN/Vm]	119,01	119,90	116,82	114,45	208,22	238,24	237,05	251,49
P_{Rk} [kN/Vm]		103,00			187,40		213,35	
δ_u [mm]	6,53	6,74	9,2	6,06	2,46	3,60	2,92	2,68
δ_{uk} [mm]		5,45			2,21		2,41	
Z_{Sta} [kN]	2,25	0,09	1,05	0,24	3,49	6,60	4,43	1,55
C [kN/cm]	4164	3793	3744	3384	6126	8763	6196	6629

2.4 Failure modes

Figure 5a shows the crack pattern (traced) on the concrete surface after a push-out test in longitudinal direction. From the outside, a crack can be seen in line with the dowel strip. This splitting crack occurred before the maximum load was reached. Further growth was prevented by the reinforcement between the dowels. Thus, the crack width is small compared to other cracks. Figure 5b shows the results of selected strain gauges from a push-out test in longitudinal direction. The strains are shown as a function of dowel force. Up to a force of 90 kN/D, the steel behavior is linear elastic. Beyond this, plastic deformations occur in the dowel strip at the measuring points at 52 and 90°. Only the strain gauges at 24° showed linear elastic behavior over the entire test period. An image of the deformed dowel strip can be seen in Figure 5c. The individual dowels show a clear rotation due to the previous shear stress. A crack in the dowel strips could not be detected in any test specimen. The strains at maximum load were also well below the ultimate strain shown in Table 1. After the end of the test, the concrete slabs were sawed open along the strip. In the saw cut in Figure 5c another crack can be seen above the dowel strip. This crack indicates a spalling of the concrete cover. On some test specimens, this was already apparent on the surface (Figure 5a, crack with w = 4 mm). Only numerical calculations and a comparison with the equations of the GDA showed that this is a special form of punching. If the material parameters used from the tests are inserted into the design equation of the GDA, a punching failure of the concrete with a load-bearing capacity of 102 kN/D becomes decisive. If this resistance is compared with the 103 kN/D from the tests, the values agree very well. Due to the very small dowel spacing of only 100 mm, the punching cones overlap, which in turn causes large-scale spalling of the concrete cover. Figure 5d visualizes this failure mode.

Selected results of the strain gauges from the push-out tests in transverse direction are shown in Figure 6. In pairs, two strain gauges opposite each other have the same color and differ in line type. Strain gauges

11 and 12 are located at the base of the dowel strip. Strain gauge 11 shows positive strain, while strain gauge 12 shows compression. The change of sign at the foot of the dowel strip indicates the bending stress due to the transmission of the shear force. Strain gauge 12 did not provide any measured values above 190 kN/D, but strain gauge 11 shows that irreversible deformations occur at the foot of the dowels even before the maximum force is reached. The image of the deformed dowel strip in Figure 7 confirms this measurement. Gauges 17 and 18 are directly opposite each other at the level of the smallest width of the dowel. Here, too, strains can be seen on one side and compressions on the other. The comparison of strain gauges 11 and 17, which are both located on the side facing the load, show that a change in sign occurs over the height of the dowel strip. While strain was measured at the base of the dowel strip, compression was measured at the smallest width of the dowel. The reverse situation can be seen on the side facing away from the load. The reason for the change in sign of the strains over the dowel height is the load-bearing mechanism of the dowel strip under transverse shear load. Due to the relative displacement of the concrete against the dowel strip, it is subjected to shear and bending at the base. The geometry of the strip allows the dowels to anchor themselves in the concrete. A supporting force was created below the head of the dowel, which causes the dowel to hang back, similar to the headed stud connectors [12]. A comparison of the amounts of strain gauges on opposite sides shows that a small tensile component was generated in addition to the stresses due to bending in the dowel strip. The strains at the smallest width of the dowel are in the range of 0,2 % when the ultimate load is reached and beyond. The plastic deformations are too small to be detected on the dowel after the end of the test (Figure 7).

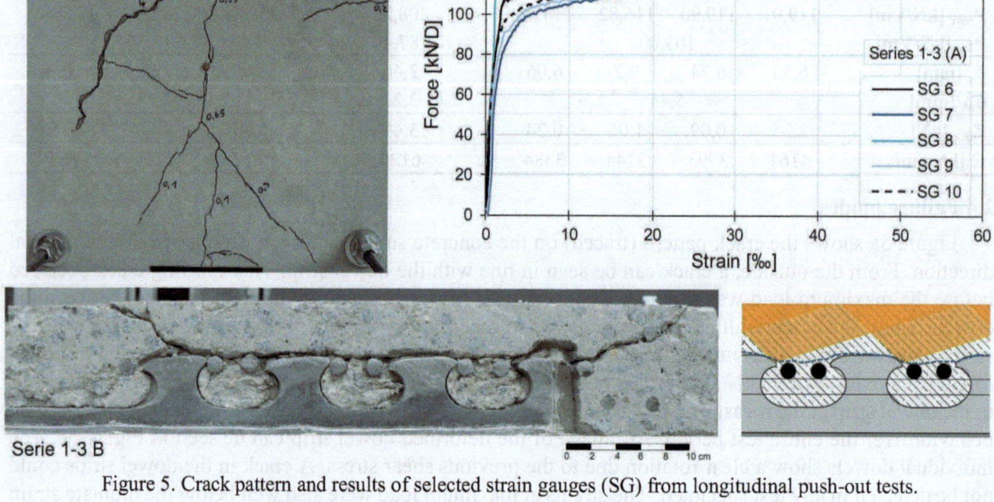

Figure 5. Crack pattern and results of selected strain gauges (SG) from longitudinal push-out tests.

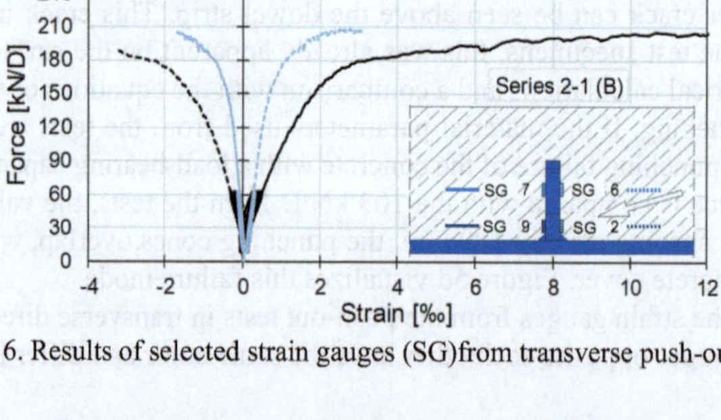

Figure 6. Results of selected strain gauges (SG)from transverse push-out tests.

Figure 7 shows the failure mode of the push-out tests in transverse direction. In the left figure, the position of the photo in the center of the concrete chord is marked in green. It shows that the concrete was pushed together in front of the base of the dowel strip. On some test specimens, the destroyed concrete came loose and was visible as a concrete wedge in front of the dowel strip after the removal of the concrete slabs (Figure 7, right). A multi-axial compressive stress condition develops on the load-facing side of the dowel strip. Under these conditions, concrete can absorb many times its uniaxial compressive strength. If the multi-axial concrete compressive strength is exceeded in this area, the concrete pulverizes and the load drop occurs in the shear force-slip curve.

Figure 7. Transverse push-out tests: Concrete compression failure mode in front of the dowel base.

3. APPLICATION LIMITS OF DUCTILE AND NON-DUCTILE CONNECTORS

In general, both EN 1994-1-1 [7] for buildings and EN1994-2 [10] for bridges require that „shear connectors shall have sufficient deformation capacity to justify any inelastic redistribution of shear assumed in design". The earlier mentioned 6 mm criterion defines when a shear connector has sufficient deformation capacity. In building construction there are composite girders with a full or partial shear connection. In order to limit the slip to a maximum in the case of a partial shear connection, a minimum degree of shear connection has been introduced. The formation of the partial shear connection is not permitted for composite bridges according to EN1994-2 [10].

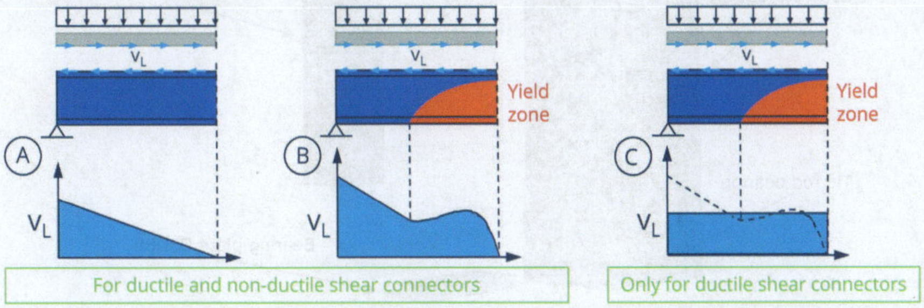

Figure 8. Longitudinal shear forces in the composite joint according to [11].

Although the 6 mm criterion with a partial shear connection is to be assigned to buildings, sufficient deformation capacity is nevertheless required in bridge constructions. Therefore, it remains to be clarified when a plastic redistribution of longitudinal shear forces is to be assumed in bridge constructions. Figure 8 shows the distribution of longitudinal shear forces using a composite girder as an example. Figure A shows the longitudinal shear force distribution according to the elasticity theory. It is affine to the shear force curve. This calculation method is permissible in both building construction and bridge construction for all cross-section classes and in all limit states. The plastic moment capacity of the beam is utilized in Figure B. The longitudinal shear force curve corresponds to the curve taking into account the nonlinear material behavior. In the area of the yield zone of the beam, the longitudinal shear forces increase. The determination according to this scheme is permissible for bridges as well as in building construction, but only for calculations in ULS and if the cross-section belongs to cross-section class 1 or 2. Figure C also exploits the nonlinear material behavior and thus the plastic load-carrying capacity of the girder. However, a plastic redistribution of the longitudinal shear forces takes place and a uniform constant course occurs. The shear connectors are arranged equidistantly for this purpose.

In bridge construction, a section-by-section equidistant distribution of the composite dowels is permissible in those areas where the design value of the acting longitudinal shear force does not exceed the shear capacity by more than 10 %. In addition, it shall be verified that the total longitudinal shear force acting in the area with equidistant arrangement is not greater than the shear capacity resulting from the total number of dowels in this area. When determining the longitudinal shear forces, a distinction is made between girders of classes 1 and 2 and those that are designed elastically or strain-limited. These rules already ensure that the requirements for ductility are far lower than in building construction when the partial composite is used. In the case of OC-slabs, the specially adapted dowel strips meet the ductility criterion of 6 mm for longitudinal shear in a good approximation, even though they are not intended for the partial composite. The design under transverse loading will be carried out without redistribution of the shear forces, so that the ductility determined in the tests is adequate.

4 NUMERICAL SIMULATIONS OF PUSH-OUT TESTS

4.1 Modeling and material models

The experimental push-out tests were simulated with a three-dimensional, non-linear FE-model. The aim was to calculate both tests with the same configurations in order to use the models for future parameter studies. The simulations were performed with the program Abaqus-Standard [13].

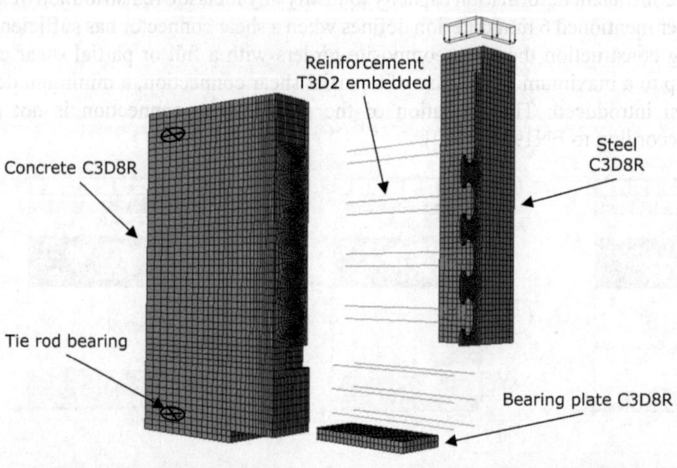

Figure 9. Modeling example on the push-out test in longitudinal direction.

Taking advantage of the double symmetry of the push-out tests, only a quarter of the specimens was modeled at a time (Figure 9). Steel and concrete were modeled using C3D8R solid elements and the reinforcement as embedded bars in the concrete. T3D2 elements, called truss elements, were used for the reinforcing bars. Consequently, they can only support axial forces. By embedding the bars in the concrete, the slip that occurs in reality between concrete and reinforcement cannot be taken into account. This saves calculation time and provides sufficiently accurate results. The aim was to make the bearing as realistic as possible, so that the model was placed on a bearing plate. Thus, the absorption of tensile forces at the support could be excluded. For the contact relationships between steel and concrete, a distinction was made between tangential and normal directions to the surface. In normal direction, a "hard contact" was defined. This means that only contact pressure can be transmitted between steel and concrete. In tangential direction, a friction contact was defined with $\mu = 0,3$ as a coefficient of friction. The load was applied displacement controlled to the welded profile in the simulation.

Material models used in the FE model					
Concrete Concrete Damaged Plasticity	uniaxial compression	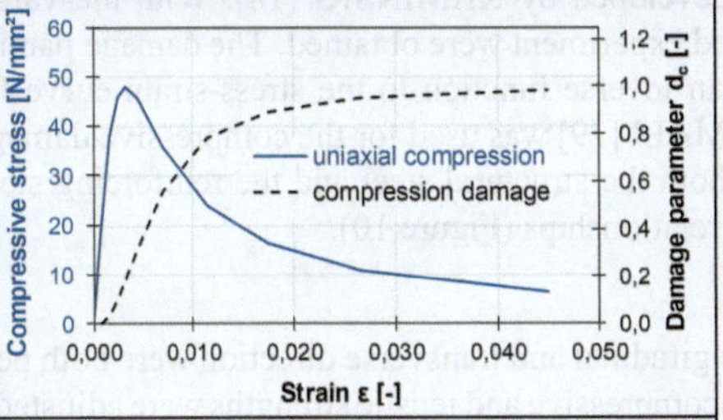	Material properties and CDP-Parameter	Compr. strength	$f_{cm} = 48$ N/mm²
				Tens. strength	$f_{ct} = 4{,}94$ N/mm²
				E-Modul	$E_{cm} = 26700$ N/mm²
				Poisson's ratio	$\nu = 0{,}2$
				Ductil. para.	$D = 1{,}0$
				Strain at f_{cm}	$\varepsilon_{c1} = 0{,}0032$
	uniaxial tension	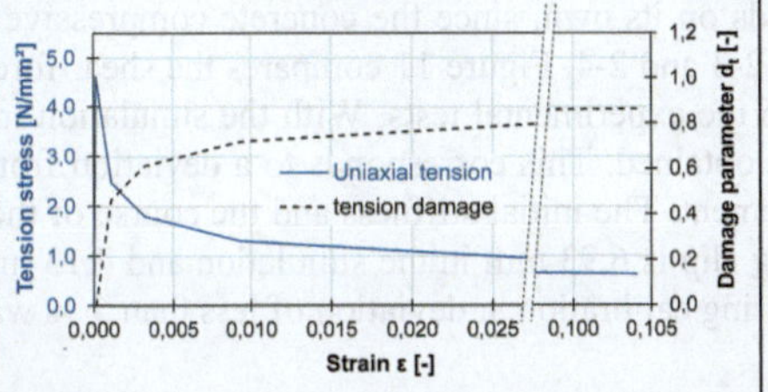		Tension/Compr. meridian	$K_c = 0{,}67$
				Eccentricity	$\varepsilon_x = 0{,}2$
				Dilation angle	$\psi = 38°$
				Strength bi-/uniaxial	$f_{b0}/f_{c0} = 1{,}16$
Steel elastic-plastic	Structural steel	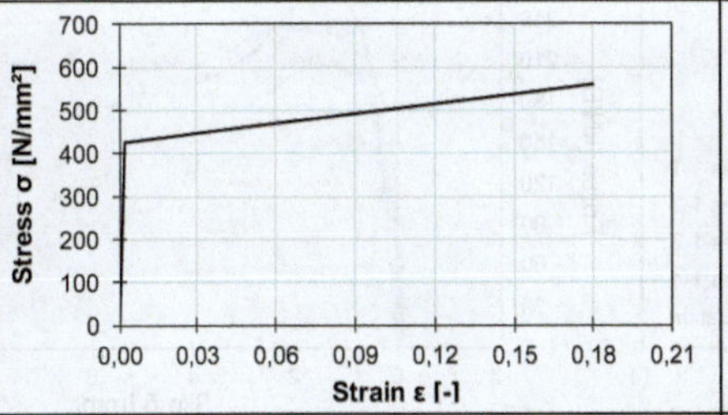	Material properties S355	Yield strength	$f_y = 425$ N/mm²
				Tensile strength	$f_u = 559$ N/mm²
				E-Modul	$E = 210000$ N/mm²
				Poisson's ratio	$\nu = 0{,}3$
	Reinforcing steel	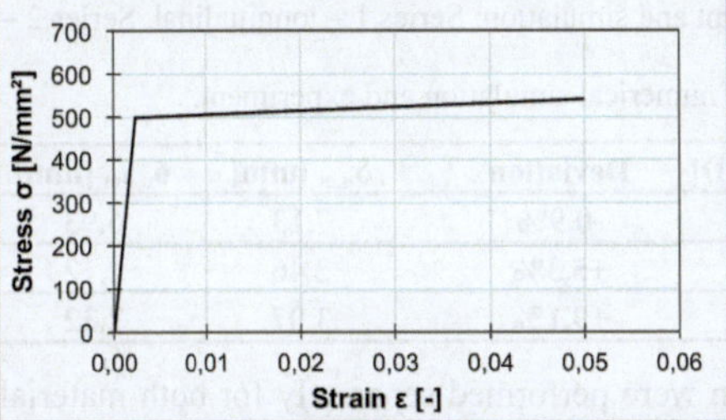	Material properties B500B	Yield strength	$f_y = 500$ N/mm²
				Tensile strength	$f_u = 542$ N/mm²
				E-Modul	$E = 210000$ N/mm²
				Poisson's ratio	$\nu = 0{,}3$

Figure 10. Material models used for concrete and steel in the FE model for the longitudinal direction.

Figure 10 shows the properties for the material models of concrete, structural steel and reinforcing steel. The concrete compressive strength, the modulus of elasticity of the concrete and the yield strength of the structural steel correspond to the material properties from the longitudinal push-out tests. The "Concrete Damaged Plasticity Model" (CDP model) implemented in Abaqus was used to describe the nonlinear material behavior of concrete. Cracks are displayed smeared in this model. Several research institutions obtained very good results with it in the investigation of composite dowel strips (e. g. [13, 14]). For the implementation of the CDP model, the behavior of concrete under uniaxial tensile and compressive loading was necessary. For the representation of the uniaxial compressive stress-strain curve, the approach according to SARGIN [16] was used. By specifying the ductility parameter D, the ductility of the concrete after exceeding the compressive strength is controlled. The parameter D can be between 0 and 1. The nonlinear behavior of concrete under uniaxial tensile loading was captured according to WANG [17] with an exponential function, which was further developed by KAMINSKI [18]. With the value 0.3 for the exponent, the best agreements of simulation and experiment were obtained. The damage parameters, which are part of the CDP model, were defined as an inverse function to the stress-strain curve for the tensile damage d_t, and the approach according to KAMALI [19] was used for the compressive damage d_c. Elastic-plastic material behavior was considered for both the structural steel and the reinforcing steel. A bilinear approximation was used for both stress-strain relationships (Figure 10).

4.2 Model verification

The simulations of the push-out tests in longitudinal and transverse direction were both performed with the same modeling. Only the uniaxial concrete compressive and tensile strengths were adjusted to the results from the accompanying tests. Series 2-1 stands on its own, since the concrete compressive strength was significantly smaller than those of series 2-2, 2-3 and 2-4. Figure 11 compares the shear force-slip curves obtained from the simulation with those from the experimental tests. With the simulation in longitudinal direction, a maximum load of 117 kN/D was obtained. This corresponds to a deviation from the tests of less than 1 % and thus shows very good agreement. The initial stiffness and the course of the shear force-slip curve also agree very well. The governing slip is 6,93 mm in the simulation and 7,13 mm in the test. The deviation is 2,8 %. As a limit criterion during calibration, a deviation of less than 5 % was considered permissible.

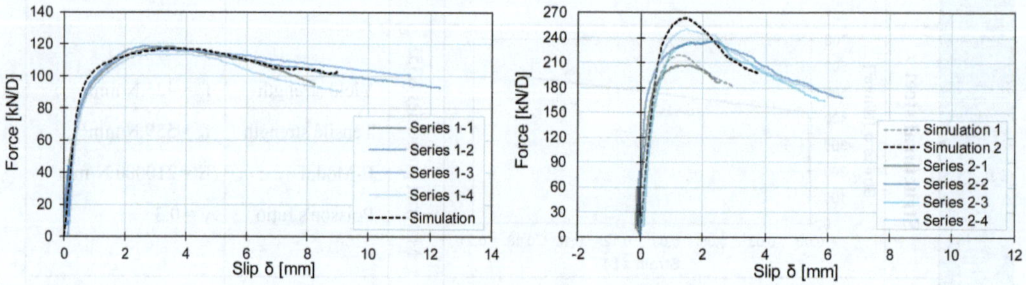

Figure 11. Shear force-slip curves from experiment and simulation: Series 1 – longitudinal, Series 2 – transverse.

Table 4: Comparison of numerical simulation and experiment.

Series	$P_{max,m}$ [kN/D]	$P_{max,sim}$ [kN/D]	Deviation	$\delta_{u,m}$ [mm]	$\delta_{u,sim}$ [mm]	Deviation
1-1 to 1-4	118	117	**-0,9%**	7,13	6,93	**-2,8%**
2-1	208	219	**+5,3%**	2,46	2,12	**-13,8%**
2-2 to 2-4	242	264	**+9,1%**	3,07	2,32	**-24,4%**

The FE calculations in transverse direction were performed separately for both material groups. The results show good agreement, but still need to be refined. The deviations of the maximum loads are 5,3 and 9,1 %. For the deformation capacity, the deviations are larger with 13,8 and 24,4 %. A comparison of the results from experiment and simulation is summarized in Table 2. Despite the deviations in the deformation capacity of the push-out tests in transverse direction, the suitability of the selected material models, discretization, and contact conditions was confirmed. The models are suitable for parameter studies.

Figure 12 shows a comparison of the crack formation of numerical simulation and test in the specimen interior of the longitudinal push-out test. The crack suggesting spalling of the concrete cover was also generated in the FE calculation. Figure 12 shows the deformations in the y-direction on the right. With a slip between concrete and steel of 3 mm, deformation occurs in the concrete exactly at the tips of the dowels. At a displacement of 5 mm there is a formation of a punching cone at each individual composite dowel. From this it can be deduced that the results are to be classified in the punching-out failure mode. At a slip of 7 mm, it can be seen what occurred at the end of the test. The concrete cover detaches completely above the transverse reinforcement due to the superposition of the punching cones.

The failure mode in the push-out tests in transverse direction is the exceeding of the multi-axial concrete compressive strength in front of the dowel base. Figure 13 shows the failure pattern from the test and the simulation. This failure mode can be captured very well with the structural mechanic calculation.

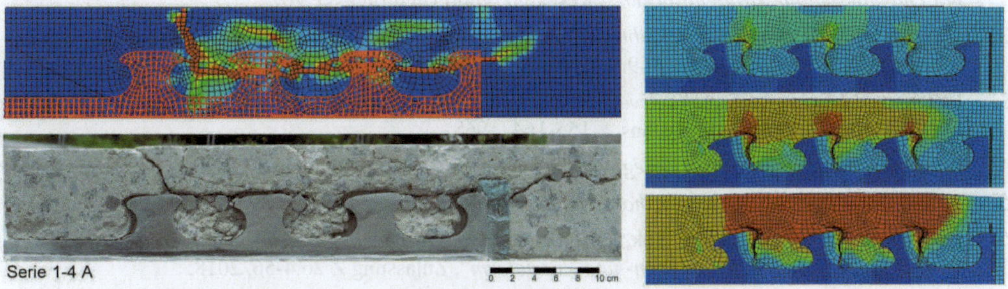

Figure 12. Saw cut and crack pattern from test and FE calculation (left), deformations from FE calculation at $\delta = 3$ mm, 5 mm and 7 mm with cant factor = 5 (right).

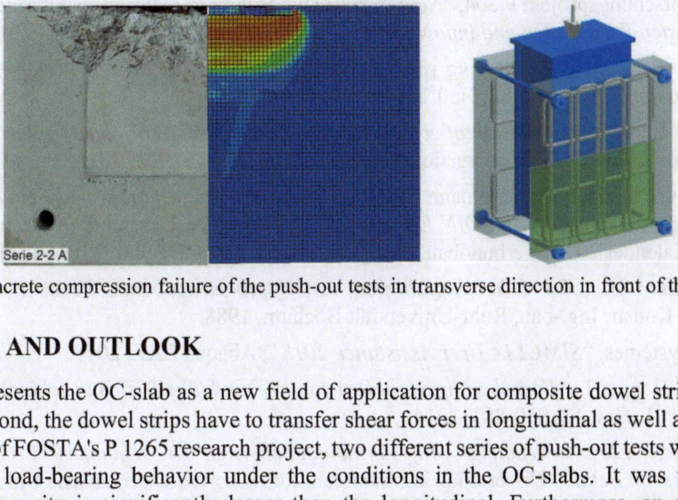

Figure 13. Concrete compression failure of the push-out tests in transverse direction in front of the dowel strip.

5 SUMMARY AND OUTLOOK

This article presents the OC-slab as a new field of application for composite dowel strips. Due to the two-dimensional bond, the dowel strips have to transfer shear forces in longitudinal as well as in transverse direction. As part of FOSTA's P 1265 research project, two different series of push-out tests were conducted to investigate the load-bearing behavior under the conditions in the OC-slabs. It was found that the transverse shear capacity is significantly larger than the longitudinal. Furthermore, an analogy of the bearing behavior of dowel strips in the transverse direction to that of headed stud connectors was established. Based on the experimental tests, numerical models were created and verified. The models serve as a basis for further parameter studies with different material and geometry configurations. Furthermore, tests are planned that not only investigate the local loading of the dowel strips, but also large-scale static and cyclic tests that map the global loading in the roadway slab.

ACKNOWLEDGMENTS AND REFERENCES

The IGF project 20255 BR / FOSTA P1265 " Wirtschaftliche und dauerhafte Orthoverbundfahrbahnplatten mit Dübelleisten für Straßenbrücken" of FOSTA – Forschungsvereinigung Stahlanwendung e. V., Düsseldorf, was funded by the German Federal Ministry for Economic Affairs and Energy via the AiF as part of the program for the promotion of Collective industrial research (IGF) on the basis of a resolution of the German Bundestag. The project was carried out at the Institute of Steel and Timber Construction of the Technische Universität Dresden. The authors would like to thank the BMWi for financial support, FOSTA for taking over the sponsorship, organization and handling, as well as all members of the committee accompanying the project for their active participation and support in the form of provision of materials, production of test specimens and support constructions for carrying out the tests.

[1] FOSTA-Forschungsprojekt P 1265, "*Wirtschaftliche und dauerhafte Orthoverbundfahrbahnplatten mit Dübelleisten für Straßenbrücken*", Research report (in progress), 2022.

[2] Leonhardt F., "*Die neue Straßenbrücke über den Rhein von Köln nach Deutz*", Die Bautechnik Jahrgang 26, Heft 7, 193-199, 1949.

[3] Fauchart J., Sfintesco D., "*The developement and use of the Robinsons composite deck in France*", IABSE Rep. Working Commissions 2, 155-164, 1968.

[4] Mansperger T., Jung R., "*Neue Wege im Verbundbrückenbau*", Stahlbau 87, Heft 8, 752–758, 2018.

[5] Jung R., Mansperger T., "*Die Orthoverbund-Fahrbahnplatte*", Stahlbau 89, Heft 2, 129-137, 2020.

[6] Deutsches Institut für Bautechnik, „*Allgemeine Bauartgenehmigung für Stahlverbundträger mit Verbunddübelleisten in Klothoiden- und Puzzleform*", Zulassung Z 26.4-56, 2018.

[7] EN 1994-1-1 Eurocode 4, "*Design of composite steel and concrete structures – Part 1-1: General rules and rules for buildings*", German version EN 1994-1-1:2004 + AC:2009, 2010.

[8] FOSTA-Forschungsprojekt P 804, "*Neue Systeme für Stahlverbundbrücken – Verbundfertigteilträger aus hochfesten Werkstoffen und innovativen Verbundmitteln*", Research report, 2014.

[9] Gündel M. et al., "*Die Bemessung von Verbunddübelleisten nach neuer Allgemeiner bauaufsichtlicher Zulassung*", Stahlbau 83, Heft 2, 112-121, 2014.

[10] EN 1994-2 Eurocode 4, "*Design of composite steel and concrete structures – Part 2: General rules and rules for bridges*", German version EN 1994-2:2005 + AC:2008, 2010.

[11] Hanswille G., Schäfer M., Bergmann M., "*Verbundtragwerke aus Stahl und Beton, Bemessung und Konstruktion – Kommentar zu DIN EN 1994-1-1*" – Ausgabe Dezember 2010. In: Kuhlmann, U., Stahlbau Kalender 2018, Verbundbau, Fertigung. Berlin Ernst & Sohn, 2018.

[12] Lungershausen H., "*Zur Schubtragfähigkeit von Kopfbolzendübeln*", Tech.-wiss. Mitteilungen / Institut für Konstr. Ing.-bau, Ruhr-Universität Bochum, 1988.

[13] Dassault Systèmes, "*SIMULIA User Assistance 2018*", Abaqus CAE, 2018.

[14] Claßen M., Hegger J., "*Verankerungsverhalten von Verbunddübelleisten in schlanken Betongurten*", Bautechnik 91, Heft 12, 869-883, 2014.

[15] Lechner T., "*Zur Anwendung von Verbunddübelleisten in schlanken Verbundträgern aus ultrahochfestem Beton*", Dissertation am Lehrstuhl für Massivbau, Technische Universität München, München, 2018.

[16] Stempniewski L., Eibl J., "*Finite Elemente im Stahlbeton*", Betonkalender Band 85, 577-647, 1996.

[17] Wang H., "*Nonlinear finite element analysis of concrete structures using new constitutive models*", Computers and Structures 79, 2781-2791, 2001.

[18] Kmiecik P., Kaminski M., "*Modelling of reinforced concrete structures and composite structures with concrete strength degradation taken into consideration*", Archives of Civil and Mechanical Engineering XI, 623-636, 2011.

[19] Kamali A. Z., "*Shear Strength of Reinforced Concrete Beams subjected to Blast Loading*", Master Thesis, Stockholm, 2012.

9TH INTERNATIONAL CONFERENCE ON COMPOSITE CONSTRUCTION IN STEEL AND CONCRETE
M. Knobloch, U. Kuhlmann, W. Kurz, M. Schäfer. (Eds.)
Online, Germany, July 27-29, 2021

LONGITUDINAL SHEAR TRANSFER IN COMPOSITE STEEL TRUSS AND CONCRETE (CSTC) BEAMS

Valentino Vigneri*, Robert Kroyer*, Stefano China, Alessio Argentoni** and Andreas Taras***

*ETH Zurich, D-BAUG, IBK, Steel and Composite Structures, 8093 Zurich, Switzerland
vigneri@ibk.baug.ethz.ch, kroyer@ibk.baug.ethz.ch, taras@ibk.baug.ethz.ch

** Tecnostrutture s.r.l - Via A. Meucci 26, 30020 Noventa di Piave (VE) Italy
schina@tecnostrutture.eu, aargentoni@tecnostrutture.eu

Keywords: Composite Beams, Shear Connection, CSTC, Steel Truss, Slim-Floor Construction.

Abstract. *The Composite Steel Truss and Concrete (CSTC) beam represents an effective and structurally economical solution which can look back to a long tradition of successful use in Italy. This type of structural element consists of a truss made of structural steel which is partly or entirely encased in in-situ cast concrete. The shear transfer mechanisms activated in these elements do not rely on dedicated shear connectors but on a mechanical interlocking between the truss and the surrounding concrete. In order to investigate the shear transfer mechanism and to support the development of appropriate design rules, 39 push-out tests have been planned in accordance with EN 1994-1-1 B.2 testing procedure and they are presented in this work. Also, a finite element (FE) model was developed to study in detail the stress and strain field of the specimens. This paper will present the findings of the experimental campaign and numerical studies on the longitudinal shear transfer in CSTC beams.*

1 INTRODUCTION

The last few decades have seen the emergence of a large variety of different structural solutions for steel-concrete composite beams, which generally make use of prefabricated elements made of structural steel that are complemented by concrete and interact with the latter through composite action. The shear connection mechanism between the components is certainly a key aspect in composite construction. Whilst some shear connection mechanisms and steel-concrete composite solutions have become common and efficient solutions for beams under high loads, the popularity of composite construction in the field of floors is still developing. This is certainly also due to a longer research history in the field of composite beams than in composite floors. However, there are several well-known alternatives, especially for applications in slim-floor beams [1]. Among the wide range of alternatives, the so-called composite steel trussed-concrete (CSTC) beams, also known as HSCTC beams, have become popular structural solutions in light industrialization in Europe, Asia, and Central and South America during the last 50 years. As the name implies, CSTC beams consist of a self-supporting system made of a structural steel truss welded to a steel plate. In addition, upper and bottom chords may be also included. The prefabricated steel elements can bear their own self-weight as well as the weight of the slab without any provisional support during the construction stage enabling a quick erection of moment resisting frames without using any formwork on the construction site. Such systems do not only have high speeds of installation and construction, but they also provide good structural performance in bending allowing to cover wide span lengths with relatively shallow depths. These advantages lead to a highly competitive and cost-effective solution which is currently widely used in the design of buildings. Among the large variety of CSTC beams currently available on the market, the current work focuses mainly on those with a sinusoidal bar welded directly on the bottom steel plate and upper chords which will be fully embedded in concrete. A typical CSTC solution with a single serpentine is displayed in Figure 1. However, the truss system can vary significantly depending on several parameters such as geometry and number of the serpentines, transversal spacing, number of upper chords.

Figure 1. Typical CSTC beam (before casting concrete) with a single sinusoidal steel truss system.

2 MOTIVATION

Unlike in typical composite beams with downstand steel profiles, in CSTC beams the longitudinal shear force is not transferred via dedicated shear connectors (e.g. headed studs or bolts). The mechanism of shear transfer is rather activated at the interface between the diagonal steel bars and concrete resulting in a stress field in either of these elements that differs from the one near shear studs and it influences the overall structural behaviour more directly. However, notwithstanding that the CSTC beams have been widely used in the last decades as structural solutions in buildings, the transfer of the longitudinal shear has not yet been investigated thoroughly with the aim of developing appropriate and reliable design rules for its verification, especially for modern types of CSTC systems. There are only a few experimental push-out campaigns that aimed to study the shear transfer mechanisms between steel elements and concrete in CSTC beams. However, as can be seen in the layout of these tests [2], the upper chords can potentially transfer part of the longitudinal shear force at the extremities which may somewhat compromise the results as the upper chords are continuous along the span length. Therefore, in order to prevent this phenomenon which may compromise the global response of the specimen, more appropriate push-out tests were recently designed by the authors and are presented in this contribution. Further to the experimental campaign described in Section 3, a finite element (FE) model was also developed in order to analyse in detail the resistance mechanisms involved in the shear transfer. Finally, the comprehensive understanding of the longitudinal shear transfer in such systems would not only guarantee the composite action of the members but it would also support the application and future development of design guidelines for CSTC beams.

3 EXPERIMENTAL CAMPAIGN

3.1 Overview and test specimens

Push-out tests represent a simple and convenient way to assess the performance of shear connectors in composite steel-concrete beams such as headed studs or bolts. Notwithstanding that the composite system considered in the current work does not make use of dedicated shear connectors such as headed studs or bolts, the same guidelines on the specimen preparation and testing procedure given in EN 1994-1-1 B.2 [3] were extended and adapted to the configurations considered in this study, as visualized in Figure 2.

The experimental campaign consists of 13 test series of 3 nominally identical specimens each for a total of 39 tests. The main mechanical and geometrical properties (Figure 3) of each series are detailed in Table 1. The width of the slabs *B* was kept 900 mm and its length was 1000 mm with H_{slab} between 290 and 335 mm depending on the configuration. Unlike "type A" specimens, in the other samples the upper part of the sinusoidal bar was cut out. The only difference between type "B" and "C" specimens lies in the fact that the latter possesses additional bottom chords welded.

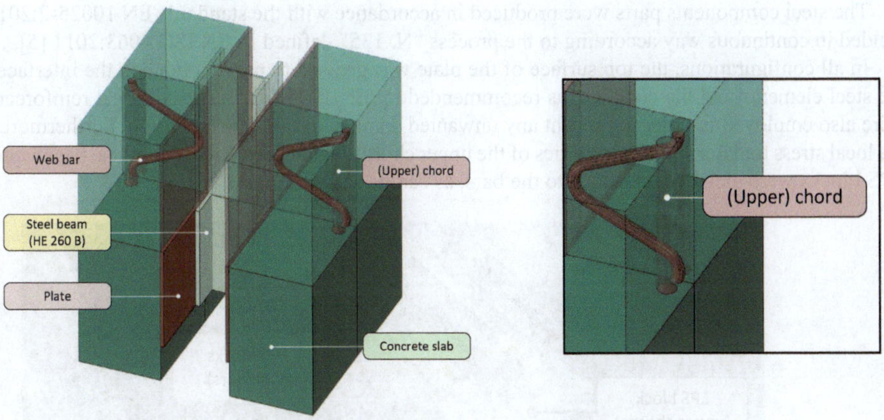

Figure 2. 3D view of a typical type "A" push-out test specimen.

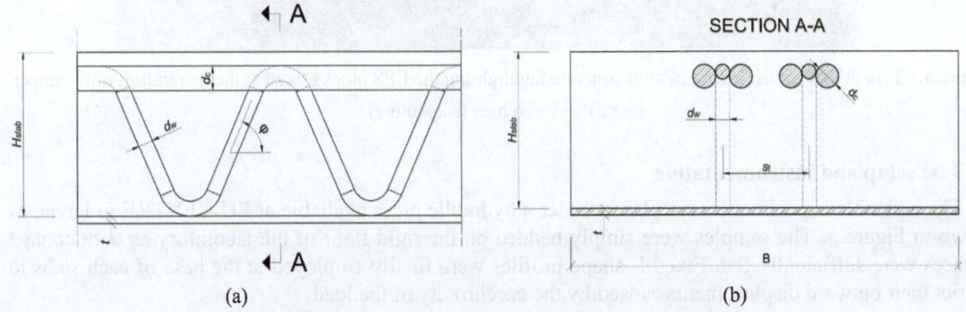

Figure 3. Main geometrical properties of the test samples.

Table 1: Data of the push-out test series.

Series	Web bar (V-shaped rebar)				Upper chord	Plate	Slab
	Diameter d_w [mm]	Number (Transv.)	Inclination angle φ [deg]	Transversal spacing s_t [mm]	Diameter d_c [mm]	Thickness t [mm]	Strength class
D32-1-C25-A	32	1	66°	-	52	15	C25/30
D32-1-C45-A	32	1	66°	-	52	15	C45/55
D32-2-C25-A	32	2	66°	Contiguous	52	15	C25/30
D32-S20-C25-A	32	2	66°	200	52	15	C25/30
D20-1-C25-A	20	1	63°	-	26	10	C25/30
D20-1-C45-A	20	1	63°	-	26	10	C45/55
D20-2-C25-A	20	2	63°	Contiguous	26	10	C25/30
D20-S125-C25-A	20	2	63°	125	52	10	C25/30
D32-1-C25-B	32	1	66°	-	-	15	C25/30
D32-2-C25-B	32	2	66°	Contiguous	-	15	C25/30
D32-3-C25-B	32	3	66°	Contiguous	-	15	C25/30
D32-1-C25-C	32	1	66°	-	52 (Bottom)	10	C25/30
D20-1-C25-C	20	1	63°	-	26 (Bottom)	10	C25/30

The steel components parts were produced in accordance with the standards EN 10025-2:2019 [4] and welded in continuous way according to the process "N. 135" defined in EN ISO 4063:2011 [5].

In all configurations, the top surface of the plate was greased to prevent bond at the interface between the steel elements and the concrete, as recommended by EN 1994-1-1 B.2. Additional reinforcement bars were also employed in order to prevent any unwanted damage during transportation. Furthermore, to avoid the local stress transfer at the extremities of the upper chords which does not occur in real beam applications, EPS blocks were included and fixed to the bars, as can be seen in Figure 4.

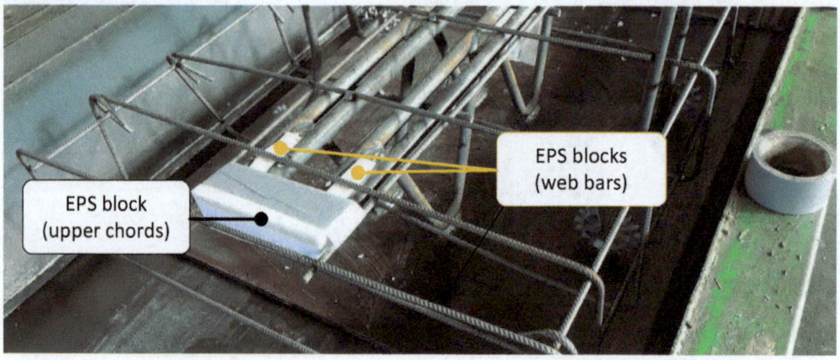

Figure 4. "Type A" specimen before casting concrete highlighting the EPS blocks fixed at the extremities of the upper chords and web bars (serpentine).

3.2 Test setup and instrumentation

The assembled specimens were placed under a hydraulic press available at EU-CENTRE in Pavia, as shown in Figure 3. The samples were simply bedded on the rigid floor of the laboratory as both contact surfaces were sufficiently flat. Fixed L-shape profiles were finally employed at the base of each slabs to restrict their outward displacements caused by the eccentricity of the load.

Each specimen was equipped with three LVDTs that measured continuously the slip between the steel beam and the concrete. Two of them detect the relative displacement at the top of each slabs while the third was positioned underneath the steel profile measuring the "absolute" vertical displacement of the steel beam. Only one specimen of each series was instrumented with additional strain gauges placed along the serpentine (before concreting) as well as on the steel plates.

Figure 5. Setup of the push-out test.

3.3 Testing procedure

As suggested by EN 1994-1-1, the load was initially imposed in increments up to 40% of the expected failure load. After that, 25 load cycles between 5% and 40% of the expected capacity were applied. However, considering the potential inaccuracies in the estimation of the peak load, the initial cycles were skipped in the first test of each series. The failure load obtained in the first test was then used to estimate the amplitude of the load cycles of the other two specimens of the series. Except for the load cycles, the whole push-out tests were conducted in displacement control mode at a constant speed of ca. 0.5 mm/min.

3.4 Initial test results

3.4.1 Load-slip curves

Only the measurements of some tests were post-processed at the time of writing this paper, with more results becoming available shortly. To understand the global response of the samples, the load-slip curves of the series D20-1-C25-A are shown in Figure 6 and are based on the measurements of the actuator force and the average slip displacements measured by the top LVDTs (01 and 02). As can be seen from the plot, the specimens show a relatively ductile behaviour comparable with headed stud shear connectors employed in solid slabs. After reaching the peak load at ca. 2-3 mm slip, the specimens were able to withstand at least 90% of its maximum resistance beyond 6 mm slip. Finally, at higher displacements (10-20 mm), a significant load reduction occurred and the test was stopped.

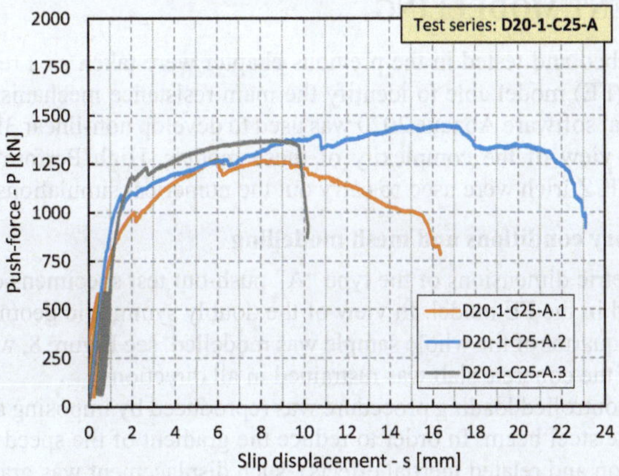

Figure 6. Experimental load-slip curves of the test series D20-1-C25-A

3.4.2 DMS measurements and failure modes

Whilst the load-slip curves provide information on the global response of the specimen, a better understanding of the force transfer between steel elements and surrounding concrete may be eased by the measurements of the strain gauges (DMS-transducers) located along the serpentine.

An interesting result was found in the strain values measured by the strain gauges 13 and 14 located in proximity of the diagonal web-plate intersection of the specimen D20-1-C45-A.1. As can be seen in Figure 7 both strain gauges 13 and 14 located at the two ends of the web bar detected similar values of compressive axial strains up to a slip displacement of approximately 1.5 mm. After this point, a sudden inversion of the strain rate was sensed by the DMS-transducer 13 reaching the positive (i.e. tensile) strains after ca. 2 mm slip. This indicates that the bar transmits axial forces as well as bending moment leading to plastic curvature observed after demounting. In parallel, the surrounding concrete undergoes severe damage with resulting loss of stiffness leading to the redistribution of the bearing forces along the serpentine. Therefore, it can be concluded that the failure occurred due to the local crushing of the concrete in combination with the plastic deformation of the diagonal web bars.

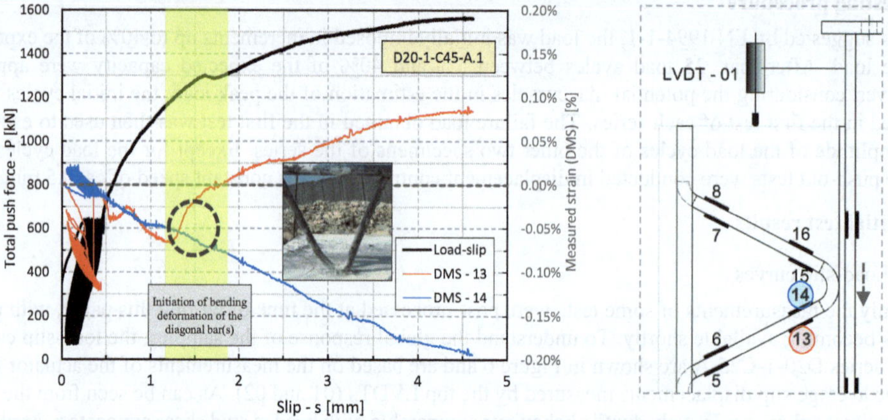

Figure 7. Values of the strain measured by DMS 13 and 14 of the specimen D20-1-C45-A.1.

4 FINITE ELEMENT MODELLING

The samples described and tested in the previous chapter were taken as a reference for developing a reliable finite element (FE) model able to identify the main resistance mechanisms involved in the shear transfer. The commercial software Abaqus 2020 was used to develop non-linear 3D FE models of the push-out test specimens. In view of the complexity of these models, High Performance Computing (HPC) clusters available at ETH Zürich were used to carry out the numerical simulations.

4.1 Geometry, boundary conditions and mesh modelling

The nominal geometric dimensions of the type "A" push-out test specimens described in the previous section were considered in the FE model. In view of the doubly symmetric geometry of the specimen and the test setup, only one quarter of the whole sample was modelled, see Figure 8, with appropriate boundary conditions. The base of the concrete slab was restrained in all directions.

The displacement-controlled loading procedure was reproduced by imposing a displacement (of 5 mm) on the top surface of the steel beam. In order to reduce the gradient of the speed over the simulation time (i.e. limit the acceleration and related inertial effects), such displacement was gradually applied according to the *SMOOTH amplitude function available in the software ABAQUS.

Due to the complex geometry of the system, solid tetrahedral elements (C3D10) were employed for modelling the steel elements as well as the surrounding concrete, as already done in previous numerical investigations on similar applications [6]. Notwithstanding the versatility and robustness of these mesh elements for such geometries, this solution is more time-consuming compared to hex elements. Therefore, further optimizations may be considered in future studies to optimize the numerical simulations. Only the steel beam was modelled through common hexahedral elements (C3D8R), as illustrated in Figure 8. The topology of the mesh was carefully optimized to avoid distorted elements, especially in proximity of the contact surfaces around the serpentine. For this reason, a finer mesh was adopted in that zone resulting in a local mean element size of ca. 6-8 mm. On the other hand, coarser mesh was employed on the rest of the specimen reaching a size of 30-40 mm.

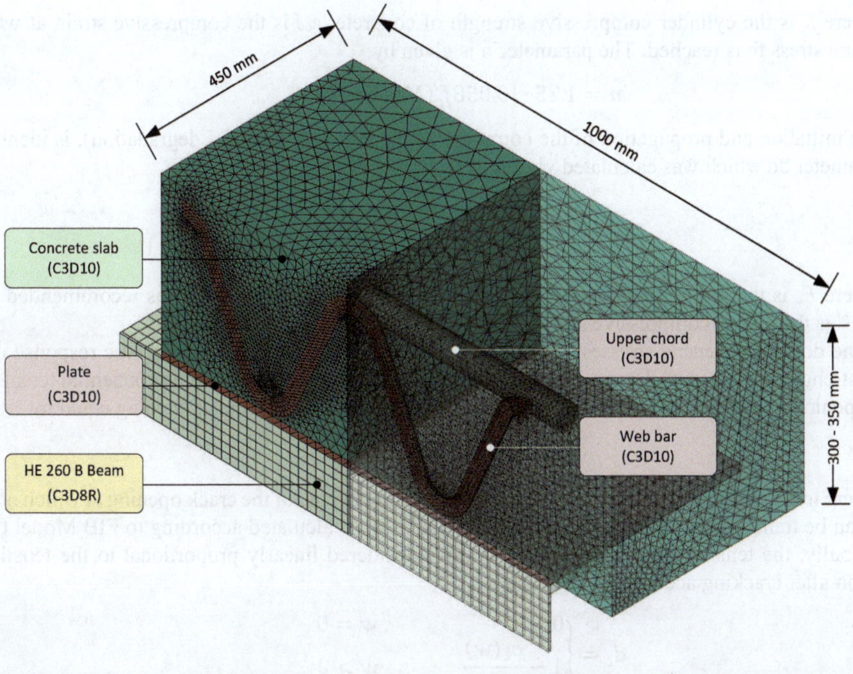

Figure 8. 3D finite element model of the push-out test specimen including mesh and different instances.

4.2 Material modelling and contact interactions

4.2.1 Concrete

The continuum approach "Concrete Damaged Plasticity" (CDP) model available in ABAQUS library [7] was used to model the behaviour of the material accounting for the initiation and propagation of the damage. The uniaxial behaviour of the material is extended to the 3D stress-strain field according to a modified Drucker-Prager model [7] where the plastic failure surface is defined through the plasticity parameters. Such variables were chosen on the basis of previous numerical studies on similar applications [8] and they are given in Table 2.

Table 2: Plasticity parameters implemented in the "Concrete Damaged Plasticity" (CDP) model. [7]

Dilation angle	Biaxial-to uniaxial strength ratio	Eccentricity parameter	Ratio of second stress invariants on tensile and compressive meridians	Viscoplasticity parameter
Ψ [deg]	f_{bc}/f_c [-]	e [-]	K_c [-]	μ [-]
38	1.16	0.1	0.67	-

The uniaxial compressive constitutive behaviour of the material was modelled in accordance with [9, 10] and the corresponding stress-strain σ_c - ε_c function can be defined as follows:

$$\sigma_c = f_c \frac{n\left(\frac{\varepsilon_c}{\varepsilon_{c0}}\right)}{(n-1) + \left(\frac{\varepsilon_c}{\varepsilon_{c0}}\right)^n} \quad (1)$$

Where f_c is the cylinder compressive strength of concrete; ε_{c0} is the compressive strain at which the maximum stress fc is reached. The parameter n is given by:

$$n = 1.25 \cdot [0.058 f_c(MPa) + 1] \tag{2}$$

The initiation and propagation of the compressive damage (i.e. stiffness degradation), is identified by the parameter dc which was calculated via Eq. (3):

$$d_c = 1 - \frac{\sigma_c E_c^{-1}}{\varepsilon_c^{pl}(1/b_c - 1) + \sigma_c E_c^{-1}} \tag{3}$$

Where E_c is the elastic modulus of the concrete, b_c was taken equal to 0.7 as recommended by [11] while ε_c^{pl} is the plastic compressive strain [7].

In the domain of tensile stresses, the material was assumed to have a linear-elastic response until the tensile strength f_{ct} is reached. Beyond this point (i.e. post-crack stage), a negative exponential tensile stress-crack opening σ_t - w function according to [12] was used, with a critical crack opening equal to:

$$w_c = 5.14\, G_f / f_t \tag{4}$$

Where w indicates the crack opening and wc is the critical value of the crack opening at which no tensile stress can be transferred. G_f is the fracture energy which was calculated according to FIB Model Code 90 [13]. Finally, the tensile damage parameter dt was considered linearly proportional to the tensile stress reduction after cracking according to Eq. (5):

$$d_t = \begin{cases} 0 & w = 0 \\ 1 - \dfrac{\sigma_t(w)}{f_t} & w > 0 \end{cases} \tag{5}$$

4.2.2 Steel elements

The elements of the truss system (i.e. plate, web bars and chords) are made of the same steel material which were modelled according to the "Von Mises" failure criterion. The fracture of the steel was neglected as it is thought that it would not occur in the configurations considered in this study. The experimental data on yield strength and ultimate strength from the tensile tests on the steel bars were used for modelling the material used in the numerical model. The corresponding key mechanical properties given in Table 3 were finally implemented in the quad-linear uniaxial stress-strain relationship presented in [8] and converted in true stress - true strain.

Table 3: Mechanical properties of the steel elements.

Elastic modulus E_s [MPa]	Poisson's ratio v_s [MPa]	Yielding strength f_y [MPa]	Ultimate strength f_u [MPa]
210000	0.3	383	530

4.2.3 Welding

Based on preliminary studies on the welding material, it was assumed that no failure occurs in the weld zone in the push-out tests considered in this study. Whilst the weld joints on the upper part, i..e between the central serpentine and the upper chords, were simply replaced by rigidly coupled *FASTENER elements [7], the geometry of the bottom weld zone was accurately reproduced, as shown in Figure 9. This enables to capture the actual local transfer of the forces between steel and concrete occurring in the real tests.

(a)

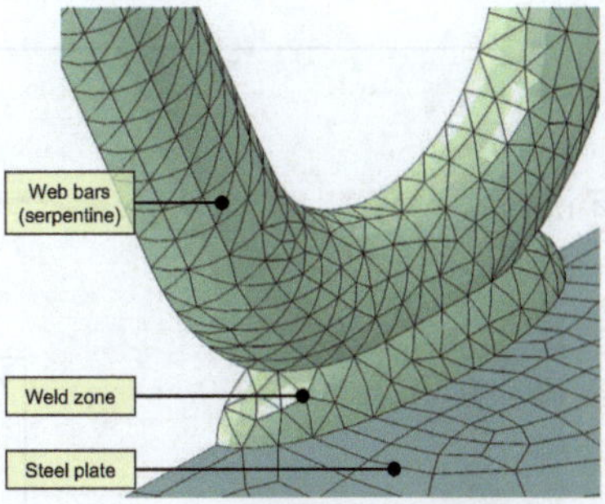

(b)

Figure 9. (a) Detail of the weld zone of the test D32-1-C25-A between the serpentine and the underlying steel plate, and (b) respective 3D finite element mesh modelling.

4.2.4 Contact formulation

The general contact formulation available in ABAQUS library was employed. It consists in a surface-to-surface contact where "hard" contact pressure-overclosure relationship was employed allowing for the separation of the surfaces under tension. The tangential behaviour was considered through the Coulomb model using a constant friction coefficient μ of 0.5. Only the plate-concrete and chord-concrete contact surfaces were considered with μ=0.1 owing to the grease spread over the steel plate and the upper chords. Additionally, in order to reproduce the application of EPS cubes shown in Figure 4, no interaction was applied between the end of the upper chords and the surrounding concrete.

4.3 Solving procedure

As thoroughly studied in similar applications for headed stud shear connectors [14, 15], a dynamic-explicit solver is recommended for reproducing the actual behaviour of push-out test specimens which involves complicated interactions. As the name implies, such approach does not enforce the static equilibrium in each node but it allows a certain amount of inertial forces. This solver procedure helps to overcome convergence issues which may arise in proximity of the complex contact surfaces (e.g. web-concrete interface). However, in order to guarantee a quasi-static simulation process, it should be checked that the dynamic effects are negligible and they do not affect the response of the system.

4.4 Results

4.4.1 Validation

The suitability of the FE model presented in the previous sections is assessed by comparing the numerically obtained results of the simulations with the experimental results. First of all, the respective load-slip curves of the test series D20-1-C-25 were plotted together with the corresponding numerical results within the first 10 mm slips. As can be seen in Figure 10(a), a good agreement was found in terms of initial stiffness, initiation of non-linear behaviour (i.e. beginning of hardening branch) as well as peak load. Furthermore, the numerical model is able to reproduce the bending curvature observed in the specimen after demounting, as shown in Figure 10(b). Based on this consideration, it can be stated that the presented FE model is suitable to simulate the mechanical behaviour of the full push-out test specimen in a consistent manner. Therefore, the model is used in the presented study to investigate in more detail the distribution of the contact forces at the steel-concrete interfaces and to identify the key resistance mechanisms activated throughout the test.

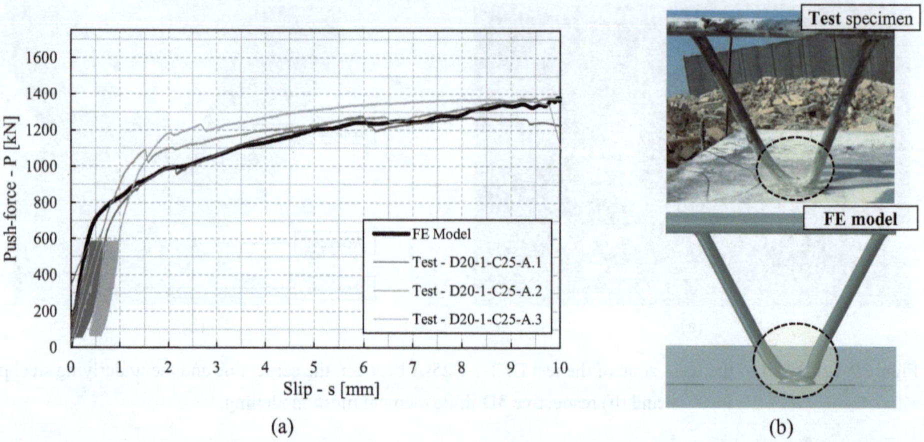

Figure 10. Experimentally and numerically obtained results in terms of (a) load-slip curve and (b) deformed shape

4.4.2 Internal forces and resistance components

Only a configuration with a single 20 mm diameter serpentine was considered in this evaluation. The feature *FREE BODY available in Abaqus was used to extract the internal forces along the serpentine between 0 mm and 10 mm slip. Considering an infinitesimal element of the diagonal dx, the tangential and perpendicular components of the contact forces along per unit of length, p and q respectively, is given by:

$$p(x) = \frac{dN(x)}{dx} \tag{6}$$

$$q(x) = \frac{dV(x)}{dx} \tag{7}$$

Where N is the axial force and V indicates the shear force along the serpentine.

For relatively small increments, Eqs. (6) and (7) can be applied discretely and the computed perpendicular and tangential contact forces per unit of length, p and q respectively, are plotted in Figure 11.

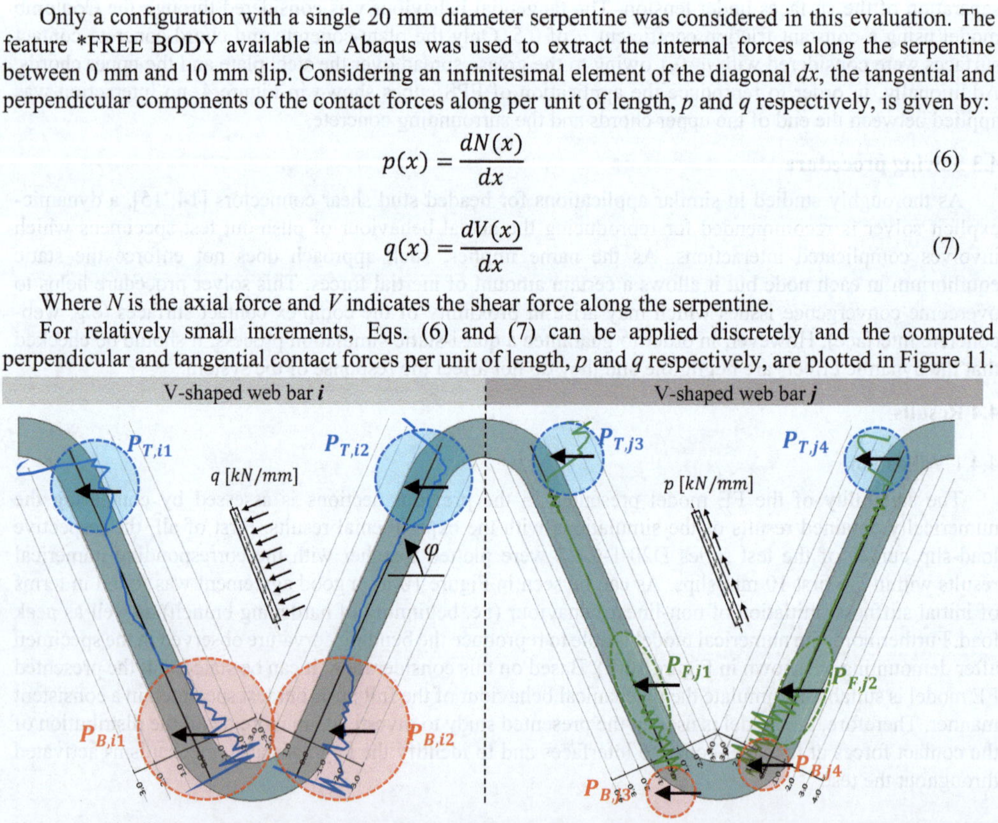

Figure 11. Distribution of the perpendicular and tangential component of the contact forces per unit of length, q and p respectively [kN/mm] and respective definition of upper, bottom and friction resistance components

With reference to Figure 11, top P_T, bottom P_B and friction P_F resistance components are given by:

$$P_B = \sum_i \left(\sum_{k=1}^{4} P_{B,ik} \right) = \sum_i \left(\sum_{k=1}^{4} \int_{B,ik} q(x)\, dx \right) \cdot \sin \varphi \tag{8}$$

$$P_T = \sum_i \left(\sum_{k=1}^{4} P_{T,ik} \right) = \sum_i \left(\sum_{k=1}^{4} \int_{T,ik} q(x)\, dx \right) \cdot \sin \varphi \tag{9}$$

$$P_F = \sum_i \left(\sum_{k=1}^{2} P_{F,ik} \right) = \sum_i \left(\sum_{k=1}^{2} \int_{F,ik} p(x)\, dx \right) \cdot \cos \varphi \tag{10}$$

It should be noted that the contact forces p arisen at the curved part of the bars are for the calculation of the top and bottom components because the forces are not perfectly tangential to the steel surface for being considered as frictional forces. Finally, the aforementioned components P_B, P_T and P_F were finally extracted from the FE model up to 10 mm slip, as shown in Figure 12.

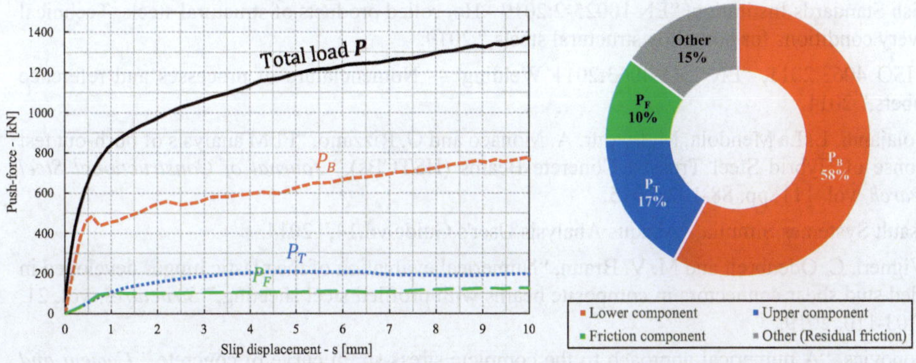

Figure 12. Total load and resistance components numerically computed from the FE model for a push-out test specimen with single 20 mm diameter serpentine

It was found that the stiffer lower component P_B covers about 58% of the entire load carried by the specimen while the upper component P_T reaches only 17%. Unlike the bottom component, P_T is fully activated only after the initiation of the concrete damage near the weld zones, at a slip of 3-6 mm. The rest of the load is transferred through the friction component P_F (ca. 10%) and residual friction mostly arising between the steel plate and the bottom surface of the slab.

5 SUMMARY AND OUTLOOK

The current work focuses on the longitudinal shear transfer in CSTC beams by means of appropriate push-out tests and with the support of numerical simulations.

The results currently available showed that the longitudinal shear resistance of these systems is mainly governed by concrete damage together with the plastic deformation of the diagonal bars which justifies the ductility observed in the test. Also, a non-linear FE model was developed and validated against test results to investigate the distribution of the forces at different displacements. From the evaluation of the numerical results, it was found that around 58% of the peak load (at ca. 5-6 mm slip) is transferred from the web bars to the surrounding concrete via high bearing stresses in proximity of the weld zone (i.e. lower component). On the other hand, the upper part of the serpentine (i.e. upper component) transfers a minor part the load at early stages and it becomes significant after the initiation of the concrete damage and respective bending moment in the diagonal bars. This resistance component reaches around 17% of the total load while only 10% is related to the friction forces developing along the serpentine. The remaining 15% of the load is transferred via residual friction at the slab-steel plate interface. Finally, the foreseen post-processing of the entire experimental campaign lays the foundations for the development of analytical equations which predict the longitudinal shear resistance of this system.

REFERENCES

[1] P. Colajanni, L. La Mendola and A. Monaco, "Stress transfer mechanism investigation in hybrid steel trussed–concretebeams by push-out tests," *Journal of Constructional Steel Research,* vol. 95, pp. 56-70, 2014.

[2] P. Colajanni, L. La Mendola and A. Monaco, "Review of Push-Out and Shear Response of Hybrid Steel-Trussed Concrete Beams. 2018, 8, 134.," *Buildings,* vol. 8, no. 10, 2018.

[3] British Standards Institution, "EN 1994-1-1:2004 Eurocode 4 - Design of composite steel and concrete structures Part 1-1: General rules and rules for buildings," London, 2004.

[4] British Standards Institution, "EN 10025-2:2019 - Hot rolled products of structural steels. Technical delivery conditions for non-alloy structural steels," 2019.

[5] EN ISO 4063:2011, "EN ISO 4063:2011 Welding — Nomenclature of processes and reference numbers," 2011.

[6] P. Colajanni, L. La Mendola, M. Latour, A. Monaco and G. Rizzano, "FEM analysis of push-out test response of Hybrid Steel Trussed Concrete Beams (HSTCBs)," *Journal of Constructional Steel Research,* vol. 111, pp. 88-102, 2015.

[7] Dassault Systèmes Simulia, "Abaqus Analysis User's Guide v6.14," 2014.

[8] V. Vigneri, C. Odenbreit and M. V. Braun, "Numerical evaluation of the plastic hinges developed in headed stud shear connectors in composite beams with profiled steel sheeting," *Structures,* vol. 21, pp. 103-110, 2019.

[9] S. Popovics, "A numerical approach to the complete stress-strain curve of concrete," *Cement and concrete research,* vol. 3, pp. 583-599, 1973.

[10] E. Thorenfeldt, A. Tomaszewicz and J. J. Jensen, "Mechanical properties of high strength concrete and application in design," in *Symposium on Utilization of High-strength Concrete*, Trondheim, 1987.

[11] V. Birtel and P. Mark, "Parameterised Finite Element Modelling of RC Beam Shear Failure," in *Abaqus users' conference*, 2006.

[12] H. A. W. Cornelissen, D. A. Hordijk and H. W. Reinhardt, "Experimental determination of crack softening characteristics of normalweight and lightweight concrete," *HERON,* 1986.

[13] FIB - Fédération internationale du béton, Model Code for Concrete Structures 2010, 2010.

[14] J. Quareshi, "Finite element modelling of steel-concrete composite structures. Ph.D. Dissertation," Leeds, 2010.

[15] V. Vigneri, "Load bearing mechanisms of headed stud shear connections in profiled steel sheeting transverse to the beam. Ph.D. Dissertation," University of Luexmbourg, Luxembourg, 2021.

[16] X. Yun and L. Gardner, "Stress-strain curves for hot-rolled steels," *Journal of Constructional Steel Research,* vol. 133, 2017.

9TH INTERNATIONAL CONFERENCE ON COMPOSITE CONSTRUCTION IN STEEL AND CONCRETE
M. Knobloch, U. Kuhlmann, W. Kurz, M. Schäfer. (Eds.)
Online, Germany, July 27-29, 2021

MECHANICAL PERFORMANCE OF INJECTED STEEL-REINFORCED RESIN CONNECTORS UNDER DIFFERENT CONFINED CONDITIONS FOR REUSABLE COMPOSITE STRUCTURES

A. Christoforidou*, F. Kavoura*, M. Pavlovic* and M. Veljkovic*

* First, Second, Third and Fourth Author's Affiliation
e-mails: A.Christoforidou@tudelft.nl, F.Kavoura@tudelft.nl,
M.Pavlovic@tudelft.nl and M.Veljkovic@tudelft.nl

Keywords: Injected steel-reinforced resin (iSRR) connections, steel-reinforced resin (SRR), steel-concrete composite floor system, demountable and reusable structures, sustainable construction, Fiber Reinforced Polymer panels.

Abstract. *A novel injection material for the development of demountable injected bolted connections in oversized holes has been recently developed by the Steel and Composite Structure research group at Delft University of Technology [1]. This injection material is a steel-reinforced resin (SRR) which consists of spherical steel particles embedded in a resin. The SRR proved to have several benefits such as the increase of the stiffness and strength of the connection and the decrease of its creep deformation when compared to injected connections with conventional epoxy resin. Research on the performance of the injected steel-reinforced resin (iSRR) has been focused on steel-to-steel, steel-to-concrete, and steel-to-FRP connections. This paper presents an overview of the recent experimental and numerical research on the mechanical properties of demountable iSRR connections and provides a comparison between different confined conditions of the injected volume.*

1 INTRODUCTION

Every structural system, either it is a building or a bridge, consists of multiple components or members. The structural segments should be successfully interconnected to make sure that a proper load transfer is accomplished. Throughout the years, due to new demands and the exploration of new materials, the traditional connection technologies have been modified or even replaced. Nowadays, the increase in the demand for demountable and reusable elements urges the application of innovative joining techniques. Additionally, the new materials that are being introduced i.e., Fiber Reinforced Polymers (FRP), steer once again the research in investigating novel connections.

A major role in the demountability and reusability of composite structures is presumed by the demountable connectors. Variety of connectors have been investigated in the literature which include bolted headed studs [2-7], bolted headed studs with embedded nuts [6, 8-11], friction-based shear connectors [12], and embedded coupler shear connectors [13-16]. Some of these demountable shear connectors use injection bolts [16-19] to establish the shear interaction.

The injection bolts are bolts filled with a two-component epoxy resin from their head and are commonly used in shear connections where slip should be avoided as an alternative for fitted [18] or pretension bolts [12]. Their general concept is described in Figure 1. After the injection and curing, the epoxy resin acts as a load-bearing element and its resistance is substantially higher than its uniaxial compression strength due to the natural confinement provided by the bolt hole [16]. The shear load in the connection is transferred through bearing and shear of the bolt. The injection bolts have been characterized as a more practical solution than the fitted bolts [17] because they can establish the shear interaction despite nominal hole clearances.

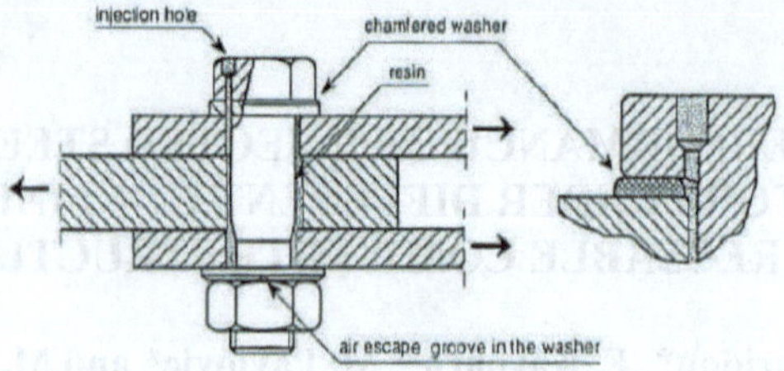

Figure 1. Cross-section of injection bolt [2].

The two-component epoxy resins used for the injection bolts should comply with the specifications of EN 1090-2 [20] regarding their viscosity, their thixotropic behaviour, and their pot time. Several epoxy resin systems have been studied in the literature [21-23] and have shown a differentiation in their mechanical properties. This paper focuses on the applicability of the Steel Reinforced Resin (SRR) proposed by Nijgh [1] which contains spherical steel particles in a resin (see Figure 2a). The SRR was tested on double lap shear connections in large oversized holes (bolt diameter + 12 mm) with a volume fraction of the steel particles of approximately 60%. The results indicated that the connection realized an increase in the instantaneous stiffness (by 71%) and a decrease in the time-dependent deformations (by 38%) in comparison with the conventional resin-injected specimens while enabling the use of (larger) oversized holes [1].

Although the initial investigation of the injected bolt with SRR was done in a steel-to-steel double lap geometry, Nijgh [16] explored its applicability on composite floor systems between concrete and steel elements. The shear connector comprises a bolt with a coupler embedded in a thick concrete element. The inserted parts of the shear connector are clamped with the injection bolt to the flange of a steel section that contains an oversized hole. In between the concrete slab and the steel beam, a steel angle profile is introduced to prevent damage to the bottom edges of the prefabricated concrete elements during the reuse [16]. An illustration of all the components of this injected steel-reinforced resin (iSRR) connection is shown in Figure 2b.

The application of the aforementioned iSRR connector on steel-concrete composite floor system proved to be very beneficial in terms of their execution efficiency, demountability and reusability which eventually leads to more sustainable structural systems [24]. These observations were supported by the fact that the iSRR connectors allow large nominal hole clearances during the execution processes which originate from geometrical and dimensional imperfections. Further, experimental and numerical work estimated the iSRR connectors' mechanical performance which is discussed in Section 2.

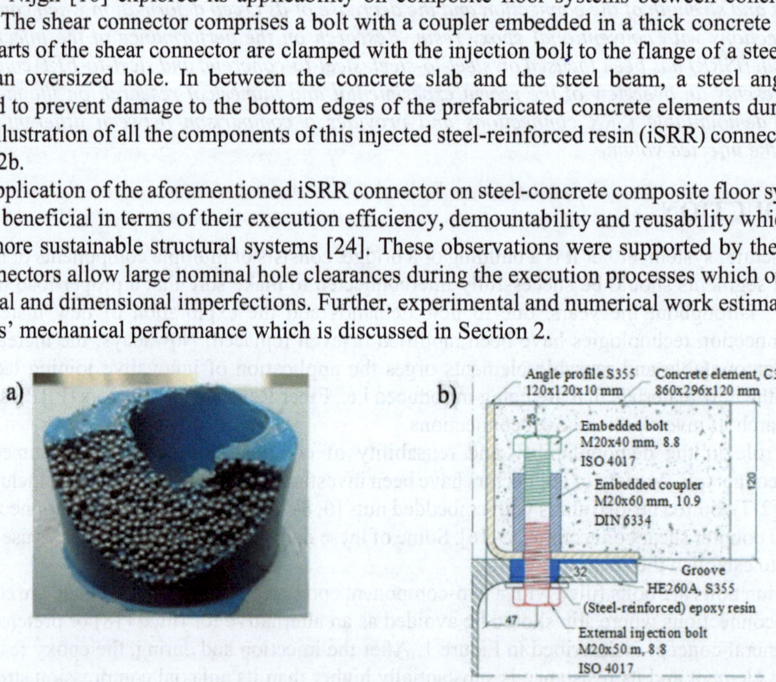

Figure 2. a) Cross-section of SRR [1] b) Demountable shear connector for composite floor systems [16].

Apart from demountability requests, new joining technologies are introduced because of new materials. In bridge renovation projects, a very competitive solution for deck replacement is the use of fatigue- and environmental-resistant FRP panels [25-28]. Over the years, the high strength-to-weight ratio of composite materials has shown excellent durability and freedom in design. In addition, they require low maintenance and short installation time [29]. The success of their application depends on the connection technology that will be applied between the FRP and the well-maintained steel support structures. Up to now, the lack of generic and feasible connection technology between FRP decks to other components of dissimilar material has hindered their implementation.

The iSRR connector that was previously investigated on steel-concrete composite floor systems has proven to be a possible way to connect also an FRP deck with a steel support structure [30-32]. Due to the nature of the FRP, slight modifications of this joining system were introduced. In this case, the iSRR connector design does not consist of an injected bolt. A threaded rod with double nuts (see Figure 3a) or a standard bolt with a coupler (see Figure 3b) is applied to connect the steel beam with the FRP deck. The iSRR material is poured into a large and almost cylindrical hole inside the FRP deck that encloses the mechanical connector.

a) b)

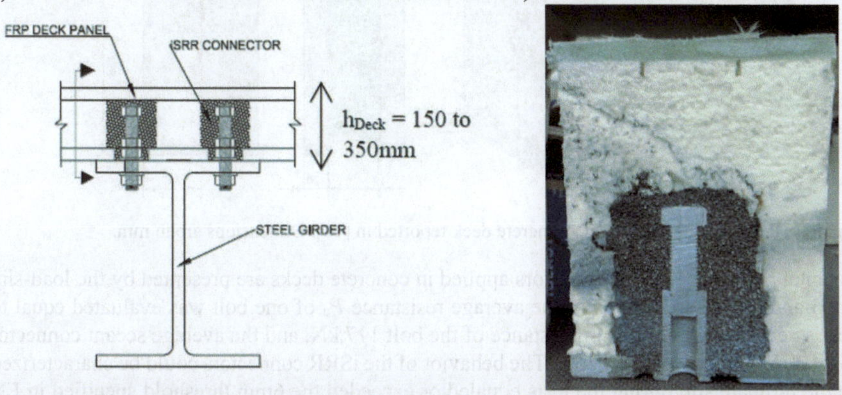

Figure 3. a) Shear connector in this study for FRP decks [32] b) Cross-section of iSRR connector inside FRP deck.

The iSRR connector has been studied and developed for being applied on steel-concrete composite buildings and on bridges with FRP decks and steel substructures [30-31, 33]. The original invention of the SRR as injection material assumes restraining the injection material in a rigid compartment formed by holes in steel plates [1]. Thus, the SRR material is considered almost fully-confined. On the other hand, for the steel-to-FPR application, the injected material is surrounded by 1) FRP panel bottom laminate that transfers the load from the connector to the laminate and 2) the foam of the sandwich web core panel which is not considered to contribute in the load transfer (see Figure 3b). For iSRR these are considered to be less rigid boundary conditions than in the case of injected steel connections and therefore the injected material is semi-confined. This paper aims to give a short overview of the mechanical performance and the failure modes of the iSRR connectors under nominally fully-confined and semi-confined conditions.

2 MECHANICAL PROPERTIES OF ISRR CONNECTORS

This section gives an overview of the push-out experiments performed on the connections for the determination of the mechanical properties of the iSRR connectors tested for applications to the steel-concrete composite floor systems and steel-FRP bridges.

2.1 iSRR connector used in composite floor systems (fully-confined condition)

Push-out experiments performed by Nijgh [16] focused on the slip resistance of the iSRR connectors (see Figure 4). The specimens consisted of four concrete elements connected with a steel flange of a steel section through eight iSRR connectors. The bolts used for the tests were M20 grade 8.8 and the hole clearance was 12 mm which is higher than the limit provided in EN 1090-2 [20] of 1-4 mm. Four iSRR push-out specimens were tested under two loading protocols following the provisions of EN 1994-1-1[34]. The first loading protocol included 25 load cycles between 5% and 40% of the expected failure load, several unloading-loading cycles of 0.5 – 1 mm increments of slip to determine the permanent relative deformation followed by loading to failure at a rate of 1 mm/minute.

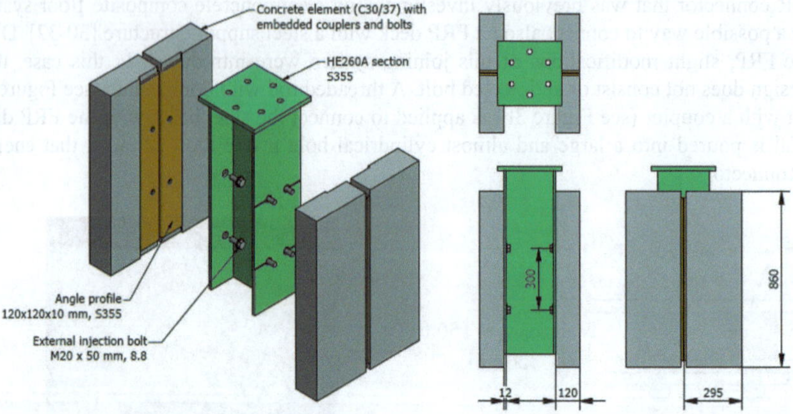

Figure 4. Push-out specimens with concrete deck reported in [16]. Dimensions are in mm.

The experimental results per iSRR connectors applied in concrete decks are presented by the load-slip curves in Figure 5 and reported in Table 1. The average resistance P_u of one bolt was evaluated equal to 118.2 kN, which is very close to the shear resistance of the bolt 117.kN, and the average secant connector stiffness at 70% of P_u [34] equal to 98 kN/mm. The behavior of the iSRR connectors could be characterized as ductile since the ultimate slip during the tests equaled or exceeded the 6mm threshold specified in EN 1994-1-1 [34]. Specimens can be divided into two groups according to the ultimate slip at failure $S_u \approx 6$mm and $S_u \approx 15$mm.

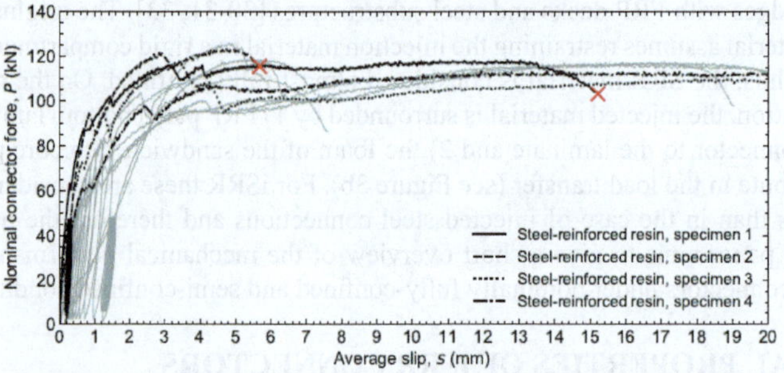

Figure 5. Push-out test results for specimens connected with iSRR connectors under confined conditions [16].

Table 1: Average push-out test results with SRR injectant in steel-concrete structures.

P_u (kN)	S_u (mm)	Secant stiffness at 0.7Pu (kN/mm)	Initial stiffness (kN/mm)
118.2	-	98.0	171.6

The shear resistance of the connectors was compared with a hand calculation model proposed by Pavlovic [35] for bolted shear connectors with single embedded nuts. The results showed a good agreement between the experiments and the hand calculation model in terms of the failure mode (bolt shear failure) and the failure load (5.6% larger than the average resistance from the experiments). Furthermore, finite element models (FEM) simulating the push-out tests were created which showed to be in line with the experimental results (see Figure 6) regarding the resistance P_u (118.2 kN from experiments and 115.6 kN from FEM), stiffness at 0,4 P_u (171.6 kN/mm from experiments and 102.7 kN/mm from FEM) and the failure mode of the connectors (bolt shear failure). FE sensitivity study showed that the large scatter of ultimate slip at failure is associated with local reduction of concrete strength as detailed in Section 3.1.

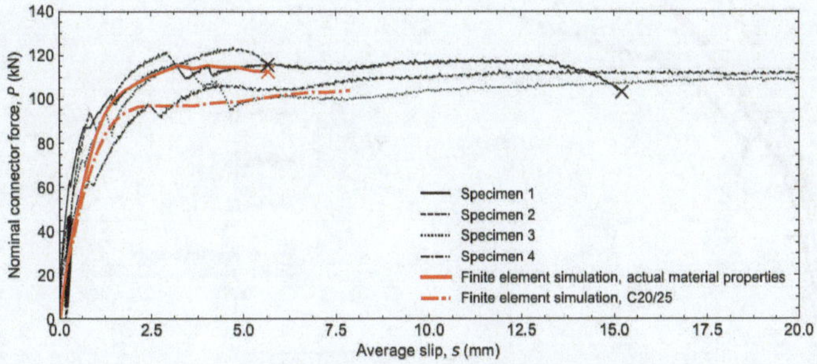

Figure 6. Comparison between experimental (black lines) and numerical (red lines) load-slip curves for push-out specimens reported in [16].

2.2 Shear connector used in FRP web core panels (semi-confined condition)

An experimental study on different types of connectors, including the novel iSRR connector, was studied in [30]. The specimens resembled push-out tests used for steel-concrete composite structures. Two FRP sandwich web core panels were connected to steel flanges of a steel section by four iSRR connectors with M20 grade 8.8 bolts as illustrated in Figure 7a. The dimensions of the SRR volume are given in Figure 7b. The initiation of the irrelevant FRP deck panel failure at boundary condition resulted in reduction of the standard eight connectors to four connectors. The same provisions of EN 1994-1-1 [34] were followed for the loading protocols. Specifically, the specimens were firstly loaded under load-controlled cycles to accommodate settling of the connectors and after 26 cycles a monotonic static load of 2mm/min was applied until failure.

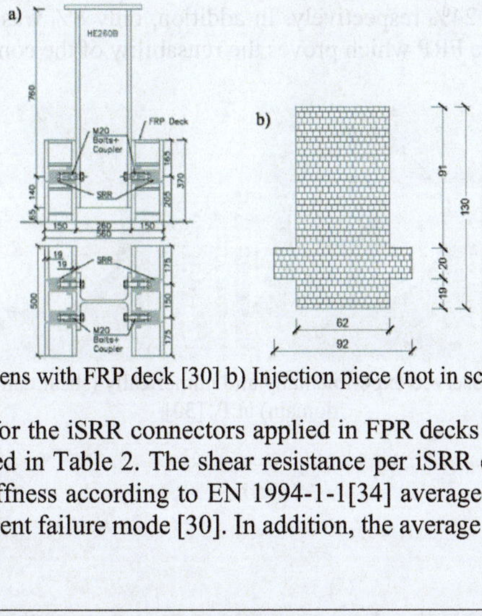

Figure 7. a) Push-out specimens with FRP deck [30] b) Injection piece (not in scale). Dimensions are in mm.

The experimental results for the iSRR connectors applied in FPR decks are presented by the load-slip curves in Figure 8 and reported in Table 2. The shear resistance per iSRR connector ranged from 109.5-130.5 kN while the secant stiffness according to EN 1994-1-1[34] averaged at 73 kN/mm excluding the specimen that failed in a different failure mode [30]. In addition, the average characteristic slip at failure S_u

could be divided once again into two groups. The first one refers to an average S_u equal to 3.2 mm characterized by bolt shear failure while the other one is equal to 8 mm due to false injection. More specifically, the reduced strength of the resin resulted in a reduced relative shear resistance between the bolt and the injection piece, and thus more ductile behaviour was observed.

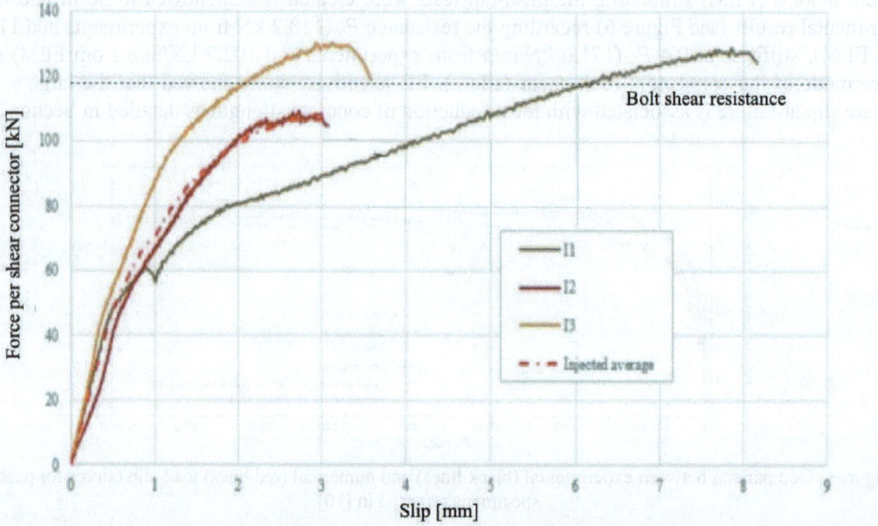

Figure 8. Experimental load-slip curves for the push-out specimens with iSRR connectors under semi-confined conditions in FRP web core deck panel [30].

Table 2: Average push-out test results with SRR injectant in steel-FRP structures with similar failure mode.

P_u (kN)	S_u (mm)	Secant stiffness at 0.7 P_u (kN/mm)	Initial stiffness (kN/mm)
120	3.2	73.0	97.9

FE models were validated and used to gain a better insight into the performance of the iSRR connector when applied in steel-FRP structures [30]. Especially for the slip development, it was concluded that the deformation of the bolt due to shear and bending and the deformation of the injected piece contributed to a slip accumulation of 34% and 24% respectively. In addition, only 8% was attributed to bearing stresses around the oversized hole of the FRP which proves the reusability of the composite panel (see Figure 9).

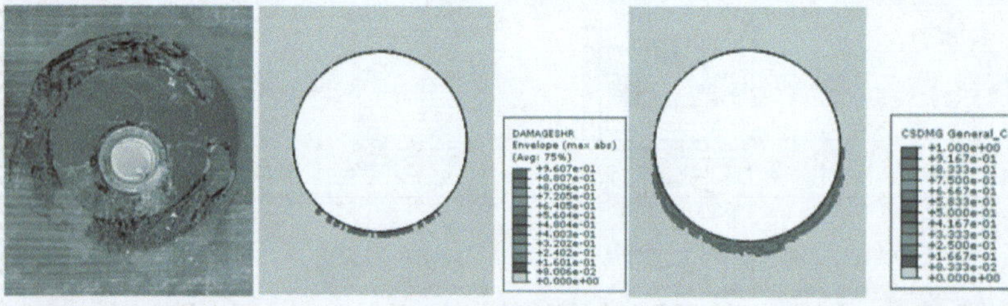

Figure 9. FRP bearing failure observed experimentally and numerically (shear damage index, damage of contact domain) at P_u [30].

3 FAILURE MODES OF ISRR CONNECTORS

The failure modes observed experimentally and numerically in the case of the fully-confined and the semi-confined connectors are different. Specifically, the fully-confined connectors failed by bolt shear failure accompanied by compressive and tensile damage occurring in the surrounding concrete [16]. The failure modes observed for the semi-confined connectors were bolt shear failure accompanied by partial or full cracks of the surrounding SRR piece in tension and debonding of the interface between the bolt and the SRR piece for bolt in axial tension. The failure modes are presented in detail in the next paragraphs.

3.1 Failure modes of fully-confined iSRR connectors

The specimens presented in Section 2.1, were divided into two groups according to the ultimate slip at failure S_u. The first group failed at relatively low slip (5,7 mm) by fracture of the bolt shortly after the maximum load was reached and by demonstrating a quasi-static linear branch at their force-slip response (see Figure 6). In this case, no damage to the concrete and the coupler was observed. The second group of specimens failed at moderate slip >=15 mm by fracture of the bolt but they demonstrated a non-linear load-slip response followed by a plateau and subsequent increase of the load which indicates a relatively high ductility. For this second group, the main failure mechanism was the bolt shear failure while significant damage was observed to the coupler (see Figure 11). In addition, compressive and tensile damage occurred in the concrete which was not visible in the surfaces of the slabs but confirmed through numerical results (see Figure 12) [16]. Consequently, the confinement of the SRR was reduced causing additional slip but the failure of the connector was not governed by the iSRR.

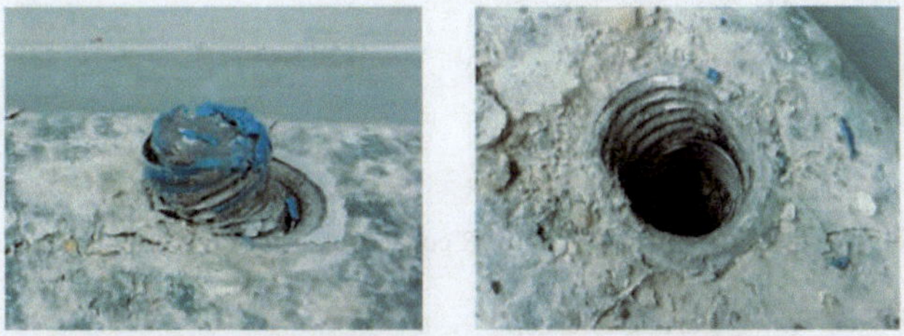

a) Coupler with part of fractured bolt **b) Ovalisation of coupler**

Figure 11. Damage of connection components for specimens with slip>15mm [16].

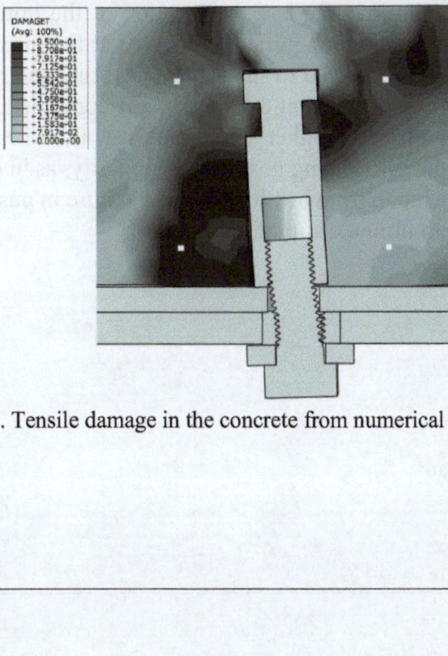

Figure 12. Tensile damage in the concrete from numerical results [16].

The numerical results showed a good agreement in comparison with the experimental results especially for the group of specimens that failed at lower slip (5,7 mm) (see Figure 6). The difference in the overall behaviour between the two groups of specimens (failure at low or moderate slip) was justified with the different modeling of the properties of the concrete strength. Specifically, it was found that the reduction of the concrete strength class led to a pronounced difference in the load-slip response. This concrete strength reduction was justified by the fact that inadequate compaction of the concrete can occur locally around the shear connector because of the reinforcement used for confinement. This fact explained the categorization of the specimens into two groups according to their slip at failure (low and moderate). Further, the numerical analysis indicated significant compressive and tensile damage in the concrete, although no obvious cracking was observed on the external faces of the specimens [16].

3.2 Failure modes of semi-confined iSRR connectors

A big part of the iSRR piece has no confinement when it is applied in an FRP sandwich web core panel. However, a smaller volume is surrounded by the FRP facing. For this application vast amount of iSRR piece is free to deform and therefore different failure modes can occur. A distinction of these failure mechanisms is made based on the type of loading that is to be expected and a visual representation of both situations is given in Figure 13.

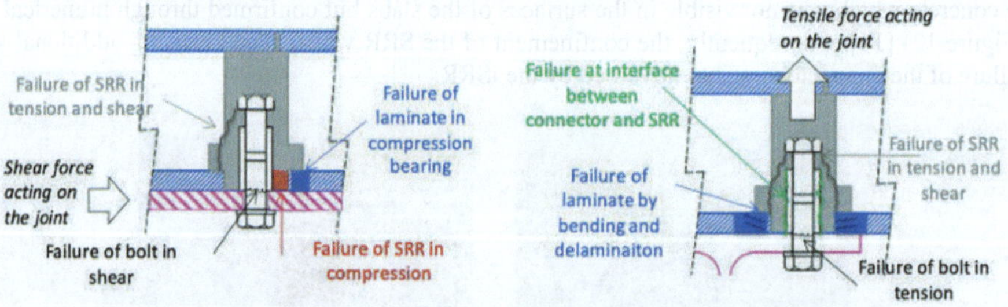

Figure 13. Possible failure modes of iSRR connection loaded in shear and tension.

3.2.1 Shear loading acting on iSRR connector

Shear forces are transferred through the bolts due to the engagement of composite action. The possible failure modes that can occur in an injected SRR joint under static load can vary. Olivier et al. explored them by performing single-lap joint experiments with 60mm hole diameter in the FRP plate and 84mm diameter for the SRR piece [32]. Due to the configuration of the set-up i.e., most of the SRR piece being uncovered, it was possible to distinguish the point in time and the location of the initiation of the different failure mechanisms. Thus, it was reported that a crack initially appeared at the interface between the SRR piece and the FRP hole (Figure 14a) which progressed to subsequent cracking and large deformations of the rod. Finally, failure of the threaded rod was detected at the shear plane combined with breaking of the injected piece as shown in Figure 14b. In these tests higher steel grade was used i.e., M20 threaded rod of grade 10.9, which led to failure mode of the iSRR being governing. Similarly, as in case of connectors embedded in concrete this switch in the failure mode from brittle bolt shear failure in push-out tests [30] slip at failure of only 3.2mm to ductile failure with ultimate slip of 8.5 mm.

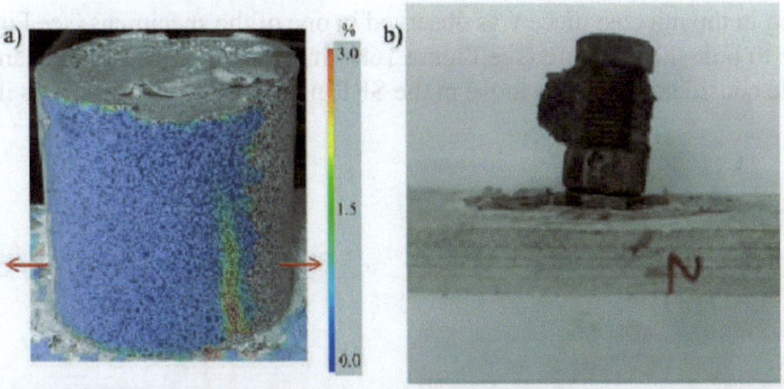

Figure 14. a) Major principal strain at 100kN loading in the iSRR piece b) Failure by crushing and shattering of SRR injection piece [32]. Arrows indicate loading direction.

3.2.2 Axial loading acting on iSRR connector

In addition to shear loads, the deck-to-girder connectors are experiencing axial loads to restrain the uplift movements and the bending of the deck. This axial loading mainly affects the resistance of the bolt but it can also generate failure on the other components. An exploration of these failure modes came from pull-out experiments performed by Csillag et. al [31]. To be more specific, mechanical connectors embedded in the middle of square sandwich panels and connecting the top FRP facing of the panels with the steel plates were tested. The test set-up is shown in Figure 15a. By pulling the embedded bolt from the fixed panel (see Figure 15b), axial forces were introduced that led to debonding of the interface between the bolt and the SRR piece at average load of 153.9 kN. Delamination of the FRP facing, failure of the SRR piece, or the bolt itself was not reported [31].

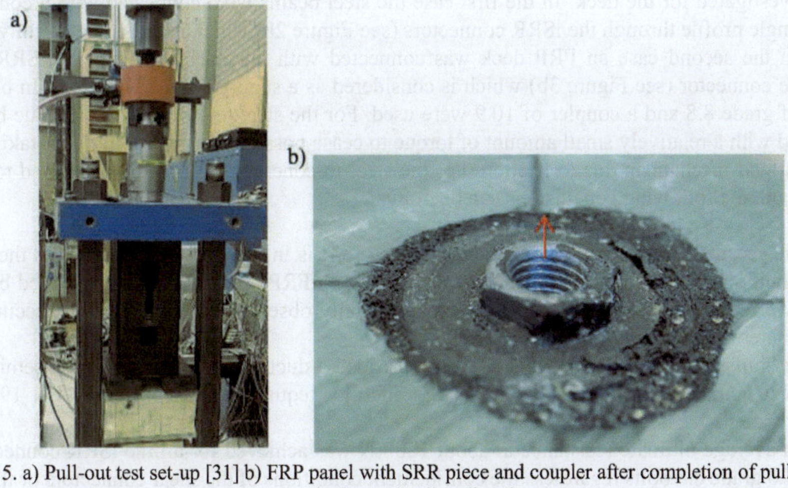

Figure 15. a) Pull-out test set-up [31] b) FRP panel with SRR piece and coupler after completion of pull-out test. Arrow indicates loading direction.

3.2.3 Combination of shear and axial loading acting on iSRR connector

The push-out experiments introduce a combination of axial and shear loading acting on the connector. Once again, the experiments conducted by Csillag & Pavlovic [30] and presented in Section 2.3 can provide sufficient information on the possible failure modes that can occur in the iSRR connection under semi-confined conditions. Because the injected piece was embedded in the FRP sandwich web core panel, it was not possible to observe the initiation and propagation of the different failure mechanics. Thus, information only for the failure mechanism that led to the failure of the specimen is available together with the FEA predictions. Cracking in the injected piece was observed in one of the specimens (see Figure 16a) while the other two failed due to bolt shear failure (see Figure 16b). In every case, negligible damage took place on the FRP facing while considerable cracks arose in the SRR piece based on the FEA as illustrated in Figure 16c [30].

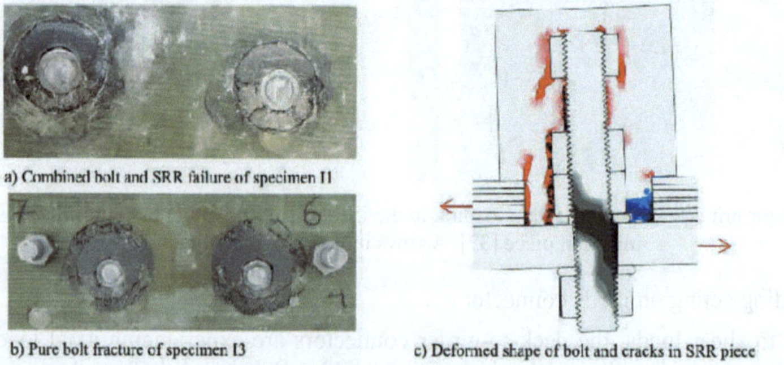

Figure 16. Failure modes obtained from push out experiments (a, b) and FEA (c) for iSRR applied in FRP decks. Arrows indicate loading direction

4 COMPARISON AND CONCLUSIONS

Push-out experiments on iSRR connectors with the injected piece under different confined conditions and different volumes were carried out in [16] and [30] and their mechanical properties and failure mechanisms are presented in this paper. The difference in confinement was a result of the different materials that were investigated for the deck. In the first case the steel beams were connected with a concrete slab and a steel angle profile through the iSRR connectors (see Figure 2b) which is considered a fully-confined condition. In the second case an FRP deck was connected with a steel beam with the iSRR material enclosing the connector (see Figure 3b) which is considered as a semi-confined condition. In both cases, M20 bolts of grade 8.8 and a coupler of 10.9 were used. For the steel-to-FRP connection, the bolts were pre-tensioned with a relatively small amount of torque to cease possible clearances but with taking care to not initiate a load-mechanism due to friction. For the steel-to-concrete solution, no preloaded took place. That being said, the following conclusions can be drawn:

1. The secant stiffness of the connector based on [34] is influenced considerably by the confined conditions of the connector. For the specimens with FRP decks and semi-preloaded bolts, 25% lower secant stiffness at 0,7 P_u was experimentally observed compared to the specimens with concrete deck and non-preloaded bolts.
2. The fully-confined shear connectors demonstrated a ductile behaviour while the semi-confined shear connectors did not accomplish the ductility requirement according to EN 1994-1-1[5] provisions.
3. An average ultimate resistance at about 120 kN was achieved for all the iSRR connectors with M20 grade 8.8 bolts regardless the confinement conditions of the shear connectors or the volume of the injected piece. This indicates that the shear resistance of the bolt is governing.
4. The fully-confined and semi-confined shear connectors showed different failure mechanisms. The failure mechanism of the fully-confined shear connectors was mainly bolt shear failure with parallel compressive and tensile damage in the concrete. In the semi-confined connectors debonding of the interface between the bolt and the SRR piece or bolt shear failure accompanied by partial or full cracks of the surrounding SRR piece took place depending on the direction of the applied force (shear, axial or interaction of both loads). Thus, the fully-confined conditions with the relatively small volume of injected material eliminate presence of cracks in the SRR.

REFERENCES

[1] Nijgh M. P., "New Materials for Injected Bolted Connections: A Feasibility Study for Demountable Connections", *MSc thesis, Delft University of Technology,* 2017.

[2] Dallam L. N., "High-strength bolt shear connectors - Pushout tests", *Journal of the America Concrete Institute, 1968.*

[3] Marshall W. T., Nelson H. M., and Banerjee H. K., "An experiment study of the use of high-strength friction grip bolts as shear connectors in composite beams", *Structural Engineer*, 1971.

[4] Hawkins N.M., "Strength in shear and tension of cast-in-place anchor bolts", *Anchorage to Concrete* 103, 233, 1987.

[5] Pavlovic M., Markovic Z., Veljkovic M., and Budevac D., "Bolted shear connectors vs. headed studs behaviour in push-out tests", *Journal of Constructional Steel Research* 88, 134, 2013.

[6] Lam D., Dai X., and Saveri E., "Behaviour of Demountable Shear Connectors in Steel-Concrete Composite Beams", *Proceedings of the 2013 Composite Construction in Steel and Concrete Conference*, American Society of Civil Engineers, 2016.

[7] Ataei A., Bradford M., and Liu X., "Sustainable composite beams and joints with deconstructable bolted shear connectors", *Proceedings of the 23rd Australasian Conference on the Mechanics of Structures and Materials (ACMSM23),* 2014.

[8] Kwon G., Hungerford B., Kayir H., Schaap B., Kyu Ju Y., Klingner R., and Engelhardt M., "Strengthening existing non-composite steel bridge girders using post-installed shear connectors", *Tech. Rep. Centre for Transportation Research of The University of Texas at Austin,* 2007.

[9] Dedic D., and Klaiber F., "High-Strength Bolts as Shear Connectors in Rehabilitation Work", *Concrete International* 6, 41, 1984.

[10] Kwon G., Engelhardt M. D., and Klingner R. E., "Behavior of post-installed shear connectors under static and fatigue loading", *Journal of Constructional Steel Research* 66, 532, 2010.

[11] Chen J., Wang W., Ding F.-X., Xiang P., Yu Y.-J., Liu X.-M., Xu F., Yang C.-Q., and Long S.-G., "Behavior of an Advanced Bolted Shear Connector in Prefabricated Steel-Concrete Composite Beams", *Materials* 12, 2958, 2019.

[12] Suwaed A.S.H. and Karavasilis T.L., "Removable shear connector for steel-concrete composite bridges", *Steel and Composite Structures* 29, 107, 2018.

[13] Yang F., Liu Y., Jiang Z., and Xin H., "Shear performance of a novel demountable steel-concrete bolted connector under static push-out tests", *Engineering Structures* 160, 133, 2018.

[14] Milosavljevic B., Milicevic V., Pavlovic M., and Spremic M., "Static behaviour of bolted shear connectors with mechanical coupler embedded in concrete", *Steel and Composite Structures* 29, 2018.

[15] Kozma A., Odenbreit C., Braun M., Veljkovic M., and Nijgh M.P., "Push-out tests on demountable shear connectors of steel-concrete composite structures", *Structures* 21, 45, 2019.

[16] Nijgh M.P., "A multi-scale approach towards reusable steel-concrete composite floor systems", *Ph.D. thesis, Delft University of Technology*, 2021.

[17] Gresnigt A. M. and Stark J. W. B., "Design of bolted connections with injection bolts", *Proceedings from the Third International Workshop on Connections in Steel Structures: Behaviour, Strength and Design*, pp. 77–87, 1995.

[18] ECCS, European recommendations for bolted connections with injection bolts, 79 ECCS, 1994.

[19] Qureshi J., and Mottram J.T., "Resin injected bolted connections: a step towards achieving slip-resistant joints in FRP bridge engineering", *Proceedings of the FRP Bridges Conference* pp. 56–66, 2012.

[20] NEN, NEN-EN 1090-2: Execution of steel structures and aluminium structures - Part 2: Technical requirements for steel structures (Nederlandse Norm(NEN), 2018.

[21] Zafari B., Qureshi J., Mottram J.T., and Rusev R. "Static and fatigue performance of resin injected bolts for a slip and fatigue resistant connection in FRP bridge engineering", *Structures* 7, 71, 2016.

[22] Koper A., "Assessment of epoxy resins for injected bolted shear connections", *MSc thesis, Delft University of Technology*, 2017.

[23] Rodrigues M., Raposa P.C., Pedrosa B., Correia J.A.F.O., Blasón S., Calvente M., Jesus A.M.P., Rebelo C., Calcada R. A. B., and Fernandez-Canteli A., "Fatigue characterisation of structural resins used in reinforcement of old bridges", *Proceedings of the XI Congresso de Construçã Metálica e mista,* 2017.

[24] Nijgh M.P, Gîrbacea I.A., and Veljkovic M., "Optimization of a composite (steel-concrete) floor system for fast execution and easy demolition", *Proceedings of the Scientific Conference on Planning, Design, Construction and Building Renewal* (iNDiS 2018), 45, 2018.

[25] Canning L. "Mount pleasant FRP bridge deck over M6 motorway", *Fourth international conference on FRP composites in civil engineering* (CICE2008): Zurich, Switzerland, 2008.

[26] Knippers J., Pelke E., Gabler M., and Berger D., "Bridges with Glass Fibre–Reinforced Polymer Decks: The Road Bridge in Friedberg, Germany", Struct Eng Int J Int Assoc Bridge Struct Eng (IABSE), 20 (4), pp. 400-404, 2010.

[27] Lee S.W., Hong K. L., "Opening the gate: construction of 300 M composite-deck bridge in Korea", *Asia-Pacific Conference on FRP in structures* (APFIS 2007), 2007.

[28] Sams M., "Broadway Bridge Case Study: Bridge Deck Application of Fiber-Reinforced Polymer", *Transportation Research Record: Journal of the Transportation Research Board*, (CD 11-S), 2005.

[29] Mara V., Hagnani R., Harryson P., "Bridge decks of fibre reinforced polymer (FRP): A sustainable solution", *Construction and Building Materials* 50, pp. 190-199, 2014.

[30] Csillag F. and Pavlović M., "Push-out behaviour of demountable injected vs. blind-bolted connectors in FRP decks", *Composite Structures* 270: 114043, 2021.

[31] Csillag F., Thie E., and Pavlović M., "Tensile and shear resistance of bolted connectors in steel-FRP hybrid beams", *Proceedings of the Belgian-Dutch IABSE Young Engineers Colloquium 2019,* YEC 2019 (pp. 50-51), 2019.

[32] Olivier G., Csillag F., Tromp L., and Pavlović M., "Conventional vs. reinforced resin injected connectors' behaviour in static, fatigue and creep experiments on slip-resistant steel-FRP joints". *Engineering Structures*, 236, 112089, 2021.

[33] Nijgh M.N., Gîrbacea I.A., and Veljkovic M., "Elastic behaviour of a tapered steel-concrete composite beam optimized for reuse", *Engineering Structures* 183, 366, 2019.

[34] EN 1994-1-1: Eurocode 4: Design of composite steel and concrete structures - Part 1-1: General rules and rules for buildings, 2011.

[35] Pavlovic M. "Resistance of bolted shear connectors in prefabricated steel-concrete composite decks", *Ph.D. thesis, University of Belgrade*, 2013.

[36] Keelor D. C., Luo Y., Earls C. J., and Yulismana W., "Service load effective compression flange width in fiber reinforced polymer deck systems acting compositely with steel stringers", *Journal of Composites for Construction*, 8(4), 289-297, 2004.